METHODS
OF
MATHEMATICAL PHYSICS

Other books available in the Cambridge Mathematical Library:

A. Baker	*Transcendental number theory*
H.F. Baker	*Abelian functions*
N. Biggs	*Algebraic graph theory, 2nd edition*
S. Chapman & T.G. Cowling	*The mathematical theory of non-uniform gases*
R. Dedekind	*Theory of algebraic integers*
G.H. Hardy	*A course of pure mathematics, 10th edition*
G.H. Hardy, J.E. Littlewood & G. Pólya	*Inequalities, 2nd edition*
D. Hilbert	*Theory of algebraic invariants*
W.V.D. Hodge & D. Pedoe	*Methods of algebraic geometry, volumes I, II & III*
R.W.H.T. Hudson	*Kummer's quartic surface*
A.E. Ingham	*The distribution of prime numbers*
H. Lamb	*Hydrodynamics*
M. Lothaire	*Combinatorics on words*
F.S. Macaulay	*The algebraic theory of modular systems*
G.N. Watson	*A treatise on the theory of Bessel functions, 2nd edition*
E.T. Whittaker	*A treatise on the analytical dynamics of particles and rigid bodies*
E.T. Whittaker & G.N. Watson	*A course of modern analysis, 4th edition*
A. Zygmund	*Trigonometric series*

METHODS OF MATHEMATICAL PHYSICS

by

SIR HAROLD JEFFREYS, M.A., D.Sc., F.R.S.
*Formerly Plumian Professor of Astronomy, University of Cambridge,
and Fellow of St John's College*

and

BERTHA SWIRLES (LADY JEFFREYS), M.A., Ph.D.
Fellow of Girton College

THIRD EDITION

CAMBRIDGE
At the University Press
1972

PUBLISHED BY THE PRESS SYNDICATE OF THE UNIVERSITY OF CAMBRIDGE
The Pitt Building, Trumpington Street, Cambridge, United Kingdom

CAMBRIDGE UNIVERSITY PRESS
The Edinburgh Building, Cambridge CB2 2RU, UK
40 West 20th Street, New York, NY 10011–4211, USA
10 Stamford Road, Oakleigh, VIC 3166, Australia
Ruiz de Alarcón 13, 28014 Madrid, Spain
Dock House, The Waterfront, Cape Town 8001, South Africa

http://www.cambridge.org

© Cambridge University Press 1946

This book is in copyright. Subject to statutory exception
and to the provisions of relevant collective licensing agreements,
no reproduction of any part may take place without
the written permission of Cambridge University Press.

First edition 1946
Second edition 1950
Third edition 1956
Reprinted 1962, 1966
First paperback edition 1972
Reprinted 1978, 1980, 1988, 1992
Reprinted 1999, 2001

Printed in the United Kingdom at the University Press, Cambridge

A catalogue record for this book is available from the British Library

ISBN 0 521 66402 0 paperback

Preface to the Third Edition

In the present edition we have made changes in Chapter 1, mainly as a result of comments by Professor A. S. Besicovitch. Some theorems are stated more explicitly, a few proofs are added, and some are shortened. We are indebted to him for an elementary proof of the theorem of bounded convergence for Riemann integrals, which appears in the notes. In Chapter 6 the proof of Poisson's equation has been improved. In Chapter 17 we have discussed the Airy integral for complex argument in more detail, and have given conditions for uniformity of approximation for asymptotic solutions of Green's type for complex argument. In Chapter 23 we have added some remarks on the analytic continuation of the solutions, and a note applies them to the parabolic cylinder functions.

We should like to express our thanks to several readers for drawing our attention to errors and misprints.

HAROLD JEFFREYS
BERTHA JEFFREYS

April 1953

Preface to the Second Edition

As a second edition of this book has been called for, we have taken the opportunity of making considerable revisions. Most of the notes at the end have been incorporated in the text. Otherwise the principal changes are as follows. In Chapter 1, the Heine-Borel theorem and Goursat's modification have been placed early, and used to derive several theorems that had been proved by separate applications of methods that could be used to prove the general theorems. In other respects, notably the theory of the Riemann integral, the theory has been given more fully. In Chapter 4 an account of block matrices has been added, and the theorem on characteristic solutions of commuting matrices has been more fully discussed. Chapter 5 (multiple integrals) has been almost completely rewritten, and now includes an account of the theory of functions of several variables, part of which was given in Chapter 11. In Chapter 9 the treatment of relaxation methods has been extended, and should now serve as an adequate introduction to the special works on the subject. Many improvements have been made in Chapters 11 and 12, including an important correction to the proof of Cauchy's theorem, a proof of the Osgood-Vitali theorem, and a complete revision of the theory of inverse functions. In Chapter 17 the conditions for the truth of Watson's lemma have been somewhat relaxed, so that they are now wide enough to cover almost all physical applications, and the method of stationary phase is more fully treated. In Chapter 24 the treatment of multipole radiation has been extended.

Where possible the proofs have been either replaced by shorter ones or generalized. Some new examples have been added.

We are indebted to numerous correspondents for pointing out errata. The two most serious corrections were given by Professor J. E. Littlewood and Dr M. L. Cartwright. We are particularly grateful for comments by Professor Littlewood (Chapters 1, 5, 11 and 12), Mr P. Hall (Chapter 4), Professor A. S. Besicovitch and Dr J. C. Burkill (Chapter 5).

HAROLD JEFFREYS
BERTHA JEFFREYS

15 *November* 1948

Preface to the First Edition

This book is intended to provide an account of those parts of pure mathematics that are most frequently needed in physics. The choice of subject-matter has been rather difficult. A book containing all methods used in different branches of physics would be impossibly long. We have generally included a method if it has applications in at least two branches, though we do not claim to have followed the rule invariably. Abundant applications to special problems are given as illustrations. We think that many students whose interests are mainly in applications have difficulty in following abstract arguments, not on account of incapacity, but because they need to 'see the point' before their interest can be aroused.

A knowledge of calculus is assumed. Some explanation of the standard of rigour and generality aimed at is desirable. We do not accept the common view that any argument is good enough if it is intended to be used by scientists. We hold that it is as necessary to science as to pure mathematics that the fundamental principles should be clearly stated and that the conclusions shall follow from them. But in science it is also necessary that the principles taken as fundamental should be as closely related to observation as possible; it matters little to pure mathematics what is taken as fundamental, but it is of primary importance to science. We maintain therefore that careful analysis is more important in science than in pure mathematics, not less. We have also found repeatedly that the easiest way to make a statement reasonably plausible is to give a rigorous proof. Some of the most important results (e.g. Cauchy's theorem) are so surprising at first sight that nothing short of a proof can make them credible. On the other hand, a pure mathematician is usually dissatisfied with a theorem until it has been stated in its most general form. The scientific applications are often limited to a few special types. We have therefore often given proofs under what a pure mathematician will consider unnecessarily restrictive conditions, but these are satisfied in most applications. Generality is a good thing, but it can be purchased at too high a price. Sometimes, if the conditions we adopt are not satisfied in a particular problem, the method of extending the theorem will be obvious; but it is sometimes very difficult, and we have not thought it worth while to make elaborate provision against cases that are seldom met. For some extensive subjects, which are important but need long discussion and are well treated in some standard book, we have thought it sufficient to give references.

We consider it especially important that scientists should have reasonably accessible statements of conditions for the truth of the theorems that they use. One often sees a statement that some result has been rigorously proved, unaccompanied by any verification that the conditions postulated in the proof are satisfied in the actual problem—and very-often they are not. This misuse of mathematics is to be found in most branches of science. On the other hand, many results are usually proved under conditions that are sufficient but not necessary, and scientists often hesitate to use them, under the mistaken belief that they are necessary. We have therefore often given proofs under more general conditions than are usually taught to scientists, where the usual sufficient conditions are often not satisfied in practice but less stringent ones are satisfied. Both troubles are due chiefly to the fact that the theorems are scattered through many books and papers, and the scientist does not know what to look for or where to look.

Preface

The book can be read consecutively, but some parts are independent of much that precedes them, and it is possible, and indeed desirable, to study different chapters concurrently. In some cases we have given special cases of a theorem before the general form where the latter involves more elaborate treatment, especially where the student is likely to meet applications to several instances of the special cases before he needs the general theorem.

We hesitated before including a chapter on the theory of functions of a real variable. This is far from a complete treatment, but fuller works are mostly longer than the theoretical physicist has time to read; and unfortunately they sometimes relegate theorems that are frequently needed to small type or unworked examples, or omit them altogether. We have aimed at giving accounts of the principal methods of the theory but not at proving every result in detail; but we think that students will benefit by filling in some of the details for themselves. If a student has difficulty in achieving the degree of abstraction needed in most of this chapter, we advise him to read as much as he can stand and then proceed to a later chapter, referring back when necessary. He will find that he has covered the whole of it before finishing Chapter 14, and that he knows both what is there and why it is there. We have not succeeded in avoiding forward references altogether, but the most serious, the proof in Chapter 12 of the theorem that an algebraic equation of degree n has n roots, used in Chapter 4, is so time-honoured that a few smaller transgressions may, we hope, be forgiven.

The notation of special functions has grown up haphazard, and is inconvenient in several respects. Quantum theorists are making wholesale changes of definition to ensure normalization, but we consider that this replaces the old complications by new ones. We have modified the usual definitions of the Legendre functions, with the result that a more symmetrical treatment becomes possible and the relation to Bessel functions becomes free from complicated numerical factors. We have returned to Heaviside's definition of the function K_n but denoted it by Kh_n. Among other advantages, this simplifies the relation to Legendre functions of the second type. We have also dropped the Γ notation for the factorial function, which seems to have no recommendations whatever.

The immediate stimulus for the book was the announcement that the second edition of *Operational Methods in Mathematical Physics* by one of us was out of print. Most of this tract has been incorporated and later developments have been added. The chapter on dispersion was somewhat out of place in the tract, as it was largely independent of the operational method, but was included because the notion of group velocity had not previously been discussed in relation to the method of steepest descents. It now finds a more natural place in a chapter on asymptotic expansions, in which some methods widely used but hitherto accessible only in scattered papers are also described. Most of *Cartesian Tensors* has also been incorporated. The applications of thermodynamics in it to hydrodynamics and elasticity would be more suitably treated in textbooks of the latter subjects.

We have not tried to give a detailed account of any branch of physics; that is a matter for the special text-books.

We are deeply indebted to many friends for their encouragement during the writing of this book. Above all we must thank Dr F. Smithies, who placed his great knowledge freely at our disposal, and generously helped in the proof reading. His suggestions have been invaluable. It is only fair to him to say that in some places we have persisted in our ways in spite of his vigorous protests. Dr J. C. P. Miller gave us special help with Chapters 9 and 23, and Mr H. Bondi with Chapter 24. We have also had valuable suggestions at various points from Professors M. H. A. Newman, A. C. Offord, L. Rosenhead and H. W. Turnbull, and from Mr A. S. Besicovitch, Miss M. L. Cartwright and Mr D. P. Dalzell.

We also thank the Universities of Cambridge, London and Manchester for permission to use examination questions as examples, and the staff of the Cambridge University Press for their care in the printing and their readiness to meet the wishes of a rather exacting pair of authors.

HAROLD JEFFREYS
BERTHA JEFFREYS

1946

The main sections of each chapter are numbered decimally at intervals of 0·01; subsections are indicated by further decimals. When the argument of a section or subsection continues that of the previous one, the numbering of the equations also continues.

Notes at the end are numbered according to the subsection referred to; references to them are indicated by a small index letter in heavy type in the text; for instance, the [a] on p. 52, in subsection 1·134, refers to note **1·134a**, which will be found on p. 692.

Sources of examples are indicated by the following abbreviations:

M. T.	Mathematical Tripos, Part II and Schedule A.
M. T., Sched. B.	Mathematical Tripos, Part III and Schedule B.
Prelim.	Preliminary Examination in Mathematics.
M/c, III.	Manchester, Final Honours in Mathematics.
I.C.	Imperial College, London.

Contents

			page
Preface			v
Chapter	1.	The Real Variable	1
	2.	Scalars and Vectors	57
	3.	Tensors	86
	4.	Matrices	114
	5.	Multiple Integrals	171
	6.	Potential Theory	199
	7.	Operational Methods	228
	8.	Physical Applications of the Operational Method	244
	9.	Numerical Methods	261
	10.	Calculus of Variations	314
	11.	Functions of a Complex Variable	333
	12.	Contour Integration and Bromwich's Integral	375
	13.	Conformal Representation	409
	14.	Fourier's Theorem	429
	15.	The Factorial and Related Functions	462
	16.	Solution of Linear Differential Equations of the Second Order	474
	17.	Asymptotic Expansions	498
	18.	The Equations of Potential, Waves, and Heat Conduction	529
	19.	Waves in One Dimension and Waves with Spherical Symmetry	546
	20.	Conduction of Heat in One and Three Dimensions	563
	21.	Bessel Functions	574
	22.	Applications of Bessel Functions	595
	23.	The Confluent Hypergeometric Function	606
	24.	Legendre Functions and Associated Functions	628
	25.	Elliptic Functions	667
Notes			691
Appendix on Notation			706
Index			711

Authors' Notes

In this *second impression* of the Third Edition, the following notes have been added: **5·051a** on differentiation under the integral sign, **10·11a** on a method used in planetary theory **23·07a** giving references for work on Coulomb wave functions. Paragraphs **10·01, 10·013** on the Calculus of Variations have been revised. Some minor corrections and addenda have been made in the text and examples.

July 1961

In this *third impression* of the Third Edition, the following notes have been added: **9·041a** on interpolation when first derivatives are given and **9·181a** on the advance in automatic computation. The treatment of orthogonal transformations in Chapter 4 has been extended and an amendment has been made to the proof of Watson's lemma in **17·03**. Further minor corrections and addenda have been made in the text and examples.

March 1966

In this *paperback edition* of the Third Edition, the following alterations have been made: **23·07** on Schrödinger's equation for the hydrogen-like atom has been revised and the note **23·07a** expanded, and in the Addenda there are references to work on Isotropic Tensors in the note **3·031a**. Further minor corrections have been made in the text.

January 1972

Chapter 1

THE REAL VARIABLE

'In dem days dey wuz monstus fon' er minners.'
JOEL CHANDLER HARRIS, *Uncle Remus*

1·01. The relation of mathematics to physics. The simplest mathematical notion is that of the number of a class. This is the property common to the class and to any class that can be matched with it by pairing off the members, one from each class, so that all members of each class are paired off and none left over. In terms of the definition we can give meanings to the fundamental operations of addition and multiplication. Consider two classes with numbers a, b and no common member. The sum of a and b is the number of the class consisting of all members of the two classes taken together. The product of a and b is the number of all possible pairs taken one from each class. We cannot always give meanings to subtraction and division, because, for instance, we cannot find a class whose number is $2-3$ or $7/5$. But it is found to be a great convenience to extend the notion of number so as to include negative numbers, ratios of numbers irrespective of whether they are positive or negative, and even irrational numbers. When this is done we can define all the four fundamental operations of arithmetic, and the result of carrying them out will always be a number within the system. We need trouble no more about whether an operation is *possible* with a particular set of numbers, since we know that it is, once we have given sufficient generality to what we mean by a number. So long as we keep to the fundamental operations we can use algebra; that is, we can prove formulae that will be correct when any numbers whatever are substituted for the symbols in them, with only one exception, namely, that we must not divide by 0.

Now the formulae may still be correct when we replace the letters in them by something other than numbers, and it is to this fact that the possibility of mathematical physics is due. It is therefore useful to know just what conditions have to be satisfied if we are to take over the rules of algebra into any subject that does not deal entirely with numbers. We may then have to find new meanings for the fundamental operations (or have them found for us) and for the sign $=$, but can still manipulate the symbols with their new meanings in the old way. A suitable set of conditions is as follows.* We say that they are to hold in a *field* F consisting of all elements of the system considered:

(1) For any a, b of F, $a+b$ and ab are uniquely determined elements of F.
(2) $b+a = a+b$. (Commutative law of addition.)
(3) $(a+b)+c = a+(b+c)$. (Associative law of addition.)
(4) $ba = ab$. (Commutative law of multiplication.)
(5) $a(bc) = (ab)c$. (Associative law of multiplication.)
(6) $a(b+c) = ab+ac$. (Distributive law.)
(7) There are two elements 0 and 1 in F, such that $a+0 = a$, $a1 = a$.
(8) For any element a of F there is an element x of F such that $a+x = 0$.
(9) For every element a of F, other than 0, there is an element y of F such that $ay = 1$.

* Stated first by Dedekind for the case where $+$ and $\times$ have their ordinary arithmetic meanings; in general by H. Weber.

It is to be noticed that the first seven rules are true if F consists only of the positive integers and 0, but the last two are false of that F, since there is no positive or zero integer x that makes $a+x = 0$ if $a = 1$, and there is no positive or zero integer y that makes $ay = 1$ if $a = 2$. The eighth rule introduces negative numbers and hence subtraction. The ninth introduces reciprocals and hence division and rational fractions. The rules are true if F consists of all rational numbers, positive or negative.

The rules mention no ordering relation: that is, they suppose a meaning attached to equality and therefore to $\neq$, but do not distinguish between greater and less. We could agree to arrange the numbers in any order, keeping the same correspondences between them according to (1), (7), (8), (9), and the rules would still be true. Algebra and pure geometry can get on to some extent without such a distinction, but higher mathematics cannot, nor can any kind of physics. A measurement is not a statement of exact equality but of equality within a certain range of error. We therefore need new rules concerning inequalities.

(10) For any a, b of F, either $a > b$, $a = b$, or $b > a$. (Law of comparability.)

(11) For given a, b of F, only one of $a > b$, $a = b$, $b > a$ can be true. (Trichotomy.)

(12) If $a > b$ and $b > c$, then $a > c$. (Transitive property.)

(13) If $a > b$, then $a+c > b+c$ for any c. (Additivity of ordering.)

(14) If $a > b$, $c > 0$, then $ac > bc$. (Multiplicativity of ordering.)

(15) If $a > b$, $b < a$. (Definition of $<$.)

The use of mathematics in science is that of a language, in which we can state relations too complicated to be described, except at inordinate length, in ordinary language. The rules satisfied by the symbols are the grammar of the language. This point of view has been developed greatly in recent years, especially by R. Carnap. But for a language to be suitable it must satisfy two conditions. It must be possible to say in it the things that we need to say; that is, it must have sufficient generality. It must also be self-consistent; that is, starting from the rules themselves it must be impossible to deduce something declared to be false by those rules. It would, for instance, be fatal to the scientific usefulness of mathematics if it was possible to prove by it that for some a and b, a is both greater and less than b. It was always taken for granted until the later nineteenth century that mathematics was consistent. But then an unexpected set of difficulties cropped up, and showed that a complete analysis of the foundations was necessary. The great *Principia Mathematica* of Whitehead and Russell showed that all the propositions asserted in mathematics concerning real numbers (not only ratios of integers, positive or negative) could be restated as propositions about the elementary notion of comparing classes by pairing their members, and demonstrable from the axioms of such comparison and others relating to pure logic. Later workers have modified some of the latter axioms, and the best choice of axioms is still a matter of discussion. Gödel and Carnap, more recently, have shown that the proposition that a given system of axioms for mathematics is consistent cannot be proved by methods using only the rules of the system. But it is found impossible to prove certain propositions that could be proved if the system was inconsistent. We have to come back to something like ordinary language after all when we want to talk *about* mathematics! This work on the boundary between logic and what we usually consider the elements of mathematics has a considerable modern literature, and it is well for physicists to know of its existence, though its detailed study is a matter for specialists.

1·02. Physical magnitudes. Generality requires that, in any particular field, the language shall contain symbols for the things that we need to talk about and for the processes that we carry out. A shepherd would be severely handicapped if he had to do his best with a language containing no words for sheep and shearing; in fact he would make such words, and that is what we habitually do in science. So long as the language is consistent it is none the worse for containing a lot of words that we do not use. A pure mathematician, working entirely on the theory of numbers, can use ordinary algebra freely in spite of the fact that he may not need to use negative numbers or fractions. For him rules (8) and (9) are just an unnecessary generality. Now in physics the fundamental notion of measurement corresponds closely to that of addition, and most physical laws are statements of proportionality, which corresponds to the notions of multiplication and division. This is the ultimate reason why mathematics is useful. Thus, for instance, we can say that if two bars are placed end to end to make one straight bar, the length of the combined bar is the sum of those of the original ones. This is not a theorem or an experimental fact; it is the definition of addition for lengths. Further, it is irrelevant which is taken first; thus the commutative law of addition holds. Again, if we unite three bars, the total length is independent of the order; hence the associative law of addition also holds. These are experimental facts established by actual comparison with other bars. These rules are enough to justify the use of scales of measurement for length, by which any length is compared with a standard one by means of a scale, every interval of which has been compared with a standard object in the process of manufacture. Quantities measurable by some process of physical addition have been called *fundamental magnitudes* by N. R. Campbell.* The most widely important ones are numbers (of classes), length, time, and mass, but physical processes of addition can also be stated for area and volume, for electric charge, potential, and current, and many other quantities.

There is a divergence of practice among physicists at the next stage. A statement that a distance is 3·7 cm. contains a number and a unit. It is often thought that algebra applies only to numbers and therefore that in the mathematical treatment the symbol used for the distance refers only to the 3·7 and not to the centimetres. The unit matters, otherwise we should find ourselves saying that 10 mm. expresses a different length from 1 cm. and that 1 cm. is the same as 1 mile; and this is contrary to physics because the only justification of using measurement at all is in the direct physical comparison by superposition. We avoid this difficulty if we say that the symbol for the length refers to the length itself and not simply to the number contained in its measure. '1 inch = 2·54 cm.' is a useful statement; either symbol, '1 inch' or '2·54 cm.', denotes the *same* length. In general theorems this procedure can always be followed. When a particular application to a measured system is made we naturally give the symbols their actual values in terms of the measures, which will include a statement of the units; but in the general theory the unit is irrelevant. The symbols will then be said to stand, not for numbers, but for *physical magnitudes*.

The alternative method would be to let the symbols stand for the numbers, but then confusion can occur, and does, between the relations between measures of the same system in different units, which are different ways of saying the same thing, and of different systems in the same units, which say different things. If, however, the numerical values in terms of special units are used for a and b in ab, their product will be the number in the

* 'Elementary' or 'Additive' might be better.

expression of ab in what is usually called the *consistent* unit for ab. The word *germane*, introduced by E. A. Guggenheim, is better because it is not inconsistent to measure distances upward in feet, horizontally in yards, and downward in fathoms; it is merely a nuisance. With adequate care this method can be used correctly, but it has several disadvantages; in particular it then leads to placing too much emphasis on the units and too little on the fundamental physical comparisons without which the units would be useless. It also suggests many comparisons that are physically meaningless, as we shall see in a moment.

If we use the notion of magnitude and retain the processes of algebra the question will at once arise, what do we mean by $a = b$ and $a+b$ if a is a length and b a time or a mass? A meaning could be attached to $a+b$, though it would be very artificial, but no physical process will give one to $a = b$. But a/b would have a meaning, being respectively a velocity or a length per unit mass.

The group of rules (10)–(14) therefore needs modification. Those up to (9) could stand, though they bring in many additions and subtractions and possibly some multiplications and divisions that we shall never have occasion to use; but in addition to the three possibilities enumerated in (10) we must admit a fourth, that a and b may not be comparable and therefore belong to different fields, and their product and ratio may belong to other fields again. This is a further disadvantage of the use of symbols to denote only the number stated in a measure, since all numbers are comparable, and the language would not exhibit the fact that it is meaningless to say that a time is greater than a density. We can then say also that if a and b are not comparable, $a+b$ is not a physical magnitude and addition does not arise. The whole field of physical magnitudes is thus divided into plots. Magnitudes in the same plot will be comparable, but their product will belong to a different plot unless at least one of them is a number.

The language needed for physics is therefore not quite the same as ordinary algebra. Since the latter is self-consistent and the statement that some magnitudes are not comparable cuts out some propositions from it and adds no new ones, the language of magnitude is also self-consistent. It will be seen that the modification corresponds to the notion of *dimensions*. Quantities of different dimensions are not comparable; also some quantities of the same dimensions are not. For instance, according to one pair of definitions in use, electric charge and magnetic pole strength have the same dimensions, and they are both additive magnitudes, but it is meaningless to add them. The field of physical magnitudes can be taken to satisfy the laws of algebra, but is classified; comparable quantities satisfy (10), and are capable of addition at least in calculation; incomparable ones do not. It should be noticed that failure of addition by a physical process is not confined to incomparable magnitudes. For instance, there is no process of combining two substances of density 1 g./cm.[3] to give one of density 2 g./cm.[3] Density is not measured directly but calculated from the additive magnitudes mass and length, and is called a *derived magnitude*. Some quantities can be both additive and derived; thus electric current measured by its magnetic effect is an additive magnitude, but regarded as the charge passing per unit time it is derived. Many derived magnitudes are ratios of two magnitudes of the same dimensions; thus we could regard the shape of a triangle as specified by two ratios, those of two sides to the third. These ratios are pure numbers and the rules of algebra can be applied to them without change.*

* A similar treatment was advocated by W. Stroud; for discussion and applications to teaching, cf. Sir J. B. Henderson, *Engineering*, 116, 1923, 409–10.

1·03. Real numbers. Most of the present chapter will be already familiar to those who have studied a good modern book on calculus, and it is not intended to compete with standard works on pure mathematics. We think, however, that some discussion here is not out of place, for several reasons. First, the latter works for the most part do not emphasize *why* the refined arguments that they give have any relevance to physics, and physicists therefore tend to believe that they are irrelevant. Secondly, they are liable to be so long that a physicist can hardly be blamed if he decides that he has not the time to work through them. Thirdly, the attention to very peculiar functions has led the subject to be regarded as the pathology of functions. The reply is that every function, except an absolute constant, is peculiar somewhere, and that by studying where a function is peculiar we can arrive at constructive results about it that would be very hard to obtain otherwise. But we are entitled to regard ourselves as general practitioners and to restrict ourselves to the kinds of peculiarities that occur in physics; rare diseases may be handed over for treatment to a specialist, in this case a professional pure mathematician.

The nature of the problem was foreshadowed in a theorem of Euclid that the ratio of the hypotenuse to one side of an isosceles right-angled triangle is not equal to any rational fraction. Euclid, it must be remembered, made no use of what we should now call numerical measures of physical magnitudes. When he said that two lines were equal he meant that one could be placed on the other so that the two ends of one coincided with the two ends of the other; this is the direct physical comparison and does not require any numerical description of the lengths. When he said that the square on the hypotenuse was twice that on a side he meant that it could be cut into pieces and that the pieces could then be put together so as to make the square on the side twice over. He was working throughout with the quantities themselves, not with the numbers that we choose to associate with them in measurement with regard to any special unit. The use of numbers for this purpose is a choice of a language. What Euclid's theorem showed was that the language of rational numbers was incapable of describing simultaneously the lengths of the side and the hypotenuse of a triangle that could easily be drawn by the rules of his geometry.

Measurement in terms of a unit is too useful a procedure to be lightly abandoned, and it could be retained, consistently with Euclid's theorem, in any of the following ways: (1) Since an infinite number of pairs of integers x, y can be found such that $x^2 + y^2 = z^2$, where z is another integer, and so that x/y is as near 1 as we like, we could suppose that the sides of a right-angled triangle satisfy $x^2 + y^2 = z^2$ exactly but that $x = y$ is not true exactly but only within the errors of measurement, and the sides are always exact multiples of some definite length. (2) We might say that x/y can be exact but $x^2 + y^2 = z^2$ is only approximate. (3) We can say that the language of rational numbers is not enough for what we need to say, and that we need a fuller language in which $x = y$ and $x^2 + y^2 = z^2$ can be both said consistently. The last alternative is the one that has been universally adopted by the admission to arithmetic of irrational numbers. It does not contradict Euclid's axioms; the first does, since he assumes that a line can have *any* length, and the second contradicts one of their best-known consequences. An experimental proof that it is right is impossible because either (1) or (2) could be true within the errors of measurement even if x, y, z were restricted to be integers. But they would be intolerably complicated, and the adoption of either would require the existence of an unknown and indeterminable standard of length such that all actual lengths are exact multiples of it, besides abandoning the simplicity of Euclid's rules without experimental reason. The universal practice in physics is to adopt alternative (3) and create a language of sufficient generality. We introduce *real numbers* and *assume* that the operations of addition, subtraction, multiplication and division can be applied to them in such a way that the same fundamental rules as for rational numbers are satisfied, and that an ordering relation satisfying rules (10)–(15) can be defined. They differ from the rationals in possessing a certain property of *completeness*, which ensures, for instance, that there is a real number $\sqrt{2}$ whose square is 2. It is not obvious that this can be done without inconsistency (and it was certainly believed for 2000 years that real numbers were meaningless*), but the 19th century investigations of Dedekind, Cantor, and others have established their workability for all practical purposes. That is enough justification for our purposes. But the logical justification involves the consideration of infinite collections. It is indeed obvious that the evaluation of $\sqrt{2}$ by root extraction or by successive approximation to a continued fraction, if taken to a finite number of steps, can never yield anything but a rational number; to give any exact meaning to $\sqrt{2}$ in numerical terms requires an infinite number. Euclid's procedure does lead in a finite number of steps to a ratio that can be identified with $\sqrt{2}$, but does not describe it in a numerical way, and the proof that his axioms are themselves consistent has so far been completed only by way of the numerical approach. The notion of $\sqrt{2}$ is accepted at school largely because we believe that a consistent system of measurement of physical objects is possible and Euclid's axioms look plausible; but we forget that the Euclidean triangle is not the real triangle, or, if we remember, we think that the real triangle is an imperfect representation of the Euclidean one. Physically the Euclidean triangle is an idealized approximation to the real one, and we cannot take it for granted that the idealization does not introduce new troubles of its own.

1·031. Nests of intervals: Dedekind section. The fundamental property of real numbers is that they can be approximated to as closely as we please by rational numbers. When we say that
$$\sqrt{2} = 1\cdot 414 \ldots,$$
we assert the following set of propositions: (1) 2 is between 1^2 and 2^2; (2) 2 is between $1\cdot 4^2$ and $1\cdot 5^2$; (3) 2 is between $1\cdot 41^2$ and $1\cdot 42^2$; (4) 2 is between $1\cdot 414^2$ and $1\cdot 415^2$; and so on to any desired accuracy. At each stage this process can be regarded as separating the decimals, to a given number of places, into two classes, those whose squares are respectively greater or less than 2. At stage 3, for instance, the squares of $1\cdot 414$, $1\cdot 413$, $1\cdot 412$ are less than 2, those of $1\cdot 415$, $1\cdot 416$, $1\cdot 417$ greater than 2. We say nothing at this stage about the fractions $1\cdot 4141$, $1\cdot 4142$, ..., $1\cdot 4149$; but at the next stage we say that 2 lies between the squares of $1\cdot 4142$ and $1\cdot 4143$. By taking a sufficient number of decimals we can make the unconsidered interval as small as we like, since we divide it by 10 at each step. Thus any decimal with a finite number of places will ultimately be classified according as its square is less or greater than 2. Now this process determines a unique infinite decimal, which we can take to *be* $\sqrt{2}$, and it can be regarded as the limit approached by the successive approximations from either side.

This process, which is capable of great extension, is an example of the definition of a

---

* Hence the name 'irrational numbers'.

real number by a *nest of rationals*. We take a succession of rationals $\{a_n\}$ and another succession $\{b_n\}$, satisfying the following conditions:

(i) $a_{n+1} \geqslant a_n$,
(ii) $b_{n+1} \leqslant b_n$,
(iii) $a_n \leqslant b_n$,

for all n, and

(iv) Given any positive rational number ϵ, a number N can be found such that

$$b_n - a_n < \epsilon \quad \text{for every } n > N.$$

Such a nest $\{a_n \mid b_n\}$ can be used as a definition of a real number. A *member* a_n, b_n of the nest consists of the set of rationals greater than or equal to a_n and less than or equal to b_n. The real number defined by the nest lies between the end-points of all its members.

A nest may turn out to define a rational number. For instance, if we consider decimals whose squares are respectively just less and just greater than 2·25 we get the nest 1, 2; 1·4, 1·6; 1·49, 1·51; 1·409, 1·501; .... The only decimal lying between the end-points of all members of the nest is 1·5, whose square is in fact 2·25. For every rational we can construct such a nest, so that the rationals themselves are real numbers.

A single real number can be defined by many different nests. For instance, instead of dividing the interval by 10 at each stage we could divide by 2, in this way generating a binary fraction or 'decimal to base 2'. It would take more than three times as many steps to get as good an approximation, but the process defines the same real number as before. Two nests $\{a_n \mid b_n\}$ and $\{\alpha_n \mid \beta_n\}$ define the same real number if and only if a_n, b_n contains α_m, β_m for sufficiently large m, and α_n, β_n contains a_m, b_m for sufficiently large m; in fact only one of these conditions need be known to hold—the other follows as a consequence.

We now come to the most important property of the real number system. We abandon the condition that a_n, b_n shall be rational and consider a nest $\{a_n \mid b_n\}$ where a_n and b_n are now real numbers. An *interval* of such a nest consists of the set of real numbers greater than or equal to a_n and less than or equal to b_n. In condition (iv) ϵ is now any positive real number. It can be proved that there is one and only one real number lying in every interval of the nest. In other words, if we apply to the real numbers the process that we have applied to the rationals, we get nothing new, but remain within the system that we have already defined. This is the property of completeness mentioned in 1·03.

Another important way of defining real numbers is by a *Dedekind section* or *cut*. If the rational numbers are divided into two classes L and R such that every member of L is less than every member of R, there is only one real number greater than or equal to every member of L and at the same time less than or equal to every member of R. If this real number is rational, then it will be either the greatest member of L or the smallest member of R. For instance, L might consist of the negative rationals together with 0 and the positive rationals whose squares are less than 2, and R of the positive rationals whose squares are greater than 2. This cut defines the real number $\sqrt{2}$.

Dedekind section arises most naturally when the numbers are classified according as they possess or do not possess a certain property. For instance, 'x has a square not greater than 2·25' defines an L class, the largest member of which is 1·5; 'x has a square less than 2·25' defines an L class with no largest member, and 1·5 is the smallest member of the R class. 'x is rational and has a square less than 2' defines

L and R classes of rationals with no largest and no smallest member respectively. 'x is real and has a square less than 2' defines an L class with no largest member and an R class with smallest member $\sqrt{2}$.

In terms of the Dedekind section, the completeness property of the real number system is equivalent to the statement that any cut in the *real* numbers defines a real number. Thus many problems that have no answer in the rational number system can be solved in terms of real numbers. We have so far considered only $\sqrt{2}$, but we are also ready for π and e when they turn up, and shall not need to search for a statement of each problem in such a form that it can be solved in rational numbers. The use of the real number system therefore avoids a lot of complications with no relevance to physics.

The methods of nested intervals and of Dedekind section are equivalent. If L and R classes exist we can form a nest of intervals, taking $a_1, a_2, \ldots$ from L and $b_1, b_2, \ldots$ from R, in such a way that the conditions required for a nest of intervals are satisfied. Conversely, if a nest exists, some rationals r will be exceeded by a_m for some m, others will not be exceeded by any a_m. These inequalities define an L and an R class and the conditions for a cut are satisfied.

If the nest (a_m, b_m) defines a positive real number x, $(1/b_m, 1/a_m)$ will define $1/x$. Then if nests $(a_m, b_m)(a'_m, b'_m)$ define x, x', $(a_m a'_m, b_m b'_m)$ will define xx'. $(-b_m, -a_m)$ will define $-x$, and whether x, x' are positive or negative, if (a_m, b_m) defines x and (a'_n, b'_n) defines x', then $(a_m + a'_m, b_m + b'_m)$ will define $x + x'$. Thus all the operations of addition, subtraction, multiplication and division are defined for the real numbers and can be shown to satisfy the fundamental rules. Full details are given by Knopp.*

Neither method proves the existence of irrational numbers, but both show that they can be used consistently and that any proposition proved by using them can be interpreted as a true proposition about rational numbers (usually, of course, much more complicated to state). In *Principia Mathematica* the aim is somewhat more ambitious: a real number is interpreted as a class of rationals (essentially the Dedekind L class) and meanings are given to the laws of algebra in terms of certain operations on these classes; and the laws so stated are proved to be true. In this sense there is an actual proof of the existence of irrationals satisfying the laws of algebra.

1·032. ϵ; indirect proofs. A peculiarity of the basic theorems about real numbers is that many of them seem incapable of direct proof. They are proved by the process known as *reductio ad absurdum*. We have to state the contradictory of the theorem and show that this itself leads to a contradiction; and then we argue that the theorem cannot be false and therefore must be true. But since most of the theorems have conclusions of the form $x = y$, their contradictories are inequalities of the form '$x < y$ or $x > y$'. Most beginners find it much more difficult to handle inequalities correctly than equalities, and of all the difficulties found in mathematical physics the greatest found by many students is in learning to approximate. That is why lower marks are obtained in problems of small oscillations in dynamics and of potentials of nearly spherical bodies than in any other part of the Mathematical Tripos. Nature does not consist entirely, or even largely, of problems designed by a Grand Examiner to come out neatly in finite terms, and whatever subject we tackle the first need is to overcome timidity about approximating. A difference between the theory of the real variable and dynamics is that in the former we are willing

* *Theory and Application of Infinite Series.*

to consider arbitrarily close approximations carried to any number of stages, whereas in the latter we only want an approximation close enough for the practical end in view. But experience in the one will tend to produce confidence in the other.

The simplest type of argument of this form is: if $x \geqslant 0$, and $x < \epsilon$, where ϵ is positive but can be chosen as small as we like, then $x = 0$. For no value of x greater than 0 can be less than *every* positive ϵ. An immediate extension is obtained by considering the *modulus* or *absolute value* of x, denoted by $|x|$ and read 'mod x'. This is equal to x when x is positive or zero, and to $-x$ if x is negative. It is therefore always $\geqslant 0$. Then if $|x| < \epsilon$ for all positive ϵ, $|x| = 0$ and therefore $x = 0$. Note that $|x| + |y| \geqslant |x+y|$, $|x-y| \geqslant |x| - |y|$.

It is necessary for this argument to use a symbol for the small quantity. If we said '$\epsilon = 0.001$', and proved that $|x| < 0.001$ by calculation, an objector might say 'you have not proved that $x = 0$; it might be 0.0001'. The symbol ϵ, to denote an *arbitrarily* small quantity, prepares us for such an objection, since by proving that $|x|$ is less than *any* ϵ we are ready to disprove any value of x, other than 0, that an objector might suggest.

The essential point is that we are concerned with processes that in the most general case could be completed only in an infinite number of steps, e.g. showing that two nests of intervals determine the same real number. We overcome this and obtain a finite proof by saying that if $a \neq b$, $|a-b|$ has a definite value M, which is not zero. If, then, we can show that $M < \epsilon$ for every positive ϵ, it follows that $M = 0$, contradicting the hypothesis, so that a and b must be equal.

1·033. Sets. A *limit-point* of a set of numbers is a number x such that for any $\epsilon > 0$ there is a member of the set, y, different from x, such that $|y-x| < \epsilon$. It follows that there are infinitely many values of y satisfying this condition. For by definition there is one; call this y_1 and take a new ϵ, say ϵ_1, less than $|y_1 - x|$. Then there must be another y of the set, say y_2, such that $0 < |y_2 - x| < \epsilon_1$. The process can evidently be continued indefinitely.*

Clearly no finite set can have a limit-point. But an infinite set also may have none; consider the set of all integers. No member has another within distance 1 of it, and no number not an integer can have more than one within distance $\frac{1}{2}$. In the set of rational numbers every member is a limit-point since there is a rational number as near as we like to any other. The same applies to the real numbers. A set may have only one limit-point; consider for instance the numbers n^{-1}, where n can be any integer. There are infinitely many within any finite distance from 0, which is therefore a limit-point; but around any other number, rational or not, we can take an interval that contains no member of the set, other than the number itself if it is a member. A limit-point of a set is not necessarily itself a member of the set. We can, for instance, make a set of rational numbers whose limit-point is $\sqrt{2}$ by taking the successive approximations to $\sqrt{2}$ by decimals, but $\sqrt{2}$ itself is not a rational number.

If all the limit-points of a set are themselves members of the set, the set is said to be *closed*. An interval $a \leqslant x \leqslant b$ as defined in 1·031 is a closed set and is called a closed interval. The corresponding open interval is $a < x < b$. We shall return to this distinction in 1·061.

1·034. *If a set has infinitely many members within a finite range $a \leqslant x \leqslant b$, then it has at least one limit-point x such that $a \leqslant x \leqslant b$.* For if we bisect the range, one half at least must

---

* It should be noticed that expressions such as 'the process can be continued indefinitely' and 'and so on' cover applications of *mathematical induction*. We shall seldom state such arguments in full, for reasons of space. The student should, however, complete some of them for himself for practice.

contain an infinite number of points of the set; bisect that half. One half again contains an infinite number, and we see that by repeating the process we can find an interval as small as we like containing an infinite number of points of the set. But this corresponds to the method of specifying a real number by a nest of intervals and therefore identifies a real number such that any small interval about it contains an infinite number of points of the set. It is therefore a limit-point of the set. This is known as the Bolzano-Weierstrass theorem.

1·035. An infinite set is *enumerable* if its members can be paired with the positive integers in such a way that to each member corresponds one and only one positive integer, and *vice versa*. Thus the squares $1^2, 2^2, \ldots, n^2, \ldots$ form an enumerable set, since to each n corresponds one n^2 and to each n^2 one n. The rational fractions between 0 and 1 form another, for they can be arranged $\frac{1}{2}, \frac{1}{3}, \frac{2}{3}, \frac{1}{4}, \frac{3}{4}, \frac{1}{5}, \frac{2}{5}, \frac{3}{5}, \frac{4}{5}, \ldots$, and the one that occurs in the nth place can be paired with n. The whole of the positive rationals form another, since they can be arranged $\frac{1}{1}, \frac{1}{2}, \frac{2}{1}, \frac{1}{3}, \frac{3}{1}, \frac{1}{4}, \frac{2}{3}, \frac{3}{2}, \frac{4}{1}, \ldots$ Here the numbers are arranged in groups, the sum of the numerator and denominator being the same for all in each group and greater by 1 than in the previous group, while those in each group are arranged in order of increasing numerator. In these two cases the comparison with the positive integers requires complete rearrangement from the natural order.

Not all infinite sets are enumerable. Far the most important exceptions are the set of all real numbers and the set of all real numbers within a given finite interval. Cantor proved that however we may try to put them into a one-one correspondence with the positive integers there will always be some omitted.

1·036. Necessary: sufficient. If two statements denoted by I and II are so related that if I is true, then II is true, we say that I is a *sufficient* condition for II and II is a *necessary* condition for I; that is, I cannot be true *unless* II is true. If II is true if and only if I is true, then I is a necessary and sufficient condition for II, and *vice versa*. In this case we may also say that I and II are *equivalent*.

In general if a necessary and sufficient condition can be stated for the truth of a given proposition several can. For instance, a necessary and sufficient condition that x, a real quantity, shall be 0 is $|x| < \epsilon$ for any assignable positive ϵ; but others are $x^2 = 0$ and $x^3 = 0$. A necessary and sufficient condition that $ax^2 - 2bx + c > 0$ for all x is that $a > 0$, $ac - b^2 > 0$; but another is that $c > 0$, $ac - b^2 > 0$.

A necessary and sufficient condition may contain superfluous information. For instance, if $ax^2 - 2bx + c > 0$ for all x, we must have $a > 0$, $c > 0$, $ac - b^2 > 0$, and conversely. Hence $a > 0$, $c > 0$, $ac > b^2$ is a necessary and sufficient condition. But if $ac > b^2$, either $a > 0$ or $c > 0$ implies the other and one of them is superfluous in the sense that it follows from the other information given. On the other hand either $a > 0$, $c > 0$, or $ac > b^2$ by itself would not guarantee that $ax^2 - 2bx + c > 0$ for all x: none of these conditions alone is sufficient. A set of necessary and sufficient conditions for the truth of a proposition is called *minimal* if the conditions left when any part of them is removed are not sufficient.

1·04. Sequences.* In considering the properties of a set we are not restricted to taking the members in any particular order. In the argument of 1·034, for instance, the

---
* Fuller discussions of sequences than are possible here will be found in K. Knopp's *Theory and Application of Infinite Series* and in Hardy's *Pure Mathematics*.

points actually in any range are determined by the specification of the set, just as, if we put some balls into a box, what balls are in the box has nothing to do with their rearrangement by shaking or sorting.

When we come to study properties essentially connected with a particular order we are dealing with *sequences*. The numbers 1, 2, 3, ... in ascending order constitute a sequence; if they were rearranged, but in such a way that we always knew where to find a particular one, they would form a different sequence but the same set. If we write s_n for the nth in a given arrangement, the property $s_{n+1} - s_n = 1$ is true for all n for the original order but for no other. In general *if s_n is completely specified when n is given, s_n may be described as a function of the positive integral variable n, and the values $s_1, s_2, ..., s_n, ...$, for successive values of n, form a sequence.* (Those who have some knowledge of series often suppose at first that the terms of a sequence are to be summed, but this is not so.) Both

$$1, \frac{1}{2}, \frac{1}{3}, ..., \frac{1}{n}, ... \qquad (1)$$

and
$$1, 2, 1, 2, 1, 2, ... \qquad (2)$$

are sequences. In the first the members are the members of an infinite set arranged in a certain order. In the second they are the members of a finite set repeated over and over again.

A sequence whose general term is s_n can be denoted by $\{s_n\}$.

1·041. Bounded, unbounded, convergent, oscillatory. Let M be an arbitrary positive number; it is possible that whatever M we take there is at least one value of s_n such that $|s_n| > M$. Such a sequence is called *unbounded*. $s_n = n$ is an obvious example, for we need only take n to be any integer greater than M. By an argument similar to that for limit-points, an unbounded sequence must have an infinite number of terms such that $|s_n|$ is greater than any assigned M.

If we can choose an M such that all $|s_n|$ are less than M, the sequence is called *bounded*. Both the sequences given at the end of 1·04 are bounded; the condition holds for both if $M = 3$.

If there is a number s such that, given any positive number ϵ, we can choose m so that for every $n > m$
$$|s_n - s| < \epsilon, \qquad (1)$$

the sequence is said to be *convergent*, and to have limit s. We then write*

$$s_n \to s \quad (n \to \infty), \qquad (2)$$

or
$$\lim_{n \to \infty} s_n = s.$$

The arrow is read 'tends to'. We can write simply
$$\lim s_n = s, \qquad (3)$$

if no ambiguity is possible. Of the above examples 1·04(1) is convergent with limit 0; we need only take $m > 1/\epsilon$. 1·04(2) is not, because whatever s and m we take, if $\epsilon < \frac{1}{2}$, there will be terms with $n > m$ such that $|s_n - s| \geq \frac{1}{2} > \epsilon$.

The most important property of a convergent sequence is that if we have a rule for calculating each term, then we can calculate the limit to any accuracy we like. Some

* J. G. Leathem, *Volume and Surface Integrals used in Physics*, 1905.

methods of approximation (cf. Chapters 9, 17) will prove that a quantity lies within a given range, but this range is not arbitrarily small; the accuracy may be enough for the application in view but is not capable of being improved indefinitely.

A sequence that is bounded but not convergent is said to *oscillate finitely*, or simply to *oscillate*. An example is 1·04 (2); another is

$$s_n = (-1)^n + \frac{1}{n}. \tag{4}$$

Unlike 1·04 (2), all s_n are different. The sequence is bounded, because $|s_n| < 2$ for every n; but it does not converge since for large n the members are alternately near to 1 and -1, and (1) cannot be satisfied if $\epsilon < \frac{1}{2}$.

If for any M there is an m such that $s_n > M$ for all $n > m$, we write

$$s_n \to \infty. \tag{5}$$

$s_n = n$ and $s_n = n^2$ are examples.

If for any M there is an m such that $s_n < -M$ for all $n > m$, we write

$$s_n \to -\infty. \tag{6}$$

$s_n = -n$ and $s_n = -n^2$ are examples.

Other types of unbounded sequences are represented by

$$s_n = (-1)^n n, \quad s_n = n \cos \tfrac{1}{2}\pi n, \quad s_n = n(1 - \cos \pi n).$$

These cannot be said to tend to anything particular, not even infinity, and are sometimes called *infinitely oscillating*. Unbounded sequences can be called *divergent*; but different writers use this term in different senses, some (e.g. Bromwich and Hardy) excluding infinitely oscillating sequences and some (e.g. Knopp) including finitely oscillating ones. A useful device is to classify sequences according as they have or have not the properties

(1) for any m, and any positive M, there is an $n > m$ such that $s_n > M$,

(2) for any m, and any positive M, there is an $n > m$ such that $s_n < -M$.

Sequences with neither property are bounded. If a sequence possesses (1) but not (2), it is *bounded below, unbounded above*, and similarly for the other two cases.

Note that no definite meaning is attached to *infinity* as such. What we do is to give meanings to all the expressions that *contain* the word *infinity* or the symbol ∞. $s_n \to \infty$ is a shorthand statement of the property of $\{s_n\}$ stated in the definition of '$s_n \to \infty$', and does not imply the existence of any real quantity denoted by ∞.

Infinity is excluded from the rules of algebra, not because there is any inconsistency in the notion of infinite numbers, but because they follow different rules. In fact the notion of an infinite set is implicit in most of our theory, since there are infinitely many values of x in any interval of x. A consistent algebra of positive infinite numbers was set up by Cantor, and has been extended by many later writers. But it is different from ordinary algebra. If a and b are positive infinite numbers we can define $a+b$ and ab uniquely; but $a+b$ need not be greater than a—in fact it is in general equal to a or to b. It is not possible to define $a-b$ and a/b uniquely. Consequently an algebra that includes both finite and infinite numbers must still distinguish between them in its rules.

1·042. *If an infinite set has a limit-point, s, then we can form a sequence from its members whose limit is s; if it has more than one limit-point we can form sequences tending to any of them.*

We have shown (beginning of 1·033) that there is an infinite number of members of the set within a given distance of a limit-point; if we take specimens in the order indicated we have a sequence with the property required.

1·043. *Any sequence formed of different members of a bounded infinite set with only one limit-point s will converge to the limit s.* It is clear that in forming a sequence from the set we have a choice at every stage; hence the number of different sequences that can be formed from the set is infinite. We have to show that they all have the same limit. For any m, the number of terms s_n of a sequence with $n > m$ is infinite. But since the set is bounded and has only one limit-point, any interval not including the limit-point can contain only a finite number of members. Hence for any ϵ only a finite number of members lie outside the range $s \pm \frac{1}{2}\epsilon$, say $s_\alpha, s_\beta, ..., s_\mu$. Let m be the greatest of $\alpha, \beta, ..., \mu$. Then for all n greater than m, $|s_n - s| \leq \frac{1}{2}\epsilon < \epsilon$, and therefore the sequence converges to s.

The result does not follow if the members of the sequence are not required to be different and some can recur infinitely often. For instance, if the set is that of the reciprocals of the integers, its only limit-point is 0; but if repetitions are allowed we can form from it the sequence

$$1, \frac{1}{2}, 1, \frac{1}{3}, 1, \frac{1}{4}, 1, ..., \frac{1}{n}, 1, ...,$$

which is oscillatory. If no member recurs more than a fixed number k times, however, the result still follows by a simple extension of the argument.

1·044. Upper and lower bounds. *A set (or sequence) bounded above has an upper bound; and one bounded below has a lower bound.* The *upper bound* of a set is a quantity M such that no member of the set exceeds M, but if ϵ is any positive quantity, however small, there is a member that exceeds $M - \epsilon$. The *lower bound* is a quantity m such that no member is less than m, but there is always one less than $m + \epsilon$.

We use the method of Dedekind section. There are quantities a such that a is exceeded by some member of the set; for we might take an a less than a known member of the set. Since the set is bounded above, there are quantities b that are not exceeded by any member of the set. Every b is greater than any a, and every quantity of the same dimensions is either an a or a b. Hence the quantities a form an L and b an R class, and determine a cut, say at M. M is a member of the R class. For if it was a member of the L class it would be exceeded by some member of the set, say K, and there would be no quantities b between M and K; hence M would not be the quantity given by the cut. Hence no member of the set exceeds M. Also $M - \epsilon$ is in the L class and therefore is exceeded by some member of the set. The corresponding result for lower bounds follows similarly.

The argument does not suppose the set infinite; but for a finite set the greatest of the set is the upper bound. For an infinite set all members may be less than the upper bound; for the set 1·04(1) the upper bound is 1 and is equal to the first term, but the lower bound is 0 and no actual member is 0.

What we call *the* upper bound is often called the *least upper bound*; and any quantity such that no member of the set exceeds it is then called *an* upper bound.

Note that if $s_n < t_n$ for all n, and $s_n \to s, t_n \to t$, then $s \leq t$, not $s < t$. Consider $s_n = 1 - 2^{-n}$, $t_n = 1 - 3^{-n}$. Here $s = t$. We may regard (s_n, t_n) as an interval whose length tends to zero, but these intervals do not constitute a nest because each is not part of its predecessor, and, in fact, the whole of each interval is on the same side of the limit.

1·0441. If $s_n \geqslant s_{n-1}$ for all n, and the sequence is bounded, then the sequence converges. Let the upper bound of s_n be s. Then for all n, $s_n \leqslant s$. But also for any ϵ there is an m such that $s_m > s - \epsilon$; and then for every $n > m$
$$s \geqslant s_n \geqslant s_m > s - \epsilon,$$
and therefore the sequence converges with limit s.

1·045. The general principle of convergence. *A necessary and sufficient condition for convergence of a sequence $\{s_n\}$ is that for any positive quantity ϵ there is an m such that for all $n \geqslant m$,*
$$|s_n - s_m| < \epsilon. \tag{1}$$

We show first that the condition is necessary. Suppose that $s_n \to s$. We have to show that m exists such that (1) is true. For any positive ω we can take m so that $|s_n - s| < \omega$ for all $n \geqslant m$. Then $|s_n - s_m| < 2\omega$ for all $n \geqslant m$. Take $\omega = \tfrac{1}{2}\epsilon$; then (1) follows.

To prove that the condition is sufficient, we notice first that the sequence is bounded, for, given any positive ω, there is an m such that $|s_n - s_m| < \omega$ for all $n \geqslant m$, and $s_1, s_2, \ldots, s_m$ are all finite. We define a_n and b_n as the lower and upper bounds respectively of s_p for $p \geqslant n$. Clearly
$$a_n \leqslant a_{n+1} \leqslant b_{n+1} \leqslant b_n.$$

Also since $|s_p - s_q| < 2\omega$ for p and q greater than m, we have that for $n \geqslant m$, $b_n - a_n \leqslant 2\omega$ since $b_n - a_n$ is the upper bound of $s_p - s_q$ for $p, q \geqslant n$. Since ω is arbitrarily small, $b_n - a_n \to 0$. The intervals (a_n, b_n) therefore form a nest, defining the real number s, say. Since
$$\left.\begin{array}{c} a_n \leqslant s_n \leqslant b_n \\ a_n \leqslant s \leqslant b_n \end{array}\right\} \text{ for all } n$$
and

we have
$$|s - s_n| \leqslant 2\omega \quad \text{for} \quad n \geqslant m,$$
that is, $s_n \to s$ as $n \to \infty$.

The device of introducing a subsidiary arbitrarily small positive quantity, usually denoted by ω, δ, or η, which is later defined as a fraction of ϵ, will be met frequently in theorems where the quantity to be proved less than ϵ is expressible as the sum of several parts.

1·05. Series. If the nth term of an infinite series is u_n, the sums
$$s_1 = u_1, \quad s_2 = u_1 + u_2, \quad s_3 = u_1 + u_2 + u_3, \quad \ldots, \quad s_n = u_1 + u_2 + \ldots + u_n, \quad \ldots,$$
constitute a sequence. If this sequence is convergent we say that the series
$$\Sigma u_n = u_1 + \ldots + u_n + \ldots,$$
where n is now made indefinitely great, is convergent; and we call the limit of s_n the sum of the series.* If $\{s_n\}$ is not convergent but finitely oscillating we shall speak of the series as finitely oscillating.

To every theorem about sequences corresponds one about series; for if $\{s_n\}$ is a sequence, and we take $u_1 = s_1$, $u_n = s_n - s_{n-1}$ for $n > 1$, $\sum_{1}^{n} u_r = s_n$.

---

* As for sequences, different definitions of *divergent* are in use; some writers restrict the term to cases where $s_n \to \infty$ or $s_n \to -\infty$, others call all non-convergent series divergent.

The geometric series is
$$\sum_{n=0}^{\infty} x^n = 1 + x + x^2 + \ldots.$$

Here, if $x \neq 1$,
$$s_n = \frac{1 - x^{n+1}}{1 - x},$$

and $|x^{n+1}|$ becomes indefinitely large with increasing n if $|x| > 1$. If $|x| < 1$, x^{n+1} tends to 0.* Hence the series is convergent if $|x| < 1$, but not if $|x| > 1$. If $x = 1$, the sum of n terms is n, and the series is not convergent. If $x = -1$ the sum of any odd number of terms is 1, but that of any even number of terms is 0. The series therefore oscillates finitely. A necessary and sufficient condition for the series to converge is therefore $|x| < 1$.

The Riemann ζ series is
$$\sum_{n=0}^{\infty} n^{-x} = 1 + \frac{1}{2^x} + \frac{1}{3^x} + \ldots.$$

First take $x > 1$. We can take the terms in batches:
$$s_n = 1 + \left(\frac{1}{2^x} + \frac{1}{3^x}\right) + \left(\frac{1}{4^x} + \frac{1}{5^x} + \frac{1}{6^x} + \frac{1}{7^x}\right) + \ldots + \left(\ldots + \frac{1}{n^x}\right)$$

and the sums in brackets after the first are respectively less than
$$\frac{2}{2^x}, \frac{4}{4^x}, \ldots, = \frac{1}{2^{x-1}}, \frac{1}{(2^{x-1})^2}, \ldots$$

Then, if $m = 2^{r-1}$, and $n \geqslant m$,
$$|s_n - s_m| < \frac{2^{-r(x-1)}}{1 - 2^{-(x-1)}},$$

which can be made $< \epsilon$ by taking r large enough. Hence the ζ series converges if $x > 1$.

If $x = 1$, we write
$$s_n = 1 + \tfrac{1}{2} + (\tfrac{1}{3} + \tfrac{1}{4}) + (\tfrac{1}{5} + \tfrac{1}{6} + \tfrac{1}{7} + \tfrac{1}{8}) + \ldots,$$

and all the sums in brackets exceed $\tfrac{1}{2}$. Hence the series is not convergent; $s_n \to \infty$. All the terms after the first are increased if $0 < x < 1$; hence again $s_n \to \infty$.

The related series for $\log 2$ is
$$1 - \tfrac{1}{2} + \tfrac{1}{3} - \tfrac{1}{4} + \ldots.$$

Here
$$s_n - s_m = \pm\left\{\left(\frac{1}{m+1} - \frac{1}{m+2}\right) + \left(\frac{1}{m+3} - \frac{1}{m+4}\right) + \ldots\right\},$$

and the sum in brackets is > 0 whether $n - m$ is even or odd. But also
$$s_n - s_m = \pm\left\{\frac{1}{m+1} - \left(\frac{1}{m+2} - \frac{1}{m+3}\right) - \left(\frac{1}{m+4} - \frac{1}{m+5}\right) - \ldots\right\},$$

and every expression in brackets ( ) is positive. Hence
$$0 < |s_n - s_m| < \frac{1}{m+1},$$

and this is less than ϵ for all $n > m$ if $(m+1) > 1/\epsilon$. Hence the series is convergent.

* Strict proofs of these apparently obvious statements will be found in Hardy's *Pure Mathematics*, pp. 134–5.

The argument can be adapted at once to show that *if $u_n > 0$, $u_n > u_{n+1}$ for all n, and $u_n \to 0$, then the series*
$$u_1 - u_2 + u_3 - u_4 + \ldots$$
is convergent.

1·051. Absolute convergence. If the series $\Sigma |u_n|$ converges, Σu_n converges; for the sum of any batch of terms u_m to u_n cannot have a modulus greater than the sum of the corresponding terms $|u_m|$ to $|u_n|$. In this case Σu_n is said to be *absolutely convergent*; if Σu_n is convergent but $\Sigma |u_n|$ is not, Σu_n is said to be *conditionally convergent*. (The word *semiconvergent* is sometimes used, but the prefix is misused, and the same word is also used for asymptotic series, which are best not regarded as infinite series at all. This word is therefore best avoided.)

We have seen that the series obtained from that for log 2 by taking all the signs positive is not convergent. Hence the series for log 2 is conditionally convergent. The geometric series, if convergent at all, is absolutely convergent.

1·052. Rearrangement of series. *The sum of an absolutely convergent series is unaltered by taking the terms in any order.* Let Σu_n be absolutely convergent, with sum s, and $\Sigma v_{n'}$ the same series, but with the terms differently arranged. It is understood that every term of either series appears in the other, but not in general in the same place. Take an arbitrary positive quantity ω and choose m so that $\sum_{n=m+1}^{\infty} |u_n| < \omega$; then the sum of the moduli of any batch of terms after the mth formed from the first series is less than ω. Take m' so that all the terms u_n up to u_m appear in the second series for values of n' less than m'. Write
$$s_m = u_1 + \ldots + u_m, \quad s'_{m'} = v_1 + \ldots + v_{m'}.$$
Then $s'_{m'} - s_m$ is the sum of a set of terms of the first series after the mth and its modulus is $< \omega$. Also if we take $n' \geqslant m'$, $s'_{n'} - s'_{m'}$ is the sum of another set of terms of the first series after the mth and therefore its modulus also is $< \omega$. Hence the second series is convergent. Let its sum be s'. Then
$$|s - s'| = |(s - s_m) - (s' - s'_{m'}) - (s'_{m'} - s_m)| < 3\omega,$$
and can therefore be proved less than any arbitrary ϵ by taking $\omega = \tfrac{1}{3}\epsilon$. Hence the two series have the same sum.

The theorem is not true of conditionally convergent series. It can be shown that if Σu_m is conditionally convergent we can rearrange it so as to make the sum anything we like. They have a precise meaning when the order of the terms is given, but not otherwise. They usually converge too slowly to be of much use for computation, but they can be used in theoretical work.

Tests for convergence based on the use of 'comparison series' are so closely related to tests for uniform convergence that we shall postpone them till we discuss the latter property (1·115, 1·117).

1·053. Double series. Similar remarks apply to *double series*, in which the general term is $u_{m,n}$. The condition of convergence is now that we can choose m, n so that for all p greater than m and all q greater than n, the sums $\sum_{r=1}^{p}\sum_{s=1}^{q} u_{r,s}, \sum_{r=1}^{m}\sum_{s=1}^{n} u_{r,s}$ differ by a quantity with modulus less than ϵ. Absolute and conditional convergence can be defined similarly, and it is again true that an absolutely convergent double series has the same sum however the terms are arranged. The proofs differ only in complexity from those for simple series.

1·06. Limits of functions: Continuity. In the most general sense, when we say that $f(x)$ is a function of x in some range of values of x we mean that for every value of x in the range one or more values of $f(x)$ exist. We can, for instance, speak of a function of x that is equal to 1 if x is rational but to 0 if x is irrational. Such a function would be fairly regarded by a physicist as pathological, and he is interested in a much narrower class of functions, roughly speaking such as can be represented by graphs.* It will usually also be required that the function shall be *single-valued*, but not necessarily. Thus for the circle

$$x^2 + y^2 = a^2,$$

we have
$$y = \pm \sqrt{(a^2 - x^2)},$$

and y is a function of x; but we get its values over the whole circle only by taking both signs for the root. A single-valued function of x in a range is one that has precisely one value for each value of x. We shall in the first place consider single-valued functions only.

The essential idea of a limit of a function is similar to that of the convergence of a sequence; for the terms of a sequence $\{s_n\}$ are the values of a function of the positive integral variable n, which is permitted to take arbitrarily large values. The new feature is that for a function $f(x)$ the variable x is not restricted to be integral; it may be permitted to take any value over an interval or even any value however large.

When the values of x form an interval we can define a *limit* of $f(\xi)$ as $\xi \to x$ as follows: if there is a quantity c such that given any positive ϵ there is a positive δ such that whenever $0 < |\xi - x| < \delta$, then $|f(\xi) - c| < \epsilon$, we say that c is the limit of $f(\xi)$ as $\xi \to x$. (We may further restrict the admissible values of ξ and, for instance, speak of the limit of $f(\xi)$ as $\xi - x \to 0$ through positive or negative values.) If also $c = f(x)$, we say that $f(\xi)$ is continuous at $\xi = x$. Then the definition of continuity may be stated as follows: *if for any positive ϵ we can choose a positive δ such that whenever $|h| < \delta$*

$$|f(a+h) - f(a)| < \epsilon$$

then $f(x)$ is said to be continuous at $x = a$. If this condition is satisfied and we take any sequence $\{h_n\}$ tending to 0, then for any δ there will be an m such that $|h_n| < \delta$ for all $n \geq m$, and then $|f(a + h_n) - f(a)| < \epsilon$. Hence for all such sequences $f(a + h_n) \to f(a)$.

Most functions met in practice are continuous, with at most a finite number of points of discontinuity. A common type of discontinuity is where $f(x+h)$, for some value of x, has one definite limit as $h \to 0$ through any set of positive values, and a different one as $h \to 0$ through any set of negative values. Such a case is called an *ordinary* or *simple discontinuity*. For instance, if

$$f(x) = 0 \quad (x < 0), \quad f(x) = 1 \quad (x > 0),$$

---

* This function is frequently used as a warning. It can be used for that purpose at once. We might try to define a pathological function as one that is neither a continuous function nor the limit of one. But nothing could be more ordinary than the function $\cos^{2n} m! \pi x$, which tends to this function when n first tends to infinity and then m does.

the limit of $f(h)$ as $h \to 0$ through any set of positive values is 1, and that as $h \to 0$ through any set of negative values is 0. This is a very common function in physical applications, since it represents, for instance, a force that begins to act on a system at a definite instant and thereafter is constant. It is usually known as the Heaviside unit function. The postage on a letter, considered as a function of weight, has simple discontinuities. The value at $x = 0$ usually does not need to be specified in experimental applications, because for an object to be visible it must have some size, and therefore if x is a position coordinate we cannot observe a quantity at an exact value of x, but only a mean value over a range. Similarly, if x is a time we cannot observe a quantity at a single moment but only over a non-zero interval. The usual tendency in pure mathematics is to insist that the function shall be specified for all values of the independent variable, but in physics it is usually enough that its integral shall be determinate. As the value of the function at a single point, provided it is finite, does not affect the integral, it is usually irrelevant to physical applications, and if a special value is assigned it is for the sake of convenience.

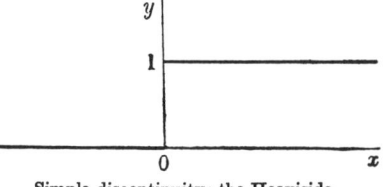

Simple discontinuity: the Heaviside unit function.

The notations, for $h > 0$,

$$\lim_{h \to 0} f(x+h) = f(x+), \quad \lim_{h \to 0} f(x-h) = f(x-),$$

are often used. Then the case we have been considering is one where

$$f(0+) \neq f(0-).$$

It may happen that $f(a+) = f(a-)$ but is not equal to $f(a)$. Such a function is said to have a *removable discontinuity*, but as $f(a)$ does not affect the integral such discontinuities are not of much importance. It is, of course, impossible to illustrate by a graph.

A limit will not exist at all if the function is unbounded in the neighbourhood of a value of x, as for $f(x) = 1/x$ near $x = 0$. For any sequence of values of x tending to 0, $f(x)$ will be unbounded. Again, if $f(x) = \sin(1/x)$, and x tends to 0 through the values $1/n\pi$, where n is an integer, the limit is 0. But if it tends to zero through the values $1/(n+\frac{1}{2})\pi$ it tends to $+1$ if n is restricted to be even and to -1 if n is restricted to be odd. This kind of misbehaviour is the most troublesome to detect when the definition of the function is at all complicated, and also it is the kind that is most easily forgotten.

The behaviour of $f(x)$ as $x \to \infty$ is even more closely analogous to that of a sequence, since in general $f(\infty)$ is not defined directly and we are concerned entirely with the limit itself, if it exists. We note only the definition and the principal criterion. *If there is a c such that for any $\epsilon > 0$ there is an X such that for all $x \geq X$ we have $|f(x) - c| < \epsilon$, then $f(x)$ is said to tend to c as $x \to \infty$.* Analogously to the general principle of convergence for series, we can show that *a necessary and sufficient condition that $f(x)$ may tend to a limit as $x \to \infty$ is that for any positive ϵ there is an X such that for all $x \geq X$, $|f(x) - f(X)| < \epsilon$.*

1·061. Continuity in an interval. $f(x)$ is said to be *continuous in an interval* if it is continuous at every point of the interval. $f(x)$ is continuous in the open interval $a < x < b$

if it is continuous for every value of x such that $a < x < b$. It is continuous in the closed interval $a \leqslant x \leqslant b$ if this condition is satisfied and also

$$f(a+) = f(a), \quad f(b-) = f(b).$$

Note that to say that $f(x)$ is continuous at $x = c$ implies that $f(c)$ is finite; for otherwise we could attach no meaning to $f(c+h) - f(c)$ at all. Similarly, if $f(x)$ is continuous in an interval it is finite at all points of the interval. We shall prove that in the latter condition it is bounded in the interval if this is closed, but not necessarily if it is open.

We denote *any* interval with end-points a, b by (a, b). When necessary we shall state explicitly whether $a < x < b$, $a \leqslant x \leqslant b$, $a < x \leqslant b$ or $a \leqslant x < b$ is to be understood.*

Note that every point x of an open interval is an interior point; that is, there are points y, z of the interval such that $a < y < x < z < b$. This is not true for a closed interval since x may then be equal to a or to b. But if a, b are finite they are limit-points of the set in $a < x < b$. Another way of expressing the distinction between closed and open intervals is to say that all limit-points of sets in a closed interval are members of the interval; those of sets in an open interval may not be, since those of some sets are the end-points. When we say that x is *within* an interval (or in later chapters within a region) we mean that it is an interior point; if we say that it is *of* a closed interval or region it may be an end or boundary point.

Functions that are continuous except at a finite number of points, where they have simple discontinuities, are called *sectionally continuous*.

A function is continuous if it is differentiable; the converse is not true, as we see from the example of $\sqrt{x}$ in $0 \leqslant x \leqslant 1$. This is continuous in the interval, including the end-points, but is not differentiable at 0.† Functions have actually been constructed that are continuous everywhere in an interval but differentiable nowhere. As a rule we shall be concerned with functions that are differentiable except possibly at isolated points, but such points are very numerous in crystal physics. There is a theorem of Weierstrass that any continuous function can be represented as closely as we like by a polynomial throughout any finite range, or by a sum of sines and cosines with suitable coefficients (cf. 14·08). Consequently, though a continuous function is not necessarily differentiable, it can be replaced with as much accuracy as we like by a function that is differentiable.

1·062. Covering theorems. We see that the property of continuity asserts that every point x of the interval (a, b) is in an interval $(x - \delta, x + \eta)$ (where δ and η may depend on x) such that (1) x is an interior point of the interval (except where $x = a$ or b, when it may be an end-point), (2) the length of the interval is not zero, (3) for every point ξ of the interval a certain property holds, in this case

$$|f(\xi) - f(x)| < \epsilon.$$

---

* Special notations are in use for open and closed intervals, and a common practice is to denote the open interval by (a, b) and the closed interval by $[a, b]$. In previous editions of this book ( ) and [ ] were used in the opposite senses.

† Nomenclature varies between different writers in such a case. However we choose x_n, positive and tending to zero, $(\sqrt{x_n} - 0)/x_n$ ultimately exceeds any given positive value. If $f(x) = x \sin(1/x)$, $f(x_n)/x_n$ can be made to tend to any limit between -1 and 1 by suitable choice of the x_n. In the latter case, $f'(x)$ is said *not to exist* at $x = 0$. For $f(x) = \sqrt{x}$, $f'(x)$ would be said by many writers to be *infinite* at $x = 0$. It is a matter of definition whether we say that $f'(x)$ does or does not exist when $\{f(x+h) - f(x)\}/h \to \infty$; we shall usually say that it does not.

We can show that in such circumstances it is possible to choose a *finite* number of intervals such that every interval satisfies such conditions and every point of (a,b) belongs to at least one of them. For different purposes we need to make the choice in somewhat different ways, and two theorems are therefore needed to show that it is possible.

1·0621. The Heine-Borel theorem. *If every point of a closed interval (a, b) is within some interval I of a family F, then there is a finite subfamily of F such that every point of (a,b) is within at least one interval of the subfamily.* We say that I covers (c, d) if every point of (c, d) is an interior point of I (i.e. not an end-point).

There may be an interval I belonging to F that covers the whole of (a, b). If so there is nothing to prove. If not, bisect (a, b). There may be a pair of intervals I_1, I_2 such that every point of $(a, \frac{1}{2}a + \frac{1}{2}b)$ is interior to I_1 and every point of $(\frac{1}{2}a + \frac{1}{2}b, b)$ to I_2. If either half is not included in an interval I, bisect that half. We say that in a finite number of steps we shall arrive at a stage where every portion of (a, b) lies within at least one interval I. For if not, the successive bisection of intervals will give a sequence of intervals, each part of the preceding one, and each half the length of the preceding one, and none of them included in an I. Such a sequence forms a nest of intervals and identifies a number x_0 common to all its members. But by hypothesis x_0 is interior to an I, say I_0, and hence there is a positive δ such that all points of $(x_0 - \delta, x_0 + \delta)$ are in I_0. Therefore all intervals of the nest whose lengths are less than δ are included in I_0, and we have a contradiction. Hence the process of bisection leads in a finite number of steps to a set of subdivisions such that every division of (a, b) is wholly interior to some I. Taking for each division an I that includes it we have the theorem.

A slight modification is often made where an end-point, say a, is an end-point of an interval of the family, say I_a, closed at a; I_a is still supposed of non-zero length δ_a. Then a is interior to the interval J_a $(a - \frac{1}{2}\delta_a, a + \frac{1}{2}\delta_a)$, and the argument applies to the set of intervals J, where J is the same as I except that I_a is replaced by J_a; every point of (a, b) is an interior point of at least one J. But then the theorem follows with the modification that a may be an end-point of I_a or b of I_b provided that I_a has a as a member and I_b has b.

The theorem gives the Bolzano-Weierstrass theorem (1·034) as a special case. If possible, let (a, b) contain no limit-point of the set. Then every point of (a, b) is in an interval I containing not more than one member of the set. Hence (a, b) can be covered by a finite set of such intervals I and therefore contains only a finite number of members of the set of points considered, contrary to hypothesis.

In the argument as we have stated it the only intervals bisected at each stage are those not already covered by an I. We could, however, equally well bisect all the intervals. For if I covers (c, d) it covers both halves of it. Hence (a, b) in the conditions stated can be divided into a finite set of *equal* intervals each covered by an I.

1·0622. The modified Heine-Borel theorem. In the Heine-Borel theorem the intervals I may be specified by any rule so long as each is of non-zero length and every point of (a, b) is an interior point of at least one of them (except that a and b may be end-points). Sometimes, however, a further restriction is made, according to which each point x of (a, b) specifies an I_x, of which x is an interior point. Then the following theorem holds. *Suppose that every point x of $a \leqslant x \leqslant b$ is within an interval I_x $(x - \delta_x, x + \eta_x)$, where $\delta_x > 0$,*

$\eta_x > 0$, except that I_a may be $a \leqslant x < a + \eta_a$ and I_b may be $b - \delta_b < x \leqslant b$; then (a, b) may be divided into a finite set of intervals such that each interval is part of the I_x defined for some point of that interval. The proof is by successive bisection as before. Assuming the theorem false, we establish the existence of a nest of intervals converging to some x_0, such that none is part of I_x for any x within that interval; but all of them less than a certain length are parts of I_{x_0}, and contain x_0, and we have a contradiction. In this case, however, it does not follow that (a, b) can be divided into *equal* intervals with the required property. If (c, d) is covered by I_x, where x is in (c, d), x can be interior to only one half of (c, d); then the other half is not necessarily covered by an I_y, when y is now restricted to be in that half.

An important application is to differentiable functions. Let $f(x)$ be differentiable at all points of $a \leqslant x \leqslant b$; this says that for any ω, for any x in (a, b), there is a positive $\delta(\omega, x)$ such that for $|h| < \delta$
$$|f(x+h) - f(x) - hf'(x)| < \omega |h|. \tag{1}$$
Then (a, b) can be divided into a finite set of intervals (x_r, x_{r+1}) such that for all x of (x_r, x_{r+1})
$$|f(x) - f(\xi_r) - (x - \xi_r) f'(\xi_r)| < \omega |x - \xi_r|,$$
ξ_r itself being a point of (x_r, x_{r+1}).

For any fixed η (1) remains true if all $\delta(\omega, x)$ are restricted to be $\leqslant \eta$. Then all $x_{r+1} - x_r$ will be $\leqslant 2\eta$.

Heine* proved that a continuous function is uniformly continuous (1·071) by what was essentially a method of Dedekind section, capable of being used to prove the general Heine-Borel theorem and so used in Lebesgue's proof. The specific use of overlapping intervals is due to Borel†, the form of the Heine-Borel theorem given here to W. H. Young.‡ The bisection method was used by Bolzano; Goursat (see 11·043) used it in an important simplification of the conditions for Cauchy's theorem, in which he recognized the effect of the restriction when each section is required to contain a point x with which the I_x covering that section is associated. He did not, however, give the general form of the modified theorem or comment on the possibility of proving the main theorem by the same method. This was first done by H. F. Baker in a note reported in title only.§

1·063. *A function continuous in $a \leqslant x \leqslant b$ is bounded in $a \leqslant x \leqslant b$.* Take an arbitrary positive ϵ. For every x of (a, b) there is an interval $I_x = (x - \delta_x, x + \eta_x)$ such that for every ξ of this interval
$$|f(\xi) - f(x)| < \epsilon. \tag{1}$$
Here $\delta_a = 0$, $\eta_b = 0$, otherwise $\delta_x, \eta_x > 0$. Therefore for every ξ_1, ξ_2 of this interval
$$|f(\xi_1) - f(\xi_2)| < 2\epsilon. \tag{2}$$
Then we can divide (a, b) into a finite number, say n, of intervals (x_r, x_{r+1}) such that for all ξ_{r1}, ξ_{r2} of each interval, including the end-points,
$$|f(\xi_{r1}) - f(\xi_{r2})| < 2\epsilon. \tag{3}$$
Hence for any x of (a, b)
$$|f(x) - f(a)| < 2n\epsilon, \tag{4}$$
and therefore $f(x)$ is bounded (above and below) in $a \leqslant x \leqslant b$.

---

* J. reine angew. Math. 74, 1871, 188.
† The I's necessarily overlap; for if I_x contained no points interior to any other, its end-points would not be interior to any I at all.
‡ Proc. Lond. Math. Soc. (1) 35, 1903, 384–8.
§ Proc. Lond. Math. Soc. (1) 35, 1903, 459.

It follows that $f(x)$ has upper and lower bounds in $(a \leqslant x \leqslant b)$. If $f(x)$ has upper bound M and lower bound m in an interval (irrespective of whether $f(x)$ is continuous) we call $M - m$ the *leap* of $f(x)$ in the interval.*

1·064. *A function continuous in a closed interval (a, b) attains in that interval its upper and lower bounds in (a, b) and every value between them.*

Let m, M be the lower and upper bounds of $f(x)$ in (a, b). Let c be any value *not* taken by $f(x)$ in (a, b). Then for any x of (a, b) there is an interval $I_x = (x - \delta_x, x + \delta_x)$, such that for all ξ common to (a, b) and I_x

$$|f(\xi) - f(x)| < \tfrac{1}{2}|f(x) - c|$$

since $f(x)$ is continuous and $|f(x) - c|$ positive. Hence at points common to (a, b) and I_x

$$|f(\xi) - c| > \tfrac{1}{2}|f(x) - c| > 0$$

and $f(\xi) - c$ has the sign of $f(x) - c$. Therefore, by the Heine-Borel theorem, (a, b) can be covered by a finite set of *overlapping* intervals I_x, say $I_{x_1}, ..., I_{x_m}$. Therefore (1) the lower bound of $|f(\xi) - c|$ in (a, b) is $\geqslant \tfrac{1}{2}|f(x_r) - c|$ for some r; none of these is zero and therefore c is not the upper or the lower bound of $f(x)$ in (a, b); (2) $f(\xi) - c$ preserves the same sign throughout (a, b) and therefore c is not between m and M.

The need for the restriction to continuous functions is made clearer by considering the function $f(x) = x$ $(0 \leqslant x < \tfrac{1}{2})$, $f(x) = 0$ $(\tfrac{1}{2} \leqslant x \leqslant 1)$. This has upper bound $\tfrac{1}{2}$, but $f(x)$ is never equal to $\tfrac{1}{2}$.

In 1·063 and 1·064 the interval must be closed. Take $f(x) = 1/x$ in $0 < x < 1$; this is continuous at every point of the interval but is unbounded. If $f(x) = x$ for $0 \leqslant x < 1$, the upper bound is 1, since for any $\eta < 1$ there is an $x < 1$ such that $f(x) > \eta$, but $f(x)$ is not equal to 1 for any $x < 1$.

1·065. Increasing and decreasing functions. A function is called *increasing* in an interval $a < x < b$ if for any x_1, x_2 such that $a < x_1 < x_2 < b$, $f(x_1) < f(x_2)$. It is called *decreasing* if, when $a < x_1 < x_2 < b$, $f(x_1) > f(x_2)$. A *non-decreasing* function is one such that $f(x_1) \leqslant f(x_2)$; similarly for a *non-increasing* function. Such functions may be constant for some parts of the interval; increasing and decreasing functions are nowhere constant. Increasing and decreasing functions are together called *monotonic*.†

1·066. Inverse functions. If $y = f(x)$ is continuous and monotonic in a closed interval (a, b) it takes once, and only once, every value between its upper and lower bounds. Hence there is a single-valued inverse function $x = g(y)$, which is also monotonic. (The condition that $f(x)$ is monotonic is necessary, for if $f(x)$ was constant in an interval, or if it was decreasing in part of the interval and increasing in another part, it could take the same value more than once, and $g(y)$ would not be single-valued.)

The inverse function $g(y)$ is continuous. This says that, for any y and a given ϵ there is a δ such that if $|\eta - y| < \delta$ then $|g(\eta) - g(y)| < \epsilon$; that is, for any x and given ϵ there is

---

* The name *oscillation* is in use. This strikes us as unfortunate because it is applied to functions that do not oscillate; when we describe a *sequence* as oscillatory there is some resemblance to what a physicist means by oscillation. *Saltus* is used by Hobson and *leap* by Newman.

† In many works what we call an increasing function is called a *strictly* increasing function, and what we call a non-decreasing one is called an increasing one. Similarly what we have called a monotonic function is often called a strictly monotonic one.

a δ such that if $|f(\xi)-f(x)|<\delta$ then $|\xi-x|<\epsilon$. To prove this we take for definiteness $f(x)$ to be increasing. Whenever $\xi_1 \leqslant x-\epsilon$ and $\xi_2 \geqslant x+\epsilon$ we have

$$f(\xi_1) \leqslant f(x-\epsilon) < f(x) < f(x+\epsilon) \leqslant f(\xi_2).$$

Then if $|f(\xi)-f(x)|$ is less than the smaller of $|f(x)-f(x-\epsilon)|$ and $|f(x)-f(x+\epsilon)|$ it follows that
$$x-\epsilon < \xi < x+\epsilon.$$

A many-valued function can often be regarded as a set of single-valued ones. Thus for any $x > 0$ there are two admissible values of $\sqrt{x}$. But if we agree to take always the positive root or always the negative one we get in either case a single-valued continuous function of x. The theorems for continuous functions will then apply to either of these separately, but having decided which to take we must not change our minds.

1·07. Uniformity of continuity. In general if we choose δ so that

$$|f(x+h)-f(x)|<\epsilon \quad \text{if} \quad |h|<\delta, \tag{1}$$

for some particular value of x, it will be found that for some other values of x and the same ϵ the inequality will not be satisfied for the same value of δ. For instance, let

$$f(x) = x^2 \quad (0 \leqslant x \leqslant 1). \tag{2}$$

If $x = 0$, (1) will be true if $\delta = \sqrt{\epsilon}$. But if $x = 1$
$$|(1-h)^2 - 1| = |2h-h^2|$$

which will not be less than ϵ if h is, say, $\tfrac{1}{2}\sqrt{\epsilon}$ and ϵ is small enough. But if we take $\delta = \tfrac{1}{2}\epsilon$, (1) will be true for all x in the range.

This brings us to the idea of *uniformity*, which we shall meet again and again. (1) specifies an inequality that is satisfied for some δ for every ϵ and x, but δ for given ϵ may depend on x, and can be written $\delta(\epsilon, x)$. A proposition (here $|f(x+h)-f(x)|<\epsilon$) is said to hold *uniformly* with regard to a variable (here x) if a condition for its truth can be stated so as not to depend on x; thus here $|h|<\delta(\epsilon)$, where $\delta(\epsilon)$ may depend on ϵ but not on x.

1·071. *A continuous function is uniformly continuous in any closed interval.* In the argument of 1·063, which applies to any value of ϵ, replace ϵ by ω and let $\tfrac{1}{2}\delta$ be the length of the shortest interval (x_r, x_{r+1}). Then any two points ξ_1, ξ_2 of (a,b) such that $|\xi_2-\xi_1|<\delta$ must belong to the same or adjacent intervals, and therefore

$$|f(\xi_2)-f(\xi_1)|<4\omega.$$

Take $\omega = \tfrac{1}{4}\epsilon$; then there is a δ such that whenever ξ_1, ξ_2 are points of (a,b) satisfying $|\xi_2-\xi_1|<\delta$
$$|f(\xi_2)-f(\xi_1)|<\epsilon.$$

1·08. Orders of magnitude. If as x tends to a limit $\phi(x)$ tends to 0 or ∞, and $f(x)/\phi(x)$ is bounded, we say that $f(x) = O\{\phi(x)\}$, or that $f(x)$ is of the same order of magnitude as $\phi(x)$. If $f(x)/\phi(x) \to 0$ as $\phi(x) \to 0$ we write $f(x) = o\{\phi(x)\}$. If $f(x)$ is bounded we can write $f(x) = O(1)$. This notation must be distinguished from the common usage in physics, where we may say that the masses of Jupiter and Saturn are of the same order of magnitude, meaning roughly that they differ by not more than a factor 10 without there being any question of a limit. In the physical sense the quantities compared must have the same dimensions. This is not necessary in the mathematical sense. x may, for instance, be a time-interval and $f(x)$ the distance travelled by a sound wave. Then $f(x) = O(x)$ because $f(x)/x$ is the velocity of sound and is supposed finite.

Note that $\qquad O(x^m) O(x^n) = O(x^{m+n}), \quad o(x^m) O(x^n) = o(x^{m+n}).$

1·09. Functions of bounded variation. If the function $f(x)$ is defined in the closed interval (a, b), and there is a number M such that

$$v = |f(x_1) - f(x_0)| + |f(x_2) - f(x_1)| + \ldots + |f(x_n) - f(x_{n-1})| \leq M,$$

for every subdivision $a = x_0 < x_1 < x_2 < \ldots < x_{n-1} < x_n = b$, $f(x)$ is said to be of *bounded variation* in the interval (a, b); and the upper bound of the sums v for all possible selections of the subdivisions is called the *total variation* of $f(x)$ in the interval.* The total variation is of interest since it is related to the condition for existence of a Stieltjes integral (1·102), and to the existence of the length of a curve, and it is useful in the theory of Fourier series and Fourier integrals.

We assume repeatedly that the sum and product of two continuous functions (and therefore of any finite number) are continuous, and that those of two functions of bounded variation are of bounded variation. The proofs are simple: for the last, notice that

$$f(x_{r+1}) g(x_{r+1}) - f(x_r) g(x_r) = \tfrac{1}{2} \{f(x_r) + f(x_{r+1})\} \{g(x_{r+1}) - g(x_r)\}$$
$$+ \tfrac{1}{2} \{g(x_r) + g(x_{r+1})\} \{f(x_{r+1}) - f(x_r)\}$$

and it follows that if M, N are the upper bounds of $|f(x)|$, $|g(x)|$, and U, V the total variations of $f(x), g(x)$ in the interval, the total variation of $f(x) g(x)$ is not greater than $MV + NU$.

Note that it is not satisfactory to define the total variation as the limit of the sum given, for there may be no limit for some ways of making the subdivision, or different ways may give different limits. Take for instance
$$f(x) = 0 \, (0 \leq x < \tfrac{1}{2}); \quad f(\tfrac{1}{2}) = 1; \quad f(x) = 0 \, (\tfrac{1}{2} < x \leq 1).$$
But the limit, if it exists, does give the total variation if the function is continuous or monotonic.

1·091. *If a function has bounded variation it need not be continuous*, or conversely. For if $f(x) = 0$ for $x \leq 0$ and $= 1$ for $x > 0$, the variation does not exceed 1 in any interval; but $f(x)$ is discontinuous. Conversely, if $f(x) = x \cos 1/x$ for $x \geq 0$, and $f(0) = 0$,

$$f\left(\frac{1}{n\pi}\right) = \frac{1}{n\pi}(-1)^n.$$

The variation between $x = \dfrac{1}{n\pi}$ and $x = \dfrac{1}{(n+1)\pi}$ is therefore at least $\dfrac{1}{n\pi} + \dfrac{1}{(n+1)\pi}$, and that between $x = 1/\pi$ and $x = 1/n\pi$

$$\geq \frac{1}{\pi}\left(1 + \frac{2}{2} + \frac{2}{3} + \ldots + \frac{2}{n-1} + \frac{1}{n}\right),$$

which tends to infinity with n. Hence $f(x)$ has not bounded variation. But $f(x)$ is seen to be continuous even at $x = 0$, since $|f(x)| \leq x, f(0) = 0$.

* Some authors use *fluctuation* instead of *variation*.

1·092. *Any function of bounded variation in (a, b) is the difference of two bounded non-decreasing functions.*

For any closed interval (a, x) we consider the sum

$$p = \Sigma \{f(x_r) - f(x_{r-1})\},$$

where $x_0 = a \leqslant x_1 \leqslant x_2 \ldots \leqslant x_{k-1} \leqslant x_k = x$

taken over all terms $f(x_r) - f(x_{r-1})$ that are positive, and

$$-n = \Sigma \{f(x_r) - f(x_{r-1})\}$$

taken over all negative terms. The upper bound of p over all possible subdivisions is the positive variation $P(a, x)$ in (a, x) and the upper bound of n is the negative variation $N(a, x)$. Let $v = p + n$; then the values of v are a bounded set since they are all $\leqslant V(a, b)$. Their upper bound $V(a, x)$ is the total variation of $f(x)$ in (a, x). Also by taking upper bounds

$$V(a, x) = P(a, x) + N(a, x).$$

Evidently $P(a, x)$, $N(a, x)$, $V(a, x)$ are all non-decreasing functions of x and are bounded in (a, b).

For any subdivision and for any fixed x

$$p - n = f(x) - f(a), \quad p + n = v.$$

Hence $\quad p = \tfrac{1}{2}v + \tfrac{1}{2}\{f(x) - f(a)\}, \quad n = \tfrac{1}{2}v - \tfrac{1}{2}\{f(x) - f(a)\}.$

Take upper bounds over all possible subdivisions; then

$$P(a, x) = \tfrac{1}{2}V(a, x) + \tfrac{1}{2}\{f(x) - f(a)\}, \quad N(a, x) = \tfrac{1}{2}V(a, x) - \tfrac{1}{2}\{f(x) - f(a)\}.$$

Hence $\quad f(x) = \{f(a) + P(a, x)\} - N(a, x),$

so that $f(x)$ is expressed as the difference of two bounded non-decreasing functions.

1·093. *All discontinuities of a function of bounded variation are simple or removable.* The characteristic feature of a simple discontinuity at $x = a$ is that $f(a-)$ and $f(a+)$ exist and are different. That of a removable discontinuity is that they exist and are equal, but not equal to $f(a)$. Suppose if possible that one of them does not exist; that is, there are two quantities M, m $(M > m)$ such that in any interval, however short, on one side of a there are points where $f(x) > M$ and points where $f(x) < m$. Let ξ_1 be a point where $f(x) > M$. Then there is a point between a and ξ_1, say ξ_2, where $f(x) < m$; then there is a ξ_3 between a and ξ_2 where $f(x) > M$, and so on. It follows that the total variation in the interval (a, ξ_1) is unbounded, and therefore $f(x)$ is not of bounded variation.

Alternatively, let $f(x)$ be of bounded variation in (a, b) and consider the positive variation $P(x)$ in (a, x). This is a bounded non-decreasing function of x and therefore has limits (not necessarily equal) as $x \to c$ (in (a, b)) through larger or smaller values. The same holds for the negative variation. Hence by subtraction $f(x)$ has limits as $x \to c$ through larger or smaller values, and therefore c is either a point of continuity or a simple or removable discontinuity.

If $f(c+)$ exists we may speak of the variation in a half-closed interval (c, d) on the right of c, meaning the variation of $g(x)$ in (c, d), where $g(c) = f(c+)$ and otherwise $g(x) = f(x)$. Then this variation tends to zero as $d \to c$. Similarly, we can define a variation in an interval on the left of c, with the same property.

Integration

1·094. Leap at a discontinuity. Let $f(x)$ be discontinuous at a but bounded in an interval including a as an interior point. Then for some positive δ, $f(x)$ has upper and lower bounds M, m in $(a-\delta, a+\delta)$. If $\delta' < \delta$, the upper bound in $(a-\delta', a+\delta')$ cannot be greater, or the lower less, than in $(a-\delta, a+\delta)$. Hence the leap in $(a-\delta, a+\delta)$ has a non-negative limit as $\delta \to 0$. If this limit is zero the function is continuous at a; if positive, we call the limit the leap of the function at a.

If $f(x) = 0$ $(x < 0)$, $f(x) = 1$ $(x \geqslant 0)$, the leap at 0 is 1. If $f(x) = 0$ $(x \neq 0)$, $f(x) = 1$ $(x = 0)$ the leap at 0 is again 1. If $f(x) = \sin 1/x$, the leap at $x = 0$ is 2, since values arbitrarily near 1 and -1 occur in any interval about 0.

1·10. Integration: Riemann, Stieltjes. Two different definitions of an integral will be used in this book.

Let $x_1, x_2, \ldots x_n$ be a set of increasing values of x between a and b, subject to all $x_{r+1} - x_r < \delta$ (we take when convenient $x_0 = a$, $x_{n+1} = b$). Take in each interval a ξ_r, so that $x_r \leqslant \xi_r \leqslant x_{r+1}$; and form the sum

$$S_n = f(\xi_0)(x_1 - a) + f(\xi_1)(x_2 - x_1) + \ldots + f(\xi_n)(b - x_n). \tag{1}$$

This sum will depend both on the values chosen for the x_r and on those for ξ_r, unless $f(x)$ is constant; but if we take a sequence of values of δ tending to zero, taking at every stage x_r and ξ_r in accordance with the inequalities, and form the sum S_n for each, these sums may tend to a limit, and this limit may be independent of the choice of the x_r and ξ_r at each stage. If so, this limit is called the *Riemann integral* and denoted by

$$\int_a^b f(x)\,dx. \tag{2}$$

It is also possible to integrate with respect to a function. If $f(x)$ and $g(x)$ are both bounded functions of x, we form the sum

$$S_n = f(\xi_0)\{g(x_1) - g(a)\} + f(\xi_1)\{g(x_2) - g(x_1)\} + \ldots + f(\xi_n)\{g(b) - g(x_n)\}, \tag{3}$$

the ξ_n being chosen as in (1). If this sum tends to a unique limit when the greatest interval of x tends to zero, the limit is called a *Stieltjes integral** and denoted by

$$\int_{x=a}^{b} f(x)\,dg(x). \tag{4}$$

The method of writing the termini needs attention because $g(x)$ may not be monotonic. It might return to its original value, but we must not write the range of integration as $g(a)$ to $g(a)$, which would apparently make the integral zero. It is x, not $g(x)$, that is required to increase steadily throughout the range.

1·101. *The Riemann integral* $\int_a^b f(x)\,dx$ *exists if and only if $f(x)$ is bounded in (a, b) and, for any positive values of ω and η, (a, b) can be divided into a finite set of intervals such that those where the leap of $f(x)$ is $\geqslant \omega$ have a total length $< \eta$.*

First, it is clearly necessary to the existence of the integral that $f(x)$ shall be bounded. For if $f(x)$ is unbounded in (a, b) there is always at least one interval (x_r, x_{r+1}) where it is

---

* T. J. Stieltjes, *Ann. d. Fac. d. Sciences*, Toulouse, 8, 1894, J., 68–75; also D. V. Widder, *The Laplace Transform*, 1941, Chapter 1; S. Pollard, *Q. J. Math.* 49, 1923, 73–138.

unbounded; and therefore the possible values of S_n formed with different choices of ξ_r in that interval are unbounded. Hence, however we choose the intervals, S_n cannot tend to a unique limit irrespective of the choice of ξ_r at each stage.

Suppose then that the upper and lower bounds of $f(x)$ in (a,b) are M, m. Suppose also that the upper and lower bounds in (x_r, x_{r+1}) are M_r, m_r, so that for any choice of ξ_r we have $m_r \leqslant f(\xi_r) \leqslant M_r$. Form the sums

$$h_n = \Sigma m_r(x_{r+1}-x_r), \quad H_n = \Sigma M_r(x_{r+1}-x_r). \tag{1}$$

These will be called the lower and upper sums for the subdivision specified by the points x_r and are the lower and upper bounds of S_n for that subdivision.

Now in any interval (x_r, x_{r+1}) there will be a value of x where $f(x) \geqslant \tfrac{3}{4}M_r + \tfrac{1}{4}m_r$ and a value where $f(x) \leqslant \tfrac{1}{4}M_r + \tfrac{3}{4}m_r$. Hence if $M_r - m_r \geqslant \omega$ it will be possible to make such choices of ξ_r that the corresponding values of $f(\xi_r)(x_{r+1}-x_r)$ differ by at least $\tfrac{1}{2}\omega(x_{r+1}-x_r)$. Then, since the choices of ξ_r in all intervals are made independently, if the intervals where $M_r - m_r \geqslant \omega$ have total length $\geqslant \eta$, where η is positive, we can make two sets of choices of the ξ_r in each interval such that the corresponding values of S_n differ by at least $\tfrac{1}{2}\omega\eta$. If then there are $\omega > 0$, $\eta > 0$ such that for any subdivision of (a,b) the total length of intervals where the leap of $f(x)$ is $\geqslant \omega$ is always at least η, S_n cannot have a unique limit. Hence the condition is necessary.

Since $m_r \leqslant M$, $M_r \geqslant m$, we have always

$$h_n \leqslant M(b-a), \quad H_n \geqslant m(b-a). \tag{2}$$

Hence the values of h_n given by all possible subdivisions have an upper bound, say h; and the values of H_n have a lower bound, say H. We show first that $h \leqslant H$.

If in any interval (x_r, x_{r+1}) we insert a further point of subdivision, say x_{r1}, and again form the lower and upper sums, the upper bound of $f(x)$ in either part may be less than M_r but cannot exceed it. Hence insertion of new points of subdivision may decrease the upper sum but cannot increase it; and similarly may increase the lower sum but cannot decrease it.

Now consider two different modes of subdivision specified by points x_r, x'_s. Let the respective sums be H_n, h_n, H'_p, h'_p. Consider the subdivision formed by taking all the points of both subdivisions together. Let the sums for it be H''_q, h''_q. It may be regarded as a subdivision of either the x_r or the x'_s set. Hence, by the last paragraph,

$$H_n \geqslant H''_q \geqslant h''_q \geqslant h'_p. \tag{3}$$

Thus it is impossible for any lower sum to exceed an upper sum, and therefore for all n, $H_n \geqslant h$ and therefore $H \geqslant h$.

Again, if we can find a method of subdivision such that $H_n - h_n < \epsilon$, it will follow that $H - h < \epsilon$; for $H_n - h_n = (H-h) + (H_n - H) + (h - h_n)$, and $H_n - H \geqslant 0$, $h - h_n \geqslant 0$. Now suppose that the intervals are classified into A intervals, where $M_r - m_r < \omega$, and B intervals, where $M_r - m_r \geqslant \omega$. In the B intervals we still have $M_r - m_r \leqslant M - m$. If α is the total length of the B intervals, we have

$$H_n - h_n < (b-a-\alpha)\omega + (M-m)\alpha. \tag{4}$$

Assume now that for any ω the total length of the B intervals can be taken arbitrarily small. Then for any positive ϵ we can take ω, α so that

$$(b-a)\omega < \tfrac{1}{2}\epsilon, \quad (M-m)\alpha < \tfrac{1}{2}\epsilon. \tag{5}$$

It follows that $0 \leqslant H - h < \epsilon$, and therefore, since H, h are independent of ϵ, that

$$H = h. \tag{6}$$

We have still to show that if we take methods of subdivision such that the length of the longest interval is δ, and we make $\delta \to 0$, then $H_n \to H$, $h_n \to h = H$. Let x_r give a set of points of subdivision satisfying (5), so that $H_n - h_n < \epsilon$. In this subdivision let the shortest interval be δ, and consider another subdivision by points x'_s such that the longest interval is less than δ. Let this give upper and lower sums H'_p, h'_p. Then any consecutive points x'_s, x'_{s+1} either belong to the same interval of the x_r set or to two adjacent ones. If the latter are both A intervals the leap is less than 2ω; if both B intervals, or if one is an A and one a B interval, it cannot exceed $M - m$. But if a B interval is of length $\mu \geqslant \delta$, the length of the x'_s intervals that have common points with it cannot exceed $\mu + 2\delta \leqslant 3\mu$. Hence the leaps of $f(x)$ in the x'_s intervals are $< 2\omega$ except possibly in a set of total length $\leqslant 3\Sigma\mu = 3\alpha$. Thus

$$H'_p - h'_p < (b - a - 3\alpha)\, 2\omega + 3(M - m)\,\alpha < \tfrac{5}{2}\epsilon. \tag{7}$$

Also $H'_p \geqslant H$, $h'_p \leqslant H$; hence

$$H'_p - H < \tfrac{5}{2}\epsilon, \quad H - h'_p < \tfrac{5}{2}\epsilon. \tag{8}$$

Since this is true for all subdivisions such that the longest interval is less than δ, the result follows.

Finally, since $h_n \leqslant S_n \leqslant H_n$, S_n also tends to H.

1·1011. The condition is due to du Bois-Reymond. It can be stated in an alternative form, which is sometimes more convenient. *A necessary and sufficient condition for the existence of $\int_a^b f(x)\,dx$ is that $f(x)$ is bounded and that for any positive ω, η the discontinuities where the leap is $\geqslant \omega$ can be enclosed in a finite set of intervals of total length $< \eta$.* Du Bois-Reymond's condition clearly implies this. Conversely, if the condition just stated is satisfied, there are no discontinuities where the leap is $\geqslant \omega$ in the remaining intervals. Then about any point in the remaining intervals there is an interval where the leap is $< \omega$. Hence, by the modified Heine-Borel theorem, the remaining intervals can be divided into a finite set such that the leap is $< \omega$ in each.

1·1012. An immediate consequence is that any continuous function has a Riemann integral; for it is bounded and has no discontinuities at all. Also any function with a finite number of finite discontinuities has a Riemann integral. The same applies to any function of bounded variation. For if, for some ω, there were an infinite number of discontinuities where the leap is greater than ω, it would not be of bounded variation.

Note that the condition does not require the number of discontinuities to be finite. Take $f(x) = 1$ when $x = 1/n$, where n is any positive integer, and otherwise zero. This is discontinuous whenever $x = 1/n$, and also at $x = 0$. But for any η the interval $(0, \tfrac{1}{2}\eta)$ contains an infinite number of discontinuities, and the remainder, with $\tfrac{1}{2}\eta \leqslant x \leqslant 1$, are finite in number and can be enclosed in intervals of total length $\tfrac{1}{2}\eta$. Thus an infinite set of discontinuities can sometimes be enclosed in a finite set of intervals of arbitrarily small total length.

If $f(x) = 0$ for x irrational and $f(x) = 1/n$ for $x = m/n$, where m/n is a proper fraction in its lowest terms, $f(x)$ is discontinuous at all rational values of x in $(0, 1)$, but continuous at all irrational values. For any irrational x_0 can be enclosed in an interval of length $1/n!$ containing no rational fraction with a denominator less than n, and therefore the values of $f(x)$ in a sufficiently small interval about x_0 will be arbitrarily small. In this case there is a discontinuity of $f(x)$ in every interval, however short. Nevertheless, it has a Riemann integral; for the number of the discontinuities where the leap of $f(x)$ exceeds ϵ is not more than the sum of the integers less than $1/\epsilon$, and is finite. The integral is, in fact, zero.

If $f(x) = 1$ for x rational and $= 0$ for x irrational, then for every x_0, rational or not, there are values of x arbitrarily near x_0 where $f(x) = 1$ and others where $f(x) = 0$. Hence every value of x is a discontinuity where the leap is 1, and those in $(0, 1)$ cannot be enclosed in any set of intervals of length < 1. In this case $H_n = 1$, $h_n = 0$, however we subdivide the interval.

Such types of irregularity are of little direct practical importance, but they have an indirect importance, since we are aiming at a considerable degree of generality and therefore need danger signals. There are other definitions of an integral, especially that of Lebesgue, which give definite values to some integrals that do not exist in Riemann's sense (including the one just mentioned); they contemplate an infinite set of subdivisions from the start. They simplify the statements and extend the generality of some later theorems appreciably. The reader is referred to the accounts given by Burkill* and Titchmarsh.† But it appears that cases where these methods are applicable and Riemann's is not are too rare in physics to repay the extra difficulty.

If $f(x)$ has a Riemann integral, $\{f(x)\}^n$ $(n > 0)$ and $|f(x)|$ have Riemann integrals over the same interval. For if $f(x)$ is bounded and the discontinuities where the leap exceeds ω can be enclosed in intervals of arbitrarily small total length, the same applies to $\{f(x)\}^n$ and $|f(x)|$. The converse is not true. Consider $f(x) = 1$ at rational values of x, $f(x) = -1$ at irrational values; $\{f(x)\}^2$ and $|f(x)|$ are integrable, $f(x)$ is not.

1·1013. 'Measure zero': 'Almost everywhere'. A set of points capable of being enclosed in intervals whose total length is arbitrarily small is said to have *measure zero*, and a proposition true except at such a set is said to be true *almost everywhere*. Any finite set of points has measure zero; so also have the integers, since we can enclose each integer n in an interval $2^{-|n|}\alpha$, where α is arbitrarily small, and $\Sigma 2^{-|n|}$ converges. So have the rational numbers in $(0, 1)$. For if p and q are integers with $p < q$ we can enclose p/q in a range of length α/q^3, where α is positive. There are $q - 1$ fractions with denominator q, excluding 0 and 1. But 0 and 1 can be enclosed in ranges α, and the other fractions in a range less than α/q^2. Summing now with regard to q we see that all rational fractions can be enclosed in ranges of total length less than $2\alpha + \alpha \sum_{2}^{\infty} q^{-2}$; the series converges and therefore the total length can be made as small as we like by a suitable choice of α. The same holds for any enumerable set.

Consider a decreasing sequence of positive quantities $\omega_1, \omega_2, \ldots$ tending to zero. If $f(x)$ has a Riemann integral the points (if any) where the leap is $\geq \omega$ can be enclosed in a finite set of intervals of arbitrarily small length; hence the discontinuities where the leap is $< \omega_{n-1}$ but $\geq \omega_n$ can be enclosed in a finite set of length $2^{-n}\eta$, and all discontinuities in a set of length η. This set of intervals is enumerable, since each can be reached in a finite

---
* J. C. Burkill, *Cambridge Mathematical Tracts*, No. 40, 1951.
† E. C. Titchmarsh, *The Theory of Functions* (1932), Chs. X, XI, XII.

number of steps from the start; hence the discontinuities of an integrable function can be enclosed in an enumerable set of intervals of arbitrarily small total length. These intervals may overlap.

1·102. Existence of Stieltjes integral. The definition in 1·10 of this type of integral allows the function $g(x)$ to be discontinuous. If $g(x)$ is non-decreasing and bounded for $a \leqslant x \leqslant b$ and if $f(x)$ is also bounded, a necessary and sufficient condition that $\int_{x=a}^{b} f(x)\, dg(x)$ shall exist is that for any ω, δ the interval can be divided into a finite number of subintervals, such that in the intervals where the leap of $f(x)$ is greater than ω the total variation of $g(x)$ is less than δ. The proof is substantially as for the Riemann integral. If $g(x)$ has bounded variation the same result follows by expressing $g(x)$ as the difference of two non-decreasing functions $\phi(x) - \psi(x)$ and considering $\int f(x)\, d\phi(x)$ and $\int f(x)\, d\psi(x)$ separately.

In particular the Stieltjes integral exists in any finite interval if $g(x)$ has bounded variation and $f(x)$ is continuous. It does not exist if $f(x)$ and $g(x)$ have a discontinuity at the same value of x, for in any interval including the discontinuity neither the leap of $f(x)$ nor the total variation of $g(x)$ is arbitrarily small. It follows that it is *not* sufficient for the existence of the Stieltjes integral that $f(x)$ and $g(x)$ shall both be of bounded variation.

We shall not give general conditions for the existence of the Stieltjes integral when $g(x)$ is not of bounded variation; we shall show that it is sufficient that $g(x)$ shall be continuous and $f(x)$ of bounded variation, but it is not sufficient that both shall be continuous.

If $a < b < c$, and we write $I(d, e) = \int_{x=d}^{e} f(x)\, dg(x)$, then if $I(a, c)$ exists both $I(a, b)$ and $I(b, c)$ exist, and their sum is $I(a, c)$. The converse is not always true. If

$$f(x) = 0 \quad (x \leqslant 0), \qquad g(x) = 1 \quad (x < 0),$$
$$ = 1 \quad (x > 0), \qquad = 0 \quad (x \geqslant 0).$$

$\int_{x=-1}^{0} f\, dg$ and $\int_{x=0}^{1} f\, dg$ both exist and are zero, but $\int_{x=-1}^{1} f\, dg$ does not exist. The converse is true with a slightly different definition of the Stieltjes integral given by Pollard.

1·103. Differentiation.

(a) If $f(x)$ is continuous and $\int_{a}^{x} f(u)\, du = F(x),$

then
$$\frac{d}{dx} F(x) = f(x),$$

and $F(x)$ is a continuous function of x.
This is almost obvious.

(b) If
$$\frac{d}{dx} F(x) = f(x),$$

and $f(x)$ is integrable, then $\int_{a}^{x} f(u)\, du = F(x) - F(a).$

Since $F(x)$ is differentiable in (a, b), we know from 1·0622 that for any positive ω, δ we can divide (a,b) into a finite set of intervals (x_r, x_{r+1}), all of lengths $\leq \delta$, each containing a point ξ_r such that for every point of (x_r, x_{r+1})

$$|F(x) - F(\xi_r) - (x - \xi_r) F'(\xi_r)| < \omega |x - \xi_r|, \tag{1}$$

and therefore
$$|F(x_{r+1}) - F(x_r) - (x_{r+1} - x_r) F'(\xi_r)| < \omega (x_{r+1} - x_r). \tag{2}$$

By addition
$$|F(b) - F(a) - \Sigma (x_{r+1} - x_r) f(\xi_r)| < \omega (b - a). \tag{3}$$

Since $f(x)$ is integrable we can, given any positive ϵ, choose δ so that the sum in (3) differs from the integral by less than ϵ. Hence

$$\left| F(b) - F(a) - \int_a^b f(x) \, dx \right| < \epsilon + \omega(b - a)$$

and therefore is zero, since ϵ and ω are arbitrarily small.

Note that it is possible for $F(x)$ to be differentiable and for its derivative not to be integrable; for example,

$$F(x) = x^2 \sin \frac{1}{x^2} \quad (0 < x \leq 1), \quad F(0) = 0.$$

The derivative exists even at $x = 0$, but is unbounded in any neighbourhood of 0.

(c) *If $f(x)$ has a Riemann integral $\int_a^b f(x) \, dx$, then $\int_a^x f(u) \, du$ exists and is continuous for all x such that $a \leq x \leq b$; and its derivative is equal to $f(x)$ except possibly at a set of measure zero, namely, the points of discontinuity of $f(x)$.*

Let x be a point of continuity of $f(x)$. Then in an interval $(x - h, x + h)$ the leap of $f(\xi)$ is ω, where $\omega \to 0$ with h. Also

$$f(x) - \omega \leq \frac{1}{h} \int_x^{x+h} f(u) \, du = \frac{1}{h} \left\{ \int_0^{x+h} f(u) \, du - \int_0^x f(u) \, du \right\} \leq f(x) + \omega.$$

Making $h \to 0$ we have

$$\frac{d}{dx} \int_0^x f(u) \, du = f(x)$$

at all points where $f(x)$ is continuous.

It follows that if
$$\int_0^x f(u) \, du = \int_0^x g(u) \, du \quad a \leq x \leq b$$

then $f(x) = g(x)$ almost everywhere in $a \leq x \leq b$, the exceptional points, if any, being at points of discontinuity of $f(x)$ or $g(x)$.

The exception in (c) is of some importance. If, for instance, $f(x) = 0$ for $x \leq 0$, and $= 1$ for $x > 0$,

$$F(x) = \int_{-1}^x f(u) \, du = 0 \quad (x \leq 0),$$
$$= x \quad (x > 0),$$

and $\dfrac{d}{dx} F(x)$ does not exist at $x = 0$. Again, if $f(x) = 0$ for $x \neq 0$ and $= 1$ for $x = 0$, $\int_a^x f(u) \, du = 0$ over any interval and has derivative 0 everywhere; but this derivative is not equal to $f(x)$ when $x = 0$.

1·1031. Integration by parts for Stieltjes integrals. We define

$$S_n = \sum_{r=1}^{n} f(\xi_r)\{g(x_r) - g(x_{r-1})\} \qquad (1)$$

with $\quad x_0 = a, \quad x_n = b, \quad x_0 \leqslant \xi_1 \leqslant x_1 \leqslant \ldots \leqslant x_{r-1} \leqslant \xi_r \leqslant x_r \ldots \leqslant x_n.$ (2)

Then $\quad S_n = f(x_n)\,g(x_n) - f(x_0)\,g(x_0) - \Sigma_n,$ (3)

where $\quad \Sigma_n = g(x_0)\{f(\xi_1) - f(x_0)\} + \sum_{r=1}^{n-1} g(x_r)\{f(\xi_{r+1}) - f(\xi_r)\} + g(x_n)\{f(x_n) - f(\xi_n)\}.$ (4)

We assume that $I = \int_{x=a}^{b} f\,dg$ exists; that is, for any $\epsilon > 0$ we can choose δ so that for all subdivisions such that the greatest subinterval is $< \delta$

$$|S_n - I| < \epsilon. \qquad (5)$$

Then for any set $a, \xi_1, \xi_2, \ldots, b$, such that $\xi_1 - a, \ldots, \xi_{r+1} - \xi_r, \ldots, b - \xi_n$ are $< \tfrac{1}{2}\delta$, x_r satisfying the inequalities (2) will also satisfy $x_r - x_{r-1} < \delta$ for all r. Hence for such a set

$$|\Sigma_n - f(b)\,g(b) + f(a)\,g(a) + I| < \epsilon, \qquad (6)$$

and therefore Σ_n tends to a limit as $\delta \to 0$, and this limit is by definition $\int g\,df$. Hence $\int g\,df$ exists and

$$\int_{x=a}^{b} g\,df = \left[fg\right]_a^b - \int_{x=a}^{b} f\,dg. \qquad (7)$$

In particular, since $\int f\,dg$ exists when f is continuous and g is of bounded variation, it also exists when f is of bounded variation and g continuous. If $g(x)$ is a Riemann integral it is both continuous and of bounded variation; hence $\int f\,dg$ exists if f has either property.

For Riemann integration the result is usually stated in the form

$$\int_a^b f(x)\,g'(x)\,dx = \left[f(x)\,g(x)\right]_a^b - \int_a^b f'(x)\,g(x)\,dx, \qquad (8)$$

thus apparently requiring both f and g to be differentiable for all $a \leqslant x \leqslant b$. If the derivatives exist and are integrable this follows immediately from (7) and 1·1032. But (7) is true under much wider conditions. Incidentally the easiest way of using (8) is to integrate g' first and rewrite (8) in the form (7).

1·1032. Change of variable in an integral. If $x = h(y) = \int_a^y g(u)\,du$, and if

$$I = \int_{y=a}^{b} f(x)\,dx, \quad J = \int_a^b f(x)\,g(y)\,dy$$

both exist, then $I = J$. Since both integrals exist by hypothesis, it is enough to prove that the partial sums tend to the same limit for some way of forming them. Take $x_r = h(y_r)$, $\xi_r = h(\eta_r)$,

$$I_n = \Sigma f(\xi_r)(x_{r+1} - x_r), \quad J_n = \Sigma f(\xi_r)\,g(\eta_r)(y_{r+1} - y_r). \qquad (1)$$

Since $g(y)$ is integrable, the intervals of y can be chosen so that those where the leap of $g(y)$ is greater than ω have total length $\leqslant \delta$, where ω, δ are arbitrarily small. Also $|g(y)|$

is bounded, say $< G$; hence if the greatest interval $y_{r+1} - y_r$ is less than λ, the greatest $|x_{r+1} - x_r|$ is less than $G\lambda$. Hence $I_n \to I$, $J_n \to J$. Now if G_r, g_r are the upper and lower bounds of $g(y)$ for $y_r \leqslant y \leqslant y_{r+1}$,

$$g_r(y_{r+1} - y_r) \leqslant x_{r+1} - x_r \leqslant G_r(y_{r+1} - y_r), \tag{2}$$

$$g_r \leqslant g(\eta_r) \leqslant G_r, \tag{3}$$

and therefore
$$|x_{r+1} - x_r - g(\eta_r)(y_{r+1} - y_r)| \leqslant (G_r - g_r)(y_{r+1} - y_r), \tag{4}$$

$$|I_n - J_n| \leqslant \Sigma f(\xi_r)(G_r - g_r)(y_{r+1} - y_r). \tag{5}$$

Let the leap of $g(y)$ in the whole interval be N; and let $G_r - g_r$ be $\leqslant \omega$ except in a set of intervals of total length δ. Then

$$|I_n - J_n| \leqslant FN\delta + \omega F(b - a - \delta), \tag{6}$$

where F is the upper bound of $|f(x)|$. This is arbitrarily small; hence

$$I_n - J_n \to 0, \quad I = J. \tag{7}$$

The usual form
$$\int_{x(a)}^{x(b)} f(x)\,dx = \int_a^b f(x)\frac{dx}{dy}\,dy, \quad \frac{dx}{dy} > 0 \tag{8}$$

is somewhat less general because it assumes dx/dy to exist everywhere. If this condition is satisfied the theorem is proved very easily by an application of Rolle's theorem (1·13). But the more general form is needed for transformation of integrals along curves, which may have corners where there is no definite tangent. (8) can be made valid if at any point c where dx/dy does not exist, we understand it to be replaced by any value between the limits, as $\delta \to 0$, of the upper and lower bounds of dx/dy in $(c - \delta, c + \delta)$.

1·104. Infinite and improper integrals. The proof of the existence of an integral breaks down if either the interval $b - a$ is infinite or the function to be integrated is unbounded in the interval. In the former case, $b - a$ is infinite and we cannot make $\omega > 0$, $(b - a)\omega < \epsilon$ by any choice of ω. In the latter the approximating sum may vary to any extent according to the point chosen to sample $f(x)$ in the subinterval where $f(x)$ is unbounded. A special device is needed in either case to give a meaning to the integral. The method used for integrals with an infinite upper limit is to use first an integral with a finite upper limit; if the integral tends to a definite limit when the upper limit tends to infinity this limit is taken as the value of the infinite integral. The need for such a device may be illustrated by the integral

$$I = \int_0^\infty \frac{\sin x}{x}\,dx.$$

According to our rule this must be interpreted as

$$\lim_{X \to \infty} \int_0^X \frac{\sin x}{x}\,dx.$$

The integral up to X exists for all X since the integrand is everywhere continuous. If we take $Y > X$, m to be the integer next greater than X/π, n to be the greatest integer less than Y/π,

$$\int_X^Y \frac{\sin x}{x}\,dx = \int_X^{m\pi} \frac{\sin x}{x}\,dx + \int_{n\pi}^Y \frac{\sin x}{x}\,dx + \sum_{r=m}^{n-1} \int_{r\pi}^{(r+1)\pi} \frac{\sin x}{x}\,dx.$$

The first of these integrals, since $|\sin x| \leq 1$ and $m\pi - X \leq \pi$, is numerically $\leq \pi/X$. Similarly, the second is numerically $\leq 1/n$. The sum consists of alternately positive and negative terms, each less in magnitude than the preceding; and we have the theorem that if $u_0 > u_1 > \ldots > u_n > 0$

$$u_0 > u_0 - u_1 + u_2 + \ldots + (-1)^n u_n > 0.$$

Hence the sum is less numerically than its first term, and

$$\left| \int_{m\pi}^{(m+1)\pi} \frac{\sin x}{x} dx \right| < \frac{1}{m}.$$

Thus
$$\left| \int_X^Y \frac{\sin x}{x} dx \right| < \frac{\pi}{X} + \frac{1}{n} + \frac{1}{m} < \frac{3\pi}{X},$$

which can be made arbitrarily small for all $Y > X$ by taking X large enough. Hence for any positive quantity ϵ, however small, we can choose X so that no matter how much we increase the upper limit beyond X we cannot change the integral by more than ϵ. Thus the integral up to X has a definite limiting value as X tends to infinity, and the infinite integral exists in the sense defined.

1·1041. Since an integral is a function of its upper terminus, we can adapt the tests for convergence of a sequence given in 1·0441 and 1·045 on the lines already mentioned in the theory of continuity (1·06). The proofs are straightforward.

If $f(x) \geq 0$ and $\int_a^X f(x) dx$ is bounded for all $X > a$, then $\int_a^X f(x) dx$ tends to a limit as $X \to \infty$.

A necessary and sufficient condition that $\int_a^X f(x) dx$ shall tend to a limit as $X \to \infty$ is that for any positive ϵ, however small, there is an A such that $\left| \int_A^X f(x) dx \right| < \epsilon$ for all $X > A$.

1·1042. The relation between infinite integrals and series is so close that the same words are convenient to express the properties:

$\int_a^\infty f(x) dx$ is *convergent* if $\lim_{X \to \infty} \int_a^X f(x) dx$ exists.

$\int_a^\infty f(x) dx$ is *unbounded* if $\left| \int_a^X f(x) dx \right|$ is unbounded as $X \to \infty$.

$\int_a^\infty f(x) dx = \infty$ if $\int_a^X f(x) dx \to \infty$ as $x \to \infty$.

$\int_a^\infty f(x) dx$ is *finitely oscillatory* if there are positive quantities ω, M such that for any X we can choose $Y_1 > X$ so that $\left| \int_X^{Y_1} f(x) dx \right| > \omega$, but cannot choose Y_2 so that $\left| \int_X^{Y_2} f(x) dx \right| > M$.

Examples of convergent integrals are

$$\int_1^\infty \frac{dx}{x^2}, \quad \int_0^\infty e^{-x} dx, \quad \int_1^\infty \frac{\sin x}{x^2} dx, \quad \int_0^\infty \frac{\sin x}{x} dx.$$

Unbounded integrals are
$$\int_1^\infty dx, \quad \int_1^\infty \frac{dx}{x}, \quad \int_0^\infty x \sin x \, dx.$$

The last of these would usually be called 'infinitely oscillating' but we have no occasion to make this distinction.

Finitely oscillatory integrals are
$$\int_0^\infty \cos x \, dx, \quad \int_0^\infty \sin x \, dx.$$

Unbounded and finitely oscillatory integrals have no definite values.

The integral $\int_a^\infty f(x) \, dx$ is called *absolutely convergent* if $\int_a^\infty |f(x)| \, dx$ is convergent. If the former integral is convergent but the latter is not, the former is called *conditionally convergent*. Of the above examples of convergent integrals, the first three are absolutely convergent, the last conditionally convergent.

1·1043. If $f(x)$ is positive and non-increasing for $x > x_0$, the integral $I = \int_{x_0}^\infty f(x) \, dx$ converges if and only if the series $\sum_{n=n_0}^\infty f(n)$ converges; where n_0 is the integer next greater than x_0. For clearly neither the series nor the integral can converge unless $f(x) \to 0$; take an integer $m > x_0$ and such that $f(m) < \epsilon$. Then

$$f(m) + f(m+1) + \ldots + f(n-1) \geqslant \int_m^X f(x) \, dx \geqslant f(m+1) + f(m+2) + \ldots + f(n-1),$$

where n is the integer next greater than X. Hence

$$\left| \int_m^X f(x) \, dx - \sum_{r=m}^{n-1} f(r) \right| \leqslant f(m) < \epsilon$$

and ϵ is arbitrarily small. Hence if either the integral or the sum tends to a definite limit the other does.

In particular $\int_1^\infty x^{-p} \, dx$ and $\sum_1^\infty x^{-p}$ both converge if and only if $p > 1$.

1·1044. Similarly, if $f(x)$ tends to infinity at some point of the range we can define an *improper integral* by first modifying the range so as to cut out an arbitrarily short interval about the infinity and then making the length of this interval tend to zero. Thus

$$\int_\epsilon^1 x^{-1/2} \, dx = \left[2x^{1/2} \right]_\epsilon^1 = 2 - 2\epsilon^{1/2}, \quad \lim_{\epsilon \to 0} \int_\epsilon^1 x^{-1/2} \, dx = 2.$$

This process is taken as the definition of $\int_0^1 x^{-1/2} \, dx$, which is not directly intelligible as it stands in terms of the definition of an integral as the limit of a sum.

The analogy between infinite and improper integrals in respect of convergence is so close that the nomenclature can be taken over unchanged.

1·1045. Change of variable may convert an ordinary integral into an infinite or improper one, but will not change its value. For instance, if for all y, $x = h(y)$ as in 1·032,

$$\int_{h(0)}^{h(y)} f(x)\,dx = \int_0^y f(x)\,g(y)\,dy, \tag{1}$$

when y and $h(y)$ are finite, $g(y) \geq 0$, then they have the same limit if either y or $h(y)$ or both tend to infinity. If $h(y) \to b$ as $y \to \infty$, and $\int_{h(0)}^b f(x)\,dx$ exists, it is the limit of the left side of (1); hence the limit is equal to the Riemann integral when this exists.

1·11. Functions of two variables. So far we have been considering sequences, which may be regarded as functions of one variable capable of taking only integral values, and functions of a continuous variable. In what follows we shall be concerned with what are essentially functions of two variables, which may be either integral or continuous. This introduces new complications when limiting processes are used, since it is not always obvious, or even true, that the same result will be obtained when the order of the limiting processes is changed. The simplest sufficient condition for the reversibility of limiting processes is provided by the following theorem on absolute convergence.

1·111. *If $f(x, y)$ is a non-decreasing function of both x and y (either or both of which may take only integral values), and*

$$\lim_{x \to \infty} f(x, y) = g(y), \quad \lim_{y \to \infty} f(x, y) = h(x), \tag{1}$$

then
$$\lim_{y \to \infty} g(y) = \lim_{x \to \infty} h(x), \tag{2}$$

in the sense that if either of the limits in (2) exists the other exists and the two are equal.

Note first that $g(y)$ is a non-decreasing function of y. For if $y_2 > y_1$

$$g(y_2) - g(y_1) = \lim_{x \to \infty} \{f(x, y_2) - f(x, y_1)\} \geq 0. \tag{3}$$

Similarly, if
$$x_2 > x_1, \quad h(x_2) \geq h(x_1). \tag{4}$$

Let $g(y)$ have a limit M. For any ϵ there is a Y such that for all $y \geq Y$

$$M \geq g(y) \geq M - \epsilon. \tag{5}$$

For all x, y, $M \geq g(y) \geq f(x, y)$. Also X exists such that for all $x > X$

$$f(x, Y) \geq g(Y) - \epsilon, \tag{6}$$

and therefore for all $y > Y$, $x > X$

$$M \geq f(x, y) \geq g(Y) - \epsilon \geq M - 2\epsilon. \tag{7}$$

Hence, if $y \to \infty$, $x > X$
$$M \geq h(x) \geq M - 2\epsilon, \tag{8}$$

and therefore, since ϵ is arbitrary, $h(x)$ also has limit M as $x \to \infty$.

We have three immediate applications. If

$$f(m, n) = \sum_{r=1}^m \sum_{s=1}^n u_{r,s}, \tag{9}$$

where $u_{r,s}$ is not negative, $f(m,n)$ is a non-decreasing function of m and n. Hence for a double series of non-negative terms

$$\sum_{r=1}^{\infty}\sum_{s=1}^{\infty} u_{r,s} = \sum_{s=1}^{\infty}\sum_{r=1}^{\infty} u_{r,s}. \tag{10}$$

If some of the $u_{r,s}$ are negative, we can write

$$v_{r,s} = |u_{r,s}| + u_{r,s}, \quad w_{r,s} = |u_{r,s}| - u_{r,s}, \tag{11}$$

and all $v_{r,s}$, $w_{r,s}$ are non-negative. If then $\Sigma\Sigma |u_{r,s}|$ exists for one order of summation we see easily that $v_{r,s}$ and $w_{r,s}$ satisfy the condition for inversion of the order of summation, and by subtraction $u_{r,s}$ does so. Hence for any absolutely convergent double series the order of summation can be inverted. As a corollary, if Σa_r and Σb_s are absolutely convergent, $\sum_r a_r \sum_s b_s = \sum_r \sum_s a_r b_s$, where the terms may be taken in any order in the sum on the right.

If $u_m(x)$ is never negative and if $\sum_{m=1}^{p} \int_0^q u_m(x)\,dx$ exists for all p, q and has limits as each of p, q tends to infinity with the other fixed, then

$$\sum_{m=1}^{\infty} \int_0^{\infty} u_m(x)\,dx = \int_0^{\infty} \{\Sigma u_m(x)\}\,dx. \tag{12}$$

If $u_m(x)$ is not always of the same sign, but $\sum_{m=1}^{p}\int_0^q |u_m(x)|\,dx$ exists and satisfies the same conditions, and if one of

$$\sum_1^{\infty}\int_0^{\infty} |u_m(x)|\,dx, \quad \int_0^{\infty}\sum_1^{\infty} |u_m(x)|\,dx$$

exists, then both the limits in (12) exist and the two are equal.

If $\phi(x,y)$ is non-negative, subject to similar conditions on the existence of the single limits,

$$\int_0^{\infty} dx \int_0^{\infty} \phi(x,y)\,dy = \int_0^{\infty} dy \int_0^{\infty} \phi(x,y)\,dx, \tag{13}$$

where we take $f(x,y)$ to be the assumed common value for the two integrals for upper termini x, y. As before, if $\phi(x,y)$ is not always of the same sign, a sufficient condition for existence and equality of the double limits is that one of them shall exist for $|\phi(x,y)|$.

1·112. Uniform convergence of sequences and series. The terms of a sequence $\{f_n(x)\}$ may be functions of a variable x. Then if the sequence converges for all values of x in an interval, its limit is a function of x, say $f(x)$. If we choose an arbitrarily small positive ϵ we shall for any x be able to choose $n(x)$ so that $|f_p(x) - f(x)| < \epsilon$ for all $p \geqslant n(x)$ because the sequence converges. In general the least value of $n(x)$ such that this is true will depend on x. But it may be possible to choose an n independent of x such that $|f_p(x) - f(x)| < \epsilon$ for all $p > n$ and for all x in the interval. If this is possible for every ϵ, $f_n(x)$ is said to be *uniformly convergent* to $f(x)$ in the interval. It can fail to be uniformly convergent if there is an x, say c, within or at the end of the interval such that if we take a succession of values of x, tending to c, the corresponding values of $n(x)$ for given ϵ tend to infinity.

As $f_n(x)$ may be the sum of the first n terms of a series, all these statements have immediate analogues for series $\Sigma u_n(x)$ over an interval of x. Thus the series Σx^n converges for all x such that $0 \leqslant x < 1$, but it is not uniformly convergent for all such x. For if we fix ϵ and then choose n so as to make

$$x^n + x^{n+1} + \ldots + x^{n+p} = \frac{x^n(1-x^{p+1})}{1-x} \qquad (1)$$

less than ϵ for all $p \geqslant 1$ we must make

$$x^n < (1-x)\epsilon, \qquad (2)$$

and therefore

$$n > \frac{\log\{(1-x)\epsilon\}}{\log x}, \qquad (3)$$

which tends to infinity as x tends to 1. This series is therefore uniformly convergent in a range $a \leqslant x \leqslant b$, where a and b are fixed quantities between 0 and $+1$, since we can choose n greater than the greater of the quantities

$$\frac{\log\{(1-a)\epsilon\}}{\log a}, \quad \frac{\log\{(1-b)\epsilon\}}{\log b},$$

and the same value of n will then do for any intermediate value of x. It is convergent for any x such that $-1 < x < 1$. But it is not uniformly convergent in the range $-1 < x < 1$ because, even though the signs $<$ exclude the possibilities that x may be actually -1 or $+1$, they permit any intermediate value, however close to 1, and however we choose n we shall always be able to find values of x such that (3) is false.

If $f_n(x) \to f(x)$ uniformly in each of a finite set of intervals (a_r, b_r) ($r = 1$ to k), then it does so uniformly in the whole set. For each interval, n_r exists such that $|f_p(x) - f(x)| < \epsilon$ for $p \geqslant n_r$ and x in (a_r, b_r). Take m equal to the greatest of the n_r; then for $p > m$, $|f_p(x) - f(x)| < \epsilon$ for x in any of the intervals.

If $f_n(x) \to f(x)$ in $a \leqslant x \leqslant b$, and $f_n(x) \to f(x)$ uniformly in $a < x < b$, then convergence is uniform in $a \leqslant x \leqslant b$. We need only apply the argument of the last paragraph to the open interval $a < x < b$ and the special points a, b, and take m equal to the largest value of n for the three. This seems to be the basis of a common statement that a sequence cannot converge uniformly in an open interval. It can, but then it also converges uniformly in the closed interval if it converges at the end-points. But if

$$f_n(0) = 2^n, \quad f_n(1) = 2^n, \quad f_n(x) = 2^{-n} \quad (0 < x < 1),$$

$\{f_n(x)\}$ is uniformly convergent in the open interval but not in the closed interval.

1·113. Continuity and integrability of uniformly convergent series. *The sum of a series of continuous functions of x, uniformly convergent in a range, is itself a continuous function of x in the range.*

The integral with regard to x of the sum of a series, uniformly convergent in a finite range of x, is the sum of the integrals of its terms, provided that the termini of the integral are in the range.

To prove the first statement, let $S(x)$ be the sum of the series. Then since the series is uniformly convergent, if ω is a positive quantity we can choose n independent of x so that if $S_n(x)$ is the sum up to $u_n(x)$

$$|S(x) - S_n(x)| < \omega, \quad |S(y) - S_n(y)| < \omega \qquad (1)$$

for all x, y in the range. But $S_n(x)$ is the sum of a finite number of continuous functions and therefore is continuous. Hence for any x we can choose δ positive but so small that

$$|S_n(y) - S_n(x)| < \omega \tag{2}$$

for all $|y - x| < \delta$. Therefore for such y

$$|S(y) - S(x)| < 3\omega, \tag{3}$$

and by taking $\omega = \tfrac{1}{3}\epsilon$ and then choosing δ in accordance with (2) we can make

$$|S(x+h) - S(x)| < \epsilon \tag{4}$$

for all h satisfying $0 \leqslant h < \delta$. Hence $S(x)$ is continuous (and therefore integrable).

To prove the second statement, we have, if

$$S(x) = S_n(x) + R_n(x), \tag{5}$$

and $|R_n(x)| < \omega$ for all x such that $a \leqslant x \leqslant b$,

$$\int_a^b S(x)\,dx = \int_a^b S_n(x)\,dx + \int_a^b R_n(x)\,dx \tag{6}$$

and

$$\left| \int_a^b R_n(x)\,dx \right| < \omega(b-a), \tag{7}$$

which is arbitrarily small. Hence by taking n large enough we can make

$$\int_a^b S_n(x)\,dx = \sum_{r=1}^n \int_a^b u_n(x)\,dx \tag{8}$$

as near as we like to $\int_a^b S(x)\,dx$. The theorem is often expressed by saying that *a uniformly convergent series of continuous functions can be integrated term by term in any finite range.*

A uniformly convergent series can also be integrated term by term if the terms are integrable but not necessarily continuous. If $S(x)$ is integrable the argument from (5) still holds. Take n so that $|R_n(x)| < \omega$. The leap of $R_n(x)$ exceeds 2ω in no interval. $S_n(x)$ is integrable. Divide (a, b) into a finite set of intervals so that the total length of those where the leap of $S_n(x)$ exceeds ω is less than δ. Then the total length of those where the leap of $S(x)$ exceeds 3ω is less than δ. ω and δ are arbitrary; hence $S(x)$ is integrable.

If $f_n(x) \to f(x)$ uniformly in (a, b), $\int_a^x f_n(\xi)\,d\xi$ tends uniformly to $\int_a^x f(\xi)\,d\xi$. For if

$$|f_n(\xi) - f(\xi)| < \omega, \quad \left| \int_a^x \{f_n(\xi) - f(\xi)\}\,d\xi \right| < \omega(x-a) \leqslant \omega(b-a).$$

1·114. Discontinuity associated with non-uniform convergence. The geometric series does not converge at either $x = 1$ or $x = -1$ and therefore does not define a value of the function at the limits; thus the question of continuity does not arise. But it is possible for a series to converge at certain values of x and yet not to be uniformly convergent in a range approaching them. The example given by Stokes, who first discussed this property, was

$$u_n(x) = \left(\frac{1}{n} - \frac{1}{n+1}\right) + 2\left(\frac{1}{(n-1)x+1} - \frac{1}{nx+1}\right). \tag{1}$$

$\Sigma u_n(x)$ converges for all x, since

$$\sum_n^{n+p} u_n(x) = \frac{1}{n} - \frac{1}{n+p+1} + 2\left(\frac{1}{(n-1)x+1} - \frac{1}{(n+p)x+1}\right). \tag{2}$$

Take $x > 0$. Then

$$\left|\sum_n^{n+p} u_n(x)\right| \leq \frac{1}{n} + \frac{2}{(n-1)x+1}, \tag{3}$$

and we can make this less than ϵ by taking n large enough. If $x = 0$, the last bracket in (2) is 0 and the sum is $< 1/n$. The series is therefore convergent for $x \geq 0$. But it is not uniformly convergent. For if the quantity on the right of (3) is greater than ϵ the quantity on the left can be made greater than ϵ by taking p large enough; and $1/n$ is always positive. If, then,

$$\frac{2}{(n-1)x+1} > \epsilon, \tag{4}$$

that is,

$$(n-1)x < \frac{2}{\epsilon} - 1, \tag{5}$$

the left of (3) will exceed ϵ for p large enough; and to make the left of (3) less than ϵ for all p we must take $n > \dfrac{2/\epsilon - 1}{x} + 1$. Hence, if we fix ϵ at the start, the appropriate values of n increase without limit as x is made smaller, and the series is not uniformly convergent. Stokes described such series as *converging with infinite slowness* near $x = 0$.

Now consider the sum of the series. We have for all x

$$\sum_1^n u_n(x) = \left(1 - \frac{1}{n+1}\right) + 2\left(\frac{1}{1} - \frac{1}{nx+1}\right),$$

and the sum of the series, if x is not zero, is 3, since the terms on the right containing n tend to zero with increasing n. Hence the limit of the sum as x tends to 0 is 3. But if we put x zero first the terms in the second bracket cancel for all n, and the sum is 1.

This example is artificial, but the functions used are quite simple, and it serves to illustrate the fact that the results of carrying out two limiting processes may be quite different according to which we do first. We have to make x tend to 0 and n to infinity. If we make x tend to 0 first and then n to infinity we get 1; if we make n tend to infinity first and then x to 0 we get 3.

1·115. Tests for uniform convergence. *A necessary and sufficient condition that $\{u_n(x)\}$ shall be uniformly convergent in an interval (a, b) is that for any ϵ we can choose m so that for all $n \geq m$, $|u_n(x) - u_m(x)| < \epsilon$ for every x of (a, b).* The proof given for simple sequences needs little alteration. (See 1·045.)

1·1151. M test. *If for all x in the interval considered $|u_n(x)| \leq v_n$, where v_n is independent of x, and the series Σv_n converges, then $\Sigma u_n(x)$ is uniformly convergent in the interval.* For we can choose n to make for all $p \geq 0$

$$\sum_n^{n+p} v_n < \epsilon$$

since Σv_n converges; and then for any x

$$\left|\sum_{n}^{n+p} u_n(x)\right| \leqslant \sum_{n}^{n+p} |u_n(x)| \leqslant \sum_{n}^{n+p} v_n < \epsilon.$$

This test is known as *Weierstrass's M test*.

The use of *comparison series* for testing ordinary convergence rests on the same principle, and we need only state the theorem. *If as $n \to \infty$, $|u_n| \leqslant v_n$, and Σv_n converges, then Σu_n converges.*

The M test is very simple to apply and we shall have numerous applications of it.

1·1152. Extension of the M test. A modification of the M test is sometimes useful even for conditionally convergent series where we cannot find a convergent series of positive terms v_n numerically greater than $u_n(x)$. Suppose that as $n \to \infty$, $u_n(x)$ tends uniformly to 0 (see below); that the terms of $\Sigma u_n(x)$ can be taken in batches of m without deranging the order, giving a series $\Sigma U_\nu(x)$; and that $|U_\nu(x)| < V_\nu$, where ΣV_ν is convergent. Then $\Sigma U_\nu(x)$ is uniformly convergent by the M test. It remains to show that in the conditions stated $\Sigma u_n(x)$ exists and is equal to $\Sigma U_\nu(x)$.

Since $u_n(x)$ tends *uniformly* to 0, for any ϵ we can choose n so that $|u_p(x)| < \epsilon$ for all $p \geqslant n$ and for all x in the range. Then if we take ν for given n so that

$$\nu m \leqslant n < (\nu+1)m,$$

$$\left|\sum_{1}^{n} u_n(x) - \sum_{1}^{\nu} U_\nu(x)\right| = |u_{m\nu+1}(x) + \ldots + u_n(x)|.$$

Take ν so that $\sum_{\nu+1}^{\infty} U_\sigma(x) < \tfrac{1}{2}\epsilon$, and so that all $|u_p(x)| < \dfrac{\epsilon}{2m}$ for $p > m\nu$. Then

$$\left|\sum_{1}^{n} u_n(x) - \sum_{1}^{\infty} U_\nu(x)\right| < \epsilon$$

for all x, and all $n > m\nu$, and $\Sigma u_n(x)$ is uniformly convergent.

This can be applied to the series

$$\sum_{1}^{\infty} (-1)^n \frac{n}{n^2 + x^2}.$$

For by taking the terms in pairs we get a series whose terms are $\leqslant$ those of $\Sigma \dfrac{1}{n(n+1)}$, and which therefore satisfies the M test. Also the general term is numerically $\leqslant 1/n$ for all x, and therefore tends uniformly to zero. Hence the series is uniformly convergent.

1·1153. Abel's lemma. Though the M test is the commonest in actual applications, series may be uniformly convergent and not satisfy it. Two more sensitive tests are based on Abel's lemma. All these tests have analogues for integrals.

If $\{v_r\}$ is a non-increasing sequence of non-negative quantities, and if the sums

$$s_p = a_1 + a_2 + \ldots a_p$$

satisfy the inequalities $h \leqslant s_p \leqslant H$ *for all p, then* $hv_1 \leqslant \sum_{1}^{n} a_p v_p \leqslant H v_1$ *for all n.* We have

$$a_1 = s_1, a_2 = s_2 - s_1, \ldots, a_n = s_n - s_{n-1};$$

$$S_n = \sum_{1}^{n} a_p v_p = s_1 v_1 + \sum_{p=2}^{n} (s_p - s_{p-1}) v_p$$

$$= s_1(v_1 - v_2) + \ldots + s_{n-1}(v_{n-1} - v_n) + s_n v_n.$$

Since all $v_p - v_{p+1} \geq 0$ and $v_n \geq 0$, the last sum will not be decreased if all the s_p are replaced by H; and therefore $S_n \leq Hv_1$. Similarly the sum will not be increased if all the s_p are replaced by h; hence $S_n \geq hv_1$.

1·1154. Abel's test. *If the series Σa_r is convergent (not necessarily absolutely) and if for all x in an interval $\{v_r(x)\}$ is a sequence of positive quantities, bounded with respect to x and r and non-increasing for given x as r increases, then $\Sigma a_r v_r(x)$ is uniformly convergent in the interval.* In Abel's lemma take

$$s_n = \sum_{p=1}^n a_{m+p}, \quad S_n(x) = \sum_{p=1}^n a_{m+p} v_{m+p}(x)$$

and take m so that $|s_n| < \omega$ for all n. Then by Abel's lemma, with

$$h = -\omega, \quad H = \omega, \quad v_1 = v_{m+1}(x) \leq M,$$
$$-\omega M \leq S_n(x) \leq \omega M$$

for all x of the interval and all $n \geq 1$. Since ω is arbitrarily small and independent of x, uniform convergence follows.*

The most important application of this theorem is to power series $\Sigma a_n x^n$. If the series converges for $x = 1$, the powers x^n, for $0 \leq x \leq 1$, satisfy the conditions imposed on $v_n(x)$; hence the series $\Sigma a_n x^n$ converges uniformly up to $x = 1$ and the limit of its sum is Σa_n. This is *Abel's theorem*. It saves a great deal of trouble; for we often get a result in the form of a power series and want to know whether the sum of the series for $x < 1$ tends in the limit to the sum for $x = 1$ when x is made to approach 1. The theorem gives us a simple answer: it does so provided the series for $x = 1$ converges.

The theorem is still true if $a_n = a_n(x)$ and $\Sigma a_n(x)$ is uniformly convergent. The proof needs no change.

1·1155. Dirichlet-Hardy test.† *If in an interval of x, $\sum_1^n a_r(x)$ is uniformly bounded with respect to n and x, and $\{v_r\}$ is a sequence of positive non-increasing quantities tending to zero, then $\Sigma a_r(x) v_r$ is uniformly convergent in the interval.* We can extend this to the case where $v_r = v_r(x)$ provided that $v_r(x) \to 0$ uniformly.

Take $S_n(x) = \sum_{p=1}^n a_{m+p} v_{m+p}$ where m is such that $v_m(x) < \omega$. Then if, for all n,

$$-M \leq \sum_{p=1}^n a_{m+p}(x) \leq M,$$

we have by Abel's lemma $\quad -M\omega \leq S_n(x) \leq M\omega$.
Uniform convergence follows.

A remarkable feature of this test is that it establishes uniform convergence without requiring any comparison series to converge. The most important applications are to series of the forms $\Sigma v_n \cos n\theta$, $\Sigma v_n \sin n\theta$. Here

$$\sum_1^n \cos n\theta = \frac{\sin(n+\tfrac12)\theta - \sin\tfrac12\theta}{2\sin\tfrac12\theta}, \quad \sum_1^n \sin n\theta = \frac{\cos\tfrac12\theta - \cos(n+\tfrac12)\theta}{2\sin\tfrac12\theta}.$$

---

* The case where $v_n(x) = x^n$ ($0 \leq x \leq 1$) was proved and used by Abel. The general form of the theorem is due to Hardy.

† Dirichlet gave a test for convergence of a series of constants, which Hardy converted into a test for uniform convergence. Hardy proposed to call the tests Abel's and Dirichlet's respectively, but the application to uniform convergence in the latter case is entirely due to Hardy.

If $\sin \tfrac12\delta$, with $\tfrac12\pi > \delta > 0$, is the smallest value of $|\sin \tfrac12\theta|$ in a range, the modulus of neither sum exceeds $\operatorname{cosec} \tfrac12\delta$, whatever n and θ may be. If then $v_1 \geqslant v_2 \geqslant \ldots \geqslant v_n \to 0$, it follows that $\Sigma v_n \cos n\theta$ and $\Sigma v_n \sin n\theta$ are uniformly convergent in any closed interval $a \leqslant \theta \leqslant b$ that contains no zero of $\sin \tfrac12\theta$; that is, excluding $\theta = 0, 2\pi, 4\pi, \ldots$.

In particular, the series
$$1 + \cos\theta + \tfrac12 \cos 2\theta + \tfrac13 \cos 3\theta + \ldots,$$
$$\sin\theta + \tfrac12 \sin 2\theta + \tfrac13 \sin 3\theta + \ldots,$$
are uniformly convergent in any range $a \leqslant \theta \leqslant b$ that excludes $0, 2\pi, \ldots$. Actually the first diverges at $\theta = 0$, the second converges everywhere, but not uniformly in any interval containing $\theta = 0$. We shall see later (ch. 14, ex. 4) that it jumps from $-\tfrac12\pi$ to $\tfrac12\pi$ as θ increases through 0, so that non-uniform convergence is associated with a discontinuity as in 1·114.

1·116. Theorem of bounded convergence. Uniformity of convergence is a sufficient condition for continuity or integrability of the sum, provided the separate terms are continuous or integrable. It is far from a necessary condition. In practice it is usually easier to test directly whether the limit function is integrable than to test for uniform convergence, and there are so many cases where the passage to the limit under the integral sign is valid without convergence being uniform that a more general rule is needed. Such a rule is as follows. It is known as the *theorem of bounded convergence*[a]. *If for all $a \leqslant x \leqslant b$, $|f_n(x)| \leqslant M$ for all n and x, if all $f_n(x)$ are integrable and if $f_n(x) \to f(x)$, where $f(x)$ is integrable, then $\int_a^b f_n(x)\,dx \to \int_a^b f(x)\,dx$.* The proof is not easy, but the result should be known.

The behaviour of $f_n(x)$ and
$$\lim_{n\to\infty} \int_0^1 f_n(x)\,dx, \quad \int_0^1 \lim_{n\to\infty} f_n(x)\,dx$$
should be studied for the cases
$$f_n(x) = xe^{-nx}, \quad f_n(x) = nxe^{-nx}, \quad f_n(x) = n^2 xe^{-nx}.$$

1·117. Useful comparison series. By far the most important comparison series are Σx^n $(0 \leqslant x < 1)$, Σn^{-s} $(s > 1)$, which we have already studied, and $\Sigma n^s a^n$ $(0 \leqslant a < 1)$. The convergence of the latter follows at once from the M test if $s < 0$. If $s \geqslant 0$, we have
$$\frac{u_{n+1}}{u_n} = \left(\frac{n+1}{n}\right)^s a = \left(1 + \frac{1}{n}\right)^s a.$$
As n increases this tends to a. Hence, since $a < 1$, we can take m large enough for this ratio to be less than b for all $n > m$, where $a < b < 1$. Then for $n > m$,
$$|u_n| < u_m b^{n-m},$$
and Σb^{n-m} is a convergent series of positive terms. Hence $\Sigma n^s a^n$ converges for $0 \leqslant a < 1$.

Comparison with the series Σn^{-s} can often be simplified. If $v_n = n^{-s}$ $(1 < s)$,
$$n\left(1 - \frac{v_n}{v_{n-1}}\right) = n\left\{1 - \left(\frac{n-1}{n}\right)^s\right\} \to s.$$
If u_n is positive for all n, and
$$n\left(1 - \frac{u_n}{u_{n-1}}\right) \to t > 1,$$

we can take $s = \frac{1}{2}(t+1)$, and then we can choose m so that for all $n > m$

$$n\left(1 - \frac{u_n}{u_{n-1}}\right) > n\left\{1 - \left(\frac{n-1}{n}\right)^s\right\};$$

and then

$$u_n < u_m \left(\frac{m}{n}\right)^s$$

and Σu_n converges. Similarly, if t exists and is less than 1, Σu_n diverges. If $t = 1$ a more sensitive test is needed, but we shall find no such case in this book.

To summarize, *if* $u_n > 0$, Σu_n *converges if either*

$$u_n/u_{n-1} \to k < 1,$$

or if

$$\frac{u_n}{u_{n-1}} \to 1, \quad n\left(1 - \frac{u_n}{u_{n-1}}\right) \to t > 1.$$

1·12. Uniform convergence of infinite integrals. If the integrand depends on x and also on another parameter y, the notion of *uniform convergence* arises as for series. We shall suppose in all cases that $\int_a^X f(x)\,dx$ exists however large X may be. This remark is needed because no meaning can be attached to the convergence of an integral, that is, to the proposition that a set of integrals with finite upper termini tend to a limit when the upper terminus tends to infinity, unless these integrals all exist. It is with the convergence of the infinite integral, assuming the existence of the finite integrals, that we are concerned in what follows. In particular, if $\int_0^X f(x)\,dx$ exists and $\int_X^Y |f(x)|\,dx < \epsilon$, for all $Y > X$, $\int_0^\infty f(x)\,dx$ converges; but the existence of $\int_0^X |f(x)|\,dx$ does not guarantee that of $\int_0^X f(x)\,dx$. If for any ϵ we can choose X so that for all Y greater than X and for all y in the range b_0 to b_1

$$\left|\int_X^Y f(x,y)\,dx\right| < \epsilon$$

the integral $\int_0^\infty f(x,y)\,dx$ is said to be uniformly convergent in the range $b_0 \leqslant y \leqslant b_1$. This property permits the reversal of the order of integration in a repeated integral even when one of the limits is infinite. By a repeated integral we mean one of the form

$$\int_{b_0}^{b_1} dy \int_{a_0}^{a_1} f(x,y)\,dx,$$

where $f(x,y)$ is to be integrated with regard to x between a_0 and a_1 and the result with regard to y from b_0 to b_1. Let us consider the integral, where $f(x,y)$ is supposed continuous with regard to both x and y,

$$I = \int_{b_0}^{b_1} dy \int_a^\infty f(x,y)\,dx = \int_{b_0}^{b_1} dy \left\{\int_a^X f(x,y)\,dx + \int_X^\infty f(x,y)\,dx\right\}. \qquad (1)$$

Now since all limits are finite

$$\int_{b_0}^{b_1} dy \int_a^X f(x,y)\,dx = \int_a^X dx \int_{b_0}^{b_1} f(x,y)\,dy. \qquad (2)$$

(The proof is simple.*) X is at our disposal; choose it so that for all $Y > X$

$$\left| \int_X^Y f(x,y)\,dx \right| < \omega. \tag{3}$$

Then the second part of (1) is numerically not greater than $(b_1 - b_0)\,\omega$; and

$$\left| I - \int_a^X dx \int_{b_0}^{b_1} f(x,y)\,dy \right| < (b_1 - b_0)\,\omega. \tag{4}$$

But ω is arbitrarily small and we can always choose X so that (3) will be satisfied. Hence

$$I = \lim_{X \to \infty} \int_a^X dx \int_{b_0}^{b_1} f(x,y)\,dy = \int_a^\infty dx \int_{b_0}^{b_1} f(x,y)\,dy, \tag{5}$$

which establishes the theorem.

This theorem can be stated in the form: *a uniformly convergent integral can be integrated under the integral sign.* It follows that an infinite integral $\int_a^\infty f(x,y)\,dx$ can be differentiated under the integral sign with regard to y provided that $\partial f/\partial y$ exists and that its integral with regard to x is uniformly convergent in the neighbourhood of the value of y under consideration. This follows immediately by putting $\partial f/\partial y$ for $f(x,y)$ in the last theorem.

An extension to uniformly convergent integrals, where $f(x,y)$ is not necessarily continuous, can be made on the lines of the argument at the end of 1·113.

1·121. M test. The commonest test for uniform convergence is the analogue of the M test for series. *If for all y such that $b_0 \leqslant y \leqslant b_1$,*

$$|f(x,y)| < g(x),$$

where $\int_a^\infty g(x)\,dx$ converges, $\int_a^\infty f(x,y)\,dx$ is uniformly and absolutely convergent in $b_0 \leqslant y \leqslant b_1$.

1·122. Abel's lemma for integrals. *If $v(x)$ is non-negative, bounded in $a \leqslant x \leqslant b$ and non-increasing with x, and if h, H are the lower and upper bounds of*

$$F(\xi) = \int_a^\xi f(x)\,dx$$

for $a \leqslant \xi \leqslant b$, then

$$hv(a) \leqslant \int_a^b f(x)\,v(x)\,dx \leqslant Hv(a).$$

Put

$$I = \int_a^b f(x)\,v(x)\,dx = \int_{x=a}^b v(x)\,dF(x)$$

$$= v(b)\,F(b) - \int_{x=a}^b F(x)\,dv(x).$$

This is valid because $F(x)$ is an integral and therefore continuous, and $v(x)$ is of bounded variation. Then, since $v(x)$ is nowhere increasing, I will not be decreased if $F(x)$ is replaced everywhere by its upper bound, or increased if it is replaced by its lower bound; then

$$h\left\{ v(b) - \int_{x=a}^b dv(x) \right\} \leqslant I \leqslant H\left\{ v(b) - \int_{x=a}^b dv(x) \right\},$$

that is,

$$hv(a) \leqslant I \leqslant Hv(a).$$

---

* For $f(x,y)$ continuous; the statement will be seen in Chapter 5 to be true under somewhat wider conditions.

It is necessary for integrals to specify that $v(x)$ is bounded; being non-negative it must have a lower bound, but it might be unbounded near $x = a$ if this is not stated separately.

1·123. Abel's test for infinite integrals. *If $\int_a^\infty f(x)\,dx$ converges (not necessarily absolutely) and if for every value of y in $b_0 \leq y \leq b_1$ the function $v(x, y)$ is non-negative, bounded for all x, y and never increasing with x, then $\int_a^\infty f(x)\,v(x, y)\,dx$ is uniformly convergent with respect to y in $b_0 \leq y \leq b_1$.*

We have $0 \leq v(x, y) \leq M$; take X so that

$$\left| \int_X^{X'} f(x)\,dx \right| < \omega \quad \text{for all } X' > X;$$

then by Abel's lemma, for $b_0 \leq y \leq b_1$

$$\left| \int_X^{X'} f(x)\,v(x, y)\,dx \right| \leq \omega M,$$

whence uniform convergence follows since ω can be taken arbitrarily small.

For instance, $\int_0^\infty \frac{\sin x}{x}\,dx$ converges; and e^{-xy} is positive, bounded and not increasing with x for $0 \leq y < \infty$. Hence $\int_0^\infty e^{-xy} \frac{\sin x}{x}\,dx$ is uniformly convergent for $y \geq 0$.

1·124. Dirichlet-Hardy test for infinite integrals. *If $\int_a^X f(x, y)\,dx$ is bounded for all $X > a$ and for $b_0 \leq y \leq b_1$, and if $v(x)$ is bounded, positive, non-increasing, and tends to zero as $x \to \infty$, then $\int_a^\infty f(x, y)\,v(x)\,dx$ is uniformly convergent for $b_0 \leq y \leq b_1$.* Here we can take X so that $v(X) < \omega$ and for all $X' > X$ there is M such that

$$-M < \int_X^{X'} f(x, y)\,dx < M.$$

Then for $b_0 \leq y \leq b_1$ and all $X' > X$

$$\left| \int_X^{X'} f(x, y)\,v(x)\,dx \right| \leq M\omega.$$

Uniform convergence follows as before.

Note that $\int_a^X f(x, y)\,dx$ is not required to tend to a limit as $X \to \infty$; it may oscillate finitely.

For instance, $$\int_0^X \sin xy\,dx = \frac{1 - \cos Xy}{y},$$

and if $|y| > \delta > 0$ this is numerically less than $2/\delta$. Also $1/x$ is positive and tends to zero with increasing x. Hence $\int_a^\infty \frac{\sin xy}{x}\,dx$, where $a > 0$, is uniformly convergent in any range such that $|y| > \delta > 0$. It is not uniformly convergent in any range that includes $y = 0$. Actually it is equal to $+\tfrac{1}{2}\pi$ for $y > 0$, $-\tfrac{1}{2}\pi$ for $y < 0$, and 0 for $y = 0$; so that, as for series, non-uniform convergence of an integral can be associated with discontinuity of its value.

Uniform convergence of an integral of a continuous function is a very useful sufficient condition for continuity of the integral and for the legitimacy of integration under the integral sign. We have had one case where non-uniform convergence is associated with discontinuity of the integral. The following example, given by Courant,* shows that it can be associated with the impossibility of reversing the order of integration. If

$$f(x, y) = (2 - xy) xy e^{-xy},$$

we find

$$\int_0^1 dx \int_0^\infty f(x, y) \, dy = 0, \quad \int_0^\infty dy \int_0^1 f(x, y) \, dx = 1.$$

We have

$$\int_0^y f(x, u) \, du = xy^2 e^{-xy}.$$

For any $x \neq 0$ this tends to 0 as $y \to \infty$; and for $x = 0$, $f(x, y) = 0$ for all y. Thus $\int_0^\infty f(x, y) \, dy$ is convergent, but it is not uniformly convergent near $x = 0$, since if η is the larger value of y that makes $xy^2 e^{-xy} = \epsilon$, $xy^2 e^{-xy} < \epsilon$ for all $y > \eta$; but η tends to infinity as $x \to 0$. In fact $\int_0^y f(x, u) \, du$ is unbounded with regard to y as $x \to 0$.

The extension of the results for integrals with respect to two variables to integrals with respect to three and more variables involves no new principles.

The following application of Dirichlet's test is sometimes useful. Let

$$I = \int_a^\infty \cos\{f(x)\} \, dx,$$

where $f'(x)$ is a positive increasing function for $x \geqslant a$, and $f'(x) \to \infty$ as $x \to \infty$. Put $f(x) = y$, $f'(x) = 1/g(y)$, $f(a) = b$. Then y is an increasing function of x, and

$$I = \int_b^\infty \cos y \, g(y) \, dy.$$

But $g(y) > 0$, and is a decreasing function with limit 0. Hence *a sufficient condition for* $\int_0^\infty \cos\{f(x)\} \, dx, \int_0^\infty \sin\{f(x)\} \, dx$ *to converge is that $f'(x)$ is an increasing function tending to ∞.* For instance

$$\int_0^\infty \cos x^2 \, dx, \quad \int_0^\infty \cos(x^3 - mx) \, dx \quad (m \text{ real})$$

converge. (For the latter, if $m > 0$, take $a > (\tfrac{1}{3}m)^{\frac{1}{2}}$.)

1·125. Integrals with upper limit tending to infinity. If $f(x, n) \to g(x)$, and $\lambda_n \to \infty$ when $n \to \infty$, we sometimes need a condition that

$$\int_a^{\lambda_n} f(x, n) \, dx \to \int_a^\infty g(x) \, dx. \tag{1}$$

The question is clearly related to that of uniform convergence; in fact we can define a function

$$h(x, n) = f(x, n) \quad (a \leqslant x \leqslant \lambda_n), \quad h(x, n) = 0 \quad (\lambda_n < x) \tag{2}$$

* *Differential and Integral Calculus*, 2, 1936, 316.

and then
$$\int_a^{\lambda_n} f(x,n)\,dx = \int_a^\infty h(x,n)\,dx. \tag{3}$$

Consequently a sufficient condition for (1) is that $h(x,n)$ shall satisfy any of the tests of 1·121, 1·123, and 1·124.

Detailed proofs of the required forms of 1·123 and 1·124, and of the analogues for series, are given by Bromwich* under the name of Tannery's theorem.

1·126. Inversion of infinite double series and repeated integrals. The theorem (due to E. H. Moore) for uniform convergence corresponding to 1·111 is as follows.

If as $y \to \infty$, $f(x,y) \to h(x)$, *and if as* $x \to \infty$, $f(x,y) \to g(y)$ *uniformly, then*
$$\lim_{x\to\infty} h(x), \quad \lim_{y\to\infty} g(y) \tag{1}$$
both exist and are equal. Take X so that $|f(x,y)-g(y)| < \omega$ for $x \geqslant X$ and all y. Then take Y so that $|f(X,y)-h(X)| < \omega$ for $y \geqslant Y$. Then for $x > X$ and $y \geqslant Y$
$$\left.\begin{array}{l} f(x,y)-h(X) = \{f(x,y)-g(y)\}-\{f(X,y)-g(y)\}+\{f(X,y)-h(X)\} \\ |f(x,y)-h(X)| < 3\omega, \quad |g(y)-h(X)| \leqslant 3\omega, \end{array}\right\} \tag{2}$$
and therefore if $y_1 > Y$,
$$|g(y_1)-g(Y)| \leqslant 6\omega. \tag{3}$$
Since ω is arbitrary, $g(y)$ has a limit F.

Hence we can take Y' so that $|g(y)-F| \leqslant \omega$ for all $y > Y'$. If Y'' is the greater of Y, Y', and $x > X$, $y > Y''$,
$$|f(x,y)-g(y)| + |g(y)-F| \leqslant 2\omega. \tag{4}$$
Let $y \to \infty$; then $|h(x)-F| \leqslant 2\omega$ and $h(x) \to F$.

There are corollaries for sums of double series, sums of infinite integrals, and repeated integrals, analogous to those of 1·111.

For series, if
$$f(m,n) = \sum_{r=1}^m \sum_{s=1}^n u(r,s), \tag{5}$$
the conditions are: as $n \to \infty$, $f(m,n)$ converges for any m, and as $m \to \infty$, $f(m,n)$ converges uniformly to $g(n)$ for all n.

If
$$f(m,x) = \sum_{r=1}^m \int_0^x u_m(\xi)\,d\xi, \tag{6}$$
the conditions are: the integrals converge as $x \to \infty$ for any m, and the sum for finite x converges uniformly for all x greater than some x_0. Alternatively, taking
$$f(m,x) = \int_0^x \left\{\sum_{r=1}^m u_m(\xi)\right\} d\xi,$$
we have the conditions: the series converges over any finite interval of ξ, and the integral converges uniformly for all m greater than m_0.

If
$$f(x,y) = \int_0^x d\xi \int_0^y \phi(\xi,\eta)\,d\eta = \int_0^y d\eta \int_0^x \phi(\xi,\eta)\,d\xi, \tag{7}$$
the conditions are: the integral with regard to η converges in any interval of ξ, and the second integral converges uniformly for all y greater than some Y.

* *Theory of Infinite Series*, 1908, 123, 438, 443.

1·13. Mean-value theorems.

We have seen that a continuous function takes its upper and lower bounds in any interval. Let $f(x)$ be continuous for $a \leqslant x \leqslant b$ and have a derivative $f'(x)$ for $a < x < b$, and let $f(a) = f(b) = 0$, but for some intermediate value ξ, $f(\xi) > 0$. Then let $x = \eta$ correspond to the upper bound of $f(x)$ in the interval. η is not equal to a or b, since $f(\eta) \geqslant f(\xi) > 0$. Now

$$f'(\eta) = \lim_{h \to 0} \frac{f(\eta+h) - f(\eta)}{h}.$$

If $h > 0$, $f(\eta+h) \leqslant f(\eta)$, and therefore $f'(\eta) \leqslant 0$. If $h < 0$, $f(\eta+h) \leqslant f(\eta)$, and therefore $f'(\eta) \geqslant 0$. These are consistent only if $f'(\eta) = 0$. $f'(x)$ is not required to be continuous.

If $f(x)$ has a lower bound < 0 in the interval, we obtain the same result by applying the argument to $-f(x)$. Hence *if $f(x)$ has a derivative in $a < x < b$, and is continuous in $a \leqslant x \leqslant b$, the derivative vanishes between any two zeros of $f(x)$*. This result is known as *Rolle's theorem*.

If c, d are any two values of x such that $f(x)$ is continuous for $c \leqslant x \leqslant d$ and has a derivative for $c < x < d$, consider the function

$$g(x) = f(x) - f(c) - \frac{x-c}{d-c}\{f(d) - f(c)\}.$$

This vanishes for $x = c$ and $x = d$. Its derivative therefore vanishes for some x between c and d, say η; and then

$$0 = g'(\eta) = f'(\eta) - \frac{f(d) - f(c)}{d-c}.$$

Thus
$$f(d) - f(c) = (d-c) f'(\eta),$$

where $c < \eta < d$. This is the *mean-value theorem* for derivatives. Geometrically it states that if we take a chord of a smooth curve, the tangent at some intermediate point is parallel to the chord.

The most important application is: *if $f(x)$ is continuous in $a \leqslant x \leqslant b$ and $f'(x) = 0$ for $a < x < b$, then $f(x)$ is constant in (a, b)*. For $f(x) - f(a) = (x-a) f'(\xi)$, where $a < \xi < x$; but $f'(\xi) = 0$ and therefore $f(x) = f(a)$. Note that it is not sufficient that $f'(x) = 0$ almost everywhere. A function is known, continuous in $(0, 1)$, and with a derivative almost everywhere zero in $(0, 1)$, but the function is not constant. But it is sufficient that $f'(x) = 0$ except at a finite set of points, where $f(x)$ is continuous. If $f(x)$ is continuous, and $F(x) = \int_a^x f(u)\,du$, then

$$\frac{dF(x)}{dx} = f(x).$$

If also $G'(x) = f(x)$, then $F(x) - G(x)$ is a continuous function with zero derivative everywhere and is therefore constant. The corresponding theorem where $f(x)$ is given only to be integrable was given in 1·103. Either can be used to justify the method of integration by first finding a function whose derivative is $f(x)$. Another consequence is useful in some cases where a derivative is required at a point $x = a$ but for some special reason is difficult to evaluate there, owing, for instance, to failure of an integral or series representing it to converge. If $f(x)$ is continuous and $f'(x)$ exists except possibly at $x = a$, we have

$$\frac{f(a+h) - f(a)}{h} = f'(a + \theta h),$$

where $0 < \theta < 1$. Now let h tend to zero; then if the left side tends to a limit this limit is $f'(a)$. But if $f'(x)$ tends to a limit when x tends to a the right side tends to this limit, and the left side, being equal to the right, also has this limit. Hence *if $f(x)$ is continuous and $f'(x)$ has a limit as x tends to a, $f'(a)$ exists and is equal to this limit.*

If $\dfrac{1}{h}\{f(a+h)-f(a)\}$ has a limit as $h \to 0$ through positive values, this limit may be called the *derivative of $f(x)$ on the right* at a and denoted by $f'(a+)$. The last argument applies equally to show that if $f(x)$ is continuous on the right at a and $f'(a+h)$ has a limit as $h \to 0$ through positive values, $f'(a+)$ exists and is equal to this limit. Similar properties hold for derivatives on the left. The statement that $f'(a)$ exists is equivalent to the statement that $f'(a+)$ and $f'(a-)$ exist and are equal.

1·131. The first mean-value theorem for integrals. If for all x such that $a \leqslant x \leqslant b$, $m \leqslant f(x) \leqslant M$,

$$m(b-a) \leqslant \int_a^b f(x)\, dx \leqslant M(b-a),$$

and therefore

$$\int_a^b f(x)\, dx = N(b-a),$$

where N is such that $m \leqslant N \leqslant M$. In particular if $f(x)$ is continuous there is a ξ in the range such that $f(\xi) = N$, and

$$\int_a^b f(x)\, dx = (b-a)f(\xi).$$

1·132. Extension of first mean-value theorem. To obtain Taylor's theorem we require a special case of an extension of the first mean-value theorem; namely, if $g(x)$ is $\geqslant 0$ for $a \leqslant x \leqslant b$, and $m \leqslant f(x) \leqslant M$,

$$m\int_a^b g(x)\, dx \leqslant \int_a^b f(x)g(x)\, dx \leqslant M\int_a^b g(x)\, dx,$$

which we can write in the form

$$\int_a^b f(x)g(x)\, dx = N\int_a^b g(x)\, dx \quad (m \leqslant N \leqslant M)$$

$$= f(\xi)\int_a^b g(x)\, dx \quad (a \leqslant \xi \leqslant b),$$

the last being true for some ξ if $f(x)$ is continuous.

1·133. Taylor's theorem. Let $f(x)$ have derivatives up to order n, and denote the nth derivative by $f^{(n)}(x)$. Consider the function

$$g(x) = f(x) - f(0) - xf'(0) - \frac{x^2}{2!}f''(0) - \ldots - \frac{x^{n-1}}{(n-1)!}f^{(n-1)}(0).$$

On differentiating $n-1$ times in succession we see that $g(x)$ and all its derivatives up to the $(n-1)$th vanish at $x = 0$. Also

$$g^{(n)}(x) = f^{(n)}(x).$$

Now consider the integral $\quad h(x) = \int_0^x \frac{(x-u)^{n-1}}{(n-1)!} f^{(n)}(u)\, du,$

where $f^{(n)}(u)$ is supposed bounded and integrable. (The reason for this choice will appear when we discuss operational methods.)

By repeated integration by parts

$$h(x) = \int_0^x \frac{(x-u)^{n-1}}{(n-1)!} df^{(n-1)}(u) = \left[\frac{(x-u)^{n-1}}{(n-1)!} f^{(n-1)}(u)\right]_0^x + \int_0^x \frac{(x-u)^{n-2}}{(n-2)!} f^{(n-1)}(u)\, du$$

$$= -\frac{x^{n-1}}{(n-1)!} f^{(n-1)}(0) + \int_0^x \frac{(x-u)^{n-2}}{(n-2)!} f^{(n-1)}(u)\, du$$

$$= -\frac{x^{n-1}}{(n-1)!} f^{(n-1)}(0) - \ldots - \frac{x^2}{2!} f''(0) - x f'(0) + \int_0^x f'(u)\, du$$

$$= g(x).$$

Hence $\quad f(x) = f(0) + x f'(0) + \frac{x^2}{2!} f''(0) + \ldots + \frac{x^{n-1}}{(n-1)!} f^{(n-1)}(0) + \int_0^x \frac{(x-u)^{n-1}}{(n-1)!} f^{(n)}(u)\, du.$

This is an exact form of Taylor's theorem. It does not require $f^{(n)}(u)$ to be continuous. Also

$$\int_0^x \frac{(x-u)^{n-1}}{(n-1)!}\, du = \frac{x^n}{n!},$$

whence if m and M are the lower and upper bounds of $f^{(n)}(u)$ for $0 \leqslant u \leqslant x$, the integral lies between $mx^n/n!$ and $Mx^n/n!$. Hence it is equal to $Nx^n/n!$, where $m \leqslant N \leqslant M$. If further $f^{(n)}(u)$ is continuous there will be a value θx ($0 < \theta < 1$) that makes it equal to N, and the integral can be written

$$\frac{x^n}{n!} f^{(n)}(\theta x).$$

This is Lagrange's form of the remainder. But the form $Nx^n/n!$ only requires the nth derivative to be integrable.

An alternative form of the remainder, due to Cauchy, is got by noticing that, if $f^{(n)}(u)$ is continuous, there is a θ such that $0 \leqslant \theta \leqslant 1$ and such that

$$\int_0^x (x-u)^{n-1} f^{(n)}(u)\, du = x(x - \theta x)^{n-1} f^{(n)}(\theta x)$$

by the first mean-value theorem for integrals; hence the remainder can be written

$$\frac{x^n}{(n-1)!}(1-\theta)^{n-1} f^{(n)}(\theta x).$$

It will be seen that these forms of Taylor's theorem do not require the convergence of an infinite series. But they are much used in proving convergence by showing that the remainder tends to 0 for large n and in estimating the error possible for a given finite number of terms.

1·134. The second mean-value theorem for integrals. We give first a simple consequence of Abel's lemma for integrals. Let $f(x)$ be integrable in (a, b) and $\phi(x)$ have bounded variation. Let $P(x)$, $N(x)$ be the positive and negative variations of $\phi(x)$ in (a, x) and let the greater of $P(b)$, $N(b)$ be ω. Then $\omega - P(x)$, $\omega - N(x)$ satisfy the conditions imposed on $v(x)$ in Abel's lemma, and if h, H are the lower and upper bounds of $\int_a^x f(t)\,dt$ for $a \leqslant x \leqslant b$,

$$\omega h \leqslant \int_a^b \{\omega - P(x)\} f(x)\,dx \leqslant \omega H, \tag{1}$$

$$\omega h \leqslant \int_a^b \{\omega - N(x)\} f(x)\,dx \leqslant \omega H. \tag{2}$$

By subtraction, $\quad \left| \int_a^b f(x)\{\phi(x) - \phi(a)\}\,dx \right| \leqslant \omega(H-h). \tag{3}$

Hence $\int_a^b f(x)\,\phi(x)\,dx$ can be replaced by $\phi(a) \int_a^b f(x)\,dx$ within a known uncertainty. This result is important especially in the theory of Fourier series and integrals.

The second mean-value theorem as usually understood means one of the following.

Bonnet's form: *if $\phi(x)$ is positive and non-increasing for $a \leqslant x \leqslant b$, there is an η such that $a \leqslant \eta \leqslant b$,*

$$\phi(a) \int_a^\eta f(x)\,dx = \int_a^b f(x)\,\phi(x)\,dx. \tag{4}$$

Du Bois-Reymond's form: *if $\phi(x)$ is monotonic for $a \leqslant x \leqslant b$, there is a ξ such that $a \leqslant \xi \leqslant b$,*

$$\int_a^b f(x)\,\phi(x)\,dx = \phi(a) \int_a^\xi f(x)\,dx + \phi(b) \int_\xi^b f(x)\,dx. \tag{5}$$

Both are easily derived from Abel's lemma; but, as Bromwich remarks,* they contain no information not contained in the lemma itself because no means is provided for estimating ξ and η; and they are less directly informative than (3) above.[a]

1·14. Infinite products. If

$$\Pi_n = (1+a_1)(1+a_2)\ldots(1+a_n), \tag{1}$$

the limit of Π_n when n tends to infinity, if it exists and is not zero, is denoted by

$$\prod_1^\infty (1+a_n), \tag{2}$$

and the infinite product is then said to converge. It is zero if any factor is zero. (In the latter case it is often said to *converge to zero* to distinguish it from such a product as

$$(1-\tfrac{1}{2})(1-\tfrac{1}{2})(1-\tfrac{1}{2})\ldots,$$

which tends to zero without any factor being zero. Such products are said to *diverge to zero*.)

The theory of convergence of infinite products is closely related to that of infinite series; in fact, if all a_n are positive, or all negative, *a necessary and sufficient condition for the convergence of* (2) *is that* Σa_n *shall converge*. We have

$$S_n = \log \Pi_n = \sum_1^n \log(1+a_r). \tag{3}$$

Clearly neither $\Pi(1+a_n)$ nor Σa_n can converge unless $a_n \to 0$. We need therefore only consider the case $a_n \to 0$. Then

$$\frac{\log(1+a_n)}{a_n} \to 1. \tag{4}$$

* *Theory of Infinite Series*, 1908, 426–7.

Hence the ratios of corresponding terms of the two series $\Sigma \log(1+a_n)$ and Σa_n are bounded and the series converge or not together. But if S_n has a limit S, Π_n has a limit e^S, and conversely if Π_n tends to a limit, not ∞ or 0, S_n tends to a limit; which proves the proposition.

If the a_n are not all of one sign and $\Sigma|a_n|$ is convergent, $\Pi(1+a_n)$ is easily shown to be convergent. If $\Sigma|a_n|$ is not convergent but $a_n \to 0$ we can choose m so that for $n > m$

$$\log(1+a_n) = a_n - \tfrac{1}{2}\lambda_n a_n^2,$$

where $c < |\lambda_n| < d$ and c, d are fixed; and therefore, if Σa_n^2 is convergent and Σa_n is conditionally convergent, $\Pi(1+a_n)$ is convergent.

The argument fails for such a product as $\Pi\left\{1 + \dfrac{(-1)^n}{\sqrt{n}}\right\}$. Here Σa_n converges but Σa_n^2 does not. This product is easily proved not convergent.

1·15. The Lipschitz condition. If

$$|f(\xi) - f(x)| \leqslant A |\xi - x|^\alpha$$

for given x and all $|\xi - x| < \delta$, where A, α are independent of ξ, and $\alpha > 0$, $f(\xi)$ is said to satisfy a Lipschitz condition of order α at $\xi = x$. If $f(\xi)$ satisfies a Lipschitz condition it is continuous at $\xi = x$, and if it satisfies one for all x in $a \leqslant x \leqslant b$ it is continuous for $a \leqslant x \leqslant b$. But even if $\alpha = 1$ the function need not be differentiable or have bounded variation. For instance, take

$$f(0) = 0, \quad f(x) = x \sin\frac{1}{x}.$$

This satisfies a Lipschitz condition of order 1 even at $x = 0$; but it is not differentiable at $x = 0$, and it has unbounded variation in any interval including $x = 0$. The function $|x|$ satisfies a Lipschitz condition of order 1 at $x = 0$, and has bounded variation in any interval, but is not differentiable at $x = 0$. For some theorems it is found that the Lipschitz condition is sufficient when continuity is not sufficient and differentiability is sufficient but not necessary.

If a Lipschitz condition of order $\alpha > 1$ is satisfied at x, clearly $f'(x) = 0$. If at every point of the interval it is satisfied for some $\alpha > 1$, $f'(x) = 0$ throughout the interval. Hence $f(x)$ is constant; consequently only the case $0 < \alpha \leqslant 1$ is of much interest.

An important case is where $f(x)$ satisfies a Lipschitz condition of order 1 uniformly in (a, b); that is, if a constant A exists such that $|f(x_2) - f(x_1)| \leqslant A|x_2 - x_1|$ for all x_1, x_2 in (a, b). Clearly $f(x)$ must be both continuous and of bounded variation. It is not necessarily differentiable at all points of (a, b), as we see from the example $f(x) = |x|$ in $(-1, 1)$. But it can be proved that $f(x)$ has a derivative almost everywhere and is the Lebesgue integral of a bounded function. This is a particular case of a rather difficult theorem due to W. H. and G. C. Young that a function of bounded variation has a derivative almost everywhere;[*] but the special case is equivalent to the proposition that a curve of finite length has a tangent almost everywhere, an elementary proof of which has been given by A. S. Besicovitch.[†]

---

[*] *Proc. Lond. Math. Soc.* (2)9, 1911, 325–35. Incidentally this is not true of all continuous functions.

[†] *J. Lond. Math. Soc.* 19, 1944, 205–7.

1·16. Cauchy's inequality.
If $a_1 \ldots a_n, b_1 \ldots b_n$ are any numbers

$$(a_1^2 + \ldots + a_n^2)(b_1^2 + \ldots + b_n^2) - (a_1 b_1 + \ldots + a_n b_n)^2 = (a_1 b_2 - a_2 b_1)^2 + \ldots.$$

This is an extension of Lagrange's identity. Then

$$\sum_1^n a_r^2 \sum_1^n b_r^2 \geqslant (\sum_1^n a_r b_r)^2.$$

This is Cauchy's inequality. It follows that if $\phi(x)$, $\psi(x)$ are two functions,

$$\int_a^b \phi^2(x)\, dx \int_a^b \psi^2(x)\, dx \geqslant \left\{ \int_a^b \phi(x)\, \psi(x)\, dx \right\}^2.$$

This is Schwarz's inequality. In particular if $\psi = 1$, $\phi(x) = |f(x)|$,

$$(b-a) \int_a^b f^2(x)\, dx \geqslant \left\{ \int_a^b |f(x)|\, dx \right\}^2.$$

Also if $n = 2$, and $b_1^2 + b_2^2 = 1$, $(a_1 b_1 + a_2 b_2)^2 \leqslant a_1^2 + a_2^2$. There are numerous applications of similar results.

EXAMPLES

1. Which, if any, of the field axioms given in 1·01 are not fulfilled by (1) the set of all positive integers, (2) the set consisting of zero and all square roots of positive integers?

2. A student (naturally not at Cambridge) was heard to say that he had found a route from his lodgings to his lectures and back that was downhill both ways. Which of the axioms does his notion of height fail to satisfy?

3. If $s_n \geqslant s_{n-1}$, $t_n \geqslant t_{n-1}$, and for every m there is an n such that $t_n > s_m$, and for every m there is a p such that $s_p > t_m$, and $\{s_n\}$ is bounded, then $\{t_n\}$ converges to the same limit as $\{s_n\}$.

4. Prove that if $a_0 = 3$, $a_{n+1} = 3 - \dfrac{2}{a_n}$, $a_n \to 2$.

Show graphically, using the curves $y = 3 - \dfrac{2}{x}$, $y = x$, that the sequence has limit 2 for all values of a_0 except $a_0 = 1$. (Infinite values of a_n are allowed.) (I.C. 1938.)

5. If $s_{n+1} = \sqrt{(2 s_n + a)}$, where s_1 and a are positive and the positive value of the square root is taken, prove that as n tends to infinity, s_n tends to the limit $1 + \sqrt{(a+1)}$. (M.T. 1940.)

6. Prove that, if $s > 1$, then

$$\sum_{\nu=1}^n \left(\frac{s}{\nu} - \frac{1}{\nu^s} \right) - s \log n$$

tends to a limit as $n \to \infty$, s remaining fixed; and that, if this limit is $\phi(s)$, then

$$0 < \phi(s) + \frac{1}{s-1} \leqslant s - 1.$$ (M.T. 1938.)

7. Show how to derive a positive root of the equation

$$x^4 + 4x - 1 = 0$$

by considering the convergence and the limit of the sequence defined by

$$x_{n+1} = \frac{1}{x_n^3 + 4}.$$

Determine this root to four places of decimals. (I.C. 1937.)

Examples

8. Solve the following difference equations

   (1) $y_{n+1} - 11y_n + y_{n-1} = 0$, where $y_0 = 0$, $y_1 = 1$.

   (2) $y_{n+1} + 5y_n + 2z_n = 0$, $z_{n+1} + 2z_n + 2y_n = 0$.   (I.C. 1941.)

9. By expressing $\sin m\theta$ and $\sinh mu$ in terms of exponentials, prove the identity
$$\sum_{m=1}^{\infty} \frac{\sin m\theta}{\sinh mu} = \sum_{n=1}^{\infty} \frac{\sin \theta}{\cosh(2n-1)u - \cos\theta} \quad (u>0,\ \theta\ \text{real}). \quad \text{(M.T. 1938.)}$$

10. If $f(x)$ is equal to 0 for irrational values of x and to 1 for rational values, prove that $xf(x)$ is continuous at $x = 0$ and nowhere else, and that $x^2 f(x)$ is differentiable at $x = 0$ and nowhere else.

11. Prove that if $a^2 \ne 1$
$$\frac{1-a^{2n}}{1-a^2} = \prod_{r=1}^{n-1}\left(1 - 2a\cos\frac{r\pi}{n} + a^2\right).$$

If a is real, prove that
$$\int_0^{\pi} \log(1 - 2a\cos x + a^2)\,dx = 0$$
if $|a| < 1$, and find its value if $|a| > 1$.   (M.T. 1939.)

12. Prove that the sum of the series
$$\frac{|x|}{1+|x|} + \frac{|x|}{(1+|x|)^2} + \frac{|x|}{(1+|x|)^3} + \ldots$$
exists for all real values of x but has a discontinuity.   (M.T. 1938.)

13. For what values of x is each of the following series uniformly convergent?

   (i) $\sum_1^{\infty} \frac{(-1)^n}{n^x}$,   (ii) $\sum_1^{\infty} \frac{1}{n}\left(1 + \tfrac{1}{2} + \ldots + \frac{1}{n}\right) \sin nx$.   (M/c, III, 1928.)

14. Prove that $\log 3 = 1 + \tfrac{1}{2} - \tfrac{2}{3} + \tfrac{1}{4} + \tfrac{1}{5} - \tfrac{2}{6} + \tfrac{1}{7} + \tfrac{1}{8} - \tfrac{2}{9} + \ldots$,

and obtain a similar series for $\log a$, where a is a positive integer. (Hint: use Abel's test for uniform convergence.)   (M/c, III, 1930.)

15. Show that for any positive value of n
$$\lim_{n\to\infty} x\left(\frac{1}{n+x} + \frac{1}{n+2x} + \ldots + \frac{1}{n+nx}\right) = \log(1+x). \quad \text{(I.C. 1938.)}$$

16. Prove that the binomial series
$$1 + \sum_{r=1}^{\infty} \frac{n(n+1)\ldots(n+r-1)}{r!} x^r$$
converges for $|x| < 1$ and is unbounded for $|x| > 1$. Prove also that (1) if $x = 1$ the series converges for $n \le 0$, and is unbounded for $n > 0$, (2) if $x = -1$ the series converges for $n < 1$, is unbounded for $n > 1$, and oscillates for $n = 1$.

17. If for $n > m$, $|u_n|^{1/n} < k < 1$, show that Σu_n converges. Apply this rule to the series given by
$$u_{2n} = 2^{-2n}, \quad u_{2n+1} = 3^{-2n-1}.$$
Will the rule of 1·117 establish the convergence of this series? If not, what extension of the rule will do so?

18. Show that the series
$$\sum_1^{\infty} \sum_1^{\infty} \frac{l+m}{(l^2+m^2)^s}$$
converges or not according as $s > \tfrac{3}{2}$ or $s \le \tfrac{3}{2}$.   (M/c, III, 1932.)

19. For the two series
$$S = \sum_1^{\infty} n^{-4}, \quad T = \sum_1^{\infty} 2^{-n},$$
find how large n must be to make the error in stopping at the nth term (1) $< 0\cdot 005$, (2) $< 0\cdot 0000005$.

20. Prove that
$$\Sigma \frac{1}{n(\log n)^p}$$
converges or tends to infinity according as $p>1$ or $p\leqslant 1$.
    Hence show that if
$$u_n>0, \quad \frac{u_n}{u_{n-1}}\to 1, \quad n\left(1-\frac{u_n}{u_{n-1}}\right)\to 1, \quad \text{and} \quad \log n\left\{n\left(1-\frac{u_n}{u_{n-1}}\right)-1\right\}\to k>1$$
the series Σu_n converges.

21. Discuss the convergence of the series
$$1+\frac{ax}{b}+\frac{a(a+2)}{b(b+1)}x^2+\frac{a(a+2)(a+4)}{b(b+1)(b+2)}x^3+\ldots$$
for positive values of a, b, x. (I.C. 1944.)

22. Show that $\quad 1+\dfrac{\alpha\beta}{2\gamma}x+\ldots+\dfrac{\alpha(\alpha+1)\ldots(\alpha+n-1)\beta(\beta+1)+(\beta+n-1)}{n!\,\gamma(\gamma+1)\ldots(\gamma+n-1)}x^n+\ldots$

converges if $0\leqslant x<1$, and if $x=1$ and $\gamma>\alpha+\beta$.

23. Prove that the product of two Riemann-integrable functions has a Riemann integral.

24. Prove the result of 1·104 by integration by parts.

25. If
$$S=\sum_{1}^{\infty}\frac{1}{n^2+1},$$
prove that $\quad \tfrac{1}{4}+\tfrac{1}{4}\pi<S<\tfrac{1}{2}+\tfrac{1}{4}\pi.$

26. If $\quad f(x)=\sum_{n=0}^{N}a_n\cos\lambda_n x\pi, \quad g(x)=\sum_{n=0}^{N}a_n\cos\left(\lambda_n x+\frac{1}{x}\right)\pi,$

where the a_n and λ_n are real, prove that if $f(x)$ or $g(x)$ has no zero in the interval $\dfrac{1}{m+1}\leqslant x\leqslant\dfrac{1}{m}$, where m is a positive integer, then the other function has at least one zero in the interior of this interval. (M.T. 1938.)

27. Prove that $\quad \displaystyle\int_1^\infty dx\int_1^\infty \frac{x-y}{(x+y)^3}\,dy=-\tfrac{1}{4}, \quad \int_1^\infty dy\int_1^\infty \frac{x-y}{(x+y)^3}\,dx=\tfrac{1}{4}.$ (M.T. 1913.)

28. Investigate the convergence of the infinite products $\Pi(1+u_n)$, where
$$(1)\ u_n=\frac{(-1)^n}{n^{1/3}}, \quad (2)\ u_n=\frac{(-1)^n}{\log(n+1)}.$$ (M/c, III, 1928.)

29. By considering $f(x)=\sqrt{x}\cos\dfrac{1}{x}$, $g(x)=\sqrt{x}\sin\dfrac{1}{x}$, show that it is not a sufficient condition for the existence of $\displaystyle\int_{x=0}^{1}f(x)\,dg(x)$ that $f(x)$ and $g(x)$ shall both be continuous.

30. If $f(x)$ has derivatives up to the $(n-1)$th for $-a<x<a$, and if $f^{(n)}(0)$ exists, prove that
$$f(x)=f(0)+\sum_{r=1}^{n}f^{(r)}(0)\frac{x^r}{r!}+o(x^n).$$

31. If $f(x)$ has a derivative $f'(x)$ for $a\leqslant x\leqslant b$, and if $f'(a)<p<f'(b)$, then there is a ξ such that $a<\xi<b$ and $f'(\xi)=p$. ($f'(\xi)$ is not assumed continuous).

32. $f_n(x)$ is non-decreasing with respect to x in (a,b) or $(-\infty,\infty)$ and uniformly bounded with respect to n; and $\lim\limits_{n\to\infty}f_n(x)=f(x)$. Prove that convergence is uniform in any interval that includes no discontinuity of $f(x)$.

Chapter 2

SCALARS AND VECTORS

'The moral of *that* is, "Take care of the sense and the sounds will take care of themselves".'
　　　　　　　　　　　　　　　　　　LEWIS CARROLL, *Alice in Wonderland*

2·01. Cartesian coordinates: summation convention. Any physical measurement is the assignment of a single magnitude. Physics may be defined as the study of the relations between magnitudes, so that from one set of measurements other sets, given the conditions of observation, can be predicted. The most elementary measures, except for simple counting, are those of distance. Now we saw that distances along the same straight line are found experimentally to be additive in a definable sense, and to satisfy the associative and commutative laws of addition. But when two distances are not along the same straight line it is found that they no longer have a unique sum unless another condition is provided; if P, Q, R are three points, it is not true that the distance PR depends only on PQ and QR. There is, however, an experimentally verifiable relation between distances along any *two* intersecting lines PQQ' and PRR', namely,

$$\frac{PQ^2 + PR^2 - QR^2}{2PQ \cdot PR} = \frac{PQ'^2 + PR'^2 - Q'R'^2}{2PQ' \cdot PR'}, \tag{1}$$

when P is not between Q and Q' nor between R and R'. This ratio (a number) is denoted by $\cos\theta$, and θ is called the angle between the lines. $\cos\theta$ is never less than -1 or greater than $+1$. Now that measurement has largely replaced Euclid's methods and the experimental treatment of 'geometry' is advocated, it is desirable that one of the first steps in teaching should be the direct verification of this law, and that it should be made the basis of the development of the subject. It is far better verified than some of the usual axioms. It makes angle a derived magnitude, and the additive property of

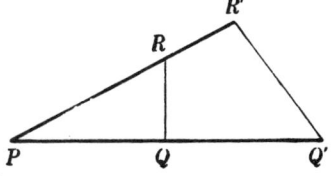

angles in a plane, taken as a postulate by Euclid, can be deduced from it. This is all to the good because a plane is a more difficult idea than a straight line. From (1) the whole Euclidean theory can be developed up to the introduction of rectangular coordinates.*

In Euclid's methods the notion of superposition plays a prominent part, and he is always speaking of the actual things compared. The tendency of modern teaching is to try to avoid superposition. But it is directly related to physical methods, and Euclid's language makes it impossible to confuse, say, a length with an area. The language of physical magnitude can say things easily that have physical meaning and would be difficult or impossible to express in Euclid's. But the attempt to reduce his system to pure mathematics removes what, for physics, are its outstanding good points.

Rectangular coordinates have the property that any distance between two points can be expressed in terms of them symmetrically by a sum of three squares of their differences.

---
* H. Jeffreys, *Scientific Inference*, Chapter 7; other considerations concerning physical magnitudes will be found in Chapters 4 and 6. It is particularly important to recognize that the establishment of scientific laws is a matter of successive approximation.

This property is shared by no other way of specifying position. The common statement that rectangular coordinates have no special physical significance is nonsense. In projective geometry the notion of distance is rejected *because* it is metrical. In physics we cannot do without it. But when it is introduced we can consider what is the shortest distance between a point and the points on a given plane (defined as the locus of points equidistant from two given points), and this leads directly to the notions of perpendiculars and rectangular coordinates.

A rectangular coordinate is a distance from a given plane; we agree to associate it with a positive sign for points on one side of the plane and a negative sign for points on the other side. Then we can speak of the displacement from P to Q as specified by three *components*, namely, the differences of the rectangular coordinates of the two points, and these differences are equal to the projections of PQ on the three axes used. As each is a distance along a given line, measured in a given direction, they have the additive property, and can be taken in any order. That is, starting from P, we can find a point P' with the same y and z coordinates as P but the x coordinate of Q; then a point P'' with the same x and z coordinates as P' and the y coordinate of Q; and finally we get to Q by varying the z coordinate. But we could adjust the coordinates in six different orders and still reach Q in three steps. This would still be true with any kind of coordinates, but with rectangular coordinates the three displacements have a special property, that each is in a given direction and has a given magnitude. (This is also true with oblique Cartesian coordinates, but these are used only in special applications and we shall not treat them till Chapter 4.) In a sense, then, we can regard parallel displacements of the same amount as equivalent. This is a particular case of the *parallelogram law*, but the latter is used in its general form only for oblique axes and we are not concerned with it at present. We can regard the equivalence as representing a physical process by supposing the displacements transferred bodily to their new starting points by means of a T-square and set-square. The process, however, is not of frequent application.

The really important property of rectangular coordinates is that they express the properties of distance equally well, and in the same form, whatever directions we take for the axes, subject only to their being mutually perpendicular. We do often want to change axes, and we require a way of inferring the coordinates with regard to one set of axes, given their values with respect to another. Now the components of a displacement with regard to the new set will be its projections on the new axes, with due attention to sign. Now if a line RS makes angles α, β, γ with the old axes, and the displacement PQ has components u, v, w with respect to them, then the projection of PQ on RS is

$$u\cos\alpha + v\cos\beta + w\cos\gamma. \tag{2}$$

This notation is cumbersome. It is usual to shorten it by denoting the three cosines by l, m, n and to call them the *direction cosines* of the line RS; then the projection becomes $lu + mv + nw$. A further shortening is achieved if we denote the axes by x_1, x_2, x_3; the components will then be denoted by u_1, u_2, u_3 and the direction cosines by l_1, l_2, l_3. Then the projection is

$$\Sigma l_i u_i \quad (i = 1, 2, 3). \tag{3}$$

The advantage of suffix notation is that the most general laws of physics have the same form for all components. Hence if we have, say, differential equations for the three coordinates of a particle it is enough to write one equation in suffix notation and let it be understood that the suffix is to take all the values 1, 2, 3 in turn. A further shortening of writing is obtained by the *summation convention*. We see that in (3) each term contains the same suffix twice, and the results are to be added. We make it a rule that in any expression in suffix notation containing a repeated suffix, that suffix is to be given all possible values and the results then added. Thus for (3) we write simply $l_i u_i$ and leave the summation to be understood from the convention.

2·02. Transformation. Now if we have two sets of rectangular axes $O123$, $O1'2'3'$ with a common origin O, we denote the two sets of coordinates by x_i, x'_j respectively. These must be regarded as two different ways of saying where P is. We denote the direction cosines of a particular x'_j axis with respect to the x_i axes by l_{ij}. Then x'_j is the projection of OP on the direction of the x'_j axis and therefore is equal to $l_{ij} x_i$. This is true whether $j = 1, 2$, or 3; hence

$$x'_j = l_{ij} x_i. \qquad (4)$$

This summarizes the three equations of the transformation, each of which has three terms on the right side.

We have not used the condition that the axes x'_j are mutually perpendicular. The condition that x'_1 and x'_2 are perpendicular is

$$l_{i1} l_{i2} = 0, \qquad (5)$$

with two similar relations for the other pairs. Again, since l_{i1} are direction cosines of the x'_1 axis referred to $O123$

$$l_{i1} l_{i1} = 1 \qquad (6)$$

with two similar relations. (We do not write $l_{ij} l_{ij}$ here because that would imply summation with regard to both i and j, which we do not intend.) Hence, though there are nine l_{ij}, they are connected by six relations of the forms (5) and (6), and we should expect that only three of them can be assigned independently. This is actually true, but this argument must not be regarded as a proof.

2·021. δ_{ik}. These six relations can be written as one. We introduce a set of numbers δ_{ik}, where i and k can each be 1, 2, or 3, such that $\delta_{ik} = 1$ if $i = k$, and $\delta_{ik} = 0$ if $i \neq k$. Then we have

$$l_{ij} l_{il} = \delta_{jl}. \qquad (7)$$

This set of quantities is called the *substitution tensor*;* in any expression containing a suffix k, not repeated, if we multiply by δ_{ik} and add, the only non-zero term is that with $k = i$, and the result is therefore to replace k by i. Note that we must distinguish between δ_{ik} with $i = k$, which is 1, and δ_{ii}, which implies a summation and is 3. Note also that in l_{il} the form of the expression itself indicates that the two l's cannot possibly mean the same thing. It is unavoidable with suffix notation that the same letter may be required in two senses, since so many letters already have special meanings, but *a suffix simply identifies an axis and can take the values* 1, 2, 3, *and the same letter on the line stands for a physical magnitude*. If this is borne in mind no confusion will be possible. The suffixes used here are $i, k, m, p, \ldots$ for the original axes, $j, l, n, q, \ldots$ for the transformed ones. o is omitted

---

* See 3·03. We do not need the definition of tensors in general at present. δ_{ik} is a particular case of the 'Kronecker δ'.

because it might be confused with a figure. Consecutive Greek or italic letters are often used in each system, but this is liable to lead to overrunning the end of the alphabet in one system or to require additional accents, which make writing more difficult.

2·022. Reverse transformation. Since the x'_j axes are an orthogonal set and the cosine of the angle between the x_i and x'_j axes is l_{ij}, we have also

$$x_i = l_{ij} x'_j, \qquad (8)$$

and since the x_i set are orthogonal

$$l_{ij} l_{kj} = \delta_{ik}. \qquad (9)$$

This set of relations is deduced algebraically from (7) in 2·073. (This in itself is a warning against complete trust in the method of counting constants. We now have 12 relations between 9 quantities.)

2·023. Velocity, acceleration, force. Definition of vector. Unit or direction vectors. Now in dynamics the equations of motion of a particle take the form

$$m\ddot{x}_i = X_i, \qquad (10)$$

when the axes are *inertial*.* (*Inertial* is better than the usual word *fixed*.) But if we take another inertial set the l_{ij} are not varying with time, and therefore

$$\dot{x}'_j = l_{ij} \dot{x}_i, \quad \ddot{x}'_j = l_{ij} \ddot{x}_i. \qquad (11)$$

Hence the components of velocity and acceleration with respect to the x'_j axes are related to the components with regard to the x_i axes as the coordinates are.

Now (10) are supposed to be true for *any* inertial axes; for different sets the force components X_i, X'_j must have different values, but we must still have

$$m\ddot{x}'_j = X'_j, \qquad (12)$$

however the axes are transformed, and quite irrespective of the actual values of X_i. But this can be true only if

$$X'_j = l_{ij} X_i. \qquad (13)$$

Thus besides displacement we find that on change of axes velocity, acceleration and force all transform according to the rule (4): force on the supposition that the equations of motion are stated in a form true for all inertial axes. Thus for a particle moving under gravity, with $O3$ upwards, we have $(\ddot{x}_1, \ddot{x}_2, \ddot{x}_3) = (0, 0, -g)$. But this is not true for other axes and the general form is $\ddot{x}_i = -gl_i$, where l_i are the direction cosines of the upward vertical.

Any three quantities A_i that transform on rotation of axes according to the rule

$$A'_j = l_{ij} A_i \qquad (14)$$

are said to be the *components of a vector* with regard to those axes. Now if we start from any set of three equations that are true for every set of axes and work out from them any consequence, using a particular set of axes $O1, O2, O3$, we could equally well have derived from the equations stated in terms of another set of axes $O1', O2', O3'$ a consequence

---

* It is not our purpose to explain in detail here how, and how far, Newtonian dynamics is based on experiment. A discussion will be found in *Scientific Inference*, Chapter 8.

differing formally from the first only in the appending of accents to all the letters. If the consequence consists of a set of three equations, and the left sides are the components of a vector, we know that the left sides in the two systems are related according to (14). But since both sets of equations are true the right sides also must transform according to (14) and be the components of a vector. Conversely if three equations assert the equality of all components of two vectors with respect to one set of axes, it follows from the fact that both sides transform according to the same rule that the set can be adapted at once to any other set of axes by merely inserting accents.

If we take any two lines whose direction cosines with regard to the x_i axes are m_i, n_i, the cosine of the angle between them is $m_i n_i$. If in particular the first line is the axis of x'_j, the cosine of the inclination to it of a line with direction cosines n_i is

$$n'_j = l_{ij} n_i, \qquad (15)$$

so that the direction cosines of a given line transform according to the rule (14) and are the components of a vector. Such a vector is often called a *unit vector*: we prefer *direction vector*, since the only application is to specify a direction.

We can speak of the component of a vector in any direction. If a line has direction cosines n_i with respect to a set of axes, the component of the vector A_i in that direction is $n_i A_i$. Now suppose that we used the x'_j system and tried to find the component of the same vector in the same direction. It would be

$$n'_j A'_j = l_{ij} n_i l_{kj} A_k = (l_{ij} l_{kj}) n_i A_k$$
$$= \delta_{ik} n_i A_k = n_i A_i, \qquad (16)$$

and therefore the component of a vector in any given direction is independent of the axes used.

This result could also be derived by taking a third set of axes, one of which is in the direction of the line n_i, and carrying out the transformation of axes first directly and then by way of the x'_j set.

A vector is often defined as an entity requiring three components for its specification, and with an additive property expressed by the parallelogram law. The latter part of the definition, however, supposes for its application that the vectors considered are to be represented by displacements with an arbitrary scale factor, usually dimensional; until this is done we do not know what we mean by the parallelogram law for that kind of vector. The introduction of the parallelogram is really an unnecessary complication, and it is better to proceed directly to the analytical statement of the required property. Again, the rule requires that we should know what we mean by addition for that kind of vector. We have natural interpretations for displacement, velocity and acceleration, and force. But we shall meet more complicated vectors such that it becomes difficult to find anything but an analytical definition of addition; and it is quite unnecessary that we should find one, since all that is required for our purposes is that our equations shall be true; if we can calculate a quantity correctly, it is not necessary for physics that we should find a separate physical interpretation for every term contained in it. Accordingly, we shall define a vector by the transformation property of its components A_i, which includes the statement that the component in a direction with direction cosines n_i is $n_i A_i$. A *scalar* is a single quantity, the same for all axes.

2·03. Single-letter notation. A vector with components A_i can be still more shortly denoted by A in heavy type. It is not really necessary, and is in fact rather difficult, to define what we mean by a vector apart from its components. In making comparisons with observation it is the components that we are concerned with, but it is often convenient to work with the single symbol. We can then define the sum of two vectors A and B to be the vector whose components are $A_i + B_i$ and denote it by $A+B$. $-B$ is the vector whose components are $-B_i$; and $A-B = A+(-B)$. We can define the product of a vector A by any scalar m to be the vector whose components are mA_i. It is obvious that with this definition vectors satisfy the commutative and associative rules of addition; $B+A$ is the vector whose components are $B_i + A_i = A_i + B_i$, and these are the components of $A+B$. Similarly, the associative rule

$$A + (B+C) = (A+B) + C$$

follows at once from the definition. We cannot significantly add a vector to a scalar, since the former is altered on change of axes and the scalar is not; but clearly mA and Am represent the same vector, and the commutative law of multiplication is satisfied. Similarly if m and n are scalars

$$(m+n)A = mA + nA = (n+m)A,$$

provided m and n have the same dimensions; otherwise addition is meaningless. Also

$$m(nA) = (mn)A.$$

The displacement from P to Q, considered as a vector, will be denoted by $\overline{PQ}$.

We have seen that a vector is completely specified by three components; but it is easy to see that this condition can be satisfied without the commutative law of addition being satisfied and therefore is not by itself a sufficient condition for the vector property. Consider the rotation of a rigid body through a finite angle about any axis through a fixed point O of the body. The natural 'sum' of two successive rotations is a single rotation, which would bring the body into the same final configuration. A single rotation is completely specified by the direction of the axis of rotation, which requires two data to specify it, and the angle turned through. Let us take a set of rectangular axes $O123$, fixed in space, and consider two successive rotations, first through $\tfrac{1}{2}\pi$ about $O1$, then through $\tfrac{1}{2}\pi$ about $O2$, both being right-handed rotations. If we take them in this order a point P with coordinates $(0, 0, 1)$ goes first to $P'(0, -1, 0)$ and remains there at the second rotation. But if we make the rotation about $O2$ first, the point goes to $P''(1, 0, 0)$ and is undisplaced by the second rotation. The order of the rotations affects the result. Now if the sum of the two rotations was obtained by the vector law this could not be so, since the commutative law of addition holds for the latter. The representation of finite rotations will be considered more fully in the next chapter.

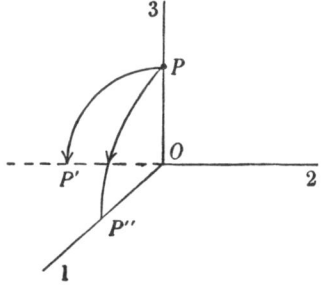

2·031. The *geometrical representation of a vector* can be shown to be possible from our definition. For if A is any vector we can multiply its components A_i by a constant c chosen so as to make them lengths. Then if $x_i = cA_i$, the projection of x on the direction l_i is,

$$l_i x_i = c l_i A_i,$$

and dividing by c we recover the component of A in the direction l_i. The addition of vectors of the same dimensions is then seen to be completely expressed by representing them by displacements on the same scale, and the parallelogram law is thus recovered for the general vector. Further properties of vectors can be inferred from those of displacements by this representation. In particular, any vector has a modulus and a direction; for if r is the length of the displacement representing it on a given scale, we have

$$r^2 = x_1^2 + x_2^2 + x_3^2 = c^2(A_1^2 + A_2^2 + A_3^2) = c^2 A^2, \tag{1}$$

where
$$A^2 = A_i A_i, \tag{2}$$

and is independent of the choice of axes. Then A (taken positive) can be called the *modulus* of A, and corresponds to the distance in a displacement. Also if $A \neq 0$, and if we write

$$x_i/r = A_i/A = m_i,$$

the m_i are the direction cosines of a definite line, which can be called the direction of A. It follows further that the projection on a line in the direction l_i of the representative displacement is $cAl_i m_i = cA\cos\theta$, where θ is the angle between the directions of l_i and m_i; and the component of A in the direction l_i is $A\cos\theta$. But the component in any given direction is independent of the axes of reference and therefore A and θ, for all l_i, are independent of the axes. Hence A is the same magnitude and m_i the same direction, whatever the axes.

It is convenient in one case to depart from the rule that we take A positive. A straight line can be represented by the equations

$$x_i = a_i + sl_i.$$

If we keep l_i the same, we can get all points on the line by allowing s to range from $-\infty$ to ∞, and this corresponds to proceeding along the line in a definite direction. If l_i is assigned it is therefore convenient to take for a line through the origin

$$x_i = xl_i,$$

thus admitting negative values of the quantity x. The distance from the origin, taken positive, will always be denoted by r. When we take

$$x_i = rl_i$$

with r positive we are considering two lines in opposite directions from the origin as different lines, with equal and opposite values of l_i. This is sometimes convenient, but not always.

2·032. Comparison of notations. The importance and use of vector notation is a matter of debate among mathematical physicists. Anything that can be said by means of A can be said by means of A_i or by writing out the components in full. If, however, the geometrical representation of a vector by a directed line segment is constantly borne in mind, to some minds the content of many physical laws is most clearly understood in vector notation. A little trouble is needed to learn to 'think in vectors'. A little is also required to acquire confidence that the compactness achieved by the summation convention does not lead to mistakes. It must, moreover, be remembered that if a physical result about a vector is obtained, it will always take three measurements to verify it, and the three components will have to be unpacked, whether it is expressed in vector or suffix notation. The unpacking from suffix notation is often easier and never harder than from vector notation. Some general theorems are more compactly expressed in the one, some in the other. In special problems judgment is needed to decide on the best moment for unpacking, and many students defer it too long when the conditions of the problem indicate one or two special directions. In elasticity and the dynamics of viscous fluids the suffix notation adapts itself far more naturally, and in the theory of relativity vector notation breaks down completely because the parallelogram law fails for velocities. Consequently some mathematical physicists hold that vector notation is pure waste of time and delays the acquisition of familiarity with the more generally useful method. What we shall do in the present chapter is to show the two methods, as far as possible, side by side. Ability to translate from either language to the other, or to the expanded form, is an absolute necessity in understanding modern physical literature. It is often useful to visualize a vector as a displacement vector, and while as a matter of definition we make a clear distinction between a general vector and a displacement vector, we shall frequently speak of a general vector in geometrical terms: e.g. the angle between two vectors A and B means strictly 'the angle between the displacement vectors representing A and B according to some specified scale'; 'two perpendicular vectors A and B' means 'two vectors A and B such that the displacements representing them are perpendicular'. The use of this analogy is unnecessary in suffix notation, analytical definitions being provided.

2·033. Null vector. A *null* or *zero* vector is one whose modulus is zero.

2·034. Direction vectors. A vector of modulus 1 (a number) in the direction of a vector A is called a unit or *direction vector* in that direction. Its components are evidently l_i, the direction cosines of the direction of A with regard to the coordinate axes. In particular we shall denote direction vectors in the directions of the axes by $e_{(1)}, e_{(2)}, e_{(3)}$ respectively; that is,

$$e_{(1)} = (1, 0, 0), \quad e_{(2)} = (0, 1, 0), \quad e_{(3)} = (0, 0, 1).$$

The use of the brackets round the suffix is to emphasize that it does not denote a component, but a particular vector. Any vector A may be written as

$$A_1 e_{(1)} + A_2 e_{(2)} + A_3 e_{(3)}.$$

Some books denote direction vectors parallel to the axes by i, j, k and write

$$A = A_x i + A_y j + A_z k.$$

2·04. Linearly dependent or coplanar vectors. If there is a relation

$$\alpha A + \beta B + \gamma C = 0 \tag{1}$$

between three vectors, where α, β, γ are real numbers (not all zero), then A, B, C are said to be linearly dependent. Geometrically this means that A, B, C can be represented by displacement vectors lying in a plane, since if $\gamma \neq 0$,

$$C = -\frac{1}{\gamma}(\alpha A + \beta B), \tag{2}$$

and therefore C is represented by a displacement vector lying in the plane of the displacements representing A and B, supposing these drawn through the same point. The vectors themselves are then said to be coplanar. We may recall at this point that parallel vectors of equal magnitude are equivalent in the system we are using. Some writers distinguish between 'free vectors' and 'localized vectors', a localized vector including, for instance, the specification both of a force and of its point of application. In our sense a localized vector is not a vector but two vectors, one to specify the force and the other to specify the point of application.

If there is no such relation as (1) the three vectors are said to be linearly independent or non-coplanar.

The corresponding results in suffix notation can be written down at once.

2·041. Expression of any vector in terms of three non-coplanar vectors. If A, B, C are any three non-coplanar vectors and D is any vector, then D can be expressed as $\alpha A + \beta B + \gamma C$, where α, β, γ are real quantities. For, let $\overline{PQ}$ represent D. Then lines through P representing A and B define a plane. Let RQ be a line through Q in the direction of C and meeting the plane defined by A and B in R. Then

$$\overline{PQ} = \overline{PR} + \overline{RQ}.$$

But $\overline{RQ}$ represents γC, where γ is a scalar; and $\overline{PR}$ is a vector coplanar with A and B, and can therefore be expressed as $\alpha A + \beta B$. Hence

$$D = \alpha A + \beta B + \gamma C.$$

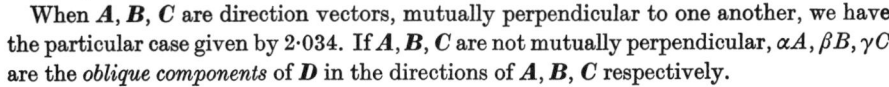

Since there is always a relation of this type between four vectors it follows that four vectors cannot be linearly independent. From the construction given it is clear that for given A, B, C (non-coplanar) and any D, the quantities α, β, γ are uniquely determined.

When A, B, C are direction vectors, mutually perpendicular to one another, we have the particular case given by 2·034. If A, B, C are not mutually perpendicular, $\alpha A, \beta B, \gamma C$ are the *oblique components* of D in the directions of A, B, C respectively.

2·05. Multiplication of vectors. We have considered multiplication of a vector by a scalar, but the meaning, if any, to be assigned to multiplication of two vectors is not immediately obvious. We can set out the nine products of the components in a square array, thus:

$$\begin{array}{ccc} A_1B_1 & A_1B_2 & A_1B_3 \\ A_2B_1 & A_2B_2 & A_2B_3 \\ A_3B_1 & A_3B_2 & A_3B_3 \end{array}$$

and we shall find that these products all reappear in Chapter 3. The two products called the *scalar* and *vector products*, which we now proceed to define, are particular combinations of these nine products, and their choice is dictated by their usefulness in physical applications.

2·06. Scalar product. This function is directly related to the fundamental and experimentally verifiable relation 2·01(1) between distances not measured along the same line. We have

$$PQ \cdot PR \cos\theta = \tfrac{1}{2}(PQ^2 + PR^2 - QR^2), \qquad (1)$$

and this is completely determined by the three distances PQ, PR, QR. Since distance is the fundamental notion of the whole subject and is the same for every frame of reference, this expression is a single quantity whose value is independent of the coordinate system:

we have called such quantities *scalars*. Now if we translate into Cartesian coordinates, if x_i denotes $\overline{PQ}$, and y_i denotes $\overline{PR}$, then $y_i - x_i$ denotes $\overline{QR}$, and

$$\tfrac{1}{2}(PQ^2 + PR^2 - QR^2) = \tfrac{1}{2}[(x_1^2 + x_2^2 + x_3^2) + (y_1^2 + y_2^2 + y_3^2)$$
$$- \{(y_1 - x_1)^2 + (y_2 - x_2)^2 + (y_3 - x_3)^2\}], \qquad (2)$$
$$= x_1 y_1 + x_2 y_2 + x_3 y_3. \qquad (3)$$

This is called the *scalar product* of the vectors $\boldsymbol{x}$ and $\boldsymbol{y}$, and in general the scalar product of $\boldsymbol{A}$ and $\boldsymbol{B}$ is defined by

$$\boldsymbol{A}.\boldsymbol{B} = A_1 B_1 + A_2 B_2 + A_3 B_3 = \sum_{i=1,2,3} A_i B_i = A_i B_i, \qquad (4)$$

using the summation convention. $\boldsymbol{A}.\boldsymbol{B}$ is equal to $AB\cos\theta$, where θ is the angle between the directions of the two vectors. We read it as 'A dot B'. The two expressions in coordinates in (2), (3) would be written by the summation convention as

$$\tfrac{1}{2}\{x_i x_i + y_i y_i - (y_i - x_i)(y_i - x_i)\} = x_i y_i, \qquad (5)$$

and the left side can be further shortened to

$$\tfrac{1}{2}\{x_i^2 + y_i^2 - (y_i - x_i)^2\}. \qquad (6)$$

We recall that x^2 would formerly have been written xx, which is what we mean by x^2; so when we see an expression like x_i^2 we interpret it as $x_i x_i$ and apply the summation convention. In terms of this convention the modulus A of a vector $\boldsymbol{A}$ is given by

$$A^2 = A_i^2.$$

This expression is the scalar product of $\boldsymbol{A}$ with itself. The saving of writing by the summation convention is enormous, so great that on the rare occasions when we do not use it we say so specially. Without it, suffix notation would have little advantage over writing everything out fully in Cartesian coordinates; with it, expressions that when fully developed would contain 9 or 81 terms can be written down and handled as easily as one. The convention remains useful in the theory of relativity and in general dynamics, since it is simply a linguistic device and does not depend on the parallelogram law.

The proof that $A_i B_i$ is independent of the axes of reference, without the use of the displacement representation, is

$$A'_j B'_j = l_{ij} l_{kj} A_i B_k = \delta_{ik} A_i B_k = A_i B_i, \qquad (7)$$

just as in deriving 2·023 (16).

Commutative law. It is clear from the definition that the order of $\boldsymbol{A}$ and $\boldsymbol{B}$ in the scalar product is irrelevant: $\boldsymbol{A}.\boldsymbol{B} = \boldsymbol{B}.\boldsymbol{A}$.

Associative law. $\boldsymbol{A}.\boldsymbol{B}$ is not a vector, and we cannot go on to form a product $\boldsymbol{A}.\boldsymbol{B}.\boldsymbol{C}$, so the associative law has no meaning. $(\boldsymbol{A}.\boldsymbol{B})\boldsymbol{C}$ means a vector in the direction of $\boldsymbol{C}$ and of $\boldsymbol{A}.\boldsymbol{B}$ times its magnitude.

Distributive law. We can, however, prove that

$$\boldsymbol{A}.(\boldsymbol{B}+\boldsymbol{C}) = \boldsymbol{A}.\boldsymbol{B} + \boldsymbol{A}.\boldsymbol{C},$$

for this follows at once from the definition. This can be immediately extended so that the scalar product of two sums of vectors can be expanded into a sum of scalar products. In particular, since

$$e_{(2)} \cdot e_{(3)} = e_{(3)} \cdot e_{(1)} = e_{(1)} \cdot e_{(2)} = 0,$$
$$e_{(1)} \cdot e_{(1)} = e_{(2)} \cdot e_{(2)} = e_{(3)} \cdot e_{(3)} = 1,$$
$$A \cdot B = (A_1 e_{(1)} + A_2 e_{(2)} + A_3 e_{(3)}) \cdot (B_1 e_{(1)} + B_2 e_{(2)} + B_3 e_{(3)})$$
$$= A_1 B_1 + A_2 B_2 + A_3 B_3.$$

We thus recover the definition (4).

Notice that when direction vectors along the axes are introduced the rectangular coordinates are regarded as scalars.

It should be noticed that if the scalar product of two displacements is zero it implies $PQ^2 + PR^2 = QR^2$; that is, either one displacement is zero or they are perpendicular. The vanishing of $A \cdot B$ does not imply that either A or B is a null vector; it implies that either they are perpendicular or one of them is a null vector. But if $A \cdot B_1$, $A \cdot B_2$, $A \cdot B_3$ are all zero, where B_1, B_2, B_3 are not coplanar, A cannot be perpendicular to all and must be null.

The cosine of the angle between two lines is the scalar product of two direction vectors along the lines; that is, if l_i, m_i are the direction cosines of the lines

$$\cos \theta = l_i m_i.$$

The component of a vector A along a line with direction cosines l_i is $l_i A_i$, which is the scalar product of A and a direction vector along the line.

2·07. Vector product. The vector product of two vectors is written $A \wedge B$ (read, A cross B) and is defined as follows by the displacement model. If $\overline{OP}$ and $\overline{OQ}$, representing A and B respectively, are not parallel they define a plane. Let $\overline{OR}$ represent a direction vector n perpendicular to this plane. Then if θ is the angle turned through from $\overline{OP}$ to $\overline{OQ}$, right-handedly about $\overline{OR}$,

$$A \wedge B = AB \sin \theta\, n.$$

It will be seen that the definition is independent of the choice of the direction of n. If we took the opposite direction as n the angle turned through right-handedly about it would be $2\pi - \theta$, and since $\sin(2\pi - \theta) = -\sin \theta$ the result for the magnitude and direction of the vector product would be the same.

In the vector product the order of the factors matters. For

$$B \wedge A = BA \sin(-\theta) n = -A \wedge B. \tag{1}$$

The commutative law is not satisfied by the vector product. We shall soon see that the associative law is also not satisfied, but we can prove that the distributive law holds, namely,

$$A \wedge (B + C) = A \wedge B + A \wedge C. \tag{2}$$

We consider first two particular cases.

(1) A is perpendicular to B and C. We notice that if A and B are perpendicular, then $A \wedge B$ is obtained from B by first multiplying B by A and then rotating it about A through a right angle. Hence the vectors $A \wedge (B + C)$, $A \wedge B$, $A \wedge C$ are represented by line segments

of lengths equal to A times the sides of the triangle representing $B+C, B, C$ respectively, and each is turned through a right angle. It follows that the vectors $A \wedge (B+C)$, $A \wedge B$, $A \wedge C$ are represented by the sides of a triangle similar to that representing $B+C, B, C$, and hence equation (2) follows for this case.

(2) A, B, C are coplanar. Then $A \wedge (B+C)$, $A \wedge B$, $A \wedge C$ are all in a direction perpendicular to the plane and the result follows from the addition formula for sines.

In the general case we assume A and B not coincident or perpendicular and write B as $B_p + B_n$, where B_p is parallel to A and B_n perpendicular to A in the plane of A and B. Then since $B_n = B \sin \theta$, it follows that

$$A \wedge B = A \wedge B_n.$$

If, similarly, $C = C_p + C_n$, then $A \wedge C = A \wedge C_n$. But

$$B + C = (B_p + C_p) + (B_n + C_n),$$

and since B_p, C_p are parallel to A and B_n, C_n perpendicular to it, it follows that $B_n + C_n$ is the component of $B+C$ perpendicular to A, and therefore

$$A \wedge (B+C) = A \wedge (B_n + C_n). \quad (3)$$

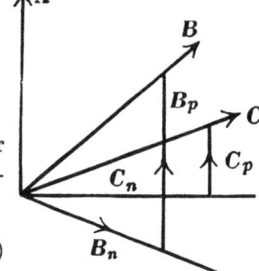

But by Case 1 this is equal to

$$A \wedge B_n + A \wedge C_n = A \wedge B + A \wedge C. \quad (4)$$

The vector product of a vector with itself or with any parallel vector is a null vector. If the vector product of any two vectors vanishes, it can be inferred that either they are parallel or one of them is a null vector. Again, it does not follow from the fact that the vector product is the null vector that one of the factors is null. But if $A \wedge B_1$ and $A \wedge B_2$ are null, and B_1, B_2 are not parallel, A must be null.

In particular, for the vectors $e_{(1)}, e_{(2)}, e_{(3)}$ we have

$$e_{(1)} \wedge e_{(1)} = e_{(2)} \wedge e_{(2)} = e_{(3)} \wedge e_{(3)} = 0, \quad (5)$$

$$e_{(2)} \wedge e_{(3)} = e_1 = -e_{(3)} \wedge e_{(2)}, \quad (6)$$

so that

$$\begin{aligned} A \wedge B &= (A_1 e_{(1)} + A_2 e_{(2)} + A_3 e_{(3)}) \wedge (B_1 e_{(1)} + B_2 e_{(2)} + B_3 e_{(3)}) \\ &= (A_2 B_3 - A_3 B_2) e_{(1)} + (A_3 B_1 - A_1 B_3) e_{(2)} + (A_1 B_2 - A_2 B_1) e_{(3)} \\ &= \begin{vmatrix} e_{(1)} & e_{(2)} & e_{(3)} \\ A_1 & A_2 & A_3 \\ B_1 & B_2 & B_3 \end{vmatrix}. \end{aligned} \quad (7)$$

A geometrical interpretation is available for the vector product of two displacements. If P is x_i and Q is y_i, the projections on the plane $x_1 = 0$ are $P_1(0, x_2, x_3)$, $Q_1(0, y_2, y_3)$, and the area of the triangle made by these two points and the origin is $\frac{1}{2}(x_2 y_3 - x_3 y_2)$. If we rotate OP_1 positively about Ox_1 the area is positive if OQ_1 is reached after a rotation less than π. It will be found that taking double the three projections in turn we have

$$x_2 y_3 - x_3 y_2, \quad x_3 y_1 - x_1 y_3, \quad x_1 y_2 - x_2 y_1. \quad (8)$$

But by a theorem of geometry these are equal to twice the magnitude of the area of the triangle OPQ multiplied by the three direction cosines of the normal to its plane, taken on the side such that the rotation of OP to OQ is in the positive sense about the normal and less than π. They are therefore the components of the vector product $x \wedge y$.

Vector area. In the definition of the vector product $x \wedge y$ the sense taken for n is irrelevant, but the right-handed rotation about n needed to bring x into the direction of y may have any value less than 2π. The statements in the last paragraph would remain true if the rotations were all taken about lines in the opposite directions to those stated and a component taken positive if the rotation about the negative direction of the corresponding axis is between π and 2π. But the signs of all components are reversed if x and y are interchanged. There are advantages in being able to speak of the triangle OPQ as having the same directed area irrespective of the labels attached to its sides; and this can be done by defining its *vector area* as

$$\tfrac{1}{2} | xy \sin \theta | n = \tfrac{1}{2} | x \wedge y | n,$$

with a particular choice of the sense of n. It will be equal to the vector product if $\sin \theta$ is positive, that is, if the rotation from OP to OQ in the positive sense about n is less than π; its sign will be reversed if n is reversed.

To make the vector area unambiguous we need a criterion for identifying the sense of n. This arises most simply in relation to a surface made up of triangles. By addition we can define a vector area for the whole surface, n being defined so that it does not cut through the surface when we pass from one face to an adjacent one. In particular, for a closed polyhedron, we can take n to be always outwards. In this case the faces with positive n_1 will have vector areas whose components are the areas of the projections of the faces on the plane $O23$ and make up the area contained in the rim of the projection. Those with negative n_1 give components whose total is the same area with the sign changed. Hence the vector area of a closed polyhedron is zero.

2·071. Analytic treatment of vector product: ϵ_{ikm}. In the analytic treatment we *define* the vector product directly as a vector with components

$$(A_2 B_3 - A_3 B_2,\ A_3 B_1 - A_1 B_3,\ A_1 B_2 - A_2 B_1).$$

It then requires proof that this set of quantities has the proper transformation properties and that its components are equal to $AB \sin \theta n_i$. The reason for introducing the vector product at all is that this set of quantities arises naturally in the discussion of the equations of dynamics, especially the motion of a rigid body and the motion of a charge under magnetic force, and in electromagnetic theory.

We consider the set of 27 numbers ϵ_{ikm} specified by the rules (1) if any two of the i, k, m are equal, $\epsilon_{ikm} = 0$; (2) if they are all different and occur in succession in the order 12312... which we call *even*, $\epsilon_{ikm} = 1$; (3) if they are all different and occur in the order 21321... which we call *odd*,* $\epsilon_{ikm} = -1$. That is,

$$\epsilon_{123} = \epsilon_{231} = \epsilon_{312} = 1, \tag{1}$$

$$\epsilon_{213} = \epsilon_{132} = \epsilon_{321} = -1, \tag{2}$$

---

* The reason for the terms *even* and *odd* is derived from the number of interchanges of suffixes needed to produce the order 123. Thus 231 can, by interchanging two suffixes once, be turned into 132 and by a second interchange to 123. But 213 is turned into 123 by a single interchange.

while any of ϵ_{111}, ϵ_{112}, ϵ_{232} and so on are zero. Now consider the sum

$$\epsilon_{ikm} A_k B_m. \qquad (3)$$

Here k and m are both repeated suffixes. For each i it is therefore the sum of nine terms. But if either $k = i$, $m = i$, or $k = m$, $\epsilon_{ikm} = 0$. The only terms that can differ from 0 are therefore the two that have k and m different from i and from each other. Thus if, for instance, $i = 1$, we must have $k = 2$ and $m = 3$ and therefore $\epsilon_{ikm} = 1$, or $k = 3$, $m = 2$, $\epsilon_{ikm} = -1$. Hence

$$\epsilon_{1km} A_k B_m = A_2 B_3 - A_3 B_2, \qquad (4)$$

and similarly for $i = 2$ and 3 we find the other two components of the vector product. Thus (3) gives a compact expression for the components of the vector product. We shall denote them at present by $(A \wedge B)_i$ to facilitate comparison with results already obtained in vector notation.

Two other important properties of ϵ_{ikm} are as follows. Clearly whatever A_i may be,

$$\epsilon_{ikm} A_k A_m = 0, \qquad (5)$$

since all the terms cancel; this formulates analytically the statement that *the vector product of a vector with itself or any parallel vector is null.* If A_i, B_i, C_i are any sets of three quantities,

$$\epsilon_{ikm} A_i B_k C_m = \begin{vmatrix} A_1 & A_2 & A_3 \\ B_1 & B_2 & B_3 \\ C_1 & C_2 & C_3 \end{vmatrix}, \qquad (6)$$

the determinant formed by the nine components. If it vanishes there are values α, β, γ such that

$$\alpha A_i + \beta B_i + \gamma C_i = 0 \qquad (7)$$

for all i, so that the vanishing of (6) is the condition for vectors A, B, C to be coplanar. If the determinant does not vanish, the equations

$$\alpha A_i + \beta B_i + \gamma C_i = D_i \qquad (8)$$

have a unique solution for any D, and we recover the result that *any vector can be expressed linearly in terms of any three non-coplanar vectors.*

We now proceed to prove *analytically* that $\epsilon_{ikm} A_k B_m$ are the components of a vector; this proof is quite independent of the argument in 2·07.

2·072. Transformation property of vector product. Take any pair of lines with direction cosines l_i, m_i. The conditions that a line with direction cosines n_i shall be perpendicular to both are, written in full,

$$l_1 n_1 + l_2 n_2 + l_3 n_3 = 0, \quad m_1 n_1 + m_2 n_2 + m_3 n_3 = 0, \qquad (1)$$

whence

$$\frac{n_1}{l_2 m_3 - l_3 m_2} = \frac{n_2}{l_3 m_1 - l_1 m_3} = \frac{n_3}{l_1 m_2 - l_2 m_1}$$

$$= \frac{(n_1^2 + n_2^2 + n_3^2)^{1/2}}{\{(l_2 m_3 - l_3 m_2)^2 + (l_3 m_1 - l_1 m_3)^2 + (l_1 m_2 - l_2 m_1)^2\}^{1/2}}. \qquad (2)$$

But the sum of squares in the denominator is, by Lagrange's identity,

$$(l_1^2 + l_2^2 + l_3^2)(m_1^2 + m_2^2 + m_3^2) - (l_1 m_1 + l_2 m_2 + l_3 m_3)^2 = 1 - \cos^2 \theta = \sin^2 \theta, \qquad (3)$$

where θ is the angle between the lines l_i, m_i. Also since n_i are direction cosines the numerator in the last expression in (2) is ± 1. Hence each of these expressions is equal to $\pm \operatorname{cosec} \theta$, and
$$n_i = \pm \operatorname{cosec} \theta \, \epsilon_{ikm} l_k m_m. \tag{4}$$
The ambiguity of sign corresponds to a choice of direction of travel along the line n_i. If we take it so that the direction of rotation from l_i to m_i through the angle θ is right-handed about n_i we see from consideration of
$$l_i = (1, 0, 0), \quad m_i = (\cos\theta, \sin\theta, 0), \quad n_i = (0, 0, 1)$$
that the positive sign must be taken. Hence
$$n_i = \operatorname{cosec} \theta \, \epsilon_{ikm} l_k m_m. \tag{5}$$
Now if l_i and m_i are given directions, the perpendicular to them is in a fixed direction independent of the axes; hence n_i, being the direction cosines of a fixed direction, transform according to the vector rule.

For two general vectors A and B we can now define two directions l_i, m_i by $A_i = Al_i$, $B_i = Bm_i$, and then $AB \sin\theta$ is a scalar since A, B, and θ are all independent of the axes. Hence $AB \sin\theta \, n_i$ is a vector. But
$$AB \sin\theta \, n_i = AB \, \epsilon_{ikm} l_k m_m = \epsilon_{ikm} A_k B_m, \tag{6}$$
which proves that for two general vectors the components of the vector product transform according to the vector rule and that their values are equal to those given by the displacement definition.

In suffix notation 2·07 (1) and (2) are obvious, and 2·07 (7) does not arise because we need never consider direction vectors along the axes.

2·073. Relations between the l_{ij}. We can now proceed to show that the relations 2·022 (9) do follow from 2·021 (7). This is really a consistency theorem. If it was not true there would be more than six independent relations between the nine direction cosines involved in a transformation of axes, and not more than two elements of the transformation could be assigned independently. But apparently we can rotate the axes by an arbitrary amount about any line, and this line itself needs two parameters to specify its direction, making three in all; and by the properties of rigid bodies the frame of the axes will remain rectangular. Hence we really have already all the information required to justify the statement that 2·022 (9) must be a consequence of 2·021 (7). But the metrical relations assumed in the argument might conceivably be mutually inconsistent, and a direct proof is desirable as a check. We take the above l_i to be the l_{i1} of the transformation, and m_i to be l_{i2}. Then since $O3'$ is perpendicular to $O1'$ and $O2'$, and the rotation from $O1'$ to $O2'$ is taken right-handedly about $O3'$ through a right angle*, $\sin\theta = 1$, and
$$l_{i3} = \epsilon_{ikm} l_{k1} l_{m2}. \tag{1}$$
Similarly,
$$l_{i1} = \epsilon_{ikm} l_{k2} l_{m3}; \quad l_{i2} = \epsilon_{ikm} l_{k3} l_{m1}. \tag{2}$$
Now this is the same as saying that in the determinant
$$L = \begin{vmatrix} l_{11} & l_{12} & l_{13} \\ l_{21} & l_{22} & l_{23} \\ l_{31} & l_{32} & l_{33} \end{vmatrix}, \tag{3}$$

---
* See note on orthogonal transformations at end of Chapter 4, p. 170.

every element is equal to its cofactor. For each of the relations (1), (2) asserts this for all the elements of a column of the determinant. If we expand in terms of the elements of the first column ($j = 1$) we therefore get l_{i1}^2, which is 1. Hence $L = 1$. But if we expand in terms of elements of the first row ($i = 1$) we get l_{1j}^2, which must therefore also be 1; similarly,

$$l_{2j}^2 = l_{3j}^2 = 1. \tag{4}$$

On the other hand, if we form $l_{ij}l_{kj}$, where i and k are unequal, we get a determinant with two rows equal and therefore zero. Hence for all i, k

$$l_{ij}l_{kj} = \delta_{ik}. \tag{5}$$

The relations (1), (2) are in the form needed for the proof of the theorem, but are stated as three separate equations. Their similarity suggests that they can be written as one; this is

$$\epsilon_{jln}l_{ij} = \epsilon_{ikm}l_{kl}l_{mn}. \tag{6}$$

The only suffixes not repeated on either side are i, l, n. (a) If n follows l in the order 1231, the only value of j that makes ϵ_{jln} different from 0 is the predecessor of l, and then $\epsilon_{jln} = 1$. Hence in this case the left side reduces to l_{ij}, where $j \neq l, n$, and this is equal to the right side by (1), (2). (b) If n precedes l in the order 1231, the left side is

$$-l_{ij}(j \neq l, n) = -\epsilon_{ikm}l_{kp}l_{ms},$$

where jps are consecutive in the order 12312; and therefore $p = n$, $s = l$, and the right side is

$$-\epsilon_{ikm}l_{kn}l_{ml} = -\epsilon_{imk}l_{mn}l_{kl}$$
$$= \epsilon_{ikm}l_{kl}l_{mn}.$$

(c) If $l = n$, both sides are unaltered if l and n are interchanged; but this interchange reverses both sides, which are therefore zero. Hence (6) is true for all values of l, n.

By multiplying (6) by l_{pl} we get

$$\epsilon_{jln}l_{ij}l_{pl} = \epsilon_{ikm}l_{kl}l_{mn}l_{pl}$$
$$= \epsilon_{ikm}l_{mn}\delta_{kp}$$
$$= \epsilon_{ipm}l_{mn},$$

and putting k for p we have
$$\epsilon_{ikm}l_{mn} = \epsilon_{jln}l_{ij}l_{kl}. \tag{7}$$

If a determinant is written as

$$\Delta = \begin{vmatrix} A_{11} & A_{12} & A_{13} \\ A_{21} & A_{22} & A_{23} \\ A_{31} & A_{32} & A_{33} \end{vmatrix}, \tag{8}$$

the first suffix referring to the row and the second to the column, it is

$$\epsilon_{jln}A_{1j}A_{2l}A_{3n} = \epsilon_{jln}A_{2j}A_{3l}A_{1n} = \epsilon_{jln}A_{3j}A_{1l}A_{2n}$$
$$= -\epsilon_{jln}A_{2j}A_{1l}A_{3n} = \text{etc.},$$

and hence
$$\epsilon_{ikm}\Delta = \epsilon_{jln}A_{ij}A_{kl}A_{mn}. \tag{9}$$

Thus a determinant whose elements are all identified by row and column suffixes can be written as an expression in one line by means of the ϵ notation. This can be extended to determinants of any order.*

2·074. The numbers $\epsilon_{ikm}\epsilon_{psm}$. One of the most important properties of ϵ_{ikm} is an identity satisfied by the 81 numbers

$$\epsilon_{ikm}\epsilon_{psm}.$$

Here we have of course to sum with regard to m, but each of i, k, p, s leads to a separate expression according as it is taken to be 1, 2, or 3. As each of these four suffixes is capable of only three values, at least two of them must be the same in each component. Evidently all components with $i = k$ or $p = s$ are zero. If $i \neq k$, there is only one value of m that makes ϵ_{ikm} different from zero, and then the only values of p and s that make ϵ_{psm} different from zero are i and k, in either order. If the orders are the same, ϵ_{ikm} and ϵ_{psm} are either both 1 or both -1, and the product is 1. If the orders are different the product is -1. Hence

$$\begin{aligned}\epsilon_{ikm}\epsilon_{psm} &= 0 & (i = k) \\ &= 0 & (p = s) \\ &= 1 & (i = p, k = s) \\ &= -1 & (i = s, k = p) \\ &= 0 & (i \neq p \text{ or } s, \text{ or } k \neq p \text{ or } s).\end{aligned}$$

Now consider the set of numbers

$$\delta_{ip}\delta_{ks} - \delta_{is}\delta_{kp}.$$

If $i = k$ or $p = s$ the components cancel. If $i \neq p$ or s, or $k \neq p$ or s, one factor of each term is 0. Hence the only non-zero components are those with i, k equal to p, s, in either order, and the members of each pair themselves unequal. But if $i = p$ and $k = s$, the first term is 1 and the second 0, and if $i = s$ and $k = p$ the first is 0 and the second 1. Hence for every possible assignment of the four non-repeated suffixes

$$\epsilon_{ikm}\epsilon_{psm} = \delta_{ip}\delta_{ks} - \delta_{is}\delta_{kp}. \tag{1}$$

We shall meet this identity again and again in different applications.

Since
$$\epsilon_{psm} = \epsilon_{mps} = \epsilon_{smp}, \tag{2}$$

for all assignments of the letters, and similarly for ϵ_{ikm}, the values of the expression on the left of (1) will not be altered by replacing ikm by kmi or mik. We can therefore provide a general rule for the signs: *take i and p to be the suffixes that follow the repeated suffix in the respective ϵ* (if the repeated suffix is the last, take the respective i or p to be the first) *factors; then δ_{ip} appears with the positive sign, and the rest of the formula can be filled in by symmetry.*

2·08. Division of vectors. This cannot be defined without ambiguity and is avoided. It is easily seen that, given a non-zero vector A and a scalar M, there is a vector B such that the scalar product $A \cdot B = M$. But the division is not unique, because we could add to B any vector perpendicular to A without affecting the scalar product. In general there is no vector B such that the vector product $A \wedge B = C$, where A and C are given vectors.

---

* Cf. Durell and Robson, *Advanced Algebra*, 1937, Chapter 16.

For $A \wedge B$ is perpendicular to A, and if C is not perpendicular to A there is no vector B that satisfies the conditions. If C, on the other hand, is perpendicular to A we could add to B any vector parallel to A without affecting the vector product, and the quotient would be ambiguous.

In the field of *quaternions*, which is an extension of vector algebra, a quotient does in general exist. Few physicists had used the quaternion method until recently, but it is now receiving some notice in quantum theory. (Cf. Chapter 4, Ex. 12.)

2·09. Triple products. The scalar product of $B \wedge C$ with another vector A, $A \cdot (B \wedge C)$ is called the triple scalar product of A, B and C. We shall show that in such a product the order of the factors is immaterial so long as the cyclic order is preserved, and the dot and the cross may be interchanged without altering the value. There are thus six possible ways of writing it.

We can also form the vector product $A \wedge (B \wedge C)$ of a vector A with the vector product of B and C.

We first examine some special cases:

(i) $B \cdot (B \wedge C)$. Clearly
$$B \cdot (B \wedge C) = 0, \tag{1}$$
since $B \wedge C$ is perpendicular to B.

(ii) $B \wedge (B \wedge C)$. $B \wedge (B \wedge C)$ is perpendicular both to B and to $B \wedge C$ and hence is in a direction perpendicular to B in the plane of B and C, obtained by rotating $B \wedge C$ right-handedly about B through a right angle. Its magnitude is B times that of $B \wedge C$, i.e. it is $B^2 C \sin \theta$. From the figure it therefore follows that

$$B \wedge (B \wedge C) = B^2 C \sin \theta \cot \theta \frac{B}{B} - B^2 C \sin \theta \operatorname{cosec} \theta \frac{C}{C} \tag{2}$$

$$= (B \cdot C) B - B^2 C. \tag{3}$$

Similarly,
$$C \wedge (B \wedge C) = -C \wedge (C \wedge B) = C^2 B - (B \cdot C) C. \tag{4}$$

(iii) $\quad (B \wedge C) \cdot (B \wedge C) = B^2 C^2 - (B \cdot C)^2. \tag{5}$

This follows immediately from the definitions.

2·091. The triple scalar product $A \cdot (B \wedge C)$. From the commutative law for the scalar product we have
$$A \cdot (B \wedge C) = (B \wedge C) \cdot A. \tag{6}$$

Further, we can write any vector A as
$$A = \alpha B + \beta C + \gamma B \wedge C, \tag{7}$$
since B, C, $B \wedge C$ are not coplanar if none of them is null. Then
$$A \cdot (B \wedge C) = \gamma (B \wedge C)^2 = \gamma [B^2 C^2 - (B \cdot C)^2]. \tag{8}$$

Also
$$A \wedge B = \beta C \wedge B - \gamma B \wedge (B \wedge C)$$
$$= \beta C \wedge B + \gamma B^2 C - \gamma (B \cdot C) B, \tag{9}$$
so that
$$C \cdot (A \wedge B) = \gamma [B^2 C^2 - (B \cdot C)^2] = A \cdot (B \wedge C). \tag{10}$$

Similarly, $$B.(C \wedge A) = A.(B \wedge C). \tag{11}$$

Hence altogether we have six equal products
$$A.B \wedge C = B.C \wedge A = C.A \wedge B = A \wedge B.C = B \wedge C.A = C \wedge A.B. \tag{12}$$

Brackets are unnecessary, since the order of forming the vector and scalar products is unambiguous. If the cyclic order is A, C, B, instead of A, B, C, the sign is changed.

Geometrical meaning. Since the modulus of $B \wedge C$ is the measure of the area of the parallelogram formed by line segments representing B and C it follows that if the angle between A and $B \wedge C$ is acute, then $A.B \wedge C$ is the measure of the volume of the parallelepiped formed by line segments representing A, B and C. If the angle is obtuse then $-A.B \wedge C$ is equal to this volume.

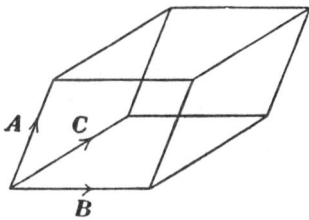

The triple scalar product $A.B \wedge C$ is sometimes written $[A, B, C]$.

2·092. The triple vector product $A \wedge (B \wedge C)$. As before we write
$$A = \alpha B + \beta C + \gamma B \wedge C. \tag{13}$$

Then
$$\begin{aligned} A \wedge (B \wedge C) &= \alpha B \wedge (B \wedge C) + \beta C \wedge (B \wedge C) \\ &= \alpha[(B.C)B - B^2 C] + \beta[C^2 B - (B.C)C] \\ &= [(\alpha B + \beta C).C]B - [(\alpha B + \beta C).B]C \\ &= (A.C)B - (A.B)C. \end{aligned} \tag{14}$$

Similarly, it may be shown that
$$(A \wedge B) \wedge C = (A.C)B - (B.C)A. \tag{15}$$

It should be noticed (1) that the associative law does not hold, (2) as an aid to remembering the signs, the term on the right-hand side in which the middle vector of the left-hand side occurs in the scalar product has the negative sign.

2·093. In suffix notation the equivalence of the various forms of the triple scalar product is obvious. For
$$A_i \epsilon_{ikm} B_k C_m = \begin{vmatrix} A_1 & A_2 & A_3 \\ B_1 & B_2 & B_3 \\ C_1 & C_2 & C_3 \end{vmatrix},$$

and the forms $A.(B \wedge C)$, $B.(C \wedge A)$, $C.(A \wedge B)$ represent the expansions of this determinant in terms of the elements of different rows. The other three are expressions of the relation $A_i B_i = B_i A_i$.

2·094. The expression for the triple vector product depends on the identity 2·074 (1) satisfied by the ϵ_{ikm}. We notice first that the use of the summation convention requires that any repeated suffix must occur *only* twice, otherwise there will be an ambiguity about the order of summation. We must therefore write the i component of the triple vector product $\mathbf{A} \wedge (\mathbf{B} \wedge \mathbf{C})$ as

$$\begin{aligned}
\epsilon_{ikm} A_k \epsilon_{mps} B_p C_s &= \epsilon_{ikm} \epsilon_{mps} A_k B_p C_s \\
&= (\delta_{ip} \delta_{ks} - \delta_{is} \delta_{kp}) A_k B_p C_s \\
&= \delta_{ks} A_k B_i C_s - \delta_{kp} A_k B_p C_i \\
&= B_i A_k C_k - C_i A_k B_k,
\end{aligned}$$

and
$$\mathbf{A} \wedge (\mathbf{B} \wedge \mathbf{C}) = (\mathbf{A} . \mathbf{C}) \mathbf{B} - (\mathbf{A} . \mathbf{B}) \mathbf{C}.$$

Also
$$(\mathbf{A} \wedge \mathbf{B}) \wedge \mathbf{C} = -\mathbf{C} \wedge (\mathbf{A} \wedge \mathbf{B}) = -(\mathbf{C} . \mathbf{B}) \mathbf{A} + (\mathbf{C} . \mathbf{A}) \mathbf{B}.$$

Mathematical physicists differ widely with respect to their ability to remember these formulae; but the formulae can always be recovered in a few lines from the identity 2·074 (1), which is much less difficult for the memory and has other applications.

2·10. Vector functions of a scalar variable. Differentiation. We shall denote a general scalar variable by t and let $\mathbf{A}(t)$ be a general vector function. We define the differential coefficient of $\mathbf{A}(t)$ with respect to t in the following way. Consider the ratio

$$\frac{\delta \mathbf{A}}{\delta t} = \frac{\mathbf{A}(t+\delta t) - \mathbf{A}(t)}{\delta t}.$$

If δt is any non-zero quantity it is clear that $\delta \mathbf{A}/\delta t$ is a vector, and if as $\delta t \to 0$ $\delta \mathbf{A}/\delta t$ tends to a limit, we define this limit as the vector $d\mathbf{A}/dt$. A formal proof that the limit is itself a vector depends simply on the theorem that the sum of the limits of two functions is equal to the limit of their sum.

The components of the vector $d\mathbf{A}/dt$ are $(dA_1/dt, dA_2/dt, dA_3/dt)$. It is important to notice that not only is the modulus of $d\mathbf{A}/dt$ in general different from that of $\mathbf{A}$, *but also its direction*. In particular the differential coefficient of a vector of constant modulus, whose direction varies, is not zero.

Differentiation of products. The rule for differentiating a product of two scalar functions is easily extended to differentiation of the product of a scalar with a vector function and to scalar and vector products. We shall simply state the results here; the proof is in every case straightforward and is left to the reader.

(1) If α is any scalar function of t, then

$$\frac{d}{dt}(\alpha \mathbf{A}) = \frac{d\alpha}{dt} \mathbf{A} + \alpha \frac{d\mathbf{A}}{dt}, \quad \frac{d}{dt}(\alpha A_i) = \frac{d\alpha}{dt} A_i + \alpha \frac{dA_i}{dt}.$$

(2) If $\mathbf{A}$ and $\mathbf{B}$ are two vector functions of t then

$$\frac{d}{dt}(\mathbf{A} . \mathbf{B}) = \frac{d\mathbf{A}}{dt} . \mathbf{B} + \mathbf{A} . \frac{d\mathbf{B}}{dt},$$

$$\frac{d}{dt}(A_i B_i) = B_i \frac{dA_i}{dt} + A_i \frac{dB_i}{dt}.$$

The order in the products is here immaterial.

(3)
$$\frac{d}{dt}(\mathbf{A} \wedge \mathbf{B}) = \frac{d\mathbf{A}}{dt} \wedge \mathbf{B} + \mathbf{A} \wedge \frac{d\mathbf{B}}{dt},$$

$$\frac{d}{dt} \epsilon_{ikm} A_k B_m = \epsilon_{ikm} \frac{dA_k}{dt} B_m + \epsilon_{ikm} A_k \frac{dB_m}{dt}.$$

Applications to particle dynamics

The order in the products must here be maintained in the vector notation. In suffix notation rearrangement of the factors is permissible.

Some special results arising from these are important:

(1) If $\boldsymbol{A}$ is a vector of *constant modulus* then

$$\frac{d}{dt}(A^2) \equiv \frac{d}{dt}(\boldsymbol{A}.\boldsymbol{A}) = 0,$$

i.e.
$$\boldsymbol{A}.\frac{d\boldsymbol{A}}{dt} = 0,$$

which shows that $d\boldsymbol{A}/dt$ is perpendicular to $\boldsymbol{A}$. Translation into suffix notation is immediate.

This can be seen geometrically. In the figure the lines representing $\boldsymbol{A}$ and $\boldsymbol{A}+\delta\boldsymbol{A}$ are of the same length. As Q approaches P the angle between OP and PQ approaches a right angle.

(2) If any vector function $\boldsymbol{A}$ is written as the product of its modulus A with the direction vector $\boldsymbol{n}$, then

$$\frac{d\boldsymbol{A}}{dt} = \frac{dA}{dt}\boldsymbol{n} + A\frac{d\boldsymbol{n}}{dt}.$$

That is, if $A_i = Al_i$, $\quad \dfrac{dA_i}{dt} = l_i \dfrac{dA}{dt} + A \dfrac{dl_i}{dt}.$

It should be noticed that $\left|\dfrac{d\boldsymbol{A}}{dt}\right|$ is not in general equal to $\left|\dfrac{dA}{dt}\right|$.

2·11. Motion of a particle under gravity with resistance varying as the velocity.

Let the origin O be at the point of projection. The resistance is assumed to act along the tangent to the path in the opposite direction to that of motion and is therefore expressed by a force vector $-m\kappa\boldsymbol{v}$, where m is the mass of the particle and κ is a constant. Let $\boldsymbol{k}$ be a direction vector in the direction of the upward vertical. Then equating the mass times acceleration of the particle to the force acting on it we have

$$m\ddot{\boldsymbol{x}} = -mg\boldsymbol{k} - m\kappa\dot{\boldsymbol{x}} \tag{1}$$

or
$$\ddot{\boldsymbol{x}} + \kappa\dot{\boldsymbol{x}} = -g\boldsymbol{k}. \tag{2}$$

We can integrate this vector equation as it stands. It may be written

$$\frac{d}{dt}(\dot{\boldsymbol{x}}e^{\kappa t}) = -ge^{\kappa t}\boldsymbol{k}, \tag{3}$$

so that
$$\dot{\boldsymbol{x}}e^{\kappa t} = -\frac{g}{\kappa}e^{\kappa t}\boldsymbol{k} + \boldsymbol{V} + \frac{g}{\kappa}\boldsymbol{k}, \tag{4}$$

if $\boldsymbol{V}$ is the velocity of projection from O at time $t=0$. Hence

$$\dot{\boldsymbol{x}} = e^{-\kappa t}\boldsymbol{V} - \frac{g}{\kappa}(1 - e^{-\kappa t})\boldsymbol{k} \tag{5}$$

and
$$\boldsymbol{x} = \frac{\boldsymbol{V}}{\kappa}(1 - e^{-\kappa t}) - \frac{gt}{\kappa}\boldsymbol{k} + \frac{g}{\kappa^2}(1 - e^{-\kappa t})\boldsymbol{k}, \tag{6}$$

since $\boldsymbol{x} = 0$ when $t = 0$.

It can be seen immediately from this equation that at time t all particles projected with speed V from O lie on a circle whose centre C is at a depth $\frac{gt}{\kappa} - \frac{g}{\kappa^2}(1-e^{-\kappa t})$ below O, for

$$\left| \mathbf{x} + \frac{gt}{\kappa}\mathbf{k} - \frac{g}{\kappa^2}(1-e^{-\kappa t})\mathbf{k} \right| = \frac{V}{\kappa}(1-e^{-\kappa t}), \qquad (7)$$

i.e.
$$|\overline{CO} + \mathbf{x}| = CP = \frac{V}{\kappa}(1-e^{-\kappa t}). \qquad (8)$$

Hence CP is equal to $\frac{V}{\kappa}(1-e^{-\kappa t})$ and is independent of the direction of projection.

Differentiating (2) with respect to the time we have

$$\dot{\mathbf{a}} + \kappa\mathbf{a} = 0, \qquad (9)$$

i.e. if the acceleration is $\mathbf{a}_0$ at time $t = 0$,

$$\mathbf{a} = \mathbf{a}_0 e^{-\kappa t}. \qquad (10)$$

Hence throughout the motion the direction of the acceleration is the same. Also if u is the horizontal component of the velocity and u_0 its initial value we have from (5)

$$u = u_0 e^{-\kappa t}, \qquad (11)$$

and if d is the distance travelled horizontally in time t, since $u = \dot{d}$,

$$d = \frac{u_0}{\kappa}(1-e^{-\kappa t}). \qquad (12)$$

Hence
$$e^{-\kappa t} = 1 - \frac{\kappa d}{u_0} \qquad (13)$$

and (10) becomes
$$\mathbf{a} = \mathbf{a}_0\left(1 - \frac{\kappa d}{u_0}\right). \qquad (14)$$

2·12. Motion of a charged particle in electric and magnetic fields at right angles. If m, $-e$ are the mass and electric charge of the particle, c the velocity of light, $\mathbf{E}, \mathbf{H}$ the electric and magnetic fields in Gaussian units, the equation of motion is

$$m\ddot{\mathbf{x}} = -e\mathbf{E} - \frac{e}{c}\dot{\mathbf{x}} \wedge \mathbf{H}, \qquad (1)$$

that is,
$$\ddot{x}_i = -\frac{e}{m}E_i - \frac{e}{mc}\epsilon_{ikm}\dot{x}_k H_m. \qquad (2)$$

Take
$$\mathbf{E} = (E, 0, 0), \quad \mathbf{H} = (0, 0, H). \qquad (3)$$

Then
$$\ddot{x}_1 = -\frac{e}{m}E - \frac{e}{mc}\dot{x}_2 H, \qquad (4)$$

$$\ddot{x}_2 = \frac{e}{mc}\dot{x}_1 H, \qquad (5)$$

$$\ddot{x}_3 = 0. \qquad (6)$$

Let the particle start from the origin with zero velocity. Then $x_3 = 0$ for all time; the motion is in a plane perpendicular to the magnetic field. Multiply (5) by i and add to (4), and put $x_1 + ix_2 = z$.* Then

$$\ddot{z} = -\frac{eE}{m} + \frac{ieH}{mc}\dot{z}.$$

Put $eH/mc = \omega$; then
$$\ddot{z} - i\omega\dot{z} = -\frac{eE}{m},$$

$$\dot{z} = -\frac{ieE}{m\omega}(1 - e^{i\omega t}),$$

since $\dot{z} = 0$ at $t = 0$; and
$$z = -\frac{ieE}{m\omega}\left(t - \frac{(e^{i\omega t} - 1)}{i\omega}\right)$$

$$= -\frac{eE}{m\omega^2}(i\omega t - e^{i\omega t} + 1),$$

$$x_1 = -\frac{eE}{m\omega^2}(1 - \cos\omega t),$$

$$x_2 = -\frac{eE}{m\omega^2}(\omega t - \sin\omega t).$$

The path is a cycloid with its cusps along the negative direction of the axis of x_2.

2·13. Small angular displacement; angular velocity. Let a particle originally at $P(x)$ receive a displacement due to a small rotation $\delta\theta$ (right-handed) about a line ON through O with direction cosines n_i. Let α be the angle between the axis of rotation and OP. Then to the first order in $\delta\theta$ the displacement of P is perpendicular to the plane of x and n and has modulus $r\sin\alpha\,\delta\theta$. Hence the displacement δx is given by

$$\delta x = \delta\theta \cdot n \wedge x + O(\delta\theta)^2, \qquad (1)$$

since $|n \wedge x| = r\sin\alpha$; or if we put

$$\boldsymbol{\delta\theta} = n\,\delta\theta, \qquad (2)$$

$$\delta x = \boldsymbol{\delta\theta} \wedge x + O(\delta\theta)^2. \qquad (3)$$

$\boldsymbol{\delta\theta}$ is a vector because $\delta\theta$ is a scalar and n_i are the direction cosines of a given line. It follows that if v is the velocity of P and $\boldsymbol{\delta\theta}/\delta t$ has a limit $\boldsymbol{\omega}$ when $\delta t \to 0$,

$$v = \lim\frac{\delta x}{\delta t} = \boldsymbol{\omega} \wedge x, \qquad (4)$$

where
$$\boldsymbol{\omega} = \omega n. \qquad (5)$$

Conversely, if the velocity $\dot{x}$ is given by an expression of the form (4) for all t, with $\boldsymbol{\omega}$ constant in magnitude and direction, we can recognize the motion as circular motion

---

* If the student is not already acquainted with the elements of the theory of the complex variable he should read the beginning of Chapter 11 at this point.

with constant velocity. For $\dot{x} \cdot \omega = 0$ and therefore the motion is in a plane perpendicular to ω. Also $\dot{x} \cdot x = 0$; and therefore r^2 is constant and the motion is on a sphere. It is therefore in a circle about ON of radius $r \sin \alpha$. Finally,

$$\dot{x} \cdot \dot{x} = (\omega \wedge x) \cdot (\omega \wedge x) = (\omega r \sin \alpha)^2, \tag{6}$$

and therefore the velocity is constant and the angular velocity ω. The sign can be checked separately.

The equation (4) can be regarded as a family of three differential equations for the x_i, namely,

$$\dot{x}_i = \epsilon_{ikm} \omega_k x_m. \tag{7}$$

The student should carry through the derivation of (6) for himself for practice in using ϵ_{ikm}. (7) can be used to illustrate a method that is often useful when one axis is specialized. We take the axis of ω as that of x_3; then $\omega = (0, 0, \omega)$ and

$$\dot{x}_1 = -\omega x_2, \quad \dot{x}_2 = \omega x_1, \quad \dot{x}_3 = 0. \tag{8}$$

Then x_3 is constant. Multiply the second equation by i and add to the first,* and put $\zeta = x_1 + i x_2$. Then

$$\dot{\zeta} = i\omega \zeta, \quad \zeta = C e^{i\beta} e^{i\omega t} \tag{9}$$

with C and β real, and the real and imaginary parts give

$$x_1 = C \cos(\omega t + \beta), \quad x_2 = C \sin(\omega t + \beta). \tag{10}$$

These equations represent uniform motion in a circle of radius C; and the solution contains three adjustable constants as it should, namely, x_3, C, and β.

We can solve equation 2·12 (1) in another way. Integrating once we have

$$m\dot{x} = -eEt - \frac{e}{c} x \wedge H. \tag{11}$$

We put
$$E = E e_{(1)} = E e_{(2)} \wedge e_{(3)}, \quad H = H e_{(3)}. \tag{12}$$

Then (11) may be rearranged as

$$\dot{x} = \frac{eH}{mc} \wedge \left(x + \frac{cEt}{H} e_{(2)} \right),$$

which we interpret immediately as follows: the particle is moving with angular velocity eH/mc about an axis parallel to $e_{(3)}$ which is itself moving with constant velocity $-\frac{cE}{H} e_{(2)}$.

2·131. In particular if i and j are two mutually perpendicular direction vectors in the $(1, 2)$ plane and, at time t, i makes an angle θ and j an angle $\frac{1}{2}\pi + \theta$ with $O1$, we have that

$$\frac{di}{dt} = \dot{\theta} j, \tag{1}$$

$$\frac{dj}{dt} = -\dot{\theta} i. \tag{2}$$

---
* Cf. footnote on p. 79.

If at time t the position vector of a particle moving in a plane is x, then the components of its velocity dx/dt and acceleration d^2x/dt^2 resolved along $O1$, $O2$ are $(\dot{x}_1, \dot{x}_2)$ and $(\ddot{x}_1, \ddot{x}_2)$ respectively. If i and j are direction vectors along x and perpendicular to it, we have

$$x = ri,$$
$$\frac{dx}{dt} = \dot{r}i + r\frac{di}{dt}$$
$$= \dot{r}i + r\dot{\theta}j, \qquad (3)$$

and
$$\frac{d^2x}{dt^2} = \ddot{r}i + \dot{r}\frac{di}{dt} + \frac{d}{dt}(r\dot{\theta})j + r\dot{\theta}\frac{dj}{dt}$$
$$= (\ddot{r} - r\dot{\theta}^2)i + \frac{1}{r}\frac{d}{dt}(r^2\dot{\theta})j. \qquad (4)$$

(3) and (4) give the components of velocity and acceleration resolved along and perpendicular to the radius vector.

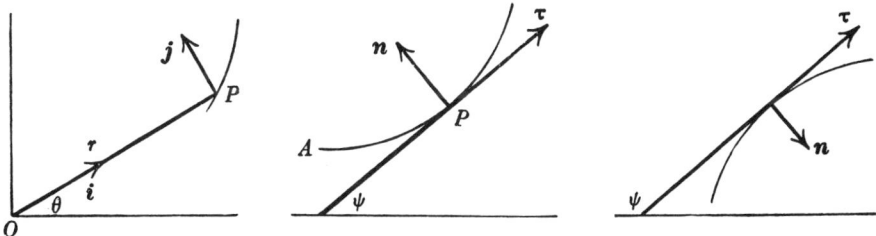

Similarly, if τ and n are direction vectors along the tangent and *inward* normal to the path of the particle, and if ψ is the angle between τ and a fixed direction chosen so that ψ is increasing going along the path in the direction of motion, τ, then

$$\frac{d\tau}{dt} = \frac{d\psi}{dt}n$$
$$= \frac{d\psi}{ds}\dot{s}n$$
$$= \frac{\dot{s}}{\rho}n,$$

where s is distance measured along the path from a fixed point and ρ is the radius of curvature (here taken essentially positive). Now the velocity vector v may be written

$$v = v\tau = \dot{s}\tau;$$

hence the acceleration dv/dt is given by

$$\frac{dv}{dt} = \dot{v}\tau + \frac{v^2}{\rho}n.$$

2·14. Angular velocity of a rigid body. Any displacement of a rigid body is equivalent to a translation, that is, a motion such that every particle receives the same displacement, followed by a rotation about an axis. By definition a rigid body is such that in any possible motion the distance between any pair of particles is unaltered. Let a particle at O go to O' in the actual displacement. First consider every particle to receive a displacement OO'; this alters no distance between particles. Let this displacement take two particles P, Q to P', Q'. Let the positions taken by P, Q in the actual displacement be P'', Q''. The metrical definition of a plane is that it is a locus of points equidistant from two given points. Then the points equidistant from P' and P'' lie on a plane, and O' is on this plane because $O'P'$ and $O'P''$ are both equal to OP. Similarly, points equidistant from Q' and Q'' lie on a plane through O'. These two planes intersect in a line, and every point R' of this line must satisfy $R'P' = R'P''$, $R'Q' = R'Q''$. Therefore, since it maintains its distance also from the particle at O', it is occupied by the same particle in the $P'Q'$ and $P''Q''$ positions.

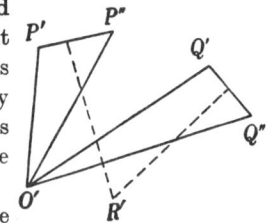

Now consider any pair of particles S, T in their three positions. Angles between planes of particles are conserved; hence the angle between the planes $O'R'S'$ and $O'R'S''$ is equal to that between $O'R'T'$ and $O'R'T''$. Therefore the $P'Q'$ position can be brought into the $P''Q''$ position by turning every particle about $O'R'$ through the same angle.

Let the position of O be $\boldsymbol{a}$, of O' $\boldsymbol{a} + \delta\boldsymbol{a}$. Let $O'R'$ have direction cosines n_i, and let the rotation about it be through a small angle $\delta\theta$. Then the displacement of $P(\boldsymbol{x})$ to P' is $\delta\boldsymbol{a}$,

and
$$\overline{P'P''} = \boldsymbol{n}\delta\theta \wedge \overline{O'P'} + O(\delta\theta)^2. \tag{1}$$

Put as before $\boldsymbol{n}\,\delta\theta = \delta\boldsymbol{\theta}$; then

$$\overline{O'P'} = \overline{OP} = \boldsymbol{x} - \boldsymbol{a}, \tag{2}$$
$$\overline{PP''} = \overline{PP'} + \overline{P'P''}$$
$$= \delta\boldsymbol{a} + \delta\boldsymbol{\theta} \wedge (\boldsymbol{x}-\boldsymbol{a}) + O(\delta\theta)^2, \tag{3}$$

which gives the displacement of a general particle of the body.

The velocity of P is then

$$\boldsymbol{v} = \lim_{\delta t \to 0} \frac{\overline{PP''}}{\delta t} = \frac{d\boldsymbol{a}}{dt} + \lim \frac{\delta\boldsymbol{\theta}}{\delta t} \wedge (\boldsymbol{x}-\boldsymbol{a})$$
$$= \dot{\boldsymbol{a}} + \boldsymbol{\omega} \wedge (\boldsymbol{x}-\boldsymbol{a}), \tag{4}$$

where
$$\boldsymbol{\omega} = \lim \frac{\delta\boldsymbol{\theta}}{\delta t}, \tag{5}$$

and is called the angular velocity of the body. Alternatively, we can write (4) as

$$v_i = \dot{a}_i + \epsilon_{ikm}\omega_k(x_m - a_m). \tag{6}$$

To check the fact that the set of velocities (6) represents a motion of a rigid body, consider the variation of the distance between two particles with coordinates x_i, y_i. We have

$$\frac{d}{dt}(y_i - x_i)^2 = 2(y_i - x_i)(\dot{y}_i - \dot{x}_i)$$
$$= 2(y_i - x_i)\,\epsilon_{ikm}\omega_k(y_m - x_m)$$
$$= 0, \tag{7}$$

which proves the result. Applying the same argument to small rotations we find that distances between particles are unchanged to the first order by a set of displacements given by (3); the second-order terms are more complicated. If we neglect them we can say that small angular displacements can be compounded by the parallelogram rule, in the sense that the sum of the displacements of a particle due to small rotations about different lines is the displacement due to the resultant of the rotations according to the rule. In particular

$$(\delta\boldsymbol{\theta} \wedge \boldsymbol{x})_i = \epsilon_{ikm}\delta\theta_k x_m, \tag{8}$$

and the components are $(x_3\delta\theta_2 - x_2\delta\theta_3, x_1\delta\theta_3 - x_3\delta\theta_1, x_2\delta\theta_1 - x_1\delta\theta_2)$ which are the sums of those due to separate rotations $(\delta\theta_1, \delta\theta_2, \delta\theta_3)$ about the axes.

2·15. Forces on a rigid body. For a particle we have the usual equations of motion, which can be written

$$m\ddot{\boldsymbol{x}} = \boldsymbol{X}, \quad m\ddot{x}_i = X_i. \tag{1}$$

Then for every particle of a rigid body these equations are true. If we simply add them and use S to denote summation over all particles we have

$$S(m\ddot{\boldsymbol{x}}) = S\boldsymbol{X}. \tag{2}$$

If we now write $Sm = M$, M will be the whole mass of the body; and if further we write

$$S(m\boldsymbol{x}) = M\bar{\boldsymbol{x}}, \tag{3}$$

$\bar{x}_i$ will be the coordinates of a point, which we shall call the centre of mass; and

$$M\ddot{\bar{\boldsymbol{x}}} = S\boldsymbol{X}, \tag{4}$$

the resultant of the forces on the particles.

Next, form the vector product of (1) with $\boldsymbol{x}$; then

$$m\boldsymbol{x} \wedge \ddot{\boldsymbol{x}} = \boldsymbol{x} \wedge \boldsymbol{X}, \quad m\epsilon_{ikm}x_k\ddot{x}_m = \epsilon_{ikm}x_k X_m. \tag{5}$$

By addition $\quad S(m\boldsymbol{x} \wedge \ddot{\boldsymbol{x}}) = S(\boldsymbol{x} \wedge \boldsymbol{X}), \quad S(m\epsilon_{ikm}x_k\ddot{x}_m) = S(\epsilon_{ikm}x_k X_m). \tag{6}$

The centre of mass of a rigid body is fixed in the body. This is usually taken for granted, but is not obvious. Let us consider r_1 the distance of the centre of mass from any given particle, originally at $\boldsymbol{x}_1$, and let any other particle of mass m_l be at $\boldsymbol{x}_l$. Then

$$(Sm_l)(\boldsymbol{x}_1 - \bar{\boldsymbol{x}}) = (Sm_l)\boldsymbol{x}_1 - S(m_l\boldsymbol{x}_l)$$
$$= Sm_l(\boldsymbol{x}_1 - \boldsymbol{x}_l), \tag{7}$$

$$(Sm_l)^2 r_1^2 = \{Sm_l(\boldsymbol{x}_1 - \boldsymbol{x}_l)\}\{Sm_{l'}(\boldsymbol{x}_1 - \boldsymbol{x}_{l'})\}$$
$$= SS'm_l m_{l'}(\boldsymbol{x}_1 - \boldsymbol{x}_l)(\boldsymbol{x}_1 - \boldsymbol{x}_{l'}), \tag{8}$$

S denoting summation with respect to l, S' with respect to l'. But

$$(\boldsymbol{x}_1 - \boldsymbol{x}_l)(\boldsymbol{x}_1 - \boldsymbol{x}_{l'}) = \tfrac{1}{2}(r_{1l}^2 + r_{1l'}^2 - r_{ll'}^2), \tag{9}$$

with an obvious notation. Hence r_1^2 is expressed entirely in terms of masses of particles and distances between them, and therefore is unaltered in any rigid-body displacement. Hence the centre of mass retains its distance from every particle of the body.

It is not enough to consider a particle originally at $\bar{\boldsymbol{x}}$, since for a hollow sphere there is no such particle.

The forces on the particles can be imagined to be separated into external forces and internal reactions. According to a principle due to d'Alembert, the internal reactions form a system in equilibrium among themselves, and their contributions to the right sides of (4) and (6) are zero. This principle is perhaps most completely understood if we regard the rigid body as the limiting case of an elastic one; but it follows at once if the body is regarded as made up of particles such that the force between any pair is along the line joining them (i.e. the acceleration components are in the ratios of the direction cosines of the line). For by Newton's third law the forces add up to zero, and if X' is the force on m_l due to $m_{l'}$, $-X'$ the reaction,

$$-x_l \wedge X' + x_{l'} \wedge X' = (x_{l'} - x_l) \wedge X' = 0,$$

if X' is along the line from x_l to $x_{l'}$. This argument is not general, because the particles might be electric or magnetic doublets, in which case the force would not be along the line joining them. But the result can still be shown to follow in a much wider class of cases subject to the condition that the distances between particles are unaltered. D'Alembert's principle is therefore an approximation valid for real solids provided the deformations can be neglected, and if they cannot it is not even strictly true that the centre of mass is fixed in the body. The reason for accepting it, however, is ultimately that experimentally it leads to the right answers.

In the right sides of (4) and (6) we need therefore consider only forces acting on the body from outside it. Further, six quantities suffice to specify the position of a rigid body, namely, the three coordinates of a given particle, and the three Euler angles specifying its orientation. These are treated in Chapter 3. But (4) and (6) form six differential equations, and six is the number required if we want to know how the body will move.

SX is the resultant force; $L = S(x \wedge X)$ is called the moment of the forces about the origin. Apart from the equations of motion the moment would have no physical interest. It should be noticed carefully that the moment of a force is $x \wedge X$, whereas the velocity due to a rotation is $\omega \wedge x$; the signs are always obvious if reference is made to a diagram, but mistakes in one or other of these expressions are common when vector notation is used throughout. If x_1 and X_2 are positive, the force is clearly tending to turn the body from $O1$ to $O2$; if ω_3 and x_1 are positive, $\dot{x}_2$ is positive. This suffices to fix the sign of one term in each component and the rest follow.

If the system is equivalent to a single force X at x, the moment is $x \wedge X = G$ and

$$G \cdot X = \epsilon_{ikm} x_k X_m X_i = 0.$$

It is therefore only in special conditions that the forces on a rigid body can be replaced by a single force.*

EXAMPLES

1. Prove that $\delta_{ik} \epsilon_{ikm} = 0$, $\epsilon_{iks} \epsilon_{mks} = 2\delta_{im}$, $\epsilon_{iks} \epsilon_{iks} = 6$.

2. If Δ denotes the determinant $\| u_{ij} \|$, prove that

$$\epsilon_{ikm} \Delta = \epsilon_{jln} u_{ij} u_{kl} u_{mn}, \quad \epsilon_{jln} \Delta = \epsilon_{ikm} u_{ij} u_{kl} u_{mn}, \quad 6\Delta = \epsilon_{ikm} \epsilon_{jln} u_{ij} u_{kl} u_{mn}.$$

3. If l_{ij} are the direction cosines of a transformation of axes prove that

$$l_{ij} = \tfrac{1}{2} \epsilon_{ikm} \epsilon_{jln} l_{kl} l_{mn}.$$

4. z is a constant unit vector, r (the position vector of a moving particle) a variable vector perpendicular to it. If the velocity at any instant is given by

$$\frac{d}{dt}(re^{kt}) = \omega z \wedge (re^{kt}), \tag{1}$$

where ω and k are constants, show that the orbit of the particle is an equiangular spiral.

---

* For reduction of a general system to two forces or a force and a couple, see H. Jeffreys, *Cartesian Tensors*, Chapter 5; C. E. Weatherburn, *Elementary Vector Analysis*, Chapter 8. These problems, however, occur only in examination questions.

Examples

A particle is moving in a plane under the action of a force to a fixed point proportional to the radial distance and a frictional resistance proportional to its velocity. Obtain the equation of motion of the particle, and by seeking solutions of the type (1) or otherwise show that the velocity at any instant is the vector sum of the velocities of the two particles describing equiangular spirals in opposite directions with equal angular velocities. (M/c, Part II, 1931.)

5. A, B, C are any three points on a sphere, centre O, of unit radius. The position vectors of A, B, C relative to O are u, v, w respectively. Show that the diameter which is perpendicular to the plane ABC cuts the sphere in the points whose position vectors are $\pm d$, where

$$[u, v, w]d \sec \theta = v \wedge w + w \wedge u + u \wedge v$$

and θ is the angle between d and u.

By considering the product $u \wedge d$, or otherwise, prove that

$$[u, v, w]\tan \theta = \pm 4 \sin \frac{a}{2} \sin \frac{b}{2} \sin \frac{c}{2},$$

where a, b, c are the sides of the spherical triangle ABC. (Prelim. 1941.)

6. If A, B, C, D are any four vectors, prove that

$$(A \wedge B).(C \wedge D) = (A.C)(B.D) - (A.D)(B.C),$$
$$(A \wedge B) \wedge (C \wedge D) = [C, D, A]B - [B, C, D]A$$
$$= [D, A, B]C - [A, B, C]D,$$

where $[A, B, C]$ denotes the triple scalar product $A.(B \wedge C)$.

Deduce the sine and cosine rules of spherical trigonometry. (Prelim. 1940.)

7. Two particles are projected simultaneously from the origin with velocities v_1, v_2 respectively and move under a constant acceleration a. Prove that if $v_1.v_2 < 0$, there is one and only one instant during the subsequent motion at which the particles subtend a right angle at the origin.

Show that at this instant the position vectors r_1, r_2 of the particles satisfy the equation

$$a^2(r_1.v_2 - r_2.v_1) + (a.r_2 - a.r_1)(a.v_1 + a.v_2) + 2v_1.v_2(a.v_2 - a.v_1) = 0.$$

(Prelim. 1940.)

8. Find an expression for the position vector r, at time t, of a particle of unit mass which moves under the action of a constant force $(n^2 + k^2)b$, together with an attractive force of magnitude $(n^2 + k^2)r$ towards the origin (where $n \neq 0$), in a medium which produces a retardation $2k$ times the speed. At time $t = 0$ the particle has velocity v and is at $r = a$.

Deduce that the triple scalar products of $r, v, a - b$ and of a, b, v are equal. (Prelim. 1941.)

9. A particle of charge e and mass m moves under the action of a uniform electric field of intensity $(0, E, 0)$ and a uniform magnetic field of intensity $(0, 0, H)$, Gaussian units being used. Prove that the motion can be regarded as the constant velocity $(Ec/H, 0, 0)$ superposed upon uniform motion in a circular helix with angular velocity $-eH/mc$ about the axis. It is to be assumed that the variation of mass with velocity is negligible.

Prove that, if the particle starts from the origin, then, whatever its initial velocity, it crosses each of the straight lines $x = 2\pi nmc^2E/eH^2, y = 0$, where $n = 1, 2, 3, \ldots$. (M.T. 1943.)

10. A particle of mass m at r is acted upon by a central force μr together with a force $e(H \wedge \dot{r})/c$, where H is a uniform magnetic field. Show that if r and $\dot{r}$ are initially perpendicular to H the particle will describe a plane curve.

Show that the particle can describe a circle about the origin under these forces, with either of two constant angular velocities. (I.C. 1942.)

11. Determine a vector $\overline{OC}$ perpendicular to $\overline{OA} = a(2, 3, 0), \overline{OB} = b(-2, 0, 1)$ such that the rotation from $\overline{OA}$ to $\overline{OB}$ is positive about $\overline{OC}$. Calculate the volume of the tetrahedron $OABC$.

Find the sides and angles of the spherical triangle ABC defined by

$$\overline{OA} = (1, 0, 0), \quad \overline{OB} = \left(\frac{1}{\sqrt{2}}, 0, \frac{1}{\sqrt{2}}\right), \quad \overline{OC} = \left(0, \frac{1}{\sqrt{2}}, -\frac{1}{\sqrt{2}}\right).$$

Chapter 3

TENSORS

> We know that intellectual food is sometimes more easily digested, if not taken in the most condensed form. It will be asked, To what extent can specialized notations be adopted with profit? To this question we reply, *only experience can tell.*
>
> F. CAJORI, *History of Mathematical Notations*, p. 77

3·01. In this chapter we develop the theory of tensors in a simple and restricted form. In many branches of physics the tensor notation in this form provides a compact mathematical expression, and familiarity with it is a preparation for the complete theory, involving the use of oblique axes, curvilinear coordinates and space of more than three dimensions; it is also an introduction to the ideas of matrix algebra. General tensor theory is indispensable as the mathematical apparatus of the theory of relativity, and matrix algebra in quantum mechanics and much of classical physics find their clearest expression in this notation. In the applications made in this chapter the physical ideas involved are simple, and practice in using the notation in this way is extremely valuable before proceeding to the applications of its complete form to theories where the physical ideas themselves are more difficult to grasp.

3·02. Transformation of coordinates. Contraction. We have defined a vector A by the transformation property

$$A'_j = l_{ij} A_i, \tag{1}$$

which is equivalent to

$$A_i = l_{ij} A'_j. \tag{2}$$

A vector can also be called a *tensor of the first order*. A scalar is a tensor of zero order.

Now if we consider the set of nine products $A_i B_k$ we notice that the scalar and vector products are particular linear combinations of these products. If we form a similar set for the components referred to new axes

$$A'_j B'_l = l_{ij} l_{kl} A_i B_k, \tag{3}$$

$$A_i B_k = l_{ij} l_{kl} A'_j B'_l. \tag{4}$$

In the same way as we use the transformation property to define a vector we now use these relations to define a tensor of the second order. A set of quantities depending on *two* directions and specified by nine components K_{ik} referred to $O123$ and K'_{jl} referred to $O1'2'3'$ forms a tensor of the second order if for all changes of axes

$$K'_{jl} = l_{ij} l_{kl} K_{ik}, \tag{5}$$

or the equivalent relation

$$K_{ik} = l_{ij} l_{kl} K'_{jl}. \tag{6}$$

The two suffixes denoting the component of K refer each to one of the coordinates of the *same* system. The direction cosines l_{ij} do *not* form a tensor since the two suffixes refer to axes of different sets.

In the square array

$$\begin{pmatrix} K_{11} & K_{12} & K_{13} \\ K_{21} & K_{22} & K_{23} \\ K_{31} & K_{32} & K_{33} \end{pmatrix}, \tag{7}$$

the components K_{11}, K_{22}, K_{33} are called the diagonal components. Their sum K_{ii} is called the *trace* or *spur* and is a scalar. For

$$K'_{jj} = l_{ij}l_{kj}K_{ik} = \delta_{ik}K_{ik} = K_{ii}. \tag{8}$$

The operation of putting two suffixes equal in a tensor and then summing is called *contraction*. The order of the tensor is reduced by 2.

The sum of two tensors K and L is defined by

$$(K+L)_{ik} = K_{ik} + L_{ik},$$

and is clearly also a tensor.

Tensors of higher orders are defined in a similar way; that is, a tensor of the nth order transforms like the product $A_i B_k C_m \ldots$ to n factors. We shall be mainly concerned with second-order tensors with some use of third and fourth order ones.

3·03. Isotropic tensors. We can show that the set of quantities δ_{ik} constitute a tensor. For if we apply the transformation (5) we get a set of quantities U'_{jl} given by

$$\begin{aligned} U'_{jl} = l_{ij}l_{kl}\delta_{ik} = l_{ij}l_{il} &= 1 \quad (j=l), \\ &= 0 \quad (j \neq l), \end{aligned} \tag{9}$$

and therefore the set δ_{ik} transforms into δ_{jl} on any rotation of axes.

Similarly, we can show that a third-order tensor with components ϵ_{ikm} referred to $O123$ has the same set of components referred to $O1'2'3'$. For on transformation we get for the jln component $l_{ij}l_{kl}l_{mn}\epsilon_{ikm}$. Expanding, we have, since i, k, m are all different in non-zero terms,

$$l_{1j}l_{2l}l_{3n} + l_{2j}l_{3l}l_{1n} + l_{3j}l_{1l}l_{2n}$$
$$- l_{2j}l_{1l}l_{3n} - l_{3j}l_{2l}l_{1n} - l_{1j}l_{3l}l_{2n}.$$

If $j = l$ all components cancel; similarly, if $j = n$ or $l = n$. If j, l, n are all different the expression is

$$\begin{vmatrix} l_{1j} & l_{1l} & l_{1n} \\ l_{2j} & l_{2l} & l_{2n} \\ l_{3j} & l_{3l} & l_{3n} \end{vmatrix},$$

which is equal to 1 if j, l, n are in even order and -1 if they are in odd order. Hence ϵ_{ikm} transforms into ϵ_{jln} under the rule for tensors of the third order.

Tensors whose components are unaltered by rotation of the axes are called *isotropic*. It can be shown* that there is no isotropic tensor of the first order, and the only ones of the second and third orders are scalar multiples of δ_{ik} and ϵ_{ikm}. There are three independent ones of the fourth order, namely,

$$\left. \begin{aligned} &\delta_{ik}\delta_{mp}, \\ &\delta_{im}\delta_{kp} + \delta_{ip}\delta_{km}, \\ &\delta_{im}\delta_{kp} - \delta_{ip}\delta_{km}. \end{aligned} \right\} \tag{10}$$

We have met the last as an alternative expression for $\epsilon_{iks}\epsilon_{mps}$. The other two appear in the derivation of the equations of motion of viscous fluids and elastic solids.

---

* H. Jeffreys, *Cartesian Tensors*, Chapter 7. See also note 3·03 a.

3·031. Isotropic tensors of order 4. If u_{ikmp} is an isotropic tensor of order 4 we have

$$u'_{jlnq} = l_{ij}l_{kl}l_{mn}l_{pq}u_{ikmp} = u_{jlnq} \tag{1}$$

for all rotations. Having regard to the fact that at least two suffixes must be equal in any component we see that the components fall into four patterns typified by u_{1111}, u_{1112}, u_{1122}, u_{1123}.

First rotate the axes about a line with direction cosines $1/\sqrt{3}$, $1/\sqrt{3}$, $1/\sqrt{3}$ so as to bring axis 1 into coincidence with the original axis 2, and so on. Since the result is a cyclic interchange of suffixes it follows from the isotropic property that

$$u_{1111} = u_{2222} = u_{3333}, \quad u_{1122} = u_{2233} = u_{3311}, \quad u_{2211} = u_{3322} = u_{1133}, \quad u_{1221} = u_{2332} = u_{3113}, \text{ etc.} \tag{2}$$

Next rotate through 90° about $O3$. Then

$$l_{12} = 1, \quad l_{21} = -1, \quad l_{33} = 1 \tag{3}$$

and the rest are zero. Take $j = 3$, $l = n = q = 1$. Then the non-zero terms are for $i = 3$, $k = m = p = 2$, and

$$u_{3111} = -u_{3222}. \tag{4}$$

Take also $j = 3$, $l = n = q = 2$. Then we must take $i = 3$, $k = m = p = 1$, and

$$u_{3222} = u_{3111}. \tag{5}$$

By similar methods it follows that all components with three suffixes equal and the other different are zero.

Similarly, we find, with

$$\begin{aligned} j = l = 1, \ n = q = 2; \ & i = k = 2, \ m = p = 1: \quad u_{1122} = u_{2211}, \\ j = n = 2, \ l = q = 3; \ & i = m = 1, \ k = p = 3: \quad u_{2323} = u_{1313}, \\ j = q = 1, \ l = n = 2; \ & i = p = 2, \ k = m = 1: \quad u_{1221} = u_{2112}, \end{aligned} \tag{6}$$

$$\begin{aligned} j = 3, \ l = n = 2, \ q = 1; \ & i = 3, \ k = m = 1, \ p = 2: \quad u_{3221} = -u_{3112}, \\ j = 3, \ l = n = 1, \ q = 2; \ & i = 3, \ k = m = 2, \ p = 1: \quad u_{3112} = u_{3221}. \end{aligned} \tag{7}$$

Hence the only non-zero components are those with the suffixes all equal or equal in pairs; and by cyclic interchange

$$u_{1111} = u_{2222} = u_{3333} = \kappa, \tag{8}$$

$$u_{1122} = u_{2211} = u_{2233} = u_{3322} = u_{3311} = u_{1133} = \lambda, \tag{9}$$

$$u_{2323} = u_{1313} = u_{3131} = u_{2121} = u_{1212} = u_{3232} = \mu, \tag{10}$$

$$u_{1221} = u_{2112} = u_{2332} = u_{3223} = u_{3113} = u_{1331} = \nu. \tag{11}$$

These relations would all be satisfied for cubic symmetry. We can now write

$$u_{ikmp} = \lambda \delta_{ik}\delta_{mp} + \mu \delta_{im}\delta_{kp} + \nu \delta_{ip}\delta_{km} + (\kappa - \lambda - \mu - \nu)v_{ikmp}, \tag{12}$$

where $v_{ikmp} = 1$ if all four suffixes are equal and otherwise zero. Now if u_{ikmp} is a tensor of order 4, $u_{ikmp}x_i y_k z_m w_p$ is a scalar, and conversely (see 3·05). This expression reduces to

$$\lambda x_i y_i z_m w_m + \mu x_i z_i y_k w_k + \nu x_i w_i y_k z_k + (\kappa - \lambda - \mu - \nu)(x_1 y_1 z_1 w_1 + \ldots). \tag{13}$$

The first three expressions are all products of scalars. But the last, if we take all the vectors the same, is $x_1^4 + x_2^4 + x_3^4$, which is not a scalar. It has cubic symmetry but not spherical symmetry. E.g. if $x_1 = x_2 = x_3 = 1/\sqrt{3}$, $x_1^4 + x_2^4 + x_3^4 = 1/3$; but if $x_1' = 1$, $x_2' = x_3' = 0$, $x_1'^4 + x_2'^4 + x_3'^4 = 1$; and this change of components would be achieved by a rotation of axes. Hence if the tensor is isotropic

$$\kappa - \lambda - \mu - \nu = 0, \tag{14}$$

and the most general isotropic tensor of order 4 is given by the first three terms of (12). It can be rewritten as the sum of three tensors of the forms 3·03 (10). See also Note 3·031*a*.

For any solid the elastic constants form a fourth-order tensor, which must be isotropic if the solid is. The v_{ikmp} term expresses an extra generality and permits the expression of the elastic properties of a cubic crystal: Young's modulus can have different values for strains along a diagonal and parallel to an edge.

3·04. Dyadic notation. It is sometimes convenient to denote a tensor of the second order by a single letter, as we do for a vector. If we multiply all components of a tensor K_{ik} by those of a vector A_m we get a tensor of the third order. But we can form from this product two different vectors by putting m equal to i or k and summing, namely, $K_{ik}A_i$ and $K_{ik}A_k$. In dyadic notation these are written $A.K$ and $K.A$ respectively, the rule for remembering which is which being that the order taken in the product is such that *summation is always taken over adjacent suffixes*; thus

$$(A.K)_k = A_i K_{ik}, \quad (K.A)_i = K_{ik} A_k. \tag{1}$$

The proof that contracting a tensor of order n gives one of order $n-2$ is similar to that of 3·02 (8) and need not be given in full. The use of heavy type can be taken as an indication that one or more suffixes are suppressed.

Similarly, we can form contracted products of two tensors K and L of the second order, namely,

$$(K.L)_{ik} = K_{im} L_{mk}, \quad (L.K)_{ik} = L_{im} K_{mk}. \tag{2}$$

Again in general the result depends on the order; this type of multiplication is not commutative. In this notation the tensor $A_i B_k$ is written as AB; the absence of the dot distinguishes the tensor from both the scalar and the vector products. Dyadic notation has analogies with matrix notation, which will be developed in the next chapter. The compression introduced by the suppression of the suffixes is compensated by the extra care that has to be taken to preserve the order, and by the fact that we sometimes do not want to contract. In particular the elastic constants of a crystal form a fourth-order tensor.

3·05. The quotient rule. If we have a set of equations

$$K_{ik} A_k = B_i, \tag{1}$$

where A_k and B_i are known to be first-order tensors, or if

$$K_{ik} T_{km} = S_{im}, \tag{2}$$

where T_{km} and S_{im} are second-order tensors, can we conversely infer that K_{ik} is a second-order tensor? The answer is that we can, provided that all the components A_k or T_{km} can be varied independently. We take the simplest case, starting from (1). We transform the axes; then we do not know how K_{ik} transforms but it must give a set K'_{jl} with nine components specified by $\partial B'_j / \partial A'_l$. Then

$$K'_{jl} A'_l = B'_j = l_{ij} B_i = l_{ij} K_{ik} A_k$$
$$= l_{ij} l_{kl} K_{ik} A'_l. \tag{3}$$

Hence $$(K'_{jl} - l_{ij} l_{kl} K_{ik}) A'_l = 0. \tag{4}$$

But if this is true for every j, when each A'_l is varied separately,

$$K'_{jl} = l_{ij} l_{kl} K_{ik}, \tag{5}$$

and K_{ik} is a second-order tensor.

An important particular case of the general theorem is that if $K_{ikm...} T_{ikm...}$ is a scalar, then if T is an arbitrary nth-order tensor K is also an nth-order tensor. In particular, the coefficients a_{ik} in a quadratic form $a_{ik} x_i x_k$, where $a_{ki} = a_{ik}$, in the coordinates of a point form a tensor of the second order.

3·06. Differentiation of scalar and vector functions of position. A scalar, or the components of a vector or a tensor, may have different values at different places, even if we consider them only at a single instant of time and do not transform the axes. Thus the different particles of a body differ in distance from the origin, but the distance of a given particle does not depend on the directions of the axes. Such a function is called a *scalar function of position*. Again, the velocity of a particle of a fluid is a vector, but in general varies with position, and can be called a *vector function of position*. The existence of such functions makes it necessary to consider their differentiation.

If ϕ is a scalar function, the set of three derivatives $\partial\phi/\partial x_i$ specifies a vector denoted by grad ϕ or $\nabla\phi$. To show that it is a vector we rotate the axes; we have

$$\frac{\partial \phi}{\partial x'_j} = \frac{\partial x_i}{\partial x'_j} \frac{\partial \phi}{\partial x_i} = l_{ij} \frac{\partial \phi}{\partial x_i}, \tag{1}$$

which proves the result. This assumes that ϕ is differentiable in three dimensions in the sense of Stolz and Young: a sufficient condition for this is that the partial derivatives of ϕ are continuous with regard to all the coordinates. Cf. 5·04.

If u_i is a vector function,

$$\frac{\partial u'_l}{\partial x'_j} = \frac{\partial x_i}{\partial x'_j} \frac{\partial u'_l}{\partial x_i} = l_{ij} \frac{\partial}{\partial x_i}(l_{kl} u_k) = l_{ij} l_{kl} \frac{\partial u_k}{\partial x_i}, \tag{2}$$

and therefore $\partial u_k/\partial x_i$ is a tensor of the second order.

It follows that $\partial u_i/\partial x_i$ is a scalar function; it is usually denoted by div u or $\nabla . u$. If, further, there is a function ϕ such that $u_i = \partial\phi/\partial x_i$,

$$\frac{\partial u_i}{\partial x_i} = \frac{\partial^2 \phi}{\partial x_i^2} = \frac{\partial^2 \phi}{\partial x_1^2} + \frac{\partial^2 \phi}{\partial x_2^2} + \frac{\partial^2 \phi}{\partial x_3^2}. \tag{3}$$

This combination of second derivatives has an importance in mathematical physics second only to differentiation with regard to the time. It is denoted by $\nabla^2\phi$.

The quantities

$$\begin{aligned}\epsilon_{ikm}\frac{\partial u_m}{\partial x_k} &= \left(\frac{\partial u_3}{\partial x_2} - \frac{\partial u_2}{\partial x_3},\ \frac{\partial u_1}{\partial x_3} - \frac{\partial u_3}{\partial x_1},\ \frac{\partial u_2}{\partial x_1} - \frac{\partial u_1}{\partial x_2}\right) \\ &= \left(\frac{\partial w}{\partial y} - \frac{\partial v}{\partial z},\ \frac{\partial u}{\partial z} - \frac{\partial w}{\partial x},\ \frac{\partial v}{\partial x} - \frac{\partial u}{\partial y}\right)\end{aligned} \tag{4}$$

in Cartesian notation, determine a vector. It is denoted by curl u. ($\nabla \wedge u$ is also used.) It will be noticed that if u is the gradient of a scalar, curl $u = 0$:

$$(\text{curl grad } \phi)_i = \epsilon_{ikm}\frac{\partial}{\partial x_k}\frac{\partial \phi}{\partial x_m} = 0. \tag{5}$$

Also if u is any vector function

$$\text{div curl } u = \frac{\partial}{\partial x_i}\epsilon_{ikm}\frac{\partial u_m}{\partial x_k} = \epsilon_{ikm}\frac{\partial^2 u_m}{\partial x_i \partial x_k} = 0, \tag{6}$$

since all the terms cancel.

A useful result is obtained by taking the curl again. We have

$$(\operatorname{curl curl} \boldsymbol{u})_i = \epsilon_{ikm}\frac{\partial}{\partial x_k}\epsilon_{mps}\frac{\partial u_s}{\partial x_p}$$

$$= (\delta_{ip}\delta_{ks} - \delta_{is}\delta_{kp})\frac{\partial^2 u_s}{\partial x_k \partial x_p}$$

$$= \frac{\partial^2 u_s}{\partial x_s \partial x_i} - \frac{\partial^2 u_i}{\partial x_k^2}$$

$$= (\operatorname{grad div} \boldsymbol{u} - \nabla^2 \boldsymbol{u})_i. \tag{7}$$

3·07. Symmetry properties. If K_{ik} is a tensor, we show that K_{ki}, obtained by interchanging rows and columns, is another. This means that if K_{ik} transforms into K'_{jl}, according to the rule

$$K'_{jl} = l_{ij}l_{kl}K_{ik}, \tag{1}$$

and we write $L_{ik} = K_{ki}$, L_{ik} will transform according to the same rule. But

$$l_{ij}l_{kl}L_{ik} = l_{ij}l_{kl}K_{ki}. \tag{2}$$

Here i and k are repeated suffixes; it is therefore immaterial which we call i and which k, and therefore they can be interchanged, giving

$$l_{kj}l_{il}K_{ik} = l_{il}l_{kj}K_{ik} = K'_{lj}, \tag{3}$$

so that the transformed set also differs from K'_{jl} in having rows and columns interchanged.

If $K_{ik} = K_{ki}$, the tensor is said to be *symmetrical*; if $K_{ik} = -K_{ki}$, it is said to be *antisymmetrical*. Again, if K_{ik} is a tensor, two others are $K_{ik}+K_{ki}$ and $K_{ik}-K_{ki}$. The first of these is unaltered if i and k are interchanged and is a *symmetrical* tensor; the second has the signs of all components reversed and is an *antisymmetrical* tensor. Since any tensor K_{ik} can be written as

$$K_{ik} = \tfrac{1}{2}(K_{ik}+K_{ki}) + \tfrac{1}{2}(K_{ik}-K_{ki}), \tag{4}$$

it can be expressed as the sum of a symmetrical and an antisymmetrical tensor.

Since $K_{ik}A_k$ and $A_i K_{ik}$ are vectors we can form their scalar products $B_i K_{ik} A_k$ and $A_i K_{ik} B_k$ with another vector $\boldsymbol{B}$. These products are not in general equal. But if K_{ik} is symmetrical they are equal, for

$$\boldsymbol{B}.\boldsymbol{K}.\boldsymbol{A} = B_i K_{ik} A_k = B_i K_{ki} A_k = A_k K_{ki} B_i = \boldsymbol{A}.\boldsymbol{K}.\boldsymbol{B}. \tag{5}$$

On the other hand, if K_{ik} is antisymmetrical the sign is reversed in the third of these expressions, and

$$\boldsymbol{B}.\boldsymbol{K}.\boldsymbol{A} = -\boldsymbol{A}.\boldsymbol{K}.\boldsymbol{B}. \tag{6}$$

It is important to notice that if K_{ik} is symmetrical and u_i a vector, every term in $K_{ik}u_i u_k$ with $i \neq k$ occurs *twice*. Thus for $i=1$, $k=2$ we get a term $K_{12}u_1 u_2$, but there is another term with $i=2$, $k=1$ equal to $K_{21}u_2 u_1$. If $K_{12} = K_{21}$ these terms are equal. Thus the expansion of

$$T = K_{ik}u_i u_k$$

is

$$K_{11}u_1^2 + 2K_{12}u_1 u_2 + K_{22}u_2^2 + 2K_{13}u_1 u_3 + \ldots,$$

and

$$\frac{\partial T}{\partial u_1} = 2(K_{11}u_1 + K_{12}u_2 + \ldots)$$

$$= 2K_{1k}u_k,$$

and in general

$$\frac{\partial T}{\partial u_i} = 2K_{ik}u_k.$$

3·071. The vector of a tensor, vec K. Consider the triad $\epsilon_{ikm}K_{km}$. This is the twice contracted product of tensors of the third and second orders and therefore is a vector; alternatively, by changing axes we have

$$\epsilon_{jln}K'_{ln} = \epsilon_{jln}l_{kl}l_{mn}K_{km} = \epsilon_{ikm}l_{ij}K_{km}. \tag{7}$$

We can therefore write this as 2 vec K, where

$$(\text{vec }K)_i = \tfrac{1}{2}\epsilon_{ikm}K_{km}; \quad \text{in components,} \quad \{\tfrac{1}{2}(K_{23}-K_{32}), \tfrac{1}{2}(K_{31}-K_{13}), \tfrac{1}{2}(K_{12}-K_{21})\}. \tag{8}$$

3·072. Relations between an antisymmetrical tensor and a vector. The components of an antisymmetrical tensor W_{ik} with $i = k$ must vanish, and since for the others $W_{ik} = -W_{ki}$ only three independent quantities need be given to specify an antisymmetrical tensor, which then takes the form

$$\begin{pmatrix} 0 & W_{12} & -W_{31} \\ -W_{12} & 0 & W_{23} \\ W_{31} & -W_{23} & 0 \end{pmatrix}. \tag{9}$$

But W_{23}, W_{31}, W_{12} are the components of vec W. We shall denote this vector by w, that is,

$$w_i = \tfrac{1}{2}\epsilon_{ikm}W_{km}. \tag{10}$$

This property, that the number of components of a vector is equal to that of the independent components of an antisymmetrical tensor, is peculiar to three dimensions. In n dimensions an antisymmetrical tensor has $\tfrac{1}{2}n(n-1)$ independent components, while a vector has n components.

It follows that the set of quantities in (9) is the same as the set

$$\begin{pmatrix} 0 & w_3 & -w_2 \\ -w_3 & 0 & w_1 \\ w_2 & -w_1 & 0 \end{pmatrix}, \tag{11}$$

that is, $W_{12} = w_3$, $W_{21} = -w_3$, and in general

$$W_{ik} = 0 \quad (i = k),$$
$$= w_m \quad (ikm \text{ in even order}),$$
$$= -w_m \quad (ikm \text{ in odd order}),$$

and therefore

$$W_{ik} = \epsilon_{ikm}w_m. \tag{12}$$

It is sometimes convenient to use the vector and sometimes the antisymmetrical tensor representation; equations (10) and (12) give the relations between them.

In particular, if we take a vector product $(w \wedge A)$,

$$(w \wedge A)_i = \epsilon_{ikm}w_k A_m = \tfrac{1}{2}\epsilon_{ikm}\epsilon_{kps}W_{ps}A_m$$
$$= \tfrac{1}{2}\epsilon_{mik}\epsilon_{psk}W_{ps}A_m$$
$$= \tfrac{1}{2}(\delta_{mp}\delta_{is} - \delta_{ms}\delta_{ip})W_{ps}A_m$$
$$= \tfrac{1}{2}(W_{mi} - W_{im})A_m$$
$$= -W_{im}A_m. \tag{13}$$

Hence we can replace vector multiplication by the vector w by multiplication by the antisymmetrical tensor $-W_{ik}$. Alternatively, we can derive the result by writing out a special component, say $i = 1$.

In physical applications it is sometimes the one and sometimes the other that appears first most naturally. In deriving the equations of angular momentum, for instance, we start from $\ddot{x}_i = X_i$; multiply the k equation by x_m, the m equation by x_k, and subtract. Then we have the nine equations

$$m(\ddot{x}_k x_m - \ddot{x}_m x_k) = X_k x_m - X_m x_k,$$

in which both sides are antisymmetrical tensors. The reason for converting these equations into the vector form is that this eliminates the three that have the form $0 = 0$ and three others that can be inferred from those retained by a change of sign.

3·08. Symmetrical tensor: principal axes. We have seen that an antisymmetrical tensor can be related to a vector. A symmetrical tensor can be related to a quadric. If K_{ik} is a symmetrical tensor with real components the equation

$$K_{ik} x_i x_k = \text{constant} \tag{1}$$

represents a central quadric with centre at the origin. Now

$$x_i = x l_i \tag{2}$$

represents a line through the origin, and the polar plane of a point on it is

$$K_{ik} l_k x_i = \text{constant.} \tag{3}$$

This plane is perpendicular to the line if

$$K_{ik} l_k = \lambda l_i, \tag{4}$$

where λ is the same for $i = 1, 2, 3$. The condition for consistency of these equations is a cubic equation for λ, and any root will in general give admissible ratios of the l_i. These will be real if λ is real, and then the line will be perpendicular to the polar plane of any point on it and in particular to the tangent plane at the point where the line meets the quadric. Such a line is a *principal axis*.

We show first that if there are two solutions corresponding to different values of λ, say λ_1 and λ_2, they give values of l_i, say l_{i1} and l_{i2} satisfying $l_{i1} l_{i2} = 0$. If λ_1 and λ_2 are real this says that the lines are perpendicular. Then

$$K_{ik} l_{k1} = \lambda_1 l_{i1}, \tag{5}$$

$$K_{ik} l_{k2} = \lambda_2 l_{i2}. \tag{6}$$

Multiply these respectively by l_{i2}, l_{i1} and contract; then

$$K_{ik} l_{k1} l_{i2} = \lambda_1 l_{i1} l_{i2}, \tag{7}$$

$$K_{ik} l_{k2} l_{i1} = \lambda_2 l_{i2} l_{i1}. \tag{8}$$

But since K_{ik} is symmetrical the expressions on the left are equal. Therefore

$$(\lambda_1 - \lambda_2) l_{i1} l_{i2} = 0, \tag{9}$$

and if $\lambda_1 \neq \lambda_2$,

$$l_{i1} l_{i2} = 0. \tag{10}$$

The condition of consistency for (4), with $l_i \neq 0$, is that the determinant $\|K_{ik} - \lambda \delta_{ik}\| = 0$; that is

$$\begin{vmatrix} K_{11} - \lambda & K_{12} & K_{13} \\ K_{21} & K_{22} - \lambda & K_{23} \\ K_{31} & K_{32} & K_{33} - \lambda \end{vmatrix} = 0. \tag{11}$$

This must have one real root; call it λ_1. Take the resulting l_{i1} as the direction cosines of a new axis of x_1', and take two axes x_2', x_3' perpendicular to this. Then if accents indicate direction cosines with regard to the new axes, $l_{11}' = 1$, $l_{21}' = l_{31}' = 0$; and by (4)

$$K_{11}' = \lambda_1, \quad K_{12}' = 0, \quad K_{13}' = 0. \tag{12}$$

The equation (11) therefore now takes the form

$$\begin{vmatrix} \lambda_1 - \lambda & 0 & 0 \\ 0 & K_{22}' - \lambda & K_{23}' \\ 0 & K_{23}' & K_{33}' - \lambda \end{vmatrix} = 0. \tag{13}$$

Hence
$$\lambda = \lambda_1 \tag{14}$$

or
$$\lambda^2 - (K_{22}' + K_{33}')\lambda + K_{22}' K_{33}' - (K_{23}')^2 = 0. \tag{15}$$

Equation (15) has real roots, since

$$(K_{22}' + K_{33}')^2 - 4\{K_{22}' K_{33}' - (K_{23}')^2\} = (K_{22}' - K_{33}')^2 + 4(K_{23}')^2 \geq 0. \tag{16}$$

If this expression is zero the roots are equal, and conversely. If the roots are different there are three real perpendicular directions satisfying (4), and they are called the *principal axes* and the values of λ the *principal values* of the tensor. When the tensor is referred to the principal axes x_j' it takes the form

$$\begin{pmatrix} \lambda_1 & 0 & 0 \\ 0 & \lambda_2 & 0 \\ 0 & 0 & \lambda_3 \end{pmatrix}, \tag{17}$$

and is said to be *reduced to diagonal form*. Then with respect to the original axes

$$K_{ik} = l_{ij} l_{kl} K_{jl}' = \lambda_1 l_{i1} l_{k1} + \lambda_2 l_{i2} l_{k2} + \lambda_3 l_{i3} l_{k3}. \tag{18}$$

If two roots are equal, take them to be λ_2; then $K_{22}' = K_{33}'$, $K_{23}' = 0$ and the quadric becomes

$$\lambda_1 x_1'^2 + \lambda_2 (x_2'^2 + x_3'^2) = \text{constant}. \tag{19}$$

It is therefore a surface of revolution and *any* line in the plane of x_2' and x_3' is a principal axis.

If all three roots are equal the quadric is a sphere.

In both these special cases it remains true that we can find three perpendicular directions satisfying (4); but we can now do it in an infinite number of ways, whereas when all the roots are unequal we can only do it in one way.

3·081. Inertia tensor of a rigid body. Consider a rigid body moving with angular velocity $\boldsymbol{\omega}$ with one point O fixed. If a particle $P(x_i)$ has mass m we have

$$\frac{d}{dt} Sm(x_k \dot{x}_m - x_m \dot{x}_k) = Sm(x_k \ddot{x}_m - x_m \ddot{x}_k) = S(x_k X_m - x_m X_k), \quad (1)$$

which is a relation between antisymmetrical tensors, expressible also in vector form

$$\frac{d}{dt} Sm \epsilon_{ikm} x_k \dot{x}_m = S \epsilon_{ikm} x_k X_m \quad (2)$$

or

$$\frac{d}{dt} Sm(\boldsymbol{x} \wedge \dot{\boldsymbol{x}}) = S(\boldsymbol{x} \wedge \boldsymbol{X}). \quad (3)$$

The expression $Sm(\boldsymbol{x} \wedge \dot{\boldsymbol{x}})$ is called the *angular momentum* of the body about O and denoted by $\boldsymbol{h}(O)$. Now since O is fixed, if $\boldsymbol{\omega}$ is the angular velocity,

$$\dot{\boldsymbol{x}} = \boldsymbol{\omega} \wedge \boldsymbol{x}, \quad (4)$$

that is,

$$\dot{x}_i = \epsilon_{ikm} \omega_k x_m, \quad (5)$$

and

$$\begin{aligned} h_i(O) &= Sm \epsilon_{ikm} x_k \epsilon_{mps} \omega_p x_s \\ &= Sm(\delta_{ip}\delta_{ks} - \delta_{ik}\delta_{ps}) x_k x_s \omega_p \\ &= Sm(x_s^2 \omega_i - x_i x_p \omega_p) \\ &= I_{ik} \omega_k, \end{aligned} \quad (6)$$

where I_{ik} is the symmetrical tensor

$$Sm(r^2 \delta_{ik} - x_i x_k). \quad (7)$$

In dyadic notation

$$\boldsymbol{h}(O) = \boldsymbol{I} \cdot \boldsymbol{\omega}. \quad (8)$$

I_{ik} is called the *inertia tensor* of the body about O. Written out in full it is

$$\begin{pmatrix} Sm(x_2^2+x_3^2) & -Smx_1x_2 & -Smx_1x_3 \\ -Smx_1x_2 & Sm(x_3^2+x_1^2) & -Smx_2x_3 \\ -Smx_1x_3 & -Smx_2x_3 & Sm(x_1^2+x_2^2) \end{pmatrix}. \quad (9)$$

The diagonal components are the moments of inertia about the axes, and the non-diagonal components are the products of inertia *multiplied by* -1. Since I_{ik} is a symmetrical tensor, axes can be found such that the products of inertia vanish and the tensor takes the form

$$\begin{pmatrix} A & 0 & 0 \\ 0 & B & 0 \\ 0 & 0 & C \end{pmatrix} \quad (10)$$

$A, B, C,$ are the principal moments of inertia at O. It is readily proved that

(1) The moment of inertia about a line with direction cosines n_i is

$$I_{ik} n_i n_k = n_i I_{ik} n_k = \boldsymbol{n} . \boldsymbol{I} . \boldsymbol{n}. \quad (11)$$

(2) The product of inertia with respect to two perpendicular lines with direction cosines n_i, n_i' is

$$-n_i I_{ik} n_k' = -\boldsymbol{n} . \boldsymbol{I} . \boldsymbol{n}'. \quad (12)$$

(3) If the centre of mass of the body is G with coordinates $\bar{x}_i$ relative to O, and if $I_{ik}(O)$ and $I_{ik}(G)$ are the inertia tensors at O and G respectively, and $Sm = M$,

$$I_{ik}(O) = I_{ik}(G) + M(\bar{x}_m^2 \delta_{ik} - \bar{x}_i \bar{x}_k). \tag{13}$$

(4) The kinetic energy of the body, moving with O fixed, is

$$\tfrac{1}{2} I_{ik}(O) \omega_i \omega_k = \tfrac{1}{2} \boldsymbol{\omega} \cdot \boldsymbol{I}(O) \cdot \boldsymbol{\omega}. \tag{14}$$

(5) The kinetic energy of the body moving in any manner is

$$\tfrac{1}{2} M V^2 + \tfrac{1}{2} I_{ik}(G) \omega_i \omega_k = \tfrac{1}{2} M \boldsymbol{V} \cdot \boldsymbol{V} + \tfrac{1}{2} \boldsymbol{\omega} \cdot \boldsymbol{I}(G) \cdot \boldsymbol{\omega}, \tag{15}$$

where V is the velocity of the centre of mass.

Since $I_{ik} \omega_k$ is a vector, its components about any set of axes can be written down at once. In particular, if we take as axes the principal axes of inertia the components of angular momentum about them are $(A\omega_1, B\omega_2, C\omega_3)$, and for a rigid body A, B, C are inindependent of time. It is this fact that makes the use of moving axes convenient.

3·09. Finite rotation of a rigid body. We have shown* that a finite rotation of a rigid body about a fixed point cannot be represented by a vector in the direction of the axis of rotation. We now show how such a rotation can be represented by a tensor.

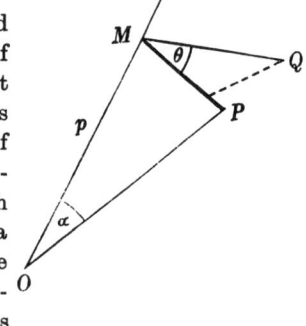

We take the origin at the point O of the body, and $O123$ is the frame of reference. Let $P(x_i)$ be a point of the body. The body is rotated through an angle θ about a line through O with direction cosines n_i, and P moves to $Q(y_i)$. Let M be the projection of P on the axis of rotation; then M is $n_k x_k n_i = p n_i$ say. The rotation displaces P through $(1 - \cos \theta) PM$ towards M, and through $PM \sin \theta$ at right angles to the plane OPM. If θ is a right-handed rotation the latter displacement is in the direction of the vector product $\boldsymbol{n} \wedge \boldsymbol{x}$. To get its magnitude, let the angle MOP be α; then the modulus of $\boldsymbol{n} \wedge \boldsymbol{x}$ is $OP \sin \alpha = PM$. Hence the second part of the displacement is $\sin \theta (\boldsymbol{n} \wedge \boldsymbol{x})$, and

$$\begin{aligned} y_i - x_i &= -(1 - \cos \theta)(x_i - p n_i) + \sin \theta (\boldsymbol{n} \wedge \boldsymbol{x})_i, \\ y_i &= \cos \theta\, x_i + (1 - \cos \theta) n_i n_k x_k + \sin \theta\, \epsilon_{ikm} n_k x_m \\ &= \{\cos \theta\, \delta_{ik} + (1 - \cos \theta) n_i n_k - \sin \theta\, \epsilon_{ikm} n_m\} x_k. \end{aligned} \tag{1}$$

The quantity in brackets { } is clearly a tensor of order 2, which we may denote by R_{ik}. It is neither symmetrical nor antisymmetrical.

If the body undergoes successive rotations represented by tensors $R_{ik}^{(1)}$, $R_{ik}^{(2)}$, $R_{ik}^{(3)}$, ..., $R_{ik}^{(n)}$, then the final position of P is given by

$$x_i^{(n)} = R_{ik_{n-1}}^{(n)} R_{k_{n-1} k_{n-2}}^{(n-1)} \ldots R_{k_1 k}^{(1)} x_k.$$

We have seen in 2·03 that the order of the rotations is important.

---

* See 2·03, p. 62, passage in small type.

3·091. When θ is small, R_{ik} reduces, to the first order, to $\delta_{ik} - \theta \epsilon_{ikm} n_m$ and

$$y - x = \mathbf{\theta} \wedge x, \tag{2}$$

where
$$\mathbf{\theta} = \theta \mathbf{n}. \tag{3}$$

The resultant of two successive small rotations $\mathbf{\theta}, \mathbf{\theta}'$ is

$$(\mathbf{\theta} + \mathbf{\theta}') \wedge x, \tag{4}$$

and in this sense a small rotation is represented by a vector. The same result was obtained in Chapter 2, but it was not possible there to give explicitly the terms of order θ^2 neglected in its derivation.

3·092. Tensor representation of angular velocity. We have shown in Chapter 2 that there is a vector $\boldsymbol{\omega}$ representing the angular velocity of a rigid body, and that the velocities $\boldsymbol{v}_P, \boldsymbol{v}_Q$ of two points are connected by the relation

$$\boldsymbol{v}_Q = \boldsymbol{v}_P + \boldsymbol{\omega} \wedge PQ. \tag{1}$$

If PQ has components x_i and we write the components of $\boldsymbol{v}_P$ and $\boldsymbol{v}_Q$ as v_i^P, v_i^Q, we have by 3·072 (13) that if

$$\Omega_{ik} = \epsilon_{ikm} \omega_m, \tag{2}$$

$$v_i^Q - v_i^P = -\Omega_{ik} x_k. \tag{3}$$

The form (3) is in a sense more general than (1). A rotation about $O3$ is the same thing as one from $O1$ towards $O2$, in three dimensions. In any number of dimensions (≥ 2) we can speak of a rotation from $O1$ towards $O2$; but it is only in three dimensions that such a rotation can be said to be *about* any particular axis. We shall consider this further in the next chapter.

3·10. General motion of a fluid. When the particles of a system are not constrained to remain at constant distances apart, the motion can no longer be specified by the velocity of one point and an angular velocity.

Let x_i be the position vector of a general particle P of the fluid and let v_i be its velocity at a given time. Then v_i is a function both of x_i and of t. The velocity $v_i + \delta v_i$ at a neighbouring point Q $(x_i + \delta x_i)$ is

$$v_i + \delta v_i = v_i + \frac{\partial v_i}{\partial x_k} \delta x_k + O(\delta x_k)^2. \tag{1}$$

To the first order in δx_k we have therefore

$$\delta v_i = \frac{1}{2} \left(\frac{\partial v_k}{\partial x_i} + \frac{\partial v_i}{\partial x_k} \right) \delta x_k - \frac{1}{2} \left(\frac{\partial v_k}{\partial x_i} - \frac{\partial v_i}{\partial x_k} \right) \delta x_k$$
$$= e_{ik} \delta x_k - \xi_{ik} \delta x_k, \tag{2}$$

say. Then e_{ik} is a symmetrical tensor and ξ_{ik} an antisymmetrical one, both of dimensions $1/t$. The part of δv_i depending on ξ_{ik} is the same as the displacement due to a rigid-body rotation with components $(\xi_{23}, \xi_{31}, \xi_{12})$. We shall see in a moment what the other part represents. Consider the rate of change of $\tfrac{1}{2} PQ^2$; this is

$$\delta x_i \delta v_i = (e_{ik} \delta x_k - \xi_{ik} \delta x_k) \delta x_i, \tag{3}$$

and the part depending on ξ_{ik} is zero because ξ_{ik} is antisymmetrical and all the terms cancel. Changes of distance between neighbouring particles therefore depend entirely on the e_{ik}. Of these, any one can be different from zero and the others zero except of course the one required not to be zero by the conditions of symmetry if $i \neq k$. Thus if $\delta v_1 = e\delta x_1$, $\delta v_2 = \delta v_3 = 0$ we have $e_{11} = e$ and all the rest 0. If $\delta v_i = (e\delta x_2, e\delta x_1, 0)$ we have $e_{12} = e_{21} = e$ and all the rest 0; and similarly for the other components by symmetry. The corresponding changes in the plane of $\delta x_3 = 0$ are illustrated.

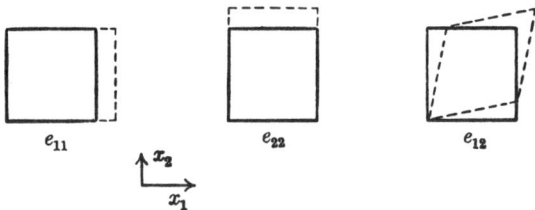

The tensor e_{ik} therefore represents the rates of change in size and shape of an element of fluid surrounding P. It is called the *rate of strain tensor*. It has three principal axes, and the changes can be reduced to three extensions along them. If the principal values are equal the rates of extension are the same in all directions, that is, the strain near P is a symmetrical expansion or compression.

In a certain sense the ξ_{ik} represent a local angular velocity; but this statement requires qualification because the e_{ik} obviously imply angular velocities, though these are in opposite senses for different parts of the element, and without further restriction an angular velocity of an element round P has no definite meaning. We consider a small element of fluid with P as its centre of mass, and suppose a small *rigid* body with the same density distribution to have the same angular momentum about P. Then we shall show that, provided the principal axes of the inertia tensor $I_{ik}(P)$ coincide with those of e_{ik}, the angular velocity of the rigid body is $\omega_i = \tfrac{1}{2}\epsilon_{ikm}\xi_{km} = \tfrac{1}{2}(\mathrm{curl}\,v)_i$.

We have, if h_i is the angular momentum of the element considered,

$$h_i = I_{ik}\omega_k = Sm\,\epsilon_{ikm}\delta x_k(v_m + \delta v_m)$$
$$= \epsilon_{ikm}v_m Sm\,\delta x_k + Sm\,\epsilon_{ikm}e_{mp}\delta x_k \delta x_p - Sm\,\epsilon_{ikm}\xi_{mp}\delta x_k \delta x_p. \qquad (4)$$

Now since P is the centre of mass of the element, $Sm\,\delta x_k = 0$. Also

$$Sm\,\delta x_k \delta x_p = -I_{kp} + \delta_{kp}Sm(\delta x_s)^2. \qquad (5)$$

Suppose now that I_{ik} and e_{ik} have the same principal axes; that is, we can take the axes so that $I_{ik} = 0$, $e_{ik} = 0$ unless $i = k$. Then

$$\epsilon_{ikm}e_{mp}I_{kp} = 0, \quad \epsilon_{ikm}e_{mp}\delta_{kp} = 0, \qquad (6)$$

since $e_{mp}I_{kp}$ and $e_{mp}\delta_{kp}$ vanish unless $m = k$ and $\epsilon_{ikm} = 0$ if $m = k$. Hence the second term on the right of (4) is

$$\{-I_{kp} + Sm(\delta x_s)^2 \delta_{kp}\}\epsilon_{ikm}e_{mp} = 0. \qquad (7)$$

If we write $\quad \xi_{mp} = \epsilon_{smp}\xi_s, \quad \xi_k = \tfrac{1}{2}\epsilon_{ikm}\xi_{km}, \quad$ (cf. § 3·072, (10) and (12))

the last term in (4) becomes

$$-\epsilon_{ikm}\epsilon_{smp}\xi_s \delta x_k \delta x_p = [(\delta x_p)^2 \delta_{ik} - \delta x_i \delta x_k]\xi_k \equiv I_{ik}\xi_k,$$

which is the angular momentum of a rigid body filling the element with inertia tensor I_{ik} and angular velocity ξ_k. Hence

$$I_{ik}\omega_k = I_{ik}\xi_k, \qquad (8)$$

and therefore $\omega_k = \xi_k$ provided $\|I_{ik}\| \neq 0$.

Conversely, if $\omega_k = \xi_k$, it follows that

$$Sm\,\epsilon_{ikm}e_{mp}\delta x_k \delta x_p = 0. \qquad (9)$$

Refer to principal axes of I_{ik}. Then $Sm\,\delta x_k \delta x_p = 0$ if $k \neq p$. Then, for example, if $i = 1$, the only non-zero terms in (9) are for $k = p = 2, m = 3$, and $k = p = 3, m = 2$; and

$$e_{33} Sm (\delta x_2)^2 - e_{23} Sm(\delta x_3)^2 = 0. \tag{10}$$

Hence either $\qquad e_{23} = 0 \quad$ or $\quad I_{22} = I_{33}$. $\hfill (11)$

Similarly, we see that $\quad e_{31} = 0$ or $I_{33} = I_{11}$, $\ e_{12} = 0$ or $I_{11} = I_{22}$. $\hfill (12)$

If then the principal axes of the element are determinate ($I_{11} \neq I_{22} \neq I_{33}$), we have $e_{ik} = 0$ for $i \neq k$, and therefore the principal axes of e_{ik} and I_{ik} coincide.

3·101. Elastic strain. The analysis of *displacement* in an elastic solid is almost identical with that of *velocity* in a fluid. It is convenient to consider the particle at $P(x_i)$ at time t to have already received a small displacement u_i, so that its undisturbed position was at $x_i - u_i$. Then if $u_i + \delta u_i$ is the displacement at $Q(x_i + \delta x_i)$, we have in just the same way

$$\delta u_i = \frac{\partial u_i}{\partial x_k} \delta x_k = \frac{1}{2}\left(\frac{\partial u_k}{\partial x_i} + \frac{\partial u_i}{\partial x_k}\right)\delta x_k - \frac{1}{2}\left(\frac{\partial u_k}{\partial x_i} - \frac{\partial u_i}{\partial x_k}\right)\delta x_k$$
$$= e_{ik}\delta x_k - \xi_{ik}\delta x_k,$$

where e_{ik} and ξ_{ik} are respectively symmetrical and antisymmetrical tensors. They can be interpreted as giving the changes of size and shape of an element, and its rotation; in a fluid, since v_i is there taken to be the velocity, the e_{ik} and ξ_{ik} there defined correspond to the rates of change of those defined here for an elastic solid.

3·102. Stress. In the interior of a substance, whether solid, liquid, or gas, there are in general reactions between the parts. The general nature of these can be seen by considering how we can apply forces to the outside of a solid with, say, one face clamped. They can be applied to any part of the accessible surface; and we can either press or pull normally on the surface or apply a tangential drag, as by friction. The notion of a state of stress extends this notion of a force across a surface to all elements of surfaces, even in the interior. If dS is a small element of surface, with its normal in the direction **n**, we speak of the reactions acting *across* dS, and representing the forces between the particles on opposite sides of it. The components of the force depend both on the size and on the direction of dS and may be written $p_i^{(n)} dS$; or, if **n** is in the direction of x_k, we denote the force by $p_{ki} dS$ and call p_{ki} the *stress components*, which are therefore *forces per unit area*. The sign is specified by taking $p_{ki} dS$ to be the reaction on the matter on the side where x_k is smaller and tending to increase its i coordinate. The force on the side where x_k is greater, tending to increase its i coordinate, will be $-p_{ki} dS$.

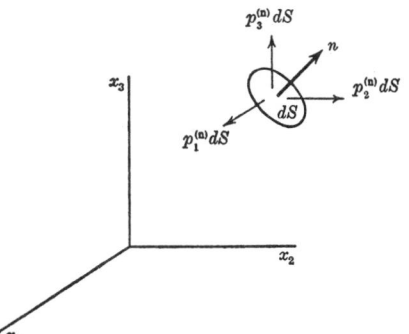

The stress components have two remarkable properties. They form a symmetrical tensor; and they have a simple linear relation to the rate of strain tensor in the case of a fluid, and to the strain tensor in an elastic solid provided the strains are small. We shall assume them to have continuous derivatives.

3·103. We first show that they form a tensor. Let a plane of x'_j constant make small intercepts of order a on the x_i axes and consider the forces on the small tetrahedron between this plane and the x_i axes. Let dS be the area of the x'_j plane forming the base of this tetrahedron. Then $p_{ji}dS$ are the forces acting on the interior across dS. Now the magnitude of the force acting across the face of constant x_k is p_{ki} times the area of that face, and the area is $l_{kj}dS$, where l_{kj} is the cosine of the angle between x_k and x'_j. It acts, however, on the matter on the positive side of $x_k = 0$ and must therefore be taken with the negative sign.

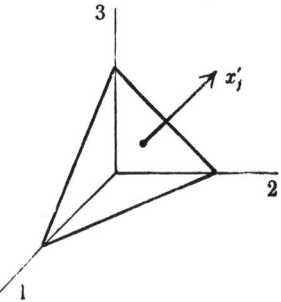

The matter inside the tetrahedron will in general be acted on by external forces such as gravity, which will be of order a^3 when a is small. These are called *body forces*. It will also have an acceleration, which we suppose always finite. Then the rate of change of momentum is also of order a^3, and the condition that the rate of change of momentum of the element is equal to the total force gives

$$(p_{ji} - l_{kj}p_{ki})dS = O(a^3).$$

This relates the forces in the direction of x_i. But we can now resolve them in the direction of x'_l; and

$$(l_{il}p_{ji} - l_{il}l_{kj}p_{ki})dS = O(a^3).$$

But $l_{il}p_{ji}dS$ is the component in the direction of x'_l of the force across a plane of x'_j constant; and therefore is the same as $p'_{jl}dS$, where p'_{jl} are the stress components with regard to the new axes. Also dS is of order a^2; hence by making a tend to zero we have

$$p'_{jl} = l_{il}l_{kj}p_{ki} = l_{ij}l_{kl}p_{ik}.$$

Therefore p_{ik} is a tensor.

3·104. Now take a small parallelepiped with centre at x_i and sides δx_i, and consider the moments about a line through the centre parallel to the axis of x_3. First ignore variations of the stress components in the region. Across the face $x_2 + \tfrac{1}{2}\delta x_2$ there is a force parallel to x_1 equal to $p_{21}\delta x_1 \delta x_3$; and this has moment $-p_{21}\delta x_1 \delta x_3(\tfrac{1}{2}\delta x_2)$. The force across the face $x_2 - \tfrac{1}{2}\delta x_2$ is equal and opposite, but as it is on the opposite side it has the same moment. The forces parallel to x_2 across the planes of x_1 constant have moment $p_{12}dx_1 dx_2 dx_3$. Evidently

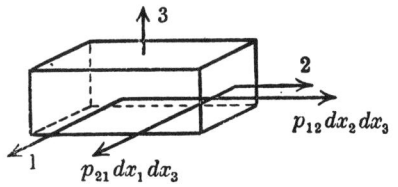

all forces arising from other stress components over the faces have no moment. Hence if the stress was uniform the moment would be

$$(p_{12} - p_{21})\delta x_1 \delta x_2 \delta x_3.$$

A little consideration will show that the change of the moment due to non-uniformity of stress is of order a^4 with the most extreme possible variability, where $\delta x_1, \delta x_2, \delta x_3$ are of order a. The body forces give a total force of order a^3, and will have a moment of order a^4.

Similarly, the total rate of change of angular momentum, so long as the acceleration is finite, is of order a^4 at most. Hence

$$(p_{12}-p_{21})\delta x_1 \delta x_2 \delta x_3 = O(a^4),$$

and by taking a sufficiently small region we show that

$$p_{12} = p_{21}.$$

By symmetry it follows that

$$p_{ik} = p_{ki},$$

and therefore p_{ik} is a symmetrical tensor.

3·105. Equations of motion. Consider again a small parallelepiped. Let ρ be the density and f_i the acceleration of the particle of matter momentarily at x_i; and let X_i be the body force per unit mass. Then $\iiint \rho f_i dx_1 dx_2 dx_3$ through the element is equal to the total force on the element. The body force contributes $\iiint \rho X_i dx_1 dx_2 dx_3$; and we have to consider what contributions arise from non-uniformity of the stress. The face $x_1 + \tfrac{1}{2}\delta x_1$ will contribute $p_{1i}\delta x_2 \delta x_3$, where p_{1i} is to be given its mean value over the face. But the opposite face contributes $-p_{1i}\delta x_2 \delta x_3$, where p_{1i} is to be given its mean value over *that* face. If the stress components are differentiable, which will in general be true, the two faces together contribute

$$\frac{\partial p_{1i}}{\partial x_1} \delta x_1 \delta x_2 \delta x_3 + O(a^4).$$

Then all six faces contribute

$$\frac{\partial p_{ki}}{\partial x_k} \delta \tau + O(a^4),$$

and by taking a sufficiently small parallelepiped we have the equations of motion

$$\rho f_i = \rho X_i + \frac{\partial p_{ki}}{\partial x_k}.$$

The above argument assumes nothing about the properties of the material except that action and reaction between neighbouring portions are equal and opposite, that all accelerations are bounded, and that the stress components are differentiable. It is equally valid for solids and liquids. Differences between the states of matter arise when we deal with the relation between stress and strain.

3·106. Stress-strain relations. 3·1061. Elastic solid. It is fairly obvious that a simple displacement or rotation of an elastic solid to a new position of equilibrium requires no change of stress; and therefore that the stress components are independent of the rotation. The fundamental relation is expressed by Hooke's law, which, in its most general form, states that as long as the strain components are small the stress is linearly related to them. This is true for the most anisotropic crystals. The usual theory of elasticity is for isotropic solids; we then assume a much more restrictive relation; one way of stating this is that the stress tensor and the strain tensor always have the same principal axes. When a tension is applied along a uniform bar, the bar extends longitudinally and contracts laterally, the changes of length of equal elements in all lateral directions being equal. This is expressed, if the axis of x_1 is taken along the bar, by

$$Ee_{11} = p_{11}, \quad Ee_{22} = Ee_{33} = -\sigma p_{11}. \tag{1}$$

E is called Young's modulus and σ Poisson's ratio; both are constants of the material. In the conditions specified all stress components other than p_{11} are zero, and the three strain components e_{23}, e_{31}, e_{12} are also zero. But then if we consider also stresses p_{22}, p_{33}, since the stress-strain relation is linear, we can add the corresponding strains and get the more general form

$$Ee_{11} = p_{11} - \sigma(p_{22} + p_{33}), \quad e_{23} = 0, \quad p_{23} = 0, \tag{2}$$

with symmetrical relations. The first of these can be written

$$Ee_{11} = (1+\sigma)p_{11} - \sigma(p_{11} + p_{22} + p_{33}), \tag{3}$$

and the whole set are summarized by

$$Ee_{ik} = (1+\sigma)p_{ik} - \sigma p_{mm}\delta_{ik}. \tag{4}$$

This set of equations is valid for the set of axes chosen, which are principal axes of the stress tensor and of the strain tensor. But if we now transform to any other set of rectangular axes, every term transforms according to the rule for second-order tensors and therefore we shall get

$$Ee'_{jl} = (1+\sigma)p'_{jl} - \sigma p'_{nn}\delta_{jl} \tag{5}$$

and

$$p'_{nn} = p_{mm}. \tag{6}$$

Hence the form (4) is not confined to principal axes and is true for any rectangular axes whatever. Its usefulness is due to the fact that in most problems of elasticity the principal axes of the stress are not in the same direction at all points and we need a form valid for all directions. It is convenient if the stresses are known and we have to find the strains from them.

In many problems, on the other hand, the stresses are unknown and the equations of motion must be regarded as differential equations for the displacements. Then we need to express the stresses in terms of the displacements, and therefore in terms of the strains. This can be done as follows. First contract the equations (4); we get

$$Ee_{mm} = (1+\sigma)p_{mm} - 3\sigma p_{mm} = (1-2\sigma)p_{mm}, \tag{7}$$

$$(1+\sigma)p_{ik} = Ee_{ik} + \frac{\sigma E}{1-2\sigma}e_{mm}\delta_{ik}, \tag{8}$$

$$p_{ik} = \lambda e_{mm}\delta_{ik} + 2\mu e_{ik}, \tag{9}$$

where

$$\lambda = \frac{\sigma E}{(1+\sigma)(1-2\sigma)}, \quad \mu = \frac{E}{2(1+\sigma)}. \tag{10}$$

λ and μ are known as Lamé's constants. λ has no special name; μ is called the *rigidity*. Evidently if a block is clamped along the plane of $x_2 = 0$ and a tangential stress p_{21} is applied over the opposite face the block will be distorted. Suppose the displacement to be $(\eta x_2, 0, 0)$. Then η is a small angle and is a measure of the *shear*. All the strain components are found to be zero except e_{12}, e_{21}, which are $\tfrac{1}{2}\eta$, $\tfrac{1}{2}\eta$; and then from (9)

$$p_{21} = \mu\eta, \tag{11}$$

so that μ is the ratio of shear stress to shear, and measures the resistance of the substance to distortion.

Another important constant arises in the case where the strain is spherically symmetrical; that is, if $e_{11} = e_{22} = e_{33}$, $e_{23} = e_{31} = e_{12} = 0$. This means that the matter is stretched in the same ratio in all directions, and the associated type of stress is called *hydrostatic*. Then (9) gives

$$p_{11} = 3\lambda e_{11} + 2\mu e_{11} = 3k e_{11}, \tag{12}$$

where
$$k = \lambda + \tfrac{2}{3}\mu. \tag{13}$$

k is called the *bulk modulus* because the relative change of volume is $3e_{11}$ to the first order, so that k is the ratio of the symmetrical stress to the change of volume. It is also known as the *incompressibility* and $1/k$ as the *compressibility*.

In terms of λ and μ

$$E = \frac{\mu(3\lambda + 2\mu)}{\lambda + \mu}, \quad \sigma = \frac{\lambda}{2(\lambda + \mu)}. \tag{14}$$

Experimentally E, k, and μ are the easiest of the elastic constants to measure directly. All have the dimensions of a stress.

An alternative method is to assert directly that if there is a universal linear relation

$$p_{ik} = c_{ikmp} e_{mp}$$

valid for all axes, c_{ikmp} is a tensor of order 4. If further its components have the same values for all axes, it is isotropic, and therefore, by 3·031,

$$c_{ikmp} = \lambda \delta_{ik}\delta_{mp} + \mu(\delta_{im}\delta_{kp} + \delta_{ip}\delta_{km}) + \nu(\delta_{im}\delta_{kp} - \delta_{ip}\delta_{km}),$$

where λ, μ, ν are scalars. Then

$$p_{ik} = \lambda \delta_{ik} e_{mm} + \mu(e_{ik} + e_{ki}) + \nu(e_{ik} - e_{ki})$$
$$= \lambda \delta_{ik} e_{mm} + 2\mu e_{ik},$$

since e_{ik} is symmetrical.

This method has the advantage that it is possible, by suitable modifications of the method of 3·031, to find out what fourth-order tensors have the symmetry properties associated with various types of crystal. Then this method can be extended to obtain the stress-strain relations for crystals.

3·1062. Fluid. In a fluid the mean stress $\tfrac{1}{3}p_{mm}$ is nearly always negative and is denoted by $-p$; p is called the *pressure*. (Contrary to what is stated in some text-books, a liquid carefully freed from dissolved gases can stand an appreciable tension; but it is true that tension seldom occurs in practice.) In a classical fluid the stress tensor is simply $-p\delta_{ik}$, and this is a good approximation in many problems relating to real fluids. The departure of the stress from this value is linearly related to the rate of strain, and if *this* is now denoted by e_{ik} it is true for a real fluid, as for an isotropic elastic solid, that the principal axes of the stress tensor are also those of the tensor e_{ik}. The required relations can therefore be written

$$p_{ik} + p\delta_{ik} = \lambda' e_{mm}\delta_{ik} + 2\mu' e_{ik}, \tag{1}$$

but we must impose the further condition that by the definition of p the tensor on the left gives zero on contraction. Hence

$$(3\lambda' + 2\mu') e_{mm} = 0 \tag{2}$$

and
$$p_{ik} = -p\delta_{ik} + 2\mu'(e_{ik} - \tfrac{1}{3}e_{mm}\delta_{ik}). \tag{3}$$

μ' is called the *viscosity*. Its dimensions are those of a stress multiplied by a time. The function multiplied by $2\mu'$ is the departure of the rate of strain from spherical symmetry.

The pressure p has important properties. It is nearly independent of the rate of strain, though it might theoretically contain a small term proportional to e_{mm}. This, however, is so small that it has no practical importance. The pressure can therefore be treated as a function of the density and temperature alone, according to the usual laws of thermal expansion and compressibility for a liquid or gas.

3·1063. The acceleration. We have left the acceleration term in the equations of motion in the form ρf_i. We need to express it, for a fluid, in terms of derivatives of the velocity; for a solid, of derivatives of the displacement. Our derivation of the equations of motion made use of a parallelepiped fixed in space. We could take an element of volume moving with the matter instead, but this would not in general remain rectangular, and the resolution of the forces would be much more difficult. But the acceleration does refer to a particular particle of the matter.

To make this explicit it is convenient temporarily to use Lagrange's way of specifying the motion. The particle at x_i at time t is supposed to have been at a_i at time t_0; then the motion of every particle is described by an equation of the form

$$x_i = g(a_1, a_2, a_3, t), \tag{1}$$

where, for a given particle, a_i is independent of t. Then the velocity and acceleration of the particle are

$$v_i = \left(\frac{\partial x_i}{\partial t}\right)_a, \quad f_i = \left(\frac{\partial^2 x_i}{\partial t^2}\right)_a, \tag{2}$$

where the suffix means that a_k is kept constant during the differentiation.

In the usual Eulerian way of specifying the motion, the velocity of a particle is regarded as a function of its position at time t instead of at time t_0. Hence if in time δt the particle moves from x_i to $x_i + \delta x_i$, its velocity will be v_i evaluated at $t + \delta t$, $x_i + \delta x_i$. Hence its acceleration is

$$\lim_{\delta t \to 0} \frac{v_i(t+\delta t, x_i+\delta x_i) - v_i(t, x_i)}{\delta t} = \lim\left\{\left(\frac{\partial v_i}{\partial t}\right)_{x_i} + \frac{\partial v_i}{\partial x_k}\left(\frac{\delta x_k}{\delta t}\right)\right\}, \tag{3}$$

and $\delta x_k/\delta t$, in the limit, is itself the velocity of the particle, v_k. Hence

$$f_i = \left(\frac{\partial v_i}{\partial t}\right)_{x_i} + v_k \frac{\partial v_i}{\partial x_k}. \tag{4}$$

The operator $\partial/\partial t + v_k \partial/\partial x_k$, which gives the time derivative of any quantity *associated with a particular particle* (i.e. a_i constant in Lagrange's specification) is usually denoted by D/Dt in English works. It is, however, simply the partial differential operator $\partial/\partial t$, with a_i kept constant instead of x_i. When, as in the Eulerian method, we suppress mention of a_i altogether, there seems to be no adequate reason against regarding the operator as an ordinary total derivative and denoting it by d/dt. The notation D/Dt is really a survival from the time when d/dt was used to denote partial differentiation.

In a fluid, therefore, the equations of motion have the form

$$\rho\left(\frac{\partial v_i}{\partial t} + v_k \frac{\partial v_i}{\partial x_k}\right) = \rho X_i + \frac{\partial p_{ki}}{\partial x_k}, \tag{5}$$

where the stress components are related to the rate of strain and the pressure according to 3·1062(3).

In a solid the conditions are somewhat different, since the initial position appears in the specification of the displacements; in fact

$$a_i = x_i - u_i. \tag{6}$$

But in nearly all problems of elasticity the displacements are small, and if we neglect their squares d/dt can be replaced by $\partial/\partial t$. Squares of the strains are also neglected in taking the stress-strain relations as linear, so that there is no loss of generality in also neglecting them in the acceleration.*

3·11. Electromagnetic stress tensor. Take the electric forces first. If K is the dielectric constant, supposed uniform, E the electric intensity, and ρ the electric charge per unit volume, the electric force per unit volume is

$$\mathbf{X} = \rho \mathbf{E} \tag{1}$$

and
$$4\pi\rho = K \operatorname{div} \mathbf{E}. \tag{2}$$

Then
$$4\pi X_i = K E_i \frac{\partial E_k}{\partial x_k}$$

$$= K \frac{\partial}{\partial x_k}(E_i E_k) - K \frac{\partial E_i}{\partial x_k} E_k$$

$$= K \frac{\partial}{\partial x_k}(E_i E_k) - K \frac{\partial E_k}{\partial x_i} E_k, \tag{3}$$

since E is the gradient of a potential; and

$$4\pi X_i = K \frac{\partial}{\partial x_k}(E_i E_k - \tfrac{1}{2} E_m^2 \delta_{ik}). \tag{4}$$

Hence the mechanical force can be regarded as derived from a stress

$$p_{ik} = \frac{K}{4\pi}(E_i E_k - \tfrac{1}{2} E_m^2 \delta_{ik}). \tag{5}$$

Now consider the force due to a magnetic field $\mathbf{H}$ on a medium carrying an electric current of density $\mathbf{j}$. The permeability μ is taken constant. Then

$$\mathbf{X} = \mu \frac{\mathbf{j}}{c} \wedge \mathbf{H}, \quad \operatorname{curl} \mathbf{H} = 4\pi \mathbf{j}/c, \tag{6}$$

$$4\pi X_i = \mu \epsilon_{ikm}(\operatorname{curl} \mathbf{H})_k H_m$$

$$= \mu \epsilon_{ikm} \epsilon_{kps} \frac{\partial H_s}{\partial x_p} H_m$$

$$= \mu(\delta_{mp}\delta_{is} - \delta_{ms}\delta_{ip}) \frac{\partial H_s}{\partial x_p} H_m$$

$$= \mu \left(\frac{\partial H_i}{\partial x_m} - \frac{\partial H_m}{\partial x_i} \right) H_m. \tag{7}$$

* The fullest discussion of the second-order terms is by F. D. Murnaghan, *Am. J. Math.* 59, 1937, 235–60.

But
$$\frac{\partial}{\partial x_m}(H_i H_m) = \frac{\partial H_i}{\partial x_m} H_m + H_i \frac{\partial H_m}{\partial x_m} \qquad (8)$$

and $\partial H_m/\partial x_m = 0$. Hence
$$4\pi X_i = \mu \frac{\partial}{\partial x_m}(H_i H_m - \tfrac{1}{2} H_k^2 \delta_{im}), \qquad (9)$$

and X_i can be derived from a stress tensor
$$p_{ik} = \frac{\mu}{4\pi}(H_i H_k - \tfrac{1}{2} H_m^2 \delta_{ik}). \qquad (10)$$

For the additional terms required when K and μ are not constant, see Abraham-Becker, *Classical Electricity and Magnetism*, pp. 104, 146.

3·12. Rate of change of a vector, when the axes are rotating. We have stated the transformation rule for vectors in a way that depends only on the mutual inclinations of the axes; so far we have not had occasion to consider what happens if these inclinations are themselves varying with the time. So long as the transformation from one set of axes to another is purely algebraic, there is no trouble; all identities depending on finding the component of a vector in a given direction will remain true even if the direction cosines themselves are varying, provided that we take all their values at the same instant. But if we have to differentiate with regard to the time we are considering different instants, and special attention to the variation of the direction cosines becomes necessary. To take a very simple case, let a particle be moving with uniform angular velocity in a circle, so that

$$x_1 = a\cos\omega t, \quad x_2 = a\sin\omega t, \quad x_3 = 0.$$

The velocity components are $(-\omega a \sin\omega t, \omega a \cos\omega t, 0)$, and the acceleration components $(-\omega^2 x_1, -\omega^2 x_2, 0)$. Now take a set of rotating axes, x_3' coinciding with x_3, while x_1' is inclined at ωt to x_1 and therefore permanently directed towards the particle. Then the coordinates in the $O1'2'3'$ system are permanently $(a, 0, 0)$, and their rates of change are $(0, 0, 0)$. But the components of the velocity *relative to* $O123$ along these axes are $(0, \omega a, 0)$ and those of the acceleration $(-\omega^2 a, 0, 0)$. *Rates of change of the coordinates with respect to sets of axes in relative rotation do not transform according to the vector rule.* We can say that the operations of differentiation with regard to t and resolution in a given direction *commute* only if the direction is fixed.*

The elementary form of the equations of motion of a particle, $m\ddot{x}_i = X_i$, requires the axes to be inertial. If we use instead a set of rotating axes x_j', the force X_i can be resolved along them by the rule for a vector, and the equations are equivalent to

$$ml_{ij}\ddot{x}_i = l_{ij} X_i = X_j', \qquad (1)$$

but the left side is not equal to $m\ddot{x}_j'$. In dealing with the motion of rigid bodies, especially, it is usually convenient to state the equations of motion referred to rotating axes,

---

* We shall meet the non-commutative property of operators repeatedly. The simplest case is where the operations are multiplication by x and differentiation with regard to x:
$$\frac{d}{dx} x f(x) = x \frac{d}{dx} f(x) + f(x)$$
which is not the same as $x \dfrac{d}{dx} f(x)$.

3·12 Rotating axes

which we can often take to be fixed in the body, and therefore we require to be able to express $l_{ij}\ddot{x}_i$ in terms of x'_j and its derivatives; similarly, if A_i are the components of displacement, velocity or angular momentum with respect to inertial axes, we need expressions for $l_{ij}dA_i/dt$ in terms of $l_{ij}A_i$ and its derivatives. We can continue to denote $l_{ij}A_i$ by A'_j. It will still be true that if $B_i = A_i$, then

$$l_{ij}B_i = l_{ij}A_i, \tag{2}$$

even though the axes of x'_j are rotating, and conversely. Then

$$A_i = l_{ij}A'_j, \tag{3}$$

$$\frac{dA_i}{dt} = l_{ij}\frac{dA'_j}{dt} + A'_j\frac{dl_{ij}}{dt}. \tag{4}$$

Now if we take a point on the x'_j axis at a fixed distance c from the origin, its coordinates with respect to the x_i axes are cl_{ij} and its velocity components are cdl_{ij}/dt. But the x'_j axes are a rigid frame; hence the velocity of a point rigidly attached to them with coordinates x_i is $-\Theta_{ik}x_k$ where Θ_{ik} represents their rate of rotation as in 3·092. Hence the x_i velocity of the point just considered is $-c\Theta_{ik}l_{kj}$, and

$$\frac{dl_{ij}}{dt} = -\Theta_{ik}l_{kj}, \tag{5}$$

$$\frac{dA_i}{dt} = l_{ij}\frac{dA'_j}{dt} - \Theta_{ik}l_{kj}A'_j, \tag{6}$$

$$l_{il}\frac{dA_i}{dt} = l_{il}l_{ij}\frac{dA'_j}{dt} - \Theta_{ik}l_{kj}l_{il}A'_j. \tag{7}$$

But $l_{il}l_{ij} = \delta_{jl}$, and $l_{il}l_{kj}\Theta_{ik}$ is Θ'_{lj}, defined as the result of transforming Θ_{ik} to the axes x'_j, x'_l according to the rule for second-order tensors. This gives the components required:

$$l_{il}\frac{dA_i}{dt} = \frac{dA'_l}{dt} - \Theta'_{lj}A'_j = \left(\frac{dA}{dt}\right)'_l + (\boldsymbol{\theta}\wedge A)_l \tag{8}$$

$$= (A'_1 - A'_2\theta'_3 + A'_3\theta'_2,\ A'_2 - A'_3\theta'_1 + A'_1\theta'_3,\ A'_3 - A'_1\theta'_2 + A'_2\theta'_1). \tag{9}$$

These are the components along x'_l of a vector whose components along x_i are dA_i/dt; and it is easy to verify that if we take a third frame of reference the components with regard to it will be the same whether we transform directly from the i frame or by way of the j frame. These components therefore satisfy the consistency rule for vectors.

Whether we regard the separate terms dA'/dt and $(\boldsymbol{\theta}\wedge A)$ as vectors is a matter of definition, and two methods are open to us. According to our original definition the transformation rule is the sole criterion for a set of three quantities to be the components of a vector, and this rule has nothing to do with whether the direction cosines occurring in the transformation are constant. If we retain this definition when the l_{il} are functions of the time, we have that since

$$\frac{dA'_l}{dt} \neq l_{il}\frac{dA_i}{dt},$$

dA/dt is not a vector if the frames considered are in relative rotation; but $dA/dt + (\boldsymbol{\theta}\wedge A)$ is one, and therefore $(\boldsymbol{\theta}\wedge A)$ is not. The fact that for compactness of writing it is written as a vector product does not make it into a vector. The test is in the transformation rule, and before we can say that a set of quantities are the components of a vector we have to show that this rule is satisfied; and in this case with this transformation rule it is not.

The alternative method is to restrict the scope of the transformation rule to axes that are not in relative rotation. To simplify matters suppose that A is a displacement from the origin. Then dA/dt is by definition a velocity, the components of which are relative to a fixed frame. We can, however, imagine a triply infinite set of fixed frames with the same origin, and at any moment the moving axes are passing through one of them, which is thereby identified. We are entitled to resolve with respect to this frame, and the velocity components with regard to it are $l_{ij}dA_i/dt$. These are the components of a vector. $\theta \wedge A$ is the velocity, referred to any of the fixed frames, of a point rigidly connected to the moving axes. Such a motion is possible, and in this sense $\theta \wedge A$ is a vector. Then $l_{ij}dA'_j/dt$ are the components, with regard to the x_i frame, of a vector; but it is not the *same* vector as dA/dt. It is the velocity of a point moving in the x_i frame with the *part* of the velocity dA/dt that is not expressed in $\theta \wedge A$.

With this interpretation the expressions in (8) are to be regarded as components of the rate of change of A, not along the moving axes, but along the fixed axes that they are instantaneously passing through.

Either of these interpretations is tenable; the important thing to realize is that they are not the same thing and they must not both be made at once, otherwise mistakes are inevitable.*

The angular velocity Θ'_{ij} is sometimes called the angular velocity of the rotating axes 'with reference to themselves', without any clear explanation of why such an angular velocity should not be identically zero. On inspection of its derivation we see that it is *the angular velocity of the axes about fixed axes instantaneously coinciding with them*. If the angular velocity about any fixed axes is known, Θ'_{ij} or its vectorial representation is found by resolving.

3·13. Applications to mechanics. Euler's angles. The position of a rigid body, of which one point O is fixed, can be specified with regard to a fixed frame of reference $Oxyz$ in the following way. We take first a marked line of particles in the body and take this as the axis $O3$ of a set of rectangular axes fixed in the body. The polar angles of this line are denoted by θ, λ. The position of the body is now known except for a possible rotation about $O3$, so that it is completely fixed if we know the angle between a marked plane of the body through $O3$ and the plane $zO3$. We call this angle χ, and the three angles θ, λ, χ are Euler's angles.

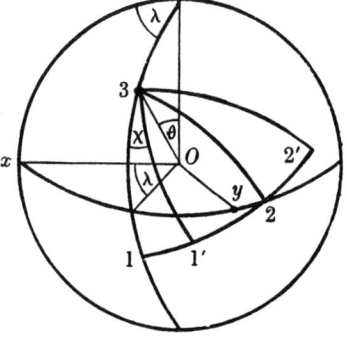

We have to express the angular velocity of the body in terms of the rates of change of these three angles. We use an intermediate frame of reference $O123$, where $O1$ is in the plane $zO3$ and at right angles to $O3$. Then $O2$ is normal to the plane $zO3$. All frames are taken to be right-handed. $O123$ is not in general fixed either in space or in the body. But if $O31'$ is a fixed

---

* A similar point arises in what is called 'covariant differentiation' in general relativity. Cf. Eddington, *Mathematical Theory of Relativity*; McConnell, *Absolute Differential Calculus*.

plane in the body making an angle χ with $O31$, we can take a third axis $O2'$ perpendicular to $O31'$, and this also is fixed in the body.

Evidently the rate of rotation of the body is specified by rates $\dot\lambda$ about Oz, $\dot\theta$ about $O2$, and $\dot\chi$ about $O3$. These can be resolved about any convenient directions. Their components with regard to $O123$ are clearly $(-\sin\theta\,\dot\lambda,\,\dot\theta,\,\dot\chi+\cos\theta\,\dot\lambda)$. Resolving again with regard to $O1'2'3$ we get the components

$$(-\sin\theta\,\dot\lambda\cos\chi+\dot\theta\sin\chi,\quad \sin\theta\,\dot\lambda\sin\chi+\dot\theta\cos\chi,\quad \dot\chi+\cos\theta\,\dot\lambda),$$

which are the components of the angular velocity of the body about axes fixed in the body.

In many actual problems the body has an axis of symmetry, which can be taken as $O3$. It is then convenient to use the frame of reference $O123$ rather than $O1'2'3$ on account of the simpler relations between the components of angular velocity and Euler's angles. There is one important difference, since we shall see that it is necessary to consider both the rotation of the body and that of the frame of reference, when the latter is not fixed in space. If we denote the angular velocity of the frame of reference by $\boldsymbol{\theta}(\theta_1,\theta_2,\theta_3)$, then if the frame is fixed in the body

$$\boldsymbol{\theta} = \boldsymbol{\omega}.$$

But if we use the frame $O123$ this has not the angular velocity component $\dot\chi$, since $O1$ always remains in the plane $zO3$. Hence for it

$$(\theta_1,\theta_2,\theta_3) = (-\sin\theta\,\dot\lambda,\,\dot\theta,\,\cos\theta\,\dot\lambda),$$

while the components of the angular velocity of the body are then

$$(\omega_1,\omega_2,\omega_3) = (-\sin\theta\,\dot\lambda,\,\dot\theta,\,\dot\chi+\cos\theta\,\dot\lambda).$$

The relations $\theta_1 = \omega_1$, $\theta_2 = \omega_2$ evidently mean that the axis $O3$ has the same angular velocity as the line of particles occupying it; that is, that $O3$ is fixed in the body. The fact that $\theta_3 \neq \omega_3$ is a reminder that $O1$ and $O2$ are not fixed in the body.

3·131. To illustrate the method, let us calculate the acceleration components of a particle in spherical polar coordinates (r,θ,λ). We take the axis $O3$ towards the particle; the axes $O1$ and $O2$ are taken as in 3·13, and the components of angular velocity of the axes are $(-\sin\theta\,\dot\lambda,\,\dot\theta,\,\cos\theta\,\dot\lambda)$. As the particle is permanently on the axis $O3$ its velocity components are $(0,0,\dot r) + (\theta_1,\theta_2,\theta_3) \wedge (0,0,r)$

$$= (r\dot\theta,\,r\sin\theta\,\dot\lambda,\,\dot r),$$

as may also be seen by inspection. Then the acceleration components are

$$(a_1,a_2,a_3) = \frac{d}{dt}(r\dot\theta,\,r\sin\theta\,\dot\lambda,\,\dot r) + (\theta_1,\theta_2,\theta_3) \wedge (r\dot\theta,\,r\sin\theta\,\dot\lambda,\,\dot r),$$

$$a_1 = \frac{d}{dt}(r\dot\theta) + \dot r\dot\theta - r\sin\theta\cos\theta\,\dot\lambda^2 = r(\ddot\theta - \sin\theta\cos\theta\,\dot\lambda^2) + 2\dot r\dot\theta,$$

$$a_2 = \frac{d}{dt}(r\sin\theta\,\dot\lambda) + \cos\theta\,\dot\lambda r\dot\theta + \sin\theta\,\dot\lambda\dot r = \frac{1}{r\sin\theta}\frac{d}{dt}(r^2\sin^2\theta\,\dot\lambda),$$

$$a_3 = \ddot r - r\sin^2\theta\,\dot\lambda^2 - r\dot\theta^2 = \ddot r - r(\dot\theta^2 + \sin^2\theta\,\dot\lambda^2).$$

3·14. Euler's dynamical equations. Let $O1'2'3'$ be taken along the principal axes of a rigid body moving with O fixed. If A, B, C are the principal moments of inertia and $\boldsymbol{\omega}$ is the angular velocity of the body about inertial axes, the angular momentum about O is $\boldsymbol{h}(O) = (A\omega_1, B\omega_2, C\omega_3)$. The rate of change of angular momentum about O is therefore

$$\frac{d\boldsymbol{h}}{dt} + \boldsymbol{\omega} \wedge \boldsymbol{h},$$

with components $(A\dot{\omega}_1 - (B-C)\omega_2\omega_3,$ etc.) about the instantaneous positions of the principal axes.

3·141. Motion of a top. The axis of symmetry is taken as $O3$; using the axes $O123$ as in 3·13, the angular velocity of the top is $(-\dot{\lambda}\sin\theta, \dot{\theta}, \dot{\chi}+\dot{\lambda}\cos\theta)$ and that of the axes is $(-\dot{\lambda}\sin\theta, \dot{\theta}, \dot{\lambda}\cos\theta)$. The moments of inertia being A, A, C, the components of angular momentum are $\{-A\dot{\lambda}\sin\theta, A\dot{\theta}, C(\dot{\chi}+\dot{\lambda}\cos\theta)\}$.

If $\boldsymbol{N}$ is the moment of the external forces

$$\frac{d\boldsymbol{h}}{dt} + \boldsymbol{\theta} \wedge \boldsymbol{h} = \boldsymbol{N}. \tag{1}$$

If the top is moving under gravity and its centre of mass is at $(0, 0, h)$,

$$\boldsymbol{N} = (0, Mgh\sin\theta, 0). \tag{2}$$

The third component of (1) gives immediately

$$\dot{\chi} + \dot{\lambda}\cos\theta = \text{const.} = n, \tag{3}$$

say. The second component gives

$$A\ddot{\theta} + Cn\dot{\lambda}\sin\theta - A\dot{\lambda}^2\sin\theta\cos\theta = Mgh\sin\theta, \tag{4}$$

so that the condition for a steady precession with $\theta = \alpha$ and $\dot{\lambda} = \Omega$ is

$$A\Omega^2\cos\alpha - Cn\Omega + Mgh = 0.$$

For the general motion the simplest method is to notice that there are two other first integrals:

(1) The angular momentum about the vertical is constant, whence

$$A\sin^2\theta\dot{\lambda} + Cn\cos\theta = \text{constant}.$$

(2) The total energy is constant, whence

$$A(\dot{\lambda}^2\sin^2\theta + \dot{\theta}^2) + C(\dot{\lambda}\cos\theta + \dot{\chi})^2 + 2Mgh\cos\theta = \text{constant}.$$

Once the expressions for the kinetic and potential energy have been found, the equations of motion can be found by using Lagrange's equations (10·07).

3·15. Tensors in two dimensions. If the only rotations of the axes permitted do not displace the axis of x_3, the possible changes of axes are more restricted than for a general rotation, and additional sets of functions are found to have the requisite transformation properties under the changes permitted. In fact if u_i ($i = 1, 2$) is a vector in two dimensions its component in a direction α is $u_1 \cos \alpha + u_2 \sin \alpha$. Now there is a vector $\boldsymbol{v}$ whose component in direction α is equal to that of $\boldsymbol{u}$ in direction $\tfrac{1}{2}\pi + \alpha$. For this is equivalent to

$$v_1 \cos \alpha + v_2 \sin \alpha = u_1 \cos (\tfrac{1}{2}\pi + \alpha) + u_2 \sin (\tfrac{1}{2}\pi + \alpha)$$
$$= -u_1 \sin \alpha + u_2 \cos \alpha, \qquad (1)$$

and therefore if

$$v_1 = u_2, \quad v_2 = -u_1, \qquad (2)$$

v_1 and v_2 are the components of a vector, which are the components of $\boldsymbol{u}$ referred to axes rotated through a right angle. In particular $(x_2, -x_1)$ are the components of a vector.

But the derivative of a vector is a tensor of the second order. Taking this vector to be $(x_2, -x_1)$ we have that

$$\eta_{ik} = \frac{\partial v_k}{\partial x_i} = \begin{pmatrix} 0 & -1 \\ 1 & 0 \end{pmatrix} \qquad (3)$$

is a tensor in two dimensions. Thus δ_{ik} is not the only isotropic tensor of order 2 in two dimensions; and any linear combination

$$\lambda \delta_{ik} + \mu \eta_{ik}, \qquad (4)$$

where λ and μ are scalars, is another.

Now consider the fourth-order tensor $\eta_{im}\eta_{kp}$. All components with $i = m$ or $k = p$ vanish, and

$$\left. \begin{array}{l} \eta_{12}\eta_{12} = \eta_{21}\eta_{21} = 1, \\ \eta_{12}\eta_{21} \phantom{= \eta_{21}\eta_{21}} = -1. \end{array} \right\} \qquad (5)$$

Now $\eta_{im}\eta_{kp} \dfrac{\partial^2 \phi}{\partial x_m \partial x_p}$ is a second-order tensor if ϕ is a scalar. But if $i = k = 1$ the only non-zero terms are for $m = p = 2$, and the component is $\partial^2\phi/\partial x_2^2$. If $i = 1, k = 2$, we must take $m = 2, p = 1$ and the component is $-\partial^2\phi/\partial x_1 \partial x_2$. Proceeding, we see that

$$\begin{pmatrix} \dfrac{\partial^2 \phi}{\partial x_2^2} & -\dfrac{\partial^2 \phi}{\partial x_1 \partial x_2} \\ -\dfrac{\partial^2 \phi}{\partial x_1 \partial x_2} & \dfrac{\partial^2 \phi}{\partial x_1^2} \end{pmatrix} \qquad (6)$$

are the components of a second-order tensor in two dimensions. This tensor has applications in elasticity, particularly in relation to the bending of thin plates and the distribution of stress between parallel planes.

3·16. Parallax. Tensor methods are sometimes useful in obtaining the formulae of spherical astronomy. Consider the parallax of the moon or a planet. Take the origin at the centre of the earth, axis 3 towards the pole, axis 2 on the observer's meridian. Let the coordinates of the planet be $x_i = r l_i$, those of the observer $\xi_i = a \lambda_i$. The distance from the observer to the planet is R, given by

$$R^2 = (x_i - \xi_i)^2 = r^2 - 2ra \lambda_k l_k + a^2, \qquad (1)$$
$$R = r - a \lambda_k l_k, \qquad (2)$$

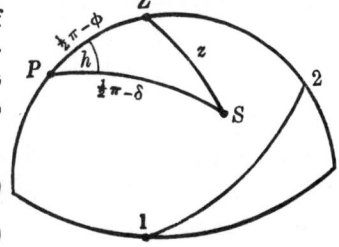

to the first order in a/r. Then the direction cosines of the line from the observer to the planet are l'_i, where

$$l'_i = \frac{x_i - \xi_i}{R} = \left(l_i - \frac{a}{r}\lambda_i\right)\left(1 + \frac{a}{r}\lambda_k l_k\right)$$

$$= l_i + P(-\lambda_i + l_i \lambda_k l_k), \tag{3}$$

where $P = a/r$, the horizontal parallax. Now the direction cosines are given in terms of the angular coordinates by

$$\begin{aligned} l_1 &= \cos\delta \sin h, & l_2 &= \cos\delta \cos h, & l_3 &= \sin\delta, \\ \lambda_1 &= 0, & \lambda_2 &= \cos\phi, & \lambda_3 &= \sin\phi, \end{aligned} \tag{4}$$

and
$$\lambda_k l_k = \cos\phi \cos\delta \cos h + \sin\phi \sin\delta. \tag{5}$$

For the parallax in declination ϖ_δ

$$\begin{aligned} l'_3 - l_3 &= \cos\delta\, \varpi_\delta \\ &= P\{-\sin\phi + \sin\delta(\cos\phi \cos\delta \cos h + \sin\phi \sin\delta)\} \\ &= P(\sin\delta \cos\delta \cos\phi \cos h - \sin\phi \cos^2\delta), \end{aligned} \tag{6}$$

whence
$$\varpi_\delta = P(\sin\delta \cos\phi \cos h - \sin\phi \cos\delta). \tag{7}$$

If h' is the observed hour angle

$$\tan h' = l'_1/l'_2, \quad \tan h = l_1/l_2, \quad \log \tan h = \log l_1 - \log l_2, \tag{8}$$

$$\frac{\sec^2 h}{\tan h}(h' - h) = \frac{l'_1 - l_1}{l_1} - \frac{l'_2 - l_2}{l_2} = P\left(-\frac{\lambda_1}{l_1} + \frac{\lambda_2}{l_2}\right)$$

$$= P\frac{\cos\phi}{\cos\delta \cos h}, \tag{9}$$

$$h' - h = P \cos\phi \sin h \sec\delta, \tag{10}$$

and ϖ_α, the parallax in right ascension, is $-(h' - h)$.

EXAMPLES

1. Show that $(\mathbf{K}.\mathbf{A}).(\mathbf{K}.\mathbf{B}) \wedge (\mathbf{K}.\mathbf{C}) = \|K\| \mathbf{A}.\mathbf{B} \wedge \mathbf{C}$.

2. Show that $\operatorname{curl}(\phi \mathbf{A}) = \phi \operatorname{curl} \mathbf{A} + \operatorname{grad} \phi \wedge \mathbf{A}$.

3. Show that $\{\operatorname{curl}(\mathbf{A} \wedge \mathbf{B})\}_i = A_i \operatorname{div} \mathbf{B} - B_i \operatorname{div} \mathbf{A} + B_k \dfrac{\partial A_i}{\partial x_k} - A_k \dfrac{\partial B_i}{\partial x_k}$.

4. If K_{ik} is a tensor, prove that

$$\begin{vmatrix} K_{22} & K_{23} \\ K_{32} & K_{33} \end{vmatrix} + \begin{vmatrix} K_{33} & K_{31} \\ K_{13} & K_{11} \end{vmatrix} + \begin{vmatrix} K_{11} & K_{12} \\ K_{21} & K_{22} \end{vmatrix} = \tfrac{1}{2} K_{ii} K_{kk} - \tfrac{1}{2} K_{ik} K_{kl}$$

and is a scalar. Relate this scalar to an invariant of the roots of the equation $\|K_{ik} - \lambda \delta_{ik}\| = 0$.

5. If the stress tensor is $2A_i A_k - \delta_{ik} A^2$, where A_i is a vector function of position, A^2 denotes $A_1^2 + A_2^2 + A_3^2$, and δ_{ik} equals unity if $i = k$ and equals zero otherwise, verify that at any point the direction of A_i is a principal direction of stress, and show that the stress at this point can be represented as a tension A^2 along the direction of A_i and a pressure A^2 normal to this direction.

Examples

6. A body is deformed by internal stress so that the particle which in the strained state is at the point (x, y, z) referred to a set of fixed rectangular axes has undergone the displacement whose components are
$$3\kappa(2x-y+z), \quad -3\kappa(x+y), \quad \kappa(3x+5z),$$
where κ is a small constant. Show that there is no rotation of any small portion of the body in the neighbourhood of (x, y, z), that one principal extension is 3κ, and determine the other two principal extensions. (Prelim. 1941.)

7. For a chain of three uniform mutually perpendicular rods each of mass m and length $2a$, show that the equation of the inertia quadric at the mass centre referred to axes parallel to the rods may be written in the form
$$\tfrac{2}{3}ma^2(5x^2 + 5y^2 + 3z^2 - 3yz + 3zx + xy) = K.$$

Deduce that one principal moment of inertia at the mass centre is ma^2 and find the others. (Prelim. 1940.)

8. Two equal uniform cones of height h are placed with their bases, of radius a, in contact. Determine the inertia tensor at the common centre of the bases, and give the ratios of the principal moments of inertia. (I.C. 1943.)

9. Determine the inertia tensor at the centre of one end of a uniform solid right prism of mass M and length $2H$, whose ends are equilateral triangles of side a. Determine the principal moments of inertia when (1) $a = H$, (2) $a = 10H$. In the second case show that the moment of inertia of the prism about any of the end edges is $\tfrac{83}{6}MH^2$. (I.C. 1940.)

10. *Motion relative to the earth.* The position vector of a particle relative to a point O at the earth's surface is $\boldsymbol{r}$; the velocity and acceleration of the particle relative to a frame at rest relative to the earth at O are $\dot{\boldsymbol{r}}$ and $\ddot{\boldsymbol{r}}$. If the particle moves under the earth's attraction and a force $\boldsymbol{F}$ per unit mass, prove that
$$\ddot{\boldsymbol{r}} + 2\boldsymbol{\omega} \wedge \dot{\boldsymbol{r}} = \boldsymbol{g}(\boldsymbol{r}) + \boldsymbol{F},$$
where $\boldsymbol{\omega}$ is the earth's angular velocity and $\boldsymbol{g}(\boldsymbol{r})$ is the acceleration relative to the frame at O of a freely falling particle instantaneously at rest relative to the frame at the point $\boldsymbol{r}$.

If a particle is projected with velocity $\boldsymbol{V}$ from O at time $t = 0$, show that to the first order in $\boldsymbol{\omega}$ its position vector $\boldsymbol{r}$ at time t is given by
$$\boldsymbol{r} = \boldsymbol{V}t + \tfrac{1}{2}\boldsymbol{g}t^2 - t^2\boldsymbol{\omega} \wedge \boldsymbol{V} - \tfrac{1}{3}t^3\boldsymbol{\omega} \wedge \boldsymbol{g},$$
where $\boldsymbol{g} = \boldsymbol{g}(O)$.

11. *Foucault's pendulum. Small oscillations.* The origin is taken at the point of suspension. $\boldsymbol{i}$ is the direction vector towards the bob of mass m. Then if T is the tension in the string, to the first order in $\boldsymbol{\omega}$
$$\ddot{\boldsymbol{r}} + 2\boldsymbol{\omega} \wedge \dot{\boldsymbol{r}} = \boldsymbol{g} - \frac{T}{m}\boldsymbol{i}.$$

Putting $\boldsymbol{g} = g\boldsymbol{z}$ and $\boldsymbol{i} = \boldsymbol{z} + \boldsymbol{\rho}$, show that to the first order in $\boldsymbol{\rho}$,
$$\ddot{\boldsymbol{\rho}} - 2\omega \sin \lambda \boldsymbol{z} \wedge \dot{\boldsymbol{\rho}} + \frac{g}{l}\boldsymbol{\rho} = 0,$$
where λ is the latitude of O, and l is the length of the string.

By taking the components of this equation in two perpendicular directions in a plane perpendicular to $\boldsymbol{z}$, use the method of 2·12 to show that the plane of oscillation rotates about the downward vertical with angular velocity $\omega \sin \lambda$.

Chapter 4

MATRICES

'All that isn't Belgrave Square is Strand and Piccadilly.'
w. s. gilbert, Utopia Limited

4·01. Introduction: definitions. In considering tensors of the second order in three dimensions we have used an abbreviated notation K_{ik} for the set of nine quantities

$$\begin{pmatrix} K_{11} & K_{12} & K_{13} \\ K_{21} & K_{22} & K_{23} \\ K_{31} & K_{32} & K_{33} \end{pmatrix},$$

When they are displayed in this way the first suffix refers to the row and the second to the column. Such a set forms a tensor if each suffix refers to one of a set of axes, of the same system of reference, and the coefficients have certain transformation properties when the axes are rotated. We have so far considered rectangular axes only, but this is a particular case of a more general tensor algebra. The generalizations take three lines: (a) to tensors of any order in n (especially four) dimensions, still referred to rectangular axes; (b) to tensors of any order in curvilinear coordinates; (c) to the study of an algebra of square arrays of quantities, which has many formal similarities to that of tensors of the second order, but is not restricted to an interpretation in terms of axes of reference.

When we take the algebraic point of view we speak of an array of quantities

$$\begin{pmatrix} K_{11} & K_{12} & \ldots & K_{1n} \\ K_{21} & K_{22} & \ldots & K_{2n} \\ \multicolumn{4}{c}{\dotfill} \\ K_{m1} & K_{m2} & \ldots & K_{mn} \end{pmatrix},$$

as a matrix of order* $m \times n$. It has m rows and n columns. We may write it as (K_{ik}), the first suffix referring to the row and the second to the column. A square matrix of order $n \times n$ is a particular case. So is a single-column matrix of m rows (order $m \times 1$) and a single-row matrix of n columns (order $1 \times n$). These three types of matrix have many applications in physical theories.

We shall denote a matrix by a single symbol, which stands for the set of $m \times n$ quantities or *elements* expressed in the matrix. For such an entity what we mean by addition, subtraction, multiplication, and division is a matter of definition. The use of heavy type is an indication that one or more suffixes are suppressed, as in vector and dyadic notation.

Addition. The sum of two matrices $\boldsymbol{a}$ and $\boldsymbol{b}$ is written as $\boldsymbol{a}+\boldsymbol{b}$ and stands for the matrix with elements $a_{ik}+b_{ik}$.

Subtraction. The matrix $-\boldsymbol{a}$ is defined as the matrix with elements $-a_{ik}$, and $\boldsymbol{a}-\boldsymbol{b}$ is defined as the matrix with elements $a_{ik}-b_{ik}$. For addition and subtraction to be significant the matrices must have the same number of rows and the same number of columns.

---

* A tensor of order 2 as written above would thus have to be spoken of as a matrix of order 3×3. The word *order* has quite distinct meanings as applied to tensors and matrices. A tensor of order greater than 2 cannot be written as a matrix.

It is clear that for addition of matrices the associative law
$$(a+b)+c = a+(b+c), \tag{1}$$
and the commutative law
$$a+b = b+a \tag{2}$$
both hold.

Multiplication. The law of multiplication is that ab is the matrix with elements $(ab)_{ik}$ given by
$$(ab)_{ik} = a_{ij}b_{jk}, \tag{3}$$
the summation convention being understood. For this to be significant j must range over the same set of values in both factors, and therefore the number of columns of a must be equal to the number of rows of b. The product is then a matrix, with as many rows as a and as many columns as b. In particular, if a is an $n \times n$ matrix, b must have n rows; but b may be a single column, and then the product matrix will also be a single column. If b is also an $n \times n$ matrix, ab will be an $n \times n$ matrix. On the other hand, if a is a single column, b must be a single row and ab is the matrix whose elements are $a_i b_k$. If a is a single row $(1 \times n)$, b must have n rows. If b is square the product is then a single row; if b has only one column the product has one row and one column, that is, it is a single quantity. The important cases for us are therefore, using single suffixes for matrices of one row or column and double ones for square matrices,

a		b		ab		In
Rows	Columns	Rows	Columns	Rows	Columns	suffixes
n	n	n	n	n	n	$a_{ij}b_{jk}$
n	n	n	1	n	1	$a_{ij}b_j$
n	1	1	n	n	n	$a_j b_k$
1	n	n	n	1	n	$a_j b_{jk}$
1	n	n	1	1	1	$a_j b_j$

The third case does not involve the use of the summation convention and is called the *outer product* of a and b. The others are all called *inner products*.

In works on algebra single-row and single-column matrices are often called vectors. This differs from the physical use of the word *vector*. In algebra the vector *is* a set of n elements and has nothing to do with any particular transformation law. In physics the vector requires *both* the elements and the assertion of a particular type of transformation law for its specification. Thus in physics we speak of the *same* vector as having different components in different systems of reference; the algebraists would call these representations themselves different vectors. We shall avoid this usage.

The commutative law of multiplication does not necessarily hold even if a and b are square. For ba must be defined as the matrix whose elements are $b_{ij}a_{jk}$, and this will be equal to $a_{ij}b_{jk}$ only in special cases; in general
$$ab \neq ba.$$

Pairs of matrices that satisfy $ab = ba$ are said to *commute*, those that satisfy $ab = -ba$ to anticommute.

The rule of multiplication is that the factors are always arranged so that repeated suffixes are *adjacent*. $a_{ij}b_{jk} = b_{jk}a_{ij}$, but this cannot be contracted to ba because the two j's are not adjacent. Explicit statement of the suffixes cannot lead to contradictions, but if they are suppressed contradictions will arise unless we have a definite rule about where the repeated suffixes are supposed to be. The rule suffices to distinguish $\boldsymbol{ba}$ from $\boldsymbol{ab}$ while maintaining the order of the factors in the explicit expression for the product.

The associative law
$$(\boldsymbol{ab})\boldsymbol{c} = \boldsymbol{a}(\boldsymbol{bc}), \tag{4}$$

and the distributive law
$$\boldsymbol{a}(\boldsymbol{b}+\boldsymbol{c}) = \boldsymbol{ab}+\boldsymbol{ac} \tag{5}$$

hold, provided the order is maintained and the operations are significant. These are easily verified by writing out the elements explicitly. For the first

$$\{(\boldsymbol{ab})\boldsymbol{c}\}_{il} = (a_{ij}b_{jk})c_{kl} = a_{ij}(b_{jk}c_{kl}) = \{\boldsymbol{a}(\boldsymbol{bc})\}_{il}. \tag{6}$$

In consequence these products can be written without brackets as $\boldsymbol{abc}$, since the position of the brackets is irrelevant. It follows that all positive powers of a given matrix commute; for $\boldsymbol{a}^2\boldsymbol{a} = \boldsymbol{aa}^2$, and $\boldsymbol{a}^m\boldsymbol{a}^n = \boldsymbol{a}^n\boldsymbol{a}^m$ follows by induction.

The *unit matrix* will be denoted by $\boldsymbol{1}$ and has components δ_{ik}, where

$$\delta_{ik} = 1 \quad (i=k), \qquad \delta_{ik} = 0 \quad (i \neq k), \tag{7}$$

and we shall write δ_{ik} for its components. It is clear that

$$\boldsymbol{1a} = \boldsymbol{a1} = \boldsymbol{a}. \tag{8}$$

The unit matrix is often denoted by $\boldsymbol{I}$, or even by 1 in cases where no ambiguity can arise.

The *null matrix* is one all of whose elements are zero.

The product of two matrices may be null without either factor being null. Thus

$$\begin{pmatrix} 1 & 0 \\ 0 & 0 \end{pmatrix} \begin{pmatrix} 0 & 0 \\ 0 & 1 \end{pmatrix} = \begin{pmatrix} 0 & 0 \\ 0 & 0 \end{pmatrix}.$$

The non-zero element of the first can be multiplied only into elements of the first row of the second, which are both zero. But if $\boldsymbol{AB} = 0$ *whatever* $\boldsymbol{B}$ may be, then $\boldsymbol{A} = 0$; similarly, if $\boldsymbol{AB} = 0$ whatever $\boldsymbol{A}$ may be, then $\boldsymbol{B} = 0$.

The *transposed matrix* of a matrix $\boldsymbol{a}$ is the matrix formed from $\boldsymbol{a}$ by interchanging its rows and columns. We shall denote it by $\tilde{\boldsymbol{a}}$ and its elements by $\tilde{a}_{ik}$. Then

$$\tilde{a}_{ik} = a_{ki}. \tag{9}$$

Since
$$(\boldsymbol{ab})_{ik} = a_{ij}b_{jk} = \tilde{a}_{ji}\tilde{b}_{kj} = \tilde{b}_{kj}\tilde{a}_{ji} = (\tilde{\boldsymbol{b}}\tilde{\boldsymbol{a}})_{ki}, \tag{10}$$

it follows that the transposed matrix of the product $\boldsymbol{ab}$, denoted by $\widetilde{\boldsymbol{ab}}$, is equal to the product $\tilde{\boldsymbol{b}}\tilde{\boldsymbol{a}}$, in this order.

Note that it would be *possible* to define $(\boldsymbol{ab})_{ik}$ to mean $a_{ij}b_{kj}$; this matrix is in fact often required, but then we should have

$$(\boldsymbol{ab}.\boldsymbol{c})_{ik} = (\boldsymbol{ab})_{ij}c_{kj} = a_{il}b_{jl}c_{kj}, \quad (\boldsymbol{a}.\boldsymbol{bc})_{ik} = a_{il}(\boldsymbol{bc})_{kl} = a_{il}b_{kj}c_{lj},$$

which are not the same. *The device of summing over adjacent suffixes is needed to make the associative law of multiplication true.* We therefore write

$$a_{ij}b_{kj}, \quad \text{not as} \quad (\boldsymbol{ab})_{ik}, \qquad \text{but as} \quad a_{ij}\tilde{b}_{jk} = (\boldsymbol{a}\tilde{\boldsymbol{b}})_{ik}.$$

Beginners sometimes find multiplication of matrices easier if they first transpose the second matrix and multiply rows into rows.

4·01   Symmetry; diagonal, adjugate, reciprocal matrices

If a is a single-column matrix, $\tilde{a}$ is a single-row matrix, and conversely. We have, if b is a square matrix,
$$(ba)_i = b_{ij}a_j = a_j b_{ij} = \tilde{a}_j \tilde{b}_{ji} = (\tilde{a}\tilde{b})_i, \tag{11}$$
so that the same rule applies as for square matrices.

The conjugate complex matrix. The elements of a matrix may be real or complex. The matrix whose elements are the conjugate complexes of those of a is denoted by a^* and its elements are a_{ik}^*.

The transposed matrix of the conjugate complex matrix, $\tilde{a}^*$, is denoted by $a^\dagger$ and its elements are
$$a_{ik}^\dagger = a_{ki}^*. \tag{12}$$

Symmetry properties. A matrix is said to be *symmetrical* if it is unaltered by interchanging rows and columns, that is,
$$a_{ik} = a_{ki} \tag{13}$$
or
$$a = \tilde{a}. \tag{14}$$
It is *antisymmetrical* or *skew-symmetrical* if the sign is changed when rows and columns are interchanged, that is,
$$a_{ik} = -a_{ki}, \tag{15}$$
$$a = -\tilde{a}. \tag{16}$$

A complex matrix is said to be *hermitian* if it is equal to its transposed conjugate complex, that is,
$$a = a^\dagger, \tag{17}$$
and *antihermitian* if
$$a = -a^\dagger. \tag{18}$$
For real matrices (17) and (18) reduce to (14) and (15) respectively. If a is hermitian, ia is antihermitian.

A *diagonal matrix* is one all of whose elements are zero except those in the leading diagonal, i.e., $a_{11}, a_{22}, \ldots, a_{nn}$. All pairs of diagonal matrices commute.

The *adjugate matrix* and the *reciprocal matrix.* If we consider the determinant formed by the elements of a square matrix a
$$\|a_{ik}\| \equiv \det a, \tag{19}$$
each element a_{ik} has a cofactor, which we may denote by A_{ik}, and
$$a_{ik}A_{jk} = (\det a)\delta_{ij}. \tag{20}$$
To preserve the rule about summing over adjacent suffixes we rewrite this by forming the *adjugate matrix* adj a,
$$(\text{adj } a)_{ik} = A_{ki}. \tag{21}$$
If a is symmetrical or hermitian, so is adj a. Further, provided that det a does not vanish, we have
$$\frac{a_{ij}A_{kj}}{\det a} = \delta_{ik} = \frac{A_{ji}a_{jk}}{\det a}. \tag{22}$$
If therefore we define the *reciprocal* a^{-1} of a matrix a by
$$(a^{-1})_{ik} = \frac{A_{ki}}{\det a}, \tag{23}$$
we have
$$a_{ij}(a^{-1})_{jk} = \delta_{ik} = (a^{-1})_{ij}a_{jk}, \tag{24}$$
or
$$aa^{-1} = a^{-1}a = 1. \tag{25}$$

The product of a matrix with its reciprocal is the unit matrix in whatever order the multiplication is carried out; provided the reciprocal exists, the order does not matter. If, however, det a vanishes, a has no reciprocal and is said to be *singular*.

Note that we must write $(a^{-1})_{ij}$, not a_{ij}^{-1}; the latter would mean the reciprocal of the ij element of a.

Division. Division by a non-singular matrix may now be defined as multiplication by its reciprocal, but the quotient depends on the order, as for the product. $a^{-1}b$ is not the same as ba^{-1}.

Reciprocal of a product. Since
$$abb^{-1}a^{-1} = a1a^{-1} = 1, \tag{26}$$
it follows that $b^{-1}a^{-1}$ is the reciprocal of ab, that is, $(ab)^{-1}$. *In forming the reciprocal of a product the order of the factors must be inverted as in forming the transpose of a product.* Written in suffixes (26) takes the form

$$a_{ij}b_{jk}(\text{adj } b)_{kl}(\text{adj } a)_{lm} = a_{ij}(\det b)\delta_{jl}(\text{adj } a)_{lm}$$
$$= a_{ij}(\text{adj } a)_{jm}(\det b)$$
$$= (\det a)(\det b)\delta_{im},$$
$$(ab)_{ik}(b^{-1}a^{-1})_{km} = 1_{im}. \tag{27}$$

Unitary and orthogonal matrices. A matrix such that
$$aa^\dagger = a^\dagger a = 1 \tag{28}$$
is called a *unitary matrix*. For such a matrix we have
$$a^{-1}aa^\dagger = a^{-1}1, \quad \text{whence} \quad a^\dagger = a^{-1}, \tag{29}$$
and hence for a unitary matrix the reciprocal is the same as the transposed conjugate complex. A matrix satisfying
$$\tilde{a} = a^{-1} \tag{30}$$
is said to be *orthogonal*. A real unitary matrix is evidently orthogonal, since then $a^\dagger = \tilde{a}$.

If a unitary matrix is also hermitian, then $a^\dagger = a$ and
$$a = a^{-1}, \tag{31}$$
from which it follows that
$$aa = a^2 = 1. \tag{32}$$

4·02. Solution of linear equations.
A set of linear equations
$$\left.\begin{array}{l}a_{11}x_1 + a_{12}x_2 + \ldots + a_{1n}x_n = y_1, \\ a_{21}x_1 + a_{22}x_2 + \ldots + a_{2n}x_n = y_2, \\ \ldots\ldots\ldots\ldots\ldots\ldots\ldots\ldots\ldots\ldots\ldots\ldots \\ a_{n1}x_1 + a_{n2}x_2 + \ldots + a_{nn}x_n = y_n,\end{array}\right\} \tag{1}$$
may be written in the abbreviated form
$$a_{ij}x_j = y_i, \tag{2}$$
where i and j run from 1 to n. If we think of x_i as a matrix with a single column (2) may be written as
$$ax = y, \tag{3}$$

where y is also a matrix with a single column. We assume at present that a is non-singular, and denote the cofactor of a_{ik} by A_{ik}.

Now since
$$A_{ik}a_{ij} = (\det a)\delta_{kj}, \tag{4}$$
we can multiply the ith equation of (2) by A_{ik} and add, and the sum will be
$$(\det a)\delta_{kj}x_j = A_{ik}y_i, \tag{5}$$
that is, if $\det a \neq 0$,
$$x_k = \frac{A_{ik}}{\det a}y_i = (a^{-1})_{ki}y_i \tag{6}$$
or
$$x = a^{-1}y, \tag{7}$$
which we might have got from (3) directly by multiplying both sides in front by a^{-1}.

Any of the sets of equations (5), (6), (7) gives the solution of the equations (1) in a compact form.*

If the matrix a is unitary (6) and (7) become respectively
$$x_k = a^\dagger_{ki}y_i, \tag{8}$$
$$x = a^\dagger y. \tag{9}$$
If a is orthogonal
$$x_k = a_{ik}y_i, \tag{10}$$
$$x = \tilde{a}y. \tag{11}$$

4·021. Multiplication of determinants. We have assumed that the reader is already familiar with the rule for multiplying determinants, but the following proof is interesting because it brings it into direct relation with the existence of solutions of homogeneous linear equations. Let a_{ij}, b_{ij} be typical elements of two determinants. Then $\|a_{ij}\| = 0$ is a necessary and sufficient condition that the equations
$$a_{ij}x_j = 0$$
have a set of solutions x_j different from 0. Multiply by b_{ik} and add; then the equations
$$b_{ik}a_{ij}x_j = 0$$
have a set of solutions different from 0, and therefore if
$$\|a_{ij}\| = 0, \quad \|b_{ik}a_{ij}\| = 0.$$
Similarly, if $\|b_{ij}\| = 0$,
$$\|a_{ik}b_{ij}\| = 0; \quad \text{and} \quad \|b_{ik}a_{ij}\| = \|a_{ik}b_{ij}\|,$$
since one of these is the transpose of the other. Hence $\|b_{ik}a_{ij}\|$ contains $\|a_{ij}\|$, $\|b_{ik}\|$ as factors; and by comparing coefficients of $a_{11}b_{11}a_{22}b_{22}\ldots a_{nn}b_{nn}$ we see that the other factor is 1.

In matrix notation
$$a_{ij}b_{ik} = \tilde{a}_{ji}b_{ik} = (\tilde{a}b)_{jk} \quad \text{and} \quad \det(\tilde{a}b) = \det\tilde{a}\det b = \det a\det b.$$

---

* In practice the numerical solution is not most easily found by this method. We give a practical method in 9·16.

4·03. Transformations. If we have a set of relations expressed by

$$y = ax, \tag{1}$$

where a is a non-singular square matrix, we can infer as in the last section that

$$x = a^{-1}y. \tag{2}$$

If the x_i are variables the matrix a can be said to effect a transformation of variables. Two specially important cases are those where a is respectively orthogonal and symmetrical.

4·031. Orthogonal transformations*. For transformations of rectangular axes we have had the rules

$$x'_j = l_{ij} x_i \tag{1}$$

and
$$x_i = l_{ij} x'_j. \tag{2}$$

The latter is already in the form of a matrix product, x_i and x'_j being matrices of one column. It can therefore be written

$$x = lx'. \tag{3}$$

But then we can suppose these equations to be solved to give the x'_j in terms of the x_i, and the solution will be

$$x' = l^{-1}x \tag{4}$$

or
$$x'_j = (l^{-1})_{ji} x_i. \tag{5}$$

Comparing this with (1) and noticing that they must be equivalent for all values of the x_i we have

$$(l^{-1})_{ji} = l_{ij}, \tag{6}$$

and therefore l is an orthogonal matrix; for

$$l^{-1} = \tilde{l}. \tag{7}$$

(1) can also be written
$$\tilde{x}' = \tilde{x}l. \tag{8}$$

We must write $\tilde{x}$, not x, because a single-column matrix cannot come first; we must therefore take the transpose of x to give a single-row matrix. Thus a rotation of axes can be expressed as an orthogonal transformation of determinant 1. Now

$$\tilde{x}x = x_i x_i \tag{9}$$

and
$$x'_j x'_j = \tilde{x} l l^{-1} x = \tilde{x} 1 x = \tilde{x} x = x_i x_i. \tag{10}$$

This verifies that the form $x_i x_i$ is invariant under an orthogonal transformation.

Conversely, if
$$x_i = l_{ij} x'_j \tag{11}$$

and
$$x_i x_i = x'_j x'_j \tag{12}$$

for all values of x'_j, we have
$$(l_{ij} x'_j)(l_{il} x'_l) = \delta_{jl} x'_j x'_l, \tag{13}$$

and therefore, since $l_{ij} l_{il}$ is a symmetrical matrix,

$$l_{ij} l_{il} = \delta_{jl}, \quad l\tilde{l} = 1. \tag{14}$$

Hence l is an orthogonal matrix.

* See note at end of Chapter 4, p. 170.

On the face of it (6) represents n^2 relations between n^2 unknowns, and we might expect only a finite number of solutions to exist, and even these might be complex. But whatever l might be, $l_{ij}l_{il}$ is symmetrical in j and l, and the sufficient condition for orthogonality therefore requires only $\tfrac{1}{2}n(n+1)$ independent relations. Thus apparently the components l_{ij} might be made to satisfy $\tfrac{1}{2}n(n-1)$ additional conditions, and this number is actually correct.

Now suppose that we keep the same axes x_i but rotate a rigid body about the origin. Imagine a set of axes x'_j to have originally coincided with the x_i ones but to be fixed in the body and therefore to be rotated with it. Then the coordinates of any particle of the body with respect to the x'_j axes are unaltered. We require its new coordinates y_i with respect to the x_i axes. If the direction cosines are denoted by l_{ij} as before we have

$$y_i = l_{ij}x'_j = l_{ij}x_j, \quad \boldsymbol{y} = l\boldsymbol{x}, \tag{15}$$

and similarly any vector $\boldsymbol{u}$ on rotation becomes $\boldsymbol{v}$, where

$$\boldsymbol{v} = l\boldsymbol{u}. \tag{16}$$

4·032. 2 × 2 orthogonal matrix in terms of one parameter. Now let l denote the orthogonal matrix

$$\begin{pmatrix} \alpha & \beta \\ \gamma & \delta \end{pmatrix}. \tag{17}$$

Then

$$ll = \begin{pmatrix} \alpha & \beta \\ \gamma & \delta \end{pmatrix} \begin{pmatrix} \alpha & \gamma \\ \beta & \delta \end{pmatrix} = \begin{pmatrix} \alpha^2+\beta^2 & \alpha\gamma+\beta\delta \\ \gamma\alpha+\delta\beta & \gamma^2+\delta^2 \end{pmatrix} = \begin{pmatrix} 1 & 0 \\ 0 & 1 \end{pmatrix}. \tag{18}$$

Then we can choose λ so that

$$\alpha = \cos\lambda, \quad \beta = \sin\lambda, \tag{19}$$

and then

$$\gamma/\delta = -\beta/\alpha = -\tan\lambda, \tag{20}$$

which, with

$$\gamma^2 + \delta^2 = 1, \tag{21}$$

gives

$$\gamma = -\sin\lambda, \quad \delta = \cos\lambda, \tag{22}$$

or

$$\gamma = \sin\lambda, \quad \delta = -\cos\lambda. \tag{23}$$

(22) gives

$$l = \begin{pmatrix} \cos\lambda & \sin\lambda \\ -\sin\lambda & \cos\lambda \end{pmatrix}, \tag{24}$$

involving one adjustable constant. $\lambda = 0$ gives $l = 1$.

(23) gives

$$l = \begin{pmatrix} \cos\lambda & \sin\lambda \\ \sin\lambda & -\cos\lambda \end{pmatrix}. \tag{25}$$

$\lambda = 0$ gives

$$l = \begin{pmatrix} 1 & 0 \\ 0 & -1 \end{pmatrix}.$$

If we regard x_1, x_2 as rectangular coordinates in two dimensions and if we substitute (24) in (15) we obtain a rotation through $-\lambda$. (23), with $\lambda = 0$, leaves x_1 unchanged but reverses the sign of x_2. In general, (25) represents a *reflexion*.

4·033. General orthogonal transformation.

In n dimensions a matrix l with
$$l_{11} = l_{22} = \cos\alpha, \quad l_{12} = -l_{21} = \sin\alpha,$$
all other diagonal elements unity and all other non-diagonal elements zero, is orthogonal of determinant $+1$. The transformation can be regarded as a rotation on the plane of the axes of x_1 and x_2, leaving coordinates in the other $n-2$ dimensions unaltered. We cannot speak of the rotation as being *about* any axis, as we can in three dimensions, but we can continue to say that it is parallel to a plane. Such a rotation, of arbitrary amount, can be made on any plane including two of the axes; hence $\frac{1}{2}n(n-1)$ independent rotations are possible in n dimensions.

The independent components of an antisymmetrical matrix have the same number. It may be shown* that every *real* orthogonal matrix can be expressed in the form
$$J(1-S)(1+S)^{-1},$$
where S is an antisymmetrical matrix and J is a matrix having either $+1$ or -1 in each diagonal place and zero elsewhere. We may note that if S is a *real* antisymmetrical matrix iS is hermitian; hence (see § 4·081) the latent roots of S are all pure imaginary and therefore $1+S$ is non-singular.

4·034. General rotation of a rigid body.

In three dimensions the matrix expressing a rotation through an angle α right-handedly about an axis with direction cosines n_i is, by 3·09,

$$\begin{pmatrix} \cos\alpha + n_1^2(1-\cos\alpha) & n_1 n_2(1-\cos\alpha) - n_3 \sin\alpha & n_1 n_3(1-\cos\alpha) + n_2 \sin\alpha \\ n_1 n_2(1-\cos\alpha) + n_3 \sin\alpha & \cos\alpha + n_2^2(1-\cos\alpha) & n_2 n_3(1-\cos\alpha) - n_1 \sin\alpha \\ n_3 n_1(1-\cos\alpha) - n_2 \sin\alpha & n_3 n_2(1-\cos\alpha) + n_1 \sin\alpha & \cos\alpha + n_3^2(1-\cos\alpha) \end{pmatrix}.$$

Let $O123$ be a set of axes and x_i the coordinates of a point of a body referred to them. Let the body be rotated into a position specified by Euler's angles, which we shall denote by θ, λ, χ. We require the final position of the particle that was at x_i before rotation. Suppose the body first rotated right-handedly through θ about $O2$; particles originally on the axes move to $1'2'3'$. The matrix for this rotation is

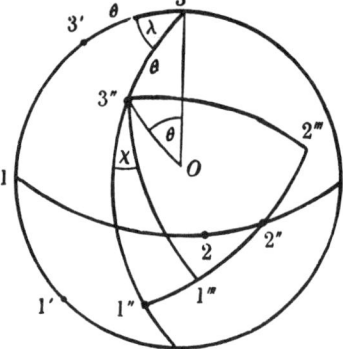

$$l_1 = \begin{pmatrix} \cos\theta & 0 & \sin\theta \\ 0 & 1 & 0 \\ -\sin\theta & 0 & \cos\theta \end{pmatrix},$$

and a particle at x moves to $y = l_1 x$. Now rotate through λ about $O3$ (not $O3'$). The particles at $1'2'3'$ move to $1''2''3''$ and the general particle moves to z, where

$$z = l_2 l_1 x, \quad l_2 = \begin{pmatrix} \cos\lambda & -\sin\lambda & 0 \\ \sin\lambda & \cos\lambda & 0 \\ 0 & 0 & 1 \end{pmatrix}, \quad l_2 l_1 = \begin{pmatrix} \cos\lambda\cos\theta & -\sin\lambda & \cos\lambda\sin\theta \\ \sin\lambda\cos\theta & \cos\lambda & \sin\lambda\sin\theta \\ -\sin\theta & 0 & \cos\theta \end{pmatrix}.$$

---
* W. L. Ferrar, *Algebra*. pp. 162–7; or see Ex. 10, p. 169.

This important matrix gives the coordinates of a particle whose position with regard to the axes $O1''2''3''$ is known, and may be regarded as the basis of most of the formulae of spherical astronomy. For a rigid body, however, the positions of all particles are not specified without a third angle; θ and λ can be chosen to specify one line of particles in the body, but those originally on $O1$ need not finish on $O1''$. We must therefore consider a further rotation χ about $O3''$. The working out of the matrix for this, followed by its multiplication into $l_2 l_1$, is not a matter to be undertaken if there is anything else to do. It can, however, be avoided. The displacement from $O123$ to $O1'''2'''3''$ is a rigid-body displacement. If the displacements λ and θ were undone, in this order, the plane $3''1'''$ would return to a position still inclined at χ to the plane 31. Hence we can allow for χ by making a rotation about $O3$ *first* and then applying the rotation $l_2 l_1$. The matrix for the former is

$$l_3 = \begin{pmatrix} \cos\chi & -\sin\chi & 0 \\ \sin\chi & \cos\chi & 0 \\ 0 & 0 & 1 \end{pmatrix},$$

and the complete rotation is given by

$$l_2 l_1 l_3 = \begin{pmatrix} \cos\lambda\cos\theta\cos\chi - \sin\lambda\sin\chi, & -\cos\lambda\cos\theta\sin\chi - \sin\lambda\cos\chi, & \cos\lambda\sin\theta \\ \sin\lambda\cos\theta\cos\chi + \cos\lambda\sin\chi, & -\sin\lambda\cos\theta\sin\chi + \cos\lambda\cos\chi, & \sin\lambda\sin\theta \\ -\sin\theta\cos\chi & \sin\theta\sin\chi, & \cos\theta \end{pmatrix}.$$

Thus, for example, a point originally at $(a, 0, 0)$ will finish at

$$a(\cos\lambda\cos\theta\cos\chi - \sin\lambda\sin\chi,\ \sin\lambda\cos\theta\cos\chi + \cos\lambda\sin\chi,\ -\sin\theta\cos\chi).$$

4·04. Symmetric matrices. We have already seen that a 3×3 real symmetric matrix has three orthogonal principal axes. If we take these as axes of reference x'_j, their direction cosines with respect to the original axes are the l_{ij} already found in 3·08, and

$$x_i = l_{ij} x'_j, \quad x = lx'. \tag{1}$$

Then
$$K_{ik} x_i x_k = l_{ij} l_{kl} K_{ik} x'_j x'_l$$
$$= K'_{jl} x'_j x'_k, \tag{2}$$

where
$$K'_{jl} = l_{ij} l_{kl} K_{ik} = \tilde{l}_{ji} K_{ik} l_{kl}, \tag{3}$$

$$K' = \tilde{l} K l. \tag{4}$$

But comparing with 3·08 (12) we see that if j and l are different $K'_{jl} = 0$; and if they are the same, equal to 1 say.

$$K'_{11} = \lambda_1 l_{i1} l_{i1} = \lambda_1. \tag{5}$$

Thus
$$K' = \begin{pmatrix} \lambda_1 & 0 & 0 \\ 0 & \lambda_2 & 0 \\ 0 & 0 & \lambda_3 \end{pmatrix}, \tag{6}$$

and the transformation (3) has reduced K to a diagonal form.

This result can be extended. A real symmetrical matrix of any order can be reduced to diagonal form by an orthogonal transformation of the form (4). A still further extension is that a hermitian matrix of any order can be reduced to diagonal form by a unitary transformation of the form $K' = l^{\dagger}Kl$.

In considering the general motion of a fluid we have seen that part of the motion in a small neighbourhood is represented by a symmetrical matrix, and that this part expresses the changes of distance between particles. Consider then the transformation expressed by a symmetrical matrix K; we take

$$y_i = K_{ij}x_j = x_j K_{ji}; \tag{7}$$

$$y = Kx; \quad \tilde{y} = \tilde{x}K. \tag{8}$$

Then
$$\tilde{y}y = \tilde{x}KKx. \tag{9}$$

For a symmetrical matrix KK is not in general equal to 1, and

$$\tilde{y}y \neq \tilde{x}x.$$

But if we refer to principal axes of K we shall have

$$y = ly', \quad x = lx', \quad \tilde{y} = \tilde{y}'\tilde{l}, \quad \tilde{x} = \tilde{x}'\tilde{l}, \tag{10}$$

$$y' = \tilde{l}y = \tilde{l}Klx' = K'x', \tag{11}$$

and
$$y'_1 = \lambda_1 x'_1, \quad y'_2 = \lambda_2 x'_2, \quad y'_3 = \lambda_3 x'_3, \tag{12}$$

since K' is in diagonal form. The displacement therefore changes the lengths of three perpendicular marked lines in the body but not their directions; all lengths parallel to each of these lines are altered in the same ratio. Such a deformation is called a *pure strain*.

We shall carry through the general reduction of an $n \times n$ hermitian matrix to the diagonal form, of which the reduction of a symmetrical matrix is a particular case. This problem occurs in many branches of mathematical physics. Before considering it, however, we must consider the solution of a set of homogeneous linear equations and some properties of determinants.

4·05. Rank of a matrix. Homogeneous linear equations. In general a set of n linear equations in n unknowns, with n constants on the right, has a unique solution. But if we take such a pair as

$$x + y = 1, \quad 2x + 2y = 3,$$

there is no solution. This fact is associated with a property of the coefficients; for whereas the pair

$$x + y = 0, \quad x - y = 0$$

has no solution other than $x = y = 0$, the pair

$$x + y = 0, \quad 2x + 2y = 0$$

has an infinite number of solutions, namely, any pair such that $y = -x$. In the former case the matrix of the coefficients is non-singular, in the latter it is singular. This is a general rule: if the constants on the right are all zero, a necessary and sufficient condition for the existence of solutions different from 0 is that the determinant of the coefficients is zero, and if the constants are not all zero and the determinant of the coefficients is zero, then either the equations are inconsistent and have no solution, or they are consistent and have an infinite number.

It is possible, moreover, for n equations to have a doubly infinite set of solutions, or indeed a set in which any number less than n of the unknowns can be assigned arbitrarily. To see how this arises it is convenient to introduce the idea of the *rank* of a matrix or a determinant. Consider the set of n equations

$$\boldsymbol{ax} = a_{ik}x_k = 0. \tag{1}$$

There is clearly always one solution $x_k = 0$ ($k = 1, 2, ..., n$).

A determinant obtained from $\|a_{ik}\|$ by suppressing m rows and m columns is called a *minor* of order $n-m$; $\|a_{ik}\|$ itself may be called the minor of order n, and any single element is a minor of order 1. If m rows and columns are suppressed in such a way that the values of i for the rows are the same as those of k for the columns, the minor is a *principal minor*; a principal minor is symmetrically placed about the leading diagonal. The principal minor in the top left-hand corner is called a *leading minor*. Thus in the determinant of order 4

$$\begin{vmatrix} a_{11} & a_{12} & a_{13} & a_{14} \\ a_{21} & a_{22} & a_{23} & a_{24} \\ a_{31} & a_{32} & a_{33} & a_{34} \\ a_{41} & a_{42} & a_{43} & a_{44} \end{vmatrix}$$

the minor enclosed by dotted lines is simply a minor of order 2. That enclosed by broken lines is a principal minor, and that enclosed by continuous lines is the leading minor of order 2.

Now it may happen that all the minors of order $r+1$ vanish (and therefore all those of order greater than $r+1$ do) while some of those of order r do not. The matrix and the determinant are then said to be *of rank r*. r is the largest integer for which it can be said 'not all minors of order r are zero'. In particular, a determinant of order n that vanishes, while not all its first minors (minors of order $n-1$) vanish, is of rank $n-1$. If the determinant $\|a_{ik}\|$ itself does not vanish, the matrix and the determinant are of rank n.*

We can see at once that unless $\boldsymbol{a}$ is of rank less than n the equations (1) have no solution other than $\boldsymbol{x} = 0$. For if $\boldsymbol{a}$ is of rank n it has a reciprocal $\boldsymbol{a}^{-1}$, and

$$\boldsymbol{a}^{-1}\boldsymbol{ax} = 0 \tag{2}$$

implies $$\boldsymbol{x} = 0. \tag{3}$$

Conversely, suppose that $\boldsymbol{a}$ is of rank $n-1$. This means that $\|a_{ik}\| = 0$, but not all the cofactors A_{ik} are 0. For definiteness suppose that A_{nn}, the leading first minor, is not zero. This can always be attained by rearranging the equations and renumbering the unknowns.

* It is possible to extend the notion of rank to $m \times n$ matrices; for instance,
$$x+y+z = 0, \quad 2x+2y+2z = 1,$$
are inconsistent for all finite x, y, z. We need not consider such cases, which have little physical interest and greatly complicate the analysis.

Then a solution of equations (1) can be obtained by putting $x_n = b$, an arbitrary constant. For we can solve the first $n-1$ equations for $x_1, x_2, ..., x_{n-1}$ in the usual way, since the determinant of the coefficients of these unknowns is A_{nn}, which by hypothesis is not zero. The solution is

$$x_k = \frac{A_{nk} b}{A_{nn}} \quad (k = 1, 2, ..., n). \tag{4}$$

We have to verify that this solution also satisfies the nth equation. Substituting, we have

$$a_{nk} x_k = \frac{b}{A_{nn}} \sum_k a_{nk} A_{nk} = \frac{b}{A_{nn}} \| a_{ik} \|, \tag{5}$$

since the sum is the expansion of $\| a_{ik} \|$ in terms of elements of the last row. But this is zero because $\| a_{ik} \| = 0$, $A_{nn} \neq 0$. Hence the nth equation is also satisfied. Thus the ratios of the x_i are unique, but the x_i can all be multiplied by an arbitrary factor.

If the cofactor A_{pq} obtained by striking out the pth row and qth column is not zero we could similarly, putting $x_q = c$, obtain the solution

$$x_k = \frac{A_{pk} c}{A_{pq}}. \tag{6}$$

But for (4) and (6) to be consistent, for all k, p, q,

$$A_{pk} A_{nq} - A_{nk} A_{pq} = 0. \tag{7}$$

The vanishing of this expression when $\| a_{ik} \|$ vanishes is a special case of Jacobi's theorem, which we shall prove later. It follows further that if a determinant and one first minor are zero, the minors either of all elements in the same row or of all in the same column are zero.

Suppose now that the matrix a is of rank r. For definiteness suppose the equations arranged so that the leading minor of order r is not zero. We must suspend the summation convention for summations that are not from 1 to n. Then the equations can be written

$$\sum_{k=1}^{r} a_{ik} x_k = - \sum_{\kappa = r+1}^{n} a_{i\kappa} x_\kappa \quad (i = 1, 2, ..., r), \tag{8}$$

$$\sum_{k=1}^{n} a_{ik} x_k = 0 \quad (i = r+1, ..., n). \tag{9}$$

For any set of values of x_κ ($\kappa = r+1$ to n) the first set has a unique solution. If in fact we denote the determinant of the coefficients on the left of (8) by α, it is non-singular and has a reciprocal; and we denote the cofactor of a_{ik} in it by α_{ik}. Then for $k = 1$ to r, (8) are satisfied if, and only if,

$$x_k = - \sum_{i=1}^{r} \frac{\alpha_{ik}}{\alpha} \sum_{\kappa=r+1}^{n} a_{i\kappa} x_\kappa. \tag{10}$$

Substitute in any of (9). We get for $i > r$

$$\sum_{k=1}^{n} a_{ik} x_k = - \sum_{k=1}^{r} \sum_{p=1}^{r} \frac{\alpha_{pk}}{\alpha} a_{ik} \sum_{\kappa=r+1}^{n} a_{p\kappa} x_\kappa + \sum_{\kappa=r+1}^{n} a_{i\kappa} x_\kappa$$

$$= \sum_{\kappa=r+1}^{n} \frac{x_\kappa}{\alpha} \left(\alpha a_{i\kappa} - \sum_{p=1}^{r} \sum_{k=1}^{r} a_{ik} \alpha_{pk} a_{p\kappa} \right). \tag{11}$$

Consider the following minor of order $r+1$, and expand in terms of the last row and column:

$$\begin{vmatrix} a_{11} & a_{12} & \ldots & a_{1r} & a_{1\kappa} \\ a_{21} & \ldots & \ldots & \ldots & a_{2\kappa} \\ \hdotsfor{5} \\ a_{r1} & a_{r2} & \ldots & a_{rr} & a_{r\kappa} \\ a_{i1} & \ldots & \ldots & a_{ir} & a_{i\kappa} \end{vmatrix}. \tag{12}$$

The coefficient of $a_{i\kappa}$ in its expansion is α. That of $a_{ik}a_{p\kappa}$ is that of $a_{i\kappa}a_{pk}$ with the sign reversed, and this is $-\alpha_{pk}$. Hence the coefficient of x_κ/α in (11) is a minor of order $r+1$ of the original determinant, which vanishes by hypothesis. Thus if the matrix a is of rank r less than n, $n-r$ of the unknowns can be assigned arbitrarily and the remainder are homogeneous linear functions of them.

4·06. Determinantal equations. We often have to consider a set of n equations

$$\lambda a_{ik} x_k = b_{ik} x_k, \tag{1}$$

where λ must be determined in such a way that solutions exist with some of the x_k different from 0. This requires that λ must be equal to some root of the equation

$$\| \lambda a_{ik} - b_{ik} \| = 0. \tag{2}$$

This will in general have n distinct roots. For each root λ_j a set of values of the x_i, say l_{ij}, can be found to satisfy (1), and any multiple of this set will be a solution. For some common types of matrices, notably symmetrical and unitary ones, the solution has special properties, which we shall study later.

If $a_{ik} = \delta_{ik}$, the left of (1) reduces to λx_i. Then (1) can be written in matrix form

$$\lambda x = bx, \tag{3}$$

and the λ are variously called the *latent roots, characteristic values, proper values,* and *eigenvalues* of the matrix b.* The equation determining them

$$\| b_{ik} - \lambda \delta_{ik} \| = 0, \tag{4}$$

is called the *characteristic equation* of the matrix. A set of x_i, say l_{ij}, satisfying (3) for a particular value of λ, say λ_j, may be called a *characteristic solution or eigensolution.*

If we take a general set of numbers X_i it may always be possible to find a set ξ_j so that

$$X_i = l_{ij} \xi_j. \tag{5}$$

The condition for this is that the matrix l_{ij} shall be non-singular. This is easily proved to be satisfied when all the λ_j are different. For if l_{ij} was singular we could take non-zero values of ξ_j so that

$$\sum_j l_{kj} \xi_j = 0 \tag{6}$$

for all k.

---

* Both λ and λ^{-1} sometimes occur in actual problems. Hilbert and Courant (*Methoden d. Math. Phys.* 1, 13) use *characteristic value* for λ defined by (3) and *eigenvalue* for $1/\lambda$.

Let r be the smallest number of the ξ_j represented in any such set of equations. Then we can arrange so that j runs from 1 to r. Then for all i

$$0 = (b_{ik} - \lambda_l \delta_{ik}) \sum_{j=1}^{r} l_{kj} \xi_j = \sum (\lambda_j - \lambda_l) l_{ij} \xi_j. \tag{7}$$

But if we take l to be any j represented in (6) the corresponding factor $\lambda_j - \lambda_l = 0$, and we have a set of relations for all i between at most $r-1$ of the $l_{ij} \xi_j$, since by hypothesis the λ_j are all different and the terms in (7) cannot all vanish. Thus the hypothesis that l_{ij} is singular leads to a contradiction.

l_{ij} is also often non-singular, but not always, when some of the roots are equal. Consider the matrices $\begin{pmatrix} 1 & 0 \\ 0 & 1 \end{pmatrix}$, $\begin{pmatrix} 1 & 1 \\ 0 & 1 \end{pmatrix}$. Each has a repeated root 1. For the first, two characteristic solutions are $(1, 0), (0, 1)$. The only characteristic solution of the second is $(1, 0)$.

Consider the matrix λ_{ij} whose *diagonal* elements are the λ_j and all others zero. Then we can write (3) as

$$b_{ik} l_{kj} = l_{ik} \lambda_{kj}. \tag{8}$$

for all i, j; and, if l is non-singular, multiplying before with l^{-1} we have

$$l^{-1} b l = \lambda. \tag{9}$$

Thus a transformation of the form (9) reduces b to diagonal form. Applied to the unit matrix it gives $l^{-1} 1 l$, which is 1. Hence such a transformation leaves the unit matrix unaltered: it is called a *collineatory* transformation and two matrices connected by such a transformation are said to be *similar*. It is not necessary to specify that l is non-singular; this is taken as understood from the fact that l^{-1} appears in the formula—if l was singular there would be no such formula.

4·061. Importance of diagonal matrices. All diagonal matrices commute with one another. If the diagonal elements of a diagonal matrix λ are λ_r, those of λ^n are λ_r^n and all others zero. If no λ_r is zero the diagonal elements of λ^{-1} are λ_r^{-1} and all others are zero. If $l^{-1} a l = \lambda$, then $a^n = (l \lambda l^{-1})^n = l \lambda^n l^{-1}$, all intermediate $l l^{-1}$ cancelling when the expression is written out in full.

This result is connected directly with the behaviour of dynamical systems satisfying linear equations of motion. If the coordinates and velocities are given at time 0, those at time τ are linear functions of them. If then we denote their values at times 0 and τ by X_i and Y_i ($i = 1$ to $2n$, where n is the number of degrees of freedom), then

$$Y_i = a_{ik} X_k, \quad Y = aX.$$

Now if 4·06(5) is satisfied by characteristic solutions of the matrix equations $ax = \lambda x$, we can write

$$X_k = l_{kj} \xi_j, \quad X = l\xi; \quad \xi = l^{-1} X,$$

and

$$Y_i = a_{ik} l_{kj} \xi_j = l_{ik} \lambda_{kj} \xi_j, \quad Y = l\lambda l^{-1} X.$$

If now Z_i are the values at time 2τ they are found from $Z = aY = a^2 X = l\lambda^2 l^{-1} X$. The process can be repeated to any multiple of τ, and therefore so long as l is not singular the solution is found. Evidently the $\lambda_{jl} (j = l)$ are the time factors $\exp(\gamma\tau)$ in the solution of the dynamical equations by the usual method.

Since in general the roots are different l is usually not singular; in some important special cases l is not singular even if some of the roots are equal, and the same method can still be used.

4·062. A matrix satisfies its own characteristic equation. If the equation

$$\|a_{ik} - \lambda \delta_{ik}\| = 0 \tag{1}$$

is expanded in powers of λ, we get an equation of degree n in λ:

$$D(\lambda) = \alpha_n + \alpha_{n-1}\lambda + \alpha_{n-2}\lambda^2 + \ldots + (-1)^n \lambda^n = 0. \tag{2}$$

The theorem means that if we substitute the matrix a for λ in each term and interpret by the rules of matrix multiplication and addition, we get a matrix $D(a)$ every element of which is 0.

We assume first that the equation (2) has n distinct roots. For each root λ_j a set of non-zero x_{ij} exists satisfying

$$\lambda_j x_{ij} = a_{ik} x_{kj}, \qquad a x_j = x_j \lambda_j, \tag{3}$$

$$a^r x_j = x_j \lambda_j^r, \tag{4}$$

since λ_j is a number. Then take $\quad y_i = \beta_j x_{ij}, \qquad y = \beta_j x_j, \tag{5}$

where β_j are n arbitrary multipliers. Then

$$D(a) y = (\alpha_n 1 + \alpha_{n-1} a + \alpha_{n-2} a^2 + \ldots + (-1)^n a^n) \beta_j x_j$$
$$= \sum_j (\alpha_n + \alpha_{n-1} \lambda_j + \alpha_{n-2} \lambda_j^2 + \ldots + (-1)^n \lambda_j^n) \beta_j x_j$$
$$= 0, \tag{6}$$

since the λ_j all satisfy (2). But the matrix of x_{ij} is non-singular, and the β_j can be chosen to make y_i anything we like; and therefore

$$D(a) = 0. \tag{7}$$

If $D(\lambda) = 0$ has multiple roots we can suppose the elements altered so as to make the roots distinct, and then apply (7). Then if the elements are made to approach their original values continuously each element of $D(a)$ also approaches its original value continuously. But it is always 0; hence the original value of each element must also be 0.

Shorter proofs exist; this one is given to illustrate the use of diagonal matrices.

4·063. Block matrices. Suppose that the elements of a matrix are grouped into blocks (squares and rectangles). These blocks themselves can be regarded as elements of a matrix of lower order. Thus from the matrix

$$\begin{pmatrix} a_{11} & a_{12} & a_{13} \\ a_{21} & a_{22} & a_{23} \\ a_{31} & a_{32} & a_{33} \end{pmatrix} \quad \text{we may derive} \quad \begin{pmatrix} A_{11} & A_{12} \\ A_{21} & A_{22} \end{pmatrix}, \tag{1}$$

where $\quad A_{11} = \begin{pmatrix} a_{11} & a_{12} \\ a_{21} & a_{22} \end{pmatrix}, \quad A_{12} = \begin{pmatrix} a_{13} \\ a_{23} \end{pmatrix}, \quad A_{21} = (a_{31}, a_{32}), \quad A_{22} = a_{33}. \tag{2}$

We shall speak of the matrix on the right in (1) as in *block form* and denote it by A. That on the left will be said to be in *expanded form*. Two matrices A, B in block form may be multiplied to give another matrix $AB = C$ in block form by the usual rule,

$$C_{ik} = \sum_j A_{ij} B_{jk}, \tag{3}$$

provided that (1) the number of block columns of A is equal to the number of block rows of B, (2) the number of columns of the matrix A_{ij} is equal to the number of rows of the matrix B_{jk}. For instance, if A, B are built up as follows, the marginal entries indicating the numbers of rows and columns, m, n, p of the components,

$$A = \begin{pmatrix} \overset{n_1}{A_{11}} & \overset{n_2}{A_{12}} \\ A_{21} & A_{22} \end{pmatrix}\begin{matrix} m_1, \\ m_2 \end{matrix} \quad B = \begin{pmatrix} \overset{p_1}{B_{11}} & \overset{p_2}{B_{12}} & \overset{p_3}{B_{13}} \\ B_{21} & B_{22} & B_{23} \end{pmatrix}\begin{matrix} n_1, \\ n_2 \end{matrix} \qquad (4)$$

where $A_{11}, A_{12}, A_{21}, A_{22}$ are respectively $m_1 \times n_1$, $m_1 \times n_2$, $m_2 \times n_1$, $m_2 \times n_2$ and $B_{11}, B_{12}, \ldots,$ are $n_1 \times p_1$, $n_1 \times p_2, \ldots,$ the product matrix is

$$C = \begin{pmatrix} \overset{p_1}{C_{11}} & \overset{p_2}{C_{12}} & \overset{p_3}{C_{13}} \\ C_{21} & C_{22} & C_{23} \end{pmatrix}\begin{matrix} m_1, \\ m_2 \end{matrix} \qquad (5)$$

where $C_{11}, C_{12}, \ldots$ are $m_1 \times p_1$, $m_1 \times p_2, \ldots$.

If a, b are simple matrices and $ab = c$, then if A, B are block forms of a, b, C is a block form of c. In other words, if we form C we can derive c by writing out all the elements of the matrix elements of C in full and removing internal brackets. We prove this for the case where A_{11}, B_{11} are $n_1 \times n_1$, A_{12}, B_{12} are $n_1 \times n_2$, A_{21}, B_{21} are $n_2 \times n_1$, and A_{22}, B_{22} are $n_2 \times n_2$. It is convenient in A_{12}, A_{22} to number the columns n_1+1 to n_1+n_2, and in A_{21}, A_{22} to number the rows n_1+1 to n_1+n_2. Then

$$\left.\begin{aligned} a_{ik} &= (A_{11})_{ik} & (i \leqslant n_1,\ k \leqslant n_1), \\ a_{ik} &= (A_{12})_{ik} & (i \leqslant n_1,\ n_1+1 \leqslant k \leqslant n_1+n_2), \\ a_{ik} &= (A_{21})_{ik} & (n_1+1 \leqslant i \leqslant n_1+n_2,\ k \leqslant n_1), \\ a_{ik} &= (A_{22})_{ik} & (n_1+1 \leqslant i \leqslant n_1+n_2,\ n_1+1 \leqslant k \leqslant n_1+n_2). \end{aligned}\right\} \qquad (6)$$

The same rules are applied to b. Now

$$c_{ik} = (ab)_{ik} = \left(\sum_{j=1}^{n_1} + \sum_{j=n_1+1}^{n_1+n_2}\right) a_{ij} b_{jk}. \qquad (7)$$

For $i, k \leqslant n_1$, this is

$$\sum_{j=1}^{n_1} (A_{11})_{ij}(B_{11})_{jk} + \sum_{j=n_1+1}^{n_1+n_2} (A_{12})_{ij}(B_{21})_{jk} = (A_{11}B_{11} + A_{12}B_{21})_{ik} = (C)_{ik}. \qquad (8)$$

The proof is similar for the other parts of the product.

We shall speak of a matrix as in *diagonal block form* if it is in block form and all the matrix elements other than the diagonal ones are null. The expanded matrix found by writing out in full and removing brackets will then have the property that all its non-zero elements are in non-overlapping squares symmetrical about the leading diagonal. A diagonal matrix is a special case.

If A, B are in diagonal block form it is clear that the product AB is also in diagonal block form.

4·064. *If A is a diagonal block matrix whose diagonal elements are $A_{rs}(s=r)$ and its expanded matrix is a, a necessary and sufficient condition that a shall be reducible to diagonal form by a collineatory transformation is that all the A_{rs} shall be so reducible; and the characteristic solutions of a are obtained from those of the A_{rs} by including zero components.*

We take the case where A is 2×2, since the extension to the case where it is $n \times n$ follows by repetition. The statement that A_{11} is reducible to diagonal form means that a non-singular L_1 exists such that $L_1^{-1} A_{11} L_1 = \lambda_1$, where λ_1 is diagonal. Then if A_{22} is similarly reducible to λ_2 by L_2,

$$\begin{pmatrix} L_1^{-1} & 0 \\ 0 & L_2^{-1} \end{pmatrix} \begin{pmatrix} A_{11} & 0 \\ 0 & A_{22} \end{pmatrix} \begin{pmatrix} L_1 & 0 \\ 0 & L_2 \end{pmatrix} = \begin{pmatrix} L_1^{-1} A_{11} L_1 & 0 \\ 0 & L_2^{-1} A_{22} L_2 \end{pmatrix} = \begin{pmatrix} \lambda_1 & 0 \\ 0 & \lambda_2 \end{pmatrix} \qquad (1)$$

The expanded matrix formed from $\begin{pmatrix} L_1 & 0 \\ 0 & L_2 \end{pmatrix}$ is non-singular because L_1, L_2 are non-singular, and

$$\begin{pmatrix} L_1^{-1} & 0 \\ 0 & L_2^{-1} \end{pmatrix} \begin{pmatrix} L_1 & 0 \\ 0 & L_2 \end{pmatrix} = \begin{pmatrix} 1 & 0 \\ 0 & 1 \end{pmatrix}. \qquad (2)$$

Hence the condition is sufficient.

Conversely, let a be reducible to diagonal form by a collineatory transformation. Let A_{11}, A_{22} be $n_1 \times n_1$ and $n_2 \times n_2$ respectively. Then the statement that a is reducible to diagonal form by a transformation $l^{-1} a l$ means that $n_1 + n_2$ linearly independent sets $l_{i(k)}$ ($i = 1$ to $n_1 + n_2$) exist such that for each set there is a λ_k satisfying

$$a_{ij} l_{j(k)} = \lambda_{(k)} l_{i(k)}, \qquad (3)$$

where the brackets round the suffix mean that we are not summing with regard to k. The λ_k need not all be different, but the statement that $l^{-1} a l$ is diagonal implies that even if some of the λ_k are equal corresponding solutions can still be chosen so as to be linearly independent.

We note first that

$$\det(a - \lambda 1) = \det(A_{11} - \lambda 1) \det(A_{22} - \lambda 1). \qquad (4)$$

Therefore every root of a is one of either A_{11} or A_{22}; and if λ is a p-fold root of A_{11} and a q-fold one of A_{22} it is a $(p+q)$-fold one of a.

We have to show that, in the given conditions, A_{11} and A_{22} have n_1 and n_2 linearly independent characteristic solutions respectively. (Neither, of course, can have more.) We number the rows and columns in A_{11} from 1 to n_1, and those in A_{22} from $n_1 + 1$ to $n_1 + n_2$. If $L_{i(k)1}$ is a characteristic solution of A_{11}, so that

$$(A_{11})_{ij} L_{j(k)1} = \lambda L_{i(k)1}, \qquad (5)$$

then if
$$\begin{aligned} l_{i(k)1} &= L_{i(k)1}, \quad (i \leqslant n_1), \\ &= 0, \quad (n_1 + 1 \leqslant i \leqslant n_1 + n_2), \end{aligned} \qquad (6)$$

we have, for $i \leqslant n_1$, $\quad a_{ij} l_{j(k)1} = (A_{11})_{ij} L_{j(k)1} = \lambda L_{i(k)1} = \lambda l_{i(k)1}, \qquad (7)$

and for $i > n_1$, $\quad a_{ij} l_{j(k)1} = 0 L_{j(k)1} + (A_{22})_{ij} 0 = \lambda l_{i(k)1}, \qquad (8)$

Hence $l_{i(k)1}$ is a characteristic solution of a with the same λ. Similarly, if $L_{i(k)2}$ is a characteristic solution of A_{22}, with $i \geqslant n_1 + 1$, there is a characteristic solution $l_{i(k)2}$ of a, given by taking

$$l_{i(k)2} = 0 \quad (i \leqslant n_1); \quad l_{i(k)2} = L_{i(k)2} \quad (n_1 + 1 \leqslant i \leqslant n_1 + n_2). \qquad (9)$$

Further, if the solutions taken for A_{11}, A_{22} respectively are linearly independent, the $l_{i(k)}$ derived from them are also linearly independent.

Suppose now that there are m_1 linearly independent characteristic solutions of A_{11} and m_2 of A_{22}, where
$$m_1 \leqslant n_1, \quad m_2 \leqslant n_2, \quad \text{but} \quad m_1 + m_2 < n_1 + n_2.$$
Then we can derive by the method just given fewer than $n_1 + n_2$ characteristic solutions of a; hence there is at least one other characteristic solution of a, say $l_{i(m_1+m_2+1)}$, satisfying, for some λ,
$$a_{ij} l_{j(m_1+m_2+1)} = \lambda l_{i(m_1+m_2+1)}. \tag{10}$$
But this is equivalent to the two sets of equations
$$\left.\begin{array}{l}(A_{11})_{ij} l_{j(m_1+m_2+1)} = \lambda l_{i(m_1+m_2+1)} \quad (i \leqslant n_1) \\ (A_{22})_{ij} l_{j(m_1+m_2+1)} = \lambda l_{i(m_1+m_2+1)} \quad (n_1+1 \leqslant i \leqslant n_1+n_2).\end{array}\right\} \tag{11}$$
The l_i by hypothesis are not all zero; hence the first set give a characteristic solution of A_{11}, or the second one of A_{22}, or both statements are true. In any case we can write
$$l_{i(m_1+m_2+1)} = \alpha l_{i(m_1+m_2+1)1} + \beta l_{i(m_1+m_2+1)2}. \tag{12}$$
But by hypothesis we have already found all the characteristic solutions of A_{11}, A_{22}; hence the $l_{i(m_1+m_2+1)1}$ can be expressed linearly in terms of the m_1 known solutions for A_{11}, and the $l_{i(m_1+m_2+1)2}$ in terms of the m_2 known solutions for A_{22}. But then it follows that $l_{i(m_1+m_2+1)}$ can be expressed linearly in terms of the solutions already obtained; this contradicts the fact that it is a new independent solution. Hence $m_1 = n_1$, $m_2 = n_2$, and the theorem is proved. In particular, if λ is a p-fold root of A_{11} and a q-fold root of A_{22}, this process yields $p+q$ independent characteristic solutions of a corresponding to that λ.

If $l_{i(k)}$ is any characteristic solution of a, the method used to get (11) will give a characteristic solution of either A_{11} or A_{22} or both. If the associated λ is a root of A_{11} but not of A_{22}, it will yield a solution of A_{11}, but a zero solution of A_{22}.

We notice also that
$$\begin{pmatrix} L_1 & 0 \\ 0 & L_2 \end{pmatrix} = \begin{pmatrix} L_1 & 0 \\ 0 & 1 \end{pmatrix} \begin{pmatrix} 1 & 0 \\ 0 & L_2 \end{pmatrix}, \tag{13}$$
and therefore the reduction of a can be carried out in stages by transforming the blocks to diagonal form in turn.

4·065. *If two matrices are both reducible to diagonal form by collineatory transformations, a necessary and sufficient condition that they shall be reducible by the same transformation is that they shall commute.*

If
$$l^{-1}al = \lambda, \quad l^{-1}bl = \mu,$$
where λ, μ are diagonal and therefore commute,
$$ab = l\lambda l^{-1} l\mu l^{-1} = l\lambda\mu l^{-1} = l\mu\lambda l^{-1} = ba.$$
Hence the condition is necessary.

Conversely, if
$$l^{-1}al = \lambda, \quad l^{-1}bl = c,$$
where λ is diagonal but c is not assumed to be so, and $ab = ba$,
$$\lambda c = l^{-1}a l l^{-1}bl = l^{-1}abl = l^{-1}bal = c\lambda.$$

Now
$$(\lambda c)_{ik} = \lambda_{ij} c_{jk} = \lambda_{ii} c_{ik},$$

where summation is not understood for i; and similarly

$$(c\lambda)_{ik} = c_{ik} \lambda_{kk},$$

where summation is not understood for k. Hence either $c_{ik} = 0$ or $\lambda_{ii} = \lambda_{kk}$. If all diagonal elements of λ are different, it follows that all non-diagonal elements of c are zero and c is in diagonal form. If some diagonal elements of λ are equal, c can be arranged in diagonal block form, the diagonal of each block corresponding to a set of equal diagonal components of λ. But by hypothesis c is reducible to diagonal form; hence by 4·064 the reduction can be carried out by reducing each block in turn, and such a transformation does not affect the corresponding elements of λ because they form a multiple of a unit matrix.

The following proof of sufficiency was communicated to us by Professor P. Hall; it does not depend on the properties of block matrices. If a, b are both $n \times n$, the statement that they can be reduced to diagonal form is equivalent to the statement that each has n linearly independent characteristic solutions; and we have to show that if $ab = ba$ there are n linearly independent solutions characteristic of both a and b. Any set of n quantities can be expressed as a linear combination of the characteristic solutions of a, since these are linearly independent; in particular, any characteristic solution x_i of b corresponding to the root μ can be so expressed, say,

$$x_i = \sum_{k=1}^{r} l_{i(k)}.$$

We may assume that $a l_{(k)} = \lambda_{(k)} l_{(k)}$ with *different* λ_k, since if two characteristic solutions $l_{i(k)}$ of a correspond to the same λ, any linear combination of them is also a characteristic solution corresponding to that λ. Then

$$a_{ij} b_{jk} x_k = a_{ij} \mu x_j = \mu \sum_{k=1}^{r} \lambda_{(k)} l_{i(k)},$$

$$b_{ij} a_{jk} x_k = b_{ij} \sum_{k=1}^{r} \lambda_{(k)} l_{j(k)},$$

and since $ab = ba$ it follows that $\sum_{k=1}^{r} \lambda_{(k)} l_{i(k)}$ is also a characteristic solution of b corresponding to the root μ. Hence so is

$$\lambda_1 x_i - \sum_{k=1}^{r} \lambda_{(k)} l_{i(k)} = (\lambda_1 - \lambda_2) l_{i(2)} + (\lambda_1 - \lambda_3) l_{i(3)} + \ldots.$$

By repeating the argument we show that $(\lambda_1 - \lambda_r)(\lambda_2 - \lambda_r) \ldots (\lambda_{r-1} - \lambda_r) l_{i(r)}$ is a characteristic solution of b and therefore $l_{i(r)}$ is one. We conclude, since the ordering of k is irrelevant, that every $l_{i(k)}$ is a characteristic solution of b as well as of a. Hence every characteristic solution of b can be expressed as a linear combination of simultaneous characteristic solutions of a and b, and therefore any set of n numbers can be so expressed.

4·066. The theorems of 4·064, 4·065 are of importance in quantum theory in the special case where a is *hermitian*. It will be shown in 4·083 that if a is hermitian it can always be reduced to diagonal form by a collineatory transformation, with l *unitary*. Then if a is hermitian and in diagonal block form the diagonal blocks are also hermitian and can be reduced similarly. Hence we do not need the general theorem of 4·064, but we do need the property expressed by 4·064(13).

4·067. The following numerical example occurs in determining the eigenvalues of the angular momenta for a certain configuration of electrons in an atom (cf. Condon and Shortley, *Theory of Atomic Spectra*, C.U.P. (1935), pp. 222–6).

Consider the three matrices:

$$A = \tfrac{1}{4}\begin{pmatrix} 3 & 0 & & & \\ 0 & 3 & & 0 & \\ \hline & & 7 & 4 & 4 \\ & 0 & 4 & 7 & 4 \\ & & 4 & 4 & 7 \end{pmatrix}, \quad B = \begin{pmatrix} 4 & -2 & & & \\ -2 & 4 & & 0 & \\ \hline & & 2 & -2 & 0 \\ & 0 & -2 & 4 & -2 \\ & & 0 & -2 & 2 \end{pmatrix},$$

$$C = \tfrac{1}{2}\begin{pmatrix} -1 & 0 & \sqrt{2} & -\sqrt{2} & 0 \\ 0 & -1 & 0 & \sqrt{2} & -\sqrt{2} \\ \hline \sqrt{2} & 0 & & & \\ -\sqrt{2} & \sqrt{2} & & 0 & \\ 0 & -\sqrt{2} & & & \end{pmatrix}.$$

All three matrices are symmetrical and all commute. The roots of A are $\tfrac{3}{4}$ (four times), $\tfrac{15}{4}$. Those of B are 2, 2, 6, 6, 0. As the matrices are symmetrical the reduction can be performed for each by an orthogonal transformation (cf. 4·081). We find, starting with B, that the orthogonal matrix

$$l = \frac{1}{\sqrt{6}}\begin{pmatrix} \sqrt{3} & 0 & \sqrt{3} & 0 & 0 \\ \sqrt{3} & 0 & -\sqrt{3} & 0 & 0 \\ 0 & \sqrt{3} & 0 & 1 & \sqrt{2} \\ 0 & 0 & 0 & -2 & \sqrt{2} \\ 0 & -\sqrt{3} & 0 & 1 & \sqrt{2} \end{pmatrix}$$

gives

$$l^{-1}Al = \tfrac{1}{4}\begin{pmatrix} 3 & 0 & 0 & 0 & 0 \\ 0 & 3 & 0 & 0 & 0 \\ 0 & 0 & 3 & 0 & 0 \\ 0 & 0 & 0 & 3 & 0 \\ 0 & 0 & 0 & 0 & 15 \end{pmatrix}, \quad l^{-1}Bl = \begin{pmatrix} 2 & 0 & 0 & 0 & 0 \\ 0 & 2 & 0 & 0 & 0 \\ 0 & 0 & 6 & 0 & 0 \\ 0 & 0 & 0 & 6 & 0 \\ 0 & 0 & 0 & 0 & 0 \end{pmatrix},$$

$$l^{-1}Cl = \tfrac{1}{2}\begin{pmatrix} -1 & \sqrt{2} & & & \\ \sqrt{2} & 0 & & 0 & & 0 \\ \hline & & -1 & \sqrt{6} & \\ & 0 & \sqrt{6} & 0 & & 0 \\ \hline & 0 & & 0 & & 0 \end{pmatrix}.$$

The first two matrices are in diagonal form, the third in diagonal block form. The separate blocks can now be transformed to diagonal form without disturbing $l^{-1}Al$ and $l^{-1}Bl$. This further transformation m, where

$$m = \frac{1}{\sqrt{15}} \begin{pmatrix} \sqrt{5} & -\sqrt{10} & 0 & 0 & 0 \\ \sqrt{10} & \sqrt{5} & & & \\ \hline & & \sqrt{6} & -3 & \\ 0 & & 3 & \sqrt{6} & 0 \\ \hline 0 & & 0 & & \sqrt{15} \end{pmatrix},$$

gives

$$(lm)^{-1}C(lm) = \tfrac{1}{2}\begin{pmatrix} 1 & 0 & 0 & 0 & 0 \\ 0 & -2 & 0 & 0 & 0 \\ 0 & 0 & 2 & 0 & 0 \\ 0 & 0 & 0 & -3 & 0 \\ 0 & 0 & 0 & 0 & 0 \end{pmatrix}.$$

4·07. Minors of the matrix of cofactors: Jacobi's theorem. Let a be a matrix of the nth order; denote its determinant by D. If M is a minor of order k, then the $(n-k)$-rowed minor obtained by striking out from D all the rows and columns represented in M is called the *complement* of M. In particular, if M is a single element a_{ik} its complementary minor is the corresponding first minor, but in this case it is often more convenient to use the *signed minor* or *cofactor* A_{ik}. Similarly, it is convenient to define the *signed complement* of M as follows. If M is an r-rowed minor of D, in which the rows $i_1, i_2, \ldots, i_r$ are represented, and the columns $k_1, k_2, \ldots, k_r$, the signed complement of M is defined by

Signed complement of $M = (-1)^{i_1+i_2+\cdots+i_r+k_1+k_2+\cdots+k_r}$ (complement of M).

For a principal minor $i_1 = k_1, \ldots, i_r = k_r$, and the signed complement is the same as the complement. The signed complement of the determinant itself is defined to be 1.

The matrix whose elements are A_{ik} will be called the matrix of cofactors. The adjugate matrix (4·01 (21)) is its transpose. Consider the determinant

$$\Delta = \begin{vmatrix} A_{11} & A_{12} & \ldots & A_{1n} \\ A_{21} & A_{22} & \ldots & A_{2n} \\ \multicolumn{4}{c}{\dotfill} \\ A_{n1} & \ldots & \ldots & A_{nn} \end{vmatrix}. \tag{1}$$

Then we can prove the following theorem about the corresponding minors of D and Δ. If M and M' are corresponding r-rowed minors of D and Δ, then

$$M' = D^{r-1} \text{ (signed complement of } M). \tag{2}$$

In particular, if $r = n$, $$\Delta = D^{n-1}. \tag{3}$$

If $r = n-1$, then if α_{ik} is the cofactor of A_{ik} in Δ, we have

$$\alpha_{ik} = D^{n-2} a_{ik}. \tag{4}$$

If $r = 2$, suppose that M and M' are obtained by striking out all but the i and m rows and the k and p columns. Then

$$\begin{vmatrix} A_{ik} & A_{ip} \\ A_{mk} & A_{mp} \end{vmatrix} = D \text{ (signed complement of } M\text{).} \tag{5}$$

We prove the general theorem first for the special case where M and M' are leading minors and $D \neq 0$. Then we can write

$$M' = \begin{vmatrix} A_{11} & A_{21} & \cdots & A_{r1} & 0 & 0 & \cdots & 0 \\ A_{12} & \cdots & \cdots & A_{r2} & 0 & 0 & \cdots & 0 \\ \cdots & & & & & & & \\ A_{1r} & \cdots & \cdots & A_{rr} & 0 & 0 & \cdots & 0 \\ A_{1,r+1} & \cdots & \cdots & A_{r,r+1} & 1 & 0 & \cdots & 0 \\ \cdots & \cdots & \cdots & \cdots & \cdots & 1 & \cdots & \cdots \\ \cdots & & & & & & & \\ A_{1,n} & \cdots & \cdots & A_{r,n} & 0 & 0 & \cdots & 1 \end{vmatrix}. \tag{6}$$

Multiplying by

$$D = \begin{vmatrix} a_{11} & a_{12} & \cdots & a_{1n} \\ a_{21} & \cdots & \cdots & a_{2n} \\ \cdots & & & \\ a_{n1} & \cdots & \cdots & a_{nn} \end{vmatrix}, \tag{7}$$

we have

$$DM' = \begin{vmatrix} D & 0 & \cdots & 0 & a_{1,r+1} & \cdots & a_{1,n} \\ 0 & D & \cdots & 0 & \cdots & \cdots & \cdots \\ \cdots & \cdots & \cdots & D & a_{r,r+1} & \cdots & a_{r,n} \\ 0 & \cdots & \cdots & 0 & a_{r+1,r+1} & \cdots & \cdots \\ \cdots & & & & & & \\ 0 & \cdots & \cdots & 0 & a_{n,r+1} & \cdots & a_{nn} \end{vmatrix}. \tag{8}$$

$$= D^r \times \text{complement of } M \text{ in } D,$$

which proves the theorem. It should be noticed that with our definition of the complement of D itself this proof holds for $r = n$.

If $D = 0$ it follows from 4·05 (7) that the minors of the elements of any two rows (or columns) of D are proportional and therefore $M' = 0$ for $r \geq 2$, and the theorem still holds. If $D = 0$ and $r = 1$, D^{r-1} is undefined, but $M' = A_{11} = $ complement of M in D.

The proof for the general case when M is not a leading minor is done by rearranging both determinants so as to bring M into the leading position and studying the changes of sign involved. Details of the proof will be found in M. Bôcher, *Introduction to Higher Algebra*, p. 32.

4·08. Quadratic and hermitian forms. A function of n quantities x_i of the form $a_{ik}x_ix_k$ with a_{ik} real is called a *quadratic form*. If y_i are another set of quantities $a_{ik}x_iy_k$ is called a *bilinear form*. In matrix notation they may be written $\tilde{x}ax$ and $\tilde{x}ay$. a_{ik} can be taken to be a symmetrical matrix in the former case. For if $a_{12}x_1x_2$ is one term, $a_{21}x_2x_1$

is another, and their sum is unaltered if we replace both a_{12} and a_{21} by $\frac{1}{2}(a_{12}+a_{21})$; and similarly for any pair of values of i and k.

By a change of variables a quadratic form can always be reduced to a sum of squares (not necessarily with positive coefficients). A simple and very useful method is to take

$$\xi_1 = x_1 + \frac{a_{12}}{a_{11}} x_2 + \ldots + \frac{a_{1n}}{a_{11}} x_n \qquad (1)$$

as a new variable. Then if we subtract $a_{11}\xi_1^2$ from $a_{ik}x_ix_k$ all terms containing x_1 cancel. Then by introducing a new variable ξ_2 we can similarly remove all terms in x_2; and in general we shall reduce the quadratic to the sum of n squares. Then

$$a_{11}x_1^2 + 2a_{12}x_1x_2 + a_{22}x_2^2 + \ldots \equiv a_{11}\xi_1^2 + \beta_2\xi_2^2 + \ldots + \beta_n\xi_n^2, \qquad (2)$$

where each ξ_r starts with x_r. The method fails if the form contains no square terms originally, for then there is no starting-point; but a simple change of variable will introduce them. Thus if $a_{ik} = 0$ when $i = k$, but $a_{12} \neq 0$, put $x_2 = x_1 + \xi_2$; then $x_1x_2 = x_1^2 + x_1\xi_2$, and we can proceed.

The features to be noticed* are

(1) The product of the coefficients on the right of (2), up to β_r, is the determinant of the coefficients on the left up to that of x_r^2.

(2) The reduction to sums of squares can be done in an infinite number of ways, of which this is only one; but however it is done the numbers of positive, negative, and zero coefficients are the same so long as the transformation is non-singular; that is, if the old variables can be expressed uniquely in terms of the new ones and vice versa.

(3) $a_{ik}x_ix_k$ is called a *positive definite* form if it is in general positive and can be zero only if all the x_i are zero; and a set of necessary and sufficient conditions for it to be positive definite is

$$a_{11} > 0, \quad \begin{vmatrix} a_{11} & a_{12} \\ a_{21} & a_{22} \end{vmatrix} > 0, \quad \begin{vmatrix} a_{11} & a_{12} & a_{13} \\ a_{21} & a_{22} & a_{23} \\ a_{31} & a_{32} & a_{33} \end{vmatrix} > 0, \quad \ldots, \quad \| a_{ik} \| > 0. \qquad (3)$$

A quadratic form can be essentially positive without being positive definite; for instance, $(x-2y)^2 = 0$ for $x = 2$, $y = 1$. For such a form $\| a_{ik} \| = 0$.

$a_{ik}x_ix_k$ is *negative definite* if $-a_{ik}x_ix_k$ is positive definite

(4) If the matrix a_{ik} is of rank r, the form is reducible to r squares and is said to be of rank r.

If a_{ik} is a hermitian matrix, $a_{ik}x_ix_k^*$ is called a *hermitian form*. It is real; for if we take a *particular* term $a_{ik}x_ix_k^*$, another particular term is got by interchanging the values of i and k and is therefore $a_{ki}x_kx_i^*$. But $a_{ki} = a_{ik}^*$ for a hermitian matrix, and therefore these two terms are conjugate complexes and their sum is real. Similarly $a_{ik}x_iy_k^*$ is called a *hermitian bilinear form*. The theory of hermitian forms is almost identical with that of quadratic forms, the only change being that in place of (2) we shall have

$$a_{ik}x_ix_k^* = a_{11}\xi_1\xi_1^* + \beta_2\xi_2\xi_2^* + \ldots + \beta_n\xi_n\xi_n^*, \qquad (4)$$

* Proofs will be found in the standard books, e.g. W. L. Ferrar's *Algebra*, Chapters 10 and 11.

where the coefficients on the right are all real; and the form will be positive definite under the same conditions as (3). The determinants in (3) are all real; for they are unaltered by transposing rows and columns; but this replaces i by $-i$ everywhere, and the imaginary part would be reversed if any determinant was complex.

4·081. Reduction of pair of hermitian forms. Any pair of real quadratic forms $a_{ik}x_ix_k$, $b_{ik}x_ix_k$ can in general be reduced simultaneously to sums of squares by transformation of variables; this is always true if one of the forms is positive definite. Any pair of hermitian forms can similarly be reduced simultaneously to the form (4) under the same conditions. The analysis for the two cases is practically the same; we shall take the hermitian case since it includes the other, and has applications in quantum theory. The case where a_{ik} is symmetrical presents itself in almost every branch of physics. The simplest is the one we have already discussed, the reduction of a second-order symmetrical tensor, with $n = 3$, to principal axes, which amounts to taking $a_{ik} = \delta_{ik}$. But without this condition we could reduce two forms in three variables to sums of squares simultaneously by a linear change of variables; geometrically this implies that any two concentric quadric surfaces have a set of three mutually conjugate diameters in common. The case of n variables arises in the theory of small oscillations in dynamics, where it is clear that if we can reduce the kinetic and potential energies simultaneously to the forms

$$2T = m_1\dot\xi_1^2 + m_2\dot\xi_2^2 + \ldots + m_n\dot\xi_n^2,$$
$$2V = m_1\lambda_1\xi_1^2 + m_2\lambda_2\xi_2^2 + \ldots + m_n\lambda_n\xi_n^2,$$

each variable will satisfy a differential equation

$$\ddot\xi_r = -\lambda_r\xi_r,$$

and their variations with time are therefore independent. The condition for stability is therefore that all the λ_r shall be positive. Since T is positive definite all the m_r are positive, and therefore V also must be positive definite for stability.

It is convenient to start with the set of n equations

$$\lambda a_{ik}x_k = b_{ik}x_k. \tag{1}$$

Then the general method of solution leads to the equation for λ

$$\| \lambda a_{ik} - b_{ik} \| = 0. \tag{2}$$

For a general real λ, $\| \lambda a_{ik} - b_{ik} \|$ is hermitian and therefore real. Hence all the coefficients of λ in its expansion in powers of λ are real. We shall assume that $a_{ik}x_ix_k^*$ is positive definite.

Let λ_1 and λ_2 be two roots of (2), and let x_{i1}, x_{i2} be corresponding non-zero values of x_i. Then

$$\lambda_1 a_{ik}x_{k1} = b_{ik}x_{k1}, \tag{3}$$
$$\lambda_2 a_{ik}x_{k2} = b_{ik}x_{k2}. \tag{4}$$

Multiply (3) by x_{i2}^* and add; multiply (4) by x_{i1}^* and add. Then

$$\lambda_1 a_{ik}x_{k1}x_{i2}^* = b_{ik}x_{k1}x_{i2}^*, \tag{5}$$
$$\lambda_2 a_{ik}x_{k2}x_{i1}^* = b_{ik}x_{k2}x_{i1}^*. \tag{6}$$

Taking the conjugate of (5) we have

$$\lambda_1^* a_{ik}^* x_{k1}^* x_{i2} = b_{ik}^* x_{k1}^* x_{i2}. \qquad (7)$$

But
$$a_{ik}^* x_{k1}^* x_{i2} = a_{ki} x_{i2} x_{k1}^* = a_{ik} x_{k2} x_{i1}^* \qquad (8)$$

by interchange of dummy suffixes. Similarly

$$b_{ik}^* x_{k1}^* x_{i2} = b_{ik} x_{k2} x_{i1}^* \qquad (9)$$

and therefore, by comparing (8) and (9) with (6),

$$(\lambda_2 - \lambda_1^*) a_{ik} x_{k2} x_{i1}^* = 0. \qquad (10)$$

First let the two solutions compared be the same; then $a_{ik} x_{k1} x_{i1}^*$ is real. If it was zero $b_{ik} x_{k1} x_{i1}^*$ would also be zero by (5). This is impossible if either of these expressions is a positive definite form. It then follows that $\lambda_1 = \lambda_1^*$ and therefore λ_1 is real. Hence all the roots of (2) are real, and for real forms the ratios of the x_{i1} are also real.

Next let λ_1 and λ_2 be different. Then since they are real $\lambda_2 \neq \lambda_1^*$, and therefore

$$a_{ik} x_{k2} x_{i1}^* = 0, \quad b_{ik} x_{k2} x_{i1}^* = 0. \qquad (11)$$

We shall call this property of the solutions orthogonality with respect to a and b and denote it by "orthogonal a".

At each simple root of (2) the determinant has a non-zero derivative with regard to λ. But this derivative is a linear function of first minors of the determinant, and therefore not all of these can be zero. It follows that at a simple root of (2) the rank of the matrix

$$c_{ik} = \lambda a_{ik} - b_{ik}$$

is $n-1$. At an r-fold root the rth derivative is not zero and is a linear function of rth minors, and therefore the rank of c_{ik} is at least $n-r$. We shall show that it is actually equal to $n-r$, but this requires the rather intricate argument of the theorem of the separation of the roots.

We first take the case where all the roots are simple. Then for any root, say λ_1, the ratios of the x_i are unique and we can write

$$x_{i1} = l_{i1} \xi_1. \qquad (12)$$

Evidently
$$\lambda_1 a_{ik} l_{k1} = b_{ik} l_{k1}, \qquad (13)$$

$$\lambda_1 a_{ik} l_{k1} l_{i1}^* = b_{ik} l_{k1} l_{i1}^*. \qquad (14)$$

$a_{ik} l_{k1} l_{i1}^*$ is real and not zero. Also if λ_1 and λ_2 are different, by (11)

$$a_{ik} l_{k1} l_{i2}^* = 0, \quad b_{ik} l_{k1} l_{i2}^* = 0. \qquad (15)$$

If we denote the various solutions by the suffix j, we can regard l_{ij} as a transforming matrix:

$$l_{ij}^* a_{ik} l_{kl} = (l^\dagger a l)_{jl}. \qquad (16)$$

It follows from (15) that all non-diagonal components of $l^\dagger a l$ and $l^\dagger b l$ are zero; hence this transformation reduces both matrices to diagonal form. If we denote the diagonal terms by A_{jl} and B_{jl} ($l = j$), B_{jl}/A_{jl} is equal to the corresponding value of λ, by (14). Again,

$$\| l_{ji}^\dagger \| \, \| a_{ik} \| \, \| l_{kl} \| = \| l_{ji}^\dagger a_{ik} l_{kl} \|, \qquad (17)$$

which is the product of diagonal components, none of which is zero. Hence l and $l^\dagger$ are non-singular matrices, and if we now write in general

$$X_i = l_{ij}\xi_j, \tag{18}$$

it will be possible to solve and find the ξ_j for any assigned values of the X_i. Then

$$a_{ik}X_kX_i^* = a_{ik}l_{kl}\xi_l l_{ij}^*\xi_j^*$$
$$= A_{jl}\xi_l\xi_j^*, \tag{19}$$
$$b_{ik}X_kX_i^* = B_{jl}\xi_l\xi_j^*. \tag{20}$$

Thus all terms with $j \neq l$ are absent from both forms.

An important special case is where $a_{ik} = \delta_{ik}$, so that the original equations reduce to

$$\lambda x_i = b_{ik}x_k.$$

In this case $\quad a_{ik}X_iX_k^* = |X_i|^2 + \ldots + |X_n^2|,$

and is a positive definite form, and

$$A_{jl} = l_{ji}^\dagger \delta_{ik} l_{kl} = l_{ji}^\dagger l_{il}$$

must be in diagonal form. But for any solution λ_j the l_{ij} are arbitrary by a constant factor. Choose this so that

$$l_{ij}^* l_{il} = 1 \quad (j = l).$$

In any case $l_{ij}^* l_{il} = 0$ $(j \neq l)$; hence

$$A_{jl} = \delta_{jl}$$

and the transformation does not alter the unit matrix; hence l is unitary. Hence any hermitian matrix b with no repeated root can be reduced to diagonal form by a transformation $l^\dagger b l$, where l is unitary; and the transformation $X = l\xi$ will reduce $b_{ik}X_iX_k^*$ to $\Sigma\lambda_j X_j X_j^*$. If b is a real symmetrical matrix l will be a real unitary matrix, that is, an orthogonal matrix.

4·082. The theorem of the separation of the roots. This resembles Sturm's theorem in the theory of equations. We still take a, b hermitian $n \times n$, and $x^\dagger a x$ positive definite, and write

$$\lambda a_{ik} - b_{ik} = c_{ik}, \quad \lambda a - b = c. \tag{1}$$

We denote $\|c_{ik}\|$ by D. We have already seen that when λ is any root of $D = 0$, D cannot be of higher rank than $n-1$, and that if λ is an r-fold root D cannot be of lower rank than $n-r$. We propose to show that the rank is actually equal to $n-r$.

Let the leading minors of D of orders n, $n-1$, ..., 1, 0 be denoted by

$$D_n(=D), \quad D_{n-1}, \quad \ldots, \quad D_1(=c_{11}), \quad D_0(=1).$$

We assume first that D has no repeated root and that no two consecutive D_r, D_{r-1} vanish for the same value of λ. We consider for any value of λ the changes of sign as we go along the finite sequence D_n to D_0.

From Jacobi's theorem we have, if C_{ik} is the cofactor of c_{ik} in D,

$$DD_{n-2} = \begin{vmatrix} C_{n-1,n-1} & C_{n-1,n} \\ C_{n,n-1} & C_{nn} \end{vmatrix} = D_{n-1}C_{n-1,n-1} - |C_{n-1,n}|^2. \qquad (2)$$

Similarly, if C_{ik}^s denotes the cofactor of c_{ik} in D_s,

$$D_s D_{s-2} = D_{s-1} C_{s-1,s-1}^s - |C_{s-1,s}^s|^2. \qquad (3)$$

If λ is a root of D_{s-1}, then since neither D_s nor D_{s-2} is zero for that value of λ, $D_s D_{s-2}$ is not zero, and must be negative. Hence when $D_{s-1} = 0$, D_s and D_{s-2} have opposite signs. Now since $a_{ik} x_i x_k^*$ is positive definite we see that as $\lambda \to +\infty$ the signs of the D_s are all positive. When $\lambda \to -\infty$ they are alternately positive and negative. Hence as λ increases from $-\infty$ to $+\infty$, n changes of sign are lost in the sequence. We can see, however, that the number of changes of sign is unaffected as λ varies, except by changes of sign of D itself. For if λ increases through a zero of D_{s-1}, D_s and D_{s-2} have opposite signs, and in any case there is one change of sign in the sequence D_s, D_{s-1}, D_{s-2}, and there is no alteration in the total number of changes of sign along the sequence D_n to D_0. But $D_0 = 1$ and never changes sign; therefore all the n changes of sign lost are lost at the beginning, that is, by λ passing through n real zeros of D.

The figure shows how the roots of the sequence of equations $D_n = 0$, $D_{n-1} = 0$, ..., must lie relative to one another. The graph of D_0 lies parallel to the λ axis. That of D_1 is a straight line passing to $+\infty$ at $\lambda = +\infty$ and to $-\infty$ at $\lambda = -\infty$. $D_1 = 0$ has one real root.

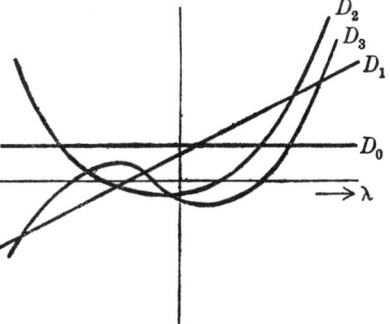

$D_2 \to +\infty$ as $\lambda \to \pm\infty$, and when $D_1 = 0$ it is negative. Therefore $D_2 = 0$ has two real roots, one less and one greater than the root of $D_1 = 0$.

$D_3 \to -\infty$ as $\lambda \to -\infty$ and to $+\infty$ as $\lambda \to +\infty$. When $D_2 = 0$ and $D_1 < 0$, D_3 must be > 0; when $D_2 = 0$ and $D_1 > 0$, $D_3 < 0$. Therefore the roots of $D_2 = 0$ separate those of $D_3 = 0$. Similarly, those of $D_s = 0$ separate those of $D_{s+1} = 0$. This proves the theorem of the separation of the roots when those of $D_n = 0$ are all different.

If D_n has an r-fold root, $\lambda = \lambda_p$, by making small changes in the b_{ik}, to b'_{ik} say, we can alter the equation so that all roots become distinct. If D_s and D_{s-1} vanish for the same λ, it will be possible to alter D_s by changing, say, b_{1s} without altering D_{s-1}. Hence if the conditions postulated are not satisfied we can make small alterations so that they are satisfied. Consider the group of r roots that coalesce at λ_p when $b'_{ik} \to b_{ik}$. Between them there are $r-1$ zeros of D_{n-1}, $r-2$ of D_{n-2}, ... 1 of D_{n-r+1}, none of D_{n-r}.

Since we could arrange the determinant with any diagonal term in the top left corner by an interchange of two rows followed by one of two columns, without disturbing its hermitian form, it follows that if λ_p is an r-fold root of $D = 0$, *all* the principal minors of orders from n down to $n-r+1$ vanish for $\lambda = \lambda_p$. The minors of order $n-r$ cannot all vanish because then λ_p would be an $(r+1)$-fold root of $D = 0$.

We have seen that if a determinant and a first minor vanish, then the first minors either of all elements in the same row or of all in the same column vanish. Consider then any principal minor of order $n-r+2$. This and any first principal minor in it vanish for $\lambda = \lambda_p$, the latter being a principal minor of order $n-r+1$ of D. Let the row and column omitted be the ith. Then either all the minors of c_{ik} or of c_{ki} vanish. But these are conjugate complexes, and if one set do the others do. Hence all minors of order $n-r+1$ of D vanish, and when $\lambda = \lambda_p$, c_{ik} is a matrix of rank $n-r$, as was to be proved.

Again, $\lambda = \lambda_p$ is at least a simple zero of all minors of order $n-r+1$ and higher orders. But the derivative of any minor is a linear function of those of the next lower order. Hence the minors of order $n-r+2$ have double zeros and those of order $n-1$, that is, the first minors of D, have at least $(r-1)$-fold zeros. We shall meet this result again when we come to the operational treatment of small oscillations.†

4·083. Extension of the orthogonal property. We have proved this property for the l_{ij} associated with any pair of different roots of the determinantal equation. If $\lambda = \lambda_j$ is an r-fold root, then since the matrix c_{ik} is of rank $n-r$ for $\lambda = \lambda_j$ we can choose r linearly independent sets of ratios of the l_{ij}. Let two of these be l_{i1} and l_{i2}. Then

$$\lambda_j a_{ik} l_{k1} = b_{ik} l_{k1}, \quad \lambda_j a_{ik} l_{k2} = b_{ik} l_{k2}, \tag{1}$$

and therefore $\qquad \lambda_j a_{ik} l_{k2} l_{i1}^* = b_{ik} l_{k2} l_{i1}^*. \tag{2}$

These expressions may be zero; if so, the solutions are already orthogonal a. If not, consider

$$m_{i2} = l_{i2} - \theta l_{i1}, \tag{3}$$

where θ is independent of i. The m_{i2} are not all zero, since l_{i1} and l_{i2} are not proportional. Then for any θ

$$\lambda_j a_{ik} m_{k2} = b_{ik} m_{k2}, \tag{4}$$

$$a_{ik} m_{k2} l_{i1}^* = a_{ik} l_{k2} l_{i1}^* - \theta a_{ik} l_{k1} l_{i1}^*, \tag{5}$$

which can be made zero by a suitable choice of θ; and then it follows from (2) that $b_{ik} m_{k2} l_{i1}^*$ is also zero.

If λ_j is an r-fold root we can choose one set of l_{ij} for it and then make all the others orthogonal a to that set by subtracting suitable multiples of it. Choosing one of the modified sets we can make all the others orthogonal a to that one, and proceed until we have made all the r sets mutually orthogonal a. Thus even if the equation for λ has multiple roots we can still choose n sets of l_{ij} with the orthogonality property, and the reduction of a_{ik} and b_{ik} to diagonal form can still be carried out (in fact in infinitely many ways). If $a_{ik} = \delta_{ik}$, l can be taken unitary as before.

If the normal coordinate corresponding to a particular solution is required we can proceed as follows. Let $x_i = l_{i1} \xi_1$ be the solution. Take a general set of x_i and determine ξ_1 so as to make $a_{ik}(x_k - l_{k1}\xi_1)(x_i^* - l_{i1}^*\xi_1^*)$ stationary for variations of ξ_1. Both real and imaginary parts may be varied, but the derivatives with regard to them can both vanish if and only if the derivatives with regard to ξ_1 and ξ_1^*, treated as independent, vanish. Hence

$$l_{i1}^* a_{ik}(x_k - l_{k1}\xi_1) = 0, \quad l_{k1} a_{ik}(x_i^* - l_{i1}^*\xi_1^*) = 0,$$

† First proved by Weierstrass, *Monatsber. d. K. Akad. d. Wiss.* Berlin, 1858; *Werke*, 1, 233–46. Several other proofs exist; one is in Lamb's *Higher Mechanics*, 1920, 222–6; another in Bromwich's *Quadratic Forms*, 1906. The last is specially interesting because it proceeds by discovering normal coordinates directly and reducing the quadratic forms to sums of squares by successive substitution. We have followed Routh's method because it is the most easily adapted to gyroscopic systems. See 4·091, 8·09.

the second of which reduces to the first on interchanging i and k and taking the conjugate. Hence
$$A_{11}\xi_1 = l_{i1}^* a_{ik} x_k.$$

If $x_k = l_{k2}\xi_2$, say, $l_{i1}^* a_{ik} x_k = 0$ by the orthogonal property; hence the ξ that is not zero in any characteristic solution can be found explicitly in terms of the x_i without the need to find all the solutions.

As an example, take
$$a = \begin{pmatrix} 3 & -1 & 1 \\ -1 & 3 & 1 \\ 1 & 1 & 9 \end{pmatrix}, \quad b = \begin{pmatrix} 1 & 3 & 12 \\ 3 & 1 & -8 \\ 12 & -8 & -2 \end{pmatrix}.$$

$\tilde{x}ax$ is positive definite and the roots of $\det(b - \lambda a) = 0$ are $2, 2, -3$.

The three equations for $\lambda = 2$ all reduce to
$$x_1 - x_2 - 2x_3 = 0,$$
and we may take solutions to be $(1, 1, 0)$, $(0, -2, 1)$. These give
$$\tilde{l}_1 a l_1 = 4, \quad \tilde{l}_1 a l_2 = -2;$$
hence a solution normal to the first is $\tilde{m}_2 = 2(l_2 + \tfrac{1}{2} l_1) = (1, -3, 2)$. The solution for $\lambda = -3$ is $\tilde{l}_3 = (-3, 1, 2)$.

Then we may write
$$x_1 = \xi_1 + \xi_2 - 3\xi_3, \quad x_2 = \xi_1 - 3\xi_2 + \xi_3, \quad x_3 = 2\xi_2 + 2\xi_3.$$

Also $\qquad \tilde{m}_2 a m_2 = 64, \quad \tilde{l}_3 a l_3 = 64,$

hence $\qquad \tilde{x} a x = 4\xi_1^2 + 64\xi_2^2 + 64\xi_3^2, \quad \tilde{x} b x = 8\xi_1^2 + 128\xi_2^2 - 192\xi_3^2.$

To illustrate what can happen if neither matrix is that of a positive definite form, we take the forms $\tfrac{1}{2}(x_1^2 - x_2^2)$, $x_1 x_2$. The equations required are
$$\lambda x_1 = x_2, \quad -\lambda x_2 = x_1,$$
and $\lambda^2 = -1$, $\lambda = \pm i$. We can therefore take
$$x_1 = \xi_1 + \xi_2, \quad x_2 = i(\xi_1 - \xi_2);$$
and
$$\tfrac{1}{2}(x_1^2 - x_2^2) = \xi_1^2 + \xi_2^2, \quad xy = i(\xi_1^2 - \xi_2^2).$$
Thus the forms are reduced to sums of squares, but not with real coefficients. Here $a_{ik} l_{k1} l_{i1}^* = 0$.

Since a hermitian matrix can always be reduced to diagonal form by a transformation $l^{-1}al$, where l is unitary, the condition required in 4·065 that the matrices shall be reducible to diagonal form by a collineatory transformation is always satisfied when the matrices are hermitian.

The peculiarity in the case of equal roots may be illustrated geometrically by the problem of finding principal axes of an ellipsoid of revolution. Any diameter in the plane of circular symmetry is a principal axis. If we arbitrarily choose two directions l_1, l_2 in this plane they will not in general be perpendicular; but we can use them to form a coplanar direction $l_2 - \theta l_1$ perpendicular to l_1.

4·084. The stationary property of λ: Rayleigh's principle. For any set of x_i we can define a quantity λ by
$$\lambda = \frac{b_{ik} x_k x_i^*}{a_{ik} x_k x_i^*} = \frac{\Sigma \lambda_j A_{jj} \xi_j \xi_j^*}{\Sigma A_{jj} \xi_j \xi_j^*}. \tag{1}$$

This always lies between the greatest and least of the λ_j since all $A_{jj} > 0$. If all the ξ_j are zero except one, the ratio reduces to the corresponding λ_j. If the others are small the change in λ is of the second order. Hence the ratio is stationary for all variations of the ξ_j and therefore of the x_i. This is **Rayleigh's principle**.

Conversely, the ratio has no other stationary values. For we have

$$\lambda a_{ik} x_k x_i^* = b_{ik} x_k x_i^*, \tag{2}$$

and if $\delta\lambda$ is of the second order we have to the first order for all changes of the x_i, both real and imaginary changes being permitted,

$$\lambda \delta(a_{ik} x_k x_i^*) = \delta(b_{ik} x_k x_i^*), \tag{3}$$

whence $\quad \lambda a_{ik} x_k = b_{ik} x_k, \quad \lambda a_{ik} x_i^* = b_{ik} x_i^*, \tag{4}$

the second of which reduces to the first on interchanging i and k and taking the conjugate complex.

The principle is often useful in numerical work. Without actually forming and solving the determinantal equation, we may be able to see roughly what ratios the x_i must have in a particular solution, usually the one with the smallest λ. If we substitute rough values for these ratios in (1) and work out λ, it will be correct within a second-order error. This may be accurate enough or it may serve as a starting point for a closer approximation to the ratios of the x_i. If the λ_j all have the same sign, the approximate value taken by λ with any ratios of the x_i is numerically greater than the smallest of them. Direct use of the principle therefore never underestimates the smallest root.†

4·09. Small oscillations. If the changes of the coordinates in a dynamical system from their values in a position of equilibrium are $x_i \ldots x_n$, the kinetic energy can be written $T = \frac{1}{2} a_{ik} \dot{x}_i \dot{x}_k$, and is a positive definite form. The change of the work function‡ can be written as $W = \frac{1}{2} b_{ik} x_i x_k$. Then Lagrange's equations of motion are

$$a_{ik} \ddot{x}_k = b_{ik} x_k. \tag{1}$$

Assume a solution of the form $\quad x_i \propto e^{\gamma t}, \tag{2}$

where γ is constant. Then $\quad \gamma^2 a_{ik} x_k = b_{ik} x_k, \tag{3}$

and we have a set of equations of the form already studied, γ^2 replacing λ. It follows that all the values of γ^2 are real. If any of them is positive the system is unstable; the condition of stability is that they shall all be negative. We wish to see what this implies about T and W. Referring back to 4·082 we see that the condition is that all the changes of sign in the sequence of determinants shall be lost as γ^2 increases from $-\infty$ to 0. Therefore the sequence

$$\| -b_{ik} \| \ldots \begin{vmatrix} -b_{11} & -b_{12} & -b_{13} \\ -b_{21} & -b_{22} & -b_{23} \\ -b_{31} & -b_{32} & -b_{33} \end{vmatrix}, \quad \begin{vmatrix} -b_{11} & -b_{12} \\ -b_{21} & -b_{22} \end{vmatrix}, \quad -b_{11}, \tag{4}$$

are all positive; and this is the condition that W is a negative definite form, or alternatively that the potential energy $V = -W$ is a positive definite form.

† A full account is given by G. Temple and W. G. Bickley, *Rayleigh's Principle*, Oxford, 1933.
‡ In most dynamical problems the work function W is slightly more convenient than the potential energy $V = -W$. The exceptions are when W is obviously a negative definite form and therefore V a positive one.

Evidently if we make a substitution
$$x_i = l_{ij}\xi_j \tag{5}$$
that reduces $a_{ik}x_ix_k$ to a sum of squares of the ξ_j it will also reduce $a_{ik}\dot x_i\dot x_k$ to sums of squares of the ξ_j. Our general argument then shows that T and W can be expressed simultaneously as sums of squares (the latter with all the coefficients negative in a stable system) even if the determinantal equation has multiple roots. Then Lagrange's method gives independent differential equations for the ξ_j with regard to the time, and they therefore all vary independently. They are known as the *normal coordinates* of the system. In practice this method of obtaining the solution is laborious, and the operational method to be described later is much easier if anything more than the free periods is required. The present method, however, explains a feature always found in the operational method when the period equation has equal roots. At present we need only give an illustration of what happens to the normal coordinates in that case. Consider a pendulum free to vibrate in any horizontal direction. If the horizontal rectangular coordinates are x_1 and x_2 we have to the second order, in the usual notation,
$$2T = m(\dot x_1^2 + \dot x_2^2), \quad 2W = -\frac{mg}{l}(x_1^2 + x_2^2).$$

Both are already expressed as sums of squares and x_1, x_2 are normal coordinates. But we could put
$$x_1 = \xi_1 \cos\alpha - \xi_2 \sin\alpha, \quad x_2 = \xi_1 \sin\alpha + \xi_2 \cos\alpha,$$
and then
$$2T = m(\dot\xi_1^2 + \dot\xi_2^2), \quad 2W = -\frac{mg}{l}(\xi_1^2 + \xi_2^2),$$
so that ξ_1, ξ_2 are also normal coordinates. In fact the coincidence of the periods implies that we can choose the normal coordinates in an infinite number of ways.

4·091. Small oscillations about steady motion. In these cases, first studied systematically by Routh, the equations of motion take the form
$$a_{ik}\ddot x_k + g_{ik}\dot x_k - b_{ik}x_k = 0, \tag{1}$$
where a_{ik} and b_{ik} are symmetrical but g_{ik} is antisymmetrical. All are real, and $a_{ik}x_ix_k$ is positive definite. A sufficient condition for stability follows if we multiply by $\dot x_i$ and add; the g_{ik} terms cancel, and
$$\frac{d}{dt}(a_{ik}\dot x_i\dot x_k - b_{ik}x_ix_k) = 0. \tag{2}$$

The first term inside the bracket is positive definite, and it follows that if $b_{ik}x_ix_k$ is negative definite it can never exceed numerically a value determined by the initial conditions; hence in these conditions the system is stable. For oscillations about equilibrium this condition is also necessary, but for oscillations about steady motion it is not, otherwise, for instance, a top could not stand up. The method of solution for given systems is straightforward, but some general features will be discussed.

If we assume $x_i \propto e^{\gamma t}$ as before we get the equation of consistency
$$D = \|a_{ik}\gamma^2 + g_{ik}\gamma - b_{ik}\| = 0. \tag{3}$$
If we denote the general element by c_{ik} the matrix c_{ik} is hermitian if γ is purely imaginary. But if γ is real the matrix is real and not symmetrical, so that our previous discussion

needs modification. Further, the general element contains terms of three degrees 0, 1, 2 in γ, so that the theory of small oscillations about steady motion is in two respects more complicated than that of what we may now call simple hermitian matrices, where terms are of degrees 0 and 1 in λ.

Changing γ to $-\gamma$ and then interchanging rows and columns leaves D unaltered; hence the values of γ occur in equal and opposite pairs. But if γ is imaginary, the ratios of the x_i are usually complex and the phases of the coordinates differ in any simple oscillation. No reduction to normal coordinates is possible.

For the stability condition we form the minors $D_n (= D)$, $D_{n-1}, ..., D_0$ as before. We still have
$$D_s D_{s-2} = D_{s-1} C^s_{s-1,s-1} - C^s_{s-1,s} C^s_{s,s-1} \tag{4}$$
If γ is pure imaginary
$$C^s_{s-1,s} C^s_{s,s-1} = |C^s_{s-1,s}|^2 > 0, \tag{5}$$
and again as γ^2 increases from $-\infty$ to 0 changes of sign can be lost only at the beginning. But if γ^2 is positive $C^s_{s-1,s}$ and $C^s_{s,s-1}$ are no longer conjugate complexes and their product need not be positive. The root-separation theorem therefore still holds so long as γ^2 is negative; if then $a_{ik}\dot{x}_i\dot{x}_k$ is a positive definite form, as it must be, and if also $b_{ik}x_i x_k$ is negative definite, it follows exactly as before that there must be n negative real values of γ^2 and the system is stable. This is the case that we derived directly from the equations of motion. But it does not follow that if $b_{ik}x_i x_k$ is not negative definite the system is unstable. For here, unlike the case of oscillations about equilibrium, a change of sign can be *gained* at the beginning of the sequence. This did not arise before because n changes of sign were lost in any case as γ^2 varied from $-\infty$ to $+\infty$, and if any had been gained at any stage they would have had to be compensated by additional losses, so that we should have had an algebraic equation of the nth degree with more than n roots. But changes of sign can be gained in a gyroscopic system. Consequently in a gyroscopic system, with $b_{ik}x_i x_k$ not negative definite, we cannot assert instability without actually investigating the period equation in detail.

Again, if the system is unstable, γ^2 may not be real. For the root-separation theorem fails for positive γ^2, and without it changes of sign may be lost when γ^2 passes through a zero of some intermediate member of the series. Thus there may be a loss of an even number of changes of sign not accounted for by real zeros of D, which must therefore have complex zeros.

Since the root-separation theorem is true for $b_{ik}x_i x_k$ negative definite it still follows that if there is an r-fold zero of D for negative γ^2 the matrix c_{ik} is of rank $n-r$ for that value and all first minors of D have $(r-1)$-fold zeros. But this is not true if $b_{ik}x_i x_k$ is not negative definite.

4·092. These considerations are illustrated by the upright top if we take the coordinates x_1, x_2 to be direction cosines of the axis with respect to two perpendicular horizontal axes. The equations of motion are, in a usual notation,
$$A\ddot{x}_1 + Cn\dot{x}_2 - Mghx_1 = 0, \quad A\ddot{x}_2 - Cn\dot{x}_1 - Mghx_2 = 0, \tag{1}$$
so that
$$b_{ik}x_i x_k = Mgh(x_1^2 + x_2^2), \tag{2}$$
and is *positive* definite.

We put x_1, x_2 proportional to $e^{\gamma t}$; then

$$(A\gamma^2 - Mgh)x_1 = -Cn\gamma x_2, \quad (A\gamma^2 - Mgh)x_2 = Cn\gamma x_1,$$
$$D_2 = (A\gamma^2 - Mgh)^2 + C^2n^2\gamma^2 = 0, \quad A\gamma^2 \pm iCn\gamma - Mgh = 0,$$
$$\gamma = \pm \tfrac{1}{2}\frac{iCn}{A} \pm \frac{i}{2}\left(\frac{C^2n^2}{A^2} - \frac{4Mgh}{A}\right)^{\frac{1}{2}}.$$
(3)

These values are all purely imaginary if $C^2n^2 > 4AMgh$. If $C^2n^2 < 4AMgh$ two roots have positive real parts and the top is unstable; but γ^2 is complex, whereas it would be real and positive for an unstable position of equilibrium. The motion of the end of the axis is an equiangular spiral, to the first order.

If $C^2n^2 = 4AMgh$ the roots become equal in pairs. Taking $\gamma = -\tfrac{1}{2}iCn/A$ we find that the matrix of the coefficients is $2Mgh\begin{pmatrix} -2 & -2i \\ 2i & -2 \end{pmatrix}$, which is of rank 1. Hence there is only one solution of the form $e^{\gamma t}$ with this γ, and similarly with the opposite sign for γ. The other solutions are of the form $te^{\gamma t}$, which do not occur at all in oscillations about equilibrium.

We have $$D_1 = A\gamma^2 - Mgh, \quad D_0 = 1,$$

and the signs run as follows, failing to reveal any zeros of D_2 for negative γ^2:

γ^2	D_2	D_1	D_0
$-\infty$	+	−	+
0	+	−	+
∞	+	+	+

If, however, we discuss the top hanging downwards, $b_{ik}x_ix_k$ is negative definite. We need only reverse the sign of g. In this case there cannot be equal roots; and the signs in the last table run:

γ^2	D_2	D_1	D_0
$-\infty$	+	−	+
0	+	+	+
∞	+	+	+

so that two changes of sign are lost between $-\infty$ and 0.

4·093. The imaginary values of γ have a stationary property for gyroscopic systems. If we take γ so that

$$\gamma^2 = \frac{b_{ik}x_kx_i^* - \gamma g_{ik}x_kx_i^*}{a_{ik}x_kx_i^*},$$

we see that the equations 4·091 (1) are equivalent to the statement that γ is stationary for both real and imaginary variations of the x_i. For any complex x_i, x_k, $g_{ik}x_kx_i^*$ is purely imaginary. We take it not zero. Put for any solution of the form $x_i = l_i e^{\gamma t}$

$$a_{ik}x_kx_i^* = P, \quad g_{ik}x_kx_i^* = iQ, \quad b_{ik}x_kx_i^* = R.$$

Then $2P\gamma = -iQ \pm (-Q^2 + 4RP)^{\frac{1}{2}}$.

If $Q^2 < 4RP$ both roots are complex, and $|\gamma^2| = R/P$. Thus if the system would be unstable in the absence of gyroscopic terms, the real part of γ is less than it would be for the same x_i if the g_{ik} were all 0. But this, by Rayleigh's principle, is less than for the actual values of the x_i in the mode without gyroscopic terms that gives the largest γ. Hence the gyroscopic effects tend to reduce instability.

If $Q^2 > 4RP$ both roots are imaginary and the system is stable. If R is a negative definite form, the system is stable for all Q, and if for oscillations with $g_{ik} = 0$ there are two roots $\pm i\sigma$, the gyroscopic effects will increase the speed of vibration for one and diminish it for the other.

4·10. Roots of unitary and orthogonal matrices. Let a_{ik} be a unitary matrix of n rows and columns. Then

$$a_{ik} a_{im}^* = \delta_{km}. \tag{1}$$

Consider the equations
$$a_{ik} x_k = \lambda x_i. \tag{2}$$

In general the roots of the determinantal equation

$$\| a_{ik} - \lambda \delta_{ik} \| = 0 \tag{3}$$

will be complex. Then the conjugate of (2) gives

$$a_{im}^* x_m^* = \lambda^* x_i^*. \tag{4}$$

Multiply (2) and (4). Then $\quad a_{ik} a_{im}^* x_k x_m^* = \lambda \lambda^* x_i x_i^*. \tag{5}$

But on account of (1), (5) reduces to
$$x_k x_k^* = \lambda \lambda^* x_i x_i^*, \tag{6}$$

and therefore $\quad \lambda \lambda^* = 1. \tag{7}$

Hence all the roots of a unitary matrix have modulus 1.

Now let λ_1 and λ_2 be two different roots of (3) and x_{i1}, x_{i2} corresponding values of the x_i. Then proceeding as before we get

$$a_{ik} a_{im}^* x_{k1} x_{m2}^* = \lambda_1 \lambda_2^* x_{i1} x_{i2}^*, \tag{8}$$

and also $\quad = x_{k1} x_{k2}^*. \tag{9}$

Hence either $\quad \lambda_1 \lambda_2^* = 1 \tag{10}$

or $\quad x_{i1} x_{i2}^* = 0. \tag{11}$

In the former case $\lambda_1 = \lambda_2$, which we have already considered. Hence the solutions have a property of complex orthogonality similar to that possessed by a set of hermitian equations.

A unitary matrix $\boldsymbol{a}$ can always be transformed to diagonal form by a transformation $l^{-1} \boldsymbol{a} l$, where l is unitary.† This holds in particular for a real unitary matrix; but l in general remains complex.

4·101. Consider now the most general 2×2 unitary matrix. As in 4·032 we write

$$l = \begin{pmatrix} a & b \\ c & d \end{pmatrix}. \tag{1}$$

† D. E. Littlewood, *The Theory of Group Characters*, 1940, 16. See also Ex. 10, p. 169.

Then
$$\mathfrak{u}^\dagger = \begin{pmatrix} aa^*+bb^* & ac^*+bd^* \\ ca^*+db^* & cc^*+dd^* \end{pmatrix} = \begin{pmatrix} 1 & 0 \\ 0 & 1 \end{pmatrix}, \tag{2}$$

and we can take
$$\left.\begin{aligned} a &= \cos\alpha\, e^{i\beta}, & b &= -\sin\alpha\, e^{i\gamma}, \\ a^* &= \cos\alpha\, e^{-i\beta}, & b^* &= -\sin\alpha\, e^{-i\gamma}, \end{aligned}\right\} \tag{3}$$

with α, β, γ real. Then
$$\left.\begin{aligned} \frac{c}{d} &= -\frac{b^*}{a^*} = \tan\alpha\, e^{i(\beta-\gamma)}, \\ \frac{c^*}{d^*} &= -\frac{b}{a} = \tan\alpha\, e^{-i(\beta-\gamma)}, \end{aligned}\right\} \tag{4}$$

and these are equivalent. Also
$$\frac{cc^*}{dd^*} = \tan^2\alpha, \quad dd^* = \cos^2\alpha, \quad cc^* = \sin^2\alpha, \tag{5}$$

and all the conditions are satisfied if
$$d = \cos\alpha\, e^{i\delta}, \quad c = \sin\alpha\, e^{i(\beta-\gamma+\delta)}. \tag{6}$$

Take
$$\beta+\delta = 2\epsilon, \quad \beta-\delta = 2\eta. \tag{7}$$

Then
$$\begin{aligned} \mathfrak{l} &= \begin{pmatrix} \cos\alpha\, e^{i(\epsilon+\eta)} & -\sin\alpha\, e^{i\gamma} \\ \sin\alpha\, e^{i(2\epsilon-\gamma)} & \cos\alpha\, e^{i(\epsilon-\eta)} \end{pmatrix} = e^{i\epsilon}\begin{pmatrix} \cos\alpha\, e^{i\eta} & -\sin\alpha\, e^{i(\gamma-\epsilon)} \\ \sin\alpha\, e^{i(\epsilon-\gamma)} & \cos\alpha\, e^{-i\eta} \end{pmatrix} \\ &= e^{i\epsilon}\begin{pmatrix} e^{i\theta} & 0 \\ 0 & e^{-i\theta} \end{pmatrix}\begin{pmatrix} \cos\alpha & -\sin\alpha \\ \sin\alpha & \cos\alpha \end{pmatrix}\begin{pmatrix} e^{i\psi} & 0 \\ 0 & e^{-i\psi} \end{pmatrix}, \end{aligned} \tag{8}$$

where
$$\theta = \tfrac{1}{2}(\eta+\gamma-\epsilon), \quad \psi = \tfrac{1}{2}(\eta-\gamma+\epsilon). \tag{9}$$

Thus whereas the most general real orthogonal 2×2 matrix can be expressed in terms of one angle, the most general unitary one needs four.

4·102. General rotation in three dimensions in terms of a 2×2 unitary transformation. Let $A_1 = (A_1, B_1, C_1)$, $A_2 = (A_2, B_2, C_2)$ be two real vectors, equal in magnitude and perpendicular, that is,

$$A_1^2+B_1^2+C_1^2 = A_2^2+B_2^2+C_2^2, \quad A_1 A_2+B_1 B_2+C_1 C_2 = 0. \tag{1}$$

Four of the real quantities $A_1, B_1, C_1, A_2, B_2, C_2$ can evidently be assigned independently. Under an orthogonal transformation both vectors preserve their magnitudes and remain perpendicular; and if under a transformation this is true for all pairs of equal perpendicular real vectors the transformation is orthogonal.

Form the 'complex vector' (a, b, c) such that

$$a = A_1+iA_2, \quad b = B_1+iB_2, \quad c = C_1+iC_2. \tag{2}$$

Then by (1)
$$a^2+b^2+c^2 = 0, \tag{3}$$

whence
$$(ib-a)(ib+a) = c^2.$$

We may call (a, b, c) a complex null vector; and x_1, x_2 exist such that

$$\left.\begin{aligned} ib-a &= x_1^2, \\ c &= x_1 x_2, \\ ib+a &= x_2^2, \end{aligned}\right\} \quad \left.\begin{aligned} a &= \tfrac{1}{2}(x_2^2-x_1^2), \\ b &= \frac{1}{2i}(x_1^2+x_2^2), \\ c &= x_1 x_2. \end{aligned}\right\} \tag{4}$$

Then x_1, x_2 are determined except for sign if (a,b,c) are given; and conversely for any assigned (complex) x_1, x_2 (4) will give (a,b,c) satisfying (3), and (1) follow on equating real and imaginary parts. Also

$$|a^2|+|b^2|+|c^2| = \tfrac{1}{2}(x_1 x_1^* + x_2 x_2^*)^2,$$

and also
$$= (A_1^2+B_1^2+C_1^2)+(A_2^2+B_2^2+C_2^2). \tag{5}$$

If x_1, x_2 undergo a linear transformation to y_1, y_2 such that

$$x_1 x_1^* + x_2 x_2^* = y_1 y_1^* + y_2 y_2^*, \tag{6}$$

a', b', c' defined by replacing x_1, x_2 by y_1, y_2 in (4) will be linear in $x_1^2, x_2^2, x_1 x_2$ and therefore in (a,b,c); and

$$|a'^2|+|b'^2|+|c'^2| = ||a^2|+|b^2|+|c^2|. \tag{7}$$

It follows that the vectors A_1', A_2' derived from a', b', c' are equal in magnitude to A_1 and A_2 and mutually perpendicular. But in general A_1' will depend on both A_1 and A_2. We want a real transformation that will represent a general rotation of A_1 into A_1', and A_2 into A_2', where A_1' is independent of A_2 and A_2' of A_1. This requires a further restriction, namely that the real and imaginary parts of a, b, c transform separately. We proceed to find in what conditions the transformation given by (6) will have this property.

Now the most general unitary transformation of two variables can be put in the form

$$y_1 = (\alpha x_1 + \beta x_2) e^{i\epsilon}, \quad y_2 = (-\beta^* x_1 + \alpha^* x_2) e^{i\epsilon}, \tag{8}$$

where ϵ is real and $\alpha\alpha^* + \beta\beta^* = 1$. Then if

$$a' = \tfrac{1}{2}(y_2^2 - y_1^2), \quad b' = \frac{1}{2i}(y_1^2 + y_2^2), \quad c' = y_1 y_2, \tag{9}$$

we find that

$$\left.\begin{aligned}
a' e^{-2i\epsilon} &= \tfrac{1}{2}(\alpha^2 + \alpha^{*2} - \beta^2 - \beta^{*2}) a + \frac{1}{2i}(\alpha^2 - \alpha^{*2} + \beta^2 - \beta^{*2}) b - (\alpha\beta + \alpha^*\beta^*) c, \\
b' e^{-2i\epsilon} &= \frac{1}{2i}(-\alpha^2 + \alpha^{*2} + \beta^2 - \beta^{*2}) a + \tfrac{1}{2}(\alpha^2 + \alpha^{*2} + \beta^2 + \beta^{*2}) b + \frac{1}{i}(\alpha\beta - \alpha^*\beta^*) c, \\
c' e^{-2i\epsilon} &= (\alpha\beta^* + \alpha^*\beta) a + \frac{1}{i}(\alpha\beta^* - \alpha^*\beta) b + (\alpha\alpha^* - \beta\beta^*) c.
\end{aligned}\right\} \tag{10}$$

The coefficients on the right are all real. Hence the condition that the real and imaginary parts of a, b, c shall transform separately is that $e^{-2i\epsilon}$ is real and therefore ± 1. In either case the transformation is orthogonal. If $\alpha = 1, \beta = 0$, it is seen at once that the determinant of the coefficients on the right is 1. For any α, β we can take $\epsilon = 0$, and the transformation, being orthogonal, must give determinant ± 1; and as it is a continuous function of α and β it must therefore always be 1. Hence if $e^{-2i\epsilon} = 1$ the transformation is a rotation. If $e^{-2i\epsilon} = -1$ it is a rotatory reflexion. (With a lavish expenditure of paper the determinant can be shown to be identically $(\alpha\alpha^* + \beta\beta^*)^3$.) Hence we can represent the most general rotation by taking $\epsilon = 0$ in (8) and (10).

In particular, take
$$\begin{pmatrix} \alpha & \beta \\ -\beta^* & \alpha^* \end{pmatrix} = \begin{pmatrix} \cos\tfrac{1}{2}\theta & -\sin\tfrac{1}{2}\theta \\ \sin\tfrac{1}{2}\theta & \cos\tfrac{1}{2}\theta \end{pmatrix}. \tag{11}$$

Then
$$A_1' = A_1 \cos\theta + C_1 \sin\theta, \quad B_1' = B_1, \quad C_1' = -A_1 \sin\theta + C_1 \cos\theta. \tag{12}$$

This represents a right-handed rotation through θ about the y axis.

If
$$\alpha = e^{-\frac{1}{2}i\psi}, \quad \beta = 0, \tag{13}$$

we have
$$A_1' = A_1 \cos\psi - B_1 \sin\psi, \quad B_1' = A_1 \sin\psi + B_1 \cos\psi, \quad C_1' = C_1, \tag{14}$$

which represents a right-handed rotation through ψ about the z axis.

If
$$\alpha = \cos\tfrac{1}{2}\phi, \quad \beta = -i\sin\tfrac{1}{2}\phi, \tag{15}$$

$$A_1' = A_1, \quad B_1' = B_1 \cos\phi - C_1 \sin\phi, \quad C_1' = B_1 \sin\phi + C_1 \cos\phi, \tag{16}$$

which represents a right-handed rotation through ϕ about the x-axis.

The most general unimodular unitary transformation can be built up from operations of this type. The transformation given by the matrix

$$\begin{pmatrix} \alpha & \beta \\ -\beta^* & \alpha^* \end{pmatrix} = \begin{pmatrix} e^{-\frac{1}{2}i\lambda} & 0 \\ 0 & e^{\frac{1}{2}i\lambda} \end{pmatrix} \begin{pmatrix} \cos\tfrac{1}{2}\theta & -\sin\tfrac{1}{2}\theta \\ \sin\tfrac{1}{2}\theta & \cos\tfrac{1}{2}\theta \end{pmatrix} \begin{pmatrix} e^{-\frac{1}{2}i\chi} & 0 \\ 0 & e^{\frac{1}{2}i\chi} \end{pmatrix}$$

$$= \begin{pmatrix} \cos\tfrac{1}{2}\theta \, e^{-\frac{1}{2}i(\lambda+\chi)} & -\sin\tfrac{1}{2}\theta \, e^{-\frac{1}{2}i(\lambda-\chi)} \\ \sin\tfrac{1}{2}\theta \, e^{\frac{1}{2}i(\lambda-\chi)} & \cos\tfrac{1}{2}\theta \, e^{\frac{1}{2}i(\lambda+\chi)} \end{pmatrix}, \tag{17}$$

leads to the general rotation in the 3-space given in 4·034. To each such rotation there correspond two 2×2 transformations.

4·11. The Pauli spin matrices. In the x, y, z space *small* rotations ϵ about the axes x, y, z respectively are given by the three matrices

$$\begin{pmatrix} 1 & 0 & 0 \\ 0 & 1 & -\epsilon \\ 0 & \epsilon & 1 \end{pmatrix}, \quad \begin{pmatrix} 1 & 0 & \epsilon \\ 0 & 1 & 0 \\ -\epsilon & 0 & 1 \end{pmatrix}, \quad \begin{pmatrix} 1 & -\epsilon & 0 \\ \epsilon & 1 & 0 \\ 0 & 0 & 1 \end{pmatrix}, \tag{18}$$

where powers of ϵ above the first are neglected. Corresponding 2×2 unitary transformations are given by the matrices

$$\begin{pmatrix} 1 & -\tfrac{1}{2}i\epsilon \\ -\tfrac{1}{2}i\epsilon & 1 \end{pmatrix}, \quad \begin{pmatrix} 1 & -\tfrac{1}{2}\epsilon \\ \tfrac{1}{2}\epsilon & 1 \end{pmatrix}, \quad \begin{pmatrix} 1-\tfrac{1}{2}i\epsilon & 0 \\ 0 & 1+\tfrac{1}{2}i\epsilon \end{pmatrix}. \tag{19}$$

If we write
$$\sigma_1 = \begin{pmatrix} 0 & 1 \\ 1 & 0 \end{pmatrix}, \quad \sigma_2 = \begin{pmatrix} 0 & -i \\ i & 0 \end{pmatrix}, \quad \sigma_3 = \begin{pmatrix} 1 & 0 \\ 0 & -1 \end{pmatrix}, \tag{20}$$

the matrices (19) are $\quad 1 - \tfrac{1}{2}i\epsilon\sigma_1, \quad 1 - \tfrac{1}{2}i\epsilon\sigma_2, \quad 1 - \tfrac{1}{2}i\epsilon\sigma_3.$

The matrices (20) are the 'Pauli matrices', which occur in the theory of electron spin in quantum mechanics.

We can verify that

$$\sigma_1^2 = \sigma_2^2 = \sigma_3^2 = 1, \quad \sigma_1\sigma_2 = -\sigma_2\sigma_1 = i\sigma_3, \quad \sigma_2\sigma_3 = -\sigma_3\sigma_2 = i\sigma_1, \quad \sigma_3\sigma_1 = -\sigma_1\sigma_3 = i\sigma_2.$$

The matrices σ_1, σ_2, σ_3 therefore *anticommute*.

4·12. The Eddington and Dirac 4×4 matrices. Consider the 4×4 matrices

$$E_1 = \begin{pmatrix} i\sigma_1 & 0 \\ 0 & i\sigma_1 \end{pmatrix}, \quad E_2 = \begin{pmatrix} i\sigma_3 & 0 \\ 0 & i\sigma_3 \end{pmatrix}, \quad E_3 = \begin{pmatrix} 0 & -\sigma_2 \\ \sigma_2 & 0 \end{pmatrix}, \quad E_4 = \begin{pmatrix} -i\sigma_2 & 0 \\ 0 & i\sigma_2 \end{pmatrix},$$

where, for example,

$$\begin{pmatrix} i\sigma_1 & 0 \\ 0 & i\sigma_1 \end{pmatrix} = \begin{pmatrix} 0 & i & 0 & 0 \\ i & 0 & 0 & 0 \\ 0 & 0 & 0 & i \\ 0 & 0 & i & 0 \end{pmatrix},$$

the elements σ being written out as indicated and the inner enclosing brackets then removed. By 4·063, multiplication may be carried out on the 2×2 matrices of which the elements are themselves 2×2 matrices, and the results then expanded into 4×4 matrices. Then

$$E_1^2 = \begin{pmatrix} -\sigma_1^2 & 0 \\ 0 & -\sigma_1^2 \end{pmatrix} = -1,$$

and similarly the squares of E_2, E_3, E_4 are -1. Also the four matrices are easily shown to anticommute. If we now write

$$iE_5 = E_1 E_2 E_3 E_4 = \begin{pmatrix} 0 & \sigma_2 \\ \sigma_2 & 0 \end{pmatrix},$$

we have
$$E_5^2 = -E_1 E_2 E_3 E_4 E_1 E_2 E_3 E_4 = E_1 E_1 E_2 E_3 E_4 E_2 E_3 E_4$$
$$= -E_2 E_2 E_3 E_4 E_3 E_4 = -E_3 E_3 E_4 E_4 = -1,$$

and $\quad iE_1 E_5 = -E_2 E_3 E_4, \quad iE_5 E_1 = E_1 E_2 E_3 E_4 E_1 = -E_1 E_1 E_2 E_3 E_4 = E_2 E_3 E_4.$

Thus E_5 anticommutes with E_1, and similarly with E_2, E_3, and E_4.

Thus we have an anticommuting pentad of square roots of -1. Written in full they are as follows, in order:

$$\begin{pmatrix} 0 & i & 0 & 0 \\ i & 0 & 0 & 0 \\ 0 & 0 & 0 & i \\ 0 & 0 & i & 0 \end{pmatrix} \begin{pmatrix} i & 0 & 0 & 0 \\ 0 & -i & 0 & 0 \\ 0 & 0 & i & 0 \\ 0 & 0 & 0 & -i \end{pmatrix} \begin{pmatrix} 0 & 0 & 0 & i \\ 0 & 0 & -i & 0 \\ 0 & -i & 0 & 0 \\ i & 0 & 0 & 0 \end{pmatrix} \begin{pmatrix} 0 & -1 & 0 & 0 \\ 1 & 0 & 0 & 0 \\ 0 & 0 & 0 & 1 \\ 0 & 0 & -1 & 0 \end{pmatrix} \begin{pmatrix} 0 & 0 & 0 & -1 \\ 0 & 0 & 1 & 0 \\ 0 & -1 & 0 & 0 \\ 1 & 0 & 0 & 0 \end{pmatrix}$$

Any matrix $E_\mu E_\nu (\mu \neq \nu)$ has square -1. Denoting it by $E_{\mu\nu}$ we find

$$E_{12} = \begin{pmatrix} i\sigma_2 & 0 \\ 0 & i\sigma_2 \end{pmatrix} = \begin{pmatrix} 0 & 1 & 0 & 0 \\ -1 & 0 & 0 & 0 \\ 0 & 0 & 0 & 1 \\ 0 & 0 & -1 & 0 \end{pmatrix}, \quad E_{13} = \begin{pmatrix} 0 & \sigma_3 \\ -\sigma_3 & 0 \end{pmatrix} = \begin{pmatrix} 0 & 0 & 1 & 0 \\ 0 & 0 & 0 & -1 \\ -1 & 0 & 0 & 0 \\ 0 & 1 & 0 & 0 \end{pmatrix},$$

$$E_{14} = \begin{pmatrix} i\sigma_3 & 0 \\ 0 & -i\sigma_3 \end{pmatrix} = \begin{pmatrix} i & 0 & 0 & 0 \\ 0 & -i & 0 & 0 \\ 0 & 0 & -i & 0 \\ 0 & 0 & 0 & i \end{pmatrix}, \quad E_{15} = \begin{pmatrix} 0 & i\sigma_3 \\ i\sigma_3 & 0 \end{pmatrix} = \begin{pmatrix} 0 & 0 & i & 0 \\ 0 & 0 & 0 & -i \\ i & 0 & 0 & 0 \\ 0 & -i & 0 & 0 \end{pmatrix},$$

$$E_{23} = \begin{pmatrix} 0 & -\sigma_1 \\ \sigma_1 & 0 \end{pmatrix} = \begin{pmatrix} 0 & 0 & 0 & -1 \\ 0 & 0 & -1 & 0 \\ 0 & 1 & 0 & 0 \\ 1 & 0 & 0 & 0 \end{pmatrix}, \quad E_{24} = \begin{pmatrix} -i\sigma_1 & 0 \\ 0 & i\sigma_1 \end{pmatrix} = \begin{pmatrix} 0 & -i & 0 & 0 \\ -i & 0 & 0 & 0 \\ 0 & 0 & 0 & i \\ 0 & 0 & i & 0 \end{pmatrix},$$

$$E_{25} = \begin{pmatrix} 0 & -i\sigma_1 \\ -i\sigma_1 & 0 \end{pmatrix} = \begin{pmatrix} 0 & 0 & 0 & -i \\ 0 & 0 & -i & 0 \\ 0 & -i & 0 & 0 \\ -i & 0 & 0 & 0 \end{pmatrix}, \quad E_{34} = \begin{pmatrix} 0 & -i1 \\ -i1 & 0 \end{pmatrix} = \begin{pmatrix} 0 & 0 & -i & 0 \\ 0 & 0 & 0 & -i \\ -i & 0 & 0 & 0 \\ 0 & -i & 0 & 0 \end{pmatrix},$$

$$E_{35} = \begin{pmatrix} iU & 0 \\ 0 & -iU \end{pmatrix} = \begin{pmatrix} i & 0 & 0 & 0 \\ 0 & i & 0 & 0 \\ 0 & 0 & -i & 0 \\ 0 & 0 & 0 & -i \end{pmatrix}, \quad E_{45} = \begin{pmatrix} 0 & -1 \\ 1 & 0 \end{pmatrix} = \begin{pmatrix} 0 & 0 & -1 & 0 \\ 0 & 0 & 0 & -1 \\ 1 & 0 & 0 & 0 \\ 0 & 1 & 0 & 0 \end{pmatrix}$$

We thus have fifteen matrix square roots of -1, to which we can add a sixteenth, $E_{16} = i1$. It can be shown that the first fifteen form six anticommuting pentads, each E being a member of two pentads.* All are antihermitian and unitary.

The matrices introduced by Dirac in the solution of the relativistic wave equation are anticommuting square roots of 1 and are connected with a pentad of Eddington's by the relations
$$\alpha_1 = iE_{25}, \quad \alpha_2 = iE_5, \quad \alpha_3 = -iE_{15}, \quad \beta = -iE_{35}, \quad \alpha_1\alpha_2\alpha_3\beta = iE_{45}.$$

4·13. Oblique axes. So far the axes of reference that we have considered have been rectangular. The position of a point could, however, be expressed either by the orthogonal projections on the axes of its displacement from the origin, or by displacements in any three non-coplanar directions that will add up vectorially to the displacement from the origin to the point. The second method corresponds to the usual resolution into oblique components; we shall see that the first also has a physical significance. The characteristic feature of rectangular axes is that the two sets of quantities are then identical. Let us see what happens with oblique axes. Suppose that we have a set of rectangular axes x_i and take three oblique axes x'_j with direction cosines l_{ij} with respect to the rectangular ones. For a reason that we shall see in a moment we denote the oblique coordinates by an index instead of a suffix, thus, x'^j. Then the rectangular coordinates of a point P will be $l_{ij}x'^j$ as before. Since l_{ij} for fixed j are the direction cosines of a single line,

$$\sum_i l_{ij}^2 = 1, \tag{1}$$

but in general for $j \neq l$
$$\sum_i l_{ij}l_{il} \neq 0, \tag{2}$$

since this is the cosine of the angle between the directions of x'^j and x'^l, which are not perpendicular. The distance of a point from the origin is r, where
$$r^2 = x_i^2 = (l_{ij}x'^j)(l_{il}x'^l)$$
$$= l_{ij}l_{il}x'^j x'^l, \tag{3}$$

---

* Eddington, *Relativity Theory of Protons and Electrons*, 1936. Eddington does not write the matrices out in full, but they are implicit in equations 3·61, p. 42. The antisymmetrical ones appear with reversed sign; this is due to the fact that in an element a_{ik} he takes i to refer to the column and k to the row, contrary to the usual convention. In *Fundamental Theory*, 1946, p. 142, he gives the matrices in detail with the present convention; his E_{31} is $-E_{13}$.

which is a quadratic form, and can be written

$$r^2 = g'_{jl} x'^j x'^l, \qquad (4)$$

where $g'_{11} = g'_{22} = g'_{33} = 1$, but $g'_{23}, g'_{31}, g'_{12}$ are not 0. If the axes x'_j are rectangular, but not otherwise, $g'_{jl} = \delta_{jl}$.

Now consider the orthogonal projection of OP on the direction of x'^j and denote it by x'_j with a suffix. It is

$$x'_j = l_{ij} x_i = l_{ij} l_{il} x'^l = g'_{jl} x'^l = \frac{\partial}{\partial x'^j}(\tfrac{1}{2} r^2), \qquad (5)$$

and

$$x'_j x'^j = g'_{jl} x'^l x'^j = r^2. \qquad (6)$$

Thus the form $x_i x_i$ correct for rectangular axes needs to be modified for oblique ones by raising one of the suffixes. We cannot now write it as x'^2_j.

The position of P can be specified equally well by the values of either x'_j or x'^j; but the actual magnitudes of the three quantities are different. The former are called the *covariant* components and the latter the *contravariant* components of the displacement OP. It is commonly thought anomalous that the adjectives are not interchanged. But the covariant component in the direction Oj has the property that for a given position of P it is independent of the directions of the other two axes; this is not true of the contravariant component, and in this respect the covariant components might appear to be the more fundamental. On the other hand, if we vary one contravariant component without altering the other two, we know directly how much and in what direction the point is displaced. This is not obvious for variation of one covariant component. Both systems therefore have their advantages and we need a means of transforming from either to the other.

We write
$$\| g'_{jl} \| = G', \qquad (7)$$

denote the cofactor of g'_{jl} in G' by G'^{jl}, and put

$$g'^{lj} = G'^{jl}/G' = g'^{jl}. \qquad (8)$$

Then g'^{jl} is the reciprocal matrix of g_{jl}, and

$$x'^l = g'^{lj} x'_j. \qquad (9)$$

This formula with its companion

$$x'_j = g'_{jl} x'^l \qquad (10)$$

are known as the formulae for raising or lowering indices. The determinant G' is seen to be fundamental. If we denote the angle between the x'_1 and x'_2 axes by α_3 and so on,

$$g'_{12} = \cos \alpha_3, \qquad (11)$$

$$G' = \begin{vmatrix} 1 & \cos \alpha_3 & \cos \alpha_2 \\ \cos \alpha_3 & 1 & \cos \alpha_1 \\ \cos \alpha_2 & \cos \alpha_1 & 1 \end{vmatrix} = 1 - \cos^2 \alpha_1 - \cos^2 \alpha_2 - \cos^2 \alpha_3 + 2 \cos \alpha_1 \cos \alpha_2 \cos \alpha_3. \qquad (12)$$

But $g'^{11}, g'^{22}, g'^{33}$ are not in general equal to 1 or to one another.

Alternatively $\qquad G' = \| l_{ij} l_{il} \| = \| \tilde{l}_{ji} \| \| l_{il} \| = \| l_{ij} \|^2. \qquad (13)$

But $\| l_{ij} \|$ is the volume of the parallelepiped with edges of unit length along the axes, or alternatively the continued scalar product of direction vectors along the axes. It could therefore be zero only if the axes were coplanar. The element of volume is therefore $G'^{1/2}dx'^1 dx'^2 dx'^3$.

If A is a general vector, the resultant of components A'^1, A'^2, A'^3 along the oblique axes, its rectangular components will be

$$A_i = l_{ij} A'^j, \tag{14}$$

and its covariant components will be

$$A'_j = g_{jl} A'^l = l_{ij} A_i. \tag{15}$$

Now let ϕ be a scalar and consider its derivatives with regard to x'^j. We have

$$\frac{\partial \phi}{\partial x'^j} = \frac{\partial x_i}{\partial x'^j} \frac{\partial \phi}{\partial x_i} = l_{ij} \frac{\partial \phi}{\partial x_i}, \tag{16}$$

and $\partial \phi / \partial x'^j$ are therefore the covariant components of grad ϕ. To get the contravariant components we must multiply by g'^{jl} and contract.

If A and B are two vectors,

$$A'_j B'^j = l_{ij} A_i B'^j = A_i B_i, \tag{17}$$

which is their scalar product. Similarly,

$$\frac{\partial A'^j}{\partial x'^j} = l_{ij} \frac{\partial A'^j}{\partial x_i} = \frac{\partial A_i}{\partial x_i}, \tag{18}$$

which is the divergence of A.

As for rectangular axes we can define tensors of any order, but the transformation rules will be different according as each index is upper or lower. We can contract with respect to a repeated index provided that it is upper in one occurrence and lower in the other, and the result will be a tensor of order lower by 2.

4·131. Crystal structure: the reciprocal lattice. A simple three-dimensional lattice is specified if three fundamental displacement vectors are given. Taking any atom as origin, there will be a similar atom at any point $n_1 a_1 + n_2 a_2 + n_3 a_3$, where a_1, a_2, a_3 are the displacements to three neighbouring atoms not in the same plane as the origin, and n_1, n_2, n_3 are integers, positive or negative. The points so specified are called lattice points. A plane through any three lattice points will include similar atoms in a repeating pattern. Its direction can be specified by its intercepts on axes through the origin in the directions of a_1, a_2, a_3; let these, divided by a suitable integer, be $a_1/h_1, a_2/h_2, a_3/h_3$, where h_1, h_2, h_3 are integers with no common factor. Then h_1, h_2, h_3 are called the indices (*Miller indices*) of the plane,* and are the same for all parallel planes.

We shall denote the volume $a_1 \cdot a_2 \wedge a_3$ of a single cell of the lattice by v_a.

The set of vectors b_1, b_2, b_3 reciprocal to the set a_1, a_2, a_3 is defined as

$$b_1 = \frac{a_2 \wedge a_3}{v_a}, \quad b_2 = \frac{a_3 \wedge a_1}{v_a}, \quad b_3 = \frac{a_1 \wedge a_2}{v_a}. \tag{1}$$

They satisfy the relations

$$a_i \cdot b_j = \delta_{ij} \tag{2}$$

and

$$v_b = b_1 \cdot b_2 \wedge b_3 = \frac{b_1 \cdot b_2 \wedge (a_1 \wedge a_2)}{v_a}$$

$$= \frac{b_1 \cdot \{(b_2 \cdot a_2) a_1 - (b_2 \cdot a_1) a_2\}}{v_a} = \frac{1}{v_a}. \tag{3}$$

---

* The suggestion for this specification was made by Grassmann and others, but first became popular through Miller's *Lehrbuch der Krystallographie* (1863).

Any vector A can be expressed by

$$A = (A \cdot b_1)a_1 + (A \cdot b_2)a_2 + (A \cdot b_3)a_3 \qquad (4)$$

$$= (A \cdot a_1)b_1 + (A \cdot a_2)b_2 + (A \cdot a_3)b_3. \qquad (5)$$

If, therefore, we build up a lattice with b_1, b_2, b_3 as fundamental vectors, the volume of a cell is the reciprocal of that in the original lattice.

Since $a_1/h_1, a_2/h_2, a_3/h_3$ are points in a lattice plane, the directions of the vectors

$$\frac{a_1}{h_1} - \frac{a_2}{h_2}, \quad \frac{a_1}{h_1} - \frac{a_3}{h_3}$$

are parallel to this plane; but both these directions are perpendicular to $h_1 b_1 + h_2 b_2 + h_3 b_3$, and this displacement in the reciprocal lattice is normal to the planes of the atomic lattice with Miller indices h_1, h_2, h_3. The equation of any of these planes can therefore be written

$$\frac{(h_1 b_1 + h_2 b_2 + h_3 b_3) \cdot x}{|h_1 b_1 + h_2 b_2 + h_3 b_3|} = p, \qquad (6)$$

and p is the perpendicular from the origin on to it. Now if x is a lattice point it is of the form $n_1 a_1 + n_2 a_2 + n_3 a_3$; then

$$(h_1 b_1 + h_2 b_2 + h_3 b_3) \cdot (n_1 a_1 + n_2 a_2 + n_3 a_3) = h_1 n_1 + h_2 n_2 + h_3 n_3. \qquad (7)$$

Now if h_1, h_2, h_3 have no common factor we can choose n_1, n_2, n_3 so that the sum on the right is 1. For, first, if h_1, h_2 have a common factor q the process of finding the highest common factor enables us to determine s_1, s_2 so that $s_1 h_1 + s_2 h_2 = q$. Similarly, if h_2, h_3 have a common factor r we can find t_1, t_2 so that $t_1 h_2 + t_2 h_3 = r$. But, by hypothesis, q and r have no common factor; therefore we can find a linear combination of these expressions, with integral coefficients, that is equal to 1. This can be taken as $h_1 n_1 + h_2 n_2 + h_3 n_3$. Evidently $h_1 n_1 + h_2 n_2 + h_3 n_3$ can be made 0 by taking $n_1 = n_2 = n_3 = 0$. Then the perpendicular distance d_h between the planes with the corresponding values of x is equal to $|h_1 b_1 + h_2 b_2 + h_3 b_3|^{-1}$, and this is the spacing of the crystal planes with Miller indices h_1, h_2, h_3.

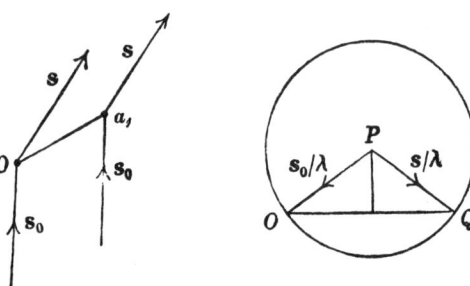

Now consider a parallel beam of X-rays falling on a crystal. Suppose that plane waves travelling in a direction s_0 fall on the atomic lattice and are scattered by the separate atoms. We want the condition that those travelling away in a direction s shall reinforce one another. The difference of path for waves scattered by two atoms separated by a_j is

$(s-s_0) \cdot a_j$, and the condition for reinforcement is that this shall be a multiple of the wave-length λ for all atoms in the region. Hence

$$(s-s_0) \cdot a_j = k_j \lambda \quad (j = 1, 2, 3), \tag{8}$$

where the k_j are integers. Then from (5)

$$s - s_0 = \lambda(k_1 b_1 + k_2 b_2 + k_3 b_3), \tag{9}$$

or, supposing k_1, k_2, k_3 to have a common factor n,

$$s - s_0 = n\lambda(h_1 b_1 + h_2 b_2 + h_3 b_3). \tag{10}$$

Now take an origin at a point O of the reciprocal lattice. The geometrical meaning of (10) is that if P is a point such that $PO = s_0/\lambda$, then if PQ is s/λ, Q must be a point of the reciprocal lattice, and all such points that lie on a sphere of radius $1/\lambda$ about P will correspond to reflexions of rays of wave-length λ. Moreover OQ is parallel to the external bisector of the angle OPQ, and hence the reflexion can be regarded as taking place at planes with Miller indices (h_1, h_2, h_3) whose normal is in the direction OQ.

If we write 2θ for the angle OPQ, so that θ is the angle of reflexion, we have

$$\frac{n}{d_h} = n \mid h_1 b_1 + h_2 b_2 + h_3 b_3 \mid = \frac{1}{\lambda} \mid s - s_0 \mid = \frac{2}{\lambda} \sin \theta \tag{11}$$

or

$$n\lambda = 2d_h \sin \theta, \tag{12}$$

which is Bragg's reflexion condition.

Further, by squaring (11) we have that

$$n^2 \lambda^2 \mid h_1 b_1 + h_2 b_2 + h_3 b_3 \mid^2 = 4 \sin^2 \theta, \tag{13}$$

and also from (10) since

$$s^2 = \mid n\lambda(h_1 b_1 + h_2 b_2 + h_3 b_3) + s_0 \mid^2 \tag{14}$$

that

$$n\lambda = -\frac{2 s_0 \cdot (h_1 b_1 + h_2 b_2 + h_3 b_3)}{\mid h_1 b_1 + h_2 b_2 + h_3 b_3 \mid^2}. \tag{15}$$

Further developments are given by P. P. Ewald.*

4·14. Curvilinear coordinates. For many problems it is convenient to specify the position of a point by three functions of the rectangular coordinates that are not constant over planes. We may, for instance, use spherical polar coordinates; then the coordinate r is constant over a sphere with centre at the origin. If we continue to denote rectangular coordinates by x_i and call the curvilinear ones x'^j, we shall have

$$dx_i = \frac{\partial x_i}{\partial x'^1} dx'^1 + \frac{\partial x_i}{\partial x'^2} dx'^2 + \frac{\partial x_i}{\partial x'^3} dx'^3 = \frac{\partial x_i}{\partial x'^j} dx'^j, \tag{1}$$

$$dx'^j = \frac{\partial x'^j}{\partial x_i} dx_i, \quad \frac{\partial \phi}{\partial x_i} = \frac{\partial x'^j}{\partial x_i} \frac{\partial \phi}{\partial x'^j}, \tag{2}$$

so that the summation convention is still applicable. But the partial derivatives are no longer constant. In (1) x_i must of course be regarded as a known function of the three x'^j, which are permitted to vary independently; in (2), x'^j is conversely regarded as a

* *Kristalle und Röntgenstrahlen*, 1923, particularly Notes 1 and 2.

function of the three x_i. The relations between the coordinates are no longer linear, but those between small changes are linear, and so are those between derivatives of a scalar. Sets of quantities that transform like dx'^j are called contravariant and those that transform like $\partial\phi/\partial x'^j$ are called covariant, as for oblique rectilinear axes. If ds is the distance between two neighbouring points we have

$$ds^2 = dx_i^2 = g'_{jl} dx'^j dx'^l, \tag{3}$$

where
$$g'_{jl} = \frac{\partial x_i}{\partial x'^j} \frac{\partial x_i}{\partial x'^l}. \tag{4}$$

Just as for oblique rectilinear coordinates we can form the reciprocal set of quantities g'^{jl} defined by

$$g'^{jl} = \frac{\partial x'^j}{\partial x^i} \frac{\partial x'^l}{\partial x^i} \tag{5}$$

and
$$g'_{jl} g'^{jn} = \frac{\partial x^i}{\partial x'^j} \frac{\partial x^i}{\partial x'^l} \frac{\partial x'^j}{\partial x^k} \frac{\partial x'^n}{\partial x^k} = \left(\frac{\partial x^i}{\partial x'^j} \frac{\partial x'^j}{\partial x^k}\right)\left(\frac{\partial x^i}{\partial x'^l} \frac{\partial x'^n}{\partial x^k}\right)$$

$$= \delta_{ik} \frac{\partial x^i}{\partial x'^l} \frac{\partial x'^n}{\partial x^k} = \frac{\partial x'^n}{\partial x^k} \frac{\partial x^k}{\partial x'^l} = \delta_{ln}. \tag{6}$$

Covariant and contravariant tensors of the second order can be defined according as they transform under further changes of coordinates like g'_{jl} or g'^{jl}. It may be verified, as an exercise in differential calculus, that if we take a third set of coordinates x''^a we get the same form for g''_{ab} by transforming first from x^i to x'^j and then to x''^a as if we transformed to x''^a directly.

All information about magnitudes of displacements for small changes of the curvilinear coordinates is summarized in the g'_{jl}. If we vary x'^1 without varying x'^2 and x'^3 the displacement will be $\{\sqrt{(g_{11})} dx'^1, 0, 0\}$. The three component displacements for separate changes of the new coordinates will not, however, in general be at right angles. The inclinations are easily found in terms of the g'_{jl} as for oblique coordinates, but are seldom required. We usually choose our coordinates so that the displacements corresponding to small changes of the x'^j separately will be at right angles; and the condition for this is

$$\frac{\partial x^i}{\partial x'^j} \frac{\partial x^i}{\partial x'^l} = 0 \quad (j \neq l), \tag{7}$$

or
$$g'_{jl} = 0 \quad (j \neq l). \tag{8}$$

If these conditions are satisfied the new coordinates are said to be *orthogonal*.

As an example let us take the x'^j to be spherical polar coordinates r, θ, λ. Then

$$x_1 = r \sin\theta \cos\lambda, \quad x_2 = r \sin\theta \sin\lambda, \quad x_3 = r \cos\theta,$$
$$dx_1 = \sin\theta \cos\lambda \, dr + r \cos\theta \cos\lambda \, d\theta - r \sin\theta \sin\lambda \, d\lambda,$$
$$dx_2 = \sin\theta \sin\lambda \, dr + r \cos\theta \sin\lambda \, d\theta + r \sin\theta \cos\lambda \, d\lambda,$$
$$dx_3 = \cos\theta \, dr \quad - r \sin\theta \, d\theta,$$

and
$$(dx_1)^2 + (dx_2)^2 + (dx_3)^2 = dr^2 + r^2 d\theta^2 + r^2 \sin^2\theta \, d\lambda^2.$$

Hence
$$g'_{11} = 1, \quad g'_{22} = r^2, \quad g'_{33} = r^2 \sin^2\theta.$$

The determinant of the transformation is $(g'_{11}g'_{22}g'_{33})^{1/2} = r^2 \sin\theta$. The inverse transformation is
$$dr = \sin\theta\cos\lambda\, dx_1 + \sin\theta\sin\lambda\, dx_2 + \cos\theta\, dx_3,$$
$$r\, d\theta = \cos\theta\cos\lambda\, dx_1 + \cos\theta\sin\lambda\, dx_2 - \sin\theta\, dx_3,$$
$$r\sin\theta\, d\lambda = -\sin\lambda\, dx_1 + \cos\lambda\, dx_2.$$

The reciprocal matrix to g'_{jl} is

$$g'^{jl} = \frac{1}{g'_{11}g'_{22}g'_{33}}\begin{pmatrix} g'_{22}g'_{33} & 0 & 0 \\ 0 & g'_{33}g'_{11} & 0 \\ 0 & 0 & g'_{11}g'_{22} \end{pmatrix} = \begin{pmatrix} 1 & 0 & 0 \\ 0 & r^{-2} & 0 \\ 0 & 0 & r^{-2}\csc^2\theta \end{pmatrix}.$$

We have here a decided difference from any rectilinear coordinates. The non-zero elements in g'_{jl} and g'^{jl} are now not merely unequal but of different dimensions. In fact in spherical polar coordinates the contravariant components of a displacement are $dr, d\theta, d\lambda$. We can define a set of covariant components by
$$g'_{jl} dx'^l = (dr, r^2 d\theta, r^2\sin^2\theta\, d\lambda).$$
But neither set are the physical components. The latter would be taken to be component *displacements* with regard to *rectangular* axes at the point, and would be $(dr, r\, d\theta, r\sin\theta\, d\lambda)$. The physical components are all lengths. In practice we are usually concerned with the physical components. We denote g_{11}, g_{22}, g_{33} by h_1^2, h_2^2, h_3^2, and small changes of the curvilinear coordinates make displacements ds_1, ds_2, ds_3; then we have for the physical components
$$ds_1 = h_1 dx'^1, \quad ds_2 = h_2 dx'^2, \quad ds_3 = h_3 dx'^3.$$

The same relations hold for components of velocity. If we have for rectangular axes a relation between vectors of the form
$$\dot{x}_i = \partial\phi/\partial x_i,$$
this transforms directly to any set of orthogonal axes, and, written in physical components, will be
$$\dot{s}_j = \frac{\partial\phi}{\partial s_j}, \quad h_1 \dot{x}'^1 = \frac{\partial\phi}{h_1 \partial x'^1}, \text{ etc.}$$

Multiplying or dividing by the corresponding h we get the same equations written as relations between covariant and contravariant components.

The derivative of a vector A_i, in curvilinear coordinates, does not in general transform like a second-order tensor, on account of the variation of g'_{ji} with position. This is the greatest complication of the tensor treatment in curvilinear coordinates. It can be overcome by a suitable modification of the derivative, but this would take us beyond the scope of this book.*

The rules for transforming coordinates will work equally well even if distances between neighbouring points cannot be put in the form $\delta_{ik} x^i x^k$ by any choice of coordinates. On a sphere, for instance, we can express the position of any point by two variables, but there is no way of choosing variables x_1, x_2 so that in all neighbourhoods on the sphere
$$ds^2 = dx_1^2 + dx_2^2.$$

* Full accounts are given by A. J. McConnell, *Applications of the Absolute Differential Calculus*, 1931; T. Levi-Cività, *The Absolute Differential Calculus*, 1927.

The theory of transformations when the quantity corresponding to distance cannot be put into Euclidean form without introducing new dimensions is the basis of Riemannian geometry and of the general theory of relativity.

4·15. Electromagnetic theory. The tensor method can be extended to four dimensions, and then forms a convenient way of stating the equations of electromagnetism. If we consider the quantity

$$ds^2 = dx_1^2 + dx_2^2 + dx_3^2 - c^2 dt^2, \tag{1}$$

where c is the velocity of light, ds taken between two neighbouring events (each specified by three position coordinates and the time) is the same for all observers even if they are in uniform relative motion. This statement is equivalent to three physical ones: (a) A particle moving with uniform velocity relative to one frame is moving with uniform velocity relative to the other. (b) Both observers attach equal values to distances at right angles to their direction of relative motion. (c) Both observers find the same value for the velocity of light, in whatever direction. (a) and (b) are taken over from Newtonian physics, (c) is an additional rule required by the Michelson-Morley experiment. Now if we write $x_4 = ict$, ds^2 reduces to the sum of four squares and can be treated like the square of a distance in Euclidean geometry, except that we now have to work in four dimensions. Since it is the same for all frames of reference the transformation from one to another is an orthogonal transformation in four dimensions.

Let us denote a new frame by accented letters and take the axes of x_2 and x_2', also x_3 and x_3', at right angles to the direction of the relative motion. Then $x_2 = x_2'$, $x_3 = x_3'$, and

$$x_1^2 + x_4^2 = x_1'^2 + x_4'^2. \tag{2}$$

This is satisfied if

$$x_1' = x_1 \cos\alpha - x_4 \sin\alpha, \quad x_4' = x_1 \sin\alpha + x_4 \cos\alpha. \tag{3}$$

The origin of the second frame has zero velocity in that frame; hence if we take $dx_1'/dx_4' = 0$, dx_1/dx_4 will be v/ic, where v is the velocity of the second origin with respect to the first frame. But this gives

$$\tan\alpha = \frac{v}{ic}, \tag{4}$$

$$x_1' = \beta x_1 + \frac{iv\beta}{c} x_4, \quad x_4' = -\frac{iv\beta}{c} x_1 + \beta x_4, \tag{5}$$

where

$$\beta = 1\bigg/\sqrt{\left(1 - \frac{v^2}{c^2}\right)}. \tag{6}$$

This transformation, due to Larmor and Lorentz, is a complex orthogonal transformation, not a unitary one; it satisfies $ll = 1$, not $ll^\dagger = 1$. For real x_1 and t it leaves x_1' and t' real. The ordinary transformations due to rotation of the axes of x_1, x_2, x_3, leaving x_4 unchanged, can of course be superposed on it. Sets of four quantities, defined for each system of reference, that are transformed into one another by the Larmor-Lorentz transformation, can be called components of four-vectors. The fundamental four-vector is x_α ($\alpha = 1, 2, 3, 4$) itself. But since ds is a scalar, $m\, dx_\alpha/ds$ is a four-vector, where m is any other scalar. If m is the intrinsic mass of a particle, supposed the same in all frames of reference, and u is the resultant velocity, it follows that $mu_\alpha/\sqrt{(1 - u^2/c^2)}$ is a four-vector, where

$$u_4 = dx_4/dt = ic.$$

The first three components correspond to those of linear momentum in Newtonian dynamics, the Newtonian mass being replaced by $m/\sqrt{(1-u^2/c^2)}$. Then if we denote the first three components by italic suffixes when treating them separately from the fourth,

$$\left(\frac{mu_a}{\sqrt{(1-u^2/c^2)}}, \frac{mic}{\sqrt{(1-u^2/c^2)}}\right) \tag{7}$$

is a four-vector.

Again, since the transformation is orthogonal, the four-dimensional volume element $dx_1 dx_2 dx_3 dx_4$ is unaltered. Hence the three-dimensional element $d\tau = dx_1 dx_2 dx_3$ transforms like $1/dx_4$, that is, like

$$\frac{ds}{dx_4} = \sqrt{\left(1-\frac{u^2}{c^2}\right)}. \tag{8}$$

If we define density as intrinsic mass per unit volume it therefore transforms like $m(1-u^2/c^2)^{-1/2}$; and a momentum per unit volume like the product of the density and the velocity. Comparing with (7) we see that if ρ is the density

$$(\rho u_a, ic\rho) \tag{9}$$

is a four-vector. In the same way, assuming that the electric charge of a particle is the same in all frames of reference we have that, if ρ is the electric charge per unit volume and j_a the current density, $(j_a, ic\rho)$ is a four-vector. It follows that

$$\operatorname{div} \mathbf{j} + ic\frac{\partial \rho}{\partial x_4} = \operatorname{div} \mathbf{j} + \frac{\partial \rho}{\partial t} \tag{10}$$

is a scalar and unaltered by transformation. It is actually zero;

$$\frac{\partial \rho}{\partial t} + \operatorname{div} \mathbf{j} = 0 \tag{11}$$

by the equation of continuity.

The pair of Maxwell's equations

$$c \operatorname{curl} \mathbf{H} - \frac{\partial \mathbf{E}}{\partial t} = 4\pi \mathbf{j}, \tag{12}$$

$$\operatorname{div} \mathbf{E} = 4\pi \rho, \tag{13}$$

now show that

$$\left(c(\operatorname{curl} \mathbf{H})_a - \frac{\partial E_a}{\partial t}, ic \operatorname{div} \mathbf{E}\right) \tag{14}$$

is a four-vector. Here $\mathbf{E}$ and $\mathbf{H}$ are the intensities of electric and magnetic field, the former in e.s.u. The four-dimensional divergence of (14) is obviously 0.

The other pair
$$\operatorname{div} \mathbf{H} = 0, \quad \operatorname{curl} \mathbf{E} + \frac{1}{c}\frac{\partial \mathbf{H}}{\partial t} = 0, \tag{15}$$

show that $\mathbf{H}$ is the curl of a three-vector $\mathbf{A}$, and that then

$$\mathbf{E} = -\frac{1}{c}\frac{\partial \mathbf{A}}{\partial t} - \operatorname{grad} \phi, \tag{16}$$

where ϕ is a scalar. $\mathbf{A}$ is arbitrary for given $\mathbf{H}$ to the extent of the gradient of any scalar, the effect of which on $\mathbf{E}$ could be compensated by a suitable change in ϕ. Hence we can impose another condition on A; we suppose that $(A, i\phi) = A_a$ is a four-vector. The condition that $\partial A_a/\partial x_a$ shall be scalar is satisfied if

$$\operatorname{div} A + \frac{1}{c}\frac{\partial \phi}{\partial t} = 0. \tag{17}$$

Then
$$iE_1 = \frac{\partial A_1}{ic\,\partial t} - i\frac{\partial \phi}{\partial x_1} = \frac{\partial A_1}{\partial x_4} - \frac{\partial A_4}{\partial x_1}, \tag{18}$$

$$H_1 = \frac{\partial A_3}{\partial x_2} - \frac{\partial A_2}{\partial x_3}, \tag{19}$$

with symmetrical relations; and iE_a and H_a are six components of an antisymmetrical tensor. We write

$$f_{\alpha\beta} = \frac{\partial A_\beta}{\partial x_\alpha} - \frac{\partial A_\alpha}{\partial x_\beta}, \tag{20}$$

$$H_1 = f_{23},\quad H_2 = f_{31},\quad H_3 = f_{12},\quad iE_1 = j_{41},\quad iE_2 = f_{42},\quad iE_3 = f_{43}, \tag{21}$$

and $f_{\alpha\beta}$, the field tensor, is

$$\begin{pmatrix} 0 & H_3 & -H_2 & -iE_1 \\ -H_3 & 0 & H_1 & -iE_2 \\ H_2 & -H_1 & 0 & -iE_3 \\ iE_1 & iE_2 & iE_3 & 0 \end{pmatrix}. \tag{22}$$

If we now write $(j_a, ic\rho) = s_a$, the pair of equations (12), (13) can be written

$$\frac{\partial f_{\alpha\beta}}{\partial x_\beta} = \frac{4\pi}{c} s_\alpha. \tag{23}$$

The pair (15) can be written

$$\frac{\partial f_{\alpha\beta}}{\partial x_\gamma} + \frac{\partial f_{\gamma\alpha}}{\partial x_\beta} + \frac{\partial f_{\beta\gamma}}{\partial x_\alpha} = 0. \tag{24}$$

If α, β, γ are all different these reduce to (15); if two of α, β, γ are equal they are identities.

The Lorentz force and the generalized stress tensor. If k is the mechanical force per unit volume,

$$k = \rho E + \frac{1}{c} j \wedge H, \tag{25}$$

which may be written as the first three components of

$$k_\alpha = \frac{1}{c} f_{\alpha\beta} s_\beta. \tag{26}$$

The fourth component defined by this is

$$k_4 = \frac{c}{i} E \cdot j, \tag{27}$$

so that k_4 is i/c times the work done by the field on the charge per unit volume per unit time. Using (23) we now have

$$k_\alpha = \frac{1}{4\pi} f_{\alpha\beta} \frac{\partial f_{\beta\gamma}}{\partial x_\gamma}. \tag{28}$$

If we define a tensor $T_{\alpha\gamma}$ by

$$T_{\alpha\gamma} = \frac{1}{4\pi}\{f_{\alpha\beta}f_{\beta\gamma} + \tfrac{1}{4}\delta_{\alpha\gamma}f_{\mu\nu}f_{\mu\nu}\}, \tag{29}$$

we find after some algebra that

$$k_\alpha = \frac{\partial T_{\alpha\gamma}}{\partial x_\gamma}. \tag{30}$$

The tensor T has the form

$$\begin{pmatrix} T_{11} & T_{12} & T_{13} & -iS_1/c \\ T_{21} & T_{22} & T_{23} & -iS_2/c \\ T_{31} & T_{32} & T_{33} & -iS_3/c \\ -iS_1/c & -iS_2/c & -iS_3/c & u \end{pmatrix}, \tag{31}$$

where the 3×3 set in the top left corner is the Maxwell stress tensor, S is the Poynting vector $\dfrac{c}{4\pi}\boldsymbol{E} \wedge \boldsymbol{H}$, and u is the energy density $\dfrac{1}{8\pi}(E^2 + H^2)$.

4·16. Probabilities in chains. We consider a system capable of several different states. At any instant there is a probability x_i that it is in a state denoted by suffix i. We consider the probability that it will be in state i at a later instant. Given that it is in state k at the first instant (i.e., if $x_k = 1$) the probability is a_{ik}. Then since it must be in some state at the second instant

$$\sum_i a_{ik} = 1, \tag{1}$$

and the total probability, for any set of values x_k, that it will be in state i at the second instant, is

$$y_i = a_{ik}x_k. \tag{2}$$

Card shuffling is the most familiar instance. The x_i will be the probabilities of the 52! possible orders of the cards at the start. The conditions of shuffling imply that for any order before a redistribution several different orders are possible after it; for a known order k at the first instant these would have probabilities a_{ik}, which must add up to 1. The total probability of order i after redistribution is then given by (2), by the usual rules for combining probabilities.

The same principle occurs in statistical mechanics. If the data are that the momenta and coordinates of a system are within specified finite (not zero) ranges at one instant, the subsequent motion is not exactly determinate even on classical mechanics, and motions differing considerably will occur according, say, to what pair of molecules encounter at the next collision.

Such probabilities can occur in chains, since the processes can be repeated. If a second rearrangement is made and the probability of state i after it is z_i, we shall have

$$z_i = a_{im}y_m, \tag{3}$$

and so on. The successive probabilities are obtained by multiplication by the same matrix.

This suggests a general treatment. The equations

$$\lambda\theta_i = a_{ik}\theta_k \tag{4}$$

are consistent if

$$\|a_{ik} - \lambda\delta_{ik}\| = 0. \tag{5}$$

Suppose that (5) has n roots λ_j, and denote the θ_i in the respective solutions by θ_{ij}. Summations with regard to j must be made explicit. Suppose further that x_i can be expressed in the form $\sum_j \alpha_j \theta_{ij}$. Then

$$y_i = \sum_k a_{ik} \sum_j \alpha_j \theta_{kj} = \sum_j \lambda_j \alpha_j \theta_{ij}$$

and the result of p applications of the operation (2) will be $\sum_j \lambda_j^p \alpha_j \theta_{ij}$, as for the class of dynamical problems described in 4·061. If in (5) we add the elements of each column we get n sums all equal to $1-\lambda$, by (1). Hence $\lambda = 1$ is always a root.

It follows that for any set a_{ik} there is a possible set of values of the x_i such that $y_i = x_i$. They need not all be equal, since we have not assumed

$$\sum_k a_{ik} = 1. \tag{6}$$

But this condition is often simply a matter of our choice of what distributions are taken together and what are considered as alternatives. In the card-shuffling problem each a_{ik} is $1/52!$, both i and k have $52!$ possible values, and the condition is satisfied. But if we treated all separately, except that we lumped together those where the ace of spades is later in the order than the ace of hearts, all $\sum_i a_{ik}$ would still be 1, but all $\sum_k a_{ik}$ would not, since for a given i, a_{ik} would be systematically greater if i corresponds to one of the combined alternatives than to one of the unaltered ones. Similarly, in the problem of collisions between molecules of a gas $\sum_k a_{ik}$ will be systematically greater if i refers to a region of larger volume than for a smaller one. But this can be remedied by taking all the regions of equal size. Thus in actual problems (6) will often be satisfied, and if not we can usually choose the alternatives in such a way that it will become satisfied.

We shall therefore assume (6), which makes a considerable simplification in the analysis. We return to (2). Let the greatest and least of the x_i be M and m. We assume that the x_i are not all equal, so that $M > m$. We assume also that no $a_{ik} = 0$. Then, using (6),

$$y_i - M = \sum_k a_{ik}(x_k - M). \tag{7}$$

Terms with $x_k = M$ contribute nothing to this sum and no term is positive. Of the others there is at least one with $x_k = m$, and therefore if α is the least of the a_{ik}, and therefore $\leqslant \tfrac{1}{2}$ if there are at least two states,

$$y_i - M < \alpha(m - M). \tag{8}$$

Similarly $\qquad y_i - m > \alpha(M - m). \tag{9}$

If then M' and m' are any two of the y_i,

$$M' < M - \alpha(M - m), \tag{10}$$

$$m' > m + \alpha(M - m), \tag{11}$$

$$|M' - m'| < (M - m)(1 - 2\alpha). \tag{12}$$

Thus the extreme range of the variables is multiplied at each step by a positive factor less than $1 - 2\alpha$, and must therefore tend to zero with a sufficient number of trials, whatever the initial probabilities of the possible distributions may be. Thus the probabilities of all the distributions tend to equality, provided only that none of the a_{ik} is zero. The theorem can be extended even if some of the a_{ik} are zero. It might be impossible, for instance, to pass directly from state 1 to state 2, but possible to pass from state 1 to state 3 and back to state 2. The analysis applies equally if we apply it to the result of taking r steps, a_{ik} now being the probability that the system would reach state i in r steps, given that it was in state k initially. Evidently it will be possible for all a_{ik} to be positive in this case when some of them are zero for one step. Hence, provided that for some finite number r of steps, and for any state k, there is a non-zero probability for every i that state i will be reached, the probabilities of all states will tend to equality whatever the initial probabilities.

It might happen that $|M'-m'|$ in (12) was 0 for all pairs. In that case uniform probability distribution would be reached in one step.

An alternative proof, using complex variable theory, is due to Fréchet. For a given λ_j let the θ_{ij} of largest modulus be such that $|\theta_{ij}| = R$. Then in the Argand diagram* the θ_{kj} are a set of points within or on a circle of radius R about the origin, and whatever the a_{ik} may be, subject to their not being negative and to their sum being 1, $a_{ik}\theta_{kj}$ cannot lie outside this circle. Therefore $|\lambda| R \leqslant R$ and all

$$|\lambda_j| \leqslant 1. \tag{13}$$

Again for each i, $(\lambda_i - a_{ii})\theta_{ij} = \sum_{k \neq i} a_{ik}\theta_{kj}$

$$|\lambda_j - a_{ii}||\theta_{ij}| \leqslant R \sum_{k \neq i} a_{ik} = R(1 - a_{ii}). \tag{14}$$

But for one θ_{ij}, $|\theta_{ij}| = R$; hence for this i,

$$|\lambda_j - a_{ii}| \leqslant 1 - a_{ii}. \tag{15}$$

If then $a_{ii} > 0$, λ_j lies within or on a circle with centre a_{ii} touching the unit circle at $+1$. Further such a circle centred on the smallest a_{ii} will include all the λ_j. Hence if all a_{ik} with $i = k$ are different from 0 it will be impossible for any λ other than 1 to have modulus 1. But this condition says simply that whatever state the system is in it is not certain to move out of that state; and the conclusion is that when p, the number of steps taken, is large enough λ^p tends to zero unless $\lambda = 1$. Hence if none of $a_{11}, a_{22}, \ldots, a_{nn}$ is zero the probabilities of the states will all tend to definite limits given by the solutions of (4) with $\lambda = 1$. It is possible to have more than one such solution. In fact we could have all the diagonal elements 1 and then $y_i = x_i$ for all i and the probabilities never change. A necessary and sufficient condition that there shall be only one solution with $\lambda = 1$ is that the matrix $a_{ik} - \delta_{ik}$ shall be of rank $n-1$.

Up to a point we can consider all possible solutions with $|\lambda| = 1$ together. In (4) let θ_1 be the θ_i with the largest modulus R. Then if any θ_k such that $a_{1k} \neq 0$ is such that $|\theta_k| < R$

$$R = |\theta_1| = |a_{1k}\theta_k| < \Sigma a_{1k} R = R,$$

which is a contradiction. The argument can then be extended to $i = 2$ if $|\theta_2| = |\theta_1|$, and so on. Then if all the a_{ik} are different from 0 the only solution with $|\lambda| = 1$ is

$$\theta_1 = \theta_2 = \ldots = \theta_n, \quad \lambda = 1,$$

* Cf. 11·04.

and therefore we have another proof that the only possible ultimate steady values of the probabilities are equal. The argument can again be extended to the case where some of the a_{ik} are zero; it can only stop if after all $|\theta_i|$ up to $|\theta_m|$ ($m < n$) it is found that all a_{ik} vanish for $k > m$ and $i \leqslant m$. But this means that it is impossible for any state with $k > m$ to pass to one with $k \leqslant m$. In this case it is easily seen by using the relations (1) and (6) again that all a_{ik} will also vanish if $k \leqslant m$ and $i > m$, and the converse process is also impossible. Then there will be an infinite number of possible limiting states, depending on the total initial probabilities in the various independent sets. Thus the case of a multiple root at $\lambda = 1$ corresponds to the case where the probabilities fall into two or more independent sets: and the limiting probability will be the same for each state of the same set.

This shows incidentally that we cannot *always* make (6) true. For we could certainly have a set of a_{ik} such that probability could pass from the states with $i \leqslant m$ to those with $i > m$ but not conversely. It is clear that in this case the whole probability tends to become concentrated in the latter set.

We have already seen that the existence of a solution with $|\lambda| = 1$ but $\lambda \neq 1$ requires that some a_{ik}, with $i = k$, is zero, and further, in order to make, for $|\theta_k| \leqslant |\theta_1|$,

$$|a_{1k}\theta_k| = |\theta_1|, \tag{16}$$

there must be some θ_k, say θ_2, such that $|\theta_2| = |\theta_1|$, and all θ_k such that $a_{1k} \neq 0$ must be equal to θ_2. Since the a_{ik} are real, $a_{11} = 0$ (as we know already) and $\theta_2 = \lambda\theta_1$.

Now we can take θ_2 in place of θ_1 and infer that there is a θ_3 equal to $\lambda\theta_2$, and that for all k such that $\theta_k \neq \theta_3$, $a_{2k} = 0$. But since only n values of k are available the process must close in a number of steps m, where $m \leqslant n$, and it follows, since all the θ_k found are different, that in every row of a_{ik} all elements are 0 except one, which must therefore be 1. Thus the matrix a_{ik} contains a minor of the form, for $m = 4$,

$$\begin{pmatrix} 0 & 1 & 0 & 0 \\ 0 & 0 & 1 & 0 \\ 0 & 0 & 0 & 1 \\ 1 & 0 & 0 & 0 \end{pmatrix},$$

and the equation $\|a_{ik} - \lambda\delta_{ik}\| = 0$ is satisfied if

$$1 - \lambda^m = 0, \tag{17}$$

the roots of which are $\exp(2r\pi i/m)$, where r is any integer from 0 to $m-1$. The form of a_{ik} shows directly that if the system starts in any state of these m, it will necessarily proceed to the others in a definite order and return to the original one in m steps.

If $m = n$, the system must describe the whole set of possible states. For $m < n$ the states not included in the cycle are independent of those in the cycle and may form other cycles or have $|\lambda| < 1$ for all roots not equal to 1.

The problem therefore resolves itself into several cases.

(1) If the a_{ik} are such that, no matter what the initial state, there is a non-zero probability that any other state will be reached in some given finite number of steps, then the probabilities of all states tend to become equal when the number of redistributions is made large. This is the *ergodic theorem*.

(2) If the a_{ik} are such that the states form sets, each member of any one of which is a possible successor of any other of that set but not of one of any other set, then the cha-

racteristic equation has 1 as a multiple root and the probabilities within each set tend to become equal; but their limiting values are not independent of the initial values.

(3) In the above cases the roots of the characteristic equation are all either 1 or have modulus less than 1. If there is a root with modulus 1 but not equal to 1, some of the states will form a cycle such that each one has a determinate successor, and the original state will be reattained after a number of steps not greater than n.

Case 1 is the one that arises in ordinary card shuffling and in the kinetic theory of gases. (The arguments usually given, due to Boltzmann and Gibbs, are fallacious.) Case 2 could arise in card shuffling if the pack was divided into two halves and these shuffled separately and finally placed together. Obviously many orders attainable by shuffling the complete pack become impossible. But the probability that the aces of spades and hearts will be together will never become independent of the probability that they were in the same half of the pack to start with. Case 3, in card shuffling, would describe a case where the 'shuffle' consisted of always removing one card from the top and putting it at the bottom —which will never give more than a cut and never a true shuffle. But it also connects any deterministic mechanics with chain probabilities, since it shows that a necessary and sufficient condition that later states shall be exactly specifiable given the state at one instant is that the time factors in the solutions shall have modulus 1.*

4·17. Integral equations. These are of several types, the common feature being the occurrence of an unknown function under the integral sign. They have a considerable literature, but can only be considered briefly in the present book. Three related types are as follows:

$$\int_0^a K(x,y)\phi(y)\,dy = f(x), \tag{1}$$

$$\phi(x) + \int_0^a K(x,y)\phi(y)\,dy = f(x), \tag{2}$$

$$\int_0^a K(x,y)\phi(y)\,dy = \lambda\phi(x). \tag{3}$$

Here the limits are fixed; $K(x,y)$ and $f(x)$ are known functions, and ϕ is a function to be found. These equations can be considered as limiting cases of matrix equations. For if we take points of subdivision at $y/a = 0, 1/n, 2/n, \ldots, (n-1)/n$, we can put

$$f(x_i) = X_i, \quad \phi(y_k) = Y_k,$$
$$K(x_i, y_k) = K_{ik}. \tag{4}$$

Then the equations are the limits of

$$\frac{a}{n}\sum_{k=0}^{n} K_{ik}Y_k = X_i, \tag{5}$$

$$Y_i + \frac{a}{n}\sum_{k=0}^{n} K_{ik}Y_k = X_i, \tag{6}$$

$$\frac{a}{n}\sum_{k=0}^{n} K_{ik}Y_k = \lambda Y_i, \tag{7}$$

---

* For applications to statistical mechanics and to quantum theory, cf. *Proc. Roy. Soc.* A, 160, 1937, 337–47; *Phil. Mag.* (7) 33, 1942, 815–31.

when $n \to \infty$. These are useful for numerical solution. With more accurate integration formulae they can be solved directly for any number of intervals up to 12 on Mallock's machine for solving simultaneous equations. But they also show that considerable similarities are to be expected between these types of integral equation and algebraic linear equations. In particular, (7) will in general be soluble only for a certain set of discrete values of λ, n in number, and therefore in the limit there will be an infinite number of solutions. There will be complications in the solutions of (5) and (6) also in cases where the determinant formed by the coefficients of the Y_k vanishes.

The function $K(x,y)$ is called the kernel of the equation. The condition $K_{ik} = K_{ki}$ for a symmetrical matrix corresponds to

$$K(y,x) = K(x,y). \tag{8}$$

If this is true the kernel is said to be symmetrical. We can define a hermitian kernel by the condition

$$K(x,y) = K^*(y,x). \tag{9}$$

Similarly, we can have an antisymmetrical kernel defined by

$$K(x,y) = -K(y,x). \tag{10}$$

Analogues of orthogonal matrices exist. In (1), a solution may be

$$\phi(y) = \int_0^a K(y,x) f(x)\, dx, \tag{11}$$

which is easily seen to correspond to the solution of a set of simultaneous equations by multiplication by the reciprocal matrix when $a\tilde{a} = 1$.

Integral equations can seldom be solved in finite terms; but there are extensive theories of their solution by successive approximation.*

EXAMPLES

1. By considering the matrices $\begin{pmatrix} 1 & 0 \\ 0 & 0 \end{pmatrix}$, $\begin{pmatrix} 0 & 1 \\ 1 & 0 \end{pmatrix}$, show that two symmetrical matrices do not necessarily commute.

2. Prove that $\begin{pmatrix} \cos\theta & -\sin\theta \\ \sin\theta & \cos\theta \end{pmatrix} = \begin{pmatrix} 1 & -\tan\tfrac{1}{2}\theta \\ \tan\tfrac{1}{2}\theta & 1 \end{pmatrix} \begin{pmatrix} 1 & \tan\tfrac{1}{2}\theta \\ -\tan\tfrac{1}{2}\theta & 1 \end{pmatrix}^{-1}$.

3. One of the quadratic forms

$$3x^2 + 2y^2 + 5z^2 + 2yz - 2zx, \quad x^2 + 2y^2 + 8yz + 12zx + 12xy$$

is positive definite. Determine which, and find a real linear transformation that reduces them to the forms $\xi^2 + \eta^2 + \zeta^2$, $4\xi^2 - 2\eta^2 - \zeta^2$. (Prelim. 1943.)

---

* E. Schmidt, *Math. Ann.* 63, 1907, 433–76; F. Smithies, *Proc. Lond. Math. Soc.* 43, 1937, 255–79; 46, 1940, 409–66; *Duke Math. J.* 8, 1941, 107–33. H. Bückner, *Die praktische Behandlung von Integralgleichungen*, Springer, 1952.

Examples

4. Find a real non-singular linear transformation that will reduce the form

$$P - \frac{2}{n-1}Q$$

to the standard form
$$y_1^2 + y_2^2 + \ldots + y_p^2 - y_{p+1}^2 - \ldots - y_{p+q}^2,$$

where
$$P \equiv x_1^2 + \ldots + x_n^2, \quad Q = \sum_{i<j} x_i x_j.$$

Determine for all real λ the rank $r = p+q$ and the signature $s = p-q$ of the forms $P + 2\lambda Q$.
(M.T. 1935.)

5. Two uniform bars AB, BC are freely pivoted at B, and supported in a straight line by springs of stiffness λ at A, B and C. The mass of each bar being m, show that the periods of the normal modes of vibration are given by $m\gamma^2/\lambda = 3, 3 \pm \sqrt{3}$. (I.C. 1940.)

6. A long chain of rods each of mass $3m$ and length $2a$, pivoted freely at the joints, lies in a straight line, and at each joint there is a spring producing a restoring force equal to $-m\lambda_0^2$ times the transverse displacement. One end of the chain is acted on by a force of period $2\pi/\lambda$, and the other is fixed. Show that if $\lambda_0/\sqrt{3} < \lambda < \lambda_0$ all rods will be equally agitated, but that for other values of λ the motion is confined to the neighbourhood of the exciting end. (This is a mechanical analogue of a radio frequency filter.)

7. A light string AB of length l has $n-1$ particles $P_1, P_2, \ldots, P_{n-1}$ of mass m/n attached to it, so that $AP_1 = P_1P_2 = \ldots = P_{n-1}B$. A and B are fixed, and the whole is under tension P.
Show that in a normal mode of period $2\pi/\gamma$ the amplitudes a_j of small transverse motions of the particles are connected by the relation

$$a_{j+1} - 2\cos\alpha \, a_j + a_{j-1} = 0 \quad (1 \leqslant j \leqslant n-1),$$

where $\cos\alpha = 1 - \gamma^2 ml/2n^2T$; find the normal frequencies.

By taking the limit as $n \to \infty$, obtain the normal frequencies of a uniform *heavy* string with the ends fixed.

8. Illustrate Rayleigh's method by considering the vibration of a uniform rod of length l and flexural rigidity EI, clamped at both ends which are free from end-thrust. Show that Rayleigh's method gives the fundamental frequency ν as $\frac{1}{2\pi}\sqrt{(504EI/\rho l^4)}$ if the approximate form of the rod at any instant is taken to be given by $y = f(x)$, where $f(x)$ is the simplest polynomial satisfying the boundary conditions. (M/c, III, 1936.)

9. If a, b are hermitian, show that $ab + ba$ and $i(ab - ba)$ are hermitian.

10. Prove that if H is a hermitian matrix, the matrix $U = (1 - iH)(1 + iH)^{-1}$ exists and is unitary; and that if U is a unitary matrix, $H = -i(1 + U)^{-1}(1 - U)$ exists and is hermitian provided that U has not a characteristic value -1.

Hence show that any characteristic solution of U is one of H and conversely, and hence that there is a unitary matrix l such that $l^\dagger U l$ is diagonal. Extend the last result to the case where U has a characteristic value 1.

11. Show that any 2×2 matrix that anticommutes with Pauli's σ_1 is of the form $a\sigma_2 + b\sigma_3$. Hence show that there are 8 linearly independent 4×4 matrices that anticommute with Eddington's E_1.

12. If a quaternion is defined by the rule

$$u = 1u_0 + iu_1 + ju_2 + ku_3,$$

where
$$i^2 = j^2 = k^2 = -1, \quad ij = k, \quad ji = -k,$$

and u_0, u_1, u_2, u_3 are real numbers, show that for any quaternions u, w there exists a quaternion v such that

$$uv = w,$$

provided that u_0, u_1, u_2, u_3 are not all zero.

Show that the Eddington matrices E_{12}, E_{23}, E_{31} satisfy the conditions for i, j, k.

Examples

13. A triangular matrix T is defined as one such that $T_{ik} = 0$ if $k > i$, or one such that $T_{ik} = 0$ if $k < i$. Show that a triangular matrix with all its diagonal elements different from zero has an inverse, and relate this result to the method of solving a set of linear simultaneous equations $a_{ik} x_k = b_i$ by successive elimination.

Show also that if $TT^{\dagger} = T^{\dagger}T$, then T is diagonal.

14. Show that any transformation of the form $B = PAQ$, where P, Q are non-singular, makes the ranks of A, B, equal.

15. Show that if H is hermitian $l^{\dagger}Hl$ is hermitian, and that if A is antihermitian $l^{\dagger}Al$ is antihermitian.

16. A is $n \times n$ and has n different latent roots, and B commutes with A. Prove that B can be expressed as a polynomial in A of degree $n - 1$.

17. $l^{-1}Al$ is diagonal, and all its diagonal elements have modulus 1; and l is unitary. Prove that A is unitary.

Note on Orthogonal Transformations.

The treatment of orthogonal transformations in three dimensional Euclidean space in Chapters 2, 3 and 4 is almost entirely restricted to those of determinant $+1$, that is to say to rotations. Those of determinant -1 represent a reflexion in a plane followed by a rotation about the normal to the plane (*a rotatory reflexion*). Since a reflexion in a plane is equivalent to an inversion in the origin (reversal of sign of all coordinates) followed by a rotation through π about the normal to the plane, we have also that any orthogonal transformation of determinant -1 represents a *rotatory inversion*.

A full discussion of the isometries of Euclidean space is given by H. S. M. Coxeter in *Introduction to Geometry*, 1961, in particular Chapters 3 and 7. The transformation matrices are given explicitly by H. Jeffreys in the *Mathematical Gazette*, XLIX, 192–4 (1965).

Chapter 5

MULTIPLE INTEGRALS

'One by one, or all at once.'

W. S. GILBERT, *The Yeomen of the Guard*

5·01. Distance; neighbourhoods; curves; regions. In this chapter we shall be concerned with functions of two or more variables, say x, y, z. These will often be rectangular coordinates of a point in the ordinary sense, but not always. We shall generally take them to be two or three in number, but extensions to more will be obvious. Some simple ideas from Euclidean geometry can be extended usefully to the cases where the variables are not rectangular coordinates. One of the chief of these is *distance*. The distance r between two points given by rectangular coordinates (x, y, z) (x', y', z') is the non-negative value that satisfies
$$r^2 = (x'-x)^2 + (y'-y)^2 + (z'-z)^2. \tag{1}$$
If the variables are not all of the same dimensions this will be meaningless, but we can extend the definition by taking
$$r^2 = \alpha(x'-x)^2 + \beta(y'-y)^2 + \gamma(z'-z)^2, \tag{2}$$
where α, β, γ are positive constants chosen to make addition possible; and then by a change of variables we can reduce this again to the form (1). The distance so defined will not in general be the Euclidean distance; for instance, x, y, z may be spherical polar coordinates, and r will then be quite different from the distance expressed in terms of these coordinates. Often x, y, z will be numbers. The reason for introducing it is that if $x' \to x$, $y' \to y$, $z' \to z$, then $r \to 0$, and conversely, so that the statement that two sets of values of the variables are near together can be adequately summarized by the single statement that the distance is small.

Note that if P, Q, R are three points with coordinates x_i, y_i, z_i (i not necessarily running from 1 to 3)
$$y_i - z_i = (y_i - x_i) - (z_i - x_i), \tag{3}$$
$$r_{QR}^2 = r_{PQ}^2 + r_{PR}^2 - 2\Sigma (y_i - x_i)(z_i - x_i). \tag{4}$$
But by Cauchy's inequality
$$\{\Sigma (y_i - x_i)(z_i - x_i)\}^2 \leq \Sigma(y_i - x_i)^2 \Sigma (z_i - x_i)^2 = r_{PQ}^2 r_{PR}^2. \tag{5}$$
Hence
$$(r_{PQ} - r_{PR})^2 \leq r_{QR}^2 \leq (r_{PQ} + r_{PR})^2, \tag{6}$$
$$|r_{PQ} - r_{PR}| \leq r_{QR} \leq r_{PQ} + r_{PR}. \tag{7}$$
This is an extension of a familiar inequality in Euclidean geometry; but it is here to be understood as a purely analytical theorem. By induction, if $P_r(x_{ri})$ are a set of points, the distance from P_1 to P_n is $\leq$ the sum of the distances from P_1 to P_2, P_2 to P_3 ..., P_{n-1} to P_n.

We shall use geometrical language freely; in particular, we shall say that a given set of values of the variables specifies a *point*. We need an analogue for more than one variable of the notion of an *interval* of one variable. This contemplates (1) that the values of the variable, if necessary after multiplying by a constant dimensional factor, are real numbers, and uses the facts: (2) between any two real numbers there is another; (3) any limit-point of a set of real numbers is also a real number. Further (4) if x_1, x_2 are points of an interval, any value between x_1 and x_2 is also a point of the interval—intervals have no gaps in them; and (5) an interval, except that of the whole set of finite real numbers, has at least one end-point, which may be characterized by the fact that within any distance of it, however small, there are some real numbers that belong to the interval and some that do not. An end-point may belong to the interval or not. (2) and (4) require only a notion of an order; we could conceivably say $x_2 > x_1$ without saying how much it is greater. But (3) and (5) do require the idea of a distance, which we express by $|x_2 - x_1|$.

The generalization of an interval of one variable will be called a *region* for two or more. We are not yet ready to define it, but we have seen that distance for one variable plays an essential part, and that distance is easily defined for any finite number of variables. It enables us to give an immediate definition of a *limit-point* in m dimensions in terms of the notion of a *neighbourhood*. A neighbourhood of a point P specified by coordinates (x, y, z) in three dimensions is the set of points $Q(x', y', z')$ such that the distance PQ, defined as the non-negative value of r satisfying

$$r^2 = (x'-x)^2 + (y'-y)^2 + (z'-z)^2,$$

is less than some positive quantity δ. (If we exclude the point P itself we say so explicitly.) If R is any set of points, P is a limit point of the set R if for any δ there is a point of R, not identical with P, whose distance from P is less than δ; or more briefly, if every neighbourhood of P contains at least one point of R other than P, and therefore an infinite number as in one dimension. A set is *closed* if all limit points of subsets of it are members of it.

For more than one variable, we might require that, in general, if all but one are kept fixed, the other should be capable of values over an interval and no others. But this leads at once to a difficulty. In two dimensions, if y is fixed, it would say that the values of x must lie in one interval; thus we should not be able to call the interior of a polygon with a re-entrant angle a region. Thus the extension of (4) to more than one dimension requires some modification of the obvious procedure of varying only one variable at a time. We say instead that if P, Q are within a region they can be connected by a curve lying wholly within the region. This rule, however, makes it necessary also to provide a definition of a curve.

The notion of end-points is replaced by that of boundary points; that of interior points presents no difficulty.

Since we shall provide an analytical definition of a curve, we might attempt to provide one of a region by saying that it is a set of points specified by an inequality, just as an interval may be specified by $a < x < b$. In fact, regions are often specified in this way, but a precaution is needed. Suppose that we took the inequality $|x| \times |x-2| + |y| < \frac{1}{2}$. This specifies two patches about $(0,0)$ and $(2,0)$, with no point in common. The inequality $|\sin x| + |y| < \frac{1}{2}$ specifies an infinite set of patches about the points $x = n\pi$, $y = 0$, no two of which have a point in common. A single inequality therefore does not necessarily provide a satisfactory specification of a region.

5·011. Curves. We have to translate into mathematical language what we mean by saying that a set of points forms a curve. We shall state the argument for three dimensions, but it applies for any finite number. The first essential is that the points occur in an order as we proceed along the curve. The second is that there are no gaps in the curve—we can travel between any two points of the curve without ever leaving the curve. We can express these conditions by saying that the variables x, y, z, for points on the curve, are continuous functions of a parameter t in an interval of t. (If the interval of t is finite and closed, x, y, z will be bounded.) The order of increasing t specifies the order of points on the curve; and the condition that x, y, z are continuous functions of t and all values of t in the interval are admissible ensures the absence of gaps. It may appear that the assumption that x, y, z are continuous functions of a continuous parameter t is not obviously true for all sets that we may wish to call curves. But the definition gives more generality than we need, not less. Curves have been defined satisfying this condition, with t bounded, that pass through every point of the closed region bounded by a unit square.* The possibility of obtaining useful results about curves arises from the further restriction, which we shall come to in a moment, that they have lengths; and if they have, the length along the curve from a fixed point to the point considered will serve as the parameter t.

A curve will have a multiple point if there are unequal values $t = c$, $t = d$, such that $(x, y, z)_{t=c} = (x, y, z)_{t=d}$. Curves with no multiple points are called *simple*.

If x, y, z are bounded on the curve and take the same values for both extreme values of t, the curve is *closed*. A simple curve that is not closed is called an *arc*. (The meaning of *closed* as applied to curves is quite unrelated to its meaning as applied to intervals, and as we shall apply it to regions.) A closed curve in two dimensions is usually called a *contour*.

5·012. The length of a curve is defined as follows. Take points $P_r(x_r, y_r, z_r)$ on the curve, in order of increasing t_r, where A is (x_0, y_0, z_0) and B is

$$(X, Y, Z) = (x_n, y_n, z_n),$$

define h_r as the positive value satisfying

$$h_r^2 = (x_{r+1} - x_r)^2 + (y_{r+1} - y_r)^2 + (z_{r+1} - z_r)^2,$$

and let
$$s_n = \sum_{r=0}^{n-1} h_r.$$

As n increases indefinitely, the largest interval of t tending to zero, s_n may tend to a limit. If it does, and if the limit is independent of the way of choosing the t_r at each stage, the limit is called the length of the curve between A and B. An equivalent statement is that the length is the upper bound (if one exists) of s_n for all ways of choosing n and the t_r; and evidently it is unaltered if t is replaced by any parameter t' that is a monotonic function of t. We denote the length of the curve between A and an arbitrary point P by s. A given value of s defines a point on the curve, and s, if it exists, can be used as a parameter in place of t. Curves with lengths are called *rectifiable*.

5·013. *A necessary and sufficient condition for a curve to have a finite length is that x, y, z all have bounded variation on the curve.* First assume that the curve has a finite length. We have
$$h_r \geqslant |x_{r+1} - x_r|, \quad s \geqslant s_n \geqslant \sum |x_{r-1} - x_r|.$$

Hence for any choice of the points P_r, $\sum |x_{r+1} - x_r|$ is not greater than the length of the curve; hence x has bounded variation on the curve. Similarly, y and z have bounded variation on the curve. Conversely, let V_x, V_y, V_z, the total variations of x, y, z on the curve, all be less than M. Then for any subdivision

$$\Sigma h_r \leqslant \Sigma \{|x_{r+1} - x_r| + |y_{r+1} - y_r| + |z_{r+1} - z_r|\} \leqslant 3M,$$

* First by Peano: cf. Hobson, *Functions of a Real Variable*, Camb. Univ. Press, 1907, 330.

and therefore Σh_r is bounded above. But adding new points of subdivision cannot reduce Σh_r; hence for any process of subdivision Σh_r tends to a limit. Uniqueness of the limit is proved as for the Riemann integral.

5·014. Hence we are led to the following definitions. An *interior point* of a set of points in m dimensions is a point P of the set such that for some δ, every point within distance δ of P is a member of the set. An *exterior point* Q is one not of the set such that for some δ, no point within distance δ of Q is a member of the set. A *boundary point* R is a point (either of the set or not) such that however we choose δ, within distance δ of R there is at least one point that is a member of the set and at least one that is not. A *region* is a set such that every pair of points of the set can be joined by a curve all of whose points, except possibly the end points, are interior points of the set. A region is *closed* if all its boundary points are members of it, *open* if all its points are interior points. It is *bounded* if for all pairs P, Q of points of it, the distances PQ have an upper bound. This upper bound is called the *diameter*. Clearly in a bounded region the values of all coordinates are bounded. As for intervals, if we refer to a region as closed we imply that it is bounded.

We take the following examples, mostly in two dimensions.

(1) The circle $x^2 + y^2 = 1$ is not a region. For if we take any circle about a point of it, it contains points that are on the circle and points that are not. Hence all points of the set are boundary points; and if we pass from one point to another along the circle, we never pass through an interior point at all.

(2) The points where $x^2 + y^2 < 1$ form an open region. The boundary is the circle
$$x^2 + y^2 = 1.$$
About any point within the circle we can draw a circle with sufficiently small radius for all points within it to satisfy $x^2 + y^2 < 1$; hence all points are interior points and the region is open. The diameter is 2.

(3) The points where $x^2 + y^2 \leqslant 1$ form a bounded closed region.

(4) The points where $0 \leqslant x \leqslant a$, $0 \leqslant y \leqslant b$ form a bounded closed region of diameter $(a^2 + b^2)^{\frac{1}{2}}$.

(5) The points where $0 \leqslant x \leqslant a$, $0 < y < b$ form a bounded region, which is neither open nor closed, since part of the boundary belongs to the region and part does not.

(6) The set of all points in a plane is an unbounded open region.

(7) The set of all points in the plane, excluding points on the line from $(-1, 0)$ to $(1, 0)$ is an unbounded open region. The line is the boundary—in this case an internal one.

(8) The set of points such that $0 < x^2 + y^2 < 1$ is a bounded open region; the origin and the circle together constitute the boundary.

(9) The set of all points interior to two circles of radius $\frac{1}{2}$ and centres $(-1, 0)$ and $(1, 0)$ is not a single region, because if we take a pair of points, one within each circle, there is no curve that connects them without passing outside both circles.

(10) Even the set of all points within and on two circles of radius 1 and centres $(-1, 0)$ and $(1, 0)$ is not a single region. For though we can pass from any point of it to any other without leaving the set, we cannot pass from $(-1, 0)$ to $(1, 0)$ in the set without passing through $(0, 0)$, which is not an interior point but a boundary point.

(11) The set of points defined by $1 \leqslant x^2 + y^2 \leqslant 4$ is a closed region, but has the peculiarity that we can pass from any point of it to any other by paths of two distinct sets (clockwise and counterclockwise) while keeping within the region, and a path of one set cannot be

deformed into one of the other without passing outside the region. Such regions are called *multiply connected*, and play a prominent part in the theory of functions of a complex variable.

(12) On the other hand, in three dimensions, if we take the region given by
$$1 \leqslant x^2+y^2+z^2 \leqslant 4,$$
though it has the same property that we may describe as having a hole in it, any path connecting two points can be continuously deformed into any other. Thus there is a quite fundamental difference between regions in two and more dimensions.

In two dimensions any simple closed curve divides the plane into two regions, an interior and an exterior one; this looks obvious, but is in fact difficult to prove,* and we shall assume it. Conversely, the boundary of a region in two dimensions may be a simple closed curve, but may consist in part or wholly of arcs or isolated points, as in examples (7), (8). In m dimensions, where $m>2$, a closed curve does not divide the space into parts; the boundary of a region is usually a region of $m-1$ dimensions, but may be of any smaller number. If $m=3$ the boundary is usually of two dimensions and is specified by taking the coordinates as functions of two parameters; such sets are called *surfaces*.

By definition every point of a region and every boundary point are limit-points of points of the region, whether the region is open or closed or neither. Conversely, any limit-point of a set of points of a region is a point of the region or a boundary point (and therefore a point of the region if the region is closed). For if not, it would be an exterior point and therefore no point of the region would be within a certain non-zero distance from it.

5·02. The Heine-Borel theorem in m dimensions. *Let D be a bounded closed set of points; let F be a family of regions I such that every point P of D is an interior point of one of the family, say I_P; then a finite subset of F exists such that the same relations hold for the I of the subset.* The proof is almost the same as in one dimension. We state it as for three dimensions. Since D is bounded it lies within a cube E of side L, with sides parallel to the axes. As before, if all points of D are interior to a single I there is nothing to prove. If not, divide E by planes parallel to all the axes into 8 cubes of side $\frac{1}{2}L$. Any cube that contains no points of D needs no further consideration. If for each cube that contains points of D these points are all interior to a single I the result holds; if not, again divide any cube that is not included in a single I into eight equal cubes. Then if the result is not established in a finite number of steps we have a nest of cubes, each part of the preceding one, with half the side, all containing at least one point of D, and none included in an I. This nest determines a limit point P_0, which must be a point of D since D is closed. But P_0 is interior to an I, say I_0; and therefore there is a δ such that all points within distance δ of P_0 lie in I_0. But for some n the diameter of the cube of side $2^{-n}L$ is less than δ and therefore this cube is contained in I_0. Thus, contrary to supposition, there is an n such that the cube of the nest of side $2^{-n}L$ is wholly interior to an I. Hence the result follows.

Note that the argument does not assume that D is a region. It might be the set of points $(1/n, 0, 0)$, where n is a positive integer, together with $(0, 0, 0)$, which is the only limit point of the set.

Note also that, though we have proceeded by dividing only cubes not yet covered by an I, the theorem follows equally if we divide all cubes at each stage; for if a single I includes a cube it also includes any of its parts. Thus there is no objection to taking all the cubes equal.

* M. H. A. Newman, *Topology of Plane Sets*, 1939, 104.

As in one dimension we may relax the conditions slightly. If P is a boundary point of D it is enough that it should also be a boundary point of a closed I_P such that all points of a neighbourhood of P on the boundary of D belong to I_P. For about each boundary point we may take a sphere and define J_P as the region consisting of all points belonging either to I_P or to this sphere (or both); and for other points of D take J_P identical with I_P. Then the conditions hold for J_P; and each boundary point is a point of I_P, though not necessarily an interior point.

An immediate consequence, as for one variable, is that *an infinite set of points in a bounded closed region has a limit-point belonging to the region.*

5·021. The modified Heine-Borel theorem. *Let D be a bounded closed region. Let every point P of D be interior to a region I_P of a family F. Then D can be divided into a finite set of regions G such that every G_P is part of the I_P associated with a point P common to G_P and D. The same relaxation for boundary points of D can be made as for the main theorem.*

The proof is the same as for the Heine-Borel theorem. But now we cannot necessarily make all the cubes equal; for if a cube is covered by an I_P corresponding to a point P interior to the cube, and is then divided, only one part can contain this P as an interior point; and then another part may not be covered by an I_Q corresponding to any Q of itself.

5·03. Functions of several variables. As for one variable, a function is defined for all points of a set. In particular we may consider its behaviour over a region.

5·031. Continuity in more than one dimension. We say that *a function $f(P) = f(x,y)$, $f(x,y,z), \ldots$ is continuous at a point P if for any ϵ there is a neighbourhood of P such that for all points Q of this neighbourhood $|f(Q)-f(P)| < \epsilon$.* Clearly if f is continuous in this sense, it is a continuous function of any coordinate when the others are kept constant. The converse is not true. Take
$$f(x,y) = \frac{xy}{x^2+y^2},$$
except when $x = y = 0$, when $f(0,0) = 0$. This is a continuous function of x for any fixed y, and of y for any fixed x. But if $x = r\cos\theta$, $y = r\sin\theta$,
$$f(x,y) = \cos\theta\sin\theta$$
which takes all values from $-\tfrac{1}{2}$ to $\tfrac{1}{2}$ irrespective of r. Hence in any neighbourhood of $(0,0)$ there are points where $f(x,y)$ differs from $f(0,0)$ by more than $\tfrac{1}{4}$.

A function is continuous in a region if for every P of the region and any ϵ there is a $\delta(P)$ (depending possibly on the position of P) such that for any Q of the region satisfying
$$r_{PQ} < \delta(P), \quad |f(Q)-f(P)| < \epsilon.$$
This does not assume that f is continuous at boundary points; the inequality stated might not hold if P is a boundary point and Q a point exterior to the region, but we are not interested in exterior points and therefore insert the words 'of the region' in the second line.

Since a neighbourhood is a region, we can infer at once from the Heine-Borel theorem that *if $f(P)$ is continuous in a bounded closed region, for any ϵ there is a δ such that if P and Q are any points of the region satisfying*
$$r_{PQ} < \delta, \quad \text{then} \quad |f(Q)-f(P)| < \epsilon.$$
The following theorems may be proved by similar methods to those used for one variable:

A function continuous in a closed region is bounded in the region, and attains its upper and lower bounds.

If $f(P)$ is continuous in a region and there are points A, B in the region where
$$f(A) < 0, \quad f(B) > 0,$$
then for any curve in the region connecting A, B there is a point P on the curve where $f(P) = 0$.

5·032. We define a distance between a point P and a set of points S as the lower bound of the distances between P and points of the set. We define a distance between two sets similarly.

The distance of a point P from a set S is a continuous function of the position of P. Denote the distance of P from the set by $\delta(P)$. Then for any ω there is a point Q of S such that
$$\delta(P) \leq PQ \leq \delta(P) + \omega.$$
Take another point P' and put $PP' = r$. Then
$$PQ - r \leq P'Q \leq PQ + r.$$
Similarly there is a Q' such that
$$\delta(P') \leq P'Q' \leq \delta(P') + \omega$$
and
$$P'Q' \leq P'Q.$$
Hence
$$\delta(P') \leq P'Q' \leq P'Q \leq PQ + r \leq \delta(P) + r + \omega.$$
Similarly
$$\delta(P) \leq \delta(P') + r + \omega;$$
hence
$$|\delta(P') - \delta(P)| \leq 2r + 2\omega.$$
Take $\omega = \tfrac{1}{4}\epsilon$; then for all $r < \tfrac{1}{4}\epsilon$ $\quad |\delta(P') - \delta(P)| < \epsilon$,
and the result follows.

It follows that *if C is a curve all points of which are interior to a region, the distance of C from the boundary of the region is positive.* For the distances of points of C from the boundary are a set of positive quantities, and their lower bound is positive or zero. But since the distance is a continuous function of position on C it takes its lower bound; the distance never being zero the lower bound is not zero.

5·033. *Any two interior points A, B of a region can be connected by a finite number of displacements, in each of which only one coordinate varies.* We state the proof as for two dimensions. By definition of a region A, B can be connected by a curve C wholly interior to the region, and this curve has a positive distance δ from the boundary. Also, since (x, y) on the curve are continuous functions of a parameter t, the distance of a point on the curve from a fixed point is a continuous function of t, and is therefore bounded. Take a square with sides parallel to the axes and large enough to include the whole of C, and subdivide it into squares of diameter less than δ. Of those intersected by the curve none meets the boundary of the region. The curve may intersect the boundary of a square many times, even an infinite number. But if $t = 0$ at A, the values of t such that the point on the curve specified by t is in the same square as A have an upper bound, which we may say defines the point P_1 where the curve last leaves the square containing A and enters another. Then for the square entered at P_1 take P_2, the last point where the curve enters another, and so on. The number of squares is finite; hence the number crossed by the curve is finite, so that it is possible to proceed from A to B by a finite number of steps each within one square. For each square we can now proceed from P_r to P_{r+1} by changing the coordinates in succession.

We see incidentally that *any two interior points of a region can be connected by a rectifiable curve.*

A consequence is: *If the partial derivatives of a function are zero everywhere in a region, the function is constant at all interior points.* If x alone is varied in a displacement, and $\partial f/\partial x = 0$ from $x = x_r$ to x_{r+1}, with y, z constant, it follows from 1·13 that $f(x_r, y) = f(x_{r+1}, y)$, and similarly for variations of y or z alone. Hence we can show in a finite number of steps that f has the same value, say a, at any two interior points. If in addition $f(x, y)$ is continuous in a closed region and P is a boundary point, for any ϵ there is an interior point Q such that $|f(Q) - f(P)| < \epsilon$, and therefore $|f(P) - a| < \epsilon$. Thus $f(x, y)$ is also equal to a at boundary points if it is continuous in the closed region.

This result is the basis of the extension to several dimensions of the method of finding an integral as the inverse of a derivative. It would not be true if we had defined a region in a way that permitted it to consist of several detached pieces.

5·04. Differentiability in more than one dimension. The derivative of a function $f(x)$ of one variable is

$$f'(x) = \lim_{h \to 0} \frac{1}{h} \{f(x+h) - f(x)\}. \tag{1}$$

This may be written
$$f(x+h) = f(x) + hf'(x) + o(h). \tag{2}$$

In several dimensions we want an approximation to the values of $f(P)$ in a neighbourhood of P, with an error again that becomes arbitrarily small compared with the distance when the distance is small enough. The kind of approximation needed is therefore, for two variables,

$$f(x+h, y+k) = f(x,y) + hf_x + kf_y + \lambda(h^2+k^2)^{\frac{1}{2}}, \tag{3}$$

where f_x, f_y are independent of h, k, and λ for given x, y is a function of h, k, tending to 0 when $h^2 + k^2 \to 0$. Taking $k = 0$, $h = 0$ in turn we have

$$\frac{\partial f}{\partial x} = f_x; \quad \frac{\partial f}{\partial y} = f_y \tag{4}$$

by definition of partial derivatives. Then we may write the condition as: *$f(x, y)$ is differentiable at (x, y) if, for any ϵ, there is a δ such that for $h^2 + k^2 < \delta^2$*

$$|f(x+h, y+k) - f(x,y) - hf_x - kf_y| \leq \epsilon(h^2+k^2)^{\frac{1}{2}}, \tag{5}$$

where f_x, f_y are independent of h and k. We shall see in a moment that this condition is more severe than the mere existence of the partial derivatives, less severe than their continuity.

If for every $P(x, y)$ of a region, a relation of the form (5) is satisfied for given ϵ by the values of f at other points $Q(x+h, y+k)$ of the region satisfying $(h^2+k^2)^{\frac{1}{2}} < \delta$, where δ may now depend on P, $f(x, y)$ is said to be differentiable in the region.

The modifications for larger numbers of variables are obvious.

This definition of differentiability of a function of several variables is due to Stolz and W. H. Young.[*] It is the basis of the theory of functions of a complex variable (cf. Chapter 11) and also has important physical applications, of which we have made use in Chapter 3, and which follow immediately from the definition. *If a function is continuous in a region, a sufficient condition for its gradient at a point to be a vector is that the function shall be differentiable there. A sufficient condition for the gradient of a function to be a vector function in a region is that the function shall be differentiable at every point of the region. Similarly, if u_i is a vector function in a region, a sufficient condition that $\partial u_i/\partial x_k$ shall be a tensor of the second order is that the components u_i shall be differentiable.*[a]

[*] *Proc. Lond. Math. Soc.* (2) 7, 1909, 157–80.

Note that the partial derivatives of a function may exist everywhere without the function being differentiable. This is true, for instance, of the function

$$f(x,y) = \frac{xy}{x^2+y^2}, \quad f(0,0) = 0.$$

Here at $(0,0)$ we have $\partial f/\partial x = \partial f/\partial y = 0$, but for small x, y

$$\frac{1}{r}\left\{f(x,y) - f(0,0) - x\frac{\partial f}{\partial x}(0,0) - y\frac{\partial f}{\partial y}(0,0)\right\} = \frac{xy}{r(x^2+y^2)},$$

which is unbounded in any neighbourhood of $(0,0)$.

Even for the function $f(x,y) = \dfrac{x^2 y}{x^2+y^2}$, $f(0,0) = 0$,

we find $\dfrac{1}{r}\left\{f(x,y) - f(0,0) - x\dfrac{\partial f}{\partial x}(0,0) - y\dfrac{\partial f}{\partial y}(0,0)\right\} = \dfrac{x^2 y}{r(x^2+y^2)} = \dfrac{x^2 y}{r^3}$

which takes values over an interval $\left(-\dfrac{2}{3\sqrt{3}}, \dfrac{2}{3\sqrt{3}}\right)$ within any neighbourhood of the origin.

5·041. *If the partial derivatives of a function are continuous in a neighbourhood of a point, the function is differentiable at the point.* For

$$f(x+h, y+k) - f(x,y) = \{f(x+h, y+k) - f(x+h, y)\} + \{f(x+h, y) - f(x,y)\}$$
$$= k\left(\frac{\partial f}{\partial y}\right)_{x+h, y+\theta k} + h\left(\frac{\partial f}{\partial x}\right)_{x+\phi h, y} \tag{1}$$

for some θ, ϕ between 0 and 1, since the derivatives exist throughout the interval considered; and since they are continuous, for any positive ω there is a δ such that if

$$h^2 + k^2 < \delta^2, \tag{2}$$

$$\left|\left(\frac{\partial f}{\partial x}\right)_{x+h, y+k} - \left(\frac{\partial f}{\partial x}\right)_{x, y}\right| < \omega, \quad \left|\left(\frac{\partial f}{\partial y}\right)_{x+h, y+k} - \left(\frac{\partial f}{\partial y}\right)_{x, y}\right| < \omega, \tag{3}$$

and therefore, since $h^2 + \theta^2 k^2 < \delta^2$, $\phi^2 h^2 < \delta^2$,

$$\left|f(x+h, y+k) - f(x,y) - h\left(\frac{\partial f}{\partial x}\right)_{x,y} - k\left(\frac{\partial f}{\partial y}\right)_{x,y}\right| < \omega|h| + \omega|k| \leq \omega\sqrt{2}\sqrt{(h^2+k^2)}. \tag{4}$$

This is the simplest sufficient condition for differentiability. To show that it is not a necessary condition, take

$$f(x,y) = x^2 \sin\frac{1}{x}; \quad f(0,0) = 0.$$

$\partial f/\partial x$ and $\partial f/\partial y$ exist everywhere and are zero at $(0,0)$; and evidently

$$\frac{1}{r}\{f(x,y) - f(0,0)\} = \frac{x^2}{r}\sin\frac{1}{x},$$

which tends to 0 with r. Hence the function is differentiable at $(0,0)$. But for $x \neq 0$

$$\frac{\partial f}{\partial x} = -\cos\frac{1}{x} + 2x\sin\frac{1}{x},$$

which is not continuous in a neighbourhood of $(0,0)$.

5·042. *Covering theorem for differentiable functions.* For variety we take the case of three variables. If $f(x,y,z)$ is differentiable throughout a bounded closed region D, for any ϵ it is possible to superpose a finite set of cubes G, such that for each G_r there is a point P_r common to G_r and D such that for any point belonging to G_r and to D

$$|f(x,y,z) - (x-x_r)f_x(P_r) - (y-y_r)f_y(P_r) - (z-z_r)f_z(P_r)| < \epsilon\{(x-x_r)^2 + (y-y_r)^2 + (z-z_r)^2\}^{\frac{1}{2}}.$$

An inequality of the form 5·04 (5) extended to three variables holds for a neighbourhood of every point of D. Hence, by the modified Heine-Borel theorem, we can superpose a finite set of cubes with the required property. The cubes will in general not be equal.

5·05. Double integrals. We define first the double integral of a function $f(x,y)$ over a rectangle $0 \leqslant x \leqslant A$, $0 \leqslant y \leqslant B$. We suppose the rectangle divided into rectangles of sides h_r, k_s all of whose diagonals are $\leqslant \delta$; in each of these we take a specimen value of $f(x,y)$, say $f(\xi_r, \eta_s)$, and form the sum $\Sigma\Sigma f(\xi_r, \eta_s) h_r k_s$. If this sum tends to a unique limit when $\delta \to 0$, this limit is the double integral

$$\int_0^A \int_0^B f(x,y)\,dx\,dy. \tag{1}$$

A necessary and sufficient condition for the existence of the double integral is that $f(x,y)$ shall be bounded and that for any ω, α the points where the leap of $f(x,y)$ is $\geqslant \omega$ can be enclosed in a finite set of rectangles whose total area is $< \alpha$. The proof is a straightforward extension of that of the corresponding theorem for the Riemann integral. It is easy to show that a rectifiable curve can be enclosed in a finite set of rectangles of arbitrarily small area. Consequently discontinuities along a finite set of rectifiable curves do not affect the existence of the integral.

A double integral, say $\iint g(x,y)\,dx\,dy$, over S, the interior of a closed plane curve C, can then be defined by taking a rectangle D enclosing C and defining $f(x,y) = g(x,y)$ in S and $f(x,y) = 0$ outside S. Then $\iint_S g(x,y)\,dx\,dy$ is defined as $\iint_D f(x,y)\,dx\,dy$ over the interior of D, provided $f(x,y)$ satisfies the condition of integrability over D.

5·051. Repeated integrals. A double integral, if it exists, is usually evaluated by integrating with respect to the variables in turn. For given x, $\int_0^B f(x,y)\,dy$ is a single integral and is a function of x. If this is then integrated with regard to x the result is the repeated integral $\int_0^A \left\{\int_0^B f(x,y)\,dy\right\} dx$; this is usually written $\int_0^A dx \int_0^B f(x,y)\,dy$, but other conventions are in use. Alternatively, we could integrate first with respect to x and then with respect to y, obtaining the repeated integral $\int_0^B dy \int_0^A f(x,y)\,dx$.

If $\int_0^A \int_0^B f(x,y)\,dx\,dy$ exists, both repeated integrals exist and are equal to the double integral. This needs to be understood in a rather extended sense. If for some x, say $x = a$, $f(a,y) = 0$ for irrational y and $= 1$ for rational y, $\int_0^B f(a,y)\,dy$ does not exist. But the line of integration can be enclosed in a rectangle of arbitrarily small area, and such irregularity for only one value of x will not affect the existence of the double integral. However, we saw in the discussion of the Riemann integral that whether the integral exists or not, so long as the function is bounded, the upper and lower integrals H, h exist. Then we define $H(x)$, $h(x)$ as the upper and lower integrals of $f(x,y)$ with respect to y for given x, and form the sums $\sum_r H(\xi_r) h_r$, $\sum_r h(\xi_r) h_r$ $(x_r \leqslant \xi_r \leqslant x_{r+1})$.

5·051 *Repeated integrals*

The existence of the double integral 5·05(1), which we shall denote by I, implies that for any ϵ a δ exists such that whenever the subdivision is into rectangles whose diagonals do not exceed δ,
$$I - \epsilon \leqslant \sum_r \sum_s m_{rs} h_r k_s \leqslant \sum_r \sum_s M_{rs} h_r k_s < I + \epsilon \tag{2}$$
where $x_{r+1} - x_r = h_r$, $y_{s+1} - y_s = k_s$, and M_{rs}, m_{rs} are the upper and lower bounds of $f(x,y)$ in the rectangle indicated by the suffixes. Take a line of constant x ($x_r \leqslant x \leqslant x_{r+1}$). In any rectangle the upper and lower bounds of $f(x,y)$ on this line lie in (m_{rs}, M_{rs}); hence
$$\sum_s m_{rs} k_s \leqslant h(x) \leqslant H(x) \leqslant \sum_s M_{rs} k_s \tag{3}$$
and if the largest k_s tends to zero
$$I - \epsilon \leqslant \Sigma n_r h_r \leqslant \Sigma N_r h_r \leqslant I + \epsilon \tag{4}$$
where by n_r we mean the lower bound of $h(x)$, by N_r the upper bound of $H(x)$ in each interval of x. Now let the largest h_r tend to zero, and take the upper bound I_1 of the first sum, the lower bound I_2 of the second for all such subdivisions. If I_1 and I_2 are equal we say that the repeated integral exists. Now
$$I - \epsilon \leqslant I_1 \leqslant I_2 \leqslant I + \epsilon. \tag{5}$$
Since I_1, I_2 are independent of ϵ, both must be equal to I. Hence $h(x)$, $H(x)$ are both integrable with respect to x and their integrals are equal to the double integral. The repeated integral can therefore be interpreted as either $\int_0^A h(x)\,dx$ or $\int_0^A H(x)\,dx$. The argument shows incidentally that if there are points where $H(x) - h(x) \geqslant \omega > 0$, these points can be enclosed in a finite set of intervals of x of arbitrarily small total length.

The converse is not true; the repeated integrals may both exist and be equal without the double integral existing. But in practice such cases are rare and in any ordinary case direct examination of the function will decide quickly whether the condition for the existence of the double integral is satisfied.

A double integral $\iint f(x,y)\,dx\,dy$ over an infinite region R can be defined by taking a sequence of regions $\{R_n\}$ such that, for any part of R, this part is included in all R_n for n greater than some m. If the double integral over R_n has a unique limit for all such sequences, this limit can be taken as the definition of the integral over R. Improper double integrals may be defined similarly. It appears, however, that unless the same process gives a unique value when $|f(x,y)|$ is substituted for $f(x,y)$ the value of the limit will depend on the shapes of the regions R_n, and consequently a non-absolutely convergent double integral has no meaning unless these are specified. For an absolutely convergent double integral inversions of limiting processes can be justified by the theorem of 1·111*

Analogous statements are true for triple and n-ple integrals. With the definition of the area of a surface that we shall give, if a surface possesses a finite area it can be enclosed in a set of parallelepipeds of arbitrarily small total volume (5·07a).

If $f(x,y) = 0$ outside a region other than a rectangle with sides parallel to the axes, the termini for y in the repeated integral, if x is kept constant, will depend on x. Thus let us suppose that the region is a quadrant of a circle defined by $x \geqslant 0$, $y \geqslant 0$, $x^2 + y^2 \leqslant a^2$. For given x, these inequalities require
$$0 \leqslant y \leqslant (a^2 - x^2)^{1/2},$$

* For further details see Hobson, *Theory of functions of a real variable*, 1907, Ch. 5; 1920, Vol. 1, Ch. 6.

and the range of integration for y is therefore 0 to $(a^2-x^2)^{1/2}$. Then x can range from 0 to a, and the integral over the quadrant is

$$\int_0^a dx \int_0^{\sqrt{(a^2-x^2)}} f(x,y)\,dy.$$

Similarly, it can be written $\int_0^a dy \int_0^{\sqrt{(a^2-y^2)}} f(x,y)\,dx.$

The region of integration always specifies a set of inequalities to be satisfied by the independent variables. From these the limits for a repeated integral can be found either by direct transformation, as in the example just considered, or with the aid of a figure. The latter is often useful, particularly when the boundary consists of several curves with different equations.

A repeated integral often arises directly, but is sometimes most easily evaluated by inverting the order of integration. When the limits are not constant, the easiest way of deciding the limits of the inverted integral is usually to find inequalities as for the double integral as an intermediate step. Thus consider

$$I = \int_0^t dx \int_0^x f(y)\,dy.$$

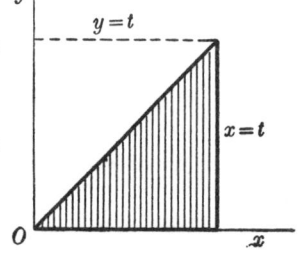

The inequalities indicated, if x and y are to be within the region of integration, are

$$0 \leqslant y \leqslant x, \quad 0 \leqslant x \leqslant t,$$

and for given y, $y \leqslant x \leqslant t$, while y, if unrestricted by x, can range from 0 to t. Then

$$I = \int_0^t dy \int_y^t f(y)\,dx = \int_0^t (t-y)f(y)\,dy,$$

so that one integration has been performed immediately. Geometrically, the integral is over the triangle whose corners are $(0, 0)$, $(t, 0)$, (t, t); and if we integrate first with regard to x it must proceed from y to t to cover the variation possible for given y within the triangle.[a]

5·052. Change of variables. Take a double integral

$$I = \iint f(x,y)\,dx\,dy$$

which is ordinarily evaluated by successive integration with respect to x and y. Let ξ and η be two differentiable functions of x and y. The curves of ξ constant and η constant will in general mark out the plane of x, y into four-cornered figures, which could be used as the elements of area in defining the double integral. We have to express the elements of area in terms of ξ and η. x and y will now be regarded as functions of ξ, η.

As we turn to the left in turning from the axis of positive x to that of positive y, we also take ξ and η so that we turn to the left in changing from the direction of increasing ξ to that of increasing η; but these directions will not in general be perpendicular. The simplest approach is to notice that as $\delta\xi$ and $\delta\eta$ tend to zero the element of area specified approximates to a parallelogram except possibly near special points, such as the centre of a circle if ξ, η are polar coordinates. To the first order in $\delta\xi$, $\delta\eta$, if one corner is (x, y) the two adjacent ones are

$$\left(x + \frac{\partial x}{\partial \xi}\delta\xi,\ y + \frac{\partial y}{\partial \xi}\delta\xi\right),\quad \left(x + \frac{\partial x}{\partial \eta}\delta\eta,\ y + \frac{\partial y}{\partial \eta}\delta\eta\right).$$

5·052   Change of variables

But the area of a parallelogram with three of its vertices at $(x_1 y_1)$, $(x_2 y_2)$, $(x_3 y_3)$ is

$$\begin{vmatrix} x_1 & y_1 & 1 \\ x_2 & y_2 & 1 \\ x_3 & y_3 & 1 \end{vmatrix} = \begin{vmatrix} x_2 - x_1 & y_2 - y_1 \\ x_3 - x_1 & y_3 - y_1 \end{vmatrix},$$

and therefore the area of the parallelogram is

$$\begin{vmatrix} \dfrac{\partial x}{\partial \xi} & \dfrac{\partial y}{\partial \xi} \\ \dfrac{\partial x}{\partial \eta} & \dfrac{\partial y}{\partial \eta} \end{vmatrix} \delta\xi \, \delta\eta.$$

The determinant is called the *Jacobian* of x, y with respect to ξ, η and denoted by $\partial(x,y)/\partial(\xi,\eta)$. Then

$$I = \iint f(x,y) \frac{\partial(x,y)}{\partial(\xi,\eta)} d\xi \, d\eta,$$

where $f(x,y)$ must be expressed as a function of ξ and η, and the range of integration is such that each element of area within the original region appears once and only once in the transformed integral.

This is the simplest way of getting the answer, but has several disadvantages. It is difficult to fix limits to the error in replacing the element bounded by curves of constant ξ and η by a parallelogram, and therefore to show that the total error tends to zero when *all* the ranges δx, δy do. Also it is difficult to generalize, for though the argument in this form is easily extended to triple integrals, since we have a convenient form for the volume of a general parallelepiped, the extension to n-ple integrals is not obvious. We get, in fact,

$$\iiint f(x,y,z) \, dx \, dy \, dz = \iiint f(x,y,z) \frac{\partial(x,y,z)}{\partial(\xi,\eta,\zeta)} d\xi \, d\eta \, d\zeta,$$

where

$$\frac{\partial(x,y,z)}{\partial(\xi,\eta,\zeta)} = \begin{vmatrix} \dfrac{\partial x}{\partial \xi} & \dfrac{\partial y}{\partial \xi} & \dfrac{\partial z}{\partial \xi} \\ \dfrac{\partial x}{\partial \eta} & \dfrac{\partial y}{\partial \eta} & \dfrac{\partial z}{\partial \eta} \\ \dfrac{\partial x}{\partial \zeta} & \dfrac{\partial y}{\partial \zeta} & \dfrac{\partial z}{\partial \zeta} \end{vmatrix}.$$

But four variables of integration occur in general relativity, and in more than three dimensions it becomes difficult to see what we mean by any generalization of the volume even of a parallelepiped, except by adopting a purely analytic definition by an integral. Then the known properties of area and volume in Euclidean geometry can no longer be used to short-circuit the direct transformation of variables. The variables x, y, z may not be Cartesian coordinates; then it becomes difficult to see what we mean by the area or volume of a region. For all these reasons it is best to proceed by direct transformation of variables, one at a time. It is convenient to use a suffix notation and to take the case of a triple integral to illustrate the method, which can clearly be extended to any order. We start with

$$I = \iiint f(x_1, x_2, x_3) \, dx_1 \, dx_2 \, dx_3, \tag{1}$$

and regard x_1, x_2, x_3 as functions of ξ_1, ξ_2, ξ_3, with continuous first partial derivatives, such that there is a one-one correspondence between points of the x_i, ξ_i regions. Then, by the mean value theorem,

$$\delta x_i = \frac{\partial x_i}{\partial \xi_j} \delta \xi_j, \tag{2}$$

where each $\partial x_i/\partial \xi_j$ is to be regarded as evaluated at some point (not necessarily all at the same point) inside the element bounded by two sets of values of the ξ_j differing by $\delta \xi_j$. We first express x_1 as a function of ξ_1, x_2 and x_3. Then, keeping x_2 and x_3 constant, we have to solve the equations

$$\delta x_1 = \frac{\partial x_1}{\partial \xi_j} \delta \xi_j, \quad 0 = \frac{\partial x_2}{\partial \xi_j} \delta \xi_j, \quad 0 = \frac{\partial x_3}{\partial \xi_j} \delta \xi_j, \tag{3}$$

and the solution is

$$\begin{vmatrix} \frac{\partial x_1}{\partial \xi_1} & \frac{\partial x_1}{\partial \xi_2} & \frac{\partial x_1}{\partial \xi_3} \\ \frac{\partial x_2}{\partial \xi_1} & \frac{\partial x_2}{\partial \xi_2} & \frac{\partial x_2}{\partial \xi_3} \\ \frac{\partial x_3}{\partial \xi_1} & \frac{\partial x_3}{\partial \xi_2} & \frac{\partial x_3}{\partial \xi_3} \end{vmatrix} \delta \xi_1 = \begin{vmatrix} \frac{\partial x_2}{\partial \xi_2} & \frac{\partial x_2}{\partial \xi_3} \\ \frac{\partial x_3}{\partial \xi_2} & \frac{\partial x_3}{\partial \xi_3} \end{vmatrix} \delta x_1. \tag{4}$$

When $\delta \xi_1 \to 0$ the determinants tend to the Jacobians evaluated at $(x_1 x_2 x_3)$; and therefore

$$I = \iiint f(x_1 x_2 x_3) \left\{ \frac{\partial(x_1, x_2, x_3)}{\partial(\xi_1, \xi_2, \xi_3)} \bigg/ \frac{\partial(x_2, x_3)}{\partial(\xi_2, \xi_3)} \right\} d\xi_1 dx_2 dx_3. \tag{5}$$

Now express x_2 as a function of ξ_2, ξ_1 and x_3; we find similarly, keeping ξ_1, x_3 constant,

$$\left. \begin{aligned} \delta x_2 &= \frac{\partial x_2}{\partial \xi_2} \delta \xi_2 + \frac{\partial x_2}{\partial \xi_3} \delta \xi_3, \\ 0 &= \frac{\partial x_3}{\partial \xi_2} \delta \xi_2 + \frac{\partial x_3}{\partial \xi_3} \delta \xi_3, \end{aligned} \right\} \tag{6}$$

$$\frac{\partial(x_2, x_3)}{\partial(\xi_2, \xi_3)} \delta \xi_2 = \frac{\partial x_3}{\partial \xi_2} \delta x_2, \tag{7}$$

$$I = \iiint f(x_1, x_2, x_3) \left\{ \frac{\partial(x_1, x_2, x_3)}{\partial(\xi_1, \xi_2, \xi_3)} \bigg/ \frac{\partial x_3}{\partial \xi_3} \right\} d\xi_1 d\xi_2 dx_3. \tag{8}$$

Finally, express x_3 as a function of ξ_3, ξ_1 and ξ_2; then keeping ξ_1 and ξ_2 constant

$$\delta x_3 = \frac{\partial x_3}{\partial \xi_3} \delta \xi_3 \tag{9}$$

and

$$I = \iiint f(x_1, x_2, x_3) \frac{\partial(x_1, x_2, x_3)}{\partial(\xi_1, \xi_2, \xi_3)} d\xi_1 d\xi_2 d\xi_3. \tag{10}$$

The successive changes of variables are justified by 1·1032 provided the integrals exist, and the result (10) can be identified as the triple integral if the latter exists. In the conditions stated, J is continuous and bounded in any finite region, so that the triple integral necessarily exists.

There is, however, a difficulty if J changes sign within the region of integration. Even for change of a single variable, say x to ξ, if $dx/d\xi$ changes sign within the range of integration, two or more values of ξ will correspond to a given x, and the integral must be written in the Stieltjes manner if the correct answer is to be reached. In the transformation of repeated integrals, if the Jacobian vanishes at a point within the region, there is a neighbourhood of the point where a given set of values of x_1, x_2, x_3 will correspond to more than one set ξ_1, ξ_2, ξ_3. If this occurs, the best procedure is usually to break up the region into subregions in each of which the Jacobian keeps a constant sign and to consider each separately.

This method shows that the first method, based on treating elements of area or volume as parallelograms or parallelepipeds, actually gives the right answer, and being justified to this extent can be used directly in many cases where it is more convenient than working out the Jacobian explicitly. The transformation is often simplified by use of the theorem

$$\frac{\partial(x_1 \ldots x_n)}{\partial(\eta_1 \ldots \eta_n)} = \frac{\partial(x_1 \ldots x_n)}{\partial(\xi_1 \ldots \xi_n)} \frac{\partial(\xi_1 \ldots \xi_n)}{\partial(\eta_1 \ldots \eta_n)}, \tag{11}$$

for
$$\frac{\partial x_i}{\partial \eta_k} = \frac{\partial x_i}{\partial \xi_j} \frac{\partial \xi_j}{\partial \eta_k}, \tag{12}$$

where x_i on the left is regarded as a function of the η_k and on the right as a function of the ξ_j. But the determinant of the expressions on the right of (12) is the product of the two determinants on the right of (11), by the rule for multiplying determinants. Alternatively, the theorem is required by consistency; for if it was false we could get a different result by transforming an integral to variables η_k directly from what we should get by transforming first to ξ_j and then to η_k.

There is a theorem that if the Jacobian of $\xi_1, \ldots, \xi_n$ with respect to $x_1, \ldots, x_n$ is zero everywhere, one of the ξ's is determined when the others are given, and they are not a suitable system of coordinates. A proof will be found in most books on calculus.

5·053. Changes of limits. Change of variables naturally implies changes of the limits. The new limits may be found either analytically, by writing the ranges in terms of inequalities, or by drawing a figure and finding the limits by examining the ranges of the new variables required to cover the latter. No general rule of transformation can be given, but the methods will be illustrated by examples.

There is an apparent inconsistency between the Jacobian transformation and the simple reversal of order in an integral with regard to two variables, for $\frac{\partial(x,y)}{\partial(y,x)} = -1$. This is explained as follows: $\frac{\partial(x,y,z)}{\partial(\xi,\eta,\zeta)}$ is unaltered by any cyclic interchange of x, y, z or of ξ, η, ζ. It is positive if, when the directions of $d\xi, d\eta, d\zeta$ are turned so that $d\xi$ and $d\eta$ lie in the plane of dx and dy, the rotation about dz from $d\xi$ to $d\eta$ being positive, $d\zeta$ makes an acute angle with positive dz. Thus the positive Jacobian means that $d\xi, d\eta, d\zeta$ form a right-handed set of directions, and if x, y, z increase throughout their ranges of integration, ξ, η, ζ will also increase; for each integration the lower limit is the smaller. If the Jacobian is negative $d\xi, d\eta, d\zeta$ form a left-handed set, but an odd number of them will decrease through the range. If we still make all the lower limits the smaller we must therefore also reverse the sign. Hence *if the lower limit is made the smaller for every variable, the Jacobian must be replaced by its modulus.* The formula for two variables is right because the limits are *not* reversed and therefore we must not use $\partial(x,y)/\partial(y,x)$ but its modulus, which is $+1$.

5·054. Polar coordinates. Take

$$I = \iint f(x,y)\, dx\, dy \quad (0 \leqslant x^2 + y^2 \leqslant a^2),$$

and transform to polar coordinates r, θ. Then
$$x = r\cos\theta, \quad y = r\sin\theta,$$
$$\frac{\partial(x,y)}{\partial(r,\theta)} = \begin{vmatrix} \cos\theta & -r\sin\theta \\ \sin\theta & r\cos\theta \end{vmatrix} = r,$$
and
$$I = \iint f(x,y)\, r\, dr\, d\theta.$$

With the usual conventions we always take $r \geq 0$; then we represent every point within the circle once by taking θ from 0 to 2π, and the limits are 0 to a for r, 0 to 2π for θ. It would be possible to allow negative values of r, so that $(-x, -y)$ would correspond to $(-r, \theta)$ with the same θ as for (x, y); but then to represent every point once θ can only range from 0 to π, and the limits become $-a$ to a for r, 0 to π for θ. Either system is equally valid. The origin is a singular point of the transformation, but gives no trouble.

5·055. Evidently if we change from rectangular coordinates x, y, z to cylindrical coordinates ϖ, λ, z we get similarly
$$\frac{\partial(x,y,z)}{\partial(\varpi,\lambda,z)} = \varpi.$$

If we now change from cylindrical to spherical polar coordinates (r, λ, θ) we must put
$$z = r\cos\theta, \quad \varpi = r\sin\theta,$$
$$\frac{\partial(\varpi,\lambda,z)}{\partial(r,\lambda,\theta)} = \frac{\partial(\varpi,z)}{\partial(r,\theta)} = -r,$$

and therefore
$$\frac{\partial(x,y,z)}{\partial(r,\lambda,\theta)} = -r\varpi = -r^2\sin\theta, \quad \frac{\partial(x,y,z)}{\partial(r,\theta,\lambda)} = r^2\sin\theta.$$

This can also be obtained directly by taking
$$x = r\sin\theta\cos\lambda, \quad y = r\sin\theta\sin\lambda, \quad z = r\cos\theta.$$

We represent the whole of the interior of a sphere of radius a by letting r range from 0 to a, θ from 0 to π, and λ from 0 to 2π. But either of the following would be equivalent:
$$-a \leq r \leq a. \qquad 0 \leq \theta \leq \pi, \quad 0 \leq \lambda \leq \pi,$$
$$0 \leq r \leq a, \quad -\pi \leq \theta \leq \pi, \quad 0 \leq \lambda \leq \pi.$$

We must *not* take the ranges as, for instance, $0 \leq r \leq a$, $-\pi \leq \theta \leq \pi$, $0 \leq \lambda \leq 2\pi$, for this would cover the sphere twice over and give twice the correct result.

5·056. As an illustration of several features in the treatment of double integrals let us take the method given in many text-books for evaluating the integral
$$I = \int_0^\infty e^{-x^2} dx.$$

It proceeds as follows:
$$I^2 = \int_0^\infty e^{-x^2} dx \times \int_0^\infty e^{-y^2} dy \tag{1}$$
$$= \int_0^\infty \int_0^\infty e^{-(x^2+y^2)} dx\, dy \tag{2}$$
$$= \int_0^\infty \int_0^{\frac{1}{2}\pi} e^{-r^2} r\, dr\, d\theta \tag{3}$$
$$= \int_0^\infty e^{-r^2} r\, dr \times \int_0^{\frac{1}{2}\pi} d\theta \tag{4}$$
$$= \tfrac{1}{2} \cdot \tfrac{1}{2}\pi; \tag{5}$$
therefore
$$I = \tfrac{1}{2}\sqrt{\pi}. \tag{6}$$

The passage from (1) to (2) requires justification; it is not obvious that the product of two single integrals can be converted into a double integral. Actually I must be understood to mean

$$\lim_{X \to \infty} \int_0^X e^{-x^2} dx = \lim_{Y \to \infty} \int_0^Y e^{-v^2} dy, \qquad (7)$$

and we can without loss of generality take $X = Y$. Then (1) reads

$$I^2 = \lim_{X \to \infty} \int_0^X e^{-x^2} dx \times \int_0^X e^{-v^2} dy. \qquad (8)$$

But in

$$\int_0^X \int_0^X e^{-(x^2+v^2)} dx dy \qquad (9)$$

the limits are finite and the integrand continuous. Hence it can be evaluated as a repeated integral and is the same as the product of integrals in (8). Hence

$$I^2 = \lim_{X \to \infty} \int_0^X \int_0^X e^{-(x^2+v^2)} dx dy. \qquad (10)$$

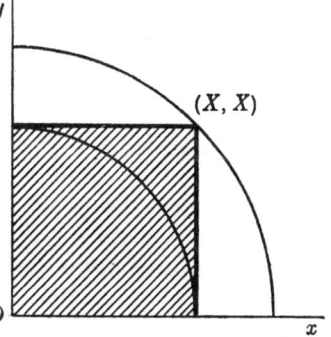

But this is not obviously transformed to (3) by change of variables; for the region of integration is a square, and (3) must be understood to mean

$$\lim_{R \to \infty} \int_0^R \int_0^{1/2\pi} e^{-r^2} r \, dr \, d\theta, \qquad (11)$$

and in this the region of integration is a quadrant. The justification is that the integrand is positive for all x, y, and the integral over the square must lie between those over quadrants of radii X and $X\sqrt{2}$. That is,

$$\int_0^X \int_0^{1/2\pi} e^{-r^2} r \, dr \, d\theta < \int_0^X \int_0^X e^{-(x^2+v^2)} dx dy < \int_0^{X\sqrt{2}} \int_0^{1/2\pi} e^{-r^2} r \, dr \, d\theta. \qquad (12)$$

When we integrate with regard to θ and then proceed to the limit, the first and third expressions tend to the same limit; hence

$$I^2 = \tfrac{1}{2}\pi \int_0^\infty e^{-r^2} r \, dr = \tfrac{1}{4}\pi. \qquad (13)$$

By an obvious transformation we find the result, often wanted later,[*]

$$(-\tfrac{1}{2})! = \int_0^\infty x^{-1/2} e^{-x} dx = \sqrt{\pi}. \qquad (14)$$

5·057. A multiple integral. Especially in the theory of probability the integral

$$I = \int\!\!\int \ldots \int \exp(-\tfrac{1}{2}W) dx_1 \ldots dx_n \qquad (1)$$

is often wanted, where

$$W = a_{ik} x_i x_k \qquad (2)$$

is a positive definite form and the limits are $-\infty$ to ∞ for all variables. If W is not positive definite the integral does not converge. We know that W can be reduced to a sum of squares by a transformation to variables $\xi_1 \ldots \xi_n$, such that the Jacobian of the transformation is unity (cf. 4·08). Then if

$$W = A_{11}\xi_1^2 + \ldots + A_{nn}\xi_n^2, \qquad (3)$$

I is the product of n integrals of the form

$$\int_{-\infty}^\infty \exp(-\tfrac{1}{2}A_{rr}\xi_r^2) d\xi_r = \sqrt{\left(\frac{2\pi}{A_{rr}}\right)}. \qquad (4)$$

[*] See Chapter 15 for the definition of the factorial function.

But
$$A_{11}A_{22}\ldots A_{nn} = \|a_{ik}\|. \tag{5}$$

Hence
$$I = \frac{(2\pi)^{\frac{1}{2}n}}{\sqrt{\|a_{ik}\|}}. \tag{6}$$

5·06. Integrals along rectifiable curves. If
$$I = \int_A^B f(x,y,z)\,ds, \tag{1}$$

s increases continuously as we pass along the curve, and I is a Riemann integral. Note that if $|f| < M$, and the length of the curve from A to B is L, $|I| < ML$.

If u, v, w are three functions of x, y, z, bounded and continuous on the curve, let
$$J = \int_{s=s_0}^{s_1} (u\,dx + v\,dy + w\,dz). \tag{2}$$

This is regarded as the sum of three Stieltjes integrals, since x, y, z are not necessarily monotonic functions of s. x is both continuous as a function of s and of bounded variation. But even if u is continuous as a function of s it may not be continuous as a function of x, for part of the curve may be perpendicular to the axis of x, say for $x = a$, and then u may change discontinuously as x passes through a because y or z does so. But it follows from (sufficient) conditions for the existence of the Stieltjes integral that J will exist if the curve can be divided into a finite number of arcs, on each of which u, v, w are either continuous or of bounded variation.

It can be shown* that a rectifiable curve has a tangent almost everywhere; hence the derivatives
$$l = dx/ds, \quad m = dy/ds, \quad n = dz/ds, \tag{3}$$

exist almost everywhere. By a change of variable we can write
$$J = \int_{s_0}^{s_1} (lu + mv + nw)\,ds = \int l_i u_i\,ds = \int \boldsymbol{l}\cdot\boldsymbol{u}\,ds. \tag{4}$$

Sufficient conditions for this change of variable are that both (2) and (4) shall exist and that the dx_i/ds are bounded, as they obviously are. In practice u_i and l_i are usually continuous except possibly at a finite number of points, and the integrals will be seen to exist on inspection.

5·07. Surface integrals: area of surfaces. Corresponding to simple closed curves and arcs are closed surfaces and surfaces with boundaries. A closed surface is one of finite diameter and with an inside and an outside, so that we can pass from any point of the surface to any other without leaving the surface, from any internal point to any other without crossing the surface, and similarly from one external point to another; but we cannot pass from an internal point to any external point without crossing the surface. If in addition any simple closed curve on the surface can be shrunk up to a point without leaving the surface, the surface is called simply connected. These conditions are satisfied by most ordinary surfaces, in particular by the boundaries of solids, but it is possible to construct an anomalous surface known as the *Klein bottle*, which has no inside and outside, and no bounding curve. It is of finite extent and we can pass from any point of

* Most simply by A. S. Besicovitch, *J. Lond. Math. Soc.* 19, 1944, 205–7.

5·07   *Surfaces*

it to any other without leaving it, but any two points not on the surface can be connected by a path that does not cross it. Such surfaces are excluded from our discussion. Any straight line that intersects a closed surface at all will intersect it in an even number of points; we shall not consider surfaces such as, in cylindrical coordinates,

$$\varpi = \sin^2 \frac{1}{z},$$

which is intersected an infinite number of times by any line $\lambda = 0$, $\varpi = a$, where $0 < a < 1$. An example of a surface that is not simply connected is an anchor ring.

Any simple closed curve on a closed simply connected surface divides it into two portions, which we shall call *caps*. The curve will be called the *rim* of either cap. Either cap separately has no inside, but usually has the further property that a simple closed curve on it and not meeting the boundary divides it into two regions separated by the curve, and any such interior curve can be contracted to a point without leaving the surface. That is, the cap must not have a hole in it. We exclude the *Möbius strip* because it is possible to draw closed curves on it that do not divide the strip into two mutually inaccessible regions. A Möbius strip can be made by taking a long rectangle of paper, giving it a half twist, and pasting the ends together. It clearly has a single boundary, and a complete longitudinal cut along it, following the original middle line, does not separate it into two pieces because the strip is still held together at the edge. Again, we can make a longitudinal cut one-third or less of the way across; this will be found to divide the surface into two pieces, the edge portion giving a strip with two edges and a complete twist in it, while the inner portion remains as a narrower Möbius strip interlocked with the edge portion. We thus see that the original edge of the strip can be deformed into a circle by continuous distortion. Further, had the strip been made of a more extensible material, the inner portion, instead of being severed, might have been stretched to maintain its connexion with the outer; hence a circle can be filled by a one-sided surface. Therefore any closed curve capable of being continuously deformed into a circle is also capable of being filled by a one-sided surface. But it is obviously also capable of being filled by a two-sided surface. Imagining the deformation now reversed so that the circle is deformed into the edge of a Möbius strip, we see that the edge of the latter can also be filled by a two-sided surface.* In what follows we shall be entirely concerned with two-sided surfaces.

Such considerations about properties of figures in space that are maintained in any continuous deformation of the figures belong to the branch of mathematics known as *topology*. Such statements as that a plane closed curve has an inside and an outside seem trivial at first sight, but are actually true only when the definition of a closed curve is made perfectly clear and even then are quite difficult to prove. Again, in the theory of the magnetic field of an electric current it is usually taken for granted that the closed circuit can be filled in by a two-sided surface; but the edge of a Möbius strip would be quite a possible form for such a circuit, and how to fill it up with a two-sided surface is far from obvious, though it can be done.

---

* For illustrations of a one-sided surface filling a circle, and a two-sided surface filling the edge of a Möbius strip, see Courant and Robbins, *What is Mathematics?* pp. 261, 388. The one-sided surface is not simply a mathematical curiosity; one will actually be formed by a soap film stretched across a boundary of suitable form. Whether the film will be stable as a two-sided or a one-sided surface is simply a matter of which has the smaller area.

Corresponding to the length of a curve it is natural to try to define the area of a curved surface by taking a set of points in the surface, connecting them up so as to form triangles, and defining the area of the surface as the limit of the sum of the areas of the triangles when these are made indefinitely small. (With plane polygons of more than three vertices, there is, of course, an extra complication; four points on a curved surface do not necessarily lie in one plane.) Unfortunately, without some further restrictions on how we are to select the points and what pairs are to be joined, this does not lead to a unique definition of the area, except when the surface is a plane. From the study of the definition of a multiple integral it is natural to require that *all* sides of the triangles should tend to zero and not simply that the areas should; but even this is not sufficient. To take an example given by H. A. Schwarz,* imagine a circular cylinder of radius 1 with its axis parallel to the z axis, divided up by plane cross-sections at interval m. Take on each section n points equally spaced about it; for each section the n points are opposite the points midway between those chosen in the adjacent sections. Then these points specify a set of isosceles triangles with their vertices in the surface, and their sides can be made arbitrarily small by taking m small enough and n large enough. But if A, B are points of a section with their cylindrical coordinates λ differing by $2\pi/n$, the midpoint of A, B is inside the cylinder by a distance $1 - \cos \pi/n$, and the z coordinate differs from that of C, the nearest point of the next section, by m. Hence the plane of the triangle is inclined to the tangent plane at C at an angle

$$\tan^{-1}\left(\frac{2}{m}\sin^2\frac{\pi}{2n}\right).$$

This is small only if mn^2 is large; if m tends to zero like n^{-1} the planes of the triangles approach the tangent planes, but not if m tends to zero like n^{-2}. Again, the area of a triangle is

$$\sin\frac{\pi}{n}\left(m^2 + 4\sin^4\frac{\pi}{2n}\right)^{1/2},$$

and that of the $\frac{2n}{m}$ triangles covering length 1 of the cylinder is

$$2n\sin\frac{\pi}{n}\left(1 + \frac{4}{m^2}\sin^4\frac{\pi}{2n}\right)^{1/2}.$$

If $n \to \infty$, this tends to 2π provided that $mn^2 \to \infty$. Thus the condition that the sum of the areas of the triangles shall approach that of the cylinder is the same as the condition that their planes shall tend to the tangent planes. If this condition is not satisfied the sum may tend to any limit greater than 2π.

Consequently the definition of the area of a curved surface is more difficult than that of the length of a curve. For a curve, so long as the direction cosines of the tangent vary continuously with position, the direction of a short chord must tend to that of the tangent, and the ratio of its length to that of the arc must tend to 1. For a surface, the corresponding approximations for triangles require a further condition, which can be taken to be that all angles of the triangles are greater than some fixed angle δ. But this introduces the further question: for what sort of surfaces is such a choice possible however short the

* *Ges. math. Abhand.* 2, 1890, 309–11.

sides? It has been shown possible* provided that the surface is bounded in all directions, and has a normal at every point, such that if $Q_1, Q_2 \ldots$ are points on the surface tending to P, the normal at Q_n tends to that at P. The last condition is often described by saying that the surface has a continuously turning normal. These conditions are somewhat unnecessarily strict, being obviously not satisfied by a cube. But since the usual formula for the area of a surface is an integral of a function of direction cosines the mere existence of the integral implies that they cannot be greatly relaxed if this formula is to be correct, and they may as well be adopted.

If the surface is given in the form
$$z = F(x,y), \tag{1}$$
where F has continuous partial derivatives F_x, F_y with regard to x and y, we can cover the projection of the surface on the x, y plane by rectangles of sides h, k so that the leaps of F_x and F_y in each rectangle are less than ϵ. Take the points where lines through the corners of these rectangles parallel to the z axis meet the surface. In each quadrilateral so determined on the surface join one diagonal. We thus find a set of triangles with all their vertices in the surface; the projection of each on the x, y plane is $\frac{1}{2}hk$. If now the vertices on the (x,y) plane are $(x_r, y_r)(x_r + h, y_r)(x_r, y_r + k)$, and we label them with suffixes 1, 2, 3, the projection on the plane of x constant is
$$\tfrac{1}{2}(y_3 - y_1)(z_2 - z_1) = \tfrac{1}{2}khF_x(x_r + \theta h, y_r), \tag{2}$$
where $0 < \theta < 1$. Similarly for $(x_r + h, y_r)(x_r, y_r + k)(x_r + h, y_r + k)$ the projection is
$$\tfrac{1}{2}khF_x(x_r + \theta' h, y_r + k).$$
Then the sum of the areas of the triangles is
$$\tfrac{1}{2}\Sigma hk\{1 + (F_x)_r^2 + (F_y)_r^2\}^{\frac{1}{2}}, \tag{3}$$
where F_x, F_y are to be evaluated in each case for two points of the rectangle in the (x, y) plane. If now we make ϵ tend to zero this sum tends to the integral
$$\iint (1 + F_x^2 + F_y^2)^{\frac{1}{2}} dx dy, \tag{4}$$
which exists in the conditions stated. But it must be remembered that this result depends on a particular way of choosing the triangles, though it is a very natural way. If we want the integral of a function $f(x, y, z)$ over the surface we can take for each triangle the value of $f(x, y, F(x, y))$ for a point of the triangle in the (x, y) plane, and if f is continuous the result will be $\iint f(x, y, z)(1 + F_x^2 + F_y^2)^{\frac{1}{2}} dx dy$, which may be briefly written $\iint f dS$.

The method assumes z single-valued. If a line parallel to the z axis meets the surface in two points it will be necessary to treat the two separately; thus the surface of a sphere must be regarded as two surfaces, one corresponding to the set of the smaller values for given x, y, the other to the larger values. Also it assumes F_x, F_y continuous and therefore bounded, since the integration is over a closed region of x, y; but for a sphere they tend to infinity and the integrals must be interpreted as improper ones by first excluding a zone

* O. D. Kellogg, *Foundations of Potential Theory*, 1929, 100–112.

about the boundary of the region and then making the width of the zone tend to zero. If F_x, F_y are discontinuous along a rectifiable curve or at a finite set of isolated points, the surface can be subdivided so as to make the discontinuities part of the boundaries of the pieces, and there is no special difficulty. If part of the surface consists of lines parallel to the axis of z, we take it in the form $x = G(y, z)$ or $y = H(z, x)$.

The integrals are usually improper, and to avoid difficulties connected with infinite series of improper integrals it is usual to require that the surface shall not be intersected an infinite number of times by any line parallel to an axis. The restrictions on the types of surface considered are therefore rather numerous, but fortunately they are satisfied by most of the surfaces we meet in practice.[a]

The surface may be given parametrically by $x = x(\xi, \eta)$ and similar equations, where x, y, z have continuous derivatives with regard to ξ and η. If we take λ, μ so that the leaps of these derivatives are always less than ϵ when $|\xi_1 - \xi_2| \leq \lambda$, $|\eta_1 - \eta_2| \leq \mu$, we can apply a similar argument to triangles specified by rectangles in the ξ, η plane of sides λ, μ. It is found that the sum now tends to

$$\iint J \, d\xi \, d\eta, \tag{5}$$

where
$$J^2 = \left\{\frac{\partial(y, z)}{\partial(\xi, \eta)}\right\}^2 + \left\{\frac{\partial(z, x)}{\partial(\xi, \eta)}\right\}^2 + \left\{\frac{\partial(x, y)}{\partial(\xi, \eta)}\right\}^2, \tag{6}$$

and that the direction cosines of the normal are given by

$$J(l, m, n) = \pm \left(\frac{\partial(y, z)}{\partial(\xi, \eta)}, \frac{\partial(z, x)}{\partial(\xi, \eta)}, \frac{\partial(x, y)}{\partial(\xi, \eta)}\right) \tag{7}$$

the sign being positive if the rotation for $d\xi$ to $d\eta$ is positive about the normal. By direct transformation of variables it can be shown that this integral is equal to that given by (4) and reduces to it when $\xi = x, \eta = y$, so that the rule for the area is invariant for different parametric representations of the surface. This is specially important because the method of triangulation adopted is specially chosen for each representation and it is therefore necessary to verify consistency.

The difficulties of a definition of area for a general surface are considerable. Lebesgue avoided the difficulty of Schwarz by a process involving the use of lower bounds of the sum of the areas of triangles approximating to the surface, thus preventing gross overestimates of the area, but Besicovitch has shown that it can lead to gross underestimates. He has invented a set of points with a finite area according to Lebesgue's definition, but with a positive volume.[*] The best general definition seems to be by C. Carathéodory.[†] All these definitions lead to the same integral formula for the areas of ordinary surfaces. Necessary and sufficient conditions for the formula to be given by Carathéodory's definition are given by Besicovitch.[‡]

In suffix notation, in transforming an integral over a surface we can interpret $l_i dS$ as

$$l_i dS = \epsilon_{ikm} \frac{\partial x_k}{\partial \xi} \frac{\partial x_m}{\partial \eta} d\xi \, d\eta, \tag{8}$$

---

[*] Q. J. Math (Oxford series), 16, 1945, 86–102.
[†] Gött. Nachr. 1914, 404–26.
[‡] Q. J. Math (Oxford series), 20, 1949, 1–7.

and
$$\epsilon_{ips} l_i dS = \frac{\partial(x_p, x_s)}{\partial(\xi, \eta)} d\xi d\eta, \tag{9}$$

subject again to the restriction that $d\xi$, $d\eta$ and the normal form a right-handed frame. If they form a left-handed frame the signs are reversed.

5·08. Green's lemma. This theorem, otherwise known as Gauss's theorem, Ostrogradsky's theorem,[a] and the divergence theorem, asserts that if V is a closed region and S its bounding surface

$$\iiint_V \left(\frac{\partial u}{\partial x} + \frac{\partial v}{\partial y} + \frac{\partial w}{\partial z}\right) dx\,dy\,dz = \iint_S (lu + mv + nw)\,dS, \tag{1}$$

provided that the triple integrals of $\partial u/\partial x$, $\partial v/\partial y$, $\partial w/\partial z$ through V exist and no straight line parallel to an axis meets S more than a fixed number of times; l, m, n are the direction cosines of the outward normal to S. It is to be understood as in 5·05 that the integral of $\partial u/\partial x$ through V means the integral through a parallelepiped D enclosing V, bounded by planes $x = X_1, X_2$, $y = Y_1, Y_2$, $z = Z_1, Z_2$, of a function $p(x,y,z)$ equal to $\partial u/\partial x$ in V and zero outside it, and the statement that the integral through V exists means that the function $p(x,y,z)$ is integrable through D.

By 5·051 the existence of the triple integral implies that the integral can be evaluated by successive integration. Then

$$\iiint_V \frac{\partial u}{\partial x} dx\,dy\,dz = \int_{Y_1}^{Y_2} \int_{Z_1}^{Z_2} dy\,dz \int_{X_1}^{X_2} p(x,y,z)\,dx. \tag{2}$$

For any pair of values of y, z that are taken by no point of V, p is zero for all x. If a line parallel to the axis of x does meet S, label the intersections 1, 2, 3, ..., in order of increasing x, so that odd suffixes will correspond to points of entry and even ones to points of exit. The number of intersections $2k$ is bounded with respect to y, z. Then along such a line

$$\int_{X_1}^{X_2} p\,dx = \int_{x_1}^{x_2} \frac{\partial u}{\partial x} dx + \int_{x_3}^{x_4} \frac{\partial u}{\partial x} dx + \dots = \sum_1^{2k} (-1)^j u_j \tag{3}$$

and
$$\iiint_V \frac{\partial u}{\partial x} dx\,dy\,dz = \int_{Y_1}^{Y_2} \int_{Z_1}^{Z_2} \sum_1^{2k} (-1)^j u_j\,dy\,dz = \sum_1^{2k} (-1)^j \iint u_j\,dy\,dz \tag{4}$$

the integrand being taken as 0 for values of y, z corresponding to no point of S.

Now from 5·07(4)
$$|l| = \left\{1 + \left(\frac{\partial x}{\partial y}\right)^2 + \left(\frac{\partial x}{\partial z}\right)^2\right\}^{-1/2} \tag{5}$$

and for any portion of S
$$\iint dS = \iint dy\,dz/|l|. \tag{6}$$

But we see by reference to a figure that at a point of exit the outward normal makes an acute angle with the x axis and therefore has an x direction cosine $l = |l|$; at a point of entry $l = -|l|$. Hence the integral on the right of (4) is $\iint lu\,dS$ taken over all points of intersection. But these points of entry and exit cover the whole of S except where a line parallel to the axis of x lies on S over an interval of x; and on such a line $l = 0$ and therefore any part of S composed of such lines makes no contribution to the integral.

We obtain similar results for $\iiint \frac{\partial v}{\partial y} dx\,dy\,dz$ and $\iiint \frac{\partial w}{\partial z} dx\,dy\,dz$ by integrating first with respect to y and z respectively, and by addition we have the theorem (1).

The expression $l\,dS$ is really to be regarded as an abbreviation for $\pm dy\,dz$, the sign being chosen according as points of a neighbourhood are points of exit or entry, so that the right side of (1) is to be regarded as the sum of three ordinary double integrals and does not involve the special difficulties of the area of a surface. In evaluation of the surface integral, when the surface is expressed parametrically, the formula 5·07(8) will be used:

$$\iint l_i u_i\, dS = \iint \epsilon_{ikm} u_i \frac{\partial x_k}{\partial \xi} \frac{\partial x_m}{\partial \eta} d\xi\, d\eta, \tag{7}$$

where $l_i, d\xi, d\eta$ form a right-handed set of directions.

The above proof gives greater generality than is often needed. Sufficient conditions usually satisfied are (1) S is bounded, has a continuously turning normal except possibly on a finite set of rectifiable curves, and is not intersected more than a fixed number of times by any straight line (2) the derivatives $\partial u/\partial x$, $\partial v/\partial y$, $\partial w/\partial z$ exist in V and are bounded, and are continuous except possibly on a finite set of surfaces of finite area. There is no objection to V having a hole in it. It may, for instance, be the region bounded by two concentric spheres. Then S consists of both the outer and the inner boundary, so that the normal at the inner boundary is taken outwards from V and therefore toward the centre.

If u, v, w are the components of a vector u_i we can write the theorem in the forms

$$\iiint_V \frac{\partial u_i}{\partial x_i} d\tau = \iint l_i u_i\, dS, \quad \iiint_V \operatorname{div} \mathbf{u}\, d\tau = \iint u_n\, dS, \tag{8}$$

u_n being the component of $\mathbf{u}$ along the outward normal. The method of proof given treats the terms separately and there would be no special advantage in using vector or tensor notation in it. The proof that we shall give later (11·053) uses a suffix notation, and is somewhat more general, but also is more difficult.

A common practice, especially in German books, is to write the integrals with only one sign of integration. This has the disadvantage that for evaluation the integrals must be written as double or triple integrals, and some confusion can arise: there are no variables of integration S and τ.

If all the components are independent of z and we apply the theorem to the region between two planes of constant z and a cylinder with its generators parallel to the z axis, we get the two-dimensional form of the theorem

$$\iint \left(\frac{\partial u}{\partial x} + \frac{\partial v}{\partial y}\right) dx\, dy = \int (lu + mv)\, ds. \tag{9}$$

For the ends contribute nothing to the surface integral, since $l = m = 0$ there, and the values of nw at corresponding points of the two ends are equal and opposite; and the length of the cylinder cancels. This result is also easily proved directly.

Replacing u by v and v by $-u$ we have

$$\iint \left(\frac{\partial v}{\partial x} - \frac{\partial u}{\partial y}\right) dx\, dy = \int (lv - mu)\, ds. \tag{10}$$

But if we proceed along the tangent in the positive direction (i.e. keeping the area on the left), the direction cosines of the tangent are $(-m, l) = (dx/ds, dy/ds)$. Hence the integral on the right is $\int (u\,dx + v\,dy)$ taken around the boundary in the positive sense. This is the two-dimensional form of *Stokes's theorem*.

5·081. Green's theorem. In Green's lemma put

$$u_i = U \frac{\partial V}{\partial x_i}.$$

Then
$$\iiint \frac{\partial}{\partial x_i}\left(U \frac{\partial V}{\partial x_i}\right) d\tau = \iint l_i U \frac{\partial V}{\partial x_i} dS,$$

that is
$$\iiint U\nabla^2 V \, d\tau + \iiint \frac{\partial U}{\partial x_i} \frac{\partial V}{\partial x_i} d\tau = \iint U \frac{\partial V}{\partial n} dS,$$

where by $\partial V/\partial n$ we understand differentiation along the outward normal. Similarly

$$\iiint V\nabla^2 U \, d\tau + \iiint \frac{\partial U}{\partial x_i} \frac{\partial V}{\partial x_i} d\tau = \iint V \frac{\partial U}{\partial n} dS,$$

and therefore
$$\iiint (U\nabla^2 V - V\nabla^2 U) \, d\tau = \iint \left(U \frac{\partial V}{\partial n} - V \frac{\partial U}{\partial n}\right) dS,$$

provided that $U\partial V/\partial x_i$ and $V\partial U/\partial x_i$ have first derivatives integrable within the region; that is, that the second derivatives of U and V exist and are integrable.

5·082. Stokes's theorem. Let C be a simple rectifiable closed curve and S a two-sided surface with C as its boundary. Let x_i be the coordinates, u_i three functions of position with continuous first derivatives. Stokes's theorem is that

$$\int_C u_i \, dx_i = \iint_S l_i \epsilon_{ikm} \frac{\partial u_m}{\partial x_k} dS, \tag{1}$$

or in vector notation
$$\int_C \boldsymbol{u} \cdot d\boldsymbol{x} = \iint_S \boldsymbol{l} \cdot \operatorname{curl} \boldsymbol{u} \cdot dS, \tag{2}$$

where l_i are the direction cosines of the normal to the surface at the element dS. The sense to be taken for the normal is given by the consideration that if C is continuously deformed and displaced so that it is described in the positive sense in a plane of z constant, and S lies in this plane, the normal becomes parallel to the positive direction of the z axis.

We suppose the surface expressed in terms of parameters ξ, η. Then

$$\iint \epsilon_{ikm} l_i \frac{\partial u_m}{\partial x_k} dS = \iint \frac{\partial u_m}{\partial x_k}\left(\frac{\partial x_k}{\partial \xi}\frac{\partial x_m}{\partial \eta} - \frac{\partial x_m}{\partial \xi}\frac{\partial x_k}{\partial \eta}\right) d\xi \, d\eta$$
$$= \iint \left(\frac{\partial u_m}{\partial \xi}\frac{\partial x_m}{\partial \eta} - \frac{\partial u_m}{\partial \eta}\frac{\partial x_m}{\partial \xi}\right) d\xi \, d\eta$$
$$= \iint \left\{\frac{\partial}{\partial \xi}\left(u_m \frac{\partial x_m}{\partial \eta}\right) - \frac{\partial}{\partial \eta}\left(u_m \frac{\partial x_m}{\partial \xi}\right)\right\} d\xi \, d\eta$$
$$= \int_C u_m \left(\frac{\partial x_m}{\partial \eta} d\eta + \frac{\partial x_m}{\partial \xi} d\xi\right), \tag{3}$$

by the two-dimensional form of Stokes's theorem. But this is just $\int_C u_m \, dx_m$.

The proof assumes that $\partial^2 x_m/\partial\xi\partial\eta$ and $\partial^2 x_m/\partial\eta\partial\xi$ exist and are equal; this can be proved on the supposition that they are continuous.*

$\boldsymbol{n}.d\boldsymbol{S}$ is often written as $d\boldsymbol{S}$, regarded as a vector with magnitude dS directed along $\boldsymbol{n}$.

5·09. Flux and circulation. If u_i is a vector, n_i the direction of the normal to a surface, $\iint n_i u_i dS$ taken over a surface is called the *flux* of u_i through the surface. In hydrodynamics, if u_i is the velocity of a fluid at any point, the flux is the volume of fluid passing through the surface per unit time; hence the name. $\int_C u_i dx_i$ taken around a circuit is called the *circulation* around the circuit. Then Green's lemma can be read: The flux of a vector through a closed surface is equal to the volume integral of its divergence through the interior. Stokes's theorem can be read: The circulation of a vector around a circuit is equal to the flux of its curl through a cap filling the circuit.

If the u_i are differentiable at P and we take any sequence of surfaces of constant form and orientation surrounding P, the diameter and volume of S_n being a_n, V_n, it follows from 5·04(5) that

$$\frac{1}{V_n}\iint l_i u_i dS_n = \left(\frac{\partial u_i}{\partial x_i}\right)_P + \frac{1}{V_n}\iiint l_i v_i r dS_n, \qquad (1)$$

where $|v_i| \to 0$ uniformly with a_n.

The last term tends to 0 with a_n, since $\iiint r dS_n = O(a_n^3) = O(V_n)$. Hence

$$\lim_{a_n \to 0} \frac{1}{V_n}\iint l_i u_i dS_n = \left(\frac{\partial u_i}{\partial x_i}\right)_P. \qquad (2)$$

This result is particularly useful when the divergence of a vector has to be expressed in terms of curvilinear orthogonal coordinates.

Vectors with zero divergence everywhere in a region are called *solenoidal* in the region. The flux of a solenoidal vector through any closed surface in the region is zero; and conversely, by the last result, if a vector has zero flux through any closed surface in a region, it is solenoidal in the region. Especially, in the flow of a fluid of density ρ, where ρ may be variable, the rate of transfer of mass through a surface is $\iint \rho l_i u_i dS$. If the mass within every surface remains constant with time, the vector ρu_i is therefore solenoidal and div$(\rho \boldsymbol{u}) = 0$. In particular, if the fluid is homogeneous and incompressible, so that the mass within a closed surface within it is necessarily constant and $\partial\rho/\partial x_i = 0$, we have simply div $\boldsymbol{u} = 0$. In many hydrodynamical problems the latter condition is satisfied. A sufficient condition for a vector to be solenoidal is evidently that it shall be the curl of another vector. We shall prove in 6·11 that this condition is necessary.

Vectors with zero circulation about every circuit in a region are called *irrotational* in the region. If $A(\xi_i)$ and $B(x_i)$ are any two points in the region and we connect them by two different paths L and L' in the region,

$$\int_L u_i dx_i = \int_{L'} u_i dx_i. \qquad (3)$$

* Cf. Gibson, *Calculus*, 221-2.

For the path along L from A to B together with the path L' from B to A form a closed circuit in the region, and if u_i is irrotational the integral around it will be zero. But in this circuit the part L' is traversed in the opposite direction; hence the result stated follows. Thus $\int_A^B u_i dx_i$ depends only on the termini and not on the intervening path, and can be expressed in the form $\phi_B - \phi_A$, where ϕ is a scalar function of position. Now take a point B' with coordinates $(x_1 + \delta x_1, x_2, x_3)$. To get from A to B' we can go to B and then on to B', and

$$\frac{1}{\delta x_1}(\phi_{B'} - \phi_B) = \frac{1}{\delta x_1}\int_B^{B'} u_i dx_i \to u_1, \tag{4}$$

whence
$$\frac{\partial}{\partial x_1}\phi_B = u_1, \tag{5}$$

and by symmetry
$$u_i = \frac{\partial \phi}{\partial x_i}.$$

Hence an irrotational vector is the gradient of a scalar. Again,

$$\frac{\partial u_i}{\partial x_k} = \frac{\partial^2 \phi}{\partial x_i \partial x_k} = \frac{\partial u_k}{\partial x_i}, \tag{6}$$

and therefore an irrotational vector has zero curl. Conversely, if a vector has zero curl everywhere in a region we can apply Stokes's theorem to show that its circulation about any circuit capable of being filled by a cap in the region is zero and therefore it is irrotational, and therefore, by the above argument, it is the gradient of a scalar.

If u_i is both solenoidal and irrotational, ϕ exists and then, since $\partial u_i/\partial x_i = 0$,

$$\frac{\partial^2 \phi}{\partial x_i^2} = \frac{\partial^2 \phi}{\partial x_1^2} + \frac{\partial^2 \phi}{\partial x_2^2} + \frac{\partial^2 \phi}{\partial x_3^2} = 0, \tag{7}$$

which is written compactly
$$\nabla^2 \phi = 0. \tag{8}$$

EXAMPLES

1. Prove that, if $-\pi < \alpha < \pi$,

$$\int_0^\infty \int_0^\infty \exp(-x^2 - 2xy\cos\alpha - y^2)\,dx\,dy = \frac{\alpha}{2\sin\alpha}.$$

2. S, T are the fixed points $(0, 0, \pm \tfrac{1}{2}R)$, and P is the variable point (x, y, z); the distances PS, PT are denoted by s, t respectively. If

$$\xi = \frac{s+t}{R}, \quad \eta = \frac{s-t}{R},$$

and ϕ is the angle between the plane PST and the plane $y = 0$, show that

$$\left|\frac{\partial(\xi, \eta, \phi)}{\partial(x, y, z)}\right| = \frac{8}{R^3(\xi^2 - \eta^2)}.$$

Hence prove that
$$\iiint \frac{1}{st} e^{-(s+t)/R}\,dx\,dy\,dz = \frac{2\pi R}{e},$$

the integral being taken over all space. (Prelim. 1942.)

3. By using the transformation
$$x+iy = (u+iv)^2$$
or otherwise, evaluate
$$\iint (x^2+y^2)^{-1/2} \, dx \, dy$$
taken over the region enclosed by arcs of the confocal parabolas
$$y^2 = 4a_r(x+a_r) \quad (r = 1, 2, 3),$$
where
$$a_1 > a_2 > 0, \quad a_3 < 0.$$

4. Prove that the integral
$$\int \frac{dS}{p}$$
taken over the surface of an ellipsoid of semi-axes a, b, c, where p is the length of the perpendicular from the centre on to the tangent plane at a point of the ellipsoid, is equal to
$$\tfrac{4}{3}\pi abc \left(\frac{1}{a^2} + \frac{1}{b^2} + \frac{1}{c^2} \right).$$

Evaluate
$$\int \frac{dS}{p^3}. \tag{M.T. 1940.}$$

5. Determine the new limits in the following integrals, when the orders of integration are reversed:
$$\int_0^1 dx \int_{x^2}^1 f(x, y) \, dy, \quad \int_\beta^{1/2\pi} d\theta \int_0^{b \operatorname{cosec} \theta} f(r, \theta) \, dr. \tag{I.C. 1940.}$$

6. Express the integral
$$\int_0^a dz \int_0^z dy \int_0^{\sqrt{(yz-y^2)}} dx$$
in spherical polar coordinates, and show that its value is $\pi a^3/24$. \hfill (I.C. 1938.)

7. If K_n is a homogeneous polynomial of positive integral degree n in the coordinates, satisfying $\nabla^2 K_n = 0$, and S is a sphere of radius a about the origin, prove that
$$\iint \left(\frac{\partial K_n}{\partial x_i} \right)^2 dS = \frac{n(2n+1)}{a^2} \iint K_n^2 \, dS.$$

8. If A and B are two vector functions of position, prove that
$$\operatorname{div}(A \wedge B) = B \cdot \operatorname{curl} A - A \cdot \operatorname{curl} B.$$
If further A and B are functions of the time and are connected by the relations
$$\frac{dA}{dt} = \operatorname{curl} B, \quad \frac{dB}{dt} = -\operatorname{curl} A,$$
and τ is a volume enclosed by a fixed surface S, prove that
$$\frac{1}{2} \frac{d}{dt} \iiint (A^2 + B^2) \, d\tau = -\iint (A \wedge B) \cdot dS. \tag{M/c, Part III, 1931.}$$

9. Prove that
$$\operatorname{div}(\phi A) = \phi \operatorname{div} A + A \cdot \operatorname{grad} \phi.$$
If $\operatorname{div} D = 4\pi\rho$, $E = -\operatorname{grad} \phi$, $D = KE$, where K is independent of E, show that, if ϕ is $O(1/r)$ at infinity
$$\frac{1}{2} \iiint \rho\phi \, d\tau = \frac{1}{8\pi} \iiint E \cdot D \, d\tau,$$
the integrals being taken through all space.

Chapter 6

POTENTIAL THEORY

'But all that moveth doth Mutation love.'
SPENSER, *The Faerie Queene*, Bk. 7

6·01. $1/r$ as a solution of $\nabla^2 \phi = 0$. Let x_i be the coordinates of a point P and r its distance from the origin. Then

$$\frac{\partial}{\partial x_i}\frac{1}{r} = \frac{d}{dr}\left(\frac{1}{r}\right)\frac{\partial r}{\partial x_i} = -\frac{x_i}{r^3}, \tag{1}$$

$$\frac{\partial^2}{\partial x_i \partial x_k}\frac{1}{r} = -\frac{\partial}{\partial x_k}\frac{x_i}{r^3} = -\frac{1}{r^3}\frac{\partial x_i}{\partial x_k} + \frac{3 x_i x_k}{r^5}$$

$$= \frac{3 x_i x_k - r^2 \delta_{ik}}{r^5}. \tag{2}$$

Now put $k = i$ and apply the summation convention; since

$$x_i x_i = r^2, \quad \delta_{ii} = 3,$$

$$\frac{\partial^2}{\partial x_i \partial x_i}\frac{1}{r} = 0, \tag{3}$$

so that $1/r$ is a solution of Laplace's equation, $\nabla^2 \phi = 0$, except at the origin.

It follows at once that if ξ_i are the coordinates of another point Q, and

$$r^2 = (x_i - \xi_i)^2$$

(summed), then

$$\frac{\partial^2}{\partial x_i^2}\frac{1}{r} = 0, \tag{4}$$

except at $x_i = \xi_i$. Note that $\dfrac{\partial}{\partial x_i}\dfrac{1}{r} = -\dfrac{\partial}{\partial \xi_i}\dfrac{1}{r}$, a result that will be needed repeatedly. Further, if we take n points $Q_1 \ldots Q_n$ and denote their coordinates by ξ_{is}, and the various distances $Q_s P$ by r_s, then

$$\nabla^2 \sum_{s=1}^{n} \frac{a_s}{r_s} = 0, \tag{5}$$

where a_s are any constants, except when any of the r_s is 0, that is, when P coincides with any of the points Q_s. Differentiation is of course understood to be with regard to the coordinates of P. Hence with this restriction any function of the form $\sum_{s=1}^{n} a_s/r_s$ is a solution of Laplace's equation.

Now the gravitational potential due to a distribution of particles is of this form. So is the electrostatic potential due to a set of point charges. Hence both satisfy Laplace's equation. This equation arises also in the hydrodynamics of an incompressible fluid. For if u_i is the velocity at $P(x_i)$ the condition that the mass within any closed surface is constant requires that u_i is a solenoidal vector; and for any circuit capable of being filled by a cap occupied wholly by fluid that has never passed near a solid boundary the circulation $\int u_i dx_i$ round it is practically zero, so that to a good approximation u_i is also an irrotational vector. In classical hydrodynamics we adopt the approximation

$$u_i = \frac{\partial \phi}{\partial x_i},$$

where ϕ is a scalar function of the coordinates, satisfying Laplace's equation, and called the *velocity potential*. The solutions of this equation therefore contain the whole of the part of hydrodynamics that neglects viscosity and treats the fluid as having a constant density independent of pressure and of any other complication such as variation of temperature or composition. These conditions are not satisfied in any real fluid, but in many actual motions of fluids about solid boundaries they are satisfied within the observational uncertainty except in parts of the fluid that have passed close to a solid; and the modern development called *boundary layer theory* deals with the latter regions, which are usually thin.

If r is the distance of P from the origin, $r_s/r \to 1$ when $r \to \infty$. Hence when r is large enough

$$r\phi = r \sum_{s=1}^{n} \frac{a_s}{r_s} \to \sum_{s=1}^{n} a_s. \tag{6}$$

Thus unless $\Sigma a_s = 0$, ϕ behaves for large r like $1/r$. If $\Sigma a_s = 0$, ϕ will decrease more rapidly than $1/r$.

Now consider the flux of the gradient of $1/r$ through a sphere of radius a about the origin. The direction cosines of the outward normal are x_i/r; hence

$$\iint l_i \frac{\partial}{\partial x_i} \frac{1}{r} dS = -\iint \frac{x_i}{r} \frac{x_i}{r^3} dS = -\iint \frac{1}{a^2} dS.$$

But the area of the sphere is $4\pi a^2$; hence

$$\iint \frac{\partial}{\partial n}\left(\frac{1}{r}\right) dS = \iint l_i \frac{\partial}{\partial x_i} \frac{1}{r} dS = -4\pi. \tag{7}$$

This can be extended to any closed surface surrounding the origin. For if we take such a surface S, and take a sphere Σ large enough to enclose S completely, we can apply Green's lemma to the region $\Sigma - S$ between S and Σ. Using $\partial/\partial \nu$ to denote differentiation along the outward normal from this region, which will be outwards from O over Σ and inwards over S, we have

$$\iiint_{\Sigma-S} \frac{\partial^2}{\partial x_i^2}\left(\frac{1}{r}\right) dx_1 dx_2 dx_3 = \iint_{S,\Sigma} \frac{\partial}{\partial \nu}\left(\frac{1}{r}\right) dS. \tag{8}$$

The left side is zero. The contribution to the right from Σ is -4π. Hence

$$\iint_S \frac{\partial}{\partial \nu}\left(\frac{1}{r}\right) dS = 4\pi. \tag{9}$$

Since $d\nu$ is here taken towards the side containing O, it follows that if we take dn outwards from O

$$\iint_S \frac{\partial}{\partial n}\left(\frac{1}{r}\right) dS = -4\pi, \tag{10}$$

which extends (7) to any surface enclosing the origin. (An easy application of Green's lemma shows that this integral is zero if S does not enclose the origin.)

In classical hydrodynamics, if the flow is radial and symmetrical about the origin, and u is the velocity at distance r from the origin, u is a function of r only. The rate of outflow through a sphere of radius r is then $4\pi r^2 u$. If the region between two such spheres of radii r_1 and r_2 is filled with fluid throughout the motion, it follows that $r_1^2 u_1 = r_2^2 u_2$; thus

$$u = \frac{m}{r^2}, \tag{11}$$

where $4\pi m$ is the volume of fluid issuing from the origin per unit time. Then

$$\phi = -\frac{m}{r}, \tag{12}$$

apart from an irrelevant constant. If m is positive, this is the velocity potential of a *source* emitting volume $4\pi m$ of fluid per unit time; if m is negative, it is that of a *sink*. That due to any set of sources and sinks is obtainable by addition as for gravitation and electrostatics.

The resemblance between these three branches of mathematical physics, to which the flow of electric current in a uniform conductor may be added, is so close that the mathematical theory common to all is most conveniently developed in one piece. With a slight modification much of the theory is also applicable to magnetism. It is known as *potential theory*.

It follows at once from (10) that if

$$\phi = \sum_{s=1}^{n} \frac{a_s}{r_s}, \tag{13}$$

$$\iint \frac{\partial \phi}{\partial n} dS = -4\pi \Sigma' a_s, \tag{14}$$

where the summation in Σ' covers all the a_s corresponding to the points Q_s that lie within S; for if Q_s lies within S, a_s/r_s contributes $-4\pi a_s$ to the sum, and if Q_s lies outside S the term contributes 0. This is a case of *Gauss's theorem*.

6·02. Continuous distributions. The application of these results to continuous distributions meets with two difficulties, one mathematical and one physical. If we identify our particles with protons and electrons, the gravitational and electrical potentials will consist of finite sums of the form just discussed. In practice, however, the number of elements in any piece of matter of ordinary size is so large that the working out of the sum would be impossible, and also there are additional forces at short distances. But just for this reason another method becomes possible. An integral is the limit of a finite sum when the number of intervals becomes very large, their total length remaining the same. Thus a sum over a large finite number of intervals is a good approximation to the integral; but, conversely, the integral is a good approximation to the sum. This suggests that instead of taking the matter or charge as concentrated in separate points we may take it as distributed continuously through the volume, in such a way as to keep approximately the same total mass or charge in any given region containing many elements. It is obviously impossible to make the totals exactly the same for every region. For if so we could take a closed surface surrounding a single element; the mass within it will tend to zero when the surface is taken small enough if the distribution is continuous, but to a finite limit for the actual distribution. The approximation is in fact good only for expressing average properties of regions containing many elements, and will not deal with individual particles. The former type of properties are usually called *molar* or *macroscopic*, the latter *molecular* or *microscopic*. (It need not be inferred from this choice of words that an actual microscope is capable of observing individual molecules!) If, for instance, a region not exceptionally long in one dimension in comparison with the others contains n elements, where n is large, in the actual distribution, and the equivalent mass or charge of $n \pm \sqrt{n}$ in the continuous one, we have the sort of approximation required. The density at each point can be identified with the ratio of the total mass in such a region to its volume, and similarly for the charge density, and both will be finite everywhere.

The physical difficulty is that if a solid consisted entirely of stationary particles, under no forces except inverse square ones, it could not be stable and would collapse to zero volume. This has been partly met either by supposing additional repulsive forces, considerable at short distances, but falling off with distance faster than r^{-2}, or by supposing the particles to be in rapid orbital motion. In the former case the additional forces must be studied separately; this method has been adopted especially by Born and Lennard-Jones in the theories of crystals and gases. In the latter the potential due to a given body, even apparently at rest, will be a rapidly varying function of the time. The quantum theory aims at combining both suggestions into a single hypothesis. The various solutions all make the forces between particles follow the inverse square law so long as the distance is considerably greater than 10^{-8} cm., and their mean values over intervals of the order of 10^{-17} sec. change little from one such interval to the next. Consequently the mathematical and physical difficulties can be met in the same way: the predictions of the inverse square law will be right provided that we apply them only to changes of mean position or momentum, during intervals of time longer than about 10^{-17} sec., of the matter within regions greater than about 10^{-8} cm. in linear extent, and when they are right they will be approximately the same as those for continuous distributions that preserve the same total mass or charge within such regions.

In our formula for ϕ we therefore replace a_s by $\rho \, d\xi_1 d\xi_2 d\xi_3$, so that ξ_i specifies a point of the distribution. We continue to use x_i for the point where ϕ is being evaluated. Then we are led to study the function

$$\phi = \gamma \iiint \frac{\rho}{R} d\xi_1 d\xi_2 d\xi_3, \tag{1}$$

where now ρ will be a function of ξ_1, ξ_2, ξ_3, the coordinates of a point Q; x_1, x_2, x_3 are coordinates of a point P, and R is the distance QP given by

$$R^2 = (x_i - \xi_i)^2. \tag{2}$$

γ is a constant with different values in different branches of physics. For shortness we write $d\xi_1 d\xi_2 d\xi_3 = d\tau$, so that $d\tau$ is an element of volume.

We suppose as before that $\rho = 0$ for points at more than a given distance from the origin. Then again when r is large

$$r\phi \to \gamma \iiint \rho \, d\tau. \tag{3}$$

If $\rho = 0$ throughout any region, $\nabla^2 \phi = 0$ in that region. For if we differentiate (1) under the integral sign we get

$$\frac{\partial^2 \phi}{\partial x_1^2} = \gamma \iiint \frac{\rho}{R^5} \{3(x_1 - \xi_1)^2 - R^2\} d\xi_1 d\xi_2 d\xi_3. \tag{4}$$

For values of R sufficiently small, Q will be in the region where $\rho = 0$. The integrand is therefore a continuous function of x_1, x_2, x_3, wherever R does not vanish, and the ranges of integration are finite. Differentiation under the integral sign is therefore justified. Then, by addition, we get as before $\nabla^2 \phi = 0$.

Important intermediate cases between discrete particles and continuous distributions are *surface and line distributions*, in which the mass or charge per unit area or unit length respectively is finite. It is again obvious that the potential satisfies Laplace's equation except possibly on the surface or line.

6·03. Uniform spherical shell. Before proceeding further with the general theory we consider a few important special cases. Take first a uniform surface density σ over a sphere of radius a. ϕ is obviously independent of direction and therefore is a function of r only, and it must satisfy Laplace's equation except actually on the sphere. Now if accents denote differentiation with respect to r, and ϕ is a function of r only

$$\frac{\partial \phi}{\partial x_i} = \frac{x_i}{r}\phi', \quad \frac{\partial^2 \phi}{\partial x_i \partial x_k} = \frac{\delta_{ik}}{r}\phi' - \frac{x_i x_k}{r^3}\phi' + \frac{x_i x_k}{r^2}\phi'', \tag{1}$$

$$\nabla^2 \phi = \frac{3}{r}\phi' - \frac{\phi'}{r} + \phi'' = \phi'' + \frac{2\phi'}{r} = \frac{d^2}{r\,dr^2}(r\phi). \tag{2}$$

Hence since $\nabla^2 \phi = 0$ everywhere except on the sphere

$$\phi = A + \frac{B}{r}, \tag{3}$$

where A and B are constants. But when r is large

$$r\phi \to \gamma \iint \sigma\, dS = 4\pi\gamma a^2 \sigma. \tag{4}$$

Therefore outside the sphere $A = 0$ and $B = 4\pi\gamma a^2 \sigma$; and*

$$\phi = \frac{4\pi\gamma a^2 \sigma}{r} \quad (r > a). \tag{5}$$

At the centre ϕ is obviously finite and equal to $4\pi\gamma a\sigma$; hence B must be 0 within the sphere and

$$\phi = 4\pi\gamma a\sigma \quad (r < a). \tag{6}$$

Thus ϕ has two distinct analytic forms inside and outside the sphere. Its radial derivative is

$$\left.\begin{aligned}\frac{\partial \phi}{\partial r} &= -\frac{4\pi\gamma a^2 \sigma}{r^2} \quad (r > a),\\ &= 0 \quad (r < a),\end{aligned}\right\} \tag{7}$$

* No stronger evidence for the need for an adequate notation could be required than that provided by the history of this formula. There is good reason to suppose that Newton delayed for about twenty years his publication of the comparison between the gravitational acceleration of the moon with gravity at the earth's surface because he was unable to prove that the attraction of a sphere at all external points was the same as that of a particle of equal mass at its centre. But the proof from Laplace's equation is so easy as to be almost trivial. Why did Newton not discover Laplace's equation? Presumably because his fluxion notation did not allow him to contemplate more than one independent variable and could not express partial derivatives.

and therefore has a discontinuity $-4\pi\gamma\sigma$ as r increases through a. ϕ itself, however, is continuous on crossing the sphere. This is a general result: the potential is continuous when a surface distribution is crossed, but the normal derivative has a discontinuity of $-4\pi\gamma\sigma$.

6·031. Uniform sphere. Next consider a uniform sphere of density ρ and radius a. We can imagine it built up of concentric spherical shells of radius α, thickness $\delta\alpha$, and therefore surface densities $\rho\delta\alpha$. Then at points outside the sphere

$$\phi = \frac{4\pi\gamma\rho}{r}\int_0^a \alpha^2 d\alpha = \frac{4}{3}\frac{\pi\gamma\rho a^3}{r} \quad (r>a), \tag{8}$$

as we should expect. For internal points it is easiest to work with $\partial\phi/\partial r$, since shells with $\alpha > r$ make no contribution to this. We have

$$\frac{\partial\phi}{\partial r} = -\gamma\int_0^r \frac{4\pi\rho\alpha^2}{r^2} d\alpha = -\tfrac{4}{3}\pi\gamma\rho r \quad (r<a). \tag{9}$$

At the centre
$$\phi = 4\pi\gamma\rho \int_0^a \alpha\, d\alpha = 2\pi\gamma\rho a^2, \tag{10}$$

and therefore
$$\phi = 2\pi\gamma\rho(a^2 - \tfrac{1}{3}r^2) \quad (r<a). \tag{11}$$

Hence both ϕ and $\partial\phi/\partial r$ are continuous on crossing the sphere, for

$$\lim_{r\to a}\frac{4}{3}\frac{\pi\gamma\rho a^3}{r} = \lim_{r\to a} 2\pi\gamma\rho(a^2-\tfrac{1}{3}r^2) = \tfrac{4}{3}\pi\gamma\rho a^2, \tag{12}$$

$$\lim_{r\to a}\left(-\frac{4}{3}\frac{\pi\gamma\rho a^3}{r^2}\right) = \lim_{r\to a}\left(-\tfrac{4}{3}\pi\gamma\rho r\right) = -\tfrac{4}{3}\pi\gamma\rho a. \tag{13}$$

But ϕ for $r<a$ does not satisfy Laplace's equation; for

$$\frac{\partial}{\partial x_i}(-\tfrac{2}{3}\pi\gamma\rho r^2) = -\tfrac{4}{3}\pi\gamma\rho x_i,$$

$$\nabla^2(-\tfrac{2}{3}\pi\gamma\rho r^2) = -4\pi\gamma\rho. \tag{14}$$

This is *Poisson's equation* and expresses a general property of the potential inside matter. It follows that when the density is discontinuous we can expect the potential and its first derivatives to be continuous but at least one second derivative to be discontinuous.

6·032. Spherical cap. Consider now the potential on the axis of a segment of a sphere of radius a and semi-angle α.

We have
$$\phi = \gamma\iint\frac{\sigma\, dS}{R} = \gamma\sigma\int_0^\alpha\int_0^{2\pi}\frac{a^2\sin\theta\, d\theta\, d\lambda}{R}, \tag{15}$$

θ and λ being spherical polar coordinates of Q. But

$$R^2 = a^2 + r^2 - 2ar\cos\theta, \tag{16}$$
$$R\, dR = ar\sin\theta\, d\theta, \tag{17}$$

for variations of Q; hence

$$\phi = 2\pi\gamma\sigma\int_{\theta=0}^{\alpha}\frac{a\, dR}{r} = 2\pi\gamma\sigma\frac{a}{r}\{(a^2+r^2-2ar\cos\alpha)^{1/2} - |r-a|\}, \tag{18}$$

and again $\partial\phi/\partial r$ has a discontinuity $-4\pi\gamma\sigma$ when r increases through a. R for $\theta = 0$ must of course be taken positive, $|r-a|$, whether $r > a$ or $r < a$, being simply the distance AP. If $\alpha = \pi$, so that the segment becomes the whole surface of a sphere, we recover the formulae (5).

If a becomes large, with $r - a = x$ and $a\sin\alpha = b$ fixed, we get the case of a circular disk of radius b. In this case

$$\phi = 2\pi\gamma\sigma\{\sqrt{(b^2+x^2)} - |x|\}, \tag{19}$$

$$\frac{\partial\phi}{\partial x} = 2\pi\gamma\sigma\left\{\frac{x}{\sqrt{(b^2+x^2)}} - \frac{x}{|x|}\right\}, \tag{20}$$

which tends to $+2\pi\gamma\sigma$ or $-2\pi\gamma\sigma$ as $x \to 0$ from negative or positive values respectively. Formulae (18) (20) are correct, of course, only when P is on the line of symmetry. But they show how the discontinuity in the normal derivative maintains the same value even though the values on one side range from zero to $2\pi\gamma\sigma$ as a varies.

6·033. Line density. As an example of a line density consider a distribution λ per unit length along the axis of z, from $\zeta = -a$ to $+a$. We have

$$\phi = \gamma\lambda \int_{-a}^{a} \frac{d\zeta}{\{x^2+y^2+(z-\zeta)^2\}^{1/2}}. \tag{21}$$

If we put
$$(x^2+y^2)^{1/2} = \varpi, \quad z - \zeta = \varpi\tan\theta, \quad R = \{x^2+y^2+(z-\zeta)^2\}^{1/2} = \varpi\sec\theta,$$

$$R_1 = \{x^2+y^2+(z-a)^2\}^{1/2} = \varpi\sec\theta_1, \quad R_2 = \{x^2+y^2+(z+a)^2\}^{1/2} = \varpi\sec\theta_2,$$

this becomes
$$\gamma\lambda \int_{\theta_1}^{\theta_2} \sec\theta\, d\theta = \gamma\lambda \log\left(\frac{\sec\theta_2 + \tan\theta_2}{\sec\theta_1 + \tan\theta_1}\right) = \gamma\lambda \log\frac{R_2 + z + a}{R_1 + z - a}$$

$$= \gamma\lambda \log\frac{R_1 + R_2 + 2a}{R_1 + R_2 - 2a} \tag{22}$$

since
$$4az = R_2^2 - R_1^2.$$

If $R_1 + R_2 \to 2a$, this tends to $+\infty$ logarithmically; but if R_1 and R_2 are both large the limit is finite. If a is very large and x, y, z not large

$$\phi \doteqdot -2\gamma\lambda \log\frac{\varpi}{2a}. \tag{23}$$

This is a solution of Laplace's equation; for

$$\frac{\partial}{\partial x}\log(x^2+y^2) = \frac{2x}{x^2+y^2}, \quad \frac{\partial^2}{\partial x^2}\log(x^2+y^2) = \frac{2(x^2+y^2) - 4x^2}{(x^2+y^2)^2} = \frac{2(y^2-x^2)}{(x^2+y^2)^2},$$

$$\frac{\partial^2}{\partial y^2}\log(x^2+y^2) = \frac{2(x^2-y^2)}{(x^2+y^2)^2}, \quad \frac{\partial^2}{\partial z^2}\log(x^2+y^2) = 0. \tag{24}$$

Thus $\log \varpi/2a$ is a two-dimensional solution of Laplace's equation. Its gradient is independent of a, which is needed only because $\log \varpi$ is, strictly, meaningless because ϖ is a length, not a number. This is the simplest possible case of another general result: near a line density λ the potential tends to infinity like $-2\gamma\lambda \log \varpi/b$, where ϖ is the shortest distance to the line and b is some fixed length, even though the line may be curved.

6·034. A closely related two-dimensional potential function, as we shall see in the theory of the complex variable, is

$$\phi = \tan^{-1}\frac{y}{x}. \tag{25}$$

For here
$$\frac{\partial \phi}{\partial x} = -\frac{y}{x^2}\frac{1}{1+y^2/x^2} = -\frac{y}{x^2+y^2}, \quad \frac{\partial^2 \phi}{\partial x^2} = \frac{2xy}{(x^2+y^2)^2},$$

$$\frac{\partial \phi}{\partial y} = \frac{1}{x}\frac{1}{1+y^2/x^2} = \frac{x}{x^2+y^2}, \quad \frac{\partial^2 \phi}{\partial y^2} = -\frac{2xy}{(x^2+y^2)^2},$$

and
$$\nabla^2 \phi = 0. \tag{26}$$

The magnitude of $\operatorname{grad} \phi$ is $1/\varpi$, as for $\phi = \log \varpi$, but its direction is along the circle $\varpi = $ constant whereas that due to $\phi = \log \varpi/b$ is radial. ϕ is not a single-valued function of x and y, but its derivatives are single-valued.

Volume densities are good approximations to those that arise in problems of gravitation, and surface densities to those of electrostatics. Line densities are more difficult to realize in the theory of attractions, but the particular potential distribution $\log(\varpi/b)$ is important in two-dimensional hydrodynamics and the flow of electricity in plane sheets. The form $\tan^{-1} y/x$ occurs in vortex motion and the magnetic field due to an electric current.

6·035. Doublet. Another type of potential of theoretical importance is that of a *doublet* or *dipole*. If

$$\phi = \gamma\left(\frac{A}{R_1} - \frac{A}{R_2}\right), \tag{27}$$

where
$$R_1^2 = (x-a)^2 + y^2 + z^2, \quad R_2^2 = (x+a)^2 + y^2 + z^2, \tag{28}$$

and we make $a \to 0$, $A \to \infty$, in such a way that $2aA \to \mu$,

$$\phi \to -\gamma\mu \frac{\partial}{\partial x}\frac{1}{r} = \gamma\mu \frac{x}{r^3}. \tag{29}$$

This is a solution of Laplace's equation except at $r = 0$, since

$$\nabla^2 \phi = -\gamma\mu \nabla^2 \frac{\partial}{\partial x}\frac{1}{r} = -\gamma\mu \frac{\partial}{\partial x}\nabla^2 \frac{1}{r} = 0. \tag{30}$$

ϕ is called the potential due to a doublet of moment μ at the origin, directed along the axis of x. For a doublet of moment μ in a direction l_i at ξ_i

$$\phi = -\gamma\mu l_i \frac{\partial}{\partial x_i}\frac{1}{R} = \gamma\mu l_i \frac{\partial}{\partial \xi_i}\frac{1}{R}, \tag{31}$$

where
$$R^2 = (x_i - \xi_i)^2. \tag{32}$$

The direction l_i is called that of the *axis* of the doublet.

The doublet field represents closely that of a small bar magnet, and occurs in many physical problems. The equipotential surfaces are closed, with the axis of the doublet as axis of symmetry, and all touch at the doublet.

6·036. A doublet shell is a distribution of doublets over a surface so that their axes are all along the outward normal to the surface. Nothing like it exists, but some general theorems make use of its properties. If μdS is the moment per element of area dS, and l_i are the direction cosines of the normal,

$$\phi = -\gamma \iint \mu l_i \frac{\partial}{\partial x_i}\left(\frac{1}{R}\right) dS = \gamma \iint \mu l_i \frac{\partial}{\partial \xi_i}\left(\frac{1}{R}\right) dS. \quad (33)$$

This is capable of a simple transformation. Take P to be the point x_i, Q to be ξ_i, and let QP make an angle χ with the normal at Q. Then

$$l_i \frac{\partial}{\partial \xi_i} \frac{1}{R} = \frac{l_i(x_i - \xi_i)}{R^3} = \frac{\cos \chi}{R^2}. \quad (34)$$

If a cone is drawn from P to the boundary of dS, and we draw around P a sphere of radius α, the area of the intercept on this sphere is

$$\alpha^2 d\omega = \alpha^2 \frac{\cos \chi}{R^2} dS. \quad (35)$$

The ratio $d\omega$ is called the element of *solid angle* subtended by dS at P, and the sum of the $\alpha^2 d\omega$ is the intercept on the sphere by a cone joining P to the boundary of S; and

$$\phi = \gamma \iint \mu \, d\omega, \quad (36)$$

a remarkably simple form. It is important to attend to the sign of $d\omega$, which is that of $\cos \chi$. In the figure $d\omega$ is positive for ϕ_P, negative for ϕ_{P_1} and ϕ_{P_2}. Evidently the potential at a great distance behaves like r^{-2} instead of r^{-1}; and if μ is constant over a closed surface $\phi = 0$ at all external points and $\phi = -4\pi\gamma\mu$ at all internal ones. Hence ϕ has a discontinuity $4\pi\gamma\mu$ on crossing the surface outwards. For a uniform plane doublet sheet ϕ jumps from $-2\pi\gamma\mu$ to $2\pi\gamma\mu$ on crossing it.

Consider a uniform doublet shell filling the half-plane $y = 0$, $x < 0$, with the axes directed towards positive y. The solid angle subtended at P is $\theta/2\pi$ times that subtended by a whole sphere about P, and therefore is 2θ, where θ is $\tan^{-1} y/x$. Hence

$$\phi = 2\gamma\mu\theta = 2\gamma\mu \tan^{-1} \frac{y}{x}. \quad (37)$$

When $\theta \to \pi$, $\phi \to 2\pi\gamma\mu$; when $\theta \to -\pi$, $\phi \to -2\pi\gamma\mu$, and therefore again there is a discontinuity $4\pi\gamma\mu$ on crossing the shell in the direction of the axes of the doublets. In this case ϕ does not tend to 0 as $r \to \infty$, but that is because the sheet does not satisfy the condition that all matter is within some given finite distance from the origin.

To sum up, if we call a surface of simple discontinuity of density a discontinuity of zero order in the density, we can call a surface density a discontinuity of order -1, and a doublet shell one of order -2. All can be regarded as limits of continuous distributions. If $\rho = \rho_0 \tanh x/a$, and $a \to 0$, $\rho \to \rho_0$ for $x > 0$ and $\to -\rho_0$ for $x < 0$. If $\rho = \frac{\sigma}{a\sqrt{\pi}} \exp\left(-\frac{x^2}{a^2}\right)$, $\rho \to 0$ as $a \to 0$ for any $x \neq 0$, but $\int_{-h}^{h} \rho \, dx \to \sigma$ however small h may be. If $\rho = \frac{2}{a^3 \sqrt{\pi}} \mu x \exp\left(-\frac{x^2}{a^2}\right)$, $\rho \to 0$ for any $x \neq 0$, $\int_{-h}^{h} \rho \, dx \to 0$, $\int_{-h}^{h} \rho x \, dx \to \mu$. The order is that of increasing irregularity in the continuous distribution needed to give the requisite properties in the limit: in the respective cases $\rho_{\max} = O(1)$, $O(a^{-1})$ and $O(a^{-2})$, where a represents the linear scale of the distribution. The corresponding orders of discontinuity in the potential, measured by the orders of the lowest discontinuous derivatives, are 2, 1, and 0. The rule can be extended to discontinuities in higher derivatives; in most practical problems the order of the discontinuity of the potential is higher by 2 than that of the density.

6·04. Potential and field inside a continuous distribution. We have now to consider more general cases. $\nabla^2 \phi = 0$ always holds when P is outside matter. When P is inside matter (which would be impossible for particle distributions), ρ does not vanish at P and the integrand tends to infinity. It is therefore necessary to define ϕ as an improper integral by first excluding a small region about P from the region of integration and then taking the limit of the integral when the diameter of the small region tends to zero. In order that the limit shall exist we shall require that its value shall be the same for every shape of the excluded region.

Lemma. The integral $\iiint \dfrac{d\tau}{R^m}$, *where R is measured from a point P in the volume of integration V, converges if $m < 3$; for all regions of the same volume the integral is largest for a sphere with centre P for $0 < m < 3$.*

Take τ to be the region inside a sphere of radius a with centre P and let τ' be a region entirely inside τ. Then

$$\iiint_{\tau-\tau'} \frac{1}{R^m} d\tau = \iiint_{\tau-\tau'} R^{2-m} d\omega \, dR$$

$$\leqslant 4\pi \int_\tau R^{2-m} dR = \frac{4\pi}{3-m} a^{3-m} \quad (m<3). \tag{1}$$

For $m < 3$, given any positive ϵ, we can take a so that $\dfrac{4\pi}{3-m} a^{3-m} < \epsilon$; then there is a sphere of radius a such that

$$\iiint_{\tau-\tau'} \frac{1}{R^m} d\tau < \epsilon \tag{2}$$

for any τ' included in τ. Hence if we exclude a cavity around P from the volume of integration the integral tends to a unique limit, independent of the shape of the cavity, as we make all its dimensions tend to zero.

Suppose now that we replace the volume of integration V by a sphere with centre P of radius b with the same volume. Then

$$\underbrace{\iiint \frac{1}{R^m} d\tau}_{\text{Sphere}} - \underbrace{\iiint \frac{1}{R^m} d\tau}_{V} = \underbrace{\iiint \frac{1}{R^m} d\tau}_{\substack{\text{Parts of sphere} \\ \text{not in } V}} - \underbrace{\iiint \frac{1}{R^m} d\tau}_{\substack{\text{Parts of } V \text{ not} \\ \text{in sphere}}}. \tag{3}$$

In the integrands on the right the volumes of integration are equal; in the first $R < b$ and in the second $R > b$ except on parts of the boundary, where $R = b$. Hence the right side is positive.

In what follows we shall assume throughout that ρ is integrable, and this implies that it is bounded.

6·04 Internal potential

We have to consider the two integrals for the potential and the intensity of field X_i,

$$\phi = \gamma \iiint \frac{\rho}{R} d\tau, \tag{4}$$

$$X_i = -\gamma \iiint \frac{\rho}{R^3}(x_i - \xi_i) d\tau. \tag{5}$$

These are both improper integrals. Consider a spherical region τ about P and a cavity τ' of any form, included in τ and enclosing P. If we consider the contribution from the region $\tau - \tau'$ and denote the upper bound of $|\rho|$ within the sphere by ρ_m we have

$$\left| \iiint_{\tau-\tau'} \frac{\rho \, d\tau}{R} \right| \leq \rho_m \iiint_{\tau-\tau'} \frac{d\tau}{R} \tag{6}$$

and

$$\left| \iiint_{\tau-\tau'} \frac{\rho}{R^3}(x_i - \xi_i) d\tau \right| \leq \rho_m \iiint_{\tau-\tau'} \frac{d\tau}{R^2}. \tag{7}$$

Hence from the lemma the integrals on the right can be made as small as we please, and those for ϕ and X_i converge to limits independent of the shape of the cavity.

X_i is obtained formally from ϕ by differentiating under the integral sign, but to show that it is actually equal to $\partial \phi / \partial x_i$ we have to show not only that the integrals exist but also that, if $P'(x_i + h l_i)$ is a point in the neighbourhood of P, then given any positive ϵ we can find h_0 such that for all $0 < h < h_0$

$$\left| \frac{\phi(P') - \phi(P)}{h} - l_i X_i \right| < \epsilon. \tag{8}$$

We divide the volume of integration into (i) the interior of a sphere τ_1 with centre P and radius a and (ii) the remainder of the region τ_0. We denote the contributions to the integrals from these two parts by suffixes 1 and 0 respectively. Let P' be any other point interior to τ_1. We shall show that (i) for any given positive ϵ the contribution of τ_1 to the left side of (8) can be made less than $\frac{1}{2}\epsilon$ for any P' by taking a sufficiently small, (ii) with a fixed we can find h_0 such that the contribution from τ_0 is also less than $\frac{1}{2}\epsilon$ for all $h < h_0 < a$.

(i) We denote distances from P' by R'. Then

$$\phi_1(P') - \phi_1(P_1) = \gamma \iiint_{\tau_1} \rho \left(\frac{1}{R'} - \frac{1}{R} \right) d\tau = \gamma \iiint_{\tau} \rho \frac{R - R'}{RR'} d\tau. \tag{9}$$

We have

$$|R - R'| \leq h < a, \tag{10}$$

$$\frac{1}{RR'} \leq \frac{1}{2}\left(\frac{1}{R^2} + \frac{1}{R'^2} \right). \tag{11}$$

Hence

$$\left| \frac{\phi_1(P') - \phi_1(P)}{h} \right| \leq \frac{\gamma \rho_m}{2} \iiint_{\tau_1} \left(\frac{1}{R^2} + \frac{1}{R'^2} \right) d\tau. \tag{12}$$

But, from the second part of the lemma, since $\iiint_{\tau_1} \frac{1}{R'^2} d\tau$ is taken over the interior of a sphere with its centre not at the origin, it is $\leq \iiint_{\tau_1} \frac{d\tau}{R^2}$ which is taken over the interior of a sphere of the same volume with its centre at the origin. Hence

$$\left| \frac{\phi_1(P') - \phi_1(P)}{h} \right| \leq \gamma \rho_m \iiint \frac{d\tau}{R^2} = 4\pi \gamma \rho_m a. \tag{13}$$

Also, since

$$|l_i(x_i - \xi_i)| \leq R, \tag{14}$$

$$|l_i X_{i1}| \leq \gamma \rho_m \iiint \frac{d\tau}{R^2} = 4\pi \gamma \rho_m a. \tag{15}$$

Hence
$$\left| \frac{\phi_1(P') - \phi_1(P)}{h} - l_i X_{i1} \right| \leq 8\pi\gamma\rho_m a, \tag{16}$$

for all P' in τ_1. Given any positive ϵ we choose a so that
$$8\pi\gamma\rho_m a < \tfrac{1}{2}\epsilon. \tag{17}$$

(ii) Since P and P' are both external points to τ_0, for a given value of a,
$$\left| \frac{\phi_0(P') - \phi_0(P)}{h} - l_i X_{i0} \right|$$
can be made less than any given positive $\tfrac{1}{2}\epsilon$ by taking $h < h_0 < a$, where h_0 may depend on a. We then have that
$$\left| \frac{\phi(P') - \phi(P)}{h} - l_i X_i \right| \leq \left| \frac{\phi_1(P') - \phi_1(P)}{h} - l_i X_{i1} \right| + \left| \frac{\phi_0(P') - \phi_0(P)}{h} - l_i X_{i0} \right| < \epsilon \tag{18}$$
for all $h < h_0$.

Therefore, since ϵ is arbitrarily small,
$$\phi(P') = \phi(P) + l_i h X_i + \omega h, \tag{19}$$
where $\omega \to 0$ with h uniformly with respect to l_i; hence ϕ is differentiable and has derivatives X_i, and
$$X_i = \frac{\partial \phi}{\partial x_i} = \gamma \iiint \rho \frac{\partial}{\partial x_i}\left(\frac{1}{R}\right) d\tau. \tag{20}$$

Now apply Green's lemma to the region between a small sphere Σ about P and the surface S of the body. We have
$$\left(\frac{\partial \phi}{\partial x_i}\right)_0 = -\gamma \iiint_{\tau_0} \rho \frac{\partial}{\partial \xi_i}\left(\frac{1}{R}\right) d\tau. \tag{21}$$

Apply Green's lemma to this integral; this will be justified, in the conditions so far adopted for the lemma, if ρ has integrable derivatives. In surface integrals we take the direction of the normal away from P; then
$$\left(\frac{\partial \phi}{\partial x_i}\right)_0 = \gamma \iiint_{\tau_0} \frac{\partial \rho}{\partial \xi_i} \frac{d\tau}{R} - \gamma \iint_S l_i \rho \frac{dS}{R} + \gamma \iint_\Sigma l_i \rho \frac{dS}{R}. \tag{22}$$

When the radius of Σ tends to 0 the last integral tends to 0; hence
$$\frac{\partial \phi}{\partial x_i} = -\gamma \iint_S l_i \rho \frac{dS}{R} + \gamma \iiint \frac{\partial \rho}{\partial \xi_i} \frac{d\tau}{R}, \tag{23}$$
the last being an improper integral through the interior of S. Hence X_i is the sum of two potential functions, one due to a surface density $-l_i \rho$ over S, the other to density $\partial \rho / \partial \xi_i$ through the interior of S. Since ρ has been supposed to have integrable derivatives, $\partial \rho / \partial \xi_i$ is integrable and bounded in the region. Hence $\partial \phi / \partial x_i$ in its turn is differentiable in the region.

Note that the argument up to (20) assumes only that ρ is integrable; ϕ exists and has first derivatives at a finite discontinuity of ρ, as for instance at a free surface. The further condition that ρ has integrable derivatives is used in the transformation leading to (23).

Note also that if a real cavity is made in the body and the distribution of ρ outside the cavity is kept unaltered, the field in the cavity will be given by X_{i0} and therefore is arbitrarily near X_i if the cavity is small enough.

6·041. Gauss's theorem. Let us consider the flux of $\partial\phi/\partial x_i$ through a closed surface S. The points x_i are taken to be on S. ρ is a function of the ξ_i only and does not involve x_i; hence

$$\iint l_i \frac{\partial \phi}{\partial x_i} dS = \gamma \iint l_i dS \iiint \rho \frac{\partial}{\partial x_i}\left(\frac{1}{R}\right) d\xi_1 d\xi_2 d\xi_3$$
$$= \gamma \iiint \rho \, d\xi_1 d\xi_2 d\xi_3 \iint l_i \frac{\partial}{\partial x_i}\left(\frac{1}{R}\right) dS, \tag{24}$$

since inversion of the order of integration is permissible provided ρ is continuous and S has a finite area (actually in somewhat wider conditions). Now in the integration over S the ξ_i are to be treated as constants. If a point Q with coordinates ξ_i lies outside S, then by Green's lemma

$$\iint l_i \frac{\partial}{\partial x_i}\left(\frac{1}{R}\right) dS = \iiint \nabla^2\left(\frac{1}{R}\right) dx_1 dx_2 dx_3 \tag{25}$$

taken through the interior of S. If Q is outside S, $\nabla^2(1/R) = 0$ at all points within S and the integral is zero. If Q is inside S, we have, as in 6·01 (7),

$$\iint l_i \frac{\partial}{\partial x_i}\left(\frac{1}{R}\right) dS = -4\pi. \tag{26}$$

Hence
$$\iint l_i \frac{\partial \phi}{\partial x_i} dS = -4\pi\gamma \iiint \rho \, d\tau, \tag{27}$$

where the range of integration for τ is through the interior of S. This extends Gauss's theorem to continuous distributions.

6·042. First treatment of Poisson's equation. Gauss's theorem provides an easy, though not satisfactory, way of getting Poisson's equation. The last integral in (27) is unaltered if we now replace ξ_i by x_i everywhere, including ρ; and the first integral is

$$\iiint \nabla^2 \phi \, dx_1 dx_2 dx_3. \tag{28}$$

Thus
$$\iiint \nabla^2 \phi \, dx_1 dx_2 dx_3 = -4\pi\gamma \iiint \rho \, dx_1 dx_2 dx_3. \tag{29}$$

This must be true for every region, and therefore

$$\nabla^2 \phi = -4\pi\gamma\rho \tag{30}$$

almost everywhere. This is *Poisson's equation*, and gives the appropriate modification of Laplace's equation inside matter.

6·043. Second treatment of Poisson's equation. This is one way of getting Poisson's equation, but is not altogether satisfactory for three reasons. (i) The use of Green's lemma to derive (28) from (27) has been completely justified only if $\partial\phi/\partial x_i$ have integrable derivatives with regard to all the coordinates; we have proved that the derivatives $\partial^2\phi/\partial x_i \partial x_k$ exist in certain conditions, but they might not be integrable. (ii) There is also a complication from points Q actually on S, but this is easily treated. For in that case the integral in (26) is easily seen to be bounded, though not equal to either 0 or -4π; and since the total volume of the points on S is 0 it does not affect the right of (27). (iii) For one variable, if for all x of a range $\int_a^x f(t)\,dt = \int_a^x g(t)\,dt$, it follows only that $f(x) = g(x)$ almost everywhere; they might differ at any set of values of x capable of being enclosed in ranges of arbitrary small total length. The proof that $f(x) = g(x)$ everywhere is easily completed if it is known that $f(x)$ and $g(x)$ are continuous, but in the present case it has not been proved that $\nabla^2\phi$ is continuous even if ρ is.

The following treatment starts from 6·04 (23). If we consider the contributions to $\partial\phi/\partial x_i$ from the exterior of a small sphere of radius a it is clear from 6·04 (21) that $\nabla^2\phi_0 = 0$. For the contributions from the interior of Σ we can apply 6·04 (23) to the interior of Σ; then

$$\left(\frac{\partial^2\phi}{\partial x_i \partial x_k}\right)_0 = -\gamma\iint_\Sigma l_i \rho \frac{\partial}{\partial x_k}\left(\frac{1}{R}\right) dS + \gamma\iiint_{\tau_1} \frac{\partial\rho}{\partial \xi_i} \frac{\partial}{\partial x_k}\left(\frac{1}{R}\right) d\tau$$

$$= \gamma\iint_\Sigma l_i \rho \frac{\partial}{\partial \xi_k}\left(\frac{1}{R}\right) dS - \gamma\iiint_{\tau_1} \frac{\partial\rho}{\partial \xi_i} \frac{\partial}{\partial \xi_k}\left(\frac{1}{R}\right) d\tau. \tag{31}$$

In this the second integral tends to zero with a. For the first, if we put $i = k$ and sum,

$$\gamma\iint_\Sigma \rho l_i \frac{\partial}{\partial \xi_i}\left(\frac{1}{R}\right) dS = -\gamma\iint \frac{\rho}{R^2} dS. \tag{32}$$

Since ρ is continuous at P the integral tends to $4\pi\rho_P$ as $a \to 0$; hence

$$\nabla^2\phi = -4\pi\gamma\rho. \tag{33}$$

We have proved this under the (sufficient) condition that ρ has integrable derivatives. This condition is not necessary. Consider a heterogeneous sphere of radius a, with ρ a function of r. The potential at an internal point is

$$\phi = \frac{4\pi\gamma}{r}\int_0^r \rho\alpha^2 d\alpha + 4\pi\gamma\int_r^a \rho\alpha\, d\alpha,$$

$$\frac{\partial\phi}{\partial x_i} = \frac{x_i}{r}\frac{d\phi}{dr} = \frac{4\pi\gamma x_i}{r}\left\{-\frac{1}{r^2}\int_0^r \rho\alpha^2 d\alpha + \frac{1}{r}\rho r^2 - \rho r\right\} = -\frac{4\pi\gamma x_i}{r^3}\int_0^r \rho\alpha^2 d\alpha,$$

$$\frac{\partial^2\phi}{\partial x_i \partial x_k} = -4\pi\gamma\left\{\left(\frac{\delta_{ik}}{r^3} - \frac{3x_i x_k}{r^5}\right)\int_0^r \rho\alpha^2 d\alpha + \frac{x_i x_k}{r^3}\frac{\rho r^2}{r}\right\},$$

$$\nabla^2\phi = -4\pi\gamma\rho,$$

with no restriction on ρ except that it is continuous at the point considered and integrable for $0 \leqslant r \leqslant a$. (See also Note **6·043a**.)

6·05. Surface distributions. Surface density. Let us suppose that over a surface S there is a bounded concentration σ per unit area; then the potential at P is

$$\phi = \gamma\iint \frac{\sigma}{R} dS. \tag{1}$$

If P is on the surface this is to be interpreted as an improper integral by excluding the interior of a curve C on the surface about P and then making C shrink up to P. Evidently Laplace's equation is satisfied at all points not on S.

Take a point O on S, and the axis of x_3 so that at every point of S within a certain non-zero distance from O, except possibly at O itself, there is a normal to S making an angle with the x_3 axis differing from a right angle by more than some fixed amount. If S is a smooth surface we need only take the axis of x_3 to be the normal at O. If S has a conical point at O (not a cusp) we can take the axis of x_3 within the angle. Then we can take a curve C about O so that σ/l_3 is bounded within C, where l_3 is the third direction cosine of the normal. Then if P is $(0, 0, x_3)$ and we take a closed curve c on S so that c is entirely within C the contribution to ϕ from the part of the surface between c and C is

$$\gamma \iint \frac{\sigma}{l_3} \frac{d\xi_1 d\xi_2}{\{\xi_1^2 + \xi_2^2 + (\xi_3 - x_3)^2\}^{1/2}} \tag{2}$$

over the part stated. If we take cylindrical coordinates ϖ, λ

$$|\sigma/l_3| < K, \tag{3}$$

$$\xi_1^2 + \xi_2^2 + (\xi_3 - x_3)^2 \geqslant \varpi^2, \tag{4}$$

and (2) has modulus $\leqslant \gamma K \iint d\varpi d\lambda \leqslant 2\pi\gamma aK, \tag{5}$

where a is the greatest value of ϖ on C. This is arbitrarily small, and therefore even for $x_3 = 0$ the improper integral has the same value irrespective of the limiting form of C. Hence ϕ has a definite value even if P is on S.

Now take C so that $4\pi\gamma aK < \frac{1}{2}\epsilon$. Then take x_3 so small that the contributions to ϕ at P and O, say ϕ and ϕ_0, from the part of S outside C differ by less than $\frac{1}{2}\epsilon$; then

$$|\phi - \phi_0| < 4\pi\gamma aK + \tfrac{1}{2}\epsilon < \epsilon. \tag{6}$$

Hence ϕ approaches ϕ_0 continuously when P approaches O from either side.

ϕ is also continuous on S. For if P is on S, within C, and at a distance r from O, the largest distance of P from any point on C is $\leqslant a + r \leqslant 2a$. Hence the contribution to $|\phi - \phi_0|$ from the interior of C is $\leqslant 6\pi\gamma aK$, by (5). We take a so that this is $\leqslant \tfrac{1}{2}\epsilon$ and then take r so small that the contributions from parts outside C differ by less than $\tfrac{1}{2}\epsilon$. Then $|\phi - \phi_0| < \epsilon$, and we can find a region on S about O such that $|\phi - \phi_0| < \epsilon$ at all points of it. Hence ϕ is continuous everywhere.

If S has an edge passing through O, or if a finite number of edges meet at O, the results follow on applying the argument to each face separately and adding.

The results still follow for a cusp if σ is bounded. These cover the cases possible for surfaces not intersected an infinite number of times by any straight line.

Referring back to 6·04 (23) we see that *at a simple discontinuity of a volume density* $\partial\phi/\partial x_i$ *is continuous*. For $\partial\phi/\partial x_i$ can be expressed as the sum of two potential functions, one due to a volume density and the other to a surface density. But both these potentials are continuous, by what we have just proved.

The normal gradient of ϕ is discontinuous on S. If dn is an element of the normal to S measured in the same sense on both sides of S, we indicate the value of $\partial/\partial n$ on the side where dn is positive as we recede from S by suffix 0, and on the other side by suffix 1. We show that in suitable conditions

$$\left(\frac{\partial\phi}{\partial n}\right)_0 - \left(\frac{\partial\phi}{\partial n}\right)_1 = -4\pi\gamma\sigma. \tag{7}$$

This result is found most easily by using Gauss's theorem, but the argument is unsatisfactory for similar reasons to those given with regard to the derivation of Poisson's equation. Direct study of the expression for $\partial\phi/\partial n$ is more satisfactory. We take a small piece of the tangent plane at a point O of the surface as a standard of comparison. We take O as origin and the tangent plane there as $x_3 = 0$. We assume that the surface has finite curvatures at and near O. Take C to be such that all points of it are at distance a from the axis of x_3, and consider the contribution to $\partial\phi/\partial x_3$ at points on the axis of x_3 from the part of S within C. This is

$$\gamma \iint \sigma \frac{\xi_3 - x_3}{R^3} dS = \gamma \iint \tau \frac{\xi_3 - x_3}{R^3} d\xi_1 d\xi_2 \tag{8}$$

for $\xi_1^2+\xi_2^2 < a^2$, where $\tau = \sigma/l_3$. We have to consider the error introduced by taking $\xi_3 = 0$. For any function whose derivative exists

$$f(\xi_3) = f(0) + \xi_3 p, \qquad (9)$$

where $p = f'(\theta \xi_3)$, $0 < \theta < 1$. But

$$\left| \frac{\partial}{\partial \xi_3} \frac{\xi_3 - x_3}{R^3} \right| = \left| \frac{1}{R^3} - \frac{3(\xi_3 - x_3)^2}{R^5} \right| \leqslant \frac{2}{R^3} \qquad (10)$$

and

$$\frac{\xi_3 - x_3}{R^3} = -\frac{x_3}{R_0^3} + \frac{2A\xi_3}{R_1^3}, \qquad (11)$$

where

$$R_0^2 = \xi_1^2 + \xi_2^2 + x_3^2 = \varpi^2 + x_3^2, \qquad (12)$$

$$|A| < 1, \qquad (13)$$

and R_1 lies between R_0 and R. Then

$$\gamma \iint_C \sigma \frac{\xi_3 - x_3}{R^3} dS = -\gamma \int_0^a \int_0^{2\pi} \frac{\tau x_3}{R_0^3} \varpi\, d\varpi\, d\lambda + \gamma \int_0^a \int_0^{2\pi} 2A\tau \frac{\xi_3}{R_1^3} \varpi\, d\varpi\, d\lambda. \qquad (14)$$

In the second integral on the right, $R_1 \geqslant \varpi$, and since the surface is assumed to have a finite curvature $\xi_3 = O(\varpi^2)$. The integral is therefore small of order a. For any ϵ we can therefore choose a so that the second integral is less than $\tfrac{1}{8}\epsilon$.

Again we have $\varpi\, d\varpi = R_0 dR_0$, since x_3 is taken constant; then

$$-\gamma \int_0^a \int_0^{2\pi} \frac{\tau x_3}{R_0^3} \varpi\, d\varpi\, d\lambda = -\gamma \int_{|x_3|}^{\sqrt{(a^2 + x_3^2)}} \int_0^{2\pi} \frac{\tau x_3}{R_0^2} dR_0\, d\lambda; \qquad (15)$$

τ is supposed continuous. This is equal to

$$-2\pi\gamma\tau_1 \left\{ \frac{x_3}{|x_3|} - \frac{x_3}{\sqrt{(a^2 + x_3^2)}} \right\}, \qquad (16)$$

where τ_1 lies between the upper and lower bounds of τ within C. Since τ is continuous we can put $\tau = \sigma + \eta$, where σ is evaluated at O and equal to τ there, and $\eta \to 0$ with a. Also the modulus of the terms in brackets is < 1. Hence if a is small enough (15) differs from

$$-2\pi\gamma\sigma \left\{ \frac{x_3}{|x_3|} - \frac{x_3}{\sqrt{(a^2 + x_3^2)}} \right\}$$

by less than $\tfrac{1}{8}\epsilon$. Next, by taking δ sufficiently small compared with a we can make the second term in the brackets as small as we like, say $< \tfrac{1}{8}\epsilon$, for all $|x_3| < \delta$. In all (8) will differ from $-2\pi\gamma\sigma x_3/|x_3|$ by less than $\tfrac{3}{8}\epsilon$. It therefore changes discontinuously by an amount arbitrarily near to $-4\pi\gamma\sigma$ when x_3 increases through 0.

Further, since the contribution to $\partial\phi/\partial x_3$ from the part of S outside C is continuous, we can choose x_3 so that its values at O and P differ by less than $\tfrac{1}{8}\epsilon$. Then for values of x_3 such that $|x_3| < \delta$, but x_3 has opposite signs, the difference between the values of $\partial\phi/\partial x_3$ on the two sides of S differs from $-4\pi\gamma\sigma$ by less than ϵ.

Here a has to be chosen *first*, so as to make the differences arising in (14) and (15) from the variations of ξ_3 and τ within C each $< \tfrac{1}{8}\epsilon$; then x_3 is chosen so that those arising from the second term in (16) and from $\tau - \sigma$ are $< \tfrac{1}{8}\epsilon$, and the contribution from the parts of S outside C differs by less than $\tfrac{1}{8}\epsilon$ from its value at O; thus we show that $\partial\phi/\partial x_3$ tends to a limit as $x_3 \to 0$ from either side, the difference between it and its limiting value on that side being less than $\tfrac{1}{2}\epsilon$; and finally the difference between the limiting values on the two sides differs from $-4\pi\gamma\sigma$ by less than ϵ, which is arbitrarily small.

The argument for the tangential derivatives proceeds similarly. We have

$$\frac{\partial\phi}{\partial x_1} = \gamma \iint \sigma \frac{\xi_1 - x_1}{R^3} dS = \gamma \iint \tau \frac{\xi_1 - x_1}{R^3} d\xi_1\, d\xi_2. \qquad (17)$$

The only serious modification of the argument is in the treatment of the integral over the circle in the tangent plane. The integrand here is an even function of x_3 and the integral must therefore have the same limit when $x_3 \to 0$ from either side if it has a limit at all. Now

$$\int \tau \frac{\xi_1 - x_1}{R^3} d\xi_1 = -\int_{\xi_1 = -\sqrt{(a^2 - \xi_2^2)}}^{\sqrt{(a^2 - \xi_2^2)}} \tau d\left(\frac{1}{R}\right) = -\left[\frac{\tau}{R}\right] + \int \frac{1}{R} \frac{\partial \tau}{\partial \xi_1} d\xi_1,$$

in which both terms are potential functions. Hence if τ is differentiable $\partial \phi / \partial x_1$ and $\partial \phi / \partial x_2$ exist at all points of $O3$ and are continuous.

These various results require conditions of different stringency on the form of the surface and the distribution of surface density. Continuity of ϕ is satisfied if σ is bounded and integrable and S meets no given line more than a fixed number of times. The normal gradient has discontinuity $-4\pi\gamma\sigma$ if σ is continuous and S has finite curvatures. The tangential derivatives of ϕ are continuous if σ has integrable derivatives and S has finite curvatures. Some of these conditions can be somewhat relaxed, but the cases where they are not satisfied and the theorems remain true are rare.

6·06. Doublet shell. We have already had

$$\phi = \gamma \iint \mu \, d\omega.$$

If, as before, we take O on the surface and a small circuit C around it, the contribution to ϕ from the part of S outside C is continuous. For the part within C, if μ is continuous, the contribution can be made as near as we like to $\gamma \mu_0 \iint d\omega$. But this increases by $4\pi\gamma\mu_0$ when the point considered passes through O.

There is a relation between this result and the discontinuity in $\partial \phi / \partial n$ for a surface density. If we displace the whole surface by $-dn$ parallel to $O3$, the change in ϕ will be to the first order in dn equal to that of a distribution of doublets of moment density $-\sigma dn$ with axes parallel to the x_3 axis, and for points close to this axis the axes will be nearly normal to the surface. This relation can be presented formally in various ways so as to show directly that the discontinuity in $\partial \phi / \partial n$ due to surface density σ is the same as that in ϕ due to doublet density $-\sigma$, but the partial integrations are tricky.

6·07. Uniqueness theorem for solutions of Laplace's equation. Let ϕ and ϕ' be two different functions satisfying Laplace's equation within a closed surface S and having continuous first and second derivatives. Then their difference ϕ'' also satisfies it. But by Green's theorem

$$\iiint \left(\frac{\partial \phi''}{\partial x_i}\right)^2 d\tau = \iiint \left\{\frac{\partial}{\partial x_i}\left(\phi'' \frac{\partial \phi''}{\partial x_i}\right) - \phi'' \nabla^2 \phi''\right\} d\tau$$

$$= \iint \phi'' \frac{\partial \phi''}{\partial n} dS. \tag{1}$$

Now if either ϕ and ϕ', or $\partial \phi / \partial n$ and $\partial \phi' / \partial n$, are equal at all points of S, this integral vanishes. But the integral on the left is not less than 0 and can vanish only if

$$\frac{\partial \phi''}{\partial x_i} = 0 \tag{2}$$

at all points within S. Hence ϕ'' is a constant and will be 0 if it is 0 on S. Hence (1) if ϕ is given over S, it is uniquely determined inside S, (2) if $\partial \phi / \partial n$ is given over S, ϕ is uniquely determined inside S except for an additive constant.

6·071. The same result is true if ϕ and ϕ' satisfy Laplace's equation outside a closed surface S, provided that they decrease sufficiently rapidly at large distances. Take a sphere of large radius A surrounding S and apply (1) to the region between S and this sphere. $d\nu$ on S will now be towards the interior of S. Now let ϕ, ϕ' be given to tend to 0 at large distances and $\partial\phi/\partial x_i$, $\partial\phi'/\partial x_i$ to be $O(1/r^2)$ when r is large. Then ϕ and ϕ' are $O(1/r)$. Then the integral over the large sphere is $O(1/A)$ and decreases indefinitely as $A \to \infty$. We therefore have

$$\iiint \left(\frac{\partial \phi''}{\partial x_i}\right)^2 d\tau = \iint \phi'' \frac{\partial \phi''}{\partial \nu} dS,$$

where the integration on the left is through all space outside of S. Hence the integral on the left vanishes if $\phi = \phi'$ or $\partial\phi/\partial n = \partial\phi'/\partial n$ at all points of S; then $\partial\phi''/\partial x_i = 0$ outside S. In this case $\phi'' \to 0$ at infinity and the difference must be zero at all points. Thus if ϕ or $\partial\phi/\partial n$ is given at all points of S and satisfies Laplace's equation outside it, while $\phi \to 0$ and $\partial\phi/\partial x_i = O(1/r^2)$ when $r \to \infty$, ϕ is uniquely determined at all points outside S.

The theorems remain true if ϕ is given over part of S and $\partial\phi/\partial n$ is given over the remainder. For, forming ϕ'' as before, the integrals over S still vanish.

6·072. Minimal theorems. Let ϕ satisfy Laplace's equation within S and let ϕ' be a function not satisfying Laplace's equation but such that $\phi' = \phi$ at all points of S. Put $\phi' - \phi = \phi''$. Then

$$\iiint \left\{ \left(\frac{\partial \phi'}{\partial x_i}\right)^2 - \left(\frac{\partial \phi}{\partial x_i}\right)^2 - \left(\frac{\partial \phi''}{\partial x_i}\right)^2 \right\} d\tau = 2 \iiint \frac{\partial \phi}{\partial x_i} \frac{\partial \phi''}{\partial x_i} d\tau$$

$$= 2 \iiint \left\{ \frac{\partial}{\partial x_i}\left(\phi'' \frac{\partial \phi}{\partial x_i}\right) - \phi'' \nabla^2 \phi \right\} d\tau$$

$$= 2 \iint \phi'' \frac{\partial \phi}{\partial n} dS = 0. \tag{1}$$

Hence
$$\iiint \left(\frac{\partial \phi'}{\partial x_i}\right)^2 d\tau \geq \iiint \left(\frac{\partial \phi}{\partial x_i}\right)^2 d\tau. \tag{2}$$

The solution of Laplace's equation therefore makes $\iiint \left(\frac{\partial \phi}{\partial x_i}\right)^2 d\tau$ a minimum subject to the given boundary conditions. The extension to the region outside S, subject to the same restrictions as in the last theorem about the behaviour of ϕ, ϕ' for large r, is simple.

6·073. A related theorem due to Kelvin is as follows. ϕ is taken to satisfy $\nabla^2 \phi = 0$ within S, and u_i is a solenoidal vector such that $l_i u_i = \partial\phi/\partial n$ over S. Put

$$u_i = \frac{\partial \phi}{\partial x_i} + v_i. \tag{3}$$

Then $l_i v_i = 0$ on S;
$$\frac{\partial v_i}{\partial x_i} = \frac{\partial u_i}{\partial x_i} - \nabla^2 \phi = 0,$$

$$\iiint \left\{ u_i^2 - \left(\frac{\partial \phi}{\partial x_i}\right)^2 - v_i^2 \right\} d\tau = 2 \iiint v_i \frac{\partial \phi}{\partial x_i} d\tau$$

$$= 2 \iiint \left\{ \frac{\partial}{\partial x_i}(\phi v_i) - \phi \frac{\partial v_i}{\partial x_i} \right\} d\tau$$

$$= 2 \iint \phi l_i v_i dS = 0,$$

and therefore
$$\iiint u_i^2 d\tau \geqslant \iiint \left(\frac{\partial \phi}{\partial x_i}\right)^2 d\tau. \tag{4}$$

Thus $\iiint u_i^2 d\tau$ is a minimum, given that u_i is a solenoidal vector within S and has a given normal component on S, if u_i is the gradient of a solution of $\nabla^2 \phi = 0$.

These theorems prove that subject to the given conditions there cannot be more than one solution. The proof that one solution exists is difficult.*

6·074. Uniqueness theorem in two dimensions. The above theorems apply equally well to two-dimensional problems for the interiors of closed curves and cylinders.

Modification is needed for external problems because in two important cases the vector u_i in two dimensions is $O(1/r)$, not $O(1/r^2)$, when r' is large. For a cylinder with a finite charge per unit length the potential at a large distance behaves like $\log r$. For two plates extending to infinity with asymptotes inclined at θ_1 and θ_2 to the x axis, and with different limiting values of ϕ on them at large distances, the conditions will be satisfied if, for large r, $\phi = A + B\theta$, which is a solution of Laplace's equation. B must be zero for the exterior of a closed curve to make ϕ single-valued; but it can be non-zero if the boundary is such that a complete circuit about the origin is impossible for large r without crossing the boundary. In both cases u_i will be $O(1/r)$. But then $\iint u_i^2 dS$ diverges since $dS = r\, dr\, d\theta$. It therefore becomes meaningless to say that if we alter u_i we shall increase this integral. The same applies if the boundary, which we shall now call C, goes to infinity along two curves with parallel asymptotes, since if ϕ tends to different limits on them the normal gradient tends to a constant and again the integral diverges. The minimal theorems therefore do not hold in two dimensions unless there is some further restriction on the behaviour of u_i at large distances.

If, however, we take the extra conditions in the uniqueness theorem to be, for a closed curve, that a, b exist so that for large r

$$\phi - a\log\frac{r}{b} \to 0, \quad \phi' - a\log\frac{r}{b} \to 0,$$

$$\frac{\partial}{\partial r}\left(\phi - a\log\frac{r}{b}\right) = O(1/r^2), \quad \frac{\partial}{\partial r}\left(\phi' - a\log\frac{r}{b}\right) = O(1/r^2),$$

we have
$$\phi'' = O(1/r), \quad \frac{\partial \phi''}{\partial r} = O(1/r^2),$$

and for a large circle
$$\iint \phi'' \frac{\partial \phi''}{\partial n} ds = O(1/r^2).$$

Hence the uniqueness theorem still holds if the behaviour of ϕ and ϕ' at large distances is suitably restricted.

Evidently, from Green's theorem, $\int (\partial \phi/\partial n)\, ds$ has the same value for C and any closed curve enclosing it. This integral can be taken as a datum in many problems, since in electrostatics it is related directly to the total charge and in hydrodynamics to the total rate of outflow, and we can restrict ourselves to variations that preserve these. But if this integral is finite and not zero the radial component must tend to zero like $1/r$, and the potential will behave like $\log r$, with a given coefficient.

---

* O. D. Kellogg, *Foundations of Potential Theory*, 236 and 277–328.

If C goes to infinity in two directions we can take ϕ, ϕ' to behave like $a\theta + b$ for large r, a and b being chosen so as to fit the limiting values of ϕ on C in the two directions. If we take
$$\phi - a\theta - b = o(1), \qquad \phi' - a\theta - b = o(1),$$
$$\frac{\partial}{\partial r}(\phi - a\theta - b) = o(1/r), \quad \frac{\partial}{\partial r}(\phi' - a\theta - b) = o(1/r),$$
then
$$\int \phi'' \frac{\partial \phi''}{\partial n} ds \to 0$$

for a large arc, and the uniqueness theorem holds. This applies equally well to a semi-infinite plate, for which we can take the limiting values of θ on the two faces to be 0 and 2π.

If C goes to infinity with two parallel asymptotes, on which ϕ has limits, we can take the external boundary to be a curve normal to both. A sufficient extra condition to ensure uniqueness is now that $\partial \phi''/\partial n \to 0$ on any curve cutting the two asymptotes normally at a large distance.

6·08. The Rayleigh-Ritz method. Minimal theorems are the basis of a useful method of numerical solution used by Rayleigh and Ritz and justified by Krylov. The usual method of solution would be to solve the partial differential equation and then combine solutions so as to satisfy the boundary conditions. But when the differential equation is equivalent to the principle that a quadratic form is stationary it is possible to choose a function f_0 that satisfies the boundary conditions and a set of functions $f_1, f_2, \ldots$ that contribute nothing to the boundary values; then

$$\phi' = f_0 + a_1 f_1 + a_2 f_2 + \ldots,$$

satisfies the boundary conditions. But if, as here, the correct solution makes $\iiint \left(\frac{\partial \phi}{\partial x_i}\right)^2 d\tau$ a minimum we can substitute ϕ' for ϕ, evaluate the various integrals numerically, and then determine $a_1, a_2, \ldots$ so as to make the resulting expression a minimum. If the functions f_r are such that any twice differentiable function can be expressed in terms of them the solution is theoretically complete, and can, with sufficient effort, give accurate numerical solutions where the formal solution of the differential equation is too complicated. An extension due to Richardson and Southwell does not even require the functions f_r to be given explicitly. In this method a rectangular network of points is taken at sufficiently close intervals, and the integral to be minimized is expressed directly in terms of the values at these points, using centred finite differences. The values of the function are then solved for directly by successive approximation. The method is laborious, but, as Southwell remarks, will always work if the computer will. Whether it is more laborious than tabulating solutions of a partial differential equation depends on special circumstances.

6·09. Green's equivalent stratum. Let ϕ satisfy Laplace's equation within a closed surface S, and let R be the distance of the point $Q(\xi_i)$ from $P(x_i)$. If P is within S, take a small sphere σ about P and apply Green's theorem to the region between σ and S. If P is not within S, apply the theorem to the interior of S directly. In either case $\nabla^2(1/R) = 0$ within the region used, and we have for P within S

$$\iint \phi \frac{\partial}{\partial n}\left(\frac{1}{R}\right) dS + \iint \phi \frac{\partial}{\partial \nu}\left(\frac{1}{R}\right) d\sigma = \iint \frac{1}{R} \frac{\partial \phi}{\partial n} dS + \iint \frac{1}{R} \frac{\partial \phi}{\partial \nu} d\sigma, \tag{1}$$

where $d\nu$ is normally outwards from the region towards P, and therefore equal to $-dR$. For P not within S the two integrals over σ do not arise, and we have simply

$$\iint \left\{ \phi \frac{\partial}{\partial n}\left(\frac{1}{R}\right) - \frac{1}{R}\frac{\partial \phi}{\partial n} \right\} dS = 0, \tag{2}$$

$\partial \phi / \partial n$ being the derivative approaching S from the inside.

For P within S, and σ of radius a, which is arbitrarily small,

$$\iint \phi \frac{\partial}{\partial \nu} \frac{1}{R} d\sigma \to \iint \frac{\phi}{a^2} a^2 d\omega = 4\pi \phi_P, \tag{3}$$

$$\iint \frac{1}{R} \frac{\partial \phi}{\partial \nu} d\sigma = O\left\{\iint a\, d\omega\right\} \to 0. \tag{4}$$

Thus
$$4\pi \phi_P = \iint \left\{ \frac{1}{R}\frac{\partial \phi}{\partial n} - \phi \frac{\partial}{\partial n}\left(\frac{1}{R}\right) \right\} dS. \tag{5}$$

This equation is a three-dimensional analogue of the theorem (11·13) in the theory of functions of a complex variable

$$f(z) = \frac{1}{2\pi i} \int_C \frac{f(t)}{t-z} dt, \tag{6}$$

when $f(z)$ is analytic within C and z is within C. It suffices to determine ϕ at all points within a surface given ϕ and its normal derivative on the surface. The term in $1/R$ is the potential due to a surface density on S, that in $\partial(1/R)/\partial n$ that of a doublet shell on S. But these distributions cannot be assigned independently; for we see from (2) that if P' is a point external to S and R' the distance QP', there is a relation

$$\iint \phi \frac{\partial}{\partial n}\left(\frac{1}{R'}\right) dS = \iint \frac{1}{R'} \frac{\partial \phi}{\partial n} dS \tag{7}$$

for every such position of P'. We should expect this since we know that if there is a solution at all, the values of either ϕ or $\partial \phi/\partial n$ over the boundary are sufficient to determine it, and therefore either will determine the other, apart possibly from an additive constant. The complex variable analogue is that either the real or the imaginary part of a function $f(z)$ over a closed contour will determine the other, apart from an additive constant, subject to $f(z)$ being analytic within the contour.

With a suitable restriction on ϕ at a large distance the result can be adapted to determine ϕ outside a surface, given ϕ and $\partial \phi/\partial n$ on the surface. For suppose that ϕ satisfies $\nabla^2 \phi = 0$ outside S and tends to 0 at a large distance like $1/r$, r being the distance from a fixed point within S. P is outside S; we draw a large sphere Σ enclosing S and P, and a small sphere σ about P as before, and apply the theorem to the region between S, σ, and Σ. We have

$$4\pi \phi_P = \iint \left\{ \frac{1}{R}\frac{\partial \phi}{\partial \nu} - \phi \frac{\partial}{\partial \nu}\left(\frac{1}{R}\right) \right\} dS + \iint \left\{ \frac{1}{R}\frac{\partial \phi}{\partial \nu} - \phi \frac{\partial}{\partial \nu}\left(\frac{1}{R}\right) \right\} d\Sigma, \tag{8}$$

where $d\nu$ is still taken outwards from the region used and therefore, on S, is from the outside. On Σ, ϕ and $1/R$ are of order $1/r$, and $\partial \phi/\partial \nu$ and $\partial(1/R)/\partial \nu$ of order $1/r^2$. Hence the integral over Σ is of order $\iint d\Sigma/r^3$ and tends to 0 when $r \to \infty$. Taking dn outwards from S,

$$4\pi \phi_P = \iint \left\{ \phi \frac{\partial}{\partial n}\left(\frac{1}{R}\right) - \frac{1}{R}\frac{\partial \phi}{\partial n} \right\} dS. \tag{9}$$

Comparing this with (2) we see that they can be consistent only if either ϕ or $\partial\phi/\partial n$ is discontinuous on crossing S, and by subtraction

$$4\pi\phi_P = \iint\left[(\phi_0-\phi_1)\frac{\partial}{\partial n}\left(\frac{1}{R}\right) - \frac{1}{R}\left\{\left(\frac{\partial\phi}{\partial n}\right)_0 - \left(\frac{\partial\phi}{\partial n}\right)_1\right\}\right]dS, \qquad (10)$$

suffixes 0 and 1 indicating limits on approaching S from the outside and inside respectively. If ϕ is continuous across S,

$$4\pi\phi_P = -\iint\frac{1}{R}\left\{\left(\frac{\partial\phi}{\partial n}\right)_0 - \left(\frac{\partial\phi}{\partial n}\right)_1\right\}dS. \qquad (11)$$

If $\partial\phi/\partial n$ is continuous,

$$4\pi\phi_P = \iint(\phi_0-\phi_1)\frac{\partial}{\partial n}\left(\frac{1}{R}\right)dS. \qquad (12)$$

The field can therefore be represented in terms of a distribution either of charge or of doublets over S. Consequently all theorems derived from the integral definition of a potential are true given only that the function satisfies Laplace's equation in the region.

Unfortunately, direct application of these theorems to find the internal or external field is seldom possible. The integrand in each of the equations, given either ϕ or $\partial\phi/\partial n$ over S, involves the other, and to find the latter usually involves the complete solution of the problem, and the potential at P will be found in the process. It can, however, be carried out in the important special cases of a sphere, a circle, and a plane. The case of the circle is treated in Chapter 14.

6·091. Solution for a sphere: external point. Let ϕ be given over a sphere of radius a, P an external point at distance r, and P' its inverse point in the sphere. Let Q be any point on the surface and take $PQ = R$, $P'Q = R'$. Then

$$4\pi\phi_P = \iint\left\{\phi_Q\frac{\partial}{\partial n}\left(\frac{1}{R}\right) - \frac{1}{R}\left(\frac{\partial\phi}{\partial n}\right)_Q\right\}dS, \qquad (1)$$

and

$$0 = \iint\left\{\phi_Q\frac{\partial}{\partial n}\left(\frac{1}{R'}\right) - \frac{1}{R'}\left(\frac{\partial\phi}{\partial n}\right)_Q\right\}dS, \qquad (2)$$

because P' is an internal point and the value taken for $\partial\phi/\partial n$ at Q is the external value. But on the sphere

$$\frac{R'}{R} = \frac{a}{r}. \qquad (3)$$

Hence we can eliminate $(\partial\phi/\partial n)_Q$; we get

$$4\pi\phi_P = \iint\phi_Q\left\{\frac{\partial}{\partial n}\left(\frac{1}{R}\right) - \frac{a}{r}\frac{\partial}{\partial n}\left(\frac{1}{R'}\right)\right\}dS. \qquad (4)$$

Denote the angle POQ by ϑ. Then

$$R^2 = a^2 + r^2 - 2ar\cos\vartheta, \quad \frac{\partial}{\partial n}\frac{1}{R} = \frac{\partial}{\partial a}\frac{1}{R} = -\frac{a - r\cos\vartheta}{R^3}, \qquad (5)$$

and similarly (keeping P' fixed in the differentiation)

$$\frac{\partial}{\partial n}\frac{1}{R'} = -\frac{a - (a^2/r)\cos\vartheta}{R'^3}. \qquad (6)$$

Hence, on substituting and simplifying,

$$4\pi\phi_P = \iint \frac{\phi}{aR^3}(r^2 - a^2)\,dS. \tag{7}$$

6·092. Internal point. Similarly, we find for the potential at the internal point P', where $OP' = r'$,

$$4\pi\phi_{P'} = \iint \phi \frac{a^2 - r'^2}{aR'^3}\,dS. \tag{8}$$

This problem, solved by Green, is sometimes called the first boundary potential problem for a sphere. The second problem is the determination of the field given the normal gradient over a sphere. It can be solved by noticing that if ϕ is a potential function, $r\partial\phi/\partial r$ is another, and the information supplied gives its surface value. The third problem is the one that occurs in the study of gravity, where the level surfaces are approximately spherical but $\partial\phi/\partial n$ is observed, not on a sphere, but over a surface where ϕ is constant. It is found that, to the first order of small departures from spherical symmetry, this information is equivalent to the values of $\frac{\partial\phi}{\partial r} + 2\frac{\phi}{r}$ over the sphere of equal volume. But $\frac{\partial}{r\partial r}(r^2\phi)$ is a potential function and the information determines it over the sphere. Its values outside the sphere are therefore determined and those of ϕ are found by an integration under the integral sign with regard to r. The problem was solved by this method by Idelson and Malkin; the original solution was by Stokes, using spherical harmonics. Cf. 24·114.

In (8) we notice that if P' is at O, $r' = 0$, $R' = a$ for all Q, and

$$4\pi\phi_0 = \frac{1}{a^2}\iint \phi\,dS.$$

Thus *the mean value of a potential function over a sphere is equal to its value at the centre.* It follows that a function cannot have a maximum or a minimum at any internal point of a region where it satisfies Laplace's equation. The extreme values must be taken on the boundary.

6·093. Solution for a plane. If we make the radius of the sphere very large we approach in the limit the solution of the corresponding problem to Green's for a plane. If z is the normal distance of P from the plane,

$$2\pi\phi_P = \iint \frac{\phi}{R^3} z\,dS. \tag{1}$$

It can be verified that the solutions obtained satisfy $\nabla^2\phi = 0$ and tend to the proper value on the boundary; the method* is similar to that used in 6·091 and 6·092. Also if

$$2\pi\phi_P = -\iint \left(\frac{\partial\phi}{\partial z}\right)_Q \frac{dS}{R}, \tag{2}$$

the solution has a derivative tending to the proper value on the boundary. The last result is of course obvious, since the surface density corresponding to a derivative $(\partial\phi/\partial z)$ over a plane is $-\frac{1}{2\pi\gamma}\frac{\partial\phi}{\partial z}$.

---

* Given in full by Poincaré, *Théorie du Potential Newtonien*, 1899, pp. 183–91.

6·10. Potential and field in a polarized medium. Consider a finite region where there is no total electric charge, but the polarization at a point Q is $\boldsymbol{P}$; that is, if $d\tau$ is a small element of volume about $Q(\xi_i)$ the dipole moment of the element is $\boldsymbol{P}d\tau$. Consider first the potential and field at $P(x_i)$ a point outside the region. Then

$$\phi = \gamma \iiint \frac{P_i(x_i - \xi_i)}{R^3} d\tau = -\gamma \iiint P_i \frac{\partial}{\partial x_i}\left(\frac{1}{R}\right) d\tau = \gamma \iiint P_i \frac{\partial}{\partial \xi_i}\left(\frac{1}{R}\right) d\tau, \tag{1}$$

$$E_i = -\frac{\partial \phi}{\partial x_i} = -\gamma \iiint \left\{ \frac{P_i}{R^3} - \frac{3P_k(x_k - \xi_k)(x_i - \xi_i)}{R^5} \right\} d\tau. \tag{2}$$

Since
$$\iiint P_i \frac{\partial}{\partial \xi_i}\left(\frac{1}{R}\right) d\tau = \iint \frac{l_i P_i}{R} dS - \iiint \frac{1}{R} \frac{\partial P_i}{\partial \xi_i} d\tau, \tag{3}$$

ϕ is the same as the potential due to a distribution of charge of density $-\operatorname{div} \boldsymbol{P}$ through the region and one of surface density $l_i P_i$, the normal component of $\boldsymbol{P}$, over the boundary of the region.

As for a continuous distribution of charge, when we wish to find the potential and field at a point P inside the region we must first suppose a small cavity made about P and study the behaviour of the integrals when the size of the cavity tends to zero.* But now the improper integrals in (2) depend on the shape of the cavity. Those in (3) do not. We denote the outer boundary by S, that of the cavity by Σ, the region inside S by V and that inside Σ by v, and therefore the region between them by $V - v$. Then the potential ϕ is given by

$$\phi = \gamma \iint \frac{l_i P_i}{R} dS + \lim_{v \to 0} \gamma \iint \frac{l_i P_i}{R} d\Sigma - \lim_{v \to 0} \gamma \iiint_{V-v} \frac{1}{R} \frac{\partial P_i}{\partial \xi_i} d\tau. \tag{4}$$

Provided P_i is bounded the second integral tends to zero, and

$$\phi = \gamma \iint \frac{l_i P_i}{R} dS - \lim_{v \to 0} \gamma \iiint_{V-v} \frac{1}{R} \frac{\partial P_i}{\partial \xi_i} d\tau, \tag{5}$$

which is the same as the potential due to a surface distribution $l_i P_i$ over S and a volume distribution $-\operatorname{div} \boldsymbol{P}$ through V. Then if $\operatorname{div} \boldsymbol{P}$ satisfies the conditions imposed on ρ in 6·04 we know by the considerations there that we may differentiate (5) under the integral sign; then

$$-\frac{\partial \phi}{\partial x_i} = \gamma \iint l_k P_k \frac{x_i - \xi_i}{R^3} dS - \lim_{v \to 0} \gamma \iiint_{V-v} \frac{\partial P_k}{\partial \xi_k} \frac{x_i - \xi_i}{R^3} d\tau. \tag{6}$$

We denote this by E_i and call it the electric intensity; it also is independent of the limiting shape of the cavity.

Again, E_i is the *field* due to the surface distribution $l_i P_i$ over S and the volume distribution $-\operatorname{div} \boldsymbol{P}$ through V. As we shall see, it is not in general the same as the limit of the field in the cavity. It follows from Poisson's equation that

$$\operatorname{div} \boldsymbol{E} = -\nabla^2 \phi = -4\pi\gamma \operatorname{div} \boldsymbol{P}, \tag{7}$$

or
$$\operatorname{div}(\boldsymbol{E} + 4\pi\gamma\boldsymbol{P}) = 0. \tag{8}$$

If in addition to the distribution of dipoles there was a charge density ρ there would be additional terms in ϕ and $\boldsymbol{E}$, and (8) would be replaced by

$$\operatorname{div}(\boldsymbol{E} + 4\pi\gamma\boldsymbol{P}) = 4\pi\gamma\rho. \tag{9}$$

* $\boldsymbol{P}$ outside the cavity remaining unaltered.

This is physically possible. We could make a hole in a block of paraffin wax, create a charge on its interior by friction, and then fill up the hole again, thus leaving a charge density in the interior of a solid. The vector $E + 4\pi\gamma P$ is called the *displacement* or *induction* and denoted by D.

Now consider the field inside v. Since P is an external point of the region $V - v$ the field is

$$F_i = \gamma \iint l_k P_k \frac{x_i - \xi_i}{R^3} dS + \gamma \iint l_k P_k \frac{x_i - \xi_i}{R^3} d\Sigma - \gamma \iiint_{V-v} \frac{\partial P_k}{\partial x_k} \frac{x_i - \xi_i}{R^3} d\tau, \quad (10)$$

which is equal, if we make the cavity tend to zero, to

$$E_i + \lim \gamma \iint l_k P_k \frac{x_i - \xi_i}{R^3} d\Sigma. \quad (11)$$

If we take the cavity to be a circular cylinder of length $2a$ and radius b with its centre at Q and its axis in the direction of P, then $l_k P_k$ at A is P and at B is $-P$ (the normal being into the cavity) and is zero over the sides, apart from small effects due to variation of P in the neighbourhood. Then the field at Q due to the surface distribution over the walls of the cavity is

$$4\pi\gamma P \left(1 - \frac{a}{(a^2 + b^2)^{1/2}} \right). \quad (12)$$

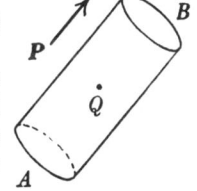

Hence if as a and b become small, $b/a \to 0$ this contribution to the field is zero. If $a/b \to 0$ the contribution is $4\pi\gamma P$. Hence E as defined by (5) is the limit of the field in a needle-shaped cavity with its axis in the direction of P, and D ($= E + 4\pi\gamma P$) is the limit of the field in a coin-shaped cavity with its axis in the direction of P.

For a spherical cavity the surface density on the wall is $-P\cos\theta$ and hence the field at the centre due to it is $\tfrac{4}{3}\pi\gamma P$.

Suppose now that there is a relation

$$P = \kappa E, \quad (13)$$

so that

$$D = E(1 + 4\pi\gamma\kappa) = KE, \quad (14)$$

where κ, and similarly K, are continuous scalar functions of position. Then κ is called the electric susceptibility and K the dielectric constant of the material, which is *isotropic*. If the relation between D and E has the form

$$D_i = K_{ik} E_k, \quad (15)$$

where K_{ik} is a tensor that is not a multiple of δ_{ik}, the medium is anisotropic.

We consider now the behaviour of D and E at the boundary of two uniform media of different dielectric constants K_1, K_2. The intensity E at any point is that due to a volume distribution div P and a surface distribution $P_{1n} - P_{2n}$, where n denotes the component in the direction of the normal drawn from the medium 1 to the medium 2. Owing to the surface distribution there is a discontinuity in the normal component of E, namely,

$$E_{2n} - E_{1n} = 4\pi\gamma(P_{1n} - P_{2n}).$$

Hence the normal component of D has the same value on both sides of the boundary.

The corresponding theory for magnetism is similar. The differences are (1) ρ is always 0, (2) κ and K (now called the permeability μ) may vary greatly with the magnetic intensity H, (3) there are permanent magnets, with fixed intensity of magnetization, so that we can write $B = H + 4\pi\gamma I$, but there is no relation between I and H.

6·11. Vector potential. Let $\mathbf{u}$ be a solenoidal vector. Then we can show that in general there is another vector $\mathbf{A}$ such that $\mathbf{u}$ is its curl. It is convenient first to call the components u, v, w and A, B, C; then we have to satisfy

$$u = \frac{\partial C}{\partial y} - \frac{\partial B}{\partial z}, \quad v = \frac{\partial A}{\partial z} - \frac{\partial C}{\partial x}, \quad w = \frac{\partial B}{\partial x} - \frac{\partial A}{\partial y}. \tag{1}$$

Take
$$C_1 = 0, \quad A_1 = \int_{z_0}^{z} v\, dz, \quad B_1 = -\int_{z_0}^{z} u\, dz, \tag{2}$$

where z_0 is independent of x, y and z, and the path of integration is parallel to the axis of z. Then the first two equations are satisfied. But

$$\frac{\partial B_1}{\partial x} - \frac{\partial A_1}{\partial y} = -\int_{z_0}^{z}\left(\frac{\partial u}{\partial x} + \frac{\partial v}{\partial y}\right)dz = \int_{z_0}^{z}\frac{\partial w}{\partial z}\,dz$$
$$= w - w(x, y, z_0). \tag{3}$$

since
$$\frac{\partial u}{\partial x} + \frac{\partial v}{\partial y} + \frac{\partial w}{\partial z} = 0. \tag{4}$$

In general $w(x, y, z_0)$ will not be zero, but it depends on x, y only and can be denoted by w_0. Now put
$$A = A_1 + A_2, \text{ etc.}; \tag{5}$$

then
$$\frac{\partial C_2}{\partial y} - \frac{\partial B_2}{\partial z} = 0, \quad \frac{\partial A_2}{\partial z} - \frac{\partial C_2}{\partial x} = 0, \quad \frac{\partial B_2}{\partial x} - \frac{\partial A_2}{\partial y} = w_0. \tag{6}$$

Take
$$B_2 = 0, \quad C_2 = 0, \quad A_2 = -\int_{y_0}^{y} w_0\, dy. \tag{7}$$

Then all the equations are satisfied. There is still a considerable arbitrariness in the solution; for we could add to (A, B, C) the gradient of any scalar whatever without affecting its curl. Conversely we can impose an additional condition on (A, B, C) to make its divergence anything we like, in particular zero. For we need only add the gradient of a scalar Φ such that $\nabla^2\Phi$ cancels the divergence of the solution we have already found; and given suitable boundary conditions Φ will be determinate. Hence, translating into tensor notation, if

$$\frac{\partial u_i}{\partial x_i} = 0, \tag{8}$$

a vector A_i exists satisfying

$$u_i = \epsilon_{ikm}\frac{\partial A_m}{\partial x_k}, \quad \frac{\partial A_i}{\partial x_i} = 0, \tag{9}$$

and A_i is unique subject to the same sort of boundary conditions as ensure uniqueness of solution in a potential problem. $\mathbf{A}$ is called the *vector potential*.

6·111. The above method is very unsymmetrical in the coordinates. A symmetrical solution can be found in some cases as follows, and is useful when curl $\mathbf{u}$ is given everywhere. Let

$$\epsilon_{ikm}\frac{\partial u_m}{\partial x_k} = \omega_i, \quad \frac{\partial u_i}{\partial x_i} = 0, \tag{10}$$

and assume
$$u_i = \epsilon_{ips}\frac{\partial A_s}{\partial x_p}, \quad \frac{\partial A_i}{\partial x_i} = 0. \tag{11}$$

Then
$$\epsilon_{ikm}\frac{\partial u_m}{\partial x_k} = \epsilon_{ikm}\epsilon_{mps}\frac{\partial^2 A_s}{\partial x_k \partial x_p}$$
$$= (\delta_{ip}\delta_{ks} - \delta_{is}\delta_{kp})\frac{\partial^2 A_s}{\partial x_k \partial x_p}$$
$$= \frac{\partial^2 A_s}{\partial x_i \partial x_s} - \frac{\partial^2 A_i}{\partial x_k^2} = -\nabla^2 A_i. \qquad (12)$$

Therefore the conditions are satisfied if
$$A_i = \frac{1}{4\pi}\iiint \frac{\omega_i}{R}d\tau \qquad (13)$$

through all space and if A_i has zero divergence. But applying Green's theorem to the region between a small sphere about x_i and a large sphere

$$\frac{\partial A_i}{\partial x_i} = -\frac{1}{4\pi}\lim\iiint \omega_i \frac{\partial}{\partial \xi_i}\left(\frac{1}{R}\right)d\tau$$
$$= -\frac{1}{4\pi}\lim\iint \frac{l_i \omega_i}{R}dS + \frac{1}{4\pi}\lim\iiint \frac{1}{R}\frac{\partial \omega_i}{\partial x_i}d\tau. \qquad (14)$$

The second integral vanishes since ω_i is the curl of a vector. The first taken over the inner sphere tends to 0 if ω_i is bounded near $\xi_i = x_i$, and over the larger sphere also tends to 0 if ω_i tends to 0 suitably at large distances. Hence (13) gives a solution of the problem. It is still not unique because to A_i we could add the gradient of any solution of Laplace's equation without affecting the relations (11).

This method fails if u_i is irrotational. For then ω_i is 0 everywhere and (13) vanishes. Its chief use is when $\omega_i = 0$ outside a filament of very small cross-section and is large inside it. This case arises in vortex motion and in the magnetic field due to an electric current. In either case the integral
$$\Omega = \int_C u_i dx_i, \qquad (15)$$

taken around a closed circuit, is zero if the circuit can be filled up by a cap not cutting the filament or wire, and has the same value for all circuits that can be filled up by caps that cut through it once. If we consider the contribution to A_i from an element between two planes separated by $d\xi_i$, and call the element of surface in a plane parallel to them dS, we have
$$d\tau = d\xi_i dS, \qquad (16)$$
$$\iint \omega_i dS = \int u_i d\xi_i = \Omega \qquad (17)$$
by Stokes's theorem; hence
$$A_i = \frac{\Omega}{4\pi}\int \frac{d\xi_i}{R}, \qquad (18)$$

taken along the length of the filament. Also
$$u_i = \epsilon_{ikm}\frac{\Omega}{4\pi}\frac{\partial}{\partial x_k}\int \frac{1}{R}d\xi_m$$
$$= \frac{\Omega}{4\pi}\int \epsilon_{ikm}\frac{(\xi_k - x_k)l_m}{R^3}ds, \quad \mathbf{u} = \frac{\Omega}{4\pi}\int \frac{(\boldsymbol{\xi}-\boldsymbol{x})\wedge d\boldsymbol{\xi}}{R^3}, \qquad (19)$$

where ds is an element of length of the filament and l_m a direction cosine of the tangent.

In hydrodynamics Ω is the circulation about the filament. In electromagnetism it is $4\pi J/c$, where J is the current in electrostatic units, and u_i is the magnetic field.

6·112. Vector potentials of point charge and doublet. Take a doublet along the axis of z, so that
$$u = \mu \operatorname{grad}(z/r^3).$$
Then in 6·11 (2) we take the lower limit for z to be $-\infty$; and
$$A_1 = -\mu \int_{-\infty}^{z} \frac{3yz}{r^5} dz = \frac{\mu y}{r^3},$$
$$B_1 = \mu \int_{-\infty}^{z} \frac{3zx}{r^5} dz = -\frac{\mu x}{r^3}$$
and from 6·11 (7)
$$A_2 = 0,$$
since w tends to 0 when $z \to -\infty$. Then a solution is $\left(\frac{\mu y}{r^3}, -\frac{\mu x}{r^3}, 0\right)$, and the divergence of this is zero. Therefore, if u_i is the gradient of a potential $\mu_k x_k/r^3$, it is also the curl of a vector potential
$$A_i = -\epsilon_{ikm} \frac{\mu_k x_m}{r^3} = -\frac{(\mu \wedge x)_i}{r^3}.$$

A similar treatment applied to the elementary vector $\operatorname{grad}(1/r)$ shows that it is the curl of the vector
$$\left\{-\frac{y}{\varpi^2}\left(1+\frac{z}{r}\right),\ \frac{x}{\varpi^2}\left(1+\frac{z}{r}\right),\ 0\right\}.$$
The part $(-y/\varpi^2, x/\varpi^2, 0)$ is the gradient of the scalar $\tan^{-1} y/x$ and is irrelevant; so we can take
$$A = \left(-\frac{yz}{r\varpi^2},\ \frac{xz}{r\varpi^2},\ 0\right).$$
This is still very unsymmetrical; partial symmetry can be given to it by taking its mean with the two other vectors obtained by cyclic interchange of the coordinates, but at the cost of producing lines of singularity along three coordinate axes instead of one. Generalization by rotation of axes merely increases the complexity.

EXAMPLES

1. If the attraction between two elements of mass m, m' is $mm'f(R)$, where R is the distance between them, show that the inward acceleration due to a uniform thin spherical shell of mass M and internal radius a, at an internal point distant x ($x < a$) from the centre, is F, where
$$F = \frac{M}{4ax^2} \int_{a-x}^{a+x} (R^2 + x^2 - a^2) f(R) \, dR.$$
If F is zero for all values of a and x, subject to the condition $x < a$, show that the only possible form of $f(R)$ is A/R^2, where A is a constant.

2. Assuming that at any point the magnetic potential of a body magnetized to intensity I is the same as that of a volume distribution of magnetic poles of density $-\operatorname{div} I$ together with a surface distribution $n_i I_i$, where n_i are the direction cosines of the outward normal, prove that (i) the field inside a sphere permanently magnetized to intensity I is of uniform intensity $-\frac{4}{3}\pi I$, (ii) the field vanishes in a spherical cavity (not necessarily concentric) made in this sphere, (iii) if a plane lamina of

uniform thickness and infinite extent is permanently magnetized to uniform intensity I the field in a spherical hole made in the lamina is of intensity $\frac{4}{3}\pi I - 4\pi(I.n)n$, where n is normal to one of the plane faces of the lamina. (M.T. 1939.)

3. O is the centre of a uniform cube of mass M, and P is an external point. The distance OP is large compared with the edge a of the cube, and the direction cosines of OP referred to three concurrent edges of the cube are l, m, n. Show how to express the gravitation potential at P in a power series in OP^{-1}, and evaluate it up to the term in OP^{-5}. (Prelim. 1936.)

4. If ϕ_1 and ϕ_2 satisfy the conditions (1) ϕ is continuous everywhere, and has continuous second derivatives except on certain surfaces, (2) $K\partial\phi/\partial n$ has a given discontinuity on crossing such surfaces, (3) ϕ tends to 0 like $1/r$, and $\partial\phi/\partial x_i$ like $1/r^2$ when r is large; and if further div$\{K\operatorname{grad}(\phi_1-\phi_2)\} = 0$ except on the surfaces of discontinuity, show that

$$\iiint K\left(\frac{\partial\phi_1}{\partial x_i}\right)^2 d\tau - \iiint K\left(\frac{\partial\phi_2}{\partial x_i}\right)^2 d\tau = \iiint K\left\{\frac{\partial(\phi_1-\phi_2)}{\partial x_i}\right\}^2 d\tau$$

taken through all space; and hence that, subject to the conditions given, ϕ is uniquely determined provided K is everywhere positive.

5. Prove that if ϕ satisfies conditions (1) and (3) above, the conditions that

$$V = \frac{1}{8\pi}\iiint K\left(\frac{\partial\phi}{\partial x_i}\right)^2 d\tau$$

shall be stationary for all small variations of ϕ are that

$$\frac{\partial}{\partial x_i}\left(K\frac{\partial\phi}{\partial x_i}\right) = 0,$$

except on the special surfaces, and that $K\partial\phi/\partial n$ shall be continuous on crossing such surfaces; and that if K is everywhere positive the stationary value of V is a minimum. Show also that if $K\partial\phi/\partial n$ has a prescribed discontinuity in crossing any such surface, V is still stationary provided ϕ is not varied on that surface.

If ϕ is the electrostatic potential in a field containing dielectrics, give the physical interpretations of the postulates and conclusions.

6. Show that if a is a constant vector

$$\operatorname{curl}\left(\frac{a\wedge r}{r^n}\right) = \frac{(2-n)a}{r^n} + \frac{nr(a.r)}{r^{n+2}},$$

and hence or otherwise find an expression for the vector potential inside an infinite straight solenoid. (M/c, Part III, 1936.)

7. $\nabla^2 u_1 = 0$, $\nabla^2 u_2 = 0$ in a closed region D, and $u_1 = u_2$ in a region D_1, which is part of D. Prove that $u_1 = u_2$ in the whole of D.

8. Deduce from Stokes's theorem that

$$\int \phi \, d\mathbf{s} = \int d\mathbf{S} \wedge \operatorname{grad} \phi$$

by considering the projection of $\int \phi \, d\mathbf{s}$ on a line of direction n.

Deduce that the mutual potential energy of two uniform magnetic shells of strengths μ, μ' is

$$-\mu\mu'\iint \frac{d\mathbf{s}.d\mathbf{s}'}{r}.$$

Chapter 7

OPERATIONAL METHODS

Even Cambridge mathematicians deserve justice.

OLIVER HEAVISIDE

7·01. Rules of arithmetic for differential operator. In a certain sense the operations of differentiation and definite integration satisfy the rules of arithmetic. For if a and b are constants

$$\frac{d}{dx}f(x) + af(x) = af(x) + \frac{d}{dx}f(x), \tag{1}$$

$$\left(\frac{d}{dx}f(x) + af(x)\right) + bf(x) = \frac{d}{dx}f(x) + \{af(x) + bf(x)\}, \tag{2}$$

$$\frac{d}{dx}af(x) = a\frac{d}{dx}f(x), \tag{3}$$

$$\frac{d}{dx}a\{bf(x)\} = \frac{d}{dx}(ab)f(x), \tag{4}$$

and if we define

$$\left(\frac{d}{dx} + a\right)f(x) = \frac{d}{dx}f(x) + af(x) = \left(a + \frac{d}{dx}\right)f(x), \tag{5}$$

which is permissible by (1), we have the forms of the distributive law

$$\frac{d}{dx}(a+b)f(x) = \frac{d}{dx}af(x) + \frac{d}{dx}bf(x), \tag{6}$$

$$b\left(\frac{d}{dx} + a\right)f(x) = b\frac{d}{dx}f(x) + abf(x). \tag{7}$$

Thus in any algebraic combination with constants, involving only addition (and therefore subtraction) and multiplication, the differential operator can be manipulated as if it were itself a numerical constant. The function $f(x)$ operated on remains on the right and the operation can be carried out on it at the end.

7·011. Operation of definite integration. The same applies to the operation of definite integration. If we write $Qf(t)$ for $\int_0^t f(\xi)\,d\xi$,

$$Qf(t) + af(t) = af(t) + Qf(t), \tag{1}$$

and therefore both can be denoted by

$$(Q+a)f(t) = (a+Q)f(t); \tag{2}$$

also

$$\{Qf(t) + af(t)\} + bf(t) = Qf(t) + \{af(t) + bf(t)\}, \tag{3}$$

$$Q\{af(t)\} = a\{Qf(t)\}, \tag{4}$$

$$Q\{abf(t)\} = (Qa)\{bf(t)\}, \tag{5}$$

$$Q\{(a+b)f(t)\} = Qaf(t) + Qbf(t), \tag{6}$$

$$a(Q+b)f(t) = aQf(t) + abf(t). \tag{7}$$

Analogous relations can be found by replacing a or b by Q and interpreting

$$Q^2 f(t) = Q\{Qf(t)\} = \int_0^t Qf(\tau)\,d\tau,$$

where $Qf(\tau)$ is the same function of τ as $Qf(t)$ is of t.

7·012. Non-commutative property of differentiation and definite integration.
The above properties of differentiation, in the hands of Boole, became the basis of a well-known method of obtaining particular integrals of linear differential equations of certain types. It is also the basis of a method of obtaining the formulae required in numerical interpolation, differentiation, and integration. The corresponding property of definite integration was apparently noticed first by J. Caqué,* and led to a number of developments in the theory of the solution of linear differential equations in the hands of Fuchs, Peano, Picard, and H. F. Baker. Heaviside attended particularly to linear differential equations with constant coefficients, which enable the maximum use to be made of the fact that the operator of integration can be combined with constants exactly as if it was a number. Of course it does not commute with variables; $Q\{tf(t)\}$ is not the same as $t\{Qf(t)\}$. But the reason why his methods worked can really be traced back to the method used by Picard in proving, for assigned conditions, the existence of solutions of differential equations. Unfortunately, though Heaviside noticed that the operators of differentiation and integration combine with constants without restriction, he did not notice that they do not commute with each other. In fact

$$\frac{d}{dt} Qf(t) = \frac{d}{dt} \int_0^t f(\xi)\,d\xi = f(t),$$

$$Q \frac{d}{dt} f(t) = \int_0^t f'(\xi)\,d\xi = f(t) - f(0)$$

which differ by $f(0)$. Heaviside obtained a considerable number of wrong results through interchanging the order of differentiation and integration, and their explanation in terms of this non-commutative property was first given by H. Jeffreys.† Heaviside was also not particularly interested in questions of convergence, and this fact so disturbed the pure mathematicians of the time that they omitted to find out in what conditions the methods could be justified, as in fact they can within their proper scope. For dynamical systems with a finite number of degrees of freedom the systematic use of the definite integration operator provides far more concise solutions than any other method gives, and its justification is complete for such problems without needing any pure mathematics more advanced than was available in Heaviside's time.

7·02. Interpretation of $Q^n f(t)$ as a single integral. If $f(t) = 1$ we have by successive integrations

$$Q1 = t, \quad Q^2 1 = \frac{t^2}{2!}, \quad Q^3 1 = \frac{t^3}{3!}, \quad \ldots, \quad Q^n 1 = \frac{t^n}{n!}. \tag{1}$$

For a general function $f(t)$ we have by definition

$$Qf(t) = \int_0^t f(\tau)\,d\tau, \tag{2}$$

$$Q^2 f(t) = Q \int_0^t f(\tau)\,d\tau = \int_0^t \left[\int_0^\xi f(\tau)\,d\tau\right] d\xi. \tag{3}$$

---
* *J. de Math.*, (2) 9, 1864, 185–222.   † *Operational Methods in Mathematical Physics*, 1927.

The integral is over the shaded region in the figure. We change the order of integration and get

$$Q^2 f(t) = \int_0^t \left[\int_\tau^t f(\tau)\, d\xi \right] d\tau = \int_0^t (t-\tau) f(\tau)\, d\tau. \tag{4}$$

Similarly

$$Q^3 f(t) = Q \int_0^t (t-\tau) f(\tau)\, d\tau = \int_0^t \left[\int_0^\xi (\xi - \tau) f(\tau)\, d\tau \right] d\xi$$

$$= \int_0^t \left[\int_\tau^t (\xi - \tau) f(\tau)\, d\xi \right] d\tau = \int_0^t \frac{(t-\tau)^2}{2!} f(\tau)\, d\tau, \tag{5}$$

and in general, by induction,

$$Q^n f(t) = \int_0^t \frac{(t-\tau)^{n-1}}{(n-1)!} f(\tau)\, d\tau. \tag{6}$$

The same may be proved as follows. $Q^n f(t)$ is the function that vanishes at $t = 0$ and has derivative equal to $Q^{n-1} f(t)$. The former condition is satisfied by (6). For the latter we differentiate under the integral sign with regard to t, and the result is

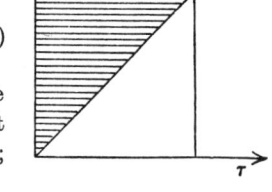

$$\int_0^t \frac{(t-\tau)^{n-2}}{(n-2)!} f(\tau)\, d\tau = Q^{n-1} f(t). \tag{7}$$

The integrand vanishes at the upper limit if $n > 1$ and therefore differentiation of the limits gives nothing. Hence if the result is right for Q^{n-1} it is right for Q^n; but it is right for $n = 1$; hence it is right for all positive integer values of n.

Thus any integral power of Q operating on a function gives a result expressible by a single integral, provided only that the function is integrable.

7·03. Series of operators. These rules permit us to express as a single integral any sum of a finite number of terms of the form

$$(a_0 + a_1 Q + a_2 Q^2 + \ldots + a_n Q^n) f(t)$$

$$= a_0 f(t) + \sum_{r=1}^n \int_0^t a_r \frac{(t-\tau)^{r-1}}{(r-1)!} f(\tau)\, d\tau$$

$$= a_0 f(t) + \int_0^t \left\{ a_1 + a_2(t-\tau) + \frac{a_3(t-\tau)^2}{2!} + \ldots + a_n \frac{(t-\tau)^{n-1}}{(n-1)!} \right\} f(\tau)\, d\tau.$$

The extension to infinite series requires special justification.

7·031. Convergence theorem. *If $\int_0^T f(t)\, dt$ exists, and if the series*

$$a_0 + a_1 z + \ldots + a_n z^n + \ldots \tag{1}$$

converges for some value of z different from zero, then the series

$$(a_0 + a_1 Q + \ldots + a_n Q^n + \ldots) f(t) \tag{2}$$

is uniformly and absolutely convergent for $0 \leqslant t \leqslant T$, and is equal to

$$a_0 f(t) + \sum_{n=1}^\infty \int_0^t a_n \frac{(t-\tau)^{n-1}}{(n-1)!} f(\tau)\, d\tau. \tag{3}$$

The two series (2) and (3) are equal term by term and therefore have the same sum if either of them converges. But since the series (1) converges for, say, $z = r$, then for any positive ρ less than $|r|$ we can find a quantity M such that for all n

$$|a_n|\rho^n < M. \tag{4}$$

Also since $Qf(t)$ is bounded for $0 \leq t \leq T$, say, there is a positive quantity C such that

$$|Qf(t)| < C \quad (0 \leq t \leq T). \tag{5}$$

Hence
$$|Q^2f(t)| < \left|\int_0^t C\,d\tau\right| < C|t|, \quad |Q^3f(t)| < \int_0^t C|t|\,d\tau < \tfrac{1}{2}C|t^2|, \tag{6}$$

and in general
$$|Q^nf(t)| < \frac{1}{(n-1)!} C|t^{n-1}|. \tag{7}$$

Hence the moduli of the terms of the series after the first are respectively less than the terms of the series

$$C\left(a_1 + a_2|t| + \tfrac{1}{2}a_3|t|^2 + \ldots + \frac{1}{(n-1)!}|t|^{n-1} + \ldots\right) \tag{8}$$

$$< \frac{CM}{\rho}\left(1 + \frac{|t|}{\rho} + \frac{1}{2}\frac{|t|^2}{\rho^2} + \ldots + \frac{1}{(n-1)!}\frac{|t|^{n-1}}{\rho^{n-1}} + \ldots\right). \tag{9}$$

But this is an exponential series and is absolutely convergent for all t. Further, the terms are not greater than those of the series obtained by replacing t by T, and this also is a convergent series of positive terms independent of t. Hence our series satisfies the M test for uniform convergence in the range $0 \leq t \leq T$; that is, in any range of t such that $Qf(t)$ is bounded.

If $f(t)$ is a continuous function, every term of the series is a continuous function; and since the sum of a uniformly convergent series of continuous functions is continuous, it follows that the sum of the series is a continuous function of t in the range $0 \leq t \leq T$. For the same reason the summation can be carried out under the integral sign: thus

$$\sum_{n=0}^{\infty} a_n Q^n f(t) = a_0 f(t) + \int_0^t f(\tau)\left(\sum_{n=1}^{\infty} a_n \frac{(t-\tau)^{n-1}}{(n-1)!}\right) d\tau. \tag{10}$$

The argument is still valid if the terms of (1) are required only to be bounded; also if $\int_0^t f(\tau)\,d\tau$ exists only as an improper integral owing to $f(\tau)$ being unbounded.

7·032. Composition of operators. Let

$$F(z) = a_0 + a_1 z + a_2 z^2 + \ldots, \tag{1}$$

$$G(z) = b_0 + b_1 z + b_2 z^2 + \ldots, \tag{2}$$

be two series that both converge for some non-zero value of $|z|$, say r. Then for any positive quantity ρ less than r, quantities M, N exist such that

$$|a_n| < M/\rho^n, \quad |b_n| < N/\rho^n. \tag{3}$$

For any z such that $|z| \leq \rho$ the product series
$$F(z)G(z) = a_0b_0 + (a_0b_1 + a_1b_0)z + (a_0b_2 + a_1b_1 + a_2b_0)z^2 + \ldots$$
$$= c_0 + c_1z + c_2z^2 + \ldots \tag{4}$$

is absolutely convergent. We shall prove that if $F(Q)$ and $G(Q)$ are the operational series obtained by replacing z by Q, and if $G(Q)$ operates on a function $\phi(t)$ satisfying the conditions of the last theorem, and if $F(Q)$ then operates on the resulting function, the result is the same as if we replaced z by Q in the product series (4) and operated on $\phi(t)$ directly. We have

$$F(Q)G(Q)\phi(t) = F(Q)(b_0 + b_1Q + b_2Q^2 + \ldots)\phi(t). \tag{5}$$

By the last theorem the series representing $G(Q)\phi(t)$ is an absolutely and uniformly convergent series, and can therefore be integrated term by term. Hence

$$Q^m(b_0 + b_1Q + b_2Q^2 + \ldots)\phi(t) = b_0Q^m\phi(t) + b_1Q^{m+1}\phi(t) + b_2Q^{m+2}\phi(t) + \ldots \tag{6}$$

and
$$F(Q)G(Q)\phi(t) = \sum_m \sum_n a_m b_n Q^{m+n}\phi(t), \tag{7}$$

where the summation with regard to n is to be carried out first. But if $|Q\phi(t)| < C$ and $m+n \geq 1$

$$|a_m b_n Q^{m+n}\phi(t)| < \frac{MNC}{(m+n-1)!}\left(\frac{T}{\rho}\right)^{m+n-1}, \tag{8}$$

and the terms are less than those of a double series of positive terms. All terms of this series with the same $m+n = k$ are equal and there are $k+1$ of them, n ranging from 0 to k. Hence their sum is

$$MNC\frac{(k+1)}{(k-1)!}\left(\frac{T}{\rho}\right)^{k-1}, \tag{9}$$

and the sum of this with regard to k is convergent. Hence the series in (7) is absolutely convergent and can be rearranged in any order without affecting its convergence or its sum. We can therefore collect all terms in Q^{m+n}; but we then have

$$F(Q)G(Q)\phi(t) = (c_0 + c_1Q + c_2Q^2 + \ldots)\phi(t), \tag{10}$$

which proves the theorem.

7·04. First order linear differential equations. Now consider the linear differential equation of the first order
$$\frac{dx}{dt} - \alpha x = \phi(t), \tag{1}$$

where α is constant and x is given to be equal to x_0 at $t = 0$. Replace t by τ and integrate both sides with regard to τ from 0 to t. We get

$$x - x_0 - \alpha Qx = Q\phi(t) \tag{2}$$

that is, $\qquad (1 - \alpha Q)x = x_0 + Q\phi(t), \tag{3}$

and the expression on the right is a continuous function of t if $\phi(t)$ is integrable. Now operate on both sides of the equation with the series $(1 + \alpha Q + \alpha^2 Q^2 + \ldots)$. By the last theorem the result of performing the operations $1 - \alpha Q$ and $1 + \alpha Q + \alpha^2 Q^2 + \ldots$ successively

7·04 *First order linear differential equations*

on x is simply x; for if we put z for Q in these series we get two series convergent for $|z| < 1/\alpha$ whose product series is 1. Hence

$$x = (1 + \alpha Q + \alpha^2 Q^2 + \ldots)\{x_0 + Q\phi(t)\}. \tag{4}$$

This gives a formal solution, which could be developed in a series of powers of α. But we know otherwise that the solution of (1) is

$$x = x_0 e^{\alpha t} + e^{\alpha t} \int_0^t e^{-\alpha \tau} \phi(\tau)\, d\tau. \tag{5}$$

Hence the expressions on the right of (4) and (5) are equal for all values of x_0; therefore

$$(1 + \alpha Q + \alpha^2 Q^2 + \ldots) x_0 = e^{\alpha t} x_0, \tag{6}$$

$$(1 + \alpha Q + \alpha^2 Q^2 + \ldots) Q\phi(t) = e^{\alpha t} Q\{e^{-\alpha t}\phi(t)\}. \tag{7}$$

We have so far given no meaning to division by an operator. But just for that reason we are entitled to do so now, provided that we can ensure consistency. If, for instance, we give a meaning to $\dfrac{1}{1-\alpha Q} g(t)$, it must be such that the operator $1 - \alpha Q$ acting on it gives $g(t)$. But we have

$$(1 - \alpha Q)(1 + \alpha Q + \alpha^2 Q^2 + \ldots) g(t) = g(t) \tag{8}$$

by the rule for the composition of operators, for all $g(t)$ such that the operations are applicable. Hence we may *define*

$$\frac{1}{1-\alpha Q} = 1 + \alpha Q + \alpha^2 Q^2 + \ldots, \tag{9}$$

and write the above interpretations in the compact form

$$\frac{1}{1-\alpha Q} 1 = e^{\alpha t}, \tag{10a}$$

$$\frac{1}{1-\alpha Q} Q\phi(t) = e^{\alpha t} Q\{e^{-\alpha t}\phi(t)\} = \int_0^t e^{\alpha(t-\tau)} \phi(\tau)\, d\tau. \tag{10b}$$

Similarly, if $F(z)$ is any function of z expansible in a power series near $z = 0$, $F(0)$ not being zero, and $G(z)$ is the reciprocal series, we can interpret $1/F(Q)$ to mean $G(Q)$. That is, the fundamental interpretation of any operator is always its expansion in positive powers of Q as if Q was a constant. Operators such as Q^{-n} and $e^{h/Q}$ are not expansible in positive powers of Q in the sense indicated; that is, the functions z^{-n} and $e^{h/z}$ are not capable of being expanded in positive powers of z. Consequently we can give them no interpretation at present; and it turns out that in the important class of physical problems that require the solution of a finite number of linear differential equations with constant coefficients such operators do not arise.

We have

$$\frac{1}{(1-\alpha Q)^2} 1 = (1 + 2\alpha Q + 3\alpha^2 Q^2 + \ldots) 1 = 1 + 2\alpha t + \tfrac{3}{2}\alpha^2 t^2 + \tfrac{4}{6}\alpha^3 t^3 + \ldots, \tag{11}$$

which is not an immediately recognizable form; but

$$\frac{Q}{(1-\alpha Q)^2} 1 = (Q + 2\alpha Q^2 + 3\alpha^2 Q^3 + \ldots + n\alpha^{n-1} Q^n \ldots) 1 = t + \alpha t^2 + \tfrac{1}{2}\alpha^2 t^3 + \ldots + \frac{\alpha^{n-1}}{(n-1)!} t^n + \ldots$$

$$= t e^{\alpha t}. \tag{12}$$

We now see that
$$\frac{1}{(1-\alpha Q)^2}1 = \frac{1}{1-\alpha Q}1 + \frac{\alpha Q}{(1-\alpha Q)^2}1 = (1+\alpha t)e^{\alpha t}, \tag{13}$$

which is the same as (11). But $(1-\alpha Q)^{-n}1$ contains more and more terms in its finite interpretation as n increases; $Q^{n-1}(1-\alpha Q)^{-n}1$ does not, for

$$\frac{Q^{n-1}}{(1-\alpha Q)^n}1 = \left(Q^{n-1} + n\alpha Q^n + \frac{n(n+1)}{2!}\alpha^2 Q^{n+1} + \ldots + \frac{n(n+1)\ldots(n+r-1)}{r!}\alpha^r Q^{n+r-1} + \ldots\right)1$$

$$= \frac{t^{n-1}}{(n-1)!} + \frac{\alpha t^n}{(n-1)!} + \frac{\alpha^2 t^{n+1}}{2!(n-1)!} + \ldots$$

$$= \frac{t^{n-1}}{(n-1)!}e^{\alpha t}, \tag{14}$$

on interpreting the separate terms by 7·02 (1). Similarly, by expansion,

$$\frac{Q^n}{(1-\alpha Q)^n}\phi(t) = \int_0^t \phi(\tau)\left\{\frac{(t-\tau)^{n-1}}{(n-1)!} + \frac{\alpha(t-\tau)^n}{(n-1)!} + \frac{\alpha^2(t-\tau)^{n+1}}{2!(n-1)!} + \ldots\right\}d\tau$$

$$= \int_0^t \phi(\tau)\frac{(t-\tau)^{n-1}}{(n-1)!}e^{\alpha(t-\tau)}d\tau. \tag{15}$$

Alternatively we may use (10b): we have

$$\frac{Q^2}{(1-\alpha Q)^2}\phi(t) = \frac{Q}{1-\alpha Q}\int_0^t e^{\alpha(t-\tau)}\phi(\tau)d\tau = \int_0^t\left[\int_0^\tau e^{\alpha(t-\tau)}e^{\alpha(\tau-\xi)}\phi(\xi)d\xi\right]d\tau$$

$$= \int_0^t\left[\int_\xi^t e^{\alpha(t-\xi)}\phi(\xi)d\tau\right]d\xi = \int_0^t \phi(\xi)(t-\xi)e^{\alpha(t-\xi)}d\xi,$$

and so on as in proving 7·02 (6).

7·05. Set of n linear differential equations of the first order. Consider the equations

$$\left.\begin{aligned} e_{11}y_1 + e_{12}y_2 + \ldots + e_{1n}y_n &= S_1, \\ e_{21}y_1 + e_{22}y_2 + \ldots + e_{2n}y_n &= S_2, \\ &\cdots\cdots\cdots\cdots\cdots\cdots\cdots\cdots \\ e_{n1}y_1 + e_{n2}y_2 + \ldots + e_{nn}y_n &= S_n, \end{aligned}\right\} \tag{1}$$

where $y_1, y_2, \ldots, y_n$ are dependent variables, t is the independent variable, S_1 to S_n are known integrable functions of t, in $0 \leqslant t \leqslant T$, and

$$e_{rs} = a_{rs}\frac{d}{dt} + b_{rs}, \tag{2}$$

where a_{rs} and b_{rs} are constants. We do not assume at present that $a_{rs} = a_{sr}$, $b_{rs} = b_{sr}$, but we do assume that the determinant

$$A = \|a_{rs}\| \tag{3}$$

is not zero. If it is, we shall be able to show later that there is a defect in the specification of the conditions.

Using the summation convention we can write the equations in the form

$$a_{rs}\frac{dy_s}{dt}+b_{rs}y_s = S_r. \tag{4}$$

Perform the operation Q on both sides of each equation; that is, replace t for a moment by an auxiliary variable τ and integrate with regard to τ from 0 to t. We get

$$a_{rs}(y_s-u_s)+b_{rs}Qy_s = QS_r, \tag{5}$$

u_s being the value of y_s at $t=0$. We rearrange this in the form

$$f_{rs}y_s = (a_{rs}+b_{rs}Q)y_s = a_{rs}u_s+QS_r. \tag{6}$$

These equations take account of both the differential equations and the initial conditions.

Now let D denote the operational determinant

$$D = \|f_{rs}\|. \tag{7}$$

If this determinant is expanded by the rules of algebra we shall obtain a polynomial in Q, in general of the nth degree. The term not containing Q is A, which by hypothesis does not vanish. Let F_{rs} be the cofactor of f_{rs} in this determinant. F_{rs} also is a polynomial in Q.

Now operate on the first equation of (1) with F_{1m}, the second with F_{2m}, and so on, and add. We have

$$F_{rm}f_{rs}y_s = F_{rm}(a_{rs}u_s+QS_r). \tag{8}$$

But $F_{rm}f_{rs} = 0$ unless $m = s$, for it is a determinant with two columns equal. If $m = s$, $F_{rm}f_{rs} = D$. Therefore

$$Dy_m = F_{rm}(a_{rs}u_s+QS_r). \tag{9}$$

The expression on the right is a bounded integrable function of t because the S_r are. Also the function $D(z)$, obtained by replacing Q by a number z, is not zero at $z = 0$ because $A \neq 0$. Hence $1/D(z)$ can, for z less than some positive quantity ρ, be expressed as a power series in z. In accordance with the rule of 7·04 we define D^{-1} as the power series in Q obtained by putting Q for z in the series for $1/D(z)$. Then operate on both sides of (9) with D^{-1}. We get

$$D^{-1}Dy_m = D^{-1}F_{rm}(a_{rs}u_s+QS_r). \tag{10}$$

But series of powers of Q can be multiplied together according to the rules of algebra; hence $D^{-1}Dy_m$ is simply y_m, and we have the formal solution

$$y_m = D^{-1}F_{rm}(a_{rs}u_s+QS_r). \tag{11}$$

The fundamental rule of interpretation is that the operators are to be multiplied out and interpreted term by term, but we shall see that they can all be reduced, at the worst, to single integrals by means of rules that we have already. The series D^{-1} operating on an integrable function always gives a convergent series; hence the result has a meaning, and must be the solution corresponding to the differential equations and the initial conditions if these have a solution at all. To show that there actually is a solution we must verify that (11) satisfies the initial conditions and the differential equations.

First, if t tends to 0 all terms containing Q tend to 0; then D^{-1} tends to A^{-1}, F_{rm} to A_{rm}, the cofactor of a_{rm} in D. Hence for $t = 0$

$$y_m = A^{-1}A_{rm}a_{rs}u_s. \tag{12}$$

But $A_{rm}a_{rs} = 0$ unless $m = s$, when it is equal to A. Hence

$$y_m = u_m,$$

and the solution satisfies the initial conditions.

Secondly,
$$a_{rs} = f_{rs} - b_{rs}Q, \tag{13}$$

and therefore the solution can be written

$$y_m = D^{-1}F_{rm}(f_{rs}u_s - b_{rs}u_s Q + QS_r). \tag{14}$$

The first term as before reduces to u_m on summation. Hence

$$y_m = u_m + D^{-1}F_{rm}Q(S_r - b_{rs}u_s), \tag{15}$$

and the last term consists of positive powers of Q operating on a known function. But

$$\frac{d}{dt}Qf(t) = \frac{d}{dt}\int_0^t f(\tau)\,d\tau = f(t). \tag{16}$$

Hence
$$\left(a_{vm}\frac{d}{dt} + b_{vm}\right)Qf(t) = (a_{vm} + b_{vm}Q)f(t) = f_{vm}f(t), \tag{17}$$

$$\left(a_{vm}\frac{d}{dt} + b_{vm}\right)y_m = b_{vm}u_m + f_{vm}D^{-1}F_{rm}(S_r - b_{rs}u_s). \tag{18}$$

But again $f_{vm}F_{rm} = 0$ unless $r = v$, when it is equal to D. Hence the last term reduces to $S_v - b_{vs}u_s$. The second term cancels the term $b_{vm}u_m$ and finally

$$\left(a_{vm}\frac{d}{dt} + b_{vm}\right)y_m = S_v, \tag{19}$$

which shows that the solution obtained satisfies the differential equations and completes the proof that the problem has a unique solution given by (11).

We now consider the case of $A = 0$. Multiply the equations (1) by the respective A_{rm} and add. Then $a_{rs}A_{rm} = 0$ even for $s = m$, since the sum is then A. Hence

$$A_{rm}b_{rs}y_s = A_{rm}S_r, \tag{20}$$

for all t, and in particular for $t = 0$. Thus the values of the y_s at $t = 0$ cannot be assigned independently. If they are assigned so as to satisfy (20) there is a fixed relation between the y_s for all time and one of the variables can be eliminated; if they do not satisfy (20) the conditions are self-contradictory. The condition that $A \neq 0$ (i.e. a_{rs} is a matrix of rank n) therefore expresses the condition that the initial values of the unknowns can be assigned independently, and will be satisfied in any properly stated problem.

7·051. The symbol p. The process of getting the operational solution (11) from (6) is the same as that of solving a set of algebraic equations in the y_s. The above procedure is the most convenient for establishing the general theorems, but the actual evaluation of the operational solution is made easier by a change of notation. We replace Q by p^{-1}; the rule that operators must be expressible in the form

$$a_0 + a_1 Q + \ldots$$

then becomes the rule that they must be expressible in the form

$$a_0 + a_1 p^{-1} + \ldots,$$

where the coefficients are such that the series

$$a_0 + a_1 z + \ldots$$

converges for some value of z different from 0. Rewriting in this notation the interpretations that we have so far obtained we have

$$\left.\begin{aligned}
p^{-1}1 &= t, \quad p^{-2}1 = \tfrac{1}{2}t^2, \quad p^{-n}1 = \frac{t^n}{n!}, \\
p^{-1}f(t) &= \int_0^t f(t)\,dt, \quad p^{-n}f(t) = \int_0^t \frac{(t-\tau)^{n-1}}{(n-1)!} f(\tau)\,d\tau, \\
\sum_{n=0}^{\infty} a_n p^{-n} f(t) &= a_0 f(t) + \int_0^t f(\tau) \left(\sum_{n=1}^{\infty} a_n \frac{(t-\tau)^{n-1}}{(n-1)!} \right) d\tau, \\
\frac{p}{p-\alpha} 1 &= e^{\alpha t}, \quad \frac{1}{p-\alpha} f(t) = \int_0^t f(\tau) e^{\alpha(t-\tau)}\,d\tau, \\
\frac{p}{(p-\alpha)^n} 1 &= \frac{t^{n-1}}{(n-1)!} e^{\alpha t}, \quad \frac{1}{(p-\alpha)^n} f(t) = \int_0^t f(\tau) \frac{(t-\tau)^{n-1}}{(n-1)!} e^{\alpha(t-\tau)}\,d\tau.
\end{aligned}\right\} \quad (21)$$

The advantage of this notation is that the operators in the last two equations expressible by a single term have p or 1 in the numerator instead of Q^{n-1} or Q^n.

We also have immediately by direct expansion

$$\frac{p^2}{p^2 + n^2} 1 = \cos nt, \quad \frac{np}{p^2 + n^2} 1 = \sin nt, \tag{22}$$

$$\frac{p^2}{p^2 - n^2} 1 = \cosh nt, \quad \frac{np}{p^2 - n^2} 1 = \sinh nt. \tag{23}$$

Returning to (6) we see that as the solution is a purely algebraic process, if we write p^{-1} for Q in each of the equations (6) and then formally multiply by p, and carry through the solution by algebra we shall arrive at the same solution, provided that we keep to the fundamental rule that operators are to be expanded in zero and negative powers of p before interpretation. But with this rule we get in place of (6)

$$(a_{rs} p + b_{rs}) y_s = p a_{rs} u_s + S_r. \tag{24}$$

These equations are called the *subsidiary equations*. They are formed from the differential equations as follows, as we see on inspection.

Write p for d/dt on the left of each equation; to the right of each equation add the result of dropping the b_{rs} on the left and replacing the y_s by their initial values. The resulting subsidiary equations are to be solved by algebra as if p was a number; and the result is to be interpreted by expanding in decreasing powers of p and interpreting p^{-1} as the operation of integrating from 0 to t.

7·052. Partial fraction rule. Since the operational solution (11) is expansible in powers of Q or p^{-1}, beginning with a constant term, the operator must be of the form $F(p)/G(p)$, where $F(p)$ is a polynomial in p of the same degree as $G(p)$ or lower. If p is replaced by a number z, $F(z)/G(z)$ is a rational function of z and can therefore be expressed in partial fractions. Each such fraction can be expressed in descending powers of z, possibly beginning with a constant, and the sum of the expansion is the expansion of

$F(z)/G(z)$. Consequently if we formally break up the operator into partial fractions and apply each separately to a given function, the sum of the results is the result of applying the expansion of the operator $F(p)/G(p)$ to the same function.

The resolution is particularly simple when $G(z)$ has only simple zeros of the form $z = \alpha$ and is not zero at $z = 0$. We have then the algebraic identity

$$\frac{F(p)}{pG(p)} = \frac{F(0)}{pG(0)} + \sum_\alpha \frac{F(\alpha)}{\alpha G'(\alpha)} \frac{1}{p-\alpha},$$

whence

$$\frac{F(p)}{G(p)} = \frac{F(0)}{G(0)} + \sum_\alpha \frac{F(\alpha)}{\alpha G'(\alpha)} \frac{p}{p-\alpha},$$

$$\frac{F(p)}{G(p)} 1 = \frac{F(0)}{G(0)} + \sum_\alpha \frac{F(\alpha)}{\alpha G'(\alpha)} e^{\alpha t}. \tag{25}$$

Hence the part of the solution that depends on the initial conditions is expressed directly in finite terms. A different form is more convenient when the function operated on is not a constant; we can write

$$\frac{F(p)}{G(p)} = \lim_{z \to \infty} \frac{F(z)}{G(z)} + \sum_\alpha \frac{F(\alpha)}{G'(\alpha)} \frac{1}{p-\alpha},$$

$$\frac{F(p)}{G(p)} S(t) = \lim_{z \to \infty} \frac{F(z)}{G(z)} S(t) + \sum_\alpha \frac{F(\alpha)}{G'(\alpha)} \int_0^t S(\tau) e^{\alpha(t-\tau)} d\tau. \tag{26}$$

The interpretation (25) is often called Heaviside's expansion theorem. But his methods involve two other expansion theorems, namely, expansion in powers of p^{-1} and in powers of e^{-ph}, where h is a constant, and in the present treatment the former is fundamental. Consequently (25) will be called the *partial fraction rule* in the present work. It can be read: *Divide by p, put into partial fractions, multiply by p and interpret.*

If there are multiple zeros of $G(p)$, or if it contains p as a factor, the expression of $F(p)/pG(p)$ in partial fractions can still be carried out, but there will be terms of the form p^{-s} or $(p-\alpha)^{-s}$, and $F(p)/G(p)$ will contain terms of the forms $p^{-(s-1)}$ or $p/(p-\alpha)^s$. These can be interpreted by means of (21). Consequently, whenever the functions S_r are integrable and the initial values of the unknowns can be assigned independently the solution can be obtained by operational methods and the result can at worst be expressed in terms of a finite number of single integrals.

A convenient way of finding the terms in $(p-\alpha)^{-s}$ may be illustrated by the following example. Take

$$F(p) = \frac{p}{(p+1)^2(p+2)}.$$

When $z \to -1$, $1/(z+2)$ tends to 1; then

$$F(p) - \frac{p}{(p+1)^2} = \frac{p}{(p+1)^2}\left(\frac{1}{p+2} - 1\right) = -\frac{p}{(p+1)(p+2)} = -\frac{p}{p+1} + \frac{p}{p+2},$$

$$F(p) 1 = \left(\frac{p}{(p+1)^2} - \frac{p}{p+1} + \frac{p}{p+2}\right) 1 = te^{-t} - e^{-t} + e^{-2t}.$$

By subtraction we can in this way reduce the highest index in the denominator by 1 at each stage. Each step checks the algebra of the previous one.

7·053. Principle of superposition. It is, however, sometimes inconvenient to have to use two different resolutions into partial fractions according as the operand is a constant or not. This can be avoided by the *principle of superposition*. We have in the p notation, if

$$F(p) = a_0 + a_1 p^{-1} + a_2 p^{-2} + \ldots,$$

$$F(p)1 = a_0 + a_1 t + a_2 \frac{t^2}{2} + \ldots + a_n \frac{t^n}{n!} + \ldots = f(t),$$

say, then

$$f'(t) = a_1 + a_2 t + \ldots + \frac{a_n t^{n-1}}{(n-1)!} + \ldots,$$

and from the third line of (21)

$$F(p)\phi(t) = a_0 \phi(t) + \int_0^t \phi(\tau) f'(t-\tau)\, d\tau. \tag{27}$$

Integrating by parts we have

$$F(p)\phi(t) = \phi(0)f(t) + \int_{\tau=0}^t f(t-\tau)\, d\phi(\tau), \tag{28}$$

since $f(0) = a_0$. Hence if we know $F(p)1$, the evaluation of $F(p)\phi(t)$ is reduced to a single integration. Using this result we can derive 7·052 (26) from (25), and need only one resolution into partial fractions.

This theorem can be interpreted physically as follows. We can regard a system as subject to disturbances represented by our $S_r(t)$. But if it was in the state of $y_s = 0$ up to time 0 and then the y_s were suddenly raised to u_s, we could represent this as due to a set of impulsive disturbances, thus virtually absorbing the initial values into QS_r at the cost of making $\int_0^\delta S_r(\tau)\, d\tau = a_{rs} u_s$ in the limit when δ is made arbitrarily small. Then the term $\phi(0)f(t)$ can be regarded as the residual effect at time t of the impulsive disturbances $\phi(0)$ at time 0. The later disturbances due to S_r or $\phi(t)$ can then be regarded as the resultant effect of numerous small disturbances $S_r d\tau$ or $d\phi(\tau)$ in time $d\tau$. Each produces its residual effect at time t, but the interval is now $t-\tau$ instead of t. Consequently their total contribution is the sum of elements of the form $f(t-\tau)\, d\phi(\tau)$, which gives the form of the integral. It is not necessary for this purpose that $\phi(t)$ should be differentiable, but if it is not the integral is not the usual Riemann integral but the extended form due to Stieltjes. (Cf. 1·10, 1·102.) This result (28) is known as the Convolution Theorem.

7·054. A third method is often most convenient when the operand itself can be expressed in the form $G(p)1 = g(t)$. Then

$$F(p)g(t) = F(p)G(p)1, \tag{29}$$

and we can proceed directly to the interpretation of the right side by the partial fraction rule.

7·06. Equations of higher order. The method is most easily extended to equations of higher order by breaking them up into equations of the first order. Thus if we have an equation of the second order such as

$$\frac{d^2 x}{dt^2} + a\frac{dx}{dt} + bx = 1 \tag{1}$$

with $x = x_0$, $dx/dt = x_1$ at $t = 0$, we introduce a new variable y given by

$$\frac{dx}{dt} - y = 0, \tag{2}$$

and the original equation can be replaced by

$$\frac{dy}{dt} + bx + ay = 1. \tag{3}$$

Then (2) and (3) are two equations of the first order, and the subsidiary equations are

$$px - y = px_0, \tag{4}$$

$$py + bx + ay = px_1 + 1. \tag{5}$$

Eliminating y by algebra we get

$$x = \frac{(p^2 + ap) x_0 + px_1 + 1}{p^2 + ap + b} 1, \tag{6}$$

which we can interpret by the partial fraction rule on putting

$$p^2 + ap + b = (p - \alpha)(p - \beta). \tag{7}$$

7·061. If we have n differential equations of the second order, possibly with variable functions on the right, we proceed in the same way. If a typical equation is

$$a_{rs} \frac{d^2 y_s}{dt^2} + b_{rs} \frac{dy_s}{dt} + c_{rs} = S_r(t), \tag{1}$$

we take

$$z_s = \frac{dy_s}{dt} \tag{2}$$

as defining a new set of variables z_s; then (1) can be written

$$a_{rs} \frac{dz_s}{dt} + b_{rs} z_s + c_{rs} y_s = S_r(t), \tag{3}$$

and instead of n equations of the second order we have now $2n$ equations of the first order. The operational method of solution will then work provided that the initial values of all the y_s and z_s can be assigned independently. If they are u_s and v_s we write the subsidiary equations

$$py_s - z_s = pu_s, \tag{4}$$

$$(a_{rs} p + b_{rs}) z_s + c_{rs} y_s = S_r(t) + a_{rs} p v_s. \tag{5}$$

The first step of solving is to eliminate z_s between these two; then

$$(a_{rs} p + b_{rs}) p (y_s - u_s) + c_{rs} y_s = S_r(t) + a_{rs} p v_s, \tag{6}$$

that is,

$$(a_{rs} p^2 + b_{rs} p + c_{rs}) y_s = S_r(t) + (a_{rs} p^2 + b_{rs} p) u_s + a_{rs} p v_s. \tag{7}$$

These can be solved for the y_s as for a set of first order equations and the same rules of interpretation apply. The allowance for the initial values of y_s and dy_s/dt is made by the terms in u_s and v_s on the right.

7·062 *Further operators*

The determinant of the coefficients of dy_s/dt and dz_s/dt in (2) and (3) is

$$\begin{vmatrix} a_{11} & a_{12} & \dots & a_{1n} & 0 & 0 & \dots & \dots \\ a_{21} & \dots & \dots & \dots & 0 & 0 & \dots & \dots \\ \dots & \dots & \dots & \dots & \dots & \dots & \dots & \dots \\ a_{n1} & \dots & \dots & \dots & a_{nn} & 0 & \dots & \dots \\ 0 & \dots & \dots & \dots & 0 & 1 & \dots & \dots \\ \dots & \dots & \dots & \dots & \dots & \dots & 1 & \dots \\ \dots & \dots & \dots & \dots & \dots & \dots & \dots & \dots \\ \dots & \dots & \dots & \dots & \dots & \dots & \dots & 1 \end{vmatrix} = \|a_{rs}\|.$$

so that solution is possible with arbitrary initial values of all y_s and dy_s/dt provided again that $\|a_{rs}\| \neq 0$.

The physical interpretation of the condition $A = \|a_{rs}\| \neq 0$ is clear in this case. The y_s may be the coordinates of a dynamical system and will satisfy differential equations of the second order with regard to the time. Then the condition that the initial values of y_s and dy_s/dt can be assigned independently amounts to saying that the coordinates chosen and their rates of change may have any initial values.

7·062. We have already had the rules

$$\frac{p^2}{p^2+n^2}1 = \cos nt, \quad \frac{np}{p^2+n^2}1 = \sin nt. \tag{1}$$

These can also be verified by the partial fraction rule. If we differentiate with respect to n the series expansions of these operators in powers of p^{-1} we get series that converge; hence

$$\frac{2n^2p^2}{(p^2+n^2)^2}1 = nt\sin nt, \tag{2}$$

$$\left\{\frac{p}{p^2+n^2} - \frac{2n^2p}{(p^2+n^2)^2}\right\}1 = t\cos nt,$$

whence

$$\frac{2n^3p}{(p^2+n^2)^2}1 = \sin nt - nt\cos nt. \tag{3}$$

7·04 (10*b*) may be written

$$e^{-\alpha t}\frac{1}{p-\alpha}f(t) = \frac{1}{p}\{e^{-\alpha t}f(t)\}. \tag{4}$$

Hence

$$e^{-\alpha t}\frac{1}{(p-\alpha)^n}f(t) = \frac{1}{p^n}\{e^{-\alpha t}f(t)\}, \tag{5}$$

$$e^{-\alpha t}F(p-\alpha)f(t) = F(p)\{e^{-\alpha t}f(t)\}. \tag{6}$$

In particular

$$e^{-\alpha t}\frac{p(p-\alpha)}{(p-\alpha)^2+\beta^2}1 = \frac{p(p+\alpha)}{p^2+\beta^2}e^{-\alpha t} = \frac{p(p+\alpha)}{p^2+\beta^2}\frac{p}{p+\alpha}1 = \frac{p^2}{(p^2+\beta^2)}1 = \cos\beta t, \tag{7}$$

and therefore

$$\frac{p(p-\alpha)}{(p-\alpha)^2+\beta^2}1 = e^{\alpha t}\cos\beta t. \tag{8}$$

Also

$$e^{-\alpha t}\frac{\beta p}{(p-\alpha)^2+\beta^2}1 = \frac{\beta(p+\alpha)}{p^2+\beta^2}e^{-\alpha t} = \frac{\beta p}{p^2+\beta^2}1 = \sin\beta t,$$

and therefore

$$\frac{\beta p}{(p-\alpha)^2+\beta^2}1 = e^{\alpha t}\sin\beta t. \tag{9}$$

These are given here as an illustration of 7·04 and 7·054. Alternatively, we can apply the partial fraction rule directly:

$$\frac{p}{p-\alpha-i\beta} = e^{(\alpha+i\beta)t} = e^{\alpha t}(\cos\beta t + i\sin\beta t),$$

and we separate real and imaginary parts.

7·07. We sometimes want the limits, if any, of $\frac{F(p)}{G(p)} 1$ and its integral as t tends to infinity. The problem of the induction balance in the next chapter is an instance. These can be simply found from the partial fraction rule. It is not necessary to consider repeated factors, since we can separate them by making small changes in the constants. The roots α must all have negative real parts, otherwise the interpretation would contain exponentials with positive indices and increase without limit, or else trigonometrical terms, which will oscillate finitely. Then

$$\frac{F(p)}{G(p)} 1 = \frac{F(0)}{G(0)} + \Sigma \frac{F(\alpha)}{\alpha G'(\alpha)} e^{\alpha t}, \tag{10}$$

and the limit as t tends to infinity is $F(0)/G(0)$. Also if the integral is to have a finite limit $F(0)/G(0)$ must be 0; then

$$\int_0^\infty \Sigma \frac{F(\alpha)}{\alpha G'(\alpha)} e^{\alpha t} dt = -\Sigma \frac{F(\alpha)}{\alpha^2 G'(\alpha)}$$

$$= \lim_{\lambda\to 0} \Sigma \frac{F(\alpha)}{\alpha G'(\alpha)} \frac{1}{\lambda-\alpha} = \lim_{\lambda\to 0} \frac{F(\lambda)}{\lambda G(\lambda)}. \tag{11}$$

Hence
$$\lim_{t\to\infty} \frac{F(p)}{G(p)} 1 = \frac{F(0)}{G(0)}, \tag{12}$$

$$\lim_{t\to\infty} p^{-1} \frac{F(p)}{G(p)} 1 = \lim_{\lambda\to 0} \frac{F(\lambda)}{\lambda G(\lambda)}, \tag{13}$$

provided that the limits on the right exist, and that all zeros of $G(p)$ have negative real parts.

7·08. In dynamical applications $G(p)$ is often an even function of p with simple zeros $\pm in$. We can separate $F(p)$ into even and odd parts, thus

$$F(p) = \tfrac{1}{2}\{F(p)+F(-p)\} + \tfrac{1}{2}\{F(p)-F(-p)\} = S(p) + pT(p), \tag{14}$$

where $S(p)$ and $T(p)$ are even functions; and then

$$\frac{F(p)}{G(p)} 1 = \frac{S(p)+pT(p)}{G(p)} 1 = \frac{S(0)}{G(0)} + \Sigma \frac{S(in)+inT(in)}{inG'(in)} e^{int}. \tag{15}$$

Taking the terms from $e^{\pm int}$ together we have

$$\frac{S(0)}{G(0)} + \Sigma \frac{2S(in)}{inG'(in)} \cos nt - \Sigma \frac{2nT(in)}{inG'(in)} \sin nt. \tag{16}$$

7·09   Solution in cosines and sines

But
$$\frac{d}{dp^2}G(p) = \frac{1}{2p}G'(p), \quad inG'(in) = -2n^2\left\{\frac{d}{dp^2}G(p^2)\right\}_{p^2=-n^2}, \quad (17)$$

$$\frac{F(p)}{G(p)}1 = \frac{S(0)}{G(0)} - \Sigma\frac{S(in)\cos nt - nT(in)\sin nt}{n^2\left[\dfrac{d}{dp^2}G(p^2)\right]_{p^2=-n^2}}. \quad (18)$$

Since S and T are polynomials in p containing only terms of even degrees this expresses the interpretation directly in real form in terms of trigonometrical functions.

7·09. The Heaviside unit function. This function $H(t)$ is defined by

$$H(t) = 0 \;\; (t<0); \quad H(t) = 1 \;\; (t>0). \quad (1)$$

Evidently $p^{-n}H(t) = p^{-n}1 \;\; (t>0); \quad p^{-n}H(t) = 0 \;\; (t<0).$ (2)

Hence if $F(p)1 = f(t)$,

$$F(p)H(t) = f(t) \;\; (t>0), \quad F(p)H(t) = 0 \;\; (t<0), \quad (3)$$

and in general $F(p)H(t) = f(t)H(t).$ (4)

We are usually interested only in positive values of t; and then it is irrelevant whether $F(p)$ is supposed to operate on 1 or on $H(t)$. Then for either $F(p)1$ or $F(p)H(t)$ it is customary to write simply $F(p)$ and leave the fact that $F(p)$ is supposed to operate on 1 or on $H(t)$ to be understood.

EXAMPLES

Solve the following differential equations with the initial conditions stated:

1. $\dfrac{d^2x}{dt^2} + 4\dfrac{dx}{dt} + 3x = 1;\quad x_0 = 3,\; x_1 = -2.$

2. $\dfrac{d^2x}{dt^2} + 5\dfrac{dx}{dt} + 6x = 12;\quad x_0 = 2,\; x_1 = 0.$

3. $\dfrac{dx}{dt} + 3x = e^{-2t};\quad x_0 = 0.$

4. $\dfrac{d^2x}{dt^2} + 4\dfrac{dx}{dt} + 4x = t^2 e^{-2t};\quad x_0 = 0,\; x_1 = 0.$

5. $\dfrac{d^4y}{dx^4} + 6\dfrac{d^3y}{dx^3} + 11\dfrac{d^2y}{dx^2} + 6\dfrac{dy}{dx} = 20 e^{-2x}\sin x,$

given that $y = 6,\; \dfrac{dy}{dx} = 0,\; \dfrac{d^2y}{dx^2} = 4,\; \dfrac{d^3y}{dx^3} = 0$ when $x = 0$.   (M/c, 1930.)

6. Solve the equations: $\dfrac{dx}{dt} + 5x + 2y = e^{-t},$

$$\dfrac{dy}{dt} + 2x + 2y = 0,$$

given that $x = 1,\; y = 0$ when $t = 0$.   (Prelim. 1945.)

7. If $F(p)1 = f(t)$

prove that $\left\{\dfrac{1}{p}F(p) - F'(p)\right\}1 = tf(t).$

Chapter 8

PHYSICAL APPLICATIONS OF THE OPERATIONAL METHOD

Cut the cackle and come to the hosses.

8·01. Charging of a condenser. An electric circuit contains a cell, a condenser and a coil with self-induction and resistance. Initially the circuit is open. It is suddenly completed; find how the charge on the plates varies with the time.

Let y be the charge on the condenser, t the time, C the capacity of the condenser, L the self-induction, R the resistance of the circuit and E the electromotive force of the cell. The current is $\dot{y}$, and the charging of the condenser produces a potential difference y/C tending to oppose the original e.m.f. Then y satisfies the differential equation

$$E - \frac{y}{C} = L\ddot{y} + R\dot{y}. \tag{1}$$

Initially y and $\dot{y}$, the current, are zero. Hence the subsidiary equation is simply

$$\left(Lp^2 + Rp + \frac{1}{C}\right)y = E, \tag{2}$$

and the operational solution is

$$y = \frac{E}{Lp^2 + Rp + 1/C} = \frac{E}{L(p+\alpha)(p+\beta)}, \tag{3}$$

say. The interpretation is, by the partial fraction rule,

$$y = \frac{E}{L\alpha\beta} + \frac{Ee^{-\alpha t}}{L(-\alpha)(-\alpha+\beta)} + \frac{Ee^{-\beta t}}{L(-\beta)(-\beta+\alpha)}$$

$$= EC + \frac{E}{L(\alpha-\beta)}\left(\frac{1}{\alpha}e^{-\alpha t} - \frac{1}{\beta}e^{-\beta t}\right). \tag{4}$$

Since $\alpha + \beta$ and $\alpha\beta$ are both positive, α and β must be either both real and positive, or else conjugate complexes with positive real parts. In either case y tends to a limit CE, as we should expect.

We notice that if the circuit contained no capacity or self-induction the differential equation would be simply

$$R\dot{y} = E. \tag{5}$$

Hence if the solution has been found for simple resistances, self-induction and capacity can be allowed for by writing $Lp + R + 1/Cp$ for R. For this reason this expression is sometimes called a resistance operator, and the operational method generally the method of resistance operators. The exponential terms in the solution become negligible after a short time, though they are important in experiments where we need to know how long it will take to approach a steady state. They are often called the *transient*.

8·02. Alternating e.m.f. applied to a coil with self-induction. Let x be the current produced. The e.m.f. is $v \cos nt$, which we can take as the real part of ve^{int}. Then we have to solve

$$L\dot{z} + Rz = ve^{int} = \frac{vp}{p-in},$$

and if z is initially zero the initial conditions contribute nothing to the subsidiary equation. Thus the operational solution is

$$z = \frac{vp}{(Lp+R)(p-in)}$$

$$= \frac{v}{L}\left(\frac{e^{int}}{in+R/L} + \frac{e^{-Rt/L}}{-R/L-in}\right)$$

$$= \frac{v(R-Lin)}{L^2n^2+R^2}(e^{int} - e^{-Rt/L}),$$

and the real part of this is

$$x = \frac{v}{L^2n^2+R^2}(R\cos nt + Ln\sin nt - Re^{-Rt/L}).$$

The first two terms give a harmonic variation, out of phase with the e.m.f. The last term gives the transient, which becomes negligible after a time of order L/R.

The harmonic part has amplitude $v/(L^2n^2+R^2)^{1/2}$, and can be written as the real part of

$$\frac{ve^{int}}{R+Lin}.$$

This is the basis of the so-called 'vector diagram', which has nothing to do with vectors, but is a special case of the geometrical representation of complex quantities usually associated with the name of Argand, though he was anticipated by Wallis and Wessel.

8·03. Discharge of a condenser in one of two mutually influencing circuits.
Suppose that we have two similar circuits, each with self-induction L and containing a condenser of capacity C, but negligible resistance, and that the condenser in one has a charge x_0 initially and the other none. The coefficient of mutual induction is M. The first circuit is closed; find the ensuing variations of the charges.

Put $CL = 1/\alpha^2$, $M = L\beta$; if x, y denote the charges on the condensers in the two circuits

$$L(\ddot{x} + \beta\ddot{y} + \alpha^2 x) = 0, \tag{1}$$

$$L(\beta\ddot{x} + \ddot{y} + \alpha^2 y) = 0. \tag{2}$$

Initially $\qquad x = x_0, \quad \dot{x} = 0, \quad y = 0, \quad \dot{y} = 0, \tag{3}$

and the subsidiary equations are

$$(p^2+\alpha^2)x + \beta p^2 y = p^2 x_0, \tag{4}$$

$$\beta p^2 x + (p^2+\alpha^2)y = \beta p^2 x_0. \tag{5}$$

On solving by algebra

$$\frac{x}{p^2(p^2+\alpha^2)-\beta^2 p^4} = \frac{y}{\beta p^2(p^2+\alpha^2)-\beta p^4} = \frac{x_0}{(p^2+\alpha^2)^2-\beta^2 p^4}, \tag{6}$$

that is, $\qquad \dfrac{x}{(1-\beta^2)p^4+p^2\alpha^2} = \dfrac{y}{\beta\alpha^2 p^2} = \dfrac{x_0}{\{(1+\beta)p^2+\alpha^2\}\{(1-\beta)p^2+\alpha^2\}}. \tag{7}$

Then
$$x = \frac{p^2(1-\beta^2)+\alpha^2}{\{(1+\beta)p^2+\alpha^2\}\{(1-\beta)p^2+\alpha^2\}}p^2 x_0$$
$$= \left\{\frac{(1+\beta)p^2}{(1+\beta)p^2+\alpha^2} + \frac{(1-\beta)p^2}{(1-\beta)p^2+\alpha^2}\right\}\tfrac{1}{2}x_0$$
$$= \left\{\cos\frac{\alpha}{\sqrt{(1+\beta)}}t + \cos\frac{\alpha}{\sqrt{(1-\beta)}}t\right\}\tfrac{1}{2}x_0, \tag{8}$$

$$y = \frac{\beta\alpha^2}{\{(1+\beta)p^2+\alpha^2\}\{(1-\beta)p^2+\alpha^2\}}p^2 x_0$$
$$= \left\{\frac{(1+\beta)p^2}{(1+\beta)p^2+\alpha^2} - \frac{(1-\beta)p^2}{(1-\beta)p^2+\alpha^2}\right\}\tfrac{1}{2}x_0$$
$$= \left\{\cos\frac{\alpha}{\sqrt{(1+\beta)}}t - \cos\frac{\alpha}{\sqrt{(1-\beta)}}t\right\}\tfrac{1}{2}x_0. \tag{9}$$

If we write
$$\frac{\alpha}{\sqrt{(1-\beta)}} = \gamma+\delta, \quad \frac{\alpha}{\sqrt{(1+\beta)}} = \gamma-\delta, \tag{10}$$

the solutions take the forms
$$x = x_0 \cos\gamma t \cos\delta t, \quad y = x_0 \sin\gamma t \sin\delta t. \tag{11}$$

If M is small, δ is small, and the disturbance consists of a rapid oscillation in period $2\pi/\gamma$, with the amplitude varying so that the oscillation is transferred from one circuit to the other in time $\pi/2\delta$. This is the case of beats due to weak coupling. A similar phenomenon is well known for two pendulums hanging on the same support, the support being not quite rigid, so that one pendulum influences the other by displacing the support. The same phenomenon of the transfer of the vibration from one pendulum to the other occurs at regular intervals.

If the coupling is strong, so that β is nearly 1, the two periods $2\pi\sqrt{(1\pm\beta)}/\alpha$ are very different, and the variation of the charge consists of a rapid oscillation superposed on a slow one of equal amplitude. The slow component has the same phase in the two circuits, the rapid one opposite phases.

8·04. Rimington's method of determining self-induction.[*] In this method the unknown inductance is placed in the first arm of a Wheatstone bridge; the fourth arm is shunted, a known capacity being placed in the shunt.

First consider the ordinary Wheatstone bridge, the resistances of the arms being $R_1 R_2 R_3 R_4$, G that of the galvanometer, b that of the battery and leads; x is the current in R_1, y that in R_2, g that through the galvanometer. Then

$$R_1 x - R_2 y + Gg = 0, \tag{1}$$
$$R_3 x - R_4 y - (R_3 + R_4 + G)g = 0 \tag{2}$$
$$b(x+y) + R_2 y + R_4(y+g) = E, \tag{3}$$

and on solving (1) and (2) we find
$$\frac{g}{R_2 R_3 - R_1 R_4} = \frac{x+y}{G(R_1+R_2+R_3+R_4)+(R_1+R_2)(R_3+R_4)}. \tag{4}$$

[*] E. C. Rimington, *Phil. Mag.* (5) 24, 1887, 54–60; Bromwich, *Phil. Mag.* (6) 37, 1919, 407–19.

If g is small compared with x and y we have nearly

$$x+y = E \bigg/ \bigg\{ b + \frac{(R_1+R_3)(R_2+R_4)}{R_1+R_2+R_3+R_4} \bigg\}, \qquad (5)$$

$$g = \frac{(x+y)(R_2R_3 - R_1R_4)}{G(R_1+R_2+R_3+R_4)}. \qquad (6)$$

The important feature of the arrangement is that $g = 0$ if $R_2R_3 = R_1R_4$, irrespective of the accuracy of the approximation (5).

According to our first result we can allow for the self-induction L in the first arm by replacing R_1 by $Lp+R_1$. Let the arrangement in the fourth arm be as shown. The resistance of the main wire is R_4, that of the shunted portion of it r. The shunt has resistance S. Then the effective resistance of the whole arm is

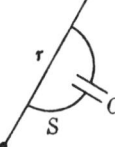

$$R_4 - r + \frac{rS}{r+S} = R_4 - \frac{r^2}{r+S}.$$

If the shunt contains a capacity C we allow for it by replacing S by $S+1/Cp$. Hence in the formula (6) for g we must replace R_1 by $Lp+R_1$, and R_4 by

$$R_4 - \frac{r^2}{r+S+1/Cp} = R_4 - \frac{r^2 Cp}{(r+S)Cp+1}. \qquad (7)$$

The result expresses the current through the galvanometer when the battery circuit is suddenly closed.

It can be shown that in actual conditions g cannot vanish for all values of the time. A sufficient condition for this would be that the modified operator $R_2R_3 - R_1R_4$ should be identically zero; then g would vanish whatever the remaining factor might represent. A little consideration will show that this condition is also necessary. This factor is modified to

$$R_2R_3 - (Lp+R_1)\bigg(R_4 - \frac{r^2 Cp}{(r+S)Cp+1}\bigg). \qquad (8)$$

Multiplying up and equating coefficients of powers of p to zero, we find

$$R_4(r+S) = r^2, \qquad (9)$$

$$-LR_4 + (R_2R_3 - R_1R_4)(r+S)C + R_1 r^2 C = 0, \qquad (10)$$

$$R_2R_3 - R_1R_4 = 0. \qquad (11)$$

From the construction of the apparatus $r \leqslant R_4$, $S \geqslant 0$. Hence (9) can hold only if $r = R_4$ and $S = 0$; the shunt wire must be attached to the ends of R_4 and have zero resistance. (11) is the usual condition for balance; and substituting in (10) we have

$$L = R_1 R_4 C. \qquad (12)$$

These conditions cannot be completely satisfied. But the changes of current on closing the battery circuit are so rapid that an ordinary galvanometer will not follow them. If the current settles down to nothing we have the usual condition for balance; but if there is a resultant flow through the galvanometer in one direction or the other it will act on the galvanometer as an impulse, and there will be a ballistic throw. The condition that g tends to 0 and that there shall be no ballistic throw are

$$g \to 0, \quad \int_0^\infty g\,dt = 0. \tag{13}$$

But these are satisfied, by 7·07, if the constant term and the term in p in the operational form for g vanish; and again, irrespective of the approximation (5), we can use (8). The formal limit when $p \to 0$ is

$$R_2 R_3 - R_1 R_4 = 0 \tag{14}$$

as before. The coefficient of p is

$$LR_4 - R_1 r^2 C, \tag{15}$$

and the vanishing of this is the condition for no ballistic throw. Condition (9), which came from the terms in p^2, no longer arises. The method is therefore first to set up the bridge in balance in the usual way, thus satisfying (14); and then to connect the shunt containing the capacity to different points in the arm R_4 so as to vary r. When the adjustment is such that there is no ballistic throw r is determined, and then (15) gives L.

8·05. The seismograph. In principle most seismographs are Euler pendulums—pendulums with supports rigidly attached to the Earth, so that when the ground moves it displaces the point of support horizontally and disturbs the pendulum. The seismograph differs from the Euler pendulum as considered in text-books of dynamics in two ways. Instead of being free to vibrate in a vertical plane, it is constrained to swing, like a gate, about an axis nearly, but not quite, vertical, so that the period is much lengthened; and fluid viscosity or electromagnetic damping is introduced to give a frictional term proportional to the relative velocity. The displacement of the mass with regard to the Earth then satisfies an equation of the form

$$\ddot{x} + 2\kappa \dot{x} + n^2 x = \lambda \dot{\xi}, \tag{1}$$

where ξ is the displacement of the ground and κ, n, λ are constants of the instrument. Some instruments, such as those of Wiechert and Wood-Anderson, are not on the principle of the Euler pendulum, but nevertheless give an equation of this form. Others are arranged to record vertical displacement of the ground; this requires a heavy mass elastically supported, and is convenient for ground movements of short period, as in seismic prospecting. For longer periods it is more difficult to design an instrument such that x will satisfy a linear differential equation, but the difficulties have been overcome in several different ways, and the differential equation is again of the form (1).

The first object of the instrument is to record as accurately as possible the time of any sudden change of the velocity of the ground. The second is that when such a change has been recorded the instrument shall return as quickly as possible to its original position so as to be ready to record any later disturbances.

Suppose first that the ground suddenly acquires a finite velocity, say unity. Then ξ jumps from 0 to 1, and therefore $\dot{x}$ from 0 to λ. The initial conditions are therefore

$$x = 0, \quad \dot{x} = \lambda, \tag{2}$$

and our subsidiary equation is

$$(p^2 + 2\kappa p + n^2) x = \lambda p \quad (t > 0). \tag{3}$$

Put
$$p^2 + 2\kappa p + n^2 = (p + \alpha)(p + \beta). \tag{4}$$

Then
$$x = \frac{\lambda p}{(p+\alpha)(p+\beta)} \quad (t>0) \tag{5}$$

$$= \frac{\lambda}{\alpha-\beta}(e^{-\beta t} - e^{-\alpha t}). \tag{6}$$

The recorded displacement x therefore begins by increasing at a finite rate λ, reaches a maximum $\lambda \left(\frac{\beta^\beta}{\alpha^\alpha}\right)^{1/(\alpha-\beta)}$ after a time $\frac{1}{\alpha-\beta}\log\frac{\alpha}{\beta}$, and then tends asymptotically to zero.

If α and β are real, and $\beta < \alpha$, the behaviour after a long time depends mainly on $e^{-\beta t}$; to confine the effects of a disturbance to as short an interval of time as possible, we should therefore make β as large as possible. But

$$\beta = \kappa - \sqrt{(\kappa^2 - n^2)} = \frac{n^2}{\kappa + \sqrt{(\kappa^2 - n^2)}}, \tag{7}$$

and for given n, β is greatest (given that it is real and therefore $\kappa \geq n$) when $\kappa = n$. This is the condition for what is called aperiodicity. The solution then reduces to

$$x = \frac{\lambda p}{(p+n)^2} = \lambda t e^{-nt} \quad (t>0). \tag{8}$$

The maximum displacement is now at time $1/n$ and is equal to λ/en.

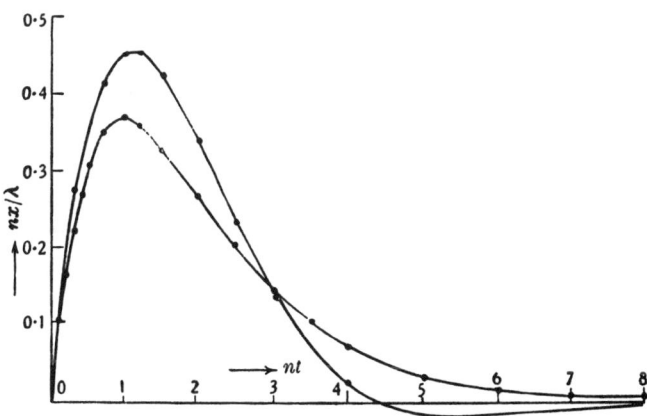

Response of aperiodic and Milne-Shaw seismographs.

If $\kappa < n$, we can put
$$n^2 - \kappa^2 = \gamma^2. \tag{9}$$
Then (6) becomes
$$x = \frac{\lambda}{\gamma} e^{-\kappa t} \sin \gamma t, \tag{10}$$

and the motion dies down more rapidly the larger κ is, in the range considered. The aperiodic state $\kappa = n$ therefore gives the least motion after a long time for given n.

In practice, however, κ is usually made rather less than n. In the Milne-Shaw instrument, for instance, κ is about $0.7n$. The motion is then oscillatory, but the ratio of the first swing to the second is $e^{\kappa \pi/\gamma}$, about 20. But x vanishes after an interval π/γ from the start, or about $4/n$, and ever afterwards is a small fraction of its first maximum. The reduced damping effect after a long time is considered less important than the quick recovery to zero after the first maximum. The time of the first maximum is $1\cdot 1/n$ from the start, as against $1/n$ for the aperiodic instrument and $1\cdot 57/n$ for the undamped one.

The Galitzin seismograph is similarly arranged, but the motion of the pendulum, instead of being recorded directly, generates by electromagnetic induction a current, which passes through a galvanometer. If x is the displacement of the pendulum, and y that of the galvanometer mirror, the differential equations are

$$\ddot{x} + 2\kappa_1 \dot{x} + n_1^2 x = \lambda \xi, \tag{11}$$

$$\ddot{y} + 2\kappa_2 \dot{y} + n_2^2 y = \mu \dot{x}, \tag{12}$$

where the reaction of the induced current on the pendulum is neglected. Again supposing the ground to start with unit velocity, we have

$$y = \frac{\lambda \mu p^2}{(p^2 + 2\kappa_1 p + n_1^2)(p^2 + 2\kappa_2 p + n_2^2)} \quad (t > 0). \tag{13}$$

Response of Galitzin seismograph.

In instruments of the original design κ and n were made the same for both the interacting systems, and both were made aperiodic, so that

$$\kappa_1 = \kappa_2 = n_1 = n_2 = n. \tag{14}$$

Then
$$y = \frac{\lambda \mu p^2}{(p+n)^4}, \quad (t > 0)$$

$$= \lambda \mu \frac{d}{dt}\left(\frac{t^3}{3!} e^{-nt}\right)$$

$$= \tfrac{1}{2}\lambda \mu (t^2 - \tfrac{1}{3} n t^3) e^{-nt}. \tag{15}$$

The indicator therefore begins to move with a finite acceleration, instead of with a finite velocity as for the pendulum. The maximum displacement follows after time $(3-\sqrt{3})/n = 1\cdot 27/n$, the mirror passes through the equilibrium position after time $3/n$, and there is a maximum displacement in the opposite direction after time $4\cdot 73/n$. The mirror then returns asymptotically to the position of equilibrium. The ratio of the two extreme displacements is $e^{2\sqrt{3}}/(2+\sqrt{3})^2 = 2\cdot 3$. In comparison with a partially damped instrument such as the Milne-Shaw, recording directly, the Galitzin machine gives the first maximum a little later, the first zero a little earlier, and the next extreme displacement is larger in comparison with the first. It will be seen from the graph that in spite of the fact that y for small t is proportional to t^2 instead of t, it begins to increase rapidly so soon that the beginning of the movement can be very accurately read.

Later modifications of the Galitzin instrument have been to abandon the relations (14) by reducing the damping and by making the galvanometer period shorter than the pendulum period. For harmonic motions of the ground this makes the magnification vary less with the forced period. It does not seem possible, however, to reduce the overswing on recovery after an impulsive change in the velocity of the ground. In some modern designs the reaction of the induced current on the pendulum can no longer be neglected.*

8·06. Resonance. A simple pendulum, originally hanging in equilibrium, is disturbed for a finite time by a force varying harmonically in a period equal to the free period of the pendulum. Find the motion after the force is removed.

The differential equation is
$$\ddot{x} + n^2 x = f \sin nt = f \frac{np}{p^2 + n^2} \quad (0 < t < T),$$
with $x = 0$, $\dot{x} = 0$ at $t = 0$. Then
$$x = f \frac{np}{(p^2 + n^2)^2} = \frac{f}{2n^2}(\sin nt - nt \cos nt).$$

The motion can therefore be regarded as a harmonic motion of continually increasing amplitude. Suppose that the disturbance acts for a time $T = r\pi/n$, where r is an integer. At the end of this time
$$x = -\frac{f}{2n^2} r\pi (-1)^r, \quad \dot{x} = 0.$$

The subsequent motion is therefore given by
$$x = -(-1)^r \frac{r\pi f}{2n^2} \cos(nt - r\pi) = -\frac{r\pi f}{2n^2} \cos nt,$$
and is therefore a harmonic motion with amplitude proportional to the duration of the disturbance.

Linear differential equations in dynamics are usually the result of neglecting the square and higher powers of the displacement. What the result shows is that the amplitude, if the forced and free periods agree, will grow until the neglected terms need to be taken into account.

8·07. *Three particles of masses m, $\tfrac{21}{20}m$, and m, in order, are attached to a light stretched string of length $4l$, dividing it into equal intervals. One of the particles of mass m is struck by a transverse impulse I. Find the subsequent motion of the middle particle.* (*Intercollegiate Examination*, 1923.)

If x_1, x_2, x_3 are the displacements of the three particles and P the tension, we find in the usual way the equations of motion
$$\left. \begin{aligned} \ddot{x}_1 &= -\lambda(2x_1 - x_2), \\ \tfrac{21}{20}\ddot{x}_2 &= -\lambda(-x_1 + 2x_2 - x_3), \\ \ddot{x}_3 &= -\lambda(-x_2 + 2x_3), \end{aligned} \right\} \quad (1)$$

---

* The theory is more fully developed by J. Rybner, *Gerlands Beitr.* 31, 1931, 259–81; 51, 1937, 375–401; 55, 1939, 303–13.

where $\lambda = P/ml$. Initially, all the displacements are zero, $\dot x_2 = \dot x_3 = 0$, $\dot x_1 = I/m$. Then the subsidiary equations are

$$\left.\begin{aligned}(p^2+2\lambda)x_1-\lambda x_2 &= pI/m,\\ -\lambda x_1+(\tfrac{21}{20}p^2+2\lambda)x_2-\lambda x_3 &= 0,\\ -\lambda x_2+(p^2+2\lambda)x_3 &= 0.\end{aligned}\right\} \qquad (2)$$

As we are asked only for the variation of x_2 we eliminate x_1 and x_3. We have

$$x_1 = \frac{\lambda}{p^2+2\lambda}x_2 + \frac{p}{p^2+2\lambda}\frac{I}{m}, \quad x_3 = \frac{\lambda}{p^2+2\lambda}x_2; \qquad (3)$$

and then

$$\left(\tfrac{21}{20}p^2+2\lambda-\frac{2\lambda^2}{p^2+2\lambda}\right)x_2 = \frac{\lambda p}{p^2+2\lambda}\frac{I}{m}, \qquad (4)$$

and on simplifying

$$(7p^2+4\lambda)(3p^2+10\lambda)x_2 = 20\lambda pI/m. \qquad (5)$$

Then

$$x_2 = \frac{20}{58}\frac{I}{m}\left(\frac{7p}{7p^2+4\lambda}-\frac{3p}{3p^2+10\lambda}\right) \qquad (6)$$

$$= \frac{10}{29}\frac{I}{m}\left(\frac{\sin\alpha t}{\alpha}-\frac{\sin\beta t}{\beta}\right), \qquad (7)$$

where

$$\alpha^2 = \tfrac{4}{7}\lambda, \quad \beta^2 = \tfrac{10}{3}\lambda. \qquad (8)$$

If we want also the motions of the other particles they can be found by using (3) and applying the partial-fraction rule. They will contain terms with the same periods as those in (7), but also terms with period $2\pi/\sqrt{(2\lambda)}$. These correspond to a third normal mode, in which the middle particle does not move. This illustrates one great advantage of the operational method. We are asked only for x_2, and the method gives it directly. In the usual method we should have to determine the amplitudes of all three normal modes separately, even though one of them is irrelevant to the question asked.

8·08. Small oscillations in dynamics. Consider a dynamical system with a Lagrangian function given by

$$2L = a_{rs}\dot x_r\dot x_s - c_{rs}x_r x_s, \qquad (1)$$

so that the equations of motion are

$$a_{rs}\ddot x_s + c_{rs}x_s = S_r, \qquad (2)$$

S_r being any generalized force component applied to x_r and not taken into account in the potential energy. If the system starts from rest and only one of the S_r differs from zero we can write

$$e_{ms} = a_{ms}p^2 + c_{ms}, \qquad (3)$$

and the subsidiary equations are

$$e_{ms}x_s = 0 \quad (m \neq r), \qquad e_{ms}x_s = S_r \quad (m = r). \qquad (4)$$

Writing Δ for the determinant of the e_{ms} and E_{rs} for the cofactor of e_{rs} in this determinant, we have the operational solution

$$x_s = \frac{E_{rs}}{\Delta}S_r. \qquad (5)$$

Notice that r is a *particular* suffix and is not summed over. Now the determinant Δ is symmetrical, so that $E_{rs} = E_{sr}$. Thus a given force S_r applied to the coordinate x_r will produce exactly the same variation in x_s as the same force would produce in x_r if it was applied to x_s. Thus we have a reciprocity theorem applicable to all non-gyroscopic and frictionless systems.

It is easy to see that friction does not affect the result if it is expressible by a dissipation function $F = \frac{1}{2}b_{rs}\dot{x}_r\dot{x}_s$. In particular, the result is true for electrical networks.

Now suppose that the force reduces to an impulse J_r at $t = 0$ and suppose that Δ has no repeated factor; we can write
$$\Delta = A\Pi(p^2+\alpha^2), \tag{6}$$
and replace S_r by pJ_r. Then
$$x_s = \frac{E_{rs}p}{A\Pi(p^2+\alpha^2)}J_r = \sum_\alpha \frac{E_{rs}(-\alpha^2)}{\Pi'(-\alpha^2)}\frac{pJ_r}{p^2+\alpha^2}, \tag{7}$$
where $E_{rs}(-\alpha^2)$ and $\Pi'(-\alpha^2)$ denote the results of putting $-\alpha^2$ for p^2 in E_{rs} and $d\Delta/dp^2$; and then
$$x_s = \sum_\alpha \frac{E_{rs}(-\alpha^2)}{\alpha\Pi'(-\alpha^2)}J_r \sin\alpha t. \tag{8}$$

The separate terms have different periods, and the terms of the same period in different coordinates constitute a normal mode of the system.

An immediate consequence is that if for some s and α, say α_1, $E_{rs}(-\alpha^2) = 0$ for all r, x_s contains no term in $\sin\alpha_1 t$ whatever impulses are applied; in other words, if x_s is the displacement of a particle of the system, that particle is at a node of the mode in question. But then if we consider an impulse J_s applied to x_s we shall have
$$x_r = \sum_\alpha \frac{E_{sr}(-\alpha^2)}{\alpha\Pi'(-\alpha^2)}J_s \sin\alpha t, \tag{9}$$
and again, since $E_{sr} = E_{rs}$, the term in $\sin\alpha_1 t$ will have zero coefficient in every coordinate. Hence we have another general reciprocity theorem; no mode can be excited by striking the system at any node of that mode. It can be shown similarly that if the initial conditions specify initial values of the coordinates but the velocities are zero, the subsequent values contain terms with factors $E_{rs}(-\alpha^2)\cos\alpha t$, and the initial displacement at a node of any mode will not contribute to the terms in that mode in the subsequent motion.

This principle, in a continuous system, provided one of the crucial tests of the existence of deep-focus earthquakes. Most earthquakes occur at depths not over about 50 km., and produce, besides the waves that travel right through the earth, two types of surface waves explained theoretically by Rayleigh and Love. These resemble waves on deep water in that the displacements die down rapidly with increasing depth and are inappreciable at depths over about a wave-length, in this case something of the order of 50–100 km. By the above principle they should not be excited appreciably by disturbances at greater depths. The late Professor H. H. Turner had inferred from the times of travel of the bodily waves that a few earthquakes originated at depths up to some 400 km., but the evidence appeared capable of other interpretations. Examination of the seismograms of these earthquakes by Stoneley, however, showed that the surface waves were absent, and this fact was not explicable by any of the other suggestions, but was just what would be expected from the reciprocity principle if the earthquakes in question originated at great depths.

8·09. Case of equal roots. In the discussion of the oscillations of dynamical systems about equilibrium the ordinary method of seeking for solutions of the form $x_s = \lambda_s e^{\gamma t}$ meets with a difficulty when the determinantal equation for γ^2 has equal roots. In the ordinary way, if we have a set of simultaneous linear differential equations for n variables, and we eliminate them in succession in favour of one, we get a differential equation for that one. If we substitute $e^{\gamma t}$ for it we shall get an equation for γ, and if there is a repeated root there will be a second solution $te^{\gamma t}$. If this happened in the theory of small oscillations it would appear that a repeated value of γ would lead to terms of the form $t\cos \kappa t$, $t\sin \kappa t$ ($\kappa = i\gamma$), and except for special initial conditions a small oscillation would grow indefinitely. This was never found to happen, and in fact if it could happen it would contradict the fundamental principle that if the potential energy is a minimum in the position of equilibrium, and the initial displacements and velocities are sufficiently small but not zero, there is a limit that no displacement can ever exceed. Laplace was puzzled, and the explanation was finally given by Routh[*] and Heaviside.[†] If the system is not dissipative and the roots are unequal, we know from 4·082 and 4·09 that the zeros of the minor of any element in the leading diagonal separate those of the original determinant, and if the determinant Δ has a factor $(p^2+\alpha^2)^k$, every first minor contains the factor $(p^2+\alpha^2)^{k-1}$. Hence when we evaluate the contribution from the initial conditions to the operational solution, namely, from 7·061(7),

$$x_m = \frac{E_{rm}}{\Delta} a_{rs}(p^2 u_s + p v_s),$$

a factor $(p^2+\alpha^2)^{k-1}$ will cancel and we are left with only a single factor $(p^2+\alpha^2)$ in the denominator. The same will happen for every repeated root, and the interpretation will contain only terms of the forms $\cos \alpha t$ and $\sin \alpha t$. Varying u_s and v_s will alter the ratios of the coefficients of these trigonometric factors for different coordinates; it will not introduce terms like $t\cos \alpha t$ or $t\sin \alpha t$.

8·10. Dissipative and gyroscopic systems. Here the root separation theorem may not hold. Then the operational solution may have a repeated factor in the denominator and terms like $te^{-\alpha t}$, $t\cos \alpha t$, $t\sin \alpha t$ may occur in the interpretation. We have had a simple instance of this for a dissipative system in the aperiodic seismograph. This will not affect stability if the undamped system is stable and non-gyroscopic, since the solutions are exponentially decreasing and will still tend to 0 with increasing t. But if a system is kept stable only by gyroscopic action, coincidence of the roots may ruin the stability. Suppose that the equations satisfied by two coordinates x_1, x_2 are

$$\ddot{x}_1 - b\dot{x}_2 + c_1 x_1 = 0, \quad \ddot{x}_2 + b\dot{x}_1 + c_2 x_2 = 0. \tag{1}$$

Assume
$$x_1 = \lambda_1 e^{\gamma t}, \quad x_2 = \lambda_2 e^{\gamma t}. \tag{2}$$

We find that γ must satisfy the determinantal equation

$$\begin{vmatrix} \gamma^2 + c_1 & -b\gamma \\ b\gamma & \gamma^2 + c_2 \end{vmatrix} = 0, \tag{3}$$

that is,
$$\gamma^4 + (c_1 + c_2 + b^2)\gamma^2 + c_1 c_2 = 0. \tag{4}$$

A necessary condition for stability is that both values of γ^2 shall be real and < 0. Hence $c_1 c_2 > 0$, and we have two cases according as c_1 and c_2 are both positive or both negative.

[*] *Stability of a given State of Motion*, 1877.   [†] *Electrical Papers*, 1, 529.

Take first the case where they are both positive. Then the system would be stable even if b was 0. The condition for equal roots is

$$(c_1+c_2+b^2)^2 = 4c_1c_2, \tag{5}$$

that is,
$$(c_1-c_2)^2 + 2b^2(c_1+c_2) + b^4 = 0. \tag{6}$$

With $c_1, c_2 > 0$ this can be satisfied *only* if $b = 0$, and then $c_1 = c_2$. Hence if (3) has equal roots and γ^2 is equal to one of them, all elements of the determinant vanish. This is what we should expect, since $c_1 x_1^2 + c_2 x_2^2$ is a positive form in this case, and the root separation theorem still holds in a gyroscopic system when the terms $c_{rs} x_r x_s$ are a positive form.

If, however, c_1 and c_2 are both negative (6) can be satisfied, provided that

$$b^2 = -(c_1+c_2) \pm \sqrt{\{(c_1+c_2)^2-(c_1-c_2)^2\}}$$
$$= -(c_1+c_2) \pm 2\sqrt{(c_1c_2)} = \{\sqrt{(-c_1)} \pm \sqrt{(-c_2)}\}^2. \tag{7}$$

Thus the determinantal equation can have equal roots in this case without b being zero. It reduces now to
$$\gamma^4 \pm 2\sqrt{(c_1c_2)}\gamma^2 + c_1c_2 = 0$$
and
$$\gamma^2 = \mp\sqrt{(c_1c_2)}. \tag{8}$$

In this case the separate elements of the determinant (3) do not vanish, though γ^2 is still real and negative. With the lower sign in (7) and (8), γ^2 would be positive and the system obviously unstable. We therefore take the negative sign in (8) and the positive sign in (7). To see what will happen to a system satisfying these conditions, with $x_1 = u_1, \dot{x}_1 = 0, x_2 = 0, \dot{x}_2 = 0$ at $t = 0$, we write

$$c_1 = -\alpha^2, \quad c_2 = -\beta^2, \quad b = \alpha+\beta, \tag{9}$$

$$\left.\begin{array}{l}(p^2-\alpha^2)x_1 - (\alpha+\beta)px_2 = p^2 u_1,\\ (\alpha+\beta)px_1 + (p^2-\beta^2)x_2 = (\alpha+\beta)pu_1.\end{array}\right\} \tag{10}$$

The operational solution is

$$x_1 = \frac{p^4 + (\alpha^2+2\alpha\beta)p^2}{(p^2+\alpha\beta)^2} u_1, \quad x_2 = -\frac{(\alpha+\beta)\alpha^2 p}{(p^2+\alpha\beta)^2} u_1, \tag{11}$$

and
$$\frac{x_1}{u_1} = \frac{p^2}{p^2+\alpha\beta} + \frac{\alpha(\alpha+\beta)p^2}{(p^2+\alpha\beta)^2} = \cos\sqrt{(\alpha\beta)}t + \frac{\alpha(\alpha+\beta)}{2\sqrt{(\alpha\beta)}} t \sin\sqrt{(\alpha\beta)}t, \tag{12}$$

$$\frac{x_2}{u_1} = -\frac{(\alpha+\beta)\alpha^2}{2(\alpha\beta)^{3/2}} \{\sin\sqrt{(\alpha\beta)}t - \sqrt{(\alpha\beta)} t \cos\sqrt{(\alpha\beta)}t\}. \tag{13}$$

Hence if a system is kept stable only by the gyroscopic terms, and the coefficients of these are such as to make the periods equal, the stability may be ruined in the sense that the amplitude of a disturbance will increase linearly with the time. This corresponds to the top with $C^2 n^2 = 4AMgh$.

8·101. Gyroscopic system with slight friction. Here we need not consider a general initial disturbance, but confine ourselves to the period equation. If the equations of motion are
$$\ddot{x}_1 + f\dot{x}_1 + c_1 x_1 - b\dot{x}_2 = 0, \quad \ddot{x}_2 + f\dot{x}_2 + c_2 x_2 + b\dot{x}_1 = 0, \tag{1}$$

where $c_1 < 0, c_2 < 0, f \geqslant 0$, and b is large enough to ensure stability when f is put equal to 0, we assume solutions proportional to $e^{\gamma t}$ and find that γ must satisfy

$$\gamma^4 + 2f\gamma^3 + \gamma^2(c_1+c_2+b^2+f^2) + f\gamma(c_1+c_2) + c_1 c_2 = 0. \tag{2}$$

Hence
$$\Sigma \gamma = -2f \leqslant 0, \quad \Sigma \frac{1}{\gamma} = -f\left(\frac{1}{c_1}+\frac{1}{c_2}\right) \geqslant 0. \tag{3}$$

Let the roots for $f = 0$ be $\pm in_1$, $\pm in_2$, with $n_1 > n_2$; and for f small and > 0 let them be $\pm in_1 - \alpha_1$, $\pm in_2 - \alpha_2$ to order f. Then we have to this order

$$\alpha_1 + \alpha_2 = f > 0,$$

$$\frac{1}{in_1 - \alpha_1} + \frac{1}{-in_1 - \alpha_1} + \frac{1}{in_2 - \alpha_2} + \frac{1}{-in_2 - \alpha_2} \doteqdot -\frac{2\alpha_1}{n_1^2} - \frac{2\alpha_2}{n_2^2} > 0.$$

These inequalities are consistent only if α_1 and α_2 have opposite signs, and in fact, since $n_1 > n_2$, $\alpha_1 > 0$, $\alpha_2 < 0$. Hence if a system is kept stable by gyroscopic action only, the effect of small friction is always to produce instability. The quicker free vibration will be damped, but the slower will increase in amplitude with time.

This feature of gyroscopic motion has considerable theoretical and practical importance. If we use the usual method to treat small oscillations about steady motion, neglecting friction, we often find that all the roots γ are purely imaginary, and infer that the system is stable. If the expression $c_{rs} x_r x_s$ is essentially $\geqslant 0$, and there is a little friction, $a_{rs}\dot{x}_r\dot{x}_s + c_{rs}x_r x_s$ will decrease, and the oscillations will be gradually damped down. Such systems are called *secularly stable*. But if the quadratic form in question is not essentially $\geqslant 0$ and the system is kept stable only by the gyroscopic terms, the slower oscillation about steady motion will gradually increase in amplitude until it can no longer be treated as small, and may lead to a complete change in the character of the motion. Such systems are called *ordinarily stable* but *secularly unstable*. The engineer tries to avoid them. They have possibly had considerable importance in the development of stellar systems, and of the solar system in particular.

8·11. Radioactive disintegration. The uranium family of elements are such that an atom of any of them, except the last, is capable of breaking up into an atom of the next and either an atom of helium (α-particle) or a free electron (β-particle). The emitted particle leaves the atom and has no effect on the later stages. The number of atoms of any element that break up in a short interval of time is proportion to the time interval and to the number of atoms of that element present.* If $u, x_1, x_2, \ldots, x_n$ are the expectations of the numbers of atoms of the various elements present at time t, they will satisfy the differential equations

$$\left.\begin{aligned}\frac{du}{dt} &= -\kappa u, \\ \frac{dx_1}{dt} &= \kappa u - \kappa_1 x_1, \\ \frac{dx_2}{dt} &= \kappa_1 x_1 - \kappa_2 x_2, \\ &\cdots\cdots\cdots\cdots\cdots \\ \frac{dx_n}{dt} &= \kappa_{n-1} x_{n-1}.\end{aligned}\right\} \tag{1}$$

* Strictly speaking, since radioactivity is a random process, this rule is true of the expectation of the number of atoms breaking up. The actual number will deviate somewhat from expectation, but we can neglect the difference if the expectation is large.

Suppose that at $t = 0$ only uranium is present; then $u = u_0$, and all the other dependent variables are zero. The subsidiary equations are

$$\begin{aligned}
(p+\kappa)u &= pu_0, \\
(p+\kappa_1)x_1 &= \kappa u, \\
(p+\kappa_2)x_2 &= \kappa_1 x_1, \\
&\cdots \\
(p+\kappa_{n-1})x_{n-1} &= \kappa_{n-2}x_{n-2}, \\
px_n &= \kappa_{n-1}x_{n-1},
\end{aligned} \quad (2)$$

and the operational solutions are written down immediately:

$$u = \frac{pu_0}{p+\kappa}, \quad x_1 = \frac{\kappa p u_0}{(p+\kappa)(p+\kappa_1)}, \quad x_2 = \frac{\kappa \kappa_1 p u_0}{(p+\kappa)(p+\kappa_1)(p+\kappa_2)}, \quad \ldots,$$

$$x_n = \frac{\kappa \kappa_1 \ldots \kappa_{n-1} u_0}{(p+\kappa)(p+\kappa_1) \ldots (p+\kappa_{n-1})}. \quad (3)$$

These are directly adapted for interpretation by the partial-fraction rule; in fact

$$\begin{aligned}
u &= u_0 e^{-\kappa t}, \quad x_1 = \frac{\kappa u_0}{\kappa_1 - \kappa}(e^{-\kappa t} - e^{-\kappa_1 t}), \\
x_2 &= \kappa \kappa_1 u_0 \left\{ \frac{1}{(\kappa_1-\kappa)(\kappa_2-\kappa)} e^{-\kappa t} + \frac{1}{(\kappa-\kappa_1)(\kappa_2-\kappa_1)} e^{-\kappa_1 t} + \frac{1}{(\kappa-\kappa_2)(\kappa_1-\kappa_2)} e^{-\kappa_2 t} \right\},
\end{aligned} \quad (4)$$

$$x_n = u_0 - \frac{\kappa_1 \ldots \kappa_{n-1} u_0}{(\kappa_1-\kappa) \ldots (\kappa_{n-1}-\kappa)} e^{-\kappa t} - \ldots. \quad (5)$$

Of all the decay constants κ is much the smallest. If the time elapsed is long enough for all the exponential factors except $e^{-\kappa t}$ to have become insignificant, the results reduce approximately to

$$u = u_0 e^{-\kappa t}, \quad x_1 = \frac{\kappa}{\kappa_1} u_0 e^{-\kappa t}, \quad x_2 = \frac{\kappa}{\kappa_2} u_0 e^{-\kappa t}, \quad \ldots, \quad (6)$$

$$x_n = u_0(1 - e^{-\kappa t}). \quad (7)$$

With the exception of the last, the quantities of the various elements decrease, retaining constant ratios to one another in the inverse ratios of their decay constants.

On the other hand, if the time elapsed is so short that unity is still a first approximation to all the exponential functions, we can proceed by expanding the operators in descending powers of p and interpreting term by term. Hence at first x_1 will increase in proportion to t, x_2 to t^2, and x_n to t^n.

In experimental work an intermediate condition often occurs. Some of the exponentials may become insignificant in the time taken by the experiment, while others are still nearly unity. We have

$$x_r = \frac{\kappa_{r-1} x_{r-1}}{p+\kappa_r} = \kappa_{r-1}(p^{-1}x_{r-1} - \kappa_r p^{-2}x_{r-1} + \ldots), \quad (8)$$

and if $\kappa_r t$ is small we can neglect the second and later terms in comparison with the first. Hence in this case

$$x_r \doteqdot \kappa_{r-1} p^{-1} x_{r-1}. \quad (9)$$

If x_{r-1} is of the form t^s, we have from (8)

$$x_r = \frac{\kappa_{r-1} s!}{(p+\kappa_r)p^s} = \frac{\kappa_{r-1} s!}{p^{s+1}} \frac{p}{p+\kappa_r} = \frac{\kappa_{r-1} s!}{p^{s+1}} e^{-\kappa_r t}. \tag{10}$$

If $\kappa_r t$ is small we can replace the exponential by 1 and confirm (8). But if it is great

$$p^{-1} e^{-\kappa_r t} = \int_0^t e^{-\kappa_r t} dt = \frac{1}{\kappa_r} + O(e^{-\kappa_r t}), \tag{11}$$

and on continuing the integrations

$$p^{-s-1} e^{-\kappa_r t} \doteqdot p^{-s} \frac{1}{\kappa_r} = \frac{1}{\kappa_r} \frac{t^s}{s!}. \tag{12}$$

Hence
$$x_r \doteqdot \frac{\kappa_{r-1}}{\kappa_r} t^s = \frac{\kappa_{r-1}}{\kappa_r} x_{r-1}. \tag{13}$$

Classifying elements into long-lived and short-lived according as $\kappa_r t$ is small or large for them, t being the duration of the experiment in question, we find that the quantity of the first long-lived disintegration product increases in proportion to t, the second to t^2, and so on. Short-lived products vary nearly in proportion to the previous long-lived one. All β-ray products are short-lived when t has ordinary values.

Radium is the third α-ray disintegration product of uranium. In rock specimens the time elapsed since formation is usually such that the relations (6) have become established. As a matter of observation the numbers of atoms of radium and uranium are found to be in the constant ratio $3 \cdot 58 \times 10^{-7}$. This determines κ/κ_3. Also the rate of break-up of radium is known directly; in fact

$$1/\kappa_3 = 2280 \text{ years.}$$

Hence
$$1/\kappa = 6 \cdot 37 \times 10^9 \text{ years.}$$

This gives the rate of disintegration of uranium itself.*

A number of specimens of uranium compounds were carefully freed from radium by Soddy, and then kept for ten years. It was found that new radium was formed, increasing like the square of the time. This would suggest that of the two elements between uranium and radium in the series one was long-lived (in comparison with ten years) and the other short-lived. Actually, however, it is known independently that both are long-lived. The first, however, is chemically inseparable from ordinary uranium, and therefore was present in the original specimens; initially, instead of $x_1 = 0$, we have

$$x_1 \doteqdot \frac{\kappa}{\kappa_1} u_0.$$

For the next element, ionium, we have

$$x_2 \doteqdot \kappa_1 p^{-1} x_1 \doteqdot \kappa u_0 t,$$

and for radium
$$x_3 = \kappa_2 p^{-1} x_2 = \tfrac{1}{2} \kappa \kappa_2 u_0 t^2.$$

* The numerical data used here have been revised in later experimental determinations, but it has also been found that the series branch and reunite to some extent, so that to take the more recent results into account would complicate the analysis without introducing any new principle.

Soddy* found that 3 kg. of uranium in 10·15 years gave 202×10^{-12} g. of radium. Hence, allowing for the difference of atomic weights,

$$x_3/u_0 = 7 \cdot 1 \times 10^{-14},$$

and, κ being known,

$$\kappa_2 = 8 \cdot 64 \times 10^{-6}/\text{year}, \quad 1/\kappa_2 = 1 \cdot 16 \times 10^5 \text{ years}.$$

This gives the rate of degeneration of ionium. Soddy gets a slightly lower value of $1/\kappa_2$ from more numerous data.

In another case that sometimes occurs in experiment the specimen has been found in nature with the various products in approximately the ratios indicated by (6), but the uranium and possibly some later products are then removed chemically, and the behaviour of the remainder is studied. A solution for this case has been given by W. F. Sedgwick.† The operational treatment, the possibility of which was suggested by A. F. Crossley, is as follows. Let x_s at $t = 0$ be 0 for $s < r$, and u_s for $s \geqslant r$, where $u_s = \dfrac{\kappa_r}{\kappa_s} u_r$, except for $r = n$.

The subsidiary equations are now

$$(p + \kappa_r) x_r = p u_r, \tag{14}$$

$$(p + \kappa_s) x_s = p u_s + \kappa_{s-1} x_{s-1} \quad (r < s < n), \tag{15}$$

$$p x_n = p u_n + \kappa_{n-1} x_{n-1}. \tag{16}$$

We have for $r < s < n$

$$(p + \kappa_s) x_s = (p + \kappa_s) u_s - \kappa_s u_s + \kappa_{s-1} x_{s-1}$$
$$= (p + \kappa_s) u_s - \kappa_{s-1}(u_{s-1} - x_{s-1}), \tag{17}$$

that is, if

$$u_s - x_s = y_s, \tag{18}$$

$$(p + \kappa_s) y_s = \kappa_{s-1} y_{s-1}, \tag{19}$$

with

$$(p + \kappa_r) y_r = \kappa_r u_r, \quad p y_n = -\kappa_{n-1} u_{n-1} + \kappa_{n-1} y_{n-1}. \tag{20}$$

The operational solutions are therefore

$$y_r = \frac{\kappa_r u_r}{p + \kappa_r}, \quad y_{r+1} = \frac{\kappa_r^2 u_r}{(p + \kappa_r)(p + \kappa_{r+1})}, \quad y_{r+2} = \frac{\kappa_r^2 \kappa_{r+1} u_r}{(p + \kappa_r)(p + \kappa_{r+1})(p + \kappa_{r+2})},$$

$$y_n = -\frac{1}{p} \kappa_r u_r + \frac{1}{p} \frac{\kappa_r^2 \kappa_{r+1} \ldots \kappa_{n-2} u_r}{(p + \kappa_r) \ldots (p + \kappa_{n-1})}. \tag{21}$$

Therefore

$$y_r = u_r(1 - e^{-\kappa_r t}), \quad x_r = u_r e^{-\kappa_r t}; \tag{22}$$

and since $\kappa_r u_r = \kappa_s u_s$ each y_s except y_n tends to u_s when t tends to infinity, by the partial fraction rule. Hence each x_s except x_n tends to 0, as would be expected;

$$x_{r+1} = u_{r+1} \left(\frac{\kappa_{r+1}}{\kappa_r(\kappa_{r+1} - \kappa_r)} e^{-\kappa_r t} + \frac{\kappa_{r+1}}{\kappa_{r+1}(\kappa_r - \kappa_{r+1})} e^{-\kappa_{r+1} t} \right)$$

$$= \frac{u_{r+1}}{\kappa_{r+1} - \kappa_r} \left(\frac{\kappa_{r+1}}{\kappa_r} e^{-\kappa_r t} - e^{-\kappa_{r+1} t} \right), \tag{23}$$

and so on.

* *Phil. Mag.* (6) 38, 1919, 483–88. † *Proc. Camb. Phil. Soc.* 38, 1942, 283.

The interesting case occurs when the duration of the experiment is such that some of the earlier exponentials have time to become small. The short-lived earlier products then disappear during the experiment. If the first not to become small is $e^{-\kappa_m t}$, we have approximately

$$x_m = \frac{\kappa_r \kappa_{r+1} \cdots \kappa_m u_m}{\kappa_m (\kappa_r - \kappa_m)(\kappa_{r+1} - \kappa_m) \cdots (\kappa_{m-1} - \kappa_m)} e^{-\kappa_m t} \doteq u_m e^{-\kappa_m t}, \tag{24}$$

so that this element decays nearly as if the others were not present; and the decay of later elements in the series will follow the same law so long as there is no intervening element with $\kappa_s \leqslant \kappa_m$. For those with longer lives, however, x_s will contain a term in $e^{-\kappa_s t}$, which may be larger.

EXAMPLES*

1. A light string of length $3l$ is stretched under tension P between two fixed points. Masses $5m$ and $8m$ are attached at the points of trisection. A small transverse velocity u is given to the particle of mass $5m$. Prove that the displacement of the other particle is

$$\frac{5}{14} \frac{u}{\alpha} \left(\sqrt{\frac{20}{3}} \sin \sqrt{\frac{3}{20}} \alpha t - \sqrt{2} \sin \frac{\alpha t}{\sqrt{2}} \right),$$

where $\alpha^2 = P/ml$. (M.T. 1929.)

2. A Galitzin seismograph is so adjusted that

$$\kappa_1 = \kappa_2 = \kappa, \quad n_1^2 = n_2^2 = 2\kappa^2.$$

Prove that the response to a unit impulsive change of velocity is

$$\frac{\lambda \mu}{2\kappa^2} (\kappa t \sin \kappa t - \sin \kappa t + \kappa t \cos \kappa t) e^{-\kappa t}.$$

3. Prove that if x is the displacement on the record given by a Galitzin seismograph, due to an impulsive change of velocity of the ground,

$$\int_0^\infty x \, dt = 0$$

whatever the constants of the instrument may be.

* Numerous examples are given by G. W. Carter, *The Simple Calculation of Electrical Transients*, 1944.

Chapter 9

NUMERICAL METHODS

I have no satisfaction in formulas unless I feel their numerical magnitude.
LORD KELVIN, *Life by Sylvanus Thompson*, p. 827

9·01. Approximation by polynomials. The characteristic feature of most numerical methods is that values of a function $f(x)$ are given for a set of distinct values of x, but not for intermediate values; for purposes of computation these are filled in on the hypothesis that $f(x)$ can be replaced by a polynomial agreeing with $f(x)$ at the places where its values are given. The simplest case is that of linear interpolation, in which only two adjacent values of $f(x)$ are taken from a table and intermediate values are calculated on the supposition that $f'(x)$ is constant in the interval. This procedure is accurate provided that $f'(x)$ changes little in the interval, but cases often arise that require allowance for higher derivatives. The use of a polynomial for fitting can never be mathematically exact unless $f(x)$ is itself a polynomial, but it can, in suitable circumstances, be as accurate as the tabulated values themselves.

9·011. Lagrange's interpolation formula.* Let $f(x)$ be given for $x = x_1, x_2, ..., x_n$. Then the function

$$g(x) = f(x_1)\frac{(x-x_2)(x-x_3)...(x-x_n)}{(x_1-x_2)(x_1-x_3)...(x_1-x_n)} + f(x_2)\frac{(x-x_1)(x-x_3)...(x-x_n)}{(x_2-x_1)(x_2-x_3)...(x_2-x_n)} + ...$$
$$+ f(x_n)\frac{(x-x_1)...(x-x_{n-1})}{(x_n-x_1)...(x_n-x_{n-1})}, \quad (1)$$

tends to $f(x_1)$ for $x = x_1$, to $f(x_2)$ for $x = x_2$, and so on. Also it is a polynomial of degree $n-1$. It is symmetrical in the sense that it is unaltered by any interchange of the suffixes; the tabulated values can therefore be taken in any order.

Most interpolation formulae can be derived from this. It is not usually convenient on account of the fact that in practice $g(x)$ will usually be determined mainly by the adjacent tabulated values, so that linear interpolation will need only a small correction; but all the arguments appear symmetrically in (1) and the contributions from all terms will need to be taken into account. Computations are made easier by using a form that makes the special dependence on neighbouring values explicit and therefore by abandoning the symmetry.

9·012. Divided differences. The values of x_r and $f(x_r)$ are first arranged in a table $x_1 ... x_n$. For any two consecutive arguments x_r and x_{r+1} we form the ratio

$$f_1(x_r x_{r+1}) = [x_r x_{r+1}] = \frac{f(x_{r+1}) - f(x_r)}{x_{r+1} - x_r}. \quad (2)$$

---

* Really due to E. Waring, *Phil. Trans.* 69 (1779), 59–67; Euler rediscovered it in 1783. Lagrange's publication was in 1795. It must however have been obvious to Newton. For historical references see Karl Pearson, *Tracts for Computers*, 2 (1920).

The notations f_1 and $[x_r x_{r+1}]$ are both in frequent use. This ratio is called the *first divided difference*. We then form

$$f_2(x_r x_{r+1} x_{r+2}) = [x_r x_{r+1} x_{r+2}] = \frac{[x_{r+1} x_{r+2}] - [x_r x_{r+1}]}{x_{r+2} - x_r}. \tag{3}$$

This is the *second divided difference*; so we proceed, the divisor at each stage being the difference of the two values of x each used in only one of the differences subtracted. Now consider a general x not equal to any of $x_1, x_2, \ldots, x_n$. The divided differences involving x exist; and by definition

$$[xx_1] = \frac{f(x_1) - f(x)}{x_1 - x}, \quad \text{whence} \quad f(x) = f(x_1) + [xx_1](x - x_1), \tag{4}$$

$$[xx_1 x_2] = \frac{[x_1 x_2] - [xx_1]}{x_2 - x}, \quad \text{whence} \quad [xx_1] = [x_1 x_2] + [xx_1 x_2](x - x_2), \tag{5}$$

$$\cdots\cdots\cdots\cdots\cdots\cdots\cdots\cdots\cdots\cdots\cdots\cdots\cdots\cdots$$

$$[xx_1 x_2 \ldots x_n] = \frac{[x_1 x_2 \ldots x_n] - [xx_1 x_2 \ldots x_{n-1}]}{x_n - x}, \tag{6}$$

whence
$$[xx_1 x_2 \ldots x_{n-1}] = [x_1 \ldots x_n] + [xx_1 x_2 \ldots x_n](x - x_n). \tag{7}$$

Substitute for $[xx_1]$ in the first identity its value given by the second; we get a three-term relation involving $[xx_1 x_2]$. Substitute for $[xx_1 x_2]$ its value given by the third identity and proceed. We have finally

$$f(x) = f(x_1) + (x - x_1)\{[x_1 x_2] + (x - x_2)\{[x_1 x_2 x_3] + \{\ldots + (x - x_{n-1})[x_1 x_2 \ldots x_n]\}\ldots\}\} + R(x), \tag{8}$$

where
$$R(x) = [xx_1 x_2 \ldots x_n](x - x_1)(x - x_2) \ldots (x - x_n). \tag{9}$$

Expanding the series we have

$$f(x) = f(x_1) + (x - x_1)[x_1 x_2] + (x - x_1)(x - x_2)[x_1 x_2 x_3] + \ldots$$
$$+ (x - x_1) \ldots (x - x_{n-1})[x_1 x_2 \ldots x_n] + R(x) \tag{10}$$

$$= P(x) + R(x), \tag{11}$$

say. This is a pure identity arising out of the definition of divided differences. Its utility depends on the value of $R(x)$, which is not known from the definitions unless $f(x)$ is; but if we can fix limits to $R(x)$ otherwise, we thereby get limits to the error involved in omitting it. Then $P(x)$ will be a polynomial of degree $n - 1$ representing $f(x)$ with assignable accuracy.

Now consider the divided differences of x^r, where r is an integer. We have

$$[xx_1] = \frac{x_1^r - x^r}{x_1 - x} = x_1^{r-1} + x_1^{r-2} x + \ldots + x^{r-1}, \tag{12}$$

which is a polynomial of degree $r - 1$. This can be extended at once to any polynomial of degree r. Therefore the rth divided difference of a polynomial of degree r is a constant and all higher ones are zero.

Now $f(x)$ is equal to Lagrange's interpolation function $g(x)$ whenever $x = x_1, x_2, \ldots, x_n$. $g(x)$ therefore has the same divided differences based on those n points as $f(x)$. Consequently if we apply (10) to $g(x)$ we shall obtain

$$g(x) = P(x) + [xx_1 \ldots x_n](x - x_1) \ldots (x - x_{n-1})(x - x_n), \tag{13}$$

the divided difference in the last term being the nth divided difference of $g(x)$. But $g(x)$ is a polynomial of degree $n-1$ and therefore its nth divided difference is 0. Hence

$$g(x) = P(x) \qquad (14)$$

for all values of x. Therefore if we define $R(x)$ as $f(x) - P(x)$, $R(x) = 0$ for $x = x_1, x_2 \ldots x_n$. This is not obvious from the occurrence of a zero factor in (9), because the divided difference is not defined when two entries coincide.

Now suppose that $f(x)$ has derivatives up to the nth in a range (a, b). Then the same applies to $R(x)$, since $g(x)$ is a polynomial. Suppose $x_1, x_2, \ldots, x_n$ arranged in ascending order. Since $g(x)$ is symmetrical this does not affect $R(x)$. Then by Rolle's theorem, since $R(x) = 0$ for $x = x_1$ and $x = x_2$, $R'(x) = 0$ for some intermediate x; similarly $R'(x) = 0$ for some x in each of the ranges x_2 to x_3, ..., x_{n-1} to x_n. Again applying Rolle's theorem we see that $R''(x) = 0$ for $n-2$ values of x between x_1 and x_n, and proceeding we have $R^{(n-1)}(x) = 0$ for one intermediate value, say $x = \xi$. But by differentiating (8) we have

$$f^{(n-1)}(x) = (n-1)! [x_1 x_2 \ldots x_n] + R^{(n-1)}(x), \qquad (15)$$

and therefore
$$f^{(n-1)}(\xi) = (n-1)! [x_1 x_2 \ldots x_n]. \qquad (16)$$

Hence there is at least one value of x between x_1 and x_n such that the $(n-1)$th divided difference is $1/(n-1)!$ times the $(n-1)$th derivative of $f(x)$. This result is true for all n; hence we can replace $n-1$ by n and infer that

$$[xx_1 \ldots x_n] = \frac{f^{(n)}(\eta)}{n!}, \qquad (17)$$

where η is within the range whose end-points are the least and greatest of x, x_1 and x_n. Then returning to (9) we have

$$R(x) = \frac{1}{n!} f^{(n)}(\eta) (x-x_1)(x-x_2) \ldots (x-x_n). \qquad (18)$$

If we can fix limits to the nth derivative in any range of n values of the argument, we can therefore fix a limit to the error introduced by using $P(x)$ instead of $f(x)$. The result of neglecting $R(x)$ in (10) is *Newton's interpolation formula*.

It is obvious from successive applications of (12) that the nth divided difference of x^n is 1. It also follows from (17). For if $f(x) = x^n$ the right of (17) is 1 for all x; and the numerators of all higher differences are 0. This fact is convenient in the fitting of power series to given values of a function. For if the nth divided differences are all found to be a_n, a_n is the coefficient of x^n. We subtract $a_n x^n$ from all tabulated values, and again form divided differences. If the arithmetic has been done correctly the differences of order $n-1$ will be constant, and this value will be the coefficient of x^{n-1}, and so on. The process is self-checking, any arithmetical mistake being detected in the next stage of the calculation.

From the form (18) it is clear that if $f^{(n)}(x)$ does not vary greatly $R(x)$ will be least if the tabular values used, x_1 to x_n, are as nearly as possible symmetrically placed about x. The formula is valid for any set of tabular values, but the error inevitable if $f^{(n)}(x)$ is not zero is much less for interpolation than for extrapolation. Similarly, we shall ordinarily get greater accuracy in interpolating in the middle of a table than near the ends. The table of differences will be set out as follows:

		f_1	f_2	f_3
x_1	$f(x_1)$			
		$[x_1x_2]$		
x_2	$f(x_2)$		$[x_1x_2x_3]$	
		$[x_2x_3]$		$[x_1x_2x_3x_4]$
x_3	$f(x_3)$		$[x_2x_3x_4]$	
		$[x_3x_4]$		$[x_2x_3x_4x_5]$
x_4	$f(x_4)$		$[x_3x_4x_5]$	
		$[x_4x_5]$		$[x_3x_4x_5x_6]$
x_5	$f(x_5)$		$[x_4x_5x_6]$	

If we wish to interpolate between x_3 and x_4 the contributions from f_3 will be least if we use the differences $[x_3x_4]$, $[x_2x_3x_4]$, and $[x_2x_3x_4x_5]$, or any other sequence that ends in the same third difference. It is necessary that the second difference used shall be one of the two used in forming the third difference, and that the first difference used shall be one of those used in forming the second difference used. Otherwise it is irrelevant what route we choose so long as we end with the same third difference; the results will always be values of the cubic polynomial that agrees with $f(x)$ at $x = x_2, x_3, x_4, x_5$. But the arithmetic is easier if we keep as nearly as possible to a horizontal line of the table.

The form (10) is usually less convenient than (8); the higher divided differences are usually small and it is troublesome to keep track of the decimal point when they are multiplied by two or more factors. Using (8) we begin at the end and calculate $(x-x_{n-1})[x_1x_2...x_n]$. We add this to $[x_1x_2...x_{n-1}]$; multiply the result by $(x-x_{n-2})$, add to $[x_1x_2...x_{n-2}]$ and so work back to the beginning. A better way still is to compute the first two terms directly; these represent the result of linear interpolation, and can be built up directly on a multiplying machine. First $f(x_1)$ is set up and transferred to the product register by one turn of the handle. Then $[x_1x_2]$ is set up and multiplied into all values of $x-x_1$ required up to and including x_2-x_1. The last should give $f(x_2)$ and check the calculation of $[x_1x_2]$. The multiplier register should be cleared before multiplication begins so that the successive values of $x-x_1$ can be read directly on it. We then write (8) in the form

$$f(x) = \{f(x_1) + (x-x_1)[x_1x_2]\} + (x-x_1)(x-x_2)\{[x_1x_2x_3] + (x-x_3)\{[x_1x_2x_3x_4] + ...\}\}.$$

The last batch of terms will be a small correction in most cases, and those in the first brackets have already been built up on the machine. It is desirable to work to one more figure than is given in the data in order to prevent accumulation of rounding-off errors.

The standard numerical methods all depend on replacing $f(x)$ by the interpolation polynomial $P(x)$, which is of degree one less than the number of data. In general $f(x) \neq P(x)$ except at the datum values, but the difference lies within assignable limits. $P(x)$ is the *smoothest* function that agrees with $f(x)$ at the required points, since $d^n P(x)/dx^n = 0$ for all x; this would not be true of any other function.

9·013. As a specimen of the method let us take some irregularly spaced values of $\sin x°$ and interpolate to multiples of $5°$. The data and the divided differences are as follows:

x	$\sin x°$	f_1	f_2	f_3
0	0·0000			
		$0·2250/13 = 0·01731$		
13	0·2250		$-0·00079/24 = -0·000033$	
		$0·1817/11 = 0·01652$		$-0·000030/37 = -0·0000008$
24	0·4067		$-0·00151/24 = -0·000063$	
		$0·1951/13 = 0·01501$		$-0·000031/41 = -0·0000008$
37	0·6018		$-0·00282/30 = -0·000094$	
		$0·2072/17 = 0·01219$		$-0·000026/43 = -0·0000006$
54	0·8090		$-0·00361/30 = -0·000120$	
		$0·1115/13 = 0·00858$		$-0·000020/42 = -0·0000005$
67	0·9205		$-0·00349/25 = -0·000140$	
		$0·0611/12 = 0·00509$		$-0·000009/36 = -0·0000002$
79	0·9816		$-0·00342/23 = -0·000149$	
		$0·0184/11 = 0·00167$		
90	1·0000			

9·013–9·02   *Equal intervals*   265

The interpolation runs as follows. The second column gives the part built up by linear interpolation between the two adjacent datum values, the third the correction for f_2 and f_3.

x	Linear formula	Correction	$\sin x°$	Correct value
5	0·08655	$- 5 \times 8(-0·000033+19\times 0·0000008) = 40 \times 0·000018 = 0·00072$	0·08727	0·0872
10	0·17310	$-10 \times 3(-0·000033+14\times 0·0000008) = 30 \times 0·000022 = 0·00066$	0·17376	0·1736
15	0·25804	$- 2 \times 9(-0·000033-15\times 0·0000008) = 18 \times 0·000045 = 0·00081$	0·25885	0·2588
20	0·34064	$- 7 \times 4(-0·000033-20\times 0·0000008) = 28 \times 0·000052 = 0·00137$	0·34201	0·3420
25	0·42171	$- 1 \times 12(-0·000063-12\times 0·0000008) = 12 \times 0·000073 = 0·00088$	0·42259	0·4226
30	0·49676	$- 6 \times 7(-0·000063-17\times 0·0000008) = 42 \times 0·000077 = 0·00323$	0·49999	0·5000
35	0·57181	$-11 \times 2(-0·000063-22\times 0·0000008) = 22 \times 0·000081 = 0·00178$	0·57359	0·5736
40	0·63837	$- 3 \times 14(-0·000094-16\times 0·0000006) = 42 \times 0·000104 = 0·00437$	0·64274	0·6428
45	0·69932	$- 8 \times 9(-0·000094-21\times 0·0000006) = 72 \times 0·000107 = 0·00770$	0·70702	0·7071
50	0·76027	$-13 \times 4(-0·000094-26\times 0·0000006) = 52 \times 0·000110 = 0·00572$	0·76599	0·7660
55	0·81758	$- 1 \times 12(-0·000120-18\times 0·0000005) = 12 \times 0·000129 = 0·00155$	0·81913	0·8192
60	0·86048	$- 6 \times 7(-0·000120-23\times 0·0000005) = 42 \times 0·000132 = 0·00554$	0·86602	0·8660
65	0·90338	$-11 \times 2(-0·000120-28\times 0·0000005) = 22 \times 0·000134 = 0·00295$	0·90633	0·9063
70	0·93577	$- 3 \times 9(-0·000140-16\times 0·0000002) = 27 \times 0·000143 = 0·00386$	0·93963	0·9397
75	0·96122	$- 8 \times 4(-0·000140-21\times 0·0000002) = 32 \times 0·000144 = 0·00461$	0·96583	0·9659
80	0·98327	$- 1 \times 10(-0·000149-13\times 0·0000002) = 10 \times 0·000152 = 0·00152$	0·98479	0·9848
85	0·99162	$- 6 \times 5(-0·000149-18\times 0·0000002) = 30 \times 0·000153 = 0·00459$	0·99621	0·9962

The results of the interpolation are given in the last column but one, and the values taken directly from the tables in the last. It will be seen that the difference only once exceeds a unit in the fourth place, and is fully accounted for by the fact that errors from neglect of the fifth decimal would run up to half a unit in the fourth place both in the datum values and in those used for comparison at the end. At 45° the contribution from f_3 amounts to 10 units in the fourth place, and an error of half a unit in the last figure of f_3 would contribute 0·7 in the fourth place of the interpolate. It is only for rather wide intervals such as these that the third difference matters in interpolation to four figures.

Since the first difference used in each case is based on the two adjacent values, the coefficient of f_2 is always negative. From 15° onwards the second difference used is that given on the same horizontal line as the *beginning* of the interval, and the extra datum value used in forming it is the one *before* the beginning of the interval. Consequently $x-x_3$, the new factor multiplying f_3, is positive. This is not possible in the first interval; the second difference used is opposite the end of the interval and involves the datum at 24°. Hence the factor multiplying f_3 is negative in this range.

An increase of accuracy is sometimes possible if the derivative of the function is known for some value in the range. Here, for instance, we know that at 90° the derivative of $\sin x$ is 0. This can be treated by extending the table one line as follows:

x	$\sin x$	f_1	f_2	f_3
67	0·9205			
		0·00509		
79	0·9816		$-0·000149$	
		0·00167		$-0·000003/23 = -0·0000001$
90	1·0000		$-0·000152$	
		0·00000		
90	1·0000			

Between 79° and 90° we can now use the formula

$$\sin x = 1·0000 - 0·000152(x-90)^2 - 0·0000001(x-90)^2(x-79).$$

9·02. Interpolation with equal intervals. The formation of divided differences is rather laborious, but cannot be avoided when the intervals of the argument and the function are both irregular. If the intervals of the argument are all equal it can be replaced by simple subtraction. Two classes of formulae are available: the *Gregory formula* and what is called the *Gregory-backwards formula*, on the one hand, and the various *central difference formulae* on the other. The Gregory formula corresponds to the method that we used near the beginning of the range in the above illustration, whereas the central difference formulae use as far as possible differences near the same horizontal line of the table. For the same reason as with divided differences the latter will be the better when they can be used, since the higher differences will be multiplied by smaller factors.

With equal intervals we form the differences as follows:

$$\Delta y_r = y_{r+1} - y_r, \quad \Delta^2 y_r = \Delta y_{r+1} - \Delta y_r = y_{r+2} - 2y_{r+1} + y_r, \quad \text{etc.}$$

each difference being formed by subtracting the two nearest entries to its left. This notation is the most convenient when the Gregory formula is being used. With central differences and backward differences other notations are more convenient, but the actual entry in each position in the table is the same:

		Forward differences			Central differences			Backward differences		
$x_0 - 2h$	y_{-2}									
$x_0 - h$	y_{-1}	Δy_{-2}	$\Delta^2 y_{-2}$		$\delta y_{-3/2}$	$\delta^2 y_{-1}$	$\delta^3 y_{-1/2}$	∇y_{-1}	$\nabla^2 y_0$	$\nabla^3 y_1$
x_0	y_0	Δy_{-1}	$\Delta^2 y_{-1}$	$\Delta^3 y_{-2}$	$\delta y_{-1/2}$	$\delta^2 y_0$	$\delta^3 y_{1/2}$	∇y_0	$\nabla^2 y_1$	$\nabla^3 y_2$
$x_0 + h$	y_1	Δy_0	$\Delta^2 y_0$	$\Delta^3 y_{-1}$	$\delta y_{1/2}$	$\delta^2 y_1$	$\delta^3 y_{3/2}$	∇y_1	$\nabla^2 y_2$	$\nabla^3 y_3$
$x_0 + 2h$	y_2	Δy_1	$\Delta^2 y_1$	$\Delta^3 y_0$	$\delta y_{3/2}$	$\delta^2 y_2$		∇y_2	$\nabla^2 y_3$	

It is evident from the mode of formation that the first divided differences in corresponding positions would be $\Delta y/h$, the second $\Delta^2 y/2h^2$, and the nth $\Delta^n y/n!h^n$. Then if we use differences based on $x_0, x_0 + h, x_0 + 2h, \ldots$ we have at once from Newton's formula

$$y(x_0 + \theta h) = y_0 + \theta h \frac{\Delta y_0}{h} + \theta h (\theta h - h) \frac{1}{2!} \frac{\Delta^2 y_0}{h^2} + \ldots$$

$$= y_0 + \theta \Delta y_0 + \frac{\theta(\theta - 1)}{2!} \Delta^2 y_0 + \ldots + \frac{\theta(\theta - 1) \ldots (\theta - n + 1)}{n!} \Delta^n y_0 + R_{n+1}, \quad (1)$$

where R_{n+1} is found from the remainder in 9·012 (18) to be

$$\frac{\theta(\theta - 1) \ldots (\theta - n)}{(n+1)!} h^{n+1} \left(\frac{d^{n+1} y}{dx^{n+1}} \right)_{x = \eta}.$$

This is Gregory's formula, discovered by James Gregory in 1670; Newton's more general formula was published in 1687. We see that it has the form of a binomial series, y_0 being preceded by the operator $(1 + \Delta)^\theta$. It has in fact an operational interpretation. If we define D as meaning d/dx, Taylor's theorem may be written

$$f(x + \alpha) = \left(1 + \alpha \frac{d}{dx} + \frac{\alpha^2}{2!} \frac{d^2}{dx^2} + \ldots \right) f(x) = e^{\alpha D} f(x). \quad (2)$$

If we also write
$$f(x + h) = E f(x) = (1 + \Delta) f(x) = e^{hD} f(x), \quad (3)$$

we have
$$f(x + \theta h) = e^{\theta h D} f(x) = (1 + \Delta)^\theta f(x). \quad (4)$$

These operators occurring in interpolation theory are fundamentally different from those of Heaviside's methods; here the fundamental operator is D, whereas in Heaviside's methods it is p^{-1}, which is not simply the inverse of D because the two do not commute. The justification of their use is therefore quite different. Expansion in powers of p^{-1} is

9·03 Gregory's extrapolation formula

justifiable in a much wider class of cases than that in powers of D. The infinite Taylor series becomes meaningless if the function operated on has no derivative above a certain finite order at some point of the range, but the p^{-1} series requires nothing more of the function than that it should be integrable. The justification here rests on the fact that in interpolation the function is replaced by a polynomial $P(x)$, with an accuracy fixed by the lowest difference neglected. The terms of the Taylor series of $P(x)$ containing D^n and higher powers are all zero if only differences to order $n-1$ are used. The operational process in powers of D is therefore valid in problems of interpolation because it is carried out only on the interpolation polynomial, not on the original function, of which the polynomial is only an approximate representation of known accuracy.

The binomial theorem can be derived from Gregory's formula. Take intervals 1 of the exponent in $(1+x)^n$ for given x; the difference table reads:

n	$f(n)$	$\Delta f(n)$	$\Delta^2 f(n)$	$\Delta^3 f(n)$
0	1			
1	$1+x$	x		
2	$(1+x)^2$	$x(1+x)$	x^2	
3	$(1+x)^3$	$x(1+x)^2$	$x^2(1+x)$	x^3

and the Gregory formula based on $f(0)$ and its differences reads

$$f(n) = 1 + nx + \frac{n(n-1)}{2!}x^2 + \ldots + \frac{n(n-1)\ldots(n-r+1)}{r!}x^r(1+\eta), \qquad (5)$$

$$0 < \eta < \frac{n-r}{r+1}(1+x)^n x \left\{\frac{\log(1+x)}{x}\right\}^{r+1},$$

which is the binomial theorem for a real fractional index.*

9·03. In the *Gregory backwards formula* the differences *ascending* diagonally from x_0 are used. We have

$$f(x_0 + \theta h) = f_0 + \theta \Delta f_{-1} + \frac{\theta(\theta+1)}{2!}\Delta^2 f_{-2} + \ldots. \qquad (6)$$

This also can be easily derived operationally. If we write

$$f(x_r) - f(x_{r-1}) = \nabla f(x_r), \qquad (7)$$

we have
$$E\nabla = \Delta = E - 1, \qquad (8)$$

whence
$$E = \frac{1}{1-\nabla}, \qquad (9)$$

$$f(x+\theta h) = E^\theta f(x) = (1-\nabla)^{-\theta} f(x) \qquad (10)$$

$$= f(x) + \theta \nabla f(x) + \frac{\theta(\theta+1)}{2!}\nabla^2 f(x) + \ldots. \qquad (11)$$

---

* It appears to be established that James Gregory knew and used Taylor's theorem as early as 1670, and therefore had another approach to the binomial theorem at that time. He apparently did not publish it on account of a mistaken belief that Newton must have found it too. Brook Taylor's publication of the theorem was in 1712, that of the so-called Maclaurin's theorem in 1742. It is incredible that between the latter two dates nobody thought of putting $a = 0$ in Taylor's theorem. Maclaurin has two better titles to fame, namely his independent discovery of the Euler-Maclaurin expansion and of the 'Maclaurin ellipsoids' in hydrodynamics. Cf. H. W. Turnbull, *James Gregory Tercentenary Volume* (1939) and *Mathematical Discoveries of Newton* (1945); Bell, *Development of Mathematics*.

This formula is really an extrapolation formula, since it can be used to infer values of the function beyond the range tabulated; of course with the usual increase of inaccuracy due to the extra range. The same applies to the Gregory formula for $\theta < 0$.

9·04. Using the central difference notation and a zigzag set of differences following a horizontal line as closely as possible we have from Newton's formula

$$f(x_0 + \theta h) = f(x_0) + \theta h \frac{\delta f_{1/2}}{h} + \frac{\theta h(\theta h - h)}{2!} \frac{\delta^2 f_0}{h^2} + \frac{\theta h(\theta h - h)(\theta h + h)}{3!} \frac{\delta^3 f_{1/2}}{h^3}$$
$$+ \frac{\theta h(\theta h - h)(\theta h + h)(\theta h - 2h)}{4!} \frac{\delta^4 f_0}{h^2} + \ldots$$
$$= f_0 + \theta \delta f_{1/2} + \frac{\theta(\theta-1)}{2!} \delta^2 f_0 + \frac{\theta(\theta-1)(\theta+1)}{3!} \delta^3 f_{1/2} + \frac{\theta(\theta-1)(\theta+1)(\theta-2)}{4!} \delta^4 f_0 + \ldots$$
$$+ \frac{\theta(\theta-1)(\theta+1) \ldots (\theta-n+1)(\theta+n-1)}{(2n-1)!} \delta^{2n-1} f_{1/2}$$
$$+ \frac{\theta(\theta-1)(\theta+1) \ldots (\theta-n+1)(\theta+n-1)(\theta-n)}{(2n)!} \delta^{2n} f_0 + \ldots \quad (12)$$

This is the *Newton-Gauss formula*. An equivalent formula can of course be obtained by using the differences $\delta f_{-1/2}$, $\delta^2 f_0$, $\delta^3 f_{-1/2}$, ..., but would be less convenient for interpolating between $\theta = 0$ and $\theta = 1$. The zigzag arrangement of the differences used makes the formula somewhat awkward to use, but this can be circumvented in three ways. We introduce a further symbol μ to indicate the mean of two adjacent elements in the same vertical column; thus

$$\mu \delta f_0 = \tfrac{1}{2}(\delta f_{-1/2} + \delta f_{1/2}), \quad \mu \delta^2 f_{1/2} = \tfrac{1}{2}(\delta^2 f_0 + \delta^2 f_1), \quad \mu \delta^3 f_0 = \tfrac{1}{2}(\delta^3 f_{-1/2} + \delta^3 f_{1/2}), \quad (13)$$

and so on. Then we can rewrite the Newton-Gauss formula as follows:

$$f(x_0 + \theta h) = f_0 + \theta(\delta f_{1/2} - \tfrac{1}{2}\delta^2 f_0) + \frac{\theta^2}{2!} \delta^2 f_0 + \frac{\theta(\theta^2-1)}{3!} (\delta^3 f_{1/2} - \tfrac{1}{2}\delta^4 f_0) + \ldots$$
$$+ \frac{\theta(\theta^2-1^2)(\theta^2-2^2) \ldots \{\theta^2-(n-1)^2\}}{(2n-1)!} (\delta^{2n-1} f_{1/2} - \tfrac{1}{2}\delta^{2n} f_0) + \ldots$$
$$= f_0 + \theta \mu \delta f_0 + \frac{\theta^2}{2!} \delta^2 f_0 + \frac{\theta(\theta^2-1)}{3!} \mu \delta^3 f_0 + \ldots$$
$$+ \frac{\theta(\theta^2-1^2)(\theta^2-2^2) \ldots \{\theta^2-(n-1)^2\}}{(2n-1)!} \mu \delta^{2n-1} f_0$$
$$+ \frac{\theta^2(\theta^2-1^2)(\theta^2-2^2) \ldots \{\theta^2-(n-1)^2\}}{(2n)!} \delta^{2n} f_0 + \ldots \quad (14)$$

This is the *Newton-Stirling* formula. We see that it involves rewriting the difference table so that all the entries lie on horizontal lines through the datum values. The even differences remain where they were, but the odd ones are replaced by means in accordance with the definition of μ.

Alternatively, if we are interpolating between $\theta = 0$ and 1, we may keep the odd differences where they are but use mean even differences centred on $\theta = \tfrac{1}{2}$. We have

$$\delta^{2n+1} f_{1/2} = \delta^{2n} f_1 - \delta^{2n} f_0, \quad (15)$$

whence $\quad \delta^{2n} f_0 = \tfrac{1}{2}\delta^{2n} f_0 + \tfrac{1}{2}(\delta^{2n} f_1 - \delta^{2n+1} f_{1/2}) = \mu \delta^{2n} f_{1/2} - \tfrac{1}{2}\delta^{2n+1} f_{1/2}, \quad (16)$

9·04 Central difference formulae

and the terms in δ^{2n}, δ^{2n+1} of the Newton-Gauss formula can be written

$$\frac{\theta(\theta^2-1^2)(\theta^2-2^2)\ldots\{\theta^2-(n-1)^2\}(\theta-n)}{(2n)!}\left\{\mu\delta^{2n}f_{1/2}-\tfrac{1}{2}\delta^{2n+1}f_{1/2}+\frac{\theta+n}{2n+1}\delta^{2n+1}f_{1/2}\right\}.$$

The last factor is $\qquad \mu\delta^{2n}f_{1/2}+\dfrac{\theta-\tfrac{1}{2}}{2n+1}\delta^{2n+1}f_{1/2}.$

Hence

$$f(x_0+\theta h)=f_0+\theta\delta f_{1/2}+\frac{\theta(\theta-1)}{2!}\mu\delta^2 f_{1/2}+\frac{\theta(\theta-\tfrac{1}{2})(\theta-1)}{3!}\delta^3 f_{1/2}+\frac{\theta(\theta-1)(\theta+1)(\theta-2)}{4!}\mu\delta^4 f_{1/2}+\ldots$$
$$+\frac{\theta(\theta^2-1^2)\ldots\{\theta^2-(n-1)^2\}(\theta-n)}{(2n)!}\left(\mu\delta^{2n}f_{1/2}+\frac{\theta-\tfrac{1}{2}}{2n+1}\delta^{2n+1}f_{1/2}\right)+\ldots. \qquad (17)$$

This is the *Newton-Bessel formula*.*

A further modification of the Newton-Gauss formula is obtained by using (15) to eliminate the odd differences. We find

$$\delta^{2n}f_0+\frac{\theta+n}{2n+1}(\delta^{2n}f_1-\delta^{2n}f_0)=\frac{\theta+n}{2n+1}\delta^{2n}f_1+\frac{n+1-\theta}{2n+1}\delta^{2n}f_0$$
$$=\frac{\theta+n}{2n+1}\delta^{2n}f_1+\frac{\phi+n}{2n+1}\delta^{2n}f_0,$$

where $\qquad \phi = 1-\theta;$

also $\qquad \theta(\theta-1)\ldots(\theta+n-1)(\theta-n)=\phi(\phi-1)\ldots(\phi+n-1)(\phi-n).$

Hence $\qquad f(x)=\left\{\phi f_0+\dfrac{\phi(\phi^2-1^2)}{3!}\delta^2 f_0+\dfrac{\phi(\phi^2-1^2)(\phi^2-2^2)}{5!}\delta^4 f_0+\ldots\right\}$
$$+\left\{\theta f_1+\frac{\theta(\theta^2-1^2)}{3!}\delta^2 f_1+\frac{\theta(\theta^2-1^2)(\theta^2-2^2)}{5!}\delta^4 f_1+\ldots\right\}. \qquad (18)$$

This is *Everett's formula*.

These three formulae all have special advantages. The Newton-Stirling formula, proceeding in terms of differences centred on x_0, has the terms in the even differences even functions of θ, those in the odd differences odd functions of θ. Hence to get values for equal and opposite values of θ we can build up the terms in the odd and even differences separately, and then the values of $f(x_0\pm\theta h)$ are found by simple addition and subtraction, that is, by three turns on a multiplying machine. It is also convenient for deriving expressions for the derivatives of the function at the datum values. The advantage of the Newton-Bessel formula is that the odd differences after the first are all multiplied by functions that vanish at $\theta=\tfrac{1}{2}$. In comparison with the Newton-Stirling formula we notice that the maximum of $|\theta(\theta^2-1)|$ for $0<\theta\leqslant\tfrac{1}{2}$ is $\tfrac{3}{8}$, but that of $\theta(\theta-1)(\theta-\tfrac{1}{2})$ is 0·048. If then we neglect the third difference the Newton-Bessel formula is seven times as accurate. In other words, if we want the error to be less than half a unit we can neglect third differences under 60 units. In practice we often need to retain second differences,

---

* These three formulae are actually all due to Newton. The second name in each case is only a label. See Karl Pearson's bibliography mentioned on p. 261. Pearson condemns the Newton-Gauss formula, and it is true that this formula is never used for computation. But its direct relation to Newton's formula makes it the easiest to prove, and the others follow from it without trouble.

and third differences are sometimes needed. But the need for them is very greatly reduced if we partly absorb them into the second differences by using a mean second difference and the Newton-Bessel formula. If we do this it is convenient to arrange the difference table slightly differently. We notice that

$$\mu\delta^2 f_{1/2} = \tfrac{1}{2}(\delta^2 f_1 + \delta^2 f_0) = \tfrac{1}{2}(\delta f_{3/2} - \delta f_{1/2} + \delta f_{1/2} - \delta f_{-1/2}) = \tfrac{1}{2}(\delta f_{3/2} - \delta f_{-1/2}), \tag{19}$$

and therefore is half the difference of the two first differences just after and just before the interval that we are to interpolate in. Consequently we need not write out the column of second differences explicitly. We rewrite Bessel's formula, accurate to second differences, in the form

$$f(x_0 + \theta h) = f_0 + \theta \delta f_{1/2} - \tfrac{1}{4}\theta(1-\theta)(\delta f_{3/2} - \delta f_{-1/2}), \tag{20}$$

and the last factor, obtained by subtracting alternate first differences, is written instead of the second difference. The function of θ is as follows, for multiples of 0·1:

θ	$-\tfrac{1}{4}\theta(1-\theta)$
0·0, 1·0	0·0000
0·1, 0·9	−0·0225
0·2, 0·8	−0·0400
0·3, 0·7	−0·0525
0·4, 0·6	−0·0600
0·5	−0·0625

A more extended table is given by Milne-Thomson and Comrie.*

Everett's formula, if second differences are kept, also takes complete account of the third differences, and in this respect is even better than the Newton-Bessel formula. The coefficients of the second and fourth differences are as follows:

θ	$-\tfrac{1}{6}\theta(1-\theta^2)$	$\tfrac{1}{120}\theta(1-\theta^2)(4-\theta^2)$
0·0	−0	0
0·1	−0·0165	0·00329
0·2	−0·0320	0·00634
0·3	−0·0455	0·00890
0·4	−0·0560	0·01075
0·5	−0·0625	0·01172
0·6	−0·0640	0·01165
0·7	−0·0595	0·01044
0·8	−0·0480	0·00806
0·9	−0·0285	0·00455
1·0	0	0

Allowing for the fact that adjacent values of $\delta^4 f$ will in general be nearly equal we see that $\delta^4 f$ can reach 20 units in a given decimal place without introducing a correction of half a unit in that place into the interpolate. Second differences exceeding 4 units require attention. If Bessel's formula is used third differences over 60 units should be retained; but it then becomes just as easy to use Everett's formula.

Another method, known as the *throw-back*, is usefully combined with Everett's formula. The coefficients of each difference in this formula keep the same sign across an interval. In particular, that of the fourth difference is $(4-\theta^2)/20$ times that of the second, and this ratio varies only in a ratio of 3 to 4. Consequently the fourth differences can be largely taken into account by a suitable modification of the second differences. It is shown by Comrie† that if, instead of using δ^2 as it stands, we use $\delta^2 - 0.184\delta^4$, the resulting error in

---

* *Standard Four-Figure Mathematical Tables.*
† *British Association Mathematical Tables*, vol. 1, Introduction.

the interpolate will not exceed half a unit if δ^4 itself does not exceed 1000 units. A similar device is extensively used in the British Association tables for modifying δ^4 to take partial account of δ^6 and so on.[a]

Many mathematical tables are now published with differences ready printed. First differences are given when the function will stand linear interpolation, but the usual arrangement of a difference table with the odd differences on intermediate lines gives rise to some trouble in printing if second and higher differences are given. Everett's formula has the great advantage at this point that it uses only even differences, so that only these need be printed, and they lie on the same line as the datum value.

9·041. Discussion of efficiency. The most convenient formulae to use with equal intervals in various circumstances are as follows, if the lowest difference neglected is to contribute less than half a unit in the last place.

(1) Near the beginning or end of the table, where a centred second difference is not available at one end of the interval, there is no alternative to the Gregory formula. (2) If the third differences do not exceed 60 units in the last place and the second differences exceed 4 units, Bessel's formula with mean second differences is far the most convenient. (3) If the third differences exceed 60 units but the fourth differences do not exceed 1000 units Everett's formula with the throw-back is adequate. (4) Larger fourth differences need explicit allowance for δ^4 and possibly higher differences; a full account is given in the introduction to the *British Association Mathematical Tables*, vol. 1.

The conditions contemplated in the third case, and still more so in the fourth, arise for functions tabulated to a large number of figures. A prohibitive number of entries would then be needed to permit even Bessel interpolation, with second differences, and there is no alternative to using intervals so long that higher differences become necessary. In such cases it is quite possible for interpolation with fourth differences to give an answer correct to ten figures when linear interpolation will not give one correct to three.

Mention should be made at this point of the use of Taylor's theorem. As it depends only on the function and its derivatives at one datum value it can hardly be called interpolation; but when the derivatives are known it will achieve a higher accuracy for a given interval with the same number of terms. In the divided difference formula knowledge of derivatives up to the nth at one datum value is equivalent to having $n+1$ data with their divided differences at that value, and the corresponding terms in the interpolation formula are those of Taylor's series. There is no way of taking such information into account with equal intervals.

In general $\mu\delta^{2n+1}f_0$ is about $f^{(2n+1)}(x_0)h^{2n+1}$, and for θ small its coefficient in the Newton-Stirling series is about $\theta\dfrac{(n!)^2}{(2n+1)!}$. The whole term is therefore about $\dfrac{(n!)^2}{(2n+1)!}h^{2n+1}\theta f^{(2n+1)}(x_0)$. The corresponding term in Taylor's series is $\dfrac{h^{2n+1}\theta^{2n+1}}{(2n+1)!}f^{(2n+1)}(x_0)$, and is much smaller for $|\theta|<1$ even for quite small values of n. Consequently if the derivatives are known there is no point in using any of the interpolation formulae; these are required when our only information about the function is derived from the tabular values themselves. (See also Note **9·041a**, p. 705).

9·042. The following example illustrates the *use of Bessel's formula*. Given $x^{1/2}$ for $x = 2, 3, 4, 5, 6$, infer values between $x = 3$ and $x = 4$. The difference table is as follows:

x	$\sqrt{x}$	Δf	$\Delta^2 f$	$2\mu\delta^2$
2·0	1·414			
3·0	1·732	+0·318	−0·050	
4·0	2·000	+0·268	−0·032	−0·082
5·0	2·236	+0·236	−0·023	−0·055
6·0	2·449	+0·213		

Inspection of the second differences shows that the third differences are under 20 in the last place; hence we can use Bessel's formula with second differences. We first interpolate linearly at intervals of 0·1. We then multiply the double mean second difference −0·082 ($= -0·050 - 0·032 = 0·236 - 0·318$) by the coefficients $-\tfrac{1}{4}\theta(1-\theta)$ from the table above, and add to the linear interpolate (note that in all formulae the coefficient of the second difference is negative):

		Correct value
3·1	1·7588 + 0·0018 = 1·7606	1·761
3·2	1·7856 + 0·0033 = 1·7889	1·789
3·3	1·8124 + 0·0043 = 1·8167	1·817
3·4	1·8392 + 0·0049 = 1·8441	1·844
3·5	1·8660 + 0·0051 = 1·8711	1·871
3·6	1·8928 + 0·0049 = 1·8977	1·897
3·7	1·9196 + 0·0043 = 1·9239	1·924
3·8	1·9464 + 0·0033 = 1·9497	1·949
3·9	1·9732 + 0·0018 = 1·9750	1·975

The error never exceeds 1 in the third figure. It is surprising at first sight that such good agreement should be possible when a constant second difference is used for interpolation, seeing that the second differences at the beginning and end of the range are nearly as 3 to 2. But the mean second difference must give agreement at the beginning, middle, and end of each interval, and the errors never have a chance to accumulate.

9·043. The following harder example illustrates the *use of Everett's formula*:

x	$\cot x°$	Δf	$\Delta^2 f$	$\Delta^3 f$	$\Delta^4 f$	$\delta^2 f$ (modified)
30	1·7321					
35	1·4281	−0·3040	+0·0677			
40	1·1918	−0·2363	+0·0445	−0·0232	+0·0096	+0·0427
45	1·0000	−0·1918	+0·0309	−0·0136	+0·0047	+0·0300
50	0·8391	−0·1609	+0·0220	−0·0089	+0·0030	+0·0214
55	0·7002	−0·1389	+0·0161	−0·0059		
60	0·5774	−0·1228				

The third differences forbid the use of Bessel's formula to second differences only, but the fourth differences can be thrown back on the second for the use of Everett's formula. Each is multiplied by 0·184 and subtracted from the second difference in the same line to give the modified second difference. This is then multiplied by the coefficients in Everett's formula and combined with the linear interpolate:

x	$\cot x°$	Correct value
41	1·15344 − 0·00205 − 0·00096 = 1·1504	1·1504
42	1·11508 − 0·00273 − 0·00168 = 1·1107	1·1106
43	1·07672 − 0·00239 − 0·00192 = 1·0724	1·0724
44	1·03836 − 0·00137 − 0·00144 = 1·0356	1·0355
45	1·00000 = 1·0000	1·0000
46	0·96782 − 0·00144 − 0·00068 = 0·9657	0·9657
47	0·93564 − 0·00192 − 0·00120 = 0·9325	0·9325
48	0·90346 − 0·00168 − 0·00137 = 0·9004	0·9004
49	0·87128 − 0·00096 − 0·00103 = 0·8693	0·8693

There are only two differences of 1 in the last place in spite of the rather large higher differences. The effect of $\delta^4 f$ is well over a unit in places, but is adequately taken into account by the throw-back with hardly any additional work. It will be noticed that one set of corrections is symmetrical about 45°. In long interpolations this halves the work with Everett's formula; there is a corresponding simplification with the Newton-Stirling formula and also with the Newton-Bessel one.

9·044. When a function is originally given at a number of irregularly and widely spaced values, as often happens when the determinations are from experiment, and a detailed table is wanted, the usual procedure would be: first interpolate by divided differences to equal intervals such that on an average about two values lie between consecutive datum values; then interpolate by Bessel's or Everett's formula to intervals such that linear interpolation is possible. It is a matter of convenience whether we proceed by stages in this way or do the whole interpolation by divided differences at once. Detailed tables of the Bessel and Everett coefficients at intervals 0·001 of θ have been published by A. J. Thompson,* E. Chappell,† Comrie‡ and L. J. Briggs and A. N. Lowan.|| The above specimen values would usually suffice for an interpolation, but if only a few values are wanted the use of these tables will give them with only one rounding-off error instead of two. On the other hand, this difficulty can be greatly reduced by carrying out the preliminary interpolations to an extra figure. It is a customary requirement of mathematical tables that the last figure given should be correct to half a unit, and in using them it is often well worth while to keep an extra figure to reduce the accumulation of rounding-off errors if the work has to be done in several stages. A lot could be said for tolerating errors up to 3 in the last place of published tables; a five-figure table with such errors is more accurate than a four-figure one with errors up to 0·5 in the last figure, and involves no more trouble in interpolation. Such a device is virtually used in the tables of Milne-Thomson and Comrie, which are printed to four decimals, but an upper dot is added at the end if the correction needed is between $+\frac{1}{6}$ and $+\frac{1}{2}$ in the last figure, and a lower dot if it is between $-\frac{1}{6}$ and $-\frac{1}{2}$. Thus if 0·0008˙ is read as 0·00083, and 0·0008. as 0·00077 the tables can be used as five-figure ones, and interpolation does not need the retention of more figures than are needed to prevent rounding-off errors in the last place with the usual four-figure ones.

It is customary to round off to the nearest integer in the last place, not to the next integer below. This prevents all the rounding-off errors from having the same sign and accumulating in a sum. When the first figure neglected is 5, one usually takes the nearest *even* integer in the place kept.

9·045. Interpolation when a derivative becomes infinite. It should be remembered that the limitation of accuracy in interpolation formulae imposed by the higher derivatives of the function is not trivial. This is obvious if the function is infinite at a point of the range, since differences that involve the value at that point are infinite and the whole method breaks down. But it is also serious if a derivative is infinite, as for, say, $x^{1/2}$ at $x = 0$. We have the following difference table:

x	$x^{1/2}$	Δ	Δ^2	Δ^3	Δ^4
0	0·000				
		+1·000			
1	1·000		−0·586		
		+0·414		+0·490	
2	1·414		−0·096		−0·444
		+0·318		+0·046	
3	1·732		−0·050		
		+0·268			
4	2·000				

---

* *Tracts for Computers*, 5, 1921; second edition 1944.   † Published privately, 1929.
‡ 'Interpolation and Allied Tables', from *Nautical Almanac*, 1937.
|| Tables of Lagrangian interpolation coefficients, W.P.A., New York, 1944.

We try to interpolate a value for $x = 0.5$ by Gregory's formula. We get

$$0.5000 + \frac{\frac{1}{2}(\frac{1}{2}-1)}{2}(-0.586) + \frac{\frac{1}{2}(\frac{1}{2}-1)(\frac{1}{2}-2)}{6}(0.490) + \frac{\frac{1}{2}(\frac{1}{2}-1)(\frac{1}{2}-2)(\frac{1}{2}-3)(-0.444)}{24} + \cdots$$

$$= 0.5000 + 0.0732 + 0.0306 + 0.0172 + \cdots = 0.6210 + \cdots.$$

The correct value is 0.7071, but we have not achieved a tolerable approach to it even with fourth differences. It is necessary to the success of the interpolation process that derivatives up to the order of the last difference retained shall exist throughout the range used in forming that difference.

This difficulty can often be circumvented by a change of $f(x)$. Thus though $\cot x°$ is infinite at $x = 0$, $x \cot x°$ has a definite limit there equal to 57.30, and its derivatives are finite. If $\cot x°$ is given for a set of values of x we can therefore interpolate $x \cot x°$ and then divide by x. Again, $\cosh^{-1}(1+x)$ behaves like $(2x)^{1/2}$ near $x = 0$; but we can interpolate its square and take the root afterwards.

9·05. Inverse interpolation: solution of equations. This process is useful when we have values of a function for equal intervals of the argument and want to know for what value the function takes a given value.* Consider the equation

$$f(x) = x^3 - 3x - 7 = 0.$$

By inspection there is a root between $+2$ and $+3$. We can begin by calculating values for $x = 2, 2.1, \ldots, 3.0$ or by calculating for $2, 2.2, \ldots, 3.0$ and then interpolating to the midpoints, where the third difference is irrelevant. We find

x	$f(x)$	$\Delta f(x)$	$\Delta^2 f(x)$	$\Delta^3 f(x)$
2·2	−2·952			
		+1·219		
2·3	−1·733		+0·138	
		+1·357		+0·006
2·4	−0·376		+0·144	
		+1·501		+0·006
2·5	+1·125		+0·150	
		+1·651		+0·006
2·6	+2·776		+0·156	
		+1·807		
2·7	+4·583			

The constancy of $\Delta^3 f(x)$ checks the arithmetic. The root is clearly about 2.425. We interpolate by Everett's formula for the highest accuracy and get

x	$f(x)$		$\Delta f(x)$	$\Delta^2 f(x)$
2·40		−0·37600		
			+0·14352	
2·41	$-0.2259 - 0.00410 - 0.00248$	$= -0.23248$		+0·00145
			+0·14497	
2·42	$-0.0758 - 0.00691 - 0.00480$	$= -0.08751$		+0·00145
			+0·14642	
2·43	$+0.0743 - 0.00857 - 0.00682$	$= +0.05891$		+0·00145
			+0·14787	
2·44	$+0.2244 - 0.00922 - 0.00840$	$= +0.20678$		+0·00147
			+0·14934	
2·45	$+0.3745 - 0.00900 - 0.00938$	$= +0.35612$		

The third differences are now irrelevant, and the root is near 2.426. We interpolate by Bessel's formula as follows:

x	$f(x)$
2·425	$-0.01430 - 0.00018 = -0.01448$
2·426	$+0.00034 - 0.00017 = +0.00017$

* Comrie, *Inverse Interpolation*.

The contributions from the second difference are nearly constant and we have

$$x = 2 \cdot 425 + 0 \cdot 001 \times \frac{0 \cdot 01448}{0 \cdot 01465} = 2 \cdot 4259884,$$

with a possible error in the last figure.

Alternatively, we could interpolate between 2·4 and 2·5 by using a mean second difference. The neglect of third differences might then make an error of 0·0001 in $f(x)$ and an error of 0·000007 in x might be expected. For most purposes such accuracy would be ample.

These methods are not restricted to algebraic equations. They would, for instance, enable us to construct a table of $\sin^{-1} x$ given a table of $\sin x$.

The following method is also useful for algebraic equations. In the above equation we put

$$x = 2 + x_1;$$

then $\quad -f(x) = 7 + 3(2 + x_1) - (8 + 12x_1 + 6x_1^2 + x_1^3) = 5 - 9x_1 - 6x_1^2 - x_1^3 = 0.$

Now put $x_1 = 0 \cdot 4 + x_2$, and the equation becomes

$$+ 0 \cdot 376 - 14 \cdot 28 x_2 - 7 \cdot 2 x_2^2 - x_2^3 = 0.$$

Write this as $\quad 14 \cdot 28 x_2 = + 0 \cdot 376 - 7 \cdot 2 x_2^2 - x_2^3$

and put $x_2 = +0 \cdot 026$ on the right, which is now $+0 \cdot 3711152$, giving a further approximation $x_2' = +0 \cdot 02598846$. Try next $x_2 = +0 \cdot 0259$; the right side becomes $+0 \cdot 3711528$, $x_2' = +0 \cdot 02599109$. Linear interpolation gives for

$$x_2 = 0 \cdot 02599, \quad x_2' = 0 \cdot 02598872,$$

and direct calculation $x_2' = 0 \cdot 02598872$. This justifies linear interpolation; and finally interpolating to $x_2 = 0 \cdot 025989$ we have $x_2' = 0 \cdot 02598875$. Then

$$x = 2 \cdot 42598875.$$

In principle this method[a] is the same as that usually known as Horner's, but it is closer to one given by Newton. Horner's contribution seems to have been the introduction of synthetic division, a useful device in its proper place, but experience of the method does not encourage the belief that the easiest way to add 3×4 is to add 4 in three separate operations. One great advantage of *not* multiplying the roots by 10 at each stage is that the coefficient of the first power of the unknown then varies little in the later stages, and it is easier to see what higher powers can be neglected consistently with the accuracy required.*

9·06. Checking by differences. When a function is given at close intervals the higher differences decrease rapidly. When the values have been found independently the formation of differences therefore gives an easy check on the arithmetic. They will not in general be zero, since the tabulated values will usually have errors up to 0·5 in the last place, with either sign. Consequently the error of Δf may reach a whole unit; while

$$\delta^2 f_0 = f_1 - 2f_0 + f_{-1}, \quad \delta^3 f_{1/2} = f_2 - 3f_1 + 3f_0 - f_{-1}, \quad \delta^4 f_0 = f_2 - 4f_1 + 6f_0 - 4f_{-1} + f_{-2}$$

* Cf. also Jeffreys, *Math. Gaz.* 27, 1943, 20; L. J. Mordell, *Nature*, 119, 1927, 42; Jeffreys, *Nature*, 119, 1927, 565.

and may reach 2, 4, and 8 units respectively. Thus we take a set of values from Bottomley's table of natural logarithms:

x	$\log x$	Δf	$\Delta^2 f$	$\Delta^3 f$	$\Delta^4 f$
7·00	1·9459				
7·01	1·9473	+14			
7·02	1·9488	+15	+1	−2	
7·03	1·9502	+14	−1	+1	+3
7·04	1·9516	+14	0	0	−1
7·05	1·9530	+14	0	0	0
7·06	1·9544	+14	0	+1	+1
7·07	1·9559	+15	+1	−2	−3
7·08	1·9573	+14	−1	+1	+3
7·09	1·9587	+14	0	0	−1
7·10	1·9601	+14	0		

and the differences call for no comment. In interpolating such a table there is not only no gain but an appreciable loss of accuracy if any attempt is made to keep differences above the first. The rounding-off error may have the same sign at two consecutive entries, and nothing can reduce it; but if it has opposite signs at two consecutive entries the errors of the linear interpolates will be intermediate and on the whole smaller than those of the tabulated values. To see the effect of this let us round off the values at intervals of 0·02 to the third figure and see what happens if we then try to keep a second difference and halve the interval.

x	$\log x$	Δf	$\Delta^2 f$
7·00	1·946		
7·02	1·949	+3	0
7·04	1·952	+3	−1
7·06	1·954	+2	+1
7·08	1·957	+3	0
7·10	1·960	+3	0
7·12	1·963	+3	0
7·14	1·966	+3	−1
7·16	1·968	+2	+1
7·18	1·971	+3	0
7·20	1·974	+3	

The linear interpolates to four figures, the Bessel corrections, and the correct values are as follows:

x	$\log x$		Correct	Error	x	$\log x$		Correct	Error
7·01	1·9475	0	1·9473	+2	7·11	1·9615	0	1·9615	0
7·03	1·9505	+1	1·9502	+3	7·13	1·9645	+1	1·9643	+2
7·05	1·9530	0	1·9530	0	7·15	1·9670	0	1·9671	−1
7·07	1·9555	−1	1·9559	−4	7·17	1·9695	−1	1·9699	−4
7·09	1·9585	0	1·9587	−2	7·19	1·9725	0	1·9727	−2

The errors given are those of the last figure in the linear interpolate and never reach 5. The Bessel corrections are at most 0·6 in this place and trivial in any case; but in all four cases where they are not zero they increase the magnitude of the error. This is a general result and applies also to higher differences; the errors of interpolated values are on the whole a little less than those of the tabular values, but the gain in accuracy on interpolation is reduced by taking account of nth differences less than 2^{n-1} in the last place.*

This statement applies to interpolation between two neighbouring values. Many books of tables print mean differences over a whole line of the table. If these are used the tendency of adjacent errors to cancel disappears, and instead each interpolated value has three nearly independent errors: rounding of the datum values; rounding of the printed

* R. A. Fisher and J. Wishart, *Proc. Camb. Phil. Soc.* 23, 1927, 912–21.

difference; and variation of the difference within a line of the table. Mean differences save some trouble if last figure errors do not matter, but should not be used if they do.

Differencing will often show up a mistake in arithmetic immediately; thus suppose that we had miscopied an entry in the table of $\log x$ as follows:

f	Δf	$\Delta^2 f$	$\Delta^3 f$	$\Delta^4 f$
1·9459				
1·9473	+14			
1·9488	+15	+1	−2	+2
1·9502	+14	−1	0	+3
1·9515	+13	−1	+3	−6
1·9530	+15	+2	−3	
1·9544	+14	−1		

The large fourth difference, though just possible with only rounding-off errors present, picks out the incorrect value of $f(x)$.

9·07. Differentiation. Newton's and Gregory's interpolation formulae can be differentiated at once; we have

$$\left[\frac{d}{dx}(f(x))\right]_{x=x_1} = [x_1 x_2] + (x_1 - x_2)[x_1 x_2 x_3] + \ldots, \qquad (1)$$

$$h\frac{d}{dx}f(x) = \Delta y_0 + \frac{2\theta - 1}{2!}\Delta^2 y_0 + \frac{3\theta^2 - 6\theta + 2}{3!}\Delta^3 y_0 + \ldots. \qquad (2)$$

The former is not often used because it yields derivatives only at the tabular values of the argument. If they are wanted for intermediate values it is easier to interpolate the function to equal intervals by divided differences and apply one of the rules for equal intervals to the interpolate. The most useful form for equal intervals is got by differentiating the Newton-Stirling formula, for in this all even differences are multiplied by θ^2 and give terms in the derivatives that vanish for the tabular values of the argument. We have

$$h\left[\frac{d}{dx}f(x)\right]_{x=x_0} = \mu\delta f_0 - \frac{1}{3!}\mu\delta^3 f_0 + \ldots + \frac{(-)^n (n!)^2}{(2n+1)!}\mu\delta^{2n+1}f_0 + \ldots. \qquad (3)$$

This is, of course, subject to the same limitation as applies to the central difference formulae for interpolation, that it cannot be used near the beginning and end of the table.

Since the tabular errors in the function will usually be of the same order of magnitude for all values of x, the accuracy of the right side of this formula is independent of the interval, but the left contains a factor h. Consequently accuracy can be increased by using a large interval, even though it may make higher differences important. Thus consider $\sin x$ at $x = 1\cdot0$, with x in radians. The table of Milne-Thomson and Comrie is at intervals of 0·001, and the first differences above and below 1·000 are $+0\cdot0006$ and $+0\cdot0005$. All that we could say from this is that the derivative is likely to be between 0·5 and 0·6. If we use intervals of 0·01 instead we have

x	$\sin x$	Δf	$\Delta^2 f$	$\Delta^3 f$	$\mu\delta$	$\mu\delta^3$
0·980	0·8305					
0·990	0·8360	+55	0			
1·000	0·8415	+55	−2	−2	0·0054	0·0000
1·010	0·8468	+53	0	+2		
1·020	0·8521	+53				

Then $0\cdot01 f' = 0\cdot0054$, $f' = 0\cdot54$. But $\mu\delta$ may be wrong by half a unit in the last place and we can say only that $0\cdot535 \leqslant f' \leqslant 0\cdot545$.

Now try intervals of 0·1:

x	$\sin x$	Δf	$\Delta^2 f$	$\Delta^3 f$	$\mu\delta$	$\mu\delta^3$
0·8	0·7174					
		+659	−77			
0·9	0·7833	+582	−85	−8		
1·0	0·8415			−4	+539·5	−6
1·1	0·8912	+497	−89			
		+408				
1·2	0·9320					

$0{\cdot}1f' = 0{\cdot}05395 + 0{\cdot}00010$, $f' = 0{\cdot}5405$ with a possible error of $0{\cdot}0005$. The correct value is $0{\cdot}5403$.

Second derivatives can be found by differentiating the Newton-Stirling formula twice and then putting $\theta = 0$. We have

$$h^2\left[\frac{d^2}{dx^2}f(x)\right]_{x=x_0} = \delta^2 f_0 - \frac{1}{4!}2\delta^4 f_0 + \ldots + \frac{(-)^{n-1}\{(n-1)!\}^2}{(2n)!}2\delta^{2n}f_0 + \ldots \qquad (4)$$

The remarks above on the need for a wide interval of course apply even more forcibly to this formula.

9·08. Integration. The simplest integration formula and one of the most useful is the *Euler-Maclaurin formula*. If $f(x)$ is differentiable we have by integration by parts

$$\int_0^1 f(x)\,dx = \left[xf(x)\right]_0^1 - \int_0^1 xf'(x)\,dx = \tfrac{1}{2}f(0) + \tfrac{1}{2}f(1) - \int_0^1 (x-\tfrac{1}{2})f'(x)\,dx. \qquad (1)$$

The first two terms give the 'trapezoidal' rule for integration; the integral expresses a correction to it. Now for $0 \leqslant x < 1$ and $r \geqslant 2$ define

$$\left.\begin{aligned} P_r(0) &= 0, \\ P_2'(x) &= x - \tfrac{1}{2}, \\ P_3'(x) &= b_2 + P_2(x), \\ P_4'(x) &= b_3 + P_3(x), \\ &\cdots\cdots\cdots\cdots\cdots \\ P_r'(x) &= b_{r-1} + P_{r-1}(x), \end{aligned}\right\} \qquad (2)$$

and choose the b_r so that $P_r(1)$ also equals 0 for all $r \geqslant 2$.

Then
$$P_r(x) = \int_0^x \{b_{r-1} + P_{r-1}(\xi)\}\,d\xi, \qquad (3)$$

$$b_r = -\int_0^1 P_r(x)\,dx. \qquad (4)$$

Then by successive integrations by parts

$$\begin{aligned}
\int_0^1 P_2'(x)f'(x)\,dx &= \left[P_2(x)f'(x)\right]_0^1 - \int_0^1 P_2(x)f''(x)\,dx \\
&= -\int_0^1 \{P_3'(x) - b_2\}f''(x)\,dx \\
&= b_2\{f'(1) - f'(0)\} - \left[P_3(x)f''(x)\right]_0^1 + \int_0^1 P_3(x)f'''(x)\,dx \\
&= b_2\{f'(1) - f'(0)\} - b_3\{f''(1) - f''(0)\} + \ldots \\
&\quad + (-)^r b_r\{f^{(r-1)}(1) - f^{(r-1)}(0)\} - (-)^r \int_0^1 P_{r+1}'(x)f^{(r)}(x)\,dx, \qquad (5)
\end{aligned}$$

since all $P_r(x)$ vanish at both limits.

Now consider the series

$$\frac{a}{e^a-1} + \tfrac{1}{2}a = \sum_{r=0}^{\infty} b_r a^r, \qquad (6)$$

$$a\frac{e^{at}-1}{e^a-1} = \sum_{r=0}^{\infty} P_r(t) a^r. \qquad (7)$$

We shall show that the b_r and $P_r(t)$ defined in this way are identical with those given by the above definitions (2). Differentiate (7) with regard to t; then

$$\frac{a^2 e^{at}}{e^a-1} = \sum_{r=0}^{\infty} P'_r(t) a^r,$$

and also
$$= \sum_{r=0}^{\infty} P_r(t) a^{r+1} + \frac{a^2}{e^a-1}$$

$$= \sum_{r=1}^{\infty} P_{r-1}(t) a^r + \sum_{r=1}^{\infty} b_{r-1} a^r - \tfrac{1}{2}a^2.$$

Equating coefficients of a^r we have for $r \geq 2$

$$P'_r(t) = P_{r-1}(t) + b_{r-1}. \qquad (8)$$

For $r = 0, 1, 2$ we expand and get

$$\frac{a(at + \tfrac{1}{2}a^2 t^2)}{a + \tfrac{1}{2}a^2} = at + \tfrac{1}{2}a^2(t^2-t) + \ldots = P_0(t) + aP_1(t) + a^2 P_2(t) + \ldots,$$

whence
$$P_0(t) = 0, \quad P_1(t) = t, \quad P_2(t) = \tfrac{1}{2}(t^2-t), \qquad (9)$$

and
$$P'_2(t) = t - \tfrac{1}{2}. \qquad (10)$$

Also if $t = 0$, the function on the left of (7) vanishes; hence all $P_r(0) = 0$. If $t = 1$ the function reduces to a; hence $P_r(1) = 0$ except for $P_1(1)$, which is 1. This proves that for $r \geq 2$ the functions defined by (2) and (6) and (7) are identical.

In (6) change a to $-a$; we have

$$-\frac{a}{e^{-a}-1} - \tfrac{1}{2}a = \frac{ae^a}{e^a-1} - \tfrac{1}{2}a = a + \frac{a}{e^a-1} - \tfrac{1}{2}a = \frac{a}{e^a-1} + \tfrac{1}{2}a.$$

Hence (6) is an even function of a, and all b_r with r odd are zero. Then (5) simplifies to

$$-\int_0^1 (x - \tfrac{1}{2}) f'(x)\, dx = -b_2\{f'(1) - f'(0)\} - b_4\{f'''(1) - f'''(0)\} - \ldots$$
$$- b_{2r}\{f^{(2r-1)}(1) - f^{(2r-1)}(0)\} + \int_0^1 P'_{2r+1}(x) f^{(2r)}(x)\, dx. \qquad (11)$$

Integrating the remainder term by parts we have

$$\left[P_{2r+1}(x) f^{(2r)}(x) \right]_0^1 - \int_0^1 P_{2r+1}(x) f^{(2r+1)}(x)\, dx,$$

in which the integrated part vanishes.

If we now apply this result to the intervals 0 to 1, 1 to 2, ..., $n-1$ to n and add, we have, from (1) and (11) the *Euler-Maclaurin formula* (the method is essentially due to W. Wirtinger,[*])

$$\int_0^n f(x)\, dx = \tfrac{1}{2}f(0) + f(1) + f(2) + \ldots + f(n-1) + \tfrac{1}{2}f(n)$$
$$- b_2\{f'(n) - f'(0)\} - b_4\{f'''(n) - f'''(0)\} - \ldots - b_{2r}\{f^{(2r-1)}(n) - f^{(2r-1)}(0)\}$$
$$- \sum_{m=0}^{n-1} \int_m^{m+1} P_{2r+1}(x-m) f^{(2r+1)}(x)\, dx. \qquad (12)$$

[*] *Acta Math.* 26, 1902, 255-60.

In its usual form this formula is stated in terms of the Bernoulli numbers* and polynomials, B_r, $\phi_r(x)$ defined by

$$b_r = B_r/r!, \quad P_r(x) = \phi_r(x)/r!. \tag{13}$$

The easiest way of calculating them is by successive applications of the relations (2). The introduction of the factorials reduces the accumulation of large denominators in the successive integrations, but slightly complicates the proof of the above theorem.

$$B_0 = 1, \quad B_2 = \tfrac{1}{6}, \quad B_4 = -\tfrac{1}{30}, \quad B_6 = +\tfrac{1}{42}, \quad B_8 = -\tfrac{1}{30}, \quad B_{10} = +\tfrac{5}{66}, \tag{14}$$

$$\begin{aligned}
&\phi_2(x) = x^2 - x, \quad \phi_3(x) = x^3 - \tfrac{3}{2}x^2 + \tfrac{1}{2}x, \quad \phi_4(x) = x^4 - 2x^3 + x^2, \\
&\phi_5(x) = x^5 - \tfrac{5}{2}x^4 + \tfrac{5}{3}x^3 - \tfrac{1}{6}x, \quad \phi_6(x) = x^6 - 3x^5 + \tfrac{5}{2}x^4 - \tfrac{1}{2}x^2, \\
&\phi_7(x) = x^7 - \tfrac{7}{2}x^6 + \tfrac{7}{2}x^5 - \tfrac{7}{6}x^3 + \tfrac{1}{6}x, \quad \phi_8(x) = x^8 - 4x^7 + \tfrac{14}{3}x^6 - \tfrac{7}{3}x^4 + \tfrac{2}{3}x^2, \\
&\phi_9(x) = x^9 - \tfrac{9}{2}x^8 + 6x^7 - \tfrac{21}{5}x^5 + 2x^3 - \tfrac{3}{10}x, \quad \phi_{10}(x) = x^{10} - 5x^9 + \tfrac{15}{2}x^8 - 7x^6 + 5x^4 - \tfrac{3}{2}x^2.
\end{aligned} \tag{15}$$

Changing x to $x_0 + \theta h$, we have

$$\int_{x_0}^{x_0+nh} f(x)\,dx = h[\tfrac{1}{2}f(x_0) + f(x_0+h) + \ldots + f\{x_0 + (n-1)h\} + \tfrac{1}{2}f(x_0+nh)]$$
$$- \frac{B_2 h^2}{2!}\{f'(x_0+nh) - f'(x_0)\} - \ldots - \frac{B_{2r} h^{2r}}{(2r)!}\{f^{(2r-1)}(x_0+nh) - f^{(2r-1)}(x_0)\}$$
$$- \sum_{m=0}^{n-1} h^{2r+2} \int_m^{m+1} P_{2r+1}(\theta - m) f^{(2r+1)}(x_0 + \theta h)\,d\theta. \tag{16}$$

Now in (7) put $t = \tfrac{1}{2}$. We have

$$a \frac{e^{\frac{1}{2}a} - 1}{e^a - 1} = \Sigma P_r(\tfrac{1}{2}) a^r.$$

Change a to $-a$ and subtract; we have

$$a = 2 \sum_{\text{odd } r} P_r(\tfrac{1}{2}) a^r,$$

and therefore $\quad P_1(\tfrac{1}{2}) = \tfrac{1}{2}, \quad P_{2r+1}(\tfrac{1}{2}) = 0, \quad (2r+1 > 1).$

We shall prove that $P_{2r+1}(t)$ has no other zeros than 0, $\tfrac{1}{2}$, and 1 for $0 \leq t \leq 1$; and $P_{2r}(t)$ has none but 0 and 1.

Suppose that $P_{2r-1}(t)$ has simple zeros at $t = 0$, $\tfrac{1}{2}$ and 1, and at no other value. Take it positive for $0 < t < \tfrac{1}{2}$. Then by (8)

$$P'_{2r}(t) = P_{2r-1}(t) \quad > 0 \quad (0 < t < \tfrac{1}{2}),$$
$$\phantom{P'_{2r}(t) = P_{2r-1}(t)} < 0 \quad (\tfrac{1}{2} < t < 1),$$

and $\quad P_{2r}(0) = P_{2r}(1) = 0.$

Hence P_{2r} has one maximum at $t = \tfrac{1}{2}$ and no other stationary value, and in $0 < t < 1$ it has the sign of $P_{2r-1}(\epsilon)$, where by ϵ we mean some number between 0 and $\tfrac{1}{2}$.

Next, $\quad P'_{2r+1}(t) = P_{2r}(t) + b_{2r}.$

---

* At least three different definitions of B_r are current; this one is that used by Milne-Thomson except for B_1, and is far the most convenient. His $B_r(x)$ is the present $\phi_r(x) + B_r$.

The formula was first given by Euler and rediscovered independently by Maclaurin a few years later. Continental authors usually refer to it as the Euler formula, but there are many Euler formulae and only one Euler-Maclaurin formula.

But as P'_{2r} has only one zero in $0 < t < 1$, $P_{2r} + b_{2r}$ can vanish for at most two values of t, and $P_{2r+1}(t)$ can have at most three zeros in $0 \leqslant t \leqslant 1$. Hence $t = 0, \frac{1}{2}, 1$ are the only zeros of $P_{2r+1}(t)$. Also $P_{2r}(\frac{1}{2})$ must be numerically greater than $-b_{2r}$, and therefore the zero of $P_{2r+1}(t)$ at $t = \frac{1}{2}$ is simple.

Also since $P_{2r}(t)$ does not change sign between 0 and 1, $P_{2r}(t)$ has the opposite sign to b_{2r}, by (4). Since $P_{2r}(0) = 0$, $P'_{2r+1}(\epsilon)$ and therefore $P_{2r+1}(\epsilon)$ have the same sign as b_{2r}, which is the opposite sign to P_{2r} and therefore the opposite sign to $P_{2r-1}(\epsilon)$. Hence $P_{2r+1}(\epsilon)$ *alternate in sign with r*. It follows that *the b_{2r} alternate in sign*.

The sign of a typical remainder term in (12) is that of

$$-\int_m^{m+1} P_{2r+1}(x-m)\{f^{(2r+1)}(x) - f^{(2r+1)}(m+\tfrac{1}{2})\}\,dx,$$

and if $f^{(2r+1)}(x)$ is monotonic in the interval, we have, since $P_{2r+1}(x)$ has opposite signs for $0 < x < \frac{1}{2}$ and $\frac{1}{2} < x < 1$, that the sign is that of $-P_{2r+1}(\epsilon)\{f^{(2r+1)}(m) - f^{2r+1}(m+\tfrac{1}{2})\}$. *In an important class of cases all the odd derivatives have the same sign.* Hence, since the $P_{2r+1}(\epsilon)$ alternate in sign, the errors due to stopping at a given value of r alternate in sign. Hence *the true value of the integral always lies between the sums of r and $r+1$ terms of the series.* The condition will be seen to be satisfied in the examples that follow.

The expansion can be derived operationally as follows. From (1)

$$\int_{x_e+mh}^{x_e+(m+1)h} f(x)\,dx = h[\tfrac{1}{2}f(x_0+mh) + \tfrac{1}{2}f\{x_0+(m+1)h\}] - h^2\int_0^1 (\theta - \tfrac{1}{2})f'(x_0+mh+\theta h)\,d\theta. \quad (17)$$

Put
$$\frac{\partial}{\partial x_0} = D. \quad (18)$$

Then the last term is
$$-h^2\int_0^1 (\theta - \tfrac{1}{2}) D e^{(m+\theta)hD} f(x_0)\,d\theta, \quad (19)$$

in which the operator D is independent of θ and therefore commutes with all functions of θ. Hence we can integrate with regard to θ as if D was a constant. Then we get on integration by parts

$$e^{mhD}\left[-\tfrac{1}{2}h(e^{hD}+1) + \frac{1}{D}(e^{hD}-1)\right]f(x_0) = e^{mhD}\left[-h + \left(\frac{1}{D} - \tfrac{1}{2}h\right)(e^{hD}-1)\right]f(x_0). \quad (20)$$

Also
$$\sum_{m=0}^{n-1} e^{mhD} = \frac{e^{nhD}-1}{e^{hD}-1}, \quad (21)$$

and therefore the sum with regard to m of the last terms in (17) is

$$-\frac{(e^{nhD}-1)}{D}\left\{\frac{hD}{e^{hD}-1} - 1 + \tfrac{1}{2}hD\right\}f(x_0) = -\frac{e^{nhD}-1}{D}\sum_{r=2}^{\infty}(hD)^r\frac{B_r}{r!}f(x_0)$$
$$= -h^2\frac{B_2}{2!}\{f'(x_0+nh) - f'(x_0)\} - h^4\frac{B_4}{4!}\{f'''(x_0+nh) - f'''(x_0)\} - \ldots. \quad (22)$$

We thus obtain the Euler-Maclaurin expansion again but without a form for the remainder after a given number of terms.

The expansion should not be interpreted as an infinite series. The theorem of 9·012 fixes an upper bound to the remainder term in interpolation formulae, and the integral of $f(x)$ is the sum of the integrals of the interpolation polynomial and this remainder term. If the remainder is small throughout the range the integral of the interpolation polynomial is within specifiable limits an approximation to that of the function. But the derivatives of the polynomial vanish exactly above a certain finite order, and the expansions are properly interpreted as the sum of a finite number of terms. The justification of the operational method in this problem therefore has nothing to do with convergence of series. It rests on the facts that (1) the operators are expansible in positive integral powers of D and therefore the terms after a certain order vanish when the operand is a polynomial, (2) since negative powers of D do not arise, the non-commutative property of differentiation and definite integration does not matter, (3) the error is the integral of the error of the interpolation polynomial, and is fixed for any finite order irrespective of questions of convergence.

As a matter of fact what usually happens is that the terms of the expansion decrease rapidly at first but afterwards increase, on account of the tendency of higher derivatives to increase if the function is not a polynomial. The most accurate value of the integral is then got by taking the sum up to the smallest term. We shall return to this matter when we come to asymptotic expansions, of which this is an example.

9·081. Consider the integral
$$\log 2 = \int_{10}^{20} \frac{dx}{x}.$$

We have $\quad f(x) = \dfrac{1}{x}, \quad f'(x) = -\dfrac{1}{x^2}, \quad f'''(x) = -\dfrac{2.3}{x^4}, \quad f^{(5)}(x) = -\dfrac{5!}{x^6}, \quad \ldots,$

and
$$\log 2 = \frac{1}{2.10} + \frac{1}{11} + \frac{1}{12} + \ldots + \frac{1}{19} + \frac{1}{2.20} - \frac{B_2}{2}\left(\frac{1}{10^2} - \frac{1}{20^2}\right) - \frac{B_4}{4}\left(\frac{1}{10^4} - \frac{1}{20^4}\right) - \frac{B_6}{6}\left(\frac{1}{10^6} - \frac{1}{20^6}\right) \ldots.$$

We arrange the calculation as follows:

0·0500000000	$-\frac{1}{12}(0·01 - 0·0025) = -0·000625$
0·0909090909	
0·0833333333	$+\frac{1}{120}(0·0001 - 0·00000625) = +0·0000008333 - 0·0000000521$
0·0769230769	
0·0714285714	$-\frac{1}{252}(0·000001)(1-\frac{1}{64}) = -0·0000000039$
0·0666666667	
0·0625000000	$+\frac{1}{240}(0·00000001)(1-\frac{1}{256}) = +0·0000000000$
0·0588235294	
0·0555555556	Total $= -0·0006242227$
0·0526315789	
0·0250000000	Hence $\quad \log 2 = +0·6931471804$
0·6937714031	

The correct result is $0·6931471805\ldots$.

9·082. Consider next Euler's constant γ defined by
$$\gamma = \lim_{n \to \infty}\left(1 + \frac{1}{2} + \frac{1}{3} + \ldots + \frac{1}{n} - \log n\right).$$

We have
$$\log n - \log 10 = \int_{10}^{n} \frac{dx}{x}$$
$$= \frac{1}{20} + \frac{1}{11} + \frac{1}{12} + \ldots + \frac{1}{n-1} + \frac{1}{2n} - \frac{1}{12}\left(\frac{1}{10^2} - \frac{1}{n^2}\right) + \frac{1}{120}\left(\frac{1}{10^4} - \frac{1}{n^4}\right) - \frac{1}{252} \cdot \frac{1}{10^6} + \ldots,$$

and
$$\lim_{n \to \infty}\left(\frac{1}{11} + \frac{1}{12} + \ldots + \frac{1}{n} - \log n + \log 10\right) = -\frac{1}{20} + \frac{1}{1200} - \frac{1}{12.10^5} + \frac{1}{252.10^6} = -0·049167496.$$

Also $1 + \tfrac{1}{2} + \ldots + \tfrac{1}{10} - \log 10 = 0\cdot 626383161$

by direct summation. Hence by addition

$$\gamma = 0\cdot 577215665.$$

This is correct to nine figures. It will be noticed that the extension from three- to nine-figure accuracy requires the computation of only two extra terms.*

9·083. Gregory's integration formula. The Euler-Maclaurin formula is the simplest and most accurate formula of numerical integration, but requires that direct calculation of the derivatives shall be possible. Higher derivatives may be mathematically complicated and it may be easier to replace them by differences. This can be done either by a formula due to Gregory or by a central difference formula.

We have seen that the correcting terms in the Euler-Maclaurin expansion can be expressed as in 9·08 (22). We also have at the beginning and end of the range respectively

$$hD = \log(1+\Delta), \quad hD = -\log(1-\nabla), \qquad (1)$$

$$-\frac{h}{e^{hD}-1} + \frac{1}{D} - \tfrac{1}{2}h = -\frac{h}{\Delta} + \frac{h}{\log(1+\Delta)} - \tfrac{1}{2}h = -\frac{h}{\nabla} - \frac{h}{\log(1-\nabla)} + \tfrac{1}{2}h. \qquad (2)$$

Two procedures are possible. We can express D in powers of Δ and ∇ and substitute in the Euler-Maclaurin formula; or we can expand the operators in (2) directly without appealing to previous knowledge of the Bernoulli numbers. Both involve rather heavy algebra. A direct attack (Gregory again!) seems easier than either method. Developing the Gregory interpolation function in powers of θ we have

$$\theta(\theta-1) = \theta^2 - \theta, \quad \theta(\theta-1)(\theta-2) = \theta^3 - 3\theta^2 + 2\theta,$$

$$\theta(\theta-1)(\theta-2)(\theta-3) = \theta^4 - 6\theta^3 + 11\theta^2 - 6\theta,$$

$$\theta(\theta-1)(\theta-2)(\theta-3)(\theta-4) = \theta^5 - 10\theta^4 + 35\theta^3 - 50\theta^2 + 24\theta,$$

$$\theta(\theta-1)\ldots(\theta-5) = \theta^6 - 15\theta^5 + 85\theta^4 - 225\theta^3 + 274\theta^2 - 120\theta,$$

$$\theta(\theta-1)\ldots(\theta-6) = \theta^7 - 21\theta^6 + 175\theta^5 - 735\theta^4 + 1624\theta^3 - 1764\theta^2 + 720\theta,$$

$$\theta(\theta-1)\ldots(\theta-7) = \theta^8 - 28\theta^7 + 322\theta^6 - 1960\theta^5 + 6769\theta^4 - 13132\theta^3 + 13068\theta^2 - 5040\theta,$$

and the respective integrals from 0 to 1 are

$$-\tfrac{1}{6}, \ +\tfrac{1}{4}, \ -\tfrac{19}{30}, \ +\tfrac{9}{4}, \ -\tfrac{863}{84}, \ +\tfrac{1375}{24}, \ -\tfrac{33953}{90}.$$

Hence

$$\frac{1}{h}\int_{x_0}^{x_0+h} f(x)\,dx = f(x_0) + \tfrac{1}{2}\Delta f(x_0) - \frac{1}{6.2!}\Delta^2 f(x_0) + \frac{1}{4.3!}\Delta^3 f(x_0) - \frac{19}{30.4!}\Delta^4 f(x_0)$$

$$+ \frac{9}{4.5!}\Delta^5 f(x_0) - \frac{863}{84.6!}\Delta^6 f(x_0) + \frac{1375}{24.7!}\Delta^7 f(x_0) - \frac{33953}{90.8!}\Delta^8 f(x_0) \ldots. \qquad (3)$$

* These and other fundamental numerical constants were computed by J. C. Adams to 272 figures (*Collected Scientific Papers* 1, 459–470).

The first two terms are $\frac{1}{2}\{f(x_0)+f(x_0+h)\}$. On adding for intervals up to x_0+nh we have therefore

$$\frac{1}{h}\int_{x_0}^{x_0+nh} f(x)\,dx = \tfrac{1}{2}f(x_0)+f(x_0+h)+\ldots+f\{x_0+(n-1)h\}+\tfrac{1}{2}f(x_0+nh)$$
$$-\frac{1}{6.2!}\{\Delta f(x_0+nh)-\Delta f(x_0)\}+\frac{1}{4.3!}\{\Delta^2 f(x_0+nh)-\Delta^2 f(x_0)\}+\ldots. \quad (4)$$

But we could equally well work in powers of ∇, and the expansion will be the same as in terms of Δ except that the signs of all even powers will be reversed. The point again is that 9·08 (22) is exact as applied to the interpolation polynomial. The relation between ∇ and Δ given in (2) is also exact as applied to this polynomial. Hence we can replace the terms in $\Delta^r(x_0+nh)$ by the equivalent expression in $\nabla^r(x_0+nh)$. The argument does *not* assume that $f(x)$ is determinate beyond x_0+nh, merely that the interpolation polynomial is, and this is true. Hence

$$\frac{1}{h}\int_{x_0}^{x_0+nh} f(x)\,dx = \tfrac{1}{2}f(x_0)+f(x_0+h)+\ldots+f\{x_0+(n-1)h\}+\tfrac{1}{2}f(x_0+nh)$$
$$-\tfrac{1}{12}\{\nabla f(x_0+nh)-\Delta f(x_0)\}-\tfrac{1}{24}\{\nabla^2 f(x_0+nh)+\Delta^2 f(x_0)\}$$
$$-\tfrac{19}{720}\{\nabla^3 f(x_0+nh)-\Delta^3 f(x_0)\}-\tfrac{3}{160}\{\nabla^4 f(x_0+nh)+\Delta^4 f(x_0)\}$$
$$-\tfrac{863}{60480}\{\nabla^5 f(x_0+nh)-\Delta^5 f(x_0)\}-\tfrac{275}{24192}\{\nabla^6 f(x_0+nh)+\Delta^6 f(x_0)\}$$
$$-\tfrac{33953}{3628800}\{\nabla^7 f(x_0+nh)-\Delta^7 f(x_0)\}\ldots \quad (5)$$

This is the Gregory formula.

9·084. Central difference formula. Similarly we can integrate the Newton-Bessel formula. Here all the terms involving odd differences give 0 on integration; the others give

$$\frac{1}{h}\int_{x_0}^{x_0+h} f(x)\,dx = \tfrac{1}{2}f(x_0)+\tfrac{1}{2}f(x_0+h)$$
$$-\frac{1}{6.2!}\mu\delta^2 f_{1/2}+\frac{11}{30.4!}\mu\delta^4 f_{1/2}-\frac{191}{84.6!}\mu\delta^6 f_{1/2}+\frac{2497}{90.8!}\mu\delta^8 f_{1/2}+\ldots. \quad (6)$$

But $\quad 2\mu\delta^{2r}f_{1/2} = (\delta^{2r}f_1+\delta^{2r}f_0) = \delta^{2r-1}f_{3/2}-\delta^{2r-1}f_{1/2}+\delta^{2r-1}f_{1/2}-\delta^{2r-1}f_{-1/2}$
$$= 2\mu\delta^{2r-1}f_1-2\mu\delta^{2r-1}f_0. \quad (7)$$

Hence

$$\frac{1}{h}\int_{x_0}^{x_0+nh} f(x)\,dx = \tfrac{1}{2}f(x_0)+f(x_0+h)+\ldots+f\{x_0+(n-1)h\}+\tfrac{1}{2}f(x_0+nh)$$
$$-\tfrac{1}{12}\{\mu\delta f(x_0+nh)-\mu\delta f(x_0)\}+\tfrac{11}{720}\{\mu\delta^3 f(x_0+nh)-\mu\delta^3 f(x_0)\}$$
$$-\tfrac{191}{60480}\{\mu\delta^5 f(x_0+nh)-\mu\delta^5 f(x_0)\}+\tfrac{2497}{3628800}\{\mu\delta^7 f(x_0+nh)-\mu\delta^7 f(x_0)\}. \quad (8)$$

Fewer terms have to be calculated with this formula than with the Gregory one, and the coefficients of the higher ones are smaller. On the other hand, the formation of the central differences requires knowledge of the function outside the range of integration, whereas the Gregory formula does not.

9·085. As an illustration take the integral

$$\tfrac{1}{6}\pi = \int_0^{1/2} \frac{dx}{\sqrt{(1-x^2)}}.$$

The square roots to 8 figures were taken from Barlow's tables.

x	$(1-x^2)^{-1/2}$	Δ	Δ^2	Δ^3	Δ^4	Δ^5	Δ^6	Δ^7
0	1·00000000		0·00250470		0·00005682		0·00000372	
0·05	1·00125235	0·00125235	0·00253311	0·00002841	0·00005868	0·00000186	0·00000376	
0·10	1·00503781	0·00378546	0·00262020	0·00008709	0·00006430	0·00000562	0·00000469	
0·15	1·01144347	0·00640566	0·00277159	0·00015139	0·00007461	0·00001031	0·00000593	0·00000260
0·20	1·02062072	0·00917725	0·00299759	0·00022600	0·00009085	0·00001624	0·00000853	
0·25	1·03279556	0·01217484	0·00331444	0·00031685	0·00011562	0·00002477	0·00001246	
0·30	1·04828484	0·01548928	0·00374691	0·00043247	0·00015285	0·00003723	0·00001952	0·00001201
0·35	1·06752103	0·01923619	0·00433223	0·00058532	0·00020960	0·00005675	0·00003153	
0·40	1·09108945	0·02356842	0·00512715	0·00079492	0·00029788	0·00008828	0·00005372	
0·45	1·11978502	0·02869557	0·00621995	0·00109280	0·00043988	0·00014200	0·00009597	0·00008660
0·50	1·15470054	0·03491552	0·00775263	0·00153268	0·00067785	0·00023797	0·00018257	0·00019081
0·55	1·19736869	0·04266815	0·00996316	0·00221053	0·00109839	0·00042054	0·00037338	
0·60	1·25000000	0·05263131	0·01327208	0·00330892	0·00189231	0·00079392		
0·65	1·31590339	0·06590339	0·01847331	0·00520123				
0·70	1·40028009	0·08437670						

We have

$$\tfrac{1}{2}f(0) + f(0·05) + \ldots + f(0·45) + \tfrac{1}{2}f(0·50) = 10·47518052.$$

Using the central difference formula we see that all odd differences vanish at $x = 0$; and the odd differences give at 0·50

$$2\mu\delta = 0·07758367, \quad 2\mu\delta^3 = 0·00374321, \quad 2\mu\delta^5 = 0·00065851, \quad 2\mu\delta^7 = 0·00027741.$$

Then the correction terms are

$$-\tfrac{1}{24}(2\mu\delta) + \tfrac{11}{1440}(2\mu\delta^3) - \tfrac{191}{120960}(2\mu\delta^5) + \tfrac{2497}{7257600}(2\mu\delta^7)$$
$$= -0·00323265 + 0·00002859 - 0·00000104 + 0·00000010$$
$$= -0·00320500.$$

Then $\quad \tfrac{1}{6}\pi = 0·05(10·47518052 - 0·00320500) = 0·5235987760,$

$$\pi = 3·1415926560.$$

The correct value is
$$\pi = 3·141592654.$$

Using the Gregory formula we find the following correction terms:

Δ	$-0·00280526$
$\Delta^2 -$	36471
$\Delta^3 -$	2654
$\Delta^4 -$	679
$\Delta^5 -$	111
$\Delta^6 -$	43
$\Delta^7 -$	9
	$-0·00320493$

and $\pi = 3·141592677$. This is inferior in accuracy to the last; the difference of 7 units in the last figure of the sum might just be due to rounding-off errors but is not likely to be. We see that the last pair of terms in the central difference formula differ by a factor of about 10; so do the last pair from odd differences in the Gregory formula. We might then expect that Δ^9 would contribute about -1 in the last place. But then we should also expect Δ^8 to contribute about -4 in this place. It is probable therefore that the difference arises because the terms in the Gregory formula decrease more slowly than those of the central difference formula.

This example is rather favourable to the Gregory formula because of the infinity of the function at $x = 1$. In finding centred differences we have to use values of the function up to 0·70, and the higher centred differences are correspondingly larger than those within the range of integration. If we had used instead the expression

$$\tfrac{1}{4}\pi = \int_0^1 \frac{dx}{1+x^2},$$

the contrast would have been more striking.*

9·09. Special rules of integration. Let $f(x)$ be given for $x = -h$, 0, and h; then by Lagrange's formula the interpolation quadratic is

$$g(x) = \left(1 - \frac{x^2}{h^2}\right)f(0) + \left(\frac{x^2}{2h^2} - \frac{x}{2h}\right)f(-h) + \left(\frac{x^2}{2h^2} + \frac{x}{2h}\right)f(h)$$

and

$$\int_{-h}^{h} g(x)\,dx = \tfrac{1}{3}h\{f(-h) + 4f(0) + f(h)\}.$$

This leads to *Simpson's rule*.† *Divide the range into a number of equal intervals; take the sum of the end ordinates, add four times the sum of the ordinates at the middles of the intervals and twice the sum of the ordinates at the junctions of the intervals, and multiply by a sixth of the interval.* It amounts to fitting a quadratic to the three values of $f(x)$ at the beginning, middle, and end of each interval. No attempt is made to maintain smoothness at the junctions between the quadratics.

The possible presence of a cubic term does not affect the rule. For the cubic could differ from the quadratic only by a function that vanishes at $-h$, 0 and h, and such a function must be of the form $Ax(h^2 - x^2)$. But the integral of this from $-h$ to h is 0.

Next, suppose $f(x)$ is given for $x = -3h, -h, h, 3h$, that is, for four equally spaced values. These determine an interpolation cubic

$$g(x) = f(-3h)\frac{(x+h)(x-h)(x-3h)}{(-2h)(-4h)(-6h)} + f(-h)\frac{(x+3h)(x-h)(x-3h)}{2h(-2h)(-4h)}$$

$$+ f(h)\frac{(x+3h)(x+h)(x-3h)}{4h \cdot 2h \cdot (-2h)} + f(3h)\frac{(x+3h)(x+h)(x-h)}{6h \cdot 4h \cdot 2h}$$

$$= -\frac{f(-3h)}{48}\left(\frac{x^3}{h^3} - \frac{3x^2}{h^2} - \frac{x}{h} + 3\right) + \frac{f(-h)}{16}\left(\frac{x^3}{h^3} - \frac{x^2}{h^2} - \frac{9x}{h} + 9\right)$$

$$- \frac{f(h)}{16}\left(\frac{x^3}{h^3} + \frac{x^2}{h^2} - \frac{9x}{h} - 9\right) + \frac{f(3h)}{48}\left(\frac{x^3}{h^3} + \frac{3x^2}{h^2} - \frac{x}{h} - 3\right)$$

* This integral is worked out by Whittaker and Robinson, *Calculus of Observations*, pp. 147–9, third differences being used in the central difference formula; there is an error of 2 in the seventh decimal of π. Ten intervals were used as here and seven decimals in the integrand.

† Simpson's publication was in 1743. It had been given earlier by Cavalieri (1639) and James Gregory (1668).

and
$$\int_{-3h}^{3h} g(x)\,dx = -\frac{h}{48}f(-3h)(-2.3^3+3.6) + \frac{h}{16}f(-h)(-\tfrac{2}{3}.3^3+9.6)$$
$$-\frac{h}{16}f(h)(\tfrac{2}{3}.3^3-9.6) + \frac{h}{48}f(3h)(2.3^3-3.6)$$
$$= \tfrac{3}{4}h\{f(-3h)+3f(-h)+3f(h)+f(3h)\}.$$

Comparing with Simpson's rule, if we call the length of the range H in both cases, this rule is
$$\int_{-\tfrac{1}{2}H}^{\tfrac{1}{2}H} g(x)\,dx = \tfrac{1}{8}H\{f(-\tfrac{1}{2}H)+3f(-\tfrac{1}{6}H)+3f(\tfrac{1}{6}H)+f(\tfrac{1}{2}H)\},$$

and is called the *three-eighths rule*. Simpson's rule for the same range is
$$\int_{-\tfrac{1}{2}H}^{\tfrac{1}{2}H} g(x)\,dx = \tfrac{1}{6}H\{f(-\tfrac{1}{2}H)+4f(0)+f(\tfrac{1}{2}H)\}.$$

Note that if $f(x)$ is constant both rules give $Hf(x)$, as they should; this is a help towards remembering the numerical coefficients. The three-eighths rule is due to Cotes. Both are correct up to cubic terms. Let us see how they work for a fourth power. Take $H = 2$, $f(x) = x^4$. The correct value is then
$$\tfrac{2}{5} = 0.4.$$
Simpson's rule gives
$$\tfrac{1}{3}(1+0+1) = 0.667.$$
The three-eighths rule gives
$$\tfrac{1}{4}(1+\tfrac{1}{27}+\tfrac{1}{27}+1) = \tfrac{14}{27} = 0.519.$$

Thus both give results in excess of the true value, the three-eighths rule being the better.[a] But both will be inferior to the rules based on interpolation formulae, which take the fourth and higher powers into account explicitly.

As an example, we apply Simpson's rule to the data of 9·085. They give
$$\tfrac{1}{6}\pi = 0.523599265,$$

which is in error by 0·00000049. The δ^3 term in the central difference formula contributed $0.05 \times 0.00002859 = 0.00000143$, the δ^5 term 0·00000005. Hence the error given by Simpson's rule is about a third of the δ^3 term in the central difference formula and about ten times the δ^5 term. A quartic term would be integrated exactly by the central difference formula including δ^3, but less accurately by Simpson's rule.

The merits of Simpson's and the three-eighths rules are that they are simple and easily remembered; but they are less accurate than either difference formula up to δ^3 if the number of intervals used is the same. They do not need the formation of differences, but differences should be formed in any case as a check on the calculation of $f(x)$, whether they are used for calculation or not. If saving of labour is a consideration it is better to save it by computing just enough values of $f(x)$ to give a good check by differences. The use of the elementary rules should be restricted to cases where there really are only three or four determined values of the function and we have to do our best with them.

A number of more complicated rules, the best-known of which is Weddle's, take partial account of higher differences; but it is not often convenient to divide the range up into an integral multiple of the number of intervals that they require. For the same reason they fail if we require the integral up to *every* tabular value of the argument. Some other rules, one of them due to Gauss, attempt to economize labour by choosing the datum points so that the integral based on them will be independent of powers *higher* than the number of datum points. But the trouble of interpolating the function to the suggested datum points is far greater than that of using a formula based on equal intervals, even if they have to be more numerous.b

9·091. L. F. Richardson's method. One device for taking approximate account of second differences is due to L. F. Richardson, and sometimes gives an answer of the same order of accuracy as Simpson's rule with very little trouble. It is based on the principle that if a result is of the form $A + B/n^2 + \ldots$, we can get an approximate value of A by using two finite values of n and extrapolating to $n = \infty$. For instance,

$$\pi = \lim_{n \to \infty} n \sin \frac{\pi}{n},$$

and $n \sin \frac{\pi}{n}$ is of the form $\pi + \frac{B}{n^2} + \frac{C}{n^4} + \ldots$.

Take $n = 4$ and 6; then neglecting C,

$$\pi + \frac{B}{16} = \frac{4}{\sqrt{2}} = 2 \cdot 828,$$

$$\pi + \frac{B}{36} = 3 \cdot 000.$$

Solving for π we have $\pi = 3 \cdot 138$,

which is a good return for this amount of work. Or take

$$\log 2 = \int_1^2 \frac{dx}{x}.$$

With one interval this gives by the trapezoidal rule

$$A + B = \tfrac{1}{2}(1 + \tfrac{1}{2}) = 0 \cdot 75.$$

With another ordinate at $x = 1 \cdot 5$, and therefore two intervals,

$$A + \tfrac{1}{4}B = \tfrac{1}{4}(1 + \tfrac{4}{3} + \tfrac{1}{2}) = 0 \cdot 7083;$$

whence $\log 2 = A = 0 \cdot 6944.$

This method is useful for estimating limits of many different types, and is not confined to integration. An extension to take account of terms in n^{-4} is also possible, and would give integrals equivalent to those containing third differences at the termini without the need to remember the coefficients in the formulae. The method is called by Richardson 'the deferred approach to the limit.'

9·092. Functions that behave like $x^{\pm 1/2}$ at a terminus. All the usual formulae for numerical integration fail when the integrand behaves like $x^{-1/2}$ at a terminus; yet the integral converges. They are also unsatisfactory when it behaves like $x^{1/2}$. Thus

$$\int_0^2 x^{1/2} dx = \tfrac{4}{3}\sqrt{2} = 1 \cdot 8856,$$

9·092 *Functions like $x^{\pm 1/2}$ at terminus*

while Simpson's rule applied to $x = 0, 1, 2$ gives $1·8047$. Similarly,

$$\int_0^3 x^{1/2}\,dx = 2\sqrt{3} = 3·4641,$$

and the three-eighths rule applied to 0, 1, 2, 3 gives $3·3655$. These rules will therefore underestimate the contributions from the intervals near the ends by 3 or 4 per cent. It is easy, however, to obtain more accurate formulae where the integrand is known to behave like $x^{-1/2}$ or $x^{1/2}$ at the end of the range, using values at $x = h$ and $2h$ or $3h$ as the data. The Gregory formula can then be used for the remainder of the range. Writing suffixes to indicate the arguments used, we find*

$$\int_0^{2h} (\alpha x^{-1/2} + \beta x^{1/2})\,dx = h(\tfrac{8}{3}\sqrt{2}\,y_1 - \tfrac{4}{3}y_2) = h(3·7712 y_1 - 1·3333 y_2), \tag{1}$$

$$\int_0^{3h} (\alpha x^{-1/2} + \beta x^{1/2})\,dx = 2h\sqrt{3}\,y_1 + 0\,y_3 = 3·4641 h y_1 + 0·0000 y_3, \tag{2}$$

$$\int_0^{3h} (\alpha x^{-1/2} + \beta x^{1/2} + \gamma x^{3/2})\,dx = h(\tfrac{14}{5}\sqrt{3}\,y_1 - \tfrac{8}{5}\sqrt{6}\,y_2 + \tfrac{12}{5}y_3)$$
$$= h(4·8497 y_1 - 3·9192 y_2 + 2·4000 y_3), \tag{3}$$

$$\int_0^{2h} (\alpha x^{1/2} + \beta x^{3/2})\,dx = h(\tfrac{16}{15}\sqrt{2}\,y_1 + \tfrac{4}{15}y_2) = h(1·5085 y_1 + 0·2667 y_2), \tag{4}$$

$$\int_0^{3h} (\alpha x^{1/2} + \beta x^{3/2})\,dx = h(\tfrac{6}{5}\sqrt{3}\,y_1 + \tfrac{4}{5}y_3) = h(2·0785 y_1 + 0·8000 y_3), \tag{5}$$

$$\int_0^{3h} (\alpha x^{1/2} + \beta x^{3/2} + \gamma x^{5/2})\,dx = h(\tfrac{6}{7}\sqrt{3}\,y_1 + \tfrac{12}{35}\sqrt{6}\,y_2 + \tfrac{16}{35}y_3)$$
$$= h(1·4846 y_1 + 0·8398 y_2 + 0·4571 y_3). \tag{6}$$

As an example let us evaluate

$$\tfrac{1}{3}\pi = \int_{1/2}^{1} \frac{dx}{\sqrt{(1-x^2)}},$$

using only two and three intervals. (1) gives with $h = \tfrac{1}{4}$

x	$(1-x^2)^{-1/2}$	
$\tfrac{1}{2}$	$1·15470$	$-1·3333 y_2 = -1·53956$
$\tfrac{3}{4}$	$1·51186$	$3·7712 y_1 = \underline{5·70153}$
		$4·16197$

Hence $\pi = 3 \times \tfrac{1}{4} \times 4·16197 = 3·1215$.

(3) gives with $h = \tfrac{1}{6}$

x	$(1-x^2)^{-1/2}$		
$\tfrac{1}{2}$	$1·1547 \times$	$2·4000$	$= +2·7713$
$\tfrac{2}{3}$	$1·3416 \times$	$-3·9192$	$= -5·2580$
$\tfrac{5}{6}$	$1·8091 \times$	$4·8497$	$= \underline{+8·7736}$
			$6·2869$

Hence $\pi = 3 \times \tfrac{1}{6} \times 6·2869 = 3·1434$.

Thus we have an error of only $0·06$ per cent from the use of only three intervals. A much closer estimate could be obtained by using the Gregory formula at intervals of $0·05$ up

* An extension is due to Bickley (Adm. Res. Ctee., Paper 9211, 1946).

to $x = 0.85$ and adding the integral from 0·85 to 1·00 obtained by means of (3). This method has been extensively used in calculations in seismology, where the integrand is of the form $\int_0^a \cosh^{-1}(1+\xi)\,dx$ and $\xi = \alpha x + O(x^2)$ for x small, or $\int_0^a \xi^{-1/2}\,dx$, where ξ has the same property.

9·093. *Graphical methods* are best avoided entirely. A rough sketch is useful to exhibit the general appearance of a function, but in any attempt at accurate work a numerical method will always give a more accurate result than a graphical one and with less trouble. From observations of the disastrous results of graphical methods in seismology it has appeared that graphical methods are liable to be less accurate than numerical ones that use only *first* differences. This seems nearly incredible at first sight, but may be due to a defect in most commercial squared paper. The spaces between the *edges* of the lines are uniform, not those between the *centres*. Consequently there is a systematic difference between the small squares adjacent to the thick lines that separate large squares and those near the middle of large squares.

9·10. Numerical solution of differential equations. The simplest method, in principle, and one that has come increasingly into prominence recently, is the direct use of Taylor's series. For a second order equation,

$$y'' = f(x, y, y'),$$

given that $y = y_0$ and $y' = y_1$ when $x = 0$, we can calculate y'' for $x = 0$ directly from the differential equation. Differentiating the equation, we have

$$y''' = \frac{\partial f}{\partial x} + \frac{\partial f}{\partial y} y' + \frac{\partial f}{\partial y'} y'',$$

in which we can substitute the value of y'' just found, and so determine y'''. Differentiating again we determine $y^{(4)}$ and so on to any order desired. The results are substituted in Taylor's series for both y and y', as follows:

$$y = y_0 + y_1 x + \frac{y''(0)}{2!} x^2 + \frac{y'''(0)}{3!} x^3 + \dots,$$

$$y' = y_1 + y''(0) x + \frac{y'''(0)}{2!} x^2 + \frac{y^{(4)}(0)}{3!} x^3 + \dots.$$

These are used to calculate y and y' up to such a value of x that the terms neglected do not affect the last figure retained. Let this be h. Then for $x = h$ we have y and y'; again using the relations found by differentiation we determine $y''(h)$, $y'''(h)$, $y^{(4)}(h)$, ... and form new Taylor series in $x - h$. These are used to find values up to $x = 2h$. An important check is obtained by summing the odd and even powers in the series separately. If we have them for $x - h = \xi$, their sum gives y for $x = h + \xi$; but their difference gives y for $x = h - \xi$, which is among the values already calculated, and the two calculations for $h - \xi$ should agree. If they do they check the whole of the formation of the derivatives and the Taylor expansion about h. By repetition we can proceed to any desired value of x.

This process was given by J. R. Airey for the solution of Emden's equation

$$y'' + \frac{2}{x} y' + y^n = 0,$$

where $y = 1$, $y' = 0$ at $x = 0$, and the calculations were done by J. C. P. Miller and D. H. Sadler.* It is quite straightforward, and on account of the fact that the numerical coefficients in Taylor's series are much smaller than in any of the finite difference formulae it enables a given range of argument to be covered in a smaller number of stages. It can be applied to a differential equation of any order; for a first-order equation, of course, only a series for y is wanted; for a third-order one, series for y, y' and y''.

As a simple example, take the equation

$$\frac{dy}{dx} = -xy,$$

with $y = 1$ at $x = 0$, the solution of which is $\exp(-\tfrac{1}{2}x^2)$. By successive differentiation and substitution for y' at each stage we get

$$y'' = (x^2 - 1)y,$$
$$y''' = (3x - x^3)y,$$
$$y^{(4)} = (3 - 6x^2 + x^4)y,$$
$$y^{(5)} = (-15x + 10x^3 - x^5)y,$$
$$y^{(6)} = (-15 + 45x^2 - 15x^4 + x^6)y.$$

The series expansion for x small is

$$y = 1 - \tfrac{1}{2}x^2 + \frac{1}{4!}3x^4 - \frac{1}{6!}15x^6 + \ldots.$$

When $x = 0.5$ the last term is 0.0003, and four terms will give four-figure accuracy in this range. We find

x	y
0.1	0.9950
0.2	0.9802
0.3	0.9560
0.4	0.9231
0.5	0.8825

We now use our general expressions to work out the derivatives at $x = 0.5$. With their factorial divisors they are:

$$y' = -0.4412, \quad y''/2! = -0.3309, \quad y'''/3! = +0.2022,$$
$$y^{(4)}/4! = +0.0575, \quad y^{(5)}/5! = -0.0462, \quad y^{(6)}/6! = -0.0057.$$

Now we use $\xi = x - 0.5$ and work out sums of even and odd powers, as follows:

ξ	Even powers	Odd powers	x	Difference	x	Sum
0.1	0.8792	−0.0439	0.4	0.9231	0.6	0.8353
0.2	0.8694	−0.0866	0.3	0.9560	0.7	0.7828
0.3	0.8532	−0.1270	0.2	0.9802	0.8	0.7262
0.4	0.8311	−0.1640	0.1	0.9951	0.9	0.6671
0.5	0.8033	−0.1968	0.0	1.0001	1.0	0.6065

The greatest difference from the values previously calculated is 0.0001 at $x = 0.1$, and this entry is the sum of six, all with rounding-off errors in the last place. An error of 0.0003 is therefore possible.

* *British Association Mathematical Tables*, vol. 2, 1932.

We next take $\xi = x - 1{\cdot}0$ and work out derivatives at $x = 1{\cdot}0$.

$$y' = -0{\cdot}6065, \quad y''/2! = 0, \quad y'''/3! = +0{\cdot}2022,$$

$$y^{(4)}/4! = -0{\cdot}0505, \quad y^{(5)}/5! = -0{\cdot}0303, \quad y^{(6)}/6! = +0{\cdot}0135.$$

In the next stage we get

ξ	Even powers	Odd powers	x	Difference	x	Sum
0·1	0·6065	−0·0604	0·9	0·6669	1·1	0·5461
0·2	0·6064	−0·1197	0·8	0·7261	1·2	0·4867
0·3	0·6061	−0·1766	0·7	0·7827	1·3	0·4295
0·4	0·6053	−0·2299	0·6	0·8352	1·4	0·3754
0·5	0·6035	−0·2788	0·5	0·8823	1·5	0·3247

The greatest discrepancy is $0{\cdot}0002$, at $x = 0{\cdot}5$ and $0{\cdot}9$. So we may proceed. There is an inevitable tendency for rounding-off errors to accumulate. Generally speaking if an entry is the sum of m rounded-off values an error of $\tfrac{1}{2}m$ in the last figure is possible, but can occur only if all have such signs as will produce errors in the same direction and all approach the extreme possible. Ordinarily their distribution is nearly random and each can be taken as $\pm \dfrac{1}{2\sqrt{3}}$ (standard error); the resultant of m will then be $\pm \tfrac{1}{2}\sqrt{(\tfrac{1}{3}m)}$, in accordance with the usual principles of the composition of random errors. Errors up to this will then be usual, and if m is large errors of $\sqrt{(\tfrac{1}{3}m)}$ or a little more may occur in about 1 entry in 20. If we proceed by series up to sixth powers the basic values for $x = 0{\cdot}5, 1{\cdot}0, \ldots$ are sums of seven terms, and if integration is carried out to 10 stages an error of 5 units may easily accumulate. This is avoided in practice by carrying out computations to one or even two places more than are needed in the final answer. Some computers go so far as to insist that computations should be carried to such a stage that it can be decided with certainty whether the rounded-off figures are 0·499 or 0·501, and round off three figures at the end in every entry to avoid the risk that 1 entry in 500 may be rounded off in the wrong direction. This is hardly worth the extra labour, but the policy of keeping one extra figure is a good one.[a]

9·11. The Adams-Bashforth method. This starts with the Gregory backwards extrapolation formula

$$f(a+\theta h) = f(a) + \theta \nabla f(a) + \frac{\theta(\theta+1)}{2!}\nabla^2 f(a) + \ldots.$$

Expanding the terms and integrating we find

$$\frac{1}{h}\int_a^{a+h} f(x)\,dx = f(a) + (\tfrac{1}{2}\nabla + \tfrac{5}{12}\nabla^2 + \tfrac{3}{8}\nabla^3 + \tfrac{251}{720}\nabla^4 + \tfrac{95}{288}\nabla^5 + \ldots)f(a).$$

Hence if
$$y' = f(x,y),$$

and we know y and f up to $x = a$, we can obtain a value for y at $x = a+h$ from f at a and its backward differences. For a given interval the terms decrease much more slowly than with Taylor series. Thus, in our example above, the sixth derivative of y at the origin is -15. For an interval 0·5 the seventh backward difference of xy will be of order $15 \times 0{\cdot}5^6$ or about 0·24, and its coefficient is about $\tfrac{1}{4}$. Hence if this method is used with the same interval there will be serious errors in the second decimal. Shorter intervals therefore become necessary. We try the same example with interval 0·1.

With a finite difference method we always need a few values for a start. Here we know that at $x = 0$, $y = 1$, $y' = 0$, $y'' = -1$. We use the first three terms of the Taylor series to compute values at ± 0.1, and form xy and its first two differences. Then for the next stage we have

$$10\Delta y = -0.0995 - \tfrac{1}{2}(0.0995) - 0 = -0.1493, \quad \Delta y = -0.0149.$$

This gives y for $x = 0.2$. We form xy, and now have differences up to ∇^3. In the next stage

$$10\Delta y = -0.1960 - \tfrac{1}{2}(0.0965) + \tfrac{5}{12}(0.0030) + \tfrac{3}{8}(0.0030) = -0.2420, \quad \Delta y = -0.0242.$$

Hence y is found for $x = 0.3$, and we proceed. It saves trouble both in writing and reading to keep only significant figures in the differences.

x	y	$-xy$	∇	∇^2	∇^3	∇^4
-0.1	0·9950	$+0.0995$				
0	1·0000	0·0000	-995	0		
0·1	0·9950	-0.0995	-995	$+30$	$+30$	-3
0·2	0·9801	-0.1960	-965	$+57$	$+27$	-0
0·3	0·9559	-0.2868	-908	$+84$	$+27$	-7
0·4	0·9230	-0.3692	-824	$+104$	$+20$	-3
0·5	0·8824	-0.4412	-720	$+121$	$+17$	-6
0·6	0·8352	-0.5011	-599	$+132$	$+11$	-6
0·7	0·7826	-0.5478	-467	$+137$	$+05$	-5
0·8	0·7260	-0.5808	-330	$+137$	$+00$	
0·9	0·6668	-0.6001	-193			
1·0	0·6064	-0.6064				

The contributions from the second and higher differences are at most in the third decimal and can be safely worked out on the slide rule. As the total contribution calculated from xy and its differences is divided by 10 there is virtually only one rounding-off error at each stage. But such an error, once made, is carried on through the rest of the calculation.[a]

9·12. Central-difference method. One difficulty of the Adams-Bashforth method is to know how to start. In the above example values of y were computed for $x = \pm 0.1$ from the first three terms of Taylor's series, and a second difference centred on $x = 0$ was thus obtained. But no higher differences could be found for early values without using more terms of Taylor's series. In this example this would be practicable; but it often happens that higher derivatives become excessively complicated in form, and their calculation is a serious undertaking. This may make the whole method of Taylor's series impracticable; for instance, though it is extensively used in the calculation of mathematical tables, astronomers prefer a quite different method for the computation of cometary perturbations. Usually, however, a few terms of Taylor's series are found and used to compute four or five values of the solution, and the rest of the work is done by finite differences.

The other difficulty is in the large coefficients of the higher differences. In the above calculation the fourth differences gave effects at each step well within the rounding-off error, but as they keep the same sign over several intervals they cannot be safely neglected. It was absolutely necessary to keep third differences. Still higher differences would be needed if greater accuracy was being attempted. But we know that the coefficients of higher differences are much less with central differences formulae than with the Gregory formulae, and it is possible to modify the calculation so as to make use of this fact. We have simply calculated, for instance, y for $x = 0.3$ by using y at $x = 0.2$ and the differences of $-xy$ running backwards from $0\cdot 2$; $-xy$ for $x = 0\cdot 3$ is then calculated from the corresponding value of y. We have made no direct use of the fact that the increment of y from $x = 0\cdot 2$ to $x = 0\cdot 3$ must be the integral of $-xy$ over the range right up to $0\cdot 3$. If some higher difference that we have neglected actually made an error in the estimated y, we should be able to check it by computing an integral from our estimates of $-xy$; the two should agree, but they will not if some high-order differences have been illegitimately neglected. When the table is complete such an integral can be found by a central-difference formula; but it can also be found as we proceed. We have

$$\frac{1}{h}\int_a^{a+h} f(x)\,dx = \tfrac{1}{2}f(a) + \tfrac{1}{2}f(a+h) - \tfrac{1}{12}\mu\delta^2 f_{a+\frac{1}{2}h} + \tfrac{11}{720}\mu\delta^4 f_{a+\frac{1}{2}h} + \ldots.$$

In this the terms in $\mu\delta^2$ and $\mu\delta^4$ will be a small correction. But we cannot use it directly to compute the integral because we do not know $\mu\delta^2 f_{a+\frac{1}{2}h}$ until f_{a+2h} is known, and we do not know $\mu\delta^4 f_{a+\frac{1}{2}h}$ until f_{a+3h} is known. We can, however, proceed by successive approximation. These terms are small in any case, much smaller than those of the same orders in the Adams-Bashforth method. We can therefore extrapolate the last difference retained one stage, add the result to the previous difference, thus extrapolating that one, and so work up to an extrapolated value of $f(a+h)$. Also by extrapolating two stages we get an extrapolated value of $\delta^2 f_{a+h}$. The fourth difference is small and has a very small coefficient, so that it can usually be neglected; if it cannot we may have to extrapolate three stages to determine it. In this way we get trial values of all the quantities needed to compute the integral, and in forming it we introduce the further small factor h. Hence the value found for y at $x = a+h$ will be very nearly right. We now use it to calculate $f(a+h)$ and form corrected differences. With these we recalculate the integral and get a much closer approximation. If we have chosen the interval suitably the change will not be more than a few units in the last place.

Returning to the same example, we write our first few values as follows:

x	y	$-xy$	Δ	Δ^2
$-0\cdot 1$	$0\cdot 9950$	$+0\cdot 0995$		
0	$1\cdot 0000$	$0\cdot 0000$	-995	0
$+0\cdot 1$	$0\cdot 9950$	$-0\cdot 0995$	-995	

At this stage we have no means of predicting the variation of Δ^2 and so our first step is to take $\delta^2 f_{0\cdot 1}$ as zero. We then have $\delta f_{0\cdot 15} = -0\cdot 0995$, $f_{0\cdot 2} = -0\cdot 0995 - 0\cdot 0995 = -0\cdot 1990$. Then

$$10\Delta y = \tfrac{1}{2}(-0\cdot 0995 - 0\cdot 1990) = -0\cdot 1492, \quad \Delta y = -0\cdot 0149.$$

We therefore enter $y_{0\cdot 2}$ as $0\cdot 9801$; this makes $-xy = -0\cdot 1960$ instead of $-0\cdot 1990$, and $\delta f_{0\cdot 15} = -0\cdot 0965$, $\delta^2 f_{0\cdot 1} = +0\cdot 0030$, $\delta^3 f_{0\cdot 05} = +0\cdot 0030$. These suggest a revised value $\delta^2 f_{0\cdot 2} = +0\cdot 0060$, and $\mu\delta^2 f_{0\cdot 15} = +0\cdot 0045$. The table, with the necessary corrections, is now

x	y	$-xy$	Δ	Δ^2	Δ^3
$-0\cdot 1$	$0\cdot 9950$	$+0\cdot 0995$			
$0\cdot 0$	$1\cdot 0000$	0	-995	0	
$+0\cdot 1$	$0\cdot 9950$	$-0\cdot 0995$	-995	$+30$	$+30$
$+0\cdot 2$		$-0\cdot 1960$	-965	$(+60)$	$(+30)$

where extrapolated values are shown in brackets. With the corrected values

$$10\Delta y = \tfrac{1}{2}(-0\cdot 0995 - 0\cdot 1960) - \tfrac{1}{12}(0\cdot 0045) = -0\cdot 1481, \quad \Delta y = -0\cdot 0148,$$

and hence $y_{0\cdot 2} = 0\cdot 9802$. We thus make only a last-figure change in y at the second approxi-

9·12   *Central-difference method*   295

mation, even though we had no information at all for our first extrapolation of Δ^2. There is no further change in $-xy$ to this accuracy. We now have

x	y	$-xy$	Δ	Δ^2	Δ^3
0	1·0000	0		0	
0·1	0·9950	−0·0995	−995	+30	+30
0·2	0·9802	−0·1960	−965	(+60)	
0·3		(−0·2865)	(−905)	(+90)	

Then
$$10\Delta y = \tfrac{1}{2}(-0.1960 - 0.2865) - \tfrac{1}{12}(0.0075) = -0.2418, \quad \Delta y = -0.0242, \quad y_{0·3} = 0.9560.$$
This gives $-xy = -0.2868$ and we correct the differences accordingly. $10\Delta y$ is changed to
$$\tfrac{1}{2}(-0.1960 - 0.2868) - 0.0006 = -0.2420,$$
and we need no further change in y. So we proceed; the final table, after the readjustments have been made, is

x	y	$-xy$	Δ	Δ^2	Δ^3
−0·1	0·9950	+0·0995	−995		
0	1·0000	0	−995	0	+30
0·1	0·9950	−0·0995	−965	+30	+27
0·2	0·9802	−0·1960	−908	+57	+27
0·3	0·9560	−0·2868	−824	+84	+20
0·4	0·9231	−0·3692	−720	+104	+16
0·5	0·8825	−0·4412	−600	+120	+13
0·6	0·8353	−0·5012	−467	+133	+4
0·7	0·7827	−0·5479	−330	+137	0
0·8	0·7261	−0·5809	−193	+137	−7
0·9	0·6669	−0·6002	−63	+130	
1·0	0·6065	−0·6065			

This method has several advantages over the Adams-Bashforth one. The third differences do not appear at all except in so far as they are taken into account in the extrapolated Δ^2. The coefficient of the second difference is only one-fifth as large. Consequently we can have greater confidence that inaccuracy has not crept in through neglect of higher differences. There is also a difference at the start. There is a difficulty in all methods about starting the integration if derivatives are troublesome to calculate. In both methods we started assuming only the first two derivatives at $x = 0$. In the Adams-Bashforth method this gave us a second difference at 0 but at no earlier value; but a third backward difference from this was needed to infer y at $x = 0.2$. This was not available and we had to proceed to 0·2, effectively, with the same quadratic formula as was used up to 0·1. If then there are terms in x^3 and x^4 that are appreciable at $x = 0.2$ but not at $x = 0.1$, the method will give an error there. This could be corrected by working backwards to $x = -0.2$ as well, but this sacrifices the direct progress, which is the outstanding good point of the method. In the central-difference method, on the other hand, we get a first approximation at $x = 0.2$, which is the same as the Adams-Bashforth value, but also have the means of correcting it, and actually introduce a last-figure change in y.

Actually with the central-difference method we do not even need a second difference at the start. Suppose that we simply start with the information that at $x = 0$, $y = 1$, $y' = 0$. Then since $y' = 0$ we can take as our trial values at ± 0.1 the *same* value of y as at $x = 0$. Then the table at this stage reads

x	y	$-xy$	Δ	Δ^2
−0·1	1·0000	+0·1000	−0·1000	
0·0	1·0000	0·0000	−0·1000	0
0·1	1·0000	−0·1000		

Then using the formula up to $x = 0.1$ we have
$$10(y_{0·1} - y_{0·0}) = \tfrac{1}{2}(0.0000 - 0.1000), \quad y_{0·1} - y_{0·0} = -0.0050,$$

giving a corrected value $y_{0\cdot1} = -0\cdot9950$. Similarly, we get a value at $x = -0\cdot1$, and without ever differentiating the differential equation we have the datum values needed for a start.

The correct values, taken from the *British Association Tables*, are as follows. The last-figure errors of the solutions by the various methods are given for comparison:

x	y	Taylor series	Adams-Bashforth	Central differences
0·0	1·0000	0	0	0
0·1	0·9950	0	0	0
0·2	0·9802	0	−1	0
0·3	0·9560	0	−1	0
0·4	0·9231	0	−1	0
0·5	0·8825	0	−1	0
0·6	0·8353	0	−1	0
0·7	0·7827	+1	−1	0
0·8	0·7261	+1	−1	0
0·9	0·6670	+1	−2	−1
1·0	0·6065	0	−1	0

As far as this example goes there is little to choose between the Taylor series* and the central differences method. The error at $x = 0\cdot2$ in the Adams-Bashforth method, which could not be corrected in that method without either working backwards a stage or finding another term of the series, is carried on throughout the work. An india-rubber is indispensable for the central-difference method.

9·121. Equations of higher order. All these methods can be directly adapted to equations of higher orders. If our equation is

$$y'' = Py' + Qy + R(x),$$

with at $x = 0$ $\quad y = y_0, \quad y' = y_1,$

we need only take $z = y'$ as a new variable and write two first-order equations

$$y' = z, \quad z' = Pz + Qy + R,$$

which we can then proceed to solve as before, the initial values of y and z being given. We naturally advance by alternate stages in the solution of the two equations, and y and z will be found with comparable accuracy. Take, for instance, the equation satisfied by the Airy integral

$$\frac{d^2y}{dx^2} = xy.$$

We treat this as two, $\quad \dfrac{dy}{dx} = z, \quad \dfrac{dz}{dx} = xy,$

and investigate the solution that makes $y = 1$, $z = 0$ when $x = 0$. Without using Taylor's series we start with the values not bracketed in the following:

x	z	xy	Δ	Δ^2	y	z	Δ	Δ^2
−0·1	(0·0050)	−0·1000				(+0·0050)		
0·0	0·0000	0·0000	+1000	0	1·0000	0	(−50)	(+100)
0·1	(+0·0050)	0·1000	+1000			(+0·0050)	(+50)	

For the first step we have $\mu\delta^2(xy) = 0$, and

$$10(z_{0\cdot0} - z_{-0\cdot1}) = -\tfrac{1}{2}(0\cdot1000 + 0) = -0\cdot0500, \quad z_{-0\cdot1} = +0\cdot0050,$$
$$10(z_{0\cdot1} - z_{0\cdot0}) = \tfrac{1}{2}(0 + 0\cdot1000) = +0\cdot0500, \quad z_{0\cdot1} = +0\cdot0050.$$

* Which proceeded five steps at a time in comparison with the others' one.

9·13 *Solution by extrapolation of $\delta^2 y$* 297

Proceeding to the right part of the table we have

$$10(y_{0\cdot 0} - y_{-0\cdot 1}) = \tfrac{1}{2}(0\cdot 0050 + 0) - 0\cdot 0008 = 0\cdot 0017, \quad y_{-0\cdot 1} = 0\cdot 9998,$$

$$10(y_{0\cdot 1} - y_{0\cdot 0}) = \tfrac{1}{2}(0\cdot 0050 + 0) - 0\cdot 0008 = 0\cdot 0017, \quad y_{0\cdot 1} = 1\cdot 0002.$$

Filling in these values we have

x	z	xy	Δ	Δ^2	y	z	Δ	Δ^2
$-0\cdot 1$	0·0050	$-0\cdot 1000$			0·9998	0·0050		
0·0	0·0000	0·0000	$+0\cdot 1000$		1·0000	0·0000	-50	
0·1	0·0050	$+0\cdot 1000$	$+0\cdot 1000$	0	1·0002	0·0050	$+50$	$+100$
0·2	(0·0200)	$(+0\cdot 2000)$	$(+0\cdot 1000)$	(0)	(1·0014)	(0·0200)	$(+150)$	$(+100)$

$\Delta(xy)$ can be extrapolated and gives $xy = +0\cdot 2000$ at $x = 0\cdot 2$. Then

$$10(z_{0\cdot 2} - z_{0\cdot 1}) = \tfrac{1}{2}(0\cdot 3000), \quad z_{0\cdot 2} = 0\cdot 0050 + 0\cdot 0150 = 0\cdot 0200.$$

We enter this on the right and form differences. Extrapolating $\Delta^2 z$ we have

$$10(y_{0\cdot 2} - y_{0\cdot 1}) = \tfrac{1}{2}(0\cdot 0250) - 0\cdot 0008 = 0\cdot 0117, \quad \Delta y = 0\cdot 0012, \quad y_{0\cdot 2} = 1\cdot 0014.$$

Multiplying this by 0·2 and returning to the left we have $xy = 0\cdot 2003$ instead of 0·2000, so we revise the differences. The change is too small to make any appreciable change in z. So we proceed, advancing a stage in each table alternately. The solution up to $x = 2\cdot 0$ is as follows:

x	z	xy	Δ	Δ^2	Δ^3	y	z	Δ	Δ^2	Δ^3
$-0\cdot 1$	0·0050	$-0\cdot 1000$				0·9998	0·0050			
0·0	0·0000	0·0000	1000			1·0000	0·0000	-50	100	
0·1	0·0050	0·1000	1000	0		1·0002	0·0050	50	100	
0·2	0·0200	0·2003	1003	3		1·0014	0·0200	150	101	
0·3	0·0451	0·3014	1011	8	10	1·0046	0·0451	251	102	
0·4	0·0804	0·4043	1029	18	15	1·0108	0·0804	353	104	
0·5	0·1261	0·5105	1062	33	18	1·0210	0·1261	457	109	
0·6	0·1827	0·6218	1113	51	24	1·0364	0·1827	566	115	
0·7	0·2508	0·7406	1188	75	27	1·0580	0·2508	681	123	
0·8	0·3312	0·8696	1290	102	34	1·0870	0·3312	804	136	
0·9	0·4252	1·0122	1426	136	41	1·1247	0·4252	940	153	17
1·0	0·5345	1·1725	1603	177	48	1·1725	0·5345	1093	169	16
1·1	0·6607	1·3553	1828	225	58	1·2321	0·6607	1262	196	27
1·2	0·8065	1·5664	2111	283	67	1·3053	0·8065	1458	228	32
1·3	0·9751	1·8125	2461	350	81	1·3942	0·9751	1686	267	39
1·4	1·1704	2·1017	2892	431	100	1·5012	1·1704	1953	315	48
1·5	1·3972	2·4440	3423	531	116	1·6293	1·3972	2268	374	59
1·6	1·6614	2·8510	4070	647	146	1·7819	1·6614	2642	445	71
1·7	1·9701	3·3373	4863	793	170	1·9631	1·9701	3087	533	88
1·8	2·3321	3·9199	5826	963	212	2·1777	2·3321	3620	639	106
1·9	2·7580	4·6200	7001	1175	260	2·4316	2·7580	4259	770	131
2·0	3·2609	5·4636	8436	1435	310	2·7318	3·2609	5029	(930)	(160)
				(1745)						

9·13. For a second-order equation with no term in y' other methods are available. If

$$y'' = f(x, y), \tag{1}$$

we have

$$\delta^2 y = (e^{Dh} - 2 + e^{-Dh})y = h^2 D^2 y + \{(2 \sinh \tfrac{1}{2} hD)^2 - h^2 D^2\} y$$

$$= h^2 f + \left\{\frac{(2 \sinh \tfrac{1}{2} hD)^2}{h^2 D^2} - 1\right\} h^2 f. \tag{2}$$

Also since

$$(2 \sinh \tfrac{1}{2} hD)^2 = \delta^2, \tag{3}$$

this is

$$h^2 f + h^2 \left\{\frac{\delta^2}{(2 \sinh^{-1} \tfrac{1}{2} \delta)^2} - 1\right\} f. \tag{4}$$

To expand the operator we have

$$\sinh^{-1} t = t - \tfrac{1}{6}t^3 + \tfrac{3}{40}t^5 - \tfrac{5}{112}t^7 + \ldots,\qquad(5)$$

$$\left(\frac{t}{\sinh^{-1} t}\right)^2 = (1 - \tfrac{1}{6}t^2 + \tfrac{3}{40}t^4 - \tfrac{5}{112}t^6 + \ldots)^{-2}$$

$$= 1 + \tfrac{1}{3}t^2 - \tfrac{1}{15}t^4 + \tfrac{31}{945}t^6 - \ldots,\qquad(6)$$

whence $\qquad h^{-2}\delta^2 y = f + \tfrac{1}{12}\delta^2 f - \tfrac{1}{240}\delta^4 f + \tfrac{31}{60480}\delta^6 f - \ldots.\qquad(7)$

The coefficient of the fourth difference in this formula is very small. Consequently it practically permits a definitive calculation of $\delta^2 y$ given f and $\delta^2 f$. Then having, say, $y(a-h)$ and $y(a)$, we have $\nabla y(a)$, and adding $\delta^2 y$ we have $\Delta y(a) = y(a+h) - y(a)$. Then $y(a+h)$ is found by addition. The complication here is that $f(a+h)$ will have to be calculated from $y(a+h)$, and until we know it we do not know $\delta^2 f(a)$. But this is easily circumvented. If, for instance, $h = 0.1$, $\tfrac{1}{12}h^2\delta^2 f$ is about $\tfrac{1}{1200}f''$, and will not affect the fourth place of decimals if f'' is less than 0·0600. Further, if we can infer f'' within 0·0600 from entries further up the table, we can use this approximate value in (7) and still get $\delta^2 y$ right within the rounding-off error. We then proceed by addition to infer $y(a+h)$; from this we calculate $f(a+h)$, and form the differences of the latter accurately. If necessary we can correct the approximate value taken for f'' and repeat the calculation. In most cases, if the interval is suitably chosen, it will rarely be found that further revision changes $\delta^2 y$ by more than a unit in the last place.

It is slightly more convenient to extrapolate $\tfrac{1}{12}\delta^2 f$ than $\delta^2 f$, since it is going to be multiplied by the small quantity h^2 in any case. But if we proceed directly to y in this way we must pay special attention to the initial conditions. If in the above example we worked to the fourth decimal, the values of y at $x = 0$ and 0.1 would only determine y' there within 0·0005, and to this extent the solution would be uncertain by 0·0005 times the solution that makes $y = 0$, $y' = 1$ at $x = 0$. In this case the solution in question reaches 3·6 at $y = 2.0$, so that for this reason alone an error of 18 in the fourth place might accumulate. This is avoided by the method already used, since this attends to y' explicitly and takes its value at $x = 0$ as a starting point. But if we use the present method we can save the situation only by keeping an extra figure in the calculation.

To treat the example of 9·121 in this way we notice that the first two terms of the Taylor series are
$$y = 1 + \tfrac{1}{6}x^3 + \ldots.$$

We write down a few values of y from this formula:

x	y	Δ	Δ^2		xy	Δ	Δ^2	$\tfrac{1}{12}\delta^2$
−0·2	0·99866				−0·19973			
−0·1	0·99983	+117	−100		−0·09998	+9975	+23	2
0·0	1·00000	+17	0		0·00000	+9998	+4	0
0·1	1·00017	+17	(+100)		+0·10002	+10002	(+23)	2
0·2	(1·00134)	(+117)	(+200)		(0·20027)	(10025)	(83)	7
0·3	(1·00451)	(317)			(0·30135)	(10108)		

We can ignore $\tfrac{1}{12}\delta^2(xy)$ in proceeding to the next stage. We have

$$100\delta^2 y_{0\cdot 1} = 0\cdot 10002 + 0, \quad \delta^2 y_{0\cdot 1} = 0\cdot 00100.$$

We enter this in its place in the differences of y, add to $\delta\Delta y_{0\cdot 05}$ to give $+0\cdot 00117$, and add this in turn to $y_{0\cdot 1}$ to give $y_{0\cdot 2} = 1\cdot 00134$. We then work out xy with this value, 0·20027, and form its differences. We now see that $\tfrac{1}{12}\delta^2(xy)$ should be 2 in the fifth decimal, but this will affect nothing in the calculation yet made, which can therefore be taken as confirmed.

9·13 *Example of method*

At the next stage we provisionally try $\tfrac{1}{12}\delta^2(xy)_{0\cdot 2} = 0\cdot 00004$; then

$$100\delta^2 y_{0\cdot 2} = 0\cdot 20027 + 0\cdot 00004, \quad \delta^2 y_{0\cdot 2} = 0\cdot 00200,$$

and we enter this and work up to $y_{0\cdot 3}$. Calculate xy as before and form its differences. We now find $\tfrac{1}{12}\delta^2(xy)_{0\cdot 2} = 0\cdot 00007$. Again the change does not affect the last figure of y.

We proceed with the calculation; $\tfrac{1}{12}\delta^2(xy)$ increases until it does begin to affect the extrapolation of y, and then to a stage when a supplementary table of its differences becomes worth while to assist its extrapolation. Thus at a later stage we have

x	y	Δ	Δ^2	xy	Δ	Δ^2	$\tfrac{1}{12}\delta^2$	Δ	Δ^2
1·4	1·50089		2106	2·10125		5298	442		16
1·5	1·62894	12805	2449	2·44341	34216	6480	540	98	21
1·6	1·78148	15254	2857	2·85037	40696	7907	659	119	
1·7	1·96259	18111	(3344)	3·33640	48603	(9642)	(804)		
1·8	(2·17714)	(21455)		(3·91885)	(58245)				

From the differences of $\tfrac{1}{12}\delta^2(xy)$ we infer that its next second difference is likely to be about 26, the next first difference therefore 145, so in the next stage we try

$$\tfrac{1}{12}\delta^2(xy) = 659 + 145 = 804.$$

Then $\qquad 100\delta^2 y_{1\cdot 7} = 3\cdot 33640 + 804 = 3\cdot 34444;$

and we enter 3344 as the second difference of y. The next steps, in order, are

$$\Delta y_{1\cdot 7} = 0\cdot 21455, \quad y_{1\cdot 8} = 2\cdot 17714, \quad (xy)_{1\cdot 8} = 3\cdot 91885,$$
$$\Delta(xy)_{1\cdot 7} = 0\cdot 58245, \quad \delta^2(xy)_{1\cdot 7} = 9642, \quad \tfrac{1}{12}\delta^2 = 804.$$

This agrees with the trial value and no change is needed. The complete table is as follows:

x	y	Δ	Δ^2	xy	Δ	Δ^2	$\tfrac{1}{12}\delta^2$	Δ	Δ^2
−0·2	0·99866			−0·19973					
−0·1	0·99983	117	−100	−0·09998	9975	23	2		
0·0	1·00000	17	0	0·00000	9998	4	0		
0·1	1·00017	17	+100	0·10002	10002	23	2		
0·2	1·00134	117	200	0·20027	10025	83	7		
0·3	1·00451	317	302	0·30135	10108	185	15		
0·4	1·01070	619	405	0·40428	10293	326	27		
0·5	1·02094	1024	511	0·51047	10619	511	43		
0·6	1·03629	1535	622	0·62177	11130	743	62		
0·7	1·05786	2157	741	0·74050	11873	1024	85	29	
0·8	1·08684	2898	871	0·86947	12897	1364	114	33	
0·9	1·12453	3769	1014	1·01208	14261	1707	147	40	
1·0	1·17236	4783	1174	1·17236	16028	2248	187	48	
1·1	1·23193	5957	1357	1·35512	18276	2820	235	57	11
1·2	1·30507	7314	1569	1·56608	21096	3503	292	68	14
1·3	1·39390	8883	1816	1·81207	24599	4319	360	82	16
1·4	1·50089	10699	2106	2·10125	28918	5298	442	98	21
1·5	1·62894	12805	2449	2·44341	34216	6480	540	119	26
1·6	1·78148	15254	2857	2·85037	40696	7907	659	145	31
1·7	1·96259	18111	3344	3·33640	48603	9642	804	176	39
1·8	2·17714	21455	3929	3·91885	58245	11756	980	215	47
1·9	2·43098	25384	4631	4·61886	70001	14339	1195	262	
2·0	2·73113	30015	(5477)	5·46226	84340		(1457)		

The solution differs at $x = 2\cdot 0$ by $0\cdot 0007$ from the previous one. This is about the difference that might be expected from the accumulation of rounding-off errors. The correct solution found from power series is $1\cdot 17230$ for $x = 1\cdot 0$, $2\cdot 73088$ for $x = 2\cdot 0$.

With regard to the neglected term in $\delta^4(xy)$, this also may accumulate. But if we sum (7) we see that the total contribution of $\delta^4(xy)$ will be about $h^2/240$ times the change of

$\delta^3(xy)$, and by inspection of the table this change is about 3000 in the fifth decimal. Hence the total effect of the δ^4 term in the range is about 1 in the sixth decimal and is correctly neglected.

9·14. The Gauss-Jackson method. The last method was used by Cowell and Crommelin in their work on the motion of Halley's comet between 1759 and 1910, and has been extensively used in the solution of Schrödinger's equation, particularly by D. R. Hartree.[*] Cowell, however, recommended a slightly different method in an appendix.[†] This is discussed further by J. Jackson,[‡] who remarks that the matter had been left in a practically perfect state by Gauss. The procedure is *to introduce a function whose second differences are f*. If we have such a function we can denote it by $\delta^{-2}f$; and then (7) can be written

$$h^{-2}\delta^2 y = \delta^2(\delta^{-2}f + \tfrac{1}{12}f - \tfrac{1}{240}\delta^2 f + \tfrac{31}{60480}\delta^4 f - \ldots), \tag{8}$$

and the two functions

$$h^{-2}y, \quad \delta^{-2}f + \tfrac{1}{12}f - \tfrac{1}{240}\delta^2 f + \tfrac{31}{60480}\delta^4 f - \ldots \tag{9}$$

have the same second differences. But a differential equation of the second order needs two adjustable constants to specify its solution; and any function of the form $A + Bx$ will give zero second difference. Consequently we are at liberty to add $A + Bx$ to $\delta^{-2}f$ and choose A and B so that the expressions (9) will be equal for two values of x. Then since the second differences are equal the functions are equal for all tabulated values of x, and

$$h^{-2}y = \delta^{-2}f + \tfrac{1}{12}f - \tfrac{1}{240}\delta^2 f + \tfrac{31}{60480}\delta^4 f - \ldots. \tag{10}$$

The advantage of this procedure is that the summation to get δ^{-2}, once we have started, can be done exactly, and each rounding-off error in the correcting terms of (10) arises only once, and, with $h = 0\cdot1$, is divided by 100 before it is passed on to the next stage of the calculation.

To make a start with the solution of the same equation as before we fit the values already found at $x = 0$ and $0\cdot1$. We have

$$100\cdot000 = \delta^{-2}f_{0\cdot0} + 0, \quad 100(1\cdot000167) = \delta^{-2}f_{0\cdot1} + \tfrac{1}{12}(0\cdot10002),$$

whence
$$\delta^{-2}f_{0\cdot1} = 100\cdot0083.$$

The calculation is now straightforward.[§] We enter the table as follows:

x	$\delta^{-2}f$	$\delta^{-1}f$	f	$100y$	x	$\delta^{-2}f$	$\delta^{-1}f$	f	$100y$
0·0	100·0000		0·00000	100·000	1·0	117·13221		1·17230	117·230
0·1	100·0083	0·0083	0·10002	100·017	1·1	123·07291	5·94070	1·35505	123·186
0·2	100·11662	0·10832	0·20027	100·133	1·2	130·36866	7·29575	1·56599	130·499
0·3	100·42521	0·30859	0·30135	100·450	1·3	139·23040	8·86174	1·81195	139·381
0·4	101·03515	0·60994	0·40428	101·069	1·4	149·90409	10·67369	2·10111	150·079
0·5	102·04937	1·01422	0·51046	102·092	1·5	162·67889	12·77480	2·44323	162·882
0·6	103·57405	1·52468	0·62178	103·626	1·6	177·89692	15·21803	2·85014	178·134
0·7	105·72051	2·14646	0·74047	105·782	1·7	195·96509	18·06817	3·33613	196·243
0·8	108·60744	2·88693	0·86944	108·680	1·8	217·36939	21·40430	3·91853	217·696
0·9	112·36381	3·75637	1·01203	112·448	1·9	242·69222	25·32283	4·61846	243·077
1·0	117·13221	4·76840	1·17230	117·230	2·0	272·63351	29·94129	5·46178	273·089

[*] *Manchester Lit. and Phil. Mem. and Proc.* 77, 1933, 91–107; D. R. Hartree and W. Hartree, *Proc. Roy. Soc.* A, 150, 1935, 9–33; 154, 1936, 588–607; 156, 1936, 45–62; 166, 1938, 450–64.
[†] *Greenwich Observations*, 1909.   [‡] *Mon. Not. R. Astr. Soc.* 84, 1924, 602–6.
[§] With no extra trouble we could find an extra figure in $\delta^{-2}f_{0\cdot1}$, and this would improve the accuracy. This has not been done, however, so as to have a fairer comparison with the other methods.

To start with, we difference the first two values of $\delta^{-2}f$ to get $\delta^{-1}f_{0·05}$. f is known at 0·1, and we add it to $\delta^{-1}f_{0·05}$ to get $\delta^{-1}f_{0·15}$. This is then added to $\delta^{-2}f_{0·1}$ to give $\delta^{-2}f_{0·2}$. Then $y_{0·2}$ is given by the equation
$$100y = 100·1166 + \tfrac{1}{12}f_{0·2}.$$

By extrapolation we try $f_{0·2} = 0·200$, and the correcting term is $+0·0167$, making $100y = 100·133$. Multiplying this by 0·2 we have $f_{0·2} = 0·20027$, and the change makes no change in the third decimal of $100y$. If there was a change at any stage we should continue the approximation till there is none. It is convenient to extrapolate $\tfrac{1}{12}f$ at each stage, and not to fill in f till the second approximation to save rewriting.

The fourth decimal in $\delta^{-1}f$ and the third in $\delta^{-2}f$ have little direct importance, but it is as easy to write them in as not, and they enable the rounding-off errors to be absorbed harmlessly in a place where they will be divided by 100 before y is calculated. The result for $x = 2·0$ is $y = 2·73089$, which is 1 unit of the fifth place from the correct value; and the amount of subsidiary calculation is less than with either of the other methods. It is not even necessary to write in the differences of f and y, since neither affect the calculation. The approximation for y, however, is of a kind such that a mistake is likely to be repeated at the next approximation, and differences should be used as a check. Occasional inspection of the second differences of f is also desirable in case their contributions should become appreciable; but they must reach $120h^{-2}$ in the last place retained for them to matter, and if they do it is less trouble to use a shorter interval. Special attention should be given to the calculation of the first two values of $\delta^{-2}f$, because an error in their difference produces an error that may increase steadily throughout the calculation. As soon as four or five values of y have been found they should be differenced to check this stage of the calculation.

The possibility of using this method depends on the absence of a term in dy/dx from the differential equation.[a] The convenience of the method is such that when such a term is present in a linear equation it is best to begin by transforming the equation so as to remove it. Astronomers, in computing perturbations, therefore largely prefer to use rectangular coordinates, even though this involves sacrificing the use of the elliptic orbit as a first approximation. The component accelerations due to the sun are included in the numerical computation at each stage and treated just like the planetary terms. This inconvenience is far more than compensated by having to deal with differential equations of the form

$$\frac{d^2 x_r}{dt^2} = f(x_1, x_2, \ldots, x_n),$$

instead of, for example, in polar coordinates,

$$\frac{d}{dt}(\sin^2 \theta \dot\lambda) = g(x_1, x_2, \ldots, x_n).$$

If $f(x, y)$ varies considerably within the range of integration it may be convenient to increase or reduce the interval. To change from interval 0·1 to 0·2 at $x = 2·0$, we should use the values already found for $y_{1·8}$ and $y_{2·0}$ to find corresponding values of $\delta_{-2}f$, and start afresh. To change from 0·1 to 0·05 would require first the interpolation of a value for $y_{1·95}$, and then the calculation of $\delta^{-2}f_{1·95}$, $\delta^{-2}f_{2·0}$. The latter will *not* be the same as for the original interval.

9·15. Estimation of an eigenvalue.

This method of solution is conveniently combined with Rayleigh's principle to give a rapidly converging series of approximations to, for instance, the period of a dynamical system. Consider for instance the oscillations of water in a narrow lake of elliptical plan. If ζ is the elevation of the water surface, u the velocity, h the depth, g gravity, and b the breadth, the equations for a small oscillation of period $2\pi/\gamma$ are

$$\frac{\partial u}{\partial t} = -g\frac{\partial \zeta}{\partial x}, \quad \frac{\partial}{\partial t}(b\zeta) = -\frac{\partial}{\partial x}(hbu). \tag{1}$$

Put
$$hbu = V; \tag{2}$$

then on elimination in favour of V we have

$$b\frac{\partial}{\partial x}\left(\frac{1}{b}\frac{\partial V}{\partial x}\right) + \kappa^2 V = 0, \tag{3}$$

where
$$\kappa^2 = \gamma^2/gh. \tag{4}$$

The boundary conditions are that $V = 0$ at the ends. We can remove the term in $\partial V/\partial x$ by the substitution

$$V = b^{1/2}U; \tag{5}$$

then
$$\frac{\partial^2 U}{\partial x^2} + \left(\kappa^2 + \frac{b''}{2b} - \frac{3b'^2}{4b^2}\right)U = 0. \tag{6}$$

With
$$b \propto (1-x^2)^{1/2}, \tag{7}$$

this is
$$\frac{\partial^2 U}{\partial x^2} = -\left(\kappa^2 - \frac{1+\tfrac{3}{2}x^2}{2(1-x^2)^2}\right)U. \tag{8}$$

It can be shown* that the two solutions near $x = \pm 1$ make U behave like $(1-x^2)^{-1/4}$ or $(1-x^2)^{5/4}$. The former would make V different from zero at $x = \pm 1$, and must therefore be excluded. The problem is to find the values of κ^2 that make it possible to avoid this solution at both ends. By symmetry U must be either an even or an odd function of x.

The mean kinetic energy over a period is given by

$$4T = \int bhu^2 dx = \int \frac{V^2}{hb}dx, \tag{9}$$

and the mean potential energy by

$$4W = \int gb\zeta^2 dx = \int \frac{g}{\gamma^2 b}\left(\frac{\partial V}{\partial x}\right)^2 dx. \tag{10}$$

Using the principle that the mean kinetic and potential energies in a period are equal we get

$$\kappa^2 = \frac{\gamma^2}{gh} = \int \frac{1}{b}\left(\frac{\partial V}{\partial x}\right)^2 dx \bigg/ \int \frac{V^2}{b}dx = \int \frac{1}{b}\left(\frac{\partial V}{\partial x}\right)^2 dx \bigg/ \int U^2 dx. \tag{11}$$

Rayleigh's principle is that any form of V satisfying the boundary conditions, but not the differential equation, will give a second-order error in κ^2 when substituted in (11).

It is fairly clear that the lowest value of κ will be such that $\partial V/\partial x$ is, on the whole, as small as possible for given mean V^2; and therefore that V keeps the same sign for all x. The next lowest value will make V change sign once, and so on.

* Cf. Chapter 16.

9·15 *Estimation of an eigenvalue* 303

For each trial value of κ^2 we first make a table of the coefficient of U in (8). We take $U = 1\cdot00000$, $\partial U/\partial x = 0$ at $x = 0$, and have for small x

$$U = 1 - \tfrac{1}{2}(\kappa^2 - \tfrac{1}{2})x^2 + \tfrac{1}{12}\{\tfrac{1}{2}(\kappa^2 - \tfrac{1}{2})^2 + \tfrac{7}{4}\}x^4 \ldots, \tag{12}$$

which gives U for $x = 0\cdot1$ and hence $\delta^{-2}f$, where f is the right side of (8).

The solution is then developed. An incorrect value of κ^2 will be shown by the solution tending to $\pm\infty$ at $x = 1$. Specimen solutions are as follows:

x	$U(\kappa^2 = 3)$	$U(\kappa^2 = 4)$	x	$U(\kappa^2 = 3)$	$U(\kappa^2 = 4)$
0·0	1·0000	1·0000	0·5	0·7129	0·6024
0·1	0·9875	0·9826	0·6	0·6028	0·4515
0·2	0·9506	0·9310	0·7	0·4862	0·2912
0·3	0·8908	0·8478	0·8	0·3718	0·1288
0·4	0·8104	0·7365	0·9	0·2781	−0·0292

With the information that the solutions near $x = 1$ should be of the form

$$A(1-x)^{5/4} + B(1-x)^{-1/4}$$

we can find A and B roughly from the last two entries in each table. We find

$$\kappa^2 = 3, \quad A = +15\cdot5, \quad B = +10\cdot7,$$
$$\kappa^2 = 4, \quad A = +17\cdot2, \quad B = -7\cdot1.$$

Hence the former solution will make U tend to $+\infty$, the second to $-\infty$, as $x \to 1$. By interpolation B should vanish with κ^2 about 3·6. A pair of trials for 3·4 and 3·6 suggested 3·56; but at this point it seemed that intervals of 0·05 instead of 0·1 would be safer in testing the behaviour near $x = 1$. The solution with this interval is as follows:

$\kappa^2 = 3\cdot56$

x	U	x	U	x	U
0	1·0000	0·35	0·8210	0·70	0·3754
0·05	0·9962	0·40	0·7693	0·75	0·3036
0·1	0·9850	0·45	0·7127	0·80	0·2327
0·15	0·9662	0·50	0·6510	0·85	0·1643
0·20	0·9401	0·55	0·5857	0·90	0·1000
0·25	0·9069	0·60	0·5175	0·95	0·0428
0·30	0·8671	0·65	0·4471		

If the solution was correct the ratio of the last two entries should be nearly $2^{5/4} = 2\cdot378$. It is actually 2·338, so that we are very close. To get a better approximation we use Rayleigh's principle. We multiply U by $(1-x^2)^{1/4}$ to get V, differentiate numerically, work out $(1-x^2)^{-1/2}(\partial V/\partial x)^2$, and integrate. As the integrand behaves like $(1-x^2)^{1/2}$ near $x = 1$ it is best to use the formulae of 9·092 from 0·90 to 1·00, and the Gregory formula up to 0·90. This gives

$$\int_0^1 (1-x^2)^{-1/2}\left(\frac{\partial V}{\partial x}\right)^2 dx = 1\cdot6287.$$

The integration of U^2 is simple except again for values beyond 0·9. For these we assume that

$$U^2 = 0\cdot0100\left(\frac{1-x}{0\cdot1}\right)^{5/2},$$

whence

$$\int_{0\cdot9}^{1\cdot0} U^2 dx = 0\cdot0004$$

and

$$\int_0^1 U^2 dx = 0\cdot4596.$$

Then $\kappa^2 = \dfrac{1 \cdot 6287}{0 \cdot 4596} = 3 \cdot 544; \quad \kappa = 1 \cdot 882.$

Previous solutions had given $\kappa = 1 \cdot 886$ by totally different methods.*

The solutions for $\kappa^2 = 3$ and $\kappa^2 = 4$ were not really necessary to the method, but are given to show how we can detect with a wrong value that the solution is not tending to zero in the way it should. Rayleigh's solution will usually give an accuracy within a few per cent with an assumed form that is even roughly near the truth, and it would have been possible to apply it at the start with $U = (1-x^2)^{5/4}$. This gives $\kappa^2 \doteqdot 3 \cdot 6$ immediately. As the error is squared at each stage it should be possible to get four-figure accuracy with at most three solutions.

Alternatively, we could assume a solution $(1-x^2)^{5/4}(1+Ax^2+Bx^4)$ and determine A, B to make κ^2, as found from (11), stationary for small variations of A, B. A similar method in principle was used by Ritz to determine the normal modes of vibration of a square plate.

To treat the second mode it is desirable to begin by subtracting from the trial solution such a multiple of the solution for the lowest mode as will make the remainder exactly orthogonal with the latter solution (cf. 6·08).†

9·16. Numerical solution of simultaneous linear equations. The methods usually given are unnecessarily laborious. We shall illustrate the solution by an example. Take the three equations

$$6 \cdot 3x - 3 \cdot 2y + 1 \cdot 0z = +7 \cdot 8, \tag{1}$$

$$-3 \cdot 2x + 8 \cdot 4y - 2 \cdot 6z = -2 \cdot 3, \tag{2}$$

$$+1 \cdot 0x - 2 \cdot 6y + 5 \cdot 7z = +8 \cdot 6. \tag{3}$$

The coefficients here form a symmetrical matrix: this is not necessary to the method, but in practice the condition is so often satisfied that we may as well take an instance of it. *Divide* the first equation by the coefficient of x, and then multiply the resulting equation by the coefficients of x in the other two. By addition or subtraction we then eliminate x and proceed. The complete solution is arranged as follows:

$$
\begin{array}{l|l}
6 \cdot 3x - 3 \cdot 2y + 1 \cdot 0z = +7 \cdot 8 & x - 0 \cdot 508y + 0 \cdot 159z = +1 \cdot 238 \\
-3 \cdot 2x + 8 \cdot 4y - 2 \cdot 6z = -2 \cdot 3 & 3 \cdot 2x - 1 \cdot 63y + 0 \cdot 51z = +3 \cdot 96 \\
+1 \cdot 0x - 2 \cdot 6y + 5 \cdot 7z = +8 \cdot 6 & x - 0 \cdot 51y + 0 \cdot 16z = +1 \cdot 24 \\
\\
6 \cdot 77y - 2 \cdot 09z = +1 \cdot 66 & y - 0 \cdot 309z = +0 \cdot 245 \\
-2 \cdot 09y + 5 \cdot 54z = +7 \cdot 36 & 2 \cdot 09y - 0 \cdot 65z = +0 \cdot 51 \\
\\
4 \cdot 89z = +7 \cdot 87 & z = +1 \cdot 609 \\
& y = +0 \cdot 245 + 0 \cdot 497 = +0 \cdot 742 \\
& x = 1 \cdot 238 + 0 \cdot 377 - 0 \cdot 256 = +1 \cdot 359
\end{array}
$$

---

* Jeffreys, *Proc. Lond. Math. Soc.* (2) 23, 1924, 463. Goldstein, *Proc. Lond. Math. Soc.* (2) 28, 1928, 95.

† For further discussion of the Ritz method see Temple and Bickley, *Rayleigh's principle*, pp. 150–2.

Check by substitution:

$$8.56 - 2.37 + 1.61 = +7.80,$$
$$-4.35 + 6.23 - 4.18 = -2.30,$$
$$+1.36 - 1.93 + 9.17 = +8.60.$$

The check shows that the solution is right to about 0·001.

When there are more than three or four equations a mistake will usually be made, and it is desirable to be able to detect it before reaching the final check by substitution. This can be done in two ways. (1) If the unknowns are $x_1, x_2, \ldots, x_n$ we can first eliminate x_1 and then x_2 by the above method. Then transpose the first two equations and eliminate first x_2 and then x_1. The whole of the coefficients in the simplified equations for $x_3, x_4, \ldots$ should be the same for both methods, and the place where any inconsistency occurs indicates at once a small set of steps where a mistake can have occurred. (2) If all the coefficients are calculated to the same number of decimals, we can take their sum and perform the same operations on the sums. It does not matter whether we reverse the sign of the term on the right so long as we always do the same. Thus

$$6.3 - 3.2 + 1.0 + 7.8 = +11.9, \quad 11.9/6.3 = +1.889,$$
$$1.000 - 0.508 + 0.159 + 1.238 = +1.889,$$

which checks the first division. Next,

$$-3.2 + 8.4 - 2.6 - 2.3 = +0.3, \quad 1.889 \times 3.2 = 6.04,$$
$$0.3 + 6.04 = 6.34, \quad 6.77 - 2.09 + 1.66 = 6.34,$$

which checks the elimination of y. In this method the check sum is written in an extra column to the right of its equation, and any mistake can be detected by adding coefficients.

An alternative method due to von Seidel can be used when the matrix of the coefficients on the left is that of a positive definite quadratic form. To illustrate it on the above set of equations we write them in the form

$$x = +1.238 + 0.508y - 0.159z, \tag{4}$$
$$y = -0.274 + 0.381x + 0.310z, \tag{5}$$
$$z = +1.509 - 0.175x + 0.456y, \tag{6}$$

and solve by successive approximation. The method requires for its rapid convergence that the coefficients of x, y, z on the right shall be fairly small; in these equations they are a little too large to show the method at its best. We first neglect y and z in (4) and take the first approximation $x = +1.238$. Substitute this on the right of (5). This gives $y = +0.198$. Substitute both these values in the right of (6); we get $z = +1.382$.

Now return to (4) and substitute $y = +0.198$, $z = +1.382$; we have now $x = +1.119$. Proceeding, we get approximations in turn as follows:

x	+1·238	1·119	1·282	1·340,	1·351
y	+0·198	0·580	0·703	0·735	0·740
z	+1·382	1·577	1·606	1·610	1·608

The change from the third approximation to the fourth is small. Apart from the formation of (4), (5), (6) the method is iterative and checks itself. It is seen at its best when the number of equations is large and the non-diagonal coefficients are all small and many of them zero.

The point of the method is that the solution is the set of values that make the function
$$S = 3 \cdot 15x^2 - 3 \cdot 2xy + 1 \cdot 0xz + 4 \cdot 2y^2 - 2 \cdot 6yz + 2 \cdot 85z^2$$
$$- 7 \cdot 8x + 2 \cdot 3y - 8 \cdot 6z$$

a minimum. The minimum exists because the quadratic terms are positive definite. Let it be Σ, and let the corresponding values of x, y, z be x_0, y_0, z_0. When we adjust any unknown we find the value that makes S a minimum given that the other unknowns have the values taken in the previous approximation. Hence the values of S, say S_n, corresponding to successive approximations form a non-increasing sequence. Also, unless all the equations are satisfied, adjustment of at least one of x, y, z will reduce S_n; hence if $S_n > \Sigma$, $S_{n+3} < S$. Put $S_n - S_{n+3} = T_n$; then, if we consider all values of x, y, z that give the same value of S_n, T_n is continuous and takes its lower bound, which therefore cannot be zero. Also, if we put $x = x_0 + x'$, $y = y_0 + y'$, $z = z_0 + z'$, $S = \Sigma + S'$, T_n is a positive definite quadratic in x'_n, y'_n, z'_n. Let the lower bound of T_n when x', y', z' vary, S_n being constant, be $\alpha S'_n$. Then $1 \geqslant \alpha > 0$. If x'_n, y'_n, z'_n are all multiplied by the same constant factor k, S'_n and T'_n are both multiplied by k^2. Thus for any set of values x'_n, y'_n, z'_n, the approximations after three more steps will give
$$S'_{n+3} = S'_n - T_n \leqslant (1-\alpha) S'_n.$$

Since $\{S'_n\}$ is non-increasing, it follows that $S'_n \to 0$. Also the successive inequalities $S' \leqslant S'_n$ specify a set of regions of x, y, z each contained in the preceding, with diameters tending to zero. Hence the values of x, y, z given by the process converge to values corresponding to $S' = 0$, that is, to the correct solution. [a]

9·17. Jury problems: ordinary differential equations. For an ordinary differential equation of the second order we may either have given values of y and dy/dx at one terminus, or a value of one of them at each of two termini. In the former case we form the numerical solution by proceeding one step at a time; this has therefore been called a *marching problem* by L. F. Richardson. The latter type of problems are called *jury problems*. For jury problems, when the equation is linear, we can make the solution depend on those of two marching problems from one terminus, which are then combined linearly so as to satisfy the condition at the other. This method fails for non-linear equations. It is possible to obtain a first approximation that satisfies the terminal conditions, and then the differential equation can be converted into a finite difference equation and used to obtain better approximation to the intermediate values. As a simple case, take the equation
$$\frac{d^2y}{dx^2} = -y, \qquad (1)$$
with $y = 0$ at $x = 0$, and $y = 1$ at $x = 1$. We know that the solution is
$$y = \frac{\sin x}{\sin (1 \text{ radian})}. \qquad (2)$$

But suppose that we did not know this. Try to interpolate a value of y at $x = 0.5$. We have, using second differences at interval 0.5,
$$\frac{d^2y}{dx^2} = 4\delta^2 y = 4(y_1 + y_0 - 2y_{0\cdot 5}) = -y_{0\cdot 5}, \qquad (3)$$

whence
$$7y_{0\cdot5} = 4y_1 + 4y_0 = 4, \tag{4}$$

$$y_{0\cdot5} = 0\cdot57. \tag{5}$$

Now interpolate to intervals $0\cdot2$ by second differences. This gives the first approximation y_1:

x	y_1	y_2	y_3	y_4	Correct y
0	0	0	0	0	0
0·2	0·245	0·239	0·238	0·237	0·236
0·4	0·467	0·468	0·466	0·464	0·463
0·6	0·667	0·672	0·674	0·673	0·671
0·8	0·845	0·851	0·853	0·854	0·853
1·0	1·000	1·000	1·000	1·000	1·000

With intervals $0\cdot2$, (3) is replaced by

$$25(y_{-h} - 2y_0 + y_h) = -y_0, \tag{6}$$

that is
$$y_0 = 0\cdot5102(y_{-h} + y_h). \tag{7}$$

With the values of y_1 at $x = 0\cdot6$ and $1\cdot0$ this gives a second approximation for y at $0\cdot8$, namely, $0\cdot851$. This, with y_1 at $0\cdot4$, gives y_2 at $0\cdot6$ equal to $0\cdot672$. We thus get the column y_2; further similar approximations give y_3 and y_4. The correct values are given in the last column. Fourth differences can be taken into account if required, but it is then necessary to continue the solution one place beyond the ends of the table.

9·18. Relaxation Methods. Partial differential equations with given boundary conditions can be treated by an extension of the method. As an example we take Laplace's equation. Suppose that a solution is expressed in the form

$$\phi = a_0 + a_1 r \cos\theta + b_1 r \sin\theta + a_2 r^2 \cos 2\theta + b_2 r^2 \sin 2\theta + \ldots + b_4 r^4 \sin 4\theta. \tag{1}$$

Suppose further that we are given the values of ϕ at the points $(1, 0)$ $(0, 1)$ $(-1, 0)$ $(0, -1)$. Denote these by $\phi_1, \phi_2, \phi_3, \phi_4$. Then we take the coefficients up to b_2 as unknowns and try to adjust them so as to make the sum agree with these values as closely as possible, judged by the sum of squares: that is, we make

$$(a_0 + a_1 + a_2 - \phi_1)^2 + (a_0 + b_1 - a_2 - \phi_2)^2 + (a_0 - a_1 + a_2 - \phi_3)^2 + (a_0 - b_1 - a_2 - \phi_4)^2 \tag{2}$$

a minimum. The condition for a minimum, with regard to a_0, is

$$\phi_0 = a_0 = \tfrac{1}{4}(\phi_1 + \phi_2 + \phi_3 + \phi_4). \tag{3}$$

Now consider the set shown in the next diagram and retain terms to b_4. (The point marked 5 is $(2, 0)$ and so on.) Forming a sum of squares similarly we find that the conditions for a minimum, so far as they contain a_0, are

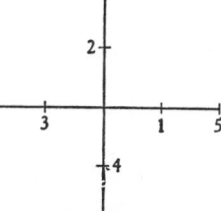

$$4a_0 + 34a_4 = \tfrac{1}{2}(\phi_1 + \ldots + \phi_8)$$

$$34a_0 + 514a_4 = \tfrac{1}{2}(\phi_1 + \phi_2 + \phi_3 + \phi_4) + 8(\phi_5 + \phi_6 + \phi_7 + \phi_8)$$

whence
$$a_0 = \tfrac{4}{15}(\phi_1 + \phi_2 + \phi_3 + \phi_4) - \tfrac{1}{60}(\phi_5 + \phi_6 + \phi_7 + \phi_8). \tag{4}$$

For the set in the next figure, we can again retain terms to b_4; but it is obvious that the estimate of a_0 depends only on the sums

$$S_1 = (\phi_1+\phi_2+\phi_3+\phi_4), \quad S_2 = (\phi_5+\phi_6+\phi_7+\phi_8) \tag{5}$$

and we shall get the same value for a_0 by taking mean values

$$\phi_1 = \phi_2 = \phi_3 = \phi_4 = \tfrac{1}{4}S_1, \quad \phi_5 = \phi_6 = \phi_7 = \phi_8 = \tfrac{1}{4}S_2. \tag{6}$$

This makes $a_1 = b_1 = a_3 = b_3 = 0$,

$$\tfrac{1}{4}S_1 = a_0 + a_4$$

$$\tfrac{1}{4}S_2 = a_0 - 4a_4.$$

Hence, irrespective of a_4,

$$a_0 = \tfrac{1}{5}S_1 + \tfrac{1}{20}S_2. \tag{7}$$

If ϕ satisfies Laplace's equation in a region, and we want its value at a point of the region given those at surrounding points of a rectangular network, (3) will give an approximation, which is simply the mean of the values at the four adjacent points. It takes account only of terms in r^2; the formulae (4) and (7), which are accurate to r^4, will be substantially more accurate.

The procedure will then be to take a trial set of values over a rectangular network so as to satisfy the boundary conditions, and to adjust them in turn.

Special attention is desirable at corners, where the appropriate expansion of ϕ will not be of the form (1). In the problem we shall consider in a moment we have the distribution shown. Take ϕ to be 0 over the boundary and to have given trial values at the points marked 2, 3, 4, 5, 6, where the values for 2 and 6, and for 3 and 5, are equal. Then the appropriate form of ϕ near the corner is

$$\phi = Ar^{2/3}\sin\tfrac{2}{3}\theta + Br^2\sin 2\theta \tag{8}$$

and the correct values will be

Point
$$\begin{aligned}
4 \quad & A \cdot 2^{1/3}\sin\tfrac{1}{2}\pi - 2B = 1\cdot 260A - 2B, \\
3, 5 \quad & A\sin\tfrac{1}{3}\pi = 0\cdot 866A, \\
2, 6 \quad & A \cdot 2^{1/3}\sin\tfrac{1}{6}\pi + 2B = 0\cdot 630A + 2B.
\end{aligned} \tag{9}$$

It will in general be impossible to find A and B so as to fit three datum values for ϕ, but we can adjust them by least squares to give the best fit as a whole and then use the solution as a smoothing function. If the trial values are ϕ_4, ϕ_3, ϕ_2 we get the minimum sum of squares of residuals at the five points by taking

$$\left.\begin{aligned}
A &= 0\cdot 3247\phi_4 + 0\cdot 4463\phi_3 + 0\cdot 3247\phi_2, \\
B &= \tfrac{1}{3}\phi_2 - \tfrac{1}{6}\phi_4.
\end{aligned}\right\} \tag{10}$$

If the adjusted values are $\phi'_4, \phi'_3, \phi'_2$,

$$\left.\begin{aligned}
\phi'_4 &= 0\cdot 742\phi_4 + 0\cdot 562\phi_3 - 0\cdot 258\phi_2, \\
\phi'_3 &= 0\cdot 281\phi_4 + 0\cdot 386\phi_3 + 0\cdot 281\phi_2, \\
\phi'_2 &= -0\cdot 129\phi_4 + 0\cdot 281\phi_3 + 0\cdot 871\phi_2.
\end{aligned}\right\} \tag{11}$$

These can be checked by taking ϕ_2, ϕ_3, ϕ_4 satisfying (9) exactly and verifying that $\phi_2', \phi_3', \phi_4'$ are equal to them respectively. The adjustment does not assume that the term in $r^{10/3} \sin \tfrac{10}{3}\theta$ is zero, but that it is small, and distributes the errors arising from its presence among $\phi_2', \phi_3', \phi_4'$; but close to the corner the smallness of this term will permit a good adjustment.

We can use (8) to halve the interval. If we take points 7, 8, 9 bisecting the lines joining the origin to 2, 3, 4, we have

$$\left. \begin{array}{l} \phi_7 = 0 \cdot 3968 A + \tfrac{1}{2} B = 0 \cdot 2955 \phi_2 + 0 \cdot 1771 \phi_3 + 0 \cdot 0455 \phi_4, \\ \phi_8 = 0 \cdot 5456 A \phantom{+ \tfrac{1}{2} B} = 0 \cdot 1771 \phi_2 + 0 \cdot 2435 \phi_3 + 0 \cdot 1771 \phi_4, \\ \phi_9 = 0 \cdot 7937 A - \tfrac{1}{2} B = 0 \cdot 0910 \phi_2 + 0 \cdot 3542 \phi_3 + 0 \cdot 3410 \phi_4. \end{array} \right\} \quad (12)$$

Even if higher terms are not negligible at 2, 3, 4, they will be much reduced at 7, 8, 9.

The process gives a rapid smoothing of departures from the true solution so far as they produce differences between the values of ϕ at adjacent points of the net. For departures that have the same sign at a block of neighbouring points the adjustment is much slower, and the process may appear to have converged sufficiently when in fact considerable errors survive. (This was true in the example given in the first edition of this book.) This is treated by Southwell by a method known as block adjustment. If we consider the sum

$$(\phi_0 - \phi_1 + \delta)^2 + (\phi_0 - \phi_2 + \delta)^2 + (\phi_0 - \phi_3 + \delta)^2 + (\phi_0 - \phi_4 + \delta)^2, \quad (13)$$

it is made a minimum for variations of δ by taking

$$\delta = \tfrac{1}{4}(\phi_1 + \phi_2 + \phi_3 + \phi_4 - 4\phi_0). \quad (14)$$

Hence $\phi_0 + \delta$ is the value of a_0 given by (3). In general the process of adjustment by (3) is equivalent to minimizing $\Sigma(\phi_r - \phi_s)^2$, where r and s indicate adjacent points of the net. Now suppose that we have a block of trial values and we wish to apply a uniform correction δ to all values within it, leaving the values outside it unaltered. Then the only terms in the sum that contain δ are those where r indicates an edge point of the block and s an outside point adjacent to it on the net, and these may be written

$$\Sigma(\phi_r - \phi_s + \delta)^2,$$

taken over all such pairs of points, say N in number; and the condition for this to be a minimum is

$$\delta = \frac{1}{N} \Sigma(\phi_s - \phi_r). \quad (15)$$

This correction is applied to the whole of the block.

In addition to the improvement made by block adjustment, we now discuss a method by which the convergence can often be made more rapid than that obtained by a naïve use of (3). If ϕ at a point of the net differs by $-\delta$ from its mean value at neighbouring points, and we simply apply a correction δ, then at the next approximation at the neighbouring points ϕ will be increased by $\tfrac{1}{4}\delta$. Thus the difference is not removed, but only divided by 4, and further corrections will be needed. We can anticipate these by making the correction in the first place $\tfrac{4}{3}\delta$ instead of δ. (See also Note 9·16a.)

For a wholly internal block (i.e. one entirely surrounded by adjustable values) the effect may be more serious. If we simply apply the correction δ given by (15) and adjust values at points adjacent to the block twice, the corrections at these may approach $\frac{1}{2}\delta$. Hence for such a block it is usually worth while to multiply by a factor greater than $\frac{4}{3}$ before applying them; $\frac{3}{2}$ is generally found satisfactory.

The normal procedure would be to tabulate the suggested values of δ at each stage, using either (3), (4), or (7). The larger values would be multiplied by $\frac{4}{3}$ and applied. This process will be varied in two ways. If it is noticed that a circuit of values at internal points need corrections nearly all of the same sign, a block correction will be evaluated for points within and on this circuit. If the problem indicates a singularity such that the solution near it is not of the form (1), it will be best to examine specially what the form will be, and to devise a method of approximation near the singularity that will be adapted to this form, as for (8).

A block may contain an internal block and it may be convenient to adjust both together. If the whole block needs a correction δ, and the treatment of the inner block indicates one of δ', the latter is really relative to the outer parts of the main block, and the whole correction needed by the inner block is $\delta + \delta'$.

9·181. As an example we consider a condenser consisting of two concentric long square prisms of sides 2 and 4, similarly situated. The inner is at potential 1, the outer at potential 0. Find the distribution of potential between them. Evidently the region consists of eight similar pieces and we need consider only one of them. As a first approximation we take the values at c, d by direct interpolation according to 9·18 (3), giving $c = d = 0.50$. For b we have (using points at 45° to the axes)

$$\phi_b = \tfrac{1}{4}(1+0+0+\phi_b); \quad \phi_b = 0.33,$$

and for a from (7)

$$\phi_a = \tfrac{4}{5} \cdot \tfrac{1}{4}(2\phi_b + 0) + \tfrac{1}{5} \cdot \tfrac{1}{4}(1+0) = 0.13 + 0.05 = 0.18.$$

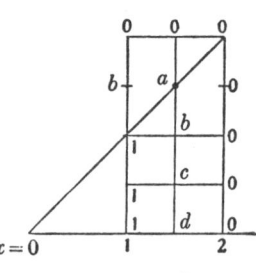

A second approximation to ϕ_b is now, from (7),

$$\phi_b = \tfrac{1}{5}(1+\phi_a+\phi_c+0) + \tfrac{1}{20}(0+\phi_b+1+0),$$

whence $\phi_b = 0.41$. Then ϕ_c and ϕ_d are corrected to 0·48 and 0·49.

At this point it becomes worth while to retain an extra figure. We continue to use (7) and get

$$\phi_a = 0.214, \quad \phi_b = 0.409, \quad \phi_c = 0.480, \quad \phi_d = 0.492.$$

This is as close as we need attempt without reducing the size of the meshes, since the values of ϕ_a and ϕ_b may be affected by the proximity of the corner. We therefore halve the size of the meshes; we interpolate to the centres of the coarse meshes by (3) for diagonal elements, and then fill up by (3) for adjacent elements. But for the three values nearest to the inner corner we also use (12); the results by this method are 0·706, 0·643, 0·475, as against 0·722, 0·660 and 0·508 by linear interpolation. The former are the better because they have taken the peculiar behaviour of the function in this region into account. The values are shown in the diagram.

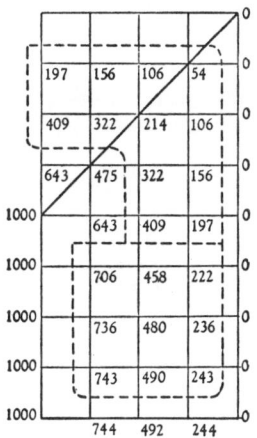
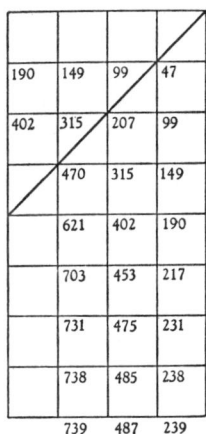

We take blocks as indicated by dotted lines. For the upper block the correction indicated is given by
$$-16\delta = 2(197+156+106+54) - 2(25+49+234+153)$$
$$\delta = -5\cdot 8;$$

but we allow partially for correlation with values at adjacent points by multiplying by $1\tfrac{1}{3}$ and therefore take $\delta = -7$.

Apply this correction and work out δ for the lower block; it is $-4\cdot 1$, which we similarly change to -5. The revised distribution, after the values near the corner have been readjusted, is as in the second diagram above.

We now work out corrections by (7) for all elements; the largest indicated are the adjacent ones where the present values are 0·402 and 0·315, reaching -9 and -8 in the third figure. We apply corrections -14 and -12 and readjust all entries in turn. The result is shown in the next diagram.

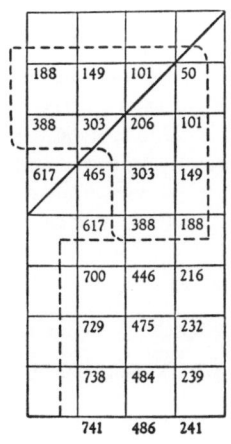
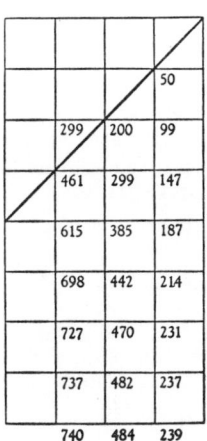

A block correction gives $\delta = -2$ for the upper block. For the lower we use a double block for the sake of symmetry (it would have been better to do this at the previous block adjustment) and get $\delta = -3$. These corrections are applied; in the next adjustment the largest correction needed is -0.004, and the solution is as in the last diagram. It should be right to about 0.002.

To get the capacity, apply Gauss's theorem to the square with the same centre and side 1.5. Using centred first differences we have the following values of $-\tfrac{1}{2}\partial\phi/\partial x$.

y	$-\tfrac{1}{2}\partial\phi/\partial x$	Δ
1.5	0.200	
1.25	0.314	0.114
1.0	0.428	0.114
0.75	0.484	0.056
0.50	0.496	0.012
0.25	0.500	0.004
0.00	0.501	0.001

By numerical integration

$$\int_0^{1.5}\left(-\frac{1}{2}\frac{\partial\phi}{\partial x}\right)dy = 0.645,$$

and the charge on the inner prism, per unit length, is

$$\frac{1}{4\pi}\times 8 \times 2 \times 0.645 = 0.821,$$

which is the capacity per unit length.

An analytical solution of the problem has been given by F. Bowman.* His result is 0.8144. D. C. Gilles, by a relaxation method, has got 0.832.† For comparison, the capacity per unit length of a condenser formed of two circular cylinders of radii 1 and 2 is

$$1/2 \log 2 = 0.721.$$

This adaptation of the method of finite differences to the solution of partial differential equations is due to L. F. Richardson, and successive approximation is always valid if the solution corresponds to making an integral a minimum. Extensions to many other types of differential equations, especially

$$\nabla^2\phi = w(x,y), \quad \nabla^4\phi = w(x,y)$$

have been given by Richardson‡ and by R. V. Southwell§ and his collaborators. A valuable introduction is given by L. Fox.∥ A method of adjusting the values near a corner, somewhat similar to that used in the above example, is due to H. Motz.¶ The method used above combines features of the methods of Richardson, Southwell and Motz. In a recent paper Bickley** gives approximations to $\nabla^2\phi$ and $\nabla^4\phi$ taking account of higher differences. One triumph of Southwell's methods is the calculation of the form of a waterfall. (See also Note 9·181a, p. 705).

* Proc. Lond. Math. Soc. (2) 39, 1935, 211–215; 41, 1936, 271–7.
† Proc. Roy. Soc. A, 193, 1948, 428.
‡ Phil. Trans. A, 210, 1910, 307–357; 226, 1927, 299–361 (with J. A. Gaunt).
§ Relaxation Methods in Engineering Science, 1940; Relaxation Methods in Theoretical Physics, 1946. ∥ Quart. J. Mech. Appl. Math. 1, 1948, 253–280.
¶ Quart. Appl. Math. 4, 1937. 371–7.
** Quart J. Mech. Appl. Math. 1, 1948, 35–42.

EXAMPLES

Practice in numerical work is the only way of learning it. The student should begin by taking selected entries from standard tables of mathematical functions and applying the methods described in this chapter.

1. If $g(x)$ is a polynomial of degree $n-1$ equal to $f(x)$ at $x_1, x_2, ..., x_n$, show that

$$\begin{vmatrix} g(x) & 1 & x & ... & x^{n-1} \\ f(x_1) & 1 & x_1 & ... & x_1^{n-1} \\ \multicolumn{5}{c}{\dotfill} \\ f(x_n) & 1 & x_n & ... & x_n^{n-1} \end{vmatrix} = 0,$$

and that Lagrange's and Newton's interpolation formulae arise from two different ways of expanding this determinant.

2. Find the general solution of the difference equation

$$\Delta^3 y_n = \Delta^2 y_n + 12 \Delta y_n.$$ (I.C. 1940.)

3. Show that when u_n is a polynomial of degree N in n

$$\sum_0^\infty u_n x^n = \frac{u_0}{1-x} + \frac{1}{1-x} \sum_1^N \left(\frac{x}{1-x}\right)^n \Delta^n u_0$$

if the series is absolutely convergent. Hence evaluate $\sum_1^\infty \frac{n^3+n}{3^n}$. (I.C. 1937.)

4. Find the real roots of the equation
$$(x-3)(x^2-1) = 1$$
to three significant figures. (I.C. 1943.)

5. Using
$$\int_1^n \log x\, dx = n \log n - n + 1,$$
apply the Euler-Maclaurin formula and show that for integral n

$$\log n! = C + (n+\tfrac{1}{2})\log n - n - \frac{B_2}{1.2n} - \frac{B_4}{3.4n^3} - \dots.$$

6. Estimate $\sum_{n=1}^{50} \frac{1}{n^2+5^2}$.

7. Prove that

$$\int_{-\frac{1}{2}}^{n+\frac{1}{2}} f(x)\, dx = f_0 + f_1 + \dots + f_n + \tfrac{1}{24}(\delta f_{n+\frac{1}{2}} - \delta f_{-\frac{1}{2}}) - \tfrac{17}{5760}(\delta^3 f_{n+\frac{1}{2}} - \delta^3 f_{-\frac{1}{2}}) + \dots$$

and check the formula by integrating $\int_{-\frac{3}{2}}^{\frac{3}{2}} x^4\, dx$.

8. Taking logarithms of
$$2n+1 = \tfrac{3}{1}.\tfrac{5}{3}.\dots.\frac{2n+1}{2n-1}$$
and using Richardson's method for $n=2$ and 3, derive a value for Euler's constant. (0·5780.)

9. A solution of the equation
$$\frac{dy}{dx} = 3x^2 + y^2$$
passes through $(0,0)$. Tabulate its values, correct to three decimal places, at intervals 0·1, over the range $0 \leqslant x \leqslant 1$. (I.C. 1936.)

10. Illustrate the method of relaxation by finding the values of x_1, x_2, x_3 that make the following function a minimum:
$$V = 10x_1^2 + 15x_2^2 + 20x_3^2 + x_1 x_2 + 2x_1 x_3 - x_1 - 2 \cdot 5 x_2 - x_3.$$ (I.C. 1939.)

Chapter 10

CALCULUS OF VARIATIONS

When change itself can give no more
'Tis easy to be true.
SIR CHARLES SEDLEY, *Reasons for Constancy*

10·01. Condition for an integral to be stationary. Suppose that we have an integral of the form

$$S = \int_{t_0}^{t_1} f\left(\frac{dx}{dt}, x, t\right) dt, \qquad (1)$$

where f is a given function; x is to be a function of t, but we have not yet specified what function. The problem of the calculus of variations is to decide what function x must be in order that S may be stationary for small variations of x. In its simplest form we can consider the determination of the shortest distance between two points. Using Cartesian coordinates and assuming that y is a differentiable function of x, with an integrable derivative, the distance along an arbitrary path is

$$s = \int_{x_0}^{x_1} \left\{\left(\frac{dy}{dx}\right)^2 + 1\right\}^{1/2} dx. \qquad (2)$$

If the ends are specified, so that $y(x_1) = y_1$, $y(x_0) = y_0$, two given quantities, we know that s is made a minimum by taking

$$\frac{y - y_0}{y_1 - y_0} = \frac{x - x_0}{x_1 - x_0}. \qquad (3)$$

This makes the path the straight line connecting (x_0, y_0) and (x_1, y_1). If we make y any other function of x we are choosing a different path, and its length will necessarily be greater than that of the straight line if the termini are kept the same. The characteristic feature of the calculus of variations, in contrast to ordinary problems of maxima and minima, is the occurrence of the unknown function or its derivative under the integral sign. To evaluate the integral (1) we must have the value of x for every value of t in the range; to make it stationary we have therefore, effectively, to determine an infinite number of values of x. The meaning of 'stationary' therefore needs definition.

We write $dx/dt = p$ and regard f as a function of the three variables x, p, t and suppose that within a certain region of these variables the second partial derivatives of f with regard to x, p, t are continuous as functions of x, p, t. We outline the argument for the case when only functions $x(t)$ with continuous second derivatives are regarded as admissible. Then if $x(t)$ and $x'(t)$ are two such slightly differing functions and

$$x'(t) = x(t) + \delta x(t), \qquad (4)$$

where
$$\delta x(t) = \alpha g(t) \qquad (5)$$

we say that $x(t)$ makes the integral stationary if

$$\frac{\partial}{\partial \alpha}\int_{t_0}^{t_1} f\left(\frac{dx'}{dt}, x', t\right) dt = 0 \tag{6}$$

when $\alpha = 0$, where $g(t)$ is any function with continuous second derivatives. At $\alpha = 0$,

$$\frac{\partial}{\partial \alpha}\int_{t_0}^{t_1} f\left\{p + \alpha\frac{dg}{dt}, x + \alpha g(t), t\right\} dt$$

$$= \int_{t_0}^{t_1} \frac{\partial}{\partial \alpha} f\left\{p + \alpha\frac{dg}{dt}, x + \alpha g, t\right\} dt = \int_{t_0}^{t_1}\left(\frac{dg}{dt}\frac{\partial f}{\partial p} + g\frac{\partial f}{\partial x}\right) dt. \tag{7}$$

Integration by parts of the first term is certainly legitimate under the given conditions and hence (7) becomes

$$\left[g\frac{\partial f}{\partial p}\right]_{t_0}^{t_1} + \int_{t_0}^{t_1} g\left(\frac{\partial f}{\partial x} - \frac{d}{dt}\frac{\partial f}{\partial p}\right) dt. \tag{8}$$

We cannot immediately infer from (8) that $\partial f/\partial p = 0$ and that

$$\phi \equiv \frac{\partial f}{\partial x} - \frac{d}{dt}\left(\frac{\partial f}{\partial p}\right) = 0 \tag{9}$$

at all intermediate values, as we could if δx were completely arbitrary. But if, for instance $\partial f/\partial p \neq 0$ at $t = t_0$ we could take

$$\begin{aligned} g(t) &= \frac{(\tau - t)^3}{(\tau - t_0)^3} \quad (t_0 \leqslant t \leqslant \tau) \\ &= 0 \quad (\tau \leqslant t \leqslant t_1), \end{aligned} \tag{10}$$

where $\tau - t_0$ may be as small as we like. Then d^2g/dt^2 is continuous and by taking $\tau - t_0$ sufficiently small we can ensure that $\partial S/\partial \alpha$ has the same sign as the integrated part. Again if there is any range, say from $t = a$ to $t = b$, where ϕ is positive we could take

$$\begin{aligned} g(t) &= 0 & (t_0 \leqslant t \leqslant a), \\ g(t) &= (t-a)^3(b-t)^3 & (a \leqslant t \leqslant b), \\ g(t) &= 0 & (b \leqslant t \leqslant t_1) \end{aligned} \tag{11}$$

and d^2g/dt^2 is continuous; with this form of δx, $\partial S/\partial \alpha$ would not vanish. Hence for equation (6) to be satisfied $\partial f/\partial p$ must vanish at both limits and $\phi = 0$ at all intermediate values. $\partial f/\partial p$ will in general involve p; hence $\phi = 0$ is ordinarily a differential equation of the second order for x.

For a fuller treatment of the theory the reader is referred to the works of Bolza and Bliss and a recent work by Pars. In particular the continuity of the second derivative of $g(t)$ need not be assumed. It will be seen from (8) that it is essential that the function under the integral sign should be integrable and our restricted conditions ensure this. Further we have considered only a one-parameter family of curves and consideration of a more general class of curves may in some cases be necessary.

The conditions of the problem may also include the condition that $\delta x = 0$ at the termini. This happens in the simple problem of finding the shortest path between two given points. For if in (2) we are given y_0 and y_1 our data forbid us to vary y at the termini, and the admissible forms of δy are all such that $\delta y_0 = \delta y_1 = 0$; but then it does not follow that $\partial f/\partial p$ vanishes at the ends, and the two terminal conditions to be satisfied by the solution are no longer $\partial f/\partial p = 0$ but that y has to take the assigned values. It will be noticed that in both cases we get two terminal conditions, the normal number that can be satisfied by the solution of an equation of the second order.

10·011 A very important case is where f does not contain t explicitly. In that case we multiply the differential equation by p:

$$\frac{\partial f}{\partial x}\frac{dx}{dt} - p\frac{d}{dt}\frac{\partial f}{\partial p} = 0. \tag{12}$$

But
$$\frac{df}{dt} = \frac{\partial f}{\partial x}\frac{dx}{dt} + \frac{\partial f}{\partial p}\frac{dp}{dt}, \quad \frac{d}{dt}\left(p\frac{\partial f}{\partial p}\right) = \frac{\partial f}{\partial p}\frac{dp}{dt} + p\frac{d}{dt}\left(\frac{\partial f}{\partial p}\right), \tag{13}$$

since $\partial f/\partial t = 0$. Therefore

$$\frac{d}{dt}\left(p\frac{\partial f}{\partial p} - f\right) = p\frac{d}{dt}\frac{\partial f}{\partial p} - \frac{\partial f}{\partial x}\frac{dx}{dt} = 0, \tag{14}$$

and a first integral of the differential equation is

$$p\frac{\partial f}{\partial p} - f = \text{constant}. \tag{15}$$

10·012. This case is exemplified by (2); writing

$$p = dy/dx, \quad f = (p^2+1)^{1/2},$$

we have
$$p\frac{\partial f}{\partial p} - f = \frac{p^2}{(p^2+1)^{1/2}} - (p^2+1)^{1/2} = -\frac{1}{(p^2+1)^{1/2}}.$$

Hence p is constant along the path, which is therefore a straight line.

10·013. A slightly more complicated problem is that of the *brachistochrone*, first propounded by John Bernoulli. Let A and B be two points connected by a smooth wire. A bead slides on the wire; its velocity at A is v_0 ($\neq 0$); what must be the form of the wire if the bead takes the shortest possible time to reach B?

Take A as origin and the axis of y downwards. Then the velocity of the bead at depth y is $\sqrt{(2gy+v_0^2)}$ and the time taken for x to reach a given value X is, with $dy/dx = p$,

$$T = \int \frac{ds}{\sqrt{(2gy+v_0^2)}} = \int_0^X \frac{(p^2+1)^{1/2}dx}{\sqrt{(2gy+v_0^2)}}.$$

For this to be stationary for variations of the path with the ends fixed, we have the first integral

$$\frac{p^2}{(p^2+1)^{1/2}\sqrt{(2gy+v_0^2)}} - \frac{(p^2+1)^{1/2}}{\sqrt{(2gy+v_0^2)}} = -\frac{1}{(p^2+1)^{1/2}\sqrt{(2gy+v_0^2)}} = \text{constant},$$

and therefore
$$\left(\frac{dy}{dx}\right)^2 + 1 = \frac{\alpha}{y+v_0^2/2g}.$$

This is integrated by the substitution $y + v_0^2/2g = \alpha \sin^2\theta$ and gives

$$x = \beta \pm \alpha(\theta - \tfrac{1}{2}\sin 2\theta).$$

When $\alpha \neq 0$ the path is therefore a cycloidal arc. Provided that $X \neq 0$, $Y + v_0^2/2g > 0$, the constants are determined from the conditions that $A(0,0)$ and $B(X, Y)$ lie on the arc and that there is no cusp between these points. The case when $v_0 = 0$ can be treated by a limiting process.

This answers the question if B is given. But suppose that we are given only that x has a given value at B, not the value of y there. Then we need the further condition at B denoted above by $\partial f/\partial p = 0$, which gives that $p = 0$ at B. Thus if we are told only that the lower terminus is in a given vertical line, the cycloid required is the one that cuts that line horizontally.

10·014. Sufficiency of conditions; maxima and minima. We have obtained necessary conditions for the integral to be stationary; in the cases just considered it is clear that they are also sufficient. To decide whether the choice makes the integral a minimum or a maximum, or merely stationary without being a maximum or a minimum for all possible variations, requires that account should be taken of the squares of the variations. In the problem of the shortest distance between two points this is simple. We can take the line joining the points as the axis of x; then

$$\int_0^X \left\{\left(\frac{dy}{dx}\right)^2 + 1\right\}^{1/2} dx \geqslant X$$

and the length of the path chosen is a minimum.

10·015. Variation of the limits. In obtaining (8) we have taken the limits t_0, t_1 as given. If they also are subject to variations Δt_0, Δt_1, S will be increased by $[f\Delta t]_{t=t_1}$ at the upper limit and decreased by $(f\Delta t)_{t=t_0}$ at the lower. The effect of allowing variation of the limits is therefore to change the integrated part to

$$\left[f\Delta t + \frac{\partial f}{\partial p}\delta x\right]_{t_0}^{t_1}. \tag{16}$$

In this expression, however, δx is the variation of x for given t, and therefore we must *not* replace t_1 by $t_1 + \Delta t_1$ in calculating δx_1. If the varied x at $t_1 + \Delta t_1$ is $x_1 + \Delta x_1$ we have

$$\Delta x_1 = \delta x_1 + p_1 \Delta t_1, \tag{17}$$

and the integrated part can be written

$$\left[\left(f - p\frac{\partial f}{\partial p}\right)\Delta t + \frac{\partial f}{\partial p}\Delta x\right]_{t_0}^{t_1}. \tag{18}$$

Take, for example, the problem of finding the shortest distance from a given point (a, b) to a given line $x\cos\alpha + y\sin\alpha = 1$. As before, the path must be a straight line. But at the intersection the possible variations of y entail corresponding variations of x, since $\Delta y_1 = -\cot\alpha \Delta x_1$. Then (18) is

$$\left\{(p^2+1)^{1/2} - \frac{p^2}{(p^2+1)^{1/2}}\right\}\Delta x_1 + \frac{p}{(p^2+1)^{1/2}}\Delta y_1 = \frac{\Delta x_1}{(p^2+1)^{1/2}}\{1 - p\cot\alpha\},$$

and vanishes if $p = \tan\alpha$.

Notice that since δy_1 is the variation of y_1 with x_1 kept constant we could conveniently write it as $(\delta y_1)_{x_1}$, in accordance with a convention used in thermodynamics; and then Δy_1 could be written $(\delta y_1)_{x_1 \cos \alpha + y_1 \sin \alpha}$. It is curious that in spite of the obvious need in partial differentiation for precise statement of what is being kept constant, such statement is not embodied in the customary notation of pure mathematics; though it is provided in thermodynamics, the theory of partial correlation in statistics, and in probability theory.*

10·02. Several dependent variables. The extension to the case where there are several dependent variables is quite straightforward. If the independent variables are $q_1, q_2, \ldots, q_n$ and we denote their derivatives with regard to t by $\dot{q}_r$, the variation of

$$S = \int_{t_0}^{t_1} f(q_1 \ldots q_n, \dot{q}_1 \ldots \dot{q}_n, t)\, dt \tag{1}$$

for small variations of the functions of t chosen for the q's is

$$\delta S = \left[\left(f - \sum_r \dot{q}_r \frac{\partial f}{\partial \dot{q}_r}\right)\Delta t + \sum_r \frac{\partial f}{\partial \dot{q}_r}\Delta q_r\right]_{t_0}^{t_1} + \int_{t_0}^{t_1} \sum_r \left\{\frac{\partial f}{\partial q_r} - \frac{d}{dt}\left(\frac{\partial f}{\partial \dot{q}_r}\right)\right\}\delta q_r\, dt, \tag{2}$$

where, if $\Delta t_1 \neq 0$,
$$(\Delta q_r)_1 = (q_r + \delta q_r)_{t_1 + \Delta t_1} - (q_r)_{t_1}, \tag{3}$$

and similarly if $\Delta t_0 \neq 0$. It follows that if the variations δq_r can be chosen independently of one another the conditions for S to be stationary are

$$\frac{\partial f}{\partial q_r} - \frac{d}{dt}\left(\frac{\partial f}{\partial \dot{q}_r}\right) = 0 \quad (r = 1, 2, \ldots, n), \tag{4}$$

$$\left(f - \sum_r \dot{q}_r \frac{\partial f}{\partial \dot{q}_r}\right)\Delta t + \sum_r \frac{\partial f}{\partial \dot{q}_r}\Delta q_r = 0, \tag{5}$$

the latter condition holding at each limit. If f does not involve t explicitly there will be a first integral as in 10·011

$$f - \sum_r \dot{q}_r \frac{\partial f}{\partial \dot{q}_r} = \text{constant}. \tag{6}$$

If $\Delta t_0 = \Delta t_1 = 0$, $\quad\quad \delta q_r\, \partial f / \partial \dot{q}_r = 0 \tag{7}$

at the limits.

10·03. Most physical applications of the calculus of variations fall under three types. (1) Determination of conditions of equilibrium from the condition that the potential energy must be stationary. (2) Fermat's principle in wave transmission, that the path is such that the time of transmission is stationary for small variations of the path. (3) Hamilton's principle in dynamics.

10·04. Fermat's principle. The examples that we have already considered can be used to illustrate Fermat's principle. If the velocity of a wave is the same at all points of the medium, the time of travel is proportional to the distance along the ray, and therefore is stationary if the ray is straight. If the velocity is proportional to $z^{1/2}$, where z is the distance from some fixed plane, the time of travel is proportional to $\int ds/\sqrt{z}$, and making

* Cf. Yule, J. Roy. Statist. Soc. 99, 1936, 770–1.

this stationary involves precisely the same analysis as the brachistochrone problem; the rays will therefore be cycloids with their cusps on $z = 0$. This example looks artificial, but actually seems to fit the propagation of explosion waves in clay.

10·041. Another interesting case is that of wave propagation when the velocity is proportional to the distance from a fixed plane. This arises in the seismic survey of the earth's outer layers; an explosion is made near the surface, and the times of arrival of the waves are recorded over a range of distance.

The velocity is $c(z_0+z)$, where c and z_0 are constant; then the time of transmission from $(0, 0)$ to (x_1, z_1) is

$$T = \int_0^{x_1} \frac{(1+p^2)^{1/2}\,dx}{c(z_0+z)},$$

with $p = dz/dx$ taken along the ray. x does not occur in the integrand; hence a first integral is

$$\frac{(1+p^2)^{1/2}}{c(z_0+z)} - \frac{p^2}{(1+p^2)^{1/2}c(z_0+z)} = \text{constant},$$

that is,
$$(1+p^2)^{1/2}(z_0+z) = \text{constant}.$$

Let the ray begin at an angle e to the surface, so that at $z = 0$, $p = \tan e$, and

$$(1+p^2)^{1/2}(z_0+z) = z_0 \sec e.$$

Then
$$x_1 = \int_0^{z_1} \frac{(z_0+z)\,dz}{\{z_0^2\sec^2 e - (z_0+z)^2\}^{1/2}} = -\left[\{z_0^2\sec^2 e - (z_0+z)^2\}^{1/2}\right]_0^{z_1}$$

$$= z_0 \tan e - \sqrt{\{z_0^2\sec^2 e - (z_0+z_1)^2\}},$$

and the ray is part of the circle

$$(x - z_0 \tan e)^2 + (z+z_0)^2 = z_0^2 \sec^2 e.$$

The deepest point of the ray is at $x = z_0 \tan e$, $z = z_0(\sec e - 1)$. The time of travel to this point is

$$\int_0^{z_*\tan e} \frac{z_0 \sec e}{c(z_0+z)^2}\,dx = \int_0^{z_*\tan e} \frac{z_0 \sec e}{c\{z_0^2\sec^2 e - (x-z_0\tan e)^2\}}\,dx$$

$$= \frac{1}{c}\left[\tanh^{-1}\frac{x - z_0 \tan e}{z_0 \sec e}\right]_0^{z_*\tan e} = \frac{1}{c}\tanh^{-1}\sin e.$$

The ray is refracted symmetrically up to the surface again; if X, T are the horizontal distance traversed and the time taken when it again reaches the surface

$$X = 2z_0 \tan e, \quad T = \frac{2}{c}\tanh^{-1}\sin e, \quad X = 2z_0 \sinh \tfrac{1}{2}cT.$$

This gives in terms of the parameter e the relation between distance and time of transmission between points on the surface.

10·05. Restricted variation: catenary. The admissible variations may be connected by some condition that makes them not independent. Consider, for instance, a uniform chain hanging from two fixed points; the position is one of minimum potential energy under gravity, and therefore if y is the height at any point $\int y\,ds$ is stationary. But the length of the chain is fixed; hence we can vary y only in such ways that $\delta \int ds = 0$. Then if λ is any constant the variations will satisfy

$$\delta \int (y-\lambda) \, ds = \delta \int (y-\lambda)(1+y'^2)^{1/2} \, dx = 0,$$

and conversely if we can find a λ such that this is satisfied for *all* variations of y, then $\int y \, ds$ is stationary for all variations that do not alter $\int ds$. The condition required has the first integral

$$\frac{y'^2}{\sqrt{(1+y'^2)}}(y-\lambda) = (y-\lambda)(1+y'^2)^{1/2} + c,$$

that is,

$$\frac{y-\lambda}{(1+y'^2)^{1/2}} = c,$$

$$x = \int \frac{c \, dy}{\{(y-\lambda)^2 - c^2\}^{1/2}} = c \cosh^{-1} \frac{y-\lambda}{c} + a,$$

$$y = \lambda + c \cosh \frac{x-a}{c}.$$

We are given the values of y for two fixed values of x, and also the length of the chain. These three data suffice to determine the three constants λ, c, a.

10·06. Hamilton's principle. Consider a system of n particles, a typical particle having mass m_r and coordinates x_{ri}. The components of force acting on it are X_{ri}. Then the equations of motion are, for each r,

$$m_r \ddot{x}_{ri} = X_{ri} \quad (r = 1, 2, \ldots, n;\ i = 1, 2, 3). \tag{1}$$

Multiply these by a set of small vectorial displacements δx_{ri}, which are arbitrary functions of the time, and add; then we have (summation with respect to r being explicit)

$$\sum_r m_r \ddot{x}_{ri} \delta x_{ri} = \sum X_{ri} \delta x_{ri} \tag{2}$$

This equation is completely equivalent to the equations of motion; for the δx_{ri} are completely arbitrary and we may therefore equate all their coefficients and recover the original equations. Now integrate between two given times; we have

$$\int_{t_0}^{t_1} \sum_r m_r \ddot{x}_{ri} \delta x_{ri} \, dt = \int_{t_0}^{t_1} \sum X_{ri} \delta x_{ri} \, dt. \tag{3}$$

Now

$$\int_{t_0}^{t_1} m_r \ddot{x}_{ri} \delta x_{ri} \, dt = \left[m_r \dot{x}_{ri} \delta x_{ri} \right]_{t_0}^{t_1} - \int_{t_0}^{t_1} m_r \dot{x}_{ri} \frac{d}{dt} \delta x_{ri} \, dt$$

$$= \left[m_r \dot{x}_{ri} \delta x_{ri} \right]_{t_0}^{t_1} - \int_{t_0}^{t_1} m_r \dot{x}_{ri} \delta \dot{x}_{ri} \, dt, \tag{4}$$

and therefore (3) is the same as

$$\left[\sum_r m_r \dot{x}_{ri} \delta x_{ri} \right]_{t_0}^{t_1} - \int_{t_0}^{t_1} \delta(\tfrac{1}{2} \sum_r m_r \dot{x}_{ri}^2) \, dt = \int_{t_0}^{t_1} \sum_r X_{ri} \delta x_{ri} \, dt. \tag{5}$$

If then the δx_{ri} are zero at the limits,

$$\delta S \equiv \int_{t_0}^{t_1} \{\delta(\tfrac{1}{2} \sum_r m_r \dot{x}_{ri}^2) + \sum_r X_{ri} \delta x_{ri}\} \, dt = 0. \tag{6}$$

This is the most general form of Hamilton's principle in classical dynamics. The function $\frac{1}{2}\sum_r m_r \dot{x}_{ri}^2$ is the kinetic energy, T. If there is a work function W, a function of the x_{ri}, and possibly of t, but not of $\dot{x}_{ri}$, such that

$$X_{ri} = \frac{\partial W}{\partial x_{ri}}, \tag{7}$$

$\sum_r X_{ri} \delta x_{ri} = \delta W$, the variation of W when the coordinates are varied by δx_{ri}; and (6) becomes

$$\delta S = \delta \int_{t_0}^{t_1} (T+W)\, dt = 0. \tag{8}$$

This is the form taken by the principle if the system is conservative.

The expressions in (6) and (8) are scalars, so that the device of introducing the variations has enabled us, for n particles, to summarize $3n$ equations of motion in one scalar equation.

10·07. Generalized coordinates and Lagrange's equations. Now consider $3n$ functions q_s of the coordinates, such that, if they are known, all the coordinates are determinate. We can then write

$$x_{ri} = x_{ri}(q_1 \ldots q_{3n}), \tag{9}$$

and call the q_s *generalized coordinates*. Then, if we use the summation convention with regard to s,

$$\delta x_{ri} = \frac{\partial x_{ri}}{\partial q_s} \delta q_s, \quad \dot{x}_{ri} = \frac{\partial x_{ri}}{\partial q_s} \dot{q}_s, \quad \frac{\partial \dot{x}_{ri}}{\partial \dot{q}_s} = \frac{\partial x_{ri}}{\partial q_s}, \tag{10}$$

$$2T = \sum_r m_r \left(\frac{\partial x_{ri}}{\partial q_s} \dot{q}_s\right)^2, \tag{11}$$

$$\sum_r X_{ri} \delta x_{ri} = \sum_r X_{ri} \frac{\partial x_{ri}}{\partial q_s} \delta q_s = Q_s \delta q_s, \tag{12}$$

and δS can be put into the form

$$\delta S = \int_{t_0}^{t_1} (\delta T + Q_s \delta q_s)\, dt, \tag{13}$$

where T and Q_s are now given functions of the q_s and $\dot{q}_s$. Then

$$\int_{t_0}^{t_1} \delta T\, dt = \left[\frac{\partial T}{\partial \dot{q}_s} \delta q_s\right]_{t_0}^{t_1} + \int_{t_0}^{t_1} \left(\frac{\partial T}{\partial q_s} - \frac{d}{dt}\frac{\partial T}{\partial \dot{q}_s}\right) \delta q_s\, dt, \tag{14}$$

and the condition that $\delta S = 0$ to the first order for all twice differentiable δq_s gives, on equating coefficients,

$$\frac{d}{dt}\frac{\partial T}{\partial \dot{q}_s} - \frac{\partial T}{\partial q_s} = Q_s. \tag{15}$$

These are *Lagrange's equations*. They are usually obtained in text-books on dynamics by direct transformation of (2); but the derivation from Hamilton's principle explains also why the left side has the characteristic form of the calculus of variations.

Now it may happen that in the actual motion certain relations between the x_{ri}, and therefore between the q_s, are specified. The most important case is where many particles belong to the same rigid body, and the coordinates can vary only in such ways that distances between particles of the same body remain unaltered. Another is when some coordinate is constrained by external force to vary in some prescribed way with time, as when a part of the system is made to move with given linear or angular velocity. Such constraints do not prevent us from considering variations δq_s such that the constraints are violated, and we can therefore still treat all the δq_s as independent and equate their respective coefficients. Then (15) remain true. But their physical interpretation is altered. Whereas in a system of free particles they are differential equations for the separate coordinates, in a system with constraints some of the q_s in the actual motion will be determinate functions of the others and of the time, and the corresponding Q_s will be the reactions needed to keep the constraints satisfied. It then becomes convenient to use one set of q_s just sufficient in number for them all to be varied without violating the constraints; and then Lagrange's equations will hold for this set, and the other q_s need not be considered unless we want to know the corresponding reactions explicitly. But some of the latter set may be given functions of the time, and in that case the x_{ri} will depend on the time as well as on the unconstrained q_s, and the time may appear explicitly in the kinetic energy. This does not affect the form of (15), but it will affect the first integrals.

For a rigid body we need six coordinates to say exactly where all its particles are. D'Alembert's principle follows at once if the body is regarded as made up of particles such that the force between any pair is along the line joining them. For the two forces of any pair add up to 0 and so do their moments about any axis. Also if x_i, y_i are the coordinates of two particles r apart and X'_i the force on the first due to the second, with resultant R', then the contribution from their reactions to $\Sigma X_{ri} \delta x_{ri}$ is

$$X'_i \delta x_i - X'_i \delta y_i = R' \frac{y_i - x_i}{r} \delta x_i + R' \frac{x_i - y_i}{r} \delta y_i$$
$$= \frac{R'}{r}(x_i - y_i)(\delta y_i - \delta x_i) = -\frac{1}{2}\frac{R'}{r}\delta(y_i - x_i)^2,$$

which equals 0 if the variations are such that the distance between the particles is unaltered. Even if the reactions between a pair of particles are not along the line joining them, so long as the internal reactions have a work function depending only on the mutual distances of the particles, without necessarily being separable into terms each depending only on the distance between two of them, it will follow that they contribute nothing to $\sum_r X_{ri} \delta x_{ri}$ whenever the δx_{ri} are such that they do not alter the mutual distances. The generalization of such a sum for an elastic solid would be minus the change of elastic strain energy, which vanishes if the distances between particles are unaltered. Without some equivalent supposition there seems to be no reason why d'Alembert's principle should be true, but in any case it is really an approximation since all real solids have some elasticity.

10·071. Non-holonomic systems. It sometimes happens that some linear relation connects the velocities but is not integrable, so that it is impossible to use it to eliminate one coordinate in favour of the others and leave the variations independent. This happens particularly in problems of rolling spheres and disks in three dimensions. Such systems are called *non-holonomic*. The method can still be adapted to their treatment by the use of undetermined multipliers. For simplicity let us suppose that there is only one such constraint, of the form

$$a_s \dot{q}_s = 0, \qquad (1)$$

where the a_s will involve the q_s. We consider variations such that

$$a_s \delta q_s = 0. \qquad (2)$$

We still have
$$\left(\frac{d}{dt}\frac{\partial T}{\partial \dot q_s} - \frac{\partial T}{\partial q_s} - Q_s\right)\delta q_s = 0, \tag{3}$$

provided the δq_s are such that the reactions do no work. This implies, in cases of rolling, that the δq_s are such that they involve no slipping; if there was slipping the tangential force would do work equal to its amount times the amount of slip. Hence the condition that (3) may be true, with unaltered Q_s, is simply (2). We do *not* assume that the varied path itself satisfies the constraints, and in general it does not; but we do assume that the δq_s do. Then for any λ

$$\left(\frac{d}{dt}\frac{\partial T}{\partial \dot q_s} - \frac{\partial T}{\partial q_s} - Q_s - \lambda a_s\right)\delta q_s = 0. \tag{4}$$

Choose a particular q_s, say q_1, where $a_1 \neq 0$, and suppose λ chosen so that the coefficient of δq_1 in (4) vanishes; that is,

$$\frac{d}{dt}\frac{\partial T}{\partial \dot q_1} - \frac{\partial T}{\partial q_1} - Q_1 - \lambda a_1 = 0. \tag{5}$$

Then we can assign all the other δq_s arbitrarily, since with any choice of their values δq_1 is determined by (3), and contributes nothing to (4) on account of (5). Then since (4) is true for all δq_s ($s = 2, 3, \ldots$), we can equate coefficients and get

$$\frac{d}{dt}\frac{\partial T}{\partial \dot q_s} - \frac{\partial T}{\partial q_s} - Q_s - \lambda a_s = 0 \quad (s \neq 1). \tag{6}$$

If there are m coordinates q_s these equations with (5) give m differential equations involving the coordinates and also the unknown λ; but we have also (1), and the equations are in general soluble.

In spite of the apparent simplicity and generality of the method of undetermined multipliers it is hardly ever used for concrete problems of non-holonomic systems. We see that the sum $\lambda a_s \delta q_s$ is added to $Q_s \delta q_s$, and therefore is the work done by the reactions in an arbitrary displacement not satisfying (2). If the multiplier is chosen so that $a_s \dot q_s$ is the velocity of slip, $a_s \delta q_s$ is the amount of slip in an arbitrary displacement, and $-\lambda$ is therefore the reaction resisting slipping. The method therefore does not avoid the explicit introduction of the reactions, but merely gives another way of determining them. It does require the explicit statement of all the coordinates, which the moving axes method often avoids. For a rolling sphere, for instance, the method of moving axes need not concern itself at all with the absolute position of any axis fixed in the sphere; it states the equations of motion directly in terms of angular velocities with respect to axes conveniently chosen with respect to the surface that the sphere is rolling on. The method of undetermined multipliers requires the introduction of three Eulerian angles and their subsequent elimination, since their actual behaviour is usually of negligible interest.

10·072. First integrals. For a conservative system, there is a work function W depending only on the q_s such that $\partial W/\partial q_s = Q_s$, and if T and W do not involve the time explicitly, we have the usual first integral 10·02 (6), namely,

$$\dot q_s \frac{\partial T}{\partial \dot q_s} - T - W = \text{constant}. \tag{1}$$

This is the energy integral $T - W = $ constant if T is quadratic in the $\dot q_s$. But a similar integral can exist even if work has to be done from outside to maintain the constraints, provided that the *other* forces have a work function, which we shall still denote by W. In this case some of the coordinates are given functions of the time, and T may not be a homogeneous quadratic in the unknown $\dot q_s$, since $\dot x_{ri}$ depends partly on the prescribed velocities. But if T and W do not involve t explicitly the integral (1) will still exist. If

$$T = T_2(\dot q_s) + T_1(\dot q_s) + T_0(\dot q_s), \tag{2}$$

where T_2, T_1, T_0 are homogeneous of degrees 2, 1, 0 in the unconstrained $\dot{q}_s$, but may involve q_s but not t, we have

$$\dot{q}_s \frac{\partial T}{\partial \dot{q}_s} = 2T_2 + T_1, \tag{3}$$

and therefore the energy integral is replaced by

$$T_2 - T_0 - W = \text{constant}. \tag{4}$$

This integral is often useful. Consider a circular wire made to rotate with given angular velocity ω about a vertical diameter. A bead is free to slide on the wire. Then the kinetic energy is

$$T = \tfrac{1}{2}ma^2(\dot{\theta}^2 + \sin^2\theta\,\omega^2) + \tfrac{1}{2}I\omega^2, \tag{5}$$

and the work function is $mga\cos\theta$. If then ω is constant T does not involve the time explicitly and we can write down at once the first integral

$$\tfrac{1}{2}ma^2(\dot{\theta}^2 - \sin^2\theta\,\omega^2) - \tfrac{1}{2}I\omega^2 - mga\cos\theta = \text{constant}. \tag{6}$$

The term $\tfrac{1}{2}I\omega^2$ is itself constant and therefore irrelevant. The function on the left is *not* the energy, which is $T - W$ and is not constant, but varies on account of the work that has to be done by the constraint to keep ω constant while θ varies. In fact if N is the couple needed we have for the rate of performance of work by it

$$N\omega = \frac{d}{dt}(T-W) = \frac{d}{dt}\{\tfrac{1}{2}ma^2(\dot{\theta}^2 + \sin^2\theta\,\omega^2) + \tfrac{1}{2}I\omega^2 - mga\cos\theta\}$$

$$= \frac{d}{dt}(ma^2\omega^2\sin^2\theta + I\omega^2) \quad \text{by (6)}$$

$$= 2ma^2\omega^2\sin\theta\cos\theta\,\dot{\theta}. \tag{7}$$

Hence
$$N = ma^2 \frac{d}{dt}(\sin^2\theta\,\omega), \tag{8}$$

which is the couple needed to maintain the angular velocity of the particle, since the angular momentum of the particle about the vertical is $ma^2\sin^2\theta\,\omega$ and therefore varies with θ when ω is kept constant.

10·073. Lagrange's equations for the top. We have

$$2T = A(\dot{\theta}^2 + \sin^2\theta\,\dot{\lambda}^2) + C(\dot{\chi} + \dot{\lambda}\cos\theta)^2,$$

$$W = -Mgh\cos\theta,$$

whence, since λ and χ occur only through their derivatives, we write down at once two first integrals

$$\frac{\partial T}{\partial \dot{\chi}} = C(\dot{\chi} + \dot{\lambda}\cos\theta) = Cn = \text{constant},$$

$$\frac{\partial T}{\partial \dot{\lambda}} = A\sin^2\theta\,\dot{\lambda} + Cn\cos\theta = \text{constant}.$$

The θ equation is

$$A(\ddot{\theta} - \sin\theta\cos\theta\,\dot{\lambda}^2) + Cn\sin\theta\,\dot{\lambda} = Mgh\sin\theta.$$

It is convenient to use the last equation rather than the energy equation in treating *small* oscillations about steady motion because $\ddot{\theta}$ is of the first order in the amplitude.

If the axis is nearly vertical the angles λ, χ are measured about nearly the same axis. Hence in treating small oscillations about the vertical it is convenient to take $\chi + \lambda = \psi$ as a new coordinate instead of χ, so that

$$2T = A(\dot\theta^2 + \sin^2\theta \dot\lambda^2) + C\{\dot\psi - \dot\lambda(1 - \cos\theta)\}^2.$$

If l, m, ν are direction cosines of the axis with regard to fixed axes, that of z being vertical,

$$\cos\theta = \nu = 1 - \tfrac{1}{2}(l^2 + m^2) + O(\theta^4),$$

$$\dot\lambda(1 - \cos\theta) = 2\dot\lambda \sin^2\tfrac{1}{2}\theta = \tfrac{1}{2}\dot\lambda\sin^2\theta + O(\theta^4) = \tfrac{1}{2}(l\dot m - m\dot l) + O(\theta^4),$$

$$\dot\theta^2 + \sin^2\theta \dot\lambda^2 = \dot l^2 + \dot m^2 + \dot\nu^2 = \dot l^2 + \dot m^2 + O(\theta^4).$$

Hence, exactly, $\qquad \dot\psi - \dot\lambda(1 - \cos\theta) = n,$

and to order θ^2
$$2T = A(\dot l^2 + \dot m^2) + C\{\dot\psi - \tfrac{1}{2}(l\dot m - m\dot l)\}^2,$$
$$2W = Mgh(l^2 + m^2),$$

whence, taking l and m as Lagrangian coordinates

$$A\ddot l + Cn\dot m = Mghl, \quad A\ddot m - Cn\dot l = Mghm,$$

which we have discussed in 4·092.

The device of taking the sum of two rotations about nearly coincident axes as a coordinate is used in this way in the theory of the motions of the planets; it makes the maximum use of the simplification introduced by the fact that the mutual inclinations of the orbits are small.

10·08. The Hamilton-Jacobi equation. Suppose that a work function W exists and that the system is holonomic, and put $T + W = L$, the Lagrangian function. We have

$$\delta \int_{t_0}^{t_1} L\,dt = \left[\frac{\partial L}{\partial \dot q_s}\delta q_s\right]_{t_0}^{t_1} + \int_{t_0}^{t_1}\left(\frac{\partial L}{\partial q_s} - \frac{d}{dt}\frac{\partial L}{\partial \dot q_s}\right)\delta q_s\,dt, \tag{1}$$

if the limits t_0, t_1 are unaltered. But if t_0, t_1 are also varied by $\Delta t_0, \Delta t_1$ and Δq_s is the variation of q_s to the new limits, the integrated part will become

$$\left[\left(L - \dot q_s \frac{\partial L}{\partial \dot q_s}\right)\Delta t + \frac{\partial L}{\partial \dot q_s}\Delta q_s\right]_{t_0}^{t_1}, \tag{2}$$

since $\Delta q_s = \delta q_s + \dot q_s \Delta t$.

We put
$$\dot q_s \frac{\partial L}{\partial \dot q_s} - L = H, \tag{3}$$

$$\frac{\partial L}{\partial \dot q_s} = p_s. \tag{4}$$

Then H is called the Hamiltonian function and p_s a generalized momentum; and

$$\delta S = \delta \int_{t_0}^{t_1} L\,dt = \left[-H\Delta t + p_s \Delta q_s\right]_{t_0}^{t_1} + \int_{t_0}^{t_1}\left(\frac{\partial L}{\partial q_s} - \frac{d}{dt}\frac{\partial L}{\partial \dot q_s}\right)\delta q_s\,dt. \tag{5}$$

Now if we suppose the q_s given at times t_0, t_1, the corresponding p_s will in general be determinate, since only one set of momenta at time t_0 will give the same set of displacements up to time t_1. Hence if S is taken along a dynamical path it will be a definite function of $t_0, t_1, (q_s)_0$, and $(q_s)_1$. It is called Hamilton's *principal function*. It is a function of the q_s at the actual t_0, t_1 however they are varied, and therefore we can replace t_1 by t; then at time t, since the last integral in (5) is zero by Lagrange's equations,

$$\frac{\partial S}{\partial t} = -H, \quad \frac{\partial S}{\partial q_s} = p_s, \tag{6}$$

but at time t_0

$$\left(\frac{\partial S}{\partial t_0}\right) = (H)_{t_0}, \quad \left(\frac{\partial S}{\partial q_{s0}}\right) = -(p_s)_{t_0}. \tag{7}$$

Now (4) can be used to eliminate $\dot{q}_s$ from H, and then H is a function of q_s, p_s and possibly t. Then

$$\frac{\partial S}{\partial t} = -H(q_s, p_s, t) = -H\left(q_s, \frac{\partial S}{\partial q_s}, t\right) \tag{8}$$

This is the Hamilton-Jacobi equation. It is a partial differential equation of the first order in the $n+1$ variables, not involving S explicitly, and its complete integral will therefore contain $n+1$ adjustable constants. From our first point of view S was a function of t only, containing $2n+1$ adjustable constants, namely, t_0 and the initial coordinates and momenta; but if the initial and final coordinates and $t-t_0$ are given the initial momenta are determinate and therefore are not adjustable; $n+1$ is therefore the right number when S is expressed as a function of t and the q_s.

If L does not contain the time explicitly, H = constant is the energy integral; if we denote this constant by h we have

$$\frac{\partial S}{\partial t} = -h, \quad \left(\frac{\partial S}{\partial t_0}\right) = h, \tag{9}$$

and therefore

$$S = -h(t-t_0) + f(q_s, q_{s0}). \tag{10}$$

Again, $\partial S/\partial q_{s0}$ is simply $-p_{s0}$, which is independent of t; and thus we have n equations expressing that the $\partial S/\partial q_{s0}$, which are functions of the q_s, q_{s0}, and possibly t, are constant throughout the motion and equal to $-p_{s0}$. Hence, given S, we have n equations to determine the q_s in terms of t and the initial conditions; the whole solution of the problem is therefore reduced to manipulation if we can determine S. This result is due to Hamilton. The difficulty in using it as it stands is that, while it is often fairly easy to obtain a complete integral of (8) involving $n+1$ constants, it is not often easy to express these constants in terms of q_{s0}; they are usually functions of both the q_{s0} and the p_{s0}. The theorem was completed by Jacobi, who showed that *any* complete integral of (8) can be used in exactly the same way. Before proving this, however, we need Hamilton's form of the equations of motion.

10·09. Hamilton's equations. L is a function of q_s, $\dot{q}_s$, and possibly t; H is a function of q_s, p_s, and possibly t. Then for arbitrary variations of q_s and $\dot{q}_s$, without varying t,

$$\delta H = \delta(\dot{q}_s p_s - L) = \dot{q}_s \delta p_s + p_s \delta \dot{q}_s - \frac{\partial L}{\partial q_s} \delta q_s - \frac{\partial L}{\partial \dot{q}_s} \delta \dot{q}_s. \tag{11}$$

But by definition $p_s = \partial L/\partial \dot{q}_s$; and therefore

$$\delta H = \dot{q}_s \delta p_s - \frac{\partial L}{\partial q_s} \delta q_s, \quad \dot{q}_s = \frac{\partial H}{\partial p_s}, \quad \frac{\partial H}{\partial q_s} = -\frac{\partial L}{\partial q_s}. \tag{12}$$

But by Lagrange's equations

$$\dot{p}_s = \frac{d}{dt}\frac{\partial L}{\partial \dot{q}_s} = \frac{\partial L}{\partial q_s} = -\frac{\partial H}{\partial q_s}, \tag{13}$$

and therefore
$$\dot{p}_s = -\frac{\partial H}{\partial q_s}, \quad \dot{q}_s = \frac{\partial H}{\partial p_s}. \tag{14}$$

These are *Hamilton's equations*. They are to be regarded as a set of $2n$ differential equations of the first order, Lagrange's being n of the second order.

Hamilton's equations can be directly related to a variational principle as follows. Take*
$$B = \int_{t_0}^{t_1} \{p_s \dot{q}_s - H(q_s, p_s)\} dt. \tag{15}$$

Then
$$\delta B = \int_{t_0}^{t_1} \left(p_s \delta \dot{q}_s + \dot{q}_s \delta p_s - \frac{\partial H}{\partial q_s} \delta q_s - \frac{\partial H}{\partial p_s} \delta p_s \right) dt$$
$$= \left[p_s \delta q_s \right]_{t_0}^{t_1} + \int_{t_0}^{t_1} \left(-\dot{p}_s \delta q_s + \dot{q}_s \delta p_s - \frac{\partial H}{\partial q_s} \delta q_s - \frac{\partial H}{\partial p_s} \delta p_s \right) dt. \tag{16}$$

The conditions that B shall be stationary when $\delta q_s = 0$ at t_0 and t_1, for all variations δq_s, δp_s at intermediate times, q_s and p_s being supposed to vary independently, are
$$\dot{p}_s = -\frac{\partial H}{\partial q_s}, \quad \dot{q}_s = \frac{\partial H}{\partial p_s}, \tag{17}$$

which are Hamilton's equations. If they are satisfied and the limits also are varied,
$$\delta B = \left[(p_s \dot{q}_s - H) \Delta t + p_s \delta q_s \right]_{t_0}^{t_1} = \left[p_s \Delta q_s - H \Delta t \right]_{t_0}^{t_1}, \tag{18}$$

which is precisely the same as δS.

The last argument does not *prove* Hamilton's equations. For if we were to define p_s as $\partial L/\partial \dot{q}_s$ in the usual way there are relations between p_s and $\dot{q}_s$, and the variations δq_s, δp_s are not independent. Hence we cannot equate their coefficients to zero. On the other hand, if we do not use a preliminary definition of p_s there is no particular reason why the integral should be stationary for variations of p_s irrespective of q_s. But we have seen that, given $p_s = \partial L/\partial \dot{q}_s$, $\dot{q}_s = \partial H/\partial p_s$ is merely a matter of differential calculus. The dynamics is contained in the other set of equations. Then in (16) the coefficients of the δp_s do vanish identically, and therefore those of δq_s can be equated to zero, leading to the other set of Hamilton's equations. The remarkable point is that, though in fact the p_s are originally defined in terms of the velocities, nevertheless if we choose to regard them as subject to independent variations, B is stationary subject to Hamilton's $2n$ equations. The variations are not independent, but B is stationary whether they are or not.

10·10. Jacobi's theorem. Let
$$S = f(q_1, \ldots, q_n; t; \alpha_1, \ldots, \alpha_n) + \alpha_{n+1} \tag{19}$$

be a complete integral of the Hamilton-Jacobi equation. In the original form the constants $\alpha_1 \ldots \alpha_n$ are the coordinates at time t_0; but Jacobi gives up this restriction and in place of $\partial S/\partial q_{s0} = -p_{s0}$ takes
$$\frac{\partial S}{\partial \alpha_r} = -\beta_r, \tag{20}$$
$$\frac{\partial S}{\partial q_s} = p_s, \tag{21}$$

* G. H. Livens, *Proc. R. S. Edinb.* 39, 1919, 113–19.

where β_r is another constant. The theorem is that if we still take these as equations to determine q_s and p_s, the resulting q_s, p_s will satisfy Hamilton's equations and therefore give a dynamical motion of the system. For, from (20),

$$0 = \frac{d}{dt}\frac{\partial S}{\partial \alpha_r} = \left(\frac{\partial}{\partial t} + \dot{q}_s \frac{\partial}{\partial q_s}\right)\frac{\partial S}{\partial \alpha_r} = \frac{\partial^2 S}{\partial \alpha_r \partial t} + \dot{q}_s \frac{\partial^2 S}{\partial \alpha_r \partial q_s}$$
$$= -\frac{\partial H}{\partial \alpha_r} + \dot{q}_s \frac{\partial p_s}{\partial \alpha_r}. \tag{22}$$

But α_r enters into H only through the fact that the p_s determined by (21) will contain α_r. Hence

$$0 = -\frac{\partial H}{\partial p_s}\frac{\partial p_s}{\partial \alpha_r} + \dot{q}_s \frac{\partial p_s}{\partial \alpha_r} = \left(\dot{q}_s - \frac{\partial H}{\partial p_s}\right)\frac{\partial p_s}{\partial \alpha_r}. \tag{23}$$

This is true for $r = 0, 1 \ldots n$; hence either

$$\dot{q}_s = \partial H/\partial p_s \quad (s = 0, 1, \ldots, n) \tag{24}$$

or
$$\frac{\partial(p_1 \ldots p_n)}{\partial(\alpha_1 \ldots \alpha_n)} = 0. \tag{25}$$

In the latter case there would be a functional relation between the p_s and the initial momenta could not be varied independently; hence (25) does not hold and therefore (24) do.

Again, from (21),

$$\frac{dp_s}{dt} = \left(\frac{\partial}{\partial t} + \dot{q}_m \frac{\partial}{\partial q_m}\right)\frac{\partial S}{\partial q_s} = -\left(\frac{\partial H}{\partial q_s}\right)_\alpha + \left(\frac{\partial H}{\partial p_m}\right)_q \left(\frac{\partial p_m}{\partial q_s}\right)_\alpha. \tag{26}$$

But
$$\left(\frac{\partial H}{\partial q_s}\right)_\alpha = \left(\frac{\partial H}{\partial q_s}\right)_p + \left(\frac{\partial H}{\partial p_m}\right)_q \left(\frac{\partial p_m}{\partial q_s}\right)_\alpha, \tag{27}$$

and therefore
$$\frac{dp_s}{dt} = -\left(\frac{\partial H}{\partial q_s}\right)_p. \tag{28}$$

Hence q_s and p_s found from (20) and (21) satisfy Hamilton's equations.

10·11. Transformation theory. Any transformation of the q_s to a new set of n coordinates q'_r with no functional relation connecting them will give a new way of stating the dynamical problem; Lagrange's equations will hold for the q'_r, and can be transformed to Hamilton's form in exactly the same way. Such a transformation is called a *point-transformation*. There is, however, a more general type called a *contact-transformation* such that the q'_r are defined as functions of both q_s and p_s, and nevertheless we can still define a set of p'_r so that q'_r and p'_r satisfy equations of Hamilton's form. Hamilton's equations in q_s, p_s are equivalent to

$$\delta B = \delta \int_{t_0}^{t_1} \{p_s \dot{q}_s - H(q_s, p_s, t)\} dt = \left[p_s \delta q_s\right]_{t_0}^{t_1}. \tag{29}$$

We require also that there shall be a function H' such that

$$\delta B' = \delta \int_{t_0}^{t_1} \{p'_r \dot{q}'_r - H'(q'_r, p'_r, t)\} dt = \left[p'_r \delta q'_r\right]_{t_0}^{t_1}. \tag{30}$$

Take a function $J(q_s, p'_r, t)$ and suppose

$$p_s = \partial J/\partial q_s, \quad q'_r = \partial J/\partial p'_r, \quad H' = H + \partial J/\partial t. \tag{31}$$

For small variations of q_s, p_s with t constant these equations can be solved to give small variations of q'_r, p'_r, and conversely, provided

$$\| \partial^2 J/\partial q_s \partial p'_r \| \neq 0. \tag{32}$$

Then
$$\delta(B'-B) = \delta \int_{t_0}^{t_1} (p'_r \dot{q}'_r - p_s \dot{q}_s - H' + H)\, dt$$
$$= \delta[p'_r q'_r] - \delta \int (q'_r \dot{p}'_r + p_s \dot{q}_s + H' - H)\, dt$$
$$= \delta[p'_r q'_r - J] \tag{33}$$

by (31); then

$$\delta(B'-B) = \delta[p'_r q'_r] - \left[\frac{\partial J}{\partial q_s} \delta q_s + \frac{\partial J}{\partial p'_r} \delta p'_r \right] = [p'_r \delta q'_r - p_s \delta q_s] \tag{34}$$

by using (31) again. Hence (29) and (30) are equivalent. But Hamilton's equations hold for q_s, p_s, H if and only if (29) is true to orders δq_s, δp_s for all small variations of the path. Hence (30) is true to orders $\delta q'_r$, $\delta p'_r$, and therefore q'_r, p'_r, H' also satisfy Hamilton's equations.

Note that if $J = q_s p'_s$, we have $q'_s = q_s$, $p'_s = p_s$; and that if $q_r = p'_r$, $p_r = q'_r$, q'_r and p'_r satisfy Hamilton's equations with $H' = -H$.

In particular let $H = K + mK_1$, where K is a Hamiltonian such that the solution of the Hamilton-Jacobi equation is known, say $S(q_s, \alpha_r, t)$, and m is small. Define

$$p'_r = \alpha_r, \quad q'_r = \beta_r = \partial S/\partial \alpha_r, \quad J = S. \tag{35}$$

β_r differs in sign from that defined by (20). If K_1 is now expressed in terms of α_r, β_r, t we shall have

$$\dot{\alpha}_r = -m\, \partial K_1/\partial \beta_r, \quad \dot{\beta}_r = m\, \partial K_1/\partial \alpha_r. \tag{36}$$

Let the integral of K_1 with regard to t as if α_r, β_r were constant be $-S_1(\alpha_r, \beta_r, t)$, and now take

$$J_1 = \alpha'_r \beta_r + mS_1(\alpha'_r, \beta_r, t). \tag{37}$$

The determinant in (32) is $1 + O(m)$.

Then
$$\beta'_r = \beta_r + m\,\partial S_1/\partial \alpha'_r, \quad \alpha_r = \alpha'_r + m\,\partial S_1/\partial \beta_r, \tag{38}$$

$$H'' = mK_1(\alpha_r, \beta_r, t) - mK_1(\alpha'_r, \beta_r, t) = O(m^2). \tag{39}$$

Thus the method is suited to rapid approximation.

In celestial mechanics six parameters are needed to specify the coordinates of a planet as functions of the time; the α_r, β_r can be used for this purpose, and are constant when there is no disturbance. When there are perturbations by other planets we can use (35) to specify α_r, β_r in terms of the coordinates and velocities, and conversely; and then the method of transformation expresses the α_r, β_r in terms of quantities as nearly constant as we like.[*a]

---

[*] The type of transformation (31) seems to have been introduced by W. F. Donkin, *Phil. Trans.* 1855, pp. 313–22. He seems to have also been the first to see that the transformations may contain t explicitly. The form (35), with extensions, is used by E. W. Brown and C. A. Shook, *Planetary Theory*, 1933.

Hamilton's equations are not much used in ordinary dynamics, because the first step in solving them would usually be to eliminate half the variables by differentiation and obtain equations of the second order. Their usefulness is in difficult problems. The Hamiltonian equations are also fundamental in statistical mechanics and the function H itself plays an important but still imperfectly understood part in quantum mechanics.

10·12. The principle of least action. We have from Hamilton's equations if H does not involve t explicitly,

$$\frac{dH}{dt} = \dot{q}_s \frac{\partial H}{\partial q_s} + \dot{p}_s \frac{\partial H}{\partial p_s} = \frac{\partial H}{\partial p_s}\frac{\partial H}{\partial q_s} - \frac{\partial H}{\partial q_s}\frac{\partial H}{\partial p_s} = 0. \tag{40}$$

Hence in any conservative holonomic system the Hamiltonian function does not vary with the time in any dynamical motion of the system. This follows also from 10·072 (1), since by definition H is the function there shown to be constant. Now we have $S = \int_{t_0}^{t_1} L\,dt$, and if the times at the limits are varied

$$\Delta S = \left[-H\Delta t + \frac{\partial L}{\partial \dot{q}_s}\Delta q_s \right]_{t_0}^{t_1} + \int_{t_0}^{t_1}\left(\frac{\partial L}{\partial q_s} - \frac{d}{dt}\frac{\partial L}{\partial \dot{q}_s}\right)\delta q_s\,dt. \tag{41}$$

Now take
$$A = S + H(t_1 - t_0); \tag{42}$$

then
$$\Delta A = \left[t\Delta H + \frac{\partial L}{\partial \dot{q}_s}\Delta q_s \right]_{t_0}^{t_1} + \int_{t_0}^{t_1}\left(\frac{\partial L}{\partial q_s} - \frac{d}{dt}\frac{\partial L}{\partial \dot{q}_s}\right)\delta q_s\,dt. \tag{43}$$

Now in deriving Lagrange's equations from Hamilton's principle we took fixed limits t_0, t_1, but allowed the q_s to vary quite arbitrarily. Thus the variations admitted permitted H to vary. But if we restrict ourselves to varied paths such that H is constant and equal to its value in the actual path, $\Delta H = 0$, and if also $\Delta q_s = 0$ and Lagrange's equations are satisfied, $\Delta A = 0$. Since $L = T + W$, $H = T - W$,

$$A = \int_{t_0}^{t_1}(L+H)\,dt = \int_{t_0}^{t_1} 2T\,dt.$$

The function A is called the *action*, and the rule just given, that in the conditions specified $\Delta A = 0$, is the *principle of least action*.* A is also called the *characteristic function* and can be made the basis of the transformation theory instead of the principal function. The principle of least action is equivalent to Hamilton's principle, but is less convenient to use. When it is spoken about, Hamilton's principle is usually intended.

10·13. Routh's modified Lagrangian function. In many dynamical problems some of the coordinates do not occur explicitly in L, only their rates of change occurring. Such coordinates are called *ignorable*, the others *palpable*. By a simple transformation it is possible to eliminate any or all of the former from the equations of motion. Let us keep the notation q_s for the coordinates that we propose to keep, but ϕ_σ for the ignorable ones that we propose to eliminate. Let

$$\partial L/\partial \dot{\phi}_\sigma = \eta_\sigma. \tag{1}$$

---
* It can be shown that A is a minimum for a dynamical path if $t_1 - t_0$ is not too large; if $t_1 - t_0$ is large A may be stationary for small variations but neither a true minimum nor a maximum.

Then by Lagrange's equations η_σ is constant throughout the motion. Now form the function
$$R = L - \phi_\sigma \eta_\sigma \tag{2}$$
and eliminate $\dot\phi_\sigma$ in favour of η_σ, not varying η_σ. We have
$$\delta R = \frac{\partial L}{\partial q_s}\delta q_s + \frac{\partial L}{\partial \dot q_s}\delta \dot q_s + \frac{\partial L}{\partial \dot\phi_\sigma}\delta\dot\phi_\sigma - \eta_\sigma \delta\dot\phi_\sigma, \tag{3}$$
and the last terms cancel. Hence when R is expressed in terms of q_s, $\dot q_s$ and η_σ, and L in terms of q_s, $\dot q_s$ and $\dot\phi_\sigma$,
$$\left(\frac{\partial R}{\partial q_s}\right)_\eta = \left(\frac{\partial L}{\partial q_s}\right)_{\dot\phi}, \quad \left(\frac{\partial R}{\partial \dot q_s}\right)_\eta = \left(\frac{\partial L}{\partial \dot q_s}\right)_{\dot\phi}. \tag{4}$$
Therefore, by Lagrange's equations,
$$\frac{d}{dt}\frac{\partial R}{\partial \dot q_s} - \frac{\partial R}{\partial q_s} = 0. \tag{5}$$

Routh's transformation, apart from a sign, is similar to Hamilton's, but is applied only to the ignorable coordinates (and not necessarily to all of them) instead of to all. Its applications are totally different, being mostly to small oscillations about steady motion. A steady motion may be defined as one such that the palpable coordinates are constant. It follows that in a steady motion $\partial R/\partial q_s = 0$, and we can expand R to the second order in departures of the q_s from their values in the steady motion and form linear differential equations for them exactly as for small oscillations about equilibrium. There is, however, one important difference. The elimination of $\dot\phi_\sigma$ in favour of η_σ usually brings in terms of the form $f(q_r)\dot q_s$, and when we approximate there will be terms in R of the form $g_{rs}q_r\dot q_s$. Now for particular values of r, s,
$$\frac{d}{dt}\left\{\frac{\partial}{\partial \dot q_s}(g_{rs}q_r\dot q_s)\right\} - \frac{\partial}{\partial q_s}(g_{rs}q_r\dot q_s) = g_{rs}\dot q_r, \tag{6}$$
$$\frac{d}{dt}\left\{\frac{\partial}{\partial \dot q_r}(g_{rs}q_r\dot q_s)\right\} - \frac{\partial}{\partial q_r}(g_{rs}q_r\dot q_s) = -g_{rs}\dot q_s, \tag{7}$$
and these terms introduce terms in the velocities into the equations of motion. These are called *gyroscopic* terms.

10·14. Variations of multiple integrals. The fundamental equations of many subjects are equivalent to statements that an integral is stationary for small variations of some function in it. The equations of static elasticity, for instance, can be expressed by a principle of stationary energy, the energy being a volume integral of a quadratic function of the strain components. In some problems the use of this principle is the nearest approach to a reliable way of getting the signs right.

We take as an illustration the derivation of *Schrödinger's wave equation for a single particle* from a variation principle. The Hamiltonian function is (apart from certain constant factors)
$$H(p_i, x_i) = \tfrac{1}{2}p_i^2 + V, \tag{1}$$
and we replace p_i by $-i\dfrac{\partial}{\partial x_i}$: V is the potential energy, supposed to be a given function of x_i. Consider the integrals
$$I = \iiint \psi H\!\left(-i\frac{\partial}{\partial x_i}, x_i\right)\psi\, d\tau, \quad J = \iiint \psi^2 d\tau, \tag{2}$$

through all space, subject to the conditions that ψ tends to zero at infinity at least as rapidly as $1/r$, and that J is given. Then the condition for I to be stationary for small variations of ψ is

$$0 = \delta \iiint \{\psi(-\tfrac{1}{2}\nabla^2\psi + V\psi) - \lambda\psi^2\}\, d\tau \qquad (3)$$

$$= \iiint \{-\delta(\tfrac{1}{2}\psi\nabla^2\psi) + 2V\psi\delta\psi - 2\lambda\psi\delta\psi\}\, d\tau. \qquad (4)$$

But

$$\iiint \delta(\psi\nabla^2\psi)\, d\tau = \iiint \{\delta\psi\nabla^2\psi + \psi\nabla^2(\delta\psi)\}\, d\tau \qquad (5)$$

and

$$\iiint (\psi\nabla^2\delta\psi - \delta\psi\nabla^2\psi)\, d\tau = \iint \left(\psi\frac{\partial\delta\psi}{\partial n} - \delta\psi\frac{\partial\psi}{\partial n}\right) dS \qquad (6)$$

by Green's theorem; and tends to 0 in the conditions stated. Hence

$$0 = 2\iiint \delta\psi(-\tfrac{1}{2}\nabla^2\psi + V\psi - \lambda\psi)\, d\tau, \qquad (7)$$

and if this is true for all small variations $\delta\psi$

$$\nabla^2\psi = 2(V-\lambda)\psi \qquad (8)$$

everywhere.

Further, we can take $\delta\psi$ proportional to ψ, and when ψ satisfies this differential equation

$$\lambda \iiint \psi^2\, d\tau = \iiint (-\tfrac{1}{2}\nabla^2\psi + V\psi)\psi\, d\tau, \qquad (9)$$

which determines λ. It is a common practice to choose the constant factor in ψ so that the integral J is equal to 1.

EXAMPLES

1. If $I = \int \left\{\left(\dfrac{dy}{dx}\right)^2 + 1\right\}^{1/2} dx$ between specified limits for x is stationary subject to $J = \int y\, dx$ having a given value, prove that the graph of y against x is an arc of a circle.

If the terminal conditions are $x = \pm a$, $y = 0$, what happens if the given value of J is greater than $\tfrac{1}{2}\pi a^2$ and I is interpreted as (1) an improper Riemann integral, (2) an integral $\displaystyle\int_{x=-a}^{a} (dx^2 + dy^2)^{1/2}$ with an appropriate generalization of the definition of a Stieltjes integral?

2. Find the curves in the (x, y) plane such that $\int \sqrt{(2E - n^2 y^2)}\, ds$ is stationary, where E and n are constants and the integral is taken between fixed end-points.

Verify that these curves are the tracks of a particle of unit mass moving with energy E under the force $(0, -n^2 y)$, taking the potential energy to be zero on the line $y = 0$. (M.T. 1944.)

3. If the velocity of waves in a sphere is $c = \alpha - \beta r^2$, where α and β are constants, prove that the paths of stationary time are circles; and if a path enters the sphere at an angle e to the surface, find the polar coordinates of the deepest point of the path and the time taken to reach it. (Wiechert.)

4. If $\int ds$ is stationary for variations of a path with fixed termini, where

$$ds^2 = g_{ik}\, dx_i\, dx_k \quad (i, k = 1, 2, 3, 4),$$

prove that

$$\frac{d}{ds}\left(g_{ik}\frac{dx_k}{ds}\right) = \frac{\partial g_{km}}{\partial x_i}\frac{\partial x_k}{\partial s}\frac{\partial x_m}{\partial s},$$

three of these equations being independent. (Riemann.)

5. If in Example 4,

$$ds^2 = c^2\left(1 - \frac{2fM}{c^2 r}\right) dt^2 - \left(1 - \frac{2fM}{c^2 r}\right)^{-1} dr^2 - r^2(d\theta^2 + \sin^2\theta\, d\lambda^2),$$

find three first integrals of the equations of motion; and if a particle moves nearly in a circle in the plane $\theta = \tfrac{1}{2}\pi$, find the apsidal angle. (Einstein.)

Chapter 11

FUNCTIONS OF A COMPLEX VARIABLE

Of fowls after their kind, and of cattle after their kind, of every creeping thing of the earth after his kind, two of every sort shall come unto thee, to keep them alive.

Genesis vi, 20

11·01. Meaning and algebra of complex numbers. There are three chief reasons why complex functions, involving a symbol i such that $i.i = -1$, are of importance in physics, which involves only real quantities directly. The first is that many physical quantities are functions ϕ, ψ of two variables x and y, where ϕ and ψ are connected by the relations
$$\frac{\partial \phi}{\partial x} = \frac{\partial \psi}{\partial y}, \quad \frac{\partial \phi}{\partial y} = -\frac{\partial \psi}{\partial x}.$$
Such pairs of functions occur, for instance, in two-dimensional problems of electrostatics, where ϕ is the potential and ψ the charge function; in two-dimensional hydrodynamics of an incompressible fluid, where ϕ is the velocity potential and ψ the stream function; and in the closely analogous problem of flow of electric current in a uniform sheet. Then ϕ and ψ are the real and imaginary parts of what is called an *analytic function* w of the complex variable $z = x+iy$. The second is that the solutions of the differential equations of physics, for certain ranges of a real variable, are usually obtained as power series; but the same power series will equally well specify the values of a function of a complex variable, and the study of the complex values is often a great help towards obtaining more compact expressions for the real ones and relating expressions by power series valid in different ranges. The third is that many integrals given in real form are most easily evaluated by relating them to complex integrals and using the powerful method of contour integration based on Cauchy's theorem.

The important property of complex numbers is that they can be defined in such a way that they satisfy the fundamental rules of algebra 1·01 (1) to (9). We first consider the consequences of applying these rules to the real numbers together with a symbol i with the property $i^2 = -1$. Since there is no real number with this property it is customary to speak of i as imaginary. If a and b are real numbers, $c = a+ib$ is called a complex number, a its real part, and b (not ib) its imaginary part. We also use the notations $\Re(c) = a$, $\Im(c) = b$ to denote the real and imaginary parts of c.

First, if $i^2 = -1$ and a, b are real, and $a = ib$, it follows that if the rules of algebra are obeyed by i,
$$a^2 = (ib)^2 = ibib = iibb = -b^2,$$
and therefore $a = b = 0$. If a real quantity is equal to an imaginary one, both are zero.

Next, if $c = a+ib$, $c' = a'+ib'$, where a, a', b, b' are real, the rules of algebra give

$$-c = -a-ib, \tag{1}$$
$$c+c' = a+a'+i(b+b'), \tag{2}$$
$$c-c' = c+(-c') = a-a'+i(b-b'), \tag{3}$$
$$cc' = aa'-bb'+i(ab'+a'b), \tag{4}$$
$$ic = -b+ia. \tag{5}$$

By (3) if $c-c' = 0$, $a = a'$ and $b = b'$. If two complex numbers are equal, their real parts are equal and their imaginary parts are equal. Hence by (2) and (4) the real and imaginary parts of the sum and product of two complex numbers are uniquely determined in terms of those of the two original numbers.

We can, however, formulate these rules without the use of the symbol i, as an algebra of pairs of real numbers.* We think now of a pair (a,b) as corresponding to $a+ib$, and to show the comparison of the notations we put

$$\gamma = (a,b); \quad c = a+ib.$$

Then a real number a corresponds to the pair $(a,0)$, and an imaginary number ib to the pair $(0,b)$, and in particular i to $(0,1)$. If we now define $-\gamma$, $\gamma \pm \gamma'$, $\gamma\gamma'$ by the rules

$$-\gamma = (-a,-b), \tag{1}$$

$$\gamma+\gamma' = (a+a', b+b'), \tag{2'}$$

$$\gamma-\gamma' = (a-a', b-b'), \tag{3'}$$

$$\gamma\gamma' = (aa'-bb', ab'+a'b), \tag{4'}$$

we have
$$i^2 = (0,1)(0,1) = (-1,0), \tag{6}$$

$$i\gamma = (0,1)(a,b) = (-b,a), \tag{7}$$

$$i.i\gamma = (-a,-b) = -\gamma; \quad i\gamma.i = -\gamma. \tag{8}$$

Thus the definitions of the components of $-\gamma$, $\gamma \pm \gamma'$, $\gamma\gamma'$ and $i\gamma$ are identical with the rules for the real and imaginary parts of $-c$, $c \pm c'$, cc', ic, and i^2 corresponds to -1. We can henceforth use c instead of γ, leaving it to be understood from the context whether we are speaking of the complex number or the pair of real numbers.

These rules are consistent with the ordinary rules of algebra for addition and multiplication of real quantities, which have been stated in Chapter 1. We have the commutative law of addition
$$c+c' = c'+c,$$
the associative law of addition
$$c+(c'+c'') = (c+c')+c'',$$
the commutative law of multiplication
$$cc' = c'c,$$
the associative law of multiplication
$$c(c'c'') = (cc')c'',$$
and the distributive law
$$c(c'+c'') = cc'+cc''.$$

* A similar idea of number pairs occurs in the theory of rational fractions. What we write as $c = a/b$ can be written as a number pair $c = (a,b)$, the rule for addition being taken to be

$$(a,b)+(a',b') = (ab'+ba', bb')$$

and the rule for multiplication $\quad (a,b)(a',b') = (aa', bb')$.

$1/(a,b)$ is defined as (b,a), and $(a,a) = (1,1)$. It may be verified that with these definitions the laws still hold.

11·01 *Algebra of complex numbers* 335

The truth of the first three and the fifth of these, when addition and multiplication are carried out according to the rules, is obvious. For the fourth we have

$$c(c'c'') = (a,b)(a'a''-b'b'', a'b''+a''b')$$
$$= (aa'a''-ab'b''-a'bb''-a''bb', ba'a''-bb'b''+aa'b''+aa''b'),$$
$$(cc')c'' = (aa'-bb', ab'+a'b)(a'',b'')$$
$$= (aa'a''-a''bb'-ab'b''-a'bb'', aa''b'+a'a''b+aa'b''-bb'b''),$$

and the explicit interpretations are identical. Hence complex numbers can be handled by algebraic methods just like real numbers.

We see at once that $(c-c')+c' = c$, so that subtraction is the inverse of addition, as in ordinary algebra. Division is a little more complicated; we write

$$\frac{1}{c} = \frac{a-ib}{a^2+b^2} = \left(\frac{a}{a^2+b^2}, -\frac{b}{a^2+b^2}\right).$$

We easily verify that with this definition $c(1/c) = 1$, so that we have defined the reciprocal of a complex number except for the case of $a = b = 0$, which we write $1/c = \infty$. Then we take for the ratio of two complex numbers c and c'

$$\frac{c}{c'} = c\left(\frac{1}{c'}\right) = \left(\frac{aa'+bb'}{a'^2+b'^2}, \frac{a'b-ab'}{a'^2+b'^2}\right),$$

and we see that this, written in the i notation, is

$$\frac{aa'+bb'+i(a'b-ab')}{a'^2+b'^2} = \frac{(a+ib)(a'-ib')}{a'^2+b'^2}.$$

We can verify that with this definition

$$\left(\frac{c}{c'}\right)c' = c'\left(\frac{c}{c'}\right) = c.$$

Also $(-b,a)(-b',a') = -(a,b)(a',b')$, which we can write $(ic)(ic') = -cc'$. (These should be proved by means of the five laws of algebra stated above.)

We have now verified that all the fundamental processes of algebra can be carried out with our number pairs, and the result will always be a number pair. All the rules can be stated in terms of real numbers, and therefore the consistency of the real number system implies that of the system of pairs of real numbers subject to these rules; and as each number pair (a,b) corresponds to a complex number $c = a+ib$ we have a consistent algebra for the complex number system.

If a, b are physical magnitudes they must have the same dimensions; the relation of complex magnitudes to complex numbers is similar to that of real magnitudes to real numbers.

We have meanings at once for $\Re(c) > \Re(c')$ and $\Im(c) > \Im(c')$; there is no meaning for $c > c'$. An important related quantity is the *modulus* or *absolute value*, which we write as $|c|$, and define by

$$|c| = |a+ib| = (a^2+b^2)^{1/2},$$

the positive root being taken. If $|a+ib| < M$, where M is some positive real number, then $|a|$ and $|b|$ separately are less than M. If $|a+ib| > M$, either $a^2 > \frac{1}{2}M^2$ or $b^2 > \frac{1}{2}M^2$. The vanishing of $|c|$ is a necessary and sufficient condition for the vanishing of a and b.

On account of this result the modulus thus defined plays the same part in the theory of the complex variable as in that of the real variable. We have always

$$|cc'| = |c||c'|.$$

Also if $|c| < \epsilon$, whatever positive value ϵ may have, then $c = 0$ and $a = b = 0$.

There are some differences in inequalities between complex and real numbers. If a and b are two real numbers we always have

$$|a^2 + b^2| \geqslant |a^2|.$$

But for complex numbers we have not necessarily

$$|c^2 + c'^2| \geqslant |c^2|.$$

For c may be 1 and $c' = i$. Then $c^2 = 1$, $c'^2 = -1$, and the left side is 0 and the right 1.

We have, however,
$$|c| + |c'| \geqslant |c+c'|,$$

for 0, c, $c+c'$ are three points in the plane and the inequality is a case of 5·01(7). Also

$$\sum_{r=1}^{n} |c_r| \geqslant \left|\sum_{r=1}^{n} c_r\right|.$$

Also if λ, μ are real
$$(\lambda a + \mu b)^2 \leqslant (\lambda^2 + \mu^2)(a^2 + b^2)$$

by Cauchy's inequality, and therefore

$$\left|\frac{\lambda a + \mu b}{a + ib}\right| \leqslant (\lambda^2 + \mu^2)^{1/2}.$$

The notion of a limit can be extended to complex numbers. If $c_n = a_n + ib_n$, and $a_n \to a$, $b_n \to b$, we say that $c_n \to c = a + ib$.

11·02. Differentiation and integration of a complex function of a real variable x.
Let ϕ, ψ be two functions that depend on a real variable x, and put

$$f = \phi + i\psi.$$

If x receives a small increment δx and ϕ and ψ corresponding increments $\delta\phi$, $\delta\psi$, we have

$$\frac{\delta f}{\delta x} = \frac{\delta \phi}{\delta x} + i\frac{\delta \psi}{\delta x}.$$

Making δx tend to limit zero we have ultimately

$$\frac{d\phi}{dx} + i\frac{d\psi}{dx},$$

which we take as the definition of df/dx. It exists only if both ϕ and ψ are differentiable at the value of x considered.

Similarly, we can define
$$\int_a^b f\,dx = \int_a^b \phi\,dx + i\int_a^b \psi\,dx,$$
provided that ϕ and ψ are integrable in the range a to b.

11·03. Functions of a complex variable. Let x and y be a pair of real variables, and express the complex pair (x, y) briefly by
$$z = x + iy. \tag{1}$$
Let ϕ and ψ be a pair of real functions of x and y, and put
$$f = \phi + i\psi. \tag{2}$$
Consider what meaning we can attach to df/dz, if any. If x and y receive small increments δx, δy,
$$\frac{\delta f}{\delta z} = \frac{\delta\phi + i\delta\psi}{\delta x + i\delta y}, \tag{3}$$
which we can always interpret by our rules.

The question is whether this always tends to the same value when δx, δy both tend to zero. We get a *necessary* pair of conditions in the following way. Take $\delta y = 0$ and then let δx tend to 0; then if the partial derivatives exist
$$\frac{\delta f}{\delta z} \to \frac{\partial\phi}{\partial x} + i\frac{\partial\psi}{\partial x}. \tag{4}$$
But if we take $\delta x = 0$ and then let δy tend to 0 we get
$$\frac{\delta f}{\delta z} \to -i\frac{\partial\phi}{\partial y} + \frac{\partial\psi}{\partial y}, \tag{5}$$
and these can be equal only if
$$\frac{\partial\phi}{\partial x} = \frac{\partial\psi}{\partial y}, \quad \frac{\partial\phi}{\partial y} = -\frac{\partial\psi}{\partial x}. \tag{6}$$
These relations are called the *Cauchy-Riemann relations*. They evidently imply
$$\frac{\partial f}{\partial y} = i\frac{\partial f}{\partial x}. \tag{7}$$
If they are not satisfied, df/dz can have no unique meaning irrespective of the limiting value of $\delta x/\delta y$, for $\delta f/\delta z$ will tend to different values according as $\delta x/\delta y$ tends to 0 or infinity. The first requirement, if df/dz is to have a meaning for all values of x, y within a range such that we can vary x and y independently, is therefore that (6) shall be true for all these values.

The second requirement for physical applications is that the components in any direction of the gradients of ϕ and ψ shall be derived from those in the x, y directions by the vector rule; that is, if
$$\delta x = h\cos\theta, \quad \delta y = h\sin\theta, \tag{8}$$
and θ is fixed while $h \to 0$,
$$\frac{\phi(x + h\cos\theta, y + h\sin\theta) - \phi(x, y)}{h} \to \phi_x\cos\theta + \phi_y\sin\theta, \tag{9}$$
where ϕ_x, ϕ_y are independent of θ; with a similar relation for ψ. If further the limit is approached uniformly with regard to θ, ϕ and ψ are differentiable as functions of two variables in the sense defined in Chapter 5.

Taking $\theta = 0$, $\theta = \tfrac{1}{2}\pi$ we have, by definition of partial derivatives,

$$\phi_x = \frac{\partial \phi}{\partial x}, \quad \phi_y = \frac{\partial \phi}{\partial y}, \quad \psi_x = \frac{\partial \psi}{\partial x}, \quad \psi_y = \frac{\partial \psi}{\partial y}. \tag{10}$$

A necessary and sufficient condition that (3) *shall have a unique limit when* δx, $\delta y \to 0$ *is that* ϕ, ψ *shall satisfy the Cauchy-Riemann relations and be differentiable.** If ϕ, ψ are differentiable we have

$$\frac{f(x + h\cos\theta, y + h\sin\theta) - f(x, y)}{h(\cos\theta + i\sin\theta)} = \frac{\phi_x \cos\theta + \phi_y \sin\theta + i\psi_x \cos\theta + i\psi_y \sin\theta}{\cos\theta + i\sin\theta} + \frac{o(h)}{h}$$

$$= \phi_x + i\psi_x + o(1), \tag{11}$$

by the Cauchy-Riemann relations; and if $h \to 0$ we have

$$\frac{\delta f}{\delta z} \to \phi_x + i\psi_x, \tag{12}$$

which is independent of θ. Hence the condition is sufficient.

To show that it is necessary, let

$$\frac{f(x + h\cos\theta, y + h\sin\theta) - f(x, y)}{h(\cos\theta + i\sin\theta)} \to u + iv, \tag{13}$$

when $h \to 0$, where u and v are real and independent of θ. Multiply by $h(\cos\theta + i\sin\theta)$ and separate real and imaginary parts; then

$$\phi(x + h\cos\theta, y + h\sin\theta) - \phi(x, y) = h(u\cos\theta - v\sin\theta) + o(h), \tag{14}$$

$$\psi(x + h\cos\theta, y + h\sin\theta) - \psi(x, y) = h(u\sin\theta + v\cos\theta) + o(h), \tag{15}$$

and therefore ϕ and ψ are differentiable; and

$$\phi_x = u = \psi_y, \quad \phi_y = -v = -\psi_x, \tag{16}$$

so that the Cauchy-Riemann relations are satisfied.

If ϕ, ψ satisfy the above conditions, and we take axes of x', y' so that

$$x' = lx + my, \quad y' = -mx + ly, \tag{17}$$

then

$$x = lx' - my', \quad y = mx' + ly', \tag{18}$$

$$\frac{\partial \phi}{\partial x'} = l\frac{\partial \phi}{\partial x} + m\frac{\partial \phi}{\partial y} = l\frac{\partial \psi}{\partial y} - m\frac{\partial \psi}{\partial x} = \frac{\partial \psi}{\partial y'}, \tag{19}$$

and similarly

$$\frac{\partial \psi}{\partial x'} = -\frac{\partial \phi}{\partial y'}. \tag{20}$$

Hence the Cauchy-Riemann relations are satisfied for axes in any direction. In particular if x' is taken along the normal to a curve, and y' along the tangent, so that $dx' = dn$, $dy' = ds$, the rotation from dn to ds being $+\tfrac{1}{2}\pi$,

$$\frac{\partial \phi}{\partial n} = \frac{\partial \psi}{\partial s}, \quad \frac{\partial \phi}{\partial s} = -\frac{\partial \psi}{\partial n}. \tag{21}$$

---

* S. Pollard, *Proc. Lond. Math. Soc.* (2) 21, 1923, 456–482.

11·03 *Existence of df/dz* 339

Just as a function of a real variable may be defined only for a certain range of the argument, or be continuous or differentiable in a certain range, a function of a complex variable is defined for ranges of x and y (the range of x for given y possibly depending on the value of y). The set of pairs of values x, y such that the function is defined for them is called a *region*. We shall give a geometrical interpretation in a moment. The essential ideas have already appeared in Chapter 5.

If within a region

$$f(z) = \phi + i\psi, \quad \lambda(z) = \mu + i\nu$$

and $f(z)$, $\lambda(z)$ are so related that for any positive ϵ there is a δ (possibly depending on z) such that for all complex h satisfying $|h| < \delta$

$$\left| \frac{f(z+h) - f(z)}{h} - \lambda(z) \right| < \epsilon, \tag{22}$$

then $f(z)$ is said to be analytic *in the region, and $\lambda(z)$ can be denoted by $f'(z)$ or df/dz.* Subject to these conditions ϕ, ψ satisfy the Cauchy-Riemann relations and are differentiable.

We can also speak of $f(z)$ as *analytic in a closed region* if there is also a unique limit when z is a boundary point and $z+h$ is restricted to be a point of the region.

We shall find that if (22) is true at all points of a region *second* derivatives of $f(z)$ exist; and therefore if it is true the first derivatives are continuous. But this takes a great deal of proof and we are not yet in a position to assume it.

Note that $z^* = x - iy$ is a differentiable function of x and y, but is not an analytic function of z because it does not satisfy the Cauchy-Riemann relations.

At present we shall consider only single-valued functions. This excludes functions like $z^{1/2}$, which require special attention to their behaviour before we can say that they are differentiable. We shall see that this difficulty can be avoided when we come to consider branch points. An analytic single-valued function in a region is also called *regular*,† *holomorphic* or *monogenic* in the region.

We notice that if ϕ and ψ have continuous second derivatives,

$$\frac{\partial^2 \psi}{\partial x \partial y} = \frac{\partial^2 \psi}{\partial y \partial x},$$

whence

$$\frac{\partial}{\partial x} \frac{\partial \phi}{\partial x} = \frac{\partial}{\partial y} \left(-\frac{\partial \phi}{\partial y} \right),$$

that is,

$$\frac{\partial^2 \phi}{\partial x^2} + \frac{\partial^2 \phi}{\partial y^2} = 0. \tag{23}$$

Similarly,

$$\frac{\partial^2 \psi}{\partial x^2} + \frac{\partial^2 \psi}{\partial y^2} = 0, \tag{24}$$

and ϕ and ψ satisfy Laplace's equation in two dimensions.

We verify easily that if $\phi_1 + i\psi_1$, $\phi_2 + i\psi_2$ are two functions of z, their product function $(\phi_1\phi_2 - \psi_1\psi_2) + i(\psi_1\phi_2 + \psi_2\phi_1)$ satisfies the relations (6) and therefore is a function of z. Since z itself is a function of z it follows that $z \cdot z$ is another and so on to all positive integral

---

† 'Regular' is the most usual term in recent mathematical works. We avoid it because we need also to speak of regular singularities of a differential equation.

powers of z. Also, since $\dfrac{d}{dz}(\phi+i\psi) = \dfrac{\partial}{\partial x}(\phi+i\psi)$, the usual formula for differentiation of a product holds, namely,

$$\frac{d}{dz}(w_1 w_2) = w_1 \frac{dw_2}{dz} + w_2 \frac{dw_1}{dz}. \tag{25}$$

The same applies to sums and quotients; hence we infer that any rational function of z is an analytic function except possibly at points where the denominator is zero.

We verify by induction that $\dfrac{d}{dz} z^n = n z^{n-1}$; alternatively

$$\frac{(z+h)^n - z^n}{(z+h) - z} = (z+h)^{n-1} + (z+h)^{n-2} z + \ldots + z^{n-1}, \tag{26}$$

which gives the result on making h tend to 0.

In general a function of the complex variable is defined originally only in a given region of x and y, and special devices are needed to give it a meaning outside that region. This does not apply to rational functions, which could be calculated for any pair of values by means of the fundamental rules. We need notice for them only that for certain values of z the division rule may fail to give a value for the function. Thus if

$$f(z) = z - 1,$$

our rule gives $\qquad \dfrac{1}{f(z)} = \dfrac{1}{x - 1 + iy} = \dfrac{x - 1 - iy}{(x-1)^2 + y^2},$

both components of which have the form 0/0 when $x = 1$, $y = 0$ and are indeterminate. But we have for any other pair of values

$$|f|\left|\frac{1}{f}\right| = 1;$$

hence if $|f|$ tends to 0 as z approaches some special value, $|1/f|$ will ultimately exceed any specified finite value; we say that it *tends to infinity*. Such a point is an example of a *pole* of the function $1/f$. We notice also that if for any reason $f(z)$ is indeterminate at $z = z_0$,

$$\frac{1}{\zeta}\{f(z_0+\zeta) - f(z_0)\}$$

is indeterminate; hence our rule for defining a derivative fails and z_0 does not belong to any region where the function is analytic.

This result suggests a difficulty in the treatment of such a function as $f = \dfrac{z+1}{z^2-1}$ at $z = -1$. Applying the rules we find that ϕ and ψ take the form 0/0. But this can be avoided; we can show that for any z other than ± 1

$$\frac{z+1}{z^2-1} = \frac{1}{z-1},$$

which has a definite value $-\tfrac{1}{2}$ at $z = -1$. We can take $f = -\tfrac{1}{2}$ at $z = -1$, and if we do so f is found to be differentiable there. Similar considerations apply to such functions as $(1/z)\sin z$ at $z = 0$; if a function analytic elsewhere tends to a unique limit at a particular point, the limit may be taken as the value of the function at the point even though the direct application of the rules gives no determinate result. This process is called *definition by continuity*.

In consequence of the rules that if ϕ^2 or $\psi^2 > M^2$, $|\phi+i\psi| > M$, and that if $|\phi+i\psi| > M$ either ϕ^2 or $\psi^2 > \frac{1}{2}M^2$, we can say that if the modulus of a function tends to infinity at a point, either its real or its imaginary part will be unbounded, and speak of the function itself as unbounded. Similarly if either the real or the imaginary part is unbounded the modulus is unbounded. We speak of a function $f(z)$ as *bounded* in a region if we can choose a positive quantity M such that at all points of the region $|f(z)| < M$.

11·04. The Argand diagram.* A geometrical representation of functions of a complex variable can be obtained by regarding x and y as rectangular coordinates of a point in a plane; this point is completely identified by the real and imaginary parts of z. Then the functions ϕ and ψ are functions of position in the plane, and satisfy Laplace's equation in two dimensions. From the point of view of pure mathematics this device is only an aid to visualization, but in physical applications it is often much more; x and y may really be coordinates, and ϕ and ψ quantities with definite physical meanings, which it is usually our task to determine as functions of x and y. Further,

$$|z| = (x^2+y^2)^{1/2} = r,$$

where r is the distance of (x, y) from the origin. There will also be an angle θ such that

$$x = r\cos\theta, \quad y = r\sin\theta,$$

and therefore $x+iy$ is equivalent to $r(\cos\theta + i\sin\theta)$.

Then θ is called the *argument*† or *phase* of z and denoted by arg z. But θ is not single-valued; we could alter it by any integral multiple of 2π and still get the same values of $\cos\theta$ and $\sin\theta$, and hence the same values of x and y for given r. It can be made single-valued by the following device. When $x > 0$, $y = 0$, we take arg $z = 0$; for any other z we take arg z to vary continuously, that is, jumps of 2π are not allowed; and we make it a rule never to cross the negative real axis. Then for x negative and y small and positive, arg z is nearly π; for x negative and y small and negative, arg z is nearly $-\pi$; and for all x, y, $-\pi \leqslant \arg z \leqslant \pi$. This makes a change approaching 2π on crossing the negative real axis, but we avoid this by not crossing it; we take arg $z = \pi$ on the negative real axis and approaching, but never attaining, $-\pi$ as we approach the negative real axis from the side of y negative. Then we write $-\pi < \arg z \leqslant \pi$. We could equally well, of course, take $-\pi \leqslant \arg z < \pi$. We shall have several other occasions to use this device of *cuts* to avoid ambiguities; they are particularly important in the use of conformal representation. The value of the argument, with this restriction, is called its *principal value*.

We sometimes write, if $z \neq 0$,

$$z/|z| = \cos\theta + i\sin\theta = \operatorname{sgn} z,$$

and more generally, if $f(z) \neq 0$,

$$f(z)/|f(z)| = \operatorname{sgn} f(z).$$

If $f(z) = 0$ we take $\operatorname{sgn} f(z) = 0$. This expression plays the part of a direction vector.

* Given first by C. Wessel (1797), J. R. Argand (1806). But J. Wallis (1673) is stated by E. T. Bell to have missed it by a hairsbreadth, if at all.

† The word *argument* is also used in the sense that if $f(x)$ is a function of x, x is called the argument of $f(x)$. It will generally be clear from the context which sense is intended, but it would be an advantage if pure mathematicians would agree to alter one of the usages. *Amplitude* is also used; this can only be considered a disaster.

If $f(z)$ is a function of z, ϕ and ψ can similarly be regarded as coordinates of a point in another plane; and the function expresses a correspondence between points in the z plane and in the $f(z)$ plane.

If $f'(z) = 0$ *throughout a region, $f(z)$ is constant in the region.* For if $f(z) = \phi + i\psi$, it follows that throughout the region

$$\frac{\partial \phi}{\partial x} = \frac{\partial \phi}{\partial y} = \frac{\partial \psi}{\partial x} = \frac{\partial \psi}{\partial y} = 0.$$

But then it follows from 5·033 that ϕ and ψ, and therefore f, are constant.

Note that if ϕ is constant in a region, so is ψ, by the Cauchy-Riemann relations, and conversely.

11·041. Continuity. A function $f(z)$ is said to be continuous at $z = z_0$ if for any positive ϵ we can choose δ so that $|f(z_0+h)-f(z_0)| < \epsilon$ for all $|h| < \delta$. It follows immediately from the definition of an analytic function that if $f(z)$ is analytic at $z = z_0$ it is also continuous there.

If $f(z)$ is analytic at all points within a given boundary it is not necessarily analytic or even continuous at a point on the boundary. For instance, if

$$f(z) = \frac{1}{z-1}$$

$f(z)$ is analytic (and therefore continuous) at every value of z such that $|z| < 1$. The boundary of the region is the circle $|z| = 1$; but the point $z = 1$ is on this circle and $f(z)$ is there discontinuous.

We shall speak of $f(z)$ as *continuous in a closed region* if, as in 5·031, for any z of the region and for any positive ϵ there is a $\delta(z, \epsilon)$ such that for every z_1 of the region satisfying

$$|z_1 - z| < \delta(z, \epsilon),$$

we have $|f(z_1) - f(z)| < \epsilon$.

11·042. Uniformity of continuity. *If $f(z)$ is continuous in a closed region, $f(z)$ is uniformly continuous; that is, for any ϵ we can choose δ so that $|f(z+h)-f(z)| < \epsilon$, for all values of z and $z+h$ of the closed region and satisfying $|h| < \delta$.* ϕ and ψ are continuous; hence by 5·031, for a given ϵ we can choose a δ so that, whenever (x_1, y_1) (x_2, y_2) are points of the region satisfying

$$(x_1-x_2)^2 + (y_1-y_2)^2 < \delta^2,$$

$$|\phi(x_1,y_1)-\phi(x_2,y_2)| < \epsilon, \quad |\psi(x_1,y_1)-\psi(x_2,y_2)| < \epsilon,$$

and therefore

$$|f(z_1)-f(z_2)| < \epsilon\sqrt{2}.$$

11·043. Goursat's lemma. *Let $f'(z)$ exist at all points of a closed region D; and let ϵ be an arbitrary small positive quantity. Then a set of squares G_r can be superposed on D, each containing a point z_r common to G_r and D such that for all other points common to G_r and D*

$$|f(z)-f(z_r)-(z-z_r)f'(z_r)| < \epsilon |z-z_r|.$$

For any Z of D there is a neighbourhood of Z such that if z is in this neighbourhood and in D

$$|f(z)-f(Z)-(z-Z)f'(Z)| < \epsilon |z-Z|.$$

This neighbourhood is a region I as in the modified Heine-Borel theorem (5·021), which therefore applies directly. A z_r may be on the boundary of a G_r provided it is not exterior to D; if it is on a boundary between two G_r it may be used for both if the inequality is satisfied. The G_r need not all be equal.

We shall call a G_r satisfying this condition an ϵ-neighbourhood of z_r. The lemma has already appeared in a more general form in 5·042, but we state the case of it that we shall need in 11·052.

11·05. Integration. We attach a meaning to the integral of a function $f(z)$ of a complex variable along a rectifiable curve L with termini z_0, Z as in 5·06(2); that is, we take as parameter the arc s measured along the curve, and consider the set of points z_r for increasing s_r. In each interval (s_r, s_{r+1}) take a point ζ_r of the curve and consider the sum, with $z_{n+1} = Z$,

$$S_n = \sum_{r=0}^{n} f(\zeta_r)(z_{r+1} - z_r) = \Sigma\{\phi(\zeta_r)(x_{r+1} - x_r) - \psi(\zeta_r)(y_{r+1} - y_r)\} \\ + i\Sigma\{\phi(\zeta_r)(y_{r+1} - y_r) + \psi(\zeta_r)(x_{r+1} - x_r)\}. \tag{1}$$

We take a positive quantity δ, and since the curve is rectifiable we can take all the intervals $s_{r+1} - s_r < \delta$ with n finite. As $\delta \to 0$, and n correspondingly to infinity, the sums in S_n define the two sums of Stieltjes integrals $\int \phi \, dx - \psi \, dy$, $\int \phi \, dy + \psi \, dx$. x, y are of bounded variation on the curve; if $f(z)$ is analytic on the curve, ϕ and ψ are continuous on the curve with regard to s. Hence the integrals exist and we can write

$$\lim_{\delta \to 0} S_n = \int_L f(z) \, dz. \tag{2}$$

We sometimes need to consider the integral of a function along a curve that forms part of the boundary of the region where the function is analytic. Then ϕ and ψ may not be continuous on the curve but the integrals exist subject to the conditions stated for Stieltjes integrals.

Note that if $|f(z)| < M$ at all points of the curve

$$|f(z_r)(z_{r+1} - z_r)| < M \, |z_{r+1} - z_r|, \tag{3}$$

$$|\Sigma f(z_r)(z_{r+1} - z_r)| < M\Sigma \, |z_{r+1} - z_r|, \tag{4}$$

and therefore, proceeding to the limit, if K is the length of the curve,

$$\left| \int_{z_0}^{Z} f(z) \, dz \right| \leq MK. \tag{5}$$

11·051. Two special integrals. Clearly

$$\int_{z_0}^{Z} dz = \lim\{(z_1 - z_0) + (z_2 - z_1) + \ldots + (Z - z_n)\} \\ = Z - z_0. \tag{1}$$

Also $\quad z_{r-1}(z_r - z_{r-1}) = \tfrac{1}{2}\{z_r^2 - z_{r-1}^2 - (z_r - z_{r-1})^2\}$
and therefore if $z_{n+1} = Z$

$$z_0(z_1 - z_0) + z_1(z_2 - z_1) + \ldots + z_n(Z - z_n) = \tfrac{1}{2}(Z^2 - z_0^2) - \tfrac{1}{2}\sum_{r=1}^{n+1}(z_r - z_{r-1})^2.$$

Take points z_r at intervals $< \omega$ along the path; then the modulus of the sum on the right is $\leqslant \omega K$, and
$$\int_{z_0}^{Z} z\,dz = \tfrac{1}{2}(Z^2 - z_0^2). \tag{2}$$

Consider now the integral of $f(z)$ round a contour C in an ϵ-neighbourhood of z_0. Let the upper bound of $|z - z_0|$ be a. Then
$$f(z) = f(z_0) + (z - z_0)f'(z_0) + (z - z_0)v, \tag{3}$$
where $|v| < \epsilon$; and
$$\int_C f(z)\,dz = \int_C \{f(z_0) + (z - z_0)f'(z_0)\}\,dz + \int_C (z - z_0)v\,dz. \tag{4}$$

The first integral on the right is zero by (1) and (2) because the termini are identical. Also
$$\left|\int_C (z - z_0)v\,dz\right| < \int a\epsilon\,ds < \epsilon a \lambda, \tag{5}$$
where λ is the length of the boundary C. Hence
$$\left|\int_C f(z)\,dz\right| < \epsilon a \lambda. \tag{6}$$

We are concerned with two types of contour: (1) a square of side b, so that
$$a \leqslant b\sqrt{2}, \quad \lambda = 4b, \quad a\lambda \leqslant 4\sqrt{2}b^2 < 6b^2;$$
(2) a square with part of its boundary replaced by a curve within the square. In the second case let the length of the curved part be μ. Then in case (1)
$$\left|\int_C f(z)\,dz\right| < 6\epsilon A, \tag{7}$$
where A is the area; and in case (2), since $\lambda \leqslant 4b + \mu$,
$$\left|\int_C f(z)\,dz\right| < 6\epsilon A + \sqrt{2}\,.\,\epsilon b\mu. \tag{8}$$

A being the whole area of the square.

11·052. Cauchy's theorem. *If $f(z)$ is analytic in the closed region bounded by C, where C is a contour of finite length, then $\int_C f(z)\,dz = 0$.*

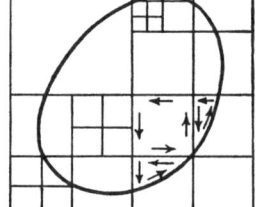

Denote C and its interior by D. We can surround C by a square E of side B. By Goursat's lemma, for any positive ω we can subdivide E into a finite set of squares G_n of side b_n, such that for every G_n containing points of D there is a point z_n common to G_n and D such that for every z common to G_n and D
$$|f(z) - f(z_n) - (z - z_n)f'(z_n)| < \omega |z - z_n|.$$
Some of the G_n will be wholly within D. Others will be intersected by C. Consider for each G_n the integral $\int f(z)\,dz$ taken round the boundary of G_n if G_n is entirely within D; if G_n is intersected by C, take the integral about the part of the boundary of G_n that lies in D together with the part of C that

lies in G_n. Then all these are closed contours satisfying the conditions of the last section of 11·051. If the area of an internal square G_n is A_n,

$$\left|\int_{G_n} f(z)\,dz\right| < 6\omega A_n.$$

If G_p is a square of side b_p overlapping C, and the part of C within it has length μ_p,

$$\left|\int_{G_p} f(z)\,dz\right| < 6\omega A_p + \sqrt{2}\,.\,\omega b_p \mu_p.$$

Take the sum of all the integrals, all circuits being described in the direction of positive rotation. Then every interior side or part of a side of a G_n traversed in forming the corresponding integral is traversed in the opposite direction in forming that for an adjacent G_n. Hence all contributions from internal sides cancel and the sum is simply the integral about C. Hence

$$\left|\int_C f(z)\,dz\right| < 6\omega(\Sigma A_n + \Sigma A_p) + 2\omega \Sigma b_p \mu_p.$$

Since the G_n, G_p are non-overlapping parts of E,

$$\Sigma A_n + \Sigma A_p \leqslant B^2.$$

Also $b_p \leqslant B$; and $\Sigma \mu_p = L$, the length of C, which is supposed finite. Hence

$$\left|\int_C f(z)\,dz\right| < \omega(6B^2 + 2BL).$$

The left side and the second factor on the right are independent of ω. Hence, since ω is arbitrarily small,

$$\int_C f(z)\,dz = 0.$$

Cauchy's theorem is the pivotal theorem of the theory of the complex variable. It often seems surprising at first sight that such a restriction on the boundary values of a function should be deducible merely from the condition that the function must have a derivative within and on the boundary. It becomes less surprising, however, when we see first that the function should really be regarded as a pair of functions and that the definition of a derivative that is being used implies two exact relations between the partial derivatives of these functions at every point where its existence is asserted.

Several mathematicians have contributed to the relaxation of the conditions for the theorem. The proof subject only to the conditions that $f(z)$ has a derivative and C has a finite length is due to Goursat.

11·053. Relation to Green's lemma. An alternative approach to Cauchy's theorem, similar to Riemann's treatment, is to use the two-dimensional form of Green's lemma. If u and v are two functions of x and y with continuous first derivatives within and on C,

$$\iint \left(\frac{\partial u}{\partial x} + \frac{\partial v}{\partial y}\right) dx\,dy = \int_C (lu + mv)\,ds, \tag{1}$$

where the integral on the left is taken through the interior of C, and l and m are the direction cosines of the outward normal to C with regard to the axes of x and y. The conditions are more restrictive than in the proof already given since they assume continuity of the separate derivatives; this is the most manageable sufficient condition for the reversal of the order of integration in the proof of Green's theorem. The direction cosines of the tangent to C, proceeding in the direction of positive rotation, are $(-m, l)$, so that for displacements along C

$$\frac{dx}{ds} = -m, \quad \frac{dy}{ds} = l \tag{2}$$

and
$$\int_C (lu + mv) \, ds = \int_C (u \, dy - v \, dx). \tag{3}$$

Now take $f(z) = \phi + i\psi$, where ϕ, ψ satisfy the Cauchy-Riemann relations; then

$$\int_C f(z) \, dz = \int_C (\phi + i\psi)(dx + i \, dy) = \int_C (\phi \, dx - \psi \, dy) + i \int_C (\psi \, dx + \phi \, dy)$$

$$= \iint \left(-\frac{\partial \phi}{\partial y} - \frac{\partial \psi}{\partial x}\right) dx \, dy + i \iint \left(-\frac{\partial \psi}{\partial y} + \frac{\partial \phi}{\partial x}\right) dx \, dy = 0. \tag{4}$$

Green's lemma can be proved in three dimensions in somewhat wider conditions than we used in 5·08, if we use methods similar to those used in Cauchy's theorem. Note first that if

$$u_i = a_i + b_{ik} x_k,$$

where a_i and b_{ik} are constants, the result follows at once. Note also that if the volume integral of div u exists the interior of S can be divided up into regions of volume $\delta\tau_n$ such that however we choose a specimen value of div u in each, at R_n, say, we cannot alter the sum $\Sigma (\text{div } u)_n \delta\tau_n$ by more than ϵ. Now if P is x_i and Q is $x_i + x_i'$, where $x_i'^2 = r^2$, and each u_i is differentiable, there is a neighbourhood of P such that

$$u_i(Q) = u_i(P) + x_k' \left(\frac{\partial u_i}{\partial x_k}\right)_P + v_i r, \tag{5}$$

where $|v_i| < \omega$ for $r < \delta$. Then over a surface G_p of volume τ_p enclosing P

$$\iint_{G_p} l_i u_i \, dS = \left(\frac{\partial u_i}{\partial x_i}\right)_P \tau_p + \iiint_{G_p} l_i v_i r \, dS. \tag{6}$$

By subdivisions of the regions $\delta\tau_n$ into $\delta'\tau_p$ and making straightforward modifications of the proof of Cauchy's theorem we can show that the sum of the values of the last term for all G_p, added for all $\delta\tau_n$, is less in modulus than $M\omega$, where M is fixed. Hence there is at least one P_{np} within each $\delta'\tau_p$ such that

$$\left| \iint l_i u_i \, dS - \sum_n \sum_p \left(\frac{\partial u_i}{\partial x_i}\right)_{P_{np}} \delta'\tau_p \right| < M\omega, \tag{7}$$

and, since the volume integral exists

$$\left| \sum_n \sum_p \left(\frac{\partial u_i}{\partial x_i}\right)_{P_{np}} \delta'\tau_p - \sum_n \left(\frac{\partial u_i}{\partial x_i}\right)_{R_n} \delta\tau_n \right| < \epsilon, \tag{8}$$

$$\left| \sum_n \left(\frac{\partial u_i}{\partial x_i}\right)_{R_n} \delta\tau_n - \iiint \frac{\partial u_i}{\partial x_i} \, d\tau \right| < \epsilon. \tag{9}$$

Hence
$$\left| \iint l_i u_i \, dS - \iiint \frac{\partial u_i}{\partial x_i} \, d\tau \right| < M\omega + 2\epsilon. \tag{10}$$

and therefore must be zero since ϵ and ω are both arbitrary. Hence sufficient conditions for Green's lemma are that u_i is differentiable on and inside S, that the volume integral of div u exists, and that S is bounded and has a finite area given as in 5·07.

Since differentiability is a sufficient condition for the derivatives of u_i to be a tensor, these conditions cover all cases where we should want to use Green's lemma.

Note that the proof must proceed in two stages, the usual rectangular subdivision being used in each. In the first we make the regions $\delta\tau_n$ small enough for the sum of $(\operatorname{div}\boldsymbol{u})_n \delta\tau_n$ to give a good approximation to the volume integral; in the second we subdivide the $\delta\tau_n$ into $\delta'\tau_p$ so that the properties of differentiable functions can be used to establish the approximation of the sum to the surface integral. The former step does not arise in Cauchy's theorem because the analogue of $\operatorname{div}\boldsymbol{u}$ is zero.

11·054. Extension of Cauchy's theorem. The theorem also holds under the following conditions: $f(z)$ is analytic within C, and continuous in the closed region bounded by C. Following Goursat,* we suppose that C is such that there is at least one internal point c such that every straight line from c intersects C just once. Then if $z' = c + \lambda(z-c)$, where $0 < \lambda < 1$, and z is on C, z' is within C, and as z travels around C, z' describes a contour C' arbitrarily near to C. Then $\int_{C'} f(z')\,dz' = 0$, by what we have proved; that is,

$$\int_C f\{c + \lambda(z-c)\}\lambda\,dz = 0; \tag{1}$$

and we can omit the constant non-zero factor λ. But by the principle of uniform continuity, for any ω we can take δ so that $|f(z) - f(z')| < \omega$ for all $|z - z'| < \delta$. But

$$|z - z'| = (1 - \lambda)|z - c| \tag{2}$$

and $|z - c|$ is bounded, say $< R$. Hence the condition is satisfied if

$$1 - \lambda = \delta/R. \tag{3}$$

Then
$$\int_C f(z)\,dz - \int_C f(z')\,dz = \int_C \{f(z) - f(z')\}\,dz \tag{4}$$

and
$$\left|\int_C \{f(z) - f(z')\}\,dz\right| < \omega L. \tag{5}$$

Hence $\left|\int_C f(z)\,dz\right| < \omega L$ and therefore is zero.

If C is such that its interior can be cut up into a finite number of regions each with a suitable internal point, the result follows by addition. There are contours C of finite length that require an infinite number of subdivisions before each region will satisfy Goursat's condition, and then the theorem remains true, but becomes much more difficult to prove.† We shall not need to consider such cases.

This extension of Cauchy's theorem appears to be necessary for some physical applications, where we wish to determine a function analytic within a contour and taking given values on the contour, though the derivative of the function at some points on the contour may not exist in the sense of the theory of functions of a complex variable, or, indeed, in that of the theory for a real variable. The further extension to cases where $f(z)$ behaves like $1/(z-z_0)$ near a point z_0 on the contour is impossible because there is no unique way of defining the integral through such a point; we shall refer later to the *principal value* of such an integral, but this is not equal to the limit of the integral around C'.

* *Cours d'Analyse*, 2, 1905, 88.
† S. Pollard, *Proc. Lond. Math. Soc.* (2), 21, 1923, 456–82; M. H. A. Newman, *Topology of Plane Sets of Points*, 1939, 154–6.

The theorem remains true if $f(z)$ is bounded but has a finite number of discontinuities on C. We cut out parts of C by drawing circular arcs of radius η about the discontinuities, and take C'' to consist of the parts of these arcs within C and of the rest of C. Then the theorem applies to show that $\int_{C''} f(z)\,dz = 0$. For any arc and the part cut out, $\int f(z)\,dz = O(\eta)$, since $f(z)$ is bounded; hence the theorem again follows. But (cf. 14·05) a simple discontinuity of $\Re f(z)$ implies that $\Im f(z)$ is unbounded.*

11·055. An important corollary follows at once. Suppose that $f(z)$ is analytic between two contours C and C', of which C encloses C', and continuous as z approaches either C or C'. Draw two lines AB, EF close together, so as to connect the two contours. Then $ABDEFGA$, described as shown, is a closed contour and $f(z)$ is analytic within it and continuous as z approaches it. Denote it by S. Then

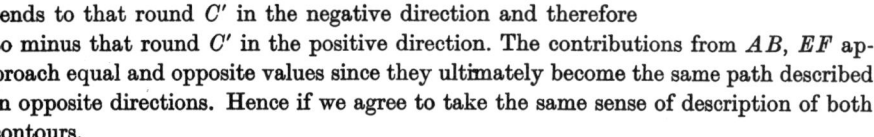

$$\int_S f(z)\,dz = 0.$$

Now let the lines AB, FE be made to approach indefinitely close together. The contribution from the part BDE tends to the integral around C in the positive direction. That from FGA tends to that round C' in the negative direction and therefore to minus that round C' in the positive direction. The contributions from AB, EF approach equal and opposite values since they ultimately become the same path described in opposite directions. Hence if we agree to take the same sense of description of both contours,

$$\int_C f(z)\,dz = \int_{C'} f(z)\,dz.$$

Hence: *if a function is analytic between two contours, and continuous on approaching them, its integral with regard to z round each contour has the same value.*

If the argument used in proving Cauchy's theorem is applied to the region between C and C', the result will be seen to follow directly. We need a separate proof in this case only because in proving Cauchy's theorem we assumed the region to be simply connected.

An immediate extension is to the case where C encloses several closed paths C', C'', ..., all external to one another. We can show similarly that the integral around C is equal to the sum of the integrals around C', C'', ..., provided that the function is analytic at all points that lie within C and outside C', C'', ... and continuous in the closed region.

11·056. Integral of an analytic function. It follows from Cauchy's theorem that if L, L' are two paths connecting z_0 and Z, and $f(z)$ is analytic on L, L' and at all points between them,

$$\int_L f(z)\,dz = \int_{L'} f(z)\,dz = F(Z),$$

---

* A general proof is given by Littlewood, *Theory of Functions*, p. 144.

say, and $F(Z)$ is single-valued. Also if $Z+\zeta$ is in the region where $f(z)$ is analytic, and $|\zeta|$ is small enough, we can take L'' to coincide with L from z_0 to Z and then to proceed to $Z+\zeta$ in a straight line. Hence

$$F(Z+\zeta)-F(Z) = \int_Z^{Z+\zeta} f(z)\,dz,$$

where the integral is along a straight line. Let ζ tend to 0; $f(z)$ will differ from $f(Z)$ by an arbitrarily small amount, and

$$\frac{1}{\zeta}\{F(Z+\zeta)-F(Z)\} \to f(Z).$$

Hence $F'(z) = f(z)$, and $F(z)$ is an analytic function of z.

This theorem should be compared with the three-dimensional one of 5·09(3).

It follows that if $G(z)$ is analytic within a region and $G'(z) = f(z)$, $F(z) - G(z)$ is constant; for it is an analytic function with zero derivative. This enables us to extend to complex integrals the method of integration by finding an indefinite integral.

11·06. Power series. The fundamental rules, if applied a finite number of times, will define a rational function of z. But other functions can be obtained by considering sums of infinite series, the most important of which are those in positive integral powers of z. Consider then the series

$$f(z) = a_0 + a_1 z + a_2 z^2 + \ldots + a_n z^n + \ldots, \tag{1}$$

where the a_n may be real or complex. Consider also the companion series

$$g(r) = b_0 + b_1 r + b_2 r^2 + \ldots + b_n r^n + \ldots, \tag{2}$$

where
$$b_n = |a_n|, \quad r = |z|. \tag{3}$$

According to the value of r, the terms $b_n r^n$ may be bounded or not. That is, we have

Case 1. There is an M such that $b_n r^n \leqslant M$ for all n.

Case 2. For any M there is an n such that $b_n r^n > M$.

The geometric series $\qquad 1 + z + z^2 + \ldots \tag{4}$

comes under Case 1 for $r \leqslant 1$, and under Case 2 for $r > 1$.

The exponential series $\qquad 1 + z + \dfrac{z^2}{2!} + \ldots + \dfrac{z^n}{n!} + \ldots \tag{5}$

comes under Case 1 for all r. For if we take $m > 2r$ and $n > m$,

$$\frac{r^n}{n!} = \frac{r^m}{m!}\frac{r^{n-m}}{(m+1)\ldots n} < \frac{r^m}{m!}(\tfrac{1}{2})^{n-m}. \tag{6}$$

Then if M is the largest of $1, r, r^2/2!\ldots r^m/m!$, it follows that $r^n/n! \leqslant M$ for all n.

The series $\qquad 1 + z + 2!z^2 + \ldots + n!z^n + \ldots$

comes under Case 2 for all $r > 0$; for if we take $m > 2/r$ and $n > m$

$$n!r^n = m!r^m \cdot (m+1) \ldots nr^{n-m} > m!r^m \cdot 2^{n-m},$$

which can be made to exceed any M by taking n large enough, and all later terms are larger still.

Thus there are series that fall within Case 1 for every value of r, others that are always within Case 2 except for $r = 0$, and others again that are within Case 1 for some values of r and Case 2 for others.

Every term of $g(r)$ increases with r unless it is 0 for all r. Hence if the series is in Case 1 for $r = r_1$ and Case 2 for $r = r_2$, we must have $r_1 < r_2$. Now suppose that values of r_1 and r_2 satisfying these conditions are found. The property 'the terms of $g(r)$ are bounded for $r < r_0$' defines a cut in the positive values of r_0, say at $r_0 = R$.

Hence if $g(r)$ belongs to Case 1 for some values of r and to Case 2 for others there is always a unique quantity R associated with the series such that all values of $r < R$ belong to Case 1 and all greater than R to Case 2. We call this the *radius of convergence* and the circle $|z| = R$ the *circle of convergence*. If $g(r)$ belongs to Case 1 for all r we can write $R = \infty$.

When $r = R$, $g(r)$ may be in either Case 1 or Case 2. Thus for (4) and the series

$$1 + z + \tfrac{1}{2}z^2 + \ldots + \frac{1}{n}z^n + \ldots$$

if $r = 1$, r^n and r^n/n are ≤ 1 for all n; and r^n and r^n/n are unbounded if $r > 1$. Thus the radius of convergence is 1 and the terms are bounded on the circle of convergence. But for the series

$$1 + 2z + 3z^2 + \ldots + (n+1)z^n + \ldots$$

the radius of convergence is again 1, but the terms are unbounded on the circle of convergence.

11·061. Absolute convergence. Suppose now that c is any positive quantity less than R. We know that for $r = c$ the terms of $g(r)$ are bounded, that is, they do not increase indefinitely; hence there is a quantity M such that

$$b_n c^n < M$$

for all n. Hence for any $r < c$ $\quad b_n r^n < M(r/c)^n$.

Further, for any m and $p \, (m < p)$,

$$\sum_{n=m+1}^{p} b_n r^n < M \left(\frac{r}{c}\right)^{m+1} \left\{ 1 + \frac{r}{c} + \ldots + \left(\frac{r}{c}\right)^{p-m-1} \right\}$$
$$= M \left(\frac{r}{c}\right)^{m+1} \frac{1 - (r/c)^{p-m}}{1 - r/c} < M \left(\frac{r}{c}\right)^{m+1} \frac{1}{1 - r/c}.$$

Hence if we choose an ϵ, however small, we can choose m_0 so that the sum of terms after the mth for $m \geq m_0$ will never exceed ϵ however many we take. The series $\Sigma b_n r^n$ therefore converges. Further

$$\left| \sum_{n=m+1}^{p} a_n z^n \right| \leq \sum_{n=m+1}^{p} |b_n z^n| = \sum_{n=m+1}^{p} b_n r^n,$$

and therefore the series $\Sigma a_n z^n$ converges for $|z| = r < c < R$, and therefore for $|z| < R$.

Within the circle of convergence the sum of the moduli of the terms is a convergent series. Such convergence is called *absolute* by analogy with the corresponding property for real series.*

---

* It is usual to take the property "the series is convergent for any z such that $|z| < r_0$" as defining R by a cut in the values of r_0; but it appears to us slightly more obvious that boundedness of the terms defines a cut, and this boundedness is used directly in proving many later theorems.

It is obvious that the series never converges if $|z| > R$.

Since $(1-r/c)^{-1}$ is bounded in any circle $|z| < d < c$, we have always
$$\left| f(z) - \sum_{n=0}^{m} a_m z^m \right| = O(z^{m+1}).$$

11·062. Uniform convergence. The definitions of uniform convergence for sequences and series of analytic functions are immediate extensions of those for functions of a real variable. If for all z in a region $f_n(z) \to f(z)$, and
$$f(z) = f_n(z) + R_n(z),$$
and if for any positive ϵ we can choose n so that for every z in the region $|R_n(z)| < \epsilon$, the sequence $\{f_n(z)\}$ is said to be uniformly convergent to $f(z)$ in the region.* If
$$f_n(z) = \phi_n + i\psi_n \to \phi + i\psi,$$
it follows that with the same choice of n
$$|\phi_n - \phi| < \epsilon, \quad |\psi_n - \psi| < \epsilon,$$
and therefore $\{\phi_n\}$, $\{\psi_n\}$ are uniformly convergent to ϕ, ψ respectively. Conversely if $\{\phi_n\}$, $\{\psi_n\}$ are uniformly convergent to ϕ, ψ, $\{\phi_n + i\psi_n\}$ is uniformly convergent to $\phi + i\psi$.

If we choose $d < c < R$ in 11·061, and m so that $\sum_{n=m+1}^{\infty} b_n d^n < \epsilon$, then for any z such that $|r| = r \le d$
$$\sum_{n=m+1}^{\infty} |b_n r^n| < \epsilon, \quad \left| \sum_{n=m+1}^{\infty} a_n z^n \right| \le \sum_{n=m+1}^{\infty} b_n r^n < \epsilon.$$

That is, we can choose m once for all, given ϵ and d, and it will do for all values of r within a range up to and including d. We thus arrive at a case of the M test for uniform convergence, extended to the complex variable. Formally we may state it as follows: *If for all values of z in a region $|u_n(z)|$ is less than v_n, which is independent of z, and the series Σv_n is convergent, then the series $\Sigma u_n(z)$ is uniformly convergent in the region.* This test is sufficient for uniform convergence, but not necessary. We see that any series satisfying it is also absolutely convergent in the region, and a series can be uniformly convergent without being absolutely convergent. We thus have the theorem: *A power series in z is uniformly convergent within and on any circle with centre $z = 0$ and lying wholly within the circle of convergence.* It may not converge uniformly, or even converge at all, on the circle of convergence itself.

It follows that if a power series has a radius of convergence R different from 0, then for any z such that $|z| \le c < R$ the sum of the series has a definite value; it therefore defines a pair of functions $\phi(x, y)$ and $\psi(x, y)$. Each of them is the sum of two uniformly convergent real series. If $g(r)$ belongs to Case 1 for all values of r, we may take any finite value for c in 11·061, and the argument proceeds as before. In that case $\Sigma a_n z^n$ defines such a pair of functions over the whole plane.

Some examples will now be given to illustrate the possible modes of behaviour of power series on the circle of convergence. We have considered the behaviour of the separate terms; we have now to consider the sums.

---

* We are not assuming at present that $f(z)$ is analytic; but cf. 11·20.

11·063. Types of power series: integral functions.

Consider again the series
$$f(z) = 1 + z + z^2 + \dots \quad (1)$$

It does not converge for any value on the circle $|z| = 1$, since all the terms have modulus 1 and the sum of n terms tends to no limit. This series is easily summed for $|z| < 1$; as for the real variable
$$f(z) = \frac{1}{1-z}. \quad (2)$$

We notice that though $f(z)$ as defined by (1) is meaningless for $|z| \geqslant 1$, the expression (2) has a definite value for any z except $z = 1$.

If we take the series
$$f(z) = z + \tfrac{1}{2}z^2 + \dots + \frac{z^n}{n} + \dots, \quad (3)$$

we find a different behaviour; as for the last, it converges for all $|z| < 1$ and diverges for all $|z| > 1$, but it also converges for all $|z| = 1$ except for $z = 1$ itself.* This series does not represent a rational function; it can be taken as the definition of $-\log(1-z)$.

The series
$$f(z) = z + \frac{z^2}{2^2} + \dots + \frac{z^n}{n^2} + \dots \quad (4)$$

has the same circle of convergence but converges even at $z = 1$. Thus we have three series with the same circle of convergence but behaving radically differently on the circle itself.

The series
$$\exp z = 1 + z + \frac{z^2}{2!} + \dots + \frac{z^n}{n!} + \dots, \quad (5)$$

on the other hand, converges for any z.

Functions definable by the same power series over the whole plane are called *integral functions*. Apart from terminating series, the exponential series is the most familiar example; closely related to it are the functions $\cosh z$, $\sinh z$, $\cos z$, $\sin z$.

The series
$$1 + 1!z + 2!z^2 + \dots + n!z^n + \dots \quad (6)$$

is not convergent for any z other than 0, however small. Such a series defines no function except for $z = 0$, and we may say that its radius of convergence is zero.

11·07. Differentiation of power series.

We have still to show that a function defined by a power series is analytic. If $f(z)$ is defined by $\Sigma a_n z^n$ we may call the series
$$a_1 + 2a_2 z + 3a_3 z^2 + \dots + na_n z^{n-1} + \dots$$

obtained by differentiating term by term, the first derived series. We can show easily that if R is the radius of the circle of convergence of the series defining $f(z)$, it is also that for the derived series. We can construct derived series of higher orders similarly.

A power series can always be integrated term by term within the circle of convergence, because it is uniformly convergent (1·113). Hence if
$$f(z) = a_0 + a_1 z + \dots + a_n z^n + \dots,$$

we have
$$\int_{z_0}^{Z} f(z)\, dz = F(Z) - F(z_0),$$

where
$$F(z) = a_0 z + \tfrac{1}{2}a_1 z^2 + \dots + \frac{1}{n+1} a_n z^{n+1} + \dots.$$

* Cf. 1·1155.

Similarly, we may start with the derived series and infer, since this has the same circle of convergence,

$$\int_{z_0}^{Z} \{a_1 + 2a_2 z + \ldots + na_n z^{n-1} + \ldots\} dz = [a_0 + a_1 z + \ldots + a_n z^n + \ldots]_{z_0}^{Z}$$
$$= f(Z) - f(z_0).$$

Differentiating with regard to Z, we have

$$a_1 + 2a_2 Z + \ldots + na_n Z^{n-1} + \ldots = df(Z)/dZ,$$

and therefore the derived series is the derivative of the original series anywhere within the circle of convergence. This shows further that a function defined by a power series has a derivative and therefore is an analytic function within the circle of convergence. The second derived series similarly represents the second derivative of $f(z)$, and since it converges the first derivative is continuous. Thus functions defined by power series satisfy the conditions used in 11·053.

11·08. Multiplication of power series. It can also be proved that if two series

$$f(z) = a_0 + a_1 z + a_2 z^2 + \ldots, \quad g(z) = b_0 + b_1 z + b_2 z^2 + \ldots,$$

both converge within any circle, then the product series

$$h(z) = c_0 + c_1 z + c_2 z^2 + \ldots,$$

where $\quad c_0 = a_0 b_0, \quad c_1 = a_0 b_1 + a_1 b_0, \quad c_2 = a_0 b_2 + a_1 b_1 + a_2 b_0, \quad \ldots,$

obtained by multiplying terms in pairs and collecting coefficients of the same power of z, converges within the same circle and is equal to $f(z) g(z)$. The proof is similar to that for absolutely convergent series of real terms.

It can also be proved that if a series is absolutely convergent it will give the same sum when the terms are taken in any order.

An immediate application of a similar argument gives

$$\exp z \exp z' = \exp(z + z')$$

for all z, z', as for the case of two real variables. Since if we write e^z for $\exp z$, z obeys the usual rules of indices, we can take this as a definition of e^z when z is complex. It should be noticed that e^z, from this point of view, is not to be regarded as the result of a process of raising e to the power z. Thus if we take $z = \frac{1}{2}$, $\exp \frac{1}{2}$ is a unique number defined as the sum of the series

$$1 + \frac{1}{2} + \frac{1}{2! \, 2^2} + \frac{1}{3! \, 2^3} + \ldots;$$

but the result of taking the square root of e might be either $\pm \exp \frac{1}{2}$. We take e^z to mean the same as $\exp z$, but other conventions are in use.

11·09. Limit-points. The definition of a limit-point in more than one dimension has been given in 5·01. We recall that any neighbourhood of a limit-point of a set contains an infinite number of members of the set and that any bounded infinite set has at least one limit-point (5·02).

11·091. *A power series cannot have $z = 0$ as a limit-point of zeros, unless it vanishes for all z.* For suppose that the series

$$f(z) = a_0 + a_1 z + a_2 z^2 + \ldots$$

has a zero sum for some non-zero z within any circle about 0. If possible, suppose that there is at least one term with a non-zero coefficient. Let the first term with a non-zero coefficient be $a_m z^m$. Then $g(z) = z^{-m} f(z) = \sum_{n=m}^{\infty} a_n z^{n-m}$ is a power series with the same circle of convergence as $f(z)$. Therefore it is uniformly convergent and therefore continuous in a neighbourhood of $z = 0$. But it is zero for some point in any neighbourhood of 0; hence $g(0) = 0$ and $a_m = 0$. Hence $f(z) = 0$ for all z.

In particular, if $f(z) = 0$ for all points within a circle about 0, however small, or even for all points on the real axis within some finite range of x, $f(z)$ is 0 for all z. But these conditions, which are those usually given in practice, are more than sufficient for the truth of the theorem: it would be enough, for instance, if $f(1/n) = 0$ for every integral n greater than 1000.

The most important application is that a function can have only one expansion in powers of z. For if, for $|z| < R$,
$$f(z) = \Sigma a_n z^n = \Sigma a'_n z^n,$$
we have
$$\Sigma (a_n - a'_n) z^n = 0$$
for all $|z| < R$. Hence $a_n = a'_n$. This justifies the method of equating coefficients of powers of z.

11·10. Taylor's theorem. Let
$$f(z) = a_0 + a_1 z + a_2 z^2 + \ldots \tag{1}$$
with radius of convergence R. Let as before c be less than R and let all $|a_n c^n|$ be less than M. Then
$$b_n = |a_n| < M/c^n. \tag{2}$$
Take z_0 such that $|z_0| < c$ and put $z = z_0 + z'$. Substitute in the series, and expand each term by the binomial theorem; we have
$$f(z_0 + z') = a_0 + a_1(z_0 + z') + a_2(z_0^2 + 2z_0 z' + z'^2) + \ldots + a_n \sum_{m=0}^{n} \frac{n!}{m! \, n-m!} z_0^{n-m} z'^m + \ldots. \tag{3}$$
Consider the companion series obtained by taking the modulus of every term; writing as before $|z| = r$, $|z_0| = r_0$, $|z'| = r'$, we have
$$g(r_0, r') = \sum_{n=0}^{\infty} \sum_{m=0}^{n} b_n \frac{n!}{m!(n-m)!} r_0^{n-m} r'^m$$
$$< M \sum_{n=0}^{\infty} \sum_{m=0}^{n} \frac{n!}{m!(n-m)!} \frac{r_0^{n-m} r'^m}{c^n}$$
$$= M \sum_{n=0}^{\infty} \frac{(r_0 + r')^n}{c^n}, \tag{4}$$
which converges if
$$r_0 + r' < c,$$
that is, within any circle with centre z_0 that does not pass beyond the circle of radius c. Hence the series (3) is absolutely convergent within such a circle. Its terms can therefore be taken in any order and will always give a convergent series with the same sum. Take them in order of ascending powers of z'. The terms independent of z' are
$$a_0 + a_1 z_0 + a_2 z_0^2 + \ldots = f(z_0), \tag{5}$$
the coefficient of z' is
$$a_1 + 2a_2 z_0 + 3a_3 z_0^2 + \ldots = f'(z_0), \tag{6}$$

and in general the coefficient of z'^m is

$$a_m + (m+1)a_{m+1}z_0 + \ldots + \frac{n!\,a_n}{m!\,(n-m)!}z_0^{n-m} + \ldots, \tag{7}$$

where $n \geqslant m$; on putting $n - m = n'$, the general term is

$$\frac{(n'+m)!}{m!\,n'!}a_{n'+m}z_0^{n'}, \tag{8}$$

where $n' \geqslant 0$. Then the coefficient of z'^m can be written

$$\frac{1}{m!}\{m(m-1)\ldots 1 . a_m + (m+1)m\ldots 2a_{m+1}z_0 + \ldots + (n+m)(n+m-1)\ldots(n+1)a_{m+n}z_0^n + \ldots\}$$

$$= \frac{f^{(m)}(z_0)}{m!}, \tag{9}$$

the bracketed index indicating the mth derivative. Hence

$$f(z) = f(z_0) + z'f'(z_0) + \frac{1}{2!}z'^2 f''(z_0) + \ldots + \frac{1}{m!}z'^m f^{(m)}(z_0) + \ldots \tag{10}$$

within any circle about z_0 that does not reach the circle of convergence of the series (1).

This is the form taken by Taylor's theorem when the variable is complex. The infinite series always converges, and there is no need for a remainder term.

11·11. Singularities. We have so far restricted ourselves to functions that are uniquely defined at each point of a region and differentiable and therefore continuous for all variations of z within the region. We have thus excluded any function if at any point of the region it is capable of taking two or more values or if it is non-differentiable; in particular, if it tends to infinity as z approaches some point in the region. We proceed now to consider what can happen in the latter cases.

A *singularity* a of a function $f(z)$ is any value of z such that we cannot choose a positive δ so that $f(z)$ is analytic and single-valued for $|z-a| < \delta$. It follows from the definition that a limit-point of singularities is a singularity.

(a) *Branch points.* Consider the function $z^{1/2}$. This is finite for all finite z, but even for z real and positive there is an ambiguity about which sign shall be taken. Suppose that we agree to take the positive root in that case. Now let us proceed in a circle about the origin in the direction of increasing argument, and let $z^{1/2}$ vary continuously. Then if we put

$$z = re^{i\theta},$$

we can take

$$z^{1/2} = r^{1/2}e^{1/2\,i\theta}.$$

We are not varying r and therefore need not vary $r^{1/2}$, and θ must vary continuously. But when θ has reached 2π, $\frac{1}{2}\theta$ has reached π, and $e^{1/2\,i\theta}$ is -1. Increasing θ further we repeat all the previous values with the opposite sign. Thus we cannot attach a single value to $z^{1/2}$ at every point if we allow θ to vary by more than 2π with r constant. But if we make it a rule that θ is never to reach 2π, we can make $z^{1/2}$ single-valued. For then, though we may make θ as near 2π as we like, the only way of getting back to the positive real axis is to make the circuit of the origin in the opposite direction, and in doing so we undo the previous variation of θ and arrive back at the original value of $z^{1/2}$. To use this device it will be necessary to exclude the possibility of a complete circuit for every value of r, and we speak of a *cut* along the positive real axis. This amounts to defining

$$z^{1/2} = r^{1/2} \exp \tfrac{1}{2} i\theta \quad (0 \leqslant \theta < 2\pi),$$

where we always take the positive value of $r^{1/2}$. We could equally well take the range of θ to be $-\pi < \theta \leqslant \pi$, or $-\pi \leqslant \theta < \pi$. In general if n is fractional we can define

$$z^n = r^n \exp ni\theta$$

with the same restrictions on θ.

A function may be single-valued and even differentiable at $z = 0$ without being analytic there; $z^{\frac{1}{2}} = 0$ at $z = 0$, and $z^{3/2} = 0$ and has a zero derivative there, no matter what sign we take first. The definition of a branch point concerns its *neighbourhood*; *a point $z = a$ is a branch point of $f(z)$ if when z moves around a in an arbitrarily small circle, not of zero radius, the value of $f(z)$ being chosen at each value of z to preserve continuity, $f(z)$ does not return to its original value.*
If
$$f(z) = (z^2 - a^2)^{1/2},$$

where a is real, there are branch points at $\pm a$. If z makes a complete circuit about any curve including both of them, $f(z)$ returns to its original value; for the square roots of $z-a$ and $z+a$ both change sign and their product has its original sign. But $f(z)$ would be reversed if we went round any circuit that included either of $\pm a$ and excluded the other. Here we can make $f(z)$ single-valued by making it a rule that we take the positive sign on the real axis when $x > a$, and never cross the real axis between $-a$ and $+a$.

When we make a cut, we select one value of the function for every point in the region, and have a single-valued function in the region. But it is discontinuous when z crosses the cut, which is therefore a line of singularities. Thus if $z^{1/2}$ is defined by $r^{1/2} \exp \tfrac{1}{2} i\theta \, (-\pi < \theta \leqslant \pi)$, where we take the positive sign for z real and positive, $z^{1/2}$ has a discontinuity $2ir^{1/2}$ when z crosses the negative real axis. If we took $z^{1/2}$ as meaning $-r^{1/2} \exp \tfrac{1}{2} i\theta$, we should get a different single-valued function, which can be called a different *branch* of $z^{1/2}$. In what follows we shall assume that all functions are single-valued or have been made so by means of a cut or cuts.

If $f(z)$ is not single-valued on a contour C, that is, if it does not return to its original value when z describes the contour, $f(z)$ varying continuously, $f(z)$ has a singularity within or on the contour. For if $f(z)$ is analytic within and on the contour we can superpose a net of squares such that $f(z)$ varies continuously when z describes any interior square or any fringing portion as in 11·052 and returns to its original value. It follows by addition that the change of $f(z)$ when z describes C is zero.

(b) *Poles.* A function $f(z)$ may be unbounded in any circle about a, however small, but be such that when we make a circuit of the point the function returns to its original value. A pole of $f(z)$ of order m is a point a such that there is a positive integer m such that for $z \neq a$,

$$f(z) = \frac{A_m}{(z-a)^m} + \frac{A_{m-1}}{(z-a)^{m-1}} + \ldots + \frac{A_1}{z-a} + g(z),$$

in a region enclosing a, where $A_m \neq 0$, and $g(z)$ is analytic at a. The terms containing negative powers of $z-a$ are called the *principal part* of the function near $z = a$. Poles of order 1 are also called *simple poles*. We do not speak of poles of non-integral order, since a would be a branch point of $(z-a)^{-n}$ if n was not an integer.

If a is a pole of order m, $(z-a)^m f(z)$ is analytic in a neighbourhood of a, and so is $1/\{f(z)-c\}$, where c is any constant.

(c) *Essential singularities.* An essential singularity is any point a, not a branch point or a pole, such that $f(z)$ is not analytic in a neighbourhood of a. $\exp(1/z)$ has an essential singularity at $z = 0$. If z is real and positive, then whatever m we choose, $z^m \exp(1/z)$ tends to infinity as z tends to 0. The behaviour of functions near essential singularities is more peculiar than near poles. If $f(z)$ has a pole at a, then for all methods of approach to a, $|f(z)|$ tends to infinity. But if z tends to 0 through negative real values $\exp(1/z)$ tends to 0. We can say, if we like, that $f(z)$ is infinite at a pole, provided that we understand that we mean by this nothing more than that, for every sequence of points z_n tending to the pole, $|f(z_n)| \to \infty$. We cannot say that it is infinite at an essential singularity because it may be possible by choosing the method of approach suitably to make the limit of $f(z)$ finite. Thus if we take the equation

$$\exp(1/z) = b,$$

where b is not 0 or ∞, it is satisfied wherever

$$z = \frac{1}{2ni\pi + \log b},$$

n being an integer, and by taking n larger and larger we can make z as near as we like to 0 while keeping $\exp(1/z)$ always equal to b. It will be seen that when a many-valued function has been replaced by a single-valued one by means of a cut, every point of the cut is an essential singularity.

(d) A formally possible type of singularity, as for the real variable, is a *removable discontinuity*. If $f(z)$ is analytic in any ring $0 < \delta < |z-a| < c$, where δ is arbitrarily small, and if $f(z) \to d$ when $z \to a$ in any manner, but $f(a) \neq d$, we call a a removable discontinuity. Such singularities have no practical importance but are mentioned for completeness. We shall always suppose that if $f(z)$ tends to a unique limit d when $z \to a$ in any manner, then $f(a) = d$.

11·111. If $f(z) = \Sigma a_n z^n$ for all z such that $|z| < R$, there is no singularity of $f(z)$ for $|z| < R$. For $\Sigma a_n z^n$ is single-valued and has a derivative everywhere within the circle of convergence.

If $f(z)$ has a line of discontinuity, the circle of convergence may overlap this line. Then $f(z)$ has no singularity within the part of the circle that includes $z = 0$. In the other part the series will converge but not be equal to $f(z)$.

11·112. Singularities at infinity. If $z = 1/\zeta$ and

$$f(z) = g(\zeta),$$

$g(\zeta)$ may either be analytic at $\zeta = 0$ or have a branch point, a pole of order m, or an essential singularity there. In these cases we say respectively that $f(z)$ is analytic at $z = \infty$, or has a branch point, a pole of order m, or an essential singularity at $z = \infty$. This extension of the definitions saves some writing.

11·113. Integrals around poles.

The equation

$$\frac{d}{dz} z^m = m z^{m-1} \tag{1}$$

is extended to cases where m is a negative integer by using the equation

$$\frac{d}{dz}(uv) = u\frac{dv}{dz} + v\frac{du}{dz} \tag{2}$$

with $u = z^m$, $v = z^{-m}$. The left side vanishes, and with $-m$ a negative integer

$$\frac{dv}{dz} = -m z^{-m-1}, \tag{3}$$

whence the result follows. Hence we have (1) for all integral m, positive or negative. Then by using (1) we have for any integral m other than 0

$$\int_{z_0}^{Z} z^{m-1} dz = \frac{1}{m}(Z^m - z_0^m). \tag{4}$$

The argument fails for $m = 0$, since the derivative of z^0 is not a multiple of z^{-1} but zero. We define provisionally

$$\log z = \int_1^z \frac{dt}{t}. \tag{5}$$

The integrand is analytic within any region that does not include the origin, and by Cauchy's theorem the integral has the same value for any two paths such that it is possible to deform one into the other without passing through the origin. Now put $|z| = r$, $\arg z = \theta$ and take the path to be from $t = 1$ along the real axis to $t = r$, and then along a circle about the origin to z. On the first part of the path t is real and positive, and

$$\int_1^r \frac{dt}{t} = \log r.$$

On the second part we put

$$t = r(\cos\lambda + i\sin\lambda),$$

and then

$$dt = r(-\sin\lambda + i\cos\lambda) d\lambda;$$

then

$$\int_r^z \frac{dt}{t} = \int_0^\theta i\, d\lambda = i\theta.$$

Hence

$$\log z = \log r + i\theta. \tag{6}$$

In particular, if we make a complete circuit about the origin

$$\int_C z^{m-1} dz = 0 \tag{7}$$

for $m \neq 0$, since z^m is single-valued; but if $m = 0$

$$\int_C \frac{dz}{z} = [\log r + i\theta] = 2\pi i, \tag{8}$$

since $\log r$ is single-valued but θ increases by 2π.

Log z is not single-valued without some restriction on the path of integration used in its definition; in other words, it has a branch point at the origin. In fact, if we maintained continuity and allowed z to make several circuits about the origin, $\log z$ would increase by $2\pi i$ for each circuit, and thus would have infinitely many possible values differing by integral multiples of $2\pi i$. Our first example, $z^{1/2}$, had only two. We can, however, make $\log z$ single-valued by means of any cut used for $z^{1/2}$. The *principal value* of $\log z$ is that such that $|\mathfrak{J}(\log z)| < \pi$, when z is not real and negative.

Now apply these rules to integrate the expression in 11·11 (b) around a circuit including the pole at $z = a$ and no other singularity. All terms in $(z-a)^{-n}$, with n different from 1, give zero. The integral of $g(z)$ is zero by Cauchy's theorem, since $g(z)$ is analytic in the region. Hence

$$\int_C f(z)\,dz = \int_C \frac{A_1 dz}{z-a} = 2\pi i A_1. \tag{9}$$

The integral of an analytic function about a pole therefore depends wholly on the coefficient of $(z-a)^{-1}$ in its principal part. This coefficient is called the *residue* of the function at the pole. The characteristic feature of the method of evaluating definite integrals known as *contour integration* is to find a contour containing no singularities other than poles, the integral around which is equal to the definite integral sought, and then to equate the integral around this contour to $2\pi i$ times the sum of the residues at poles within the contour. There is no simple analogous rule for branch points and essential singularities, though one can be found for the latter in some cases by means of Laurent's theorem, which will be proved later.

11·114. Relation of the exponential and logarithmic functions. By direct multiplication we have

$$\exp(\log z) = \exp(\log r) \times \exp i\theta$$

(since both factors are absolutely convergent series and the same argument applies as for real numbers)

$$= r(\cos\theta + i\sin\theta)$$

$$= x + iy = z. \tag{10}$$

We can make use of the exponential and logarithmic functions to define z^n for irrational and even complex indices; we take

$$z^n = \exp(n\log z). \tag{11}$$

This is single-valued and analytic in any region such that $\log z$ is. The verification that it has a derivative equal to nz^{n-1} may be left to the student.

If a is real and positive

$$a^z = \exp(z\log a) = \exp(x\log a)\exp(iy\log a),$$

$$|a^z| = \exp(x\log a) = a^x.$$

11·12. Isolated and non-isolated essential singularities. An essential singularity a may be *isolated* or not. If it is isolated we can take a circle about a, with a radius not zero, such that a is the only singularity within it. Thus $\operatorname{cosec}(1/z)$ has a pole whenever $z = 1/n\pi$ and n is an integer, and all these singularities are isolated in the sense that we can draw a circle about each small enough to contain no other. But they have a limit-point at $z = 0$, and we see easily that the function is indeterminate and has no derivative at this point, nor has any function of the form $z^m \operatorname{cosec}(1/z)$. Hence $z = 0$ is an essential singularity, and it is not isolated since there is another singularity within any circle about it, however small. On the other hand, $\exp \operatorname{cosec}(1/z)$ has isolated essential singularities at all points $z = 1/n\pi$ and a non-isolated one at $z = 0$. Functions can be constructed with non-isolated essential singularities at all points of a curve. The most important cases are those of single-valued functions that have been derived from many-valued ones by introducing cuts.

If $f(z)$ is analytic and single-valued in a region except for poles or essential singularities, and the singularities have a limit-point in the region, the limit-point is a non-isolated essential singularity. For if the limit-point is $z = 0$, any neighbourhood of $z = 0$ contains singularities of $f(z)$, and therefore $z = 0$ is a singularity; and it is not a pole because, for any positive integral m, $z^m f(z)$ has singularities in any neighbourhood of $z = 0$.

11·13. Cauchy's integral. Let $f(z)$ be analytic and single-valued within a contour C and continuous in the closed region. Then if z is within C,

$$f(z) = \frac{1}{2\pi i} \int_C \frac{f(t)}{t-z} dt. \tag{1}$$

If $f(t)$ is expressible as a power series in $t-z$ this is obvious because by Taylor's theorem the residue of $f(t)/(t-z)$ at $t = z$ is $f(z)$. But we have not yet proved that an analytic function can be expressed as a power series and this theorem is one step towards proving it.

The only singularity within the contour is the simple pole at $t = z$. If we take a small circle C' about z there is no singularity between C and C', and the integrals about C and C' are equal by the corollary to Cauchy's theorem. On C', since $f(t)$ is differentiable at $t = z$,

$$f(t) = f(z) + (t-z)\{f'(z) + v(t)\}, \tag{2}$$

where $v(t) \to 0$ with $t-z$. Then

$$\int_C \frac{f(t)}{t-z} dt = \int_{C'} \left\{\frac{f(z)}{t-z} + f'(z) + v(t)\right\} dt. \tag{3}$$

The first term gives $2\pi i f(z)$, the second 0, and the third tends to 0 as the radius of C' tends to 0. But the left side is independent of the radius of C'; therefore it is equal to $2\pi i f(z)$, which proves the proposition.

It follows that
$$f'(z) = \lim_{\zeta \to 0} \frac{1}{\zeta} \frac{1}{2\pi i} \int_C \left(\frac{f(t)}{t-z-\zeta} - \frac{f(t)}{t-z}\right) dt$$
$$= \lim_{\zeta \to 0} \frac{1}{2\pi i} \int_C \frac{f(t)}{(t-z)(t-z-\zeta)} dt$$
$$= \frac{1}{2\pi i} \int_C \frac{f(t)}{(t-z)^2} dt \tag{4}$$

since $|t-z|$ has a positive lower bound on C. We can find similarly

$$\frac{f^{(n)}(z)}{n!} = \frac{1}{2\pi i} \int_C \frac{f(t)}{(t-z)^{n+1}} dt, \tag{5}$$

and all these integrals exist. Hence Cauchy's integral can be differentiated under the integral sign.

This result is particularly important because in the argument leading up to Cauchy's theorem we deliberately avoided assuming that ϕ and ψ had continuous derivatives; we assumed only that they were differentiable and satisfied the Cauchy-Riemann relations. But starting from this assumption we have now shown that the first derivatives in these conditions are differentiable and therefore are continuous after all. We thus reach the end of a long story.

11·14. Relation of Cauchy's integral to power series. Now

$$\frac{1}{t-z} = \frac{1}{t} + \frac{z}{t^2} + \ldots + \frac{z^m}{t^{m+1}} + \frac{z^{m+1}}{t^{m+1}(t-z)}. \tag{6}$$

Hence, if C of 11·13 is a circle, $|t| = c$, where $|z| < c$ and $f(t)$ is bounded on C,

$$f(z) = \frac{1}{2\pi i} \sum_{n=0}^{m} \int_C \frac{z^n}{t^{n+1}} f(t)\, dt + R_m(z), \tag{7}$$

where
$$R_m(z) = \frac{1}{2\pi i} \int_C \frac{z^{m+1} f(t)}{t^{m+1}(t-z)}\, dt. \tag{8}$$

Since z is *within* C, $|t-z|$ has a lower bound $\rho > 0$ when t lies on C. Hence

$$|R_m(z)| \leq \frac{1}{2\pi} \left|\frac{z}{c}\right|^{m+1} \frac{M}{\rho} 2\pi c = \frac{Mc}{\rho} \left|\frac{z}{c}\right|^{m+1}, \tag{9}$$

where M is the upper bound of $|f(t)|$ on C. Hence $|R_m(z)| \to 0$, and

$$f(z) = \frac{1}{2\pi i} \sum_{n=0}^{\infty} z^n \int_C \frac{f(t)}{t^{n+1}}\, dt, \tag{10}$$

for all z within C. Hence, in the conditions stated, $f(z)$ *has a convergent expansion in a power series, whose radius of convergence is not less than c*; and comparing with (5) we see that this series is the Taylor series for $f(z)$ about $z = 0$. Consequently all results proved for functions defined by power series are true for analytic functions in general.

11·141. Cauchy's inequality. If we now write

$$f(z) = \Sigma a_n z^n \tag{11}$$

then for $|z| \leq c$
$$|a_n z^n| = \frac{1}{2\pi} \left| z^n \int_C \frac{f(t)}{t^{n+1}} dt \right| \leq \frac{1}{2\pi} c^n \frac{M}{c^{n+1}} 2\pi c = M. \tag{12}$$

Hence *within and on a circle about $z = 0$ no term of the power series expansion of $f(z)$ has a modulus greater than the maximum modulus of $f(z)$ on C*. This is *Cauchy's inequality*.

If R is the distance of 0 from the nearest singularity or from the nearest point of the boundary of the region, whichever is smaller, then by 11·14 the series $\Sigma a_n z^n$ converges and is equal to $f(z)$ for all z such that $|z| < R$. Further, R is the greatest value such that this is true whenever $|z| < R$. For if not, let it be true whenever $|z| < R'$, where $R' > R$. Then either (1) the circle of convergence of $\Sigma a_n z^n$ contains a singularity of the function given by the sum of the series, which is impossible by 11·111, (2) for part of the circle the sum of the series is not $f(z)$, or (3) the circle of convergence extends beyond the boundary of the region where $f(z)$ is defined.

If $f(z)$ is defined uniquely for all z except possibly for isolated singularities the theorem takes the simpler form: the circle of convergence of a power series passes through the singularity of the function nearest to its centre.

Also, by comparison with 11·091 and 11·14, *if two functions are analytic for $|z| < R$, a necessary and sufficient condition that they shall be equal for $|z| < R$ is that they have the same expansion in powers of z for $|z| < c$ for some $c < R$; and both functions will then be equal to the sum of the series for $|z| < R$.*

11·142. Liouville's theorem; integral functions for large $|z|$. If $f(z)$ is analytic over the whole plane, we can take C to be a circle of arbitrarily large radius c. Then if $f(z)$ is bounded over the whole plane, say $|f(z)| \leq M$,

$$|a_n| \leq \frac{M}{c^n},$$

which is arbitrarily small for $n > 0$. Hence

$$f(z) = a_0,$$

and therefore *a function bounded and analytic over the whole plane is a constant.* This is *Liouville's theorem*.

Conversely if a function is analytic over the whole plane (an integral function) it is unbounded for large $|z|$ unless it is constant.

11·15. Analytic continuation. We have seen from Taylor's theorem that if a function $f(z)$ is given by a power series in z, it can be represented also as a power series in $z - z_0$, where z_0 is any point within the original circle of convergence, and this series will converge within any circle about z_0 that does not pass beyond the original circle of convergence. It may, however, converge within a circle that does pass beyond the original circle of convergence. Take the function

$$f(z) = 1 + z + z^2 + \ldots,$$

and put $z_0 = \tfrac{1}{2}i$. $f(z)$ is already known to be equal to $1/(1-z)$ for $|z| < 1$, and its derived series express the functions

$$\frac{1!}{(1-z)^2}, \quad \frac{2!}{(1-z)^3}, \quad \frac{3!}{(1-z)^4}, \quad \ldots$$

Hence the Taylor expansion of $f(z)$ in powers of $z' = z - \tfrac{1}{2}i$ is

$$\frac{1}{1-\tfrac{1}{2}i} + \frac{z'}{(1-\tfrac{1}{2}i)^2} + \frac{z'^2}{(1-\tfrac{1}{2}i)^3} + \ldots.$$

By Taylor's theorem we know that this series must converge and be equal to the original function if $|z'| < \tfrac{1}{2}$, since i is the point of the circle $|z| = 1$ nearest to $\tfrac{1}{2}i$. But we see by inspection that it actually converges if $|z'| < |1 - \tfrac{1}{2}i| = \tfrac{1}{2}\sqrt{5}$. This is what we should expect since by 11·141 the circle of convergence must pass through the singularity of the function nearest to the origin used. But the series considered might have represented no function already known; in that case the new Taylor series would define values of an analytic function over a range of z where no function is defined by the original series.

Then we may be able to extend the range of definition further by taking a new Taylor series about a point in the new region. This process is called *analytic continuation*, and is fundamental in Weierstrass's theory of functions, which takes the power series as the fundamental definition of an analytic function.* Weierstrass's definition has the merit of being *constructive*; that is, we can assign any coefficients we like and always get a power series, the convergence and continuation of which we can proceed to study. Some functions arise naturally as power series, as, for instance, when we solve a differential equation by series. Some of the proofs, however, are more difficult with Weierstrass's approach than with Cauchy's. With the latter, however, we achieve little until we actually find functions satisfying the conditions stated for a function to be analytic; and what we find is that the most general analytic functions *have* power-series expansions. Hence the two theories are completely equivalent: but in practice our initial information about a function sometimes shows that it satisfies Weierstrass's condition, sometimes Cauchy's, and therefore, strictly speaking, both developments are necessary to a complete theory. In practice, however, when continuation is required the direct use of Taylor series is laborious and seldom used.

In the Weierstrass theory a function is defined at the outset *only* within the circle of convergence of the original power series. The function is then defined by the values given by the series together with all its continuations, which may pass branch points on opposite sides and thus give more than one value of $f(z)$ for given z. But if $f(z)$ is a rational function, not with a pole at the origin, the Weierstrass method would define it initially only within a circle extending to the pole of smallest modulus, and the process of continuation is needed before we can calculate it anywhere outside that circle. The method that we have adopted, on the other hand, enables us to calculate it directly from the fundamental rules for any value of z except the poles. Similarly, definitions by definite integrals are often directly applicable over larger regions than power series.

The introduction of artificial barriers to replace many-valued functions by single-valued ones is an awkward feature of the present method, since the same object would usually be attained by many different cuts. It is possible either with the Weierstrass or the Cauchy method to dispense altogether with the use of cuts and to consider the function as a whole; we can choose one possible value of $f(z)$ at $z = z_0$, say, and consider how $f(z)$ behaves if z varies continuously from z_0 to z_1, $f(z)$ varying continuously, or we can use the method of continuation by Taylor series. The value found for $f(z_1)$ will then depend on the route chosen if $f(z)$ is many-valued, and the route therefore must be specified. This is done most systematically by the Riemann surface method, which replaces the z plane by a number of sheets winding into one another at the branch points. In this theory it is always true that there is a singularity on the circle of convergence, and the possibility of the series converging but not being equal to the function in part of the circle does not arise; the sum of the series is always one value of the function. The general treatment is beyond the scope of this book. We have however a number of cases, especially in Chapters 21 and 25, where special attention has to be given to different paths of integration in presence of one or more branch points. Fortunately in these cases the distinctions between the paths are fairly simple.

* Harkness and Morley, *Theory of Functions*; Hurwitz and Courant, *Allgemeine Funktionentheorie*.

The treatment of functions with branch points is particularly important in the use of conformal representation in Chapter 13, but there the principle of one-one correspondence between points of the two planes compared makes the introduction of cuts unavoidable.

11·151. *If two functions $f_1(z)$, $f_2(z)$ are analytic in a region D and equal in a region D' within D, they are equal everywhere in D.* Take z_0 to be any point in D', and Z any other point in D. Then $f_1(z) = f_2(z)$ within any circle about z_0 that does not reach a singularity or the boundary of D'. Now suppose z_0 and Z connected by a curve of finite length in D not reaching the boundary of D. The distances between points on the curve and the boundary of D have a positive lower bound δ. Hence we can choose points $z_1, z_2, ..., z_{n-1}$, $z_n = Z$ on the curve such that $|z_r - z_{r-1}| < \delta$, and n is finite. z_1 lies within the circle of convergence of the series representing $f_1(z)$ and $f_2(z)$ in powers of $z - z_0$; hence both functions have the same Taylor series in $z - z_1$, and z_2 is within its circle of convergence. Proceeding we can show in a finite number of steps that both functions have the same Taylor series in $z - z_{n-1}$, and its circle of convergence includes Z. Hence $f_1(Z) = f_2(Z)$.

It is not necessary to the argument that the functions should be known to be equal at every point of a region. It is enough that they should be equal at, for instance, an infinite number of points within a square, or even along a finite stretch of a straight line. For we can establish the existence of a limit-point z_0 by the method of successive bisection, and it is in a region where $f_1(z) - f_2(z)$ is analytic. Take it as origin and apply the argument of 11·091, and it follows that $h(z) = f_1(z) - f_2(z)$ is everywhere zero for $|z - z_0| < \delta$.

It is astonishing that so much can be inferred from a knowledge of the values of the function in a limited region, but we must remember, as for Cauchy's theorem, the severe restrictions on the possible behaviour of the function imposed by the condition that it is analytic. We shall see under Fourier's theorem that it is sometimes possible to extend the definition of a function outside the original range in a quite different way by assuming different properties outside the range.

The condition that the set of points where $h(z) = 0$ must have a limit-point is essential. If, for instance, two functions are known to agree whenever z is an integer, their difference vanishes at an infinite number of points, but these have no limit-point, and the functions could differ by any multiple of $\sin 2\pi z$. The condition that $h(z)$ must also be analytic at the limit-point is also essential; for instance, if $h(z) = 0$ when $z = 1/n$ it could be any multiple of $\sin \pi/z$. The additional information that $h(z)$ is analytic at $z = 0$ removes this possibility.

We have seen in 11·141 that any function $f(z)$ defined by a power series must have at least one singularity on the circle of convergence. The process of continuation may lead to definitions all round a singularity a. If the result at a given z depends on whether we pass a on the side where $\arg z < \arg a$ or $\arg z > \arg a$, a is a branch point. If it is independent of the route, a is a pole or an essential singularity.

A type of application that we shall often meet is to the solution of a differential equation of the form
$$F(z, \zeta'', \zeta', \zeta) = 0,$$
where F is such that if ζ is an analytic function of z in a region D so also is F. We may find by some special method that $\zeta = f(z)$, where $f(z)$ is analytic, satisfies this equation in a certain region D' of z, included in D. Then $F\{z, f''(z), f'(z), f(z)\}$ is an analytic function

of z, identically zero in D'. It follows by continuation that this function is zero over the whole region D, and therefore $f(z)$ satisfies the differential equation in D.

11·152. It follows from the modified Heine-Borel theorem that if $f(z)$, a function of the complex variable, has a power series expansion about every point of a closed region, a finite set of squares can be found, covering the whole region, so that every square, or the part of it that is not outside the region, lies wholly within the circle of convergence of the expansion of $f(z)$ about some point of the square.

11·16. Two other theorems with some resemblance to Cauchy's inequality may be proved here. First, *the real or the imaginary part of an analytic function has no maximum or minimum within any open region*. For, if it has, let us take the maximum to be at $z = 0$ and the contour C of 11·13 to be a circle about 0. Then

$$f(0) = \frac{1}{2\pi i}\int_C f(t)\frac{dt}{t} = \frac{1}{2\pi}\int_0^{2\pi} f(ce^{i\theta})\,d\theta,$$

where we have put $t = ce^{i\theta}$ on C. Hence $f(0)$ is equal to the mean value of $f(z) = \phi + i\psi$ on any circle about 0 and therefore neither ϕ nor ψ can be greater at 0 than at every point on the circle; similarly, they cannot be less than at every point of the circle. The extreme values of ϕ and ψ in any region must therefore be taken on the boundary. If ϕ is constant on C, then by the uniqueness theorem (6·074) ϕ is constant within C; then by the Cauchy-Riemann relations ψ is constant within C. Hence $f(z)$ is constant within C, and by analytic continuation (11·15) $f(z)$ is constant over the whole region. ϕ and ψ can have stationary points, but if they are maxima for some directions of displacement they are minima for displacements in some other directions, as we can see by considering the power-series expansions.

The three-dimensional analogue is that if $\nabla^2 \phi = 0$ within and on a sphere, ϕ at the centre is the mean of the values over the sphere. (Cf. 6·092.)

11·161. Maximum modulus principle. *If $f(z)$ is analytic in a region, $|f(z)|$ has no maximum within the region; and if $|f(z)| \leq M$ at all points of the boundary, $|f(z)| < M$ at all points of the interior unless $f(z)$ is constant.* Let z_0 be a point within the region, put $z = z_0 + t$, and take a small circle C, $t = ce^{i\theta}$, within the region. Then, if asterisks denote conjugate complexes,

$$f(z) = \phi + i\psi = a_0 + \sum_1^\infty a_n t^n; \quad f^* = \phi - i\psi = a_0^* + \sum_1^\infty a_n^* t^{*n}; \quad |f(z)|^2 = ff^*$$

and the series are absolutely convergent on C. If m, n are unequal integers,

$$\int_0^{2\pi} t^m t^{*n} d\theta = c^{m+n}\int_0^{2\pi} e^{(m-n)i\theta}d\theta = 0$$

and therefore
$$\frac{1}{2\pi}\int_0^{2\pi} ff^* d\theta = \frac{1}{2\pi}\int_0^{2\pi}\left(a_0 a_0^* + \sum_1^\infty a_n a_n^* t^n t^{*n}\right)d\theta$$
$$= |a_0|^2 + \sum_1^\infty |a_n|^2 c^{2n} \geq |a_0|^2 = |f(z_0)|^2.$$

Hence $|f(z_0)|^2$ is not greater than the mean value of $|f(z)|^2$ on C, equality holding only if $f(z)$ is constant. If $f(z)$ is not constant it follows that for any z_0 not on the boundary there are points z arbitrarily near z_0 such that $|f(z)| > |f(z_0)|$, and the upper bound of $|f(z)|$ can be taken only on the boundary.

Alternatively, if $\arg f(z_0) = \alpha$, take $g(z) = e^{-i\alpha} f(z)$. Then $\Re g(z_0) = |f(z_0)|, \Im g(z_0) = 0$. If we take any circle C about z_0 as centre, and $\Re g(z)$ is not constant, then, by 11·16, $\Re g(z) > |f(z_0)|$ at some point z_1 of C. Hence $|f(z_1)| \geqslant |g(z_1)| > |f(z_0)|$.

By applying the same arguments to $1/f(z)$, we see that $|f(z)|$ cannot have a minimum at an internal point if $1/f(z)$ is analytic.

The maximum modulus principle has considerable mathematical and also physical importance. It can be stated alternatively: if $|f(z)| = M$ on a closed contour, and at some point within the contour $|f(z)| > M$, then $f(z)$ has a singularity within the contour. If at some point within the contour $|f(z)| < M$, $1/f(z)$ has a singularity within the contour and $f(z)$ will have a zero or an essential singularity.

It shows also that in two-dimensional electrostatics the maximum electric intensity, and in hydrodynamics the maximum velocity, occur on the boundary provided there are no singularities within it. These statements can be extended to three dimensions.

A curve of constant $|f(z)|$ may have a node z_1. If so, we have that for two distinct directions through z_1, $\dfrac{d}{ds}|f(z)|^2 = 0$. But $|f(z)|^2$ is differentiable;* hence $\dfrac{d}{ds}|f(z)|^2 = 0$ for any direction through z_1 and $f'(z_1) = 0$. (The case $|f(z_1)| = 0$ can be excluded. For if there was a continuous curve with $|f(z)| = |f(z_1)| = 0$, z_1 would be a limit-point of zeros of $f(z)$, which would therefore be zero everywhere.) Conversely, if $f'(z_1) = 0$, $\dfrac{d}{ds}|f(z)|^2 = 0$ for every direction through z_1, and a curve of constant $|f(z)|$ through z_1 has a node if $f(z)$ is not constant everywhere.

If the curve $|f(z)| = M$ passes through a node and has a loop within the region, the loop may be regarded as a closed curve by itself and the maximum modulus principle applies to it.

11·162. Schwarz's lemma. If $f(z)$ is analytic and $|f(z)| < M$ for $|z| \leqslant c$, and $f(0) = 0$, $f(z)/z$ is analytic for $|z| \leqslant c$ and $|f(z)/z| < M/c$ for $|z| = c$. Hence, for $|z| \leqslant c$, $|f(z)/z| < M/c$, that is, $|f(z)| < \dfrac{M}{c}|z|$. This lemma is due to Schwarz.

11·17. Laurent's theorem. Let $f(z)$ be analytic and single-valued between and on two circles C and C', with centre O, C' being interior to C. Take z between them and draw a small circle C'' around z. Then $\dfrac{f(t)}{t-z}$ is a function of t with no singularity in the region bounded by C, C' and C'' or on the bounding curves. Hence

$$\int_{C''} \frac{f(t)}{t-z} dt = \int_C \frac{f(t)}{t-z} dt - \int_{C'} \frac{f(t)}{t-z} dt,$$

all contours being taken in the positive sense. But as for Cauchy's integral the integral on the left is $2\pi i f(z)$. To evaluate that round C, we can expand in powers of z/t since $|t| > |z|$; hence it is

$$\sum_{n=0}^{\infty} z^n \int_C \frac{f(t)}{t^{n+1}} dt.$$

* As a function of x and y, in the sense of Stolz and Young.

On C', $|z| > |t|$, and we can expand in powers of t/z; the corresponding integral is

$$-\sum_{n=1}^{\infty} z^{-n} \int_{C'} f(t) t^{n-1} dt.$$

Altogether $\quad f(z) = \dfrac{1}{2\pi i} \left[\sum_{n=0}^{\infty} z^n \int_C \dfrac{f(t)}{t^{n+1}} dt + \sum_{n=1}^{\infty} z^{-n} \int_{C'} t^{n-1} f(t) dt \right].$

This shows that *in any region between two concentric circles that is free from singularities a function can be expanded in a power series including negative powers.*

11·171. Several important consequences follow. First suppose that $f(z)$ has a singular point at the origin and no other within C. Then the integrals around C' are independent of the radius of C', by 11·055. If possible let m be an integer such that $z^m f(z)$ is bounded within some circle D within C', except possibly at the origin; that is, a, M exist so that

$$|z^m f(z)| < M \quad (0 < |z| < a).$$

Then if $n \geqslant m+1$ and the radius of D is $b < a$

$$\left| \int_{C'} t^{n-1} f(t) dt \right| = \left| \int_D t^{n-1} f(t) dt \right| < 2\pi M |t|^{n-m-1} |t| \quad (|t| = b)$$
$$= 2\pi M b^{n-m},$$

which is arbitrarily small since we can take b as small as we like. But since the integral is independent of b it must be zero. Therefore if $z^m f(z)$ is bounded within any given radius from the origin (except possibly at the origin itself) the Laurent expansion contains no terms in z^{-n} with $n > m$. If m is the lowest integer (> 0) that makes this true, $f(z)$ has a pole of order m at the origin. If it is true for $m = 0$, the whole of the negative powers disappear and we recover Cauchy's expansion. We can prove nothing about the value of $f(0)$ by using Laurent's theorem alone, since $z = 0$ is always excluded from the region. But if $m = 0$ and $f(z)$ is continuous at $z = 0$, $f(0)$ must be equal to the constant term in the Cauchy expansion and therefore $f(z)$ is analytic at $z = 0$. We can therefore say that if $m = 0$ and $f(z)$ is continuous, $f(z)$ is analytic at the origin. Hence *if m is a positive integer and $z^m f(z)$ is bounded in the neighbourhood of the origin, but $z^{m-1} f(z)$ is not, $f(z)$ has a pole of order m at the origin; and if $f(z)$ is bounded and continuous it is analytic.*

Now this is the converse of the definition of a pole of order m. Therefore *a necessary and sufficient condition for an isolated singularity at $z = 0$ of a single-valued function to be an essential singularity is that $z^m f(z)$ shall be unbounded near the origin for all positive integral values of m.*[a]

11·172. Behaviour near an isolated essential singularity. Again, *if $f(z)$ has a pole of order m, $1/f(z)$ has a zero of order m, and conversely.* The condition that $f(z)$ is not analytic at $z = a$, but $1/f(z)$ is, is in fact often taken as the definition of a pole. Similarly, $f(z) - c$ has a pole of order m, where c is any constant, and therefore $1/\{f(z) - c\}$ has a zero of order m. Conversely, if the last function has a zero of order m, $f(z) - c$ and therefore $f(z)$ have a pole of order m. If $1/\{f(z) - c\}$ is analytic at 0, and not zero, $f(z)$ is analytic. Hence if $f(z)$ has an essential singularity at 0, $1/\{f(z) - c\}$ also has an essential singularity at 0. For if the latter function was analytic $f(z)$ could have at worst a pole, and if $1/\{f(z) - c\}$ had a pole $f(z)$ would be analytic. But if $1/\{f(z) - c\}$ has an isolated essential singularity at 0, it is unbounded near 0; hence *in any circle about an isolated essential singularity $f(z)$ must somewhere approach arbitrarily close to any finite value.* It has been proved by Picard that it actually takes every value an infinite number of times,* except possibly for one value that it never takes (for example 0 for $f(z) = e^{1/z}$). We can, of course, in no circumstances speak of the value at the singularity itself, since this is undefinable directly and a definition by any limiting process will depend on the method of approach.

11·173. When an expansion of Laurent's form exists about an essential singularity, say

$$f(z) = \sum_{n=0}^{\infty} a_n z^n + \sum_{m=1}^{\infty} a_{-m} z^{-m},$$

we have
$$\int_D f(z)\,dz = 2\pi i a_{-1},$$

where D is any contour lying between C and C' and surrounding the origin. Thus the method of residues is applicable even to functions with essential singularities provided that the conditions for Laurent's theorem are satisfied.

11·174. It should be noted that, if there are several singularities within C', the expansion will alter for any change of C' that makes it pass over a singularity. Thus $\exp(1/z)$ has a single expansion in negative powers valid everywhere except at $z = 0$. But $\operatorname{cosec}(1/z)$ and $\exp(\operatorname{cosec} 1/z)$ can be expanded only in a zone from $|z| = 1/(n+1)\pi$ to $|z| = 1/n\pi$, and will require different expansions for every value of n. No power series, even including negative powers, can hold in the neighbourhood of an unisolated essential singularity.

11·175. *If $f(z)$ has no singularities except a finite number of poles, $\alpha_1, \alpha_2, \ldots$ and is bounded at infinity, $f(z)$ is the sum of the principal parts at the poles together with a constant.* Take small circles $C_1, C_2, \ldots$ about the poles and a contour C large enough to include all the poles; then if z is within C and outside $C_1, C_2, \ldots$.

$$f(z) = \frac{1}{2\pi i}\int_C \frac{f(t)}{t-z}dt - \frac{1}{2\pi i}\sum_r \int_{C_r} \frac{f(t)}{t-z}dt.$$

The first integral is independent of C so long as C contains all the poles of $f(t)/(t-z)$. Hence we can take C arbitrarily large. Then since $f(t)$ is bounded for large t the integral is a finite constant. As in the proof of Laurent's theorem the integral about each C_r gives the principal part at the pole α_r. This proves the theorem.

It follows that a function with no singularities except poles, and bounded at infinity, is a rational function. For we have only to bring the principal parts to a common denominator, and $f(z)$ is the ratio of two polynomials, the numerator being of degree not higher than the denominator.

11·176. Fourier's theorem. A form of this theorem can be derived from Laurent's theorem. Under the conditions for the latter all the $f(t)\,t^{n-1}$ and $f(t)\,t^{-n-1}$ are analytic in the region used, and the paths C and C' can therefore be replaced by a circle D, $|t| = |z|$. Put $z = re^{i\theta}$, $t = re^{i\phi}$; then

$$z^{-n}\int_D f(t)\,t^{n-1}dt = i\int_0^{2\pi} f(re^{i\phi})\,e^{ni(\phi-\theta)}d\phi,$$

$$z^n\int_D f(t)\,t^{-n-1}dt = i\int_0^{2\pi} f(re^{i\phi})\,e^{-ni(\phi-\theta)}d\phi,$$

$$f(z) = \frac{1}{2\pi}\int_0^{2\pi} f(re^{i\phi})\,d\phi + \frac{1}{\pi}\Sigma\int_0^{2\pi} f(re^{i\phi})\cos n(\phi-\theta)\,d\phi.$$

---

* The proof is surprisingly difficult to complete. Cf. Titchmarsh, *Theory of Functions*, p. 283; Landau, *Darstellung...Funktiontheorie*, 1929, Ch. 7.

This is one of the simplest ways of finding the Fourier expansion; but it assumes that $f(t)$ is analytic within a zone $|t_1| < |t| < |t_2|$, where $|t_1| < |z| < |t_2|$, and this condition is not usually satisfied in the cases where we need the expansion. We shall obtain the expansion under more general conditions in Chapter 14.

11·18. Functions with no continuation. It might be supposed that, since the most familiar power series can be continued by Taylor's series beyond their original circles of convergence, this can always be done. This is not so. Consider the series

$$f(z) = 1 + z + z^2 + z^6 + \ldots + z^{n!} + \ldots.$$

This has radius of convergence 1, and hence there is at least one singularity with modulus 1. As we approach $z = 1$ from inside, every term tends to 1 and the sum to infinity; and since a point where $f(z)$ tends to infinity for at least one direction of approach must be a singularity, $z = 1$ is a singularity of the function.

Now within any arc of the circle, however short, there are points where $\arg z$ is a rational fraction of 2π. Put then

$$z = r \exp 2\pi i \frac{m}{n} \quad (r < 1),$$

where m and n are integers. Then

$$z^{n!} = r^{n!} \exp 2\pi i m(n-1)! = r^{n!},$$

$$z^{(n+1)!} = (z^{n!})^{n+1} = (r^{n!})^{n+1} = r^{(n+1)!},$$

and so on; and the series becomes

$$f(z) = (1 + z + \ldots + z^{(n-1)!}) + r^{n!} + r^{(n+1)!} + \ldots.$$

As r approaches 1 the first part tends to a finite limit and the rest to infinity. Hence there is a singularity at $z = \exp(2\pi i m/n)$ and therefore in every arc of the circle. No Taylor expansion about an internal point of the circle can converge beyond the nearest singularity, and this is at the nearest point on the circle (since a limit-point of singularities is a singularity). Hence no continuation is possible.

A still more peculiar case is Osgood's series

$$f(z) = \Sigma \frac{z^{a^n+2}}{(a^n+1)(a^n+2)},$$

where a is an integer greater than 1. The circle of convergence is again $|z| = 1$, but not only does the series converge at all points of the circle, but its first derived series does. Yet we know from general considerations that there must be at least one singularity on the circle. We call this z_0 and consider $f(z_1)$ and $f(z_2)$, where

$$|z_1| = |z_2| = r < 1, \quad \arg z_1 = \arg z_0, \quad \arg z_2 = \arg z_0 + 2\pi k/a^m,$$

k being an integer. As for the last series, the terms up to that with $n = m-1$ differ in the two cases, but the later ones are all identical except for the factor $(z_2/z_1)^2$. Consequently if we form the Taylor series about z_1 and z_2 and consider points on the respective radii there will be differences in the early terms, but the later ones, which determine the convergence, are in a constant ratio. Hence the Taylor series in $z - z_2$ is not convergent if $|z - z_2| > 1 - r$. Since m may be taken arbitrarily large we have again the result that there must be a singularity in every arc of the circle of convergence, in spite of the apparently good behaviour of the function there. We may compare the fact that $z^2 \log z$ and its first derivative vanish at $z = 0$, which is nevertheless a branch point of the function; but it is surprising that such behaviour should be possible over an entire circumference. The reason why the function must be considered non-analytic at $|z| = 1$ is that it is impossible to say what we mean by its derivative as $|z|$ approaches 1 *from outside* because we cannot say what $f(z)$ is outside.

These cases are artificial, but are given here to show that it is quite possible that an analytic function may be defined in part of the plane and completely indefinable outside it. Every point of the circle of convergence is a singularity.

11·19. Abel's theorem. *If a power series $\Sigma a_n z^n$ is convergent at a point z_0 on its circle of convergence, it is uniformly convergent on the radius up to and including z_0, and its sum approaches the sum $\Sigma a_n z_0^n$ in the limit.* It is not necessary that $\Sigma a_n z_0^n$ should be absolutely convergent. The theorem follows by applying that of 1·1154 to the real and imaginary parts of the sum separately. The theorem is important because it often provides a way of summing conditionally convergent series that would otherwise be unmanageable. Thus consider the series
$$\log(1+z) = z - \tfrac{1}{2}z^2 + \tfrac{1}{3}z^3 - \ldots,$$
which converges on the circle of convergence except at $z = -1$. Hence by Abel's theorem we can put $z = e^{i\theta}$ directly, and get
$$\log(1+e^{i\theta}) = e^{i\theta} - \tfrac{1}{2}e^{2i\theta} + \tfrac{1}{3}e^{3i\theta} - \ldots,$$
and also
$$= \log\{2\cos\tfrac{1}{2}\theta e^{\frac{1}{2}i\theta}\}$$
$$= \log(2\cos\tfrac{1}{2}\theta) + \tfrac{1}{2}i\theta.$$

Separating real and imaginary parts we have
$$\left.\begin{array}{l}\cos\theta - \tfrac{1}{2}\cos 2\theta + \tfrac{1}{3}\cos 3\theta - \ldots = \log(2\cos\tfrac{1}{2}\theta),\\ \sin\theta - \tfrac{1}{2}\sin 2\theta + \tfrac{1}{3}\sin 3\theta - \ldots = \tfrac{1}{2}\theta,\end{array}\right\} \quad (-\pi < \theta < \pi).$$

This device of inserting powers of $r\,(<1)$ in the coefficients of a series to make it absolutely convergent is now usually known as *Abel summation*. It had previously been used by Euler, and is therefore also called *Euler summation*; but the latter name is now usually given to another method also due to Euler.

The method can even be used to suggest a meaning for a series in a region where it does not converge. For instance, we might obtain the series
$$a\cos\theta - \tfrac{1}{2}a^2\cos 2\theta + \tfrac{1}{3}a^3\cos 3\theta - \ldots$$
as the solution of some problem, but in the conditions of that problem $a > 1$ and the series has no definite meaning as it stands. Nevertheless, we may try the result of taking $a < 1$, when the series becomes
$$\Re\{\log(1+ae^{i\theta})\} = \log|1+ae^{i\theta}|$$
$$= \tfrac{1}{2}\log(1+2a\cos\theta+a^2).$$

This suggests a meaning even when $a > 1$, and a suggested form for the answer is often a great help towards obtaining a valid proof. In this case the justification would be completed if we knew that the function required was the real part of an analytic function of the complex variable $ae^{i\theta}$ for all a; for then we could sum within the circle of convergence and apply the result outside by analytic continuation.

11·20. Morera's theorem. Suppose that for all paths L within a certain closed region the integral of a continuous single-valued complex function, not assumed analytic, between given termini $z_0 = (x_0, y_0)$, $Z = (X, Y)$ has the same value, and therefore depends only on the termini. We write

$$\int_L \{\phi(x,y) + i\psi(x,y)\} dz = F(X, Y).$$

Also
$$\int_{L'} \{\phi(x,y) + i\psi(x,y)\} dz = F(X+\xi, Y+\eta),$$

where L' is a path connecting z_0 with $Z+\zeta$, where ζ is small enough for the straight line connecting Z with $Z+\zeta$ to lie wholly within the region. Then L' can be taken to lie entirely within the region. Since $F(X+\xi, Y+\eta)$ is unaltered by changes of the path L' so long as the ends remain the same and it continues to lie within the region, we can take L' to coincide with L from z_0 to Z and then proceed as a straight line to $Z+\zeta$. Then if $\arg \zeta = \theta$ we have for variations with argument θ

$$\lim_{|\zeta| \to 0} \frac{1}{\zeta} \{F(X+\xi, Y+\eta) - F(X, Y)\}$$
$$= \lim_{|\zeta| \to 0} \frac{1}{\zeta} \int_Z^{Z+\zeta} (\phi + i\psi) dz$$
$$= \phi(X, Y) + i\psi(X, Y);$$

and this is independent of $\arg \zeta$. Hence $F(X, Y)$ has a derivative in the sense of 11·03 and is an analytic function of the complex variable Z. Hence $\phi + i\psi$ also is an analytic function.

This theorem is a converse of Cauchy's.

One importance of this theorem is that it shows that the uniqueness of a complex integral for variations of the path involves the same sort of restrictions on the real and imaginary parts of the integrand as the uniqueness of the derivative for variations of direction.

Another is that it provides an easy proof of the proposition that a uniformly convergent series of analytic single-valued functions is an analytic single-valued function. So far we have proved this only for power series. Let

$$S(z) = \sum_{n=0}^{\infty} u_n(z),$$

where the $u_n(z)$ are analytic single-valued functions of z. If the series is uniformly convergent in a region it can be integrated term by term along any path in the region; hence

$$\int_L S(z) dz = \sum_{n=0}^{\infty} \int_L u_n(z) dz.$$

But since the $u_n(z)$ are analytic and single-valued in the region their integrals depend only on the termini. Hence the integral on the left depends only on the termini. Therefore, by Morera's theorem, $S(z)$ is an analytic single-valued function in the region.

11·21. The Osgood-Vitali theorem.

A much more powerful condition for the limit of a sequence of analytic functions to be itself an analytic function was given by W. F. Osgood* and extended by G. Vitali† and R. Jentzsch. *Let $\{f_n(z)\}$ be a sequence of functions, each analytic in a region D; let $|f_n(z)| \leqslant M$ for every n and every z in D; and let $f_n(z)$ tend to a limit, as $n \to \infty$, at a set of points having a limit-point inside D. Then in any region interior to D, $f_n(z)$ tends uniformly to a limit, and the limit is an analytic function of z.*

We take the limit-point to be $z = 0$. All $f_n(z)$ have power series expansions valid for $|z| \leqslant c$, where c is any quantity less than the distance of 0 from the boundary of D; that is,

$$f_n(z) = \sum_{p=0}^{\infty} a_{n,p} z^p.$$

If $|f_n(z)| \leqslant M$ for $|z| = c$, then $|f_n(z)| \leqslant M$, $|a_{n,p} z^p| \leqslant M$ for $|z| \leqslant c$. Suppose that $a_{n,m}$ ($m \geqslant 0$) does not tend to a limit as $n \to \infty$ but that $a_{n,p}$ does so for all $p < m$, if any. Consider

$$g_n(z) = z^{-m}\{f_n(z) - \sum_{p=0}^{m-1} a_{n,p} z^p\}.$$

On $|z| = c$, $|g_n(z)| \leqslant (m+1) M/c^m$. Hence $g_n(z)$ is uniformly bounded with respect to n and z, and $g_n(0) = a_{n,m}$. Then $g_n(z) - g_n(0)$ is uniformly bounded ($\leqslant (m+2) M/c^m$), and is zero for $z = 0$; hence for $|z| \leqslant c$, by Schwarz's lemma (11·162)

$$|g_n(z) - a_{n,m}| \leqslant (m+2) M |z|/c^{m+1}.$$

Now if $q > n$

$$|a_{n,m} - a_{q,m}| \leqslant |g_n(z) - a_{n,m}| + |g_q(z) - a_{q,m}| + |g_n(z) - g_q(z)|.$$

Take r so that $(m+2) Mr/c^{m+1} < \omega$, and z' such that $0 < |z'| < r$ and such that $f_n(z')$ and therefore $g_n(z')$ has a limit; take n so that for all $q > n$

$$|g_n(z') - g_q(z')| < \omega.$$

Then for all $q > n$ $\qquad |a_{n,m} - a_{q,m}| < 3\omega,$

and therefore $a_{n,m}$ has a limit, contrary to hypothesis. Hence $a_{n,m}$ tends to a limit a_m for every m. Also since for all n, $|a_{n,p}| \leqslant M/c^p$, we have $|a_p| \leqslant M/c^p$, and the series $\Sigma a_p z^p$ converges in any circle $|z| < c$ and defines an analytic function $f(z)$. Also convergence of $\Sigma a_{n,p} z^p$ is uniform with respect to both n and z for $|z| \leqslant d < c$. Hence for $|z| < d$

$$\lim_{n \to \infty} \Sigma a_{n,p} z^p = \Sigma a_p z^p = f(z),$$

and $f_n(z)$ tends uniformly to $f(z)$.

Now take D' to be any region interior to D and including $z = 0$. Let δ be the distance of the boundary of D' from that of D. Then D' can be covered by a finite set of overlapping circles of radius $c < \delta$, in such a way that no circle meets the boundary of D, and the centre of every circle lies within at least one other. Let the centre of a circle C_1 lie within C_0, where C_0 has centre 0. Then the conditions hold in the common part of C_0 and C_1 and therefore in the whole of C_1; and by repetition they hold in every circle of the set. Hence $f_n(z)$ tends to an analytic limit function $f(z)$ in the whole of D', and convergence is uniform because the set of circles is finite.

---

* *Annals of Mathematics*, (2), 3, 1902, 25–34.
† *Annali di Matematica*, (3), 10, 1904, 65–82.

The most difficult part of any application of the last part of 11·20 is to show that the series (or the sequence) considered is uniformly convergent. The Osgood-Vitali theorem establishes this in general subject only to the existence of a limit at an infinite set of points with a limit-point in the region and to the sequence being bounded. In practice the existence is usually established at all points of some region enclosed in D, which is a more stringent condition than that assumed by Osgood, and still more than that assumed by Vitali.

The limit-function is not necessarily analytic on the boundary of D.

EXAMPLES

1. Prove that if we tried to define an algebra of number pairs by 11·01 (1′), (2′), (3′) and
$$\gamma\gamma' = (aa' + bb', ab' - ba'),$$
the commutative law of multiplication would not be satisfied.

2. If z_n is the $(n+16)$th root of unity, of least positive argument, and w_n is the nth power of z_n, prove that for $n = 1, 2, \ldots$ the points w_n proceed anticlockwise once round the unit circle. Determine how many are situated in each of the four quadrants. Also evaluate $|w_n - z_n|$ as a function of n and determine for what n its value is largest. (I.C. 1942.)

3. Prove that if a sequence of complex numbers z_n tends to a finite limit $c (\neq 0)$, then $1/z_n \to 1/c$. Prove also that if $f(z)$ is analytic and $f(0) \neq 0$, then $1/f(z)$ is analytic at $z = 0$.

4. Show that, if r_0 and A are positive real numbers, and
$$r_{n+1} + \frac{1}{r_n} = 2A,$$
then the condition $A \geqslant 1$ is necessary for the convergence of the sequence $\{r_n\}$; show that it is also sufficient in the case $r_0 > 1$, by verifying that $r_n > 1$ for every n, and
$$|r_n - c| \leqslant \left|\frac{r_0 - c}{c^n}\right|$$
for a suitable $c > 1$. (I.C. 1939.)

5. Given that the series $\Sigma a_n z^n$ has the radius of convergence 2, find the radii of convergence of
$$\Sigma a_n^k z^n, \quad \Sigma a_n z^{nk}, \quad \Sigma a_n z^{n^2}, \quad \Sigma (a_1^k + a_2^k + \ldots + a_n^k) z^n,$$
where k is a fixed positive integer, and in the fourth series the numbers a_n are positive. (M.T. 1942.)

6. By considering the function $\exp\{kf(z)\}$ for suitable constants k, or otherwise, show that
(i) if u is bounded for all z then $f(z)$ is constant,
(ii) if $u < v$ for all z then $f(z)$ is constant,
where $f(z) = u + iv$ is an analytic function of z and u and v are real. (Prelim. Exam. 1943.)

7. What are the radii of convergence of the expansions in powers of z of the following functions?

(i) $\dfrac{z}{e^z - 1}$, (ii) $\log \dfrac{\sin z}{z}$, (iii) $\exp\{(1 - z^2)^{1/2}\}$. (M/c III, 1928.)

8. Derive an alternative proof of the maximum modulus principle from the fact that $\log|f(z)| = \Re \log f(z)$ in a suitably defined region.

9. If $f(z) = a_0 + a_m z^m + R(z)$, where a_m is the first non-zero coefficient after a_0 in the expansion, prove that c exists such that for $|z| < c$, $|R(z)| < \frac{1}{2}|a_m z^m|$, and hence prove the maximum modulus principle.

10. Prove that an analytic function whose only singularity is a pole at infinity is a polynomial.

11. If
$$\Omega(\kappa) = \int_{-\infty}^{\infty} e^{\kappa x} f(x)\, dx,$$
where x is real and $f(x)$ is real and ≥ 0, exists for every complex value of κ (including $\kappa = 0$), prove that $\Omega(\kappa)$ is an integral function.

12. If $f(z)$ is analytic in a region and not constant, prove that the values of z where $|f(z)| = M$ do not form an arc with an end interior to the region.

13. Comment on the following proof of Cauchy's theorem.

Put
$$I(\lambda) = \int_C f(\lambda z)\, dz.$$

Then
$$\frac{dI}{d\lambda} = \int_C z f'(\lambda z)\, dz = \int \frac{z}{\lambda} df(\lambda z) = -\frac{1}{\lambda}\int_C f(\lambda z)\, dz = -\frac{I}{\lambda}.$$

Therefore $I = A/\lambda$, and by making $\lambda \to 0$ we have $A = 0$. Finally put $\lambda = 1$.

14. If $\Sigma u_n(z)$ is uniformly convergent on a rectifiable contour C, and $u_n(z)$ is analytic within C and continuous in the closed region, prove that $\Sigma u_n(z)$ is uniformly convergent in any region interior to C.

15. Prove that the real and imaginary parts of $f(z)$, where $f(z)$ is analytic within C, have bounded variation on any rectifiable path within C.

16. If
$$f(a) = \frac{1}{2\pi i} \int_C \frac{\phi(z)}{z-a}\, dz$$
where C is the circle $|z| = 1$, find $f(a)$, (i) when $\phi(z) = 1/z$, (ii) when $\phi(z) = \Re(z)$. (M.T. 1940.)

17. Prove that the radius of convergence of $\Sigma a_n z^n$ is the smallest limit-point of the set $|a_n|^{-1/n}$.

Chapter 12

CONTOUR INTEGRATION AND BROMWICH'S INTEGRAL

'Go round about, Peer Gynt!'
IBSEN, *Peer Gynt*

12·01. Description of method. This method of evaluating definite integrals is based directly on Cauchy's theorem. We have had instances of it already in the proofs of Cauchy's inequality and Laurent's theorem. A simple example is the following. Take

$$I = \int_0^\pi \frac{d\theta}{a - b\cos\theta}, \qquad (1)$$

where a and b are real and $a > b > 0$. The integrand is an even function of θ, and therefore

$$I = \frac{1}{2}\int_0^{2\pi} \frac{d\theta}{a - b\cos\theta} = \int_0^{2\pi} \frac{e^{i\theta}d\theta}{2ae^{i\theta} - b(e^{2i\theta} + 1)}. \qquad (2)$$

Put
$$e^{i\theta} = z. \qquad (3)$$

As θ increases from 0 to 2π, z moves round the circle $|z| = 1$. Then

$$I = -\frac{1}{i}\int_C \frac{dz}{bz^2 - 2az + b}, \qquad (4)$$

where the path of integration is around the unit circle. But this is a closed contour and the integral is therefore equal to $2\pi i$ times the sum of the residues at any poles within it. There are two poles, namely, the zeros of the denominator, and their product is 1; write

$$b\alpha = a - \sqrt{(a^2 - b^2)}, \quad b/\alpha = a + \sqrt{(a^2 - b^2)}, \qquad (5)$$

Then α is within the unit circle and $1/\alpha$ outside. Then

$$I = -\frac{1}{ib}\int_C \frac{dz}{(z - \alpha)(z - 1/\alpha)}. \qquad (6)$$

Near α the integrand has the form

$$\frac{1}{\alpha - 1/\alpha}\left(\frac{1}{z - \alpha} + \text{terms analytic at } z = \alpha\right), \qquad (7)$$

and the residue is therefore $(\alpha - 1/\alpha)^{-1}$. Hence

$$I = -\frac{2\pi i}{ib(\alpha - 1/\alpha)} = \frac{\pi}{\sqrt{(a^2 - b^2)}}. \qquad (8)$$

12·011. Now consider the rather more complicated integral

$$I = \int_0^\pi \frac{d\theta}{(a - b\cos\theta)^2} \qquad (1)$$

under the same conditions.

With the same transformation

$$I = \frac{2}{i} \int_C \frac{z\,dz}{(bz^2 - 2az + b)^2} \tag{2}$$

$$= \frac{2}{ib^2} \int_C \frac{z\,dz}{(z-\alpha)^2 (z - 1/\alpha)^2} \tag{3}$$

again around the unit circle. But now $z = \alpha$ is a double pole and we must expand to get the coefficient of $(z-\alpha)^{-1}$. Put

$$z - \alpha = z', \tag{4}$$

$$\frac{z}{(1/\alpha - z)^2} = \alpha\left(1 + \frac{z'}{\alpha}\right)\left(\frac{1}{\alpha} - \alpha - z'\right)^{-2}$$

$$= \alpha\left(\frac{1}{\alpha} - \alpha\right)^{-2}\left\{1 + z'\left(\frac{1}{\alpha} + \frac{2}{1/\alpha - \alpha}\right) + \ldots\right\} \tag{5}$$

and the coefficient of z' is

$$\left(\frac{1}{\alpha} - \alpha\right)^{-3}\left(\frac{1}{\alpha} - \alpha + 2\alpha\right) = \left(\frac{1}{\alpha} - \alpha\right)^{-3}\left(\frac{1}{\alpha} + \alpha\right). \tag{6}$$

This is the coefficient of z'^{-1} when the integrand is developed in powers of z', and therefore is the residue. Hence

$$I = 2\pi i \frac{2}{ib^2}\left(\frac{1}{\alpha} - \alpha\right)^{-3}\left(\frac{1}{\alpha} + \alpha\right)$$

$$= \frac{4\pi}{b^2}\left\{\frac{2(a^2 - b^2)^{1/2}}{b}\right\}^{-3}\frac{2a}{b}$$

$$= \frac{\pi a}{(a^2 - b^2)^{3/2}}. \tag{7}$$

12·02. The case when $b > a$ may be used to illustrate the notion of the *principal value of an integral*. Referring to 12·01(5) we see that in this case the poles $\{a \pm i\sqrt{(b^2 - a^2)}\}/b$ are complex and have modulus 1; the integrand therefore is unbounded on the suggested contour, just as it is near $\cos\theta = a/b$ in the original integral. In such cases the integral is strictly meaningless, but a related integral sometimes occurs in practice, though its use always needs special justification. For real variables, if $f(x)$ has a simple pole at $x = a$ we cut out a range on the path from $a - h$ to $a + h$ and form the integral over the remainder. If this has a limit when h tends to 0 we call the limit the principal value of the integral. The simplest case is that of the logarithm. If a and b are real and have the same sign, we can choose a real path from a to b without passing through the origin, and then

$$\int_a^b \frac{dx}{x} = \log\frac{b}{a}.$$

But if a is negative and b positive the integral diverges at $x = 0$. But we can still define

$$P\int_a^b \frac{dx}{x} = \lim_{h \to 0}\left[\int_a^{-h} \frac{dx}{x} + \int_h^b \frac{dx}{x}\right]$$

$$= \lim_{h \to 0}\left(\log\frac{h}{-a} + \log\frac{b}{h}\right) = \log\frac{b}{-a} = \log\frac{b}{|a|}.$$

This integral is called the *principal value* and is unambiguous. The condition that we must use the same h on both sides is essential; it will be seen that by taking the first integral from a to $-h$ and the second from k to b, we could make the limit anything we like by making h and k tend to zero in a suitable ratio. Principal values are often written as ordinary integrals, but this practice should be avoided. We see that if the complex variable was used we could complete the path by a semicircle from $-h$ to $+h$ about the origin, either above or below the real axis. The former, being described in the negative direction, would give a contribution $-\pi i$; the latter, $+\pi i$. According to the path permitted by any cuts made in the complex plane we should therefore have in this case

$$\int_a^b \frac{dz}{z} = \log\frac{b}{|a|} \pm i\pi.$$

The principal value is the mean of these alternatives.

Similarly, if a path in the complex plane passes through a simple pole a we can define a principal value of the integral along the path by cutting out the part of the path within a small circle of radius h about a and then making h tend to 0. If we change the variable z to ζ, and $dz/d\zeta$ is finite and not zero at the pole, the same device will define an integral in the ζ plane, and the two will be equal. For if the circle in the z plane cuts the path at $a-k$ and $a+k'$, where $|k| = |k'| = h_1$ and that in the ζ plane cuts the path at $\alpha-\kappa$ and $\alpha+\kappa'$, then if k and k' tend to 0 so that $k/k' \to 1$, κ and κ' will also tend to 0 so that $\kappa/\kappa' \to 1$.

In the case of the integral 12·01 (1), if $b > a$,

$$P\int_0^\pi \frac{d\theta}{a-b\cos\theta} = -\frac{1}{i} P\int_C \frac{dz}{b(z-\alpha)(z-1/\alpha)},$$

where we must cut out parts of the path within small circles about the two poles and then make the radii tend to 0. But we can still complete the contour by adding small semicircles about the poles and inside the unit circle. There is no singularity within this contour and the integral about it is 0. The two arcs tend to semicircles, and as they are described in the negative sense the integral on each tends to $-\pi i$ times the residue. But the residues are, at α and $1/\alpha$ respectively,

$$-\frac{1}{i}\frac{1}{\alpha-1/\alpha}, \quad -\frac{1}{i}\frac{1}{1/\alpha-\alpha},$$

which are equal and opposite. Hence the integrals around the arcs together tend to 0, and therefore the principal value of the integral around the unit circle is 0.

This device for defining a principal value succeeds only at simple poles.

12·03. If $f(z) = g(z)/h(z)$, where $g(z)$ is analytic and not zero at $z = a$, while $h(z)$ has a simple zero there, we can write

$$f(z) = \frac{g(a)}{(z-a)\,h'(a)} + \phi(z),$$

where $\phi(z)$ is analytic at a; then the residue is obtained immediately as $g(a)/h'(a)$. For multiple poles it is usually necessary to carry out the expansion as far as the term in $(z-a)^{-1}$, and this may be troublesome.

12·04. Let $f(z)$ be analytic in a region except for poles, and consider the function $f'(z)/f(z)$. If a is a pole of order m,
$$f(z) = A(z-a)^{-m}\{1+\phi(z)\}, \tag{1}$$
where $\phi(z)$ is analytic near a and is zero at a; and
$$f'(z) = -mA(z-a)^{-m-1}\{1+\phi(z)\}+A(z-a)^{-m}\phi'(z). \tag{2}$$
Hence near a
$$\frac{f'(z)}{f(z)} = -\frac{m}{z-a}+\psi(z), \tag{3}$$
where $\psi(z)$ is analytic at a. If b is a zero of order n, we can write
$$f(z) = B(z-b)^n\{1+\chi(z)\}, \tag{4}$$
and get similarly
$$\frac{f'(z)}{f(z)} = \frac{n}{z-b}+\omega(z). \tag{5}$$

Hence the function $f'(z)/f(z)$ is analytic in the region except for simple poles at the poles and zeros of $f(z)$; and the residue at a pole of order m is $-m$, and that at a zero of order n is n. If then we take
$$\frac{1}{2\pi i}\int_C \frac{f'(z)}{f(z)}dz \tag{6}$$
around any contour C in the region not passing through a pole or a zero of $f(z)$, its value is $\Sigma n - \Sigma m$, the excess of the number of zeros of $f(z)$ within C over the number of poles, multiple poles and zeros being counted multiply. We notice that (6) can also be written $\frac{1}{2\pi i}[\log f(z)]_C$, the brackets indicating the change of $\log f(z)$ when z completes the circuit C, or as $\frac{1}{2\pi}[\arg f(z)]_C$.

Similarly if $h(z)$ is analytic in the region,
$$\frac{1}{2\pi i}\int_C h(z)\frac{f'(z)}{f(z)}dz = \Sigma nh(b) - \Sigma mh(a). \tag{7}$$
In particular if $h(z) = z$ the integral is the excess of the sum of the values of z at the zeros within C over the sum at the poles.

12·041. If $f(z) = g(z) + h(z)$, where $f(z)$ and $g(z)$ are analytic on C, and if at all points on C $|h(z)| < |g(z)|$, then
$$[\arg f(z)]_C = [\arg g(z)]_C. \tag{1}$$
Put
$$f(z) = (1+k)g(z), \tag{2}$$
where $|k| < 1$ for all z on C. Then
$$\arg f(z) - \arg g(z) = \arg(1+k). \tag{3}$$
But $1+k$ has a positive real part; hence $-\tfrac{1}{2}\pi < \arg(1+k) < \tfrac{1}{2}\pi$.

Now the changes of $\arg f(z)$ and $\arg g(z)$ on describing C are integral multiples of 2π. But since $|\arg f(z) - \arg g(z)|$ is always less than $\tfrac{1}{2}\pi$ the difference of their changes is less than π and therefore must be zero. This is *Rouché's theorem*.

12·042. This theorem provides a proof of the theorem used in algebra, but incapable of being proved by purely algebraic methods, that an *algebraic equation of degree n has n roots*. Let the equation be
$$f(z) = a_0 z^n + a_1 z^{n-1} + \ldots + a_n,$$
where $a_0 \neq 0$. We can find positive quantities $r_1, r_2, \ldots$ so that
$$|a_0|r_1 > |a_1|, \quad |a_0|r_2^2 > |a_2|, \quad \ldots, \quad |a_0|r_n^n > |a_n|.$$
Take
$$r = r_1 + r_2 + \ldots + r_n.$$
Then at all points of the circle $|z| = r$,
$$|a_0 z^n| > |a_1 z^{n-1} + a_2 z^{n-2} + \ldots + a_n|.$$
Hence, by 12·041, if we proceed around this circle the changes of argument of $f(z)$ and $a_0 z^n$ are equal. But $\arg z^n$ increases by $2n\pi$; hence $\arg f(z)$ increases by $2n\pi$. But $f(z)$ has no poles; hence it has n zeros within the circle. It clearly has none outside.

12·043. Beginners sometimes argue that since
$$\sin z = z - \frac{z^3}{3!} + \frac{z^5}{5!} - \ldots = 0$$
is an equation of infinite degree, it must have an infinite number of roots. The result is correct, but the argument would apply equally to
$$\exp z = 1 + z + \frac{z^2}{2!} + \ldots = 0,$$
which has no roots at all. The reason why the method of 12·042 breaks down for equations of infinite degree is that there is no term of highest degree to make a starting point, nor are there any r and n such that the nth term has a greater modulus than all others for all $|z| > r$, so that there is no comparison function satisfying the conditions of 12·041. If we apply 12·04 directly to $\exp z$ we have
$$\frac{1}{2\pi i} \int_C \frac{f'(z)}{f(z)} dz = \frac{1}{2\pi i} \int_C dz = 0$$
whatever path we choose; and therefore $\exp z$ has no zeros and no poles.

12·05. Inverse functions. Suppose that we have an equation
$$\zeta = f(z) \qquad (1)$$
giving ζ as an analytic function of z within a certain region of z. Then conversely we may regard z as a function of ζ, say $z = p(\zeta)$, at a certain set of points, namely the values of $f(z)$ taken for values of z in the region of z. The question is whether $p(\zeta)$ is an analytic function of ζ, and in the first place whether it is single-valued over a region of ζ. If $f(z_1) = f(z_2)$, and we take a path from z_1 to z_2 within the region, ζ describes a contour, since by hypothesis $f(z)$ is analytic and therefore bounded on the path. Then when ζ describes this contour z does not return to its original value; hence $p(\zeta)$ has a branch point within the contour, by 11·11(a). For instance, $f(z) = z^2$ takes the same value for $z = a$ and $-a$. If $z = ae^{i\theta}$ with a constant and θ increasing from 0 to π, $\arg \zeta$ increases by 2π but z does

not return to its original value. Hence, since a is arbitrarily small, $\zeta = 0$ is a branch point of $p(\zeta) = \zeta^{1/2}$. Again, if $f(z) = e^z$, and if y varies from $-\pi$ to π, x remaining equal to a constant a, $|\zeta|$ remains constant at e^a and $\arg \zeta$ increases by 2π, so that ζ describes the circle $|\zeta| = e^a$. Since a may be indefinitely large and negative, every circle about $\zeta = 0$ contains a branch point of the inverse function $\log \zeta$, and therefore $\log \zeta$ has a branch point at $\zeta = 0$. Hence if $p(\zeta)$ is to be single-valued we must be prepared to make such cuts in the z plane that $f(z)$ never takes the same value twice in any portion. Functions that take no value more than once in a region are called *simple* or *schlicht* in the region.

Even if this is done it is not obvious that the set of values of ζ determined by (1) constitute a region; this requires a theorem, which we shall proceed to prove, that in certain conditions every value of ζ over a neighbourhood of ζ is a value of $f(z)$.

12·051. *If $f(z)$ is analytic at $z = 0$, and if $f(0) = 0$, $f'(0) \neq 0$, there is a region of the ζ plane about $\zeta = 0$ such that the equation $f(z) = \zeta$ has one and only one solution $p(\zeta)$ that is analytic in ζ and tends to 0 when ζ tends to 0. If $f'(0) = 0$, $\zeta = 0$ is a branch point of $p(\zeta)$.*

Since $f(z)$ is analytic at $z = 0$ (and therefore in a neighbourhood of $z = 0$ by definition (11·11)) there is an R such that

$$f(z) = a_1 z + a_2 z^2 + \ldots \quad (z < R). \tag{1}$$

Take $a_1 \neq 0$. $a_2 + a_3 z + \ldots$ is bounded in any circle $|z| \leq c < R$. Put

$$f(z) = a_1 z + h(z). \tag{2}$$

Then, for some M, $|h(z)| < M|z^2|$; take d so that $0 < Md < \tfrac{1}{2}|a_1|$. Then for $0 < |z| \leq d$

$$|f(z) - a_1 z| < M|z^2| \leq |z| Md < \tfrac{1}{2}|a_1 z|. \tag{3}$$

For given ζ the number of roots of $f(z) = \zeta$ within the circle C ($|z| = d$) is

$$\frac{1}{2\pi i} \int_C \frac{f'(t)}{f(t) - \zeta} dt = \frac{1}{2\pi} [\arg \{f(t) - \zeta\}]_C$$

$$= \frac{1}{2\pi} [\arg \{a_1 z + h(z) - \zeta\}]_C. \tag{4}$$

Let $|\zeta| < \tfrac{1}{2}|a_1 d|$. Since $|h(z)| < \tfrac{1}{2}|a_1 d|$ at all points of C, $|a_1 z| > |h(z) - \zeta|$ on C, and the number of roots of $f(z) = \zeta$ within C is the same as those of $a_1 z = 0$, that is, 1. If $\zeta = 0$ the root is $z = 0$. Hence z is a single-valued function of ζ, determinate for all ζ satisfying $|\zeta| < \tfrac{1}{2}|a_1|d$.

Now consider the integral
$$\int_C \frac{tf'(t)}{f(t) - \zeta} dt. \tag{5}$$

The only pole of the integrand is where $f(t) = \zeta$, and the residue is the value of t at this point, that is, z. Hence the integral is $2\pi i z$. The integral is an analytic function of ζ; hence for $|\zeta| < |\tfrac{1}{2} a_1 d|$, z is an analytic function of ζ, equal to 0 when $\zeta = 0$.

If $a_1 = 0$, let $a_n (n > 1)$ be the first non-zero coefficient in (1). Applying the necessary modifications to the argument we find that there is a δ such that for any ζ satisfying $0 < |\zeta| < \delta$, there will be n values of z, tending to 0 with ζ, that make $f(z) = \zeta$, and therefore z is not a single-valued function of ζ in any neighbourhood of $\zeta = 0$. This completes the proof of the theorem.

If $f(z) = a_0$ for $z = z_0$, corresponding results hold, as we see by considering $f(z) - a_0$ and $z - z_0$.

12·052. If $f'(0) \neq 0$, z is expressed as an analytic function $p(\zeta)$ over a certain neighbourhood, in which $f\{p(\zeta)\} = \zeta$. If $p(\zeta)$ can be extended by continuation, this relation will hold over the whole region attainable by continuation, and therefore the continuation remains the inverse function. It is desirable to say something about the positions of its singularities, but a complete account is not possible.

We have seen that if $f'(z_0) = 0$, $f(z_0)$ is a branch point of $p(\zeta)$, as for $f(z) = z^2$ near $z = 0$.

Also a value ζ_0 not taken by $f(z)$ obviously cannot be part of a region where $p(\zeta)$ is analytic. For instance, if $f(z) = e^z$, the excluded value zero is a branch point of $\log \zeta$.

In any neighbourhood of a pole or an isolated essential singularity of $f(z)$, $|f(z)|$ is unbounded. Then we may consider
$$\zeta' = 1/f(z).$$

If $f(z)$ has a simple pole at z_0, $d\zeta'/dz$ is not zero; hence in a neighbourhood of z_0, z is an analytic function of ζ' and therefore of $1/\zeta$. Then $p(\zeta)$ is analytic, tending to z_0, for values of ζ outside a sufficiently large circle. But if $f(z)$ has a multiple pole, $p(\zeta)$ outside any circle will be many-valued. If $f(z)$ has an isolated essential singularity at z_0, any value except possibly one is taken infinitely many times. Hence a function with such a singularity cannot have a single-valued inverse unless we introduce cuts. If $f(z)$ never takes a value ζ_0, then ζ_0 is a branch point of $p(\zeta)$. If $f(z)$ takes all values in a neighbourhood of z_0 there is at least a branch point at $\zeta = \infty$, since values of $f(z)$ of arbitrarily large modulus occur infinitely many times.

General rules for the neighbourhood of unisolated essential singularities would be extremely difficult to state.

If near a branch point of $f(z)$
$$\zeta = f(z) = z^{1/m}(a_0 + a_1 z + \ldots),$$
with m a positive integer, we can put $z^{1/m} = Z$. Then Z, and therefore z, have not branch points at $\zeta = 0$. In this case, though we need a cut to make $f(z)$ single-valued, nevertheless if we ignore the cut and allow z to proceed many times around 0, $f(z)$ cannot repeat a value until m circuits have been completed, and then repeats the initial value only for the initial value of z.

If however
$$f(z) = z^{m/n}(a_0 + a_1 z + \ldots)$$
where m, n are positive integers > 1 and prime to each other, then the inverse function has a branch point at $\zeta = 0$.

More complicated behaviour of $f(z)$ near a branch point will necessitate special treatment.

If z tends to infinity, $f(z)$ may or may not tend to a limit, and the limit, if it exists for different paths, may or may not have the same value for all paths. Then for a corresponding path for ζ, $p(\zeta)$ tends to infinity, and the limit of $f(z)$ corresponds to a singularity of $p(\zeta)$. The easiest treatment is to put $z' = 1/z$, $f(z) = g(z')$, and to consider the inversion of $\zeta = g(z')$ near $\zeta = g(0)$.

12·053. Lagrange's expansion. This provides a formal development of the power series of expansion of the inverse function, and an extension gives one of an analytic function of the inverse function. In the integral 12·051(5) we can replace the circle C by a contour D within the region where $f(z)$ is analytic, and D can be taken as large as we like subject to $f(z)$ not taking the same value twice within or on it. Then in 12·051 (5) put $f(t) = \tau$ and take $|\zeta|$ less than the smallest value of $|\tau|$ on D. Then if Δ is the corresponding path in the τ plane

$$z = p(\zeta) = \frac{1}{2\pi i} \int_\Delta \frac{t\, d\tau}{\tau - \zeta}$$

$$= \frac{1}{2\pi i} \int_\Delta t \sum_{n=0}^\infty \frac{\zeta^n}{\tau^{n+1}} d\tau = \sum_{n=0}^\infty a_n \zeta^n. \tag{1}$$

Here $a_0 = 0$, and for $n \geqslant 1$

$$a_n = \frac{1}{2\pi i} \int_\Delta \frac{t\, d\tau}{\tau^{n+1}} = \frac{1}{2\pi i} \int_D \frac{tf'(t)\, dt}{\{f(t)\}^{n+1}}$$

$$= \frac{1}{2\pi i} \left[-\frac{t}{n\{f(t)\}^n} \right]_D + \frac{1}{2\pi i n} \int_D \frac{dt}{\{f(t)\}^n}. \tag{2}$$

The integrated part vanishes because $f(t)$ is single-valued. Hence a_n is $1/n$ times the residue of $\{f(t)\}^{-n}$ at $t = 0$. This form is originally due to Jacobi. Specification of the radius of convergence of the series may be difficult; by 11·141 it is at least equal to the value of $|f(z)|$ at the singularity of $p(\zeta)$ of smallest modulus, but this rule needs care, partly on account of the treatment of cuts. If we take curves C_m enclosing $z = 0$ such that

$$|f(z)| = m > 0,$$

such curves are simple for m small enough; and if $m' < m$, C_m will enclose $C_{m'}$. If we increase m, C_m may reach a point $z = b$ where $f'(z) = 0$. Then C_m has a node at b, and for larger m will open at b and form a bulge, and values of $f(z)$ near $f(b)$ will be taken twice or more. Then $|f(b)|$ will be the radius of convergence. But this does not exclude the possibility that $|f(z)|$ may have smaller values at other stationary points. Consider for instance

$$\zeta = f(z) = z(z-1)^2.$$

This is stationary at $z = \frac{1}{3}$ and $z = 1$. At the latter point $f(z) = 0$, but the radius of convergence of the solution of $f(z) = \zeta$ that vanishes with ζ is $|f(\frac{1}{3})| = \frac{4}{27}$. If we try to continue this solution to the neighbourhood of $z = 1$ we must cross the closed curve about $z = 0$ specified by $|\zeta| = \frac{4}{27}$, and the solutions $z = 1 \pm \zeta^{1/2} + \ldots$ that make ζ small near $z = 1$ correspond to different branches of the inverse function.

The coefficients can be evaluated by putting

$$f(z) = \frac{z}{\phi(z)}, \tag{3}$$

where $\phi(z)$ will not be zero within D. Then

$$a_n = \frac{1}{2\pi i n} \int_D \frac{\{\phi(t)\}^n\, dt}{t^n}. \tag{4}$$

But from the formulae for Taylor coefficients in 11·13

$$\frac{1}{2\pi i}\int_D \frac{g(t)\,dt}{t^n} = \frac{g^{(n-1)}(0)}{(n-1)!} \quad (n>0), \tag{5}$$

whence
$$a_n = \frac{1}{n!}\left[\frac{d^{n-1}}{dz^{n-1}}\{\phi(z)\}^n\right]_{z=0} \tag{6}$$

and
$$z = \sum_{n=1}^{\infty} \frac{1}{n!}\left[\frac{d^{n-1}}{dz^{n-1}}\{\phi(z)\}^n\right]_{z=0} \zeta^n. \tag{7}$$

This is *Lagrange's expansion*. We have taken $z=0$ as starting point. If we have $f(z) = \alpha$ when $z = a$, and

$$\zeta - \alpha = \frac{z-a}{\phi(z)}, \tag{8}$$

we have the more general form

$$z - a = \sum_{n=1}^{\infty} \frac{1}{n!}\left[\frac{d^{n-1}}{dz^{n-1}}\{\phi(z)\}^n\right]_{z=a} (\zeta-\alpha)^n \tag{9}$$

convergent within a circle in the ζ plane such that

$$\frac{d}{dz}\frac{z-a}{\phi(z)} = \frac{\phi(z)-(z-a)\phi'(z)}{\{\phi(z)\}^2} \tag{10}$$

never vanishes within it.

It is unusual for the general coefficient in the inverted series to be expressible in a convenient form.* But if the radius of convergence is known it guarantees the existence of the expansion within a definite region, and extends to analytic functions the theorem of inversion of a monotonic function given in 1·066.

If $h(z)$ is another function of z analytic in the region it is a simple extension to find its expansion in terms of ζ. We have only to put $h(t)$ for t in (1), and

$$h(z) - h(a) = \sum_{n=1}^{\infty} c_n (\zeta-\alpha)^n, \tag{11}$$

where
$$c_n = \frac{1}{n!}\left[\frac{d^{n-1}}{dz^{n-1}}[\{\phi(z)\}^n h'(z)]\right]_{z=a}. \tag{12}$$

c_n is $1/n$ times the residue of $\{f(z)-f(a)\}^{-n} h'(z)$ at $z = a$.

12·06. Mittag-Leffler's theorem. Suppose that $f(z)$ has an infinite number of poles and no other singularities. We wish to know whether it is possible to extend the result of 11·175 and say that $f(z)$ differs from the sum of the principal parts at the poles by a constant. We cannot assume the extra condition used in 11·175 that $f(z)$ is bounded for all $|z| > R$, where R is fixed, for $f(z)$ is infinite at some points outside any given contour. But it may be possible to choose a set of contours $C_1, C_2, \ldots$ such that on every

* They are given in detail up to the coefficient of z^{12} in terms of the coefficients in the series expansion of $f(z)$ by W. E. Bleieck, *Phil. Mag.* (7) 33, 1942, 637–8; cf. also W. G. Bickley and J. C. P. Miller, *Phil. Mag.* (7) 34, 1943, 35–6.

C_m, $|f(z)| < M$, where M is independent of m and $|z| \geq R_m$, where $R_m \to \infty$, and such that $\int_{C_m} \left|\frac{dz}{z}\right| < 2\pi\lambda$, where λ is fixed; that is, we choose the contours to pass *between* the poles and ultimately to become indefinitely large. Then if $t = z$ is not a pole of $f(t)$, and C_m encloses z,

$$f(z) = \frac{1}{2\pi i} \int_{C_m} \frac{f(t)\, dt}{t-z} + P_m(z), \qquad (1)$$

where $P_m(z)$ is the sum of the principal parts at all poles α within C_m. But also, if $f(t)$ is analytic at $t = 0$,

$$\frac{1}{2\pi i} \int_{C_m} \frac{f(t)\, dt}{t-z} = \frac{1}{2\pi i} \int_{C_m} \frac{f(t)}{t}\, dt + \frac{1}{2\pi i} \int_{C_m} \frac{zf(t)}{t(t-z)}\, dt, \qquad (2)$$

and the first term on the right is $f(0) - P_m(0)$, by (1). Then

$$f(z) = f(0) + \{P_m(z) - P_m(0)\} + \frac{1}{2\pi i} \int_{C_m} \frac{zf(t)}{t(t-z)}\, dt. \qquad (3)$$

As we take C_m larger and larger we include more and more poles and add more and more terms to the sum, and then the integral gives a remainder term. If it tends to zero the sum will therefore converge as $R_m \to \infty$ and be equal to $f(z)$. But this is true; for

$$\left|\int_{C_m} \frac{zf(t)}{t(t-z)}\, dt\right| < \frac{M|z|\, 2\pi\lambda}{R_m - |z|} \to 0. \qquad (4)$$

Then $\qquad\qquad f(z) = f(0) + \lim\{P_m(z) - P_m(0)\}. \qquad (5)$

If a function analytic at the origin has no singularities other than poles for finite z, and if we can choose a sequence of contours C_m about $z = 0$ tending to infinity, such that $|f(z)|$ never exceeds a given quantity M on any of these contours and $\int |\, dz/z\,|$ is uniformly bounded on them, then (5) holds; where $P_m(z)$ is the sum of the principal parts of $f(z)$ at all poles α within C_m. If there is a pole at $z = 0$ we can replace $f(0)$ by the negative powers and the constant term in the Laurent expansion of $f(z)$ about $z = 0$.

It may not be legitimate to write (5) in the form

$$f(z) = \{f(0) - \lim P_m(0)\} + \lim P_m(z);$$

for (5) may converge without either $\{P_m(0)\}$ or $\{P_m(z)\}$ converging.

12·061. As an example, take $\quad f(z) = \operatorname{cosec} z - 1/z. \qquad (1)$

This is analytic at $z = 0$, since we can define

$$f(0) = \lim_{z \to 0} \frac{z - \sin z}{z \sin z} = 0$$

and then $\qquad f'(0) = \lim_{z \to 0} \frac{f(z) - f(0)}{z} = \lim \frac{z - \sin z}{z^2 \sin z} = \frac{1}{6}.$

It has simple poles at $z = \pm n\pi$ for all integral n, and

$$\operatorname{cosec}(n\pi+z') - \frac{1}{n\pi+z'} = \frac{1}{z'\cos n\pi} + \text{analytic terms};$$

thus the residue at $n\pi$ is $(-1)^n$.

Now
$$\sin(x+iy) = \sin x \cosh y + i \cos x \sinh y,$$
$$|\sin(x+iy)|^2 = \sin^2 x \cosh^2 y + \cos^2 x \sinh^2 y$$
$$= \sinh^2 y + \sin^2 x.$$

It is easiest to see that $f(z)$ is bounded on a suitable series of contours by taking squares C_m with their sides $x = \pm(m+\tfrac{1}{2})\pi$, $y = \pm(m+\tfrac{1}{2})\pi$, where m is an integer. $\int |dz/z|$ around C_m is $8\sinh^{-1}1$. On the sides parallel to the y axis

$$|\sin z| \geq |\sin(m+\tfrac{1}{2})\pi| = 1,$$

and on those parallel to the x axis

$$|\sin z| \geq |\sinh(m+\tfrac{1}{2})\pi| \geq |(m+\tfrac{1}{2})\pi|.$$

Hence $f(z)$ is bounded on C_m for all m and the conditions for Mittag-Leffler's theorem are satisfied. Hence

$$\operatorname{cosec} z - \frac{1}{z} = \Sigma(-1)^n \left(\frac{1}{z-n\pi} + \frac{1}{n\pi} \right), \qquad (2)$$

the summation being over all positive and negative integers n, not including 0. If we combine equal and opposite values of n,

$$\operatorname{cosec} z = \frac{1}{z} - \sum_{1}^{\infty} (-1)^n \frac{2z}{n^2\pi^2 - z^2}. \qquad (3)$$

12·062. Again, take
$$f(z) = \cot z - \frac{1}{z}. \qquad (4)$$

The sum given by the formula is, since all residues are 1,

$$\Sigma\left(\frac{1}{z-n\pi} + \frac{1}{n\pi}\right) = -\sum_{n=1}^{\infty} \frac{2z}{n^2\pi^2 - z^2}. \qquad (5)$$

To justify it we take the same series of squares C_m. We have

$$\tan z = \frac{\tan x + i \tanh y}{1 - i \tan x \tanh y},$$

$$|\cot z|^2 = \frac{1 + \tan^2 x \tanh^2 y}{\tan^2 x + \tanh^2 y}$$

$$= \tanh^2 y + \frac{1 - \tanh^4 y}{\tan^2 x + \tanh^2 y}.$$

When $x = \pm(m+\tfrac{1}{2})\pi$ this is equal to $\tanh^2 y$ and less than 1. When y is large, $\tanh^2 y$ is nearly 1, and $|\cot z|^2$ is nearly 1 for all x. Hence the conditions are satisfied and (5) is true.

12·063. By transformation or by similar arguments (not necessarily with the same choice of squares) we can obtain the following:

$$\tan z = 8z \sum_1^\infty \frac{1}{(2n-1)^2 \pi^2 - 4z^2}, \qquad (6)$$

$$\sec z = 4 \sum_1^\infty \frac{(-1)^{n-1}(2n-1)\pi}{(2n-1)^2 \pi^2 - 4z^2}, \qquad (7)$$

$$\coth z = \frac{1}{z} + 2z \sum_1^\infty \frac{1}{n^2\pi^2 + z^2}, \qquad (8)$$

$$\operatorname{cosech} z = \frac{1}{z} + 2z \sum \frac{(-1)^n}{n^2\pi^2 + z^2}, \qquad (9)$$

$$\operatorname{sech} z = 4 \sum_1^\infty \frac{(-1)^{n-1}(2n-1)\pi}{(2n-1)^2 \pi^2 + 4z^2}, \qquad (10)$$

$$\tanh z = 8z \sum_1^\infty \frac{1}{(2n-1)^2 \pi^2 + 4z^2}. \qquad (11)$$

These series are all uniformly convergent in any bounded closed region excluding all poles. Thus if r is the largest value of $|z|$ in a range and we take n in the cotangent series greater than r/π, the nth and succeeding terms can be written

$$-2r \left(\frac{1}{n^2\pi^2 - z^2} \frac{z}{r} + \frac{1}{(n+1)^2 \pi^2 - z^2} \frac{z}{r} + \dots \right),$$

and the moduli are less than the terms of the absolutely convergent series

$$2r \left(\frac{1}{n^2\pi^2 - r^2} + \frac{1}{(n+1)^2 \pi^2 - r^2} + \dots \right).$$

For the series for $\sec z$ and $\operatorname{sech} z$, which are not absolutely convergent, we take the terms in pairs and apply the extension of the M test in 1·1152. Hence we can integrate all these series term by term. Then

$$\left[\log \frac{\sin z}{z} \right]_0^z = \int_0^z \left(\cot z - \frac{1}{z} \right) dz$$

$$= \int_0^z \left(\frac{2z}{z^2 - \pi^2} + \frac{2z}{z^2 - 4\pi^2} + \dots \right) dz$$

$$= \Sigma \log \frac{z^2 - n^2\pi^2}{-n^2\pi^2} = \Sigma \log \left(1 - \frac{z^2}{n^2\pi^2} \right),$$

and therefore

$$\sin z = z \prod_{n=1}^\infty \left(1 - \frac{z^2}{n^2\pi^2} \right). \qquad (12)$$

Similarly

$$\left[\log \sec z \right]_0^z = \int_0^z \tan z \, dz = -\Sigma \log \left(1 - \frac{4z^2}{(2n-1)^2 \pi^2} \right),$$

$$\cos z = \prod_{s=1}^\infty \left(1 - \frac{4z^2}{(2n-1)^2 \pi^2} \right). \qquad (13)$$

12·07. The Bernoulli numbers and polynomials.

The method of residues gives a convenient way of connecting these with trigonometrical functions. If we write 9·08(6) as

$$\frac{1}{e^a-1}+\frac{1}{2}-\frac{1}{a}=\sum_{r=1}^{\infty}b_r a^{r-1}, \qquad (1)$$

and regard a as a complex variable, the function on the left has simple poles of residue 1 at all the points $a=\pm 2ni\pi$, 0 excepted, and is bounded on circles of radius $(2n+1)\pi$ centred on the origin. It vanishes for $a=0$. Hence it is equal to

$$\sum_{n=-\infty}^{\infty'}\left(\frac{1}{a-2ni\pi}-\frac{1}{2ni\pi}\right)=\sum_{n=1}^{\infty}\frac{2a}{4n^2\pi^2+a^2} \qquad (2)$$

$$=\sum_{n=1}^{\infty}\sum_{r=1}^{\infty}\frac{2a}{4n^2\pi^2}\left(\frac{-a^2}{4n^2\pi^2}\right)^{r-1} \quad (|a|<2\pi), \qquad (3)$$

the accent meaning that $n=0$ is excluded from the sum. Comparing with (1) we have

$$b_{2r}=(-1)^{r-1}\sum_{n=1}^{\infty}\frac{2}{(4n^2\pi^2)^r}, \quad b_{2r+1}=0. \qquad (4)$$

In particular
$$b_2=\frac{1}{2\pi^2}\left(1+\frac{1}{2^2}+\frac{1}{3^2}+\ldots\right), \qquad (5)$$

$$b_4=-\frac{1}{8\pi^4}\left(1+\frac{1}{2^4}+\frac{1}{3^4}+\ldots\right), \qquad (6)$$

and therefore, since $B_r=r!\,b_r$,

$$1+\frac{1}{2^2}+\frac{1}{3^2}+\ldots=2\pi^2 b_2=\pi^2 B_2=\frac{\pi^2}{6}, \qquad (7)$$

$$1+\frac{1}{2^4}+\frac{1}{3^4}+\ldots=-8\pi^4 b_4=-\tfrac{1}{3}\pi^4 B_4=\frac{\pi^4}{90}, \qquad (8)$$

$$1+\frac{1}{2^6}+\frac{1}{3^6}+\ldots=32\pi^6 b_6=\tfrac{2}{45}\pi^6 B_6=\frac{\pi^6}{945}. \qquad (9)$$

On account of the smallness of the second term we have, very nearly, for large r,

$$b_{2r}=(-1)^{r-1}\frac{2}{2^{2r}\pi^{2r}}; \quad \text{and} \quad B_{2r}=(-1)^{r-1}\frac{2(2r)!}{2^{2r}\pi^{2r}}. \qquad (10)$$

Now combining 9·08(6), (7) we can write

$$\frac{e^{at}}{e^a-1}+\frac{1}{2}-\frac{1}{a}=\Sigma\{P_r(t)+b_r\}a^{r-1}. \qquad (11)$$

This function is bounded on circles of radius $(2n+1)\pi$ for $0<t<1$ and reduces to t at $a=0$. Hence it is equal to

$$t+\sum_{n=-\infty}^{\infty'}e^{2n\pi it}\left(\frac{1}{a-2n\pi i}+\frac{1}{2n\pi i}\right)$$

$$=t+\sum_{n=1}^{\infty}\left(\frac{2a\cos 2n\pi t}{4n^2\pi^2+a^2}+\frac{a^2}{n\pi(4n^2\pi^2+a^2)}\sin 2n\pi t\right)$$

$$=t+\sum_{n=1}^{\infty}\sum_{r=1}^{\infty}\frac{2a\cos 2n\pi t}{4n^2\pi^2}\left(\frac{-a^2}{4n^2\pi^2}\right)^{r-1}$$

$$-\sum_{n=1}^{\infty}\sum_{r=1}^{\infty}\frac{1}{n\pi}\sin 2n\pi t\left(\frac{-a^2}{4n^2\pi^2}\right)^r. \qquad (12)$$

Comparing with (11) we have

$$P_{2r}(t) + b_{2r} = \sum_{n=1}^{\infty} (-1)^{r-1} \frac{2\cos 2n\pi t}{(4n^2\pi^2)^r}, \tag{13}$$

$$P_{2r-1}(t) = (-1)^r \sum_{n=1}^{\infty} \frac{\sin 2n\pi t}{n\pi(4n^2\pi^2)^{r-1}} \quad (r > 1). \tag{14}$$

When $t = 0$ the right of (13) reduces to b_{2r}, and $P_{2r}(0) = 0$ as we should expect. In particular

$$P_2(t) + b_2 = \frac{1}{2\pi^2}\left(\cos 2\pi t + \frac{1}{2^2}\cos 4\pi t + \frac{1}{3^2}\cos 6\pi t + \ldots\right), \tag{15}$$

$$P_3(t) = \frac{1}{4\pi^3}\left(\sin 2\pi t + \frac{1}{2^3}\sin 4\pi t + \frac{1}{3^3}\sin 6\pi t + \ldots\right), \tag{16}$$

$$P_4(t) + b_4 = -\frac{1}{8\pi^4}\left(\cos 2\pi t + \frac{1}{2^4}\cos 4\pi t + \frac{1}{3^4}\cos 6\pi t + \ldots\right). \tag{17}$$

As for the numbers b_r, these reduce with increasing accuracy to the first terms for large r.

12·08. Operational methods and contour integration. Complex integrals can be used to interpret many operational solutions of problems relating to systems with an infinite number of degrees of freedom. We shall first show their relation to the simple operators p^{-n} of Chapter 7.

Consider the integral

$$\frac{1}{2\pi i}\int_C \frac{e^{zt}}{z^{n+1}}\,dz, \tag{1}$$

where C is a closed path in the z plane enclosing the origin, n is a positive integer, and t is independent of z. On expanding the exponential function we see that the coefficient of z^{-1} is $t^n/n!$; hence this is the residue of the integrand at the origin, and there are no other poles within C. Hence the integral is equal to $t^n/n!$.

It follows that

$$\frac{1}{2\pi i}\int_C \frac{e^{zt}}{z}\left(a_0 + \frac{a_1}{z} + \ldots + \frac{a_n}{z^n}\right)dz = a_0 + a_1 t + \ldots + \frac{a_n t^n}{n!}. \tag{2}$$

Now let

$$F(z) = a_0 + \sum_{n=1}^{\infty} a_n z^{-n} \tag{3}$$

converge when $|z| = R$; it will be uniformly convergent for all greater values of $|z|$. Then in the integral

$$f(t) = \frac{1}{2\pi i}\int_C \frac{e^{zt}}{z} F(z)\,dz, \tag{4}$$

where C is now a contour in the region where $|z| > R$, the integrand is a uniformly convergent series and we can integrate term by term. Hence

$$\frac{1}{2\pi i}\int_C \frac{e^{zt}}{z} F(z)\,dz = a_0 + \sum_{n=1}^{\infty} \frac{a_n t^n}{n!}. \tag{5}$$

This may be compared with the rule for interpreting the operational expression

$$F(p)\,1 = a_0 + \sum_{n=1}^{\infty} \frac{a_n t^n}{n!} = (a_0 + \Sigma a_n p^{-n})\,1, \tag{6}$$

which was shown to be valid provided that a positive quantity r exists such that the series $\sum_{n=0}^{\infty} a_n \zeta^n$ converges for $|\zeta| < r$. Changing ζ to $1/z$ and r to $1/R$ we see that the condition for the validity of (5) is that R exists such that $\sum_{n=0}^{\infty} a_n z^{-n}$ converges when $|z| \geq R$. But this is precisely the condition assumed in deriving (5), and we can now express it by saying that $F(z)$ is analytic at infinity. Hence if $F(p)$ satisfies this condition

$$F(p)\,1 = \frac{1}{2\pi i} \int_C \frac{e^{zt}}{z} F(z)\,dz. \tag{7}$$

Hence the result of applying any operator so far defined to unity can be immediately translated into a contour integral.

Several of our other rules can be derived at once. Thus if $F(z)$ and $G(z)$ are both expressible in power series in $1/z$, converging when $|z| \geq R$, they converge absolutely on C and can be multiplied to give another series in $1/z$, which is uniformly convergent. Hence

$$F(p)\,G(p)\,1 = \frac{1}{2\pi i} \int_C \frac{e^{tz}}{z} F(z)\,G(z)\,dz, \tag{8}$$

since on expansion we see that the two sides are identical.

The partial fraction rule also follows. For if $F(z)/G(z)$ is the ratio of two polynomials, $G(z)$ being of the same degree as $F(z)$ or higher,

$$\frac{F(z)}{zG(z)} = \Sigma P(z-\alpha), \tag{9}$$

where $P(z-\alpha)$ is the principal part of $F(z)/zG(z)$ at the pole $z = \alpha$; $\alpha = 0$ being in general one pole. There is no constant term since both sides tend to zero at $z = \infty$. Then

$$\frac{1}{2\pi i} \int_C \frac{F(z)}{zG(z)} e^{tz}\,dz = \frac{1}{2\pi i} \int_C \{\Sigma P(z-\alpha)\} e^{tz}\,dz. \tag{10}$$

The pole at $z = 0$ gives a finite series of powers of t; while the general term in the contribution from $z = \alpha$ will be

$$\frac{1}{2\pi i} \int_C \frac{A_m e^{tz}}{(z-\alpha)^m}\,dz = \frac{A_m}{2\pi i} e^{\alpha t} \int_{C'} \frac{e^{t\zeta}\,d\zeta}{\zeta^m}, \tag{11}$$

where $\zeta = z - \alpha$, and the transformed contour C' now encloses $\zeta = 0$, since C encloses $z = \alpha$. But this is $A_m t^{m-1} e^{\alpha t}/(m-1)!$, which is the interpretation given by the partial fraction rule.

12·081. The contour integral interpretation, however, will not give directly the result of applying an operator to any function other than a constant. We can verify consistency if the operand $g(t)$ is of the form $G(p)\,1$, where $G(p)$ is an operator within the meaning of the definition; for then our rule of 7·054 gives

$$F(p)\,g(t) = F(p)\,G(p)\,1, \tag{1}$$

which can be consistently interpreted by the rule for the composition of operators. But this is not general. For we have seen that if $G(p)$ fulfils the condition that $G(z)$ is expansible in a power series in $1/z$ for $|z| > R$, $g(t)$ is an integral function of t. We cannot therefore apply (1) directly if, for instance, the force on a dynamical system is of the form $A(a+t)^{-1}$, even if a is positive, much less to any case where $g(t)$ is not an analytic function of t for real positive values of t. But we can appeal to the principle of superposition

$$F(p)g(t) = g(0)f(t) + \int_0^t f(t-\tau)\,dg(\tau), \qquad (2)$$

which will provide an interpretation if only $F(p)$ satisfies the fundamental rule. So far, therefore, the contour integral interpretation is less general than the operational one, which will always give a solution in terms of a single integral when the problem depends on a finite number of linear differential equations with constant coefficients.

12·082. Limits of operators. Many physical problems are stated in terms of partial differential equations, which may be regarded as the formal result of having a large number n of degrees of freedom and making n tend to infinity. In such cases for any finite n the operator $F_n(p)$ will satisfy the conditions; but the limit of $F_n(z)$ may not. In particular it may not be bounded for large $|z|$, and there is then no suitable contour C. A simple case capable of being treated purely operationally is the following. Consider the equations

$$\frac{dx_r}{dt} + \frac{n}{h}x_r = \frac{n}{h}x_{r-1},$$

where all $x_r = 0$ $(r \geqslant 1)$ at $t = 0$, and $x_0 = g(t)$. Then for $r \geqslant 1$

$$x_r = \left(1 + \frac{ph}{n}\right)^{-r} g(t).$$

Take $r = n$ and consider

$$F_n(p)g(t) = \left(\frac{ph}{n} + 1\right)^{-n} g(t), \qquad (1)$$

where t and h are positive. $F_n(z)$ has a convergent expansion in powers of $1/z$ for all $|z| > n/h$. We can therefore evaluate this as

$$\left(\frac{n}{h}\right)^n \left(p + \frac{n}{h}\right)^{-n} g(t) = \left(\frac{n}{h}\right)^n \int_0^t g(\tau) \frac{(t-\tau)^{n-1}}{(n-1)!} e^{-n(t-\tau)/h}\,d\tau$$

$$= \left(\frac{n}{h}\right)^n \int_0^t g(t-\tau) \frac{\tau^{n-1}}{(n-1)!} e^{-n\tau/h}\,d\tau \qquad (2)$$

$$= \int_0^{nt/h} g\!\left(t - \frac{hv}{n}\right) \frac{v^{n-1}}{(n-1)!} e^{-v}\,dv. \qquad (3)$$

When n is large the integrand, apart from the g factor, has a sharp maximum near $v = n-1$. Hence if t is positive and greater than h, and $g(t-\tau)$ is continuous, the integral approximates as n increases to

$$\int_0^\infty g(t-h) \frac{v^{n-1}}{(n-1)!} e^{-v}\,dv = g(t-h). \qquad (4)$$

Now when $n \to \infty$, $F_n(z) \to \exp(-zh)$. When an operator $F_n(p)$ is such that as $n \to \infty$, $F_n(z) \to F(z)$, we shall speak of $F(p)$ as the *formal limit* of $F_n(p)$. Then we have in this sense

$$\exp(-ph)g(t) = g(t-h), \qquad (5)$$

provided t and h are positive and $t > h$. We therefore obtain an interpretation of $\exp(-ph)$, although $\exp(-zh)$ is not expansible in powers of z^{-1} and $\exp(-ph)$ is therefore not defined in our fundamental rules.

If $h > t$ (3) still holds, but the maximum of the factor $v^{n-1}e^{-v}$ is at $v = n-1$, which for $n > (1-t/h)^{-1}$ is outside the range of integration. The integral can then be shown easily to tend to zero.

Hence the result (5) depends essentially on the condition $t > h$; if $t < h$

$$\exp(-ph)g(t) = 0. \tag{6}$$

Thus this example of a formal limit of an operator leads to an intelligible result; but even if $g(t)$ is an analytic function the result of the operation is not an analytic function, since its values for $t < h$ are not the analytic continuation of $g(t-h)$ for $t > h$.

It is a common occurrence in mathematics for the limit of an infinite number of operations to give something fundamentally different from the separate operations. If $\sqrt{2}$ is expressed as a decimal, the digits up to any finite stage express a rational fraction, but the complete decimal is not rational. In the theory of the complex variable, any finite number of applications of the fundamental rules can give only rational functions, with no singularities other than poles; but an infinite series and its continuations can define a function with essential singularities and branch points. Hence there is no occasion for surprise at the interpretation of $\exp(-ph)$. But we must verify whether it commutes with p^{-1}. We have, writing $\exp(-ph)$ formally as e^{-ph},

$$e^{-ph}p^{-1}g(t) = e^{-ph}\int_0^t g(\tau)\,d\tau = \int_0^{t-h} g(\tau)\,d\tau, \tag{7}$$

$$p^{-1}e^{-ph}g(t) = \int_0^t g(\tau-h)\,d\tau = \int_{-h}^{t-h} g(\tau)\,d\tau, \tag{8}$$

so that the commutative rule holds if and only if $g(t) = 0$ for negative values of t; and if it is, *all* our operations on it will give 0 for negative t.

If we take $\exp(ph)$ with $h > 0$ we are not led to the interpretation

$$\exp(ph)g(t) = g(t+h) \tag{9}$$

nor to any interpretation whatever. We have in this case

$$\left(1 - \frac{ph}{n}\right)^{-n} g(t) = \left(-\frac{n}{h}\right)^n \int_0^t g(t-\tau) \frac{\tau^{n-1}}{(n-1)!} e^{n\tau/h}\,d\tau, \tag{10}$$

which can be shown* to tend to no limit as $n \to \infty$. We shall illustrate this by a special case later (12·10). Consequently $\exp(ph)$ is not an operator admissible in the system.

These results are intelligible physically. We ordinarily regard the state of a physical system at time t as determined by its state at time 0 and the disturbances that have affected it between times 0 and t. For systems with a finite number of degrees of freedom the operational method makes direct use of this principle. But if $\exp(-ph)g(t)$ was equal to $g(t-h)$ for $h > t$ we should be saying that the state at time t is determined by disturbances before time 0 and not taken into account in the specification of the state at time 0. If $\exp(ph)g(t)$ was equal to $g(t+h)$ for $h > 0$ we should be saying that a system can be influenced by disturbances that have not yet happened. Our results are therefore just what we should expect physically. The problem is to express them in our mathematical language. It is necessary to do so because when we treat the vibrations of strings and other problems of

* *Proc. Camb. Phil. Soc.* 36, 1940, 274.

sound by the operational method the operator $\exp(-ph)$ makes an appearance. But it will not obey the commutation rule unless we restrict the operand to be 0 for $t<0$, and though this makes no change in the solution for $t>0$ it implies that the contour integral as defined above is not quite what we need. But a modification of the path of integration gives a result with the required property.

The particular operators investigated above are far from the only ones that have $\exp(\pm ph)$ as their formal limits. A criterion of consistency is therefore needed if $\exp(-ph)$ is to be satisfactorily defined. The failure to arrive at a satisfactory definition of $\exp(ph)$ can be understood from the form of $F_n(p)$ in this case. It would correspond to a family of n differential equations of the form

$$\frac{dx_r}{dt} - \frac{n}{h} x_r = -\frac{n}{h} x_{r-1} \quad (r \geq 1), \quad x_0 = g(t),$$

with all the variables zero at $t = 0$. But the complementary functions of this set of equations have the form $t^m \exp(nt/h)$ and become meaningless if n tends to infinity. This will happen for any mode of approach to a limit such that at least one complementary function has the form $\exp(\alpha_n t)$, where the real part of α_n tends to $+\infty$ with n. We may speak of such sequences of systems as *tending to infinite instability*. Hence a necessary condition for a consistent definition of an operator by a limiting process is that the approach shall be through systems with no complementary function of this type. It is hard to invent reasonable sequences of physical systems that tend to infinite instability, but the principle that it is not approached in the limit turns out to play an important part in the justification of the application of the operational method to the solution of partial differential equations.

12·09. Bromwich's integral. The device introduced by Bromwich[*] is to replace the path C of 12·08 by one from $c - i\infty$ to $c + i\infty$, where c is real and positive and the path is so chosen that all singularities of $F(z)$ are to the left of it. We denote this path by L. It should be recalled that for the complex variable, as for the real variable, the usual proof of the existence of an integral fails if the path has an infinite length. The integral is then defined by first considering termini at a finite distance, for which the existence of the integral can be proved; if the integral tends to a limit as the termini approach infinity the limit is taken as the definition of the integral over an infinite range.

Now let $F(z)$ be a rational function bounded at infinity, and therefore such that its numerator is of the same or lower degree than the denominator. We shall show that for t real

$$\frac{1}{2\pi i} \int_L \frac{e^{tz}}{z} F(z) \, dz = \frac{1}{2\pi i} \int_C \frac{e^{tz}}{z} F(z) \, dz = f(t) \quad (t > 0) \tag{1}$$

$$= 0 \quad (t < 0). \tag{2}$$

12·091. Jordan's lemma. For this we need a form of the theorem known as *Jordan's lemma*. We first replace $F(z)/z$ by $\phi(z)$ and impose on $\phi(z)$ the weaker condition that for any ω we can choose r so that for any $|z| \geq r$ and $\Re(z) \leq c$ ($c > 0$), $|\phi(z)| < \omega$. Then for $t > 0$ we first take AB a finite stretch of L as shown and complete the contour by a rectangle

---

[*] *Proc. Lond. Math. Soc.* (2), 15, 1916, 401–448. A similar device was introduced at the same time by K. W. Wagner, *Archiv. f. Electrotechnik*, 4, 1916, 159–193.

$ABCD$, the corners of which are $c-iX$, $c+iX$, $-X+iX$, $-X-iX$. We can choose X, so that $|\phi(z)| < \omega$ at all points of BC, CD, and DA.

Then
$$\left|\int_{BC} e^{tz}\phi(z)\,dz\right| = \left|\int_{c}^{-X} e^{t(x+iX)}\phi(z)\,dx\right| < \omega \int_{-c}^{X} e^{-tx}dx < \frac{\omega}{t}(e^{tc}-e^{-tX}). \tag{1}$$

Similarly
$$\left|\int_{DA} e^{tz}\phi(z)\,dz\right| < \frac{\omega}{t}(e^{tc}-e^{-tX}). \tag{2}$$

Also
$$\left|\int_{CD} e^{tz}\phi(z)\,dz\right| = \left|\int_{X}^{-X} e^{t(-X+iy)}\phi(z)\,i\,dy\right| < 2\omega X\,e^{-tX}. \tag{3}$$

Thus the integrals along BC, CD, DA are together less than

$$2\omega\left(\frac{1}{t}e^{ct} + Xe^{-tX}\right). \tag{4}$$

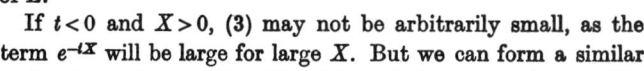

If then we choose an arbitrary ϵ we can choose $\omega < \frac{1}{4}t\epsilon e^{-ct}$, and then X so that $2\omega Xe^{-tX} < \frac{1}{2}\epsilon$; and then the integrals together are $<\epsilon$. Therefore $\int_L e^{tz}\phi(z)\,dz$ converges and is equal to the contour integral around all singularities on the negative side of L.

If $t<0$ and $X>0$, (3) may not be arbitrarily small, as the term e^{-tX} will be large for large X. But we can form a similar rectangle on the positive side of L, and the appropriate modifications of the argument show that the integral on L is equal to a contour integral including all singularities to the positive side of L, provided now that we can choose r so that for any $|z| \geqslant r$ and $\Re(z) \geqslant c$, $|\phi(z)| < \omega$.

The argument still holds if $\phi(z)$ has an infinite number of singularities on the imaginary axis. For we are not restricted to vary the path continuously, and if we can find a sequence of paths tending to infinity such that $|\phi(z)| \to 0$ on them the theorem will still follow, as in the proof of Mittag-Leffler's theorem.

12·092. Heaviside's unit function. $F(z)/z$ of 12·09 satisfies the conditions imposed on $\phi(z)$ in the proof of this lemma; hence (1) and (2) follow, and the use of the path L instead of the closed contour gives the same interpretations for positive t but zero for negative t, since there are no singularities to the right of L. In particular if we take $F(z) = 1$ the integral is 1 for positive t and 0 for negative t. We define as in 7·09

$$H(t) = 1 \quad (t>0), \qquad H(t) = 0 \quad (t<0), \tag{1}$$

and call $H(t)$ the Heaviside unit function, since it occurs, written as **1**, in many places in Heaviside's writings. He also wrote $p\mathbf{1}$ for a function whose integral is $H(t)$ and called this the impulse function. It is identical with the Dirac δ-function. For the theory of such functions see I. M. Gel'fand and G. E. Shilov, *Generalized Functions*, Vol. I, 1964 and M. J. Lighthill, *An Introduction to Fourier Analysis and Generalised Functions*, 1958.

The unit function is discontinuous at $t = 0$. If we take the principal value of the integral we get $\frac{1}{2}$, but there appear to be no occasions when this is used.

If $F(z)$ is analytic and bounded to the right of L, and the integral exists, we define

$$F(p)H(t) = \frac{1}{2\pi i}\int_L F(z)\frac{e^{tz}}{z}\,dz = f(t)H(t). \tag{2}$$

12·093. Integration and differentiation of Bromwich's integral. In
$$F(p)H(t) = f(t)H(t) = \frac{1}{2\pi i}\int_L F(z)e^{tz}\frac{dz}{z}, \tag{1}$$
let us integrate under the integral sign from 0 to t; we have
$$p^{-1}f(t)H(t) = \frac{1}{2\pi i}\int_L F(z)(e^{tz}-1)\frac{dz}{z^2}. \tag{2}$$
But if $F(z)/z$ is analytic and bounded to the right of L we can write
$$\frac{1}{2\pi i}\int_L F(z)\frac{dz}{z^2} = \frac{1}{2\pi i}\int_{L'} F(z)\frac{dz}{z^2}, \tag{3}$$
where L' lies entirely within the region where $|F(z)/z| < \omega$, and ω is arbitrarily small. Hence this integral is 0 and
$$p^{-1}f(t)H(t) = \frac{1}{2\pi i}\int_L F(z)e^{tz}\frac{dz}{z^2}, \tag{4}$$
which is our interpretation of $p^{-1}F(p)H(t)$. Hence the integral is consistent with the interpretation of p^{-1} as meaning definite integration from 0 to t.

If we differentiate (1) under the integral sign we get
$$\frac{d}{dt}f(t)H(t) = \frac{1}{2\pi i}\int_L F(z)e^{tz}dz = pF(p)H(t), \tag{5}$$
provided now that $F(z)$, and not merely $F(z)/z$, satisfies the conditions of Jordan's lemma: if it does not the integral is meaningless. But if it does, take $t = 0$ and use the path L'; then
$$\lim_{t\to 0}f(t)H(t) = \frac{1}{2\pi i}\int_{L'}\frac{F(z)}{z}dz = 0. \tag{6}$$
Hence the interpretation
$$\frac{d}{dt}f(t)H(t) = pF(p)H(t) \tag{7}$$
is valid provided that $f(0) = 0$, but not otherwise. Our restriction on the identification of p with differentiation therefore reappears in relation to Bromwich's integral.

12·10. If we interpret $\exp(-ph)F(p)H(t)$ or $F(p)\exp(-ph)H(t)$ by Bromwich's rule we get
$$\frac{1}{2\pi i}\int_L F(z)\frac{e^{t-h}}{z}dz = f(z-h)H(t-h), \tag{1}$$
which agrees with the interpretation obtained by treating $\exp(-ph)$ as the limit of an operator. If we apply the rule to $\exp(ph)$ with h positive we get $f(t+h)H(t+h)$, whereas our limiting process showed that this operator is not uniquely definable. In one sense this is trivial because this operator never occurs in practice, but it suggests that further precautions are needed. Our limiting process gives an explanation. We tried to define $\exp(ph)$ as the formal limit of $(1-ph/n)^{-n}$; but then the $F(z)$ of the Bromwich integral is $(1-zh/n)^{-n}$ and has a pole at $z = n/h$. Hence, however we draw the path L, there will be values of n such that the integrand has a pole to the right of it; and the limit of the operation on $f(t)$ is not to be identified with the Bromwich integral along any path, since it is essential to the definition that all singularities shall be on the negative side. These conditions will clearly arise in any case where there is a tendency to infinite instability.

To see what happens in this case, we have from 12·082 (10)

$$f_n(t) H(t) = \left(1 - \frac{ph}{n}\right)^{-n} H(t) = \frac{1}{2\pi i} \int_{L_n} \left(1 - \frac{zh}{n}\right)^{-n} e^{zt} \frac{dz}{z}. \tag{2}$$

For $t > 0$ the pole at $z = 0$ contributes 1; a contour C about $z = n/h$ contributes

$$\frac{1}{2\pi i} \int_C \left(-\frac{hz'}{n}\right)^{-n} e^{n t/h} e^{z't} \frac{dz'}{n/h + z'}, \tag{3}$$

where $z = n/h + z'$. The residue of $\frac{z'^{-n} e^{z't}}{1 + hz'/n}$ at $z' = 0$ is $\frac{t^{n-1}}{(n-1)!}\left(1 + O\left(\frac{h}{t}\right)\right)$. Hence for large t, $f_n(t)$ behaves like

$$1 - \left(-\frac{n}{h}\right)^{n-1} e^{n t/h} \frac{t^{n-1}}{(n-1)!} = 1 + u_n(t). \tag{4}$$

Then
$$\frac{u_{n+1}(t)}{u_n(t)} = -e^{t/h} \frac{t}{h} \frac{(n+1)^n}{n^n} \to -\frac{t}{h} \exp\left(\frac{t}{h} + 1\right) \tag{5}$$

which is < -1 for sufficiently large t. Hence $u_n(t)$, and therefore $f_n(t)$, oscillate infinitely for sufficiently large t as n increases.

It will be true that
$$F(p, n) H(t) = \frac{1}{2\pi i} \int_L F(z, n) e^{zt} \frac{dz}{z} \tag{6}$$

for L drawn so that all singularities of $F(z, n)$ are on the negative side; it will also be true that

$$\lim_{n \to \infty} F(p, n) H(t) = \lim_{n \to \infty} \frac{1}{2\pi i} \int_{L_n} F(z, n) e^{zt} \frac{dz}{z}, \tag{7}$$

provided that these limits exist, L_n being chosen for each n in accordance with the above rule; but if α_n is the largest real part of z at any pole of $F(z, n)$ and $\alpha_n \to \infty$ we cannot necessarily invert the order of integration and make $n \to \infty$. The result would be

$$\frac{1}{2\pi i} \int_L \left\{\lim_{n \to \infty} F(z, n)\right\} e^{zt} \frac{dz}{z}, \tag{8}$$

but this assumes a fixed path L, and we cannot draw any fixed L so that all poles of $F(z, n)$ are on its negative side for all n. Further, the limit (8) may exist, but (7) cannot for all initial conditions, since the function will in general ultimately increase like $e^{\alpha_n t}$, which cannot tend to any limit as $\alpha_n \to \infty$.

On the other hand, if the systems do not tend to infinite instability, inversion is possible in fairly wide conditions. For then we can choose k greater than the real part of any pole of any $F_n(z)$; and then we can take the c of the Bromwich integral greater than k, and use the same path L for all n. A proof has been given by D. P. Dalzell* (cf. 12·101). Consequently the operational method for continuous systems is now justified. We regard the differential equation as a formal limit of a set of ordinary differential equations in a finite number of variables, which can be solved by the operational method; and the formal limit of the operational solution is the solution for the continuous system when the number of variables is made infinite, and can be evaluated by means of Bromwich's integral. The formal limit of the operational solution can be found directly by forming the subsidiary equation from the partial differential equation and solving with the given boundary conditions.

* *Proc. Camb. Phil. Soc.* 36, 1940, 276–9·

This approach seems more satisfactory than the usual one, which takes as a fundamental requirement that the partial differential equation shall be satisfied. In, for instance, the motion of a stretched string, the equations of motion are rightly applied to a finite length of the string since this has a finite mass. But if we proceed to the limit and speak of the acceleration at a given point of the string we are differentiating a function that may have no derivative. If, for instance, the initial conditions are that the string is drawn aside a distance η at $x = \xi$, the string being straight at intervening points, the partial differential equation is meaningless at $x = \xi$, which is the only place where the initial acceleration is not zero. But if we replace the string by a set of particles uniformly spaced and with total mass equal to that of the string, we get an unambiguous solution, and the limit of this when the number of particles is made indefinitely large satisfies the equation for any finite length of the string, which is as much as we have any right to expect. A further consideration is that the actual string has an atomic structure and is really composed of a finite number of particles. The physical problem, therefore, is not the motion of a continuous string. The interest of the latter is simply that of an approximation, which for many purposes is good enough; but any question about the validity of a solution should be directed towards its accuracy for the discrete system and not towards the validity of the process used for solving the partial differential equation, since this equation is itself under suspicion as an expression of the physical facts. The process of making δx tend to zero continuously is intrinsically meaningless for an actual string, because there are no particles to give the displacement a meaning within a certain spacing. Similarly in thermal problems it is meaningless to specify an absolute temperature within, say, 1 part in 1000, unless something of the order of a million molecules are considered, and the differential equation of heat flow must also be regarded as a somewhat faulty idealization.

12·101. Dalzell's theorem. Let $F(z, n)$, an analytic function of z, satisfy the following conditions:

(i) *For each n a finite R_n exists such that for $|z| > R_n$*

$$F(z, n) = a_{0,n} + \sum_{1}^{\infty} \frac{a_{r,n}}{z^r}.$$

(ii) *For all $\Re(z) > k$, $F(z, n)$ has a limit $F(z)$ as $n \to \infty$.*

(iii) *Positive values of M, k exist such that for all n, and for all $\Re(z) > k$, $|F(z, n)| < M$.*

(iv) *In any finite interval of t, if $c > k$, possibly excluding a finite set of fixed intervals of arbitrarily small total length δ,*

$$\left| \int_c^{c+iY} F(z, n) e^{zt} dz \right| < N$$

for any real Y, where N may depend on δ but not on n or t. We show that (v)

$$F(p, n) H(t) \to F(p) H(t)$$

almost everywhere; and (vi)

$$\frac{1}{p} F(p, n) H(t) \to \frac{1}{p} F(p) H(t)$$

for all t.

By the Osgood-Vitali theorem (11·21) it follows from conditions (ii) and (iii) that in any bounded region of the half plane $\Re(z) > k$, $F(z, n) \to F(z)$ uniformly, and $F(z)$ is an analytic function of z in the half plane, possibly with a singularity at infinity. Clearly also $|F(z)| \leq M$ everywhere in the half plane.

The function
$$f_n(t) = F(p,n)H(t) = \frac{1}{2\pi i}\int_L F(z,n)e^{zt}\frac{dz}{z}, \tag{1}$$

where $\Re(z) = c > k$ on L, exists for all $t \neq 0$ by (i), is 0 for $t < 0$ by (i) or by (iii) and Jordan's lemma, and is a continuous function of t for $t > 0$, by (i). Also if $z = c + i\eta$,
$$\frac{1}{z} = \frac{c - i\eta}{c^2 + \eta^2}, \tag{2}$$

and $\Re(1/z)$ and $\Im(1/z)$ for $|\eta| > c$ are monotonic functions tending to zero as $|\eta| \to \infty$. Hence by Dirichlet's test, in any interval where (iv) is satisfied, for any ω we can choose Y so that
$$\left|\frac{1}{2\pi i}\left(\int_{c-i\infty}^{c-iY} + \int_{c+iY}^{c+i\infty}\right)F(z,n)e^{zt}\frac{dz}{z}\right| < \omega \tag{3}$$

for all n; further $F(z,n)e^{zt}/z \to F(z)e^{zt}/z$ uniformly in $-Y < \Im(z) < Y$. Therefore
$$F(p,n)H(t) \to \frac{1}{2\pi i}\int_L F(z)e^{zt}\frac{dz}{z} = F(p)H(t) = f(t), \tag{4}$$

say; and $f(t)$, being the limit of a uniformly convergent sequence of continuous functions, is continuous in any such interval.

Next,
$$p^{-1}f_n(t) = \int_0^t f_n(\tau)d\tau = \frac{1}{2\pi i}\int_L F(z,n)e^{zt}\frac{dz}{z^2} \tag{5}$$

by 12·093, since $F(z,n)$ has no singularities for $\Re(z) > k$; and
$$\left|\frac{1}{2\pi i}\left(\int_{c-i\infty}^{c-iY} + \int_{c+iY}^{c+i\infty}\right)F(z,n)e^{zt}\frac{dz}{z^2}\right| \leq \frac{1}{\pi}\frac{Me^{ct}}{\sqrt{(c^2+Y^2)}} < \omega \tag{6}$$

if $Y > Me^{ct}/\pi\omega$. Also $F(z,n)e^{zt}/z^2 \to F(z)e^{zt}/z^2$ uniformly for $-Y < \Im(z) < Y$; and therefore in any interval of t
$$p^{-1}f_n(t) \to \frac{1}{2\pi i}\int_L F(z)e^{zt}\frac{dz}{z^2}, \tag{7}$$

and the right side is a continuous function of t.

But in any interval where the sequence $\{f_n(t)\}$ is uniformly convergent
$$\int_{t_0}^{t_1} f(t)dt = \frac{1}{2\pi i}\int_L F(z)(e^{zt_1} - e^{zt_0})\frac{dz}{z^2} = \lim\left[p^{-1}f_n(t)\right]_{t_0}^{t_1}, \tag{8}$$

and therefore the right side of (7) has derivative $f(t)$ except possibly at a finite set of values of t in any interval of t, and it is continuous at these values. It can therefore be denoted by $p^{-1}f(t)$.

Of the conditions stated, (i) is implied by the general principle that we proceed through a sequence of discrete systems, and (ii) is obviously necessary. (iii) expresses the necessary condition that the sequence of systems does not tend to infinite instability. A less severe condition could be sufficient, but not much less. None of the functions
$$\frac{z^2}{n^{1/2}(z^2+n^2)}, \quad \frac{z^2}{z^2+n^2}, \quad \frac{nz^2}{z^2+n^2}$$

is uniformly bounded in the half plane, as we see by taking $z = in + c$. For the first, both (v) and (vi) are true; for the second, (v) is false and (vi) true; for the third, both are false. The exclusion of a set of measure δ from (iv) allows us to deal with a sequence that would tend, say, to sech ph. Such sequences arise in wave problems, and the discontinuities are an essential feature of the solutions.

Dalzell's argument is different but rests on substantially the same physical principles. It would be tempting to argue that as $p^{-1}F(p)H(t)$ is the limit of a uniformly convergent series of analytic functions of t, expressible by power series, $p^{-1}F(p)H(t)$ is an analytic function of t and therefore has a derivative everywhere. This is not so; for in general the integral (1) does not exist for complex t, and the limiting process does not lead to a definition of $p^{-1}f(t)$ off the real axis.

12·11. Text-books on differential equations describe a method of finding particular integrals by expansion in powers of $D = d/dt$. This is valid if the function operated on is a polynomial, when the series terminates. Its relation to the expansion in powers of p^{-1} is as follows. We have

$$F(p)t^m = F(p)\frac{m!}{p^m}1 = m!\,\Sigma p\frac{F(p)}{p^{m+1}}1, \tag{1}$$

where we take the principal parts of $F(z)/z^{m+1}$ at all poles and then replace z by p. One such pole in general is $z = 0$. If then near $z = 0$

$$F(z) = \alpha_0 + \alpha_1 z + \ldots + \alpha_m z^m + O(z^{m+1}), \tag{2}$$

the corresponding terms in $F(p)t^m$ are

$$m!\left(\frac{\alpha_0}{p^m} + \frac{\alpha_1}{p^{m-1}} + \ldots + \alpha_m\right)1 = \alpha_0 t^m + m\alpha_1 t^{m-1} + \ldots + m!\,\alpha_m = F(D)t^m, \tag{3}$$

which is the particular integral found by the usual method. The principal parts at the other poles α give terms with factors $\exp(\alpha t)$, which in the usual method are part of the complementary function. In the Heaviside method they have such coefficients that the function and its derivatives up to the $(n-1)$th vanish at $t = 0$ when the expansion of $F(p)$ in negative powers of p begins with a term in p^{-n}.

It follows that
$$F(D)t^m = \frac{1}{2\pi i}\int_C \frac{m!\,F(z)}{z^{m+1}}e^{zt}dz, \tag{4}$$

where C is a contour surrounding the origin but *not* enclosing any other pole of $F(z)$; while

$$F(p)t^m H(t) = \frac{1}{2\pi i}\int_L \frac{m!\,F(z)}{z^{m+1}}e^{zt}dz. \tag{5}$$

The difference between the two methods can therefore be stated as a difference in the path of integration for the same function when the operand is a polynomial.

If, however, the operand $g(t) = t^m$ is a positive fractional power of t, the interpretation of $F(p)g(t)$ remains significant for all positive t. That of $F(D)g(t)$ does not; for

$$(\alpha_0 + \alpha_1 D + \ldots)t^m = \alpha_0 t^m + m\alpha_1 t^{m-1} + \ldots + \alpha_r m(m-1)\ldots(m-r+1)t^{m-r} + \ldots,$$

and if $\Sigma \alpha_r z^r$ converges like a geometric progression this series never converges.

Most of the non-convergent series obtained by Heaviside were due to the fact that he never clearly distinguished p from D.

12·12. Special contour integrals. The unit function can be transformed to give an integral that is fundamental in the theory of Fourier series.

We can replace L by a path from $-i\infty$ to $-i\delta$, and $i\delta$ to $i\infty$, connecting them by a small semicircle of radius δ about 0, on the positive side. The first two paths give the principal value of a complex integral;

$$\frac{1}{2\pi i}\left[\int_{-i\infty}^{-i\delta} + \int_{i\delta}^{i\infty} \frac{e^{zt}}{z}dz\right] \to \frac{1}{2\pi i}P\int_{-\infty}^{\infty} e^{iyt}\frac{dy}{y} \qquad (1)$$

$$= \frac{1}{\pi}\int_0^{\infty} \sin yt \frac{dy}{y}. \qquad (2)$$

The small semicircle, in the limit, gives $\tfrac{1}{2}$. Hence

$$\frac{1}{2} + \frac{1}{\pi}\int_0^{\infty} \sin yt \frac{dy}{y} = H(t); \qquad (3)$$

$$\int_0^{\infty} \sin yt \frac{dy}{y} = \tfrac{1}{2}\pi \quad (t>0), \qquad (4)$$

$$= -\tfrac{1}{2}\pi \quad (t<0). \qquad (5)$$

If $t = 0$ the integral is zero, so that we can write it in general as $\tfrac{1}{2}\pi \operatorname{sgn} t$.

12·121. Another similar integral is, for real λ,

$$I = \int_0^{\infty} \frac{\cos \lambda x}{1+x^2}dx = \frac{1}{2}\int_{-\infty}^{\infty} \frac{e^{i\lambda z}}{1+z^2}dz. \qquad (6)$$

We complete the contour for $\lambda > 0$ by a large rectangle on the upper side of the real axis. The contour contains a pole at $z = i$; and the residue is $e^{-\lambda}/2i$. Thus

$$I = \tfrac{1}{2} \cdot 2\pi i \frac{e^{-\lambda}}{2i} = \tfrac{1}{2}\pi e^{-\lambda}. \qquad (7)$$

For the integral as stated this is the value if λ is taken positive. But (6) would be equally correct if we took the negative value of λ. In that case however we should have to complete the contour by a large rectangle *below* the real axis, and it would now contain the pole at $z = -i$, described in the negative direction. The residue at this pole is $-e^{\lambda}/2i$; hence we get

$$I = \tfrac{1}{2}\pi e^{\lambda}, \qquad (8)$$

which is the same as (7) since we are now using the negative value of λ.

12·122. This integral can also be written

$$\left.\begin{aligned}\int_{-\infty}^{\infty} \frac{e^{\kappa x}dx}{\pi(1+x^2)} &= e^{i\kappa} \quad (\Im(\kappa) > 0),\\ &= e^{-i\kappa} \quad (\Im(\kappa) < 0),\end{aligned}\right\} \qquad (9)$$

where κ is purely imaginary. In this form it appears in probability theory as the characteristic function of a certain probability law.*

* Cf. Jeffreys, *Theory of Probability*, p. 76.

12·123. Take again

$$\int_0^\infty \frac{\cos ax - \cos bx}{x^2}\,dx = \tfrac{1}{2}\Re P\int_{-\infty}^\infty \frac{e^{iaz}-e^{ibz}}{z^2}\,dz. \tag{10}$$

For $b > a > 0$ we take a large rectangle as in 12·121, indented by a small semicircle about 0; there is a simple pole of residue $i(a-b)$ at the origin, and the integral is

$$\tfrac{1}{2}\Re\{(\pi i)\,i(a-b)\} = \tfrac{1}{2}\pi(b-a). \tag{11}$$

12·124. For real a we know that

$$\int_0^\infty e^{-\tfrac{1}{2}az^2}\,dz = \left(\frac{\pi}{2a}\right)^{1/2}. \tag{1}$$

But if we take a path going to infinity in any direction such that $-\tfrac{1}{4}\pi < \arg z < \tfrac{1}{4}\pi$ the integral will be the same; for we can complete the contour by an arc at a large distance, on which $|z\exp(-\tfrac{1}{2}az^2)|$ can be made arbitrarily small. Putting

$$z = r\exp i\alpha,$$

we have
$$\int_0^\infty e^{-\tfrac{1}{2}ar^2 \exp 2i\alpha}\,dr = \left(\frac{\pi}{2a}\right)^{1/2} e^{-i\alpha}, \tag{2}$$

whence, separating real and imaginary parts,

$$\left.\begin{array}{l}\displaystyle\int_0^\infty e^{-\tfrac{1}{2}ar^2\cos 2\alpha}\cos(\tfrac{1}{2}ar^2\sin 2\alpha)\,dr = \left(\dfrac{\pi}{2a}\right)^{1/2}\cos\alpha \\[2mm] \displaystyle\int_0^\infty e^{-\tfrac{1}{2}ar^2\cos 2\alpha}\sin(\tfrac{1}{2}ar^2\sin 2\alpha)\,dr = \left(\dfrac{\pi}{2\alpha}\right)^{1/2}\sin\alpha\end{array}\right\} -\tfrac{1}{4}\pi < \alpha < \tfrac{1}{4}\pi. \tag{3}$$

It can be shown that at the limit $\alpha = \tfrac{1}{4}\pi$ the integral still converges and is continuous for $|\alpha| \leqslant \tfrac{1}{4}\pi$. (Cf. 1·124.) Hence

$$\int_0^\infty e^{-\tfrac{1}{2}iar^2}\,dr = \left(\frac{\pi}{2a}\right)^{1/2}\frac{1-i}{\sqrt{2}},\quad \int_0^\infty \cos(\tfrac{1}{2}ar^2)\,dr = \frac{1}{2}\left(\frac{\pi}{a}\right)^{1/2},\quad \int_0^\infty \sin(\tfrac{1}{2}ar^2)\,dr = \frac{1}{2}\left(\frac{\pi}{a}\right)^{1/2}. \tag{4}$$

These integrals are the basis of the methods of steepest descents and stationary phase for the approximate evaluation of complex integrals, used especially in the theory of dispersion of waves and the theory of probability, including statistical mechanics.

12·125. The theory of the factorial function makes use of the integral

$$I = \int_0^\infty \frac{z^a\,dz}{(1+z)^2}, \tag{1}$$

where a may be fractional or complex but $1 > \Re(a) > -1$. Then there is a branch point at $z = 0$. Take

$$\int \frac{(ze^{i\pi})^a}{(1+z)^2}\,dz = 0, \tag{2}$$

around the path shown, where $\arg(ze^{i\pi})$ is taken as zero for z between -1 and 0. The large arcs contribute 0 in the limit since $\Re(a) < 1$. The pole at $z = -1$ contributes $(-2\pi i)(-a)$ in the limit, since the residue is $-a$, and the paths from and to $-\infty$ give contributions cancelling in the limit, since -1 is not a branch point. The small circle about 0 gives 0 in the limit since $\Re(a) > -1$. Thus we are concerned only with the integrals between 0 and $+\infty$. But these are

$$(e^{-i\pi a} - e^{i\pi a}) \int_0^\infty \frac{z^a dz}{(1+z)^2} = -2i \sin \pi a \int_0^\infty \frac{z^a dz}{(1+z)^2},$$

since $\arg(ze^{i\pi})$ is $-\pi$ on the upper line and $+\pi$ on the lower. Therefore

$$\int_0^\infty \frac{z^a dz}{(1+z)^2} = \frac{\pi a}{\sin \pi a}. \tag{3}$$

12·126. Next consider the integral,* possibly for fractional m,

$$p^m H(t) = \frac{1}{2\pi i} \int_L z^{m-1} e^{zt} dz, \tag{1}$$

which converges if $m < 1$. If $m > 1$ the integral along L does not exist. But if we modify the path to L' so that L' has asymptotes in the third and second quadrants, not parallel to the imaginary axis, the integral will converge for t real and positive without restriction on m. We therefore consider the integral along L'. Now this is equivalent to an integral along M, which consists of a small circle about the origin and two lines from and to $-\infty$. In the first place we take $m > 0$; then the integral about the small circle tends to 0 when the circle is made arbitrarily small. We take z to be $\mu e^{i\pi}$ and $\mu e^{-i\pi}$ on CD and AB respectively. Then on CD

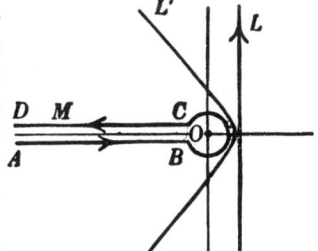

$$z^{m-1} e^{zt} dz = e^{m\pi i} \mu^{m-1} e^{-\mu t} d\mu \tag{2}$$

and on AB similarly it is $e^{-m\pi i} \mu^{m-1} e^{-\mu t} d\mu$. Hence

$$\frac{1}{2\pi i} \int_M z^{m-1} e^{zt} dz = \frac{1}{2\pi i} \int_\infty^0 e^{-m\pi i} \mu^{m-1} e^{-\mu t} d\mu + \frac{1}{2\pi i} \int_0^\infty e^{m\pi i} \mu^{m-1} e^{-\mu t} d\mu$$

$$= \frac{1}{\pi} \sin m\pi \int_0^\infty \mu^{m-1} e^{-\mu t} d\mu$$

$$= \frac{\sin m\pi}{\pi} \frac{(m-1)!}{t^m}. \tag{3}$$

But by a known identity (15·02)

$$\frac{\sin m\pi}{m\pi} = \frac{1}{m!(-m)!} \tag{4}$$

and hence

$$\frac{1}{2\pi i} \int_M z^{m-1} e^{zt} dz = \frac{1}{t^m(-m)!}, \tag{5}$$

for $0 < m$. But both sides of this equation are analytic functions of m; hence it is true for all m. (The singularity at $z = 0$ does not affect this statement. It is proved for $0 < m$ when the circle is arbitrarily small; it is therefore true for $0 < m$ for any size

---

* Chapter 15 should be read before this section. 12·13 and later sections are independent of Chapter 15.

of the circle; and with a non-zero radius of the circle the question of divergence at 0 does not arise.) Changing m to $-n$ we have therefore for all n, if $t > 0$,

$$p^{-n} H(t) = \frac{t^n}{n!}, \tag{6}$$

if the left side is determined by an integral on L' or M; and this is also true for the path L if $n > -1$. In particular, since

$$(-\tfrac{1}{2})! = \sqrt{\pi}, \quad (\tfrac{1}{2})! = \tfrac{1}{2}\sqrt{\pi}, \quad (\tfrac{3}{2})! = \tfrac{3}{2} \cdot \tfrac{1}{2}\sqrt{\pi}, \quad (-\tfrac{3}{2})! = -2\sqrt{\pi}, \quad (-\tfrac{5}{2})! = \tfrac{2}{3} \cdot \tfrac{2}{1}\sqrt{\pi}, \quad \ldots$$

$$p^{1/2} H(t) = \frac{1}{\sqrt{(\pi t)}}, \quad p^{-1/2} H(t) = 2\sqrt{\frac{t}{\pi}}, \quad p^{-3/2} H(t) = \frac{2}{1} \cdot \frac{2}{3} \frac{t^{3/2}}{\sqrt{\pi}}, \quad p^{-5/2} H(t) = \frac{2}{1} \cdot \frac{2}{3} \cdot \frac{2}{5} \frac{t^{5/2}}{\sqrt{\pi}}$$

and in a special sense
$$p^{3/2} H(t) = -\frac{1}{2} \frac{t^{-3/2}}{\sqrt{\pi}}, \quad p^{5/2} H(t) = \frac{1}{2} \cdot \frac{3}{2} \frac{t^{-5/2}}{\sqrt{\pi}}.$$

If n is a negative integer $n!$ is infinite and $p^{-n} H(t) = 0$.

Operators with $n \leqslant -1$ are not valid in the strict sense that they can be interpreted by integrals on the Bromwich path. They require modification of the path to L' or M to give them a meaning. They do not arise directly in the solution of physical problems. It often happens, however, that operational solution leads to an integral valid on the Bromwich path, but the easiest way of evaluating it is first to modify the path to L' or M (subject to there being no singularities between the paths) and then to expand in ascending powers of z on L' or M. The integrals on L corresponding to $n > -1$ will then be as stated in (6); but if $n \leqslant -1$ this formula should be regarded simply as an aid to memory, as a shorthand for

$$\frac{1}{2\pi i} \int_M e^{zt} z^{-n-1} dz = \frac{t^n}{n!} \quad (t > 0). \tag{7}$$

If we differentiate this under the integral sign with regard to n we get another uniformly convergent integral. If $n > -1$ it converges on L. Then we have

$$\frac{1}{2\pi i} \int_M e^{zt} z^{-n-1} \log z \, dz = -\frac{t^n \log t}{n!} + \frac{t^n}{n!} \frac{d}{dn} \log n!, \tag{8}$$

or, in operational form,
$$p^{-n} \log p \, H(t) = \left(\frac{d}{dn} \log n! - \log t\right) \frac{t^n}{n!} H(t). \tag{9}$$

In particular, if $n = 0$, $\quad \log p \, H(t) = (-\gamma - \log t) H(t), \tag{10}$

where γ is Euler's constant (cf. 15·04).

Specially important operators are $p^{1/2} \exp(-a p^{1/2})$ and $\exp(-a p^{1/2})$, where a is a positive constant. Interpreted by integrals on L they are significant, since on L the argument of $z^{1/2}$ is between $\pm \tfrac{1}{4}\pi$ and the integrand decreases exponentially at the limits. Then

$$p^{1/2} \exp(-a p^{1/2}) H(t) = \frac{1}{2\pi i} \int_L e^{zt - a z^{1/2}} z^{-1/2} dz, \tag{11}$$

$$\exp(-a p^{1/2}) H(t) = \frac{1}{2\pi i} \int_L e^{zt - a z^{1/2}} \frac{dz}{z}. \tag{12}$$

These can be evaluated in two ways, both of which have many other applications. First, we notice that the integrals are unaffected for $t>0$ if we use the paths L' or M. Also $\exp(-az^{1/2})$ can be expanded in an absolutely convergent series. Hence we can reverse the order of integration and summation (1·111) and write

$$p^{1/2}\exp(-ap^{1/2})H(t) = \frac{1}{2\pi i}\int_{L'} e^{zt} z^{-1/2}\left(\sum_{n=0}^{\infty}(-1)^n \frac{a^n z^{1/2 n}}{n!}\right)dz$$

$$= \frac{1}{2\pi i}\sum_{n=0}^{\infty}(-1)^n \int_{L'} a^n e^{zt}\frac{z^{1/2(n-1)}}{n!}dz$$

$$= \Sigma(-1)^n a^n \frac{p^{1/2(n+1)}}{n!}H(t) = \Sigma(-1)^n a^n \frac{t^{-1/2(n+1)}}{n!\{-\frac{1}{2}(n+1)\}!}H(t) \quad (13)$$

in the sense of (7); then all terms with odd n vanish, and if $n = 2m$

$$n!\{-\tfrac{1}{2}(n+1)\}! = (2m)!\{-\tfrac{1}{2}(2m+1)\}! = \frac{1.2\ldots 2m\sqrt{\pi}}{(-\tfrac{1}{2})(-\tfrac{3}{2})\ldots(-\tfrac{1}{2}(2m-1))}$$

$$= (-2)^m 2.4\ldots 2m\sqrt{\pi} = (-4)^m m!\sqrt{\pi}. \quad (14)$$

Then for $t>0$

$$p^{1/2}\exp(-ap^{1/2})H(t) = \frac{1}{\sqrt{\pi}}\sum_{m=0}^{\infty}(-)^m \frac{a^{2m} t^{-m-1/2}}{4^m m!} = \frac{1}{\sqrt{(\pi t)}}\exp\left(-\frac{a^2}{4t}\right). \quad (15)$$

The effect of removing the factor $p^{1/2}$ is that it is now the terms with even n that vanish, with the exception of the first; we have now

$$\exp(-ap^{1/2})H(t) = \sum_{n=0}^{\infty}(-)^n a^n \frac{p^{1/2 n}}{n!} = 1 - \sum_{m=0}^{\infty}\frac{a^{2m+1} t^{-m-1/2}}{(2m+1!)(-m-\tfrac{1}{2})!}$$

$$= 1 - \frac{2}{\sqrt{\pi}}\sum_{m=0}^{\infty}\left\{\frac{a}{2\sqrt{t}} - \frac{1}{3}\left(\frac{a}{2\sqrt{t}}\right)^3 + \frac{1}{2!5}\left(\frac{a}{2\sqrt{t}}\right)^5 - \ldots\right\}. \quad (16)$$

If we define

$$\operatorname{erf} x = \frac{2}{\sqrt{\pi}}\int_0^x e^{-u^2}du, \quad (17)$$

we have

$$\operatorname{erf} x = \frac{2}{\sqrt{\pi}}\left(x - \tfrac{1}{3}x^3 + \frac{1}{2!5}x^5 - \ldots\right), \quad (18)$$

and hence

$$\exp(-ap^{1/2})H(t) = 1 - \operatorname{erf}\frac{a}{2\sqrt{t}}. \quad (19)$$

Alternatively, return to (11) and write

$$zt - az^{1/2} = t\left(z^{1/2} - \frac{a}{2t}\right)^2 - \frac{a^2}{4t}. \quad (20)$$

With a and $t>0$, $a/2t$ is positive and we can take a path for $z^{1/2}$ through it parallel to the imaginary axis. Put then

$$z^{1/2} - \frac{a}{2t} = i\zeta;$$

then

$$p^{1/2}\exp(-ap^{1/2})H(t) = \frac{1}{\pi}\int_{-\infty}^{\infty}\exp\left(-\frac{a^2}{4t} - t\zeta^2\right)d\zeta$$

$$= \frac{1}{\sqrt{(\pi t)}}\exp\left(-\frac{a^2}{4t}\right). \quad (21)$$

Also if (12) is differentiated with regard to a we get (11) with the sign changed. Hence

$$\exp(-ap^{1/2}) H(t) = -\int_0^a \frac{1}{\sqrt{(\pi t)}} \exp\left(-\frac{a^2}{4t}\right) da + C. \tag{22}$$

But if $a = 0$ (12) is simply $H(t)$. Hence

$$\exp(-ap^{1/2}) H(t) = 1 - \int_0^a \frac{1}{\sqrt{(\pi t)}} \exp\left(-\frac{a^2}{4t}\right) da$$

$$= 1 - \operatorname{erf} \frac{a}{2\sqrt{t}}. \tag{23}$$

12·13. The generalized principle of superposition. The operators just interpreted do not satisfy our original rule that $F(z)$ shall be expansible in powers of z^{-1} when $|z|$ is large enough. The question is whether the interpretations will still satisfy the principle of superposition. Let

$$f(t) = \frac{1}{2\pi i} \int_L e^{zt} \frac{F(z)}{z} dz. \tag{1}$$

Then it can be shown that under suitable restrictions

$$F(u) = u \int_0^\infty e^{-ut} f(t) dt. \tag{2}$$

Take $\Re(u)$ greater than c; $F(z)$ is analytic for $\Re(z) \geq c$; and let the integral of $F(z)/z^2$ around an infinite semicircle to the right of L be zero. Then if we may invert the order of integration

$$u \int_0^\infty e^{-ut} f(t) dt = u \int_0^\infty e^{-ut} \frac{1}{2\pi i} dt \int_{c-i\infty}^{c+i\infty} e^{zt} \frac{F(z)}{z} dz = \frac{1}{2\pi i} \int_{c-i\infty}^{c+i\infty} \frac{F(z)}{z} dz \int_0^\infty u e^{-(u-z)t} dt$$

$$= \frac{1}{2\pi i} \int_{c-i\infty}^{c+i\infty} \frac{u F(z)}{z(u-z)} dz. \tag{3}$$

But $\Re(u)$ is greater than c; hence if we deform the path L into an infinite semicircle to the right of L we pass over the pole at u, and over no other singularity. But by hypothesis the integral along this semicircle is zero. Hence the last integral is $-2\pi i$ times the residue at this pole, and therefore* $u \int_0^\infty e^{-ut} f(t) dt = F(u)$. The justification of the inversion of the order of integration will be carried out later when we come to the Fourier-Mellin theorem, which is really the converse of this.

Now suppose that we have two operators $F(p)$ and $G(p)$ satisfying the conditions just imposed on $F(p)$. Then $g(t-\tau) = 0$ if $\tau > t$, and

$$\int_0^t f(\tau) g(t-\tau) d\tau = \int_0^\infty f(\tau) g(t-\tau) d\tau = \frac{1}{2\pi i} \int_0^\infty f(\tau) d\tau \int_L e^{z(t-\tau)} G(z) \frac{dz}{z}. \tag{4}$$

But

$$\int_0^\infty f(\tau) e^{-z\tau} d\tau = F(z)/z; \tag{5}$$

hence

$$\int_0^t f(\tau) g(t-\tau) d\tau = \frac{1}{2\pi i} \int_L e^{zt} F(z) G(z) \frac{dz}{z^2} = \frac{F(p) G(p)}{p} H(t). \tag{6}$$

* This argument is due to S. Goldstein, *Proc. Lond. Math. Soc.* (2) 34, 1931, 104.

If we differentiate we have

$$F(p)\,G(p)\,H(t) = \int_0^t f(\tau)\,g'(t-\tau)\,d\tau + f(t)\,g(0), \tag{7}$$

which is our previous rule of superposition, extended to a much wider range of conditions. Thus again the result of the successive application of two operators whose interpretations are known can be reduced to a single integral.

12·14. Integral equations of Abel's and Poisson's types. If

$$\int_0^x \phi(\xi)f(x-\xi)\,d\xi = g(x), \tag{1}$$

where f and g are given functions, and we wish to determine ϕ so that (1) will be true for all x, we introduce the operational expressions indicated by capital letters and apply the principle of superposition. Then (1) is equivalent to

$$\frac{1}{p}\Phi(p)\,F(p) = G(p), \tag{2}$$

$$\Phi(p) = \frac{pG(p)}{F(p)}, \tag{3}$$

whence $\phi(x)$ is found by substituting in the Bromwich integral and interpreting. This is Abel's type of integral equation.

Poisson's type
$$\phi(x) + \int_0^x \phi(\xi)f(x-\xi)\,d\xi = g(x) \tag{4}$$

is treated similarly. We get

$$\Phi(p) + \frac{1}{p}\Phi(p)\,F(p) = G(p), \tag{5}$$

$$\Phi(p) = \frac{pG(p)}{p+F(p)}, \tag{6}$$

whence the solution follows. Examples are given by Goldstein.*

12·15. The staircase, parapet, and saw-tooth functions. Consider the series

$$(e^{-ph} + e^{-2ph} + e^{-3ph} + \ldots)H(t) = H(t-h) + H(t-2h) + \ldots + H(t-nh) + \ldots \tag{1}$$

If $[t/h]$ denotes the integral part of t/h, all terms with $n < t/h$ are 1, all with $n > t/h$ are 0, and the number of the former set is $[t/h]$. Hence the function is equal to $[t/h]$. It is equal to 0 for $0 < t < h$, and rises discontinuously by 1 at $t = h, 2h, 3h, \ldots$. It behaves like the date, given as a function of time. Now if we translate this by Bromwich's integral p is replaced by z, whose real part is the positive quantity c; then the series is a convergent geometrical series, and

$$e^{-zh} + e^{-2zh} + \ldots = \frac{e^{-zh}}{1-e^{-zh}} = \frac{e^{-\frac{1}{2}zh}}{2\sinh\tfrac{1}{2}zh}.$$

Hence
$$\left[\frac{t}{h}\right] = \frac{1}{2\pi i}\int_L \frac{e^{z(t-\frac{1}{2}h)}}{2\sinh\tfrac{1}{2}zh}\frac{dz}{z} = \frac{e^{-\frac{1}{2}ph}}{2\sinh\tfrac{1}{2}ph}H(t). \tag{2}$$

* *Journ. Lond. Math. Soc.* 6, 1931, 262–8.

Similarly the series
$$(\tfrac{1}{2} - e^{-ph} + e^{-2ph} - e^{-3ph} + \ldots) H(t) \qquad (3)$$

is equal to $+\tfrac{1}{2}$ if $[t/h]$ is 0 or an even integer, $[-\tfrac{1}{2}]$ if $[t/h]$ is an odd integer, and may therefore be written as $\tfrac{1}{2}(-1)^{[t/h]}$. The series also can be summed like a geometrical progression and gives

$$\left(\frac{1}{2}\frac{1-e^{-ph}}{1+e^{-ph}}\right) H(t) = (\tfrac{1}{2}\tanh \tfrac{1}{2}ph) H(t) = \tfrac{1}{2}(-1)^{[t/h]} \qquad (t>0). \qquad (4)$$

These two functions are fundamental in the operational treatment of waves.

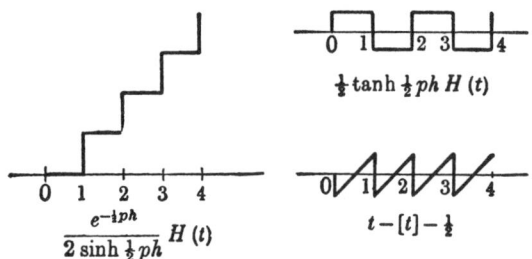

The average of t/h for t between nh and $nh+h$ is $n + \tfrac{1}{2} = \left[\tfrac{t}{h}\right] + \tfrac{1}{2}$. Then the function $\tfrac{t}{h} - \left[\tfrac{t}{h}\right] - \tfrac{1}{2}$ is $-\tfrac{1}{2}$ at $t=0$, increases uniformly to $\tfrac{1}{2}$ at $t=h$, drops discontinuously to $-\tfrac{1}{2}$, and then repeats itself periodically. Its operational expression is

$$\left(\frac{1}{ph} - \frac{1}{e^{ph}-1} - \frac{1}{2}\right) H(t). \qquad (5)$$

Now the first stage of the Euler-Maclaurin formula of integration could be written

$$\int_0^{nh} f(t)\,dt = \tfrac{1}{2}f(0) + f(h) + f(2h) + \ldots + f\{(n-1)h\} + \tfrac{1}{2}f(nh) - \int_0^{nh}\left[\tfrac{t}{h} - \left[\tfrac{t}{h}\right] - \tfrac{1}{2}\right] f'(t)\,dt, \qquad (6)$$

and the operator

$$\frac{hD}{e^{hD}-1} + \tfrac{1}{2}hD - 1 \qquad (7)$$

occurred in the operational derivation of the Euler-Maclaurin series. The similarity of the two expressions suggests an intimate relation, but the operator (5) acts on $H(t)$ and cannot be developed in ascending powers of p, whereas (7) must be developed in ascending powers of D. A relation probably exists, but it does not seem likely to be simple enough to replace the development of the Euler-Maclaurin formula by integration by parts.

12·16. Frullani's integrals. Consider

$$I = \int_0^\infty \{f(ax) - f(bx)\}\frac{dx}{x},$$

$f(x)$ satisfying certain conditions that will appear as we proceed. Break up the range of integration into 0 to δ and δ to ∞, where δ is arbitrarily small. Then if the integral I converges at the lower limit the integral over 0 to δ is arbitrarily small. Then

$$I = \lim_{\delta \to 0} \int_\delta^\infty \{f(ax) - f(bx)\}\frac{dx}{x}$$

$$= \lim \left\{\int_{a\delta}^\infty f(u)\frac{du}{u} - \int_{b\delta}^\infty f(u)\frac{du}{u}\right\},$$

provided the two integrals converge; and this is

$$\int_{a\delta}^{b\delta} f(u) \frac{du}{u}.$$

But if $f(u)$ tends continuously to a finite limit as $u \to 0$ this approaches

$$f(0+) \int_{a\delta}^{b\delta} \frac{du}{u} = f(0+) \log \frac{b}{a}.$$

Then
$$I = f(0+) \log \frac{b}{a}.$$

In particular
$$\int_0^\infty (\cos ax - \cos bx) \frac{dx}{x} = \int_0^\infty (e^{-ax} - e^{-bx}) \frac{dx}{x} = \log \frac{b}{a}.$$

Neither integral can be treated by contour integration. The first integrand is an odd function and the integral obtained by changing the range to $(-\infty, \infty)$ vanishes identically.

The same method can be extended to many integrals of odd functions. Take, for instance,

$$I = \int_0^\infty \frac{\sin^3 x}{x^2} dx = \frac{1}{4} \int_0^\infty \frac{3 \sin x - \sin 3x}{x^2} dx$$

$$= \lim_{\delta \to 0} \frac{1}{4} \int_\delta^\infty (3 \sin x - \sin 3x) \frac{dx}{x^2}$$

$$= \tfrac{1}{4} \lim \left\{ \int_\delta^\infty 3 \sin x \frac{dx}{x^2} - \int_{3\delta}^\infty 3 \sin u \frac{du}{u^2} \right\} = \tfrac{3}{4} \lim \int_\delta^{3\delta} \sin x \frac{dx}{x^2}$$

$$= \tfrac{3}{4} \log 3.$$

EXAMPLES

1. Evaluate the integral
$$\int_0^\infty \frac{x^4 dx}{1+x^6}.$$
(M.T. 1943.)

2. Show that
$$\int_0^\infty \frac{\sin^2 x}{x^2} dx = \int_0^\infty \frac{\sin x}{x} dx, \quad \int_0^\infty \frac{\sin^2 x}{x^2(1+x^2)} dx = \tfrac{1}{4}\pi(1+e^{-2}).$$

3. Find the principal value of
$$\int_0^\infty \frac{\tan x}{x} dx.$$

4. Prove that as X tends to infinity
$$\int_0^X \frac{\cos ax - \cos bx}{x^2} dx = \tfrac{1}{2}\pi(b-a) + O\left(\frac{1}{X^2}\right) \quad (a>0, b>0).$$
(M.T. 1938.)

5. If B_r are Bernoulli's numbers, prove that
$$\frac{B_{r-1}}{1!(r-1)!} + \frac{B_{r-2}}{2!(r-2)!} + \ldots + \frac{B_2}{(r-2)!2!} - \frac{1}{2(r-1)!} + \frac{1}{r!} = 0.$$

6. Prove that
$$x \cot x = 1 - B_2 \frac{(2x)^2}{2!} + B_4 \frac{(2x)^4}{4!} - \ldots.$$

7. a and b are real and positive and unequal constants. Find
$$\int_0^\infty \frac{dx}{(x^2+a^2)^2(x^2+b^2)^2}.$$
Is the result valid also when $a = b$?
(M.T. 1935.)

8. Prove from Liouville's theorem, by considering $z^n/f(z)$, that any polynomial $f(z)$ of positive degree n has at least one zero. (M.T. 1935.)

9. Show that the roots of the equation
$$z^5 - 12z^2 + 14 = 0$$
lie between the circles $|z| = 1$ and $|z| = \frac{5}{2}$. How many of the roots lie inside the circle $|z| = 2$? (M.T. 1939.)

10. If $f(z) = (z-a_1)\ldots(z-a_n)$, prove that all the zeros of $f'(z)$ lie within any polygon with no re-entrant angle that includes all the zeros of $f(z)$.

11. If $|f(z)|$ is constant on a closed contour on which $f'(z)$ nowhere vanishes, show that if $f(z)$ is analytic within the contour it has one more zero than $f'(z)$ within the contour. (Macdonald.)

12. If
$$\phi(z) = b_0 + b_1 z + \ldots,$$
where $b_1 \neq 0$, the root of $\phi(z) = 0$ nearest the origin is the coefficient of $\dfrac{1}{t}$ in the Laurent expansion of
$$-\log \frac{\phi(t)}{t} = -\log\left\{b_1 + \frac{b_0}{t} + (b_2 t + \ldots)\right\}.$$
(W. R. Andress, *Math. Gaz.* 27, 1943, 92.)

13. Show that
$$\int_{-\infty}^{\infty} \frac{\sin(x+a)\sin(x-a)}{x^2-a^2}\,dx = \frac{\pi}{2a}\sin 2a.$$

14. If $f(z)/z$ is uniformly bounded on a set of contours defined as in Mittag-Leffler's theorem, tending to infinity and $f(z)$ has simple poles α_r of residue β_r, prove that
$$f(z) = f(0) + zf'(0) + \Sigma\left(\frac{\beta_r}{z-\alpha_r} + \frac{\beta_r(z+\alpha_r)}{\alpha_r^2}\right).$$

15. Evaluate the integral
$$\int_0^{\infty} \frac{\sin mx}{x(x^2+a^2)^2}\,dx \quad (m>0).$$

16. By integrating the function $\dfrac{e^{az}}{1+e^z}$ $(0<a<1)$ around a rectangle of breadth 2π, prove that
$$\int_{-\infty}^{\infty} \frac{e^{ax}}{1+e^x}\,dx = \frac{\pi}{\sin a\pi},$$
and derive an alternative proof that $z!(-z)! = \pi z \operatorname{cosec} \pi z$.

17. Prove that for $h > 0$
$$\int_0^{\infty} \sin \kappa x\, e^{-\kappa h}\frac{d\kappa}{\kappa} = \tan^{-1}\frac{x}{h}.$$

18. Prove that
$$\int_0^{\infty} \frac{\log t}{t^2-1}\,dt = 2\int_0^{\infty} \frac{u\,du}{e^u - e^{-u}} = \tfrac{1}{4}\pi^2.$$
(I.C. 1944.)

19. If for small $|z|$
$$\sec z = 1 + \Sigma a_{2r} z^{2r}$$
prove that
$$1 - \frac{1}{3^{2r+1}} + \frac{1}{5^{2r+1}} - \ldots = \frac{\pi^{2r+1}}{2^{2r+2}} a_{2r}.$$

Chapter 13

CONFORMAL REPRESENTATION

Plus ça change, plus c'est la même chose.
ALPHONSE KARR, *Les Guêpes*, 1849

13·01. Conditions for conformal mapping. Let ζ and z be two complex variables related by

$$z = f(\zeta) \tag{1}$$

where $f(\zeta)$ is an analytic function. Then if ζ moves along a curve in the plane of ξ, η, z will move along a curve in the plane of x, y. Every value of ζ that $f(\zeta)$ is defined for will identify a point in the x, y plane, and conversely if the inverse function exists for a value of z a value of ζ will be identified. Thus (1) can be regarded as a *transformation*, enabling us to map at least a part of the ξ, η plane on the x, y plane and conversely.

The importance of this type of transformation rests on the facts that it is, in general, *continuous* and *conformal*. For, in the first place, take γ a small complex number and consider $z + c = f(\zeta + \gamma)$. We have

$$c = f(\zeta + \gamma) - f(\zeta) = \gamma \{f'(\zeta) + v\},$$

where v tends to zero with γ. Hence if $f'(\zeta)$ exists both the real and imaginary parts of c tend to zero when those of γ do. Thus the transformation from ζ to z is continuous. Conversely, that from z to ζ is continuous provided that $d\zeta/dz = 1/f'(\zeta)$ exists. But in the neighbourhood of a point where $f'(\zeta)$ or its reciprocal does not exist the transformation or its inverse may not be continuous.

Now take two small complex numbers γ and γ' and consider the behaviour of

$$\frac{c'}{c} = \frac{f(\zeta + \gamma') - f(\zeta)}{f(\zeta + \gamma) - f(\zeta)} = \frac{\gamma'}{\gamma} \frac{f'(\zeta) + v'}{f'(\zeta) + v},$$

where now v tends to zero with γ and v' with γ'. Then when γ and γ' tend to zero, in any specified ratio, c'/c tends to the same ratio, provided again that $f'(\zeta)$ has a definite value different from zero. If it were zero, the limit of c'/c would of course depend on the limit of the ratio of v'/v, about which we are not yet in a position to say anything. Apart from these special cases, then, we have

$$\lim \frac{c'}{c} = \lim \frac{\gamma'}{\gamma},$$

which is the same as the pair of relations

$$\lim \left|\frac{c'}{c}\right| = \lim \left|\frac{\gamma'}{\gamma}\right|, \quad \lim(\arg c' - \arg c) = \lim(\arg \gamma' - \arg \gamma).$$

That is, if we take a small triangle in the ζ plane its vertices will determine those of a small triangle in the z plane, and if we take a sequence of triangles of the same shape in the ζ plane, the size tending to zero, the corresponding triangles in the z plane will tend to the same shape, always providing $f'(\zeta) \neq 0$. This is what we mean by saying that the transformation is *conformal*. For any two intersecting curves the angle of intersection is unaltered by the transformation, and to pass from one to the other we must travel around the point of intersection in the same sense. In particular since the axes of ξ and η are at right angles, two curves in the z plane on which ξ and η are respectively constant cut at right angles. Corresponding to straight lines parallel to the ξ axis we have a family of curves in the z plane, and corresponding to straight lines parallel to the η axis a family of curves intersecting the first family at right angles. We shall refer to these as the families respectively of η constant and ξ constant (ξ and η respectively of course varying along the same curve).

The transformation is therefore conformal and continuous except possibly at places where $f'(\zeta)$ or $1/f'(\zeta)$ does not exist. We have therefore to consider what happens at singularities and stationary points of $f(\zeta)$. First take a branch point. Since $f(\zeta)$ does not return to its original value when ζ makes a circuit about the point, more than one point in the z plane will correspond to a given point in the ζ plane unless we restrict the region in the ζ plane in such a way as to prevent such circuits. Similarly for stationary points, if m is an integer > 1

$$f(\zeta) = f(\zeta_0) + \alpha(\zeta - \zeta_0)^m + \ldots,$$

ζ will not be uniquely determined as a function of z. For a map to be useful it is clearly necessary that there shall be a one-one correspondence between the place on the ground and that on the map. Simple poles of $f(\zeta)$ do no harm in themselves. But if $f(\zeta)$ has a multiple pole or an isolated essential singularity anywhere ζ, for given z, will take a given value more than once in the neighbourhood. This applies equally if the singularity is at infinity. Apart therefore from unisolated essential singularities, which would be difficult to treat either in general or in particular, the transformation can be one-one over the whole plane only if $f(\zeta)$ is a rational function with at most simple poles and no stationary points, and behaving at infinity like either ζ, a constant, or $1/\zeta$. This limits us to functions of the form

$$f(\zeta) = a + b\zeta + \sum_{r=1}^{n} \frac{a_r}{\zeta - b_r}, \quad f'(\zeta) = b - \sum_{r=1}^{n} \frac{a_r}{(\zeta - b_r)^2}.$$

$f'(\zeta)$ has n double poles in a large circle C. But

$$\frac{1}{2\pi i} \int_C \frac{f''(\zeta)}{f'(\zeta)} d\zeta = \frac{1}{2\pi i} \int_C \left\{ 2 \Sigma \frac{a_r}{(\zeta - b_r)^3} \Big/ \left(b - \Sigma \frac{a_r}{(\zeta - b_r)^2} \right) \right\} d\zeta$$

$$= 0 \quad \text{if} \quad b \neq 0$$

$$= -2 \quad \text{if} \quad b = 0, \; \Sigma a_r \neq 0.$$

Hence $f'(\zeta)$ has $2n$ zeros if $b \neq 0$, $2n - 2$ if $b = 0$ and $\Sigma a_r \neq 0$. If $b = 0$ and $\Sigma a_r = 0$, $f(\zeta) - a$ behaves for large ζ like ζ^{-m}, where $m \geq 2$, and the transformation from z to ζ is not single-valued. Hence for ζ to be a single-valued function of z, either

$$b \neq 0, \; n = 0, \quad f(\zeta) = a + b\zeta.$$

or
$$b = 0, \; n = 1, \quad f(\zeta) = a + \frac{a_1}{\zeta - b_1}.$$

These are therefore the only transformations that need no cuts to give a one-one correspondence over the whole planes and do not involve unisolated essential singularities. The first is trivial, being merely a combination of a displacement and a change of scale. The second is interesting. We shall consider it in a simplified form.

It will be found that successive application of transformations of these forms yields only a transformation of the same form with different constants.

In other cases cuts must be made so that zeros of $f'(\zeta)$ and all singularities of $f(\zeta)$ lie on the boundary or outside it. When they are made the departure from conformality at the exceptional points is found to provide a means of transforming simple boundaries in the ζ plane into an extraordinary variety in the z plane and thereby arriving at solutions of apparently very complicated problems.

We shall speak of any point in the z plane or the ζ plane, such that $f'(\zeta)$ has a limit other than 0 and ∞ when the point is approached, as an ordinary point of the transformation; and any point such that $f'(\zeta) \to 0$, $|f'(\zeta)| \to \infty$, or $f'(\zeta)$ has no definite limit on approaching the point, as a singular point of the transformation. That is, we call a point a singular point of the transformation irrespective of whether z has a singularity when considered as a function of ζ, or ζ has one when considered as a function of z, just as in considering multiple integrals we called a transformation singular if the Jacobian tended to either 0 or ∞.

13·02. Transformations: scale factor. If now $F(\zeta)$ is an analytic function

$$z = f(\zeta), \quad w = F(\zeta) = \phi + i\psi,$$

w is also an analytic function of z. Therefore when ϕ and ψ are expressed in terms of x and y they satisfy Laplace's equation in two dimensions. If one of them is constant over a curve in the ζ plane, it will be constant over the transformed curve in the z plane. Therefore if we can find a solution of Laplace's equation in a region of the ξ, η plane, which is constant over a given boundary in that plane, then the same function expressed in terms of x, y will be a solution of Laplace's equation in the transformed figure and will be constant over the transformed boundary. Consequently the method of conformal representation is capable of solving a great variety of problems; any analytic function will yield a new set.

If $d\sigma$ is an element of area in the ζ plane and dS the corresponding one in the z plane,

$$dS = \frac{\partial(x,y)}{\partial(\xi,\eta)} d\sigma = \left|\frac{dz}{d\zeta}\right|^2 d\sigma,$$

by the Cauchy-Riemann relations: $|dz/d\zeta|$ is called the *modulus* or scale factor of the transformation.

If ζ describes a curve, points near the z curve will be on the left or right of the z curve according as the corresponding points in the ζ plane are to the left or right of the ζ curve, in consequence of the fact that rotations retain the same sense on transformation. Also the normal distances are multiplied by $|dz/d\zeta|$.

If w has a logarithmic singularity at an ordinary point ζ_0 of the ζ plane,

$$w = A \log(\zeta - \zeta_0) + g(\zeta)$$
$$= A \log \frac{z - z_0}{f'(\zeta)} + G(z),$$

where $g(\zeta)$ and $G(z)$ are analytic; hence w in the z plane also has a logarithmic singularity with the same coefficient. If ζ_0 is a singularity of the transformation the coefficient will be different, or the singularity of w may not be logarithmic at all. We must however always arrange for such singularities to be on the boundary or outside the region, and special attention must be paid to them to make sure that the physical conditions correspond.

13·03. Interpretations of complex potential in electrostatics and hydrodynamics.

Hydrodynamics.
$$w = \phi + i\psi,$$
where ϕ is the velocity potential and ψ the stream function. Then
$$\frac{dw}{dz} = \frac{\partial \phi}{\partial x} + i\frac{\partial \psi}{\partial x} = \frac{\partial \phi}{\partial x} - i\frac{\partial \phi}{\partial y} = u - iv,$$
where u, v are the components of velocity, and
$$\left|\frac{dw}{dz}\right| = (u^2 + v^2)^{1/2} = q,$$
the resultant velocity. If we take a curve joining P and Q

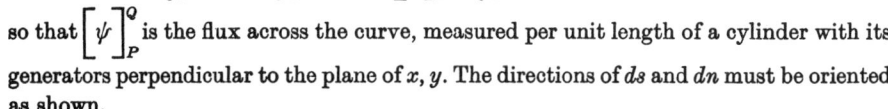

$$\int_P^Q \frac{\partial \phi}{\partial n} ds = \int_P^Q \frac{\partial \psi}{\partial s} ds = \left[\psi\right]_P^Q = \int_P^Q u\, dy - v\, dx,$$
so that $\left[\psi\right]_P^Q$ is the flux across the curve, measured per unit length of a cylinder with its generators perpendicular to the plane of x, y. The directions of ds and dn must be oriented as shown.

A line source at z_0 that emits fluid at a rate $2\pi m$ per unit length has the complex potential
$$m \log(z - z_0).$$

Electrostatics.
$$w = \phi + i\psi,$$
where ϕ is the electrostatic potential and ψ the charge function. Then
$$\frac{dw}{dz} = -(E_x - iE_y), \quad \left|\frac{dw}{dz}\right| = (E_x^2 + E_y^2)^{1/2},$$
E_x, E_y being the components of electric field intensity. If a closed conductor connects P, Q and dn is the normal outward from the conductor, the surface density σ is $-\frac{1}{4\pi}\frac{\partial \phi}{\partial n}$; and the charge on PQ is
$$-\frac{1}{4\pi}\int \frac{\partial \psi}{\partial s} ds = -\frac{1}{4\pi}\left[\psi\right]_P^Q$$
per unit length perpendicular to the plane. Moreover, if $z = f(\zeta)$ and if P', Q' and the curve joining them in the ζ plane correspond to P, Q and PQ, ϕ and ψ are the same for both planes and the charge on $P'Q'$ per unit length perpendicular to the plane is the same as that on PQ, although the surface densities at corresponding points are different. Hence the capacity of a conductor in two dimensions is invariant under conformal transformation.

The sign is most conveniently checked by physical considerations.

A line charge at z_0 of density e per unit length has the complex potential
$$-2e \log(z - z_0).$$

Current electricity

We consider steady currents flowing in a sheet of thickness t and specific conductivity σ; if ϕ is the electrostatic potential and $\boldsymbol{j}$ the current density, $\boldsymbol{E}$ the field,
$$\boldsymbol{j} = \sigma \boldsymbol{E} = -\sigma \operatorname{grad} \phi.$$

13·04   Use of $z = A/\zeta$

The flow of current across an electrode between two points P and Q is

$$-\sigma t \int_P^Q \frac{\partial \phi}{\partial n} ds = -\sigma t \int_P^Q \frac{\partial \psi}{\partial s} ds$$

$$= -\sigma t \left[\psi\right]_P^Q.$$

The complex potential corresponding to an electrode of small circular cross-section, leading in a current J at z_0, is

$$-\frac{J}{2\pi\sigma t} \log(z - z_0).$$

If there is a source of fluid or current, or an electric charge at a point where the transformation is conformal, then there is an equal source or charge at the corresponding point in the transformed plane.

Suppose there is a source on the boundary at a point where it has a corner of angle α. If we consider the source to be entirely inside the boundary all the flow will go into the region. If on the other hand it is a point source actually at the corner only a fraction $\alpha/2\pi$ will go into the region. If at this point the transformation is not conformal and the angle between the corresponding lines is β we must on this second supposition place a source α/β times the original one at the corresponding point in the transformed plane.

We proceed to consider some special transformations.

13·04. $z = A/\zeta$ (A real). This gives

$$z = A\frac{\xi - i\eta}{\xi^2 + \eta^2}, \quad x = \frac{A\xi}{\xi^2 + \eta^2}, \quad y = -\frac{A\eta}{\xi^2 + \eta^2}, \quad \xi = \frac{Ax}{x^2 + y^2}, \quad \eta = -\frac{Ay}{x^2 + y^2}.$$

It therefore resembles an inversion, since $|z||\zeta| = |A|$, a constant. But unlike an inversion, it does not make the directions of (x, y) and (ξ, η) from their respective origins coincident. If ζ describes a circle about the origin, z also describes one, but in the opposite sense, the outside of each becoming the inside of the other. But it still follows that, in general, circles transform into circles, unless they pass through the origin, when they transform into straight lines.

As an example, consider a circular cylinder of radius a lying on its side on a plane. We take the axes as shown. Incompressible fluid is flowing parallel to the axis of x with velocity U at a great distance. The boundary condition is that the stream function ψ is constant over the solid boundary and therefore has the same value over the plane and the circle. For large $|z|$ the complex potential w must approximate to Uz.

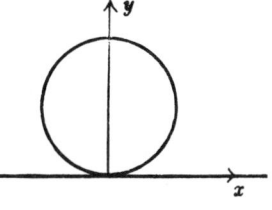

Putting $z = A/\zeta$, we see that the axis of x transforms into the ξ-axis. The circle will transform into a line parallel to the axis of ξ, which is identified by taking $z = 2ia$, which gives $\zeta = -iA/2a$. It is convenient to take A so that this is equal to $\frac{1}{2}i\pi$, and therefore

$$A = -\pi a.$$

Then

$$z = -\frac{\pi a}{\zeta}.$$

For z large w approximates to $Uz = -\pi aU/\zeta$, which represents a doublet source at the origin. We must add extra terms to make w purely real on the boundaries $\eta = 0$ and $\eta = \frac{1}{2}i\pi$. For $\eta = 0$ this is already satisfied, but for $\eta = \frac{1}{2}i\pi$ it is not, but can be adjusted by adding a term $-\pi aU/(\zeta - i\pi)$. This is analytic when $0 \leq \eta \leq \frac{1}{2}\pi$, and therefore admissible, but it disturbs the condition over $\eta = 0$ and a further term $-\pi aU/(\zeta + i\pi)$ is needed. Proceeding in this way, and using 12·06, we have

$$w = -\pi aU\left(\frac{1}{\zeta} + \frac{1}{\zeta - i\pi} + \frac{1}{\zeta + i\pi} + \frac{1}{\zeta - 2i\pi} + \frac{1}{\zeta + 2i\pi} + \ldots\right)$$

$$= -\pi aU \coth \zeta$$

$$= \pi aU \coth \frac{\pi a}{z}.$$

To verify that this satisfies the conditions, notice first that it is real for $z = x$ and approximates to Uz for $|z|$ large, as it should. On the circle the radius vector is $2a\sin\theta$, and

$$z = 2a\sin\theta\, e^{i\theta},$$

$$\zeta = -\tfrac{1}{2}\pi \operatorname{cosec}\theta\, e^{-i\theta} = -\tfrac{1}{2}\pi(\cot\theta - i),$$

$$w = -\pi aU \coth(\tfrac{1}{2}\pi i - \tfrac{1}{2}\pi\cot\theta)$$

$$= \pi aU \tanh(\tfrac{1}{2}\pi\cot\theta),$$

and this is real.

We are interested mainly in the velocity q and the pressure. We have

$$q = \left|\frac{dw}{dz}\right| = \left|\frac{U\pi^2 a^2}{z^2 \sinh^2 \pi a/z}\right|.$$

When $z \to 0$ through real values $q \to 0$; thus the velocity vanishes on the line of contact. At the top point $z = 2ia$, $\zeta = \frac{1}{2}i\pi$,

$$q = |\tfrac{1}{4}\pi^2 U \operatorname{cosec}^2 \tfrac{1}{2}\pi| = \tfrac{1}{4}\pi^2 U.$$

The velocity at the top of the cylinder is therefore about 2·5 times that at a large distance. The pressure at the top is correspondingly lower. This implies that a stream tends to lift objects lying on the bottom and is connected with the theory of the transport of sediments.*

13·05. $z = \zeta^n$ (n real, $n \neq 1$). The transformation is not conformal at the origin. Since $|z| = |\zeta|^n$, $\arg z = n \arg \zeta$, lines through the origin in one figure correspond to lines through the origin in the other, and circles about the origin to circles about the origin. This transformation is therefore useful for problems relating to *plane boundaries meeting in a line*. If the ζ figure is a pair of lines meeting at π, and therefore including the upper half of the ζ plane, the lines in the z plane will therefore meet at $n\pi$. n cannot be greater than 2 for this ζ figure because $\arg z$ would then exceed 2π for some values of ζ and the representation would not be unique. The transformation covers planes meeting at angles $\leq 2\pi$.

* For the resultant force on the cylinder, cf. *Proc. Camb. Phil. Soc.* 25, 1929, 272–6.

Consider a line source of fluid at $z = z_0$, between planes meeting at $n\pi$. The complex potential due to the source is

$$w_0 = -\frac{m}{2\pi}\log(z-z_0),$$

and that in the ζ figure is accordingly $-\frac{m}{2\pi}\log(\zeta-\zeta_0)$, since z_0 is not a singularity of the transformation. The stream function is

$$\psi_0 = -\frac{m}{2\pi}\tan^{-1}\frac{\eta-\eta_0}{\xi-\xi_0}.$$

We require the complete stream function to be constant when $\eta = 0$. ψ_0 does not satisfy this condition, but if we add on the result of changing η_0 to $-\eta_0$ it will. Hence

$$w = -\frac{m}{2\pi}\log(\zeta-\xi_0-i\eta_0)(\zeta-\xi_0+i\eta_0)$$

satisfies all the conditions. Therefore

$$w = -\frac{m}{2\pi}\log(z^{1/n}-z_0^{1/n})(z^{1/n}-z_0^{*1/n})$$

is the solution, where $z_0^* = x_0 - iy_0$.

When n is the reciprocal of an integer this result can be found by the method of images. Otherwise the latter method fails because it leads to images, and therefore new singularities, within the region of flow.

The conjugate velocity vector is

$$u - iv = \frac{dw}{dz} = \frac{dw}{d\zeta}\bigg/\frac{dz}{d\zeta}.$$

$dw/d\zeta$ is analytic at $z = 0$; but $dz/d\zeta$ behaves like $\zeta^{n-1} = z^{1-1/n}$, which tends to 0 if $n > 1$ and to ∞ if $n < 1$. Hence the velocity will tend to 0 or infinity at the corner according as $n < 1$ or $n > 1$, that is, according as the angle occupied by the fluid is less or greater than π. In the latter case the classical theory fails to give a good approximation to the motion of a real fluid; it breaks down near a projecting corner for reasons connected with viscosity.[a]

13·06. $z = ae^\zeta$ (a real). Since $\zeta = \log(z/a)$, ζ is a many-valued function of z and $z = 0$ is a branch point. We have

$$x = ae^\xi\cos\eta, \quad y = ae^\xi\sin\eta$$

and therefore every pair of values of x, y except $(0, 0)$ is obtained by letting η vary over a range 2π and ξ from $-\infty$ to $+\infty$. The whole x, y plane is therefore mapped on *an infinite strip* between $\eta = 0$ and $\eta = 2\pi$, or if more convenient for the particular problem, $-\pi < \eta \leqslant \pi$. Half the plane is mapped on a strip of width π. The circle $|z| = a$ corresponds to the line $\xi = 0$, $0 \leqslant \eta < 2\pi$; the interior to the *semi-infinite strip* to the left of the η axis and the exterior to the right of the η axis. In the z plane the curves of constant ξ are circles with the origin as centre, and the curves of constant η are straight lines through the origin.

13·07. Coaxal circles. $\zeta = c\log\dfrac{z-a}{z+a}$. For simplicity we take a and c real. Then

$$\xi + i\eta = c\left[\log\frac{r_1}{r_2} + i(\theta_1 - \theta_2)\right].$$

Here $\theta_1 - \theta_2$ is not unique and to make ξ, η one-valued we must prevent it from varying by more than 2π. This can be done either by making cuts along the real axis from $-\infty$ to $-a$, and from $+a$ to $+\infty$, or by making one from $-a$ to $+a$. We take the former case. Then θ_1 and θ_2 satisfy

$$0 \leq \theta_1 < 2\pi, \quad -\pi < \theta_2 \leq \pi.$$

The family of curves η = constant are a set of coaxal circles through $A(+a)$ and $B(-a)$, while the family ξ = constant is the orthogonal set of coaxal circles with A and B as the limit points.

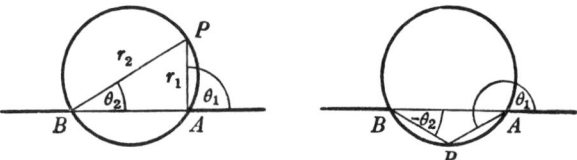

For the part of the circle C above the real axis

$$\theta_1 - \theta_2 = APB = \alpha,$$

the angle in the segment APB. For the lower part

$$(2\pi - \theta_1) + \theta_2 = \pi - \alpha,$$

that is,

$$\theta_1 - \theta_2 = \pi + \alpha.$$

Thus $\theta_1 - \theta_2$ changes discontinuously by $\pm \pi$ when P passes through A or B. For $0 < \alpha < \pi$ it is always equal to the angle APB on the side that faces *downwards*.

As we take all such circles with α from 0 to π, η goes from 0 to $2c\pi$. Straight lines $\eta = c\alpha$ with $0 < \alpha < \pi$ correspond to arcs of circles in the upper half of the z plane, while those with $\pi < \alpha < 2\pi$ correspond to arcs in the lower half of the z plane. $\eta = c\pi$ gives the straight line BA. $\eta \to 0$ for points approaching the x axis outside BA from above, $\eta \to 2c\pi$ for points approaching it from below.

Now take the orthogonal system
$$\log \frac{r_1}{r_2} = \frac{\xi}{c}$$

For any point P take the circle APB and draw the tangent to it, cutting the axis of x in K. The triangles KPA, KBP are similar, and

$$\frac{r_1}{r_2} = \frac{AP}{BP} = \frac{PK}{BK} = \frac{AK}{PK}.$$

Then
$$\frac{AK}{BK} = \left(\frac{r_1}{r_2}\right)^2 = e^{2\xi/c}.$$

Therefore for given ξ, K is a fixed point. Also

$$KP^2 = KA \cdot KB,$$

and therefore is constant. Hence the points P with given ξ lie on a circle. If r_1/r_2 is small the circle is a small one enclosing A; if it is large we get a small circle enclosing B. Thus the curves of ξ constant are the coaxal circles with limiting points A, B orthogonal to the intersecting circles of constant η.

13·071 Coaxal cylinders

Analytically,
$$\frac{r_1^2}{r_2^2} = \frac{(x-a)^2 + y^2}{(x+a)^2 + y^2} = e^{2\xi/c},$$

$$\frac{x^2 + y^2 + a^2}{2ax} = -\frac{e^{2\xi/c} + 1}{e^{2\xi/c} - 1} = -\coth\frac{\xi}{c},$$

$$x^2 + y^2 + 2ax \coth\frac{\xi}{c} + a^2 = 0.$$

Thus there is a one-one correspondence between the whole of the z plane and the infinite strip of width $2c\pi$ parallel to the ξ axis on the ζ plane.

This transformation is extremely useful in solving problems on coaxal cylinders. For example:

13·071. *A long hollow metal cylinder, whose equation is $r = a$, is divided into two parts by the plane $\sin\theta = 0$. The parts are slightly separated and charged to potentials V_1, V_2, and two thin sheets of metal at potential V_3 occupy the regions $r > a$, $\theta = 0$ and $r > a$, $\theta = \pi$. By means of the transformation $\zeta = \log(z-a) - \log(z+a)$, or otherwise, show that the surface density at a point on the inner surface of the cylinder is*

$$\frac{V_1 - V_2}{4\pi^2 a \sin\theta},$$

and find the surface density on the outside of the cylinder and on both sides of the planes.

By means of the transformation the upper side of $-\infty B$ and $A\infty$ becomes $\eta = 0$, and the lower sides become $\eta = 2\pi$. BPA becomes $\eta = \tfrac{1}{2}\pi$ and BQA becomes $\eta = \tfrac{3}{2}\pi$.

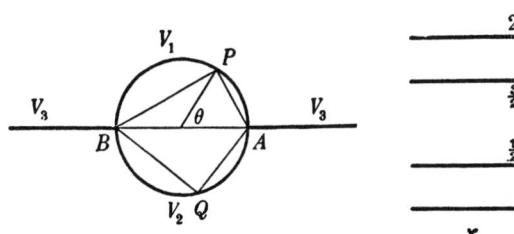

The problem is thus reduced to one of parallel plate condensers. Considering the inner condenser between $\eta = \tfrac{1}{2}\pi$ and $\tfrac{3}{2}\pi$, we have the surface densities in the ζ plane as

$$\pm (V_1 - V_2)/4\pi.\pi.$$

Hence that at a point on the inner surface of the cylinder in the z plane is

$$\pm \frac{V_1 - V_2}{4\pi^2}\left|\frac{d\zeta}{dz}\right|.$$

But
$$\frac{d\zeta}{dz} = \frac{1}{z-a} - \frac{1}{z+a} = \frac{2a}{(z-a)(z+a)},$$

$$\left|\frac{d\zeta}{dz}\right| = \frac{2a}{r_1 r_2} = \frac{2a}{4a^2 \sin\tfrac{1}{2}\theta \cos\tfrac{1}{2}\theta} = \frac{1}{a\sin\theta}.$$

Hence the surface density is $\dfrac{V_1-V_2}{4\pi^2 a \sin\theta}$.

Similarly, that on the outside of the cylinder is

$$\dfrac{V_1-V_3}{2\pi^2 a \sin\theta}, \quad \dfrac{V_2-V_3}{2\pi^2 a \sin\theta};$$

on the upper side of the planes it is $-\dfrac{V_1-V_3}{\pi^2}\dfrac{a}{x^2-a^2}$,

and on the lower side $-\dfrac{V_2-V_3}{\pi^2}\dfrac{a}{x^2-a^2}$.

13·072. Capacity of a condenser formed by two non-concentric circles. Let two circles of the system, with $c=1$, have radii α and β ($\alpha>\beta$) and let the distance between their centres be d. They can be written

$$(x+a\coth\xi)^2+y^2 = a^2\operatorname{cosech}^2\xi,$$

and therefore
$$\alpha = a\operatorname{cosech}\xi_1, \quad \beta = a\operatorname{cosech}\xi_2,$$

$$d = a(\coth\xi_1 - \coth\xi_2)$$

$$= a\dfrac{\sinh(\xi_2-\xi_1)}{\sinh\xi_1 \sinh\xi_2}.$$

Thus
$$\sinh(\xi_2-\xi_1) = \dfrac{da}{\alpha\beta}.$$

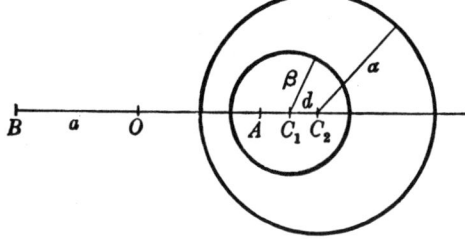
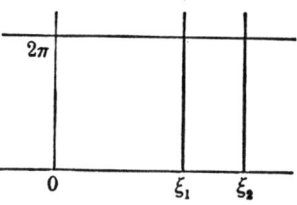

If $OC_1 = X$, $OC_2 = X+d$, then since a is the length of the tangent from O to all the circles

$$a^2 = X^2 - \beta^2 = (X+d)^2 - \alpha^2.$$

Hence
$$2dX = \alpha^2 - \beta^2 - d^2,$$

$$a^2 = \dfrac{(\alpha^2-\beta^2-d^2)^2}{4d^2} - \beta^2,$$

$$\cosh^2(\xi_2-\xi_1) = \dfrac{a^2 d^2}{\alpha^2\beta^2} + 1$$

$$= \dfrac{(\alpha^2-\beta^2-d^2)^2 - 4d^2\beta^2 + 4\alpha^2\beta^2}{4\alpha^2\beta^2},$$

$$\cosh(\xi_2-\xi_1) = \dfrac{\alpha^2+\beta^2-d^2}{2\alpha\beta}.$$

If now we have a condenser formed by two conducting cylinders in the z plane it becomes a parallel plate condenser in the ζ plane, the separation being $\xi_2 - \xi_1$. The capacity per unit length is

$$\frac{2\pi}{4\pi|\xi_2 - \xi_1|} = \frac{1}{2\cosh^{-1}\{(\alpha^2 + \beta^2 - d^2)/2\alpha\beta\}},$$

and this therefore gives the capacity in the z plane.

13·073. Another form of the transformation is

$$\frac{z-a}{z+a} = e^{\zeta/c},$$

that is,
$$\frac{z}{a} = \frac{e^{\zeta/c}+1}{1-e^{\zeta/c}} = -\coth\frac{\zeta}{2c}.$$

The restriction of a and c to be real is unnecessary; a more general form is

$$\zeta = c\log\frac{z-z_1}{z-z_2}.$$

Taking c as a pure imaginary is equivalent to turning the axes in the ζ plane through a right angle.

The transformation is related to some of the most fundamental problems in two dimensions. Thus, in electrostatics, if there are charges λ per unit length along two parallel wires the complex potential is

$$w = -\lambda\log\frac{z-a}{z+a},$$

and it follows at once that the equipotentials are coaxal circular cylinders with the wires as limiting lines, and the lines of force are the orthogonal set of circles. Similarly, in hydrodynamics, for parallel lines of sources and sinks of equal and opposite strength, the coaxal cylinders about the lines give the surfaces of equal ϕ, the orthogonal surfaces the stream lines. Analogous relations for the flow due to a pair of line vortices, and the magnetic field due to a pair of electric currents, will suggest themselves.

13·08. Confocal conics. $z = c\cosh\zeta$ (c real).

$$x+iy = c(\cosh\xi\cos\eta + i\sinh\xi\sin\eta),$$
$$x = c\cosh\xi\cos\eta, \quad y = c\sinh\xi\sin\eta.$$

Then the curves of ξ constant are

$$\frac{x^2}{c^2\cosh^2\xi} + \frac{y^2}{c^2\sinh^2\xi} = 1.$$

These are ellipses with foci at $(\pm c, 0)$, axes $c\cosh\xi$, $c\sinh\xi$. η is then the eccentric angle.

If $\eta = $ constant

$$\frac{x^2}{c^2\cos^2\eta} - \frac{y^2}{c^2\sin^2\eta} = 1.$$

These are hyperbolas with the same foci.

It must be noticed, however, that a particular value of η does not give the whole hyperbola. If $0 \leq \eta < \tfrac{1}{2}\pi$ and ξ ranges from $-\infty$ to ∞, we get the right branch; if $\tfrac{1}{2}\pi < \eta \leq \pi$ we get the left. On the other hand $+\xi$ and $-\xi$, where $0 \leq \eta \leq 2\pi$, give the same ellipse.

Problems relating to elliptic boundaries are much more frequent than those for hyperbolic boundaries. We therefore make the restriction $\xi \geq 0$, and then get the whole ellipse by taking $0 \leqslant \eta < 2\pi$. But now with ξ positive we do not even get the whole of one branch of the hyperbola for a particular value of η; in fact if η is an acute angle the parts of the hyperbola in the four quadrants are given by η, $\pi - \eta$, $\pi + \eta$, $2\pi - \eta$.

With the restrictions made, the semi-infinite strip of width 2π in the ζ plane transforms into the whole of the z plane. The function $\cosh^{-1}(z/c)$ is many-valued, but we have made it single-valued by making a cut in the z plane from $(-c, 0)$ to $(\infty, 0)$. The upper half of the z plane transforms into the semi-infinite strip $\xi > 0$, $0 \leqslant \eta \leqslant \pi$, and the lower half into $\xi > 0$, $\pi \leqslant \eta < 2\pi$. With $\xi = 0$, z travels from $(c, 0)$ to $(-c, 0)$ and back as η increases from 0 to 2π. The transformation is not conformal at $z = \pm c$, and the angle π between adjoining parts of the x axis from these points transforms into an angle $\tfrac{1}{2}\pi$ at the corresponding points in the ζ plane.

Suppose now that we have to solve $\nabla^2 \phi = 0$ in a region bounded by an ellipse of the system. The solution must have period 2π in order that ϕ may be single-valued; for if not we should be in the situation of having a waistcoat whose buttons did not come opposite to the buttonholes when it was put on. This condition is satisfied by any linear combination of $e^{n\zeta}$ and $e^{-n\zeta}$ for integral n. If we are concerned with a region containing the line of foci a further condition is needed. The cut in this case is not a physical barrier but a mathematical device, and must introduce no discontinuities in the physical problem. Not only ϕ, therefore, but $\partial\phi/\partial x$ and $\partial\phi/\partial y$ must be continuous on crossing the line of foci and thereby changing η into $2\pi - \eta$. But the normal to the line of foci, on either side, is the direction of increasing ξ, which is that of increasing y on the upper side and decreasing y on the lower. In fact

$$\frac{\partial \phi}{\partial \xi} = \frac{\partial \phi}{\partial x}\frac{\partial x}{\partial \xi} + \frac{\partial \phi}{\partial y}\frac{\partial y}{\partial \eta} = c\left(\sinh\xi\cos\eta\frac{\partial\phi}{\partial x} + \cosh\xi\sin\eta\frac{\partial\phi}{\partial y}\right) \to c\sin\eta\frac{\partial\phi}{\partial y},$$

$$\frac{\partial \phi}{\partial \eta} = \frac{\partial \phi}{\partial x}\frac{\partial x}{\partial \eta} + \frac{\partial \phi}{\partial y}\frac{\partial y}{\partial \eta} = c\left(-\cosh\xi\sin\eta\frac{\partial\phi}{\partial x} + \sinh\xi\cos\eta\frac{\partial\phi}{\partial y}\right) \to -c\sin\eta\frac{\partial\phi}{\partial x},$$

and if $\partial\phi/\partial x$, $\partial\phi/\partial y$ are to have the same values for $\xi = 0$ on replacing η by $2\pi - \eta$, $\partial\phi/\partial\xi$ and $\partial\phi/\partial\eta$ must either be 0 or reverse their signs. The admissible solutions are therefore $\cosh n\xi \cos n\eta$, for which $\partial\phi/\partial\xi = 0$ at $\xi = 0$ and $\partial\phi/\partial\eta$ reverses its sign, and $\sinh n\xi \sin n\eta$, for which $\partial\phi/\partial\eta = 0$ and $\partial\phi/\partial\xi$ reverses its sign. The solutions $\cosh n\xi \sin n\eta$ and $\sinh n\xi \cos n\eta$ are not admissible for a complete ellipse. They would enter if the line of foci was occupied by a barrier, for then $\partial\phi/\partial x$, $\partial\phi/\partial y$ need not be continuous across it.

For external problems the data will include information about the behaviour of ϕ for large ξ. The disturbance due to the presence of the elliptic boundary must then tend to zero for large ξ, and the admissible solutions are of the forms $e^{-n\xi}(\cos n\eta, \sin n\eta)$.

For the scale of the transformation we have

$$\frac{dz}{d\zeta} = c\sinh\zeta,$$

$$\left|\frac{dz}{d\zeta}\right|^2 = c^2(\sinh^2\xi\cos^2\eta + \cosh^2\xi\sin^2\eta)$$
$$= c^2(\cosh^2\xi - \cos^2\eta) = \tfrac{1}{2}c^2(\cosh 2\xi - \cos 2\eta).$$

Also
$$|z^2| = c^2(\cosh^2\xi\cos^2\eta + \sinh^2\xi\sin^2\eta)$$
$$= c^2(\cosh^2\xi - \sin^2\eta) = \tfrac{1}{2}c^2(\cosh 2\xi + \cos 2\eta).$$

13·081. Conducting elliptic cylinder in a uniform field of force parallel to the major axis.
The potential of the undisturbed field is

$$-Fx = -Fc\cosh\xi\cos\eta + \text{constant}.$$

Suppose the conductor to be given by $\xi = \alpha$. Then the extra term due to the presence of the conductor must be of the form $Ae^{-\xi}\cos\eta$, for this tends to 0 as $\xi\to\infty$, and it must cancel the variation of Fx over the surface $\xi = \alpha$. Thus

$$Ae^{-\alpha} - Fc\cosh\alpha = 0.$$

Hence
$$\phi = \text{constant} - Fc\cosh\xi\cos\eta + Fce^{\alpha-\xi}\cosh\alpha\cos\eta$$
$$= \text{constant} - Fce^{\alpha}\sinh(\xi-\alpha)\cos\eta.$$

The corresponding ψ is
$$\psi = \text{constant} - Fce^{\alpha}\cosh(\xi-\alpha)\sin\eta.$$

The charge induced on the part of the cylinder on the positive side of the y axis is

$$-\frac{1}{4\pi}[\psi_{\eta=\frac{1}{2}\pi} - \psi_{\eta=\frac{3}{2}\pi}] = \frac{1}{2\pi}Fce^{\alpha} = \frac{1}{2\pi}F(a+b).$$

13·082. Elliptic cylinder of dielectric constant K in a uniform field parallel to the major axis.
Let ϕ_1 be the potential inside the cylinder and ϕ_0 that outside. The only admissible forms are, apart from a constant,

$$\phi_0 = -Fc\cosh\xi\cos\eta + Ae^{-\xi}\cos\eta, \quad \phi_1 = B\cosh\xi\cos\eta.$$

The boundary conditions are $\phi_0 = \phi_1$, $K\dfrac{\partial\phi_1}{\partial\xi} = \dfrac{\partial\phi_0}{\partial\xi}$ at $\xi = \alpha$, since $\partial\xi/\partial n$ is continuous.
Hence
$$Ae^{-\alpha} - Fc\cosh\alpha = B\cosh\alpha, \quad -Ae^{-\alpha} - Fc\sinh\alpha = KB\sinh\alpha.$$

Then $\quad A = Fc\dfrac{K-1}{\cosh\alpha + K\sinh\alpha}\cosh\alpha\sinh\alpha\, e^{\alpha}, \quad B = -Fc\dfrac{e^{\alpha}}{\cosh\alpha + K\sinh\alpha}.$

13·083. Rotating elliptic cylinder filled with fluid.
Let ω be the angular velocity. On the boundary the velocity components of the cylinder are $(-\omega y, \omega x)$. But if ψ is the stream function

$$\frac{\partial\psi}{\partial y} = u, \quad \frac{\partial\psi}{\partial x} = -v,$$

and therefore on the boundary

$$\psi = \int(u\,dy - v\,dx) = -\tfrac{1}{2}\omega|z|^2$$
$$= -\tfrac{1}{4}\omega c^2(\cosh 2\alpha + \cos 2\eta).$$

The first term is an irrelevant constant, and we can take

$$\psi = -\tfrac{1}{4}\omega c^2\frac{\cosh 2\xi}{\cosh 2\alpha}\cos 2\eta, \quad \phi = \tfrac{1}{4}\omega c^2\frac{\sinh 2\xi}{\cosh 2\alpha}\sin 2\eta = \frac{\omega}{\cosh 2\alpha}xy.$$

Thus at points in the interior $\quad (u,v) = \dfrac{\omega}{\cosh 2\alpha}(y,x).$

We have also, since $\quad \tanh\alpha = \dfrac{b}{a}, \quad \cosh 2\alpha = \dfrac{a^2+b^2}{a^2-b^2}.$

Thus, as we should expect, there is no motion inside if the cylinder is circular. In case there is any doubt of the truth of the result, one may take a teacup with a floating leaf in it on the side next one's mouth. On turning the cup round the leaf is found to be still on the side next the mouth. The result is of course true only for fluids of small viscosity.

13·09. Generalized Joukowsky transformations. Suppose that we are given the equation of a closed curve in the z' plane in the form $f(x', y') = 0$. Then we cannot at once give an answer to the question: what is the transformation that will transform the closed curve to a circle $|z| = a$ and the region outside the curve to the outside of the circle? We have rather to examine various transformations and see what curve in the z' plane corresponds to $|z| = a$. It has, however, been found that many of the transformations important in mathematical physics belong to a very general class, and we first consider this generally before proceeding to special cases. Let

$$z' = z \sum_{r=0}^{\infty} \frac{a_r}{z^r}, \tag{1}$$

where the a_r may be complex and $a_0 \neq 0$. In practice we usually take $a_0 = 1$. Suppose further that the series converges for $|z| \geq a$ and that dz'/dz has no zeros outside the circle C defined by $|z| = a$. Then the transformation is conformal for all z outside the circle. When z travels round the circle C, z' will describe a closed curve C'. If dz'/dz has a simple zero on C, the curve C' will have a cusp at the corresponding point. Further, a large circle in the z plane with centre at the origin corresponds to a large closed curve, approximately circular, in the z' plane and conversely. If we proceed inwards from a point on the large circle in the z plane along a straight line to the origin, the curve in the z' plane corresponding to the part of the line outside C will have a unique tangent at each point. z is therefore also a single-valued function of z' and will be expressible by a power series of the form

$$z = z' \sum_{r=0}^{\infty} \frac{b_r}{z'^r}$$

if z' is sufficiently large.

With the further transformation $z = ae^{\zeta}$ we have that the part of the z' plane outside C' is represented on the semi-infinite strip $\xi \geq 0$, $0 \leq \eta < 2\pi$ of the ζ plane. Hence if a closed curve in the z' plane can be represented parametrically by

$$x' + iy' = ae^{i\eta} \sum_{r=0}^{\infty} \frac{a_r}{a^r} e^{-ri\eta} \quad (a \text{ real}, a_r \text{ complex}, a_0 \neq 0), \tag{2}$$

we can at once infer that it is the curve corresponding to $\xi = 0$ of the family

$$z' = ae^{\zeta} \sum_{r=0}^{\infty} \frac{a_r}{a^r} e^{-r\zeta} = z \sum_{r=0}^{\infty} \frac{a_r}{z^r}. \tag{3}$$

Particular cases

13·091. Ellipse. Take

$$z' = \frac{1}{2}\left\{\frac{a+b}{a}z + \frac{a(a-b)}{z}\right\}. \tag{4}$$

With $z = ae^{i\eta}$ (the circle C) we have

$$z' = x' + iy' = \tfrac{1}{2}(a+b)e^{i\eta} + \tfrac{1}{2}(a-b)e^{-i\eta}$$
$$= a\cos\eta + ib\sin\eta, \tag{5}$$

and the curve C' is an ellipse with semi-axes a, b, and η is the eccentric angle. This transformation is also often taken in the form

$$z' = z + \frac{a^2}{z}, \tag{6}$$

so as not to alter the scale at infinity; then to the circle $|z| = a$ correspond the two sides of a line from $(2a, 0)$ to $(-2a, 0)$ and back[a]. This can also be written

$$\frac{z'+2a}{z'-2a} = \left(\frac{z+a}{z-a}\right)^2. \tag{7}$$

13·092. Joukowsky aerofoils. Take

$$\frac{dz'}{dz} = \prod_{r=1}^{n}\left(1-\frac{z_r}{z}\right), \quad \Sigma z_r = 0, \quad |z_1| = a, \quad |z_r|_{r \neq 1} < a. \tag{8}$$

Then C' has a cusp at z_1'; for near it we have

$$\frac{dz'}{dz} = g(z)(z-z_1), \tag{9}$$

where $g(z)$ is analytic and not zero at $z = z_1$, and therefore

$$z' - z_1' = \tfrac{1}{2}g(z_1)(z-z_1)^2 + \ldots. \tag{10}$$

Hence as z travels along the circle and passes through z_1, z' approaches z_1' and then recedes along a curve with the same tangent.

If $n = 2$, $z_1 = a$, $z_2 = -a$, $\quad \dfrac{dz'}{dz} = 1 - \dfrac{a^2}{z^2}, \quad z' = z + \dfrac{a^2}{z},$

and we recover (6). But if now instead of the circle C we take a slightly larger circle passing through $+a$ but a little beyond $-a$ and transform it, we shall get an Indian club-shaped figure with a cusp at $z' = 2a$ and a rounded end at z' a little less than $-2a$. If further we take the centre of this circle a little off the axis of x the z' figure will not be symmetrical about the x' axis. Thus two terms are enough to give a fair representation of the form of an actual aeroplane wing. The most serious departure from actuality is that all Joukowsky aerofoils have a cusp at the trailing edge, whereas the actual angle is not a cusp. This can be remedied by a further modification due to Glauert, as follows.

13·093. Region bounded by circular arcs. We take now

$$\frac{z' - (2-n)a\cos\beta}{z' + (2-n)a\cos\beta} = \left(\frac{z - ae^{-i\beta}}{z + ae^{i\beta}}\right)^{2-n},$$

where n and β are small and positive. This is evidently the result of applying two transformations of the form 13·07 with different values of c, but now we are interested in the external region. We write

$$\arg(z - ae^{-i\beta}) = \theta_1 \qquad \arg(z' - (2-n)a\cos\beta) = \theta_1'$$
$$\arg(z + ae^{i\beta}) = \theta_2 \qquad \arg(z' + (2-n)a\cos\beta) = \theta_2'$$

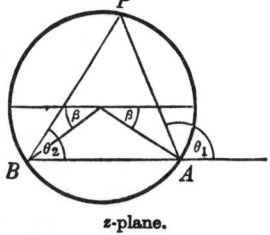

z-plane.

where $\theta_1, \theta_2, \theta_1', \theta_2'$ are defined to be zero when z is on BA produced, and vary continuously as z travels on a curve outside the circle.

Then
$$\theta_1' - \theta_2' = (2-n)(\theta_1 - \theta_2),$$

for points z outside C. At P, $\theta_1 - \theta_2 = \tfrac{1}{2}\pi - \beta$. If P transforms to P', P' is on a circular arc through $\pm(2-n)a\cos\beta$, with angle $(2-n)(\tfrac{1}{2}\pi - \beta)$ at the circumference. But if P proceeds to near B and then travels round a small semicircle about B, θ_2 increases by π and therefore $\theta_1' - \theta_2'$ becomes

$$(2-n)(\tfrac{1}{2}\pi - \beta) - (2-n)\pi = -(2-n)(\tfrac{1}{2}\pi + \beta).$$

This is negative; we add 2π to give a positive angle, which will be the angle facing downwards as in 13·07. Then the lower arc gives an arc in the z' plane containing an angle

$$2\pi - (2-n)(\tfrac{1}{2}\pi + \beta) = (2+n)\tfrac{1}{2}\pi - (2-n)\beta.$$

If this is less than π the lower arc in the z' plane will be concave *downwards*. The z' figure consists of two circular arcs intersecting at an angle $n\pi$.

If instead of C we take a circle through A but passing a little beyond B we get a rounded leading edge. This is Glauert's transformation. When z is large we find the first few terms of the series development to be

$$z' = z + ia\sin\beta + \frac{(1-n)(3-n)}{3}\cos^2\beta \cdot \frac{a^2}{z} + \ldots.$$

13·094. Closed polygonal boundary. Consider

$$\frac{dz'}{dz} = A\prod_{r=1}^{n}\left(1 - \frac{z_r}{z}\right)^{\alpha_r/\pi}, \tag{1}$$

where A is constant, and $|z_r| = a$ for all r. Take

$$z = ae^{i\eta}, \quad z_r = ae^{i\eta_r}. \tag{2}$$

Then
$$\frac{dz'}{d\eta} = Be^{i\eta(1 - 1/2\Sigma\alpha_r/\pi)}\prod_{r=1}^{n}\{\sin\tfrac{1}{2}(\eta - \eta_r)\}^{\alpha_r/\pi}, \tag{3}$$

where B is another constant. Therefore if $\Sigma\alpha_r = 2\pi$, $\arg(dz'/d\eta)$ is constant. Thus as z describes the arc between z_r and z_{r+1}, z' proceeds along a straight line. On passing half way round z_r, however, $\eta - \eta_r$ changes sign and $\arg(dz'/d\eta)$ changes by $\pm\alpha_r$. The curve C' is therefore a polygon with external angles $\pm\alpha_r$.

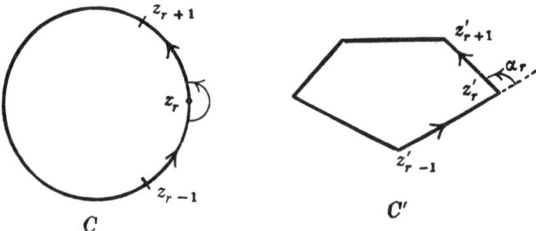

We take first the external problem. Then z can travel positively about z_r, and $\arg(z - z_r)$ *increases* by π; thus the curves are oriented as shown.

The condition $\Sigma\alpha_r = 2\pi$ is satisfied by the external angles of a closed polygon. If (1) is developed in powers of $1/z$, it is seen to contain a term in $\Sigma\alpha_r z_r/z$, and therefore z' will not be single-valued unless we have also $\Sigma\alpha_r z_r = 0$. Subject therefore to this condition and $\Sigma\alpha_r = 2\pi$ the outside of a circle C is transformed into the outside of a polygon.

The same transformation will *not* do for the *inside* of a polygon. For if we took the inside of the circle, then we should have to pass round z_r in the negative direction, and arg $(dz'/d\eta)$ would decrease by α_r. Thus we should not get the inside of the same polygon, but the outside of its mirror image.

On the other hand, (1) determines z' as a function of z, which could be continued into the interior of C by paths between any pair z_r, z_{r+1}. But then we should have to make a cut along the whole of C except between these two in order to make z' single-valued; then this cut would have to be traversed internally and we get the same result as in the last paragraph.

13·095. But consider the transformation

$$z' = \sum_{r=1}^{\infty} b_r z^r. \tag{4}$$

(There is no loss of generality in dropping a constant b_0 on the right.) Suppose that the series converges for $|z| \leq a$ and that all zeros of dz'/dz lie outside or on this circle. As before, C corresponds to a closed curve C' in the z' plane, but now a small circle about

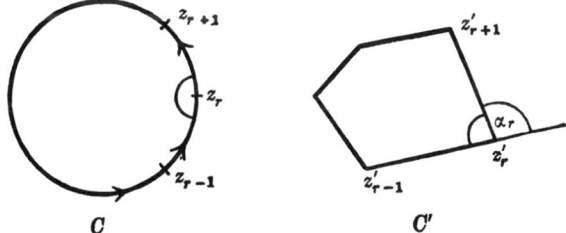

$z = 0$ corresponds to a small curve approximating to a circle. Thus the interior of C is mapped on to the interior of C'. The transformation of this type for a polygon is known and is of considerable interest. Take

$$\frac{dz'}{dz} = A \prod_{r=1}^{n} \left(1 - \frac{z}{z_r}\right)^{-\alpha_r/\pi}, \tag{5}$$

where A is constant and $|z_r| = a$. As before we get

$$\frac{dz'}{d\eta} = iae^{i\eta} A e^{-\frac{1}{2}i\Sigma\alpha_r(\eta-\eta_r)/\pi} \prod \{-2i \sin \tfrac{1}{2}(\eta-\eta_r)\}^{-\alpha_r/\pi}, \tag{6}$$

and arg $(dz'/d\eta)$ is constant for each arc provided $\Sigma \alpha_r = 2\pi$. But now as z describes a semicircle about z_r on the *inside* arg $(z-z_r)$ *decreases* by π and arg $(dz'/d\eta)$ increases by α_r. Thus the α_r are again the external angles, but the interior of C corresponds to the interior of C'.

13·096. These transformations are due to various writers.*

The transformation of the outside of a circle into the outside of a general closed curve can be seen to be unique. For we can imagine the curve C' to be occupied by a conductor carrying a given charge per unit length. Then the external field is determinate and with a

---

* W. G. Bickley, *Phil. Trans.* A, 228, 1929, 235–74; R. M. Morris, *Proc. Camb. Phil. Soc.* 33, 1937, 474–84; *Math. Ann.* 116, 1939, 374–400.

suitable adjustment of the charge we can arrange for the potential to behave like $\log r'$ and the charge function like θ' for large r. Then

$$\phi + i\psi \doteq \log z' \quad (r' \text{ large}).$$

Now if $z' = f(z)$ transforms a circle C into the curve C', $\log z$ is a function of z' whose real part is constant over C' and therefore must be equal to ϕ provided that $z' - z \to 0$ for $|z|$ large. Hence
$$\log z = \phi + i\psi, \quad z = \exp(\phi + i\psi),$$
which is uniquely determined. In fact the external transformation is determined if the potential problem is solved, and conversely. The outstanding advantage of external transformations of this type is that they turn the outside of a closed curve into the outside of another, leaving the scale and orientation at infinity unaltered. Hence if the form of the complex potential is known at large distances it can be adapted to the transformed problem by simply writing z for z'. In the internal problem if ϕ is constant over a closed curve and $\nabla^2 \phi = 0$ in the interior, ϕ is constant in the interior and tells us nothing about z. A variable ϕ can be arranged by taking $\phi = \log r' + \phi'$ within C', where $r' = |z' - c|$ and c is within C', $\nabla^2 \phi' = 0$ within C', and $\phi = 0$ on C'. Then if $z = \exp(\phi + i\psi)$, $z \doteq z' - c$ for $z' - c$ small, and $|z| = 1$ when z' is on C'. Hence the transformation represents C' and its interior on the unit circle and its interior, an arbitrary point c within C' corresponding to $z = 0$.

In both problems the existence of a solution of the potential problem is physically plausible; the analytical proof is difficult.

13·10. Another class of transformations, closely related but somewhat better known, is due to Schwarz and Christoffel. We take, keeping z' for the transformed figure,

$$\frac{dz'}{dt} = A \Pi (t - t_r)^{-\alpha_r/\pi}, \tag{1}$$

where the points t_r lie along the real axis, which is taken as the path for t. The region taken is the upper half of the t plane. Then when t travels, necessarily in the negative sense, around t_r, $\arg(dz'/dt)$ increases by α_r. If $\Sigma \alpha_r = 2\pi$ we get the interior of a closed polygon; if it is -2π we do not get the exterior, because when t describes a large semicircle $\arg z'$ will increase by 3π. If $\Sigma \alpha_r = \pi$ two sides are parallel and extend to infinity.

To see the relation to 13·095 we put

$$z + ia = \frac{2a^2}{t - ia}, \quad z = -ia\frac{t + ia}{t - ia}. \tag{2}$$

We get
$$\frac{dz'}{dt} = -\frac{2a^2 A}{(t-ia)^2} \Pi \left\{ \frac{2ia(t-t_r)}{(t-ia)(t_r+ia)} \right\}^{-\alpha_r/\pi}$$
$$= C \Pi (t - t_r)^{-\alpha_r/\pi}, \tag{3}$$

since $\Sigma \alpha_r = 2\pi$. This is of exactly the same form as (1); the only difference is that the t_r are on the real axis and the z_r on the circle.

The external transformation 13·094 (1) behaves differently. We first put $z = a^2/\zeta$ to transform the outside of the circle into the inside, and then take ζ instead of z in (2). We thus get

$$z = ia\frac{t - ia}{t + ia}, \tag{4}$$

$$\frac{dz'}{dt} = -\frac{2a^2 A}{(t+ia)^2} \Pi \left(\frac{2ia(t-t_r)}{(t-ia)(t_r+ia)} \right)^{\alpha_r/\pi}. \tag{5}$$

13·10 Schwarz-Christoffel transformation

With $\Sigma \alpha_r = 2\pi$ this has double poles at $t = \pm ia$. Consequently there is a singularity in the upper half of the t plane and z' will not be single-valued as a function of t unless the residue vanishes. Thus we again require an extra condition as for the transformation into the outside of a circle. The presence of a singularity not on the path of integration probably makes this transformation less useful for external problems than the transformation into a circle.

To transform the inside or outside of a given polygon into the upper half of the t plane, we take the α_r equal to the external angles, but still have to find the t_r to make the lengths of the sides right. This is always possible, but the proof that it is possible is difficult, except for a triangle.

In the Schwarz-Christoffel transformation for a given polygon the points t_r for three corners can be chosen arbitrarily. For if we take, with $\alpha\delta - \gamma\beta \neq 0$,

$$t = \frac{\alpha s + \beta}{\gamma s + \delta}, \quad \frac{dt}{ds} = \frac{\alpha\delta - \gamma\beta}{(\gamma s + \delta)^2}, \quad t - t_r = \frac{(\alpha\delta - \gamma\beta)(s - s_r)}{(\gamma s + \delta)(\gamma s_r + \delta)}, \quad \frac{dz'}{ds} = \frac{D}{(\gamma s + \delta)^2} \Pi \left(\frac{s - s_r}{\gamma s + \delta}\right)^{-\alpha_r/\pi},$$

and the factors $\gamma s + \delta$ cancel, leaving the form unaltered. But $\alpha, \beta, \gamma, \delta$ can be chosen to put three of the s_r anywhere we like. It is usually convenient to take them at some of the values $0, \pm 1$, and ∞. Evidently the same will be true of the internal transformation into a circle. If the polygon has more than three vertices the choice of the values of s_r for three of them will fix those for the others.

The external transformation is unique and no similar simplification is possible. The relation $\Sigma \alpha_r z_r = 0$ is equivalent to two relations between the η_r, and a factor of modulus 1 in all the z_r will be cancelled by another in the factor A.

A theory of the extension to curved boundaries is given by J. G. Leathem.*

EXAMPLES

1. A straight slit of width $2a$ and of great length is cut in a large conducting sheet. Show that, when the sheet is charged, the field in the neighbourhood of the slit, not too near the ends, can be determined by a complex transformation of the form $w = c(z^2 - a^2)^{1/2}$. Show that the surface density σ at a point distant x from the central line of the slit varies according to the law

$$\sigma = \sigma_0(1 - a^2/x^2)^{-1/2},$$

and that the equation of the equipotential surface of potential V is

$$x^2 y^2 = \left(\frac{V}{4\pi\sigma_0}\right)^2 \left\{x^2 - y^2 - a^2 + \left(\frac{V}{4\pi\sigma_0}\right)^2\right\}. \tag{M.T. 1943.}$$

2. Show that the resistance between two circular electrodes of equal radius b in an infinite plane uniform sheet of material of two-dimensional conductivity σ is approximately

$$\frac{1}{\pi\sigma} \log \frac{c}{b},$$

when the distance c between the electrodes is large compared with b.

The lines OA, OB of unlimited length form the boundary of a conducting sheet, which occupies the angle between them. At P on OA and Q on OB semicircular electrodes of radius b are let into

* *Phil. Trans.* A, 215, 1915, 439–87.

the sheet, b being small compared with PQ. The angle $AOB = \alpha$, and $OP = OQ = a$. Show that the resistance between the electrodes is approximately

$$\frac{2}{\pi\sigma}\log\frac{2\alpha a}{\pi b}.$$ (M.T. 1941.)

3. Electric charge is distributed with density e along the line $x = y = 0$, and the regions defined by $|y| \geqslant a$ are occupied by conducting matter at zero potential. Verify that in the space between the conductors the potential ϕ is given by

$$\phi + i\psi = -2e\log\tanh\frac{\pi z}{4a}, \quad z = x+iy.$$

Prove that the charge induced upon unit length of the strip of surface defined by $y = a$, $0 < x < b$ is

$$-\frac{e}{\pi}\tan^{-1}\tanh\frac{\pi b}{4a}.$$ (Prelim. 1937.)

4. Show that by means of the formulae

$$\frac{\pi z}{2a} = \sqrt{(t^2-1)} - \cosh^{-1}t + \tfrac{1}{2}\pi i,$$

$$\frac{\pi w}{2a} = Ut,$$

the solution can be found for the problem of the flow on one side of a stepped boundary consisting of $y = -a$ for $x < 0$, $y = a$ for $x > 0$, and $x = 0$ for $-a < y < a$.

Show that if the pressure at infinity is zero the force on unit width of the transverse portion is zero.

5. Sketch the transformed curves of the axes in the z plane and of the trisectors of the angles between them, under the transformation

$$w = -\frac{1}{z+i}.$$ (I.C. 1942.)

6. By considering the transformation $w = \exp z^2$ in the area A defined by $-1 \leqslant xy \leqslant 1$, $x^2-y^2 \leqslant 1$, $x \geqslant 0$, show that

$$\iint_A (x^2+y^2)e^{2(x^2-y^2)}dx\,dy = \tfrac{1}{2}e^2.$$ (I.C. 1936.)

7. Find a system of curves orthogonal to the curves of the family $x^3 - 3xy^2 = c$.

Chapter 14

FOURIER'S THEOREM

'I must go in and out.'
BERNARD SHAW, *Heartbreak House*

14·01. Harmonics fitted to n equally spaced values. Let the values of $f(x)$ be specified for n equally spaced values of x, namely,

$$x_r = \frac{2r\pi}{n} = r\lambda \quad (r = 0, 1, \ldots, n-1). \tag{1}$$

Denote $f(x_r)$ briefly by f_r; we wish to determine coefficients C_s so that

$$f_r = \sum_{s=0}^{n-1} C_s e^{irs\lambda}. \tag{2}$$

These are n equations in n unknowns. Multiply the rth equation by $\exp(-irm\lambda)$, where m is one of $0, 1, \ldots, n-1$, and add for all values of r. We get

$$\sum_{r=0}^{n-1} f_r e^{-irm\lambda} = \sum_{r=0}^{n-1} \sum_{s=0}^{n-1} C_s e^{ir(s-m)\lambda}. \tag{3}$$

Now if $s \neq m$,
$$\sum_{r=0}^{n-1} e^{ir(s-m)\lambda} = \frac{1 - e^{in(s-m)\lambda}}{1 - e^{i(s-m)\lambda}} = 0, \tag{4}$$

since $n\lambda = 2\pi$; if $s = m$ each term is 1 and the sum is n. Hence

$$\sum_{r=0}^{n-1} f_r e^{-irm\lambda} = nC_m. \tag{5}$$

To show that these satisfy the original equations, we have, replacing m by s,

$$\sum_{s=0}^{n-1} C_s e^{its\lambda} = \frac{1}{n} \sum_{s=0}^{n-1} \sum_{r=0}^{n-1} f_r e^{i(t-r)s\lambda} \tag{6}$$

and the sum of $e^{i(t-r)s\lambda}$ with regard to s is 0 unless $r = t$, when it is n. Hence the sum (6) is f_t, and (5) is the solution.

For $s = 0$, C_s is simply $\frac{1}{n}\sum_{t=0}^{n-1} f_t$. If m is a value of s and not zero, $n - m$ is another. Hence apart from $s = 0$ we can take the terms of (2) in pairs; and

$$C_s e^{its\lambda} + C_{n-s} e^{it(n-s)\lambda} = \frac{1}{n}\sum_{r=0}^{n-1} f_r \{e^{i(t-r)s\lambda} + e^{i(t-r)(n-s)\lambda}\}$$

$$= \frac{2}{n}\sum_{r=0}^{n-1} f_r \cos(t-r)s\lambda. \tag{7}$$

If n is even, $s = \tfrac{1}{2}n$ is one possible value and occurs only once; for this

$$C_s e^{its\lambda} = \frac{1}{n}\sum_{r=0}^{n-1} f_r e^{\frac{1}{2}ni(t-r)\lambda} = \frac{1}{n}\sum_{r=0}^{n-1} f_r \cos \tfrac{1}{2}n(t-r)\lambda, \tag{8}$$

since $\sin \tfrac{1}{2}n(t-r)\lambda = \sin(t-r)\pi = 0.$

Thus
$$f_t = A_0 + \Sigma A_s \cos st\lambda + \Sigma B_s \sin st\lambda, \tag{9}$$

where the summations include $s = 1, 2, \ldots$ up to $\tfrac{1}{2}n$ or $\tfrac{1}{2}(n-1)$ according as n is even or odd; and

$$A_0 = \frac{1}{n} \sum_{r=0}^{n-1} f_r, \tag{10}$$

$$A_s = \frac{2}{n} \sum_{r=0}^{n-1} f_r \cos sr\lambda, \quad B_s = \frac{2}{n} \sum_{r=0}^{n-1} f_r \sin sr\lambda \quad (0 < s < \tfrac{1}{2}n), \tag{11}$$

and if n is even
$$A_{\frac{1}{2}n} = \frac{1}{n} \sum_{r=0}^{n-1} f_r \cos \tfrac{1}{2} nr\lambda = \frac{1}{n} \sum_{r=0}^{n-1} (-1)^r f_r, \tag{12}$$

the sine term vanishing since $\sin r\pi = 0$.

An alternative method is to assume the form (9) directly and evaluate Σf_r, $\Sigma f_r \cos rs\lambda$, $\Sigma f_r \sin rs\lambda$. The summations are, however, slightly more difficult by this method.

In this way we represent the n values exactly as the sums of a constant and $n-1$ trigonometrical terms. The method is known as *harmonic analysis* and is extensively used in the study of observational data. In meteorology, for instance, the pressure, temperature, humidity and so on are recorded at intervals of an hour. From the hourly values a harmonic representation can be found for each day, including terms of periods 24, 12, 8, 6, 24/5, ... hours down to 2 hours. But for the last the data will determine only a cosine term, since the corresponding sine term vanishes at all the times where there are observations; to find it we should need a shorter interval between consecutive observations. If we extend (9) to fractional values of t we can regard it as an interpolation function. Analyses can be carried out for all days over an interval and the results compared to see whether the harmonic terms repeat themselves and can therefore be made the basis of inferences over longer intervals; for instance, the 24-hourly period in temperature is obvious, but its amplitude and phase are found by harmonic analysis, while diurnal and semidiurnal periods in the pressure are noteworthy features of the climate in many regions.

Useful two-figure tables for harmonic analysis have been published by H. H. Turner[*] for 9 to 21 intervals. His r is the present $r+1$.

14·02. Fourier series. Now suppose that $f(x)$ is given for all values of x from 0 to 2π. We can increase n indefinitely and thereby make our interpolation function agree with $f(x)$ at more and more points, and the interval is $2\pi/n$, while $r\lambda = x$. Then if the coefficients tend to definite limits these will be

$$A_0 = \frac{1}{2\pi} \int_0^{2\pi} f(x)\,dx, \quad A_s = \frac{1}{\pi} \int_0^{2\pi} f(x) \cos sx\,dx, \quad B_s = \frac{1}{\pi} \int_0^{2\pi} f(x) \sin sx\,dx, \tag{1}$$

provided that the integrals exist, since the method of subdivision used is only one of the ways that must give the same limit if the integrals exist. It is sufficient that $f(x)$ itself shall be integrable; this will imply the existence of all the other integrals.

Let $g_n(x)$ be the interpolation function obtained as in 14·01 for n intervals. We should expect that, when n increases indefinitely, the interpolation function tends to a limit $f(x)$ for every value of x. Unfortunately this is difficult to prove directly, and is not even always true. Clearly if $f(x)$ is one of the peculiar functions that interest pure mathe-

---

[*] *Tables for facilitating the use of Harmonic Analysis*, 1913.

maticians, such as one that is zero for all rational fractions of 2π and 1 for all irrational ones, then since we sample it only at points of the former set the coefficients in $g_n(x)$ will always be 0 and the limiting series will vanish for all x; it will therefore disagree with $f(x)$ at every irrational fraction of 2π. The integrals (1) then do not exist in the Riemann sense in this case, but they do exist in the Lebesgue sense, and then the series represents the function at irrational values of $x/2\pi$ and not at rational ones. Even if $f(x)$ is continuous we cannot infer that $g_n(x) \to f(x)$ unless we can also show that the limit exists, and this is not always true.*

It is easiest to proceed to direct study of the series of sines and cosines

$$A_0 + \Sigma(A_n \cos nx + B_n \sin nx),$$

where A_n and B_n are defined by (1). This series is called the *Fourier series* of $f(x)$. We recall that if $f(x)$ is of bounded variation in an interval (a, b), it need not be continuous, but $f(x-)$ and $f(x+)$ exist for every x within the interval, and $f(a+)$ and $f(b-)$ also exist. Then we shall show

(1) *If $f(x)$ is of bounded variation in $(0, 2\pi)$, the Fourier series of $f(x)$ converges to*

$$\tfrac{1}{2}\{f(x-)+f(x+)\}$$

for every interior point of the interval, and at the end points to $\tfrac{1}{2}\{f(0+)+f(2\pi-)\}$.

(2) A_n, B_n *tend to zero as n increases, at least as fast as $1/n$.*

(3) *If $f(x)$ is also continuous at all points of the interval, the sum of the Fourier series is equal to $f(x)$ at all points of the interval.*

(4) *If $f(x)$ is differentiable at all points of the interval and $f'(x)$ has bounded variation, the coefficients decrease at least as fast as n^{-2}, and similarly for higher derivatives.*

When several derivatives have bounded variation the rapid decrease of the early terms makes the series useful for computation. We have had an example in the case of the Bernoulli polynomials, for which the expansions given in 12·07 are the Fourier expansions. For $P_4(t)$ the fourth term of the expansion has an amplitude $1/256$ of that of the first, so that a few terms of the series give a good idea of the general appearance of the function and can even be used for computation if three-figure accuracy is sufficient.

The Fourier series also has important applications in potential theory; these are shared by what is called the *allied series*† $\sum_{n=1}^{\infty}(A_n \sin nx - B_n \cos nx)$. Somewhat more severe conditions are needed for the convergence of the allied series.

To prove the above statements we need a lemma.

14·03. Riemann's lemma. *If $\phi(x)$ is non-decreasing and bounded in the range a to b, and λ is large,*

$$\int_a^b \phi(x) \cos \lambda x\, dx \quad \text{and} \quad \int_a^b \phi(x) \sin \lambda x\, dx \quad \text{are} \quad O\!\left(\frac{1}{\lambda}\right).$$

For
$$\int_a^b \phi(x) \cos \lambda x\, dx = \left[\frac{\phi(x)}{\lambda}\sin \lambda x\right]_a^b - \frac{1}{\lambda}\int_{x=a}^b \sin \lambda x\, d\phi(x),$$

---

* If $f(x)$ satisfies a Lipschitz condition the statements suggested can be proved: cf. D. Jackson, *The Theory of Approximation*, 1930, 130: A. C. Offord, *Duke Math. J.* 6, 1940, 505–10.

† We use here the name introduced by W. H. Young. The term *conjugate series* is also used.

the integral on the right being a Stieltjes integral if $\phi(x)$ changes discontinuously at any value of x. All elements $\delta\phi(x)$ are $\geqslant 0$, and $|\sin \lambda x| \leqslant 1$; hence if $|\phi(x)| < A$

$$\left|\int_a^b \phi(x) \cos \lambda x\, dx\right| \leqslant \frac{2}{\lambda} A + \frac{2}{\lambda} A = \frac{4A}{\lambda}.$$

Also, similarly,
$$\left|\int_a^b \phi(x) \sin \lambda x\, dx\right| \leqslant \frac{4A}{\lambda}.$$

If we change $\phi(x)$ to $-\phi(x)$ the result is seen to be true also if $\phi(x)$ is non-increasing in the interval.

We know that if $f(x)$ is of bounded variation in the range a to b, it can be expressed as the sum of a non-increasing and a non-decreasing function, both bounded in the interval. Hence *for any function of bounded variation*

$$\int_a^b f(x) \cos \lambda x\, dx = O\left(\frac{1}{\lambda}\right), \quad \int_a^b f(x) \sin \lambda x\, dx = O\left(\frac{1}{\lambda}\right).$$

14·04. Summation of Fourier and allied series. We can write

$$A_n \cos nx + B_n \sin nx = \frac{1}{\pi} \int_0^{2\pi} f(t) \cos n(t-x)\, dt, \tag{1}$$

$$A_n \sin nx - B_n \cos nx = -\frac{1}{\pi} \int_0^{2\pi} f(t) \sin n(t-x)\, dt. \tag{2}$$

We denote the sums of the Fourier series and the allied series, up to the terms in nx, by $S_n(x)$ and $T_n(x)$, and take $f(x)$ to be of bounded variation in $(0, 2\pi)$. Then

$$S_n(x) + iT_n(x) = \frac{1}{\pi}\int_0^{2\pi} f(t)\{\tfrac{1}{2} + e^{-i(t-x)} + \ldots + e^{-ni(t-x)}\}\, dt$$

$$= \frac{1}{\pi}\int_0^{2\pi} f(t)\left(-\frac{1}{2} + \frac{1 - e^{-(n+1)i(t-x)}}{1 - e^{-i(t-x)}}\right) dt$$

$$= \frac{1}{2\pi}\int_0^{2\pi} f(t)\left\{\frac{\sin(n+\tfrac{1}{2})(t-x)}{\sin\tfrac{1}{2}(t-x)} - i\frac{\cos\tfrac{1}{2}(t-x) - \cos(n+\tfrac{1}{2})(t-x)}{\sin\tfrac{1}{2}(t-x)}\right\} dt. \tag{3}$$

This reduces $S_n(x)$ and $T_n(x)$ separately to single integrals. We have to study their behaviour as n tends to infinity. We notice first that the integrands are finite at all points, even at $t = x$. For $S_n(x)$, first exclude an arbitrarily small range $x - \delta < t < x + \delta$. Then in the remaining ranges $f(t)\operatorname{cosec}\tfrac{1}{2}(t-x)$ is bounded, and has bounded variation if $f(t)$ has. Therefore by Riemann's lemma the contribution to $S_n(x)$ from these ranges tends to 0 with increasing n. Also in $x - \delta < t < x + \delta$, $f(t)\left\{\dfrac{1}{\tfrac{1}{2}(t-x)} - \dfrac{1}{\sin\tfrac{1}{2}(t-x)}\right\}$ has bounded variation. Hence

$$S_n(x) - \frac{1}{2\pi}\int_{x-\delta}^{x+\delta} f(t)\frac{\sin(n+\tfrac{1}{2})(t-x)}{\tfrac{1}{2}(t-x)}\, dt \to 0. \tag{4}$$

Since δ is arbitrarily small it follows that that if the series converges its sum depends wholly on the values of $f(t)$ near $t = x$. Also, on putting

$$(n+\tfrac{1}{2})(t-x) = u, \tag{5}$$

$$S_n(x) - \frac{1}{\pi}\int_{-(n+1/2)\delta}^{(n+1/2)\delta} f\left(x + \frac{u}{n+\tfrac{1}{2}}\right)\frac{\sin u}{u}\, du \to 0. \tag{6}$$

As $n \to \infty$ the limits tend to $\pm \infty$. But as $t \to x$ from larger or smaller values, $f(t) \to f(x+)$ or $f(x-)$. By our choice of δ we can make the variation of the values of $f(t)$ on each side of x as small as we like, by 1·093. Hence, by du Bois-Reymond's form of the second mean value theorem (1·134), putting

$$f(u) = \frac{\sin u}{u}, \quad \phi(u) = f\left(x + \frac{u}{n+\tfrac{1}{2}}\right),$$

we can make the integral as near as we like to

$$\int_0^\infty f(x+)\frac{\sin u}{u} du + \int_0^\infty f(x-)\frac{\sin u}{u} du = \tfrac{1}{2}\pi\{f(x+)+f(x-)\}, \tag{7}$$

and therefore

$$S_n(x) \to \tfrac{1}{2}\{f(x+)+f(x-)\}. \tag{8}$$

The series is therefore convergent at any point x such that $f(t)$ tends to definite limits as t approaches x on each side. If these limits are the same and equal to $f(x)$, the sum is $f(x)$. If they are different, as when $f(t)$ has a finite jump at x, the sum is the mean of the limits. If $x = 0$ or 2π, (4) must be modified; making the appropriate changes we find

$$S_n(0) = S_n(2\pi) \to \tfrac{1}{2}\{f(0+)+f(2\pi-)\}.$$

Now consider

$$T_n(x) = -\frac{1}{2\pi}\int_0^{2\pi} f(t)\frac{\cos\tfrac{1}{2}(t-x)-\cos(n+\tfrac{1}{2})(t-x)}{\sin\tfrac{1}{2}(t-x)} dt. \tag{9}$$

This is convergent, but the convergence is only saved by the term in $\cos(n+\tfrac{1}{2})(t-x)$, which we wish to treat as a remainder term. We can, however, proceed as follows. First exclude a range of length 2δ about x as before; outside this the part depending on $\cos(n+\tfrac{1}{2})(t-x)$ tends to 0. Then $T_n(x)$ has the same limit as

$$-\frac{1}{2\pi}\left[\int_0^{x-\delta} + \int_{x+\delta}^{2\pi}\right] f(t)\cot\tfrac{1}{2}(t-x)\,dt - \frac{1}{2\pi}\int_{x-\delta}^{x+\delta} f(t)\frac{\cos\tfrac{1}{2}(t-x)-\cos(n+\tfrac{1}{2})(t-x)}{\sin\tfrac{1}{2}(t-x)} dt, \tag{10}$$

and the last portion

$$= -\frac{1}{2\pi}\int_0^\delta \{f(x+v)-f(x-v)\}\frac{\cos\tfrac{1}{2}v - \cos(n+\tfrac{1}{2})v}{\sin\tfrac{1}{2}v} dv. \tag{11}$$

Now if $\{f(x+v)-f(x-v)\}\operatorname{cosec}\tfrac{1}{2}v$ is of bounded variation for $0 \leqslant v \leqslant \delta$ we can again apply Riemann's lemma and say that the term in $\cos(n+\tfrac{1}{2})v$ tends to 0. But then, if the upper bound of this expression is M,

$$\left|\int_0^\delta \{f(x+v)-f(x-v)\}\cot\tfrac{1}{2}v\,dv\right| \leqslant \int_0^\delta M\cos\tfrac{1}{2}v\,dv = 2M\sin\tfrac{1}{2}\delta, \tag{12}$$

which we could make arbitrarily small at the outset by a suitable choice of δ. The condition assumed is true if $f(t)$ is differentiable on each side of x; the inference would remain valid if the derivative was itself discontinuous at x. Then we have the result that if $f(t)$ is differentiable on each side of x, the allied series is convergent and its sum is

$$-\frac{1}{2\pi}\lim_{\delta \to 0}\left[\int_0^{x-\delta} + \int_{x+\delta}^{2\pi}\right] f(t)\cot\tfrac{1}{2}(t-x)\,dt = -\frac{1}{2\pi} P\int_0^{2\pi} f(t)\cot\tfrac{1}{2}(t-x)\,dt, \tag{13}$$

P denoting the principal value.

14·041. The Lipschitz condition. If for some positive M and α, and $|v| \leq \delta$, $|f(x+v)-f(x)| \leq M|v|^\alpha$, it will still be true that (12) can be made arbitrarily small by a suitable choice of δ. With this condition the truth of (8) follows immediately from (6) and the fact that $\int_0^\infty \sin u \, du/u$ converges.

If a function satisfies this condition with $\alpha = 1$, and some M independent of x, it also follows immediately that the function is of bounded variation; for the total variation is the upper bound of

$$|f(x_1)-f(0)|+|f(x_2)-f(x_1)|+\ldots+|f(2\pi)-f(x_n)| \leq M\Sigma|x_r-x_{r-1}| = 2\pi M.$$

Hence a sufficient condition for the Fourier series to converge to $f(x)$ and the allied series to (13) in $(0, 2\pi)$ is that $f(x)$ shall satisfy a Lipschitz condition of order 1 uniformly; this takes in its stride the condition that the variation shall be bounded.

14·05. Complex theory. We replace x by θ and t by χ, and regard θ as the argument of a complex variable z, of modulus a. We shall show that in suitable conditions the Fourier series has a simple relation to the solution of a potential problem, where the potential ϕ is given over the circle $|z| = a$ and is required in the interior. We have seen from Laurent's theorem that the supposition that $\phi+i\psi$ is analytic in a zone containing the circle leads to the Fourier expansion. We shall now see that the condition is not necessary.

The Fourier and allied series of $f(\theta)$ can be written together

$$S+iT = \frac{1}{2\pi}\int_0^{2\pi} f(\chi)\,d\chi + \frac{1}{\pi}\sum_{n=1}^\infty \int_0^{2\pi} f(\chi)e^{ni(\theta-\chi)}\,d\chi. \tag{1}$$

Put

$$t = ae^{i\chi}, \quad z = re^{i\theta} \quad (r<a), \tag{2}$$

and consider the series

$$\phi+i\psi = \frac{1}{2\pi}\int_0^{2\pi} f(\chi)\,d\chi + \frac{1}{\pi}\sum_{n=1}^\infty \int_0^{2\pi} f(\chi)\left(\frac{z}{t}\right)^n d\chi. \tag{3}$$

When $r \to a$ the terms reduce to those of (1). Since the series is uniformly convergent in any closed interval of $r<a$,

$$\phi+i\psi = \frac{1}{\pi}\int_0^{2\pi} f(\chi)\left\{\frac{1}{2}+\sum_1^\infty \left(\frac{z}{t}\right)^n\right\}d\chi$$

$$= \frac{1}{2\pi}\int_0^{2\pi} f(\chi)\frac{t+z}{t-z}\,d\chi = \frac{1}{2\pi i}\int_C f(\chi)\frac{t+z}{t-z}\frac{dt}{t}, \tag{4}$$

where the integral in the last expression is taken around the circle $|t| = a$. This function is defined subject simply to $f(\chi)$ being integrable, and provides us with a precise starting point. We wish to study its properties, and in particular to see whether, when $|z| \to a$, $\phi \to f(\theta)$. Put $\chi-\theta = \vartheta$ and separate the real and imaginary parts. We get

$$\phi = \frac{1}{2\pi}\int_0^{2\pi} \frac{a^2-r^2}{a^2-2ar\cos\vartheta+r^2} f(\theta+\vartheta)\,d\vartheta, \tag{5}$$

$$\psi = -\frac{1}{\pi}\int_0^{2\pi} \frac{ar\sin\vartheta}{a^2-2ar\cos\vartheta+r^2} f(\theta+\vartheta)\,d\vartheta. \tag{6}$$

As they are the real and imaginary parts of a function of the complex variable z, these functions are solutions of Laplace's equation in two dimensions everywhere within the circle, and we shall expect ϕ to be the potential and ψ the charge function, given that ϕ takes the values $f(\theta)$ on the boundary and that there is no charge inside. We therefore proceed to examine the behaviour of ϕ and ψ when r tends to a. Suppose that when χ is near θ, $f(\chi)$ tends to limits $f(\theta-)$ and $f(\theta+)$ as χ tends to θ through smaller and larger values respectively. Then for any ω we can choose a quantity δ so that

$$\left. \begin{array}{l} |f(\theta+\vartheta)-f(\theta+)| < \omega \quad (0 < \vartheta \leqslant \delta), \\ |f(\theta+\vartheta)-f(\theta-)| < \omega \quad (-\delta \leqslant \vartheta < 0). \end{array} \right\} \tag{7}$$

Then
$$2\pi\phi = \left\{\int_\delta^{2\pi-\delta} + \int_{-\delta}^0 + \int_0^\delta\right\} \frac{a^2-r^2}{a^2-2ar\cos\vartheta+r^2} f(\theta+\vartheta)\,d\vartheta. \tag{8}$$

When $r \to a$ the first integral tends to 0. For the third,

$$\int_0^\delta \{f(\theta+)-\omega\} \frac{a^2-r^2}{a^2-2ar\cos\vartheta+r^2}\,d\vartheta \leqslant \int_0^\delta f(\theta+\vartheta) \frac{a^2-r^2}{a^2-2ar\cos\vartheta+r^2}\,d\vartheta$$

$$\leqslant \int_0^\delta \{f(\theta+)+\omega\} \frac{a^2-r^2}{a^2-2ar\cos\vartheta+r^2}\,d\vartheta. \tag{9}$$

Now, if $\tan\tfrac{1}{2}\vartheta = u$,

$$\int_0^\delta \frac{a^2-r^2}{a^2-2ar\cos\vartheta+r^2}\,d\vartheta = \int_0^{\tan 1/2\delta} \frac{(a^2-r^2)\,2du}{(a-r)^2+(a+r)^2 u^2} = 2\tan^{-1}\!\left(\frac{a+r}{a-r}\tan\tfrac{1}{2}\delta\right). \tag{10}$$

which tends to π as $r \to a$. Hence as $r \to a$ we can make the second integral in (9) lie between $\pi\{f(\theta+)\pm\omega\}$; and by the type of argument already familiar in potential theory ϕ has a limit as $r \to a$ equal to
$$\lim_{r \to a} \phi = \tfrac{1}{2}\{f(\theta-)+f(\theta+)\}. \tag{11}$$

We break up the range for ψ similarly. The range δ to $2\pi-\delta$ gives

$$-\frac{1}{2\pi}\int_\delta^{2\pi-\delta} f(\theta+\vartheta)\cot\tfrac{1}{2}\vartheta\,d\vartheta, \tag{12}$$

and the range $-\delta$ to δ gives

$$-\frac{1}{\pi}\int_{\vartheta=-\delta}^\delta 4f(\theta+\vartheta)\frac{aru\,du}{(1+u^2)\{(a-r)^2+(a+r)^2 u^2\}}$$
$$= \frac{1}{\pi}\int_0^{\tan 1/2\delta} \{f(\theta+\vartheta)-f(\theta-\vartheta)\}\left\{\frac{u}{1+u^2} - \frac{(a+r)^2 u}{(a-r)^2+(a+r)^2 u^2}\right\}du. \tag{13}$$

The first part on the right of (13) is independent of r. The second is less numerically than

$$\frac{1}{\pi}\int_0^{\tan 1/2\delta} \{f(\theta+\vartheta)-f(\theta-\vartheta)\}\frac{du}{u}, \tag{14}$$

which tends to 0 with δ if $f(\theta)$ is differentiable at $\vartheta = 0$. (Actually a Lipschitz condition would do; it would be enough that $\{f(\theta+\vartheta)-f(\theta-\vartheta)\}u^{-\alpha}$ should be bounded, where α is any positive number.) Then both parts tend to 0 with δ, and

$$\psi \to -\frac{1}{2\pi} P \int_0^{2\pi} f(\theta+\vartheta)\cot\tfrac{1}{2}\vartheta\,d\vartheta, \tag{15}$$

which was found for the sum of the allied series on the circle in 14·04. If $f(\theta+\vartheta)-f(\theta-\vartheta)$ behaves like $1/(-\log|u|)$ the integral (14) will not tend to 0 and the principal value will not exist, but such cases do not seem physically important.

Thus the procedure of introducing the factors $(r/a)^n$ and letting r tend to a gives the same values for the series as direct summation does. By Abel's theorem this would be expected, since we know that if a power series converges at a point on the circle of convergence its sum there is the limit of the sum on approaching that point from within. But the present results are true under somewhat wider conditions, for the argument merely assumes that $f(\theta)$ is integrable, not that it has bounded variation. It may therefore give a meaning to the series on the circle even though the latter may not converge. Our theorem then takes the form:

If $f(\theta)$ is integrable, and

$$A_0 = \frac{1}{2\pi}\int_0^{2\pi} f(\theta)\,d\theta, \quad A_n = \frac{1}{\pi}\int_0^{2\pi} f(\theta)\cos n\theta\,d\theta, \quad B_n = \frac{1}{\pi}\int_0^{2\pi} f(\theta)\sin n\theta\,d\theta,$$

then the Fourier series
$$\phi = A_0 + \sum_{n=1}^{\infty}(A_n\cos n\theta + B_n\sin n\theta)$$

and the allied series
$$\psi = \sum_{n=1}^{\infty}(A_n\sin n\theta - B_n\cos n\theta)$$

if summed by Abel's method, give

$$\phi = \tfrac{1}{2}\lim_{\delta,\eta\to 0}\{f(\theta+\delta)+f(\theta-\eta)\},$$

where $\delta,\eta > 0$, at any value of θ where this limit exists, and

$$\psi = -\frac{1}{2\pi}P\int_0^{2\pi} f(\chi)\cot\tfrac{1}{2}(\chi-\theta)\,d\chi$$

for any value of θ where $f(\theta)$ satisfies a Lipschitz condition.

The use of Abel's theorem can be justified immediately in a large class of cases. The conditions in 14·04 are sufficient for convergence on the circle.

Without previous knowledge of the properties of $f(\theta)$, something can still be said about the convergence if the coefficients in the series are known. By a theorem of Tauber the Fourier series converges if the Abel sum exists and nA_n and nB_n tend to 0; by an extension due to Littlewood it is sufficient that they should be bounded.

So far as this theorem relates to ϕ it bears the name of Fourier.* Study of the allied series is modern.† The present form of the theorem seems to be the one with the most direct physical applications. It is known that a Fourier series is always summable, even if not convergent, by a method of Cesàro known as (C, 1), which is less drastic than Abel summation, at all points where $f(\theta+)$ and $f(\theta-)$ exist; but in potential problems the trigonometrical factors are associated with powers of r in such a way as to make Abel summation arise naturally, and then the functions ϕ and ψ are determined at all internal points by the integrals (5) and (6), given the values of ϕ over the surface. It would be

---

* For no very obvious reason. The problem of the vibrating string, with twice differentiable initial displacement, was solved by d'Alembert and Euler in 1747. D. Bernoulli got the solution as a sine series in 1753. If the solution is unique the two forms must be equivalent and the Fourier sine theorem follows. Fourier, in his *Analytical Theory of Heat*, 1822, gave an alleged proof, which is a mathematical nightmare. The book is an excellent work on heat conduction. The first proof of the theorem under reasonably general conditions was due to Dirichlet in 1829.

† See Hardy and Littlewood, *Proc. Lond. Math. Soc.* (2) 24, 1925, 211–240.

equally valid, since ϕ is the function allied to ψ, to take as $f(\theta)$ the values of ψ over the surface provided ψ is differentiable; then (5) will give ψ at internal points and (6) will give $-\phi$. If there are local concentrations of charge, ψ will have finite discontinuities and ϕ will not be determinate there, since the principal value of the integral giving it will not exist there, and ϕ will tend to infinity logarithmically as we approach a discontinuity in ψ. If ϕ is discontinuous at a point it would correspond in the electrostatic problem to a doublet with its axis tangential, and the charge locally will be indeterminate. Hence the special cases that arise in the summations correspond to physical difficulties also.

The introduction of Abel summation is due to Poisson.

We have seen that a function analytic within a contour can be determined entirely in terms of its values on the contour. The real and imaginary parts, however, satisfy Laplace's equations separately, and therefore by general potential theory each is determined by its values on the contour. Further, by Cauchy's theorem the real and imaginary parts are connected even on the contour. The present result shows what this relation is when the contour is a circle. An additive constant could be included in either ϕ or ψ without upsetting the Cauchy-Riemann relations. We have arbitrarily taken the constant term in ψ as zero. This is physically unimportant because we are not usually much concerned with the absolute values of ϕ and ψ, but only with their differences from place to place, so that the constant is usually irrelevant.

14·051. The cosine and sine series. Suppose that $f(x)$ is such that $f(2\pi - x) = f(x)$. Then

$$A_0 = \frac{1}{2\pi} \int_0^{2\pi} f(x)\, dx = \frac{1}{\pi} \int_0^{\pi} f(x)\, dx, \tag{1}$$

$$A_n = \frac{1}{\pi} \int_0^{2\pi} f(x) \cos nx\, dx = \frac{2}{\pi} \int_0^{\pi} f(x) \cos nx\, dx, \tag{2}$$

$$B_n = \frac{1}{\pi} \int_0^{2\pi} f(x) \sin nx\, dx = 0. \tag{3}$$

Hence the Fourier series

$$A_0 + \Sigma A_n \cos nx \tag{4}$$

will represent $f(x)$ from 0 to π, and $f(2\pi - x)$ from π to 2π.

On the other hand, if $f(2\pi - x) = -f(x)$,

$$A_0 = 0, \quad A_n = 0, \tag{5}$$

$$B_n = \frac{2}{\pi} \int_0^{\pi} f(x) \sin nx\, dx, \tag{6}$$

and the series $\Sigma B_n \sin nx$, with B_n determined by (6), will represent $f(x)$ from 0 to π, and $-f(2\pi - x)$ from π to 2π. We thus have two representations of the same function valid from 0 to π; but they represent different functions from π to 2π. This property is very different from that possessed by power series. It can be regarded as a method of continuation; but if, for instance, $f(x) = x$ from $x = 0$ to π, the analytic continuation is x in $\pi < x < 2\pi$, the continuation by (4) is $2\pi - x$, and that by (5) is $-(2\pi - x)$. All are correct in their proper places, but the decision between them depends on the particular problem, which will itself indicate if the function is analytic, or symmetrical or antisymmetrical about $x = \pi$. If $f(-x) = f(x) = f(2\pi - x)$ and $f(\pi - x) = -f(x)$, so that the function is even but is anti-symmetrical about $\tfrac{1}{2}\pi$,

$$A_0 = \frac{1}{2\pi}\left\{\int_{-\pi}^{-1/2\pi} + \int_{-1/2\pi}^{0} + \int_{0}^{1/2\pi} + \int_{1/2\pi}^{\pi} f(x)\,dx\right\}, \tag{7}$$

in which the first and second integrals cancel, also the third and fourth,

$$A_n = \frac{2}{\pi}\int_0^{\pi} f(x)\cos nx\,dx = 0 \quad \text{for } n \text{ even}, \tag{8}$$

$$= \frac{4}{\pi}\int_0^{1/2\pi} f(x)\cos nx\,dx \quad \text{for } n \text{ odd}, \tag{9}$$

$$B_n = 0. \tag{10}$$

Hence
$$f(x) = \sum_{m=0}^{\infty} \frac{4}{\pi}\cos(2m+1)x \int_0^{1/2\pi} f(x)\cos(2m+1)x\,dx. \tag{11}$$

Examples. Take $f(\theta) = 1$ for $0 < \theta < \pi$; then the constant term in the cosine series is 1 and the rest are 0, and the function is everywhere 1; as it should be since the function represented by the cosine satisfies $f(2\pi - \theta) = f(\theta) = f(2\pi + \theta)$. But for the sine series we have

$$B_n = \frac{2}{\pi}\int_0^{\pi}\sin n\theta\,d\theta = \frac{2}{\pi n}(1 - \cos n\pi) = 0 \quad \text{or} \quad \frac{4}{\pi n}, \tag{12}$$

according as n is even or odd. Hence

$$1 = \frac{4}{\pi}(\sin\theta + \tfrac{1}{3}\sin 3\theta + \tfrac{1}{5}\sin 5\theta + \ldots) \quad (0 < \theta < \pi). \tag{13}$$

To carry out Abel summation on the series we have

$$S = \lim_{r\to 1}\frac{2}{\pi i}\{(re^{i\theta} + \tfrac{1}{3}r^3 e^{3i\theta} + \ldots) - (re^{-i\theta} + \tfrac{1}{3}r^3 e^{-3i\theta} + \ldots)\} \tag{14}$$

$$= \lim_{r\to 1}\frac{2}{\pi i}\{\tanh^{-1}(re^{i\theta}) - \tanh^{-1}(re^{-i\theta})\}. \tag{15}$$

But
$$\tanh(x - y) = \frac{\tanh x - \tanh y}{1 - \tanh x \tanh y}, \tag{16}$$

and
$$S = \lim_{r\to 1}\frac{2}{\pi i}\tanh^{-1}\frac{2ir\sin\theta}{1 - r^2}$$

$$= \lim_{r\to 1}\frac{2}{\pi}\tan^{-1}\frac{2r\sin\theta}{1 - r^2}. \tag{17}$$

$\tan^{-1} z$ being many-valued we must take the value that tends to 0 with r, since the series on the right of (14) do. Then if $\sin\theta > 0$ the limit is 1, which verifies the result. But if $\sin\theta < 0$ the limit is -1, as we should expect since the sine series represents a function antisymmetrical about π.

If $\theta = 0$ or π, the series vanishes. This agrees with our result that at a point of discontinuity the sum of the series is the mean of the limits, here $+1$ and -1, on opposite sides of it.

The allied series is $\psi = -\dfrac{4}{\pi}(\cos\theta + \tfrac{1}{3}\cos 3\theta + ...)$,

and
$$(\phi + i\psi)_{r<1} = -\frac{4i}{\pi}\left(re^{i\theta} + \frac{r^3}{3}e^{3i\theta} + ...\right) = -\frac{4i}{\pi}\tanh^{-1} re^{i\theta} = -\frac{2i}{\pi}\log\frac{1+re^{i\theta}}{1-re^{i\theta}}$$

$$= -\frac{2i}{\pi}\log\frac{1+r\cos\theta + ir\sin\theta}{1-r\cos\theta - ir\sin\theta}$$

$$= -\frac{2i}{\pi}\left(\frac{1}{2}\log\frac{(1+r\cos\theta)^2 + r^2\sin^2\theta}{(1-r\cos\theta)^2 + r^2\sin^2\theta} + i\tan^{-1}\frac{r\sin\theta}{1+r\cos\theta} + i\tan^{-1}\frac{r\sin\theta}{1-r\cos\theta}\right)$$

$$\to -\frac{2i}{\pi}\left\{\frac{1}{2}\log\frac{1+\cos\theta}{1-\cos\theta} + i\frac{\theta}{2} + i(\tfrac{1}{2}\pi - \tfrac{1}{2}\theta)\right\} \quad 0 < \sin\theta < 1$$

$$= 1 - \frac{2i}{\pi}\log\cot\tfrac{1}{2}\theta.$$

We recover the previous value of ϕ; and
$$\psi = -\frac{2}{\pi}\log\cot\tfrac{1}{2}\theta,$$

which is infinite at the points of discontinuity, as expected. Changing the sign of θ does not alter ψ, which is therefore in general $-\dfrac{2}{\pi}\log|\cot\tfrac{1}{2}\theta|$.

The points of discontinuity of ϕ are seen to correspond to the points $z = re^{i\theta} = \pm 1$, where $\phi + i\psi$ has branch points. The behaviour of ϕ and ψ at such points is connected with the fact that near $z = 0$, the real part of $i\log z$ changes by $\pm\pi$ as we go half-way round, but the imaginary part is $\log r$, which tends to infinity logarithmically.

Next, take
$$f(\theta) = \theta(\pi - \theta) \quad 0 < \theta < \pi.$$

For the cosine series
$$A_0 = \frac{1}{\pi}\int_0^\pi \theta(\pi-\theta)\,d\theta = \tfrac{1}{6}\pi^2,$$

$$A_n = \frac{2}{\pi}\int_0^\pi \theta(\pi-\theta)\cos n\theta\,d\theta = -\frac{2}{n^2}\{1 + (-1)^n\},$$

$$f(\theta) = \tfrac{1}{6}\pi^2 - \cos 2\theta - \frac{\cos 4\theta}{4} - \frac{\cos 6\theta}{9} - ....$$

Notice that since $f(\theta) \to 0$ at $\theta = 0$,
$$\tfrac{1}{6}\pi^2 = 1 + \frac{1}{2^2} + \frac{1}{3^2} + ...,$$

a relation that we have found in considering the Bernoulli numbers.

The sine series is
$$\theta(\pi - \theta) = \frac{8}{\pi}\left(\sin\theta + \frac{\sin 3\theta}{3^3} + \frac{\sin 5\theta}{5^3} + ...\right).$$

Notice the much more rapid convergence of the sine series. Three terms give an accuracy of under 0·3 % of the maximum, whereas three of the cosine series give errors reaching over 4 % of the maximum. The opposite feature was found for the function $f(\theta) = 1$, where one term of the cosine series was enough but convergence of the sine series was slow.

Generally speaking if the function tends to zero at a terminus, convergence is faster for the series all of whose terms vanish there; if its derivative is zero, it is better to use the series whose terms have zero derivatives there.

Note that if we put $\theta = \tfrac{1}{2}\pi$ in the sine series we get

$$\pi^3 = 32\left(1 - \frac{1}{3^3} + \ldots\right).$$

The allied series is
$$-\frac{8}{\pi}\left(\cos\theta + \frac{\cos 3\theta}{3^3} + \ldots\right),$$

which converges for all values of θ. There are, however, singularities on the unit circle, otherwise the combined series
$$z + \frac{z^3}{3^3} + \frac{z^5}{5^3} + \ldots$$

would converge for some values of $|z| > 1$, and it does not. From the fact that the sine series represents $\pi\theta + \theta^2$ for $-\pi < \theta < 0$ we see that the first derivative with regard to θ is continuous, the second discontinuous, and may suspect that $\phi + i\psi$ contains terms of the form $(z-1)^2 \log(z-1)$, and analogously $(z+1)^2 \log(z+1)$.

14·06. Integration of Fourier series. We know that a uniformly convergent series of continuous functions represents a continuous function and also that it is integrable term by term. The terms in a Fourier series are continuous functions, but we have seen that they can add up to a discontinuous function. Hence the series in such cases are not uniformly convergent. It may therefore be asked whether the other characteristic property of uniformly convergent series, that of being integrable term by term, also fails for such series. The answer is that it does not. *A Fourier series can always be integrated term by term, not even needing to be convergent, and gives the integral of its defining function.* Let $f(x)$ have the Fourier series in the range 0 to 2π

$$a_0 + \Sigma\,(a_n \cos nx + b_n \sin nx). \tag{1}$$

Then
$$F(x) = \int_0^x f(t)\,dt \tag{2}$$

exists, because the fact that $f(x)$ has a Fourier series implies that it is integrable. Also if $f(x)$ is integrable $F(x)$ is continuous; and if $f(x)$ is bounded $F(x)$ has bounded variation. Then $F(x) - a_0 x$ has a convergent Fourier series, say

$$F(x) - a_0 x = A_0 + \Sigma\,(A_n \cos nx + B_n \sin nx) \quad (0 \leqslant x \leqslant 2\pi), \tag{3}$$

and
$$\pi A_n = \int_0^{2\pi} \{F(x) - a_0 x\} \cos nx\,dx = \left[\frac{1}{n}\{F(x) - a_0 x\}\sin nx\right]_0^{2\pi} - \frac{1}{n}\int_0^{2\pi} \sin nx\{f(x) - a_0\}\,dx$$
$$= -\frac{1}{n}\int_0^{2\pi} f(x) \sin nx\,dx = -\frac{\pi b_n}{n} \quad (n \neq 0), \tag{4}$$

$$\pi B_n = \int_0^{2\pi} \{F(x) - a_0 x\} \sin nx\,dx = \left[-\frac{1}{n}\{F(x) - a_0 x\}\cos nx\right]_0^{2\pi} + \frac{1}{n}\int_0^{2\pi} \cos nx\{f(x) - a_0\}\,dx$$
$$= -\frac{1}{n}\{F(2\pi) - 2\pi a_0\} + \frac{\pi a_n}{n}, \tag{5}$$

and the first term is 0 by the definition of a_0 as $\dfrac{1}{2\pi}\int_0^{2\pi} f(x)\,dx$. Hence

$$F(x) - a_0 x = A_0 + \Sigma\left(\dfrac{a_n}{n}\sin nx - \dfrac{b_n}{n}\cos nx\right). \tag{6}$$

But $F(0) = 0$; hence

$$F(x) = a_0 x + \Sigma\left\{\dfrac{a_n}{n}\sin nx + \dfrac{b_n}{n}(1 - \cos nx)\right\}, \tag{7}$$

which is the result of integrating the Fourier series of $f(x)$ term by term.

Even if $f(x)$ is unbounded, but if $f(x)$ and $|f(x)|$ have improper Riemann integrals, the theorem remains true.

14·061. Differentiation of Fourier series. The corresponding proposition for differentiation is true provided that we understand that the differentiation is to be carried out within the original circle; this is valid because $\phi + i\psi$ is analytic with regard to z_0 within the circle; and the argument may be repeated to show that the limit of the derivative as z_0 approaches the circle is the derivative on the circle, and is equal to the derived series on the circle at any point where the latter converges. But what usually happens is that the first or some later derived series converges nowhere on the circle, even though the derivative itself may exist at almost all points. Thus the function equal to 1 for $0 < \theta < \pi$ and to -1 for $\pi < \theta < 2\pi$ gives the derived series

$$\dfrac{4}{\pi}(\cos\theta + \cos 3\theta + \cos 5\theta + \ldots),$$

which has no obvious meaning, though it is summable by Abel's method; yet the derivative of the function exists and is 0 except at $\theta = \pi$. In a special sense we may still speak of the Fourier series of such a derivative, for even if the simple definition of an integral fails we often define an 'improper' integral as the limit of a sequence of integrals that exclude the exceptional point. But this process can be applied to such a derivative as the last; we take ranges that do not include the value π where the derivative does not exist, and then let a terminus tend to π. In this case all the Fourier coefficients calculated in this way will be 0, and the Fourier series will be 0, thus agreeing with the derivative at every point but one. Let us suppose, then, that $f(x)$ has a derivative except at isolated points, and we want a Fourier series that will represent this derivative at all other points. We take for $0 < x < \pi$

$$f(x) = A_0 + \Sigma A_n \cos nx \tag{1}$$

and assume that $f'(x)$ satisfies the conditions for having a Fourier expansion except possibly at 0 and π. Then suppose

$$f'(x) = \Sigma b_n \sin nx. \tag{2}$$

We have
$$b_n = \dfrac{2}{\pi}\int_0^\pi f'(x)\sin nx\,dx = \dfrac{2}{\pi}\Big[f(x)\sin nx\Big]_0^\pi - \dfrac{2n}{\pi}\int f(x)\cos nx\,dx$$
$$= -nA_n. \tag{3}$$

Thus, if $f'(x)$ has a Fourier sine expansion for $0 < x < \pi$, it can be found by differentiating the cosine expansion of $f(x)$ term by term.

If for $0 < x < \pi$
$$f(x) = \Sigma B_n \sin nx, \tag{4}$$
we assume
$$f'(x) = a_0 + \Sigma a_n \cos nx. \tag{5}$$
Then
$$a_0 = \frac{1}{\pi}\int_0^\pi f'(x)\,dx = \frac{1}{\pi}\{f(\pi) - f(0)\}, \tag{6}$$

$$a_n = \frac{2}{\pi}\int_0^\pi f'(x)\cos nx\,dx = \frac{2}{\pi}\Big[f(x)\cos nx\Big]_{0+}^{\pi-} + \frac{2n}{\pi}\int_0^\pi f(x)\sin nx\,dx$$
$$= \frac{2}{\pi}\{(-1)^n f(\pi-) - f(0+)\} + nB_n. \tag{7}$$

Thus the Fourier series of the derivative of a sine series is not to be found by term-by-term differentiation unless $f(\pi-)$ is equal to $f(0+)$. But if we have the limits of $f(x)$ as x tends to 0 and to π we can find the correct coefficients. If we put
$$f(0+) + f(\pi-) = A,$$
$$f(0+) - f(\pi-) = B,$$
we can write
$$a_0 = -B/\pi,$$
$$a_n = nB_n - 2A/\pi \quad n \text{ odd},$$
$$a_n = nB_n - 2B/\pi \quad n \text{ even},$$
and (5) will be correct except possibly at 0 and π.

14·062. Fluid heated below. This result can sometimes be used in the numerical solution of differential equations. When a thin layer of liquid is heated below, it does not become unstable immediately, viscosity and heat conduction together tending to annul any differences of velocity and temperature. But when the temperature gradient is large enough ascending currents form in some places and descending ones in others, giving a cellular pattern. The temperature is no longer constant over horizontal surfaces, being of the form $\beta z + Z\cos lx\cos my$ in the simplest type of solution, where z is the height, x and y the horizontal coordinates, and Z a function of z. It is found that when the instability first arises Z must satisfy a differential equation of the dimensionless form
$$\left(\frac{d^2}{d\xi^2} - b^2\right)^3 Z + \mu b^2 Z = 0, \tag{1}$$
where ξ is proportional to z and μ to β, the undisturbed temperature gradient. The boundary conditions for two perfectly conducting solid boundaries are
$$Z = 0, \quad Z'' = 0, \quad Z''' - b^2 Z' = 0 \quad (\xi = 0, \pi). \tag{2}$$
We want a value of μ such that the differential equation can be satisfied subject to these six boundary conditions with Z not zero everywhere. Since $Z = 0$ at the boundaries it is natural to assume a sine series
$$Z = \Sigma A_n \sin n\xi. \tag{3}$$
Then
$$Z' = \Sigma nA_n \cos n\xi \tag{4}$$
by 14·061 (7); and then by 14·061 (3)
$$Z'' = -\Sigma n^2 A_n \sin n\xi. \tag{5}$$
But $Z'' = 0$ at the termini. Hence
$$Z''' = -\Sigma n^3 A_n \cos n\xi, \tag{6}$$
$$Z^{(4)} = \Sigma n^4 A_n \sin n\xi. \tag{7}$$

There is no restriction on $Z^{(4)}$ at the boundaries. We therefore put

$$\frac{2}{\pi}\{Z^{(4)}(0)+Z^{(4)}(\pi)\} = A, \tag{8}$$

$$\frac{2}{\pi}\{Z^{(4)}(0)-Z^{(4)}(\pi)\} = B, \tag{9}$$

and
$$Z^{(5)} = -\frac{B}{2}+\Sigma\{n^5 A_n-(A,B)\}\cos n\xi, \tag{10}$$

A or B being taken according as n is odd or even.
Finally,
$$Z^{(6)} = -\Sigma(n^6 A_n-(A,B)n\}\sin n\xi \tag{11}$$

and
$$\left(\frac{d^2}{d\xi^2}-b^2\right)^3 Z+\mu b^2 Z = -\Sigma(n^2+b^2)^3 A_n\sin n\xi +\Sigma\{(A,B)n+\mu b^2 A_n\}\sin n\xi = 0. \tag{12}$$

This is the Fourier series of the zero function on the left and therefore the coefficients of all terms must vanish; thus

$$\left.\begin{array}{rl}\{(n^2+b^2)^3-\mu b^2\} A_n = nA & n\text{ odd}\\ = nB & n\text{ even}\end{array}\right\} \tag{13}$$

Also we have the third pair of boundary conditions

$$Z'''-b^2 Z' = 0, \tag{14}$$

whence
$$\Sigma(n^2+b^2)nA_n = 0, \tag{15}$$

$$\Sigma(-)^n(n^2+b^2)nA_n = 0. \tag{16}$$

Hence the sums of terms with n odd and even vanish separately; and by substitution

$$\Sigma\frac{n^2(n^2+b^2)}{(n^2+b^2)^3-\mu b^2}A = 0 \quad n\text{ odd}, \tag{17}$$

$$\Sigma\frac{n^2(n^2+b^2)}{(n^2+b^2)^3-\mu b^2}B = 0 \quad n\text{ even}. \tag{18}$$

There are thus two distinct types of solution, those depending on odd values of n being symmetrical about the median plane $\xi = \frac{1}{2}\pi$, the others antisymmetrical. We wish to find the least value of μ for each type. Computation is convenient for the following reason. The variation of any term due to a change of μ is by a factor decreasing like n^{-6}, and the terms themselves decrease like n^{-2}. Hence, if we can compute the series obtained by putting $\mu = 0$, the correcting series depending on μ will consist of terms that decrease like n^{-8}, and will therefore be very rapidly convergent at the start. Taking the odd solution first, we have

$$\tanh \tfrac{1}{2}\pi b = \frac{4b}{\pi}\left\{\frac{1}{b^2+1}+\frac{1}{b^2+9}+\ldots\right\}, \tag{19}$$

whence
$$\sum_{n\text{ odd}}\frac{n^2}{(n^2+b^2)^2} = \tfrac{1}{8}\pi\left(\frac{d}{db}+\frac{1}{b}\right)\tanh \tfrac{1}{2}\pi b$$

$$= \tfrac{1}{8}\pi\left(\frac{1}{b}\tanh \tfrac{1}{2}\pi b+\tfrac{1}{2}\pi\operatorname{sech}^2\tfrac{1}{2}\pi b\right) = T(b) \text{ say}. \tag{20}$$

This can be computed directly as a function of b; and by subtraction

$$\Sigma\frac{n^2\mu b^2}{(n^2+b^2)^2\{(n^2+b^2)^3-\mu b^2\}}+T(b) = 0. \tag{21}$$

The second term in the series being small we can write this as

$$\frac{\mu b^2}{(1+b^2)^2\{(1+b^2)^3 - \mu b^2\}} + T(b) + K = 0, \tag{22}$$

where
$$K = \sum_{n=3,5,\ldots} \frac{n^2 \mu b^2}{(n^2+b^2)^2\{(n^2+b^2)^3 - \mu b^2\}} \tag{23}$$

or
$$\mu b^2 = \frac{(1+b^2)^3}{1 - 1^2/(1^2+b^2)^2\{T(b)+K\}}. \tag{24}$$

We first neglect K, assume a series of trial values of b, and work out μ for each. The results are

b	μ	Δ	Δ^2	Δ^3	Corrected μ
0·90	17·836				
0·95	17·652	−184	+141		17·587
1·00	17·609	− 43	+125	−16	17·537
1·05	17·691	+ 82	+118	− 7	17·613
1·10	17·891	+200			

We now work out the first two terms of K for $b = 0.95$ and 1.00, using our approximate values of μ. They are 0·0014 and 0·0016 respectively, $T(b)$ being 0·5177 and 0·4581. Allowing for them in (22) and interpolating for the minimum we have $\mu = 17.536$ at $b = 0.995$. The parameters a and λ used in other treatments are $\pi b = 3.14$ and $\pi^4 \mu = 1708.2$ respectively. Southwell and Pellew, solving directly in complex exponential functions, get $\lambda = 1707.8$.

The lowest mode with even n can be found similarly; starting with

$$\sum_{n \text{ even}} \frac{b^2}{b^2+n^2} = \tfrac{1}{4}\pi \coth \tfrac{1}{2}\pi b - \frac{1}{2b}, \tag{25}$$

we get
$$T(b) = \sum \frac{n^2}{(n^2+b^2)^2} = \tfrac{1}{8}\pi\left(\frac{d}{db}+\frac{1}{b}\right)\coth \tfrac{1}{2}\pi b = \tfrac{1}{8}\pi\left(\frac{1}{b}\coth \tfrac{1}{2}\pi b - \tfrac{1}{2}\pi \operatorname{cosech}^2 \tfrac{1}{2}\pi b\right), \tag{26}$$

and, proceeding as before, we find
$$\mu b^2 = \frac{(4+b^2)^3}{1 - 4/(4+b^2)^2\{T(b)+K\}}, \tag{27}$$

where
$$K = \frac{16\mu b^2}{(16+b^2)^2\{(16+b^2)^3 - \mu b^2\}} + \frac{36\mu b^2}{(36+b^2)^2\{(36+b^2)^3 - \mu b^2\}} + \cdots. \tag{28}$$

Solutions neglecting K are

b	μ	b	μ
1·5	186·9	1·7	182·8
1·55	185·1	1·75	183·0
1·6	183·9	1·8	183·5
1·65	183·0	1·85	184·4

For $b = 1.70$ the correcting terms K are 0·0040, while $T(b)$ is 0·2213. The corrected μ is then 180·8.

The interest of even n, pointed out by Southwell and Pellew, is in the fact that in these solutions the conditions $Z = Z'' = 0$ are satisfied at $\xi = \tfrac{1}{2}\pi$, with the further condition $Z^{(4)} - b^2 Z'' = 0$, which is the condition for a free surface, replacing $Z''' - b^2 Z' = 0$. Hence this solution is the solution for a liquid with a perfectly conducting rigid boundary at the bottom, and a depth half that used in the first case. We get b and μ for a layer with the same depth as in the first case by multiplying by 2 and 16 respectively. Restoring also the factor π^4 we have $\lambda = 1100.6$. Southwell and Pellew get $\lambda = 1100$.

The problem has a considerable literature.* The possibility of using the rules for finding the Fourier series for the derivatives of a function with no singularities within the range was suggested by Dr S. Goldstein. It will be seen that the rapid convergence depends on the fact that the terms of the series Σn^{-8} decrease extremely rapidly at the beginning, and on the possibility of combining the parts not involving μ into a known function of b. It is quite convenient to use considering that the solution depends on a sixth order equation with two adjustable parameters.

---

* Jeffreys, *Phil. Mag.* (7) 2, 1926, 833–44; *Proc. Roy. Soc.* A, 118, 1928, 195–208; A. Pellew and R. V. Southwell, *Proc. Roy. Soc.* A, 176, 1940, 312–43.

14·07. The Gibbs phenomenon. This is a peculiarity of the sum of a finite number of terms of a Fourier series when the function has a simple discontinuity. It is sufficient to consider the function that is equal to 1 for $0 < x < \pi$ and to -1 when $\pi < x < 2\pi$. From 14·04 (3) the sum of the terms up to those in nx is

$$S_n(x) = \frac{1}{2\pi} \int_0^\pi \frac{\sin(n+\tfrac{1}{2})(t-x)}{\sin\tfrac{1}{2}(t-x)} dt - \frac{1}{2\pi} \int_\pi^{2\pi} \frac{\sin(n+\tfrac{1}{2})(t-x)}{\sin\tfrac{1}{2}(t-x)} dt$$

$$= \frac{1}{2\pi} \int_0^\pi \left\{ \frac{\sin(n+\tfrac{1}{2})(t-x)}{\sin\tfrac{1}{2}(t-x)} - \frac{\sin(n+\tfrac{1}{2})(t+x)}{\sin\tfrac{1}{2}(t+x)} \right\} dt$$

$$= \frac{1}{2\pi} \int_{-x}^{\pi-x} \frac{\sin(n+\tfrac{1}{2})\theta}{\sin\tfrac{1}{2}\theta} d\theta - \frac{1}{2\pi} \int_x^{\pi+x} \frac{\sin(n+\tfrac{1}{2})\theta}{\sin\tfrac{1}{2}\theta} d\theta$$

$$= \frac{1}{2\pi} \left\{ \int_{-x}^x - \int_{\pi-x}^{\pi+x} \right\} \frac{\sin(n+\tfrac{1}{2})\theta}{\sin\tfrac{1}{2}\theta} d\theta.$$

We may regard the first integral as representing the effects of the discontinuity at 0, the second of that at π. For x small the second will be small, since $\sin\tfrac{1}{2}\theta$ is about 1 when θ is near π. For the first we write

$$n+\tfrac{1}{2} = m, \quad m\theta = \xi,$$

$$S_n \doteqdot \frac{1}{\pi} \int_0^{mx} \frac{\sin\xi\, d\xi}{m\sin\xi/2m}.$$

This is 0 when $x = 0$, and increases till $mx = \pi$. The maximum value then is

$$\frac{1}{\pi} \int_0^\pi \frac{\sin\xi\, d\xi}{m\sin\xi/2m} \doteqdot \frac{2}{\pi} \int_0^\pi \frac{\sin\xi}{\xi} d\xi$$

when m is large. If the upper limit was ∞ the integral would be $\tfrac{1}{2}\pi$ and thus give the limit 1, as we should expect. But

$$\int_0^\infty \frac{\sin\xi}{\xi} d\xi = \int_0^\pi \frac{\sin\xi}{\xi} d\xi + \int_\pi^{3\pi} \frac{\sin\xi}{\xi} d\xi + \int_{3\pi}^{5\pi} \frac{\sin\xi}{\xi} d\xi + \ldots,$$

and every term after the first is negative. Hence

$$\frac{2}{\pi} \int_0^\pi \frac{\sin\xi}{\xi} d\xi > 1.$$

It is actually about 1·179.[*] Hence near the discontinuity at $x = 0$ the sum of a finite number of terms of the Fourier series overshoots the mark appreciably. Increasing the number of terms does not remove this peculiarity; it merely shifts it nearer to the discontinuity.

The explanation is easy. The sum of a finite number of terms of the series is a continuous function, and the difference between it and $f(x)$ is orthogonal to all the trigonometric terms up to $\cos nx$ and $\sin nx$. But for some distance from 0, $S_n(x)$ is less than $f(x)$ because $f(x)$ jumps to 1 immediately and $S_n(x)$ does not. This difference will make a negative contribution to $\int \{S_n(x) - f(x)\} \sin mx\, dx$ ($m \leqslant n$), which must be compensated by a positive contribution somewhere else. But it can be compensated approximately for all m not too large by having $S_n(x) > f(x)$ in an adjacent range.

[*] The numerical value is given wrongly in several books.

This series could hardly be used for computation, and the Gibbs phenomenon is only another warning that Fourier series are not of much use for direct computation unless the coefficients decrease at least as fast as n^{-3}.

14·08. Weierstrass's theorem on approximations by polynomials. *If a function is continuous in any finite interval, at the ends of which it has the same value, a finite number of harmonic terms can be found such that their sum differs from the function by less than ϵ at every point of the interval; and a polynomial can be found with the same property.*

The proof of Fourier's theorem given in 14·04 assumed the function to have bounded variation; in a certain sense we shall see that this assumption is unnecessary. Evidently by a linear transformation of the independent variable we can make the interval 0 to 2π, and the function will also be continuous with regard to this variable. Then the conditions on $f(x)$ are that it is continuous for $0 \leqslant x \leqslant 2\pi$ and $f(0) = f(2\pi)$. Now, since a continuous function is uniformly continuous, for a given positive ω we can choose a set of points of subdivision $0, x_1, x_2, ..., x_m, 2\pi$ such that the upper and lower bounds of $f(x)$ in every interval differ by less than ω. In each interval x_r to x_{r+1} take the linear function that agrees with $f(x)$ at x_r and x_{r+1}. Then this function differs from $f(x)$ by not more than ω, since it always lies between the upper and lower bounds of $f(x)$ in the interval. We thus have a function $g(x)$ defined for each interval, continuous at all points, including the points of subdivision, and of bounded variation ($\leqslant (m+1)\omega$) between 0 and 2π. It therefore can be expressed as a Fourier series, and it nowhere differs from $f(x)$ by more than ω. The introduction of $g(x)$ cuts out small but rapid fluctuations such as those of $x \sin 1/x$ near $x = 0$, which could make $f(x)$ have infinite total variation without being discontinuous.

Now consider the contributions to the Fourier coefficients from the range x_r to x_{r+1}. We have in this range

$$g(x) = \frac{(x-x_r)f(x_{r+1}) + (x_{r+1}-x)f(x_r)}{x_{r+1}-x_r} = ax+b \quad \text{say};$$

$$\frac{1}{2\pi}\int_{x_r}^{x_{r+1}} g(x)\,dx = \frac{1}{4\pi}(x_{r+1}-x_r)\{f(x_r)+f(x_{r+1})\},$$

$$\frac{1}{\pi}\int_{x_r}^{x_{r+1}} g(x)\cos nx\,dx = \frac{1}{n\pi}\left[(ax+b)\sin nx\right]_{x_r}^{x_{r+1}} + \frac{a}{n^2\pi}\left[\cos nx\right]_{x_r}^{x_{r+1}}$$

$$= \frac{1}{n\pi}\{f(x_{r+1})\sin nx_{r+1} - f(x_r)\sin nx_r\} + \frac{f(x_{r+1})-f(x_r)}{(x_{r+1}-x_r)n^2\pi}(\cos nx_{r+1} - \cos nx_r),$$

$$\frac{1}{\pi}\int_{x_r}^{x_{r+1}} g(x)\sin nx\,dx$$

$$= -\frac{1}{n\pi}\{f(x_{r+1})\cos nx_{r+1} - f(x_r)\cos nx_r\} + \frac{f(x_{r+1})-f(x_r)}{(x_{r+1}-x_r)n^2\pi}(\sin nx_{r+1} - \sin nx_r).$$

When contributions for different intervals are added, the terms in the curled brackets all cancel, those from 2π cancelling those from 0. Hence

$$g(x) = \sum_{r=0}^{m} \frac{1}{4\pi}(x_{r+1}-x_r)\{f(x_{r+1})+f(x_r)\}$$
$$+ \sum_{n=1}^{\infty} \sum_{r=0}^{m} \frac{f(x_{r+1})-f(x_r)}{(x_{r+1}-x_r)n^2\pi}(\cos n(x-x_{r+1}) - \cos n(x-x_r)),$$

where $x_0 = 0$, $x_{m+1} = 2\pi$. The terms after the constant are all less than some constant multiple of $1/n^2$. Hence the series is uniformly convergent. It is therefore possible to choose n so that for every x the sum of terms up to those in nx differs from $g(x)$ by less than ω and therefore from $f(x)$ by less than 2ω.

We thus have a finite set of harmonic terms. But $\cos nx$ and $\sin nx$ can be expanded in uniformly convergent power series. We expand each harmonic term up to such a power of x, say x^s, that the total error committed in neglecting terms after x^s in all terms is less than ω. Then collecting terms in like powers of x we have a polynomial in x of degree s that nowhere differs from $f(x)$ by as much as 3ω.

For a given ϵ, we can take $\omega = \tfrac{1}{3}\epsilon$; then the harmonic series up to terms in nx nowhere differs from $f(x)$ by as much as $\tfrac{2}{3}\epsilon$, and the polynomial nowhere by ϵ. This proves the theorem.

For the polynomial approximation it is unnecessary that $f(0) = f(2\pi)$. For we can take

$$f(x) = \frac{1}{2\pi}\{xf(2\pi) + (2\pi - x)f(0)\} + h(x),$$

in which the first expression is a polynomial and the second satisfies the conditions imposed in the main theorem.

This theorem is important partly because it makes it possible to replace functions that are continuous but have not bounded variation by functions of two of the simplest possible types, with a known limit of error. But it will also replace a continuous function that is not differentiable by one that has derivatives of all orders, and if this is done many proofs can be simplified.[a]

The Fourier series for $f(x)$ and $g(x)$ are obviously different, but will agree closely in the early terms. The polynomial expression will not in general be identical with the interpolation polynomial found from $f(0), f(x_1), \ldots, f(2\pi)$ by divided differences, since the neglected terms will not vanish exactly at the values in question, and to obtain the requisite accuracy it may be necessary to keep *more* than $m+1$ terms if $f(x)$ fluctuates rapidly. In the latter case numerical interpolation might fail owing to large higher derivatives.

14·081. Extension of Weierstrass's approximation theorem. *If $\int_0^{2\pi} f(x)\,dx$ exists, the upper and lower bounds of $f(x)$ differing by M, then for any ϵ, δ a finite number of harmonic terms can be found such that their sum differs from $f(x)$ by less than ϵ at every point of the interval, except possibly within a set of subintervals of total length δ, within which the sum nowhere differs from $f(x)$ by more than $M + \epsilon$; and a polynomial can be found with the same property.*

We use du Bois-Reymond's necessary and sufficient condition (1·1011) for the existence of the Riemann integral. For an arbitrary ω we can enclose all discontinuities of $f(x)$ where the leap is $\geq \omega$ within a finite number m of subintervals of total length δ, each discontinuity being at an interior point of the subinterval, unless it is 0 or 2π, when we take the discontinuity as an end point. Call these intervals G. As in 14·08 we divide the remainder of the interval $(0, 2\pi)$ into n subintervals in each of which the leap of the function is less than ω, and construct a continuous function $g(x)$ by linear interpolation. $g(x)$ is of bounded variation, $\leq n\omega + mM$. In the m subintervals G, $|g(x) - f(x)| \leq M$; in the rest,

$$|g(x) - f(x)| < \omega.$$

Then we can find a sum of a finite number of harmonic terms, nowhere differing from $g(x)$ by more than ω, and a polynomial nowhere differing from $g(x)$ by more than 2ω; the sum of harmonic terms therefore differs from $f(x)$ nowhere by more than $M+2\omega$, and except in the subintervals G it nowhere differs from $f(x)$ by as much as 2ω. For the polynomial we need only replace 2ω by 3ω. Taking $\omega = \tfrac{1}{2}\epsilon$ or $\tfrac{1}{3}\epsilon$ (which is independent of δ) we have the required results.

Corollary. Let T be a sum of harmonic terms with the property stated in the above theorem, and consider the integral

$$I = \int_0^{2\pi} \{f(x) - T\}^2 \, dx.$$

This exists if $f(x)$ has a Riemann integral. Then

$$I < (M+\epsilon)^2 \delta + (2\pi - \delta)\epsilon^2.$$

If η is an arbitrary positive quantity, we can choose ϵ so that $2\pi\epsilon^2 < \tfrac{1}{2}\eta$, and then δ so that $(M+\epsilon)^2\delta < \tfrac{1}{2}\eta$; then $I < \eta$. Therefore if $f(x)$ has a Riemann integral we can find a sum T of a finite number of harmonic terms such that the integral I is arbitrarily small.

14·09. Approximation by least squares: Parseval's theorem. Let $S'_n(x)$ be any finite sum of the form

$$S'_n(x) = A'_0 + \sum_{r=1}^{n}(A'_r \cos rx + B'_r \sin rx). \tag{1}$$

and let $$S_n(x) = A_0 + \sum_{r=1}^{n}(A_r \cos rx + B_r \sin rx),$$

where A_r, B_r are the Fourier coefficients of $f(x)$. Then if $f(x)$ is integrable

$$\int_0^{2\pi}\{f(x)-S'_n(x)\}^2\,dx = \int_0^{2\pi}\{f(x)\}^2\,dx - 2\int_0^{2\pi} f(x)\,S'_n(x)\,dx + \int_0^{2\pi}\{S'_n(x)\}^2\,dx. \tag{2}$$

Now all products of the form $\cos rx \sin sx$ have zero integrals; so have all of the forms $\cos rx \cos sx$ and $\sin rx \sin sx$ with $r \neq s$. Hence

$$\int_0^{2\pi}\{S'_n(x)\}^2\,dx = 2\pi A'^2_0 + \pi\sum_{r=1}^{n}(A'^2_r + B'^2_r). \tag{3}$$

Also $$\int_0^{2\pi} f(x)\,dx = 2\pi A_0, \quad \int_0^{2\pi} f(x)\cos rx\,dx = \pi A_r, \quad \int_0^{2\pi} f(x)\sin rx\,dx = \pi B_r, \tag{4}$$

and therefore $$\int_0^{2\pi} f(x)\,S'_n(x)\,dx = 2\pi A_0 A'_0 + \pi\sum_{r=1}^{n}(A_r A'_r + B_r B'_r) \tag{5}$$

and $$\int_0^{2\pi}\{f(x)-S'_n(x)\}^2\,dx = \int_0^{2\pi}\{f(x)\}^2\,dx + 2\pi(A'^2_0 - 2A_0 A')$$
$$+ \pi\sum_{r=1}^{n}(A'^2_r - 2A_r A'_r + B'^2_r - 2B_r B'_r). \tag{6}$$

But $$A'^2_r - 2A_r A'_r = (A'_r - A_r)^2 - A^2_r, \quad B'^2_r - 2B_r B'_r = (B'_r - B_r)^2 - B^2_r \tag{7}$$

and $$\int_0^{2\pi}\{f(x)-S'_n(x)\}^2\,dx = \int_0^{2\pi}\{f(x)\}^2\,dx - 2\pi A^2_0 - \pi\sum_{r=1}^{n}(A^2_r + B^2_r)$$
$$+ 2\pi(A'_0 - A_0)^2 + \pi\sum_{r=1}^{n}\{(A'_r - A_r)^2 + (B'_r - B_r)^2\}. \tag{8}$$

But the terms involving A'_r and B'_r in the last expression are all $\geqslant 0$ and vanish when $A'_r = A_r$, $B'_r = B_r$. Hence: *if we measure the discrepancy between the function $f(x)$ and a trigonometrical expression of the form* (1) *by the integral of the square of the difference between them, the discrepancy is least if the coefficients are taken to be the Fourier coefficients up to A_n, B_n.* This result may be compared with that of our original method of determining the coefficients so that the trigonometrical expression would agree exactly with $f(x)$ at equally spaced points. Here we do not attempt to give an exact fit at any specified points, but aim at the best fit with the function as a whole, measuring the discrepancy as a whole by the integral of its square. We find that the best fit is the Fourier expansion up to any order we choose. Results of this type are found for all expressions in terms of orthogonal functions, and are not confined to Fourier expansions.

Again, let $S'_n(x)$ be a function T defined as in the corollary to 14·081, so that for an arbitrarily small ϵ

$$\int_0^{2\pi} \{f(x) - S'_n(x)\}^2 dx < \epsilon. \tag{9}$$

Hence

$$\int_0^{2\pi} \{f(x) - S_n(x)\}^2 dx < \epsilon \tag{10}$$

and therefore

$$\left| \int_0^{2\pi} \{f(x)\}^2 dx - 2\pi A_0^2 - \pi \sum_1^n (A_r^2 + B_r^2) \right| < \epsilon. \tag{11}$$

Hence

$$A_0^2 + \tfrac{1}{2} \sum_{r=1}^n (A_r^2 + B_r^2) \to \frac{1}{2\pi} \int_0^{2\pi} \{f(x)\}^2 dx. \tag{12}$$

This is *Parseval's theorem.*

Note that if $f(x)$ has a Riemann integral and is otherwise unrestricted, it may be of unbounded variation, and its discontinuities may not be simple; then its Fourier series may not converge or even be summable by Abel's method for some values of x.

If $f(x)$ and $\{f(x)\}^2$ are integrable, even by improper integrals, (8) still follows, and

$$\int_0^{2\pi} \{f(x) - S_n(x)\}^2 dx = \int_0^{2\pi} \{f(x)\}^2 dx - 2\pi A_0^2 - \pi \sum_{r=1}^n (A_r^2 + B_r^2). \tag{13}$$

The left side is not negative, and by hypothesis $\int_0^{2\pi} \{f(x)\}^2 dx$ is finite. Hence the sum of positive terms $\sum_{r=1}^n (A_r^2 + B_r^2)$ cannot increase without limit as n increases. Thus: *if a function and its square have improper integrals, the sum of the squares of its Fourier coefficients is convergent.*

As a corollary, if $f(x)$ and $\{f(x)\}^2$ are integrable from 0 to 2π, *even by improper integrals,* $\int_0^{2\pi} f(x) \cos nx\, dx$ and $\int_0^{2\pi} f(x) \sin nx\, dx$ tend to zero. This fact is made the basis of some modern treatments of Fourier's theorem and the trigonometric interpolation polynomials.*

14·10. Harmonic analysis: correction for averaging. In a modification of the problem of 14·01 that often occurs in practice the data are not the actual values of $f(x)$ at $x_r = r\lambda$, but means of $f(x)$ over ranges centred on x_r. In studying the diurnal variation of atmospheric pressure, for instance, it would be natural to read the barometer at hourly

---

* Dunham Jackson, *The Theory of Approximation,* 1930; *Fourier Series and Orthogonal Polynomials,* 1941.

intervals, and we should then have the conditions of 14·01. But in studying wind the data would be the runs of the anemometer in each hour, and therefore the mean wind for each hour. The same formulae could be fitted to such data, but would not give the best determination of the harmonic components of the wind itself. In fact if we take the mean of e^{isx} over the range $(r-\tfrac{1}{2})\lambda$ to $(r+\tfrac{1}{2})\lambda$, it is

$$\frac{1}{\lambda}\int_{(r-\frac{1}{2})\lambda}^{(r+\frac{1}{2})\lambda} e^{isx}\,dx = \frac{2\sin\tfrac{1}{2}s\lambda}{s\lambda} e^{isr\lambda}$$

and harmonic analysis applied to the mean values will underestimate the coefficients by a factor $(2/s\lambda)\sin\tfrac{1}{2}s\lambda$. The coefficients found from mean values must therefore be divided by this quantity to give the harmonic development of $f(x)$.

14·101. Empirical periodicities: the periodogram. The method of 14·01 fits a set of harmonic functions exactly to a finite number of values of a function. If the solution is to be used outside the original range (e.g. for prediction) it will be periodic, all terms being periodic in the interval used for analysis. But the function may be a periodic one with a period that is not a submultiple of the interval used, and we may wish to determine its amplitude and phase and possibly even the period itself. In the theory of the tides, for instance, we know from general theory that there must be harmonic components with periods calculable from the rates of the earth's rotation and revolution and the moon's orbital period, but the amplitudes and phases cannot be calculated on account of the complicated form of the ocean. What can be done, however, is to instal a tide gauge in each harbour where predictions are required. This records the tide height at regular intervals, often hourly. The amplitude and phase associated with each period can then be determined so as to fit the observations, and once these are known they provide a basis of prediction. A year's observations suffice to make predictions as far ahead as we like (unless of course the harbour silts up). This semi-empirical method, the periods being taken from astronomy and the amplitudes and phases from direct observation, was introduced by Sir G. H. Darwin and is systematically carried out at the Liverpool Tidal Institute under Prof. J. Proudman and Dr A. T. Doodson.

If only one period was concerned the calculation would be simple; the coefficients in an expression of the form $a + A\cos\gamma t + B\sin\gamma t$ could be found from only three observations, though more would be combined by the method of least squares for greater accuracy. Actually the tide contains 7 lunar and 7 solar components of incommensurable periods, and a method of successive approximation has to be used. The interval is chosen so as to be as nearly as possible a multiple of the *two* periods with the largest amplitudes, and the coefficients are estimated from the formulae

$$A\Sigma\cos^2\gamma t = \Sigma f(t)\cos\gamma t, \quad B\Sigma\sin^2\gamma t = \Sigma f(t)\sin\gamma t,$$

summation being over the times of observation. With this choice of interval the terms in $f(t)$ arising from one pair of components will produce a negligible effect on the estimates of those from the other pair. The contributions from the largest terms are then subtracted from the observed values, and then the residuals are analysed for further terms. Since these will not in general repeat themselves exactly in the original interval it may be necessary, after determining all the terms that should be possible theoretically, to return to the start and determine corrections to the largest terms from the residuals.

14·101 Detection of unknown periods

When a period is suspected but not already known, the estimation is much more difficult. Suppose that in 14·01
$$f(x) = C \exp i(\gamma x + \alpha),$$
where C and α are real and γ is not an integer. If we work out C_m from 14·01 (5) we get
$$nC_m = Ce^{i\alpha} \sum_{r=0}^{n-1} \exp\left\{\frac{2\pi i r}{n}(\gamma - m)\right\}$$
$$= Ce^{i\alpha} \frac{1 - \exp\{2\pi i(\gamma - m)\}}{1 - \exp\{2\pi i(\gamma - m)/n\}}$$
$$= C \frac{\sin \pi(\gamma - m)}{\sin \pi(\gamma - m)/n} \exp\left\{i\alpha + \pi i\left(1 - \frac{1}{n}\right)(\gamma - m)\right\}.$$

Evidently $|C_m|$ is largest when m is as near as possible to γ. If m is taken to be the integral part of γ, and therefore $m+1$ to be the smallest integer greater than γ, the corresponding terms in the harmonic analysis will be revealed by having the largest coefficients, and their phases at $x = 0$ will be nearly opposite. Taking the real part and now taking $f(x)$ to be $C\cos(\gamma x + \alpha)$ we shall have for the largest terms
$$C \frac{\sin \pi(\gamma - m)}{\sin \pi(\gamma - m)/n} \cos\left\{mx + \alpha + \pi\left(1 - \frac{1}{n}\right)(\gamma - m)\right\}$$
$$+ C \frac{\sin \pi(m+1-\gamma)}{\sin \pi(m+1-\gamma)/n} \cos\left\{(m+1)x + \alpha + \pi\left(1 - \frac{1}{n}\right)(\gamma - m) - \pi\left(1 - \frac{1}{n}\right)\right\}.$$

The ratio of the amplitudes then gives an equation for γ, and C is determined. The phase of either term then determines α.

This kind of analysis is most used for the detection of natural periodicities, of which perhaps the best-known is the sunspot period. In practice it is complicated by irregular disturbances, so that the actual variation is not simple harmonic though the greater part of it may be represented by a few harmonic terms. If there are n observed values y_r with mean A_0, let
$$\Sigma (y_r - A_0)^2 = (n-1)s^2.$$
But any harmonic coefficient A_m or B_m ($m \geq 1$) would contribute $\frac{1}{2}A_m^2$ or $\frac{1}{2}B_m^2$ to the mean square, and there are $n-1$ of them. (It is easy to verify that all product terms cancel.) Thus the average of $\frac{1}{2}A_m^2$ or $\frac{1}{2}B_m^2$ is $s^2/(n-1)$. This would be true even if the y_r were a wholly random set. Consequently a set of harmonic coefficients by itself gives no evidence that the periods can be used for prediction unless some of the amplitudes have squares much greater than $4s^2/(n-1)$. In the method just described it is best to rely on the largest coefficients because these are less affected proportionally by any irregular variation that may be present.

The best way of estimating the uncertainty introduced by the irregular variation depends on the circumstances.* One that often succeeds is to divide the range up into equal stretches and do a separate analysis for each. If each stretch contains n observations the phase of a harmonic with period n/γ will increase by $2\pi\gamma$ from one stretch to the next, and if we take the terms in $\cos mx$ and $\sin mx$ in the analysis together the phase will

* Cf. Jeffreys, *Theory of Probability*, pp. 291–5; M.N.R.A.S. 100, 1940, 139–55; *Gerlands Beitr. z. Geophysik*, 53, 1938, 111–39.

increase by $2\pi(\gamma-m)$. Hence the determinations of the phase for several stretches give a set of linear equations for γ and α, which can be solved by the method of least squares, and the residuals lead to an estimate of uncertainty. Without some such precaution periodicities found by harmonic analysis and not predicted by previous theoretical considerations should be mistrusted, as many complications are capable of giving spurious periods; not more than a tenth of those that have been asserted will bear a proper statistical examination.

14·11. Fourier's integral theorem: preliminary discussion. In Fourier's theorem in the form

$$f(x) = \frac{1}{2\pi}\int_{-\pi}^{\pi} f(u)\,du + \sum_{n=1}^{\infty}\frac{1}{\pi}\int_{-\pi}^{\pi} f(u)\cos n(u-x)\,dx \tag{1}$$

(subject to certain restrictions already stated) put

$$x = X/T, \quad u = U/T, \quad f(x) = F(X). \tag{2}$$

Then $$F(X) = \frac{1}{2\pi T}\int_{-\pi T}^{\pi T} F(U)\,dU + \sum_{n=1}^{\infty}\frac{1}{\pi T}\int_{-\pi T}^{\pi T} F(U)\cos\frac{n}{T}(U-X)\,dU. \tag{3}$$

This extends the theorem to an arbitrary interval. Now take T very large, and put

$$n/T = \kappa. \tag{4}$$

Then the values of κ for consecutive terms differ by $1/T$, and the first term will tend to zero if $\int_{-\infty}^{\infty} F(U)\,dU$ converges. Hence we may expect that

$$F(X) = \frac{1}{\pi}\int_{0}^{\infty} d\kappa \int_{-\infty}^{\infty} F(U)\cos\kappa(U-X)\,dU \tag{5}$$

for values of X where $F(X)$ is continuous, and that at simple discontinuities of $F(X)$ the repeated integral will be equal to $\frac{1}{2}\{F(X+)+F(X-)\}$. This is Fourier's integral theorem. But we may also expect that the occurrence of a repeated infinite integral will introduce new problems of convergence. The seriousness of these may be seen from the example

$$F(X) = \cos\alpha X. \tag{6}$$

Here the integral with regard to U is infinite for $\kappa = \alpha$ and indeterminate for all other values of κ, and the repeated integral is meaningless. This breakdown in the simplest possible case may serve as a warning against the common belief among physicists that Fourier's integral theorem is easy.

Even if $$F(X) = \frac{1}{X}\sin\alpha X, \tag{7}$$

although $\int_{-\infty}^{\infty} F(X)\,dX$ exists, the integral with regard to U is infinite for $\kappa = \alpha$. Convergence of $\int_{-\infty}^{\infty} F(X)\,dX$ is therefore not a sufficient condition for the truth of the theorem.

If, however, $\int_{-\infty}^{\infty} |F(X)|\,dX$ exists, the integral with regard to U will converge uniformly for all κ; and as for the series theorem, if the repeated integral is to have a definite value for all X, we shall expect to need a further condition, such as that $F(X)$ has bounded variation, and therefore that its discontinuities, if any, are simple.

14·111. A related potential problem.

We have seen (6·093) that if a potential function ϕ is given over a plane and x is the normal distance of a point P from the plane, all charges being either on the plane or on the side of it remote from P, and if ϕ satisfies some suitable condition at large distances,

$$2\pi\phi_P = \iint \frac{\phi_Q}{R^3} x\, dS. \qquad (1)$$

If Q is $(0, \eta, \zeta)$, and P is (x, y, z) and if ϕ_Q is a function of η only, say $f(\eta)$,

$$2\pi\phi_P = \int_{-\infty}^{\infty} x\, d\eta \int_{-\infty}^{\infty} \frac{f(\eta)}{\{x^2 + (\eta-y)^2 + (\zeta-z)^2\}^{3/2}} d\zeta$$

$$= 2\int_{-\infty}^{\infty} \frac{xf(\eta)}{x^2 + (\eta-y)^2} d\eta. \qquad (2)$$

Also

$$\int_0^{\infty} e^{-\kappa x} \cos\kappa(\eta-y)\, d\kappa = \frac{x}{x^2 + (\eta-y)^2} \qquad (3)$$

and therefore (2) is equivalent to

$$\phi_P = \frac{1}{\pi} \int_{-\infty}^{\infty} f(\eta)\, d\eta \int_0^{\infty} e^{-\kappa x} \cos\kappa(\eta-y)\, d\kappa. \qquad (4)$$

If we reverse the order of integration and then put $x = 0$ we get Fourier's integral theorem again. But we clearly cannot put $x = 0$ before reversing the order of integration. On the other hand if $x > 0$, 14·111 (2) exists in much wider conditions than 14·11 (5); it exists if $f(\eta)$ is integrable over any finite interval and if Y exists such that $\int_Y^{\infty} \frac{f(\eta)}{\eta^2} d\eta$ and $\int_{-\infty}^{-Y} \frac{f(\eta)}{\eta^2} d\eta$ converge. In fact $f(\eta)$ might be $\eta\cos\alpha\eta$ or $\eta\sin\alpha\eta$, so far as we can see at present. If then (2) tends to a limit $f(y)$ when x tends to 0, we shall have a much more general theorem.

This approach is of further interest because

$$\frac{x}{x^2 + (\eta-y)^2} = \Re\frac{1}{x + i(y-\eta)}, \quad e^{-\kappa x}\cos\kappa(\eta-y) = \Re\exp\{-\kappa(x + iy - i\eta)\} \qquad (5)$$

and therefore there will be, in suitable conditions, a function ψ_P allied to ϕ_P so that $\phi_P + i\psi_P$ is an analytic function of the complex variable $x + iy$, and if ϕ_P is a potential ψ_P is the corresponding charge function. We therefore consider also the function

$$\psi_P = \frac{1}{\pi}\int_{-\infty}^{\infty} \frac{(\eta-y)f(\eta)}{x^2 + (\eta-y)^2} d\eta = \frac{1}{\pi}\int_{-\infty}^{\infty} f(\eta)\, d\eta \int_0^{\infty} e^{-\kappa x} \sin\kappa(\eta-y)\, d\kappa. \qquad (6)$$

Clearly ϕ_P may exist in cases where ψ_P does not. The analogue of 14·11 (5) will be the repeated integral

$$\frac{1}{\pi}\int_0^{\infty} d\kappa \int_{-\infty}^{\infty} f(\eta) \sin\kappa(\eta-y)\, d\eta. \qquad (7)$$

The allied function was introduced by Titchmarsh;[*] besides its applications in two-dimensional potential theory, it is needed in the discussion of the linear response of a recording instrument (14·15).

[*] *Proc. Lond. Math. Soc.* (2) 1926, 109–30. But cf. Lamb, *Phil. Trans.* A, 203, 1904, 26.

14·112. Proof of Fourier's integral theorem. *If $f(y)$ is of bounded variation, and $|f(y)|$ is integrable from $-\infty$ to ∞, then*

$$\frac{1}{\pi}\int_0^\infty d\kappa \int_{-\infty}^\infty f(\eta)\cos\kappa(\eta-y)\,d\eta = \tfrac{1}{2}\{f(y+)+f(y-)\}.$$

In the conditions stated $\int_{-\infty}^\infty f(\eta)\cos\kappa(\eta-y)\,d\eta$ is uniformly convergent by the M test, and therefore for any finite A

$$I_A = \int_0^A d\kappa \int_{-\infty}^\infty f(\eta)\cos\kappa(\eta-y)\,d\eta = \int_{-\infty}^\infty f(\eta)\,d\eta \int_0^A \cos\kappa(\eta-y)\,d\kappa$$

$$= \int_{-\infty}^\infty f(\eta)\frac{\sin A(\eta-y)}{\eta-y}\,d\eta. \tag{1}$$

The rest of the argument is substantially the same as for the summation of a Fourier series. We write

$$I_A = \int_0^\delta \{f(y+\theta)+f(y-\theta)\}\frac{\sin A\theta}{\theta}\,d\theta + \int_\delta^\infty \{f(y+\theta)+f(y-\theta)\}\frac{\sin A\theta}{\theta}\,d\theta, \tag{2}$$

where δ is independent of A. Then since $\dfrac{f(y\pm\theta)}{\theta}$ is of bounded variation in $\delta \leqslant \theta < \infty$ the second integral tends to zero as $A \to \infty$, by Riemann's lemma; and the first tends to $\tfrac{1}{2}\pi\{f(y+)+f(y-)\}$, as in the proof for the series theorem. Hence the theorem follows.

The corresponding theorem for the allied integral is as follows. *If $f(y)$ satisfies the same conditions as in Fourier's integral theorem, then at any y where a Lipschitz condition is satisfied*

$$\frac{1}{\pi}\int_0^\infty d\kappa \int_{-\infty}^\infty f(\eta)\sin\kappa(\eta-y)\,d\eta = \frac{1}{\pi}P\int_{-\infty}^\infty f(\eta)\frac{d\eta}{\eta-y}.$$

We have similarly

$$J_A = \int_0^A d\kappa \int_{-\infty}^\infty f(\eta)\sin\kappa(\eta-y)\,d\eta = \int_{-\infty}^\infty f(\eta)\,d\eta \int_0^A \sin\kappa(\eta-y)\,d\kappa,$$

$$= \int_{-\infty}^\infty f(\eta)\frac{1-\cos A(\eta-y)}{\eta-y}\,d\eta. \tag{3}$$

By the Lipschitz condition, for some M and α (possibly depending on y), where $0 < \alpha \leqslant 1$,

$$|f(\eta)-f(y)| \leqslant M|\eta-y|^\alpha. \tag{4}$$

Then

$$J_A = \int_0^\delta \{f(y+\theta)-f(y-\theta)\}\frac{1-\cos A\theta}{\theta}\,d\theta + \int_\delta^\infty \{f(y+\theta)-f(y-\theta)\}\frac{1-\cos A\theta}{\theta}\,d\theta. \tag{5}$$

The first integral has modulus less than $4M\delta^\alpha/\alpha$; choose δ so that this is less than ω. This choice of δ is independent of A. Then as $A \to \infty$,

$$\int_\delta^\infty \{f(y+\theta)-f(y-\theta)\}\frac{\cos A\theta}{\theta}\,d\theta \to 0, \tag{6}$$

and therefore by choosing A suitably we can make

$$\left| J_A - \int_\delta^\infty \{f(y+\theta)-f(y-\theta)\}\frac{d\theta}{\theta} \right| < 2\omega. \tag{7}$$

Hence
$$\frac{1}{\pi}\int_0^\infty d\kappa \int_{-\infty}^\infty f(\eta) \sin\kappa(\eta-y)\,d\eta = \frac{1}{\pi}\lim_{\delta\to 0}\int_\delta^\infty \{f(y+\theta)-f(y-\theta)\}\frac{d\theta}{\theta}$$
$$= \frac{1}{\pi} P\int_{-\infty}^\infty f(\eta)\frac{d\eta}{\eta-y}. \tag{8}$$

In particular this is true if $f(\eta)$ is differentiable at $\eta = y$.

14·113. Extension of Fourier's integral theorem. *If $f(y)$ is bounded for all y and integrable in any finite range of y, the integral*

$$\phi(x,y) = \frac{1}{\pi}\int_{-\infty}^\infty d\eta \int_0^\infty f(\eta)\,e^{-\kappa x}\cos\kappa(\eta-y)\,d\kappa$$

is a solution of Laplace's equation for $x > 0$; and when x tends to 0 the integral tends to $\tfrac{1}{2}\{f(y+)+f(y-)\}$ for any value of y such that the latter expression exists.

The integral is equal to
$$\phi(x,y) = \frac{1}{\pi}\int_{-\infty}^\infty f(\eta)\frac{x\,d\eta}{x^2+(\eta-y)^2}. \tag{1}$$

This is uniformly convergent, by the M test, in any interval of x and y such that $x \geqslant c > 0$. The integrals obtained by differentiating once and twice under the integral sign are also uniformly convergent, and therefore differentiation under the integral sign is permissible. Carrying out the differentiation we show that

$$\frac{\partial^2 \phi(x,y)}{\partial x^2} + \frac{\partial^2 \phi(x,y)}{\partial y^2} = 0 \quad (x > c > 0). \tag{2}$$

Next, put
$$\eta = y + x\tan\theta. \tag{3}$$

Then
$$\phi(x,y) = \frac{1}{\pi}\int_{-\tfrac{1}{2}\pi}^{\tfrac{1}{2}\pi} f(y+x\tan\theta)\,d\theta. \tag{4}$$

It follows immediately that if $f(y)$ has upper and lower bounds M and m, then for all positive x, $m \leqslant \phi(x,y) \leqslant M$.

Suppose that $f(\eta)$ is continuous on both sides of $\eta = y$, but not necessarily at y. We can choose a range δ of η such that for $0 < x\tan\theta \leqslant \delta$

$$|f(y+x\tan\theta)-f(y+)| < \omega, \quad |f(y-x\tan\theta)-f(y-)| < \omega, \tag{5}$$

where ω is arbitrarily small; and

$$\phi(x,y) = \frac{1}{\pi}\int_0^{\tan^{-1}\delta/x} \{f(y+x\tan\theta)+f(y-x\tan\theta)\}\,d\theta$$
$$+ \frac{1}{\pi}\int_{\tan^{-1}\delta/x}^{\tfrac{1}{2}\pi} \{f(y+x\tan\theta)+f(y-x\tan\theta)\}\,d\theta. \tag{6}$$

The first term is
$$\frac{1}{\pi}[\{f(y+)+f(y-)\}+2\lambda\omega]\tan^{-1}\frac{\delta}{x}, \tag{7}$$

where $|\lambda| \leqslant 1$; and when $x \to 0$ this tends to

$$\tfrac{1}{2}\{f(y+)+f(y-)\} \tag{8}$$

with an error $\leqslant \lambda\omega$, which is arbitrarily small.

The second term in (6) tends to 0 if $f(y)$ is bounded, since the lower limit tends to $\tfrac12\pi$. Hence
$$\lim_{x \to 0} \phi(x,y) = \tfrac12\{f(y+) + f(y-)\}. \tag{9}$$

The corresponding theorem for the allied function is: *if $f(y)$ is bounded for all y, and integrable over any finite range of y, and if, for some positive Y, $\int_Y^\infty \frac{f(y)}{y} dy$ and $\int_{-\infty}^{-Y} \frac{f(y)}{y} dy$ converge, then for $x > 0$ the integral*
$$\psi(x,y) = \frac{1}{\pi} \int_{-\infty}^\infty d\eta \int_0^\infty f(\eta) e^{-\kappa x} \sin \kappa (\eta - y) d\kappa$$
is the imaginary part of an analytic function of $x + iy$, of which $\phi(x,y)$ is the real part; and when $x \to 0$
$$\psi(x,y) \to \frac{1}{\pi} P \int_{-\infty}^\infty f(\eta) \frac{d\eta}{\eta - y}$$
at any value of y where $f(\eta)$ satisfies a Lipschitz condition.

We have for $x > 0$
$$\psi(x,y) = \frac{1}{\pi} \int_{-\infty}^\infty f(\eta) \frac{\eta - y}{x^2 + (\eta - y)^2} d\eta \tag{10}$$
which is convergent in the conditions stated; and
$$\phi(x,y) + i\psi(x,y) = \frac{1}{\pi} \int_{-\infty}^\infty f(\eta) \frac{x + i(\eta - y)}{x^2 + (\eta - y)^2} d\eta$$
$$= \frac{1}{\pi} \int_{-\infty}^\infty f(\eta) \frac{d\eta}{x + iy - i\eta}, \tag{11}$$
which is an analytic function of $x + iy$.

Choose a small quantity δ such that
$$\int_0^\delta |f(y + \theta) - f(y - \theta)| \frac{d\theta}{\theta} < \omega; \tag{12}$$
this can be done because $f(\eta)$ satisfies a Lipschitz condition. Then
$$\left| \int_0^\delta \{f(y+\theta) - f(y-\theta)\} \frac{\theta}{x^2 + \theta^2} d\theta \right| < \omega \tag{13}$$
and
$$\int_\delta^\infty \{f(y+\theta) - f(y-\theta)\} \frac{\theta}{x^2 + \theta^2} d\theta \to \int_\delta^\infty \{f(y+\theta) - f(y-\theta)\} \frac{d\theta}{\theta} \tag{14}$$
since the conditions of Abel's test for uniform convergence are satisfied for all $x < \delta$. The theorem follows immediately.

The conditions stated for these two theorems can be appreciably relaxed. If
$$f(y) = y \cos \alpha y \quad \text{or} \quad y \sin \alpha y$$
it is easy to show by contour integration that the results for $\phi(x,y)$ remain true. But the most important extensions are probably to the cases where $f(y)$ tends to finite limits as $y \to \pm \infty$. If $f(y) = A$, we get $\phi(x,y) = A$, and the allied function is a constant by the Cauchy-Riemann relations. If $f(y) = A \ (y > 0), f(y) = -A \ (y < 0)$,
$$\phi(x,y) = \frac{2A}{\pi} \tan^{-1} \frac{y}{x}$$
and the allied function is $-\frac{2A}{\pi} \log (x^2 + y^2)^{1/2}$.

Hence if $f(y) \to C(y \to \infty)$, $f(y) \to D(y \to -\infty)$, and $\int_Y^\infty \frac{f(y)-C}{y} dy$, $\int_{-\infty}^{-Y} \frac{f(y)-D}{y} dy$ converge for some positive Y, we can still find an allied function by applying the method to

$$g(y) = f(y) - C \quad (y > 0)$$
$$= f(y) - D \quad (y < 0)$$

and adding $\tfrac{1}{2}(C+D) + \frac{C-D}{\pi} \tan^{-1}\frac{y}{x}$ to ϕ, $-\frac{C-D}{\pi} \log(x^2+y^2)^{1/2}$ to ψ.

14·12. The cosine and sine transforms. If $f(y)$ is given from $y = 0$ to ∞ and if $f(-y) = f(y)$, we have, subject to convergence conditions as in 14·112,

$$f(y) = \frac{2}{\pi} \int_0^\infty d\kappa \int_0^\infty f(\eta) \cos \kappa\eta \cos \kappa y \, d\eta. \tag{1}$$

Similarly if $f(-y) = -f(y)$,

$$f(y) = \frac{2}{\pi} \int_0^\infty d\kappa \int_0^\infty f(\eta) \sin \kappa\eta \sin \kappa y \, d\eta. \tag{2}$$

These rules can be stated as follows. If

$$g(\kappa) = \sqrt{\left(\frac{2}{\pi}\right)} \int_0^\infty f(\eta) \cos \kappa\eta \, d\eta, \quad \text{then} \quad f(y) = \sqrt{\left(\frac{2}{\pi}\right)} \int_0^\infty g(\kappa) \cos \kappa y \, d\kappa. \tag{3}$$

If

$$h(\kappa) = \sqrt{\left(\frac{2}{\pi}\right)} \int_0^\infty f(\eta) \sin \kappa\eta \, d\eta, \quad \text{then} \quad f(\eta) = \sqrt{\left(\frac{2}{\pi}\right)} \int_0^\infty h(\kappa) \sin \kappa y \, d\kappa. \tag{4}$$

These results show a complete symmetry between $f(y)$ and its two transforms $g(\kappa)$ and $h(\kappa)$. The latter are known as the Fourier cosine and sine transforms; they constitute two of the earliest solutions of integral equations, and were given as such by Laplace. Fourier's integral theorem in its general form can similarly be written: if

$$v(\kappa, y) = \int_{-\infty}^\infty f(\eta) \cos \kappa(\eta - y) \, d\eta, \quad \text{then} \quad f(y) = \frac{1}{\pi} \int_0^\infty v(\kappa, y) \, d\kappa, \tag{5}$$

or, more symmetrically, if

$$k(\kappa) = \frac{1}{\sqrt{(2\pi)}} \int_{-\infty}^\infty f(\eta) e^{-i\kappa\eta} \, d\eta,$$

then

$$f(y) = \frac{1}{\sqrt{(2\pi)}} \int_{-\infty}^\infty k(\kappa) e^{i\kappa y} \, d\kappa.$$

14·121. Parseval's theorem for integrals. By (3),

$$\int_0^\infty f_1(y) f_2(y) \, dy = \sqrt{\left(\frac{2}{\pi}\right)} \int_0^\infty f_1(y) \, dy \int_0^\infty g_2(\kappa) \cos \kappa y \, d\kappa$$
$$= \int_0^\infty g_1(\kappa) g_2(\kappa) \, d\kappa, \tag{6}$$

on replacing y by η and reversing the order of integration. Similarly

$$\int_0^\infty f_1(y) f_2(y) \, dy = \int_0^\infty h_1(\kappa) h_2(\kappa) \, d\kappa. \tag{7}$$

If $f_1(y) = f_2(y) = f(y)$ we have

$$\int_0^\infty f^2(y)\,dy = \int_0^\infty g^2(\kappa)\,d\kappa = \int_0^\infty h^2(\kappa)\,d\kappa. \tag{8}$$

This is the analogue for integrals of Parseval's theorem for series. For a vibrating dynamical system both theorems have the physical interpretation that the total energy of the system is the sum of the energies in the normal modes.

14·13. The Fourier-Mellin theorem.
If $H(t)$ is the Heaviside unit function the sum

$$f(\tau_1)\{H(t-\tau_1)-H(t-\tau_2)\} + f(\tau_2)\{H(t-\tau_2)-H(t-\tau_3)\} + \ldots + f(\tau_r)\{H(t-\tau_r)-H(t-\tau_{r+1})\}, \tag{1}$$

where $\tau_1 < \tau_2 < \ldots < \tau_{r+1}$, is equal to $f(\tau_1)$ for $\tau_1 < t < \tau_2$, $f(\tau_2)$ for $\tau_2 < t < \tau_3$ and so on. Proceeding to the limit when the values become indefinitely close, if $f(t)$ is continuous, we get the Stieltjes integral

$$f(t) = -\int_{\tau=-\infty}^\infty f(\tau)\,dH(t-\tau). \tag{2}$$

We can suppose that $f(t) = 0$ for $t < 0$. Now substitute for $H(t-\tau)$ in terms of Bromwich's integral.

$$f(t) = -\int_{\tau=0}^\infty f(\tau)\,d\,\frac{1}{2\pi i}\int_{c-i\infty}^{c+i\infty} e^{z(t-\tau)}\frac{dz}{z} = \frac{1}{2\pi i}\int_{c-i\infty}^{c+i\infty}\frac{e^{zt}}{z}dz\int_0^\infty zf(\tau)e^{-z\tau}d\tau \tag{3}$$

assuming that we may invert the order of integration and then differentiate under the integral sign. Then if for some $c > 0$, and $\Re(z) = c$,

$$\int_0^\infty zf(\tau)e^{-z\tau}\,d\tau = F(z), \tag{4}$$

$$F(p)H(t) = f(t)H(t). \tag{5}$$

This gives a rule for deriving an operator for any function $f(t)$ such that the integral (4) exists for $\Re(z) \geq c$. (4) is called the *Laplace transform* of $f(t)$; this name is often given to $F(z)/z$.

To justify this, put $z = c + iy$. Then the second integral in (3) is

$$\frac{1}{2\pi}\int_{-\infty}^\infty e^{ct}\,dy\int_0^\infty f(\tau)e^{-c\tau}e^{iy(t-\tau)}\,d\tau = \frac{e^{ct}}{2\pi}\int_{-\infty}^\infty dy\int_0^\infty f(\tau)e^{-c\tau}\cos y(t-\tau)\,d\tau \tag{6}$$

and (3) is equivalent to

$$f(t)e^{-ct} = \frac{1}{\pi}\int_0^\infty dy\int_0^\infty f(\tau)e^{-c\tau}\cos y(t-\tau)\,d\tau. \tag{7}$$

But this is Fourier's integral theorem applied to the function that is 0 for $t < 0$ and equal to $f(t)e^{-ct}$ for $t > 0$. If this function satisfies the conditions for the theorem for some positive value of c, the operator corresponding to $f(t)$ can therefore be found by (4). The result (7) is true for all c large enough for the integral (4) to converge absolutely, that is, for $\int_0^\infty |f(\tau)|\,e^{-c\tau}\,d\tau$ to converge. $F(z)$ found by this method will be analytic for $\Re(z) > c$, but will not in general satisfy the rule of our fundamental operators that it shall have an expansion in negative powers of z for all z with sufficiently large modulus. Consequently it has been held by some recent writers that the operational method should be abandoned, or indeed that it has been already abandoned, and replaced by the Fourier-

Mellin theorem.* This, however, seems to be a fundamental mistake in method. The operational method considers the state of a system given at time 0, and obtains a solution for its state at time t depending at no point on any information except about its state at time 0 and the disturbances acting between time 0 and t. The equation (4) is meaningless unless we know the function at all times up to infinity. It is true that the values at times later than t will often not affect the state at time t. But if $f(t) = \exp(at^2)$ with a positive the integral diverges whatever c may be; and we are never in a position in an actual experiment to verify that this does not happen for t large enough. Even without such disturbances, it is a common occurrence for second-order terms neglected in the equations to rise into importance if the time is long enough, and it is not justifiable to adopt a procedure that assumes the equations to be linear for all time; the whole theory of equipartition of energy rests on this fact. The operational method avoids all such complications by making the procedure depend directly on the data up to time t and on nothing else. The Fourier-Mellin procedure makes the validity of the solution depend on the superfluous extra condition that future disturbances are not of such a character as to make the integral in (4) diverge for every c. It is a valuable method in its proper place, as has particularly been shown by van der Pol, since when we know that the solution of a differential equation satisfies the conditions for the applicability of the method it can often be obtained in a form immediately adapted to contour integration. The use of p as a notation for a complex variable in this method is however to be deprecated. The method is not a substitute for the operational method, since each is valid in conditions where the other is not, and nothing but confusion can arise from mixing the notations. The usual symbol z for a complex variable is available, and so is Bromwich's λ, and there is no occasion to use the preoccupied p.

The practice of denoting what we have called $F(z)/z$ by $F(z)$ is to be especially deprecated. This function differs from $f(t)$ in dimensions, and gives rise to needless trouble in checking. The point has been made explicitly by McLachlan.

14·14. Harmonic oscillation of finite duration. Let
$$f(t) = \cos \gamma t \quad (-\tfrac{1}{2}T < t < \tfrac{1}{2}T),$$
$$= 0 \quad (t < -\tfrac{1}{2}T,\; t > \tfrac{1}{2}T).$$

We wish to express $f(t)$ as a Fourier integral. We find
$$\int_{-\infty}^{\infty} f(\tau) \cos \kappa(\tau - t)\, d\tau = \int_{-\tfrac{1}{2}T}^{\tfrac{1}{2}T} \cos \gamma\tau \cos \kappa(t-\tau)\, d\tau$$
$$= \left(\frac{\sin \tfrac{1}{2}(\gamma - \kappa)T}{\gamma - \kappa} + \frac{\sin \tfrac{1}{2}(\gamma + \kappa)T}{\gamma + \kappa} \right) \cos \kappa t$$

and
$$f(t) = \frac{1}{\pi} \int_0^{\infty} \left(\frac{\sin \tfrac{1}{2}(\gamma - \kappa)T}{\gamma - \kappa} + \frac{\sin \tfrac{1}{2}(\gamma + \kappa)T}{\gamma + \kappa} \right) \cos \kappa t\, d\kappa.$$

---

* From remarks of some enthusiasts one might infer that if asked to solve $\dfrac{dx}{dt} = \cos t$, they proceed as follows: first form the Laplace transform of $\cos t$; multiply by $1/z$; substitute the result for $F(z)/z$ in the Bromwich integral, and evaluate by contour integration.

If $f(t)$ is unbounded near $t = 0$, but has an improper Riemann integral, the operational method needs no special justification. The proof of the Fourier-Mellin theorem, on the other hand, meets with new difficulties when the function is unbounded. See also **14·13 a**.

Thus the Fourier representation of a harmonic oscillation of finite duration includes an infinite range of frequencies. If we take γ positive the second term will be small compared with the first for values of κ near γ; the first has a maximum equal to $\frac{1}{2}T$ for $\kappa = \gamma$, and vanishes first for $\kappa - \gamma = \pm 2\pi/T$. There is a pair of minima at $\gamma \pm 3\pi/T$, about -0.2 of the first maximum. Thus the larger amplitudes are concentrated about $\kappa = \gamma$ provided that γT is large; but for any finite T there will be a finite range of κ and γ such that the amplitude is not negligible, and this will be shown, in the optical case, by a broadening of the spectral line.

14·15. Response of a recording instrument. Instruments have usually to be tested to find out whether their actual behaviour is in accordance with that intended. A common method of testing an instrument for recording vibrations, such as a seismograph, is to apply known harmonic disturbances of different periods. For each period the lag and the magnification are recorded when they have become steady. The instrument to be useful must be stable and damped; its response may satisfy a differential equation of the second order, but need not do so, but if solutions of the form $e^{\lambda t}$ are possible in the absence of disturbance, all the values of λ have negative real parts. If a disturbance is $\cos \gamma t$, the response is $\mu \cos(\gamma t - \epsilon)$, where μ and ϵ will be functions of γ. This may be written by saying that the response to $e^{i\gamma t} + e^{-i\gamma t}$ is $\mu(e^{i\gamma t - i\epsilon} + e^{-i\gamma t + i\epsilon})$ or that for all γ, $\mu e^{-i\epsilon}$ is a function of γ, μ being an even and ϵ an odd function of γ. If we take $i\gamma = \lambda$, $\mu e^{-i\epsilon}$ is an analytic function of λ when λ has a positive real part, and the two functions $\mu \cos \epsilon$ and $-\mu \sin \epsilon$ are related in the same way as the electrostatic potential and charge function. Hence by 14·112(8),

$$-\mu \sin \epsilon = \frac{1}{\pi} \int_0^\infty \{(\mu \cos \epsilon)_{\gamma+\eta} - (\mu \cos \epsilon)_{\gamma-\eta}\} \frac{d\eta}{\eta},$$

and since $i\mu e^{-i\epsilon}$ is another analytic function

$$\mu \cos \epsilon = \frac{1}{\pi} \int_0^\infty \{(\mu \sin \epsilon)_{\gamma+\eta} - (\mu \sin \epsilon)_{\gamma-\eta}\} \frac{d\eta}{\eta}.$$

These are relations between observed quantities and can be used either as a check on the hypothesis of linear response or to use the observed values of μ and ϵ to improve each other's accuracy. As a rule $\mu \to 0$ with γ (very long period). For γ very large μ may tend to a finite limit μ_0 as for a damped pendulum, or to 0 as for a Galitzin seismograph. It can be shown* that if the disturbance is $H(t)$ the response is 0 for $t < 0$, and for $t > 0$ it is

$$\frac{1}{\pi} \int_0^\infty \mu \sin(\gamma t - \epsilon) \frac{d\gamma}{\gamma} = \mu_0 H(t) + \frac{1}{\pi} \int_0^\infty \{(\mu \cos \epsilon - \mu_0) \sin \gamma t + \mu \sin \epsilon (1 - \cos \gamma t)\} \frac{d\gamma}{\gamma}.$$

EXAMPLES

1. Show that if $0 \leqslant x \leqslant \pi$,

$$\sin x = \frac{2}{\pi} - \frac{4}{\pi} \left(\frac{\cos 2x}{1.3} + \frac{\cos 4x}{3.5} + \frac{\cos 6x}{5.7} + \ldots \right).$$

What function does the series represent in the range $-\pi \leqslant x \leqslant 0$? (Prelim. 1936.)

2. Show that the series $\Sigma a^n \cos b^n \pi x$, where $|a| < 1$, represents a continuous function of x, but that its derivative, if any, has no Fourier expansion in $0 \leqslant x \leqslant 1$ if b is an integer and $|ab| > 1$. What

---

* Jeffreys, *Phil. Mag.* (7) 30, 1940, 165–7. A term $-\mu_0 \sin \gamma t$ has been omitted from equation (29) of the paper.

can be inferred about the derivative from this result? (Actually this series, though continuous for all real x, can be proved by other methods to have no derivative for any value of x in the conditions stated. Cf. Hardy, *Trans. Amer. Math. Soc.* 17, 1916, 301–25.)

3. Find the cosine and sine expansions of $\cos \tfrac{1}{2}x$ and $\sin \tfrac{1}{2}x$ in the range $0 < x < \pi$, and verify that they satisfy the rules for differentiating Fourier series.

4. Prove that for $0 < \theta \leqslant \pi$
$$\sin\theta + \tfrac{1}{2}\sin 2\theta + \tfrac{1}{3}\sin 3\theta + \ldots = \tfrac{1}{2}\pi - \tfrac{1}{2}\theta.$$

5. If $f(n)$ is given for every positive or negative integral value of n, prove that (1) subject to suitable convergence conditions the function
$$g(x) = \frac{1}{\pi}\sum_{n=-\infty}^{\infty} f(n)\,\frac{\sin \pi(x-n)}{x-n} = \frac{1}{\pi}\int_0^\pi \left\{\sum_{-\infty}^{\infty} \cos\mu(x-n) f(n)\right\} d\mu$$
is equal to $f(n)$ for $x = n$,
$$(2) \qquad \int_{-\infty}^{\infty} g^2(x)\,dx = \sum_{-\infty}^{\infty} f^2(n).$$

(Whittaker, *Proc. R.S. Edin.* 35, 1915, 181–94.)

6. If
$$f(x) = A_0 + \Sigma A_n \cos nx + \Sigma B_n \sin nx \quad (0<x<2\pi),$$
$$f'(x) = a_0 + \Sigma a_n \cos nx + \Sigma b_n \sin nx \quad (0<x<2\pi),$$
$$A = f(2\pi-) - f(0+),$$
show that
$$a_0 = \frac{A}{2\pi}, \quad a_n = \frac{A}{\pi} + nB_n, \quad b_n = -nA_n.$$

7. If
$$f(x) = A_0 + \Sigma(A_n \cos nx + B_n \sin nx),$$
$$g(x) = C_0 + \Sigma(C_n \cos nx + D_n \sin nx),$$
prove that
$$\frac{1}{2\pi}\int_0^{2\pi} f(x)g(x)\,dx = A_0 C_0 + \tfrac{1}{2}\Sigma(A_n C_n + B_n D_n),$$
under conditions to be stated.

8. Prove that if $g(x)$ is the function fitted to $f(x)$ as in 14.01, then according as n is odd or even,
$$g(x) = \frac{1}{n}\sum_{r=0}^{n-1} f\left(\frac{2r\pi}{n}\right) \frac{\sin \tfrac{1}{2}n\left(x - \frac{2r\pi}{n}\right)}{\sin \tfrac{1}{2}\left(x - \frac{2r\pi}{n}\right)}\left(1,\ \cos\tfrac{1}{2}\left(x - \frac{2r\pi}{n}\right)\right),$$
and explain how this could have been inferred from the theory of the complex variable.

9. Find a series of sines and cosines that will represent e^x in the range $-\pi < x < \pi$, and draw the graph of the sum of the series outside this range. Deduce that
$$\coth \pi = \frac{1}{\pi} + \frac{2}{\pi}\left(\frac{1}{1+1^2} + \frac{1}{1+2^2} + \frac{1}{1+3^2} + \ldots\right). \qquad \text{(I.C. 1939.)}$$

10. Find a solution of the integral equation
$$\int_0^\infty g(x)\cos\alpha x\,dx = f(\alpha),$$
where
$$f(\alpha) = 1 - \tfrac{1}{2}\alpha^2 \quad (0 < \alpha < 1)$$
$$= 0 \qquad (\alpha > 1). \qquad \text{(I.C. 1943.)}$$

Chapter 15

THE FACTORIAL AND RELATED FUNCTIONS

'There are three hundred and seventy-two competent renderings of a single verse of one of the more cryptic Odes, and it has been aptly claimed that even the appearance of a giraffe must be capable of some rational explanation.' ERNEST BRAMAH, *The Moon of Much Gladness*.

15·01. We define the *factorial function** by

$$z! = \int_0^\infty u^z e^{-u} du, \qquad (1)$$

where $\Re(z) > -1$. Convergence is uniform in any region $\Re(z) \geq -1 + \delta$, where $\delta > 0$.
It follows immediately by integration by parts that

$$(z+1)! = (z+1)z!, \qquad (2)$$

and hence for z a positive integer, since

$$0! = 1, \qquad (3)$$

$$z! = 1.2.3....z. \qquad (4)$$

Also $z!$ is an analytic function because it has a derivative

$$\frac{d}{dz} z! = \int_0^\infty u^z \log u . e^{-u} du, \qquad (5)$$

which converges uniformly in any closed region of z for $\Re(z) > -1$.

Now since $z!$ is analytic we may expect that it will have an analytic continuation into the region where the integral (1) is not convergent. In fact if $\Re(z) < -1$ we can choose an integer n greater than $\Re(-z)$, and define

$$z! = \frac{(z+n)!}{(z+n)(z+n-1)...(z+1)}. \qquad (6)$$

By the relation (2) this is true for $\Re(z) > -1$. For $\Re(z) < -1$ the expression on the right is the ratio of two analytic functions and therefore is an analytic function for all $\Re(z) > -n$, except for poles where z is a negative integer. For its meaning to be unique it must be independent of the choice of n, subject to $\Re(n+z) > -1$; but if m is an integer greater than n,

$$\frac{(z+m)!}{(z+m)(z+m-1)...(z+1)} \bigg/ \frac{(z+n)!}{(z+n)(z+n-1)...(z+1)} = \frac{(z+m)!}{(z+n)!(z+n+1)...(z+m)} = 1, \qquad (7)$$

---

* This function is usually denoted by $\Gamma(z+1)$ in mathematical writings except when z is a positive integer. The Γ notation seems to have arisen through some idea that $n!$ is defined in the first place only for n a positive integer, and that when we extend the domain of application we need a new notation. This is contrary to usual mathematical practice. Ordinarily if a definition is applicable only in a restricted domain, and another is found that is applicable in a wider domain and is equivalent to the first in the restricted domain, the second is taken as giving a meaning to the old term in the wider domain. x^2 is originally defined only for x a positive integer and then extended in turn to rational, negative, irrational and complex numbers. $\sin x$ is originally defined for real argument and extended to complex numbers by means of power series. Weierstrass's theory of analytic continuation rests on the same principle. There is no reason why the factorial function should receive exceptional treatment in this respect. The extra 1 in the definition of $\Gamma(z)$ is a minor but continual nuisance. As the factorial notation has been adopted in the *British Association Tables*, which are the fullest tables of the function, there seems to be no reason why the Γ notation should be perpetuated. Gauss's $\Pi(z)$ is often used and is equivalent to $z!$; but it is liable to confusion with the general notation for a product.

and therefore the definition (6) is unique. It therefore must be the analytic continuation of $z!$ and can be taken as completing the definition. It follows at once that $z!$ has simple poles at all negative integral values of z, and no other singularities for finite z.

We have already had (5·056)
$$(-\tfrac{1}{2})! = \sqrt{\pi}, \tag{8}$$
whence
$$(\tfrac{1}{2})! = \tfrac{1}{2}\sqrt{\pi}, \quad (\tfrac{3}{2})! = \tfrac{3}{2}\cdot\tfrac{1}{2}\sqrt{\pi}, \quad \ldots, \tag{9}$$
$$(-\tfrac{3}{2})! = \frac{\sqrt{\pi}}{-\tfrac{1}{2}} = -2\sqrt{\pi}, \quad (-\tfrac{5}{2})! = \tfrac{2}{3}\cdot\tfrac{2}{1}\sqrt{\pi}, \quad \ldots. \tag{10}$$

15·02. The Beta function. For $\Re(m), \Re(n) > -1$ put $\Re(m) = \mu$, $\Re(n) = \nu$; then

$$m!\,n! = \int_0^\infty e^{-x} x^m\, dx \int_0^\infty e^{-y} y^n\, dy$$
$$= \lim_{A\to\infty} \int_0^A \int_0^A e^{-(x+y)} x^m y^n\, dx\, dy. \tag{1}$$

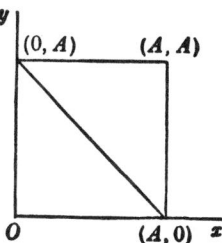

We take the triangles $(0,0)\,(A,0)\,(0,A)$ and $(A,0)\,(0,A)\,(A,A)$ separately. For the former, put $x+y=z$ and eliminate y; then $z \leq A$, and the range of x is from 0 to z, since $y \geq 0$. The integral over the triangle $(0,0)\,(A,0)\,(0,A)$ is then

$$\int_0^A dz \int_0^z e^{-z} x^m (z-x)^n\, dx, \tag{2}$$

and putting $x = tz$ we have

$$\int_0^A dz \int_0^1 z^{m+n+1} e^{-z} t^m (1-t)^n\, dt = \int_0^1 t^m (1-t)^n\, dt \int_0^A z^{m+n+1} e^{-z}\, dz$$
$$\to (m+n+1)! \int_0^1 t^m (1-t)^n\, dt. \tag{3}$$

The integral over the triangle $(A,0)\,(A,A)\,(0,A)$ of $x^m y^n$ has modulus

$$< \int_0^A \int_0^A x^\mu y^\nu\, dx\, dy = \frac{A^{\mu+\nu+2}}{(\mu+1)(\nu+1)},$$

and that of $e^{-x} x^m y^n$ has modulus

$$< \frac{e^{-A} A^{\mu+\nu+2}}{(\mu+1)(\nu+1)} \to 0. \tag{4}$$

Hence for $\Re(m) > -1$, $\Re(n) > -1$

$$\int_0^1 t^m (1-t)^n\, dt = \frac{m!\,n!}{(m+n+1)!}. \tag{5}$$

The integral is usually denoted by $B(m+1, n+1)$, but it is somewhat easier to manipulate in the present form.

In (5) put
$$t = \frac{u}{1+u};$$
then
$$\int_0^\infty \frac{u^m\, du}{(1+u)^{m+n+2}} = \frac{m!\,n!}{(m+n+1)!}. \tag{6}$$

Again, for $-1 < \Re(z) < 1$, putting $m = z$, $n = -z$, we have

$$\int_0^\infty \frac{u^z}{(1+u)^2} du = z!(-z)!. \tag{7}$$

But this integral is $\pi z \operatorname{cosec} \pi z$, by 12·125; hence

$$z!(-z)! = \frac{\pi z}{\sin \pi z}, \tag{8}$$

which is extended at once to all z other than positive or negative integers by continuation. It is more convenient than the corresponding formula in the Γ notation, and easier to remember on account of the obvious check at $z = 0$.

Again, for $-1 < \Re(z)$,
$$\frac{z!z!}{(2z+1)!} = \int_0^1 t^z(1-t)^z dt. \tag{9}$$

Put $2t = 1+s$; the integral is

$$2^{-2z-1} \int_{-1}^1 (1-s^2)^z ds = 2^{-2z} \int_0^1 (1-s^2)^z ds$$

$$= 2^{-2z-1} \int_0^1 (1-u)^z u^{-1/2} du$$

$$= 2^{-2z-1} \frac{z!(-\tfrac{1}{2})!}{(z+\tfrac{1}{2})!},$$

whence $\qquad z!(z+\tfrac{1}{2})! = 2^{-2z-1} \pi^{1/2} (2z+1)!,$

that is, $\qquad z!(z-\tfrac{1}{2})! = 2^{-2z} \pi^{1/2} (2z)!. \quad (\Re(z) > -\tfrac{1}{2}) \tag{10}$

This is generalized by continuation. If z is a positive integer the proof is simple;

$$(2z)! = (2z)(2z-2) \ldots 2 \cdot (2z-1)(2z-3) \ldots 1$$

$$= 2^{2z} z!(z-\tfrac{1}{2})!/(-\tfrac{1}{2})!. \tag{11}$$

15·03. Gauss's definition.
In (5) replace m by z and t by u/n; then

$$\frac{z!n!}{(n+z+1)!} = n^{-z-1} \int_0^n u^z \left(1 - \frac{u}{n}\right)^n du. \quad (\Re(z) > -1) \tag{1}$$

If $n \to \infty$ the last integral tends to

$$\int_0^\infty u^z e^{-u} du = z! \tag{2}$$

since $(1-u/n)^n < e^{-u}$ for $0 < u \leqslant n$ and the M test is satisfied.

Hence for fixed z and n tending to infinity

$$\frac{(n+z+1)!}{n! \, n^{z+1}} \to 1, \tag{3}$$

irrespective of z. Since also $(n+z+1)/n \to 1$ we can write this as

$$\frac{(n+z)!}{n! \, n^z} \to 1. \tag{4}$$

This is an equality for all positive n if $z = -1$ or $z = 0$.

Substitute (4) in 15·01 (6); then since we can take n as large an integer as we like

$$z! = \lim_{n \to \infty} \frac{n!\, n^z}{(z+1)(z+2)\cdots(z+n)} \quad (\Re(z) > -1), \tag{5}$$

which provides a definition of $z!$ independent of the integral, since $n!$ can be defined directly by the product 15·01 (4). (5) was taken by Gauss as the definition of the function and denoted by $\Pi(z)$. We have not yet proved that (5) agrees with 15·01 (6) for $\Re(z) < -1$.

But
$$\Pi(z) = \lim_{n \to \infty} n^z \left\{(1+z)\left(1+\frac{z}{2}\right)\cdots\left(1+\frac{z}{n}\right)\right\}^{-1}, \tag{6}$$

$$\log \Pi(z) = \lim \left[z\left(\log\frac{2}{1} + \log\frac{3}{2} + \ldots + \log\frac{n}{n-1}\right) - \sum_{m=1}^{n} \log\left(1+\frac{z}{m}\right)\right]$$

$$= \sum_{m=1}^{\infty}\left\{z \log \frac{m+1}{m} - \log\left(1+\frac{z}{m}\right)\right\}. \tag{7}$$

But
$$z \log \frac{m+1}{m} - \log\left(1+\frac{z}{m}\right) = z\left(\frac{1}{m} - \frac{1}{2m^2} + \ldots\right) - \frac{z}{m} + \frac{z^2}{2m^2} - \ldots$$

$$= O\left(\frac{1}{m^2}\right). \tag{8}$$

The series for $\log \Pi(z)$ is therefore convergent for all z. Also $\frac{d}{dz}\log \Pi(z)$ is a uniformly convergent series of analytic functions in any closed region excluding negative integers, and therefore $\Pi(z)$ is analytic in any region that excludes negative integers. It must therefore be identical with $z!$ according to our extended definition. We can therefore remove the restriction $\Re(z) > -1$ in (5). Gauss's definition has the advantage that it is immediately intelligible for all z except the obvious poles; but in practice the function usually arises through the integral.

We verify at once from (5) that

$$\frac{\Pi(z)}{\Pi(z-1)} = \lim \frac{n^z}{n^{z-1}} \frac{z}{z+n} = z, \tag{9}$$

giving a direct proof of the inductive relation from Gauss's definition. Also

$$z!(-z)! = \lim \frac{n!\, n!}{(1-z^2)(2^2-z^2)\cdots(n^2-z^2)}$$

$$= \lim (1-z^2)\left(1-\frac{z^2}{2^2}\right)\left(1-\frac{z^2}{3^2}\right)\cdots\left(1-\frac{z^2}{n^2}\right)$$

$$= \frac{\pi z}{\sin \pi z}, \tag{10}$$

in agreement with 15·02 (8).

15·04. The digamma (F) and trigamma (F') functions. These are

$$F(z) = \frac{d}{dz}\log z! = \lim_{n \to \infty}\left(\log n - \frac{1}{z+1} - \ldots - \frac{1}{z+n}\right), \tag{1}$$

$$F'(z) = \frac{d^2}{dz^2}\log z! = \frac{1}{(z+1)^2} + \frac{1}{(z+2)^2} + \ldots + \frac{1}{(z+n)^2} + \ldots \tag{2}$$

If in (1) we put $z = 0$ we have

$$F(0) = \lim_{n \to \infty} \left(\log n - 1 - \tfrac{1}{2} - \ldots - \frac{1}{n} \right) = -\gamma, \qquad (3)$$

where γ is Euler's constant, the numerical value of which is $0{\cdot}577215665\ldots$. Hence

$$\left(\frac{d}{dz} z! \right)_{z=0} = -\gamma. \qquad (4)$$

These functions, for $0 \leqslant z \leqslant 1{\cdot}0$ and $z \geqslant 10{\cdot}0$, are tabulated in the *British Association Tables*. A table covering the range $0 \leqslant z \leqslant 20{\cdot}0$ is given by E. Pairman.* The digamma function, also denoted by $\psi(z+1)$, has important applications in statistics and in the solution of differential equations. Webster denotes it by $\Psi(z)$, which avoids the difficulty of distinguishing F from F in writing; but the latter does not seem to lead to confusion in practice.

Note that
$$F(z) = F(z+n) - \frac{1}{z+n} - \ldots - \frac{1}{z+1}. \qquad (5)$$

15·05. Asymptotic formulae. The series for $\log z!$, $F(z)$ and $F'(z)$ can be summed by the Euler-Maclaurin formula (9·08). Taking (2) we have for z not real and negative

$$\frac{1}{z} = \int_0^\infty \frac{dx}{(z+x)^2} = \frac{1}{2z^2} + \frac{1}{(z+1)^2} + \frac{1}{(z+2)^2} + \ldots$$
$$- b_2 \cdot \frac{2}{z^3} - b_4 \frac{2 \cdot 3 \cdot 4}{z^5} - \ldots - \sum_{m=0}^\infty \int_m^{m+1} \frac{P_{2r+1}(x-m)(2r+2)!}{(z+x)^{2r+3}} dx. \qquad (1)$$

Hence
$$F'(z) = \frac{1}{z} - \frac{1}{2z^2} + \frac{B_2}{z^3} + \frac{B_4}{z^5} + \ldots + \frac{B_{2r}}{z^{2r+1}} + \sum_{m=0}^\infty \int_m^{m+1} \frac{(2r+2)\phi_{2r+1}(x-m)}{(z+x)^{2r+3}} dx. \qquad (2)$$

For z real and positive all odd derivatives of $(z+x)^{-2}$ have the same sign; hence the function then lies between the sums of r and $r+1$ terms of the series. The inequality is more difficult to state if z is complex, and obviously breaks down altogether if z is real and negative. Similarly

$$F(z) = \log z + \frac{1}{2z} - \frac{B_2}{2z^2} - \frac{B_4}{4z^4} - \ldots - \frac{B_{2r}}{2rz^{2r}} - \sum_{m=0}^\infty \int_m^{m+1} \frac{\phi_{2r+1}(x-m)}{(z+x)^{2r+2}} dx, \qquad (3)$$

and by integration
$$\log z! = C + (z+\tfrac{1}{2}) \log z - z + \frac{B_2}{2z} + \frac{B_4}{3 \cdot 4z^3} + \ldots, \qquad (4)$$

again in the sense that the function on the left, for real positive z, lies between the sums of r and $r+1$ terms of the series. We have to determine the constant C. Take z large and use 15·02 (10) in the form

$$\log z! + \log(z-\tfrac{1}{2})! = -2z \log 2 + \tfrac{1}{2} \log \pi + \log(2z)!. \qquad (5)$$

Substituting from (4) and simplifying we find that the terms that do not tend to zero for large z give
$$C = \tfrac{1}{2} \log 2\pi \qquad (6)$$

Hence for large z
$$\log z! = \tfrac{1}{2} \log 2\pi + (z+\tfrac{1}{2}) \log z - z + \frac{1}{12z} - \frac{1}{360z^3} + \frac{1}{1260z^5} - \frac{1}{1680z^7} + \frac{1}{1188z^9} - \ldots. \qquad (7)$$

* *Tracts for Computers*, no. 1.

This is *Stirling's formula*, and is in continual use for approximations to high factorials. It should not be regarded as an infinite series for any given z; if so regarded the terms oscillate infinitely. For we have seen (12·07) that for large r

$$B_{2r} \doteq (-1)^{r-1} \frac{2(2r)!}{(2\pi)^{2r}},$$

and the general term of (7) is therefore approximately

$$(-1)^{r-1} \frac{2(2r-2)!}{(2\pi)^{2r} z^{2r-1}}. \tag{8}$$

Thus for any z the terms are unbounded above and below. Nevertheless the decrease of the early terms is so rapid that the series is of the greatest use for computation. Even for $z = 1$ the terms of (7) up to z^{-5} and z^{-7} lead to

$$1 \cdot 83730 < \log 2\pi < 1 \cdot 83849.$$

The correct value is 1·83788, which is almost midway between the two approximations, which themselves differ by only 0·00119; a truly remarkable result to be obtained on the hypothesis that 1 is a large number. Legendre computed $z!$ for small z by using Stirling's series for large z and working back by successive division. The series was proved divergent by Bayes.* The first proof of the asymptotic property is due to Cauchy (1843).

Stirling† actually gave the following series, with $\zeta = z + \tfrac{1}{2}$ and z an integer:

$$\log z! = \tfrac{1}{2}\log 2\pi + \zeta\log\zeta - \zeta - \frac{1}{2.12\zeta} + \frac{7}{8.360\zeta^3} - \frac{31}{32.1260\zeta^5} + \frac{127}{128.1680\zeta^7} - \ldots \tag{9}$$

The form (7) usually quoted as Stirling's series was found by de Moivre in consequence of a letter from Stirling. The coefficients in (9) are somewhat smaller but not so easy to calculate. The ingenuity needed to obtain the series with the mathematical resources then available, not even the general definition of $z!$ being known, is astonishing. The usual form is most accurately described as de Moivre's form of Stirling's series, since all the principles used in finding it were given by Stirling. His method consisted in expressing $\log n$ as a difference $f(n+1) - f(n)$.

If we return to 9·08 we see that the formula can be written

$$\log z! = (z+\tfrac{1}{2})\log z - z + \tfrac{1}{2}\log 2\pi - \int_0^\infty \frac{t - [t] - \tfrac{1}{2}}{z+t}\,dt, \tag{10}$$

$[t]$ meaning the integral part of t. The last term is the Bourguet-Stieltjes form of the remainder‡—actually entirely due to Stieltjes; Bourguet found the corresponding expression in a Fourier series, which Stieltjes saw to be that of $t - [t] - \tfrac{1}{2}$.

15·06. It may be verified that

$$\int_0^{1/2\pi} \cos^m\theta \sin^n\theta\,d\theta = \frac{1}{2}\int_0^1 t^{1/2(n-1)}(1-t)^{1/2(m-1)}\,dt$$

$$= \frac{1}{2}\frac{\{\tfrac{1}{2}(m-1)\}!\{\tfrac{1}{2}(n-1)\}!}{\{\tfrac{1}{2}(m+n)\}!}.$$

When m and n are positive integers, we can substitute the explicit formulae for the factorials and derive the usual elementary formulae for the integrals.

* *Phil. Trans.* 53, 1763, 269–271. † *Methodus Differentialis*, 1730, Prop. 28.
‡ *J. des Math.* (4), 5, 1889, 425–44.

15·07. Wallis's formula for π.

We have

$$\int_0^{1/2\pi} \sin^{2m-1}\theta\, d\theta > \int_0^{1/2\pi} \sin^{2m}\theta\, d\theta > \int_0^{1/2\pi} \sin^{2m+1}\theta\, d\theta,$$

and therefore for integral m

$$\frac{(2m-2)(2m-4)\ldots 2}{(2m-1)(2m-3)\ldots 1} > \frac{(2m-1)(2m-3)\ldots 1}{2m(2m-2)\ldots 2}\frac{\pi}{2} > \frac{2m(2m-2)\ldots 2}{(2m+1)(2m-1)\ldots 1},$$

that is,

$$\frac{2m(2m-2)^2\ldots 2^2}{(2m-1)^2(2m-3)^2\ldots 1^2} > \tfrac{1}{2}\pi > \frac{(2m)^2(2m-2)^2\ldots 2^2}{(2m+1)(2m-1)^2\ldots 1^2}.$$

The ratio of the extreme members of this inequality is $\dfrac{2m}{2m+1}$, and therefore tends to 1 for large m. Hence

$$\tfrac{1}{2}\pi = \lim_{m\to\infty}\frac{(2m)^2(2m-2)^2\ldots 2^2}{(2m+1)(2m-1)^2\ldots 1^2}.$$

This is Wallis's formula, of historical interest as the first expression of $\tfrac{1}{2}\pi$ as the limit of an algebraic function. It can be used to obtain the constant in Stirling's formula as follows, and even to give the first term of the formula. Inserting the factor

$$(2m)^2(2m-2)^2\ldots 2^2$$

in both numerator and denominator we get

$$\tfrac{1}{2}\pi = \lim_{m\to\infty}\frac{2^{4m}(m!)^4}{(2m+1)(2m!)^2}. \tag{1}$$

If m is large

$$\log(2m)! - \log m! = \log(m+1) + \ldots + \log 2m = \int_{m+1/2}^{2m+1/2}\log x\, dx + O\!\left(\frac{1}{m}\right)$$

$$= (2m+\tfrac{1}{2})\log(2m+\tfrac{1}{2}) - (m+\tfrac{1}{2})\log(m+\tfrac{1}{2}) - m + O\!\left(\frac{1}{m}\right). \tag{2}$$

Use this to eliminate $\log(2m!)$ from (1) and drop small terms; then

$$\log\tfrac{1}{2}\pi = \lim\Big\{2\log m! + 4m\log 2$$
$$- (4m+1)\!\left(\log 2m + \frac{1}{4m}\right) + (2m+1)\!\left(\log m + \frac{1}{2m}\right) + 2m - \log 2m\Big\}$$

$$= \lim 2\{\log m! - (m+\tfrac{1}{2})\log m + m - \log 2\}, \tag{3}$$

whence

$$\lim_{m\to\infty}\frac{m!}{\sqrt{(2\pi)}\, m^{m+1/2}e^{-m}} = 1. \tag{4}$$

This method of course succeeds only when m is a positive integer. In view of the amazing ingenuity of many mathematical writers of the period, it is surprising that the first term was not found in this way until Stirling got the complete expansion.

15·08. Dirichlet integrals.

Take

$$I = \iint x^m y^n\, dx\, dy \quad (m, n > -1) \tag{1}$$

over the range of integration $x \geq 0$, $y \geq 0$, $x+y \leq a$. Substituting $y = z - x$ we have

$$I = \int_0^a\!\int_0^z x^m(z-x)^n\, dz\, dx, \tag{2}$$

since, if x exceeded z, y would be <0. Put $x = zt$; then

$$I = \int_0^a \int_0^1 z^{m+n+1} t^m (1-t)^n \, dz \, dt = \frac{a^{m+n+2} m! n!}{(m+n+2)!}. \tag{3}$$

If we vary a while keeping m and n fixed

$$\frac{dI}{da} = a^{m+n+1} \frac{m! n!}{(m+n+1)!}. \tag{4}$$

Now take
$$I_2 = \iiint x^m y^n z^p \, dx \, dy \, dz \tag{5}$$

over the range $x \geq 0, y \geq 0, z \geq 0, x+y+z \leq a$. Then x and y are restricted so that

$$0 \leq x+y \leq a-z;$$

and integrating with regard to them we have

$$I_2 = \int_0^a \frac{(a-z)^{m+n+2} m! n!}{(m+n+2)!} z^p \, dz = \frac{a^{m+n+p+3} m! n! p!}{(m+n+p+3)!} \tag{6}$$

by (3). In general if

$$I = \iint \ldots \int x_1^{m_1} x_2^{m_2} \ldots x_r^{m_r} \, dx_1 \, dx_2 \ldots dx_r, \tag{7}$$

where all $x_s \geq 0$ and $\Sigma x_s \leq a$,

$$I = a^{\Sigma m + r} \frac{m_1! m_2! \ldots m_r!}{(\Sigma m + r)!}. \tag{8}$$

If we take $m = n = 0$ in (1), and $a = 1$, we have the area of a triangle whose corners are at $(0,0)$, $(1,0)$, $(0,1)$; and the result is $\tfrac{1}{2}$. If we take $m = n = p = 0$ in (5), and $a = 1$, we have the volume of a tetrahedron whose vertices are at $(0,0,0)$, $(1,0,0)$, $(0,1,0)$, $(0,0,1)$, and the result is $\tfrac{1}{6}$.

The integrals are easily generalized to evaluate

$$J = \iint \ldots \int f(\Sigma x_s) x_1^{m_1} x_2^{m_2} \ldots x_r^{m_r} \, dx_1 \, dx_2 \ldots dx_r, \tag{9}$$

under the same restrictions. For from (8)

$$\frac{dI}{da} = a^{\Sigma m + r - 1} \frac{m_1! m_2! \ldots m_r!}{(\Sigma m + r - 1)!} \tag{10}$$

and
$$\frac{dJ}{da} = f(a) \frac{dI}{da}. \tag{11}$$

Hence
$$J = \int_0^a f(s) s^{\Sigma m + r - 1} \, ds \, \frac{m_1! m_2! \ldots m_r!}{(\Sigma m + r - 1)!}. \tag{12}$$

A proof in greater detail is given by L. J. Mordell.*

One of the most important applications of these integrals is in the theory of statistics, where we often want integrals of the form

$$L = \iint \ldots \int f(\Sigma x_s^2) \, dx_1 \, dx_2 \ldots dx_r, \tag{13}$$

subject to $\Sigma x_r^2 \leq a^2$. Put
$$x_s = \xi_s^{1/2}, \tag{14}$$

* Edin. Math. Soc., Math. Notes 34, 1944, 15–17.

and take all $\xi_s \geq 0$. Then

$$L = \int\int \ldots \int f(\Sigma \xi)\, 2^{-r} \xi_1^{-1/2} \xi_2^{-1/2} \ldots \xi_r^{-1/2} d\xi_1 \ldots d\xi_r$$

$$= 2^{-r} \int_0^{a^2} f(s)\, s^{1/2 r - 1} ds \, \frac{\pi^{1/2 r}}{(\tfrac{1}{2} r - 1)!}. \tag{15}$$

This is the integral through a generalized quadrant; if we allow all the variables to have negative signs we must multiply by 2^r.

In particular take $f(s)$ unity; then we get the volume of a generalized r-dimensional sphere of radius a

$$L = \frac{\pi^{1/2 r}}{(\tfrac{1}{2} r)!} a^r. \tag{16}$$

For $r = 1, 2, 3$ this has the values $2a$, πa^2, $\tfrac{4}{3}\pi a^3$, corresponding to the length of a line, the area of a circle, and the volume of a sphere.

15·09. The exponential and related integrals. These are related to a case of the incomplete factorial function, which we shall denote by

$$\mathrm{ei}(z) = \int_z^\infty \frac{e^{-t}}{t} dt. \tag{1}$$

The integral also converges if the upper limit is $\infty \exp i\alpha$, and is independent of α, so long as $-\tfrac{1}{2}\pi \leq \alpha \leq \tfrac{1}{2}\pi$. Now in

$$\int_0^z (1 - e^{-t}) \frac{dt}{t}, \tag{2}$$

the integrand is an integral function; hence by expansion and integration this integral is equal to

$$z - \frac{z^2}{2 \cdot 2!} + \frac{z^3}{3 \cdot 3!} - \ldots, \tag{3}$$

and therefore

$$\int_z^Z \frac{e^{-t}}{t} dt = \int_z^Z \frac{dt}{t} - \int_z^Z \frac{1 - e^{-t}}{t} dt$$

$$= \left[\log z - \left(z - \frac{z^2}{2 \cdot 2!} + \frac{z^3}{3 \cdot 3!} - \ldots \right) \right]_z^Z. \tag{4}$$

The left side tends to a definite limit as $Z \to \infty$; hence the right side does the same. Then

$$\int_z^\infty e^{-t} \frac{dt}{t} = C - \log z + z - \frac{z^2}{2 \cdot 2!} + \frac{z^3}{3 \cdot 3!} - \ldots, \tag{5}$$

where C is a constant. To identify it we have

$$\frac{dz!}{dz} = \int_0^\infty e^{-t} t^z \log t\, dt, \tag{6}$$

whence, if γ is Euler's constant,

$$-\gamma = F(0) = \int_0^\infty e^{-t} \log t\, dt. \tag{7}$$

Now

$$\int_z^\infty e^{-t} \log t\, dt = \left[-e^{-t} \log t \right]_z^\infty + \int_z^\infty e^{-t} \frac{dt}{t}$$

$$= e^{-z} \log z + \mathrm{ei}(z)$$

$$= C - (1 - e^{-z}) \log z + O(z), \tag{8}$$

and also

$$= -\gamma - \int_0^z e^{-t} \log t\, dt.$$

Cosine and sine integrals

Hence, making z tend to 0,
$$C = -\gamma, \tag{9}$$

and
$$\mathrm{ei}(z) = -\gamma - \log z + z - \frac{z^2}{2.2!} + \frac{z^3}{3.3!} - \frac{z^4}{4.4!} + \dots \tag{10}$$

If z is real and negative, and we take the principal value,
$$P\,\mathrm{ei}(-x) = -\gamma - \log x - x - \frac{x^2}{2.2!} - \frac{x^3}{3.3!} - \dots, \tag{11}$$

which is denoted by $-\mathrm{Ei}(x)$ in published tables.

An asymptotic expansion* for large z follows at once from 17·01, with $n = 1$.
$$\mathrm{ei}(z) = e^{-z}\left(\frac{1}{z} - \frac{1}{z^2} + \frac{2!}{z^3} - \frac{3!}{z^4} + \dots + (-)^r \frac{r!}{z^{r+1}}\right) - (-)^r (r+1)! \int_z^\infty e^{-t} t^{-r-2}\,dt. \tag{12}$$

If $z = iy$, where y is real and positive,
$$\mathrm{ei}(iy) = \int_{iy}^\infty e^{-t}\frac{dt}{t} = \int_y^\infty e^{-iu}\frac{du}{u} = \int_y^\infty \frac{\cos u}{u}\,du - i\int_y^\infty \frac{\sin u}{u}\,du$$

and also
$$= -\gamma - \log y - \tfrac{1}{2}\pi i + i\left(y - \frac{y^3}{3.3!} + \frac{y^5}{5.5!} + \dots\right) + \left(\frac{y^2}{2.2!} - \frac{y^4}{4.4!} + \dots\right).$$

Hence
$$\int_y^\infty \frac{\cos u}{u}\,du = -\gamma - \log y + \frac{y^2}{2.2!} - \frac{y^4}{4.4!} + \dots, \tag{13}$$

$$\int_y^\infty \frac{\sin u}{u}\,du = \tfrac{1}{2}\pi - y + \frac{y^3}{3.3!} - \frac{y^5}{5.5!} + \dots. \tag{14}$$

The asymptotic formulae follow from
$$\mathrm{ei}(iy) \sim e^{-iy}\left(-\frac{i}{y} + \frac{1}{y^2} + \frac{i.2!}{y^3} - \frac{1.3!}{y^4} + \dots\right),$$

whence
$$\int_y^\infty \frac{\cos u}{u}\,du \sim \cos y\left(\frac{1!}{y^2} - \frac{3!}{y^4} + \dots\right) - \sin y\left(\frac{1}{y} - \frac{2!}{y^3} + \frac{4!}{y^5} - \dots\right), \tag{15}$$

$$\int_y^\infty \frac{\sin u}{u}\,du \sim \cos y\left(\frac{1}{y} - \frac{2!}{y^3} + \frac{4!}{y^5} + \dots\right) + \sin y\left(\frac{1}{y^2} - \frac{3!}{y^4} + \dots\right). \tag{16}$$

The tabulated functions $\mathrm{Ci}(y)$ and $\mathrm{Si}(y)$ are
$$\mathrm{Ci}(y) = -\int_y^\infty \frac{\cos u}{u}\,du, \tag{17}$$

$$\mathrm{Si}(y) = \tfrac{1}{2}\pi - \int_y^\infty \frac{\sin u}{u}\,du = \int_0^y \frac{\sin u}{u}\,du. \tag{18}$$

If in (1) we put
$$z = -\log\zeta, \quad t = -\log u,$$

we have
$$\mathrm{ei}(-\log\zeta) = -\int_0^\zeta \frac{du}{\log u}, \tag{19}$$

which is denoted by $-\mathrm{li}(\zeta)$, the principal value being taken if ζ is real and greater than 1.

* This section should be read after the beginning of Chapter 17.

The integral
$$I = \int_0^\infty \frac{\sin xt}{1+t^2} dt$$
cannot be evaluated directly by the method used for the corresponding integral in $\cos xt$, since the integrand is an odd function of t. We can, however, write it as
$$\Im \int_0^\infty \frac{e^{ixt}}{1+t^2} dt,$$
and replace the path by one from 0 to $i\infty$, with a semicircle about $t = i$. The residue at this pole is $e^{-x}/2i$, and the semicircle therefore contributes nothing to the imaginary part of the integral. Then for $x > 0$

$$\begin{aligned}
I &= \Im P \int_0^{i\infty} \frac{e^{ixt}}{1+t^2} dt = \Im P \int_0^\infty \frac{e^{-xu}}{1-u^2} i\, du \\
&= -P \int_0^\infty \tfrac{1}{2} e^{-xu} \left(\frac{1}{u-1} - \frac{1}{u+1}\right) du \\
&= -\tfrac{1}{2} P \int_{-1}^\infty e^{-x-xv} \frac{dv}{v} + \tfrac{1}{2} \int_1^\infty e^{x-xv} \frac{dv}{v} \\
&= -\tfrac{1}{2} e^{-x} P \operatorname{ei}(-x) + \tfrac{1}{2} e^x \operatorname{ei}(x) \\
&= \cosh x \left(x + \frac{x^3}{3.3!} + \ldots\right) - \sinh x \left(\gamma + \log x + \frac{x^2}{2.2!} + \ldots\right).
\end{aligned}$$

By repeated integration by parts it may be shown that
$$I \sim \frac{1}{x} + \frac{2!}{x^3} + \frac{4!}{x^5} + \ldots.$$

EXAMPLES

1. Find the positive number α such that the product $1.2^2.3^3 \ldots n^n$ is large or small in comparison with $n^{\alpha' n^2}$ according as α' is less or greater than α. (M.T. 1943.)

2. Prove that the residue of $z!$ at $z = -n$, where n is a positive integer, is $(-1)^{n-1}/(n-1)!$.

3. Prove that if n is a positive integer
$$z!\left(z - \frac{1}{n}\right)! \ldots \left(z - \frac{n-1}{n}\right)! = (2\pi)^{\frac{1}{2}(n-1)} n^{-nz-\frac{1}{2}} (nz)!.$$ (Euler.)

(Show that the ratio of the two sides is unaltered by changing z to $z+1$; and that it is of the form $1 + O(1/z)$ for z large.)

4. Obtain the same result from Gauss's definition.

5. Prove that for $\Re(z) > -1$
$$\begin{aligned}
F(z) &= \lim_{n\to\infty} \left(\log n - \int_0^\infty e^{-tz} \frac{e^{-t} - e^{-(n+1)t}}{1 - e^{-t}} dt\right) \\
&= \int_0^\infty \left(\frac{e^{-t}}{t} - \frac{e^{-(z+1)t}}{1 - e^{-t}}\right) dt.
\end{aligned}$$ (Gauss.)

(Cf. Frullani's integrals.)

6. If p^{-1} denotes integration from 0 to x, prove from the Fourier-Mellin theorem that
$$\operatorname{ei}(x) H(x) = \log(1+p). H(x)$$
and hence that
$$\operatorname{ei}(x) + \log x + \gamma = x - \frac{x^2}{2.2!} + \frac{x^3}{3.3!} - \ldots.$$

Examples

7. Prove that
$$\int_0^\infty \frac{u^m du}{(Au^2+B)^{\frac{1}{2}(m+n+2)}} = \frac{\{\tfrac{1}{2}(m-1)\}!\{\tfrac{1}{2}(n-1)\}}{\{\tfrac{1}{2}(m+n)\}!} \cdot \frac{1}{2A^{\frac{1}{2}(m+1)}B^{\frac{1}{2}(n+1)}} \cdot \quad (m>-1,\ n>-1)$$

8. Prove that
$$\int_0^\infty \sigma^{-n} e^{-k/\sigma^2} d\sigma = \frac{(\tfrac{1}{2}n-\tfrac{3}{2})!}{2 \cdot k^{\frac{1}{2}n-\frac{1}{2}}} \quad (n>1).$$

9. Prove that for positive real n
$$\frac{1}{n} - \log\frac{n}{n-1} < 0, \quad \frac{1}{n} - \log\frac{n+1}{n} > 0,$$
and hence show directly that the limit defining Euler's constant exists.

10. If
$$f(x) = \int_0^x \frac{u(\xi)}{(x-\xi)^\mu} d\xi,$$
where $f(0) = 0$, $0 < \mu < 1$, and $f'(x)$ is continuous, prove that
$$u(x) = \frac{\sin \mu \pi}{\pi} \frac{d}{dx} \int_0^x \frac{f(\eta)\, d\eta}{(x-\eta)^{1-\mu}}. \qquad\qquad \text{(Abel.)}$$

11. Prove that
$$F'(x) \sim \frac{1}{2}\left(\frac{1}{x}+\frac{1}{x+1}\right) - \Sigma B_{2r}\left(\frac{1}{x^{2r}} - \frac{1}{(x+1)^{2r}}\right),$$
and derive corresponding expressions for $F(x)$ and $\log x!$. $\qquad$ (A. Lodge.)

12. If $\int_0^t f(\tau)\, d\tau$ exists, $\alpha,\ \beta \geqslant 0$, but not necessarily integers, and $p^{-\alpha} f(t) H(t)$ is defined by
$$p^{-\alpha} f(t) H(t) = \int_{\tau=0-}^t f(t-\tau)\, d\left(\frac{\tau^\alpha}{\alpha!} H(\tau)\right)$$
show that
$$p^0 f(t) H(t) = f(t) H(t)$$
$$p^{-\alpha} p^{-\beta} f(t) H(t) = p^{-\beta} p^{-\alpha} f(t) H(t) = p^{-\alpha-\beta} f(t) H(t)$$
and that for $h > 0$
$$p^{-\alpha} e^{-ph} g(t) = e^{-ph} p^{-\alpha} g(t)$$
where $g(t)$ is any integrable function zero for negative t.

13. Using 15·05(7) and 15·02(11), derive 15·05(9).

14. Using Wallis's formula and 15·03(4), show that Stirling's formula can be extended to complex z for $\Re(z)$ large and positive.

15. If $C(x)$ and $S(x)$ are Fresnel's integrals, defined for real x by
$$C(x) + iS(x) = \int_0^x e^{it^2} dt$$
prove that for large positive x
$$C(x) = \frac{1}{2}\sqrt{\frac{\pi}{2}} - P\cos x^2 + Q\sin x^2$$
$$S(x) = \frac{1}{2}\sqrt{\frac{\pi}{2}} - P\sin x^2 - Q\cos x^2$$
where
$$P(x) \sim \frac{1}{2}\left(\frac{1}{2x^3} - \frac{1 \cdot 3 \cdot 5}{2^3 x^7} + \ldots\right), \quad Q(x) \sim \frac{1}{2}\left(\frac{1}{x} - \frac{1 \cdot 3}{2^2 x^5} + \frac{1 \cdot 3 \cdot 5 \cdot 7}{2^4 x^9} \ldots\right).$$

Chapter 16

SOLUTION OF LINEAR DIFFERENTIAL EQUATIONS OF THE SECOND ORDER

A merry road, a mazy road, and such as we did tread
The night we went to Birmingham by way of Beachy Head.
G. K. CHESTERTON, *The Flying Inn*

16·01. When the coefficients in a differential equation are variable the chief methods of solution are as follows:

(1) Direct numerical solution (Chapter 9). This is often laborious, but in many cases it is the only method.

(2) Solution by power series.

(3) Solution by substitution of definite or contour integrals.

(4) Asymptotic solutions (Chapter 17). These can be obtained by several methods. Direct transformation of the differential equation will often yield solutions in the form of asymptotic series; also a solution as a definite or contour integral can be approximated to by the method of steepest descents.

16·02. Singular points of a differential equation. Any second order linear equation can be expressed in the form
$$\frac{d^2y}{dx^2} = f\left(x, y, \frac{dy}{dx}\right).$$

If y and dy/dx are given at $x = x_0$, the differential equation in general determines d^2y/dx^2 at $x = x_0$. Differentiating the differential equation we can determine d^3y/dx^3 at $x = x_0$, and so on; the terms of a Taylor series for y can thus be found in turn, and if the series has a non-zero radius of convergence a solution exists. If y and dy/dx can take any pair of values at $x = x_0$ without making d^2y/dx^2 infinite we call x_0 an *ordinary point* of the equation; if not, we call it a *singular point*. For instance, if
$$\frac{d^2y}{dx^2} = -y,$$
and $y = y_0$, $dy/dx = y_1$ at $x = x_0$, there is a solution
$$y = y_0 \cos(x - x_0) + y_1 \sin(x - x_0)$$
whatever x_0, y_0, y_1 may be; hence all values of x are ordinary points of this differential equation. But even with the first order equation
$$x\frac{dy}{dx} = y$$
we cannot assign y arbitrarily at $x = 0$; for if we take y anything but 0 at $x = 0$, dy/dx will be infinite and we cannot form the Taylor series. Similarly for
$$x^2\frac{d^2y}{dx^2} - 2x\frac{dy}{dx} + 2y = 0,$$

if y is not 0 at $x = 0$ either dy/dx or d^2y/dx^2 is infinite and we cannot form the Taylor series. The value $x = 0$ is a singular point of these two equations.

An important property of linear equations is that their singular points are fixed. With such a simple equation as

$$\frac{dy}{dx} = \frac{1}{1-y^2},$$

dy/dx is infinite where $y = \pm 1$, and this is *not* at the same value of x, irrespective of the value of y at $x = 0$. In fact the solution is $y - \tfrac{1}{3}y^3 = x + \alpha$, and $y = 1$ where $x = \tfrac{2}{3} - \alpha = \tfrac{2}{3} - y(0) + \tfrac{1}{3}[y(0)]^3$. It is this variability of the position of the singularities that leads to most of the additional difficulty of the general theory of non-linear equations.

16·03. Existence of solutions about ordinary points. Consider the equation

$$\frac{d^2y}{dx^2} + f(x)\frac{dy}{dx} + g(x)y = 0, \qquad (1)$$

where $f(x)$ and $g(x)$ are analytic functions of x (which we may take to be a complex variable for extra generality) within a region including $x = 0$. When $x = 0$ let $y = y_0$, $dy/dx = y_1$. For them to be assignable arbitrarily $f(x)$ and $g(x)$ must be bounded near $x = 0$. Define

$$Q\phi(x) = \int_0^x \phi(t)\,dt. \qquad (2)$$

Perform the operation Q on the differential equation, assuming it possible, and integrate the second term by parts. Then

$$\frac{dy}{dx} - y_1 + yf(x) - y_0 f(0) - Q\{yf'(x)\} + Q\{g(x)y\} = 0, \qquad (3)$$

which we can write

$$\frac{dy}{dx} = y_1 + y_0 f(0) - yf(x) + Q\{h(x)y\}. \qquad (4)$$

Integrating again we have

$$y = y_0 + y_1 x + y_0 f(0) x - Q\{yf(x)\} + Q^2\{h(x)y\}. \qquad (5)$$

Put

$$y_0 + y_1 x + y_0 f(0) x = u_1, \qquad (6)$$

$$- Q\{f(x)u_1\} + Q^2\{h(x)u_1\} = u_2, \qquad (7)$$

$$- Q\{f(x)u_r\} + Q^2\{h(x)u_r\} = u_{r+1}, \qquad (8)$$

and take a straight path from 0 to x. Suppose that on the path $|f(x)| < S$, $|h(x)| < T$, $|u_1| < C$. Now if $|\phi(x)| < A|x|^r$ on the path

$$|Q\phi(x)| < \int_0^{|x|} A|x|^r\,dx < \frac{A}{r+1}|x|^{r+1}; \qquad (9)$$

also there is a quantity U such that $S + \tfrac{1}{2}T|x| < U$ for all points of the path. Then

$$|u_2| = |-Q\{f(x)u_1\} + Q^2\{h(x)u_1\}| < CS|x| + \frac{1}{2!}CT|x|^2 < CU|x|, \qquad (10)$$

$$|u_3| < C\left(\tfrac{1}{2}SU|x|^2 + \frac{1}{2\cdot 3}TU|x|^3\right) < \frac{1}{2!}CU^2|x|^2, \qquad (11)$$

and in general

$$|u_{r+1}| < \frac{1}{r!}CU^r|x|^r. \qquad (12)$$

The series
$$u_1 + u_2 + u_3 + \ldots, \tag{13}$$
is therefore absolutely convergent; and if the inequalities hold for $|x| < R$ and $\arg x$ within a given range, the series is also uniformly convergent by the M test. The separate terms u_r are analytic functions of x, since u_1 is an analytic function and by definition each later term is the integral of an analytic function. But a uniformly convergent series of analytic functions is analytic; hence the series represents an analytic function. The proof that it satisfies the differential equation and the conditions imposed at $x = 0$ is straightforward.

The existence of a solution is therefore proved for straight paths of integration. By Cauchy's theorem it will be correct for any other path with the same termini provided that the new path can be deformed into a straight line without passing over any singularity of the functions integrated. But the only possible singularities of the functions u_r are those of $f(x)$ and $g(x)$. The solution is therefore analytic over the whole plane except possibly at the singular points of the equation.

The method is substantially that of Picard, foreshadowed in unpublished work of Cauchy and in the paper of Caqué already mentioned in Chapter 7.*

The process can be extended to give continuations for the solution around the singular points. Suppose for instance that there is a singularity at $x = 1$, and no other. Then we could proceed from $x = 0$ to $x = 2+i$, transform to $2+i$ as a new origin, and infer a value of y for $x = 2$, which cannot be reached by a straight path from $x = 0$. But we should not necessarily get the same value in this way as if we proceeded by way of $x = 2-i$. It is possible, in fact, for $f(x)$ and $g(x)$ to have poles at a point and therefore to be single-valued near it, but y and its first derivative cannot be assigned arbitrarily and the admissible solutions will in general have branch points.

16·04. Power series solutions. This argument proves the existence of a solution. The method is seldom convenient for obtaining it explicitly, though it can sometimes be usefully combined with numerical integration. It does, however, show that the solution is expressible by a power series within a circle reaching to the nearest singularity of the equation. Since $f(x)$ and $g(x)$ are also so expressible we can substitute directly in the differential equation and equate coefficients. Then the coefficients in the solution can be found in turn.

Thus if we take the equation satisfied by the *Airy integral*
$$\frac{d^2 y}{dx^2} = xy, \tag{1}$$
we assume
$$y = a_0 + a_1 x + a_2 x^2 + a_3 x^3 + \ldots.$$
Then by substitution
$$2a_2 + 3 \cdot 2 a_3 x + 4 \cdot 3 a_4 x^2 + r(r-1) a_r x^{r-2} + \ldots = a_0 x + a_1 x^2 + \ldots + a_{r-3} x^{r-2} + \ldots,$$
and
$$a_2 = 0, \quad a_3 = \frac{a_0}{2 \cdot 3}, \quad a_4 = \frac{a_1}{3 \cdot 4}, \quad \ldots, \quad a_r = \frac{a_{r-3}}{r(r-1)}.$$
Then
$$y = a_0 \left(1 + \frac{x^3}{2 \cdot 3} + \frac{x^6}{2 \cdot 3 \cdot 5 \cdot 6} + \ldots \right) + a_1 \left(x + \frac{x^4}{3 \cdot 4} + \frac{x^7}{3 \cdot 4 \cdot 6 \cdot 7} + \ldots \right). \tag{2}$$

This is obviously an integral function, as would be expected since the differential equation has no singular point.

* For historical references see *Encycl. d. math. Wiss.* II, 1i, 198–200.

Take now *Legendre's equation*

$$(1-x^2)\frac{d^2y}{dx^2} - 2x\frac{dy}{dx} + n(n+1)y = 0. \tag{3}$$

The singular points are at $x = \pm 1$. Try

$$y = a_0 + a_1 x + \ldots + a_r x^r + \ldots.$$

The constant terms and linear terms give

$$2.1 a_2 + n(n+1) a_0 = 0, \quad 3.2 a_3 + (n+2)(n-1) a_1 = 0,$$

and in general

$$(r+1)(r+2) a_{r+2} = -(n-r)(n+r+1) a_r. \tag{4}$$

Hence

$$y = a_0 \left(1 - \frac{n(n+1)}{2!} x^2 + \frac{(n-2)n(n+1)(n+3)}{4!} x^4 - \frac{(n-4)(n-2)n(n+1)(n+3)(n+5)}{6!} x^6 + \ldots \right)$$
$$+ a_1 \left(x - \frac{(n-1)(n+2)}{3!} x^3 + \frac{(n-3)(n-1)(n+2)(n+4)}{5!} x^5 - \ldots \right). \tag{5}$$

We see from (4) that the radii of convergence of both series are in general 1. But if n is an even positive integer the series of even powers terminates and reduces to a polynomial; if it is an odd positive integer the odd series reduces to a polynomial. In either case the other solution is still an infinite series and must have a singularity for $|x| = 1$; but there will not be even one solution analytic at both $+1$ and -1 unless n is an integer.

In both these examples there is a two-term recurrence relation between the coefficients. This is fortunately true of many of the solutions of the differential equations of physics, and the series solutions are then easy. If the recurrence relation involves three coefficients the solution is difficult, and if more it is practically impossible to obtain more than a few terms explicitly.

16·05. Solution near an isolated singularity. Let $f(x)$ and $g(x)$ have a singularity at $x = 0$ but be single-valued in S, a region about 0, and have no other singularity in S. Let $x = a$ be an ordinary point in this region. Then the differential equation has two independent analytic solutions in a region about a, the first terms being 1 and $x - a$ respectively; denote these by $Y_1(x)$ and $Y_2(x)$. They have analytic continuations throughout S except at 0; let continuation be carried out along a closed path about 0, returning to a. Let the continuations of Y_1 and Y_2 be Z_1, Z_2. These will not in general be equal to Y_1, Y_2 for given x, because $x = 0$ may be a branch point of the solutions. But Z_1, Z_2 will be solutions of the differential equation and must therefore be linear functions of Y_1, Y_2; thus for all x in S

$$\left. \begin{array}{l} Z_1 = a_{11} Y_1 + a_{12} Y_2, \\ Z_2 = a_{21} Y_1 + a_{22} Y_2. \end{array} \right\} \tag{1}$$

The matrix a_{ik} is non-singular. For if it were singular, there would be an identically zero linear form $\alpha Z_1 + \beta Z_2$ with α, β not both zero; on reversing the process of continuation we could then show that $\alpha Y_1 + \beta Y_2$ is identically zero, which is impossible. Hence a_{ik} has non-zero eigenvalues λ_1, λ_2.

First let these eigenvalues be different. Then we can find two linear combinations of Y_1 and Y_2, say W_1 and W_2, such that their continuations are $\lambda_1 W_1$ and $\lambda_2 W_2$. But the continuation of x^s is $x^s \exp 2\pi i s$. Hence if s_1 and s_2 are such that

$$2\pi i s_1 = \log \lambda_1, \quad 2\pi i s_2 = \log \lambda_2, \tag{2}$$

the functions
$$x^{-s_1}W_1, \quad x^{-s_2}W_2 \tag{3}$$
are single-valued in S. They have therefore convergent Laurent expansions in S and therefore within a circle about 0 extending to the nearest singularity of $f(x)$ or $g(x)$.

If the eigenvalues of a_{ik} are equal, it will be possible to reduce a_{ik} to a triangular matrix but not in general to a diagonal one; then we can find a W_1 such that its continuation is λW_1, but for any other linear combination W_2 the continuation will be $\lambda W_2 + \mu W_1$. Then there will be a solution W_1 such that $x^{-s}W_1$ is single-valued in S.

For the second solution, if
$$2\pi i s = \log \lambda, \quad 2\pi i t = \frac{\mu}{\lambda}, \tag{4}$$
$W_2 - W_1 t \log x$ will have continuation $\lambda(W_2 - W_1 t \log x)$, and therefore there is in general a second solution W_2 such that
$$x^{-s}(W_2 - W_1 t \log x) \tag{5}$$
has a convergent Laurent expansion with the same region of validity. But W_2 may be identically equal to $W_1 t \log x$. We could, for instance, have a differential equation with solutions $x, x \log x$. In any case, however, the solutions can be written as one of the pairs
$$x^{s_1}z_1, \quad x^{s_2}z_2; \quad x^s z_1, x^s(z_2 - z_1 t \log x); \quad x^s z_1, x^s z_1 \log x; \tag{6}$$
where z_1, z_2 are single-valued and analytic in S except possibly at $x = 0$, and therefore have convergent Laurent expansions in S valid in a circle about 0 extending to the nearest singularity of $f(x)$ or $g(x)$. The third case may be regarded as a particular case of the second when $z_2 = 0, t = -1$.

It may happen that even if the eigenvalues of a_{ik} are equal, it can be reduced to diagonal form. In this case $\mu = 0$, and there are solutions W_1, W_2 such that $x^{-s}W_1, x^{-s}W_2$ are single-valued in the neighbourhood.

The condition that $f(x), g(x)$ are single-valued in S is necessary to this argument. For if either of them did not return to its original value when x described a circle about 0, the relations (1) would not be true.

This argument shows that solutions of the types indicated exist near a pole or an isolated essential singularity of $f(x)$ or $g(x)$. It leaves open the possibility that z_1 and z_2 may themselves have singularities at $x = 0$, which may be essential singularities even if $f(x)$ and $g(x)$ have only poles.

16·051. Regular singularities. The Laurent expansions of the functions z_1, z_2 in 16·05 may each contain only a finite number of negative powers; if so, each function has a pole or an ordinary point at $x = 0$. If this is so the singularity of the differential equation is said to be *regular*. Then, since an alteration of s_1, s_2 or s in 16·05 by an integer does not affect the corresponding λ, we can rewrite the solutions in one of the forms
$$y_1 = x^{s_1}z_1, \quad y_2 = x^{s_2}z_2 \tag{1}$$
or
$$y_1 = x^s z_1, \quad y_2 = x^{s+c}z_2 + t y_1 \log x, \tag{2}$$
where z_1, z_2 are now required to be analytic and not zero at $x = 0$. In (2) c must be an integer. In (1), if $s_1 = s_2$, we can subtract a multiple of y_1 from y_2 so as to cancel the first term. The remainder will not vanish identically because y_2 is not proportional to y_1, and will be a solution independent of y_1 starting with a power different from s_1. Hence there is no loss of generality in taking $s_1 \neq s_2$. In (2), similarly, if $c = 0$ we can subtract a multiple

of y_1 to cancel the first term in x^sz_2; thus there is a solution with c a positive integer. In this case the whole of z_2 may disappear and we are left with the special case where $y_2 = y_1 \log x$. Thus in (2) we can take $c \neq 0$ unless $z_2 = 0$.

If $f(x)$ or $g(x)$ has an isolated singularity, not a branch-point, at $x = 0$, a necessary and sufficient condition for it to be a regular singularity of $y'' + f(x) y' + g(x) y = 0$ is that $xf(x)$ and $x^2g(x)$ shall be analytic at $x = 0$. We show first that the condition is necessary. Since $y = A_1 y_1 + A_2 y_2$ satisfies the differential equation for all constant A_1, A_2, we must have

$$f(x) = -\frac{y_1 y_2'' - y_1'' y_2}{y_1 y_2' - y_1' y_2}; \quad g(x) = \frac{y_1' y_2'' - y_1'' y_2'}{y_1 y_2' - y_1' y_2} \tag{3}$$

and also, since y_1 is a solution,

$$g(x) = -\frac{y_1'' + f(x) y_1'}{y_1}. \tag{4}$$

From (1) we find by direct substitution

$$f(x) = -\frac{s_1 + s_2 - 1}{x} + O(1); \quad g(x) = \frac{s_1 s_2}{x^2} + O\left(\frac{1}{x}\right). \tag{5}$$

From (2), if c is positive or if $z_2 = 0$,

$$f(x) = -\frac{2s-1}{x} + O(1); \quad g(x) = \frac{s^2}{x^2} + O\left(\frac{1}{x}\right). \tag{6}$$

From (2), if c is negative,

$$f(x) = -\frac{2s+c-1}{x} + O(1); \quad g(x) = \frac{s(s+c)}{x^2} + O\left(\frac{1}{x}\right). \tag{7}$$

In all cases $f(x)$ and $g(x)$ contain no logarithm. Hence the condition stated is necessary. Conversely, let $xf(x)$, $x^2g(x)$ be analytic at $x = 0$, and have no singularity for

$$|x| \leq |x_0| = r_0.$$

Make a cut from 0 to $-\infty \operatorname{sgn} x_0$. Then any point $x = re^{i\alpha}$ (r, α real) within the circle $|x| = r_0$ and not on the cut can be reached from x_0 by a straight path from x_0 to $x_1 = r_0 e^{i\alpha}$, and then another from $r_0 e^{i\alpha}$ to x, and also by a straight path from x_0 to x directly, and the two approaches, by the argument of 16·03, will give the same result. Now y and dy/dx are bounded on $|x| = r_0$. (For if arg $x -$ arg $x_0 > \tfrac{1}{2}\pi$ we can proceed from x_0 to $x_0 e^{\tfrac{1}{2}\pi i}$ and then to x, by straight paths, on which $r \geq r_0/\sqrt{2}$.) Define

$$Q\phi(r) = \int_{r_0}^{r} \phi(\rho) d\rho. \tag{8}$$

Since $r < r_0$ this is negative for positive ϕ, but $Q^2\phi(r)$ is then positive. Now if $|f(x)| \leq A/r$, $|g(x)| \leq B/r^2$ for $r \leq r_0$, we can apply the method of 16·03 to the integration from r_0 to r; and the moduli of the successive terms are $\leq$ those of the solution by the same method of

$$\frac{d^2z}{dr^2} + \frac{A}{r}\frac{dz}{dr} - \frac{B}{r^2}z = 0, \tag{9}$$

given that at $x = x_1$, z and dz/dr have the values of y and $-|\partial y/\partial r|$. But this equation has the solution

$$z = Cr^{t_1} + Dr^{t_2}, \tag{10}$$

where t_1, t_2 are the roots of $\quad t(t-1) + At - B = 0$. \hfill (11)

Since B is positive the roots are real and cannot be equal; hence the solution of (9) always has the form (10), and if t_1 is the smaller root $r^{-t_1}z$ is bounded* for $0 < r \leq r_0$. But $|z| \geq |y|$,

* Of course it is not analytic, being defined only for a special path.

and C, D are clearly bounded for variations of α (including on the cut, from whichever side we approach it). Hence $x^{-1}y$ is bounded for $|x| < r_0$. But this implies that the functions z_1, z_2 of (1), (2) have not essential singularities at 0. Hence the differential equation has a regular singularity at $x = 0$.

If $y = Ax + Be^{1/x}$,
$$\frac{d^2y}{dx^2} + \frac{2x}{x^3(x+1)}\left(x\frac{dy}{dx} - y\right) = 0,$$
so that both $xf(x)$ and $x^2g(x)$ are non-analytic at $x = 0$. If
$$y = Ae^{1/x} + Be^{-1/x},$$
$$\frac{d^2y}{dx^2} + \frac{2}{x}\frac{dy}{dx} - \frac{y}{x^4} = 0,$$
so that $xf(x)$ is analytic but $x^2g(x)$ is not.

16·052. Singularities at infinity. If in
$$\frac{d^2y}{dx^2} + f(x)\frac{dy}{dx} + g(x)y = 0 \tag{1}$$
we put $x = 1/\xi$, we get
$$\xi^4\frac{d^2y}{d\xi^2} + \{2\xi^3 - \xi^2 f(x)\}\frac{dy}{d\xi} + g(x)y = 0. \tag{2}$$

We call infinity an ordinary point of (1) if the functions
$$\frac{2\xi - f(x)}{\xi^2} = 2x - x^2 f(x), \quad \frac{g(x)}{\xi^4} = x^4 g(x) \tag{3}$$
are analytic at $\xi = 0$, that is, at $x = \infty$. We call it a regular singularity if they are respectively $O(1/\xi) = O(x)$ and $O(1/\xi^2) = O(x^2)$.

It is impossible for all points, including infinity, to be ordinary points of (1). For then $f(x)$ and $g(x)$ must be analytic over the whole plane and therefore integral functions, and for $|x|$ large
$$2x - x^2 f(x) = O(1), \quad x^4 g(x) = O(1).$$
These functions are bounded over the whole plane and therefore are constants, say a and b. Therefore
$$f(x) = -\frac{a}{x^2} + \frac{2}{x}, \quad g(x) = \frac{b}{x^4}$$
and $x = 0$ is not an ordinary point of (1) even if $a = b = 0$.

It is possible for all points except 0 and ∞ to be ordinary points of (1), and for 0 and ∞ to be regular singularities. In these conditions for $|x|$ small
$$f(x) = O\left(\frac{1}{x}\right), \quad g(x) = O\left(\frac{1}{x^2}\right),$$
and for $|x|$ large, $\quad 2x - x^2 f(x) = O(x), \quad x^4 g(x) = O(x^2).$

Hence $xf(x)$ and $x^2g(x)$ are bounded over the whole plane, and are constants. Then (1) reduces to the form
$$\frac{d^2y}{dx^2} + \frac{a}{x}\frac{dy}{dx} + \frac{b}{x^2}y = 0$$
which can be solved in terms of elementary functions.

16·06. Solutions near a regular singularity.

In 16·051 (1) (2) the functions z_1, z_2 are analytic in a neighbourhood of 0. Hence all the terms of the differential equation can be expressed by convergent power series, and we can proceed by substituting power series for y and equating coefficients. We rewrite the differential equation in the form

$$Dy = x^2 \frac{d^2y}{dx^2} + (p_0 + p_1 x + \ldots) x \frac{dy}{dx} + (q_0 + q_1 x + \ldots) y = 0. \tag{1}$$

Put
$$y = x^s (a_0 + a_1 x + \ldots) \quad (a_0 \neq 0) \tag{2}$$

and equate powers of x^{s+r}. We find

$$\{(s+r)(s+r-1) + p_0(s+r) + q_0\} a_r = \text{terms in } a_0 \text{ to } a_{r-1}. \tag{3}$$

For $r = 0$, we have, since $a_0 \neq 0$,

$$s(s-1) + p_0 s + q_0 = 0 \tag{4}$$

which is called the *indicial equation*. Let its roots be s_1, s_2. For each root, provided the coefficient in (3) never vanishes for $r > 0$, we can determine the a_r in turn, a_0 being arbitrary. The series will converge within a circle extending to the nearest singularity of the differential equation other than $x = 0$*, and a linear combination of the two solutions will give the general solution. There is an obvious exceptional case if $s_1 = s_2$. There may be another if they differ by an integer; if they are α and $\alpha - k$ ($k > 0$), a series solution is obtained as usual for the larger root. But for the smaller the coefficient of a_r in (3) will vanish when $r = k$, and with an arbitrary a_0, a_k and higher terms will in general be infinite. The series can then be rendered finite only by taking a_k finite and all earlier coefficients zero. But then the series reduces to that given by the root α, and we have still found only one solution.

It may happen, however, that for an arbitrary a_0 the right side of the equation for a_k also vanishes. Then the equation holds for any value of a_k, and the solution starting with $x^{\alpha-k}$ is arbitrary by a multiple of the solution that starts with x^α. We therefore get two solutions in power series in this case. This corresponds to the case where a_{ik} of 16·05 is reducible to diagonal form in spite of having equal eigenvalues.

It remains to examine the nature of the other solution in the cases where only one series solution exists. Denote this solution by

$$w(x) = x^\alpha (1 + a_1 x + \ldots), \tag{5}$$

and put
$$y = w(x) z. \tag{6}$$

The terms in z cancel because $w(x)$ is a solution of the equation, and

$$\frac{z''}{z'} = -\frac{2w'}{w} - \frac{p_0}{x} - p_1 - p_2 x \ldots, \tag{7}$$

whence
$$\log z' = C - 2 \log w - p_0 \log x - p_1 x - \tfrac{1}{2} p_2 x^2 - \ldots,$$

$$z' = \frac{A}{x^{p_0} w^2} \phi(x),$$

$$z = A \int^x x^{-p_0} w^{-2} \phi(x) \, dx, \tag{8}$$

where C and A are arbitrary, and $\phi(x)$ is analytic, and equal to 1 for $x = 0$. The first term in the expansion of $x^{-p_0} w^{-2}$ is $x^{-p_0 - 2\alpha}$. But if the other root of the indicial equation is $\alpha - k$,

$$2\alpha - k = -p_0 + 1, \tag{9}$$

* For a direct proof see Whittaker and Watson, *Modern Analysis*, 1915, p. 193.

and $-p_0-2\alpha = -k-1$. If k is not a positive integer or zero we therefore get a series of powers on integration, starting with x^{-k}. If $k = 0$,

$$z = A \int^x \frac{1}{x}(1+r_1 x + r_2 x^2 + \ldots)\,dx$$
$$= A(\log x + r_1 x + \tfrac{1}{2}r^2 x^2 + \ldots) + \text{constant}, \tag{10}$$
and
$$y = \{A(\log x + r_1 x + \tfrac{1}{2}r_2 x^2 + \ldots) + B\}w(x) \tag{11}$$

so that the solution is a case of 16·051 (2) with $c > 0$ unless all of $r_1, r_2, \ldots$ are zero, when it is a case of 16·051 (2) with $z_2 = 0$.

If k is an integer > 0,

$$z = A \int^x x^{-k-1}(1+r_1 x + r_2 x^2 + \ldots)\,dx$$
$$= B + A\left\{-\frac{1}{kx^k} - \frac{r_1}{(k-1)x^{k-1}} - \ldots + r_k \log x + \ldots\right\}. \tag{12}$$

The second solution therefore contains a part $w(x) \log x$ unless $r_k = 0$. The rest is a power series beginning with a term in $x^{\alpha-k}$, and $\alpha - k$ is the other root of the indicial equation.

The solution is then of the form of 16·051 (2) with $y_1 = w(x)$ and $c = -k$ with k positive.

The forms of the indicial equation given by 16·051 (5) (6) (7) lead respectively to the pairs of indices s_1, s_2; s, s; $s, s+c$. Hence the correspondence of the types of solution in the two methods of approach is as follows.

16·051 (1). Roots of indicial equation different; if they differ by an integer, 16·06 (12) with $r_k = 0$.

16·051 (2) with $c > 0$ or with $z_2 = 0$. Roots of indicial equation equal. One solution contains a logarithm. $z_2 = 0$ corresponds to all $r_1, r_2, \ldots$, of 16·06 (11) being zero.

16·051 (2) with $c < 0$. Roots of indicial equation differ by an integer; 16·06 (12) with $r_k \neq 0$. One solution contains a logarithm.

The roots of the indicial equation are usually real. For if

$$s = u + iv$$
$$x^s = \exp\{(u+iv)\log x\} = x^u\{\cos(v\log x) + i\sin(v\log x)\}$$

and the complicated behaviour of the last factor near $x = 0$ is usually forbidden by some physical consideration.

The importance of the case where the roots differ by an integer and a logarithm does not arise far exceeds what might be expected from the fact that it apparently implies two coincidences. The reason is that it is only in the case where s_1, s_2 are both integers and a logarithm does not arise that both solutions can be free from a branch-point at $x = 0$.

16·07. The second solution may be found explicitly, once we know its form, by either of two methods. We can assume

$$y = w(x)\log x + \sum_{r=0}^{\infty} b_r x^{\alpha-k+r}, \tag{13}$$

and determine the coefficients by substitution; or we can use *the method of Frobenius*. This consists in substituting in the differential equation a series

$$y = x^s(1 + a_1 x + a_2 x^2 + \ldots), \tag{14}$$

and equating coefficients as before, *except for the lowest power of x*. Then for any s, such that no $s+r$ for integral $r \geq 0$ satisfies the indicial equation, all the coefficients are determined in terms of a_0 and s exactly as before. But y no longer satisfies the differential equation; in fact

$$Dy = \{s(s-1) + p_0 s + q_0\} x^s. \tag{15}$$

Now if the roots are different and do not differ by an integer, and we make s tend to either of them, the coefficient tends to 0 and y therefore tends to a solution of the differential equation. We thus recover the previous solutions.

If the roots are both equal to α, the coefficient is $(s-\alpha)^2$. Hence we have when $s \to \alpha$ both

$$Dy \to 0, \quad \frac{\partial}{\partial s} Dy \to 0. \tag{16}$$

But s does not appear explicitly in D; hence

$$D \frac{\partial y}{\partial s} = \frac{\partial}{\partial s} Dy \to 0, \tag{17}$$

and therefore both y and $\dfrac{\partial y}{\partial s}$ tend to solutions of the differential equation. Thus we get two independent solutions.

If the roots are α and $\alpha - k$, the coefficient is $(s-\alpha)(s-\alpha+k)$;

$$Dy = (s-\alpha)(s-\alpha+k) x^s. \tag{18}$$

When $s \to \alpha$, y tends to the previous solution. But $\dfrac{\partial}{\partial s} Dy$ does not tend to 0 when s tends to either root. This can be remedied by multiplying by $s-\alpha+k$, and we get a second solution

$$\left[\frac{\partial}{\partial s} \{(s-\alpha+k) y\} \right]_{s=\alpha-k}. \tag{19}$$

The factor $s-\alpha+k$ cancels the vanishing factor in the denominators when $r \geqslant k$ and in general we get an infinite series.

It appears that in this case we could get a third solution

$$\left[\frac{\partial}{\partial s} \{(s-\alpha) y\} \right]_{s=\alpha}$$

by the same method. But no denominator in y tends to 0 when $s \to \alpha$, and this expression is simply $(y)_{s=\alpha}$ again, so that we recover the first solution.

In spite of its apparent simplicity the method of Frobenius is seldom used, for the following reason. In general we require the solution of a differential equation to satisfy certain conditions such as being single-valued and continuous in a region, or tending to zero at infinity. The solution containing a logarithm usually fails to satisfy the former condition and therefore we often need only the solution with the higher index, which is a straightforward series. On the other hand even if both solutions are power series, they usually tend to infinity with x and to get one that tends to zero we must take a combination of them in a special ratio. The solution given by the method of Frobenius is not in general this combination, and it can be discovered only by other methods, which lead to determinations of the coefficients on the way. We shall find instances of such methods in relation to Bessel and Legendre functions and the confluent hypergeometric function.

Legendre's equation

In Legendre's equation put $x = 1+z$. Then

$$(2z+z^2)\frac{d^2y}{dz^2} + 2(1+z)\frac{dy}{dz} - n(n+1)y = 0.$$

The indicial equation is $\quad 2s(s-1) + 2s = 0,$

and both roots are zero. Using the method of Frobenius, put

$$y_c = z^c(1 + a_1 z + \ldots + a_r z^r + \ldots).$$

The lowest power of z merely repeats the indicial equation. The terms in z^r give

$$2(r+1+c)^2 a_{r+1} = -\{(r+c)(r+1+c) - n(n+1)\}a_r$$
$$= -(r+c-n)(r+c+n+1)a_r,$$

and

$$y_c = z^c\left\{1 - \frac{(c-n)(c+n+1)}{(c+1)^2}\frac{z}{2} + \frac{(c-n)(c-n+1)(c+n+1)(c+n+2)}{(c+1)^2(c+2)^2}\left(\frac{z}{2}\right)^2 - \ldots\right\}.$$

When $c \to 0$ we get a solution

$$w(z) = 1 + \frac{n(n+1)}{1^2}\frac{z}{2} + \frac{(n-1)n(n+1)(n+2)}{1^2 \cdot 2^2}\left(\frac{z}{2}\right)^2 + \ldots,$$

which could have been found directly. Its radius of convergence is 2. This solution is denoted by $P_n(x)$. If n is an integer the series terminates and is therefore a multiple of the terminating series found already (16·04).

The second solution can be taken as $\partial y_c/\partial c$, when $c \to 0$. We have

$$\frac{\partial}{\partial c} z^{c+r} = z^r \frac{\partial}{\partial c}\{\exp(c \log z)\} \to z^r \log z,$$

and the second solution is

$$P_n(x) \log z - \frac{(-n)(n+1)}{1^2}\left(\frac{1}{-n} + \frac{1}{n+1} - \frac{2}{1}\right)\frac{z}{2}$$
$$+ \frac{(-n)(-n+1)(n+1)(n+2)}{1^2 \cdot 2^2}\left(\frac{1}{-n} + \frac{1}{-n+1} + \frac{1}{n+1} + \frac{1}{n+2} - \frac{2}{1} - \frac{2}{2}\right)\left(\frac{z}{2}\right)^2 - \ldots.$$

When n is a positive integer the terminating series is also analytic at $x = -1$, but there will be a second solution containing $\log(x+1)$. It can be shown that the second solution can be taken as

$$q_n(x) = \frac{1}{\pi}P_n(x)\log\frac{x+1}{x-1} - f_{n-1}(x),$$

where f_{n-1} consists of the positive and zero powers of x in the expansion of $\frac{1}{\pi}P_n(x)\log\frac{x+1}{x-1}$ in *descending* powers of x.

The properties of the second solution indicate at once a way of excluding it from a large class of physical problems without actually needing to evaluate it. The most important applications of the equation concern a sphere, x being the sine of the latitude. In such problems we usually know that the solution or its derivative is finite at the poles. But when both indices at the poles are zero the second solution obviously does not satisfy either condition. Hence the only admissible solution is the first, and this must be analytic at both ± 1. The equation has no other singularity, and therefore the admissible solution must be an integral function. Of the solutions found in 16·04 the one in powers of x that terminates is therefore an admissible solution; the infinite series with radius of convergence 1 is not. Consequently we can assert immediately that the terminating solution is the one required. Even when the differential equation has other singularities we can often identify the physically important solution by mere inspection of the indicial equation and the radii of convergence.

16·08. Three-term recurrence relations. The following example from tidal theory illustrates both this principle and a method of obtaining solutions when the recurrence relation involves three coefficients. In the free symmetrical oscillations of water on a rotating globe, ζ, the elevation of the surface, satisfies the equation

$$\frac{d}{d\mu}\left(\frac{1-\mu^2}{f^2-\mu^2}\frac{d\zeta}{d\mu}\right)+\beta\zeta=0,$$

where μ is the sine of the latitude, β a positive constant depending on the depth, the rate of rotation, and the radius of the earth, and f is proportional to the speed of the oscillation to be determined. ζ and $\frac{1}{f^2-\mu^2}\frac{d\zeta}{d\mu}$ must be finite for all latitudes. The equation has singularities at $\mu=\pm 1$, $\mu=\pm f$. Suppose that ζ is expanded in powers of $1-\mu$, starting with $(1-\mu)^n$. Then the term of lowest degree in $(1-\mu)$ is $\frac{2n^2}{f^2-1}(1-\mu)^{n-1}$, and the indicial equation is $n^2=0$. Consequently one solution contains terms in $\log(1-\mu)$ and is inadmissible, since it would make $d\zeta/d\mu$ infinite at the north pole. Similarly one solution would contain terms in $\log(1+\mu)$ and be inadmissible at the south pole. The determination of the periods then reduces to finding values of f that give solutions finite at both poles.

It is easily shown that if $f \neq 0$ the roots of the indicial equation at $\mu=\pm f$ are 0 and 2; if $f=0$ they are 0 and 3. On direct examination, however, it is found that the second solution near $\mu=f$ does not involve a logarithm, and both solutions are therefore analytic at $\mu=f$. Hence if we obtain a solution in powers of μ it must be analytic at all the singularities of the differential equation and therefore be an integral function. Hence the ratio of the coefficients of consecutive terms tends to 0. If it tended to ± 1 we should know at once that the solution becomes logarithmically infinite at the poles. It is not even strictly necessary to examine the singularities at $\mu=\pm f$. For if $|f|<1$ and the solution was not analytic at $\pm f$ the radius of convergence would be $|f|<1$, and if both solutions were found to have radii of convergence ≥ 1 it would follow at once that they are analytic at $\mu=\pm f$.

The method of solution adopted by Laplace and many later writers is to take

$$\frac{1}{\mu^2-f^2}\frac{d\zeta}{d\mu}=a_0+a_1\mu+a_2\mu^2+\ldots, \tag{1}$$

whence by integration

$$\zeta=A-f^2a_0\mu-\tfrac{1}{2}f^2a_1\mu^2+\sum_{r=2}^{\infty}\frac{1}{r+1}(a_{r-2}-f^2a_r)\mu^{r+1}, \tag{2}$$

and also

$$\frac{1-\mu^2}{f^2-\mu^2}\frac{d\zeta}{d\mu}=-a_0-a_1\mu-\sum_{r=2}^{\infty}(a_r-a_{r-2})\mu^r. \tag{3}$$

Hence

$$-a_1-\sum_{r=2}^{\infty}r(a_r-a_{r-2})\mu^{r-1}+\beta\left\{A-f^2a_0\mu-\tfrac{1}{2}f^2a_1\mu^2+\sum_{r=2}^{\infty}\frac{1}{r+1}(a_{r-2}-f^2a_r)\mu^{r+1}\right\}=0. \tag{4}$$

Here we can equate coefficients; the terms in μ^{r+1} give for $r \geqslant 2$

$$-(r+2)(a_{r+2}-a_r)+\frac{\beta}{r+1}(a_{r-2}-f^2 a_r)=0, \quad (5)$$

and if

$$\frac{a_{r+2}}{a_r}=N_r, \quad (6)$$

$$N_r = 1 - \frac{\beta f^2}{(r+1)(r+2)} + \frac{\beta}{(r+1)(r+2) N_{r-2}}. \quad (7)$$

We have a three-term relation between coefficients. Now if r is large and N_{r-2} not small of order r^{-2}, N_r will be nearly 1, N_{r+2} still more nearly 1, and so on. Hence the radius of convergence of the series (1), and therefore of the series (2), will be 1, which is what we must avoid. The only escape is that N_{r-2} must be small of order r^{-2}; but then the solution is an integral function. Incidentally the possibility of a singularity at $\mu = \pm f$ is disposed of. The question that remains is whether we can choose N_r so that all N_r determined by successive applications of (7) will be small of order r^{-2}. Now (7) can be rewritten in the form

$$N_{r-2} = -\frac{\dfrac{\beta}{(r+1)(r+2)}}{1 - \dfrac{\beta f^2}{(r+1)(r+2)} - N_r}, \quad (8)$$

and N_{r-2} is of the order required if N_r is. We can therefore express N_{r-2} as a convergent continued fraction; the information that ζ is an integral function is equivalent to the statement that all N_r are small enough for the continued fractions expressing the ratios of successive coefficients to converge.

The solution is either an odd or an even function of μ. If we take the even solution and pick out the terms in μ^2 from (4) we have

$$-3(a_3-a_1)-\tfrac{1}{2}\beta f^2 a_1 = 0, \quad (9)$$

that is,

$$N_1 = 1 - \frac{\beta f^2}{2.3}. \quad (10)$$

But also

$$N_1 = -\frac{\dfrac{\beta}{4.5}}{1-\dfrac{\beta f^2}{4.5}+} \frac{\dfrac{\beta}{6.7}}{1-\dfrac{\beta f^2}{6.7}+} \frac{\dfrac{\beta}{8.9}}{1-\dfrac{\beta f^2}{8.9}+} \cdots, \quad (11)$$

and equating these two expressions we have the required equation for f^2. The method of solution is by successive approximation; a trial value of f^2 is substituted in (11), the last denominator retained being such that the term in f^2 is small, and N_1 is worked out from (11). Then (10) gives a second approximation to f^2 and we repeat the process. As a check an extra denominator can be retained, and if the result agrees with the previous one to the desired accuracy it can be taken as correct.

16·09. This method of treating solutions with three-term recurrence relations between the coefficients is not confined to power series. An important example is *Mathieu's equation*

$$\frac{d^2 y}{dx^2} + (4\alpha - 16q \cos 2x) y = 0, \quad (1)$$

where q is real and α is a function of q chosen so that one solution has period 2π. The equation arose first in the study of the oscillations of an elliptic membrane; it also occurs in the oscillations of water in an elliptic lake and has generalizations to wave motion in any periodic structure such as a crystal. Periodic coefficients occur also in Hill's method of treating the moon's motion and in the motion of a pendulum on a vibrating support, but the solutions then have not necessarily the same period as the coefficient.

The periodic solutions of Mathieu's equation are all either symmetrical or antisymmetrical about 0 and $\tfrac{1}{2}\pi$; that is, they have the same symmetry properties as $\cos x$, $\cos 2x$, $\sin x$, $\sin 2x$, to which they reduce when $q = 0$ and $4\alpha = 1$ or 4. We assume them to be expressible by convergent Fourier series. Take for instance

$$y = A_0 + \Sigma A_{2r} \cos 2rx. \tag{2}$$

Then
$$-\sum_{r=0}^{\infty}(2r)^2 A_{2r}\cos 2rx + 4\alpha \sum A_{2r} \cos 2rx - 8q \sum A_{2r} \cos(2r-2)x$$
$$-8q \sum A_{2r} \cos(2r+2)x = 0. \tag{3}$$

Equating coefficients we have for $r \geqslant 2$

$$2q A_{2r-2} + (r^2 - \alpha) A_{2r} + 2q A_{2r+2} = 0, \tag{4}$$

and if
$$N_r = A_{2r+2}/A_{2r}, \tag{5}$$

$$2q N_r = -(r^2 - \alpha) - \frac{2q}{N_{r-1}}. \tag{6}$$

Hence if N_{r-1} is not approximately $-2q/r^2$, N_r will be large of order r^2, and the series will diverge exponentially. This contradicts the hypothesis. Hence we rewrite (6) as

$$N_{r-1} = -\cfrac{1}{\cfrac{r^2 - \alpha}{2q} + N_r}, \tag{7}$$

and proceed as for the tidal equation.

For the constant terms we have
$$\alpha A_0 - 2q A_2 = 0, \tag{8}$$

and for the terms in $\cos 2x$, since $\cos(-2x) = \cos 2x$,
$$4q A_0 + (1-\alpha) A_2 + 2q A_4 = 0. \tag{9}$$

Eliminating A_0 we have
$$\left(\frac{8q^2}{\alpha} + 1 - \alpha\right) A_2 + 2q A_4 = 0, \tag{10}$$

whence
$$N_1 = \frac{A_4}{A_2} = -\frac{8q^2/\alpha + 1 - \alpha}{2q}, \tag{11}$$

and also, from (7),
$$N_1 = -\cfrac{1}{\cfrac{4-\alpha}{2q} - \cfrac{1}{\cfrac{9-\alpha}{2q} - \cfrac{1}{\cfrac{16-\alpha}{2q} - \cdots}}} \tag{12}$$

Equating these two expressions we have an equation for α for given q. Analogous methods can be applied to the functions expressible by odd cosines, odd sines, and even sines.*

* Goldstein, *Camb. Phil. Trans.* 23, 1927, 303–36.

16·091. A three-term recurrence relation also occurs in the solution of the equations that arise in the solution of Schrödinger's equation for the hydrogen molecular ion. The equation is separable in spheroidal coordinates (cf. 18·063) and for a Σ-state the equations

are
$$\frac{d}{d\nu}\left[(\nu^2-1)\frac{dX}{d\nu}\right]+[A+2R\nu-p^2\nu^2]X = 0$$

and
$$\frac{d}{d\mu}\left[(1-\mu^2)\frac{dY}{d\mu}\right]+[-A+p^2\mu^2]Y = 0,$$

where A, R and p are constants.

These equations have a considerable literature. The reader is referred to a paper by W. G. Baber and H. R. Hassé,* where other references are given.

16·092. Infinite determinants. Differential equations, especially with periodic coefficients, can often be treated by a method introduced in the theory of the moon's motion by G. W. Hill, and developed by E. W. Brown. We illustrate it by an example. A light rod of length l stands on a support, and carries a mass m at its upper end. The point of support is constrained to vibrate vertically, its displacement being $a \cos nt$; a/l is very small but an^2/g is not necessarily small, so that n^2 is large compared with g/l. If θ is the inclination of the rod to the upward vertical we easily find the equation of motion, to the first order in a,

$$l\ddot\theta = (g-an^2\cos nt)\theta. \tag{1}$$

This is of Mathieu's type, but we do not assume the motion to be periodic in time $2\pi/n$. On the contrary, we try

$$\theta = e^{i\gamma t}\sum_{-\infty}^{\infty} b_m e^{imnt}, \quad \ddot\theta = -\sum(\gamma+mn)^2 b_m e^{i(\gamma+mn)t}, \tag{2}$$

$$(g-an^2\cos nt)\theta = e^{i\gamma t}\Sigma\{gb_m - \tfrac{1}{2}an^2(b_{m+1}+b_{m-1})\}e^{imnt}, \tag{3}$$

and (1) will be satisfied if, for all integral m,

$$\{(\gamma+mn)^2 l+g\}b_m = \tfrac{1}{2}an^2(b_{m+1}+b_{m-1}). \tag{4}$$

The equation for γ can then be written

$$\begin{vmatrix} \cdots & & & & & \\ -\tfrac{1}{2}an^2 & (\gamma-2n)^2l+g & -\tfrac{1}{2}an^2 & 0 & \cdots & \cdots \\ 0 & -\tfrac{1}{2}an^2 & (\gamma-n)^2l+g & -\tfrac{1}{2}an^2 & 0 & 0 \\ 0 & 0 & -\tfrac{1}{2}an^2 & \gamma^2 l+g & -\tfrac{1}{2}an^2 & 0 \\ 0 & 0 & 0 & -\tfrac{1}{2}an^2 & (\gamma+n)^2 l+g & -\tfrac{1}{2}an^2 \\ & & & & \cdots & \end{vmatrix} = 0. \tag{5}$$

The convergence of such infinite determinants is a matter for discussion,† but often irrelevant because, even if the determinant does not converge, it can usually be made to do so by multiplying rows and columns by suitable non-zero factors, which will affect neither the roots nor the method of solution. It is obvious that if γ is a root, $\gamma+rn$ is another, where

---

* *Proc. Camb. Phil. Soc.* 31, 1935, 564–81.

† Cf. Whittaker and Watson, *Modern Analysis*, 1915, pp. 36, 407–10; Riesz, *Equations linéaires à une infinité d'inconnues*, 1913.

r is any integer; but this also is irrelevant because it simply makes the diagonal element with the smallest coefficient of l come at $m = -r$, and the solution is unaltered. A solution can be found by the continued fraction method, taking $M_m = b_m/b_{m-1}$ for $m > 0$ and $N_m = b_m/b_{m+1}$ for $m < 0$.

For $n = 0$ we know that γ is purely imaginary. Let us see whether it can become real if an^2/g is large; if it can, the system becomes stable. The transition will be at $\gamma = 0$, and then, taking $m = 0, 1, -1$ in turn, we have from (4)

$$b_0 = \frac{1}{2}\frac{an^2}{g}(b_1 + b_{-1}), \tag{6}$$

$$(n^2l+g)b_1 = \tfrac{1}{2}an^2(b_0+b_2); \quad (n^2l+g)b_{-1} = \tfrac{1}{2}an^2(b_0+b_{-2}). \tag{7}$$

Then it is possible to have $b_0 > b_1 > b_2 \ldots, b_0 > b_{-1} > b_{-2} \ldots$ if l is large compared with a; and then, neglecting b_2 and b_{-2}, we have

$$1 = \frac{2(\tfrac{1}{2}an^2)^2}{g(n^2l+g)}, \tag{8}$$

that is, $$a^2n^4 - 2gln^2 - 2g^2 = 0. \tag{9}$$

The positive value of n^2 is given by

$$a^2n^2 = gl + (g^2l^2 + 2g^2a^2)^{\frac{1}{2}} \doteqdot 2gl, \tag{10}$$

so that the maximum velocity of the point of support, an, is the velocity due to a fall through a distance l. Then an^2/g is large, as expected, and we have, nearly,

$$b_1 = b_{-1} = \frac{1}{2}\frac{a}{l}b_0, \quad b_2 = b_{-2} = \frac{1}{8}\frac{a}{l}b_1, \ldots, \tag{11}$$

and the coefficients decrease rapidly with increasing $|m|$. Hence the inverted pendulum can be made stable by giving a sufficiently rapid small vertical vibration to the point of support.

16·10. Solution by complex integrals. This method consists in substituting for y in the differential equation an integral of the form $\int T e^{xt} dt$, where T is a function of t, a complex variable, and the limits are fixed. Integration by parts may then yield a differential equation for T. Take for instance Bessel's equation

$$x\frac{d}{dx}\left(x\frac{dy}{dx}\right) + (x^2 - n^2)y = 0. \tag{1}$$

With the suggested substitution this becomes

$$\int_A^B T\left\{x\frac{d}{dx}\left(x\frac{d}{dx}\right) + (x^2 - n^2)\right\} e^{xt} dt = 0, \tag{2}$$

that is, $$\int_A^B T(x^2t^2 + xt + x^2 - n^2) e^{xt} dt = 0. \tag{3}$$

The problem is to find a form of T such that this will hold for all values of x, at least within a certain range. At present we take x real and positive. We integrate by parts twice, using

$$xe^{xt} dt = d(e^{xt}), \tag{4}$$

and get

$$\left[e^{xt}\{xt^2T - t^2T' - tT + xT - T'\}\right]_A^B + \int_A^B e^{xt}\{(t^2+1)T'' + 3tT' + (1-n^2)T\} dt = 0. \tag{5}$$

This will hold if (1) the integrated part vanishes, which will be true if, for instance, the function is single-valued and the path closed, or if it is open and the real parts of A and B are infinite and negative, and the factor involving T behaves like a power of t when $|t|$ is large, and if (2)

$$(t^2+1)\,T'' + 3tT' + T = n^2 T. \tag{6}$$

Put
$$T = u/\sqrt{(1+t^2)}. \tag{7}$$

Then
$$\frac{d}{dt}\left\{\sqrt{(t^2+1)}\frac{du}{dt}\right\} = \frac{n^2 u}{\sqrt{(t^2+1)}}, \tag{8}$$

and if
$$\frac{dt}{\sqrt{(t^2+1)}} = dv, \tag{9}$$

$$\frac{d^2 u}{dv^2} = n^2 u. \tag{10}$$

Then
$$v = \log\{t + \sqrt{(t^2+1)}\}, \tag{11}$$

$$u = \{t + \sqrt{(t^2+1)}\}^{\pm n}, \tag{12}$$

and possible forms for y are

$$\int_A^B \frac{e^{xt}\,dt}{\sqrt{(t^2+1)}\{t+\sqrt{(t^2+1)}\}^n}, \quad \int_A^B \frac{e^{xt}\,dt}{\sqrt{(t^2+1)}\{t+\sqrt{(t^2+1)}\}^{-n}}, \tag{13}$$

where A and B lie at infinity in the third and second quadrants respectively. This condition is satisfied by the path M of 12·126.

Solution by series gives, except when n is zero or an integer (taken positive), the two solutions
$$J_n(x) = \sum_{r=0}^{\infty}(-)^r\frac{(\tfrac{1}{2}x)^{n+2r}}{r!(n+r)!}, \quad J_{-n}(x) = \sum_{r=0}^{\infty}(-)^r\frac{(\tfrac{1}{2}x)^{-n+2r}}{r!(-n+r)!}. \tag{14}$$

These are independent even if n is half an odd integer, though the roots of the indicial equation then differ by an integer. We compare them with the complex integral solutions by expanding in descending powers of t. The first integral has for its first term

$$\int_M \frac{e^{xt}\,dt}{t^{n+1}} = 2\pi i \frac{x^n}{n!},$$

and the second similarly $2\pi i \dfrac{x^{-n}}{(-n)!}$. Both give series in ascending powers of x. Hence

$$J_n(x) = \frac{1}{2\pi i}\int_M \frac{e^{xt}\,dt}{\sqrt{(t^2+1)}\{t+\sqrt{(t^2+1)}\}^n}, \tag{15}$$

$$J_{-n}(x) = \frac{1}{2\pi i}\int_M \frac{\{t+\sqrt{(t^2+1)}\}^n}{\sqrt{(t^2+1)}}e^{xt}\,dt. \tag{16}$$

It follows that if $n \geqslant 0$, $x > 0$, $J_n(x)H(x)$ has the operational form*

$$\frac{p}{\sqrt{(p^2+1)}\{p+\sqrt{(p^2+1)}\}^n}H(x), \tag{17}$$

but $J_{-n}(x)$ has no operational form if $n > 1$; for the expansion of the corresponding operator would start with p^n, which we cannot interpret by the fundamental rule, and if

* Originally obtained by the method of 21·01: H. Jeffreys, *Operational Methods in Mathematical Physics*, 1927.

we apply the Bromwich integral it diverges. The modification of Bromwich's path to make $\Re(t)$ tend to $-\infty$ at the ends is essential to make the integral for $J_{-n}(x)$ converge.

Numerous interesting applications of similar methods have been given by B. van der Pol, (15) and (16) in particular being found.* But he has assumed from the start that the function to be found has an image derivable by the equation

$$F(z) = \int_0^\infty zf(x)e^{-zx}dx,$$

which is meaningless unless convergence conditions are satisfied. It is significant if $f(x) = J_n(x)$, but not if $f(x) = J_{-n}(x)$ with $n > 1$. Yet his process yields the above differential equation for T, and he gives the two solutions, one of which cannot be evaluated by the Bromwich integral, which he states as a fundamental rule at the outset. The matter is further confused by his use of the expression 'operational solution'. The method used is not operational, his p being defined as a complex variable and not as an operator. The fact is that the operational method and the method of substituting contour integrals, though they have a certain formal similarity and though there is a domain where they are both applicable, can be used in different regions outside that domain, and cannot be interchanged indiscriminately. To give a meaning to the integral for $J_{-n}(x)$ it is necessary to modify the Bromwich path by making its ends proceed to infinity in directions between the imaginary axis and the negative real axis. But such a modification in problems of small oscillations would give the wrong result in every case where there are infinitely many real periods tending to zero. In fact van der Pol, in getting an expression for $J_{-n}(x)$, has used an equipment insufficiently strong to catch his fish; but the fish has jumped on to dry land beside him.

If x is complex the series solutions are single-valued if we take $-\pi < \arg x \leqslant \pi$. If n is integral we need no restriction on $\arg x$, but we shall see in Chapter 21 that $J_{-n}(x)$ and $J_n(x)$ are not then independent. The complex integrals, however, need to be modified in such a way that $\Re(xt) \to -\infty$ at both ends of the path. We can ensure continuity by making $\arg t$ at the ends vary continuously with $\arg x$, and the integrals will then always agree with the series.

The integrands have branch points at $t = \pm i$. The integral along any loop passing around one of these and passing to infinity so that $\Re(xt) \to -\infty$ will give a solution of the equation. Two other solutions can be found in this way, and are the important Hankel functions. They are of course not independent of the two solutions $J_n(x)$ and $J_{-n}(x)$, which can for some purposes conveniently be expressed in terms of them (cf. 21·02).

16·11. Conversion of series into integrals. A series can sometimes be converted into an integral directly by means of the rule

$$\frac{1}{2\pi i}\int_M \frac{e^{zx}}{z^{r+1}}dz = \frac{x^r}{r!},$$

where for x real and positive M is a path with termini at infinity in the third and second quadrants and cutting the positive real axis. If we apply this to the series for $J_n(x)$ we get

$$\frac{1}{2\pi i}\int_M \frac{e^{zx}}{z}\sum_{r=0}^\infty (-)^r \frac{(n+2r)!}{r!(n+r)!}\frac{1}{(2z)^{n+2r}}dz, \tag{18}$$

* Phil. Mag. (7) 8, 1929, 861–98; 13, 1932, 537–77.

which must be identical with (15); a direct method of summing the series is given in Ex. 9, p. 496. Alternatively, we may consider, in operational form,

$$(\tfrac{1}{2}x)^n J_n(x) H(x) = \sum_{r=0}^{\infty} (-)^r \frac{(\tfrac{1}{2}x)^{2n+2r}}{r!(n+r)!} = \sum (-)^r \frac{(2n+2r)!}{r!(n+r)!} \left(\frac{1}{2p}\right)^{2n+2r} H(x). \qquad (19)$$

But we have had
$$z!(z-\tfrac{1}{2})! = 2^{-2z} \pi^{1/2} (2z)!, \qquad (20)$$
whence

$$(\tfrac{1}{2}x)^n J_n(x) H(x) = \sum (-)^r \frac{(n+r-\tfrac{1}{2})!}{\pi^{1/2} r!} \left(\frac{1}{p}\right)^{2n+2r} H(x) = \frac{(n-\tfrac{1}{2})!}{\sqrt{\pi}\, p^{2n}} \left(1 + \frac{1}{p^2}\right)^{-(n+1/2)} H(x)$$

$$= \frac{(n-\tfrac{1}{2})!}{\sqrt{\pi}} \frac{p}{(p^2+1)^{n+1/2}} H(x) = \frac{1}{2\pi i} \frac{(n-\tfrac{1}{2})!}{\sqrt{\pi}} \int_M \frac{e^{tx}}{(t^2+1)^{n+1/2}} dt. \qquad (21)$$

In particular
$$(\tfrac{1}{2}x)^{1/2} J_{1/2}(x) H(x) = \frac{1}{\sqrt{\pi}} \frac{p}{p^2+1} H(x) = \frac{\sin x}{\sqrt{\pi}} H(x).$$

By continuation,
$$J_{1/2}(x) = \sqrt{\left(\frac{2}{\pi x}\right)} \sin x. \qquad (22)$$

This method fails for $J_{-1/2}(x)$, since $(n-\tfrac{1}{2})!$ is then infinite and the expression (21) takes the form $\infty \times 0$. But direct examination of the series for it shows that

$$J_{-1/2}(x) = \sqrt{\left(\frac{2}{\pi x}\right)} \cos x. \qquad (23)$$

16·12. The Wronskian. If $y_1, y_2 \ldots y_n$ are functions of x, with $(n-1)$th derivatives in an open interval, the determinant

$$W = \begin{vmatrix} y_1 & y_2 & \cdots & y_n \\ y_1^{(1)} & y_2^{(1)} & \cdots & y_n^{(1)} \\ \cdots \cdots \cdots \cdots \cdots \cdots \cdots \\ y_1^{(n-1)} & y_2^{(n-1)} & \cdots & y_n^{(n-1)} \end{vmatrix} \qquad (1)$$

is called their *Wronskian*. If there are constants $A_1, \ldots, A_n$ not all zero such that $\sum_{r=1}^{n} A_r y_r = 0$ throughout the interval, then $W = 0$. This follows at once if we differentiate the given relation $n-1$ times and eliminate the A_r between the derived relations. Conversely, *if $W = 0$ throughout the interval and the minor of at least one of $y_r^{(n-1)}$ nowhere vanishes, there are constants A_r such that $\sum_{r=1}^{n} A_r y_r = 0$ throughout the interval.* Suppose that the minor of $y_n^{(n-1)}$ nowhere vanishes. Then the $n-1$ equations

$$\sum_{r=1}^{n-1} B_r y_r^{(s)} + y_n^{(s)} = 0 \quad (s = 0, 1, \ldots, n-2) \qquad (2)$$

yield a set of quantities B_r for each x, and since $W = 0$ (2) holds also for $s = n-1$. Further, the B_r are differentiable, since $y_r^{(n-2)}$ is. Differentiate each of (2) and subtract the next. Then
$$\sum_{r=1}^{n-1} B_r' y_r^{(s)} = 0 \quad (s = 0, 1, \ldots, n-2). \qquad (3)$$

But since the determinant of the $y_r^{(s)}$ nowhere vanishes, it follows that all B_r' are zero.

The need of the condition that the Wronskian of some set of $n-1$ of the functions is nowhere zero may be illustrated by the example

$$y_1 = 0 \quad (-1 < x \leqslant 0), \qquad y_1 = \exp(-1/x^2) \quad (0 < x < 1),$$
$$y_2 = \exp(-1/x^2) \quad (-1 < x < 0), \qquad y_2 = 0 \quad (0 \leqslant x < 1).$$

Here $y_1 y_2' - y_1' y_2 = 0$ throughout the interval, but there are no non-zero A_1, A_2 such that $A_1 y_1 + A_2 y_2 = 0$ everywhere.

If the y_r are given to be analytic functions and there is a region where the first minor of some element of the last row is nowhere zero, the existence of a linear relation in this region follows, and can then be extended by analytic continuation to all points of the common part of the regions of definition of the y_r.

16·13. If $y_1, y_2, \ldots, y_n$ are solutions of a differential equation

$$y^{(n)} + \sum_{r=1}^{n} f_{n-r}(x) y^{(n-r)} = 0 \tag{1}$$

with no linear relation with constant coefficients connecting them, the derivative of their Wronskian is seen to be obtained from W by replacing $y_r^{(n-1)}$ by $y_r^{(n)}$. Substituting for $y_r^{(n)}$ from the differential equation and subtracting suitable multiples of the other rows from the last we find

$$\frac{dW}{dx} = -f_{n-1}(x) W, \tag{2}$$

$$W = A \exp\left\{-\int^x f_{n-1}(u) du\right\}, \tag{3}$$

where A is constant. Hence if there is an x where $W \neq 0$, W does not vanish anywhere. If $f_{n-1}(x) = 0$, W is constant. For a second-order equation with one known solution y_1 this leads to an easily soluble first-order equation for y_2, equivalent to the elementary solution by putting $y = y_1 z$.

16·14. Variation of parameters. Let the differential equation be

$$\frac{d^2y}{dx^2} + f(x)\frac{dy}{dx} + g(x) y = S, \tag{1}$$

and suppose that we have two independent solutions of the equation when $S = 0$. Denote these by y_1 and y_2. Then if $y = Ay_1 + By_2$, where A and B are constants, we have the most general solution with $S = 0$. The method of variation of parameters consists in making A and B variable functions of x and choosing them so that the equation can be satisfied for a general S. We take, then,

$$y = P(x) y_1 + Q(x) y_2; \tag{2}$$

then
$$y' = P' y_1 + Q' y_2 + P y_1' + Q y_2'. \tag{3}$$

As we have introduced two new functions P and Q we are entitled to assume one relation between them; we take
$$P' y_1 + Q' y_2 = 0. \tag{4}$$

Then
$$y'' = P y_1'' + Q y_2'' + P' y_1' + Q' y_2', \tag{5}$$

and substituting in (1) we have

$$\{P y_1'' + f(x) P y_1' + g(x) P y_1\} + \{Q y_2'' + f(x) Q y_2' + g(x) Q y_2\} + P' y_1' + Q' y_2' = S. \tag{6}$$

The terms in brackets cancel because y_1 and y_2 satisfy the equation with $S = 0$; hence we have two equations to determine P' and Q' in terms of S. Then

$$P' = \frac{Sy_2}{y_1'y_2 - y_2'y_1}, \quad Q' = -\frac{Sy_1}{y_1'y_2 - y_2'y_1}, \tag{7}$$

which are definite if y_1 and y_2 are linearly independent. Hence

$$y = Cy_1 + Dy_2 + \int_a^x \frac{y_1(x) y_2(\xi) - y_2(x) y_1(\xi)}{y_1'(\xi) y_2(\xi) - y_2'(\xi) y_1(\xi)} S(\xi) d\xi, \tag{8}$$

where C and D are constant. a may be taken arbitrarily, but a change of its value only adds multiples of $y_1(x)$ and $y_2(x)$ to the solution, and therefore is equivalent to altering C and D.

It is easy to verify that (8) satisfies (1). We have by differentiation

$$y' = Cy_1' + Dy_2' + \int_a^x \frac{y_1'(x) y_2(\xi) - y_2'(x) y_1(\xi)}{y_1'(\xi) y_2(\xi) - y_2'(\xi) y_1(\xi)} S(\xi) d\xi \tag{9}$$

(differentiation of the limit yielding nothing because the integrand vanishes there);

$$y'' = Cy_1'' + Dy_2'' + \int_a^x \frac{y_1''(x) y_2(\xi) - y_2''(x) y_1(\xi)}{y_1'(\xi) y_2(\xi) - y_2'(\xi) y_1(\xi)} S(\xi) d\xi + S(x), \tag{10}$$

since the integrand in (9) is $S(x)$ when $\xi = x$. Substituting in (1) now gives an identity, by the definitions of y_1, y_2. Further, since $y_1 y_2' - y_1' y_2 \neq 0$, the constants C, D can be chosen to make y and y' take any prescribed values at $x = a$, and therefore (8) is the most general solution. In comparison with the method where only one solution y_1 is known and we assume $y = y_1 u$, the further integrations with this method are usually much easier.

Take in particular
$$y'' + n^2 y = S,$$
with $y_1 = \cos nx$, $y_2 = \sin nx$. Then
$$y_1' y_2 - y_2' y_1 = -n,$$

$$y = -\frac{1}{n} \cos nx \int^x S(\xi) \sin n\xi \, d\xi - \frac{1}{n} \sin nx \int^x S(\xi) \cos n\xi \, d\xi$$
$$= -\frac{1}{n} \int_0^x S(\xi) \sin n(\xi - x) \, d\xi + A \cos nx + B \sin nx,$$

which is easily shown to satisfy all the conditions.

This method, in an extended form, is the basis of the methods used for the calculation of planetary orbits. Without the disturbance due to other planets, the motion of any planet would be an ellipse, specified by six constants determined by the position and velocity components at one instant. To allow for perturbations these constants are taken as variables; that is, at any moment we can speak of the instantaneous ellipse as the orbit that the planet would describe if the perturbations were removed and the planet moved under the solar attraction alone, with the initial position and velocity that it has at that instant. Perturbations will prevent this from being always the same orbit, but the changes are slow and determinate, and from them the position of the planet at a future instant can be inferred.

It is not necessary for the validity of the method that S shall be a known function of x. If S also involves y (8) will still be true, though not immediately informative since we do not know what values to take for S in the integral without knowing y first. In such a case (8) becomes an *integral equation* for y, since the unknown function y occurs under the sign of integration. We consider equations of the form

$$\frac{d^2y}{dx^2} + g(x)y = h(x)y \tag{11}$$

and take y_1, y_2 to be solutions of $y'' + g(x)y = 0$ such that $y_1'y_2 - y_2'y_1 = 1$. Then

$$y = A_1 y_1 + A_2 y_2 + \int_a^x \{y_1(x)y_2(\xi) - y_2(x)y_1(\xi)\} h(\xi) y(\xi) d\xi$$

$$= f(x) + \int_a^x K(x,\xi) y(\xi) d\xi \tag{12}$$

with
$$K(x,\xi) = \{y_1(x)y_2(\xi) - y_2(x)y_1(\xi)\} h(\xi). \tag{13}$$

This can be solved by an iterative method. First suppose $K(x,\xi)$ to be bounded in an interval $a \leqslant \xi \leqslant x \leqslant b$. Put $|K| < M$. Let the upper bound of $|f(x)|$ in the interval be N. Substitute $f(\xi)$ for $y(\xi)$ in the integral in (12); then if the integral is $f_1(x)$ we have

$$|f_1(x)| \leqslant MN(x-a). \tag{14}$$

Now substitute $f_1(x)$ in the integral in turn; if the integral is then $f_2(x)$ we have

$$|f_2(x)| \leqslant M^2 N \frac{(x-a)^2}{2!}. \tag{15}$$

Thus we find
$$y = f(x) + f_1(x) + f_2(x) + \ldots, \tag{16}$$

which converges like the exponential series $N \exp\{M(x-a)\}$. Substitution then shows that the sum actually satisfies (12). This is a practical method of solution if y_1 and y_2 are known and $h(x)$ is small.

If y_1, y_2, h are analytic in a bounded region of x, including $x = a$, and for any x of the region there is a straight line in the region connecting a and x, we can take this straight line as the path of integration, and a similar argument holds. The solution in the region is then a uniformly convergent series of analytic functions and therefore analytic.

The method still succeeds if y is required to attain given values at two fixed values of x; for A_1 and A_2 can be redetermined at each stage so that each approximation satisfies the terminal conditions exactly. Generally speaking it determines a convergent series solution, but leaves some liberty of choice of the stage when it is convenient to turn to arithmetic. It is, however, usually longer than direct numerical solution by finite differences.

16·15. Green's function. This is a method closely related to the last, but directly applicable to problems where a solution is required to take definite values at fixed termini. It can be extended to two and more dimensions, but has too many ramifications to discuss here. It has much theoretical importance because it enables a differential equation, with suitable boundary conditions, to be converted into an integral equation. Accounts are given by Courant-Hilbert,[*] Webster[†] and Temple & Bickley.[‡] The problems of 6·091, 6·092, 6·093 and 14·05 are particular cases of it.

---

[*] *Methoden der Mathematischen Physik*, 1, 1924, 273–99.
[†] *Partial Differential Equations of Mathematical Physics*, pp. 109–42, 222–38.
[‡] *Rayleigh's principle*, 1933.

EXAMPLES

1. Find the general series solution of
$$y'' - xy' + 2y = 0.$$
Show that one solution is a polynomial, and deduce the other solution in finite terms. (I.C. 1942.)

2. If
$$y = \frac{\sin n\theta}{\cos \tfrac{1}{2}\theta}, \quad x = \sin \tfrac{1}{2}\theta,$$
prove that
$$(1-x^2)\frac{d^2y}{dx^2} - 3x\frac{dy}{dx} + (4n^2 - 1)y = 0,$$
and hence that
$$y = 2n\left(x + \frac{4(1^2-n^2)}{3!}x^3 + \frac{4^2(1^2-n^2)(2^2-n^2)}{5!}x^5 + \ldots\right)$$
and
$$\frac{\cos n\theta}{\cos \tfrac{1}{2}\theta} = 1 + \frac{1^2 - 4n^2}{2!}x^2 + \frac{(1^2-4n^2)(3^2-4n^2)}{4!}x^4 + \ldots.$$
Explain why these series have radius of convergence 1 if $2n$ is not an integer.

3. If
$$(1-x^3)\frac{d^2y}{dx^2} - 6x^2\frac{dy}{dx} - 6xy = 0$$
and if $y = 1$, $y' = 0$ when $x = 0$, find a series for y. Sum the series and verify that the sum satisfies the differential equation. (I.C. 1936.)

4. Prove that if $x = \sin \tfrac{1}{2}\theta$, $y = \cos n\theta$,
$$(1-x^2)y'' - xy' + 4n^2 y = 0.$$
Hence prove that
$$\cos n\theta = 1 - \frac{n^2}{2!}(2\sin\tfrac{1}{2}\theta)^2 + \frac{n^2(n^2-1^2)}{4!}(2\sin\tfrac{1}{2}\theta)^4 - \ldots,$$
$$\frac{\sin n\theta}{\cos \tfrac{1}{2}\theta} = 2n\sin\tfrac{1}{2}\theta - \frac{n(n^2-1^2)}{3!}(2\sin\tfrac{1}{2}\theta)^3 + \ldots.$$

5. What is the least number of steps required in the continuation in 16·05 when the continuation is carried out (1) by power series, (2) by the method of 16·03, the successive origins being equally distant from the singularity?

6. If a linear differential equation of the second order has $0, 1, \infty$ as regular singularities, and no other singularities, show that it is reducible to the form
$$x(1-x)y'' + (a+bx)y' + \frac{c + dx + ex^2}{x(1-x)}y = 0.$$
Show also that if at each of $0, 1$ one solution has index 0, the last term reduces to $-ey$.

7. Solve completely the equation
$$x^2(1+x^2)y'' - 2y = 2x^4. \qquad \text{(M.T. 1936.)}$$

8. If
$$y'' + Py' + Qy = 0,$$
where P and $Q - q_1/x$ are analytic at $x = 0$, where $q_1 \neq 0$, prove that one solution of the equation always involves a logarithm.

9. If
$$y = \frac{x}{\{x + \sqrt{(x^2-1)}\}^n \sqrt{(x^2-1)}}$$
and $x = 1/\xi$, prove by putting $x = \cosh u$ that
$$\xi^2(1-\xi^2)\frac{d^2y}{d\xi^2} + (\xi - 4\xi^3)\frac{dy}{d\xi} - (2\xi^2 + n^2)y = 0.$$
Hence prove that
$$y = \sum_{r=0}^{\infty} \frac{(n+2r)!}{r!(n+r)!}\left(\frac{1}{2x}\right)^{n+2r} \quad (|x| > 1).$$

Examples

10. Show by the method of Frobenius that two solutions of

$$x\frac{d^2y}{dx^2} + y = 0$$

are
$$y_1 = x - \frac{x^2}{1.2} + \frac{x^3}{1.2^2.3} - \ldots + \frac{(-1)^m x^{m+1}}{1.2^2.3^2\ldots(m-1)^2 m}\ldots$$

and
$$y_2 = 1 + \frac{x}{1^2} - \ldots - \frac{(-1)^m x^m}{1^2.2^2\ldots(m-1)^2 m}\left(\frac{2}{1} + \frac{2}{2} + \ldots + \frac{2}{m-1} + \frac{1}{m}\right) - y_1 \log x.$$

11. Show that two solutions of
$$\frac{d^2y}{dx^2} - \frac{dy}{dx} + \frac{y}{x} = 0$$

are x and
$$1 - x\log x - \frac{x^2}{2!} - \ldots - \frac{x^m}{(m-1)m!}\ldots$$

12. If
$$y'' + \chi(x)y = 0,$$

where $\chi(x)$ is an integrable function with period 2π, prove that the solution has in general the form

$$y = C_1 e^{\lambda x}\phi_1(x) + C_2 e^{-\lambda x}\phi_2(x),$$

where $\phi_1(x)$ and $\phi_2(x)$ have period 2π; and that in the exceptional case

$$y = D_1\psi_1(x) + xD_2\psi_2(x),$$

where ψ_1 and ψ_2 have period 2π.

13. If
$$\frac{d}{dx}\left(\frac{f(x)}{x-x_0}\frac{dy}{dx}\right) + g(x)y = 0$$

where $f(x)$, $g(x)$ are analytic in a neighbourhood of x_0, and $f(x_0) \neq 0$, prove that the roots of the indicial equation at x_0 always differ by an integer but that the solution never contains a logarithm.

14. If the Wronskian of two solutions y_1, y_2 of a second-order linear equation is W, prove by direct transformation that the solution

$$y_3 = (A_1 y_1 + A_2 y_2)\int_c^x \frac{W\,dx}{(A_1 y_1 + A_2 y_2)^2}$$

is a linear combination of y_1 and y_2.

If $W = 1$ everywhere and $|y_1|$ is large compared with $x^{1/2+\epsilon}$ when x is large, ϵ being positive, show that there is a second solution y_2 that is not large compared with $x^{1/2+\epsilon}$, and that taking $A_1 = 1$, $c = \infty$ in the above expression leads to $y_3 = -y_2$ for any A_2.

Chapter 17

ASYMPTOTIC EXPANSIONS

Up the airy mountain
Down the rushy glen.

WILLIAM ALLINGHAM

17·01. Nature of asymptotic expansions: incomplete factorial function.
The incomplete factorial function is defined by the integral

$$I = \int_x^\infty e^{-t} t^{-n} dt, \tag{1}$$

where x and n are positive. Integrate by parts; we have

$$\begin{aligned}I &= e^{-x}x^{-n} - n\int_x^\infty e^{-t}t^{-n-1}dt \\ &= e^{-x}\{x^{-n} - nx^{-n-1} + n(n+1)x^{-n-2} \ldots + (-)^r n(n+1)\ldots(n+r-1)x^{-n-r}\} \\ &\quad -(-)^r n(n+1)\ldots(n+r)\int_x^\infty e^{-t}t^{-n-r-1}dt.\end{aligned} \tag{2}$$

This is exact. Now the integral in the last term is always positive, but its coefficient alternates in sign with successive values of r. Consequently the error committed in neglecting it alternates in sign, and therefore I always lies between the sums of r and $r+1$ terms of the series. But the ratio of the term in x^{-n-r} to the preceding one is $-(n+r-1)/x$. If, then, x is large compared with n, the terms will decrease to a minimum and then increase again. If we stop at the smallest term but one we shall know that the error is less than the next term, which will be a small fraction of the sum. Thus we can get a good approximation to the value of the integral. Nevertheless the terms for a general r are the terms of an infinitely oscillating series. The properties are similar to those we have already found for several approximations based on the Euler-Maclaurin expansion.

Such a series is called an *asymptotic expansion*[*]. It is really not correctly regarded as an infinite series at all, and some confusion has arisen from the expression 'use of divergent series' in relation to such expansions. It is to be regarded as the sum of a finite number of terms, stopping either at the smallest but one or at some earlier one when we have already achieved as much accuracy as we want. In suitable circumstances the accuracy may be very high. But unlike a convergent series, which will theoretically always give as much accuracy as we want if we take enough terms, an asymptotic series is definitely limited in accuracy; if we take more than a certain number of terms we increase the error again. The terms of convergent series often decrease from the start, as for the series

$$1 + \tfrac{1}{3} + \tfrac{1}{9} + \tfrac{1}{27} + \ldots, \quad 1 - \tfrac{1}{3} + \tfrac{1}{9} - \tfrac{1}{27} + \ldots.$$

For the second of these the error committed in stopping at any term is less in magnitude than the first term neglected; for the first it is less than the last term retained. But in the series

$$1 + 100 + \frac{10^4}{2!} + \frac{10^6}{3!} + \ldots,$$

---

[*] The subject matter of this chapter has been treated more fully by H. Jeffreys in *Asymptotic Approximations*, 1962.

though it is convergent, and accordingly defines a particular number quite precisely, the terms increase up to the hundredth, and enormous labour would be needed to sum it by direct computation.* For calculation an important property of a series is rapid decrease of the early terms, and successive sums can be regarded as successive approximations. This property may not be possessed by a convergent series, and may be possessed by a divergent one. But successive approximation is a necessary feature of scientific work, and is used at the stage of most calculations when the results are reduced to quantitative answers. We seldom aim at exact answers; what is desirable is to have some idea of the accuracy of the answers we do get, and this is given in a most convenient form by such a series as (2).

The Euler-Maclaurin formula is in general asymptotic. If the function operated on is a polynomial the series terminates and there is no more to be said. But if the function contains a fractional or negative power of the argument the higher derivatives acquire a pair of factors of the form $(n-2r)(n-2r-1)/x^2$ at each step; while the coefficients b_{2r} decrease only like $(2\pi)^{-2r}$. Hence, however small the interval used may be, the terms will ultimately increase indefinitely on account of the accumulation of factorials in the numerator. Consequently the series, if regarded as an infinite series, is usually divergent. Yet the high apparent accuracy obtained by using it can be justified by the method of 9·08.

17·02. Poincaré's definition. The usual definition of an asymptotic approximation, due to Poincaré, is that if $f(z)$ is an analytic function, $S_n(z)$ is the sum of the terms up to $A_n z^{-n}$ of the series

$$S(z) = A_0 + \frac{A_1}{z} + \ldots + \frac{A_n}{z^n} + \ldots, \tag{1}$$

and if $R_n(z) = f(z) - S_n(z)$, the series is called an asymptotic expansion of $f(z)$ within a given interval of $\arg z$ if for every n

$$\lim_{|z| \to \infty} z^n R_n(z) = 0. \tag{2}$$

We write
$$f(z) \sim S(z). \tag{3}$$

A power series in $1/z$ that converges for $|z| > R$ satisfies this definition of an asymptotic expansion. For there is an M such that the remainder after the term in z^{-n} has a modulus less than $\frac{M|z|^{-n}}{|z|/R-1}$, for all values of $\arg z$.

17·021. Asymptotic series can be multiplied unconditionally. For if

$$S_n(z) = A_0 + \frac{A_1}{z} + \ldots + \frac{A_n}{z^n}, \tag{1}$$

$$T_n(z) = B_0 + \frac{B_1}{z} + \ldots + \frac{B_n}{z^n}, \tag{2}$$

are asymptotic representations of $f(z)$, $g(z)$, we can choose z so that

$$|z^n \{f(z) - S_n(z)\}|, \quad |z^n \{g(z) - T_n(z)\}|$$

are arbitrarily small.

---

* Actually, of course, we should work out $100 \log_{10} e$ and then evaluate by means of a table of logarithms to the base 10. When a multiplying machine is available two uses remain for logarithms to base 10; to work out high powers and logarithms to base e of large numbers.

Also
$$S_n(z) T_n(z) = C_0 + \frac{C_1}{z} + \dots + \frac{C_n}{z^n} + o(z^{-n}) = U_n(z) + o(z^{-n}), \tag{3}$$

where
$$C_m = A_0 B_m + A_1 B_{m-1} + \dots + A_m B_0, \tag{4}$$

and $z^n\{f(z)g(z) - U_n(z)\}$ is the sum of three terms that tend to zero with $1/z$ for all n.

17·022. An asymptotic expansion can also be integrated unconditionally. If for $\alpha \leqslant \arg z \leqslant \beta$, $|z| > R$, $f(z)$ is analytic, and
$$|z^n\{f(z) - S_n(z)\}| < \omega,$$
and z_1 satisfies the inequalities, take a path from z_1 to infinity with constant $\arg z$. Then
$$\left| \int_{z_1}^{\infty} \{f(z) - S_n(z)\} dz \right| < \int_{|z_1|}^{\infty} \frac{\omega}{r^n} dr = \frac{\omega}{(n-1)|z_1|^{n-1}},$$
that is,
$$\left| z_1^{n-1} \left\{ \int_{z_1}^{\infty} f(z) dz - \int_{z_1}^{\infty} S_n(z) dz \right\} \right| < \frac{\omega}{n-1},$$
so that the term by term integration of $S_n(z)$ gives an asymptotic expansion of $\int_{z_1}^{\infty} f(z) dz$.

If z_1 and z_2 have the same modulus and we take a circular arc about the origin to connect them, this arc is of length $L < 2\pi |z_1|$, and
$$\left| \int_{z_1}^{z_2} \{f(z) - S_n(z)\} dz \right| < \frac{\omega L}{|z_1|^n} < \frac{2\pi \omega}{|z_1|^{n-1}},$$
and the same result holds. Since $f(z)$ and $S_n(z)$ have no singularity in the region we can connect any two points in the region by a path partly of constant $\arg z$ and partly of constant $|z|$ without altering the integral, and the result follows.

17·023. Asymptotic expansions are unique. For if we have for all $|z| > R$, $\alpha \leqslant \arg z \leqslant \beta$
$$\lim_{z \to \infty} z^n \left\{ f(z) - A_0 - \frac{A_1}{z} - \dots - \frac{A_n}{z^n} \right\} = 0, \quad \lim_{z \to \infty} z^n \left\{ f(z) - B_0 - \frac{B_1}{z} - \dots - \frac{B_n}{z^n} \right\} = 0, \tag{1}$$
we should have
$$\lim_{z \to \infty} z^n \left\{ A_0 - B_0 + \frac{A_1 - B_1}{z} + \dots + \frac{A_n - B_n}{z^n} \right\} = 0 \tag{2}$$
and therefore
$$A_0 = B_0, \quad A_1 = B_1, \quad \dots, \quad A_n = B_n. \tag{3}$$

It follows that $f(z)$ *can have an asymptotic expansion of the form* 17·02 (1) *for all values of* $\arg z$ *only if the series converges*. For if 17·02 (2) held for all values of $\arg z$ we could choose quantities M, R such that $|z^n R_n(z)| < M$ for all $|z| > R$; and then $z^n R_n(z)$ would have a convergent expansion in powers of $1/z$ by Cauchy's inequality. Hence $f(z)$ would have a convergent expansion with the asymptotic property. But then it follows that the only asymptotic expansion of $f(z)$ is the convergent series.

17·024. The converse is not true; the same expansion in a given region may be an asymptotic expansion of several functions provided that their differences $f(z) - g(z)$ satisfy for every n
$$\lim_{z \to \infty} z^n \{f(z) - g(z)\} = 0.$$

This could happen, for instance, if the range of argument was $-\tfrac{1}{4}\pi$ to $\tfrac{1}{4}\pi$ and $f(z) - g(z) = e^{-z}$. Poincaré's definition therefore does not fix bounds to the error for a given z, and these are usually found by special methods.

17·03. Watson's lemma.

Two of the most important methods of obtaining asymptotic expansions are the method of steepest descents, due to Debye, and that of stationary phase, due to Kelvin. They are largely, but not completely, equivalent. We need first a form of a lemma due to G. N. Watson.* Consider the integral along the real axis

$$I = \int_0^Z e^{-az} z^m f(z)\, dz \tag{1}$$

where $f(z)$ is analytic on the path and not zero at $z = 0$; Z is independent of a, and may be infinite; a is large, real and positive; and I exists for some a, say α. Hence $m > -1$. Then within the circle of convergence of the series expansion of $f(z)$

$$f(z) = a_0 + a_1 z + \ldots + a_{n-1} z^{n-1} + R_n(z). \tag{2}$$

where $R_n(z)/z^n$ tends to a finite limit as $z \to 0$. Take a fixed A in $(0, Z)$ within the circle of convergence; then

$$I = \left[\int_0^A + \int_A^Z\right] e^{-az} z^m f(z)\, dz. \tag{3}$$

In $(0, A)$ the function

$$g(z) = \{f(z) - (a_0 + a_1 z + \ldots + a_{n-1} z^{n-1})\} z^{-n} \tag{4}$$

is bounded. Let the upper bound of its modulus be M. Then

$$I_A = \int_0^A e^{-az} z^m f(z)\, dz$$

$$= \int_0^A e^{-az} z^m (a_0 + a_1 z + \ldots + a_{n-1} z^{n-1})\, dz$$

$$+ \int_0^A e^{-az} z^m \theta M z^n dz, \tag{5}$$

where $|\theta| < 1$. Put $z = A(1+u)$; then $(1+u)^{m+r} \leq e^{(m+r)u}$ $(m+r > 0)$, ≤ 1 $(m+r \leq 0)$,

$$\int_A^\infty e^{-az} z^{m+r} dz < \int_0^\infty A^{m+r+1} e^{-Aa - aAu} e^{(m+r)u} du = A^{m+r+1} e^{-Aa}/(Aa - m - r) \quad (m+r > 0)$$

$$< A^{m+r} e^{-aA}/a \quad (m+r \leq 0) \tag{6}$$

$$I_A = \sum_{r=0}^{n-1} a_r \frac{(m+r)!}{a^{m+r+1}} + O(e^{-aA}) + \theta \frac{M(m+n)!}{a^{m+n+1}}. \tag{7}$$

For $a = \alpha$ let the upper bound of $\left|\int_A^X e^{-az} z^m f(z)\, dz\right|$ for X in (A, Z) be N. Then from Abel's lemma for integrals (since $e^{-(a-\alpha)z}$ is a positive decreasing function of z)

$$\left|\int_A^Z e^{-az} z^m f(z)\, dz\right| \leq 2 e^{-(a-\alpha)A} N, \tag{8}$$

---

* *Proc. Lond. Math. Soc.* (2), (17), 1918, 133; *Theory of Bessel Functions*, p. 236.

and therefore
$$\left| I - \sum_{r=0}^{n-1} a_r \frac{(m+r)!}{a^{m+r+1}} \right| < \frac{M(m+n)!}{a^{m+n+1}} + Ke^{-aA}, \tag{9}$$

where K is independent of a. If we multiply by a^{m+n} and make $a \to \infty$ the right side tends to 0. Hence
$$\int_0^Z e^{-az} z^m f(z)\, dz \sim \sum_{r=0}^{\infty} \frac{(m+r)!\, a_r}{a^{m+r+1}}. \tag{10}$$

There is no reason against the upper limit Z being infinite, since the argument from Abel's lemma still applies so long as the integral converges for some α. It is also permissible for $f(z)$ to be unbounded at some point or points so long as the improper integral exists.

Watson took $f(z)$ bounded on the whole of the real axis and Z infinite. These conditions are often satisfied, but not always, and the slight extension we have made seems worth while.

The series is ultimately divergent if $f(z)$ has a singularity at a finite distance R from the origin. For if $R' > R$ there exists a k such that $|a_r| > k/R'^r$ for an infinite number of values of r, and the terms in (10), for any given a, are unbounded with respect to r.

The lemma proves the existence of an asymptotic expansion in Poincaré's sense, and determines the coefficients. It does not provide an estimate of the error for given a in stopping at a given term, since we have not assigned a value to M. An idea of the accuracy of the sum up to the smallest term can be got by a method related to what J. R. Airey called 'use of convergence factors'.* The principle is illustrated most easily by our first example. The integral in the remainder term was
$$\int_x^\infty e^{-t} t^{-n-r-1}\, dt,$$

and if $n+r = x$ the next term is numerically equal to the last kept. But if we take the logarithmic derivative of the integrand we get
$$\frac{d}{dt}\{-t - (n+r+1)\log t\} = -1 - \frac{n+r+1}{t},$$

and the two terms are nearly equal when $t = x$. Hence the integral is nearly
$$e^{-x} x^{-n-r-1} \int_0^\infty e^{-2u}\, du = \tfrac{1}{2} e^{-x} x^{-n-r-1},$$

and the remainder term is nearly
$$(-)^{r+1} \tfrac{1}{2} n(n+1) \ldots (n+r)\, e^{-x} x^{-n-r-1},$$

which is *half* the next term in the expansion. Thus a very substantial improvement will be made if the asymptotic series is computed up to the smallest term, and half the next term is added. Greater accuracy still is obtainable by expanding $\{t + (n+r+1)\log t\}$ to higher powers of $(t-x)/x$. We shall return to this point in Chapter 23.

* *Phil. Mag.* (7), 24, 1937, 521–552.

The case we chiefly need is
$$I = \int_{-A}^{B} e^{-\frac{1}{2}b^2 z^2} f(z)\, dz, \tag{11}$$

where A, B are positive and independent of b, and $f(z)$ still has an expansion given by (2) near $z = 0$. We put $z^2 = \zeta$,
$$I = \frac{1}{2}\int_0^{A^2} e^{-\frac{1}{2}b^2 \zeta} f(-\zeta^{1/2})\, \zeta^{-1/2}\, d\zeta + \frac{1}{2}\int_0^{B^2} e^{-\frac{1}{2}b^2 \zeta} f(\zeta^{1/2})\, \zeta^{-1/2}\, d\zeta. \tag{12}$$

The odd powers cancel and
$$I \sim \sqrt{(2\pi)}\left(\frac{a_0}{b} + \frac{a_2}{b^3} + 1.3\frac{a_4}{b^5} + \ldots + 1.3 \ldots (2n-1)\frac{a_{2n}}{b^{2n+1}}\right). \tag{13}$$

17·04. Method of steepest descents. This is due to Debye, and is applied to the approximate evaluation of integrals of the form
$$I = \int_A^B \chi(z)\, e^{tf(z)}\, dz, \tag{1}$$

where t is large, real and positive, and $f(z)$ is analytic. We write
$$f(z) = \phi + i\psi, \tag{2}$$

separating its real and imaginary parts. ϕ and ψ both satisfy Laplace's equation, and the integrand will be large where ϕ is algebraically large. The transformation from x, y to ϕ, ψ will be non-singular in a region containing no singularities or zeros of $f'(z)$. In such a region we can pass from A to B by a finite number of steps along lines of ϕ or ψ constant. Put $f(z) = \zeta$. Then
$$I = \int_{z=A}^{B} e^{t\zeta} \frac{\chi(z)}{f'(z)}\, d\zeta = \int_{z=A}^{B} e^{t\zeta} g(\zeta)\, d\zeta. \tag{3}$$

Suppose first that $g(\zeta)$ is analytic in the region. Then $g(\zeta)$ has a bounded derivative in the region and its real and imaginary parts separately will be of bounded variation on a finite path of ϕ or ψ constant. We can then apply the inequalities derived for integrals in **1·134 a**. If the path from A to B is one of ψ constant, and $\phi_A > \phi_B$, the path from A to B is called one of *steepest descent*. Then
$$I = \int_{z=A}^{B} e^{t\zeta_A} e^{-t(\phi_A - \phi)} g(\zeta)\, d\phi. \tag{4}$$

If the path is one of ϕ constant
$$I = i\int_{z=A}^{B} e^{t\zeta_A} e^{it(\psi - \psi_A)} g(\zeta)\, d\psi, \tag{5}$$
$$|e^{-t\zeta_A} I| < 2\sqrt{2}\{|g(B)| + V(B)\}/t, \tag{6}$$

where $V(B)$ is the greater of the total variations of $\Re g(\zeta)$ and $\Im g(\zeta)$ on the path. In either case
$$I = O(e^{t\phi_A}/t) \tag{7}$$

subject to $\phi_A \geq \phi_B$. For infinite paths it is necessary to verify directly whether $V(B)$ is finite.

Since $\phi_A - \phi$ is a real variable and $g(\zeta)$ is analytic, we find for (4) by Watson's lemma,
$$I \sim -e^{t\zeta_A}\left\{\frac{g(A)}{t} + \sum_1^\infty \frac{(-1)^r}{t^{r+1}}\left(\frac{\partial^r}{\partial \zeta^r} g(\zeta)\right)_{z=A}\right\}. \tag{8}$$

If a line of ϕ constant connects A, B we can in general find a path of constant ϕ from A' to B' such that $\phi_{A'} = \phi_{B'} < \phi_A$, and $\psi_{A'} = \psi_A$, $\psi_{B'} = \psi_B$. Then I is equivalent to integrals along AA', $A'B'$, $B'B$, where ϕ is constant on $A'B'$ and ψ is constant on AA' and $B'B$. Then by (7) the integral along $A'B'$ is $O(e^{t\phi_{A'}}/t)$. Those along AA' and $B'B$ can be approximated to by (8), and for all r, as $t \to \infty$,

$$\left| t^{r+1} e^{t(\phi_{A'} - \phi_A)} \right| \to 0. \tag{9}$$

Hence $A'B'$ contributes nothing to the asymptotic expansion of I (subject as before to $\Re g(\zeta)$ and $\Im g(\zeta)$ being of bounded variation on $A'B'$).

It follows that detailed tracing of the paths of constant ψ is seldom necessary; the asymptotic expansion is wholly determined by the behaviour of the integrand near the points where ϕ is greatest.

It often happens, however, that on any path (in the z plane) from A to B there are points where ϕ exceeds both ϕ_A and ϕ_B. In this case ϕ has a maximum at an interior point of the path. Suppose that the part of the path that passes through this point is one of constant ψ (it cannot be one of constant ϕ). Then if ds, dn are elements of length along and normal to the path we have at this point $\partial \psi/\partial s = 0$, $\partial \phi/\partial s = 0$, and therefore, by the Cauchy-Riemann relations, $\partial \phi/\partial n = 0$, $\partial \psi/\partial n = 0$. The point is therefore one where $f'(z) = 0$, and $g(\zeta)$ as defined in (3) is no longer analytic. Such points are known as *saddle-points* or *cols*. Another approach is to consider the maxima of ϕ on all paths connecting A and B. If the maximum is to be made as small as possible by a suitable choice of path, we must have $\partial \phi/\partial n = 0$ at it; but we also have $\partial \phi/\partial s = 0$ since ϕ is a maximum; and therefore $f'(z) = 0$.

Lines of constant ψ are called *lines of steepest descent*, because on them the direction at any point is such that $|d\phi/ds|$ is as great as possible. If θ is the inclination of the path to the axis of x,

$$\frac{\partial \phi}{\partial s} = \cos\theta \frac{\partial \phi}{\partial x} + \sin\theta \frac{\partial \phi}{\partial y}, \tag{10}$$

and if this is to be stationary for variations of θ

$$0 = -\sin\theta \frac{\partial \phi}{\partial x} + \cos\theta \frac{\partial \phi}{\partial y} = -\sin\theta \frac{\partial \psi}{\partial y} - \cos\theta \frac{\partial \psi}{\partial x} = -\frac{\partial \psi}{\partial s}, \tag{11}$$

which is satisfied on a path of constant ψ.

In these cases, therefore, it is convenient to make part of the path of integration consist of a line of steepest descent through a saddle-point so that the larger values of ϕ are concentrated in as short an interval of the path as possible.

ϕ cannot be a maximum for all variations of x and y from a point. Through a saddle-point z_0 there will be two or more curves of constant ϕ, separating the neighbourhood into sectors. Those where ϕ is less than at z_0 are called *valleys*, those where it is greater than at z_0 *hills*. If then A and B lie in different valleys specified by a saddle-point z_0, the best path will be of the form ACz_0DB, where $\phi_C = \phi_A$, $\phi_D = \phi_B$, and Cz_0D is a path of constant ψ through z_0. In this case the approximation (8) fails because $g(z_0)$ would be infinite, but this difficulty is overcome by a method described below. The contributions from AC and DB are negligible compared with that from Cz_0D, and this is the most striking feature of the method.

Isolated singularities of $g(\zeta)$, if not actually on the path, do not in general affect the approximation, since all that matters is the upper bound of its modulus or the total variations of its real and imaginary parts on the actual path. If the path from A to B is not one of constant ϕ or ψ and is replaced by one consisting of segments of ϕ or ψ constant, there may be a singularity z_1 of $g(\zeta)$ between the original path and the adopted one, but an integral about z_1 will contain a factor $\exp t\phi(z_1)$, which is exponentially small compared with $\exp t\phi(z_0)$.

Lines of steepest descent terminate only at singular points of $f(z)$ or at infinity.

If z_0 is a saddle-point and $f''(z_0) \neq 0$, $f(z)$ near it can be expanded in the form

$$f(z) = f(z_0) + \tfrac{1}{2}(z-z_0)^2 f''(z_0) + \dots, \tag{12}$$

and the direction of the path will be such that $(z-z_0)^2 f''(z_0)$ is real and negative. If then we put

$$f(z) - f(z_0) = -\tfrac{1}{2}\zeta^2, \tag{13}$$

and change the variable to ζ, the integral takes the form considered in Watson's lemma and the existence of an asymptotic expansion in negative powers of $t^{1/2}$ can be inferred. In practice, however, the inversion of series is usually troublesome, and if terms after the first are required they are usually found in some other way. For many purposes, however, the first term is sufficient, and can be obtained easily. We have

$$\begin{aligned} I &= e^{tf(z_0)} \int \chi(z) e^{-1/2 t \zeta^2} dz \\ &= e^{tf(z_0)} \int e^{-1/2 t \zeta^2} \chi(z) \frac{dz}{d\zeta} d\zeta. \end{aligned} \tag{14}$$

But if we write for values of z on the path, with r real and small

$$z - z_0 = r e^{i\alpha}, \tag{15}$$

$$\zeta^2 = -f''(z_0) r^2 e^{2i\alpha},$$

$$\zeta = \pm r |f''(z_0)|^{1/2}, \tag{16}$$

since $f''(z_0) e^{2i\alpha}$ is real and negative. Then

$$\frac{d\zeta}{dz} = \pm e^{-i\alpha} |f''(z_0)|^{1/2}. \tag{17}$$

In the range $(-\pi, \pi)$ there are two possible choices for α, and they differ by π. In any application of the method we have at this point to make an inspection of the behaviour of ϕ and ψ over the complex plane in order to decide the sense in which the path goes through the saddle-point. If we select the value of α that makes r positive at points on the path after passing through z_0, we shall have to take the positive sign in (16), as ζ goes from $-\infty$ to $+\infty$ on the path. Then by Watson's lemma the integral is given asymptotically by

$$I \sim \frac{\chi(z_0) e^{tf(z_0)} \sqrt{(2\pi)} e^{i\alpha}}{|tf''(z_0)|^{1/2}}. \tag{18}$$

Since $t^{-n} \exp\{tf(z_0)\}$ will be large for all n if $f(z_0)$ has a positive real part, we should strictly write (18) as

$$e^{-tf(z_0)} I \sim \frac{\sqrt{(2\pi)} \chi(z_0)}{|tf''(z_0)|^{1/2}} e^{i\alpha} \tag{19}$$

in order that Poincaré's definition shall be applicable. We shall, however, use the form (18) for convenience, with the understanding that where exponential factors are present in the approximation such a transposition is needed before the definition is applied.

The method restricts us to paths of steepest descent and traverses on lines of constant ϕ; and if the termini are such that ϕ must have a maximum at an interior point, the path of steepest descent is taken through a saddle-point, thereby making this maximum as small as possible. It may happen that with these restrictions any path deformable into AB will pass through two or more saddle-points. If so, each will make its contribution to the integral. The largest contribution will be from the one where ϕ is largest.

A path of constant ψ through z_0 may pass through a second saddle-point z_1. Then, if ϕ has a maximum at z_0, it will have a minimum at z_1 and will increase again on the smooth continuation of the path past z_1. Hence the line of steepest descent in such a case turns abruptly through a right angle at z_1. The contribution from the neighbourhood of z_1 will be exponentially small compared with that from z_0, but this case is interesting in relation to the Stokes phenomenon, which we shall examine later.

We have assumed $f''(z_0) \neq 0$. If the first non-vanishing derivative is $f^{(n)}(z_0)$ $(n > 2)$, three or more valleys meet at z_0, and it will be necessary to examine which pair of them contains paths leading to the termini. The argument from Watson's lemma needs straightforward modifications.

17·05. Paths of constant ϕ; method of stationary phase. If the path AB of 17·04 (1) is one of constant ϕ, and χ/ψ' is of bounded variation, we have seen that the integral is $O(e^{t\phi}/t)$, and an approximation can be found by using a path $AA'B'B$. If there is one saddle-point between A and B this argument fails, because if $\partial \psi/\partial s$ changes sign, so does $\partial \phi/\partial n$; then if $\phi_{A'} < \phi_A$, $\phi_{B'} < \phi_B$, A' and B' lie on opposite sides of AB, and ϕ on $A'B'$ cannot be uniformly less than on AB. We can, however, proceed as follows. Suppose that the path of constant ϕ is along the axis of x increasing, so that the saddle-point is x_0, $\psi'(x_0) = 0$, $\psi''(x_0) \neq 0$. Then on AB, near x_0,

$$f(z) = f(x_0) + \tfrac{1}{2} i \psi''(x_0)(x - x_0)^2 + O(x - x_0)^3, \qquad (1)$$

whence, since $f(z)$ is supposed analytic,

$$f(z) = f(x_0) + \tfrac{1}{2} i \psi''(x_0)(z - z_0)^2 + O(z - z_0)^3. \qquad (2)$$

If $\psi''(x_0) > 0$, α of 17·04 is $+\tfrac{1}{4}\pi$; if $\psi''(x_0) < 0$, $\alpha = -\tfrac{1}{4}\pi$. Then the integral along a line of steepest descent through x_0

$$= \chi(x_0) \sqrt{\frac{2\pi}{t|\psi''(x_0)|}} \exp\{t\phi(x_0) + it\psi(x_0) + \tfrac{1}{4}\pi i \operatorname{sgn} \psi''(x_0)\} + O\left(\frac{1}{t} \exp t\phi(x_0)\right). \qquad (3)$$

If there is a second saddle-point at x_1 within the range of integration, $\psi''(x_1)$ will have the opposite sign to $\psi''(x_0)$; since ϕ is constant the contributions will be of comparable magnitude.

If ψ is not given to be the imaginary part of an analytic function, it is still possible to find an approximation to $\int e^{it\psi} \chi(x) dx$ by methods similar to those used in proving Fourier's theorem. In any interval where χ/ψ' is of bounded variation it follows as before that $I = O(1/t)$. If $\psi'(x_0) = 0$, near x_0 we put, for $\psi''(x_0) > 0$,

$$\psi(x) - \psi(x_0) = \tfrac{1}{2} u^2 \doteq \tfrac{1}{2} \psi''(x_0)(x - x_0)^2. \qquad (4)$$

Take $u > 0$ when $x > x_0$. If δ is fixed and positive, and

$$I_\delta = \int_{x = x_0 - \delta}^{x_0 + \delta} \frac{\chi(x)}{\psi'(x)} \exp it\{\psi(x_0) + \tfrac{1}{2} u^2\} u\, du, \qquad (5)$$

we have

$$\psi'(x) \doteq \psi''(x_0)(x - x_0) \doteq \{\psi''(x_0)\}^{1/2} u. \qquad (6)$$

Then

$$\frac{\chi(x) u}{\psi'(x)} = a_0 + \theta u; \quad a_0 = \frac{\chi(x_0)}{\{\psi''(x_0)\}^{1/2}}, \qquad (7)$$

where $\theta u \to 0$ with u. If further θ is of bounded variation in $(-\delta, \delta)$,

$$\int_{x=x_0-\delta}^{x_0+\delta} \theta u \exp itu^2 \, du = O\left(\frac{1}{t}\right), \tag{8}$$

$$\int_{x=x_0-\delta}^{x_0+\delta} \exp itu^2 \, du \sim \sqrt{\left(\frac{2\pi}{t}\right)} \exp \tfrac{1}{4}\pi i. \tag{9}$$

Then $\quad I \sim I_\delta + O\left(\dfrac{1}{t}\right) \sim \dfrac{\chi(x_0)}{|\psi''(x_0)|^{1/2}} \sqrt{\left(\dfrac{2\pi}{t}\right)} \exp\{it\psi(x_0) + \tfrac{1}{4}\pi i\} + O\left(\dfrac{1}{t}\right). \tag{10}$

If $\psi''(x_0) < 0$, we must write $-u^2$ for u^2 in (4). The effect in (5) is to replace $\psi'(x)$ by $-\psi'(x)$ and u^2 by $-u^2$, and in (10) we have $-\tfrac{1}{4}\pi i$ for $\tfrac{1}{4}\pi i$. The result therefore agrees with (3) to this order with $\phi = 0$.

The principle of the method is due to Stokes and Kelvin, who argued that in a wave problem the contributions from the parts of the range of integration near a point of stationary phase will be nearly in the same phase and add up, whereas those from other parts will interfere. It is not so easy, however, to find higher terms by this method.

17·06. Stirling's formula by steepest descents. The simplest application of Watson's lemma is to the factorial function

$$z! = \int_0^\infty u^z e^{-u} \, du \quad (\Re(z) > -1). \tag{1}$$

Put $u = zv$. Then (z^z being interpreted as $\exp(z \log z)$ with $|\arg z| < \pi$),

$$z! = z^{z+1} \int_0^\infty \exp\{z(\log v - v)\} \, dv. \tag{2}$$

We shall take
$$z = re^{i\theta}$$
with θ fixed, and attempt an approximation for given θ with r large. Then (2) will be written

$$z! = z^{z+1} \int_0^\infty \exp[r\{e^{i\theta}(\log v - v)\}] \, dv \tag{3}$$

so that $\quad f(v) = e^{i\theta}(\log v - v), \quad f'(v) = e^{i\theta}\left(\dfrac{1}{v} - 1\right), \quad f''(v) = -\dfrac{e^{i\theta}}{v^2}. \tag{4}$

Then $\log v - v$ is analytic if we make a cut from 0 to $-\infty$. The only saddle-point is at $v = 1$, where $f(v) = -e^{i\theta}$, $f''(v) = -e^{i\theta}$.

If v is real (> 0), $\Re f(v) = \cos\theta(\log v - v)$, which never exceeds its value at $v = 1$ provided $\cos\theta$ is positive. Hence the real axis lies in two valleys reaching 0 and ∞. $\Im f(v)$ is not constant on the real axis except for $\theta = 0$. The direction of the line of steepest descent through $v = 1$ is given by putting $v - 1 = w$, with $-f''(1) w^2 = e^{i\theta} w^2$ real and positive, so that $\arg w = -\tfrac{1}{2}\theta$ or $\pi - \tfrac{1}{2}\theta$. The sense to be taken is seen as follows. If we take $\delta > 0$,

$$v_1 = 1 - \delta e^{-\frac{1}{2}i\theta}, \quad v_2 = 1 + \delta e^{-\frac{1}{2}i\theta},$$

a path consisting of straight lines from 0 to v_1, 1, v_2, ∞ in turn lies wholly in valleys except at the saddle-point, so that the direction is always from left to right through the saddle-point. Then for large r

$$z! \sim z^{z+1} \exp(-re^{i\theta}) \sqrt{\left(\frac{2\pi}{r}\right)} e^{-\frac{1}{2}i\theta} \tag{5}$$

$$= \sqrt{(2\pi)} \, z^{z+\frac{1}{2}} e^{-z} \quad \cos\theta > 0, \tag{6}$$

where the value of $z^{\frac{1}{2}}$ with a positive real part is to be taken. This is the first term of Stirling's expansion. Note that if $\cos\theta > 0$, $\Re(z) > -1$ is satisfied, so that the latter condition becomes superfluous.

17·07. The Airy integral.

Consider the integrals

$$f(z) = \frac{1}{2\pi i} \int e^{tz - 1/3 t^3} dt \tag{1}$$

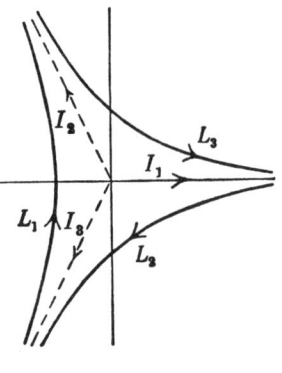

taken along any of the three paths shown in the figure. They converge exponentially provided that the real part of t^3 tends to $+\infty$ at the termini; thus the three termini can in the first place be conveniently written $+\infty$, $\infty \exp(\frac{2}{3}\pi i)$, $\infty \exp(-\frac{2}{3}\pi i)$. We have

$$\frac{d^2}{dz^2} f(z) - z f(z) = \frac{1}{2\pi i} \int e^{tz - 1/3 t^3} (t^2 - z) dt$$

$$= -\frac{1}{2\pi i} \int e^{tz - 1/3 t^3} d(tz - \tfrac{1}{3}t^3)$$

$$= 0, \tag{2}$$

since $\exp(tz - \tfrac{1}{3}t^3)$ tends to 0 at both limits in each case. Hence these three integrals are solutions of the differential equation

$$\frac{d^2 Z}{dz^2} - zZ = 0. \tag{3}$$

Since this equation is of the second order it can have only two linearly independent solutions, and there must be a linear relation between the integrals. But if we take the integral around any contour in the positive sense around the origin it will vanish since there is no singularity of the integrand at a finite distance. Hence, with the senses indicated in the diagram, the sum of the three integrals is 0.

First take the path L_1 and define

$$\text{Ai}(z) = \frac{1}{2\pi i} \int_{L_1} e^{tz - 1/3 t^3} dt. \tag{4}$$

This is the Airy integral, one form of which was studied first by Airy in relation to diffraction near a caustic surface. It may be proved (cf. 1·123) that it still converges if z is real and L_1 is reduced to the imaginary axis. If we put $t = is$ it reduces to

$$\text{Ai}(z) = \frac{1}{2\pi} \int_{-\infty}^{\infty} e^{i(sz + 1/3 s^3)} ds = \frac{1}{\pi} \int_{0}^{\infty} \cos(sz + \tfrac{1}{3}s^3) ds, \tag{5}$$

which, apart from some constant factors, is the form used by Airy.

Alternatively, take the integrals from 0 to ∞, $\infty \exp \tfrac{2}{3}\pi i$, $\infty \exp(-\tfrac{2}{3}\pi i)$ and denote them respectively by I_1, I_2, I_3.

$$I_2(z) = \frac{1}{2\pi i} \int_0^{\infty \exp 2/3 \pi i} e^{tz - 1/3 t^3} dt = \frac{1}{2\pi i} \exp \tfrac{2}{3}\pi i \int_0^{\infty} \exp(uze^{2/3 \pi i} - \tfrac{1}{3}u^3) du$$

$$= \exp(\tfrac{2}{3}\pi i) I_1(z \exp \tfrac{2}{3}\pi i), \tag{6}$$

$$I_3(z) = \exp(-\tfrac{2}{3}\pi i) I_1\{z \exp(-\tfrac{2}{3}\pi i)\}, \tag{7}$$

and

$$\text{Ai}(z) = I_2(z) - I_3(z). \tag{8}$$

17·07   *A second solution*

Also $\quad \dfrac{1}{2\pi i}\int_{L_1} = -I_1(z)+I_3(z), \quad \dfrac{1}{2\pi i}\int_{L_4} = -I_2(z)+I_1(z).$ (9)

The three solutions are thus expressed in terms of the single function $I_1(z)$. Now

$$I_1(z) = \frac{1}{2\pi i}\int_0^\infty e^{tz-\tfrac{1}{3}t^3}dt = \frac{1}{2\pi i}\int_0^\infty e^{-\tfrac{1}{3}t^3}\sum_{r=0}^\infty \frac{(tz)^r}{r!}dt$$

$$= \frac{1}{2\pi i}\int_0^\infty e^{-v}\sum 3^{-\tfrac{2}{3}}\frac{(3^{\tfrac{1}{3}}z)^r}{r!}v^{\tfrac{1}{3}r-\tfrac{2}{3}}dv$$

$$= \frac{1}{2\pi i}\cdot 3^{-\tfrac{2}{3}}\sum_{r=0}^\infty \frac{(3^{\tfrac{1}{3}}z)^r}{r!}(\tfrac{1}{3}r-\tfrac{2}{3})!. \tag{10}$$

Hence $\quad \text{Ai}(z) = \dfrac{1}{2\pi i}\exp(\tfrac{2}{3}\pi i)\, 3^{-\tfrac{2}{3}}\sum_{r=0}^\infty \dfrac{(3^{\tfrac{1}{3}}z)^r}{r!}(\tfrac{1}{3}r-\tfrac{2}{3})!\exp\tfrac{2}{3}\pi ir$

$\qquad\qquad -\dfrac{1}{2\pi i}\exp(-\tfrac{2}{3}\pi i)\, 3^{-\tfrac{2}{3}}\sum_{r=0}^\infty \dfrac{(3^{\tfrac{1}{3}}z)^r}{r!}(\tfrac{1}{3}r-\tfrac{2}{3})!\exp(-\tfrac{2}{3}\pi ir)$

$\qquad = \dfrac{1}{\pi}3^{-\tfrac{2}{3}}\sum_{r=0}^\infty \dfrac{(3^{\tfrac{1}{3}}z)^r}{r!}(\tfrac{1}{3}r-\tfrac{2}{3})!\sin\{\tfrac{2}{3}\pi(r+1)\}$

$\qquad = \dfrac{1}{\pi}3^{-\tfrac{2}{3}}\sin\tfrac{2}{3}\pi(-\tfrac{2}{3})!\left(1+\dfrac{z^3}{2.3}+\dfrac{z^6}{2.3.5.6}+\dfrac{z^9}{2.3.5.6.8.9}+\cdots\right)$

$\qquad +\dfrac{1}{\pi}3^{-\tfrac{1}{3}}\sin\tfrac{4}{3}\pi(-\tfrac{1}{3})!\left(z+\dfrac{z^4}{3.4}+\dfrac{z^7}{3.4.6.7}+\dfrac{z^{10}}{3.4.6.7.9.10}+\cdots\right). \tag{11}$

This is the sum of two convergent series for all z and is real for real z. Each separately satisfies the differential equation (3). Denote the numerical coefficients outside the brackets by α and $-\beta$ and write

$$\text{Ai}(z) = \alpha y_1(z) - \beta y_2(z). \tag{12}$$

Using the relation $\qquad z!(-z)! = \dfrac{\pi z}{\sin \pi z},$

we find $\qquad \alpha = 3^{-\tfrac{2}{3}}/(-\tfrac{1}{3})!, \quad \beta = 3^{-\tfrac{1}{3}}/(-\tfrac{2}{3})!.$

For another solution of (3), real for real z, take

$$\text{Bi}(z) = \frac{1}{2\pi}\left(\int_{L_2}-\int_{L_4}\right) = -iI_2 - iI_3 + 2iI_1$$

$$= \frac{1}{2\pi}3^{-\tfrac{2}{3}}\sum\frac{(3^{\tfrac{1}{3}}z)^r}{r!}(\tfrac{1}{3}r-\tfrac{2}{3})![2-\exp\tfrac{2}{3}(r+1)i\pi-\exp\{-\tfrac{2}{3}(r+1)i\pi\}]$$

$$= \frac{1}{\pi}3^{-\tfrac{2}{3}}\sum\frac{(3^{\tfrac{1}{3}}z)^r}{r!}(\tfrac{1}{3}r-\tfrac{2}{3})!\{1-\cos\tfrac{2}{3}(r+1)\pi\}$$

$$= \sqrt{3}\,\alpha y_1 + \sqrt{3}\,\beta y_2. \tag{13}$$

Reducing I_2 and I_3 to integrals along the imaginary axis we have also for real z

$$\text{Bi}(z) = \frac{1}{\pi}\int_0^\infty \{e^{tz-\frac{1}{3}t^3} + \sin(tz + \tfrac{1}{3}t^3)\}\,dt. \tag{14}$$

The series expansions converge too slowly to be convenient for computation if $|z|$ is more than about 3. We try the method of steepest descents.

We therefore study the behaviour of the paths of steepest descent through the saddle-points when z has a general complex value. We write $z = r\exp(i\theta)$, $tz - \tfrac{1}{3}t^3 = f(t)$. Then the saddle-points are P_1 ($t = t_1 = r^{1/2}\exp(\tfrac{1}{2}i\theta)$) and P_2 ($t = t_2 = -r^{1/2}\exp(\tfrac{1}{2}i\theta)$). t_1 has a positive and t_2 a negative real part if $-\pi < \theta < \pi$. We consider $\theta \geqslant 0$. We have

$$\Re f(t_2) = -\tfrac{2}{3}r^{3/2}\cos\tfrac{3}{2}\theta, \quad \Im f(t_2) = -\tfrac{2}{3}r^{3/2}\sin\tfrac{3}{2}\theta.$$

The path of steepest descent S_1 through P_1 makes an angle $\pi - \tfrac{1}{4}\theta$ and the path of steepest descent S_2 through P_2 makes an angle $\tfrac{1}{2}\pi - \tfrac{1}{4}\theta$ with the x axis.

We consider first the approximation to Ai(z), for which the path is L_1. There are the following cases:

(i) $0 \leqslant \theta < \tfrac{1}{3}\pi$. S_2 goes from $\infty\exp(-\tfrac{2}{3}\pi i)$ through P_2 to $\infty\exp(\tfrac{2}{3}\pi i)$, keeping P_1 always on the right; it has the form of L_1 and since $\Re f(t_2) < 0$, we have that Ai(z) is exponentially *decreasing*.

(ii) $\theta = \tfrac{1}{3}\pi$. The path S_2 still has the form of L_1 but $\Re f(t_2) = 0$ and Ai(z) becomes *oscillatory*.

(iii) $\tfrac{1}{3}\pi < \theta < \tfrac{2}{3}\pi$. The path S_2 goes from $\infty\exp(-\tfrac{2}{3}\pi i)$ through P_2, cuts the asymptote going to $\infty\exp(\tfrac{2}{3}\pi i)$ and approaches it from above. It still keeps P_1 always on the right and is of the form of L_1. $\Re f(t_2)$ is positive and Ai(z) is exponentially *increasing*.

(iv) $\theta = \tfrac{2}{3}\pi$. The path S_2 is a straight line from $\infty\exp(-\tfrac{2}{3}\pi i)$ through P_2 and P_1 to $\infty\exp(\tfrac{1}{3}\pi i)$. A path L_1 follows this as far as P_1 and then turns through a right angle along S_1 and goes to $\infty\exp(\tfrac{2}{3}\pi i)$. $\Re f(t_1)$ is negative and the contribution from the part of the path near P_1 is small compared with that from the neighbourhood of P_2. Ai(z) is exponentially increasing.

(v) $\tfrac{2}{3}\pi < \theta < \pi$. The path S_2 now goes from $\infty\exp(-\tfrac{2}{3}\pi i)$ to ∞ and to complete a path equivalent to L_1 we have to add a path S_1 from ∞ through P_1 to $\infty\exp(\tfrac{2}{3}\pi i)$. The contribution from the neighbourhood of P_1 is exponentially decreasing while that from the neighbourhood of P_2 is exponentially increasing, and the latter determines the behaviour of Ai(z).

(vi) $\theta = \pi$. The path L_1 is made up of the same two parts as in (v), but now $\Re f(t_1) = \Re f(t_2) = 0$ and the contributions from the two saddle-points are of comparable magnitude. Ai(z) is *oscillatory*.

The first few terms of the asymptotic expansions obtained in this way for $-\pi < \arg z < \pi$ and $\arg z = \pi$ are given in (20) and (22).

The treatment of Bi(z) is a little more complicated because two paths, L_2 and L_3, are involved. We have, however, putting $t = \tau\exp(-\tfrac{2}{3}\pi i)$

$$\frac{1}{2\pi i}\int_{L_2} e^{tz-\frac{1}{3}t^3}\,dt = \frac{1}{2\pi i}\exp(-\tfrac{2}{3}\pi i)\int_{L_1} e^{\tau z\exp(-\frac{2}{3}\pi i) - \frac{1}{3}\tau^3}\,d\tau$$

$$= \exp(-\tfrac{2}{3}\pi i)\,\text{Ai}(z\exp(-\tfrac{2}{3}\pi i)). \tag{15}$$

Similarly
$$\frac{1}{2\pi i}\int_{L_3} e^{tz-\frac{1}{3}t^3} = \exp(\tfrac{2}{3}\pi i)\,\text{Ai}(z\exp(\tfrac{2}{3}\pi i)). \tag{16}$$

Then (20) will yield a valid approximation for both (15) and (16) in the range
$$-\tfrac{1}{3}\pi < \arg z < \tfrac{1}{3}\pi.$$
In this range, therefore, we have the expansion given in (21).

If this were valid outside the given range it would suggest that Bi(z) is exponentially small for some values of arg z. This is not so; in fact Bi(z) is exponentially large for all values of arg z, except for some that make it oscillatory. To see this and to obtain the asymptotic expansion for other values of arg z we make use of the identities

$$\text{Ai}(ze^{2/3 k\pi i}) = e^{1/3 k\pi i}\left(\cos \tfrac{1}{3}k\pi\, \text{Ai}(z) - \tfrac{i}{\sqrt{3}}\sin \tfrac{1}{3}k\pi\, \text{Bi}(z)\right), \tag{17}$$

$$\text{Bi}(ze^{2/3 k\pi i}) = e^{1/3 k\pi i}(-\sqrt{3}\,i \sin \tfrac{1}{3}k\pi\, \text{Ai}(z) + \cos \tfrac{1}{3}k\pi\, \text{Bi}(z)), \tag{18}$$

$$\tfrac{i}{\sqrt{3}}\sin \tfrac{1}{3}k\pi\, \text{Bi}(z) = e^{1/3 k\pi i}\,\text{Ai}(ze^{-2/3 k\pi i}) - \cos \tfrac{1}{3}k\pi\, \text{Ai}(z), \tag{19}$$

where k is a positive or negative integer.

From the last of these with $k = \pm 1$ we have that, if $|\arg z| = \tfrac{1}{3}\pi$ or π, Bi(z) is oscillatory, and that otherwise one of the terms on the right is exponentially increasing. In particular if $\arg z = \pi$ and $z = \zeta e^{\pi i}$ we have the expansions (22), (23).

We now summarize the results obtained:

$$\text{Ai}(z) \sim \frac{1}{2\sqrt{\pi}} z^{-1/4} \exp(-\tfrac{2}{3}z^{3/2})\left(1 - \frac{1.5}{1!\,48}z^{-3/2} + \frac{1.7.5.11}{2!\,48^2}z^{-3} - \frac{1.7.13.5.11.17}{3!\,48^3}z^{-9/2} + \ldots\right) \tag{20}$$

for $-\pi < \arg z < \pi$;

$$\text{Bi}(z) \sim \frac{1}{\sqrt{\pi}} z^{-1/4} \exp(\tfrac{2}{3}z^{3/2})\left(1 + \frac{1.5}{1!\,48}z^{-3/2} + \frac{1.7.5.11}{2!\,48^2}z^{-3} + \frac{1.7.13.5.11.17}{3!\,48^3}z^{-9/2} + \ldots\right) \tag{21}$$

for $-\tfrac{1}{3}\pi < \arg z < \tfrac{1}{3}\pi$.

When $\arg z = \pi$ we put $z = \zeta e^{\pi i}$ and obtain

$$\text{Ai}(z) = \frac{1}{\sqrt{\pi}} \zeta^{-1/4}\{P(\zeta)\sin(\tfrac{2}{3}\zeta^{3/2} + \tfrac{1}{4}\pi) - Q(\zeta)\cos(\tfrac{2}{3}\zeta^{3/2} + \tfrac{1}{4}\pi)\}, \tag{22}$$

$$\text{Bi}(z) = \frac{1}{\sqrt{\pi}} \zeta^{-1/4}\{P(\zeta)\cos(\tfrac{2}{3}\zeta^{3/2} + \tfrac{1}{4}\pi) + Q(\zeta)\sin(\tfrac{2}{3}\zeta^{3/2} + \tfrac{1}{4}\pi)\}, \tag{23}$$

where
$$P(\zeta) \sim 1 - \frac{1.7.5.11}{2!\,48^2}\zeta^{-3} + \frac{1.7.13.19.5.11.17.23}{4!\,48^4}\zeta^{-6} - \ldots, \tag{24}$$

$$Q(\zeta) \sim \frac{1.5}{1!\,48}\zeta^{-3/2} - \frac{1.7.13.5.11.17}{3!\,48^3}\zeta^{-9/2} + \ldots. \tag{25}$$

The second terms are about a tenth of the first even at $|z| = 1$.

The particular functions Ai(z) and Bi(z) were chosen as the fundamental pair so that one would decrease exponentially along the positive real axis and so that on the negative real axis they would have similar amplitudes for large z but differ by $\tfrac{1}{2}\pi$ in phase.[a]

A linear combination of the series on the right in (20) and (21) is an asymptotic approximation to some solution of the equation, but not to the same solution for all values of arg z. This phenomenon, discovered by Stokes, is known as the discontinuity of arbitrary constants in asymptotic approximations. It is an example of the theorem of 17·023 that an asymptotic expansion, if not convergent, cannot be valid for all values of arg z.

17·08. Dispersion: wave-velocity and group-velocity. In a continuous dynamical system capable of propagating waves along the x axis, let wave-length $2\pi/\kappa$ be associated with period $2\pi/\gamma$. If the original disturbance ζ_0 is unity for $-h<x<h$ and otherwise zero, we can write it as $H(x+h)-H(x-h)$, or in Fourier form

$$\zeta_0 = \frac{2}{\pi}\int_0^\infty \sin\kappa h \cos\kappa x \frac{d\kappa}{\kappa}. \tag{1}$$

The original rate of change can also be expressed as a Fourier integral; let us take it, however, to be zero. We then have the problem of waves spreading out from an initially disturbed region. For instance, the system may be a long canal and the original disturbance an elevation or depression of the water surface by some solid striking it. The elevation for a later time will then be

$$\zeta = \frac{2}{\pi}\int_0^\infty \sin\kappa h \cos\kappa x \cos\gamma t \frac{d\kappa}{\kappa}. \tag{2}$$

We neglect possible complications due to reflexion at the ends, if any; that is, we treat the problem as determined entirely by the initial disturbance. The integrand in (2) can be broken up as follows:

$$\zeta = \frac{1}{2\pi}\int_0^\infty \{\sin(\gamma t - \kappa x + \kappa h) - \sin(\gamma t - \kappa x - \kappa h) + \sin(\gamma t + \kappa x + \kappa h) - \sin(\gamma t + \kappa x - \kappa h)\}\frac{d\kappa}{\kappa}. \tag{3}$$

If γ was proportional to κ we should have a system that propagates waves of all lengths with the same velocity, and then the first and second terms would represent waves travelling towards x positive, the third and fourth towards x negative. Those represented by the first and fourth terms would appear to have started from $x = h$, the others from $x = -h$. We are concerned here with cases where γ is not proportional to κ; in other words the wave-velocity depends on the wave-length. We take γ to be an odd function of κ, real for κ real. The treatments of the four terms are all similar, and we may confine attention to the first. We then take

$$\zeta_1 = \frac{1}{4\pi i}\int_0^\infty \{e^{i(\gamma t - \kappa x + \kappa h)} - e^{-i(\gamma t - \kappa x + \kappa h)}\}\frac{d\kappa}{\kappa}$$

$$= \frac{1}{4\pi i}P\int_{-\infty}^\infty e^{i(\gamma t - \kappa x + \kappa h)}\frac{d\kappa}{\kappa}. \tag{4}$$

Here h appears only in the combination $x-h$; we may therefore omit it, as if the waves started at $x = 0$, and restore it later if required. We can also temporarily omit the suffix in ζ_1. In the applications x and t are both large; but we can find an approximation for large t with x/t fixed. Then

$$f(\kappa) = i(\gamma - \kappa x/t) \tag{5}$$

and at a saddle-point $\quad f'(\kappa) = i(\gamma' - x/t) = 0; \quad f''(\kappa) = i\gamma'', \tag{6}$

accents denoting differentiation with regard to κ. It thus appears that a fundamental part in the method is played by $d\gamma/d\kappa$; this is called the *group-velocity*. The wave-velocity, at which all waves would travel if only one wave-length was present, is γ/κ; but we see that the Fourier representation of a local disturbance automatically introduces all possible wave-lengths, and it remains to be seen whether the wave-velocity reappears explicitly. We shall find that it does. The relation between them can be written, if we put $\gamma/\kappa = c$, $d\gamma/d\kappa = C$,

$$C = \frac{d}{d\kappa}(\kappa c) = \kappa\frac{dc}{d\kappa} + c = c - \lambda\frac{dc}{d\lambda}, \tag{7}$$

if we introduce the wave-length $\lambda = 2\pi/\kappa$ instead of κ; but the results are far more easily stated in terms of κ. (5) is an equation determining κ as a function of x/t; denote this value by κ_0. We assume at present that it is real. Since γ' is an even function of κ, if κ_0 is one (positive) root $-\kappa_0$ is another. We also assume that γ is not infinite for any real κ, so that the integrand in (4) has no singularities on the real axis except the pole at $\kappa = 0$, which is irrelevant because the other three parts of (3) will also have poles at 0 with residues ± 1, and their effects will all cancel when they are combined, whatever path of integration we choose.

Then $f''(\kappa_0)$ is purely imaginary, and γ'' will have opposite signs at κ_0 and $-\kappa_0$. If γ'' is positive at κ_0, the path of steepest descent at κ_0 will cross the real axis in the direction $\frac{1}{4}\pi$, that at $-\kappa_0$ in the direction $-\frac{1}{4}\pi$. If γ'' is negative these relations will be reversed. Then the first term in the asymptotic expansion of the contribution from the passage through κ_0, in the first case, is

$$\frac{1}{4\pi i} \frac{\sqrt{(2\pi)}}{\kappa_0} \frac{\exp i(\gamma_0 t - \kappa_0 x)}{\{t\gamma''(\kappa_0)\}^{1/2}} e^{1/4\pi i}, \tag{8}$$

and the passage through $-\kappa_0$ gives

$$-\frac{1}{4\pi i} \frac{\sqrt{(2\pi)}}{\kappa_0} \frac{\exp\{-i(\gamma_0 t - \kappa_0 x)\}}{\{t\gamma''(\kappa_0)\}^{1/2}} e^{-1/4\pi i}, \tag{9}$$

on attending to the reversals of sign but remembering that it is $|\gamma''(-\kappa_0)|$ that appears in the denominator, the argument being looked after by the last factor. Combining the two we have

$$\zeta \sim \frac{1}{\sqrt{(2\pi)}\sqrt{(t\gamma_0'')}\,\kappa_0} \sin(\gamma_0 t - \kappa_0 x + \tfrac{1}{4}\pi). \tag{10}$$

If γ_0'' is negative we still take $|\gamma_0''|$ in the denominator but reverse the signs of the exponents $\pm \frac{1}{4}\pi i$. Then in both cases

$$\zeta \sim \frac{1}{\sqrt{(2\pi)}\sqrt{|t\gamma_0''|}\,\kappa_0} \sin(\gamma_0 t - \kappa_0 x + \tfrac{1}{4}\pi \,\mathrm{sgn}\,\gamma''(\kappa_0)). \tag{11}$$

There is therefore a difference of phase of $\frac{1}{2}\pi$ according as the group-velocity increases or decreases with κ.

Now consider the angle $\qquad \theta = \gamma_0 t - \kappa_0 x, \tag{12}$

and see how it varies with x and t, remembering that κ_0 and therefore γ_0 are functions of x/t determined by (5).

$$\frac{\partial \theta}{\partial x} = t\frac{d\gamma_0}{d\kappa_0}\frac{\partial \kappa_0}{\partial x} - \kappa_0 - x\frac{\partial \kappa_0}{\partial x}. \tag{13}$$

But $t\dfrac{d\gamma_0}{d\kappa_0} - x = 0$ by (6); hence the terms in $\partial \kappa_0/\partial x$ cancel, and $\dfrac{\partial \theta}{\partial x} = -\kappa_0$. Hence the phase varies by 2π in a distance $2\pi/\kappa_0$; and therefore $2\pi/\kappa_0$ *is the wave-length of waves passing a point at distance x near time t.* Similarly

$$\frac{\partial \theta}{\partial t} = \gamma_0 + t\frac{d\gamma_0}{d\kappa_0}\frac{\partial \kappa_0}{\partial t} - x\frac{\partial \kappa_0}{\partial t} = \gamma_0. \tag{14}$$

Hence $2\pi/\gamma_0$ *is the period of waves passing a point at distance x near time t.* Then γ_0/κ_0 *is the velocity of waves passing a point at distance x at time t*, in the sense that if we move forward with this velocity we shall keep in the same position relative to the nearest crest and trough. Thus the velocity of an individual wave is the wave-velocity appropriate to its length. But we cannot infer from this that each wave travels with a constant velocity. For if γ is not proportional to κ, C is not equal to γ/κ; therefore if we take a point travelling outwards with velocity c it will come to a place where the ratio x/t is different from what it was at the place and time when we started, C will be different and therefore the local values of γ, κ, and c will also be different. We can think of an individual wave, and it travels with velocity c appropriate to its length at the moment; but as the wave goes on its period, length, and velocity will all change. If on the other hand we travel out so as to keep x/t the same, we always keep to the same value of C, and therefore of κ, γ and c; but we do not keep to the same wave because $c \neq C$. That is, *periods, wave-lengths and wave-velocities are propagated with the group-velocity; individual waves travel with the local wave-velocity, but change their periods, lengths, and velocities as they travel.*

Energy also is propagated with the group-velocity, in a certain sense. Let us consider two points starting at $x = 0$ at time 0 and moving with velocities C_1, C_2. The energy corresponding to the displacements between these points at time t can be taken to be proportional to

$$\int_{C_1 t}^{C_2 t} \zeta^2 dx \sim \int_{C_1 t}^{C_2 t} \frac{1}{2\pi(t\gamma_0'')\kappa_0^2} \sin^2(\gamma_0 t - \kappa_0 x \pm \tfrac{1}{4}\pi) dx, \qquad (15)$$

provided that there are several waves between $x = C_1 t$ and $C_2 t$. But then we can take a mean value of $\sin^2(\gamma_0 t - \kappa_0 x \pm \tfrac{1}{4}\pi)$ over each wave, since the individual waves are nearly harmonic; and then this expression is nearly

$$\int_{C_1 t}^{C_2 t} \frac{1}{4\pi t \gamma_0'' \kappa_0^2} dx = \int_{C_1}^{C_2} \frac{dC}{4\pi \gamma_0'' \kappa_0^2}, \qquad (16)$$

on putting $x = Ct$. But γ_0'' and κ_0^2 are functions of C alone; hence *the energy between two points starting at the origin and moving out with constant speeds is independent of the time*; and these speeds are the local group-velocities.

All these results are approximations subject to the condition that we can safely reduce the asymptotic expansion of ζ to its first term. This is usually satisfied; we shall see later that if it is not the local motion is not even approximately simple harmonic in the neighbourhood. Since we are effectively expanding in negative powers of t the approximation will always improve as the wave train advances.

In (15), the wave-lengths between the points $x = C_1 t$ and $C_2 t$ remain constant; but the distance increases; hence the number of waves increases in proportion to t. The original disturbance may at first give a solitary wave, but as it travels it develops into a train, which becomes longer and longer.

There may be no real value of κ that satisfies (6) for some values of x/t. In that case the best procedure is to seek for saddle-points off the real axis. Then the exponent in (4) will have a real part increasing numerically with the time. In a physical system, if this was positive, we should ultimately have a steadily increasing energy. Since this is impossible the real part of the exponent must be negative. Hence complex roots of (6) correspond to places where there is little disturbance and what there is is not approximately simple-harmonic. There may be a minimum or a maximum real group-velocity; in the former case there will be little movement within a certain distance of the origin, increasing with the time, in the latter beyond a certain distance. Since a minimum or a maximum group-velocity implies the vanishing of γ'', and therefore of the denominator of our approximate solution, we can infer that near the front or rear of such a train the amplitudes will be large; but if we are too near it we can no longer use the method in its present form, since it depends essentially on $\gamma_0'' t$ being large enough to overwhelm later terms of the series. Further consideration of this point requires attention to the cubic terms in the exponent, which introduce the Airy integral.

17·09. Dispersion of water waves. All the features described in 17·07 and 17·08 are exemplified in the theory of water waves when capillarity and gravity are both taken into account. The wave-velocity is given as a function of κ by

$$c^2 = \left(\frac{g}{\kappa} + T'\kappa\right) \tanh \kappa H, \qquad (1)$$

where the surface tension is $T'\rho$, ρ being the density, and H is the depth of the water. If we choose the solution that reduces to $(gH)^{1/2}$ when κ is small, c is a single-valued function of κ. Then

$$\gamma^2 = (g\kappa + T'\kappa^3) \tanh \kappa H, \qquad (2)$$

and γ behaves like $(gH)^{1/2}\kappa$ for κ small and like $(T'\kappa^3)^{1/2}$ for κ large. The group-velocity C therefore tends to $(gH)^{1/2}$ for κ small and to $+\infty$ for κ large. Taking the second approximation for κ small we have

$$\gamma = (gH)^{1/2}\kappa\left(1 - \tfrac{1}{6}\kappa^2 H^2 + \frac{1}{2}\frac{T'\kappa^2}{g}\right), \quad C = (gH)^{1/2}\left(1 - \tfrac{1}{2}\kappa^2 H^2 + \frac{3}{2}\frac{T'\kappa^2}{g}\right). \qquad (3)$$

If then $T'/g < \tfrac{1}{3}H^2$, C will be less than $(gH)^{1/2}$ when κ is small. This will be satisfied if the depth is more than about 0·5 cm.; and for smaller depths it would be necessary to take viscosity into account. We shall assume that H is considerably more than 1 cm. Then as κ increases from 0, C begins by decreasing; but it increases again for κ large and therefore has a minimum for an intermediate value of κ. The existence of a minimum wave-velocity for water waves is well known, that of a minimum group-velocity less so, but it has a considerable influence on their dispersion. We find that at a given time there is smooth water near the origin, up to the distance that can have been reached by waves travelling with the minimum group velocity. Further out any given distance will be associated with a group-velocity, but this will be associated with two possible wave-lengths, the shorter controlled mainly by capillarity and the longer by gravity. Beyond a distance $(gH)^{1/2}t$ there will be no gravity waves, but the capillary waves will still be possible.

If the depth is large we can take $\tanh \kappa H = 1$ for all but the longest waves; then

$$\gamma = (g\kappa + T'\kappa^3)^{1/2}, \qquad (4)$$

$$C = \frac{g + 3T'\kappa^2}{2(g\kappa + T'\kappa^3)^{1/2}}, \qquad (5)$$

$$\frac{dC}{d\kappa} = \frac{3T'\kappa}{(g\kappa + T'\kappa^3)^{1/2}} - \frac{(g + 3T'\kappa^2)^2}{4(g\kappa + T'\kappa^3)^{3/2}}, \qquad (6)$$

and the minimum group-velocity corresponds to

$$\kappa^2 = \frac{g}{T'}\left(\frac{2}{\sqrt{3}} - 1\right). \tag{7}$$

If we introduce the minimum wave-velocity c_0 given by

$$c_0^2 = 2(gT')^{1/2}, \tag{8}$$

and the corresponding wave-length given by

$$\kappa_0^2 T' = g, \tag{9}$$

we have for minimum group-velocity

$$\kappa = 0{\cdot}393\kappa_0, \quad C = 0{\cdot}768c_0, \quad c = 1{\cdot}212c_0, \tag{10}$$

$$\frac{d^2C}{d\kappa^2} = 0{\cdot}787\frac{c_0^5}{g^2}. \tag{11}$$

On water the least wave-velocity is 23 cm./sec., and the corresponding wave-length 1·7 cm. Hence the minimum group-velocity is 18 cm./sec., the corresponding wave-velocity 28 cm./sec., and the wave-length 4·3 cm. The approximation $\tanh \kappa H = 1$ is therefore justified for waves with the minimum group-velocity or shorter waves provided that the depth is more than about 5 cm.

We shall take first the case of very deep water and waves such that x/t is small compared with $(gH)^{1/2}$ but large compared with the minimum group-velocity. Then we can write

$$\gamma = (g\kappa)^{1/2}, \quad c = (g/\kappa)^{1/2}, \quad C = \tfrac{1}{2}(g/\kappa)^{1/2}, \quad dC/d\kappa = -\tfrac{1}{4}(g/\kappa^3)^{1/2}, \tag{12}$$

and from 17·08 (11) (κ_0 defined as in 17·08)

$$\zeta \sim \frac{2}{\sqrt{(2\pi t)}\,(g/\kappa_0^3)^{1/4}\kappa_0} \sin(\gamma_0 t - \kappa_0 x - \tfrac{1}{4}\pi), \tag{13}$$

where

$$\frac{x}{t} = C = \frac{1}{2}\left(\frac{g}{\kappa_0}\right)^{1/2}. \tag{14}$$

Thus

$$\zeta \sim \sqrt{\left(\frac{2}{\pi t}\right)} g^{-1/4}\kappa_0^{-1/4} \sin(\gamma_0 t - \kappa_0 x - \tfrac{1}{4}\pi), \tag{15}$$

and the amplitude decreases towards the rear of the train like $x^{1/2}$.

This result, however, is modified if we return to 17·08 (3); for the train that we have estimated is a single one, supposed to have started from $x = 0$. There will actually be two superposed trains, one having started from $x = h$ and the other from $x = -h$. If the wave-length exceeds $2h$, therefore, we should restore these, and then the full solution will be approximately $-2h\,\partial/\partial x$ of what we have just found. It will therefore increase towards the rear of the train like $x^{-3/2}$. The behaviour of the train therefore depends greatly on the form of the initial disturbance. The splash of a brick, for instance, will give $2h$ about 10 cm. Waves shorter than this will have their smaller amplitudes towards the rear of the train, longer ones their larger amplitudes. Hence there will be a wave-length associated with a maximum amplitude and determined mostly by the scale of the initial disturbance. But a rain-drop or a rising fish gives an initial disturbance with a scale comparable with 1 cm., and the amplitude of the gravity waves will increase all the way to the rear of the train.

17·09 *Capillary waves; long waves*

Next, take capillary waves. We have now

$$\gamma = (T'\kappa^3)^{1/2}, \quad c = (T'\kappa)^{1/2}, \quad C = \tfrac{3}{2}(T'\kappa)^{1/2}, \quad dC/d\kappa = \tfrac{3}{4}(T'/\kappa)^{1/2}, \tag{16}$$

$$\zeta \sim \frac{1}{\sqrt{(\tfrac{3}{2}\pi t)}\,(T'/\kappa_0)^{1/4}\,\kappa_0} \sin(\gamma_0 t - \kappa_0 x + \tfrac{1}{4}\pi), \tag{17}$$

where

$$\tfrac{3}{2}(T'\kappa_0)^{1/2} = \frac{x}{t}. \tag{18}$$

The amplitude therefore is roughly proportional to $x^{-3/2}$ at a given time. Thus even without the effect of viscosity, which is considerable for these short waves, their amplitudes will fall off rapidly towards the front.

Two exceptional cases arise, namely for group-velocities near $(gH)^{1/2}$ and those near the minimum. The standard formula found in the method of steepest descents depends on $f''(z)$ being large enough for us to be able to replace $\exp tf(z)$ by $\exp tf(z_0)\exp\{\tfrac{1}{2}tf''(z_0)(z-z_0)^2\}$ until it is small compared with its value at the saddle-point. This will be possible for t large enough if $f''(z_0)$ is not zero. If $f''(z_0) = 0$, the behaviour of the integrand will depend on the terms in $f'''(z_0)(z-z_0)^3$, and the method is not applicable in its simplest form. Near a value of z that gives $f''(z) = 0$ the approximation will not be good unless t is much larger than would suffice to give a good approximation elsewhere. We can, however, obtain useful solutions in terms of the Airy integral.

Take first the longest waves. Since the horizontal extent is then of dominating importance we allow for the rear wave by applying the operator $-2h\,\partial/\partial x$ to 17·08 (4); then from 17·09 (3)

$$\zeta \sim \frac{h}{2\pi}\int_{-\infty}^{\infty} e^{i(\gamma t - \kappa x)}\,d\kappa$$

$$\sim \frac{h}{\pi}\int_0^{\infty} \cos\{(gH)^{1/2}\kappa t(1-\tfrac{1}{6}\kappa^2 H^2) - \kappa x\}\,d\kappa$$

$$= \frac{h}{\pi}\int_0^{\infty} \cos\{\kappa x - (gH)^{1/2}\kappa t + \tfrac{1}{6}(gH)^{1/2} tH^2\kappa^3\}\,d\kappa. \tag{19}$$

Put

$$(gH)^{1/2} = \alpha, \quad z^3 = \frac{2(x-\alpha t)^3}{\alpha tH^2}, \quad \kappa = \frac{zs}{x-\alpha t}. \tag{20}$$

Then

$$\zeta \sim \frac{hz}{\pi(x-\alpha t)}\int_0^{\infty}\cos(sz + \tfrac{1}{3}s^3)\,ds = h\left\{\frac{2}{\alpha tH^2}\right\}^{1/3} \mathrm{Ai}\left\{\frac{2(x-\alpha t)^3}{\alpha tH^2}\right\}^{1/3}. \tag{21}$$

At points where $x = \alpha t$, the amplitude decreases with time like $t^{-1/3}$ instead of $t^{-1/2}$ as at places where the group-velocity behaves ordinarily. The front of the gravity wave therefore becomes more and more the most prominent feature of the disturbance. The Airy integral decreases towards positive argument, and the disturbance at values of x greater than αt falls off rapidly and smoothly. The maximum elevation is a little behind the place where $x = \alpha t$, and is followed by a train of waves becoming smaller and shorter.

If x/t is near the minimum group-velocity and h much less than the corresponding wavelength,

$$\zeta \doteqdot \frac{h}{2\pi}\int_{-\infty}^{\infty} e^{i(\gamma t - \kappa x)}\,d\kappa = \frac{h}{\pi}\int_0^{\infty}\cos(\gamma t - \kappa x)\,d\kappa. \tag{22}$$

Let suffix m indicate values corresponding to the stationary group-velocity. For the positive value κ_m, put

$$\kappa = \kappa_m + \kappa_1, \quad \gamma \doteqdot \kappa_m c_m + C_m \kappa_1 + \tfrac{1}{6}C_m''\kappa_1^3. \tag{23}$$

Then
$$\gamma t - \kappa x \doteq \kappa_m(c_m t - x) + \kappa_1(C_m t - x) + \tfrac{1}{6}\kappa_1^3 C_m'' t. \tag{24}$$

The saddle-points are near $\kappa_1 = \pm \{2(x/t - C_m)/C_m''\}^{-\frac{1}{2}}$. Suppose that x/t is such that this is small, and take a circle $|\kappa_1|$ = constant such that the saddle-points lie within it. The error in neglecting the parts of the path of integration outside the circle is $O(1/t)$, and we can deform the path inside the circle so that it passes through $\kappa_1 = 0$. Then

$$\zeta \sim \frac{h}{2\pi}\cos \kappa_m(c_m t - x)\int_{-\infty}^{\infty}\cos\{\kappa_1(C_m t - x) + \tfrac{1}{6}\kappa_1^3 C_m'' t\}d\kappa_1$$
$$-\frac{h}{2\pi}\sin \kappa_m(c_m t - x)\int_{-\infty}^{\infty}\sin\{\kappa_1(C_m t - x) + \tfrac{1}{6}\kappa_1^3 C_m'' t\}d\kappa_1. \tag{25}$$

The second integral is zero; and

$$\zeta \sim 2h\left(\frac{2}{C_m'' t}\right)^{1/3}\cos \kappa_m(c_m t - x)\,\mathrm{Ai}\left\{\left(\frac{2}{C_m'' t}\right)^{1/3}(C_m t - x)\right\}. \tag{26}$$

The errors due to neglecting C_m''' and $C_m^{(4)}$ within an arc of κ_1 such that $\kappa_1^3 C_m'' t$ has become large are easily shown to be of the order of $t^{-2/3}$ of the main term; hence the error of (26) is $O(1/t)$. For large t the last factor varies much more slowly with x than the cosine, on account of the small factor $t^{-1/3}$ in its argument. Hence the variation can be described as a series of waves of length $2\pi/\kappa_m$ and period $2\pi/\kappa_m c_m$, but with the amplitude falling off exponentially for $x < C_m t$ and oscillating slowly for $x > C_m t$. The most conspicuous feature is the ring of waves with the length corresponding to the minimum group-velocity, surrounding a circle of smooth water.

17·10. Interrupted harmonic wave train. In the direct measurement of the velocity of light or sound a continuous train of waves, which can be treated as harmonic, is interrupted at regular intervals by a toothed wheel or a revolving mirror, giving a series of flashes. What is observed is the time such that the flash returning after reflexion at a distance is blocked by the revolving mechanism near the eye or ear. Thus the experiment does not depend directly on the time of travel of individual waves, but on that of variations of intensity. The simplest statement of the phenomenon would be to regard the train as having beats, so that the disturbance can be expressed in the form

$$\zeta = \cos \gamma_0 t \cos \tfrac{1}{2}\pi t/k = \tfrac{1}{2}\cos\left(\gamma_0 - \frac{\pi}{2k}\right)t + \tfrac{1}{2}\cos\left(\gamma_0 + \frac{\pi}{2k}\right)t. \tag{1}$$

The disturbance has period $2\pi/\gamma_0$ but its amplitude vanishes whenever t is an odd multiple of k. In the next beat the phase is reversed. Now suppose that the wave-length $2\pi/\kappa_0$ for period $2\pi/\gamma_0$ is given by $\kappa_0 = \gamma_0/c_0$ and for neighbouring periods by

$$\kappa_0 + \frac{d\kappa}{d\gamma}\delta\gamma = \kappa_0 + (1/C)\,\delta\gamma.$$

Then the disturbance at distance x is

$$\zeta = \tfrac{1}{2}\cos\left\{\left(\gamma_0 - \frac{\pi}{2k}\right)t - \left(\kappa_0 - \frac{\pi}{2kC}\right)x\right\} + \tfrac{1}{2}\cos\left\{\left(\gamma_0 + \frac{\pi}{2k}\right)t - \left(\kappa_0 + \frac{\pi}{2kC}\right)x\right\}$$
$$= \cos(\gamma_0 t - \kappa_0 x)\cos\frac{\pi}{2k}\left(t - \frac{x}{C}\right). \tag{2}$$

Since $\gamma_0 = \kappa_0 c_0$, the first factor shows that the phases travel with the wave-velocity. But the second factor shows that the beats travel with the group-velocity $d\gamma/d\kappa = C$. Hence experiments of the type in question determine the group-velocity associated with period $2\pi/\gamma_0$, not the wave-velocity.

17·11. Refraction of a pulse. Suppose that there is an instantaneous disturbance at a point O within one medium and that it is partly transmitted into a second medium where the velocity of propagation is different. Both media are dispersive, that is, the velocity depends on the wave-length. Suppose that the disturbance in the second medium is observed at a point P; this may be regarded as the resultant of a set of partial disturbances coming by way of various points Q on the interface. Then the time factor in the disturbance at P is

$$\exp i(\gamma t - \kappa_1 s_1 - \kappa_2 s_2), \qquad (1)$$

where s_1 and s_2 are the distances of Q from O and P, and κ_1 and κ_2 correspond to γ in the two media. Let Q be at a distance s measured along the boundary from some fixed point. Then the chief part of the disturbance at P, since γ and s are both variables, will be found by making this factor stationary with regard to both, t and the position of P being kept fixed. This gives

$$\kappa_1 \frac{ds_1}{ds} + \kappa_2 \frac{ds_2}{ds} = 0, \qquad (2)$$

$$t - s_1 \frac{\partial \kappa_1}{\partial \gamma} - s_2 \frac{\partial \kappa_2}{\partial \gamma} = 0. \qquad (3)$$

Introducing the wave-velocity and the group-velocity for each medium we see that these are equivalent to

$$\frac{1}{c_1}\frac{ds_1}{ds} + \frac{1}{c_2}\frac{ds_2}{ds} = 0, \qquad (4)$$

$$\frac{s_1}{C_1} + \frac{s_2}{C_2} = t. \qquad (5)$$

The first of these is simply the usual law of refraction; the effective values of s are around the value that makes the time from O to P stationary for a point travelling with the wave-velocity. Thus the directions of travel of the waves are determined by the wave-velocity. The second relation, however, shows that the dominant period is determined by the group-velocities; for this is the equation corresponding to $x/t = C$ determining the dominant period in simple dispersion. Thus refraction in dispersive media is rather complicated, the lines of constant phase being in general inclined to those of constant period.*

17·12. Asymptotic solutions of differential equations. These are of several types. If

$$\frac{d^2y}{dx^2} + f(x)\frac{dy}{dx} + g(x) y = 0 \qquad (1)$$

and $f(x), g(x)$ are analytic and bounded as x tends to infinity, we can write the equation as

$$\frac{d^2y}{dx^2} + \left(a_0 + \frac{a_1}{x} + \ldots\right)\frac{dy}{dx} + \left(b_0 + \frac{b_1}{x} + \ldots\right) y = 0. \qquad (2)$$

* R. Stoneley, *Proc. Camb. Phil. Soc.* 31, 1935, 360–7.

The functions
$$z_1 = e^{\lambda_1 x} x^{\sigma_1}, \quad z_2 = e^{\lambda_2 x} x^{\sigma_2} \tag{3}$$
satisfy the differential equation
$$z'' - \left(\lambda_1 + \lambda_2 + \frac{\sigma_1 + \sigma_2}{x} + \frac{\sigma_1 - \sigma_2}{x^2\{\lambda_2 - \lambda_1 + (\sigma_2 - \sigma_1)/x\}}\right) z'$$
$$+ \left[\left(\lambda_1 + \frac{\sigma_1}{x}\right)\left(\lambda_2 + \frac{\sigma_2}{x}\right) - \frac{\lambda_1 \sigma_2 + \lambda_2 \sigma_1}{x^2\{\lambda_2 - \lambda_1 + (\sigma_2 - \sigma_1)/x\}}\right] z = 0. \tag{4}$$

The coefficients agree to terms in $1/x$ with those in (2) provided that
$$\lambda_1 + \lambda_2 = -a_0, \quad \lambda_1 \lambda_2 = b_0, \quad \lambda_2 - \lambda_1 \neq 0, \tag{5}$$
$$\sigma_1 + \sigma_2 = -a_1, \quad \lambda_1 \sigma_2 + \lambda_2 \sigma_1 = b_1. \tag{6}$$

The first two of these equations determine λ_1, λ_2, which will be different unless $b_0 = \tfrac{1}{4}a_0^2$. Except in this case σ_1 and σ_2 are determinate. Convergent solutions can then be determined by the method of variation of parameters, but contain incomplete factorial functions. Substituting asymptotic expansions for these we get forms for y_1, y_2 such that y_1/z_1 and y_2/z_2 are given asymptotically by series in descending powers of x.

The result was given by E. L. Ince,* by a rather different method. He did not attend explicitly to the case where the method leads to $\lambda_1 = \lambda_2$. The solution then takes a different form. We can suppose a factor $e^{\lambda x}$ removed from the solutions, so that in (2) a_0 and b_0 are zero. In this case the functions
$$z_1, z_2 = e^{\pm \mu x^{\frac{1}{2}}} x^\sigma \tag{7}$$

satisfy
$$z'' - 2(\sigma - \tfrac{1}{4})\frac{z'}{x} - \left(\frac{\mu^2}{4x} - \frac{\sigma^2 + \tfrac{1}{2}\sigma}{x^2}\right) z = 0, \tag{8}$$

which agrees with (2) to terms in $1/x$ if
$$2(\sigma - \tfrac{1}{4}) = -a_1, \quad \tfrac{1}{4}\mu^2 = -b_1. \tag{9}$$

There are solutions of (2) asymptotically equal to z_1, z_2 multiplied by series in descending powers of $x^{\frac{1}{2}}$, provided $b_1 \neq 0$.

If further $b_1 = 0$, (2) has a regular singularity at infinity, and there are two convergent series solutions unless the indices are equal or differ by an integer, when one solution may contain a logarithm.

Hence if $f(x), g(x)$ tend to a_0, b_0 as x tends to infinity, we can in general substitute for y a series
$$e^{\lambda x} x^\sigma \left(1 + \frac{\gamma_1}{x} + \frac{\gamma_2}{x^2} + \ldots\right) \tag{10}$$

and determine λ, σ and the coefficients in turn, and the series so obtained will either converge or be asymptotic to two solutions of the differential equation. In the exceptional case where the equation for λ has equal roots, there are solutions of the form
$$e^{\pm \mu x^{\frac{1}{2}}} x^\sigma \left(1 + \frac{\gamma_1}{x^{\frac{1}{2}}} + \frac{\gamma_2}{x} + \ldots\right) \tag{11}$$

with similar properties. We shall refer to such solutions (i.e., in descending powers of x) as of Stokes's type; though the first expansions of this type seem to have been given by Jacobi for the Bessel functions.

* *Ordinary Differential Equations*, 1927, 169–71.

17·121. Write Bessel's equation in the form

$$\frac{d^2y}{dx^2} + \frac{1}{x}\frac{dy}{dx} + \left(1 - \frac{n^2}{x^2}\right)y = 0.$$

This is in the form required, with $\lambda = \pm i$. Put

$$y = e^{ix}u.$$

Then

$$u'' + \left(2i + \frac{1}{x}\right)u' + \left(\frac{i}{x} - \frac{n^2}{x^2}\right)u = 0.$$

Now put

$$u = x^\sigma v,$$

and choose σ so that the term in v/x is zero. We find that $\sigma = -\tfrac{1}{2}$, and

$$v'' + 2iv' - (n^2 - \tfrac{1}{4})\frac{v}{x^2} = 0.$$

Substitute

$$v = 1 + \frac{a_1}{x} + \frac{a_2}{x^2} + \ldots;$$

we find the recurrence relation

$$\{r(r+1) - (n^2 - \tfrac{1}{4})\}a_r = 2i(r+1)a_{r+1}.$$

Hence

$$v \sim 1 + \frac{\tfrac{1}{4} - n^2}{2ix} + \frac{(\tfrac{1}{4} - n^2)(\tfrac{9}{4} - n^2)}{(2i)^2\, 2!\, x^2} + \frac{(\tfrac{1}{4} - n^2)(\tfrac{9}{4} - n^2)(\tfrac{25}{4} - n^2)}{(2i)^3\, 3!\, x^3} + \ldots$$

$$= 1 - \frac{(\tfrac{1}{4} - n^2)(\tfrac{9}{4} - n^2)}{2!\,(2x)^2} + \frac{(\tfrac{1}{4} - n^2)(\tfrac{9}{4} - n^2)(\tfrac{25}{4} - n^2)(\tfrac{49}{4} - n^2)}{4!\,(2x)^4} - \ldots$$

$$- i\left\{\frac{\tfrac{1}{4} - n^2}{2x} - \frac{(\tfrac{1}{4} - n^2)(\tfrac{9}{4} - n^2)(\tfrac{25}{4} - n^2)}{3!\,(2x)^3} + \ldots\right\}$$

$$= U - iV \text{ say.}$$

Thus we have found an asymptotic solution

$$y_1 \sim x^{-1/2}e^{ix}(U - iV) = x^{-1/2}(U\cos x + V\sin x) + ix^{-1/2}(U\sin x - V\cos x).$$

The real and imaginary parts separately will be asymptotic solutions.

The choice of coefficients to make the solution correspond to $J_n(x)$ can be made by considering the first term of the asymptotic expansion found from the complex integral solution. We shall postpone this until we consider the most convenient companion function to $J_n(x)$.

If n is small and x large the terms begin by decreasing rapidly. Thus for $n = 0$ and $x = 10$ the term in x^{-2} is

$$\frac{9}{16.2.(20)^2} = \frac{9}{12800}$$

of the first. But the terms of the ascending power series do not begin to decrease till the fifth, and many more are needed to give an accuracy of 1 in 1000. The asymptotic series is therefore of great use.

It is not useful unless $2x$ is considerably greater than n^2.

If n is half an odd integer both series terminate and the solutions are expressed in finite terms. In particular if $n = \tfrac{1}{2}$, $U = 1$, $V = 0$, and a pair of solutions is $x^{-1/2}\cos x$, $x^{-1/2}\sin x$.

17·122. Sometimes $f(x)$ and $g(x)$ are not conveniently expressed in power series. The following method is then useful. Put

$$y = uv, \tag{1}$$

and determine v so that the coefficient of u' is 0. This gives

$$\frac{2v'}{v} + f = 0, \tag{2}$$

$$v = \exp\left[-\tfrac{1}{2}\int^x f(x)\,dx\right], \tag{3}$$

the constant factor being irrelevant. Then the equation for u takes the form

$$u'' = \chi(x)u. \tag{4}$$

We assume $\chi(x)$ large but χ'/χ small in the interval considered. Put

$$u = \exp\left[\int^x \eta\,dx\right]. \tag{5}$$

Then
$$\frac{u'}{u} = \eta, \quad \frac{u''}{u} = \left(\frac{u'}{u}\right)^2 + \eta' = \eta' + \eta^2.$$

and
$$\eta^2 + \eta' = \chi(x). \tag{6}$$

This equation is remarkably simple in appearance, but is non-linear. However, if $\chi(x)$ varies slowly we can take, approximately,

$$\eta = \chi^{1/2}(x); \tag{7}$$

and in the second approximation η' will be small. Then

$$\eta^2 \doteq \chi(x) - \tfrac{1}{2}\frac{\chi'}{\chi^{1/2}}, \quad \eta \doteq \chi^{1/2}\left(1 - \tfrac{1}{4}\frac{\chi'}{\chi^{3/2}}\right) = \chi^{1/2} - \tfrac{1}{4}\frac{\chi'}{\chi},$$

$$u \doteq \chi^{-1/4}\exp\left[\int^x \chi^{1/2}\,dx\right]. \tag{8}$$

The easiest way of examining the accuracy is to attempt a third approximation; redetermining η' from (8) and substituting in (6) we find

$$\eta \doteq \chi^{1/2} - \tfrac{1}{4}\frac{\chi'}{\chi} - \frac{1}{8\chi^{1/2}}\frac{d}{dx}\frac{\chi'}{\chi}. \tag{9}$$

If A is the smallest value of $\chi^{1/2}$ in the range considered the integral of the last term will be of order $\dfrac{1}{8A}\left[\dfrac{\chi'}{\chi}\right]$. If therefore χ'/χ varies by a moderate factor in the range an appreciable error will accumulate unless χ is everywhere large.*

17·123. This method can frequently be used to suggest the first term in an asymptotic expansion. If we take the equation for the Airy integral

$$\frac{d^2y}{dx^2} = xy,$$

---

* Cf. W. E. Milne, *Trans. Amer. Math. Soc.* 31, 1929, 907–18; E. C. Titchmarsh, *Journ. Lond. Math. Soc.* 19, 1945, 66–8.

we have at once $\chi = x$, $\int_0^x \chi^{1/2} dx = \tfrac{2}{3} x^{3/2}$, and there are asymptotic solutions beginning with

$$x^{-1/4} \exp(\pm \tfrac{2}{3} x^{3/2}).$$

The exponent is not a multiple of x, so that the solution is not of the form considered in 17·12; but we could transform to $\tfrac{2}{3} x^{3/2}$ as a new variable and proceed.

17·13. A somewhat similar solution can be obtained if $\chi(x)$ depends both on x and on an additional parameter h, and

$$y'' = \chi(x) y = (h^2 \chi_0 + h \chi_1 + \chi_2) y, \tag{1}$$

where h is large. For any h the equation will have a pair of solutions, valid in any range and not only when x is large. We want to know how these solutions behave when h is large. We therefore want an expansion in descending powers of h. Put

$$y = \left(\frac{d\xi}{dx}\right)^{-1/2} z; \tag{2}$$

the differential equation becomes

$$\xi'^2 \frac{d^2 z}{d\xi^2} = \left(h^2 \chi_0 + h \chi_1 + \chi_2 + \frac{\xi'''}{2\xi'} - \frac{3}{4} \frac{\xi''^2}{\xi'^2}\right) z. \tag{3}$$

Take

$$\xi = \int_0^x \left(\chi_0 + \frac{\chi_1}{h} + \frac{\psi_2}{h^2}\right)^{1/2} dx, \tag{4}$$

where ψ_2 is to some extent at our disposal. Then the differential equation takes the form

$$\frac{d^2 z}{d\xi^2} - h^2 z = g(\xi, h) z; \tag{5}$$

g is expansible in descending powers of h. Substituting

$$z = e^{h\xi} \left(1 + \frac{f_1(\xi)}{h} + \frac{f_2(\xi)}{h^2} + \ldots\right) \tag{6}$$

and equating coefficients of $h^2, h, 1, \ldots$ in turn, we can find the functions $f_1, f_2, \ldots$ one by one. Similarly we find a formal solution starting with $\exp(-h\xi)$. There are advantages in choosing ψ_2 so that $|g(\xi, h)|$ will be as small as possible.

In practice we are usually concerned with only the first approximation. We take h real and positive. Even for this the proof that the solutions are asymptotic requires restrictions on $g(\xi, h)$ and on the region. In particular the region may be unbounded and we want conditions that there may be solutions Z_1, Z_2 satisfying

$$Z_1 \exp(-h\xi) = 1 + M_1/h; \quad Z_2 \exp(h\xi) = 1 + M_2/h, \tag{7}$$

where M_1, M_2 are bounded for all ξ of the region and for all $h \geq h_0 > 0$. Suitable conditions on the region, E, are (1) E contains at least one point of the real axis (2) if ξ is in E, then all points $\eta = \Re(\xi) + i\theta \Im(\xi)$, $0 \leq \theta \leq 1$, are also in E (and consequently the upper and lower bounds of $\Re(\xi)$, which may be infinite, are taken for points on the real axis); and on $g(\xi, h)$ (3) that on any paths in E along the real or parallel to the imaginary axis,

$\int_{\xi_1}^{\xi_2} |g(t,h)\,dt| < M$, with M independent of ξ_1, ξ_2 and h for $h \geqslant h_0$. The method is to notice that the solutions are solutions of the integral equations

$$Z_1(\xi) = e^{h\xi} + \int_A^\xi \frac{1}{2h} \{e^{h(\xi-t)} - e^{-h(\xi-t)}\} g(t,h) Z_1(t)\,dt, \tag{8}$$

$$Z_2(\xi) = e^{-h\xi} - \int_\xi^B \frac{1}{2h} \{e^{h(\xi-t)} - e^{-h(\xi-t)}\} g(t,h) Z_2(t)\,dt, \tag{9}$$

where A, B are the lower and upper bounds of $\Re(\xi)$, and the path in (8) is from A to $\Re(\xi)$ along the real axis and from $\Re(\xi)$ to ξ parallel to the imaginary axis; for (9) it is similarly from ξ to $\Re(\xi)$ and then to B. It can be shown that substitution of successive approximations gives corrections $k_r(\xi, h)$ times the first term, such that $|k_r| < (2M/h)^r$. Thus the method leads to absolutely convergent solutions, and the error of the first term is uniformly less than $4M/h$ of the first term for $h \geqslant 4M$. Attention to the form of k_r shows that it has an asymptotic expansion in powers of $1/h$, and Z_1, Z_2 can be rearranged in the form (6).

Condition (2) is important because the proof assumes that in (8) the upper bound on the path of $|e^{2ht}|$, and in (9) that of $|e^{-2ht}|$, are taken at $t = \xi$. This would not be true if the condition was not satisfied. If possible ψ_2 should be chosen so that condition (3) is satisfied. For instance, if

$$y'' = (h^2 + a)y, \tag{10}$$

exact solutions are $\exp\{\pm(h^2+a)^{1/2}x\}$. If in (4) we took $\psi_2 = a$, the first terms of the approximate solutions are identical with the exact solutions. If we took $\psi_2 = 0$ the first terms would be $\exp(\pm hx)$. The ratios of these to the exact solutions are approximately $\exp(\pm ax/2h)$, in which the indices are unbounded in an infinite interval of x. This is a case where it is best to take $\psi_2 = \chi_2$.

On the other hand take the equation, suggested originally by Prof. G. H. Hardy,

$$y'' - (h^2 \theta'^2 + h\theta'')y = 0. \tag{11}$$

Exact solutions are

$$y_1 = \exp h\theta; \quad y_2 = 2h e^{h\theta} \int_x^\infty e^{-2h\theta(t)}\,dt$$

$$= \frac{1}{\theta'} e^{-h\theta} - \frac{1}{2h}\frac{\theta''}{\theta'^3} \exp(-h\theta) + O\left(\frac{1}{h^2}\right). \tag{12}$$

on integration by parts.

If we take $\theta = (\log x)^2$ and $\psi_2 = \chi_2 = 0$, we find

$$\chi^{1/2} = 2h \frac{\log x}{x}\left\{1 - \frac{1}{4h}\frac{\log x - 1}{(\log x)^2} - \frac{1}{32h^2}\frac{(\log x - 1)^2}{(\log x)^2} + \ldots\right\}. \tag{13}$$

The first two terms lead to

$$y_1 = \exp h(\log x)^2, \tag{14}$$

$$y_2 = \frac{x}{\log x} \exp\{-h(\log x)^2\}, \tag{15}$$

which agree with the first terms of the exact solutions. But the third term introduces factors

$$x^{\pm(\log x - 4)/32h}(\log x)^{-1/16h}, \tag{16}$$

which vary by indefinitely large factors in an infinite range of x; that is, the approximation given by the first three terms is not uniform. Closer investigation of (5) shows that taking $\psi_2 = \chi_2$ makes $\int^\infty g(\xi, h)\, d\xi$ diverge. But if we take simply

$$h\xi^{1/2} = h\theta' + \frac{\theta''}{2\theta'} \tag{17}$$

corresponding to
$$\psi_2 = \frac{\theta''^2}{4\theta'^2}, \tag{18}$$

the integral converges absolutely and uniformly, and the approximation does not contain the objectionable factor introduced by the third term of (13).

Approximations of this type have a long history. The first with an application to a general χ_0 seems to have been given by G. Green,* who used an equivalent process in showing that for tidal waves in a canal of slowly varying section the energy is transmitted without loss by reflexion. We shall therefore describe them as of *Green's type*. The approximation for negative h^2 was given simultaneously and justified by J. Liouville.† The transition from physical optics, depending on a second-order partial differential equation in three dimensions, to geometrical optics, depending on a first-order one, really involves the same principle. A special application to Bessel functions had been given still earlier, by G. Carlini in 1817.‡ There are numerous applications in wave mechanics, the chief perhaps being in the proof that classical mechanics is the limiting case of quantum mechanics when the energy is large. The present form is due to Jeffreys.§

17·131. If χ_0 has a zero in the region, ξ' defined by 17·13 (4) tends to 0 there as $h \to \infty$ and the approximation fails. For a simple zero this can be treated by choosing ξ so that the differential equation reduces approximately to

$$\frac{d^2z}{d\xi^2} = h^2 \xi z \tag{1}$$

solutions of which are $z_1 = \text{Bi}\,(h^{2/3}\xi)$, $z_2 = \text{Ai}\,(h^{2/3}\xi)$. Let $\chi = \chi_0 + \chi_1/h + \psi_2/h^2$ vanish at $x = -\alpha$ (where α will be $O(1/h)$) and take $\chi'_0(0) > 0$. We can take

$$\xi'^2 \xi = \chi; \quad \tfrac{2}{3}\xi^{3/2} = \int_{-\alpha}^{x} \left(\chi_0 + \frac{\chi_1}{h} + \frac{\psi_2}{h^2} \right)^{1/2} dx, \tag{2}$$

and the transformation is non-singular at $\xi = 0$. Then the differential equation takes the form
$$\frac{d^2z}{d\xi^2} - h^2 \xi z = g(\xi, h)\, z, \tag{3}$$

and solutions can be developed from integral equations as for the simple case of 17·13, the successive corrections again decreasing as fast as the terms of a geometrical progression in $1/h$. Conditions are needed on the region of validity E. The most useful set appears to be that E includes a segment of the real axis from 0 to B and segments of straight lines from 0 to $B_1 e^{2/3\pi i}$ and from 0 to $B_2 e^{-2/3\pi i}$; that any ξ not on these lines can be connected to a point ξ_0 on one of these segments by an arc $\rho^{3/2} \cos \tfrac{3}{2}\theta = \text{constant}$, where $t = \rho e^{i\theta}$; the path of integration consists of segments of these three lines together with one of these

---

* *Camb. Phil. Trans.* 6, 1837, 457–62; *Papers*, p. 225. Cf. Lamb, *Hydrodynamics*, 1932, p. 274.
† *J. des Math.* 2, 1837, 22–5.   ‡ Watson, *Theory of Bessel Functions*, p. 6.
§ *Proc. Lond. Math. Soc.* (2) 23, 1924, 428–36. Other references in Jeffreys, *Phil. Mag.* (7) 33, 1942, 451–6, *Asymptotic Approximations*, 1962.

arcs; on any of the paths 0 to B, 0 to ξ_0 and ξ_0 to ξ the integral $\int |t^{-1/2}g(t,h)| |dt|$ is uniformly bounded.*

When $|h^{2/3}\xi|$ is large we can use the first terms of the asymptotic approximations to Ai and Bi, with further errors of order $1/h\xi^{3/2}$.

We use Ai and Bi to connect solutions of the forms of those of 17·13 valid on opposite sides of $\xi = 0$. We have

$$y = \xi'^{-1/2}z = \chi^{-1/4}\xi^{1/4}z. \tag{4}$$

For $\xi > 0$ put $M = \int_{-\alpha}^{x} h\chi^{1/2} dx = \tfrac{2}{3}h\xi^{3/2}$; a pair of solutions are

$$y_1 = \sqrt{\pi}\, h^{1/6}\xi'^{-1/2}\, \mathrm{Bi}\,(h^{2/3}\xi)\{1+O(1/h)\} = \chi^{-1/4}\exp M\{1+O(1/h)\}, \tag{5}$$

$$y_2 = \sqrt{\pi}\, h^{1/6}\xi'^{-1/2}\, \mathrm{Ai}\,(h^{2/3}\xi)\{1+O(1/h)\} = \tfrac{1}{2}\chi^{-1/4}\exp(-M)\{1+O(1/h)\}. \tag{6}$$

For $\xi < 0$, put $\xi = -\zeta$, $L = \int_{\alpha}^{-x} h(-\chi)^{1/2} d(-x)$; then

$$y_1 = |\chi|^{-1/4}\{\cos(L+\tfrac{1}{4}\pi)+O(1/h)\}, \tag{7}$$

$$y_2 = |\chi|^{-1/4}\{\sin(L+\tfrac{1}{4}\pi)+O(1/h)\}, \tag{8}$$

where χ can be replaced by χ_0 with a further error of order $1/h$.

It has been pointed out by Langer† that care is needed in the use of (5), (6), (7) and (8) to establish correspondences between solutions on opposite sides of a zero of χ_0. We have, if A and B are constants, a general solution $Ay_1 + By_2$, with asymptotic expressions

$$A\chi^{-1/4}\exp M\{1+O(1/h)\} + \tfrac{1}{2}B\chi^{-1/4}\exp(-M)\{1+O(1/h)\}, \tag{9}$$

$$A|\chi|^{-1/4}\{\cos(L+\tfrac{1}{4}\pi)+O(1/h)\} + B|\chi|^{-1/4}\{\sin(L+\tfrac{1}{4}\pi)+O(1/h)\}. \tag{10}$$

If the solution for $\xi > 0$ is exponentially small, it follows that $A = 0$ and hence the solution must be, to $O(1/h)$, equal to $\tfrac{1}{2}B\chi^{-1/4}\exp(-M)$ and $B|\chi|^{-1/4}\sin(L+\tfrac{1}{4}\pi)$ on the respective sides. If, however, A is not known to be zero but merely to be of order B/h, it will not affect the validity of the approximation on the oscillatory side; but the A term will be much larger than the B term for large M, and the approximation on the exponential side is ruined. Similarly if the approximation is $|\chi|^{-1/4}\cos(L+\tfrac{1}{4}\pi)$ on the oscillatory side it must be $\chi^{-1/4}\exp M$ on the exponential side, but the converse does not follow. In using the formulae (9) and (10) confusion may be avoided if explicit attention is paid to the error terms throughout the work.‡

The device for crossing a zero of χ was first given by Rayleigh§ and extended by R. Gans.‖ It has been rediscovered by several later writers. Langer points out that when χ has two zeros the equation can be transformed to one of the forms

$$\frac{d^2y}{d\xi^2} = \pm\{h^2(\xi^2-1)+g(\xi,h)\}y$$

with approximate solutions treated in 23·08, which can be used in a similar way to the solutions of $y'' = h^2\xi y$ in the case of one zero.

---

* For these conditions and those imposed in 17·13, see H. Jeffreys, *Proc. Camb. Phil. Soc.* 49, 1953, 601–11.

† Especially *Bull. Amer. Math. Soc.* 1934, 545–82; *Trans. Amer. Math. Soc.* 67, 1949, 461–90.

‡ H. Jeffreys, *Proc. Camb. Phil. Soc.* 52, 1956, 61–6. § *Proc. Roy. Soc.* A, 86, 1912, 207–26.

‖ *Ann. d. Phys.* 47, 1915, 709–36.

17·132. As an example of the method, take Bessel's equation for large order

$$x\frac{d}{dx}\left(x\frac{dy}{dx}\right) + (x^2 - n^2)y = 0, \tag{1}$$

and put
$$x = ne^{-z}. \tag{2}$$

Then
$$\frac{d^2y}{dz^2} = n^2(1 - e^{-2z})y, \tag{3}$$

the oscillatory side being $x > n$ or $z < 0$. Then for $x < n$

$$M = n\int_0^z (1 - e^{-2z})^{1/2}\,dz = n(\theta - \tanh\theta), \tag{4}$$

where
$$x = n\,\mathrm{sech}\,\theta. \tag{5}$$

For $x > n$,
$$L = n\int_0^\zeta (e^{2\zeta} - 1)^{1/2}\,d\zeta = n(\tan u - u), \tag{6}$$

where $x = n\sec u$. Then there is a solution equal, for $x > n$, to

$$2\tan^{-1/2}u\,[\sin\{n(\tan u - u) + \tfrac{1}{4}\pi\} + O(1/n)]$$

and for $x < n$ to

$$\tanh^{-1/2}\theta\,\exp\{-n(\theta - \tanh\theta)\}\{1 + O(1/n)\}$$
$$= x^n\left(\frac{n^2 - x^2}{n^2}\right)^{-1/4}\{n + (n^2 - x^2)^{1/2}\}^{-n}\exp(n^2 - x^2)^{1/2}\{1 + O(1/n)\}. \tag{7}$$

When x is small this is approximately
$$x^n(2n)^{-n}e^n. \tag{8}$$

But the first term in the expansion of $J_n(x)$ is $\dfrac{1}{n!}(\tfrac{1}{2}x)^n$, and if we approximate to the factorial by Stirling's formula we see that our solution is a representation of $(2\pi n)^{1/2}J_n(x)$.

Another solution for $x > n$, is

$$2\tan^{-1/2}u\,[\cos\{n(\tan u - u) + \tfrac{1}{4}\pi\} + O(1/n)]$$

and for $x < n$

$$2\tanh^{-1/2}\theta\,\exp\{n(\theta - \tanh\theta)\}\{1 + O(1/n)\}$$
$$= 2x^{-n}\left(\frac{n^2 - x^2}{n^2}\right)^{-1/4}\{n + (n^2 - x^2)^{1/2}\}^n\exp\{-(n^2 - x^2)^{1/2}\}\{1 + O(1/n)\}, \tag{9}$$

which is a representation of the second solution of Bessel's equation denoted by $-(2\pi n)^{1/2}Y_n(x)$ (21·02).

The errors of these approximations are of order $1/n$. They have one great advantage over those in descending power series, that they can be used to fix the adjustable constants so that the solution will represent the same function as the ascending power series (the gap near the zero of χ_0 being filled in, if necessary, by direct use of the Airy integral). The corresponding adjustment for descending series usually requires the indirect method of complex integral representation, if such a representation exists; and if it does not, we may be reduced to numerical comparison in some range where both the convergent and the asymptotic expansions can be computed directly.

Applications of the method to Mathieu functions are given by H. Jeffreys, *Proc. Lond. Math. Soc.* (2), 23, 1924, 437–76; to the transparency of a potential barrier in wave mechanics by B. Jeffreys, *Proc. Camb. Phil. Soc.* 38, 1942, 401–5, and 52, 1956, 273–9. Solutions of the differential equation treated in **23·08a** can also be used in this problem.

Extensions to cases where χ vanishes to higher orders than the first are given by S. Goldstein.*

A method, based on a suitable change of the independent variable, applicable to non-linear ordinary and partial differential equations, is given by M. J. Lighthill.†

---

* *Proc. Lond. Math. Soc.* (2) 28), 1928, 81–90. † *Phil. Mag.* (7) 40, 1949, 1179–1201.

EXAMPLES

1. If $n = x-h$, where $0 < h < 1$, $x > 0$, prove that

$$\int_x^\infty t^{-n}e^{-t}dt = x^{-n}e^{-x}\int_0^\infty \left(1+\frac{v}{x}\right)^{-x+h} e^{-v}dv$$

$$\sim x^{-n}e^{-x}\left\{\tfrac{1}{2}+(\tfrac{1}{8}+\tfrac{1}{4}h)\frac{1}{x}+(-\tfrac{1}{32}+\tfrac{1}{16}h+\tfrac{1}{8}h^2)\frac{1}{x^2}+\ldots\right\},$$

and indicate the use of such an expansion in improving the approximation to the incomplete factorial function given by the asymptotic expansion of 17·01. (Bickley and Miller.)

2. A solution of the equation

$$y'' + 256ye^{4x} = 0$$

is zero at $x = 0$. Determine approximately where its other zeros lie.
If in addition $y' = 1$ at $x = 0$, find the position and magnitude of the first maximum for $x > 0$. (I.C. 1936.)

3. Show how to approximate to the solution of the equation

$$\frac{d^2y}{dx^2} - \frac{y}{X^4} = 0, \tag{1}$$

where X, a function of x, is such that X'' is small in comparison with $1/X^3$. Hence deduce the exact solution when $X'' = 0$.

Compare the solution with that of

$$\frac{d^2y}{dx^2} + \frac{y}{X^4} = 0. \tag{2}$$

(I.C. 1937.)

4. If

$$x\frac{d}{dx}\left(x\frac{dy}{dx}\right) + (n^2 - x^2)y = 0,$$

where n is large and $y \to 0$ as $x \to \infty$, prove that y is a multiple of a function given asymptotically by

$$\left(\frac{x^2}{n^2}-1\right)^{-1/4}\exp\left\{-n\left(\frac{x^2}{n^2}-1\right)^{1/2}+n\sec^{-1}\frac{x}{n}\right\}, \quad (x > n), \quad 2\left(1-\frac{x^2}{n^2}\right)^{-1/4}\sin n\{\log\tan(\tfrac{1}{4}\pi+\tfrac{1}{2}\psi)-\sin\psi+\tfrac{1}{4}\pi\},$$

$(x < n)$, where $\cos\psi = x/n$.

5. If

$$x\frac{d}{dx}\left(x\frac{dy}{dx}\right) + (n^2+x^2)y = 0,$$

where n is large, prove that

$$y \sim \left(1+\frac{x^2}{n^2}\right)^{-1/4}\genfrac{}{}{0pt}{}{\cos}{\sin}\, n(\sec v + \log\tan \tfrac{1}{2}v),$$

where $v = \tan^{-1}x/n$.

6. If x is large and positive, prove that

$$\int_0^x \mathrm{Ai}(x)\,dx \sim \frac{1}{3} - \frac{1}{2\sqrt{\pi}}x^{-3/4}e^{-\tfrac{2}{3}x^{3/2}},$$

and if $x = -\xi$, where ξ is large and positive, prove that

$$\int_0^x \mathrm{Ai}(x)\,dx \sim -\frac{2}{3} + \frac{1}{\sqrt{\pi}}\xi^{-3/4}\cos(\tfrac{2}{3}\xi^{3/2}+\tfrac{1}{4}\pi).$$

7. Prove that

$$\mathrm{Ai}(z)\mathrm{Bi}'(z) - \mathrm{Ai}'(z)\mathrm{Bi}(z) = \frac{1}{\pi}$$

verifying that the same constant is obtained by taking z small, z large and positive, or $-z$ large and positive.

8. If $\tfrac{1}{2}\pi > |\arg z| > \delta > 0$, prove that Stirling's formula can be extended to $(-z)!$.

(Use $z!(-z)! = \pi z \operatorname{cosec} \pi z$.)

Chapter 18

THE EQUATIONS OF POTENTIAL, WAVES, AND HEAT CONDUCTION

Divide et impera.
LOUIS XI.

18·01. The gravitational potential in free space satisfies Laplace's equation

$$\nabla^2 \phi = \frac{\partial^2 \phi}{\partial x^2} + \frac{\partial^2 \phi}{\partial y^2} + \frac{\partial^2 \phi}{\partial z^2} = 0. \tag{1}$$

The same equation is satisfied in free space by the electric and magnetic potentials if the field is steady, and by the velocity potential in incompressible fluid. In a uniform compressible fluid the velocity potential satisfies

$$\nabla^2 \phi = \frac{1}{c^2} \frac{\partial^2 \phi}{\partial t^2}, \tag{2}$$

where c is the velocity of sound, provided that the velocities are small. For elastic waves in a uniform solid the same equation is satisfied by the scalar and vector potentials, which give the longitudinal and transverse displacements respectively, with two different values of c. The same equation is satisfied by the field components in electromagnetic waves, c being then the velocity of light. In a uniform material at rest the temperature satisfies

$$\nabla^2 \phi = \frac{1}{h^2} \frac{\partial \phi}{\partial t}, \tag{3}$$

where h^2 is the thermometric conductivity, defined as the thermal conductivity divided by the heat capacity per unit volume. An equation of the same form is satisfied in diffusion if ϕ is the concentration of the diffusing material.

Evidently in a steady state these two equations both reduce to Laplace's equation. The equations of vibration of water in a lake or channel of uniform depth, and that of vibration of a membrane, are the wave equation without the term $\partial^2 \phi / \partial z^2$.

These three equations are so widely applicable that they are often called 'the differential equations of physics'. They do not include the wave equation of quantum mechanics, but even for this more complex equation the study of these equations is a necessary introduction.

The possibility of solving them depends chiefly on the fact that they are *separable* in several systems of coordinates; that is, they have solutions that break up into factors, each factor being a function of one coordinate or t. We try to choose the coordinate system so that one coordinate is constant over a surface where a given boundary condition has to be satisfied by ϕ. Taking, for instance, the wave equation in rectangular coordinates, we try a form of solution

$$\phi = XYZT, \tag{4}$$

where X is a function of x only, and so on. Substitute in (2) and divide by $XYZT$. Then

$$\frac{1}{X}\frac{d^2X}{dx^2}+\frac{1}{Y}\frac{d^2Y}{dy^2}+\frac{1}{Z}\frac{d^2Z}{dz^2}=\frac{1}{c^2T}\frac{d^2T}{dt^2}. \tag{5}$$

Each term is a function of only one of the independent variables. Hence if the equation is to hold for all values of x, y, z, t each term must be constant, and every expression of the form

$$A\exp\{i(lx+my+nz-\gamma t)\}, \tag{6}$$

where A, l, m, n, γ are constants and

$$l^2+m^2+n^2 = \gamma^2/c^2 \tag{7}$$

is a solution; so is any sum of expressions of this form. The complex exponentials can evidently be replaced by cosines and sines. Now we know from Fourier's theorem that in a bounded region the values of ϕ and $\partial\phi/\partial t$ at $t=0$ can in general be expressed in a series of products of sines and cosines of lx, my, nz; and then we derive the complete solution by associating with each term its proper factor in γt. As an example, consider a rectangular membrane whose corners are at $(0,0)$, $(a,0)$, $(0,b)$, (a,b). Since the margin is supposed to be fixed we require a solution that vanishes wherever $x=0$ or a, or $y=0$ or b. Thus the admissible terms will contain factors $\sin\frac{l\pi x}{a}\sin\frac{m\pi y}{b}$, where l and m are now integers; cosines will not vanish on the edges. The solution will then be

$$\phi = \sum_{l=1}^{\infty}\sum_{m=1}^{\infty}\sin\frac{l\pi x}{a}\sin\frac{m\pi y}{b}(A_{l,m}\cos\gamma t+B_{l,m}\sin\gamma t),$$

with
$$\frac{\gamma^2}{c^2} = \pi^2\left(\frac{l^2}{a^2}+\frac{m^2}{b^2}\right).$$

Then (assuming the possibility of differentiating the series)

$$\phi_{t=0} = \sum\sum A_{l,m}\sin\frac{l\pi x}{a}\sin\frac{m\pi y}{b},$$

$$\left(\frac{\partial\phi}{\partial t}\right)_{t=0} = \sum\sum \gamma B_{l,m}\sin\frac{l\pi x}{a}\sin\frac{m\pi y}{b}.$$

If we know the initial values of ϕ and $\partial\phi/\partial t$, we can expand them in double Fourier sine series, and comparison of coefficients determines $A_{l,m}$ and $B_{l,m}$. Thus the solution is found.

For the oscillations of water in a shallow rectangular lake of uniform depth we have the same differential equation to be satisfied by ζ, the vertical displacement of the free surface. The boundary conditions are different. It is now the velocity normal to the boundary that must vanish there, and this is proportional to $\partial\zeta/\partial n$. Hence at $x=0$ and $x=a$, $\partial\zeta/\partial x=0$; at $y=0$ and $y=b$, $\partial\zeta/\partial y=0$. The appropriate solutions satisfying the boundary conditions are now products of cosines instead of sines. Hence ζ and $\partial\zeta/\partial t$ for $t=0$ must be expanded in double Fourier cosine instead of sine series; the time factors are applied as before. Evidently if $A_{0,0}$ is not 0 it means that the variation of ζ is about a mean different from 0, that is, that the origin of ζ has not been taken at the undisturbed level of the free surface. The mean of $(\partial\phi/\partial t)_{t=0}$ over the rectangle must be 0, for if not it would imply that the total quantity of water was varying; hence $B_{0,0}=0$.

The condition that the initial ϕ and $\partial\phi/\partial t$ can be expanded in Fourier series (in one, two or three variables according as the region is one-, two- or three-dimensional) will be satisfied by most functions that occur in practice.

18·011. Equations of elliptic, parabolic, and hyperbolic types. We notice that even for wave propagation in one dimension we can get solutions of

$$\frac{\partial^2 \phi}{\partial x^2} = \frac{\partial^2 \phi}{c^2 \partial t^2}$$

that vanish for all time at $x = a, b$ and are not identically zero, and such solutions can also vanish for $a \leqslant x \leqslant b$ if $t = 0$ or $2(b-a)/c$. This property is to be contrasted with that of solutions of Laplace's equation in two dimensions

$$\frac{\partial^2 \phi}{\partial x^2} + \frac{\partial^2 \phi}{\partial y^2} = 0.$$

Here if $\phi = 0$ for all y at $x = 0$ and $x = a$, and for all x at $y = 0$ and $y = b$, ϕ must be 0 for all x, y within the rectangle. The difference can be traced to the fact that, if we take all terms to the same side of the equation, the signs are the same for Laplace's equation and opposite for the wave equation. More generally, if the terms in second derivatives in a differential equation are $a\dfrac{\partial^2 \phi}{\partial x^2} + 2h\dfrac{\partial^2 \phi}{\partial x \partial y} + b\dfrac{\partial^2 \phi}{\partial y^2}$, the conditions that the solution can be made to satisfy at the boundaries are quite different according as $ab - h^2$ is positive or negative. In the former case if ϕ is given on a closed curve it is determined everywhere within it, and the equation is said to be of *elliptic type*. In the latter the equation is said to be of *hyperbolic type* and no such conclusion follows. We shall not enter into the general theory, which is given in full by Webster.* a, b, h need not be constants, but $ab - h^2$ may then change sign within the region. This condition occurs in the motion of projectiles at velocities above that of sound and is then connected with the formation of a shock wave.†

18·012. The equation of heat conduction in one dimension is intermediate in character ($ab - h^2 = 0$) and is said to be of *parabolic type*. But more generally, since it is of the first order with regard to t, we cannot assign the values of ϕ and $\partial\phi/\partial t$ at $t = 0$ independently. The extension to more dimensions is straightforward and we shall only illustrate from the one-dimensional case. If

$$\frac{\partial \phi}{\partial t} = h^2 \frac{\partial^2 \phi}{\partial x^2}$$

and $\phi = 0$ at $x = 0$ and $x = a$, a solution is

$$\phi = \sin\frac{n\pi x}{a} \exp\left(-\frac{h^2 n^2 \pi^2 t}{a^2}\right).$$

Hence if ϕ at $t = 0$ is expanded in a Fourier series we get the solution

$$\phi = \sum_{1}^{\infty} A_n \sin\frac{n\pi x}{a} \exp\left(-\frac{h^2 n^2 \pi^2 t}{a^2}\right).$$

---

* *Partial Differential Equations of Mathematical Physics*, Chap. 6.
† G. I. Taylor and J. W. Maccoll, *Proc. Roy. Soc.* A, 139, 1933, 278–311.

The characteristic feature is that all the exponents are negative; if the temperature is kept zero at two points it will tend to zero everywhere as the time increases. If the series converges for $t = 0, \phi$ and all its derivatives of whatever order with regard to both x and t exist and will be represented by convergent series for $t > 0$, however small t may be. The universal tendency of heat conduction is to smooth out differences of temperature. This, of course, is another way of stating the second law of thermodynamics, but the equation of heat conduction provides a time scale for the smoothing and the second law does not. When ϕ satisfies the wave equation we can say that when t varies the inequalities of ϕ are not reduced but merely transferred to other places.

It is possible to have periodic solutions of the heat equation, but only if there is a periodic source of heat, either internal or at the boundary.

Another peculiarity of the equation of conduction is that there is a severe restriction on the distributions of temperature that can be the successors of any previous distributions. We have seen that Fourier series in practice usually converge like Σn^{-s}, where s is a small integer. In a favourable case like the series of 14·05, it converges like $\Sigma (r/a)^n$, with $r < a$. Now suppose that ϕ was expressible, at time $-\tau$ ($\tau > 0$) by a Fourier series, however slowly convergent. Then the terms at time 0 would decrease like $\exp(-h^2 n^2 \pi^2 \tau/a^2)$, which is a faster decrease than for any geometric series. If, then, the Fourier series of ϕ at $t = 0$ does not satisfy this condition for some $\tau > 0$ it cannot be the successor of any previous distribution unless there has been disturbance from outside.

18·013. Whittaker's general solutions. The test of whether we have obtained the most general solution of a partial differential equation is not a mere matter of counting adjustable constants, as it is for ordinary differential equations. We shall only quote types of general solution that have been given by Whittaker. A general solution of Laplace's equation is

$$\phi = \int_{-\pi}^{\pi} f(z + ix \cos u + iy \sin u, u) \, du,$$

where $f(\zeta, u)$ is an arbitrary function; a general solution of the wave equation is

$$\phi = \int_{-\pi}^{\pi} \int_{-\pi}^{\pi} f(x \sin u \cos v + y \sin u \sin v + z \cos u + ct, u, v) \, du \, dv.$$

These solutions are capable of a wide range of transformation; numerous applications are given by Whittaker and Watson.

18·02. Curvilinear coordinates. For other forms of boundary solutions of the form 18·01(4) are seldom obvious, though Whittaker's general solutions can often be adapted to other forms of boundary on transformation of coordinates. We can, alternatively, write

$$\phi = f(x, y, z) \, T, \tag{1}$$

and then

$$\frac{1}{f} \nabla^2 f = \left(0, \; \frac{1}{c^2 T} \frac{\partial^2 T}{\partial t^2}, \; \frac{h^2}{T} \frac{\partial T}{\partial t}\right), \tag{2}$$

according as the equation considered is Laplace's equation, the wave equation, or the equation of heat conduction. Both sides must be constant, and the equations are all reduced to the form

$$\nabla^2 f = -\kappa^2 f, \tag{3}$$

to be solved in accordance with the boundary conditions. Now the wave equation in a continuous medium can be regarded as the limit of the set of equations of motion for a set of particles forming a stable close lattice. Then if we take $T = e^{\gamma t}$ all possible values of γ^2 for normal modes are real and negative. It follows at once that with the same boundary conditions on $f(x,y,z)$ all values of γ for free flow of heat are real and negative. The determination of the solutions for given initial conditions therefore reduces to finding an expansion of a general f in terms of the characteristic functions of the equation (3). The time factor is thus treated separately in any case and we have simply to consider (3), where κ^2 will be 0 in potential problems. The time factor will in general not be written explicitly since it can always be restored at the end. This equation is separable in several other coordinate systems besides Cartesian ones. In general curvilinear orthogonal coordinates ξ_1, ξ_2, ξ_3 are used, the elements of length ds_1, ds_2, ds_3 corresponding to small changes $d\xi_1, d\xi_2, d\xi_3$ being $h_1 d\xi_1, h_2 d\xi_2, h_3 d\xi_3$.

Now by Green's lemma
$$\iiint \nabla^2 \phi \, d\tau = \iint \frac{\partial \phi}{\partial n} dS. \tag{4}$$

Apply this result to a small volume bounded by the surfaces
$$\xi_1 = \xi_{10} \pm \tfrac{1}{2}\delta\xi_1, \quad \xi_2 = \xi_{20} \pm \tfrac{1}{2}\delta\xi_2, \quad \xi_3 = \xi_{30} \pm \tfrac{1}{2}\delta\xi_3.$$

On a surface of constant ξ_1, if n is in the direction of increasing ξ_1,
$$\frac{\partial \phi}{\partial n} = \frac{\partial \phi}{h_1 \partial \xi_1}, \tag{5}$$

and the area of the element given by ranges $\delta\xi_2, \delta\xi_3$ is $h_2 h_3 \delta\xi_2 \delta\xi_3$. Then $\iint \frac{\partial \phi}{\partial n} dS$ over such an element is $\iint \frac{h_2 h_3}{h_1} \frac{\partial \phi}{\partial \xi_1} d\xi_2 d\xi_3$. The two surfaces given by $\xi_1 = \xi_{10} \pm \tfrac{1}{2}\delta\xi_1$ will together contribute

$$\frac{\partial}{\partial \xi_1} \left(\frac{h_2 h_3}{h_1} \frac{\partial \phi}{\partial \xi_1} \right) \delta\xi_1 \delta\xi_2 \delta\xi_3 + o(\delta\xi_1 \delta\xi_2 \delta\xi_3),$$

since on the surface with the larger ξ_1 the outward normal is in the direction of increasing ξ_1 and on the opposite one it is in the direction of decreasing ξ_1. Then the integral on the right of (4) is the sum of these expressions for the three pairs of opposite faces. The element of volume is $h_1 h_2 h_3 \delta\xi_1 \delta\xi_2 \delta\xi_3 + o(\delta\xi_1 \delta\xi_2 \delta\xi_3)$. Hence

$$h_1 h_2 h_3 \nabla^2 \phi = \frac{\partial}{\partial \xi_1}\left(\frac{h_2 h_3}{h_1}\frac{\partial \phi}{\partial \xi_1}\right) + \frac{\partial}{\partial \xi_2}\left(\frac{h_3 h_1}{h_2}\frac{\partial \phi}{\partial \xi_2}\right) + \frac{\partial}{\partial \xi_3}\left(\frac{h_1 h_2}{h_3}\frac{\partial \phi}{\partial \xi_3}\right) \tag{6}$$

almost everywhere, and everywhere if both sides are continuous. The latter condition will usually be obviously satisfied by the solutions we shall obtain, and the equation to be solved is [a]

$$\frac{1}{h_1 h_2 h_3}\left\{\frac{\partial}{\partial \xi_1}\left(\frac{h_2 h_3}{h_1}\frac{\partial \phi}{\partial \xi_1}\right) + \frac{\partial}{\partial \xi_2}\left(\frac{h_3 h_1}{h_2}\frac{\partial \phi}{\partial \xi_2}\right) + \frac{\partial}{\partial \xi_3}\left(\frac{h_1 h_2}{h_3}\frac{\partial \phi}{\partial \xi_3}\right)\right\} = -\kappa^2 \phi. \tag{7}$$

The transformation is particularly simple if the coordinate ξ_3 is z. This choice of ξ_3 is convenient if the boundary is a cylinder of any form of section. Then $h_3 = 1$, and ξ_1, ξ_2 are functions of x, y only. Now if
$$\xi_1 + i\xi_2 = f(x+iy), \tag{8}$$

the orthogonality relations are satisfied on account of the Cauchy-Riemann relations, and

$$h_1^2 = h_2^2 = \left|\frac{d(x+iy)}{d(\xi_1+i\xi_2)}\right|^2 = h^2, \text{ say.} \tag{9}$$

Then
$$\frac{1}{h^2}\left(\frac{\partial^2 \phi}{\partial \xi_1^2}+\frac{\partial^2 \phi}{\partial \xi_2^2}+\frac{\partial}{\partial z}h^2\frac{\partial \phi}{\partial z}\right) = -\kappa^2\phi, \tag{10}$$

or
$$\frac{\partial^2 \phi}{\partial \xi_1^2}+\frac{\partial^2 \phi}{\partial \xi_2^2}+h^2\frac{\partial^2 \phi}{\partial z^2} = -\kappa^2 h^2 \phi. \tag{11}$$

In particular if ϕ is independent of z and t the equation reduces to Laplace's equation with regard to ξ_1 and ξ_2.

18·03. Cylindrical coordinates. If

$$\varpi^2 = x^2+y^2, \quad \lambda = \tan^{-1}y/x, \tag{1}$$

$$\log \varpi + i\lambda = \log(x+iy), \tag{2}$$

and we can take

$$\xi_1 = \log \varpi, \quad \xi_2 = \lambda, \quad x+iy = e^{\xi_1+i\xi_2}, \quad h^2 = |x+iy|^2 = \varpi^2. \tag{3}$$

Then
$$\frac{\partial}{\partial \log \varpi}\frac{\partial \phi}{\partial \log \varpi}+\frac{\partial^2 \phi}{\partial \lambda^2} = -\varpi^2\left(\frac{\partial^2 \phi}{\partial z^2}+\kappa^2\phi\right), \tag{4}$$

that is,
$$\varpi\frac{\partial}{\partial \varpi}\left(\varpi\frac{\partial \phi}{\partial \varpi}\right)+\frac{\partial^2 \phi}{\partial \lambda^2}+\varpi^2\left(\frac{\partial^2 \phi}{\partial z^2}+\kappa^2\phi\right) = 0. \tag{5}$$

If
$$\phi = P\Lambda Z, \tag{6}$$

where P, Λ, Z are functions of ϖ, λ, z respectively,

$$\frac{1}{P}\varpi\frac{d}{d\varpi}\left(\varpi\frac{dP}{d\varpi}\right)+\frac{1}{\Lambda}\frac{d^2\Lambda}{d\lambda^2}+\varpi^2\left(\frac{1}{Z}\frac{d^2Z}{dz^2}+\kappa^2\right) = 0. \tag{7}$$

The second term is a function of λ only, the others do not involve λ; take it as $-n^2$. Then

$$\frac{1}{\varpi P}\frac{d}{d\varpi}\left(\varpi\frac{dP}{d\varpi}\right)+\frac{1}{Z}\frac{d^2Z}{dz^2}+\kappa^2-\frac{n^2}{\varpi^2} = 0. \tag{8}$$

The second term involves z only, the others are independent of z. Take it to be $-\mu^2$. Then

$$\varpi\frac{d}{d\varpi}\left(\varpi\frac{dP}{d\varpi}\right)+\{(\kappa^2-\mu^2)\varpi^2-n^2\}P = 0. \tag{9}$$

Finally, put $(\kappa^2-\mu^2)^{1/2}\varpi = \xi$; then

$$\xi\frac{d}{d\xi}\left(\xi\frac{dP}{d\xi}\right)+(\xi^2-n^2)P = 0, \tag{10}$$

which is *Bessel's equation*.

This transformation has a singularity at the origin; that is, λ is not a single-valued function of x and y if we are permitted to make a complete circuit about the origin. But ϕ must be single-valued, and therefore Λ must be. This can be ensured if n is an integer, for $\cos n\lambda$ and $\sin n\lambda$ are then single-valued functions of x and y, but not otherwise. If the solution is to hold within a complete circle n must therefore be an integer, which without

loss of generality can be taken positive. But if it is to hold only within a sector $\alpha \leqslant \lambda \leqslant \beta$ there will in general be two boundary conditions at $\lambda = \alpha$ and $\lambda = \beta$. Suppose for instance that ϕ is to vanish at all points of the boundary. Then we are limited to solutions of the form
$$\Lambda = \sin n(\lambda - \alpha),$$
where n is restricted to satisfy
$$\sin n(\beta - \alpha) = 0,$$
and again only a discrete set of values of n are possible. It is a general feature of transformations with a singularity within the region considered that the single-valuedness of the solution introduces what is virtually an extra boundary condition.

Similarly if ϕ is required to be finite at the origin we are limited to the solutions $J_n(\xi)$; for the other solutions $Y_n(\xi)$ of Bessel's equation are infinite at the origin. If further ϕ is to vanish over the circle $\varpi = a$ we must have $J_n\{(\kappa^2 - \mu^2)^{1/2} a\} = 0$. J_n vanishes for an infinite number of values of the argument, and the boundary conditions determine a set of possible values of $\kappa^2 - \mu^2$. Further, if the solution is to hold within a cylinder of finite length there will be a restriction on the possible values of μ. If the solution is to hold only between two circles the other solution of Bessel's equation will also be required if the boundary conditions are to be satisfied, and the new boundary condition will determine the ratio of the coefficients.

18·04. Parabolic cylinder coordinates. Take
$$\xi_1 + i\xi_2 = (x+iy)^{1/2}, \quad x = \xi_1^2 - \xi_2^2, \quad y = 2\xi_1\xi_2. \tag{1}$$
If ξ_1 is constant, we can eliminate ξ_2 and get
$$x = \xi_1^2 - \frac{y^2}{4\xi_1^2}, \tag{2}$$
a set of parabolas with a common focus at the origin and proceeding towards negative x. If ξ_2 is constant we get similarly
$$x = \frac{y^2}{4\xi_2^2} - \xi_2^2, \tag{3}$$
a set of confocal parabolas with their axes towards $x = +\infty$. Then
$$h^2 = 4(\xi_1^2 + \xi_2^2). \tag{4}$$
If also
$$\frac{\partial^2 \phi}{\partial z^2} = -\mu^2 \phi, \tag{5}$$
there will be solutions of the form
$$\phi = X_1 X_2 Z, \tag{6}$$
if
$$\frac{d^2 X_1}{d\xi_1^2} + \{4(\kappa^2 - \mu^2)\xi_1^2 - \alpha\} X_1 = 0, \tag{7}$$
$$\frac{d^2 X_2}{d\xi_2^2} + \{4(\kappa^2 - \mu^2)\xi_2^2 + \alpha\} X_2 = 0, \tag{8}$$
where α is a constant. The substitution $\xi_1 = i\eta$ turns the first equation into the second, so that solution of problems relating to parabolic boundaries requires the solution of the same differential equation for purely real and purely imaginary argument. If $\kappa^2 > \mu^2$ solutions will be oscillating for both ξ_1 and ξ_2 if they are large enough.

This equation arises in the theory of tidal oscillations in parabolic bays. It is also the equation of the harmonic oscillator in wave mechanics. We can provide that ξ_1 and ξ_2 shall be single-valued by never crossing the axis of the parabolas on one side of the origin.

18·05. Elliptic cylinder coordinates. Take

$$x+iy = c\cosh(\xi+i\eta), \quad x = c\cosh\xi\cos\eta, \quad y = c\sinh\xi\sin\eta. \tag{1}$$

The curves of ξ constant are the ellipses

$$\frac{x^2}{c^2\cosh^2\xi} + \frac{y^2}{c^2\sinh^2\xi} = 1, \tag{2}$$

and those of η constant are the confocal hyperbolas

$$\frac{x^2}{c^2\cos^2\eta} - \frac{y^2}{c^2\sin^2\eta} = 1. \tag{3}$$

If we take ξ always ≥ 0 we describe a complete confocal ellipse by increasing η from 0 to 2π. We have also

$$h^2 = |c\sinh(\xi+i\eta)|^2 = c^2(\sinh^2\xi\cos^2\eta + \cosh^2\xi\sin^2\eta)$$
$$= \tfrac{1}{2}c^2(\cosh 2\xi - \cos 2\eta), \tag{4}$$

and if again

$$\frac{1}{\phi}\frac{\partial^2\phi}{\partial z^2} = -\mu^2, \tag{5}$$

$$\frac{\partial^2\phi}{\partial\xi^2} + \frac{\partial^2\phi}{\partial\eta^2} + \tfrac{1}{2}c^2(\cosh 2\xi - \cos 2\eta)(\kappa^2-\mu^2)\phi = 0. \tag{6}$$

The standard solutions can therefore be taken as

$$V = XYZ, \tag{7}$$

where X and Y satisfy

$$\frac{d^2X}{d\xi^2} - \{R - \tfrac{1}{2}c^2(\kappa^2-\mu^2)\cosh 2\xi\}X = 0, \tag{8}$$

$$\frac{d^2Y}{d\eta^2} + \{R - \tfrac{1}{2}c^2(\kappa^2-\mu^2)\cos 2\eta\}Y = 0. \tag{9}$$

We write

$$\tfrac{1}{2}c^2(\kappa^2-\mu^2) = 16q. \tag{10}$$

The second of these is *Mathieu's equation*. R is a constant, to be determined in such a way that the solution has period 2π. Evidently if $q = 0$, R must be the square of an integer. Since the coefficient of Y is an even function, there will be one even and one odd solution for any pair of values of R and q. It has been shown by Ince and others that not more than one of these can be periodic except for $q = 0$, and the datum that one is periodic determines a discrete set of values of R. The even solutions are denoted by $ce_n(\eta, q)$, the odd ones by $se_n(\eta, q)$. Changing ξ to $i\theta$ in (8) reproduces (9).

Since we always take ξ positive the only way of comparing values on opposite sides of the line joining the foci is to change η to $2\pi - \eta$. $\partial\phi/\partial x$ and $\partial\phi/\partial y$ are to be continuous across this line. But on $\xi = 0$

$$Y\frac{dX}{d\xi} = \frac{\partial\phi}{\partial\xi} = \frac{\partial\phi}{\partial x}\frac{\partial x}{\partial\xi} + \frac{\partial\phi}{\partial y}\frac{\partial y}{\partial\xi} = \frac{\partial\phi}{\partial x}c\sinh\xi\cos\eta + \frac{\partial\phi}{\partial y}c\cosh\xi\sin\eta = c\sin\eta\frac{\partial\phi}{\partial y}, \tag{11}$$

$$X\frac{dY}{d\eta} = \frac{\partial\phi}{\partial\eta} = -\frac{\partial\phi}{\partial x}c\cosh\xi\sin\eta + \frac{\partial\phi}{\partial y}c\sinh\xi\cos\eta = -c\sin\eta\frac{\partial\phi}{\partial x}. \tag{12}$$

Hence at $\xi = 0$, $\dfrac{1}{\sin\eta} Y \dfrac{dX}{d\xi}$ and $\dfrac{1}{\sin\eta} X \dfrac{dY}{d\eta}$ are unaltered on changing η to $2\pi - \eta$. If Y is an even function of η, $Y/\sin\eta$ changes sign, and the conditions require that $dX/d\xi = 0$. Therefore X is an even function of ξ. Secondly, if Y is $\mathrm{se}_n(\eta, q)$, an odd function, $dY/d\eta$ does not change sign and therefore $X = 0$. Hence X is an odd function. The possible types of solution (in real form) are therefore

$$XY = \mathrm{ce}_n(i\xi, q)\,\mathrm{ce}_n(\eta, q), \quad XY = -i\,\mathrm{se}_n(i\xi, q)\,\mathrm{se}_n(\eta, q). \tag{13}$$

Mathieu's equation has mathematical interest because it is the simplest second-order equation with a periodic coefficient. The restriction that the solutions also must have period 2π is required by the physical conditions in problems of membranes and tidal oscillations. There are, however, many problems of vibration where the restoring force contains a periodic coefficient, and these can be treated by similar methods to those used for the solution of Mathieu's equation.[a]

18·06. Spherical and spheroidal coordinates. The special feature here is that one coordinate is constant over the surface, respectively, of a sphere or an ellipsoid of revolution. In either case there is an axis of symmetry (for the sphere, of course, an infinite number) and one coordinate $\xi_3 = \lambda$ can be taken to be the azimuth about it and the other two to be orthogonal coordinates in planes of λ constant. If we also take cylindrical coordinates ϖ, λ, z we can simplify the analysis by making use of the fact that

$$\varpi^s(\cos s\lambda + i\sin s\lambda) = (x+iy)^s \tag{1}$$

gives a pair of solutions of Laplace's equation. In spherical coordinates, then, we know at once that there is a family of solutions $r^s \sin^s\theta(\cos s\lambda, \sin s\lambda)$. Now if we denote one of these solutions by M we are led to look for other solutions containing it as a factor, say FM, where F is independent of λ. Then

$$\nabla^2(FM) = M\nabla^2 F + 2\dfrac{\partial F}{\partial x_i}\dfrac{\partial M}{\partial x_i} + F\nabla^2 M, \tag{2}$$

and the last term is 0. Also the second term is unaltered by rotation of the axes; therefore

$$\dfrac{\partial F}{\partial x_i}\dfrac{\partial M}{\partial x_i} = \dfrac{\partial F}{\partial s_1}\dfrac{\partial M}{\partial s_1} + \dfrac{\partial F}{\partial s_2}\dfrac{\partial M}{\partial s_2} + \dfrac{\partial F}{\partial s_3}\dfrac{\partial M}{\partial s_3}. \tag{3}$$

But $ds_3 = \varpi\,d\lambda$, and $\partial F/\partial s_3 = 0$. Hence

$$\nabla^2(FM) = M\nabla^2 F + 2\left(\dfrac{\partial F}{\partial s_1}\dfrac{\partial M}{\partial s_1} + \dfrac{\partial F}{\partial s_2}\dfrac{\partial M}{\partial s_2}\right). \tag{4}$$

18·061. Spherical polar coordinates.

$$ds_1 = dr, \quad ds_2 = r\,d\theta, \quad ds_3 = r\sin\theta\,d\lambda, \tag{5}$$

and

$$\nabla^2\phi = \dfrac{1}{r^2\sin\theta}\left\{\dfrac{\partial}{\partial r}\left(r^2\sin\theta\dfrac{\partial\phi}{\partial r}\right) + \dfrac{\partial}{\partial\theta}\left(\sin\theta\dfrac{\partial\phi}{\partial\theta}\right) + \dfrac{\partial}{\partial\lambda}\left(\dfrac{\partial\phi}{\sin\theta\,\partial\lambda}\right)\right\}$$

$$= \dfrac{\partial}{r^2\partial r}\left(r^2\dfrac{\partial\phi}{\partial r}\right) + \dfrac{1}{r^2\sin\theta}\dfrac{\partial}{\partial\theta}\left(\sin\theta\dfrac{\partial\phi}{\partial\theta}\right) + \dfrac{\partial^2\phi}{r^2\sin^2\theta\,\partial\lambda^2}. \tag{6}$$

Since Laplace's equation is homogeneous in r we try

$$\phi = FM, \quad F = r^{n-s}\Theta, \tag{7}$$

where Θ is a function of θ only. Then

$$\nabla^2 F = (n-s)(n-s+1)r^{n-s-2}\Theta + r^{n-s-2}\frac{\partial}{\sin\theta\,\partial\theta}\left(\sin\theta\frac{\partial\Theta}{\partial\theta}\right). \tag{8}$$

Also
$$\frac{\partial F}{\partial r}\frac{\partial M}{\partial r}+\frac{\partial F}{r\partial\theta}\frac{\partial M}{r\partial\theta}=r^{n-s-2}\left\{s(n-s)\Theta M+sM\cot\theta\frac{d\Theta}{d\theta}\right\}. \tag{9}$$

Hence in spherical polar coordinates

$$\nabla^2(FM) = Mr^{n-s-2}\left[\frac{d}{\sin\theta\,d\theta}\left(\sin\theta\frac{d\Theta}{d\theta}\right)+2s\cot\theta\frac{d\Theta}{d\theta}+(n-s)(n+s+1)\Theta\right]. \tag{10}$$

If $\nabla^2\phi = 0$ this gives a differential equation satisfied by Θ; we put $\cos\theta = \mu$, and it becomes

$$\frac{d}{d\mu}\left\{(1-\mu^2)\frac{d\Theta}{d\mu}\right\}-2s\mu\frac{d\Theta}{d\mu}+(n-s)(n+s+1)\Theta = 0. \tag{11}$$

When $s = 0$ this is *Legendre's equation*.

We denote the solutions by Θ_1, Θ_2. The solutions of Laplace's equation in spherical polar coordinates are therefore of the form

$$\varpi^s(\cos s\lambda, \sin s\lambda)r^{n-s}(\Theta_1,\Theta_2) = r^n\sin^s\theta(\Theta_1,\Theta_2)(\cos s\lambda, \sin s\lambda). \tag{12}$$

We have not assumed n positive, and we see that (11) is unaltered if we replace n by $-n-1$; hence there will be another set of solutions with r^{-n-1} instead of r^n, the other factors remaining the same. If we express Θ in a power series in μ there are two solutions, one an even and the other an odd function of μ. If n and s are integers and $n-s$ is even, we shall see that the even series terminates; if $n-s$ is odd the odd series terminates. For the terminating series, with an appropriate constant factor, we write

$$\sin^s\theta . \Theta_1 = p_n^s(\mu), \tag{13}$$

and for the other
$$\sin^s\theta . \Theta_2 = q_n^s(\mu). \tag{14}$$

There is an exceptional case when $s = 0$; for if

$$\frac{\partial^2 \Lambda}{\partial\lambda^2} = 0.\Lambda$$

the general solution is $\quad\Lambda = A + B\lambda. \tag{15}$

But if ϕ is to return to its original value when λ is increased by 2π, B must be 0. Hence for $s = 0$ the λ factor is a constant. (11) can be written

Differentiate this: $\quad(1-\mu^2)\Theta'' - 2(s+1)\mu\Theta' + (n-s)(n+s+1)\Theta = 0. \tag{16}$

$$(1-\mu^2)\Theta''' - 2(s+2)\mu\Theta'' + (n-s-1)(n+s+2)\Theta' = 0, \tag{17}$$

which is the same equation with Θ' for Θ and $s+1$ for s. This differentiation does not assume s positive. Hence if G satisfies the equation

$$(1-\mu^2)G'' + 2n\mu G' = 0 \tag{18}$$

obtained by putting $s+1 = -n$ in (16),

$$\Theta = \frac{d^{n+s+1}}{d\mu^{n+s+1}}G \tag{19}$$

will satisfy (16). But (18) gives $\quad G' \propto (\mu^2-1)^n. \tag{20}$

Hence
$$\Theta \propto \frac{d^{n+s}}{d\mu^{n+s}}(\mu^2-1)^n, \tag{21}$$

and a family of solutions of Laplace's equation in spherical polar coordinates is
$$r^{n-s}(x+iy)^s \frac{d^{n+s}}{d\mu^{n+s}}(\mu^2-1)^n, \tag{22}$$

that is,
$$r^n \sin^s\theta(\cos s\phi, \sin s\phi) \frac{d^{n+s}}{d\mu^{n+s}}(\mu^2-1)^n. \tag{23}$$

We have obtained only one solution of (16). The nature of the second may be seen by taking $s = n$; the solution that we have found will be constant. The other is
$$\Theta' \propto (\mu^2-1)^{-(n+1)}, \tag{24}$$

and its $(n-s+1)$th integral, with $s \leqslant n$, will be infinite of order $\sin^{-2s}\theta$ at $\mu = \pm 1$ and contain a logarithm. Hence this solution is inadmissible for a complete sphere. It has other applications, however, and will be considered further under *Legendre functions*.

Postponing the choice of the constant factors for $s \neq 0$, we denote the solutions by $r^n p_n^s(\mu) (\cos s\lambda, \sin s\lambda)$, $r^n q_n^s(\mu) (\cos s\lambda, \sin s\lambda)$. For $s = 0$ the constant is chosen so that $p_n^0(1) = 1$.

We now return to
$$\nabla^2\phi = -\kappa^2\phi, \tag{25}$$

and put
$$\phi = RS_n(\theta,\lambda), \tag{26}$$

where $r^n S_n(\theta,\lambda)$ is a solution of $\nabla^2\phi = 0$. Then
$$\frac{1}{r^2}\frac{d}{dr}\left(r^2\frac{dR}{dr}\right) - n(n+1)\frac{R}{r^2} = -\kappa^2 R. \tag{27}$$

Put
$$R = r^{-1/2}K. \tag{28}$$

Then
$$r^2\frac{d^2K}{dr^2} + r\frac{dK}{dr} + \{\kappa^2 r^2 - (n+\tfrac{1}{2})^2\}K = 0, \tag{29}$$

and
$$K = AJ_{n+1/2}(\kappa r) + BY_{n+1/2}(\kappa r). \tag{30}$$

The required solutions are therefore
$$\phi = r^{-1/2}\{J_{n+1/2}(\kappa r), Y_{n+1/2}(\kappa r)\}\{p_n^s(\cos\theta), q_n^s(\cos\theta)\}(\cos s\lambda, \sin s\lambda). \tag{31}$$

The Bessel functions of order half an odd integer are expressible in finite terms.

18·062. Oblate spheroidal coordinates. We replace the coordinates ϖ, z by ξ, η according to
$$\varpi = c\cosh\xi\cos\eta, \quad z = c\sinh\xi\sin\eta, \tag{1}$$

and again seek for solutions of the form
$$\phi = FM = F(\xi,\eta)\varpi^s(\cos s\lambda, \sin s\lambda). \tag{2}$$

Then
$$ds_1 = h\,d\xi, \quad ds_2 = h\,d\eta, \quad ds_3 = \varpi\,d\lambda, \tag{3}$$

where
$$h^2 = c^2(\cosh^2\xi - \cos^2\eta). \tag{4}$$

Then
$$\nabla^2\phi = \frac{1}{h^2\varpi}\left\{\frac{\partial}{\partial\xi}\left(\varpi\frac{\partial\phi}{\partial\xi}\right)+\frac{\partial}{\partial\eta}\left(\varpi\frac{\partial\phi}{\partial\eta}\right)+\frac{\partial}{\partial\lambda}\left(\frac{h^2}{\varpi}\frac{\partial\phi}{\partial\lambda}\right)\right\}, \tag{5}$$

$$\nabla^2 F = \frac{1}{h^2}\left\{\frac{\partial^2 F}{\partial\xi^2}+\frac{\partial^2 F}{\partial\eta^2}+\frac{1}{\varpi}\frac{\partial\varpi}{\partial\xi}\frac{\partial F}{\partial\xi}+\frac{1}{\varpi}\frac{\partial\varpi}{\partial\eta}\frac{\partial F}{\partial\eta}\right\}, \tag{6}$$

$$= \frac{1}{h^2}\left\{\frac{\partial^2 F}{\partial\xi^2}+\frac{\partial^2 F}{\partial\eta^2}+\tanh\xi\frac{\partial F}{\partial\xi}-\tan\eta\frac{\partial F}{\partial\eta}\right\}, \tag{7}$$

$$\frac{\partial F}{\partial s_1}\frac{\partial M}{\partial s_1}+\frac{\partial F}{\partial s_2}\frac{\partial M}{\partial s_2} = \frac{sM}{h^2}\left(\tanh\xi\frac{\partial F}{\partial\xi}-\tan\eta\frac{\partial F}{\partial\eta}\right), \tag{8}$$

$$\nabla^2\phi = \frac{M}{h^2}\left\{\frac{\partial^2 F}{\partial\xi^2}+\frac{\partial^2 F}{\partial\eta^2}+(2s+1)\tanh\xi\frac{\partial F}{\partial\xi}-(2s+1)\tan\eta\frac{\partial F}{\partial\eta}\right\}. \tag{9}$$

And if
$$F = X(\xi)\,Y(\eta), \quad \nabla^2\phi = -\kappa^2\phi, \tag{10}$$

$$\frac{1}{X}\left\{\frac{d^2X}{d\xi^2}+(2s+1)\tanh\xi\frac{dX}{d\xi}\right\}+\frac{1}{Y}\left\{\frac{d^2Y}{d\eta^2}-(2s+1)\tan\eta\frac{dY}{d\eta}\right\} = -\kappa^2 c^2(\cosh^2\xi-\cos^2\eta), \tag{11}$$

whence
$$\frac{d^2X}{d\xi^2}+(2s+1)\tanh\xi\frac{dX}{d\xi}+(\kappa^2 c^2\cosh^2\xi - R)\,X = 0, \tag{12}$$

$$\frac{d^2Y}{d\eta^2}-(2s+1)\tan\eta\frac{dY}{d\eta}-(\kappa^2 c^2\cos^2\eta - R)\,Y = 0, \tag{13}$$

where R is constant.

In the second of these put $\sin\eta = \mu$; we get

$$(1-\mu^2)\frac{d^2Y}{d\mu^2}-2(s+1)\mu\frac{dY}{d\mu}+\{R-\kappa^2 c^2(1-\mu^2)\}\,Y = 0, \tag{14}$$

which, apart from the term in κ^2, is the equation satisfied by the function Θ for spherical polar coordinates, with $R = (n-s)(n+s+1)$. In the first, similarly, put $i\sinh\xi = \nu$; then

$$(1-\nu^2)\frac{d^2X}{d\nu^2}-2(s+1)\nu\frac{dX}{d\nu}+\{R-\kappa^2 c^2(1-\nu^2)\}\,X = 0, \tag{15}$$

which is the same equation. Hence for $\kappa^2 = 0$

$$X\cosh^s\xi = A p_n^s(i\sinh\xi)+B q_n^s(i\sinh\xi), \tag{16}$$

$$Y\cos^s\eta = C p_n^s(\sin\eta)+D q_n^s(\sin\eta). \tag{17}$$

These give the solutions of Laplace's equation. The presence of the terms in κ^2 considerably increases the complexity of the solutions of the other equations.

18·063. Prolate spheroidal coordinates. Take

$$\varpi = c\sinh\xi\sin\eta, \quad z = c\cosh\xi\cos\eta, \tag{1}$$

and proceed as before. We get

$$\frac{d^2X}{d\xi^2}+(2s+1)\coth\xi\frac{dX}{d\xi}+(\kappa^2 c^2\sinh^2\xi - R)\,X = 0, \tag{2}$$

$$\frac{d^2Y}{d\eta^2}+(2s+1)\cot\eta\frac{dY}{d\eta}+(\kappa^2 c^2\sin^2\eta + R)\,Y = 0. \tag{3}$$

In the second of these put $\cos\eta = \mu$; then

$$(1-\mu^2)\frac{d^2Y}{d\mu^2} - 2(s+1)\mu\frac{dY}{d\mu} + \{\kappa^2 c^2(1-\mu^2) + R\}Y = 0, \tag{4}$$

and in the first put $\cosh\xi = \nu$; then

$$(\nu^2-1)\frac{d^2X}{d\nu^2} + 2(s+1)\nu\frac{dX}{d\nu} + \{\kappa^2 c^2(\nu^2-1) - R\}X = 0. \tag{5}$$

In this case, if r_1 and r_2 are the distances of a point from the foci $(0, 0, \pm c)$

$$\nu = \frac{r_1+r_2}{2c}, \quad \mu = \frac{r_1-r_2}{2c}. \tag{6}$$

With
$$R = (n-s)(n+s+1), \quad \kappa = 0, \tag{7}$$

$$X\sinh^s\xi = Ap_n^s(\cosh\xi) + Bq_n^s(\cosh\xi), \tag{8}$$

$$Y\sin^s\eta = Cp_n^s(\cos\eta) + Dq_n^s(\cos\eta). \tag{9}$$

Both here and in 18·062 the q_n^s solution is inadmissible within an ellipsoid if $\xi = 0$ is a possible value. In both cases $h^2 = 0$ where $\xi = \eta = 0$, and it is found on examination that the gradient of $p_n^s q_n^s$ would tend to infinity for these values.* Hence *within* an oblate ellipsoid of revolution the solutions are $p_n^s(i\sinh\xi)p_n^s(\sin\eta)$, and within a prolate one they are $p_n^s(\cosh\xi)p_n^s(\cos\eta)$. But in problems relating to the outside of an ellipsoid ξ may be indefinitely large. If the solution is not to tend to infinity at a large distance p_n^s is inadmissible, but q_n^s becomes admissible because ξ cannot be 0 in the region considered and $q_n^s(\nu)$ can be defined, except for $-1 < \nu < 1$, so that it tends to 0 for large $|\nu|$. The solutions outside will be $q_n^s(i\sinh\xi)p_n^s(\sin\eta)$ and $q_n^s(\cosh\xi)p_n^s(\cos\eta)$ respectively.

18·07. The equation is also separable in general ellipsoidal coordinates, and leads to Lamé functions, which are an extension of Mathieu functions.†

18·08. Orthogonality relations. If ϕ and ϕ' are any two functions with continuous second derivatives in a region,

$$\iiint (\phi\nabla^2\phi' - \phi'\nabla^2\phi)\,d\tau = \iint \left(\phi\frac{\partial\phi'}{\partial n} - \phi'\frac{\partial\phi}{\partial n}\right) dS, \tag{1}$$

the first integral being through the region and the second over its boundary, dn being in the direction of the outward normal. If ϕ and ϕ' satisfy Laplace's equation in the region,

$$\iint \phi\frac{\partial\phi'}{\partial n}\,dS = \iint \phi'\frac{\partial\phi}{\partial n}\,dS. \tag{2}$$

Take the surface S to be a sphere, and take

$$\phi = r^m S_m(\theta,\lambda), \quad \phi' = r^n S_n(\theta,\lambda). \tag{3}$$

---

* For details see Hobson, *Spherical and Ellipsoidal Harmonics*, 1931, 422.
† Webster, *Partial Differential Equations of Mathematical Physics*, pp. 331–42.

Then the equation reduces to

$$m \iint r^{m+n-1} S_m S_n \, dS = n \iint r^{m+n-1} S_m S_n \, dS, \qquad (4)$$

and r is constant over the boundary. Hence if $m \neq n$,

$$\iint S_m S_n \sin\theta \, d\theta \, d\phi = 0. \qquad (5)$$

We express this verbally by saying that *any two surface harmonics of different degrees are orthogonal*. It is the analogue for a sphere of $\int_0^{2\pi} \cos m\theta \cos n\theta \, d\theta = 0$ and its companion relations when $m \neq n$.

If ϕ and ϕ', instead of satisfying Laplace's equation, satisfy

$$\nabla^2 \phi = -\kappa^2 \phi, \quad \nabla^2 \phi' = -\kappa'^2 \phi' \qquad (6)$$

and both ϕ and ϕ' satisfy a boundary condition on S of the form

$$a \frac{\partial \phi}{\partial n} + b\phi = 0, \qquad (7)$$

where a and b are constants, the same for both solutions, it is clear that if either a or b is 0 both $\iint \phi \frac{\partial \phi'}{\partial n} dS$ and $\iint \phi' \frac{\partial \phi}{\partial n} dS$ are 0. If neither of a and b is zero

$$\iint \left(\phi \frac{\partial \phi'}{\partial n} - \phi' \frac{\partial \phi}{\partial n} \right) dS = -\frac{b}{a} \iint (\phi\phi' - \phi'\phi) \, dS = 0. \qquad (8)$$

Hence
$$\iiint (\phi \nabla^2 \phi' - \phi' \nabla^2 \phi) \, d\tau = 0, \qquad (9)$$

that is,
$$(\kappa^2 - \kappa'^2) \iiint \phi\phi' \, d\tau = 0. \qquad (10)$$

Hence either $\kappa^2 = \kappa'^2$ or
$$\iiint \phi\phi' \, d\tau = 0. \qquad (11)$$

This gives a further set of orthogonality relations. Thus for a circular cylinder typical solutions are $J_m(\kappa\varpi)\cos m\lambda$, $J_n(\kappa'\varpi)\cos n\lambda$, where κ and κ' may be chosen so that on the boundary $\varpi = a$

$$J_m(\kappa a) = 0, \quad J_n(\kappa' a) = 0, \quad \text{or} \quad J'_m(\kappa a) = 0, \quad J'_n(\kappa' a) = 0. \qquad (12)$$

Then if $\kappa \neq \kappa'$,
$$\int_0^a \int_0^{2\pi} J_m(\kappa\varpi) J_n(\kappa'\varpi) \cos m\lambda \cos n\lambda \, \varpi \, d\varpi \, d\lambda = 0. \qquad (13)$$

This is satisfied if $m \neq n$; but if $m = n$ we must have

$$\int_0^a \varpi J_m(\kappa\varpi) J_m(\kappa'\varpi) \, d\varpi = 0, \qquad (14)$$

if κ and κ' are two different roots of $J_m(\kappa a) = 0$, $J'_m(\kappa a) = 0$, or of any equation of the form $\alpha J_m(\kappa a) + \beta \kappa J'_m(\kappa a) = 0$.

Similarly, by applying the argument to a sphere, for which the typical solution valid in the interior is $r^{-1/2} J_{n+1/2}(\kappa r) P_n^s(\cos\theta) \cos s\lambda$, we get

$$\int_0^a r J_{m+1/2}(\kappa r) J_{m+1/2}(\kappa' r) \, dr = 0, \qquad (15)$$

where κ and κ' are different roots of $J_{m+\frac{1}{2}}(\kappa a) = 0$, $J'_{m+\frac{1}{2}}(\kappa a) = 0$, or of

$$\alpha \frac{J_{m+\frac{1}{2}}(\kappa a)}{a^{\frac{1}{2}}} + \beta \frac{\partial}{\partial a} \frac{J_{m+\frac{1}{2}}(\kappa a)}{a^{\frac{1}{2}}} = 0.$$

The orthogonality relations determine at once the coefficients in an expansion in terms of the characteristic solution, provided such an expansion exists. For if the solutions are $\phi_0, \phi_1 \ldots \phi_n, \ldots$ (in general a three-dimensional set) and the function $f(x, y, z)$ is assumed to be

$$f(x, y, z) = \Sigma a_m \phi_m, \tag{16}$$

then on account of the orthogonality relations

$$\iiint f(x, y, z) \phi_n d\tau = \iiint (\Sigma a_m \phi_m) \phi_n d\tau = a_n \iiint \phi_n^2 d\tau \tag{17}$$

whence a_n is determined. The proof of the existence of an expansion of the form (16), however, is long and rather difficult. Two methods are used, the Sturm-Liouville method based on direct study of the differential equation,* and the method of Green's functions, which uses the theory of integral equations.† It may be said that the conditions required are similar to those for Fourier's series theorem. The integral equations method is very beautiful, but unfortunately too long for this work.

18·09. Potential at external points: MacCullagh's formula. Let O be an origin taken at a point of a distribution, $P(x_i = rl_i)$ an external point, $Q(x'_i = r'l'_i)$ an internal point. Put $PQ = R$. Then the potential at P due to a distribution of density or electric charge is

$$\phi = \gamma \iiint \rho \frac{d\tau}{R} + \gamma \iiint \sigma \frac{dS}{R}. \tag{1}$$

Now when $r' < r$, if θ is the angle POQ,

$$\frac{1}{R} = \frac{1}{\sqrt{(r^2 - 2rr' \cos\theta + r'^2)}} = \frac{1}{r} + \frac{r' \cos\theta}{r^2} + \frac{r'^2(3\cos^2\theta - 1)}{2r^3} + \ldots \tag{2}$$

on expansion. We need take only the volume integral as the modifications for the surface integral are obvious. Since

$$\cos\theta = l_i l'_i, \tag{3}$$

$$\phi = \gamma \iiint \rho \frac{d\tau}{r} + \gamma \iiint \rho r' \frac{l'_i l_i}{r^2} d\tau + \gamma \iiint \rho r'^2 \frac{3 l'_i l_k l'_k - 1}{2r^3} d\tau + \ldots. \tag{4}$$

Now

$$\iiint \rho \, d\tau = M, \tag{5}$$

the total mass or charge. The coefficient of $\gamma l_i / r^2$ is

$$\iiint \rho x'_i \, d\tau = M\overline{x_i}, \tag{6}$$

which we can make zero by taking the origin at the centre of mass or of charge, provided that M is not zero. (It does not appear to be noticed as a rule that in electrostatics a charged body has a charge centre in complete analogy with the centre of mass.) The coefficient of γr^{-3} is

$$\iiint \tfrac{1}{2}\rho(3 l_i l_k x'_i x'_k - r'^2) \, d\tau = \tfrac{1}{2} l_i l_k \iiint \rho(3 x'_i x'_k - r'^2 \delta_{ik}) \, d\tau. \tag{7}$$

---

* Ince, *Ordinary Differential Equations*.
† Erhard Schmidt, *Math. Ann.* 63, 1907, 433–76; A. Kneser, ibid. pp. 477–524; F. Smithies, *Proc. Lond. Math. Soc.* 43, 1937, 255–79.

Now if I_{ik} is the inertia tensor at O

$$I_{ik} = \iiint \rho(r'^2 \delta_{ik} - x'_i x'_k)\, d\tau, \qquad (8)$$

and (7) may be written

$$-\tfrac{3}{2} l_i l_k I_{ik} + \iiint \rho r'^2 d\tau.$$

But $I_{ik} l_i l_k$ is the moment of inertia about OP. If we denote this by I and the principal moments of inertia at O by A, B, C

$$\phi = \gamma \frac{M}{r} + \frac{\gamma}{2r^3}(A+B+C-3I) + O\left(\frac{1}{r^4}\right). \qquad (9)$$

This is known as *MacCullagh's formula*.

With an obvious extension the work function due to two gravitating bodies whose centres of mass are r apart is

$$\gamma \left(\frac{MM'}{r} + \frac{M'}{2r^3}(A+B+C-3I) + \frac{M}{2r^3}(A'+B'+C'-3I') \right) + O\left(\frac{1}{r^4}\right). \qquad (10)$$

MacCullagh's formula is also correct right down to the surface of a gravitating solid provided that this surface is an ellipsoid, the squares of whose departures from a sphere can be neglected. The justification in the latter case, however, is quite different from the one just given, and requires the theory of expansions in spherical harmonics over a sphere (24·06).

In the case $M = 0$ it is clear that the first term of (4) is 0 and that no change of origin can alter the values of the second, since the effect of any change of origin on the coefficient would be multiplied by the zero factor M. This case arises in magnetism and in some problems of dielectrics. In spherical polar coordinates the term in $1/r^2$ is

$$\frac{\gamma}{r^2} \iiint \rho r' \{\cos\theta \cos\theta' + \sin\theta \sin\theta' \cos(\phi - \phi')\}\, d\tau.$$

That in $1/r^3$ has the same form as before but the moments must of course be taken about the same point as was used as origin for the term in $1/r^2$.

EXAMPLES

1. If $f(\zeta)$ is an analytic function of ζ in a region R including the origin and a segment of the real axis, and if z, ϖ, λ are cylindrical polar coordinates, prove that the integral

$$\int_0^{2\pi} f(z + i\varpi \cos\alpha)\, d\alpha$$

is a potential function in the corresponding region of the variables z and ϖ.

By taking $f(\zeta)$ to be $\tan^{-1}(a/\zeta)$, or otherwise, verify that the free distribution of electricity over the conducting circular disk $z = 0$, $\varpi \leqslant a$ has a surface density proportional to $(a^2 - \varpi^2)^{-1/2}$; and show that the capacity of the disk is $2a/\pi$. (M.T. 1935.)

2. Six equal point charges e are situated on the axes at equal distances a from the origin, forming the vertices of a regular octahedron. It is required to expand the potential due to them in the neighbourhood of the origin. Prove:

(i) that the expression for the potential must be invariant for changes of sign and for interchanges of the coordinates (x, y, z);

(ii) that therefore the lowest terms in the expansion can be written $A + Br^2 + Cr^4 + D(x^4 + y^4 + z^4)$, where $r^2 = x^2 + y^2 + z^2$;

(iii) that in order to satisfy Laplace's equation, $B = 0$, $C = -\tfrac{3}{5}D$.

By calculating the potential at $(x, 0, 0)$, where $x \ll a$, show that D is positive and find its value.

Examples   545

If instead of six equal charges there are eight, situated at the corners of a cube whose edges, each of length $2b$, are parallel to the axes, and whose centre is at the origin, obtain the corresponding expansion of the potential near the origin, showing that in this case D is negative.

(M/c, Part II, 1938.)

3. Show that the mutual potential energy of two small magnets of strengths μ_1, μ_2 whose centres are at the points r_1, r_2 is

$$\frac{\mu_1 \cdot \mu_2}{|r_1 - r_2|^3} - \frac{3[\mu_1 \cdot (r_1 - r_2)][\mu_2 \cdot (r_1 - r_2)]}{|r_1 - r_2|^5}.$$

4. Obtain the solution of the equation of heat conduction satisfying the conditions

$$V = 0 \quad (x = 0, \; x = a),$$
$$V = x^2(a^2 - x^2) \quad (0 \leqslant x \leqslant a, \; t = 0),$$
$$V \to 0 \quad (t \to \infty),$$

in the form
$$V = 2a^2 \sum_{n=1}^{\infty} \left[(-1)^{n+1} \left(\frac{5}{n^3 \pi^3} - \frac{12}{n^5 \pi^5} \right) - \left(\frac{1}{n^3 \pi^3} + \frac{12}{n^5 \pi^5} \right) \right] e^{-n^2 \pi^2 h^2 t / a^2} \sin \frac{n \pi x}{a}.$$
(I.C. 1944.)

5. Obtain a solution of the equation

$$\frac{\partial^2 z}{\partial x^2} + \frac{\partial^2 z}{\partial y^2} = z$$

in the form $z = f(x) g(y)$, such that $z = 0$ when $x = 0$ and $y = 0$, and $\partial z / \partial x = 0$ when $x = \pi$.

6. A rectangle has its sides of lengths a and b maintained at temperatures 0 and 1 respectively. Find the steady temperature at any point, and show that at the centre of the rectangle its value is

$$\frac{4}{\pi} \sum_{n=0}^{\infty} \frac{(-1)^n}{2n+1} \operatorname{sech}(2n+1) \frac{\pi a}{2b}.$$
(I.C. 1944.)

7. A circular membrane of radius a is fixed at the edge and under uniform tension P. Show that in a symmetrical normal displacement the potential energy is

$$V = \tfrac{1}{2} P \int_0^a \left(\frac{\partial z}{\partial r} \right)^2 2\pi r \, dr.$$

Assuming
$$z = C \left(1 - \frac{r^2}{a^2} \right) \left(1 + \beta \frac{r^2}{a^2} \right)$$

and suitably adjusting β, obtain an estimate of the longest period. (I.C. 1940.)

8. A plane area of heat-conducting material is bounded by an ellipse and its major axis. The curved boundary is maintained at temperature V_1 and the straight boundary at V_2. Find the steady temperature at any point within the area. (I.C. 1940.)

9. A conducting solid is charged to potential ϕ_s. Prove that there is a point of the solid such that if r is the distance from this point the potential at a large distance is

$$\phi = \frac{c \phi_s}{r} + O\left(\frac{1}{r^3} \right),$$

where c is the electrostatic capacity of the solid.

Two conductors of capacities c_1, c_2, a long distance apart, have charges e_1, e_2 and potentials ϕ_1, ϕ_2. Prove that the potential energy V is given by

$$2V = \frac{e_1^2}{c_1} + \frac{e_2^2}{c_2} + \frac{2 e_1 e_2}{r} + O\left(\frac{1}{r^3} \right)$$
$$= \left(1 + \frac{c_1 c_2}{r^2} \right)(c_1 \phi_1^2 + c_2 \phi_2^2) - \frac{2 c_1 c_2 \phi_1 \phi_2}{r} + O\left(\frac{1}{r^3} \right),$$

where r is the distance between the charge centres of the conductors.

Chapter 19

WAVES IN ONE DIMENSION AND WAVES WITH SPHERICAL SYMMETRY

19·01. Vibrating string: d'Alembert's solution. In a large class of physical problems we meet with the differential equation

$$\frac{\partial^2 y}{\partial t^2} = c^2 \frac{\partial^2 y}{\partial x^2}, \tag{1}$$

where t is the time, x the distance from a fixed point or a fixed plane, and c is a known velocity. The general solution of this equation was given by d'Alembert. We take as new variables

$$u = x - ct, \quad v = x + ct, \tag{2}$$

and then by transforming the differential equation we get

$$\frac{\partial^2 y}{\partial u \, \partial v} = 0. \tag{3}$$

It follows that $\partial y/\partial v$ is independent of u, and therefore a function of v only; and integrating again we see that y must be of the form

$$y = f(u) + g(v) = f(x - ct) + g(x + ct). \tag{4}$$

Further, any functions f and g substituted in this equation will give a solution of (1) provided that they are twice differentiable.

Consider a uniform string under tension P, with mass ρ per unit length, and suppose it displaced transversely so that the displacement y and its gradient $\partial y/\partial x$ are small at all points. Then to the first order of small quantities the transverse component of the tension is $P \partial y/\partial x$ and is communicating transverse momentum to the part of the string to the left of x at a rate $P \partial y/\partial x$. Hence

$$\frac{\partial}{\partial t} \int_0^x \rho \dot{y} \, dx = P \left[\frac{\partial y}{\partial x} \right]_0^x, \tag{5}$$

and on differentiating with regard to x, assuming this permissible,

$$\rho \frac{\partial^2 y}{\partial t^2} = P \frac{\partial^2 y}{\partial x^2}, \tag{6}$$

which is of the form (1), with $c^2 = P/\rho$. But (5), which is physically the more fundamental equation, only assumes that y is differentiable once with regard to both x and t and $\int \dot{y} \, dx$ differentiable with regard to t. Now substitute $y = f(x - ct)$ in (5), assuming that $f(x - ct)$ has an integrable derivative. We have

$$\frac{\partial}{\partial t} \int_0^x \rho \frac{\partial}{\partial t} f(x - ct) \, dx = -\frac{\partial}{\partial t} \int_0^x \rho c f'(x - ct) \, dx$$

$$= -\frac{\partial}{\partial t} \rho c \{ f(x - ct) - f(-ct) \}$$

$$= \rho c^2 \{ f'(x - ct) - f'(-ct) \}$$

$$= P \left[\frac{\partial}{\partial x} f(x - ct) \right]_0^x. \tag{7}$$

Hence (5) is satisfied by a once differentiable function $f(x-ct)$, and similarly by $g(x+ct)$; and (4) therefore satisfies the physical conditions without the need to suppose that y is twice differentiable. The point is of some importance because (4) is habitually taken as a solution of wave problems where the second derivatives required in (1) do not exist. The differentiation with regard to x needed to give (1) or (6) is merely a mathematical device to present the problem in a tractable form, which suggests a solution but cannot prove that it is right in all the cases where we want to use it. The proof that the suggested solution satisfies the mechanical conditions when only the first derivatives exist requires the further argument leading to (7).

It may be noticed that the restriction that the first derivatives must exist at all points is still a little unnecessarily severe. The derivatives may have finite discontinuities for some values of x or t. Inspecting the proof of (7) again we see that it still holds provided that neither $x-ct$ nor $-ct$ is a point of discontinuity of the derivative of $f(x-ct)$. Even at points where the derivative is discontinuous, the value of $f(x-ct)$ can be filled in from the fact that $f(x-ct)$ is continuous. Hence there is a solution of the form (4) if, for instance, the string is displaced into a number of straight segments. In this case the second derivatives in (1) are zero except at the points where the slope changes discontinuously, where the second derivative does not exist, so that (4) becomes meaningless; but the mechanical problem still has a definite solution.

We require forms of the two functions $f(x-ct)$ and $g(x+ct)$ that will correspond to assigned values of y and $\dot{y}$ when $t=0$. Now if initially $y=\phi(x)$ it is clear that

$$\tfrac{1}{2}\{\phi(x-ct)+\phi(x+ct)\}$$

reduces to $\phi(x)$ at $t=0$ for all values of x, and its first derivative with regard to t is zero. Also $\psi(x+ct)-\psi(x-ct)=0$ at $t=0$ and

$$\frac{1}{2}\frac{\partial}{\partial t}\{\psi(x+ct)-\psi(x-ct)\} = \tfrac{1}{2}c\{\psi'(x+ct)+\psi'(x-ct)\} \to c\psi'(x).$$

Hence if at $t=0$, $\dot{y}=\chi(x)$, and we take

$$\psi'(x) = \frac{1}{c}\chi(x), \quad \psi(x) = \frac{1}{c}\int^x \chi(x)\,dx,$$

$\tfrac{1}{2}\{\psi(x+ct)-\psi(x-ct)\}$ will contribute nothing to y at $t=0$ and will give the correct value of $\dot{y}$; and

$$\psi(x+ct)-\psi(x-ct) = \frac{1}{2c}\int_{x-ct}^{x+ct}\chi(x)\,dx.$$

The solution that makes $y=\phi(x)$, $\dot{y}=\chi(x)$ at $t=0$ is therefore

$$y = \tfrac{1}{2}\{\phi(x-ct)+\phi(x+ct)\} + \frac{1}{2c}\int_{x-ct}^{x+ct}\chi(x)\,dx. \qquad (8)$$

This is d'Alembert's solution. It is the most general solution possible, but as it stands it is applicable for indefinitely large t only to a string of infinite length in both directions and subject to no external forces. For if the string extends from $x=0$ to $x=l$, our data can be only the values of y and $\dot{y}$ for $t=0$ and $x=0$ to l, and for any positive t, however small, there will be values of x between 0 and l such that $x-ct$ is negative and values between 0 and l such that $x+ct>l$. But for a string of finite length there will in general be forces at the ends, which will, for instance, keep the ends fixed. The situation is, therefore, that to obtain the position of the string for positive t we need values of the functions ϕ and χ outside the range originally given: but we have instead information for all time about the displacements at the ends. To make the problem precise we take the case where the ends are fixed; then if we can choose ϕ and ψ outside the given range so that y as given by (8) vanishes at $x = 0$ and l for all t, we satisfy the equation of motion within the string and also the end conditions, and therefore have a solution of the problem. Mathematically stated, the two functions ϕ and ψ are arbitrary outside the range $0 \leqslant x \leqslant l$, and we choose them to make the solution satisfy the conditions at the ends. Physically, we imagine the finite string replaced by an infinite one, and choose the initial displacements and velocities of the latter so that the ends will not move; and we expect that within the range $0 \leqslant x \leqslant l$ the solution will be the same as that for a finite string constrained by forces at the ends that prevent the ends from moving. In the one case the forces come from reactions with the supports, in the other from the tensions in the outlying parts of the string. The conditions on ϕ and χ are seen from (8) to be that both must be antisymmetrical about both $x = 0$ and $x = l$; that is,

$$\left.\begin{aligned}\phi(-x) &= -\phi(x), & \phi(2l-x) &= -\phi(x),\\ \chi(-x) &= -\chi(x), & \chi(2l-x) &= -\chi(x).\end{aligned}\right\} \tag{9}$$

For fixed ends these equations give at once the complete solution. When the ends are not fixed, however, as when one end carries a massive particle but is not fixed, the extrapolations are much less obvious and the easiest method of solution is the operational one.

If we take the term $f(x-ct)$ by itself, we see that it is unaltered if we increase t by τ and x by $c\tau$. Hence the part of the displacement represented by this term can be regarded as travelling with constant velocity c in the direction of increasing x; the term can therefore be considered as representing a *progressive wave*. Similarly the term $g(x+ct)$ can be regarded as representing a progressive wave travelling with velocity c in the direction of decreasing x.

19·02. Operational solution for string with fixed ends. Let us consider in detail the case of a string of length l, originally drawn aside a distance η at the point $x = b$, so that initially it lies in two straight pieces, and then released. Then if y_0 is the initial displacement,

$$\left.\begin{aligned} y_0 &= \eta x/b & (0 \leqslant x \leqslant b),\\ y_0 &= \eta(l-x)/(l-b) & (b \leqslant x \leqslant l), \end{aligned}\right\} \tag{1}$$

and the initial velocity is zero. Hence the subsidiary equation is

$$p^2 y - c^2 \frac{\partial^2 y}{\partial x^2} = p^2 y_0, \tag{2}$$

and we solve as if p was a constant, subject to the condition that y must always vanish at $x = 0$ and $x = l$. Operations are supposed performed on $H(t)$ unless the contrary is stated. The solution is

$$\left.\begin{aligned} y &= \eta \frac{x}{b} + A \sinh\frac{px}{c} \sinh\frac{p}{c}(l-b) & (0 \leqslant x \leqslant b),\\ y &= \eta \frac{l-x}{l-b} + A \sinh\frac{pb}{c} \sinh\frac{p}{c}(l-x) & (b \leqslant x \leqslant l). \end{aligned}\right\} \tag{3}$$

19·02 String with fixed ends

The constant A must be the same in both these expressions in order that y may be continuous at $x = b$. Also a discontinuity in $\partial y/\partial x$ at this point would imply an infinite acceleration, which cannot persist. There may be discontinuities of $\partial y/\partial x$ at special instants, but these will give impulsive changes in velocity. We therefore choose A so that $\partial u/\partial x$ will in general be continuous at $x = b$; this gives

$$\eta\left(\frac{1}{b}+\frac{1}{l-b}\right)+\frac{pA}{c}\left\{\cosh\frac{pb}{c}\sinh\frac{p}{c}(l-b)+\sinh\frac{pb}{c}\cosh\frac{p}{c}(l-b)\right\} = 0, \tag{4}$$

whence
$$A = -\frac{c}{p}\frac{l}{b(l-b)}\eta\operatorname{cosech}\frac{pl}{c}. \tag{5}$$

Then
$$\begin{aligned} y &= \frac{\eta l}{b(l-b)}\left\{\frac{l-b}{l}x - \frac{c}{p}\frac{\sinh px/c\,\sinh p(l-b)/c}{\sinh pl/c}\right\} && (0 \leqslant x \leqslant b), \\ y &= \frac{\eta l}{b(l-b)}\left\{\frac{b}{l}(l-x) - \frac{c}{p}\frac{\sinh pb/c\,\sinh p(l-x)/c}{\sinh pl/c}\right\} && (b \leqslant x \leqslant l). \end{aligned} \tag{6}$$

This can be interpreted at once by the partial fraction rule. Taking the first expression, we see that if we replace p by a constant z and then make z tend to 0 the second term just cancels the first; hence there is no term in the interpretation independent of t. There are poles at $zl/c = n\pi i$, where n is any integer, positive or negative, but not zero, and the partial fraction rule gives for $0 \leqslant x \leqslant b$,

$$y = -\frac{\eta l}{b(l-b)}\sum_{n=-\infty}^{\infty'}\frac{l}{n\pi i}\frac{i\sin\dfrac{n\pi x}{l}\cdot i\sin\dfrac{n\pi(l-b)}{l}}{\dfrac{n\pi i c}{l}\cdot\dfrac{l}{c}\cosh n\pi i}e^{n\pi ict/l}$$

$$= \frac{2\eta l^2}{b(l-b)}\sum_{1}^{\infty}\frac{1}{n^2\pi^2}\sin\frac{n\pi x}{l}\sin\frac{n\pi b}{l}\cos\frac{n\pi ct}{l}. \tag{7}$$

The solution for $b \leqslant x \leqslant l$ leads to exactly the same expression.

Every term of this satisfies the differential equation

$$\frac{\partial^2 y}{\partial t^2} = c^2\frac{\partial^2 y}{\partial x^2}. \tag{8}$$

The separate terms can therefore be regarded as each representing an oscillation in period $2l/nc$, the displacements for all values of x varying proportionately. As for systems with a finite number of degrees of freedom, the partial fraction rule leads to the analysis of the motion into normal modes. The motion corresponding to a given harmonic factor in the time is called a *standing wave*. We see that any standing wave can be replaced by a pair of progressive waves; for

$$2\sin\frac{n\pi x}{l}\cos\frac{n\pi ct}{l} = \sin\frac{n\pi}{l}(x+ct) + \sin\frac{n\pi}{l}(x-ct). \tag{9}$$

Similarly, any progressive wave can be replaced by a pair of standing waves with phases $\tfrac{1}{2}\pi$ apart; thus

$$\sin\frac{n\pi}{l}(x+ct) = \sin\frac{n\pi x}{l}\cos\frac{n\pi ct}{l} + \cos\frac{n\pi x}{l}\sin\frac{n\pi ct}{l}. \tag{10}$$

The natural way to try to prove that the solution found satisfies the required conditions would be to differentiate term by term and substitute in (8). But in this case that method does not work, because if we differentiate twice we get a series whose terms oscillate finitely as n increases. But if we differentiate once we get a convergent series which can be substituted in 19·01 (5), and the verification that 19·01 (5) is satisfied presents no particular difficulty. Alternatively, we can break up each term as in (9), and notice that the series is in this way converted into the sum of a function of $x-ct$ and one of $x+ct$, each separately being once differentiable, and therefore it satisfies the equation of motion. It also satisfies the end conditions. For each term vanishes at $x = 0$ and $x = l$, and the series is uniformly convergent because the modulus of the general term is $\leqslant n^{-2}$, and therefore the sum tends to 0 as $x \to 0$ or l.

The series (7) converges too slowly to be of much use for actual calculation of the displacement. Another method of evaluation is as follows. We start with Bromwich's integral:

$$-\frac{c}{p}\frac{\sinh px/c \sinh p(l-b)/c}{\sinh pl/c} H(t) = -\frac{c}{2\pi i}\int_L \frac{\sinh zx/c \sinh z(l-b)/c}{z^2 \sinh zl/c} e^{zt}\,dz, \qquad (11)$$

where $\Re(z) = k > 0$ on the line L. But then $|e^{-2zl/c}| < 1$ on L and at all points to the right of it, and we can expand cosech zl/c in a convergent geometric series $2e^{-zl/c}\sum_{n=0}^{\infty} e^{-2nzl/c}$. The order of integration and summation can be inverted, and we have

$$-\frac{c}{4\pi i}\sum_{n=0}^{\infty}\int_L e^{z(x-b)/c}(1-e^{-2zx/c})(1-e^{-2z(l-b)/c})e^{-2nlz/c} e^{zt}\,\frac{dz}{z^2}$$

$$= -\tfrac{1}{2}c\sum_{n=0}^{\infty} e^{p(x-b)/c}(1-e^{-2px/c})(1-e^{-2p(l-b)/c})e^{-2pnl/c}\frac{1}{p}H(t), \qquad (12)$$

which is the result of expanding (11) directly as if p was a constant with a positive real part. But

$$\frac{c}{p}H(t) = 0 \quad (t<0),$$
$$\phantom{\frac{c}{p}H(t)} = ct \quad (t>0), \qquad (13)$$

and

$$e^{-ph/c}\frac{c}{p}H(t) = 0 \quad (ct<h),$$
$$\phantom{e^{-ph/c}\frac{c}{p}H(t)} = ct-h \quad (ct>h). \qquad (14)$$

The first term of y in (3) is $\eta x/b$ for $x \leqslant b$. All terms of (12) are zero until $ct = b-x$, so that

$$y = \eta x/b \quad (0<ct<b-x). \qquad (15)$$

When $ct = b-x$ the first term in (12) begins to differ from 0; it is

$$-\tfrac{1}{2}e^{-p(b-x)/c} ct\, H(t) = -\tfrac{1}{2}(ct-b+x), \qquad (16)$$

and

$$y = \frac{\eta x}{b} - \frac{1}{2}\frac{\eta l}{b(l-b)}(ct-b+x) = \frac{\eta l}{b(l-b)}\left\{\left(\frac{1}{2}-\frac{b}{l}\right)x - \tfrac{1}{2}(ct-b)\right\}. \qquad (17)$$

In this stage $\partial y/\partial x$ is the mean of η/b and $-\eta/(l-b)$, the original slopes of the two parts of the string. It begins when a wave travelling with velocity c from b has had time to reach x. It will continue until the term in $2px/c$ or $2p(l-b)/c$ no longer vanishes. The former yields the expression

$$\tfrac{1}{2}e^{-p(b+x)/c}ct\, H(t) = \tfrac{1}{2}(ct-b-x) \quad (b+x<ct), \qquad (18)$$

and, adding this to the previous contributions, we find

$$y = -\frac{\eta x}{l-b}. \tag{19}$$

The beginning of this stage corresponds to the time it would take a wave to travel from b to 0 and be reflected back to x. The part of the string reached by this reflected wave is therefore parallel to the original position of the part where $b < x < l$.

When $ct = (b-x) + 2(l-b)$ a wave reflected at $x = l$ arrives, and afterwards

$$y = \frac{\eta l}{b(l-b)}\left\{\left(\frac{1}{2}-\frac{b}{l}\right)x + \tfrac{1}{2}(ct+b) - l\right\}. \tag{20}$$

This holds until $ct = b + x + 2(l-b)$; in the next stage

$$y = \frac{\eta l}{b(l-b)} x\left(1 - \frac{b}{l}\right) = \frac{\eta x}{b}. \tag{21}$$

When $ct = 2l$, the whole of the string is back in its original position; the term in $e^{-2pl/c}$ then begins to affect the motion, and the whole process repeats itself. We see that at any instant the string is in three straight pieces. The two end pieces are parallel to the two portions of the string in its original position, and are at rest. For the middle portion the gradient $\partial y/\partial x$ is the mean of those for the end portions, and the transverse velocity is $\pm \dfrac{1}{2}\dfrac{\eta lc}{b(l-b)}$. The middle portion is always either extending or withdrawing at each end with velocity c.

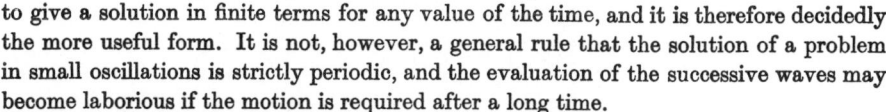

The partial fraction rule and the expansion in negative exponentials are alternative ways of evaluating the Bromwich integral. The former in general analyses it into normal modes, the latter into progressive waves. In the problem just considered the exact periodicity enables the wave expansion to give a solution in finite terms for any value of the time, and it is therefore decidedly the more useful form. It is not, however, a general rule that the solution of a problem in small oscillations is strictly periodic, and the evaluation of the successive waves may become laborious if the motion is required after a long time.

19·03. Solution for a general initial disturbance. We take $y = \phi(x)$, $\dot{y} = 0$ at $t = 0$; then the subsidiary equation is

$$p^2 y - c^2 \frac{\partial^2 y}{\partial x^2} = p^2 \phi(x), \tag{1}$$

and we want a solution that vanishes at $x = 0$ and l. Using the method of variation of parameters, we assume that the solution is

$$y = A \cosh\frac{px}{c} + B \sinh\frac{px}{c}, \tag{2}$$

where A and B are functions of x subject to

$$A'\cosh\frac{px}{c} + B'\sinh\frac{px}{c} = 0. \tag{3}$$

Substituting in (1) we find
$$A'\sinh\frac{px}{c} + B'\cosh\frac{px}{c} = -\frac{p}{c}\phi(x). \tag{4}$$

Hence
$$A' = \frac{p}{c}\phi(x)\sinh\frac{px}{c}, \quad B' = -\frac{p}{c}\phi(x)\cosh\frac{px}{c}. \tag{5}$$

For $x = 0$, $y = A(0)$; hence, since $y = 0$ at $x = 0$,

$$A = \int_0^x \frac{p}{c}\phi(\xi)\sinh\frac{p\xi}{c}\,d\xi.$$

Also, since $y = 0$ at $x = l$,
$$B(l)\sinh\frac{pl}{c} = -A(l)\cosh\frac{pl}{c},$$

which gives $B(l)$; and then $B(x)$ is determined since we have B' from (5). We have

$$B(x) = -\coth\frac{pl}{c}\int_0^l \frac{p}{c}\phi(\xi)\sinh\frac{p\xi}{c}\,d\xi + \int_x^l \frac{p}{c}\phi(\xi)\cosh\frac{p\xi}{c}\,d\xi.$$

Substituting for A and B in (2) we have

$$y = \int_0^x \frac{p}{c}\phi(\xi)\sinh\frac{p\xi}{c}\frac{\sinh p(l-x)/c}{\sinh pl/c}\,d\xi + \int_x^l \frac{p}{c}\phi(\xi)\sinh\frac{px}{c}\frac{\sinh p(l-\xi)/c}{\sinh pl/c}\,d\xi \tag{6}$$

$$= \int_{\xi=0}^x \phi(\xi)\frac{\sinh p(l-x)/c}{\sinh pl/c}\,d\cosh\frac{p\xi}{c} - \int_{\xi=x}^l \phi(\xi)\frac{\sinh px/c}{\sinh pl/c}\,d\cosh\frac{p(l-\xi)}{c}. \tag{7}$$

By the partial fraction rule

$$\frac{\sinh p(l-x)/c \cosh p\xi/c}{\sinh pl/c} = \frac{l-x}{l} - 2\sum_{n=1}^{\infty}\frac{1}{n\pi}\sin\frac{n\pi x}{l}\cos\frac{n\pi\xi}{l}\cos\frac{n\pi ct}{l}, \tag{8}$$

$$\frac{\sinh px/c \cosh p(l-\xi)/c}{\sinh pl/c} = \frac{x}{l} + 2\sum_{n=1}^{\infty}\frac{1}{n\pi}\sin\frac{n\pi x}{l}\cos\frac{n\pi\xi}{l}\cos\frac{n\pi ct}{l}. \tag{9}$$

The first terms are independent of ξ and contribute nothing to the integrals; and

$$y = -2\int_{\xi=0}^l \phi(\xi)\,d\left(\sum_{n=1}^{\infty}\frac{1}{n\pi}\sin\frac{n\pi x}{l}\cos\frac{n\pi ct}{l}\cos\frac{n\pi\xi}{l}\right). \tag{10}$$

The series represents a function with finite discontinuities and a Stieltjes integral is required.

If we invert the order of integration and summation and then differentiate $\cos n\pi\xi/l$ in the separate terms, we get

$$y = \frac{2}{l}\sum_{n=1}^{\infty}\int_0^l \phi(\xi)\sin\frac{n\pi x}{l}\sin\frac{n\pi\xi}{l}\cos\frac{n\pi ct}{l}\,d\xi. \tag{11}$$

This is Fourier's solution; putting $t = 0$ we get the sine series

$$\phi(x) = \frac{2}{l}\sum_{n=1}^{\infty}\int_0^l \phi(\xi)\sin\frac{n\pi x}{l}\sin\frac{n\pi\xi}{l}\,d\xi. \tag{12}$$

To get a wave expansion we have

$$\frac{\sinh p(l-x)/c \, \cosh p\xi/c}{\sinh pl/c} = \tfrac{1}{2}e^{-p(x-\xi)/c}(1-e^{-2p(l-x)/c})(1+e^{-2p\xi/c})\sum_{n=0}^{\infty} e^{-2npl/c} \quad (\xi < x), \tag{13}$$

$$\frac{\sinh px/c \, \cosh p(l-\xi)/c}{\sinh pl/c} = \tfrac{1}{2}e^{-p(\xi-x)/c}(1-e^{-2px/c})(1+e^{-2p(l-\xi)/c})\sum_{n=0}^{\infty} e^{-2npl/c} \quad (\xi > x). \tag{14}$$

All the exponents are negative multiples of p; this always happens. Also

$$e^{-p(x-\xi)/c}(1-e^{-2p(l-x)/c})(1+e^{-2p\xi/c})H(t) = H\left(t - \frac{x-\xi}{c}\right) - H\left(t - \frac{2l-x-\xi}{c}\right)$$
$$+ H\left(t - \frac{x+\xi}{c}\right) - H\left(t - \frac{2l-x+\xi}{c}\right), \tag{15}$$

which is constant except for jumps of ± 1 when the arguments of the unit functions pass through 0. Then the terms arising from $n = 0$ give

$$y = \tfrac{1}{2}\int_{\xi=0}^{x} \phi(\xi)\,d\left\{H\left(t-\frac{x-\xi}{c}\right) - H\left(t-\frac{2l-x-\xi}{c}\right) + H\left(t-\frac{x+\xi}{c}\right) - H\left(t-\frac{2l-x+\xi}{c}\right)\right\}$$
$$+ \tfrac{1}{2}\int_{\xi=x}^{l} \phi(\xi)\,d\left\{-H\left(t-\frac{\xi-x}{c}\right) + H\left(t-\frac{\xi+x}{c}\right) - H\left(t-\frac{2l-x-\xi}{c}\right) + H\left(t-\frac{2l+x-\xi}{c}\right)\right\}$$
$$= \tfrac{1}{2}\{\phi(x-ct) - \phi(2l-x-ct) - \phi(ct-x) + \phi(ct+x-2l)\}$$
$$+ \tfrac{1}{2}\{\phi(ct+x) - \phi(ct-x) - \phi(2l-x-ct) + \phi(2l+x-ct)\}, \tag{16}$$

where only those terms are to be taken such that the values of ξ that make the arguments of the corresponding unit functions vanish lie within the ranges of the respective integrals. The first term in the first line is seen to represent the direct wave from points between O and x, the third, which begins at time x/c, the wave reflected at $x = 0$, and the other two the reflexions of these at $x = l$. Corresponding relations hold for the terms in the second line except that they give the contributions from waves starting between x and l. It will be seen that the solution holds up to time $2l/c$, by which time all the terms have disappeared on account of their arguments passing out of the ranges permitted; but then the terms from $n = 1$ in (13) enter and repeat the entire motion. At time $4l/c$ they also have all disappeared but those from $n = 2$ enter, and so on indefinitely.

In the foregoing cases the motion repeats itself exactly at regular intervals. In the following it does not.

19·04. *A uniform heavy string of length $2l$ is fixed at the ends. A particle of mass m is attached to the middle of the string. Initially the string is straight and under tension P. A transverse impulse J is given to the particle. Find the subsequent motion of the particle.*[*]
We take x zero at the middle of the string. By symmetry we need consider only the range of values $0 \leqslant x \leqslant l$. Call the displacement of the particle η. When $t = 0$, y and $\partial y/\partial t$ are zero except at $x = 0$. The subsidiary equation for the string therefore needs no additional terms for the initial conditions. The conditions that when $x = 0$, $y = \eta$, and when $x = l$, $y = 0$, give, therefore,

$$y = \eta \frac{\sinh p(l-x)/c}{\sinh pl/c}. \tag{1}$$

[*] Cf. Rayleigh, *Theory of Sound*, 1, 1894, 204.

The equation of motion of the particle, taking account of the equal tensions in the string on both sides of it, is

$$m\frac{\partial^2 \eta}{\partial t^2} = 2P\left(\frac{\partial y}{\partial x}\right)_{x=0}. \tag{2}$$

At $t = 0$, $\eta = 0$, $m\dot\eta = J$. Hence the subsidiary equation for η is

$$mp^2\eta = 2P\left(\frac{\partial y}{\partial x}\right)_{x=0} + pJ$$

$$= -2P\eta\frac{p}{c}\coth pl/c + pJ, \tag{3}$$

and therefore

$$\eta = \frac{Jc}{mpc + 2P\coth pl/c}. \tag{4}$$

If ρ is the line density of the string, $P = \rho c^2$, and the mass of the string is $2\rho l$. Put

$$\frac{2\rho l}{m} = k, \quad 2P = \frac{kmc^2}{l}. \tag{5}$$

Then

$$\eta = \frac{lJ/mc}{(pl/c) + k\coth pl/c}. \tag{6}$$

To interpret by the partial fraction rule, we recollect that the system is a stable one without dissipation, and therefore all zeros of the denominator are purely imaginary. With $pl/c = i\omega$, ω satisfies

$$\omega = k\cot\omega. \tag{7}$$

There is a root between every two consecutive multiples of π, positive or negative, and the roots occur in pairs of equal magnitude. Then

$$\eta = \frac{lJ}{mc}\Sigma\frac{l}{i\omega c(1 + k\csc^2\omega)(l/c)}e^{i\omega ct/l}$$

$$= \frac{2lJ}{mc}\Sigma\frac{1}{\omega(1 + k\csc^2\omega)}\sin\frac{\omega ct}{l}, \tag{8}$$

the second summation being only over positive values of ω.

If a root of (7) is $n\pi + \lambda$, where $\lambda < \pi$,

$$(n\pi + \lambda)\tan\lambda = k, \tag{9}$$

and $\lambda = k/n\pi$ approximately. Then the series converges like Σn^{-3}. Four or five terms should therefore be enough to give 1 % accuracy. For higher accuracy the labour would be great.

For t not too great an exact solution can be found easily by the wave expansion. It is convenient to change the unit of time to l/c, the time taken for a wave to travel half the length of the string. We also replace J/m by V. Then

$$\eta = \frac{V}{p + k\coth p} = \frac{V(1 - e^{-2p})}{(p + k) - (p - k)e^{-2p}}$$

$$= V\frac{1 - e^{-2p}}{p + k}\left\{1 + \frac{p-k}{p+k}e^{-2p} + \left(\frac{p-k}{p+k}\right)^2 e^{-4p} + \ldots\right\}$$

$$= \frac{V}{p+k}\left(1 - \frac{2k}{p+k}e^{-2p} - \frac{2k(p-k)}{(p+k)^2}e^{-4p} - \ldots\right). \tag{10}$$

The first term is zero for $t<0$, and for $t>0$ it is equal to

$$\frac{V}{k}(1-e^{-kt}) \quad (t>0). \tag{11}$$

After $t=2$ the second term no longer vanishes. We have

$$\frac{k}{(p+k)^2} = \frac{1}{k} - \frac{p}{k(p+k)} - \frac{p}{(p+k)^2} = \frac{1}{k} - \frac{1}{k}e^{-kt} - te^{-kt} \tag{12}$$

and the second term therefore contributes

$$-\frac{2V}{k}\{1-e^{-k(t-2)} - k(t-2)e^{-k(t-2)}\} \quad (t>2). \tag{13}$$

The third term is zero for $t<4$; for $t>4$ it is easily found to be

$$\frac{2V}{k}[1-\{1+k(t-4)+k^2(t-4)^2\}e^{-k(t-4)}]. \tag{14}$$

The process may be extended to determine the motion up to any time desired. The entry of a new term into the solution corresponds to the arrival of a new pair of waves reflected at the ends.

19·05. *A uniform heavy bar is hanging vertically from one end, and a mass m is suddenly attached to the lower end. Find how the tension at the upper end varies with the time.**

For a light bar it is easy to see that the added mass will perform harmonic oscillations about the position of equilibrium; when it reaches its lowest position the extra tension is therefore twice the weight. This feature accounts for the danger of suddenly attaching a load that a system might be well able to support if the load was added gradually.

For a heavy bar, if x be the distance from the upper end, y the longitudinal displacement, y satisfies

$$\rho \frac{\partial^2 y}{\partial t^2} - E \frac{\partial^2 y}{\partial x^2} = F, \tag{1}$$

where ρ is the density, E Young's modulus, and F the external force per unit volume, in this case ρg. Put $E/\rho = c^2$, and let the displacement of a particle under the tension before the weight is attached be y_0. Then

$$-c^2 \frac{\partial^2 y_0}{\partial x^2} = g. \tag{2}$$

When $x=0$, $y_0=0$; and when $x=l$, the length of the bar, $\partial y_0/\partial x = 0$. Hence

$$y_0 = \frac{glx}{c^2}\left(1-\frac{1}{2}\frac{x}{l}\right). \tag{3}$$

After the weight is attached we still have

$$\frac{\partial^2 y}{\partial t^2} - c^2 \frac{\partial^2 y}{\partial x^2} = g = -c^2 \frac{\partial^2 y_0}{\partial x^2}, \tag{4}$$

and when $t=0$, $y=y_0$, $\dot{y}=0$. Hence the subsidiary equation is

$$p^2 y - c^2 \frac{\partial^2 y}{\partial x^2} = p^2 y_0 - c^2 \frac{\partial^2 y_0}{\partial x^2}, \tag{5}$$

* Love, *Elasticity*, § 283.

and the solution that vanishes with x is given by

$$y - y_0 = A \sinh px/c, \tag{6}$$

where A is independent of x.

If ϖ is the cross section of the bar, the equation of motion of the mass m is

$$m \frac{\partial^2 y}{\partial t^2} = mg - E\varpi \frac{\partial y}{\partial x}, \tag{7}$$

the derivatives being evaluated at $x = l$. The subsidiary equation is

$$mp^2 y = mg + mp^2 y_0 - E\varpi \frac{\partial y}{\partial x}, \tag{8}$$

and on substitution for y from (6)

$$\left(p^2 \sinh \frac{pl}{c} + \frac{E\varpi p}{mc} \cosh \frac{pl}{c}\right) A = g. \tag{9}$$

The tensile stress at the upper end is

$$g\rho l + \frac{EpA}{c} = g\rho l + \frac{Emg}{E\varpi \cosh pl/c + mcp \sinh pl/c}. \tag{10}$$

If k is the ratio of the mass of the weight to that of the bar,

$$k = m/\rho\varpi l, \quad mc/E\varpi = kl/c, \tag{11}$$

and the stress is

$$\frac{gm}{\varpi}\left[\frac{1}{k} + \frac{1}{\cosh pl/c + k(pl/c)\sinh pl/c}\right]. \tag{12}$$

We see that $gm/\varpi k$ is the stress due to the weight of the bar alone, and gm/ϖ is the statical stress due to the added load. To evaluate the actual stress we expand the operator in powers of $e^{-pl/c}$. We take l/c for the new unit of time; then

$$\frac{1}{kp \sinh p + \cosh p} = \frac{2e^{-p}}{(kp+1) - (kp-1)e^{-2p}}$$

$$= \frac{2e^{-p}}{kp+1}\left[1 + \frac{kp-1}{kp+1}e^{-2p} + \left(\frac{kp-1}{kp+1}\right)^2 e^{-4p} + \ldots\right]. \tag{13}$$

The first term vanishes to time unity, and afterwards is equal to

$$2(1 - e^{-(t-1)/k}). \tag{14}$$

This increases steadily up to time 3, when the next term enters. Again,

$$\frac{kp-1}{(kp+1)^2} = -1 + \frac{kp}{kp+1} + \frac{2kp}{(kp+1)^2}$$

$$= -1 + e^{-t/k} + 2(t/k)e^{-t/k}, \tag{15}$$

and the first two terms of (13), for $t > 3$, are equal to

$$2e^{-(t-3)/k}\{1 + (2/k)(t-3) - e^{-2/k}\}. \tag{16}$$

This has a maximum when

$$1 + e^{-2/k} = \frac{2}{k}(t-3). \tag{17}$$

This has a root less than 5 if

$$4/k > 1 + e^{-2/k}, \tag{18}$$

which is an equality if $k = 2 \cdot 7$. Thus if $k = 1$ or 2 the maximum stress will occur before $t = 5$. If $k = 1$ it is when $t = 3 \cdot 568$, and is equal to $3 \cdot 266 gm/\varpi$, that is, $1 \cdot 633$ times the statical stress. If $k = 2$ the corresponding results are $t = 4 \cdot 368$, $2 \cdot 520 gm/\varpi$, and $1 \cdot 680$ times the statical stress.

The third term enters at $t = 5$, and afterwards is equal to

$$2[1 - e^{-(t-5)/k} - 2\{(t-5)/k\}^2 e^{-(t-5)/k}]. \tag{19}$$

If $k = 4$, the maximum stress is when $t = 6 \cdot 183$, and is equal to $2 \cdot 29 gm/\varpi$. The statical stress is $1 \cdot 25 gm/\varpi$, so that the ratio is $1 \cdot 83$.

This solution and that of 19·04 are due to Bromwich.

19·06. Periodic disturbance at an internal point. It is sometimes argued that if a periodic motion is enforced at an internal point of a system of finite size, it represents a continual supply of energy, and the disturbance will ultimately exceed any bound, irrespective of any question of resonance.* It is interesting to examine what will actually happen in a simple case where these conditions are satisfied. First, consider a string of length $2l$, originally at rest, and suppose that for $t > 0$ the middle point is made to vibrate harmonically, the displacement being $\sin nt$. The operational solution expressing this displacement at $x = 0$ and zero displacement at $x = l$ is

$$y = \frac{np}{p^2 + n^2} \frac{\sinh p(l-x)/c}{\sinh pl/c} \tag{1}$$

$$= e^{-px/c}(1 - e^{-2p(l-x)/c})(1 + e^{-2pl/c} + e^{-4pl/c} + \ldots) \sin nt\, H(t)$$

$$= \sin n\left(t - \frac{x}{c}\right) - \sin n\left(t - \frac{2l-x}{c}\right) + \sin n\left(t - \frac{2l+x}{c}\right) - \sin n\left(t - \frac{4l-x}{c}\right) + \ldots, \tag{2}$$

where we are to include only those terms whose arguments are positive. At any instant the motion therefore consists of a number of superposed harmonic waves of the same period, their number increasing indefinitely with the time. At first sight this suggests that the disturbance may grow indefinitely; on the other hand it is possible that this may be prevented by successive waves interfering. This can be tested by evaluating (1) by the partial fraction rule. The pole at in contributes

$$\frac{in^2}{in \cdot 2in} \frac{\sin n(l-x)/c}{\sin nl/c} e^{int} = -\tfrac{1}{2} i \frac{\sin n(l-x)/c}{\sin nl/c} e^{int}, \tag{3}$$

and with that at $-in$ gives

$$\frac{\sin n(l-x)/c}{\sin nl/c} \sin nt. \tag{4}$$

The pole at $pl/c = ri\pi$ (r an integer) gives

$$\frac{nri\pi c/l}{n^2 - r^2\pi^2 c^2/l^2} \frac{i \sin r\pi(l-x)/l}{ri\pi \cos r\pi} e^{ri\pi ct/l} = i\frac{nc/l}{n^2 - r^2\pi^2 c^2/l^2}(-1)^r \sin\frac{r\pi(l-x)}{l} e^{ri\pi ct/l}, \tag{5}$$

and altogether

$$y = \frac{\sin n(l-x)/c}{\sin nl/c} \sin nt - 2\sum_{r=1}^{\infty}(-1)^r \frac{nc/l}{n^2 - r^2\pi^2 c^2/l^2}\sin\frac{r\pi(l-x)}{l}\sin\frac{r\pi ct}{l}. \tag{6}$$

* Cf. H. M. Macdonald, *Proc. Roy. Soc.* A, 98, 1921, 409–11.

The series is absolutely convergent so long as n is not an exact multiple of $\pi c/l$, and its sum for any x and t is less than the result of replacing the sines by 1 and taking the moduli of all the terms. Hence y does not increase beyond limit. If we take the kinetic or the potential energy we again get a series that converges like Σr^{-2}, and the energy never passes a certain value. The rate of supply of energy is in fact proportional to the product $\left(\frac{\partial y}{\partial x}\right)\left(\frac{\partial y}{\partial t}\right)$ evaluated at $x = 0$, and one factor consists of sines, the other of cosines, of multiples of the time. The rate of supply of energy is sometimes positive, sometimes negative. The mean rate of supply of energy over a long time tends to zero.

In the problem just treated the displacement is prescribed to vary finitely at one point, and it might be thought that it is this condition that prevents indefinite growth at any point. We therefore consider also the case where the transverse force, not the displacement, is prescribed to be $\sin nt$ for $t > 0$. Denoting the displacement at $x = 0$ by y_0, we have now

$$y = y_0 \frac{\sinh p(l-x)/c}{\sinh pl/c}, \tag{7}$$

$$\sin nt = \frac{np}{p^2+n^2} = -2P\left(\frac{\partial y}{\partial x}\right)_{x=0} = y_0 \frac{p}{c} 2P \coth pl/c, \tag{8}$$

and $\quad y = \frac{nc}{2P(p^2+n^2)} \frac{\sinh p(l-x)/c}{\cosh pl/c} \tag{9}$

$$= \frac{c}{2Pn} e^{-px/c}(1 - e^{-2p(l-x)/c})(1 - e^{-2pl/c} + e^{-4pl/c} - \ldots)(1 - \cos nt)$$

$$= \frac{c}{2Pn}\left[\left\{1 - \cos n\left(t - \frac{x}{c}\right)\right\} - \left\{1 - \cos n\left(t - \frac{2l-x}{c}\right)\right\} - \left\{1 - \cos n\left(t - \frac{2l+x}{c}\right)\right\} + \ldots\right], \tag{10}$$

where again only the terms with positive arguments are to be included. If we evaluate (9) by the partial fraction rule we get

$$y = \frac{c}{2Pn} \sin nt \frac{\sin n(l-x)/c}{\cos nl/c} + \sum_{r=0}^{\infty} \frac{nc}{P\{n^2 - (r+\frac{1}{2})^2 \pi^2 c^2/l^2\}(r+\frac{1}{2})\pi} \cos(r+\tfrac{1}{2})\frac{\pi x}{l} \sin(r+\tfrac{1}{2})\frac{\pi ct}{l}. \tag{11}$$

The series in this case converges like Σr^{-3} and again there is an upper bound to the displacement. The force at $x = 0$ is now $\sin nt$ and the velocity is a series of cosines, so that again after a long time as much energy comes to be taken out as is put in.

The system is supposed to be of finite extent. If we make l tend to infinity we get an infinitely long string, which is hardly a practical possibility, but the same analysis would apply fairly well to a gas in a long tube open at both ends. Simply letting l increase we see that the waves arising from terms with arguments containing l take longer and longer to return, and the solution for any given x will reduce to the first term so long as $2l - x < ct$, and therefore for a longer range of time the larger l is. Proceeding to the limit we see that the disturbance consists of a wave of fixed amplitude extending for a greater and greater distance from the origin; the energy therefore does increase indefinitely. It is therefore possible for a finite force to produce an indefinitely large amount of energy in an infinite system provided that it is made to act long enough. It will be noticed that in this case the operational solution reduces to the first term in the wave expansion and could be found by writing the subsidiary equation as

$$p^2 y - c^2 \frac{\partial^2 y}{\partial x^2} = 0,$$

and taking the solution as $Ae^{-px/c}$, rejecting the solution $Be^{px/c}$ on the ground that it would represent a wave travelling inwards and therefore a source of energy at a large distance.

It has been supposed in (6) that n is not an exact multiple of $\pi c/l$, in (11) that n is not of the form $(s+\tfrac{1}{2})\pi c/l$, where s is an integer. In the special cases where this restriction is not satisfied we have resonance and the disturbance will grow. The modification of the solution to take account of the double poles is straightforward, but not of any special interest since indefinite increase of the disturbance in the case of resonance is formally possible even for a system with one degree of freedom. The main conclusion is that a harmonic force of limited amount will never impart more than a given amount of energy to a system, however long it acts, unless either the system is of infinite extent or the period of the force agrees exactly with a free period of the system. In the latter case the solution will ultimately need modification to take account of neglected higher powers of the displacement.

19·07. Problems of spherical symmetry. The equation of propagation of sound in three dimensions is

$$\frac{\partial^2 \phi}{\partial t^2} = c^2 \nabla^2 \phi, \tag{1}$$

where ϕ is the velocity potential; the velocity components are

$$u_i = \frac{\partial \phi}{\partial x_i}, \tag{2}$$

and the pressure is

$$P = -\rho \frac{\partial \phi}{\partial t}. \tag{3}$$

(Capital P is used because we want p for the Heaviside operator.) Now if ϕ is a function of r and t only, where r is the distance from a fixed point,

$$\nabla^2 \phi = \frac{1}{r^2}\frac{\partial}{\partial r}\left(r^2 \frac{\partial \phi}{\partial r}\right) = \frac{\partial^2 \phi}{\partial r^2} + \frac{2}{r}\frac{\partial \phi}{\partial r} = \frac{1}{r}\frac{\partial^2}{\partial r^2}(r\phi). \tag{4}$$

Hence

$$\frac{\partial^2}{c^2 \partial t^2}(r\phi) = \frac{\partial^2}{\partial r^2}(r\phi), \tag{5}$$

and this has the same form as the equation of vibration of a string or transmission of sound in one dimension, the dependent variable being now $r\phi$. Hence a general solution is

$$r\phi = f(r-ct) + g(r+ct). \tag{6}$$

The first term will represent a disturbance travelling outwards, the second one travelling inwards.

19·08. Explosion wave. Consider a spherical region of high pressure, surrounded by an infinitely extended region of uniform pressure. The boundary between them is solid, and the whole is at rest. Suddenly the boundary is annihilated; find the subsequent motion. We suppose the motion small enough for squares of the displacements to be neglected. At all points, for $t > 0$, 19·07 (1) holds. We take P to be the excess of the pressure above the undisturbed pressure outside the sphere. As there is initially no motion, ϕ is constant everywhere, and may be taken as zero. The excess pressure is given by

$$P = -\rho\, \partial\phi/\partial t. \tag{1}$$

This is initially a positive constant P_0 when $r < a$, and 0 when $r > a$. Then we can take the subsidiary equations to be

$$\left(\frac{\partial^2}{\partial r^2} - \frac{p^2}{c^2}\right)(r\phi) = \frac{P_0}{\rho c^2} pr \quad (r < a)$$
$$= 0 \quad (r > a). \tag{2}$$

The pressure must remain finite at the centre, and the disturbance for $r > a$ cannot include any wave travelling inwards. Then

$$r\phi = -\frac{P_0 r}{\rho p} + A \sinh\frac{pr}{c} \quad (r < a),$$
$$= B e^{-pr/c} \quad (r > a). \tag{3}$$

The pressure and the radial velocity must be continuous at $r = a$; hence ϕ and $\partial\phi/\partial r$ must be continuous. These give

$$-\frac{P_0 a}{\rho p} + A \sinh\frac{pa}{c} = B e^{-pa/c}, \tag{4}$$

$$-\frac{P_0}{\rho p} + \frac{p}{c} A \cosh\frac{pa}{c} = -\frac{p}{c} B e^{-pa/c}, \tag{5}$$

whence $\quad A = \dfrac{P_0}{\rho p^2}(c + ap)\, e^{-pa/c}, \quad B = \dfrac{P_0}{2\rho p^2}(c - ap)\, e^{pa/c} - \dfrac{P_0}{2\rho p^2}(c + ap)\, e^{-pa/c}. \tag{6}$

Thus outside the original sphere

$$\frac{\rho}{P_0} r\phi = \left[\frac{1}{2p^2}(c - ap)\, e^{-p(r-a)/c} - \frac{1}{2p^2}(c + ap)\, e^{-p(r+a)/c}\right] H(t). \tag{7}$$

The associated pressure change is

$$P = -\frac{P_0}{2r}\left[\left(\frac{c}{p} - a\right) e^{-p(r-a)/c} - \left(\frac{c}{p} + a\right) e^{-p(r+a)/c}\right] H(t)$$
$$= -\frac{P_0}{2r}\left[e^{-p(r-a)/c}(ct - a) - e^{-p(r+a)/c}(ct + a)\right] H(t)$$
$$= -\frac{P_0}{2r}\left[\left\{c\left(t - \frac{r-a}{c}\right) - a\right\} H\left(t - \frac{r-a}{c}\right) - \left\{c\left(t - \frac{r+a}{c}\right) + a\right\} H\left(t - \frac{r+a}{c}\right)\right]$$
$$= -\frac{P_0}{2r}\left[(ct - r) H\left(t - \frac{r-a}{c}\right) - (ct - r) H\left(t - \frac{r+a}{c}\right)\right]. \tag{8}$$

For given $r > a$, therefore, the pressure change is zero up to time $(r-a)/c$, when the first wave from the compressed region arrives, and after $(r+a)/c$, when the wave from the most distant point passes. At intermediate times it is equal to $P_0(r - ct)/2r$. This is equal to

$P_0 a/2r$ just after the first wave arrives, $-P_0 a/2r$ when the last passes, and varies linearly with the time in between. The compression in front of the shock is associated with an equal rarefaction in the rear.*

Within the sphere the pressure is

$$\frac{P_0 p}{r}\left[\frac{r}{p}-\frac{1}{p^2}(c+ap)\sinh\frac{pr}{c}e^{-pa/c}\right] = P_0\left[1-\frac{1}{2r}\left(\frac{c}{p}+a\right)(e^{-p(a-r)/c}-e^{-p(a+r)/c})\right]. \quad (9)$$

This is equal to P_0 up to time $(a-r)/c$, then drops suddenly to $P_0(1-a/2r)$, decreases linearly with the time till it reaches $-P_0 a/2r$ at time $(a+r)/c$, and then rises suddenly to zero. The infinity in the pressure at the centre is only instantaneous, for the time the disturbance lasts at a given place is $2r/c$, which vanishes at the centre. It is due to the simultaneous arrival of elementary waves from all points on the surface; at other points the waves from different parts arrive at different times, giving a finite disturbance of pressure over a non-zero interval. If $r < \frac{1}{2}a$, the pressure becomes negative immediately on the arrival of the disturbance. Strictly, the occurrence of an infinity in the solution means that squares of the disturbances cannot be neglected within a certain range of r and t; but this range will be smaller the smaller P_0 is.

The behaviour of the velocity at distant points is similar to that of the pressure. If u is the radial velocity,

$$u = \partial\phi/\partial r = -p\phi/c - \phi/r.$$

If r is great the first term is simply pP/c, and its behaviour is inferred immediately. It is proportional to $1/r$, the second to $1/r^2$. The first term gives no total outward displacement, the outward movement during the stage of increased pressure being just cancelled by the inward movement during the stage of decreased pressure. The second term, however, gives a small velocity which vanishes at the beginning and end of the shock, and reaches a positive maximum at time r/c. It produces a total radial displacement of order a/r times the maximum given by the first term; this represents the fact that the matter originally compressed expands till it reaches normal pressure, and the surrounding matter moves outwards to make room for it.

The corresponding problem in one dimension would be that of an excess pressure P_0 within a length $2a$ of an infinite tube and suddenly released. Two waves of excess pressure $\frac{1}{2}P_0$ would travel in opposite directions; the reduction of pressure in the rear of the disturbance that we have found in three dimensions has no counterpart in one. We shall see when we deal with applications of Bessel functions that it has a very striking counterpart in two dimensions.

Disturbances that do not affect a given place until some definite instant are often conveniently called *pulses*. The sound wave from an explosion is an example; so are flashes of light from a rotating mirror and the elastic waves sent out by an earthquake.

19·09. Diverging waves produced by a sphere oscillating radially.† Suppose that a sphere of radius a begins at time 0 to oscillate radially in period $2\pi/n$. We require the motion of the air outside it.

Initially all is at rest; hence

$$r\phi = Ae^{-pr/c}. \quad (1)$$

* Cf. Stokes, *Phil. Mag.* 34, 1849, 52.
† Love, *Proc. Lond. Math. Soc.* (2) 2, 1904, 88; Bromwich, ibid. (2) 15, 1916, 431.

When $r = a$ the outward displacement is, say, $\frac{1}{n}\sin nt$ when $t > 0$, and the outward velocity $\cos nt$. Hence

$$A\left[\frac{\partial}{\partial r}\left(\frac{1}{r}e^{-pr/c}\right)\right]_{r=a} = \frac{p^2}{p^2+n^2}, \tag{2}$$

$$r\phi = -\frac{p^2 a^2}{(p^2+n^2)(1+pa/c)}\exp\left\{-\frac{p}{c}(r-a)\right\}$$

$$= -\frac{c^2 a^2}{c^2+a^2 n^2}\exp\left\{-\frac{p}{c}(r-a)\right\}\left\{\cos nt + \frac{na}{c}\sin nt - \exp\left(-\frac{ct}{a}\right)\right\}$$

$$= -\frac{c^2 a^2}{c^2+a^2 n^2}\left[\cos n\left(t-\frac{r-a}{c}\right) + \frac{na}{c}\sin n\left(t-\frac{r-a}{c}\right) - \exp\left(-\frac{ct-r+a}{a}\right)\right], \tag{3}$$

when $ct > r-a$.

The solution has a periodic part with a period equal to that of the given disturbance, together with a part dying down with the time at a rate independent of n, but involving the size of the sphere. As there is no corresponding term in the problem of the explosion we may regard it as the result of the constraint introduced by the prescription of a definite motion of the sphere. Its effect on the velocity or the pressure is to that of the second term in a ratio comparable with $(c/na)^2$.

EXAMPLES

1. A string of length $3l$ and line density ρ is under a tension $P = \rho c^2$ and fixed at its ends. Two particles of mass m are attached to the points of trisection. A transverse impulse J is given to one particle. Show that the operational solution for the displacement of the other particle is

$$\frac{PJ}{c}\frac{\sinh pl/c}{(mp\sinh pl/c + 2P/c\cosh pl/c)^2 - P^2/c^2},$$

and find the explicit solution up to time $3l/c$.

2. A heavy uniform string of length $3l$ and line density ρ is fixed at the ends, and a particle of mass m is attached at a distance l from one end. The tension is ρc^2. A transverse velocity v is given to the particle. Show that the displacement of the particle is

$$\eta = \frac{mv\sinh pl/c\, \sinh 2pl/c}{mp\sinh pl/c\, \sinh 2pl/c + \rho c\sinh 3pl/c}$$

and evaluate η up to time $4l/c$. (M.T. Sched. B, 1927.)

3. A closed pipe of length l contains air whose density is slightly greater than that of the outside air in the ratio $1+s_0:1$. Everything being at rest, the disk closing one end of the pipe is suddenly drawn aside. Show that after a time t the velocity potential is

$$\phi = -\frac{8lcs_0}{\pi^2}\sum_{r=0}^{\infty}\frac{(-1)^r}{(2r+1)^2}\cos\frac{(2r+1)\pi x}{2l}\sin\frac{(2r+1)\pi ct}{2l},$$

the origin being taken at the permanently closed end and c being the velocity of sound.
(M/c, Part III, 1932.)

4. Find the motion produced in the conditions of Ex. 3 except that the pipe is a narrow cone instead of a cylinder.

Chapter 20

CONDUCTION OF HEAT IN ONE AND THREE DIMENSIONS

Then cold, and hot, and moist, and dry,
In order to their stations leap.
    JOHN DRYDEN, *Song for St Cecilia's day*

20·01. Equation of heat conduction. The rate of transmission of heat across a surface by conduction is equal to $-k\partial V/\partial n$ per unit area, where V is the temperature, k a constant of the material called the thermal conductivity, and dn an element of the normal to the surface. Hence we can show easily that in a uniform material the rate of flow of heat into an element of volume $dx\,dy\,dz$ is $k\nabla^2 V\,dx\,dy\,dz$. But the quantity of heat required to produce a rise of temperature dV in unit mass is $c\,dV$, where c is the specific heat, and therefore that needed to produce a rise dV in unit volume is $\rho c\,dV$, where ρ is the density.* Hence V satisfies the equation

$$\frac{\partial}{\partial t}(\rho c V) = k\nabla^2 V. \tag{1}$$

If we put
$$k/\rho c = h^2, \tag{2}$$

h^2 is called the thermometric conductivity, and the equation becomes

$$\frac{\partial V}{\partial t} = h^2 \nabla^2 V. \tag{3}$$

In addition there may be some internal source of heat. If this would raise the temperature by P per unit time if it stayed where it was generated, a term P must be added to the right of (3). Chemical and radioactive changes are the chief producers of heat at internal points.

In applying the operational method of solution it is usually convenient to write $h^2 q^2$ for p. The operational solutions are then functions of q; but q must be expressed again in terms of p before interpreting.

20·02. Rod cooled at one end. Consider first a uniform rod, with its sides thermally insulated, and initially at temperature S. At time 0 the end $x = 0$ is cooled to temperature zero, and afterwards maintained at that temperature. The end $x = l$ is kept at temperature S. Find the variation of temperature at other points of the rod.

The problem being one-dimensional, the equation of heat conduction is

$$\frac{\partial V}{\partial t} - h^2 \frac{\partial^2 V}{\partial x^2} = 0, \tag{1}$$

while at time 0, $V = S$. Hence the subsidiary equation is

$$pV - h^2 \frac{\partial^2 V}{\partial x^2} = pS \tag{2}$$

or
$$\frac{\partial^2 V}{\partial x^2} - q^2 V = -q^2 S. \tag{3}$$

---

* Thermodynamic effects require some modification of this statement, since there will be in general a thermal expansion, and some of the heat is used in doing work against the pressure of the surrounding material. The correction required is serious for a gas, but not important for a solid or liquid. Cf. Jeffreys, *Cartesian Tensors*, Chapter 8, or *Proc. Camb. Phil. Soc.* 26, 1930, 101–6.

The end conditions are that $V = 0$ at $x = 0, t > 0$; $V = S$ at $x = l$. Hence

$$V = S\left(1 - \frac{\sinh q(l-x)}{\sinh ql}\right). \tag{4}$$

The operator is an even function of q and therefore a single-valued function of p. The poles are where $ql = \pm in\pi$, that is, $p = -h^2n^2\pi^2/l^2$, where n is any integer. But the negative values of $\Im(q)$ give the same values of p as the positive ones, and therefore when we apply the partial fraction rule we need consider only the positive and zero values. The part arising from $p = 0$ is

$$S\left(1 - \frac{l-x}{l}\right) = \frac{Sx}{l}. \tag{5}$$

The general term is

$$-S\frac{\sinh in\pi(l-x)/l}{(-h^2n^2\pi^2/l^2)(\cosh in\pi)(l^2/2h^2in\pi)} e^{-n^2\pi^2h^2t/l^2} = S\frac{2}{n\pi}\sin\frac{n\pi x}{l}e^{-n^2\pi^2h^2t/l^2}, \tag{6}$$

and the complete solution is

$$V = S\left[\frac{x}{l} + \sum_{n=1}^{\infty} \frac{2}{n\pi}\sin\frac{n\pi x}{l}e^{-n^2\pi^2h^2t/l^2}\right]. \tag{7}$$

If we use the Bromwich integral we have

$$V = \frac{S}{2\pi i}\int_L e^{zt}\left(1 - \frac{\sinh \zeta(l-x)}{\sinh \zeta l}\right)\frac{dz}{z}, \tag{8}$$

where $h^2\zeta^2 = z$. The integrand is a single-valued function of z with poles at 0 and $-n^2h^2\pi^2/l^2$. It is immaterial whether we define ζ to be real and positive or real and negative when z is real and positive. We take it positive. The path can be deformed as shown; and then into a loop about $z = 0$ and small circles about all the negative poles. The residue at 0 is x/l. For the negative poles, $\zeta = +in\pi/l$, and

$$z\frac{d}{dz}\sinh \zeta l = \tfrac{1}{2}\zeta\frac{d}{d\zeta}\sinh \zeta l = \tfrac{1}{2}\zeta l \cosh \zeta l = \tfrac{1}{2}in\pi(-1)^n.$$

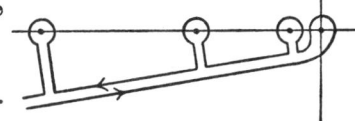

Hence the residues are

$$-e^{-n^2h^2\pi^2t/l^2}\frac{\sinh in\pi(l-x)/l}{\tfrac{1}{2}in\pi(-1)^n} = \frac{2}{n\pi}\sin\frac{n\pi x}{l}e^{-n^2\pi^2h^2t/l^2}.$$

The values $\zeta = -in\pi/l$ do not arise, since at the negative poles $\arg \zeta = \tfrac{1}{2}\arg z = \tfrac{1}{2}\pi i$ when the path is taken as shown. Hence we recover (7). The path might also be taken in either of the following two ways:

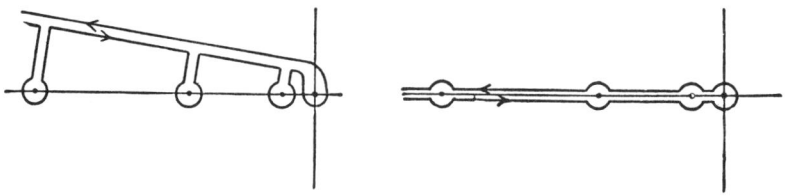

It is easy to show that either of these leads to the same result.

The series (7) is rapidly convergent if $\pi h t^{1/2}/l$ is moderate; if it is 1 the second exponential in the sum is $e^{-4} = 0{\cdot}018$, and the next $e^{-9} = 0{\cdot}0001$. At this stage there is clearly considerable cooling for half the length of the rod.

If $\pi h t^{1/2}/l$ is small the convergence is slow. In this case we can adopt a form of the expansion method applied to waves.* We write (4) in the form

$$V = S[1 - e^{-qx}(1 - e^{-2q(l-x)})(1 + e^{-2ql} + e^{-4ql} + \ldots)]. \tag{9}$$

For if we interpret this as an integral along the path L, the argument of ζ is between $\pm \tfrac{1}{4}\pi$ at all points of the path, and the series converges uniformly and absolutely. Integration term by term is therefore justifiable, and we may interpret term by term. Now

$$qx = xp^{1/2}/h, \tag{10}$$

and by 12·126 (19),
$$e^{-qx}H(t) = 1 - \operatorname{erf}\frac{x}{2ht^{1/2}} \quad (t > 0). \tag{11}$$

Hence
$$V = S\left[\operatorname{erf}\frac{x}{2ht^{1/2}} + \left(1 - \operatorname{erf}\frac{2l-x}{2ht^{1/2}}\right) - \left(1 - \operatorname{erf}\frac{2l+x}{2ht^{1/2}}\right) + \ldots\right]. \tag{12}$$

When w is great, $1 - \operatorname{erf} w$ is small compared with e^{-w^2}. If then $x/2ht^{1/2}$ is moderate, but $l/2ht^{1/2}$ large, this series is rapidly convergent, and can in most cases be reduced to its first term. This solution is therefore convenient whenever (7) is not.

The separate terms of (9) do appear to depend on which sign we take for ζ when z is real and positive, whereas the original integral does not. But if we took the negative sign the series would diverge on L. There would, however, be a convergent expansion on L with positive signs in the exponents. But since $\Re(\zeta)$ is now negative on L we are led again to precisely the same series. The choice of the positive sign for $\Re(\zeta)$ is convenient, but the negative sign would lead to the same answers if the work is done correctly.

The temperature gradient at $x = 0$ is, for large l,

$$S\left[\frac{\partial}{\partial x}(1 - e^{-qx})\right]_{x=0} = Sq = \frac{S}{h\sqrt{(\pi t)}}. \tag{13}$$

This equation played an important part in Kelvin's estimate of the age of the Earth. Neglecting the curvature of the Earth, he treated the cooling problem as one of one-dimensional flow, S being the melting point; then measures of the temperature gradient at the surface led by (13) to an estimate of t. Knowledge not available in his time has led to considerable change in the result.

20·03. One-dimensional flow of heat in a region infinite in both directions. First suppose that at time 0 the distribution of temperature is given by

$$V = H(x). \tag{1}$$

We have seen that the function
$$e^{-qx}H(t) = 1 - \operatorname{erf}\frac{x}{2ht^{1/2}} \quad (t > 0) \tag{2}$$

satisfies the differential equation
$$\frac{\partial y}{\partial t} - h^2 \frac{\partial^2 y}{\partial x^2} = 0 \tag{3}$$

---

* Heaviside, *Electromagnetic Theory*, 2, 69–79, 287–8.

for positive values of x; and it will also satisfy it for negative x, since both terms in (3) are odd functions of x. Also when t tends to zero this function tends to zero for all positive values of x, and to 2 for all negative values. It follows that the function

$$\tfrac{1}{2}(2-e^{-qx}) = \frac{1}{2}\left[1+\operatorname{erf}\frac{x}{2ht^{1/2}}\right] \tag{4}$$

satisfies the differential equation for all values of x and all positive t; and when t tends to zero it tends to 0 for negative x and to 1 for positive x; and therefore to $H(x)$. Hence this function gives the solution for positive t if $V = H(x)$ when $t = 0$.

Suppose now that the initial distribution of temperature is

$$V = f(x) = \int_{\xi=-\infty}^{\infty} f(\xi)\,dH(\xi-x). \tag{5}$$

The solution that reduces to $H(\xi-x)$ when $t = 0$ is

$$\frac{1}{2}\left[1+\operatorname{erf}\frac{\xi-x}{2ht^{1/2}}\right]. \tag{6}$$

Hence, by the principle of superposition, the solution of the more general problem is

$$V = \int_{\xi=-\infty}^{\infty} \tfrac{1}{2}f(\xi)\,d\left[1+\operatorname{erf}\frac{\xi-x}{2ht^{1/2}}\right]$$

$$= \frac{1}{\sqrt{\pi}}\int_{-\infty}^{\infty} f(\xi)\,e^{-(\xi-x)^2/4h^2t}\frac{d\xi}{2ht^{1/2}}. \tag{7}$$

Put now
$$\xi = x + 2ht^{1/2}\lambda; \tag{8}$$

then
$$V = \frac{1}{\sqrt{\pi}}\int_{-\infty}^{\infty} f(x+2ht^{1/2}\lambda)\,e^{-\lambda^2}\,d\lambda. \tag{9}$$

This is the general solution obtained by Fourier.

20·04. Imperfect cooling at the free end of a one-dimensional region. With the initial conditions of 20·02, let us suppose that the end $x = l$ is maintained at temperature S as before, but that the end $x = 0$ is not effectively cooled to temperature 0. Instead we suppose that it radiates away heat at a rate proportional to its temperature. At the same time heat is conducted to the end at a rate $k\,\partial V/\partial x$ per unit area. These effects must balance if the temperature at the surface is to vary continuously, so that instead of having $V = 0$ at the end as before we shall have a relation of the form

$$\frac{\partial V}{\partial x} - \alpha V = 0 \quad (x = 0). \tag{1}$$

The operational solution is again

$$V = S\{1 - A\sinh q(l-x)\}H(t), \tag{2}$$

where A has now to be determined to satisfy (1); then

$$qA\cosh ql - \alpha(1 - A\sinh ql) = 0, \tag{3}$$

and
$$V = S\left\{1 - \frac{\alpha \sinh q(l-x)}{q\cosh ql + \alpha \sinh ql}\right\}H(t). \tag{4}$$

The roots in p are real and negative, and we can proceed to an interpretation in partial fractions as usual. Or, using the expansion in 'waves', we have

$$V = S\left[1 - \frac{\alpha e^{-qx}}{q+\alpha}(1 - e^{-2q(l-x)})\left(1 - \frac{q-\alpha}{q+\alpha}e^{-2ql} + \ldots\right)\right]. \tag{5}$$

If the length is great enough to make the terms involving e^{-2ql} inappreciable, we can reduce this to its first two terms, thus

$$V = S\left(1 - \frac{\alpha e^{-qx}}{q+\alpha}\right). \tag{6}$$

If α is great, the solution reduces to that of 20·02: this is to be expected, for (1) then implies that $V = 0$ when $x = 0$, which is the boundary condition adopted in 20·02. If α is small, V reduces to S; the reason is that this implies that there is no loss of heat from the end, and therefore the temperature does not change anywhere. For intermediate values of α we proceed as follows. If

$$y = \frac{\alpha e^{-qx}}{q+\alpha} H(t), \tag{7}$$

Bromwich's rule gives

$$y = \frac{1}{2\pi i}\int_L \frac{\alpha h}{z^{1/2}+\alpha h}\exp\left(zt - \frac{z^{1/2}x}{h}\right)\frac{dz}{z}. \tag{8}$$

Put $z = \zeta^2$. The path for ζ is a curve from $Re^{-\frac{1}{4}\pi i}$ to $Re^{\frac{1}{4}\pi i}$, where R is great, passing the origin on the positive side. Denote this path by N. Then

$$y = \frac{1}{\pi i}\int_N \frac{\alpha h}{\zeta(\zeta+\alpha h)}\exp\left(\zeta^2 t - \frac{x\zeta}{h}\right)d\zeta$$

$$= \frac{1}{\pi i}\int_N \left(\frac{1}{\zeta} - \frac{1}{\zeta+\alpha h}\right)\exp\left(\zeta^2 t - \frac{x\zeta}{h}\right)d\zeta. \tag{9}$$

But

$$\frac{1}{\pi i}\int_N \frac{1}{\zeta}\exp\left(\zeta^2 t - \frac{x\zeta}{h}\right)d\zeta = \frac{1}{2\pi i}\int_L \exp\left(zt - \frac{z^{1/2}x}{h}\right)\frac{dz}{z}$$

$$= e^{-qx}H(t) = 1 - \operatorname{erf}\frac{x}{2ht^{1/2}} \quad (t > 0). \tag{10}$$

For the second integral in (9) we put $\zeta + \alpha h = \mu$; and this part of y is

$$-\frac{1}{\pi i}\int_N \frac{1}{\mu}\exp\left\{\mu^2 t - \mu\left(2\alpha ht + \frac{x}{h}\right)\right\}\exp(\alpha^2 h^2 t + \alpha x)\,d\mu$$

$$= -\exp(\alpha^2 h^2 t + \alpha x)\left(1 - \operatorname{erf}\frac{x + 2\alpha h^2 t}{2ht^{1/2}}\right). \tag{11}$$

Hence

$$y = 1 - \operatorname{erf}\xi - \exp(\gamma^2 + \alpha x)\{1 - \operatorname{erf}(\xi + \gamma)\}, \tag{12}$$

where $\xi = x/2ht^{1/2}$, $\gamma = \alpha h t^{1/2}$; and

$$V = S[\operatorname{erf}\xi + \exp(\gamma^2 + \alpha x)\{1 - \operatorname{erf}(\xi + \gamma)\}]. \tag{13}$$

This is the same as Riemann's solution.*

The temperature at $x = 0$ is $S\exp\gamma^2(1 - \operatorname{erf}\gamma)$, whence the gradient at the end follows by (1). For t small this has a convergent expansion in negative powers of t and therefore

* Riemann-Weber, *Partielle Differentialgleichungen*, 2, 1912, 95–8.

falls continuously. The temperature fall at $x = 0$ is not instantaneous, nor the gradient momentarily infinite, as in 20·02. For great values of t we can use the asymptotic expansion for erf γ:

$$V_{x=0} \sim \frac{S}{\alpha h \sqrt{(\pi t)}} \left[1 - \frac{1}{2} \frac{1}{\alpha^2 h^2 t} + \frac{1 \cdot 3}{2 \cdot 2} \left(\frac{1}{\alpha^2 h^2 t}\right)^2 - \cdots\right]. \tag{14}$$

This is equivalent to one found by Heaviside.*

20·05. A long rod is fastened at the end $x = 0$, the other end $x = l$ being free. Initially it is at temperature 0, but at time 0 the clamped end is raised to temperature S and kept there. Each part of the rod loses heat by radiation and convection at a rate proportional to its temperature.†

The differential equation is now

$$\frac{\partial V}{\partial t} = h^2 \frac{\partial^2 V}{\partial x^2} - \alpha^2 V, \tag{1}$$

where α is a constant. Put

$$p + \alpha^2 = h^2 r^2, \tag{2}$$

and write the equation

$$\frac{\partial^2 V}{\partial x^2} = r^2 V. \tag{3}$$

The solution for a long rod is

$$V = S e^{-rx} = \frac{S}{2\pi i} \int_L \exp\left\{zt - \frac{(z+\alpha^2)^{1/2} x}{h}\right\} \frac{dz}{z}. \tag{4}$$

Put $z + \alpha^2 = \zeta^2$; then

$$V = \frac{S}{\pi i} \int_N \exp\left\{(\zeta^2 - \alpha^2)t - \frac{\zeta x}{h}\right\} \frac{\zeta\, d\zeta}{\zeta^2 - \alpha^2}$$

$$= \frac{S}{2\pi i} \int_N \exp\left\{(\zeta^2 - \alpha^2)t - \frac{\zeta x}{h}\right\} \left\{\frac{1}{\zeta - \alpha} + \frac{1}{\zeta + \alpha}\right\} d\zeta. \tag{5}$$

But if

$$\zeta = \alpha + \mu, \tag{6}$$

the term in $1/(\zeta - \alpha)$ becomes

$$\frac{S}{2\pi i} \int_N \exp\left\{\mu^2 t - \mu\left(\frac{x}{h} - 2\alpha t\right)\right\} \exp\left(-\frac{\alpha x}{h}\right) \frac{d\mu}{\mu} = \tfrac{1}{2} S \exp\left(-\frac{\alpha x}{h}\right) \left(1 - \operatorname{erf}\frac{x - 2\alpha h t}{2 h t^{1/2}}\right), \tag{7}$$

as in 20·04 (11). The complete solution is

$$V = \tfrac{1}{2} S \left[\exp\left(-\frac{\alpha x}{h}\right)\left(1 - \operatorname{erf}\frac{x - 2\alpha h t}{2 h t^{1/2}}\right) + \exp\left(\frac{\alpha x}{h}\right)\left(1 - \operatorname{erf}\frac{x + 2\alpha h t}{2 h t^{1/2}}\right)\right]. \tag{8}$$

If $\alpha t^{1/2}$ is small the error functions are practically unity so long as $x/2ht^{1/2}$ is large, and V is very small. V can be comparable with S for small values of the time only if $x/2ht^{1/2}$ is not large, and then $\alpha x/h$ is small. Then the solution is practically $S(1 - \operatorname{erf} x/2ht^{1/2})$, which is the solution in the absence of radiation from the sides.

If $\alpha t^{1/2}$ is large and $x/2ht^{1/2}$ not large, the first error function is nearly -1, and the second $+1$, and $V \doteqdot S e^{-\alpha x/h}$. This is the solution for a steady state, and will hold so long as $\alpha t^{1/2} - x/2ht^{1/2}$ is large and positive, even if $x/2ht^{1/2}$ is itself large. The steady state is therefore reached approximately, for given x, when both $t > \alpha^{-2}$, and $t > x/2\alpha h$.

* *Electromagnetic Theory*, 2, 15.
† Ingen-Hausz's experiment: cf. Edser, *Heat for Advanced Students*, 1908, p. 424.

20·06. Cooling of the Earth

If there are several rods of different conductivities but similar surfaces, so that α is the same for all, the values of x where V attains a given steady value will be proportional to h.

20·06. The cooling of the Earth. Cooling in the Earth since it first became solid has not had time to become appreciable except at depths small compared with the radius. It is therefore legitimate to neglect the effects of curvature and treat the problem as one-dimensional. Radiation from the outer surface must have soon reduced the temperature to that maintained by solar radiation, so that we may suppose the surface temperature to be constant and adopt it as our zero of temperature. The chief difference from the problem of 20·02 is that we must allow for the heating effect of radioactivity in the outer layers. Suppose first that the quantity P defined in 20·01 is equal to a constant A down to a depth H and zero below that depth. Take the initial temperature to be $S + mx$ $(x > 0)$, where m is a constant. Then the subsidiary equation is

$$\frac{\partial^2 V}{\partial x^2} - q^2 V = -\frac{A}{h^2} - q^2(S + mx) \quad (0 < x < H) \tag{1}$$

$$= -q^2(S + mx) \quad (H < x), \tag{2}$$

and the solutions are

$$V = \frac{A}{q^2 h^2} + S + mx + Be^{-qx} + Ce^{qx} \quad (0 < x < H) \tag{3}$$

$$= S + mx + De^{-qx} \quad (H < x). \tag{4}$$

A term in e^{qx} is not required in the solution for great depths, because it would imply that the temperature dropped suddenly by a finite amount in consequence of a disturbance at the surface. V must vanish at $x = 0$, and V and $\partial V/\partial x$ are continuous at $x = H$. Hence

$$\left.\begin{array}{l} B + C + S + A/p = 0, \\ Be^{-qH} + Ce^{qH} + A/p = De^{-qH}, \\ Be^{-qH} - Ce^{qH} = De^{-qH}. \end{array}\right\} \tag{5}$$

Solving and substituting in (3), (4) we find

$$V = S(1 - e^{-qx}) + mx + \frac{A}{p}\{1 - e^{-qx} - e^{-qH} \sinh qx\} \quad (0 < x < H), \tag{6}$$

$$V = S(1 - e^{-qx}) + mx + \frac{A}{p}(\cosh qH - 1)e^{-qx} \quad (H < x). \tag{7}$$

The solutions and their derivatives with regard to x involve operators of the forms $q^{-1}e^{-qx}$, $q^{-2}e^{-qx}$. These can be evaluated by using Bromwich's integral and integrating by parts; or we can start with

$$e^{-qx} = 1 - \operatorname{erf}\frac{x}{2ht^{1/2}}, \tag{8}$$

and if

$$\Phi_1(u) = \int_u^\infty (1 - \operatorname{erf} v)\,dv, \quad \Phi_2(u) = \int_u^\infty \Phi_1(v)\,dv, \tag{9}$$

$$q^{-1}e^{-qx} = 2ht^{1/2}\Phi_1\!\left(\frac{x}{2ht^{1/2}}\right), \quad q^{-2}e^{-qx} = 4h^2 t \Phi_2\!\left(\frac{x}{2ht^{1/2}}\right). \tag{10}$$

The explicit forms of Φ_1 and Φ_2 are

$$\Phi_1(u) = \frac{1}{\sqrt{\pi}} e^{-u^2} - u(1-\operatorname{erf} u), \tag{11}$$

$$\Phi_2(u) = (\tfrac{1}{4}+\tfrac{1}{2}u^2)(1-\operatorname{erf} u) - \frac{1}{2\sqrt{\pi}} u e^{-u^2}. \tag{12}$$

The functions have been partly tabulated by Jeffreys and Hartree (cf. 23·08). They are, however, intimately related to the functions Hh_1 and Hh_2 treated later and tabulated in the *British Association Tables*; it would be useful to have the latter tabulated to four figures at a short enough interval to permit linear interpolation.

In the actual problem a considerable simplification arises from the fact that H is small compared with $2ht^{1/2}$. On this account we can expand the solutions in powers of H and retain only the earlier terms (the path for the Bromwich integral being modified to M). Then for the surface temperature gradient we have

$$\begin{aligned}\left(\frac{\partial V}{\partial x}\right)_{x=0} &= Sq+m+\frac{A}{p}(q-qe^{-qH}) \doteq Sq+m+\frac{A}{h^2q^2}q(qH-\tfrac{1}{2}q^2H^2)+\ldots \\ &= Sq+m+\frac{AH}{h^2}(1-\tfrac{1}{2}qH)+\ldots \\ &= \left(S-\frac{1}{2}\frac{AH^2}{h^2}\right)\frac{1}{h\sqrt{(\pi t)}}+m+\frac{AH}{h^2}+\ldots \end{aligned} \tag{13}$$

and for the temperature at depths greater than H

$$\begin{aligned}V &= S(1-e^{-qx})+mx+\frac{A}{h^2q^2}\tfrac{1}{2}q^2H^2 e^{-qx} \\ &\doteq mx+\left(S-\frac{1}{2}\frac{AH^2}{h^2}\right)\operatorname{erf}\frac{x}{2ht^{1/2}}+\frac{AH^2}{2h^2}. \end{aligned} \tag{14}$$

The age of the earth is now known to be of the order of 2×10^9 years; with this value the term AH/h^2 in (13) accounts for about $\tfrac{2}{3}$ of the observed temperature gradient.

An alternative possibility is that the radioactive generation of heat, instead of being confined to a uniform surface layer, may decrease exponentially with depth. The subsidiary equation becomes

$$\frac{\partial^2 V}{\partial x^2} - q^2 V = -\frac{A}{h^2} e^{-\alpha x} - q^2(S+mx) \tag{15}$$

at all depths. We already know the part of the solution contributed by $S+mx$. The remainder is

$$W = \frac{A}{h^2}\frac{e^{-\alpha x}-e^{-qx}}{q^2-\alpha^2}. \tag{16}$$

But

$$\frac{e^{-\alpha x}}{h^2(q^2-\alpha^2)} = \frac{e^{-\alpha x}}{p-h^2\alpha^2} = \frac{1}{h^2\alpha^2}e^{-\alpha x}(e^{h^2\alpha^2 t}-1), \tag{17}$$

and

$$\begin{aligned}-\frac{\alpha^2 e^{-qx}}{q^2-\alpha^2} &= \frac{1}{2}\left(\frac{\alpha}{q+\alpha}-\frac{\alpha}{q-\alpha}\right)e^{-qx} \\ &= \tfrac{1}{2}[e^{-qx}-\exp(\gamma^2+\alpha x)\{1-\operatorname{erf}(\xi+\gamma)\}] \\ &\quad + \tfrac{1}{2}[e^{-qx}-\exp(\gamma^2-\alpha x)\{1-\operatorname{erf}(\xi-\gamma)\}], \end{aligned} \tag{18, 19}$$

where

$$\xi = x/2ht^{1/2},\quad \gamma = \alpha h t^{1/2}. \tag{20}$$

Hence, collecting the terms,

$$W = \frac{A}{h^2\alpha^2}[1 - \mathrm{erf}\,\xi - e^{-\alpha x} - \tfrac{1}{2}\exp(\gamma^2 + \alpha x)\{1 - \mathrm{erf}(\xi+\gamma)\} + \tfrac{1}{2}\exp(\gamma^2 - \alpha x)\{1 + \mathrm{erf}(\xi-\gamma)\}], \quad (21)$$

which is the same as the solution obtained by Ingersoll and Zobel.*

The contribution of radioactivity to the temperature gradient at the surface is

$$\left(\frac{\partial W}{\partial x}\right)_{x=0} = \frac{A}{h^2}\frac{1}{q+\alpha} = \frac{A}{h^2\alpha}\{1 - \exp\gamma^2(1 - \mathrm{erf}\,\gamma)\}. \quad (22)$$

When γ is great, as it actually is,

$$\left(\frac{\partial W}{\partial x}\right)_{x=0} \doteqdot \frac{A}{h^2\alpha}\left(1 - \frac{1}{\gamma\sqrt{\pi}}\right) = \frac{A}{h^2\alpha}\left(1 - \frac{1}{\alpha h\sqrt{(\pi t)}}\right). \quad (23)$$

To reconcile the various data it is necessary that the radioactivity must be practically confined to the outermost 20 km.

20·07. *A spherical thermometer bulb is initially at a uniform temperature equal to that of its surroundings. The temperature of the air decreases with height, and the thermometer is carried upwards in a balloon at such a rate that the temperature at the outside of the glass varies linearly with the time. Find how the mean temperature of the mercury varies.†*

The temperature within the bulb satisfies the equation

$$\frac{\partial V}{\partial t} = h^2 \nabla^2 V = \frac{h^2}{r}\frac{\partial^2}{\partial r^2}(rV), \quad (1)$$

and the subsidiary equation is

$$\frac{\partial^2}{\partial r^2}(rV) = q^2 rV. \quad (2)$$

The solution finite at the centre is $V = (A/r)\sinh qr$, where A is a function of t. The temperature at the outer surface of the glass is Gt, where G is a constant. But the glass has only a finite conductivity, and the surface condition at the outside of the mercury is nearly

$$\frac{\partial V}{\partial r} = K(Gt - V), \quad (3)$$

K depending on the ratio of the two conductivities and the thickness of the glass. Then

$$A = \frac{a^2 KG/p}{Ka\sinh qa + qa\cosh qa - \sinh qa}, \quad (4)$$

where a is the inner radius of the glass.

The mean temperature of the mercury is

$$V_0 = \frac{3}{a^3}\int_0^a r^2 V\,dr = \frac{3A}{a^3}\int_0^a r\sinh qr\,dr$$

$$= \frac{3KG}{apq^2}\frac{qa\cosh qa - \sinh qa}{Ka\sinh qa + qa\cosh qa - \sinh qa}. \quad (5)$$

* *Mathematical Theory of Heat Conduction*, Ginn, 1913.
† A. R. McLeod, *Phil. Mag.* (6) 37, 1919, 134; Bromwich, *ibid.* (6) 37, 1919, 407–19.

In applying the partial fraction rule, we notice that near $p = 0$

$$V_0 = \frac{3KG}{apq^2} \frac{(qa)^3(\tfrac{1}{3} + \tfrac{1}{30}q^2a^2)}{qa(Ka + \tfrac{1}{6}Kq^2a^3 + \tfrac{1}{3}q^2a^2)}$$

$$= G\left\{\frac{1}{p} - \left(\frac{1}{15}\frac{a^2}{h^2} + \frac{1}{3}\frac{a}{h^2K}\right)\right\}$$

$$= G\left\{t - \frac{a^2}{h^2}\left(\frac{1}{15} + \frac{1}{3aK}\right)\right\}. \tag{6}$$

Thus there is a systematic lag in the temperature of the mercury in comparison with that of the air, which must be allowed for in the measurement of upper-air temperatures.

The other zeros of the denominator give exponential contributions, which are evaluated in Bromwich's paper.

20·08. Periodic supply of heat. Reference was made in 18·012 to the possibility of periodic solutions of the equation of heat conduction when there is a periodic source of heat. As an example, we consider a one-dimensional region where the temperature at $x = 0$ is given to be $S\cos\gamma t$. Regard this as the real part of $S\exp i\gamma t$. Then

$$h^2 \frac{\partial^2 V}{\partial x^2} = \frac{\partial V}{\partial t} = i\gamma V,$$

and the solution tending to zero for large x is

$$V = Se^{-\lambda x + i\gamma t},$$

where $\qquad h^2\lambda^2 = i\gamma, \quad \lambda = \frac{1+i}{h}\sqrt{(\tfrac{1}{2}\gamma)} = (1+i)\kappa.$

Then $\qquad V = S\exp\{i\gamma t - (1+i)\kappa x\},$

and taking the real part $\qquad V = Se^{-\kappa x}\cos(\gamma t - \kappa x).$

The variation is periodic but its amplitude falls off exponentially with depth, and the phase is continually retarded. At a depth π/κ the phase is opposite to what it is at the end and the amplitude is reduced in the ratio 1 to $e^{-\pi}$. The changes are more rapid with depth the shorter the period.

This is observed to occur for the diurnal and annual variations of temperature in the ground, the former being inappreciable at depths more than about 1 and the latter about 18 metres. It is important in meteorology, because the ocean is turbulent and heat is transferred to much greater depths by mixing. This, even more than the difference of specific heats, accounts for the greater ability of the ocean than the land to store the heat it receives during the summer, and to warm the air passing over it in the winter.

EXAMPLES

1. A uniform conducting sphere of radius a and thermometric conductivity h^2 is initially at temperature 0. Heat is supplied uniformly throughout the sphere in such a way that the temperature would rise at a rate P if there was no conduction. The outside is maintained at temperature 0. Show that at any subsequent time the temperature at any point is

$$V = \frac{1}{6}\frac{P}{h^2}(a^2 - r^2) + \Sigma \frac{2P}{h^2 r}\left(\frac{a}{n\pi}\right)^3(-1)^n \sin\frac{n\pi r}{a}\exp(-h^2 n^2 \pi^2 t/a^2).$$

If t is small show that an approximation to $-\partial V/\partial r$ at the outside is

$$P\left(\frac{2}{h}\sqrt{\frac{t}{\pi}} - \frac{t}{a}\right).$$ (M.T. Sched. B, 1926.)

2. A uniform sphere, originally at temperature S, cools from the surface. The temperature gradient at the surface is $-\kappa$ times the temperature there. Prove that after a long time the temperature at the surface is nearly

$$S\exp\left\{\frac{-3\kappa h^2 t}{a}\right\}.$$ (M.T. Sched. B, 1928.)

3. A long uniform string is stretched so as to propagate waves with velocity c. There is a resistance to transverse movement capable of producing a retardation equal to λ times the velocity. One end is suddenly drawn aside a distance y_0. Prove that (i) there is no motion at distance x from that end until time x/c, (ii) the slope at the disturbed end is asymptotically

$$-\frac{y_0}{c}\left(\frac{\lambda}{\pi t}\right)^{1/2}\left(1 - \frac{1}{4\lambda t} - \frac{3}{32\lambda^2 t^2} - \ldots\right).$$ (M.T. Sched. B, 1928.)

4. A sphere of radius a with initial temperature V_0 is surrounded by an infinite medium of the same material, and with initial temperature 0. Prove that the temperature at distance $r(>a)$ from the centre at time t is

$$V = \frac{V_0}{\sqrt{\pi}}\left\{\int_c^b e^{-\lambda^2}d\lambda - \frac{h\sqrt{t}}{r}(e^{-b^2} - e^{-c^2})\right\},$$

where
$$b = \frac{r-a}{2h\sqrt{t}}, \quad c = \frac{r+a}{2h\sqrt{t}}.$$ (I.C. 1942.)

5. Determine the solution z, valid in the range $0 < x < \pi, t > 0$, of the equation

$$\frac{\partial z}{\partial t} = \frac{\partial^2 z}{\partial x^2},$$

such that $z = 0$ for $x = 0$ and $x = \pi$, and $z = x$ for $t = 0$. (I.C. 1941.)

Chapter 21

BESSEL FUNCTIONS

'Mine is a long and a sad tale!' said the Mouse, turning to Alice, and sighing.
'It *is* a long tail, certainly', said Alice, looking down with wonder at the Mouse's tail; 'but why do you call it sad?'

LEWIS CARROLL, *Alice's Adventures in Wonderland*

21·01. Definitions of $J_{\pm n}(x)$, $I_{\pm n}(x)$. We have already had (16·10, 18·03) the functions

$$J_n(x) = \sum_{r=0}^{\infty} (-1)^r \frac{(\tfrac{1}{2}x)^{n+2r}}{r!(n+r)!}, \quad J_{-n}(x) = \sum_{r=0}^{\infty} (-1)^r \frac{(\tfrac{1}{2}x)^{-n+2r}}{r!(-n+r)!}. \tag{1}$$

The corresponding series with all the signs taken positive are

$$I_n(x) = \sum_{r=0}^{\infty} \frac{(\tfrac{1}{2}x)^{n+2r}}{r!(n+r)!}, \quad I_{-n}(x) = \sum_{r=0}^{\infty} \frac{(\tfrac{1}{2}x)^{-n+2r}}{r!(-n+r)!}. \tag{2}$$

Clearly
$$I_n(x) = e^{-\tfrac{1}{2}n\pi i} J_n(xe^{\tfrac{1}{2}\pi i}) = e^{\tfrac{1}{2}n\pi i} J_n(xe^{-\tfrac{1}{2}\pi i}) \tag{3}$$

$$I_{-n}(x) = e^{\tfrac{1}{2}n\pi i} J_{-n}(xe^{\tfrac{1}{2}\pi i}) = e^{-\tfrac{1}{2}n\pi i} J_{-n}(xe^{-\tfrac{1}{2}\pi i}). \tag{4}$$

The differential equation satisfied by J_n and J_{-n} is

$$x\frac{d}{dx}\left(x\frac{dy}{dx}\right) + (x^2 - n^2)y = 0, \tag{5}$$

and that satisfied by I_n and I_{-n} is

$$x\frac{d}{dx}\left(x\frac{dy}{dx}\right) - (x^2 + n^2)y = 0. \tag{6}$$

I_n and I_{-n} are often called Bessel functions of imaginary argument.

21·011. Complex integrals: operational forms. Convenient complex integrals for these functions can be found by starting with the operator in terms of t real and positive,

$$p^{-n} \exp(ap^{-1}) = p^{-n} \sum_{r=0}^{\infty} \frac{a^r}{r!p^r} = \sum \frac{a^r}{r!p^{n+r}} \tag{7}$$

$$p^{-n} \exp(ap^{-1}) H(t) = a^{-\tfrac{1}{2}n} t^{\tfrac{1}{2}n} \sum_{r=0}^{\infty} \frac{a^{\tfrac{1}{2}n+r} t^{\tfrac{1}{2}n+r}}{r!(n+r)!} H(t) = \left(\frac{t}{a}\right)^{\tfrac{1}{2}n} I_n\{2\sqrt{(at)}\} H(t). \tag{8}$$

Hence for $n > -1$, $t > 0$,

$$\left(\frac{t}{a}\right)^{\tfrac{1}{2}n} I_n\{2\sqrt{(at)}\} = \frac{1}{2\pi i} \int_L \exp\left(zt + \frac{a}{z}\right) \frac{dz}{z^{n+1}}. \tag{9}$$

If we now modify the path to M, which has termini at $\Re(z) = -\infty$ and crosses the positive real axis, the integral is significant for all n and can be used also to express $I_{-n}\{2\sqrt{(at)}\}$. Again, we may put

$$zt = u. \tag{10}$$

Then (9) becomes
$$\frac{1}{2\pi i} t^n \int_M \exp\left(u + \frac{at}{u}\right) \frac{du}{u^{n+1}} \quad (11)$$

and
$$I_n\{2\sqrt{(at)}\} = \frac{1}{2\pi i} (at)^{1/2 n} \int_M \exp\left(u + \frac{at}{u}\right) \frac{du}{u^{n+1}} \quad (12)$$

which is valid without restriction on t and n.

If we now put $t = \tfrac{1}{4}x^2$, $a = 1$, we have
$$I_n(x) = \frac{1}{2\pi i} (\tfrac{1}{2}x)^n \int_M \exp\left(u + \frac{1}{4}\frac{x^2}{u}\right) \frac{du}{u^{n+1}} \quad (13)$$

and with $a = -1$,
$$J_n(x) = \frac{1}{2\pi i} (\tfrac{1}{2}x)^n \int_M \exp\left(u - \frac{1}{4}\frac{x^2}{u}\right) \frac{du}{u^{n+1}} \quad (14)$$

as is obvious from comparison of the series (1) and (2).

Another interesting form is got by putting
$$u = \tfrac{1}{2}x\lambda; \quad (15)$$

then
$$I_n(x) = \frac{1}{2\pi i} \int_M \exp\left\{\tfrac{1}{2}x\left(\lambda + \frac{1}{\lambda}\right)\right\} \frac{d\lambda}{\lambda^{n+1}}, \quad (16)$$

$$J_n(x) = \frac{1}{2\pi i} \int_M \exp\left\{\tfrac{1}{2}x\left(\lambda - \frac{1}{\lambda}\right)\right\} \frac{d\lambda}{\lambda^{n+1}} \quad (17)$$

valid if the termini of M are where $\Re(x\lambda) = -\infty$. These are Schläfli's integrals. Now in (16) put
$$\lambda + \frac{1}{\lambda} = 2\mu; \quad (18)$$

then
$$\lambda = \mu + (\mu^2 - 1)^{1/2}, \quad (19)$$

the positive sign being taken for μ real and > 1. Then
$$I_n(x) = \frac{1}{2\pi i} \int_M e^{\mu x} \frac{d\mu}{(\mu^2 - 1)^{1/2} \{\mu + (\mu^2 - 1)^{1/2}\}^n} \quad (20)$$

and similarly
$$J_n(x) = \frac{1}{2\pi i} \int_M e^{\mu x} \frac{d\mu}{(\mu^2 + 1)^{1/2} \{\mu + (\mu^2 + 1)^{1/2}\}^n}. \quad (21)$$

Thus we have the operational forms for $I_n(x)$ and $J_n(x)$ for $n > -1$,
$$I_n(x) H(x) = \frac{p}{(p^2 - 1)^{1/2} \{p + (p^2 - 1)^{1/2}\}^n} H(x), \quad (22)$$

$$J_n(x) H(x) = \frac{p}{(p^2 + 1)^{1/2} \{p + (p^2 + 1)^{1/2}\}^n} H(x). \quad (23)$$

21·02. The Hankel functions $\mathrm{Hs}_n(x)$, $\mathrm{Hi}_n(x)$; $Y_n(x)$. Now integrals of the forms (13), (14) will also satisfy the differential equations if one terminus is taken at $u = 0$, provided that the approach to 0 is in such a direction that $\dfrac{x^2}{u} \to -\infty$ for $I_n(x)$ and to $+\infty$ for $J_n(x)$.

This may be verified directly by the method of 16·10. It is convenient again to take x real, and to use (16), (17). Then $\lambda+1/\lambda$ is stationary at ± 1 and $\lambda-1/\lambda$ at $\pm i$. These are saddle-points and we may take the paths through them as follows, for $I_n(x)$

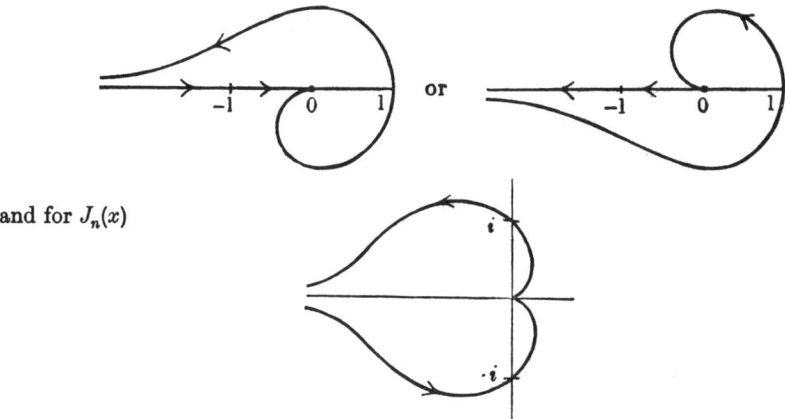

and for $J_n(x)$

In either case the sum of the integrals along the two paths, in the directions indicated, is $I_n(x)$ or $J_n(x)$, as the case may be, since the paths are together deformable into M.

In the latter figure we denote the integral along the upper path by $\tfrac12 \mathrm{Hs}_n(x)$ and that along the lower by $\tfrac12 \mathrm{Hi}_n(x)$.* Then

$$2J_n(x) = \mathrm{Hs}_n(x) + \mathrm{Hi}_n(x). \tag{24}$$

Since the dominant parts of the integrands near the respective saddle-points are $\exp(\pm ix)$, we see that the three functions are related in the same sort of way as $\cos x$, $\exp ix$, and $\exp(-ix)$, and if we also take

$$2iY_n(x) = \mathrm{Hs}_n(x) - \mathrm{Hi}_n(x) \tag{25}$$

$Y_n(x)$ will be analogous to $\sin x$. Also

$$\mathrm{Hs}_n(x) = \frac{1}{\pi i}\int_{0,i}^{-\infty} \exp\left\{\tfrac12 x\left(\lambda-\frac{1}{\lambda}\right)\right\} \frac{d\lambda}{\lambda^{n+1}}, \tag{26}$$

$$\mathrm{Hi}_n(x) = \frac{1}{\pi i}\int_{-\infty,-i}^{0} \exp\left\{\tfrac12 x\left(\lambda-\frac{1}{\lambda}\right)\right\} \frac{d\lambda}{\lambda^{n+1}}. \tag{27}$$

The way of writing the limits in (26) means that the path goes from 0 to $-\infty$ by way of i. These integrals, being analytic functions of x, will also be solutions of the differential equation for complex x such that $\Re(x) > 0$.

The paths are transformed into themselves by the substitution

$$\lambda = -1/u. \tag{28}$$

---

* These functions were introduced by Nielsen and denoted by $H_1^n(x)$, $H_2^n(x)$. Watson denotes them by $H_n^{(1)}(x)$ and $H_n^{(2)}(x)$. The former notation has the disadvantage that the same n is a suffix in $J_n(x)$; the latter is awkward in printing and writing, and almost impossible on a typewriter. We use $\mathrm{Hs}_n(x)$ and $\mathrm{Hi}_n(x)$, the 's' and 'i' meaning 'superior' and 'inferior' in accordance with the paths taken for the Schläfli integrals. Other notations are in use.

But if we are to maintain the rule that arguments are to be taken as 0 (not 2π) for quantities on the positive real axis we must take respectively,

$$\text{for (26)}, \quad \lambda = e^{i\pi}/u; \quad \text{for (27)}, \quad \lambda = e^{-i\pi}/u \tag{29}$$

and then
$$\operatorname{Hs}_n(x) = \frac{1}{\pi i} e^{-n\pi i} \int_{0,i}^{-\infty} \exp \tfrac{1}{2} x \left(u - \frac{1}{u}\right) u^{n-1} du = e^{-n\pi i} \operatorname{Hs}_{-n}(x), \tag{30}$$

$$\operatorname{Hi}_n(x) = e^{n\pi i} \operatorname{Hi}_{-n}(x). \tag{31}$$

But we have also
$$2J_{-n}(x) = \operatorname{Hs}_{-n}(x) + \operatorname{Hi}_{-n}(x) = e^{n\pi i} \operatorname{Hs}_n(x) + e^{-n\pi i} \operatorname{Hi}_n(x), \tag{32}$$

and therefore

$$i \sin n\pi \, \operatorname{Hs}_n(x) = J_{-n}(x) - e^{-n\pi i} J_n(x), \quad i \sin n\pi \, \operatorname{Hi}_n(x) = e^{n\pi i} J_n(x) - J_{-n}(x). \tag{33}$$

All functions in the last two relations being analytic in both x and n, they can be taken as definitions of $\operatorname{Hs}_n(x)$ and $\operatorname{Hi}_n(x)$ except when $\sin n\pi = 0$. In all cases they will be equal to integrals derivable from (26) and (27) by continuous modification of the path in such a way that $\Re(x\lambda) \to -\infty$ at one terminus. Also from (25)

$$Y_n(x) = \frac{J_n(x) \cos n\pi - J_{-n}(x)}{\sin n\pi}. \tag{34}$$

When n is a positive integer $J_{-n}(x)$ reduces to $(-1)^n J_n(x)$; but when n approaches a positive integer this expression for Y_n tends to a definite limit, which we can then take as the definition of $Y_n(x)$. We take then*

$$Y_n(x) = \lim_{\epsilon \to 0} \frac{J_{n+\epsilon}(x) \cos(n+\epsilon)\pi - J_{-n-\epsilon}(x)}{\sin(n+\epsilon)\pi}$$

$$= \lim \frac{J_{n+\epsilon}(x) - (-1)^n J_{-n-\epsilon}(x)}{\pi \epsilon}$$

$$= \frac{1}{\pi}\left(\frac{\partial}{\partial n} J_n(x) - (-1)^n \frac{\partial}{\partial n} J_{-n}(x)\right). \tag{35}$$

The terms in (1) fall into two classes according as there is an infinite factorial in the denominator or not. The general term of $J_n(x)$ is

$$u_{n,r}(x) = (-1)^r \frac{(\tfrac{1}{2}x)^{n+2r}}{r!\,(n+r)!}, \tag{36}$$

and
$$\frac{\partial}{\partial n} u_{n,r}(x) = (-1)^r \frac{(\tfrac{1}{2}x)^{n+2r}}{r!\,(n+r)!} \{\log \tfrac{1}{2} x - F(n+r)\}, \tag{37}$$

where F is the digamma function. A similar result holds for the terms of $J_{-n}(x)$ with $r \geqslant n$. But for J_{-n} with $r < n$ we have

$$u_{-n-\epsilon,r} = (-1)^r \frac{(\tfrac{1}{2}x)^{-n-\epsilon+2r}}{r!\,(-n-\epsilon+r)!} = (-1)^r \frac{(\tfrac{1}{2}x)^{-n-\epsilon+2r}}{r!\,(-\epsilon)!}(-\epsilon)(-\epsilon-1)\ldots(-\epsilon-n+r+1), \tag{38}$$

$$\frac{\partial u_{-n,r}}{\partial n} = (-1)^r \frac{(\tfrac{1}{2}x)^{-n+2r}}{r!\,0!}(-1)^{n-r-2}(n-r-1)! = (-1)^n \frac{(n-r-1)!}{r!}(\tfrac{1}{2}x)^{-n+2r}. \tag{39}$$

---

* An astonishing variety of notations exists in different accounts of Bessel functions. Watson's and our Y_n is called N_n by Jahnke and Emde, $-G_n$ by Heaviside, and $\pm 2K_n/\pi$ by various other writers. $G_n(x)$ of Gray, Mathews and MacRobert is $\tfrac{1}{2}i\pi \operatorname{Hs}_n(x)$. G_n has also been used in other senses. For details of notation for this and other Bessel functions see *Mathematical Tables and other Aids to Computation*, 1, 1944, 207–308.

Hence
$$Y_n(x) = \frac{1}{\pi}\sum_{r=0}^{\infty}(-1)^r\frac{(\tfrac{1}{2}x)^{n+2r}}{r!(n+r)!}\{\log\tfrac{1}{2}x - F(n+r)\}$$
$$+\frac{1}{\pi}(-1)^n\sum_{r=n}^{\infty}(-1)^r\frac{(\tfrac{1}{2}x)^{-n+2r}}{r!(-n+r)!}\{\log\tfrac{1}{2}x - F(-n+r)\}$$
$$-\frac{1}{\pi}\sum_{r=0}^{n-1}\frac{(n-r-1)!}{r!}(\tfrac{1}{2}x)^{-n+2r}. \tag{40}$$

Put $r = n+s$ in the second series; it becomes
$$\frac{1}{\pi}\sum_{s=0}^{\infty}(-1)^s\frac{(\tfrac{1}{2}x)^{n+2s}}{s!(n+s)!}\{\log\tfrac{1}{2}x - F(s)\}, \tag{41}$$

and
$$Y_n(x) = \frac{2}{\pi}J_n(x)\log\tfrac{1}{2}x - \frac{1}{\pi}\sum_{r=0}^{\infty}(-1)^r\frac{(\tfrac{1}{2}x)^{n+2r}}{r!(n+r)!}\{F(r) + F(n+r)\}$$
$$-\frac{1}{\pi}\sum_{r=0}^{n-1}\frac{(n-r-1)!}{r!}(\tfrac{1}{2}x)^{-n+2r} \tag{42}$$

for n a positive integer or 0. There is always a singularity of Y_n at the origin, and it therefore cannot arise in any solution that holds at the centre of a circle. It can, however, arise in solutions that hold between two concentric circles.

21·021. Integrals on real paths for $J_n(x)$, $Y_n(x)$. Another integral expression for $\text{Hs}_n(x)$ can be got by specifying the path of (26) definitely as follows.

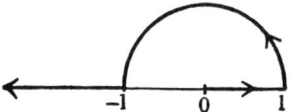

On the stretch 0 to $+1$ put $\lambda = e^{-u}$; then this part is
$$\frac{1}{\pi i}\int_0^{\infty}\exp\{\tfrac{1}{2}x(e^{-u}-e^u)\}e^{nu}du = \frac{1}{\pi i}\int_0^{\infty}\exp(-x\sinh u + nu)du. \tag{43}$$

On the semicircle $+1$ to -1 put $\lambda = \exp(i\theta)$; then we have
$$\frac{1}{\pi}\int_0^{\pi}\exp(ix\sin\theta)\exp(-ni\theta)d\theta. \tag{44}$$

From -1 to $-\infty$ put $\lambda = \exp(u+i\pi)$; this part is
$$\frac{1}{\pi i}\int_0^{\infty}\exp(-x\sinh u)\exp(-nu-n\pi i)du, \tag{45}$$

and in all
$$\text{Hs}_n(x) = \frac{1}{\pi}\int_0^{\pi}\exp(ix\sin\theta - ni\theta)d\theta + \frac{1}{\pi i}\int_0^{\infty}\exp(-x\sinh u)\{\exp nu + \exp(-nu-n\pi i)\}du. \tag{46}$$

Since when x is real Hs_n and Hi_n are conjugate complexes
$$\text{Hi}_n(x) = \frac{1}{\pi}\int_0^{\pi}\exp(-ix\sin\theta + ni\theta)d\theta - \frac{1}{\pi i}\int_0^{\infty}\exp(-x\sinh u)\{\exp nu + \exp(-nu+n\pi i)\}du \tag{47}$$

and
$$J_n(x) = \frac{1}{\pi}\int_0^{\pi}\cos(x\sin\theta - n\theta)d\theta - \frac{1}{\pi}\int_0^{\infty}\exp(-x\sinh u - nu)\sin n\pi\, du, \tag{48}$$
$$Y_n(x) = \frac{1}{\pi}\int_0^{\pi}\sin(x\sin\theta - n\theta)d\theta - \frac{1}{\pi}\int_0^{\infty}\exp(-x\sinh u)\{\exp nu + \exp(-nu)\cos n\pi\}du \tag{49}$$

valid for $\Re(x) > 0$; the first is also valid for $\Re(x) = 0$ if $\Re(n) > 0$. If n is a positive integer $J_n(x)$ reduces to the first integral, which is known as Bessel's integral. It occurs in the expression of the radius vector in planetary motion in terms of the eccentric angle.

21·022. Integrals on real paths for $I_n(x)$; the second solution $\mathrm{Kh}_n(x)$. We can get similar expressions for the integrals along the paths for $I_n(x)$ by taking x to be complex with a positive real part and choosing the appropriate path; or we can proceed directly as follows. Using (16) we take the path in the λ plane

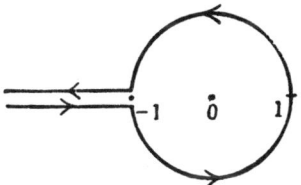

On the circle put
$$\lambda = e^{i\theta}. \tag{50}$$
Then its contribution to $I_n(x)$ is
$$\frac{1}{2\pi}\int_{-\pi}^{\pi} \exp(x\cos\theta - ni\theta)\,d\theta = \frac{1}{\pi}\int_0^{\pi} \exp(x\cos\theta)\cos n\theta\,d\theta. \tag{51}$$

From $-\infty$ to -1 and -1 to $-\infty$ put
$$\lambda = \exp(u - i\pi), \quad \lambda = \exp(u + i\pi). \tag{52}$$
The contributions are
$$-\frac{1}{2\pi i}\int_0^{\infty} \exp(-x\cosh u - nu)\,e^{ni\pi}\,du + \frac{1}{2\pi i}\int_0^{\infty} \exp(-x\cosh u - nu)\,e^{-ni\pi}\,du$$
$$= -\frac{\sin n\pi}{\pi}\int_0^{\infty} \exp(-x\cosh u - nu)\,du. \tag{53}$$

Hence
$$I_n(x) = \frac{1}{\pi}\int_0^{\pi}\exp(x\cos\theta)\cos n\theta\,d\theta - \frac{\sin n\pi}{\pi}\int_0^{\infty}\exp(-x\cosh u - nu)\,du. \tag{54}$$

Also
$$I_{-n}(x) = \frac{1}{\pi}\int_0^{\pi}\exp(x\cos\theta)\cos n\theta\,d\theta + \frac{\sin n\pi}{\pi}\int_0^{\infty}\exp(-x\cosh u + nu)\,du \tag{55}$$

and
$$I_{-n}(x) - I_n(x) = \frac{2\sin n\pi}{\pi}\int_0^{\infty}\exp(-x\cosh u)\cosh nu\,du \tag{56}$$
$$= \sin n\pi\,\mathrm{Kh}_n(x) \tag{57}$$

say.* Then Kh_n is significant even when n is an integer, subject to $\Re(x) > 0$.

Again, take the integral from $-\infty$ to 0 by itself, with $\arg\lambda = -\pi$. Then the part from $-\infty$ to -1 gives $-\dfrac{1}{2\pi i}e^{ni\pi}\displaystyle\int_0^{\infty}\exp(-x\cosh u - nu)\,du$ as before. From -1 to 0 put
$$\lambda = \exp(-u - i\pi).$$

---

* This function is that used by Heaviside; Watson, following Macdonald, takes $K_n(x)$ equal to $\tfrac{1}{2}\pi$ times this, though he recognizes explicitly that the factor complicates the relation to $\mathrm{Hs}_n(x)$ and $\mathrm{Hi}_n(x)$. It also complicates the relation to the Legendre function $q_n^*(\mu)$. Published tables refer to Macdonald's function; but the occasions for using the relations between K_n, Hs_n, and Hi_n are so numerous that the most convenient procedure would be to divide the published tables by $\tfrac{1}{2}\pi$. We write Heaviside's function as $\mathrm{Kh}_n(x)$ to distinguish it from Macdonald's.

This gives
$$-\frac{1}{2\pi i}e^{n i \pi}\int_0^\infty \exp(-x\cosh u + nu)\,du$$
and the whole integral is
$$-\frac{1}{\pi i}e^{n i \pi}\int_0^\infty \exp(-x\cosh u)\cosh nu\,du = -\frac{e^{n i \pi}}{2i}\mathrm{Kh}_n(x). \tag{58}$$

$\mathrm{Kh}_n(x)$ can be transformed back to the Schläfli form; we have
$$\mathrm{Kh}_n(x) = \frac{2}{\pi}\int_0^\infty \exp(-x\cosh v)\cosh nv\,dv = \frac{1}{\pi}\int_{-\infty}^\infty \exp(-x\cosh v + nv)\,dv \quad (\Re(x) > 0). \tag{59}$$

Put $e^v = \lambda$; then
$$\mathrm{Kh}_n(x) = \frac{1}{\pi}\int_0^\infty \exp\left\{-\tfrac{1}{2}x\left(\lambda + \frac{1}{\lambda}\right)\right\}\lambda^{n-1}\,d\lambda, \quad (\Re(x) > 0); \tag{60}$$
and if $\tfrac{1}{2}x\lambda = u$,
$$\mathrm{Kh}_n(x) = \frac{1}{\pi}(\tfrac{1}{2}x)^{-n}\int_0^\infty \exp\left(-u - \frac{1}{4}\frac{x^2}{u}\right)u^{n-1}\,du, \quad |\arg x| < \tfrac{1}{4}\pi. \tag{61}$$

Clearly $\mathrm{Kh}_n(x) = \mathrm{Kh}_{-n}(x)$, so that we can always take $n \geqslant 0$.

By continuity the same forms will hold if $\Re(x) = 0$ or $\Re(x^2) = 0$ provided that the integrals converge. In particular (61) is true if $\Re(x^2) = 0$ and $n > 0$.

Kh_n can be expressed directly in terms of Hs_n or Hi_n. For, using our original relations between the I and J functions, we have
$$\sin n\pi\,\mathrm{Kh}_n(x) = I_{-n}(x) - I_n(x) \tag{62}$$
$$= e^{\frac{1}{2}n\pi i}J_{-n}(xe^{\frac{1}{2}\pi i}) - e^{-\frac{1}{2}n\pi i}J_n(xe^{\frac{1}{2}\pi i}) \tag{63}$$
$$= i\sin n\pi e^{\frac{1}{2}n\pi i}\mathrm{Hs}_n(xe^{\frac{1}{2}\pi i}), \tag{64}$$
$$\sin n\pi\,\mathrm{Kh}_n(x) = e^{-\frac{1}{2}n\pi i}J_{-n}(xe^{-\frac{1}{2}\pi i}) - e^{\frac{1}{2}n\pi i}J_n(xe^{-\frac{1}{2}\pi i}) \tag{65}$$
$$= -i\sin n\pi e^{-\frac{1}{2}n\pi i}\mathrm{Hi}_n(xe^{-\frac{1}{2}\pi i}). \tag{66}$$
Hence
$$\mathrm{Kh}_n(x) = ie^{\frac{1}{2}n\pi i}\mathrm{Hs}_n(xe^{\frac{1}{2}\pi i}) = -ie^{-\frac{1}{2}n\pi i}\mathrm{Hi}_n(xe^{-\frac{1}{2}\pi i}). \tag{67}$$

In consequence of these relations all the Bessel functions for given n can be expressed in terms of the single function $\mathrm{Kh}_n(x)$. When n is a positive integer or 0 we can represent $\mathrm{Kh}_n(x)$ by
$$\mathrm{Kh}_n(x) = \lim_{\epsilon \to 0}\frac{I_{-n-\epsilon}(x) - I_{n+\epsilon}(x)}{\sin\pi(n+\epsilon)} = \frac{1}{\pi}(-1)^n\frac{\partial}{\partial n}\{I_{-n}(x) - I_n(x)\} \tag{68}$$
$$= (-1)^{n+1}\frac{2}{\pi}I_n(x)\log(\tfrac{1}{2}x) + (-1)^n\frac{1}{\pi}\sum_{r=0}^\infty \frac{(\tfrac{1}{2}x)^{n+2r}}{r!(n+r)!}\{F(r) + F(n+r)\}$$
$$+ \frac{1}{\pi}\sum_{r=0}^{n-1}(-1)^r\frac{(n-r-1)!}{r!}(\tfrac{1}{2}x)^{-n+2r}, \tag{69}$$
and in particular
$$\mathrm{Kh}_0(x) = -\frac{2}{\pi}I_0(x)\log(\tfrac{1}{2}x) + \frac{2}{\pi}\sum_{r=0}^\infty \frac{(\tfrac{1}{2}x)^{2r}}{r!r!}F(r). \tag{70}$$

21·03. Further complex integrals for $J_n(x)$. Other integral representations of the Bessel functions can be found by putting in the equation for $J_n(x)$
$$y = x^{-n}u. \tag{71}$$
Then
$$xu'' - (2n-1)u' + xu = 0. \tag{72}$$

Substitute
$$u = \int e^{zx} Z \, dz \tag{73}$$
along a path to be determined; then

$$\begin{aligned} xu'' - (2n-1)u' + xu &= \int \{xz^2 - (2n-1)z + x\} Z e^{zx} \, dz \\ &= [(z^2+1) Z e^{zx}] - \int e^{zx} \{(z^2+1) Z' + (2n+1) z Z\} \, dz, \end{aligned} \tag{74}$$

and if the integrand is to vanish
$$\frac{Z'}{Z} = -\frac{(2n+1)z}{z^2+1}, \tag{75}$$

$$Z \propto (z^2+1)^{-(n+1/2)}. \tag{76}$$

Then a solution will be
$$u = \frac{1}{2\pi i} \int_M \frac{e^{zx}}{(z^2+1)^{n+1/2}} \, dz, \tag{77}$$

where $\Re(zx) \to -\infty$ at the ends of the path. If $n > -\frac{1}{2}$ the integral will converge on the standard Bromwich path, and we have the operational form $p/(p^2+1)^{n+1/2}$. The first term in the expansion is $p^{-2n} = x^{2n}/(2n)!$. This identifies the solution as a constant multiple of $x^n J_n(x)$, and indeed it is $\frac{2^n n!}{(2n)!} x^n J_n(x)$, as we can verify by direct expansion of the denominator in descending powers of z. Then using the multiplication formula for $(2n)!$

$$\frac{p}{(p^2+1)^{n+1/2}} H(x) = \frac{\pi^{1/2}}{2^n (n-\tfrac{1}{2})!} x^n J_n(x) H(x). \tag{78}$$

This result was found in this way by van der Pol. Apparently, though the equation (72) is of the second order and must have another solution $x^n J_{-n}(x)$, we have found only one solution. But the integrand in (77) has branch points at $\pm i$. If we take a figure of eight contour surrounding them, or a loop from $-\infty$ about either, we shall obtain other solutions. Analogous integrals are used by Watson in Chapter 6 of his book.

If $n+\tfrac{1}{2}$ is an integer the integrand is single-valued and the solutions will be expressible in finite terms, as we have already seen in considering the asymptotic expansions for this case. The present form has one advantage over that used by Watson: he gets a factor in the integrand that would be $(z^2+1)^{n-1/2}$ in the present notation, and if $n-\tfrac{1}{2}$ was a positive integer or zero the integrals would vanish. This complication is avoided by having the factor $(z^2+1)^{-n-1/2}$. On the other hand his form is more manageable when the path is reduced to a loop about a branch point.

21·04. Recurrence formulae. Returning to

$$J_n(x) = \frac{1}{2\pi i} (\tfrac{1}{2}x)^n \int_M \exp\left(u - \frac{1}{4}\frac{x^2}{u}\right) \frac{du}{u^{n+1}} \tag{1}$$

and differentiating, we have

$$\begin{aligned} J_n'(x) &= \frac{1}{2\pi i} \tfrac{1}{2} n (\tfrac{1}{2}x)^{n-1} \int_M \exp\left(u - \frac{1}{4}\frac{x^2}{u}\right) \frac{du}{u^{n+1}} \\ &\quad - \frac{1}{2\pi i} (\tfrac{1}{2}x)^n \int_M \exp\left(u - \frac{1}{4}\frac{x^2}{u}\right) \tfrac{1}{2} x \frac{du}{u^{n+2}} \\ &= \frac{n}{x} J_n(x) - J_{n+1}(x). \end{aligned} \tag{2}$$

Differentiation of
$$J_n(x) = \frac{1}{2\pi i}\int_M \exp\left\{\tfrac{1}{2}x\left(\lambda-\tfrac{1}{\lambda}\right)\right\}\frac{d\lambda}{\lambda^{n+1}} \tag{3}$$

gives
$$J_n'(x) = \frac{1}{2\pi i}\int_M \exp\left\{\tfrac{1}{2}x\left(\lambda-\tfrac{1}{\lambda}\right)\right\}\tfrac{1}{2}\left(\tfrac{1}{\lambda^n}-\tfrac{1}{\lambda^{n+2}}\right)d\lambda$$
$$= \tfrac{1}{2}J_{n-1}(x)-\tfrac{1}{2}J_{n+1}(x). \tag{4}$$

By subtraction
$$J_{n-1}(x)+J_{n+1}(x) = \frac{2n}{x}J_n(x). \tag{5}$$

Also from (4) and (5)
$$J_n'(x) = J_{n-1}(x) - \frac{n}{x}J_n(x). \tag{6}$$

These differentiations have been carried out on the integrals over the path M, which give $J_n(x)$. But they could equally be done on the paths used for $\mathrm{Hs}_n(x)$ and $\mathrm{Hi}_n(x)$, which therefore satisfy the same recurrence relations; and then by the definition of $Y_n(x)$ this also will satisfy them.

The corresponding relations for $I_n(x)$ and $\mathrm{Kh}_n(x)$ are somewhat different, on account of the difference of sign of $1/u$ and $1/\lambda$ in the exponent. We find

$$I_n'(x) = \frac{n}{x}I_n(x)+I_{n+1}(x) = -\frac{n}{x}I_n(x)+I_{n-1}(x) = \tfrac{1}{2}I_{n-1}(x)+\tfrac{1}{2}I_{n+1}(x), \tag{7}$$

$$I_{n-1}(x)-I_{n+1}(x) = \frac{2n}{x}I_n(x), \tag{8}$$

$$\mathrm{Kh}_n'(x) = \frac{n}{x}\mathrm{Kh}_n(x)-\mathrm{Kh}_{n+1}(x) = -\frac{n}{x}\mathrm{Kh}_n(x)-\mathrm{Kh}_{n-1}(x) = -\tfrac{1}{2}\mathrm{Kh}_{n-1}(x)-\tfrac{1}{2}\mathrm{Kh}_{n+1}(x), \tag{9}$$

$$\mathrm{Kh}_{n-1}(x)-\mathrm{Kh}_{n+1}(x) = -\frac{2n}{x}\mathrm{Kh}_n(x). \tag{10}$$

In consequence of these relations it is possible, given any Bessel function for $n = 0$ and $n = 1$, to build up the same function for any integral n.

Particularly important cases are

$$J_0'(x) = -J_1(x), \quad \frac{d}{dx}\{xJ_1(x)\} = xJ_0(x); \tag{11}$$

$$I_0'(x) = I_1(x), \quad \frac{d}{dx}\{xI_1(x)\} = xI_0(x); \tag{12}$$

$$\mathrm{Kh}_0'(x) = -\mathrm{Kh}_1(x), \quad \frac{d}{dx}\{x\mathrm{Kh}_1(x)\} = -x\mathrm{Kh}_0(x). \tag{13}$$

These occur in hydrodynamical problems relating to cylinders, where if the radial velocity depends on J_0 the azimuthal velocity depends on J_1, and conversely.

21·05. Asymptotic formulae of Stokes's type. These are most easily obtained for $\mathrm{Hs}_n(x)$, $\mathrm{Hi}_n(x)$, and $\mathrm{Kh}_n(x)$, on account of the fact that the natural paths of integration to use for them pass through only one saddle-point. The forms 21·02 (26) (27) are perhaps the most convenient for getting the first term, since with them the positions of the saddle-points of the exponential factor are independent of x. Writing

$$f(\lambda) = \tfrac{1}{2}x\left(\lambda-\tfrac{1}{\lambda}\right), \quad f''(i) = -x/i^3 = -ix, \tag{1}$$

the integrand for $\mathrm{Hs}_n(x)$ at $\lambda = i$ is

$$\frac{1}{\pi i}\exp(ix)\exp\{-\tfrac{1}{2}(n+1)\pi i\} \tag{2}$$

and the path of steepest descent, for x real and positive, is at $\tfrac{3}{4}\pi$ to the positive real axis. Then

$$\mathrm{Hs}_n(x) \sim \frac{1}{\pi i}\exp i\{x-\tfrac{1}{2}n\pi-\tfrac{1}{2}\pi\}\left(\frac{2\pi}{x}\right)^{1/2}\exp(\tfrac{3}{4}\pi i)$$

$$= \left(\frac{2}{\pi x}\right)^{1/2}\exp\{i(x-\tfrac{1}{2}n\pi-\tfrac{1}{4}\pi)\}, \tag{3}$$

and similarly

$$\mathrm{Hi}_n(x) \sim \left(\frac{2}{\pi x}\right)^{1/2}\exp\{-i(x-\tfrac{1}{2}n\pi-\tfrac{1}{4}\pi)\}. \tag{4}$$

Hence

$$J_n(x) \sim \left(\frac{2}{\pi x}\right)^{1/2}\cos(x-\tfrac{1}{2}n\pi-\tfrac{1}{4}\pi), \tag{5}$$

$$Y_n(x) \sim \left(\frac{2}{\pi x}\right)^{1/2}\sin(x-\tfrac{1}{2}n\pi-\tfrac{1}{4}\pi). \tag{6}$$

These determine the coefficients of the various first terms. The rest of the expansions can be determined from some of the integrals, but are most easily found from the differential equation as in 17·121. We have

$$\mathrm{Hs}_n(x) = \left(\frac{2}{\pi x}\right)^{1/2}\exp i(x-\tfrac{1}{2}n\pi-\tfrac{1}{4}\pi)(U-iV), \tag{7}$$

$$\mathrm{Hi}_n(x) = \left(\frac{2}{\pi x}\right)^{1/2}\exp\{-i(x-\tfrac{1}{2}n\pi-\tfrac{1}{4}\pi)\}(U+iV), \tag{8}$$

$$J_n(x) = \left(\frac{2}{\pi x}\right)^{1/2}\{U\cos(x-\tfrac{1}{2}n\pi-\tfrac{1}{4}\pi)+V\sin(x-\tfrac{1}{2}n\pi-\tfrac{1}{4}\pi)\}, \tag{9}$$

$$Y_n(x) = \left(\frac{2}{\pi x}\right)^{1/2}\{U\sin(x-\tfrac{1}{2}n\pi-\tfrac{1}{4}\pi)-V\cos(x-\tfrac{1}{2}n\pi-\tfrac{1}{4}\pi)\}, \tag{10}$$

where

$$U \sim 1 - \frac{(1-4n^2)(9-4n^2)}{2!(8x)^2}+\frac{(1-4n^2)(9-4n^2)(25-4n^2)(49-4n^2)}{4!(8x)^4}-\ldots, \tag{11}$$

$$V \sim \frac{1-4n^2}{8x} - \frac{(1-4n^2)(9-4n^2)(25-4n^2)}{3!(8x)^3}+\ldots. \tag{12}$$

These are Stokes's expansions. We have taken x as real and positive, but we see that this condition can be relaxed. For with a complex x we still obtain the correct integrand at $\lambda = i$ by direct substitution. The direction of the path is rotated by $-\tfrac{1}{2}\arg x$, and this is allowed for by the factor $x^{-1/2}$. Yet there is a limit to the range permitted to $\arg x$. For, let us suppose it increased continuously by 2π. $J_n(x)$ is multiplied by $\exp(2ni\pi)$ and returns to its original value if n is a positive integer. But each term of (9) is multiplied by $\exp(-i\pi)$. We know already that an asymptotic expansion cannot be correct for all values of $\arg x$ unless it converges; we see here that to permit unlimited variation of $\arg x$ would lead to seriously wrong results. To trace the origin of the change we notice that as $\arg x$ increases, if we keep to paths of steepest descent the path near each saddle-point rotates negatively by half as much as $\arg x$ increases, and further $\arg \lambda$ must decrease at $|\lambda| = \infty$ and increase at $|\lambda|$ small, by the same amount as $\arg x$ increases.

Let arg x increase from 0 to 2π. Then the paths of steepest descent are exactly as at the start, but the directions of travel through the saddle-points are reversed, *and therefore they are traversed in opposite directions to their original ones* if we try to maintain the continuity of the approximation. But this is clearly wrong. For with n a positive integer we must get back to the same value of $J_n(x)$, and $\mathrm{Hs}_n(x)$ is always an integral from 0 to $\infty \exp(i\pi - i\arg x)$, and $\mathrm{Hi}_n(x)$ one from $\infty \exp(-i\pi - i\arg x)$ to 0, the two together constituting a loop in the positive sense about the origin. The reversal in sign must come from a failure of the asymptotic expressions to represent $\mathrm{Hs}_n(x)$ and $\mathrm{Hi}_n(x)$ over the range $0 \leqslant \arg x \leqslant 2\pi$, and the reason is that at some value of $\arg x$ continuous deformation of the steepest descents path near a saddle-point makes it change from one going from the origin to infinity instead of from infinity to the origin. Thus let $\arg x = \tfrac{1}{2}\pi$. The path for $\mathrm{Hs}_n(x)$ is straight along the imaginary axis, and there is no trouble. But that for $\mathrm{Hi}_n(x)$ is as shown, a dotted line indicating the path for $\arg x$ a little less than $\tfrac{1}{2}\pi$. If $\arg x$ is a little greater than $\tfrac{1}{2}\pi$, we could take a path in a suitable direction from infinity to i, round the circle, and from i to the origin again in a suitable way, and $\mathrm{Hi}_n(x)$ would be given correctly. But such a path would not be the steepest descents path (the other dotted line), which, if it is going from left to right near $-i$, would cross the circle from inside and go from 0 to infinity. We can see that the difference of the integrals along the two paths is $2\mathrm{Hs}_n(x)$. This is irrelevant to the definition of the asymptotic expansion; for every term in $\mathrm{Hs}_n(x)$ contains an exponentially small factor when $0 < \arg x < \pi$,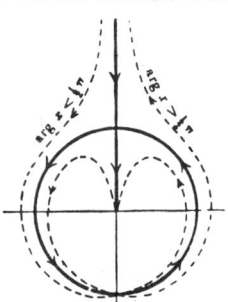
and the whole series, when $|x|$ is large enough, will be small compared with any given term of the series for $\mathrm{Hi}_n(x)$. The expansions are therefore both valid when $0 \leqslant \arg x < \pi$; but at $\arg x = \pi$ the moduli of the first terms become equal, and the new portion of Hi_n will become the larger. Hence the asymptotic expansion of $\mathrm{Hi}_n(x)$ is not valid when $\arg x > \pi$. That of $\mathrm{Hs}_n(x)$ will similarly begin to represent a function with a multiple of $\mathrm{Hi}_n(x)$ when $\arg x = \tfrac{3}{2}\pi$, and the new part will become the larger when $\arg x = 2\pi$. A little further consideration shows that the range of validity of the asymptotic expansions is $-\pi < \arg x < 2\pi$ for $\mathrm{Hs}_n(x)$, $-2\pi < \arg x < \pi$ for $\mathrm{Hi}_n(x)$. The expansions of $J_n(x)$ and $Y_n(x)$ based on them are therefore valid for $-\pi < \arg x < \pi$, but of course not at $\arg x = \pm \pi$.

21·051. Asymptotic formulae for $I_n(x)$ and $\mathrm{Kh}_n(x)$ can be derived from these by change of argument, using (3), (4), or directly from the integrals 21·022 (51), (60). The later terms are found from the differential equation.

$$I_n(x) \sim \frac{e^x}{\sqrt{(2\pi x)}} \left\{ 1 + \frac{1-4n^2}{1!\,8x} + \frac{(1-4n^2)(9-4n^2)}{2!\,(8x)^2} + \ldots \right\}, \tag{13}$$

$$\mathrm{Kh}_n(x) \sim \sqrt{\left(\frac{2}{\pi x}\right)} e^{-x} \left\{ 1 - \frac{1-4n^2}{1!\,8x} + \frac{(1-4n^2)(9-4n^2)}{2!\,(8x)^2} - \ldots \right\}. \tag{14}$$

It can be shown that for real argument the error in stopping at any term of the series U and V has the same sign as the first term neglected, and therefore the function lies between the sums of r and $r+1$ terms of the series if the number of terms retained is so large that consecutive terms alternate in sign.* The same is true of $\mathrm{Kh}_n(x)$, but the corresponding inequality for $I_n(x)$ is more complicated.

* Watson, *Theory of Bessel Functions*, p. 209.

21·052. Interpretations of $\mathrm{Kh}_0(p\varpi/c)$, $p\mathrm{Kh}_0(p\varpi/c)$, $\mathrm{Kh}_0(q\varpi)$.

The physical applications of the functions can be distinguished by the exponential factors in the asymptotic expansions. Thus where we should write for a wave travelling in the positive direction of the axis of x, in one dimension,

$$\cos\kappa(ct-x) = \Re\exp i\kappa(ct-x),$$

we should write for a symmetrical wave in two dimensions $\Re e^{i\kappa ct}\mathrm{Hi}_0(\kappa\varpi)$, the phase of which, when $\kappa\varpi$ is large, will travel outwards with velocity c. The function $D_0(x)$ used in Lamb's *Hydrodynamics* is $-i\mathrm{Hi}_0(x)$. The function $\mathrm{Hi}_n(x)$ is therefore specially convenient for treating spreading harmonic waves.

The finiteness of $I_n(x)$ at the origin makes it the suitable function to be used for problems dealing with the interior of a circle; but since $I_n(x)$ tends to infinity with x the proper function of imaginary argument to use outside a circle is $\mathrm{Kh}_n(x)$. $\mathrm{Kh}_0(pa)$, where p is the Heaviside operator, occurs regularly in problems of the spreading of cylindrical disturbances and plays a part similar to $\exp(-ph)$ in one dimension. In its simplest form we have

$$\mathrm{Kh}_0(x) = \frac{2}{\pi}\int_0^\infty \exp(-x\cosh u)\,du. \tag{1}$$

Then
$$\mathrm{Kh}_0\!\left(\frac{p\varpi}{c}\right)H(t) = \frac{1}{2\pi i}\frac{2}{\pi}\int_0^\infty\!\int_L \exp\!\left(zt - \frac{z\varpi}{c}\cosh u\right)du\,\frac{dz}{z}$$

$$= \frac{2}{\pi}\int_0^\infty H\!\left(t - \frac{\varpi}{c}\cosh u\right)du, \tag{2}$$

$$= 0\left(t < \frac{\varpi}{c}\right); \quad = \frac{2}{\pi}\int_0^{\cosh^{-1} ct/\varpi} du \quad \left(t > \frac{\varpi}{c}\right)$$

and therefore in general
$$= \frac{2}{\pi}\cosh^{-1}\frac{ct}{\varpi}\cdot H\!\left(t - \frac{\varpi}{c}\right). \tag{3}$$

The operator $\mathrm{Kh}_0(p\varpi/c)$ therefore gives a disturbance beginning at time ϖ/c and spreading out with velocity c; its magnitude at a given place will ultimately increase indefinitely with time. Differentiating, however, we get a more usual operator

$$p\mathrm{Kh}_0(p\varpi/c)\,H(t) = \frac{2}{\pi}\frac{c}{(c^2t^2-\varpi^2)^{1/2}}H\!\left(t-\frac{\varpi}{c}\right). \tag{4}$$

These show at once a characteristic feature of waves in two dimensions; unlike waves in one or three dimensions, there is no sudden end to a two-dimensional disturbance, but an indefinitely prolonged trail.

Again, if
$$h > 0,\ p = h^2q^2, \tag{5}$$

$$\mathrm{Kh}_0(q\varpi)\,H(t) = \frac{1}{\pi}\int_0^\infty \exp\!\left(-u - \frac{1}{4}\frac{q^2\varpi^2}{u}\right)\frac{du}{u}H(t)$$

$$= \frac{1}{\pi}\int_0^\infty \exp\!\left(-u - \frac{1}{4}\frac{p\varpi^2}{h^2 u}\right)\frac{du}{u}H(t)$$

$$= \frac{1}{\pi}\int_0^\infty \frac{e^{-u}}{u}H\!\left(t - \frac{\varpi^2}{4h^2 u}\right)du$$

$$= \frac{1}{\pi}\int_{\varpi^2/4h^2 t}^\infty \frac{e^{-u}}{u}du\,H(t) = \frac{1}{\pi}\mathrm{ei}\!\left(\frac{\varpi^2}{4h^2 t}\right)H(t). \tag{6}$$

Differentiating with respect to t we have for $t > 0$

$$p\mathrm{Kh}_0(q\varpi)H(t) = \frac{1}{\pi} \frac{e^{-\varpi^2/4h^2t}}{\varpi^2/4h^2t} \frac{\varpi^2}{4h^2t^2} = \frac{e^{-\varpi^2/4h^2t}}{\pi t}, \qquad (7)$$

and also
$$q\varpi\mathrm{Kh}_1(q\varpi)H(t) = -q\varpi\mathrm{Kh}_0'(q\varpi) = -\varpi\frac{\partial}{\partial\varpi}\mathrm{Kh}_0(q\varpi)$$

$$= \frac{1}{\pi}\frac{e^{-\varpi^2/4h^2t}}{\varpi^2/4h^2t}\frac{2\varpi^2}{4h^2t} = \frac{2}{\pi}e^{-\varpi^2/4h^2t}. \qquad (8)$$

These operators, which occur in problems of diffusion about circular cylinders, are special cases of some that lead to the confluent hypergeometric function.

21·06. Functions of large order: approximations of Green's type. Asymptotic approximations for large n have been found (17·132), apart from constant factors, by direct study of the differential equations. The constant factors can be identified by comparison with the Stokes expansions for x much larger than n. The approximations can be, and originally were, found by Debye by the method of steepest descents. We illustrate this by means of $I_n(x)$ and $\mathrm{Kh}_n(x)$ for x real and positive. We have from 21·022 (57),

$$\mathrm{Kh}_n(x) = \frac{2}{\pi}\int_0^\infty \exp(-x\cosh u)\cosh nu\, du$$

$$\doteqdot \frac{1}{\pi}\int_0^\infty \exp(-x\cosh u + nu)\, du = \frac{1}{\pi}\int_0^\infty \exp\{f(u)\}\, du \qquad (1)$$

since the term in $\exp(-nu)$ decreases steadily with increasing u and is easily shown to be negligible in comparison with that in $\exp(nu)$ when n is large. The path of steepest descent is the real axis. The integrand is a maximum when

$$f'(u) = n - x\sinh u = 0; \qquad (2)$$

and then $\sinh u = n/x$, $f''(u) = -x\cosh u$, $f(u) = n\sinh^{-1}\frac{n}{x} - (x^2+n^2)^{1/2}$,

$$\mathrm{Kh}_n(x) \sim \sqrt{\left(\frac{2}{\pi}\right)} x^{-n}(n^2+x^2)^{-1/4}\{n+(n^2+x^2)^{1/2}\}^n \exp\{-(n^2+x^2)^{1/2}\}. \qquad (3)$$

For $I_n(x)$ we can use 21·011 (20),

$$I_n(x) = \frac{1}{2\pi i}\int_M \exp\{f(\mu)\}\frac{d\mu}{(\mu^2-1)^{1/2}}, \qquad (4)$$

where
$$f(\mu) = \mu x - n\log\{\mu + (\mu^2-1)^{1/2}\}. \qquad (5)$$

There is a saddle-point where
$$f'(\mu) = x - \frac{n}{(\mu^2-1)^{1/2}} = 0; \qquad (6)$$

and then
$$\mu = \left(1+\frac{n^2}{x^2}\right)^{1/2}, \quad f''(\mu) = \frac{n\mu}{(\mu^2-1)^{3/2}} = \frac{x^2}{n}\left(\frac{x^2}{n^2}+1\right)^{1/2},$$

$$f(\mu) = (x^2+n^2)^{1/2} - n\log\left\{\left(1+\frac{n^2}{x^2}\right)^{1/2}+\frac{n}{x}\right\}. \qquad (7)$$

Since $f''(\mu)$ is positive the path of steepest descent is parallel to the imaginary axis, and we find

$$I_n(x) \sim \frac{1}{\sqrt{(2\pi)}} x^n (n^2+x^2)^{-1/4} \{n+(n^2+x^2)^{1/2}\}^{-n} \exp(n^2+x^2)^{1/2}. \tag{8}$$

The corresponding approximations to $J_n(x)$ and $Y_n(x)$ are, for $x < n$,

$$J_n(x) \sim \frac{1}{\sqrt{(2\pi)}} x^n (n^2-x^2)^{-1/4} \{n+(n^2-x^2)^{1/2}\}^{-n} \exp(n^2-x^2)^{1/2}, \tag{9}$$

$$Y_n(x) \sim -\sqrt{\left(\frac{2}{\pi}\right)} x^{-n} (n^2-x^2)^{-1/4} \{n+(n^2-x^2)^{1/2}\}^n \exp\{-(n^2-x^2)^{1/2}\}, \tag{10}$$

and for $x > n$

$$J_n(x) \sim \sqrt{\left(\frac{2}{\pi}\right)} (x^2-n^2)^{-1/4} \sin\{n(\tan v - v) + \tfrac{1}{4}\pi\}, \tag{11}$$

$$Y_n(x) \sim -\sqrt{\left(\frac{2}{\pi}\right)} (x^2-n^2)^{-1/4} \cos\{n(\tan v - v) + \tfrac{1}{4}\pi\}, \tag{12}$$

where $\sec v = x/n$.

Later terms in approximations of this type have been obtained and are given by Bickley[*] to order n^{-11}, but the recurrence relation is more complicated than for the Stokes expansions. The first term for $x > n$ actually gives quite a good approximation down to the first zero of $J_n(x)$.

21·07. Applications of the Wronskian. Write Bessel's equation in the form

$$y'' + \frac{y'}{x} + \left(1 - \frac{n^2}{x^2}\right) y = 0. \tag{1}$$

Then if the Wronskian of any two solutions y_1, y_2 is taken

$$W(y_1, y_2) = A \exp\left(-\int^x \frac{dx}{x}\right) = \frac{A}{x}. \tag{2}$$

In particular
$$J'_n(x) J_{-n}(x) - J'_{-n}(x) J_n(x) = \frac{A}{x}. \tag{3}$$

The constant can be fixed by considering the first terms.

$$J_n(x) = \frac{(\tfrac{1}{2}x)^n}{n!} + \ldots, \quad J_{-n}(x) = \frac{(\tfrac{1}{2}x)^{-n}}{(-n)!} + \ldots, \quad J'_n(x) = \frac{\tfrac{1}{2}(\tfrac{1}{2}x)^{n-1}}{(n-1)!} + \ldots, \quad J'_{-n}(x) = \frac{\tfrac{1}{2}(\tfrac{1}{2}x)^{-n-1}}{(-n-1)!} + \ldots$$

and
$$A = \frac{1}{(n-1)!(-n)!} - \frac{1}{(-n-1)!n!} = \frac{2n}{n!(-n)!} = \frac{2 \sin n\pi}{\pi}. \tag{4}$$

It follows at once that for n not an integer $J_n(x)$ and $J_{-n}(x)$ are two independent solutions, but become proportional when n is an integer. Also

$$J_n(x) Y'_n(x) - J'_n(x) Y_n(x) = \frac{2}{\pi x}, \tag{5}$$

$$I'_n(x) \mathrm{Kh}_n(x) - I_n(x) \mathrm{Kh}'_n(x) = \frac{2}{\pi x}. \tag{6}$$

The factors in (5), (6) also follow easily by considering the asymptotic approximations for x large (21·051). (The existence of asymptotic expansions of the derivatives follows at once from the recurrence relations.)

[*] *Phil. Mag.* (7) 34, 1943, 37–49.

21·08. Functions of order half an odd integer. We have seen from the asymptotic expansions (17·121) that these can be expressed in finite terms. This can be done compactly by means of the recurrence relations. We take in each case the relation that contains the function of order $n+1$, that of order n, and the derivative of the latter. Thus multiplying the relation for $J_{n+1}(x)$ (21·04 (2)) by x^{-n}

$$x^{-n}J_{n+1}(x) = -x^{-n}J'_n(x) + nx^{-n-1}J_n(x) = -\frac{d}{dx}\{x^{-n}J_n(x)\}, \tag{1}$$

$$x^{-n-1}J_{n+1}(x) = -\frac{d}{x\,dx}\{x^{-n}J_n(x)\}, \tag{2}$$

and by induction
$$J_{n+m}(x) = (-1)^m x^{n+m}\left(\frac{d}{x\,dx}\right)^m \{x^{-n}J_n(x)\}. \tag{3}$$

In particular
$$J_{1/2}(x) = \left(\frac{2}{\pi x}\right)^{1/2} \sin x, \tag{4}$$

and therefore
$$J_{m+1/2}(x) = \left(\frac{2}{\pi}\right)^{1/2} (-1)^m x^{m+1/2} \left(\frac{d}{x\,dx}\right)^m \frac{\sin x}{x}. \tag{5}$$

Since we know the exponential factors for the Hankel functions, we infer at once

$$\mathrm{Hs}_{m+1/2}(x) = \left(\frac{2}{\pi}\right)^{1/2} (-1)^m x^{m+1/2} \left(\frac{d}{x\,dx}\right)^m \left(-\frac{ie^{ix}}{x}\right), \tag{6}$$

$$\mathrm{Hi}_{m+1/2}(x) = \left(\frac{2}{\pi}\right)^{1/2} (-1)^m x^{m+1/2} \left(\frac{d}{x\,dx}\right)^m \left(\frac{ie^{-ix}}{x}\right), \tag{7}$$

$$Y_{m+1/2}(x) = \left(\frac{2}{\pi}\right)^{1/2} (-1)^m x^{m+1/2} \left(\frac{d}{x\,dx}\right)^m \left(-\frac{\cos x}{x}\right). \tag{8}$$

Since $I_{1/2}(x)$ consists of the same terms as $J_{1/2}(x)$ with all signs taken positive we have immediately

$$I_{1/2}(x) = \left(\frac{2}{\pi x}\right)^{1/2} \sinh x, \tag{9}$$

and from the recurrence relation (21·04 (7))

$$x^{-n-1}I_{n+1}(x) = \frac{d}{x\,dx}\{x^{-n}I_n(x)\}, \tag{10}$$

whence
$$I_{m+1/2}(x) = \left(\frac{2}{\pi}\right)^{1/2} x^{m+1/2} \left(\frac{d}{x\,dx}\right)^m \frac{\sinh x}{x}. \tag{11}$$

Also
$$\mathrm{Kh}_{1/2}(x) = \left(\frac{2}{\pi x}\right)^{1/2} e^{-x}, \tag{12}$$

$$\mathrm{Kh}_{m+1/2}(x) = \left(\frac{2}{\pi}\right)^{1/2} (-1)^m x^{m+1/2} \left(\frac{d}{x\,dx}\right)^m \frac{e^{-x}}{x}. \tag{13}$$

21·09. The functions ber, bei, kher, khei.* These are defined by

$$\mathrm{ber}_n(x) \pm i\,\mathrm{bei}_n(x) = J_n(xe^{\pm 3/4 \pi i}) = e^{\pm 1/2 n \pi i} I_n(xe^{\pm 1/4 \pi i}),$$

$$\mathrm{kher}_n(x) \pm i\,\mathrm{khei}_n(x) = e^{\pm 1/2 n \pi i} \mathrm{Kh}_n(xe^{\pm 1/4 \pi i}) = i\,\mathrm{Hs}_n(xe^{\pm 3/4 \pi i}).$$

* A. Russell, *Phil. Mag.* (6), 17, 1909, 524–552; C. S. Whitehead, *Quart. Journ. Math.*, 42, 1911, 316–42; H. G. Savidge, *Phil. Mag.* (6), 19, 1910, 49–58.

They occur especially in problems of periodic heat flow and slow periodic motion of a viscous fluid with cylindrical boundaries. $\ker_n(x)$ and $\mathrm{khei}_n(x)$ are $2/\pi$ times the tabulated functions $\ker_n(x)$, $\mathrm{kei}_n(x)$. The properties of the functions are easily inferred from those of $I_n(x)$ and $\mathrm{Kh}_n(x)$ for complex argument.

If $|x|$ is small,
$$\mathrm{Kh}_0(x) \doteqdot -\frac{2}{\pi}\log(\tfrac{1}{2}x),$$
whence
$$\mathrm{kher}_0(x) \doteqdot -\frac{2}{\pi}\log\tfrac{1}{2}|x|,\quad \mathrm{khei}_0(x) \doteqdot -\tfrac{1}{2}(x>0),\ \doteqdot \tfrac{1}{2}(x<0).$$

If $p^{-1}f(t)$ denotes $\int_0^t f(t)\,dt$,
$$p^{-n}e^{t/p}H(t) = t^{\frac{1}{2}n}e^{-\frac{3}{2}n\pi i}\{\mathrm{ber}_n 2\sqrt{t} + i\,\mathrm{bei}_n 2\sqrt{t}\}H(t).$$

21·10. Expansions and definite integrals. It follows immediately from Schläfli's integral and Laurent's theorem that if n is an integer, positive or negative, $J_n(x)$ is the coefficient of λ^n in the expansion of $\exp\tfrac{1}{2}x\left(\lambda - \dfrac{1}{\lambda}\right)$ in positive and negative powers of λ. This can also be shown directly by multiplication of series. For this reason the Bessel functions of integral order are often called *Bessel coefficients*. We therefore have

$$\exp\tfrac{1}{2}x\left(\lambda - \frac{1}{\lambda}\right) = \sum_{n=-\infty}^{\infty} J_n(x)\lambda^n \tag{1}$$

without restriction on x and λ. Put $\lambda = \exp i\theta$. Then

$$\exp(ix\sin\theta) = \sum_{n=-\infty}^{\infty} J_n(x) e^{ni\theta}. \tag{2}$$

Multiply by $\exp(-in\theta)$ and integrate from $-\pi$ to π; then

$$\int_{-\pi}^{\pi} \exp i(x\sin\theta - n\theta)\,d\theta = 2\pi J_n(x), \tag{3}$$

and
$$J_n(x) = \frac{1}{\pi}\int_0^{\pi} \cos(x\sin\theta - n\theta)\,d\theta, \tag{4}$$

a particular case of the result for general n.

If $n=0$ we can replace θ by $\theta - \tfrac{1}{2}\pi$ in (4), and

$$J_0(x) = \frac{1}{2\pi}\int_{-\pi}^{\pi} \cos(x\cos\theta)\,d\theta = \frac{1}{2\pi}\int_{-\pi}^{\pi} \exp(ix\cos\theta)\,d\theta. \tag{5}$$

Now consider
$$I = \int_0^{\infty} e^{-ax} J_0(bx)\,dx \quad (a>0,\ a,\ b\ \text{real}) \tag{6}$$

$$= \frac{1}{2\pi}\int_0^{\infty} e^{-ax}dx \int_{-\pi}^{\pi} \exp(ibx\sin\theta)\,d\theta.$$

Integrate first with regard to x; then

$$I = \frac{1}{2\pi}\int_{-\pi}^{\pi}\frac{d\theta}{a - ib\sin\theta} = \frac{1}{2\pi}\int \frac{e^{i\theta}d\theta}{ae^{i\theta} - \tfrac{1}{2}be^{2i\theta} + \tfrac{1}{2}b} = -\frac{1}{\pi i}\int\frac{dz}{bz^2 - 2az - b} \tag{7}$$

taken round the unit circle. The poles are at
$$bz = a \pm \sqrt{(a^2 + b^2)}.$$

For real a, b these are real, and one is inside and the other outside the circle. The integral is then found to be $(a^2+b^2)^{-1/2}$. With a change of notation

$$\int_0^\infty e^{-\kappa z} J_0(\kappa \varpi)\, d\kappa = \frac{1}{(\varpi^2+z^2)^{1/2}}, \tag{8}$$

with $z > 0$, z, ϖ real. We can regard ϖ, z as cylindrical coordinates; then the right side is simply r^{-1}, and we have expressed the fundamental solution of Laplace's equation in terms of solutions of the equation in cylindrical coordinates. This is valid for $z > 0$; for $z < 0$ we evidently must take the exponent as $+\kappa z$.

Now we should expect that a solution of Laplace's equation in cylindrical coordinates (ϖ, λ, z) can be expressed in another way. For subject to convergence conditions, if we keep ϖ, λ constant, the solution ϕ can be expressed in terms of $\cos \alpha z$ and $\sin \alpha z$ as a Fourier integral $\int f(\alpha, \varpi, \lambda) \cos \alpha z\, d\alpha + \int g(\alpha, \varpi, \lambda) \sin \alpha z\, d\alpha$, and the same will apply to $\nabla^2 \phi$. But if $\nabla^2 \phi = 0$ for all z, f and g must satisfy

$$\frac{1}{\varpi} \frac{\partial}{\partial \varpi}\left(\varpi \frac{\partial f}{\partial \varpi}\right) + \frac{1}{\varpi^2} \frac{\partial^2 f}{\partial \lambda^2} - \alpha^2 f = 0$$

and must therefore be of the form $\{AI_n(\alpha \varpi) + B\mathrm{Kh}_n(\alpha \varpi)\}(\cos n\lambda, \sin n\lambda)$. Such an expression has obvious drawbacks, since $I_n(\alpha \varpi)$ tends to ∞ as $\varpi \to \infty$, and $\mathrm{Kh}_n(\alpha \varpi)$ to infinity as $\varpi \to 0$. But for a distribution with a singularity at $\varpi = 0$ and tending to 0 at infinity we may expect the Kh_n solution to be admissible. It can be obtained as follows. We may think of the problem as a potential one, with a line density proportional to $\cos \alpha z$ along the axis of z. Then the potential on $z = 0$ due to such a distribution will be

$$\int_{-\infty}^{\infty} \frac{\cos \alpha \zeta}{\sqrt{(\zeta^2+\varpi^2)}}\, d\zeta = 2\Re \int_0^\infty \frac{e^{i\alpha \zeta}}{\sqrt{(\zeta^2+\varpi^2)}}\, d\zeta = 2\Re \int_0^{i\infty} \frac{e^{i\alpha \zeta}}{\sqrt{(\zeta^2+\varpi^2)}}\, d\zeta \tag{9}$$

with an indentation in the path about $\zeta = i\varpi$. Put $\zeta = i\kappa$ and then $\kappa = \varpi \cosh v$. Then we have, the integral up to $\kappa = \varpi$ being purely imaginary,

$$2\Re \int_0^\infty e^{-\alpha \kappa} \frac{d\kappa}{\sqrt{(\kappa^2 - \varpi^2)}} = 2 \int_\varpi^\infty e^{-\alpha \kappa} \frac{d\kappa}{\sqrt{(\kappa^2 - \varpi^2)}} = 2\int_0^\infty e^{-\alpha \varpi \cosh v}\, dv = \pi \mathrm{Kh}_0(\alpha \varpi), \tag{10}$$

from 21·022 (59).

Alternatively, the potential close to the axis must be $-2\log \varpi$ times the line density. But for ϖ small $\mathrm{Kh}_0(\alpha \varpi)$ behaves like $-\frac{2}{\pi} \log(\tfrac{1}{2}\alpha \varpi)$, and therefore $\pi \mathrm{Kh}_0(\alpha \varpi) \cos \alpha \zeta$ will behave like $-2\log \varpi \cos \alpha \zeta$. Then the potential due to a line density $\cos \alpha z$ will be

$$\pi \mathrm{Kh}_0(\alpha \varpi) \cos \alpha z.$$

Now take the density to be $\frac{1}{2h}$ from $z = -h$ to $z = h$ and otherwise zero. Then we can express it as a Fourier integral $\int_0^\infty f(\alpha) \cos \alpha z\, d\alpha$, with

$$f(\alpha) = \frac{1}{2\pi h} \int_{-h}^{h} \cos \alpha \zeta\, d\zeta = \frac{1}{\pi \alpha h} \sin \alpha h. \tag{11}$$

This is therefore the factor to be associated with $\cos \alpha z$ in the expression of a line density uniform for $-h < z < h$, and the limit function is $1/\pi$ when h is small and the distribution reduces to a unit mass or charge. Then we have

$$\frac{1}{r} = \int_0^\infty \mathrm{Kh}_0(\alpha \varpi) \cos \alpha z \, d\alpha \quad (\varpi > 0). \tag{12}$$

The integral converges at both limits except when $\varpi = 0$.

It is instructive to see how the apparently quite different expressions (8) and (12) can be connected directly. Starting with (8) we write it as

$$\frac{1}{r} = \frac{1}{2} \int_0^\infty e^{-\kappa z} \{\mathrm{Hs}_0(\kappa \varpi) + \mathrm{Hi}_0(\kappa \varpi)\} \, d\kappa. \tag{13}$$

The two parts must be treated separately since $\mathrm{Hs}_0(x) \to 0$ at $x = +i\infty$, $\mathrm{Hi}_0(x) \to 0$ at $x = -i\infty$. Then

$$\frac{1}{r} = \frac{1}{2} \int_0^{i\infty} e^{-\kappa z} \mathrm{Hs}_0(\kappa \varpi) \, d\kappa + \frac{1}{2} \int_0^{-i\infty} e^{-\kappa z} \mathrm{Hi}_0(\kappa \varpi) \, d\kappa$$

$$= \frac{1}{2} \int_0^\infty e^{-i\alpha z} i \mathrm{Hs}_0(i\alpha \varpi) \, d\alpha - \frac{1}{2} \int_0^\infty e^{i\alpha z} i \mathrm{Hi}_0(-i\alpha \varpi) \, d\alpha$$

$$= \frac{1}{2} \int_0^\infty e^{-i\alpha z} \mathrm{Kh}_0(\alpha \varpi) \, d\alpha + \frac{1}{2} \int_0^\infty e^{i\alpha z} \mathrm{Kh}_0(\alpha \varpi) \, d\alpha$$

$$= \int_0^\infty \mathrm{Kh}_0(\alpha \varpi) \cos \alpha z \, d\alpha, \tag{14}$$

by using the relations 21·022 (67) between Kh_0 and the Hankel functions. This type of transformation is frequently used in the treatment of waves over plane boundaries, and it is well to have it in its simplest possible application. A modification for spherical boundaries is the basis of much work on the propagation of electromagnetic waves over a sphere.

21·101. Fourier-Bessel integral. Subject to conditions similar to those for Fourier's integral theorem a function of position over a plane can be expressed in terms of Bessel functions. If $\phi(\rho, \chi)$ is the potential at Q, whose cylindrical coordinates are $(\rho, \chi, 0)$, and P is (ϖ, λ, z), where $z > 0$,

$$\phi_P = \frac{1}{2\pi} \iint \phi(\rho, \chi) \frac{z}{R^3} dS = -\frac{1}{2\pi} \frac{\partial}{\partial z} \iint \phi(\rho, \chi) \frac{dS}{R}$$

$$= -\frac{1}{2\pi} \frac{\partial}{\partial z} \iiint_0^\infty \phi(\rho, \chi) e^{-\kappa z} J_0(\kappa q) \, dS \, d\kappa, \tag{1}$$

where q is the projection of QP on the plane $z = 0$, and therefore

$$q^2 = \rho^2 + \varpi^2 - 2\rho\varpi \cos(\chi - \lambda). \tag{2}$$

We wish to have an expression in terms of $e^{-\kappa z} J_n(\kappa \varpi) (\cos n\lambda, \sin n\lambda)$, the typical solutions of Laplace's equation in cylindrical coordinates. The reduction of $J_0(\kappa q)$ to this form is unexpectedly difficult and was apparently discovered by Neumann and Heine as a limiting case of the corresponding result in spherical polar coordinates. The expansion required is

$$J_0(\kappa q) = J_0(\kappa \varpi) J_0(\kappa \rho) + 2 \sum_{n=1}^\infty J_n(\kappa \varpi) J_n(\kappa \rho) \cos n(\chi - \lambda). \tag{3}$$

To verify it, we substitute Schläfli integrals for the Bessel functions and choose the paths so that the variables of integration have moduli >1 at all points of the paths. Then

$$J_n(\kappa\varpi)J_n(\kappa\rho)\cos n(\chi-\lambda) = -\frac{1}{4\pi^2}\int_M e^{\frac{1}{2}\kappa\varpi(\alpha-1/\alpha)}\frac{d\alpha}{\alpha^{n+1}}\int_M e^{\frac{1}{2}\kappa\rho(\beta-1/\beta)}\frac{d\beta}{\beta^{n+1}}\cos n(\chi-\lambda) \quad (4)$$

and

$$1+2\sum_{n=1}^{\infty}\frac{\cos n(\chi-\lambda)}{\alpha^n\beta^n} = \frac{\alpha^2\beta^2-1}{\alpha^2\beta^2-2\alpha\beta\cos(\chi-\lambda)+1}. \quad (5)$$

Then the series is

$$S = -\frac{1}{4\pi^2}\int_M\int_M \exp\tfrac{1}{2}\left\{\kappa\varpi\left(\alpha-\frac{1}{\alpha}\right)+\kappa\rho\left(\beta-\frac{1}{\beta}\right)\right\}\frac{\alpha^2\beta^2-1}{\alpha^2\beta^2-2\alpha\beta\cos(\chi-\lambda)+1}\frac{d\alpha\,d\beta}{\alpha\beta}, \quad (6)$$

and since the integrand is single-valued we can replace the paths by any closed contours such that $|\alpha|>1$, $|\beta|>1$. Write

$$\chi-\lambda=\vartheta,\quad \alpha=\sigma/\beta, \quad (7)$$

where $|\sigma|=|\alpha\beta|>1$. Then the index of the exponential is

$$\tfrac{1}{2}\kappa\left\{\varpi\left(\frac{\sigma}{\beta}-\frac{\beta}{\sigma}\right)+\rho\left(\beta-\frac{1}{\beta}\right)\right\} = \tfrac{1}{2}\kappa\left\{\left(\rho-\frac{\varpi}{\sigma}\right)\beta-(\rho-\varpi\sigma)\frac{1}{\beta}\right\}$$

$$= \tfrac{1}{2}\kappa\left\{\rho^2-\rho\varpi\left(\sigma+\frac{1}{\sigma}\right)+\varpi^2\right\}^{1/2}\left\{\left(\frac{\rho-\varpi/\sigma}{\rho-\varpi\sigma}\right)^{1/2}\beta-\left(\frac{\rho-\varpi\sigma}{\rho-\varpi/\sigma}\right)^{1/2}\frac{1}{\beta}\right\} \quad (8)$$

$$= \psi,\text{ say,}$$

and

$$S = -\frac{1}{4\pi^2}\int_C\int_C (\exp\psi)\frac{\sigma^2-1}{\sigma^2-2\sigma\cos\vartheta+1}\frac{d\sigma\,d\beta}{\sigma\beta}. \quad (9)$$

We can now integrate with regard to β and get

$$S = \frac{1}{2\pi i}\int_C J_0\left[\kappa\left\{\rho^2-\rho\varpi\left(\sigma+\frac{1}{\sigma}\right)+\varpi^2\right\}^{1/2}\right]\frac{\sigma-1/\sigma}{\sigma-2\cos\vartheta+1/\sigma}\frac{d\sigma}{\sigma}. \quad (10)$$

The path can be taken to be any circle of radius >1 so as to enclose the poles at $\sigma=\exp(\pm i\vartheta)$. Now if we put $\sigma=1/\sigma'$, and study the changes in sign, the form is unaltered, but the new path is a circle c of radius less than 1 traversed in the negative direction. Thus we have also, taking c now in the positive sense,

$$S = -\frac{1}{2\pi i}\int_c J_0\left[\kappa\left\{\rho^2-\rho\varpi\left(\sigma+\frac{1}{\sigma}\right)+\varpi^2\right\}^{1/2}\right]\frac{\sigma-1/\sigma}{\sigma-2\cos\vartheta+1/\sigma}\frac{d\sigma}{\sigma}, \quad (11)$$

and by addition $2S$ is simply the sum of the residues at the two poles. This is evaluated immediately and gives

$$S = J_0\{\kappa(\rho^2-2\varpi\rho\cos\vartheta+\varpi^2)^{1/2}\} = J_0(\kappa q) \quad (12)$$

as was to be shown. Hence

$$\phi_P = -\frac{1}{2\pi}\frac{\partial}{\partial z}\iiint \phi(\rho,\chi)e^{-\kappa z}\left\{J_0(\kappa\varpi)J_0(\kappa\rho)+2\sum_{n=1}^{\infty}J_n(\kappa\varpi)J_n(\kappa\rho)\cos n(\chi-\lambda)\right\}\rho\,d\rho\,d\chi\,d\kappa. \quad (13)$$

The differentiation with regard to z can be done under the integral sign, and we have the expansion required. The reversal of the order of summation and integration is permissible in the same sort of conditions as for Fourier's integral theorem (applied, of course, in two dimensions). The proof of convergence for the limiting integral over the surface obtained by putting $z = 0$ at the start requires additional conditions, as for Fourier's theorems; if it does converge, it is equal to $\phi(\varpi, \lambda, 0)$. What is called the Fourier-Bessel expansion theorem is then

$$\phi(\varpi, \lambda) = \frac{1}{2\pi} \int_0^\infty \int_0^{2\pi} \int_0^\infty \phi(\rho, \chi) J_0(\kappa\varpi) J_0(\kappa\rho) \kappa \, d\rho \, d\chi \, d\kappa$$

$$+ \sum_{n=1}^\infty \frac{1}{\pi} \int_0^\infty \int_0^{2\pi} \int_0^\infty \phi(\rho, \chi) J_n(\kappa\varpi) J_n(\kappa\rho) \cos n(\chi - \lambda) \kappa \, d\rho \, d\chi \, d\kappa. \quad (14)$$

It should be noticed that when ρ is large, $J_n(\kappa\rho)$ is of order $\rho^{-1/2}$, and the absolute convergence of the integrals requires that $\phi(\rho, \chi) \to 0$ faster than $\rho^{-3/2}$.

21·102. Expansion between concentric circles. If u and v are functions of x satisfying

$$x^2 u'' + xu' + (\lambda^2 x^2 - l^2) u = 0, \quad (1)$$

$$x^2 v'' + xv' + (\mu^2 x^2 - m^2) v = 0, \quad (2)$$

we multiply by v/x, u/x respectively and subtract; then

$$x(u''v - uv'') + (u'v - uv') + \left\{(\lambda^2 - \mu^2) x - \frac{l^2 - m^2}{x}\right\} uv = 0, \quad (3)$$

and by integration

$$\int \left\{(\lambda^2 - \mu^2) x - \frac{l^2 - m^2}{x}\right\} uv \, dx = -[x(u'v - uv')]. \quad (4)$$

The boundary conditions are usually such that the terms on the right vanish there. If then $\lambda = \mu$, $l \neq m$,

$$\int uv \frac{dx}{x} = 0, \quad (5)$$

and if $\lambda \neq \mu$, $l = m$,

$$\int x uv \, dx = 0, \quad (6)$$

the limits being any values of x where the terms on the right of (4) vanish. In particular if the limits for x are 0 and a, and λ and μ are two different quantities such that

$$J_m(\lambda a) = J_m(\mu a) = 0, \quad \text{or} \quad J'_m(\lambda a) = J'_m(\mu a) = 0,$$

then

$$\int_0^a x J_m(\lambda x) J_m(\mu x) \, dx = 0. \quad (7)$$

This might have been expected from the general orthogonality relations inferred from Green's theorem. To determine the coefficients in the expansion of a given function we need also the integral of xv^2, where v is any solution of (2). Multiplying the equation by v' we have

$$0 = x^2 v'v'' + xv'^2 + (\mu^2 x^2 - m^2) vv' = \frac{d}{dx}\{\tfrac{1}{2} x^2 v'^2 + \tfrac{1}{2}(\mu^2 x^2 - m^2) v^2\} - \mu^2 xv^2,$$

and therefore, between any limits,

$$\int xv^2 \, dx = \frac{1}{2\mu^2} [x^2 v'^2 + (\mu^2 x^2 - m^2) v^2]. \quad (8)$$

EXAMPLES

1. If
$$x^2\frac{d^2y}{dx^2}+nx\frac{dy}{dx}+(b+cx^{2m})y = 0$$

and
$$\xi = x^m, \quad y = x^{-\frac{1}{2}(n-1)}\eta,$$

prove that
$$\eta = AJ_\mu\left(\frac{c^{\frac{1}{2}}\xi}{m}\right)+BJ_{-\mu}\left(\frac{c^{\frac{1}{2}}\xi}{m}\right),$$

where
$$\mu^2 m^2 = \tfrac{1}{4}(n-1)^2 - b.$$
(Lommel.)

2. If
$$\frac{d^2y}{dx^2} = x^{2m-2}y,$$

prove that
$$y = x^{\frac{1}{2}}\left\{AI_{1/2m}\left(\frac{x^m}{m}\right)+BI_{-1/2m}\left(\frac{x^m}{m}\right)\right\}$$

and hence show that the Airy integral is a multiple of $x^{\frac{1}{2}}\text{Kh}_{\frac{1}{2}}(\tfrac{2}{3}x^{\frac{3}{2}})$.
(Nicholson.)

3. Express Bi(x) in terms of Bessel functions of order $\pm\tfrac{1}{3}$.
(Miller.)

Chapter 22

APPLICATIONS OF BESSEL FUNCTIONS

22·01. The majority of applications of Bessel functions are to vibrations of systems with symmetry about an axis; the z coordinate usually either varies little, as in tidal waves on circular sheets of water, or the dependent variable is independent of z. Even if it involves z, Bessel functions usually provide the best treatment if the boundaries are planes of constant z. Bessel functions of order half an odd integer, in combination with Legendre functions, arise in problems of vibration for spherical boundaries. They also occur in various one-dimensional problems, notably the oscillations of a light string loaded with heavy particles at regular intervals, and the transmission of electric waves in a submarine cable.

22·02. Cylindrical pulse. Consider the explosion problem of 19·08, with the modification that the original excess pressure P_0 is within a cylinder of radius a instead of a sphere. With analogous initial conditions the subsidiary equation is

$$\frac{1}{\varpi}\frac{\partial}{\partial \varpi}\left(\varpi\frac{\partial \phi}{\partial \varpi}\right) - \frac{p^2}{c^2}\phi = \frac{P_0 p}{\rho c^2} \quad (\varpi < a), \\ = 0 \quad (\varpi > a). \tag{1}$$

The complementary functions are the Bessel functions of order zero, $I_0(p\varpi/c)$ and $\mathrm{Kh}_0(p\varpi/c)$. The latter is inadmissible within the cylinder because it is infinite when $\varpi = 0$. The former cannot occur outside it. For the interpretation is to be an integral through values of the variable with positive real parts, and when ϖ is great the asymptotic expansion of $I_0(z\varpi/c)$ contains $\exp(z\varpi/c)$ as a factor. Hence the solution would give a pulse travelling inwards. The solution is therefore

$$\phi = -\frac{P_0}{\rho p} + A I_0\left(\frac{p\varpi}{c}\right) \quad (0 \leqslant \varpi \leqslant a), \\ = B\mathrm{Kh}_0\left(\frac{p\varpi}{c}\right) \quad (a \leqslant \varpi). \tag{2}$$

Also $\partial\phi/\partial t$ and $\partial\phi/\partial \varpi$ must be continuous at $\varpi = a$. Hence

$$A I_0\left(\frac{pa}{c}\right) - \frac{P_0}{\rho p} = B\mathrm{Kh}_0\left(\frac{pa}{c}\right), \tag{3}$$

$$A I_0'\left(\frac{pa}{c}\right) = B\mathrm{Kh}_0'\left(\frac{pa}{c}\right). \tag{4}$$

We have the identity

$$I_0'(x)\mathrm{Kh}_0(x) - I_0(x)\mathrm{Kh}_0'(x) = 2/\pi x, \tag{5}$$

from 21·07 (6).

Hence for $\varpi \geqslant a$

$$\frac{\rho\phi}{P_0} = -\frac{\pi a}{2c} I_0'\left(\frac{pa}{c}\right)\mathrm{Kh}_0\left(\frac{p\varpi}{c}\right). \tag{6}$$

But
$$I_0(x) = \frac{1}{\pi}\int_0^\pi \exp(x\cos\theta)\,d\theta, \tag{7}$$

$$\mathrm{Kh}_0(x) = \frac{2}{\pi}\int_0^\infty \exp(-x\cosh v)\,dv, \tag{8}$$

and therefore
$$\frac{\rho\phi}{P_0} = -\frac{1}{\pi}\frac{a}{c}\int_0^\pi\int_0^\infty \cos\theta\exp\left(\frac{pa}{c}\cos\theta - \frac{p\varpi}{c}\cosh v\right)d\theta\,dv$$

$$= -\frac{1}{\pi}\frac{a}{c}\int_0^\pi\int_0^\infty \cos\theta\, H\left(t + \frac{a}{c}\cos\theta - \frac{\varpi}{c}\cosh v\right)d\theta\,dv. \tag{9}$$

Since
$$t + \frac{a}{c}\cos\theta - \frac{\varpi}{c}\cosh v \leqslant t + \frac{a}{c} - \frac{\varpi}{c}, \tag{10}$$

ϕ vanishes at any place up to time $(\varpi - a)/c$; and if we integrate first with regard to v, we can replace the upper limit by $\cosh^{-1}\{(ct + a\cos\theta)/\varpi\}$ and the unit function by 1, provided $ct + a\cos\theta > \varpi$. This will be true at least for $\theta = 0$ if $ct > \varpi - a$. Hence

$$\frac{\rho\phi}{P_0} = -\frac{a}{\pi c}\int_0^\pi \cos\theta\cosh^{-1}\left(\frac{ct + a\cos\theta}{\varpi}\right)d\theta \quad (ct > \varpi - a). \tag{11}$$

If $ct > \varpi + a$, $(ct + a\cos\theta)/\varpi > 1$ for all θ, and the upper limit is π. If $\varpi - a < ct < \varpi + a$, the upper limit is $\cos^{-1}(\varpi - ct)/a$. The disturbance can therefore be divided into three stages, according as $ct < \varpi - a$, $\varpi - a < ct < \varpi + a$, and $\varpi + a < ct$. In the first stage $\phi = 0$ and we have a cylindrical pulse travelling outwards with velocity c.

We are interested chiefly in the pressure. This is given by

$$\frac{P}{P_0} = \frac{a}{\pi}\int_0^{} \frac{\cos\theta\,d\theta}{\{(ct + a\cos\theta)^2 - \varpi^2\}^{1/2}}. \tag{12}$$

Put for $\varpi - a < ct < \varpi + a$,

$$ct + a - \varpi = 2b, \quad ct + a\cos\theta - \varpi = 2b\cos^2\psi \tag{13}$$

and suppose b small. Then soon after the arrival of the pulse

$$\frac{P}{P_0} \doteq \frac{1}{2}\sqrt{\left(\frac{a}{\varpi}\right)}\left[1 - \frac{3}{4}\frac{b}{a} - \frac{1}{4}\frac{b}{\varpi}\right]. \tag{14}$$

The increase of pressure on arrival is therefore $\tfrac{1}{2}P_0(a/\varpi)^{1/2}$, as against $\tfrac{1}{2}P_0$ in the corresponding one-dimensional problem and $\tfrac{1}{2}P_0 a/\varpi$ in the three-dimensional one. The decrease with time is at first proportionately slower than in the three-dimensional problem, and P is still positive when $ct = \varpi$. But it tends to $-\infty$ at $ct = \varpi + a$ and returns to finite negative values for greater values of t. The approximate value near $ct = \varpi + a$ is*

$$\frac{P}{P_0} \doteq -\frac{1}{2\pi}\sqrt{\left(\frac{a}{\varpi}\right)}\left\{\log\frac{32a}{|\varpi + a - ct|} - 4\right\}, \tag{15}$$

and when $ct - \varpi$ is large compared with a is

$$\frac{P}{P_0} = -\frac{1}{2}\frac{a^2 ct}{(c^2 t^2 - \varpi^2)^{3/2}}. \tag{16}$$

* Jeffreys, Proc. Camb. Phil. Soc. 39, 1943, 48–51.

This infinite disturbance of pressure does not imply infinite energy, since the infinity is only logarithmic; and it will in any case be modified by the inclusion of second-order terms in the hydrodynamical equations. The instantaneous release of the whole of the surface of an infinite cylinder would be difficult to arrange physically, but an approximation to it would be possible if the interior was filled with an explosive mixture and the velocity of the wave of combustion in it was several times the velocity of sound in cool air. The indefinitely prolonged tail of the disturbance is characteristic of two-dimensional propagation. It occurs also for a point source between two parallel plates and in the formation of elastic waves in a solid;* surface waves are formed by diffraction at the boundary, spread out in two dimensions, and give at any place only an asymptotic return to the original position, in spite of the fact that the original disturbance may be of finite extent in all three directions.

22·03. Light string with concentrated loads. We have seen that the operational method is universally valid for the treatment of a properly specified finite set of linear equations, and does not need the use of Bromwich's integral. We have suggested that continuous systems are best regarded physically as derived from discrete systems by a limiting process, and the solutions found for them as the limits of the solutions for the discrete systems. It is desirable, therefore, to have a concrete example showing how the kind of operator that arises for continuous systems can also arise as the limit of a sequence of operators applicable to discrete systems. One such example is provided by the uniform stretched string under tension $P = \rho c^2$, with mass ρ per unit length. If we replace this by a light string under tension P, with particles of mass ρl at intervals l, we have a discrete system with the same average mass per unit length, and we can approach the uniform string as a limit by taking l indefinitely small. The equation of motion of a particle is

$$\ddot{y}_r = -\frac{c^2}{l^2}(2y_r - y_{r-1} - y_{r+1}), \tag{1}$$

which reduces, on putting $x = rl$ and letting $l \to 0$ with x fixed, to

$$\frac{\partial^2 y}{\partial t^2} = c^2 \frac{\partial^2 y}{\partial x^2}. \tag{2}$$

Suppose that the system starts from rest, that the particle with $r = m$ is kept fixed, and that y_0 is made to vary with the time in a prescribed manner. Then the subsidiary equations are

$$\left(p^2 + \frac{2c^2}{l^2}\right) y_r = \frac{c^2}{l^2}(y_{r-1} + y_{r+1}) \quad (0 < r < m). \tag{3}$$

These can be solved formally by putting

$$y_r = A e^{r\lambda}; \tag{4}$$

then

$$\frac{p^2 l^2}{c^2} + 2 = e^{-\lambda} + e^{\lambda} = 2\cosh\lambda \tag{5}$$

and there are two equal and opposite real values of λ for real p.

* Lamb, *Phil. Trans.* A, 203, 1904, 1–42.

Then if
$$\sinh \tfrac{1}{2}\lambda = \frac{pl}{2c} \tag{6}$$

(5) is satisfied; and if we now take the positive root, and take
$$y_r = \frac{\sinh(m-r)\lambda}{\sinh m\lambda} y_0, \tag{7}$$

all the conditions are satisfied. But $\sinh s\lambda/\sinh \lambda$ is a polynomial in $\sinh \tfrac{1}{2}\lambda$ of degree $2(s-1)$. The operator is therefore a rational function of p, and its expansion in descending powers of p starts with $(c/lp)^{2r}$. It follows that the further a particle is from the disturbed end the more gradually it will begin to move. But also we can expand in exponentials

$$\frac{\sinh(m-r)\lambda}{\sinh m\lambda} = e^{-r\lambda}(1 - e^{-2(m-r)\lambda})(1 + e^{-2m\lambda} + e^{-4m\lambda} + \ldots). \tag{8}$$

For if we replace p by z and λ by ζ, with $\Re(z) > 0$, then $\Re(\zeta) > 0$, and $|e^{-\zeta}| < 1$. The first term in y_r is then
$$w_r = \left\{ \frac{pl}{2c} + \left(\frac{p^2 l^2}{4c^2} + 1 \right)^{1/2} \right\}^{-2r} y_0, \tag{9}$$

which again is expansible in negative powers of p and satisfies our fundamental rules. Further, if $lr = x$ and $l \to 0$, it tends formally to $e^{-px/c}$, the operator characteristic of waves in a uniform string. It does not lead to an interpretation of $e^{px/c}$ by a limiting process because if we change the sign of r or p we get an expression that is not expansible in negative powers of p.

The physical string, however, has a molecular structure, and we are concerned to know how closely the solution for the continuous string approximates to that for the actual string. For this purpose we take $y_0 = H(t)$; we want to see whether $w_r \to H(t - x/c)$ when $rl = x$ is fixed and l is small. Then
$$w_r = \frac{1}{2\pi i} \int_L e^{zt} \left\{ \left(1 + \frac{z^2 l^2}{4c^2} \right)^{1/2} + \frac{zl}{2c} \right\}^{-2r} \frac{dz}{z}, \tag{10}$$

and we use the method of steepest descents. Put
$$\phi(z) = zt - 2r \log \left\{ \left(1 + \frac{z^2 l^2}{4c^2} \right)^{1/2} + \frac{zl}{2c} \right\}; \tag{11}$$

then
$$\phi'(z) = t - \frac{rl}{c\sqrt{\{1 + z^2 l^2/4c^2\}}}, \tag{12}$$

$$\phi''(z) = \frac{rzl^3}{4c^3(1 + z^2 l^2/4c^2)^{3/2}}. \tag{13}$$

If $rl = x$, $x/ct = \xi$, the saddle-points are at $z = \pm (2c/l)(\xi^2 - 1)^{1/2}$ and therefore are on the real or the imaginary axis according as ξ is greater or less than 1.

If $\xi > 1$ we find
$$w_r \sim \frac{1}{2} \left(\frac{l}{\pi ct} \right)^{1/2} \frac{\xi}{(\xi^2 - 1)^{3/4}} \exp \left[-\frac{2ct}{l} \{\xi \cosh^{-1} \xi - (\xi^2 - 1)^{1/2}\} \right], \tag{14}$$

which tends exponentially to 0 for given ξ if $l \to 0$. If $\xi - 1$ is small, put $\xi = \cosh u$, $v = ctu^3/l$ and consider values of ξ such that v is large. We find that if

$$v = 6, \ w_r \doteqdot 0\cdot 0021, \ \xi \doteqdot 1 + \tfrac{1}{2}(vl/ct)^{2/3}.$$

Hence if ct/l is large w_r is negligible if x exceeds ct by an amount that increases only like $t^{1/3}$, and tends to zero with l. If $l = 10^{-8}$ cm., $ct = 10$ cm., $v = 6$, we have $\xi - 1 \doteqdot 1.7 \times 10^{-6}$. With ordinary magnitudes, therefore, the motion is negligible at distances so little in advance of the ideal pulse that the continuous system gives as good an approximation as we should ever need.

If $x < ct$, the lines of steepest descent through the saddle-points proceed from and to $-\infty$; obviously they cannot approach $+\infty$, since the integrand tends to infinity there. Hence they are not together equivalent to the path L, since the pole at $z = 0$ lies in between. We must therefore include a loop from $-\infty$ about the origin, and this makes a contribution 1 to w_r. We find then that for $x < ct$

$$w_r \sim 1 - \left(\frac{l}{\pi ct}\right)^{1/2} \frac{\xi}{(1-\xi^2)^{3/4}} \cos\left[\frac{2ct}{l}\{(1-\xi^2)^{1/2} - \xi \cos^{-1}\xi\} + \tfrac{1}{4}\pi\right]. \tag{15}$$

When $l \to 0$ this tends to 1, as we expected; but the correction term tends to 0 only like $l^{1/2}$ instead of exponentially, its wave-length tending to zero with l. However, with $l/ct = 10^{-9}$, $1 - \xi \geqslant 5 \times 10^{-5}$, the term never exceeds 1.7×10^{-2} and diminishes with decreasing ξ. The sharp front followed by a constant displacement is therefore a good approximation.

The change of phase of the correction term from one particle to the next is small if ξ is only a little less than 1, but approaches π if ξ is small. We therefore have dispersion. The disturbance can be regarded as including all possible wave-lengths $\geqslant 2l$; the longest have group-velocity c, but the shortest group-velocity 0. This can be seen by returning to (10) and writing the first two factors as $\exp i(\gamma t - \kappa x)$, with

$$z = i\gamma, \quad rl = x, \quad \frac{\gamma l}{2c} = \sin\theta, \quad \kappa = 2\theta/l. \tag{16}$$

The wave velocity is then

$$\frac{\gamma}{\kappa} = \frac{c\sin\theta}{\theta}, \tag{17}$$

and values of θ between 0 and $\tfrac{1}{2}\pi$ are admissible. (Larger values would give the same displacement where x ranges through exact multiples of l and are therefore irrelevant.) The wave velocity therefore ranges from c to $2c/\pi$, and the group velocity is

$$\frac{d\gamma}{d\kappa} = c\cos\theta, \tag{18}$$

which ranges from c to 0.

Suppose now that the string extends on both sides of the particle specified by $r = 0$, and that instead of the motion of this particle being prescribed it is given an initial displacement u and then released. Its subsidiary equation is

$$\left(p^2 + \frac{2c^2}{l^2}\right) y_0 - \frac{c^2}{l^2}(y_1 + y_{-1}) = p^2 u, \tag{19}$$

and if the time is short enough for waves reflected at the ends not to have arrived

$$y_1 = y_{-1} = e^{-\lambda} y_0, \quad y_0 = \frac{pul/2c}{\cosh\tfrac{1}{2}\lambda}, \quad w_r = \frac{pul/2c}{(1+p^2l^2/4c^2)^{1/2}}\left\{\frac{pl}{2c} + \left(1 + \frac{p^2l^2}{4c^2}\right)^{1/2}\right\}^{-2r}. \tag{20}$$

$$= u J_{2r}\left(\frac{2ct}{l}\right), \tag{21}$$

from 21·01 (23).

A Bessel function takes its maximum value when the argument slightly exceeds the order, so that we see at once that the greatest displacement travels out with velocity c, and from 21·05(5) that when ct is large compared with rl the motion of a particle is oscillatory, consecutive particles differing in phase by amounts approaching π. The rapid variation of the phase of the movement suggests an analogy with heat conduction, but it is quite systematic, and the essential property of heat conduction is that the variation is not systematic. Consecutive particles in random motion are as likely to be in the same phase as in opposite phases. A little further examination shows that the analogy breaks down in another respect.* If we consider random initial displacements and velocities given to all particles in a finite length of the string, the energy is found to spread out so that the length of the string that contains a given fraction of the initial energy increases in proportion to t. In heat conduction the length in question would increase like $t^{1/2}$.

There is a considerable change if there is any irregularity in the structure of the system itself. Let us assume a harmonic wave train coming from negative x, but that the particle specified by suffix 0 has mass $\rho l(1+a)$ instead of ρl, where a may be small. There will be a reflected wave; we therefore take

$$y = \exp i(\gamma t - \kappa x) + A \exp i(\gamma t + \kappa x) \quad (x/l \leq 0),$$
$$y = B \exp i(\gamma t - \kappa x) \quad (x/l \geq 0). \tag{22}$$

Then
$$1 + A = B. \tag{23}$$

Also the equation of motion for this particle is

$$\left\{-(1+a)\gamma^2 + \frac{2c^2}{l^2}\right\} y_0 = \frac{c^2}{l^2}(y_{-1} + y_{+1}), \tag{24}$$

and substituting from (22) and (23) we find, with $\kappa l = 2\theta$,

$$B = \frac{1}{1 + ia \tan \theta}. \tag{25}$$

If θ is small, corresponding to long waves, B is practically 1 and there is nearly perfect transmission. But if θ is nearly $\frac{1}{2}\pi$, corresponding to the shortest waves possible, there will be nearly perfect reflexion even if a is small. Thus even a slight irregularity of structure will practically destroy the tail of a wave train. If there is a disturbance between two such irregularities much of the energy will be reflected several times before it gets past either, and a number of minor irregularities will give an irregular motion closely resembling thermal agitation, with a slow leakage resembling conduction. In an actual solid we have a three-dimensional form of the same problem, the irregularities arising from random motions of electrons even in a crystal and from local departures from a regular pattern in a glass.

22·04. Diffusion as a limit. It is strictly meaningless to speak of the temperature at a point, since the temperature expresses the mean energy of random motion of a number of particles; if we speak of the absolute temperature as specified within a factor of 10^{-3} we must be considering something of the order of 10^6 particles. In the strict mathematical sense, therefore, the space derivatives of the temperature do not exist. But if l is sufficiently large for the difference of temperature between two places l apart to be

* Jeffreys, *Proc. Camb. Phil. Soc.* 23, 1927, 775.

considerably more than its uncertainty we can convert the equation of conduction into a finite difference equation, the one-dimensional form of which is

$$\frac{\partial y_r}{\partial t} = \frac{h^2}{l^2}(y_{r-1} - 2y_r + y_{r+1}). \tag{1}$$

We take $y_0 = H(t)$ and solve operationally. The analysis is very similar to that at the beginning of the last section, p replacing p^2, and we get as solution for the inward diffusion corresponding to (9) of 22·03

$$w_r = \left\{\left(1 + \frac{pl^2}{4h^2}\right)^{1/2} + \frac{p^{1/2}l}{2h}\right\}^{-2r} = p^{-r}\left\{\left(\frac{l^2}{4h^2} + \frac{1}{p}\right)^{1/2} + \frac{l}{2h}\right\}^{-2r} \tag{2}$$

which is expansible in powers of p^{-1}. If l is made to tend to 0 while rl tends to x, w_r tends formally to $e^{-p^{1/2}x/h}$, that is, to our e^{-qx}. Thus we have obtained the latter operator as the limit of one expressible as a power series in p^{-1}. The difference $(w_1 - w_0)/l$ yields a derivation of $p^{1/2}$, namely

$$p^{1/2} = h \lim_{l \to 0} \frac{1}{l}\left[1 - \left\{\left(1 + \frac{pl^2}{4h^2}\right)^{1/2} + \frac{p^{1/2}l}{2h}\right\}^{-2}\right]. \tag{3}$$

Now
$$w_r = \frac{1}{2\pi i}\int_L \left\{\left(1 + \frac{zl^2}{4h^2}\right)^{1/2} + \frac{z^{1/2}l}{2h}\right\}^{-2r} e^{zt}\frac{dz}{z}. \tag{4}$$

By Dalzell's theorem (12·101) we can reverse the order of integration and passage to the limit; then

$$\lim_{l \to 0} w = \frac{1}{2\pi i}\int_L \exp\left(zt - \frac{z^{1/2}x}{h}\right)\frac{dz}{z} = 1 - \operatorname{erf}\frac{x}{2ht^{1/2}}, \tag{5}$$

from 12·126.

When l is small but not zero the saddle-point is slightly displaced, but we may take the path to be the path of steepest descent for the integral in (5). On this path we can expand in ascending powers of l, for the main contribution comes from the neighbourhood of the saddle-point, and an expansion exists if l is small. Then

$$w \sim \frac{1}{2\pi i}\int_L \left(1 + \frac{z^{3/2}xl^2}{24h^3}\right)\exp\left(zt - \frac{z^{1/2}x}{h}\right)\frac{dz}{z}$$

$$= \left(1 - \frac{xl^2}{24}\frac{\partial^3}{\partial x^3}\right)\left(1 - \operatorname{erf}\frac{x}{2ht^{1/2}}\right)$$

$$= \left(1 - \operatorname{erf}\frac{x}{2ht^{1/2}}\right) - \frac{xl^2}{48h^3\sqrt{(\pi t^3)}}\left(1 - \frac{x^2}{2h^2 t}\right)\exp\left(-\frac{x^2}{4h^2 t}\right). \tag{6}$$

If $x/2ht^{1/2}$ is large both terms are small irrespective of l. The important values of x are of the order of $2ht^{1/2}$ or less. At these the last term is of the order of $l^2/48h^2 t$ of the first, so that it can be neglected if $t \gg l^2/48h^2$.

If we take $l = 10^{-5}$ cm., $h = 0·1$ c.g.s. (a value for a bad conductor), the critical value of t is 2×10^{-10} sec. Thus for a short time after the initial disturbance of temperature the usual solution is invalid because the temperature itself is meaningless; but this time is very short and would be shorter for better conductors. The approach through the finite system therefore confirms the results obtained by treating the continuous case directly and answers the logical objection to that method.

22·05. The submarine cable.

This is a uniform conductor with self-induction, capacity, resistance, and leakage. At distance x from the end suppose that the charge per unit length is y, the potential ϕ, and the current J. Then the following equations hold:

$$y = k\phi, \tag{1}$$

$$\frac{ldJ}{dt} + rJ = -\frac{\partial \phi}{\partial x}, \tag{2}$$

$$\frac{\partial y}{\partial t} = -\frac{\partial J}{\partial x} - s\phi. \tag{3}$$

l, k and r are the self-induction, capacity and resistance per unit length. The leakage is such that a potential ϕ produces a current $s\phi$ per unit length leaking away through the insulation. Up to $t = 0$, y, J, and ϕ are zero; afterwards ϕ is raised to ϕ_0, which may be a function of t, at $x = 0$. Then the subsidiary equations are

$$(lp+r) J = -\frac{\partial \phi}{\partial x}, \quad (kp+s)\phi = -\frac{\partial J}{\partial x}, \tag{4}$$

whence
$$\frac{\partial^2 \phi}{\partial x^2} = (lp+r)(kp+s)\phi. \tag{5}$$

Put
$$(lp+r)(kp+s) = q^2 = lk\{(p+\rho)^2 - \sigma^2\}. \tag{6}$$

Then the operational solution, neglecting reflexion from the far end, is

$$\phi = e^{-qx}\phi_0. \tag{7}$$

If self-induction and leakage are negligible we have $l = 0$, $s = 0$, $q^2 = krp$. Then the solution has the same form as for conduction of heat. This condition occurs in ordinary telegraph wires. If in this case $\phi_0 = H(t)$,

$$\phi = 1 - \operatorname{erf} \frac{x}{2}\left(\frac{kr}{t}\right)^{1/2}. \tag{8}$$

For given x, $\phi \to 1$ for large t, and the approach is quicker if kr is small. For fairly short lines the time needed is short enough for successive signals to be transmitted without overlapping. But for long ones, and especially for submarine cables, the time needed to built up the requisite potential at the receiving end is long enough to interfere seriously with the practicable speed of signalling. Reduction of kr means thicker conductors and therefore prohibitive cost. The modification introduced by Heaviside was the deliberate introduction of self-induction and leakage far above what the simple cable possessed. The principle of the self-induction can be seen from the rough analogy of a projectile thrown through air. If it has negligible mass the resistance damps down the motion quickly and it does not travel far. But a heavier projectile, though it needs more effort to give it the same velocity, keeps its velocity better and travels further. Self-induction acts in much the same way as inertia. The effect of leakage is less obvious, but will appear in a moment. We put
$$lk = 1/c^2. \tag{9}$$

Since $\rho + \sigma = r/l$ and $\rho - \sigma = s/k$ are both positive, ρ is positive. Then $0 < |\sigma| < \rho$.

If $\sigma = 0$ we have simply

$$\phi = \exp\left\{-(p+\rho)\frac{x}{c}\right\}\phi_0$$

$$= e^{-\rho x/c}\phi_0\left(t - \frac{x}{c}\right). \tag{10}$$

Submarine cable

Hence the variation of ϕ with time at distance x is an exact copy of that at the transmitter except for the delay x/c and the constant attenuation factor $e^{-\rho x/c}$. The velocity c is very high and the attenuation can be compensated by amplifiers at the receiving end. This arrangement is the distortionless cable, and is achieved if $ls = kr$.

If σ is not zero, we take $\phi_0 = H(t)$ and the solution is

$$\phi = \exp\left[-\frac{x}{c}\{(p+\rho)^2 - \sigma^2\}^{1/2}\right] = \frac{1}{2\pi i}\int_L \exp\left[zt - \frac{x}{c}\{(z+\rho)^2 - \sigma^2\}^{1/2}\right]\frac{dz}{z}. \quad (11)$$

The current is

$$J = -\frac{1}{l(p+\rho+\sigma)}\frac{\partial\phi}{\partial x} = \frac{1}{lc}\frac{p+\rho-\sigma}{\{(p+\rho)^2-\sigma^2\}^{1/2}}\exp\left[-\frac{x}{c}\{(p+\rho)^2-\sigma^2\}^{1/2}\right]. \quad (12)$$

First omit the term in $\rho - \sigma$ from the first factor. What remains is

$$I = \frac{1}{lc}\frac{1}{2\pi i}\int_L \frac{\exp\left[-\frac{x}{c}\{(z+\rho)^2 - \sigma^2\}^{1/2} + zt\right]dz}{\{(z+\rho)^2 - \sigma^2\}^{1/2}}$$

$$= \frac{1}{2\pi i lc}e^{-\rho t}\int_L \frac{\exp\left[\zeta t - \frac{x}{c}(\zeta^2 - \sigma^2)^{1/2}\right]d\zeta}{(\zeta^2 - \sigma^2)^{1/2}}. \quad (13)$$

Put $\zeta = \frac{1}{2}\sigma(u + 1/u)$; then

$$I = \frac{1}{2\pi i lc}e^{-\rho t}\int_L \exp\tfrac{1}{2}\sigma\left\{ut + \frac{t}{u} - \frac{xu}{c} + \frac{x}{uc}\right\}\frac{du}{u}$$

$$= \frac{1}{lc}e^{-\rho t}I_0\left\{\sigma\sqrt{\left(t^2 - \frac{x^2}{c^2}\right)}\right\} \quad (t - x/c > 0). \quad (14)$$

The part omitted follows by integration; we have

$$J = \frac{1}{lc}e^{-\rho t}I_0\left\{\sigma\sqrt{\left(t^2 - \frac{x^2}{c^2}\right)}\right\} + \frac{\rho - \sigma}{lc}\int_{x/c}^t e^{-\rho t}I_0\left\{\sigma\sqrt{\left(t^2 - \frac{x^2}{c^2}\right)}\right\}dt \quad (t > x/c). \quad (15)$$

For ϕ, notice first that

$$\exp\left[-\frac{x}{c}\{(p+\rho)^2 - \sigma^2\}^{1/2}\right] = e^{-(p+\rho)x/c}\exp\left\{\frac{\sigma^2 x}{2(p+\rho)c} + \ldots\right\} \quad (16)$$

the second exponential being in descending powers of $p + \rho$; hence ϕ is zero up to time x/c and then jumps to $e^{-\rho x/c}$, afterwards varying continuously. Secondly, we can take the termini to have real part $-\infty$, and then differentiate with regard to t; then

$$\frac{\partial\phi}{\partial t} = \frac{1}{2\pi i}\int_M \exp\left\{zt - \frac{x}{c}\{(z+\rho)^2 - \sigma^2\}^{1/2}\right\}dz$$

$$= \frac{1}{2\pi i}e^{-\rho t}\int_M \exp\left[\zeta t - \frac{x}{c}(\zeta^2 - \sigma^2)^{1/2}\right]d\zeta$$

$$= \frac{1}{2\pi i}e^{-\rho t}\int_M \exp\tfrac{1}{2}\sigma\left\{u\left(t - \frac{x}{c}\right) + \frac{1}{u}\left(t + \frac{x}{c}\right)\right\}\tfrac{1}{2}\sigma\left(1 - \frac{1}{u^2}\right)du$$

$$= \tfrac{1}{2}\sigma e^{-\rho t}\left[\left(\frac{t+x/c}{t-x/c}\right)^{1/2}I_{-1}\left\{\sigma\sqrt{\left(t^2 - \frac{x^2}{c^2}\right)}\right\} - \left(\frac{t-x/c}{t+x/c}\right)^{1/2}I_1\left\{\sigma\sqrt{\left(t^2 - \frac{x^2}{c^2}\right)}\right\}\right]$$

$$= \frac{\sigma e^{-\rho t}x/c}{(t^2 - x^2/c^2)^{1/2}}I_1\left\{\sigma\sqrt{\left(t^2 - \frac{x^2}{c^2}\right)}\right\}. \quad (17)$$

But we know ϕ just after $t = x/c$; hence

$$\phi = e^{-\rho x/c} + \int_{x/c}^{t} \frac{\sigma x/c}{(t^2 - x^2/c^2)^{1/2}} e^{-\rho t} I_1 \left\{ \sigma \sqrt{\left(t^2 - \frac{x^2}{c^2}\right)} \right\} dt \quad (t > x/c). \tag{18}$$

22·06. Line source of heat: line vortex. The equation of heat conduction, for the case of symmetry about a straight line, is

$$\frac{\partial V}{\partial t} - \frac{h^2}{\varpi} \frac{\partial}{\partial \varpi} \left(\varpi \frac{\partial V}{\partial \varpi} \right) = 0. \tag{1}$$

The same equation is satisfied by the vorticity ζ in a viscous liquid when the motion is in circles about an axis,* h^2 being replaced by ν, the kinematic viscosity. Take first a concentration of heat κ per unit length along the axis at $t = 0$; the operational solution will be

$$V = A \mathrm{Kh}_0(q\varpi). \tag{2}$$

A being a function of t. If ρc is the heat capacity per unit volume the excess heat within distance ϖ of the axis, per unit distance parallel to the axis, is,

$$2\pi \int_0^\varpi \rho c V \varpi \, d\varpi = 2\pi A \int_0^\varpi \rho c \varpi \mathrm{Kh}_0(q\varpi) \, d\varpi$$
$$= -2\pi A \rho c q^{-2} \left[q\varpi \mathrm{Kh}_1(q\varpi) \right]_0^\varpi. \tag{3}$$

When $t \to 0$ this must tend to κ for all $\varpi > 0$; but

$$\lim_{t \to 0} \frac{1}{q^2} [q\varpi \mathrm{Kh}_1(q\varpi)] = \lim_{q \to \infty} \frac{1}{q^2} \{q\varpi \mathrm{Kh}_1(q\varpi)\} = 0 \tag{4}$$

and

$$\lim_{\varpi \to 0} q\varpi \mathrm{Kh}_1(q\varpi) = \frac{2}{\pi}. \tag{5}$$

Hence

$$\kappa = 4 A \rho c q^{-2} \tag{6}$$

and

$$V = \frac{\kappa}{4\rho c} q^2 \mathrm{Kh}_0(q\varpi) = \frac{\kappa}{4\rho c} \frac{e^{-\varpi^2/4h^2 t}}{\pi h^2 t}. \tag{7}$$

Similarly if κ is the original circulation about the axis and we put $p = \nu q^2$ we get (2) for ζ instead of V, and the circulation is $2\pi \int_0^\varpi \zeta \varpi \, d\varpi$. Proceeding similarly we find†

$$\zeta = \frac{\kappa}{4\pi \nu t} e^{-\varpi^2/4\nu t}.$$

The circulation is $\kappa(1 - e^{-\varpi^2/4\nu t})$ and the velocity is therefore

$$\frac{\kappa}{2\pi\varpi}(1 - e^{-\varpi^2/4\nu t}).$$

---

* Lamb, *Hydrodynamics*, 1932, p. 591.
† Goldstein, *Proc. Lond. Math. Soc.* (2) 34, 1932, 62.

EXAMPLES

1. An elevation of the form $\sin \kappa x$ on the surface of a highly viscous fluid can be shown to decrease with time like $\exp(-Agt/\nu\kappa)$. Show that if the initial elevation is $\pm \frac{1}{2}$ according as $x \gtrless 0$, the elevation at any later time is

$$\Im \operatorname{Kh}_0 \{2e^{-\frac{1}{4}\pi i}\sqrt{(Agxt/\nu)}\}$$

and that the discontinuity at $x = 0$ persists.

2. A unit e.m.f. is applied at $t = 0$ to the sending end of a non-inductive line of resistance R, capacity C, and leakage G per unit length. Show that the current I at the sending end at time t is given by

$$I = \left(\frac{C}{\pi R}\right)^{1/2} \left\{\frac{e^{-\lambda t}}{\sqrt{t}} + \lambda \int_0^t \frac{e^{-\lambda \tau}}{\sqrt{\tau}} d\tau\right\},$$

where $\lambda = G/C$. (I.C. 1943.)

3. The curved surface and the base ($z = 0$) of the cylinder $\varpi = a$ are maintained at zero temperature. The other plane end $z = b$ is maintained at temperature T. Prove that the steady distribution of temperature is given by

$$V = \frac{2T}{a^2} \sum_{n=1}^{\infty} \frac{1}{\mu_n^2} \frac{\sinh \mu_n z}{\sinh \mu_n b} \frac{J_0(\mu_n \varpi)}{J_1(\mu_n a)},$$

where $\mu_1, \mu_2 \ldots$ are the zeros of $J_0(\mu a)$. (I.C. 1939.)

4. A uniform chain of length l and weight w hangs freely from one end, and makes small oscillations. Calculate the length of the simple pendulum equivalent to the slowest mode of vibration. (I.C. 1938.)

Chapter 23

THE CONFLUENT HYPERGEOMETRIC FUNCTION

> 'All changes trying, he will take the form
> Of ev'ry reptile on the earth, will seem
> A river now, and now devouring fire;
> But hold him ye, and grasp him still the more.'
>
> HOMER, *Odyssey* (Cowper's translation)

23·01. The *hypergeometric function* is defined in general by the series

$$F(a,b;c;z) = 1 + \frac{a \cdot b}{1 \cdot c}z + \frac{a(a+1)b(b+1)}{2!\,c(c+1)}z^2 + \frac{a(a+1)(a+2)b(b+1)(b+2)}{3!\,c(c+1)(c+2)}z^3 + \ldots \quad (1)$$

and satisfies the differential equation

$$z(1-z)\frac{d^2u}{dz^2} + \{c - (a+b+1)z\}\frac{du}{dz} - abu = 0. \quad (2)$$

A second solution is $\quad z^{1-c}F(a-c+1, b-c+1; 2-c; z).\quad (3)$

It can be shown by direct transformation that any second order differential equation with three regular singularities (one of which may be at infinity) and no other singularities can be reduced to this form, and that all the solutions about them can be expressed in terms of the hypergeometric function. The function has a large literature and several well known functions can be expressed in terms of it. We notice at once that if $b = c$ it reduces to the binomial series, and if $a = b = 1, c = 2$, it gives the series for $-z^{-1}\log(1-z)$. The series expressing the Legendre functions in terms of argument $1-x$ is also of this type.

If c is not an integer both series are significant for $|z| < 1$. There are several complications of the types discussed in Chapter 16 if c is an integer, positive, zero, or negative.

If we put
$$x = \frac{1}{z}$$

the singularities of the transformed equation are at ∞, 1, 0; and if we put

$$x = 1 - z$$

the singularities are at 1, 0, ∞. By successive applications of these transformations we can express the equation in terms of any of the independent variables

$$z, \quad \frac{1}{z}, \quad 1-z, \quad \frac{1}{1-z}, \quad \frac{z-1}{z}, \quad \frac{z}{z-1},$$

and in each case it retains its hypergeometric form, and there are two solutions expressible in terms of hypergeometric functions of the variable used. The equation therefore has 12 solutions of the forms (1), (3), any of which can be expressed in terms of two fundamental solutions. All have radius of convergence 1. Twelve more can be obtained, of the types $(1-z)^{c-a-b}F(c-a, c-b; c; z)$ and $z^{1-c}(1-z)^{c-a-b}F(1-a, 1-b; 2-c; z)$. Each of these is equal to one of the original twelve.

23·02. Series and differential equation for the confluent hypergeometric function.
If we put $bz = x$ and then let b tend to infinity the function becomes the series

$$u_1 = {}_1F_1(\alpha, \gamma, x) = M(\alpha, \gamma, x) = 1 + \frac{\alpha}{\gamma}x + \frac{\alpha(\alpha+1)}{2!\gamma(\gamma+1)}x^2 + \dots . \tag{1}$$

In this notation the above hypergeometric series would be ${}_2F_1(a, b; c; z)$, the first suffix denoting the number of factorials in the numerator of the general term, the second the number, apart from $n!$, in the denominator. The notation can be extended to series containing any number of factorials in the general term; such series are known as generalized hypergeometric functions. The Bessel functions come under the type ${}_0F_1$, and can evidently be derived by putting $\alpha x = y$ and then making α tend to infinity.

Evidently ${}_1F_1(\alpha, \gamma, x)$ is an integral function. It satisfies the differential equation

$$x\frac{d^2u}{dx^2} + (\gamma - x)\frac{du}{dx} - \alpha u = 0. \tag{2}$$

Another solution is found to be

$$u_2 = x^{1-\gamma}{}_1F_1(1+\alpha-\gamma, 2-\gamma, x) = x^{1-\gamma} + \sum_{r=1}^{\infty}\frac{(1+\alpha-\gamma)\dots(r+\alpha-\gamma)}{r!(2-\gamma)\dots(r+1-\gamma)}x^{r+1-\gamma} \tag{3}$$

except possibly when γ is an integer. Hence there are two independent series solutions except possibly if γ is an integer (positive, zero, or negative). If $\gamma = 1$, (1) and (3) are identical.

If γ is an integer $\geqslant 2$ all terms of (3) from $r = \gamma - 1$ onwards have a zero factor in the denominator, and (3) will not be a valid form of solution unless there is also a zero factor in the numerator, that is, unless α is an integer such that $1 \leqslant \alpha \leqslant \gamma - 1$.

If γ is zero or a negative integer, and $r = 1 - \gamma$, all terms of (1) from x^r onwards have vanishing denominators, and (1) will not be a valid form of solution unless α is an integer such that $\gamma \leqslant \alpha \leqslant 0$.

Hence for certain special values of α there are two series solutions even if γ is an integer different from 1. A terminating series can be found in each case.

If γ is not an integer, and α is a negative integer or zero, (1) terminates; if $\alpha - \gamma$ is a negative integer, (3) terminates.

If $\gamma > 0$, (1) is always significant. We shall see that (1) can also be expressed in terms of a complex integral, and that all solutions of (2) can be expressed in terms of the integral used for (1), with suitable changes of the termini, just as all solutions of Bessel's equation can be expressed by changes of the termini in the complex integrals used for $J_n(x)$, where $\Re(n) \geqslant 0$.

Since the function depends on three variables it is practically beyond the reach of tabulation in general. If a function of one variable takes a page to tabulate, one of two variables will take a book, one of three variables an ordinary sized room of bookshelves, and one of four variables a large library. Consequently the theory of this function, and still more of the hypergeometric function, is mainly a matter of general propositions with detailed application to a few special cases.

23·03. Complex integral solutions. Complex integrals representing the functions can be obtained at once by operational methods. We write

$$z^{\gamma-1}{}_1F_1(\alpha,\gamma,x) = \sum_{r=0}^{\infty} \frac{\alpha(\alpha+1)\ldots(\alpha+r-1)(\gamma-1)!}{r!} \frac{x^{\gamma+r-1}}{(\gamma+r-1)!}$$

$$= (\gamma-1)!\sum \frac{\alpha(\alpha+1)\ldots(\alpha+r-1)}{r!} p^{-(\gamma+r-1)} = (\gamma-1)!\left(1-\frac{1}{p}\right)^{-\alpha} p^{1-\gamma} \quad (4)$$

$$= \frac{(\gamma-1)!}{2\pi i}\int_L \frac{e^{zx}dz}{z^\gamma(1-1/z)^\alpha} = \frac{(\gamma-1)!}{2\pi i}\int_L \frac{e^{zx}dz}{z^{\gamma-\alpha}(z-1)^\alpha} \quad (5)$$

and similarly

$$x^{1-\gamma}{}_1F_1(1+\alpha-\gamma, 2-\gamma, x) = (1-\gamma)!\left(1-\frac{1}{p}\right)^{-(1+\alpha-\gamma)} p^{\gamma-1} \quad (6)$$

$$= \frac{(1-\gamma)!}{2\pi i}\int_L \frac{e^{zx}z^{\alpha-1}dz}{(z-1)^{1+\alpha-\gamma}}. \quad (7)$$

These are to be understood in the first place as valid for x real and positive; and then we have by analytic continuation, putting $zx = \lambda$,

$$_1F_1(\alpha,\gamma,x) = \frac{(\gamma-1)!}{2\pi i}\int_M \frac{e^\lambda \lambda^{\alpha-\gamma}}{(\lambda-x)^\alpha} d\lambda, \quad (8)$$

$$_1F_1(1+\alpha-\gamma, 2-\gamma, x) = \frac{(1-\gamma)!}{2\pi i}\int_M \frac{e^\lambda \lambda^{\alpha-1}}{(\lambda-x)^{1+\alpha-\gamma}} d\lambda \quad (9)$$

valid for all x. If the factorial factor is omitted, both are intelligible even if γ is an integer. If we put $\lambda = \kappa + x$ in (8) we get

$$u_1 = {}_1F_1(\alpha,\gamma,x) = \frac{(\gamma-1)!}{2\pi i}e^x\int_M \frac{e^\kappa \kappa^{-\alpha}}{(\kappa+x)^{\gamma-\alpha}} d\kappa = e^x\,{}_1F_1(\gamma-\alpha,\gamma,-x), \quad (10)$$

and similarly $\quad u_2 = x^{1-\gamma}{}_1F_1(1+\alpha-\gamma, 2-\gamma, x) = x^{1-\gamma}e^x\,{}_1F_1(1-\alpha, 2-\gamma, -x). \quad (11)$

(10) is very useful if $\gamma-\alpha$ is an integer ≤ 0, since the last factor is then a polynomial. Similarly (11) is useful if $1-\alpha$ is an integer ≤ 0. Hence if either $\gamma-\alpha$ or α is an integer, irrespective of sign, one of the series solutions reduces to an elementary function. Many recurrence relations exist, analogous to those for the Bessel functions.*

Also
$$\frac{d}{dx}{}_1F_1(\alpha,\gamma,x) = \frac{(\gamma-1)!\alpha}{2\pi i}\int_M \frac{e^\lambda \lambda^{\alpha-\gamma}}{(\lambda-x)^{\alpha+1}} d\lambda$$

$$= \frac{(\gamma-1)!\alpha}{\gamma!}{}_1F_1(\alpha+1,\gamma+1,x)$$

$$= \frac{\alpha}{\gamma}{}_1F_1(\alpha+1,\gamma+1,x), \quad (12)$$

as is also obvious by differentiation of the series.

If we put $\alpha x = y$ and then let α tend to infinity, the integrals tend to the forms of Schläfli's for the Bessel functions, apart from some simple factors.

* B.A. *Reports* (Committee for the Calculation of Mathematical Tables), 1926. Tables also in 1927 *Report*.

23·04. Asymptotic formulae of Stokes's type.

23·03 (8) represents an analytic function for all values of x, real or complex, but cuts will be needed if α and γ are non-integral. If x has a positive imaginary part we can replace the path M by a pair of loops about the branch points, and we can show that each loop separately gives a solution of the differential equation. Take any integral of the form

$$I = \int \frac{e^\lambda \lambda^{\alpha-\gamma}}{(\lambda-x)^\alpha} d\lambda, \tag{1}$$

where the limits are independent of x. Then

$$x\frac{d^2I}{dx^2} + (\gamma-x)\frac{dI}{dx} - \alpha I = \int e^\lambda \lambda^{\alpha-\gamma} \left\{ \frac{\alpha(\alpha+1)x}{(\lambda-x)^{\alpha+2}} + \frac{\alpha(\gamma-x)}{(\lambda-x)^{\alpha+1}} - \frac{\alpha}{(\lambda-x)^\alpha} \right\} d\lambda. \tag{2}$$

In the numerators replace x by $\lambda - (\lambda-x)$ and rearrange in powers of $\lambda-x$; then

$$\frac{x}{(\lambda-x)^\alpha} = \frac{\lambda}{(\lambda-x)^\alpha} - \frac{1}{(\lambda-x)^{\alpha-1}}, \tag{3}$$

$$\frac{\alpha(\alpha+1)x}{(\lambda-x)^{\alpha+2}} + \frac{\alpha(\gamma-x)}{(\lambda-x)^{\alpha+1}} - \frac{\alpha}{(\lambda-x)^\alpha} = \frac{\alpha(\alpha+1)\lambda}{(\lambda-x)^{\alpha+2}} - \frac{\alpha(\alpha+1)}{(\lambda-x)^{\alpha+1}} - \frac{\alpha\lambda}{(\lambda-x)^{\alpha+1}} + \frac{\alpha\gamma}{(\lambda-x)^{\alpha+1}}, \tag{4}$$

$$\int e^\lambda \lambda^{\alpha-\gamma+1} \frac{\alpha(\alpha+1)}{(\lambda-x)^{\alpha+2}} d\lambda = -\int e^\lambda \lambda^{\alpha-\gamma+1} \alpha d\frac{1}{(\lambda-x)^{\alpha+1}}$$
$$= -\left[e^\lambda \frac{\alpha \lambda^{\alpha-\gamma+1}}{(\lambda-x)^{\alpha+1}} \right] + \int \frac{\alpha e^\lambda}{(\lambda-x)^{\alpha+1}} \{\lambda^{\alpha-\gamma+1} + (\alpha-\gamma+1)\lambda^{\alpha-\gamma}\} d\lambda \tag{5}$$

and the integral cancels the integrals arising from the last three terms of (4). Hence (1) is a solution of the differential equation provided that the integrated part of (5) vanishes, and this will be satisfied for any path such that $\Re(\lambda) = -\infty$ at the ends. Hence integrals on the paths M_1 and M_2 in the figure give separate solutions of the differential equation.

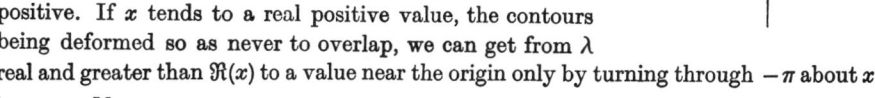

On M_1 we must attend specially to the phase of $\lambda-x$. λ^α is taken real and positive on the positive real axis, and therefore $(\lambda-x)^\alpha$, when $\lambda-x$ is real and positive, is also real and positive. If x tends to a real positive value, the contours being deformed so as never to overlap, we can get from λ real and greater than $\Re(x)$ to a value near the origin only by turning through $-\pi$ about x; hence on M_1

$$(\lambda-x)^\alpha = (x-\lambda)^\alpha e^{-\alpha\pi i}.$$

On M_1 we can now expand in descending powers of x;

$$I_1 = \int_{M_1} \frac{e^\lambda \lambda^{\alpha-\gamma}}{(\lambda-x)^\alpha} d\lambda = \int_{M_1} e^\lambda \lambda^{\alpha-\gamma} e^{\alpha i \pi} x^{-\alpha} \left(1-\frac{\lambda}{x}\right)^{-\alpha} d\lambda$$

$$\sim x^{-\alpha} e^{i\alpha\pi} \int_{M_1} e^\lambda \left(\lambda^{\alpha-\gamma} + \frac{\alpha \lambda^{\alpha-\gamma+1}}{x} + \frac{\alpha(\alpha+1)\lambda^{\alpha-\gamma+2}}{2!x^2} + \ldots \right) d\lambda$$

$$= x^{-\alpha} e^{i\alpha\pi} 2\pi i \left(\frac{1}{(\gamma-\alpha-1)!} + \frac{\alpha}{(\gamma-\alpha-2)!x} + \frac{\alpha(\alpha+1)}{(\gamma-\alpha-3)!2!x^2} + \ldots \right)$$

$$= \frac{2\pi i x^{-\alpha} e^{i\alpha\pi}}{(\gamma-\alpha-1)!} \left(1 + \frac{\alpha(\gamma-\alpha-1)}{x} + \frac{\alpha(\alpha+1)(\gamma-\alpha-1)(\gamma-\alpha-2)}{2!x^2} + \ldots \right). \tag{6}$$

On M_2 we put $\lambda = x+\kappa$; then

$$I_2 = e^x \int_{-\infty, 0+}^{-\infty} \frac{e^\kappa (x+\kappa)^{\alpha-\gamma}}{\kappa^\alpha} d\kappa$$

$$\sim e^x x^{\alpha-\gamma} \int_{-\infty, 0+}^{-\infty} e^\kappa \kappa^{-\alpha} \left(1 + \frac{(\alpha-\gamma)\kappa}{x} + \frac{(\alpha-\gamma)(\alpha-\gamma-1)\kappa^2}{2! x^2} + \ldots\right) d\kappa$$

$$= 2\pi i e^x x^{\alpha-\gamma} \left(\frac{1}{(\alpha-1)!} + \frac{\alpha-\gamma}{(\alpha-2)! x} + \frac{(\alpha-\gamma)(\alpha-\gamma-1)}{(\alpha-3)! 2! x^2} + \ldots\right)$$

$$= \frac{2\pi i x^{\alpha-\gamma} e^x}{(\alpha-1)!} \left(1 + \frac{(\alpha-1)(\alpha-\gamma)}{x} + \frac{(\alpha-1)(\alpha-2)(\alpha-\gamma)(\alpha-\gamma-1)}{2! x^2} + \ldots\right). \quad (7)$$

If $\Re(x)$ is large and positive any term of (7) is large compared with I_1, and in Poincaré's sense I_1 can be neglected. With actual values of $\Re(x)$, however, I_1 may be comparable with some terms of the I_2 series and is then worth retaining. Then for $\Im(x) > 0$

$$_1F_1(\alpha, \gamma, x) \sim \frac{(\gamma-1)!}{(\alpha-1)!} x^{\alpha-\gamma} e^x \left(1 + \frac{(\alpha-1)(\alpha-\gamma)}{x} + \frac{(\alpha-1)(\alpha-2)(\alpha-\gamma)(\alpha-\gamma-1)}{2! x^2} + \ldots\right)$$

$$+ \frac{(\gamma-1)!}{(\gamma-\alpha-1)!} x^{-\alpha} e^{\alpha i \pi} \left(1 + \frac{\alpha(\gamma-\alpha-1)}{x} + \frac{\alpha(\alpha+1)(\gamma-\alpha-1)(\gamma-\alpha-2)}{2! x^2} + \ldots\right) \quad (8)$$

If $\Im(x) < 0$ the same holds provided that M_2 still lies above M_1; but if we draw M_2, as would be more natural, so as to lie below M_1 the factor $e^{\alpha i \pi}$ must be replaced by $e^{-\alpha i \pi}$. This is another instance of the discontinuity of constants in asymptotic expansions.

It follows that, again for $\Im(x) > 0$,

$$x^{1-\gamma} {}_1F_1(1+\alpha-\gamma, 2-\gamma, x) = J_2 + J_1 \sim \frac{(1-\gamma)!}{(\alpha-\gamma)!} x^{\alpha-\gamma} e^x \left(1 + \frac{(\alpha-\gamma)(\alpha-1)}{x} + \ldots\right)$$

$$+ \frac{(1-\gamma)!}{(-\alpha)!} e^{(1+\alpha-\gamma)i\pi} x^{-\alpha} \left(1 + \frac{(1+\alpha-\gamma)(-\alpha)}{x} + \ldots\right) \quad (9)$$

and the two series in (8) and (9) are identical, but their coefficients are in different ratios. If $\Re(x)$ is large, however, the portion arising from M_1 is negligible in comparison with that from M_2, and the two solutions are almost proportional. If $\Re(x)$ is small, and especially when x is purely imaginary, we must keep both series. (For $\Im(x) < 0$, reverse i.)

It follows that for varying x, J_1 and J_2 are constant multiples of I_1 and I_2, subject to the same expansions remaining valid. For all four functions are solutions of the differential equation, and I_1 and I_2 are not proportional. Hence J_1 and J_2 can be linearly expressed in terms of I_1 and I_2. But if $J_1 = AI_1 + BI_2$, and $B \neq 0$, J_1 will increase like e^x as $\Re(x) \to \infty$, and it does not. Hence $B = 0$. Similarly by making $\Re(x) \to -\infty$ we show that J_2 is a constant multiple of I_2.

This result makes it possible to express I_1, I_2, J_1, J_2 in terms of the series solutions u_1, u_2 when these exist. For if we write

$$\frac{(\gamma-1)!}{2\pi i} \int_{M_1} \frac{e^\lambda \lambda^{\alpha-\gamma}}{(\lambda-x)^\alpha} d\lambda = S_1, \quad \frac{(\gamma-1)!}{2\pi i} \int_{M_2} \frac{e^\lambda \lambda^{\alpha-\gamma}}{(\lambda-x)^\alpha} d\lambda = S_2, \quad (10)$$

$$\frac{(1-\gamma)!}{2\pi i} x^{1-\gamma} \int_{M_1} \frac{e^\lambda \lambda^{\alpha-1}}{(\lambda-x)^{\alpha-\gamma+1}} d\lambda = J_1, \quad \frac{(1-\gamma)!}{2\pi i} x^{1-\gamma} \int_{M_2} \frac{e^\lambda \lambda^{\alpha-1}}{(\lambda-x)^{\alpha-\gamma+1}} d\lambda = J_2, \quad (11)$$

we have
$$u_1 = S_1 + S_2, \tag{12}$$

$$u_2 = J_1 + J_2 = \frac{(1-\gamma)!}{(\gamma-1)!}\left\{\frac{(\gamma-\alpha-1)!}{(-\alpha)!}e^{(1-\gamma)i\pi}S_1 + \frac{(\alpha-1)!}{(\alpha-\gamma)!}S_2\right\}, \tag{13}$$

whence
$$\left\{\frac{(\gamma-\alpha-1)!}{(-\alpha)!}e^{-\gamma i\pi} + \frac{(\alpha-1)!}{(\alpha-\gamma)!}\right\}S_1 = -\frac{(\gamma-1)!}{(1-\gamma)!}u_2 + \frac{(\alpha-1)!}{(\alpha-\gamma)!}u_1. \tag{14}$$

We make use repeatedly of the identity
$$\sin\alpha\pi = \frac{\pi}{(\alpha-1)!(-\alpha)!}. \tag{15}$$

Then the coefficient of S_1 on the left of (14) is
$$\frac{\pi\sin\gamma\pi\, e^{-i\alpha\pi}}{(-\alpha)!(\alpha-\gamma)!\sin\alpha\pi\sin(\gamma-\alpha)\pi}. \tag{16}$$

Now if $\lambda = -\mu$
$$S_1 = -\frac{(\gamma-1)!}{\pi}\sin(\alpha-\gamma)\pi\, e^{i\alpha\pi}\int_0^\infty \frac{e^{-\mu}\mu^{\alpha-\gamma}}{(x+\mu)^\alpha}d\mu \quad (\Re(\alpha-\gamma) > -1), \tag{17}$$

$$J_1 = -\frac{(1-\gamma)!}{\pi}x^{1-\gamma}\sin\alpha\pi\, e^{(\alpha-\gamma)i\pi}\int_0^\infty \frac{e^{-\mu}\mu^{\alpha-1}}{(x+\mu)^{\alpha-\gamma+1}}d\mu \quad (\Re(\alpha) > 0), \tag{18}$$

and from the latter,
$$S_1 = \frac{(\gamma-1)!\, e^{i\alpha\pi}}{(\gamma-\alpha-1)!(\alpha-1)!}x^{1-\gamma}\int_0^\infty \frac{e^{-\mu}\mu^{\alpha-1}}{(x+\mu)^{\alpha-\gamma+1}}d\mu \quad (\Re(\alpha) > 0). \tag{19}$$

We write
$$S_1 = e^{i\alpha\pi}\frac{(\gamma-1)!}{(\gamma-\alpha-1)!}U(\alpha,\gamma,x); \tag{20}$$

then for $0 \leq \arg x < \pi$ (replacing i by $-i$ for $-\pi < \arg x \leq 0$)
$$U(\alpha,\gamma,x) = \frac{e^{-i\alpha\pi}}{2\pi i}(\gamma-\alpha-1)!\int_{M_1}\frac{e^\lambda \lambda^{\alpha-\gamma}}{(\lambda-x)^\alpha}d\lambda \quad \text{from (20)}, \tag{21}$$

$$= \frac{e^{i(\gamma-\alpha-1)\pi}}{2\pi i}(-\alpha)!\, x^{1-\gamma}\int_{M_1}\frac{e^\lambda \lambda^{\alpha-1}}{(\lambda-x)^{\alpha-\gamma+1}}d\lambda \quad \text{from (11), (13) and (20)}, \tag{22}$$

$$= \frac{1}{(\alpha-\gamma)!}\int_0^\infty \frac{e^{-\mu}\mu^{\alpha-\gamma}}{(x+\mu)^\alpha}d\mu \quad (\Re(\alpha-\gamma) > -1) \quad \text{from (17)}, \tag{23}$$

$$= \frac{1}{(\alpha-1)!}x^{1-\gamma}\int_0^\infty \frac{e^{-\mu}\mu^{\alpha-1}}{(x+\mu)^{\alpha-\gamma+1}}d\mu \quad (\Re(\alpha) > 0) \quad \text{from (19)}, \tag{24}$$

$$= \frac{(-\gamma)!}{(\alpha-\gamma)!}u_1 + \frac{(\gamma-2)!}{(\alpha-1)!}u_2 \quad (\gamma \text{ non-integral}) \quad \text{from (14) and (16)} \tag{25}$$

$$\sim x^{-\alpha}\left(1 + \frac{\alpha(\gamma-\alpha-1)}{x} + \frac{\alpha(\alpha+1)(\gamma-\alpha-1)(\gamma-\alpha-2)}{2!\, x^2} + \ldots\right) \quad \text{from (6)}. \tag{26}$$

(26) is a terminating series if $\alpha-1$ or $\alpha-\gamma$ is a negative integer, and is then an exact solution. In either of these cases one of the series solutions in (25) is multiplied by zero, and U reduces to a multiple of the other.

If $\gamma-\alpha-1$ or $-\alpha$ is a negative integer, (21) or (22) is defined by continuity.

A full discussion of the functions was first given by Barnes, who converted the series into complex integrals involving $z!$, where z is a complex variable of integration, the terms of the series being the residues of a function at the poles of a factorial. We shall also give another method due to Goldstein.

23·05. Goldstein's operational expression.* We now consider a different treatment, in which the operational form for a confluent hypergeometric function is obtained in terms of Bessel functions, and the result 23·04 (23) is again derived. In reducing the series to operational expressions we have introduced such a power of x that the operator takes an algebraic form, the γ factors being removed from the denominators. We could, however, remove the α factors from the numerators. We first replace x by $-a^2/4t$, and if p is now the operator corresponding to t we have

$$p^n = \frac{t^{-n}}{(-n)!}, \quad p^{n+1} = \frac{t^{-n-1}}{(-n-1)!} = -\frac{nt^{-n-1}}{(-n)!}, \tag{1}$$

$$\begin{aligned}
{}_1F_1\!\left(\alpha,\gamma,-\frac{a^2}{4t}\right) &= 1 + \frac{(-\alpha)!}{(-\alpha-1)!\,1!\,\gamma}\frac{a^2}{4t} + \frac{(-\alpha)!}{(-\alpha-2)!\,2!\,\gamma(\gamma+1)}\left(\frac{a^2}{4t}\right)^2 + \ldots \tag{2} \\
&= (-\alpha)!\left(\frac{4t}{a^2}\right)^\alpha\left\{\frac{1}{(-\alpha)!}\left(\frac{a^2}{4t}\right)^\alpha + \frac{1}{(-\alpha-1)!\,1!\,\gamma}\left(\frac{a^2}{4t}\right)^{\alpha+1} + \ldots\right\} \\
&= (-\alpha)!\left(\frac{4t}{a^2}\right)^\alpha\left\{\left(\frac{a^2 p}{4}\right)^\alpha + \frac{1}{1!\,\gamma}\left(\frac{a^2 p}{4}\right)^{\alpha+1} + \ldots\right\} \\
&= (-\alpha)!\,(\gamma-1)!\,(\tfrac{1}{2}a)^{1-\gamma}t^\alpha p^{\alpha-1/2\gamma+1/2}I_{\gamma-1}(ap^{1/2}). \tag{3}
\end{aligned}$$

This is exact for all t with a positive real part. For if we interpret $I_{\gamma-1}(ap^{1/2})$ by means of a complex integral on the path M it gives a factor of order $\exp(az^{1/2})$, which is overwhelmed by the factor $\exp(tz)$ when t has a positive real part and $z \to -\infty$. Then using 23·03 (3) and (10), we find

$$\begin{aligned}
{}_1F_1\!\left(\alpha,\gamma,\frac{a^2}{4t}\right) &= e^{a^2/4t}{}_1F_1\!\left(\gamma-\alpha,\gamma,-\frac{a^2}{4t}\right) \\
&= e^{a^2/4t}(\alpha-\gamma)!\,(\gamma-1)!\,(\tfrac{1}{2}a)^{1-\gamma}t^{\gamma-\alpha}p^{1/2\gamma-\alpha+1/2}I_{\gamma-1}(ap^{1/2}). \tag{4}
\end{aligned}$$

Similarly, using 23·03 (11),

$$\left(\frac{a^2}{4t}\right)^{1-\gamma}{}_1F_1\!\left(1+\alpha-\gamma,2-\gamma,\frac{a^2}{4t}\right) = e^{a^2/4t}(\alpha-1)!\,(1-\gamma)!\,(\tfrac{1}{2}a)^{1-\gamma}t^{\gamma-\alpha}p^{1/2\gamma-\alpha+1/2}I_{1-\gamma}(ap^{1/2}). \tag{5}$$

The constant factors are in the same ratio as in 23·04 (14). Also if a is large $I_{\gamma-1}(az^{1/2})$ and $I_{1-\gamma}(az^{1/2})$ are nearly equal, and a suitable solution with a different behaviour at infinity will be

$$U = \pi e^{a^2/4t}(\tfrac{1}{2}a)^{1-\gamma}t^{\gamma-\alpha}p^{1/2\gamma-\alpha+1/2}\mathrm{Kh}_{1-\gamma}(ap^{1/2}). \tag{6}$$

To investigate this we use the expression

$$\mathrm{Kh}_n(x) = \frac{1}{\pi}(\tfrac{1}{2}x)^{-n}\int_0^\infty \exp\!\left(-u-\frac{x^2}{4u}\right)u^{n-1}du \quad |\arg x| < \tfrac{1}{4}\pi. \tag{7}$$

* *Proc. Lond. Math. Soc.* (2) 34, 1931, 103–25.

If we interpret (6) by using the path L, U converges since

$$\mathrm{Kh}_n(z) = O(z^{-\frac{1}{2}}e^{-z}),$$

and on this path the relation $\Re(a^2z) > 0$ is satisfied. Then

$$\mathrm{Kh}_{1-\gamma}(ap^{\frac{1}{2}}) = \frac{1}{\pi}(\tfrac{1}{2}ap^{\frac{1}{2}})^{\gamma-1}\int_0^\infty \exp\left(-u - \frac{a^2p}{4u}\right)u^{-\gamma}du$$

$$= \frac{1}{\pi}(\tfrac{1}{2}ap^{\frac{1}{2}})^{\gamma-1}\int_0^\infty e^{-u}u^{-\gamma}H\left(t - \frac{a^2}{4u}\right)du$$

$$= \frac{1}{\pi}(\tfrac{1}{2}ap^{\frac{1}{2}})^{\gamma-1}\int_{a^2/4t}^\infty e^{-u}u^{-\gamma}du, \tag{8}$$

$$\pi p^{\frac{1}{2}\gamma-\alpha+\frac{1}{2}}\mathrm{Kh}_{1-\gamma}(ap^{\frac{1}{2}}) = (\tfrac{1}{2}a)^{\gamma-1}p^{\gamma-\alpha}\int_{a^2/4t}^\infty e^{-u}u^{-\gamma}du. \tag{9}$$

But

$$\frac{1}{p}F(p)g(t) = \int_0^t f(t-\tau)g(\tau)d\tau \tag{10}$$

and

$$p^{\gamma-\alpha+1}H(t) = \frac{t^{\alpha-\gamma-1}}{(\alpha-\gamma-1)!} \quad (\Re(\alpha-\gamma) > -1). \tag{11}$$

Hence (9) is

$$(\tfrac{1}{2}a)^{\gamma-1}\int_0^t \frac{(t-\tau)^{\alpha-\gamma-1}}{(\alpha-\gamma-1)!}d\tau \int_{a^2/4\tau}^\infty e^{-u}u^{-\gamma}du. \tag{12}$$

The integration is over the shaded region in the diagram. Reversing the order we have for the limits $a^2/4u < \tau < t$, $a^2/4t < u < \infty$; and we have

$$(\tfrac{1}{2}a)^{\gamma-1}\int_{a^2/4t}^\infty e^{-u}u^{-\gamma}du \int_{a^2/4u}^t \frac{(t-\tau)^{\alpha-\gamma-1}}{(\alpha-\gamma-1)!}d\tau$$

$$= \frac{(\tfrac{1}{2}a)^{\gamma-1}}{(\alpha-\gamma)!}\int_{a^2/4t}^\infty e^{-u}u^{-\gamma}\left(t - \frac{a^2}{4u}\right)^{\alpha-\gamma}du$$

$$= \frac{(\tfrac{1}{2}a)^{\gamma-1}}{(\alpha-\gamma)!}t^{\alpha-\gamma}\int_{a^2/4t}^\infty e^{-u}u^{-\alpha}\left(u - \frac{a^2}{4t}\right)^{\alpha-\gamma}du$$

$$= \frac{(\tfrac{1}{2}a)^{\gamma-1}}{(\alpha-\gamma)!}t^{\alpha-\gamma}e^{-a^2/4t}\int_0^\infty e^{-v}v^{\alpha-\gamma}\left(v + \frac{a^2}{4t}\right)^{-\alpha}dv. \tag{13}$$

Hence

$$U = \frac{1}{(\alpha-\gamma)!}\int_0^\infty e^{-v}v^{\alpha-\gamma}\left(v + \frac{a^2}{4t}\right)^{-\alpha}dv \quad (\Re(\alpha-\gamma) > -1) \tag{14}$$

is the solution required. Also, using the definition of Kh_n in terms of I_{-n} and I_n, we have

$$U(\alpha, \gamma, x) = \frac{1}{(\alpha-\gamma)!}\int_0^\infty e^{-v}v^{\alpha-\gamma}(v+x)^{-\alpha}dv \tag{15}$$

$$= -\pi\operatorname{cosec}\gamma\pi\left\{\frac{1}{(\alpha-1)!(1-\gamma)!}u_2 - \frac{1}{(\alpha-\gamma)!(\gamma-1)!}u_1\right\}$$

$$= \left\{\frac{(-\gamma)!}{(\alpha-\gamma)!}u_1 + \frac{(\gamma-2)!}{(\alpha-1)!}u_2\right\}. \tag{16}$$

The numerical factors can be checked by taking $x = 0$. This result is identical with 23·04 (25).

Now

$$\mathrm{Kh}_{-n}(z) = \mathrm{Kh}_n(z) \tag{17}$$

and therefore also

$$U\left(\alpha, \gamma, \frac{a^2}{4t}\right) = \pi e^{a^2/4t}(\tfrac{1}{2}a)^{1-\gamma} t^{\gamma-\alpha} p^{\frac{1}{2}\gamma-\alpha+\frac{1}{2}}\mathrm{Kh}_{\gamma-1}(ap^{1/2})$$

$$= \frac{1}{(\alpha-1)!}\left(\frac{a^2}{4t}\right)^{1-\gamma} e^{a^2/4t} \int_{a^2/4t}^{\infty} e^{-u} u^{\gamma-\alpha-1}\left(u - \frac{a^2}{4t}\right)^{\alpha-1} du, \tag{18}$$

by similar methods; then

$$U(\alpha, \gamma, x) = \frac{1}{(\alpha-1)!} x^{1-\gamma} e^x \int_x^{\infty} e^{-u} u^{\gamma-\alpha-1}(u-x)^{\alpha-1} du$$

$$= \frac{1}{(\alpha-1)!} x^{1-\gamma} \int_0^{\infty} e^{-v} v^{\alpha-1}(v+x)^{\gamma-\alpha-1} dv \quad (\Re(\alpha) > 0). \tag{19}$$

The two forms (15), (19) were found in 23·04 by using the two loop integrals I_1 and J_1.

Since 23·04 (21) (22) are analytic functions of α and γ for all values, they can be taken as providing definitions of U for unrestricted α and γ; and then (16), being true for a continuous range of values of α and γ, will be true for all α, γ, the right side being defined by continuity when γ is integral. They become ambiguous if x is real and negative, so that a cut along the negative real axis is needed for x, but a similar cut is needed to define $x^{1-\gamma}$ if γ is not an integer, so that this involves no loss of generality.

The integrals are inconvenient for finding the convergent series expansions directly; for if we try to expand a power of $u-x$ in ascending powers of x the series will diverge for $|u|<|x|$ and it is impossible to find a path passing between the singularities such that the series converges at all points of it.

If we take the loop M_2 for either integral we find without much trouble that it is a multiple of $e^x U(\gamma-\alpha, \gamma, z)$, where $z = xe^{\pm i\pi}$. Owing to the restriction $|\arg z|<\pi$ (cf. 23·04 (21)) we take the lower sign when $\Im(x)>0$ and the upper sign when $\Im(x)<0$. For $\Im(x)>0$ we make M_2 lie entirely above the real axis and find a solution

$$V(\alpha, \gamma, x) = \frac{(\alpha-1)!}{2\pi i} e^x \int_{-\infty, 0+}^{-\infty} e^{\kappa}(x+\kappa)^{\alpha-\gamma} \kappa^{-\alpha} d\kappa$$

$$= \frac{(\alpha-\gamma)!}{2\pi i} e^x x^{1-\gamma} \int_{-\infty, 0+}^{-\infty} e^{\kappa}(x+\kappa)^{\alpha-1} \kappa^{\gamma-\alpha-1} d\kappa$$

$$\sim e^x x^{\alpha-\gamma}, \tag{20}$$

and

$$u_1 = {}_1F_1(\alpha, \gamma, x) = \frac{(\gamma-1)!}{(\alpha-1)!} V + \frac{(\gamma-1)!}{(\gamma-\alpha-1)!} e^{\alpha i\pi} U \quad (\Im(x)>0), \tag{21}$$

$$u_2 = x^{1-\gamma} {}_1F_1(1+\alpha-\gamma, 2-\gamma, x) = \frac{(1-\gamma)!}{(\alpha-\gamma)!} V + \frac{(1-\gamma)!}{(-\alpha)!} e^{(1+\alpha-\gamma)i\pi} U \quad (\Im(x)>0) \tag{22}$$

whence

$$V(\alpha, \gamma, x) = \frac{\pi}{\sin \gamma \pi}\left(\frac{e^{\alpha i\pi}}{(\gamma-\alpha-1)!(1-\gamma)!} u_2 + \frac{e^{(\alpha-\gamma)i\pi}}{(-\alpha)!(\gamma-1)!} u_1\right). \tag{23}$$

For $\Im(x) < 0$ we make M_2 lie entirely below the real axis; the result is a different solution $W(\alpha, \gamma, x)$, which is equal to the expression on the right of (23) with i replaced by $-i$. For $\Im(x) < 0$, $W \sim e^x x^{\alpha-\gamma}$ and the analytic continuation of $V(\alpha, \gamma, x)$ is

$$W(\alpha, \gamma, x) - \frac{2\pi i U(\alpha, \gamma, x)}{(\gamma - \alpha - 1)!(-\alpha)!}.$$

The interesting property of V, apart from its simple asymptotic expansion, is that it tends exponentially to 0 as $\Re(x) \to -\infty$. Further, V as determined by either of the integrals (20) is a solution of the differential equation, independent of U, even when γ is an integer and the two series solutions coalesce.

We have also seen from 23·03 (10) and (11) that one solution is expressible in finite terms if either $\gamma - \alpha$ or $1 - \alpha$ is an integer. In particular

$$_1F_1(\alpha, \alpha, x) = e^x, \quad _1F_1(0, \gamma, x) = 1, \quad _1F_1(-1, \gamma, x) = 1 - x/\gamma.$$

23·051. Convergent expansion of $U(\alpha, \gamma, x)$ when γ is a positive integer. Put $\gamma = m + c$, where m is a positive integer, and let c tend to zero.

$$U(\alpha, m+c, x) = \frac{(-m-c)!}{(\alpha-m-c)!} {}_1F_1(\alpha, m+c, x)$$
$$+ \frac{(m+c-2)!}{(\alpha-1)!} x^{1-m-c} {}_1F_1(1+\alpha-m-c, 2-m-c, x).$$

When $c \to 0$ the terms up to x^{m-2} in the second series give negative powers of x and have no counterparts in the first series. The corresponding terms in U tend to

$$U_1 = \frac{(m-2)!}{(\alpha-1)!} x^{1-m} G(x),$$

where $G(x)$ consists of the expansion of ${}_1F_1(1+\alpha-m, 2-m, x)$ up to the term in x^{m-2}. Next, take the term in x^r in the first series with that in x^{r+m-1} in the second. With their proper multipliers they give

$$u_r = \frac{(-m-c)!}{(\alpha-m-c)!} \frac{(\alpha+r-1)!}{(\alpha-1)! r!} \frac{(m+c-1)!}{(m+c+r-1)!} x^r$$

$$+ \frac{(m+c-2)!}{(\alpha-1)!} \frac{(r+\alpha-1-c)!(1-m-c)!}{(\alpha-m-c)!(r+m-1)!(r-c)!} x^{r-c}$$

$$= \frac{\pi}{\sin(m+c)\pi} \frac{1}{(\alpha-1)!(\alpha-m-c)!} \left(\frac{(\alpha+r-1)!}{(m+r-1+c)! r!} x^r - \frac{(\alpha+r-1-c)!}{(m+r-1)!(r-c)!} x^{r-c} \right)$$

$$\to (-1)^m \frac{x^r}{(\alpha-1)!(\alpha-m)!}$$

$$\times \left[\frac{(\alpha+r-1)!}{(m+r-1)! r!} \log x + \frac{(\alpha+r-1)!}{(m+r-1)! r!} \{-F(m+r-1) + F(\alpha+r-1) - F(r)\} \right]$$

(from 15·04)

$$= \frac{(-1)^m}{(m-1)!(\alpha-m)!} \frac{(\alpha+r-1)!}{r!(\alpha-1)!} \frac{(m-1)!}{(m+r-1)!} x^r [\log x - F(m+r-1) - F(r) + F(\alpha+r-1)].$$

But
$$-F(m+r-1)-F(r)+F(\alpha+r-1)$$
$$= -F(m-1)-F(0)+F(\alpha-1)-\frac{1}{m}-\frac{1}{m+1}\cdots-\frac{1}{m+r-1}$$
$$-\frac{1}{1}-\frac{1}{2}\cdots-\frac{1}{r}+\frac{1}{\alpha}+\frac{1}{\alpha+1}\cdots+\frac{1}{\alpha+r-1}.$$

Hence $U_2 = \sum_{r=0}^{\infty} u_r = \frac{(-1)^m}{(m-1)!(\alpha-m)!}\,_1F_1(\alpha,m,x)\{\log x - F(m-1) - F(0) + F(\alpha-1)\} + U_3$

where
$$U_3 = \sum_{r=1}^{\infty} \frac{(-1)^m}{(m-1)!(\alpha-m)!} \frac{\alpha(\alpha+1)\cdots(\alpha+r-1)}{r!\,m(m+1)\cdots(m+r-1)} x^r$$
$$\times \left(\frac{1}{\alpha}+\frac{1}{\alpha+1}\cdots+\frac{1}{\alpha+r-1}-\frac{1}{1}-\cdots-\frac{1}{r}-\frac{1}{m}-\frac{1}{m+1}-\cdots-\frac{1}{m+r-1}\right).$$

and
$$U(\alpha,m,x) = U_1 + U_2.$$

This solution is due to Stoneley.* A function partly tabulated by H. A. Webb and J. R. Airey† omits the terms U_1.

If $m = 1$, the terms U_1 do not arise.

If α is a positive integer less than m, U reduces to U_1. In this case, and also if α is a negative or zero integer, U is given exactly by 23·04 (26).

For an application see 23·07 a. If x is 0 or a negative integer, a convergent expression for U may be obtained by similar methods.

23·06. Whittaker's transformation. If in the original differential equation
$$xu'' + (\gamma - x)u' - \alpha u = 0 \tag{1}$$
we put
$$u = ve^{\frac{1}{2}x} \tag{2}$$
we get
$$v'' + \frac{\gamma}{x}v' + \left(-\tfrac{1}{4} + \frac{\gamma - 2\alpha}{2x}\right)v = 0, \tag{3}$$
and the further substitution
$$v = x^{-\frac{1}{2}\gamma}w \tag{4}$$
gives
$$w'' + \left(-\tfrac{1}{4} + \frac{(\tfrac{1}{2}\gamma - \alpha)}{x} + \frac{\gamma(2-\gamma)}{4x^2}\right)w = 0. \tag{5}$$
Putting
$$w = x^{\frac{1}{2}}y \tag{6}$$
we have the further form
$$x\frac{d}{dx}\left(x\frac{dy}{dx}\right) - \{\tfrac{1}{4}x^2 - (\tfrac{1}{2}\gamma - \alpha)x + \tfrac{1}{4}(\gamma - 1)^2\}y = 0. \tag{7}$$

(7) is interesting because if $\alpha = \tfrac{1}{2}\gamma$ it reduces to a form of Bessel's equation, and solutions are $I_{\pm\frac{1}{2}(\gamma-1)}(\tfrac{1}{2}x)$. Then
$$_1F_1(\alpha, 2\alpha, x) = (\alpha - \tfrac{1}{2})!\,e^{\frac{1}{2}x}\left(\frac{4}{x}\right)^{\alpha - \frac{1}{2}} I_{\alpha - \frac{1}{2}}(\tfrac{1}{2}x). \tag{8}$$

When the term in x is not zero, however, it can produce a profound change in the character of the solutions. Evidently $\mathrm{Kh}_{\frac{1}{2}(\gamma-1)}(\tfrac{1}{2}x)$ will be a further solution of (7) if $\alpha = \tfrac{1}{2}\gamma$ and corresponds to the solution U; but if this condition is not satisfied it is possible to find α and γ such that U is a multiple of one of the series solutions. The Bessel analogue would be that Kh_n could be proportional to either I_n or $I_{-n}!$.

* M.N.R.A.S. Geophys. Suppl. 3, 1934, 226–8. See also D. R. Hartree, Proc. Camb. Phil. Soc. 24, 1928, 426–37 for the corresponding expansion of Whittaker's function.
† Phil. Mag. (6), 36, 1918, 129–41.

(5) is Whittaker's form, and is written by him as

$$w'' + \left(-\tfrac{1}{4} + \frac{k}{x} + \frac{\tfrac{1}{4} - m^2}{x^2}\right)w = 0 \qquad (9)$$

so that
$$k = \tfrac{1}{2}\gamma - \alpha, \quad \pm m = \tfrac{1}{2}(1-\gamma). \qquad (10)$$

This notation simplifies the writing of the differential equation slightly. But if we try to solve (3), (5) or (7) for general γ, α by a series we get a three term relation between the coefficients, and it appears that in any useful series α and γ will enter explicitly into the solutions. We shall therefore write solutions of (5) as

$$e^{-\tfrac{1}{2}x} x^{\tfrac{1}{2}\gamma} \{ {}_1F_1(\alpha, \gamma, x); \ x^{1-\gamma} {}_1F_1(1+\alpha-\gamma, 2-\gamma, x); \ U(\alpha, \gamma, x\}. \qquad (11)$$

while those of (7) need an extra factor $x^{-\tfrac{1}{2}}$ and therefore, unless $\gamma = 1$, one of the series solutions tends to infinity at the origin.

An alternative form comes by putting $x = 2\mu z$; then

$$\frac{d^2w}{dz^2} + \left(-\mu^2 + (\gamma - 2\alpha)\frac{\mu}{z} + \frac{\gamma(2-\gamma)}{4z^2}\right)w = 0. \qquad (12)$$

In many applications μ is ± 1 or $\pm i$.

A case of special interest is where a solution of (5) is required to tend to zero at $x = +\infty$, and also to be small compared with $x^{\tfrac{1}{2}}$ for x small. The former condition requires that the solution shall be that in U. But U is a linear combination of two solutions of (1), of which one is bounded and not zero near $x = 0$, and the other behaves like $x^{1-\gamma}$, except for $\gamma = 1$, when the second solution behaves like $\log x$. The corresponding solutions of (5) will behave like $x^{\tfrac{1}{2}\gamma}$ and $x^{1-\tfrac{1}{2}\gamma}$ or $x^{\tfrac{1}{2}}\log x$. Thus if $\gamma \geq 1$ only the solution of (1) bounded near the origin is admissible, and if $\gamma < 1$ only the one that behaves like $x^{1-\gamma}$. In the former case the solution required is ${}_1F_1(\alpha, \gamma, x)$. But this increases like e^x for large x unless $\alpha - 1$ is a negative integer, when we take the series to end with the term in $x^{-\alpha}$. In the latter case the solution is $x^{1-\gamma} {}_1F_1(1+\alpha-\gamma, 2-\gamma, x)$, which increases like e^x unless $\alpha - \gamma$ is a negative integer, when we take the series to end with the term in $x^{\gamma-\alpha-1}$. In either case the admissible solution of (5) is a terminating series multiplied by $e^{-\tfrac{1}{2}x}$.

The function U is therefore of great physical importance. Whittaker's solution takes the form

$$W_{k,m}(x) = \frac{x^{-m+\tfrac{1}{2}} e^{-\tfrac{1}{2}x}}{(m-k-\tfrac{1}{2})!} \int_0^\infty e^{-t} t^{m-k-\tfrac{1}{2}} (t+x)^{m+k-\tfrac{1}{2}} dt$$

$$= e^{-\tfrac{1}{2}x} x^{\tfrac{1}{2}\gamma} U(\alpha, \gamma, x), \qquad (13)$$

$$\gamma = 1 \pm 2m, \quad \alpha = \tfrac{1}{2}\gamma - k. \qquad (14)$$

We shall write
$$W_{k,m}(x) = W(\alpha, \gamma, x) \qquad (15)$$

when the α, γ notation is being used. The differential equation is unaltered if x and k are replaced by $-x$ and $-k$; hence another solution is $W_{-k,m}(-x)$. The asymptotic expansion of $W_{k,m}(x)$ is

$$W_{k,m}(x) \sim x^k e^{-\tfrac{1}{2}x} \left\{ 1 + \frac{m^2 - (k-\tfrac{1}{2})^2}{1!\,x} + \frac{\{m^2 - (k-\tfrac{1}{2})^2\}\{m^2 - (k-\tfrac{3}{2})^2\}}{2!\,x^2} + \ldots \right\} \qquad (16)$$

which terminates if $m+k-\tfrac{1}{2}$ is an integer ≥ 0 or if $m-k+\tfrac{1}{2}$ is an integer ≤ 0. Also

$$W_{-k,m}(-x) \sim (-x)^k e^{\tfrac{1}{2}x} \left\{ 1 - \frac{m^2 - (k+\tfrac{1}{2})^2}{1!\,x} + \frac{\{m^2 - (k+\tfrac{1}{2})^2\}\{m^2 - (k+\tfrac{3}{2})^2\}}{2!\,x^2} - \ldots \right\} \qquad (17)$$

which terminates if $m+k+\tfrac{1}{2}$ is an integer ≤ 0 or $m-k-\tfrac{1}{2}$ an integer ≥ 0.

23·07. Schrödinger's equation for the hydrogen-like atom.[a] The radial wave function R satisfies a differential equation reducible to the form

$$\frac{d^2R}{dr^2} + \frac{2}{r}\frac{dR}{dr} + \left(2E + \frac{2Z}{r} - \frac{l(l+1)}{r^2}\right)R = 0, \tag{1}$$

where l is a positive integer or zero, and Z is positive. R is required to be bounded for $0 \leqslant r < \infty$. If we put

$$R = P/r \tag{2}$$

we get

$$\frac{d^2P}{dr^2} + \left(2E + \frac{2Z}{r} - \frac{l(l+1)}{r^2}\right)P = 0. \tag{3}$$

If E is negative (bound electron) we put $2E = -\tfrac{1}{4}\kappa^2$, $\kappa r = \rho$, $2Z = \kappa\nu$ and then find

$$\frac{d^2P}{d\rho^2} + \left(-\frac{1}{4} + \frac{\nu}{\rho} - \frac{l(l+1)}{\rho^2}\right)P = 0 \tag{4}$$

which is in Whittaker's form with

$$\gamma = 2(l+1), \quad \alpha = l+1-\nu. \tag{5}$$

The solutions have indices $l+1$ and $-l$ at $\rho = 0$. The latter is excluded, and the solution required is

$$P = e^{-\tfrac{1}{2}\rho}\rho^{l+1}\,{}_1F_1(l+1-\nu,\, 2l+2,\, \rho). \tag{6}$$

When ρ is large this will be large like $\exp(\tfrac{1}{2}\rho)$ unless the series terminates (compare 23·04 (8)), that is, unless $1+l-\nu$ is zero or a negative integer. Hence

$$\nu = l+1+s = n, \tag{7}$$

where s is an integer $\geqslant 0$. n is called the principal quantum number. Then

$$E = -\tfrac{1}{8}\kappa^2 = -\frac{1}{2}\frac{Z^2}{n^2} \quad (n = l+1, l+2, \ldots), \tag{8}$$

$$R \propto \rho^l e^{-\tfrac{1}{2}\rho}\,{}_1F_1(-s,\, 2l+2,\, \rho). \tag{9}$$

The polynomials in this case have a compact operational expression. From 23·03 (4), if γ is an integer $\geqslant 1$,

$$x^{\gamma-1}\,{}_1F_1(-s, \gamma, x) = (\gamma-1)!\left(1-\frac{1}{p}\right)^s p^{1-\gamma} = (\gamma-1)!\,(p-1)^s p^{1-\gamma-s}. \tag{10}$$

But

$$F(p-\alpha)\,1 = e^{\alpha x}\, F(p)\, e^{-\alpha x} \tag{11}$$

and therefore

$$x^{\gamma-1}\,{}_1F_1(-s, \gamma, x) = (\gamma-1)!\, e^x \frac{p^s}{(p+1)^{\gamma+s-1}} e^{-x}$$

$$= (\gamma-1)!\, e^x \frac{p^{s+1}}{(p+1)^{\gamma+s}} = (\gamma-1)!\, e^x \left(\frac{d}{dx}\right)^s \frac{x^{\gamma+s-1}}{(\gamma+s-1)!} e^{-x}. \tag{12}$$

The polynomials

$$L_s(x) = e^x \left(\frac{d}{dx}\right)^s (x^s e^{-x}) \tag{13}$$

are called the *Laguerre polynomials*.* Now

$$\rho^l e^{-\frac{1}{2}\rho} {}_1F_1(-s, 2l+2, \rho) = \rho^l e^{-\frac{1}{2}\rho} \rho^{-2l-1}(2l+1)! \, e^{\rho} \left(\frac{d}{d\rho}\right)^s \frac{\rho^{2l+s+1}}{(2l+s+1)!} e^{-\rho}$$

$$= (2l+1)! \, \rho^{-l-1} e^{\frac{1}{2}\rho} \left(\frac{d}{d\rho}\right)^s \frac{\rho^{2l+s+1}}{(2l+s+1)!} e^{-\rho}. \tag{14}$$

Writing $\dfrac{d}{d\rho} \equiv D$

$$L_s(\rho) = (D-1)^s \rho^s \tag{15}$$

and (14) may be written

$$\frac{(2l+1)!}{(2l+s+1)!} \rho^{-l-1} e^{-\frac{1}{2}\rho} (D-1)^s \rho^{2l+s+1}. \tag{16}$$

If L^{2l+1}_{2l+s+1} denotes the $(2l+1)$th derivative of the $(2l+s+1)$th Laguerre polynomial

$$L^{2l+1}_{2l+s+1} = D^{2l+1}(D-1)^{2l+s+1}\rho^{2l+s+1}. \tag{17}$$

Comparison of coefficients shows that (16) is equal to

$$-\frac{s!(2l+1)!}{[(2l+s+1)!]^2} \rho^l e^{-\frac{1}{2}\rho} L^{2l+1}_{2l+s+1}. \tag{18}$$

If E is positive (free electron) we have to replace κ by $i\kappa$; but for large r the exponential factor will now be $\exp(i\kappa r)$ and will not tend to infinity for r large and real. The differential equation now reads, with $E = \frac{1}{8}\kappa^2$, $i\kappa r = \rho$,

$$\frac{d^2 P}{d\rho^2} + \left(-\frac{1}{4} - \frac{2iZ}{\kappa\rho} - \frac{l(l+1)}{\rho^2}\right) P = 0. \tag{19}$$

The solution can now be written

$$P = e^{-\frac{1}{2}i\kappa r}(\kappa r)^{l+1} {}_1F_1\left(1+l+\frac{2iZ}{\kappa}, 2l+2, i\kappa r\right). \tag{20}$$

Here α is complex and $i\kappa r$ is purely imaginary. Hence in the asymptotic expansion for large r we must keep both the series in 23·04 (8). We have

$$P \sim e^{-\frac{1}{2}i\kappa r}(\kappa r)^{l+1}\left\{\frac{(2l+1)!}{(l+2iZ/\kappa)!}(i\kappa r)^{-l-1+2iZ/\kappa} e^{i\kappa r} + \frac{(2l+1)!}{(l-2iZ/\kappa)!}(i\kappa r)^{-l-1-2iZ/\kappa} e^{i\pi(l+1+2iZ/\kappa)}\right\} \tag{21}$$

and $i\kappa r$ must be interpreted as $\kappa r \exp(\frac{1}{2}\pi i)$. Then

$$P \sim (2l+1)! \, e^{-\pi Z/\kappa} \left\{\frac{(\kappa r)^{2iZ/\kappa}}{(l+2iZ/\kappa)!} e^{\frac{1}{2}i\kappa r - \frac{1}{2}\pi i(l+1)} + \frac{(\kappa r)^{-2iZ/\kappa}}{(l-2iz/\kappa)!} e^{-\frac{1}{2}i\kappa r + \frac{1}{2}\pi i(l+1)}\right\} \tag{22}$$

$$= \frac{2(2l+1)! \, e^{-\pi Z/\kappa}}{|(l+2iZ/\kappa)!|} \cos\left(\frac{1}{2}\kappa r - \frac{1}{2}\pi(l+1) + \frac{2Z}{\kappa}\log(\kappa r) - \arg(l+2iZ/\kappa)!\right), \tag{23}$$

in which the constant factor is independent of r.†

(20) can be put in another form. It is equivalent to

$$e^{-\frac{1}{2}i\kappa r}(\kappa r)^{l+1} \frac{(2l+1)!}{2\pi i} \int_M e^{\lambda} \frac{\lambda^{2iZ/\kappa-l-1}}{(\lambda-i\kappa r)^{2iZ/\kappa+l+1}} d\lambda. \tag{24}$$

---
* Cf. Courant-Hilbert, *Methoden der mathematischen Physik*, 1, 1924, 77–9; E. Schrödinger, *Ann. d. Physik*, (4), 80, 1926, 437–90.
† Bethe, *Handb. d. Physik*, 24/1, p. 289. His k is the present $\frac{1}{2}\kappa$.

Now (19) is unaltered by writing $-l-1$ for l. This suggests considering the expression

$$e^{-\frac{1}{2}i\kappa r}(\kappa r)^{-l}\frac{(2l+1)!}{2\pi i}\int_M e^\lambda \lambda^{2iZ/\kappa+l}(\lambda-i\kappa r)^{-2iZ/\kappa+l}d\lambda. \qquad (25)$$

Expanding the integrand in descending powers of λ and integrating, we find that (25) is equal to

$$(-i)^{2l+1}e^{-\frac{1}{2}i\kappa r}(\kappa r)^{l+1}\frac{\left(-\frac{2iZ}{\kappa}+l\right)!}{\left(-\frac{2iZ}{\kappa}-l-1\right)!} {}_1F_1\left(1+l+\frac{2iZ}{\kappa},2l+2,i\kappa r\right) \qquad (26)$$

so that (20) differs from (25) by a constant real factor. It should be noted that the solution (20) is real.

23·08. The parabolic cylinder, Hermite, and Hh functions. Consider the equation

$$\frac{d^2y}{dx^2}+(A-Bx^2)y = 0 \qquad (1)$$

where A and B are real. Clearly all solutions are integral functions of x. Put $\xi = x^2$, $y = \xi^{-1/4}z$. Then

$$\frac{d^2z}{d\xi^2}+\left(-\tfrac{1}{4}B+\frac{A}{4\xi}+\frac{3}{16\xi^2}\right)z = 0 \qquad (2)$$

which is in the extended Whittaker form 23·06 (12), with

$$\mu = \tfrac{1}{2}\sqrt{B}, \quad \gamma = \tfrac{1}{2}, \quad \alpha = \tfrac{1}{4}\left(1-\frac{A}{\sqrt{B}}\right) \qquad (3)$$

and solutions of (1) are

$$\exp\left(-\tfrac{1}{2}B^{1/2}x^2\right)\{{}_1F_1(\alpha,\tfrac{1}{2},B^{1/2}x^2);\, x\,{}_1F_1(\tfrac{1}{2}+\alpha,\tfrac{3}{2},B^{1/2}x^2);\, U(\alpha,\tfrac{1}{2},B^{1/2}x^2)\} \qquad (4)$$

with three others obtained from these by changing the sign of $\sqrt{B}$. Naturally not more than two of the six solutions can be independent, and in particular if we take the positive sign for $\sqrt{B}$ and put

$$\alpha' = \tfrac{1}{4}\left(1+\frac{A}{\sqrt{B}}\right), \qquad (5)$$

$$\exp\left(-\tfrac{1}{2}B^{1/2}x^2\right){}_1F_1(\alpha,\tfrac{1}{2},B^{1/2}x^2) = \exp\left(\tfrac{1}{2}B^{1/2}x^2\right){}_1F_1(\alpha',\tfrac{1}{2},-B^{1/2}x^2), \qquad (6)$$

$$x\exp\left(-\tfrac{1}{2}B^{1/2}x^2\right){}_1F_1(\tfrac{1}{2}+\alpha,\tfrac{3}{2},B^{1/2}x^2) = x\exp\left(\tfrac{1}{2}B^{1/2}x^2\right){}_1F_1(\tfrac{1}{2}+\alpha',\tfrac{3}{2},-B^{1/2}x^2). \qquad (7)$$

For the expressions in (6) are even solutions, and those in (7) are odd solutions; and the ratio of the two sides of each equation tends to 1 when x tends to 0. The same follows from 23·03 (10).

Three specially important cases are distinguished according to the signs of A and B. First, the equations 18·04 (7), (8) satisfied by the parabolic cylinder functions are of the form (1). For their solutions to be oscillatory for large ξ_1 or large ξ_2, as defined in 18·04, we must have $\kappa^2 > \mu^2$, and therefore B is negative. This case, in tidal theory, would correspond to a prescribed harmonic motion at a long distance from a parabolic cape, or at the mouth of a long parabolic bay. Hence in this case the parabolic cylinder functions are expressible in terms of confluent hypergeometric functions of imaginary argument.

With A real, α will be complex; but the solutions will be real as in the case of the wave equation for free electrons near a nucleus. The same differential equation arises in wave mechanics for a linear repulsive field.

B is positive in the case of the harmonic oscillator in wave mechanics, and in tidal theory or acoustics in the case of a local disturbance due to a parabolic projection. In the latter type of problem we have as a limiting case diffraction by a semi-infinite screen, which is one of the few diffraction problems that have been worked out exactly.* If we require a solution to tend to zero when x tends to infinity in each direction, no linear combination of the odd and even solutions can satisfy the conditions unless one of the coefficients vanishes, and then the solution with a non-zero coefficient must reduce to a polynomial multiplied by an exponential. Hence 2α is an integer ≤ 0. This is the case of the Schrödinger wave equation of the harmonic oscillator.

One solution also reduces to an elementary function if 2α is a positive integer. For then, since $\alpha + \alpha' = \frac{1}{2}$, one of the series on the right of (6), (7) reduces to a polynomial. But in this case, since it is multiplied by $\exp(\frac{1}{2} B^{1/2} x^2)$, the elementary solution will usually be forbidden. We can however still find a solution based on U that tends to zero as $x \to +\infty$. This will not be an elementary function, but the functions that occur in problems of heat flow in one and three dimensions, and the Hh_n functions for $n \geq 0$, which have importance in statistics, are of this type.[a]

These solutions are conveniently derived by using the function U and its operational expression. Since

$$p^{1/2} \exp(-ap^{1/2}) = \frac{1}{\sqrt{(\pi t)}} e^{-a^2/4t} \tag{8}$$

and

$$\operatorname{Kh}_{1/2}(x) = \left(\frac{2}{\pi x}\right)^{1/2} e^{-x} \quad \text{from 21·08 (13)}, \tag{9}$$

$$p^{n+1/4} \operatorname{Kh}_{1/2}(ap^{1/2}) = p^{n+1/4} \left(\frac{2}{\pi a p^{1/2}}\right)^{1/2} e^{-ap^{1/2}}$$

$$= \left(\frac{2}{\pi a}\right)^{1/2} p^n e^{-ap^{1/2}}. \tag{10}$$

But

$$\frac{\partial}{\partial a} e^{-ap^{1/2}} = -p^{1/2} e^{-ap^{1/2}}, \tag{11}$$

$$\int_a^\infty e^{-ap^{1/2}} da = p^{-1/2} e^{-ap^{1/2}}, \tag{12}$$

and therefore if $2m$ is an integer ≥ 0,

$$p^{m+1/2} e^{-ap^{1/2}} = (-1)^{2m} \left(\frac{\partial}{\partial a}\right)^{2m} \frac{1}{\sqrt{(\pi t)}} e^{-a^2/4t}, \tag{13}$$

and if $2m$ is an integer ≤ 0,

$$p^{m+1/2} e^{-ap^{1/2}} = \left(\int_a^\infty da\right)^{-2m} \frac{1}{\sqrt{(\pi t)}} e^{-a^2/4t}. \tag{14}$$

Now

$$U\left(\alpha, \tfrac{1}{2}, \frac{a^2}{4t}\right) = \pi e^{a^2/4t} (\tfrac{1}{2} a)^{1/2} t^{1/2-\alpha} p^{3/4-\alpha} \operatorname{Kh}_{1/2}(ap^{1/2}) \quad \text{from 23·05 (6)},$$

$$= \pi^{1/2} e^{a^2/4t} t^{1/2-\alpha} p^{1/2-\alpha} e^{-ap^{1/2}}. \tag{15}$$

* Lamb, *Hydrodynamics*, 1932, p. 538.

Put $\alpha = -m$; then, if 2α is an integer ≤ 0,

$$U\left(\alpha, \tfrac{1}{2}, \frac{a^2}{4t}\right) = \pi^{1/2} e^{a^2/4t} t^{1/2-\alpha} (-1)^{2\alpha} \left(\frac{\partial}{\partial a}\right)^{-2\alpha} \frac{1}{\sqrt{(\pi t)}} e^{-a^2/4t}$$

$$= t^{-\alpha} e^{a^2/4t} (-1)^{2\alpha} \left(\frac{\partial}{\partial a}\right)^{-2\alpha} e^{-a^2/4t}, \tag{16}$$

$$U(\alpha, \tfrac{1}{2}, B^{1/2} x^2) = (-1)^{2\alpha} (2 B^{1/4})^{2\alpha} e^{B^{1/2} x^2} \left(\frac{\partial}{\partial x}\right)^{-2\alpha} e^{-B^{1/2} x^2}. \tag{17}$$

These solutions are known as the *Hermite polynomials*; and their products by $\exp(-\tfrac{1}{2} B^{1/2} x^2)$ satisfy (1). For a given α, where 2α is an integer ≤ 0, (17) is the only solution that tends to zero at either $x = \infty$ or $x = -\infty$; it does so at both. If 2α is not an integer ≤ 0 there is *no* solution that tends to zero at both $x = \infty$ and $x = -\infty$.

If 2α is an integer > 0,

$$U\left(\alpha, \tfrac{1}{2}, \frac{a^2}{4t}\right) = \pi^{1/2} e^{a^2/4t} t^{1/2-\alpha} \left(\int_a^\infty da\right)^{2\alpha} \frac{1}{\sqrt{(\pi t)}} e^{-a^2/4t}, \tag{18}$$

$$U(\alpha, \tfrac{1}{2}, B^{1/2} x^2) = (2 B^{1/4})^{2\alpha} e^{B^{1/2} x^2} \left(\int_x^\infty dx\right)^{2\alpha} e^{-B^{1/2} x^2}. \tag{19}$$

Thus the solutions can be built up by successive integration or differentiation according to the sign of α. The forms (16), (18) are convenient as they stand in problems of diffusion (including heat conduction). For other purposes it is probably best to make use of the tables given in the *British Association Tables*, vol. 1; though a supplementary table of the commoner functions at closer intervals and to four or five figures would be very useful.

A table of the related functions $\operatorname{ierf} x = \int_x^\infty (1 - \operatorname{erf} w) \, dw$, $\operatorname{iierf} x = \int_x^\infty \operatorname{ierf} x \, dw$ to four figures is given by Hartree* (cf. 20·06 (9)).

23·081. For n an integer ≥ 0,

$$\operatorname{Hh}_n(x) = \int_0^\infty \frac{t^n}{n!} \exp\{-\tfrac{1}{2}(t+x)^2\} \, dt = \int_x^\infty \frac{(u-x)^n}{n!} e^{-1/2 u^2} \, du, \tag{1}$$

and for n a negative integer

$$\operatorname{Hh}_n(x) = (-1)^{n-1} \left(\frac{d}{dx}\right)^{-n-1} e^{-1/2 x^2}. \tag{2}$$

In particular

$$\operatorname{Hh}_0(x) = \int_x^\infty e^{-1/2 u^2} \, du, \quad \operatorname{Hh}_0(-\infty) = \sqrt{(2\pi)}, \tag{3}$$

$$\operatorname{Hh}_{-1}(x) = e^{-1/2 x^2}. \tag{4}$$

Clearly for all n

$$\frac{d}{dx} \operatorname{Hh}_n(x) = -\operatorname{Hh}_{n-1}(x) \tag{5}$$

and it is easy to show that $\operatorname{Hh}_n(x)$ satisfies the equation

$$\frac{d^2 y}{dx^2} + x \frac{dy}{dx} - ny = 0. \tag{6}$$

* Mem. Proc. Manchester Lit. Phil. Soc. 80, 1936, 85–102.

The functions satisfy the recurrence relation
$$(n+1)\,\mathrm{Hh}_{n+1}(x) + x\,\mathrm{Hh}_n(x) - \mathrm{Hh}_{n-1}(x) = 0. \tag{7}$$

If we put
$$y = e^{-\frac{1}{4}x^2} z, \tag{8}$$

we find
$$\frac{d^2 z}{dx^2} - (n + \tfrac{1}{2} + \tfrac{1}{4}x^2)\, z = 0 \tag{9}$$

for all integral n, positive or negative. This is of the form of 23·08 (1), which is reduced to it by taking
$$x = (4B)^{-\frac{1}{4}}\eta. \tag{10}$$

Then the solution of 23·08 (1) that concerns us, in the cases that we have investigated, can be written compactly as
$$e^{\frac{1}{4}\eta^2} \mathrm{Hh}_n(\eta) \tag{11}$$

for any integral n, positive or negative, and
$$A = -(2n+1)\sqrt{B}. \tag{12}$$

The function $D_n(x)$ is defined by
$$D_n(x) = e^{\frac{1}{4}x^2}\mathrm{Hh}_{-n-1}(x) = e^{\frac{1}{4}x^2}(-1)^n \frac{d^n}{dx^n} e^{-\frac{1}{2}x^2} \tag{13}$$

for n positive. It clearly has n real zeros. It also has an orthogonal property, as we should expect because it arises in the solution of the wave equation. In fact
$$\int_{-\infty}^{\infty} D_m(x) D_n(x)\, dx = \int_{-\infty}^{\infty} e^{\frac{1}{2}x^2}(-1)^{m+n} \left(\frac{d^m}{dx^m} e^{-\frac{1}{2}x^2}\right)\left(\frac{d^n}{dx^n} e^{-\frac{1}{2}x^2}\right) dx$$
$$= (-1)^n \int_{-\infty}^{\infty} e^{-\frac{1}{2}x^2} \frac{d^m}{dx^m}\left(e^{\frac{1}{2}x^2} \frac{d^n}{dx^n} e^{-\frac{1}{2}x^2}\right) dx. \tag{14}$$

If $m \neq n$, we can take m to be the greater. Then the function differentiated m times is a polynomial of degree n, and the derivative is zero. Hence the integral is zero unless $m = n$. If $m = n$ the term of highest degree is $(-1)^n x^n$, and
$$\int_{-\infty}^{\infty} \{D_n(x)\}^2\, dx = \int_{-\infty}^{\infty} n!\, e^{-\frac{1}{2}x^2}\, dx = \sqrt{(2\pi)}\, n!. \tag{15}$$

The Hh_n functions are all positive when $n \geq 0$ and there is no question of orthogonality. In terms of the series solutions of (6)
$$\mathrm{Hh}_n(x) = \sqrt{(\tfrac{1}{2}\pi)}\, 2^{-\frac{1}{2}n} \sum_{m=0}^{\infty} (-1)^m \frac{2^{\frac{1}{2}m} x^m}{m!\,\{\tfrac{1}{2}(n-m)\}!}$$
$$= \frac{\sqrt{(\tfrac{1}{2}\pi)}}{2^{\frac{1}{2}n}(\tfrac{1}{2}n)!}\left(1 + \frac{nx^2}{2!} + \frac{n(n-2)x^4}{4!} + \ldots\right)$$
$$- \frac{\sqrt{(\tfrac{1}{2}\pi)}}{2^{\frac{1}{2}n-\frac{1}{2}}(\tfrac{1}{2}n-\tfrac{1}{2})!}\, x\left(1 + \frac{(n-1)x^2}{3!} + \frac{(n-1)(n-3)x^4}{5!} + \ldots\right). \tag{16}$$

23·082. Asymptotic approximation to $\mathrm{Hh}_n(x)$ for x large, $n > 0$. From 23·081 (1)

$$\mathrm{Hh}_n(x) = \frac{e^{-\frac{1}{2}x^2}}{n!} \int_0^\infty t^n e^{-tx-\frac{1}{2}t^2} dt \qquad (1)$$

$$= \frac{e^{-\frac{1}{2}x^2}}{n!} \int_0^\infty e^{-tx} \Sigma (-1)^m \frac{t^{2m+n}}{2^m m!} dt \quad (\Re(x) > 0), \qquad (2)$$

$$\sim \frac{e^{-\frac{1}{2}x^2}}{n!} \sum_{m=0}^\infty (-1)^m \frac{(2m+n)!}{2^m m!} \frac{1}{x^{2m+n+1}}. \qquad (3)$$

$$\mathrm{Hh}_n(-x) + (-1)^n \mathrm{Hh}_n(x) = \frac{1}{n!} \int_0^\infty t^n \left(e^{-\frac{1}{2}(t-x)^2} + (-1)^n e^{-\frac{1}{2}(t+x)^2} \right) dt$$

$$= \frac{1}{n!} \int_{-\infty}^\infty t^n e^{-\frac{1}{2}(t-x)^2} dt$$

$$= \frac{1}{n!} \int_{-\infty}^\infty (x+u)^n e^{-\frac{1}{2}u^2} du$$

$$= \sum_{m=0}^{\prime} \frac{2^{\frac{1}{2}(m+1)}(\frac{1}{2}m - \frac{1}{2})!}{m!(n-m)!} x^{n-m}, \qquad (4)$$

summation being over even values of $m \leqslant n$. The latter expression is exact but is given with the asymptotic expansion because it is in descending powers. It is useful in estimating $\mathrm{Hh}_n(-x)$ for large x, since $\mathrm{Hh}_n(x)$ is then small.

23·083. Asymptotic approximation to $\mathrm{Hh}_n(x)$ for n large, > 0, x moderate. $\mathrm{Hh}_n(x)$ is very small when x is more than about 3, but n may be large. Then if

$$\phi(t) = n \log t - tx - \tfrac{1}{2}t^2, \qquad (5)$$

$$\phi'(t) = \frac{n}{t} - x - t, \qquad (6)$$

$$\phi''(t) = -\frac{n}{t^2} - 1. \qquad (7)$$

Then for n large and x moderate, $\phi'(t) = 0$ gives $t = n^{\frac{1}{2}} - x$ as a first approximation, $\phi''(t) = -2$. Then

$$\phi(n^{\frac{1}{2}}) = \tfrac{1}{2} n \log n - n^{\frac{1}{2}} x - \tfrac{1}{2} n + \tfrac{1}{4} x^2 \qquad (8)$$

and $$\mathrm{Hh}_n(x) \sim \frac{e^{-\frac{1}{4}x^2}}{n!} \sqrt{\pi} \, n^{\frac{1}{2}n} e^{-\frac{1}{2}n - x\sqrt{n}}. \qquad (9)$$

Taking $x = 0$ and applying Stirling's formula we recover the first term of the ascending series to order $1/n$; the approximation is therefore checked.

23·084. Asymptotic approximation to $D_n(x)$ for large n. This is most easily found from the differential equation 23·081 (9) with $-n-1$ in place of n, which is

$$\frac{d^2 z}{dx^2} + (n + \tfrac{1}{2} - \tfrac{1}{4} x^2) z = 0. \qquad (1)$$

Put $$n + \tfrac{1}{2} = m, \quad x = 2\sqrt{m}\,\xi. \qquad (2)$$

Then
$$\frac{d^2 z}{d\xi^2} + 4m^2(1-\xi^2) z = 0. \tag{3}$$

The solutions are of exponential type for $\xi > 1$ or $\xi < -1$ and oscillating for $\xi^2 < 1$; and

$$2m \int_\xi^1 \sqrt{(1-\xi^2)}\, d\xi = m(\tfrac{1}{2}\pi - \theta - \tfrac{1}{2}\sin 2\theta) \quad (\xi = \sin\theta), \tag{4}$$

$$2m \int_1^\xi \sqrt{(\xi^2 - 1)}\, d\xi = m(\tfrac{1}{2}\sinh 2u - u) \quad (\xi = \cosh u). \tag{5}$$

Then a solution decreasing exponentially as $\xi \to +\infty$ is

$$\frac{1}{(\xi^2 - 1)^{1/4}} \exp\{-m\xi(\xi^2 - 1)^{1/2}\}\{\xi + \sqrt{(\xi^2 - 1)}\}^m \longleftrightarrow \frac{2}{(1-\xi^2)^{1/4}} \sin\{m(\tfrac{1}{2}\pi - \theta - \tfrac{1}{2}\sin 2\theta) + \tfrac{1}{4}\pi\}. \tag{6}$$

The constant factor is determined from the fact that when x is large $D_n(x)$ is asymptotically $x^n \exp(-\tfrac{1}{4}x^2)$; it is thus found to be

$$2^{-1/2} m^{1/2 n} e^{-1/2 m} \doteq 2^{-1/2} n^{1/2 n} e^{-1/2 n}. \tag{7}$$

A check is obtained by taking $x = 0$ and therefore $\theta = 0$. Evidently $D_n(0) = 0$ for odd n; for even n the right side of (6) is ± 2. But for even n

$$\pm D_n(0) = \frac{d^n}{dx^n} \frac{(\tfrac{1}{2}x^2)^{1/2 n}}{(\tfrac{1}{2}n)!} = \frac{(\tfrac{1}{2})^{1/2 n} n!}{(\tfrac{1}{2}n)!} \sim \sqrt{2}\, n^{1/2 n} e^{-1/2 n}. \tag{8}$$

Hence to obtain an approximation to $D_n(x)$ with an error $O(1/n)$ we must multiply both sides of (6) by

$$2^{-1/2} (n/e)^{1/2 n}. \tag{9}$$

23·09. Accuracy of steepest descents approximations. This approximation to $\mathrm{Hh}_n(x)$, when n is large, suggests a way of estimating the error in stopping at the smallest term in an asymptotic expansion found by steepest descents. In the expression

$$I = \int_{-\infty}^\infty e^{-1/2 a^2 z^2} f(z)\, dz \tag{1}$$

put $f(z) + f(-z) = g(z)$; then $g(z)$ is an even function, and

$$I = \int_0^\infty e^{-1/2 a^2 z^2} g(z)\, dz. \tag{2}$$

Integrate $2n$ times by parts; then, since odd $g^{(m)}(z)$ vanish at $z = 0$,

$$I = \frac{1}{a} g(0)\, \mathrm{Hh}_0(0) + \frac{g''(0)}{a^3} \mathrm{Hh}_2(0) + \ldots + \frac{g^{(2n)}(0)}{a^{2n+1}} \mathrm{Hh}_{2n}(0) + R_{2n}, \tag{3}$$

where $\displaystyle R_{2n} = \frac{1}{a^{2n+1}} \int_0^\infty \mathrm{Hh}_{2n}(az)\, g^{(2n+1)}(z)\, dz = \frac{1}{a^{2n+2}} \int_0^\infty \mathrm{Hh}_{2n+1}(az)\, g^{(2n+2)}(z)\, dz.$ (4)

Now let the singularity of smallest modulus of $f(z)$ be at $re^{i\alpha}$; then $g(z)$ has singularities at $\pm re^{i\alpha}$, near which we suppose that $g(z)$ behaves like a negative power or a logarithm.

The contributions to $g(z)$ from these will have the forms

$$\frac{A}{(re^{i\alpha}-z)^m}+\frac{A}{(re^{i\alpha}+z)^m} \quad \text{or} \quad A\log(r^2 e^{2i\alpha}-z^2) \tag{5}$$

where m is independent of a and n. We assume that a is large enough for the smallest term to be at a large value of n; now

$$g^{(2n)}(z) \doteqdot \frac{(m+2n-1)!}{(m-1)!}A\left(\frac{1}{(re^{i\alpha}-z)^{m+2n}}+\frac{1}{(re^{i\alpha}+z)^{m+2n}}\right)$$

$$\doteqdot \frac{(2n)!\,M}{r^{2n}e^{2ni\alpha}}\cosh\frac{2nz}{re^{i\alpha}} \quad \text{for } z \text{ small,} \tag{6}$$

in the sense that M can be chosen so that the ratio of $g^{(2n)}(z)$ to an expression of this form tends to 1 when n is large. Also if u_{2n} is the general term of (3)

$$\frac{u_{2n+2}}{u_{2n}} \doteqdot \frac{g^{(2n+2)}(0)}{a^2 g^{(2n)}(0)}\frac{\mathrm{Hh}_{2n+2}(0)}{\mathrm{Hh}_{2n}(0)} \doteqdot \frac{1}{a^2}\frac{(2n+1)(2n+2)}{r^2 e^{2i\alpha}}\frac{1}{2(n+1)} \doteqdot \frac{2n}{a^2 r^2 e^{2i\alpha}}. \tag{7}$$

The smallest term is therefore specified by

$$2n = [r^2 a^2], \tag{8}$$

and $\quad R_{2n} \doteqdot \dfrac{u_{2n}(2n+1)(2n+2)}{r^2 a\,\mathrm{Hh}_{2n}(0)\,e^{2i\alpha}} \displaystyle\int_0^\infty \mathrm{Hh}_{2n+1}(az)\cosh\dfrac{2nz}{re^{i\alpha}}\,dz$

$$\doteqdot u_{2n}a(2n)^{1/2}e^{-2i\alpha}\int_0^\infty \exp\left(-\sqrt{(2n)}\,az-\tfrac{1}{4}a^2 z^2\right)\cosh\left(\sqrt{(2n)}\,aze^{-i\alpha}\right)dz. \tag{9}$$

If n is large and a not small, the integrand becomes small before the term in $a^2 z^2$ becomes important, and R_{2n} reduces to*

$$R_{2n} \doteqdot u_{2n}e^{-2i\alpha}\int_0^\infty e^{-x}\cosh(xe^{-i\alpha})\,dx$$

$$= \frac{u_{2n}e^{-2i\alpha}}{1-e^{-2i\alpha}}, \tag{10}$$

$$u_{2n}+R_{2n} \doteqdot \frac{u_{2n}}{1-e^{-2i\alpha}} = (\tfrac{1}{2}-\tfrac{1}{2}i\cot\alpha)u_{2n}. \tag{11}$$

The simple rule found for the incomplete factorial function that we should take half the smallest term is therefore true for steepest descents approximations only if $\alpha = \pm\tfrac{1}{2}\pi$, and this will be shown by successive terms being precisely opposite in phase. R_{2n} will exceed u_{2n} in modulus if $|\cot\alpha| > \sqrt{3}$.

This discussion is rough, but serves two purposes. Most discussions of the error in stopping at a given term of an asymptotic approximation treat special functions and take the general term. But the method has a wide generality and should be capable of a more general treatment. We see that the fact that the early terms give a rapid decrease in the error is due to the rapid decrease of the early $\mathrm{Hh}_n(az)$ with increasing n or az when a is large; the early derivatives of $g(z)$ can be treated as approximately constant in the range where the Hh_n factor is appreciable. But when we approximate many times by integration

---

* See Airey, *Phil. Mag.* (7), 24, 1937, 526.

by parts, though the Hh_n factor decreases, the $g^{(2n)}(z)$ factor varies more and more rapidly with z until when $2n = r^2a^2$ its variation becomes as important as that of $\mathrm{Hh}_{2n}(az)$, and nothing is to be gained by further attempts to approximate on these lines. This explanation is probably familiar to many pure mathematicians, but possibly they have not succeeded in stating it with the precision that they like; but a physicist likes to see even a rough discussion that brings out the point.

The approximate formula (11) makes it possible to allow for the remainder by simple inspection of the terms near the smallest, α being determined directly by comparison of the phases. It breaks down if $\alpha = 0$, that is, if all the terms have the same sign. This might be expected, since it means that the path of integration passes through a singularity. Difficulties have been found, for instance, with the estimation of the remainder in $I_n(x)$ with x real; but the path of steepest descent passes twice through the subsidiary saddle-point at -1, and the integrand is infinite there. The situation is saved to some extent by the fact that the improper integral exists, but it is necessary to break the range of integration up if integration by parts is to be used. Our approximation (9) is extremely crude in this case, but if we put $\alpha = 0$ we get

$$R_{2n} \doteqdot \tfrac{1}{2} u_{2n} (2n)^{1/2} \int_0^\infty \exp(-\tfrac{1}{2}x^2)\,dx = \tfrac{1}{2}\sqrt{(n\pi)}\,u_{2n} \tag{12}$$

and the factor $\sqrt{n}$ is a warning that the size of the smallest term is no safe guide to the accuracy in such a case.

Similar considerations are applicable to the integral

$$I = \int_0^\infty e^{-az} f(z)\,dz = 2\int_0^\infty e^{-a\zeta^2} f(\zeta^2)\,\zeta\,d\zeta. \tag{13}$$

The same methods will apply except that, $\zeta f(\zeta^2)$ being an odd function, it will be the terms in even derivatives that vanish. (11) will still hold with $2n-1$ for $2n$.

For a fuller treatment of the subject of this chapter, the reader is referred to *Confluent Hypergeometric Functions* by L. J. Slater (Cambridge University Press, 1960).

EXAMPLES

1. Prove the recurrence relations

$$x\,_1F_1(\alpha+1,\gamma+1,x) = \gamma[\,_1F_1(\alpha+1,\gamma,x) - \,_1F_1(\alpha,\gamma,x)],$$

$$\alpha\,_1F_1(\alpha+1,\gamma+1,x) = (\alpha-\gamma)\,_1F_1(\alpha,\gamma+1,x) + \gamma\,_1F_1(\alpha,\gamma,x),$$

$$(\alpha+x)\,_1F_1(\alpha+1,\gamma+1,x) = (\alpha-\gamma)\,_1F_1(\alpha,\gamma+1,x) + \gamma\,_1F_1(\alpha+1,\gamma,x),$$

$$\alpha\gamma\,_1F_1(\alpha+1,\gamma,x) = \gamma(\alpha+x)\,_1F_1(\alpha,\gamma,x) - x(\gamma-\alpha)\,_1F_1(\alpha,\gamma+1,x),$$

$$\alpha\,_1F_1(\alpha+1,\gamma,x) = (x+2\alpha-\gamma)\,_1F_1(\alpha,\gamma,x) + (\gamma-\alpha)\,_1F_1(\alpha-1,\gamma,x),$$

$$(\gamma-\alpha)x\,_1F_1(\alpha,\gamma+1,x) = \gamma(x+\gamma-1)\,_1F_1(\alpha,\gamma,x) + \gamma(1-\gamma)\,_1F_1(\alpha,\gamma-1,x).$$

(*B.A. Report*, 1926.)

2. Prove that

$$\exp(-\tfrac{1}{4}x^2 + xt - \tfrac{1}{2}t^2) = \sum_{n=0}^\infty \frac{t^n}{n!} D_n(x).$$

Hence prove that

$$\int_{-\infty}^\infty e^{i\kappa x} D_n(x)\,dx = 2\sqrt{\pi}\,i^n D_n(2\kappa).$$

3. Prove that for real a, b, c the hypergeometric series converges at $z = 1$ if $c > a+b$ and diverges if $c < a+b$; and that it converges at $z = -1$ if $c+1 > a+b$.

Chapter 24

LEGENDRE FUNCTIONS AND ASSOCIATED FUNCTIONS

> 'You boil it in sawdust; you salt it in glue;
> You condense it with locusts and tape;
> Still keeping one principal object in view,
> To preserve its symmetrical shape.'
>
> LEWIS CARROLL, *The Hunting of the Snark*

24·01. Associated Legendre functions. We have seen that the solutions of the potential, wave, and sound equations for spherical boundaries depend on the solution of the equation

$$(1-\mu^2)\frac{d^2\Theta}{d\mu^2} - 2(s+1)\mu\frac{d\Theta}{d\mu} + (n-s)(n+s+1)\Theta = 0, \tag{1}$$

where $-1 \leqslant \mu \leqslant 1$. Potential problems concerning the outside of spheroids depend on the same differential equation, with $\mu > 1$ for prolate spheroids and μ purely imaginary for oblate ones; in these problems we require the solution that tends to 0 when μ tends to ∞. Except when the contrary is stated, we shall take n, s to be integers, $n \geqslant s \geqslant 0$. Then one solution has index 0 at $\mu = \pm 1$ and the other index $-s$. The latter may contain a logarithm, and in any case will be infinite. Since one solution is an odd and the other an even function of μ, the solution with index 0 at $\mu = 1$ will also have index 0 at $\mu = -1$, and will be the solution needed for problems of spherical boundaries. It can have no other singularities and is therefore an integral function. Series solutions can be found easily. They are given explicitly for $s = 0$ in 16·04. An expression in finite terms for the solutions analytic at $\mu = \pm 1$ is found by the method of 18·061 to be

$$\Theta_n^s = \frac{d^{n+s}}{d\mu^{n+s}}(\mu^2-1)^n. \tag{2}$$

We can build up the solutions with singularities at ± 1 from those found by taking $s = n$ in (1). We have

$$(1-\mu^2)\frac{d^2}{d\mu^2}\Theta_n^n - 2(n+1)\mu\frac{d}{d\mu}\Theta_n^n = 0. \tag{3}$$

One solution of this is a constant, and successive integrations will build up polynomial solutions for smaller values of s; but these are already given by (2) without the need of special attention to fix the constants of integration. The other solution is given by

$$\frac{d}{d\mu}\Theta_n^n = -\frac{1}{(\mu^2-1)^{n+1}}, \tag{4}$$

and if we choose the constants so that the solutions tend to 0 at $\mu = \infty$ they will be

$$\Theta_n^s = \frac{1}{(n-s)!}\int_\mu^\infty \frac{(\mu-u)^{n-s}}{(u^2-1)^{n+1}}du, \tag{5}$$

where the path does not cross the real axis between -1 and 1.

The indicial equation for μ large has roots $n-s$ and $-n-s-1$; the former corresponds to (2), and the latter shows that we are justified in assuming for all s the existence of a solution vanishing at infinity. It is easy to verify that (5) satisfies

$$\frac{d}{d\mu}\Theta_n^s = \Theta_n^{s+1} \tag{6}$$

and therefore it must be the second solution required. If we take $n = s = 0$ it becomes

$$\Theta^0 = \tfrac{1}{2}\log\frac{\mu+1}{\mu-1}, \tag{7}$$

so that there are logarithmic singularities at $\mu = \pm 1$, and a special convention will be needed to give a definite value to the function for $-1 \leqslant \mu \leqslant 1$. Except in this case we can agree to take the value given by (5) for μ real and greater than 1 and define the function by continuation, excluding the real values of μ between -1 and $+1$ by a cut. For $-1 < \mu < 1$ we shall see that a modified definition is possible, but in any case the second solution will become infinite at $\mu = \pm 1$.

24·02. Solutions of Laplace's equation in spherical polar coordinates will then be

$$\phi = (r^n, r^{-n-1})\sin^s\theta \cdot \Theta \cdot (\cos s\lambda, \sin s\lambda) \tag{8}$$

where Θ is given by either (2) or (5), with an appropriate constant factor; for a complete sphere the single-valuedness of ϕ will require that for $s = 0$ the term in λ, which replaces $\sin s\lambda$ in this case, will not occur, and the finiteness of ϕ at $\theta = 0$ or π will exclude the solution (5). We shall see that any function satisfying Laplace's equation inside or outside a sphere can be expressed in terms of spherical harmonics, Θ being taken as in (2).

24·03. Potential in a cavity. Consider a closed surface not surrounding any matter. Within it a potential function exists satisfying Laplace's equation. We can draw a sphere about any point in this region and lying wholly in the region; and within such a sphere the potential is given in terms of its values on the sphere by Green's integral 6·092 (8). Now if we take $r < a$,

$$\frac{a^2-r^2}{R^3} = \frac{a^2-r^2}{(a^2-2ar\cos\vartheta+r^2)^{3/2}}.$$

If we for a moment regard r as a complex variable, this function and any of its derivatives have singularities only at $r = ae^{\pm i\vartheta}$, and therefore have expansions in power series in r, uniformly and absolutely convergent with regard to r for $|r| \leqslant c < a$. But the terms in r^n are the sum of terms of the form $(r^2)^m(r\cos\vartheta)^{n-2m}$, and $r\cos\vartheta$ is a linear function of x, y, z. Hence for any ϕ integrable over $r = a$, the terms in r^n can be expressed as a homogeneous polynomial in x, y, z; and the series is uniformly convergent with regard to ϑ for $|r| \leqslant c$. Hence ϕ within the cavity can be expressed as the sum of a series of homogeneous polynomials in x, y, z, valid for $r < a$, and any derivative of ϕ of any order with regard to x, y, z can also be so expressed. Hence sufficiently near the centre we can write ϕ as the sum of a series of homogeneous polynomials in x, y, z

$$\phi = \phi_0 + \phi_1 + \phi_2 + \ldots + \phi_n + \ldots, \tag{1}$$

ϕ_n being of the nth degree. Since $\nabla^2\phi = 0$ for all points in the sphere, $\nabla^2\phi_n = 0$, by equating terms of equal degree. Thus ϕ can be expressed as a series of polynomials each satisfying Laplace's equation. Obvious solutions are

$$\phi_0 = 1; \quad \phi_1 = x, y, z; \quad \phi_2 = xy, yz, zx, x^2 - y^2, 2z^2 - x^2 - y^2.$$

Now when $z = 0$ let $\quad \phi_n = g(x,y), \quad \dfrac{\partial \phi_n}{\partial z} = h(x,y).$ \hfill (2)

ϕ_n can be expressed as a terminating Taylor series in powers of z. But

$$\frac{\partial^{2r}}{\partial z^{2r}}\phi_n = (-1)^r\left(\frac{\partial^2}{\partial x^2} + \frac{\partial^2}{\partial y^2}\right)^r \phi_n, \quad \frac{\partial^{2r+1}}{\partial z^{2r+1}}\phi_n = (-1)^r\left(\frac{\partial^2}{\partial x^2} + \frac{\partial^2}{\partial y^2}\right)^r \frac{\partial \phi_n}{\partial z} \quad (3)$$

and hence the derivatives with regard to z at $z = 0$ can all be found by differentiating $g(x,y)$ and $h(x,y)$. If we write

$$\frac{\partial^2}{\partial x^2} + \frac{\partial^2}{\partial y^2} = \nabla_1^2, \quad (4)$$

$$\phi_n = \Sigma(-1)^m \nabla_1^{2m} g(x,y) \frac{z^{2m}}{(2m)!} + \Sigma(-1)^m \nabla_1^{2m} h(x,y) \frac{z^{2m+1}}{(2m+1)!}, \quad (5)$$

and ϕ_n is completely determined given $g(x,y)$ and $h(x,y)$. But $g(x,y)$ is a polynomial of degree n and therefore contains $n+1$ terms; $h(x,y)$ is of degree $n-1$ and contains n terms. If we substitute (5) in $\nabla^2\phi_n$, with arbitrary coefficients in $g(x,y)$ and $h(x,y)$, we find that $\nabla^2\phi_n = 0$. Hence exactly $2n+1$ coefficients can be assigned independently, and ϕ_n can be expressed in terms of $2n+1$ linearly independent polynomials.

If we take $s = 0, 1, \dots, n$ in

$$r^n \sin^s\theta \frac{d^{n+s}}{d\mu^{n+s}}(\mu^2 - 1)^n (\cos s\lambda, \sin s\lambda) \quad (6)$$

we have $2n+1$ solutions, which are clearly independent since none of $\cos s\lambda$ and $\sin s\lambda$ can be linearly expressed in terms of the λ factors for other values of s. The solutions are expressible as polynomials in x, y, z. For $r^s \sin^s\theta(\cos s\lambda, \sin s\lambda)$ are the real and imaginary parts of $(x+iy)^s$ and therefore are polynomials. Also $\dfrac{d^{n+s}}{d\mu^{n+s}}(\mu^2 - 1)^n$ is a polynomial in μ of degree $n-s$ of the form $\Sigma A_m \mu^{n-s-2m}$; and

$$r^{n-s}\mu^{n-s-2m} = (r\mu)^{n-s-2m} r^{2m} = z^{n-s-2m}(x^2 + y^2 + z^2)^m.$$

Hence the solutions (6) are $2n+1$ linearly independent polynomials. Any polynomial of degree n in x, y, z that satisfies $\nabla^2\phi = 0$ can therefore be expressed linearly in terms of them.

24·04. Solid and surface harmonics: explicit forms for $p_n^s(\mu)e^{is\lambda}$. Laplace's equation can be written

$$\frac{\partial}{\partial r}\left(r^2 \frac{\partial \phi}{\partial r}\right) + \frac{\partial}{\sin\theta\, \partial\theta}\left(\sin\theta \frac{\partial \phi}{\partial \theta}\right) + \frac{\partial^2 \phi}{\sin^2\theta\, \partial\lambda^2} = 0.$$

If $\phi = r^n S_n(\theta,\lambda)$ the first term is $n(n+1)\phi$, and the coefficient is unaltered by changing n into $-n-1$. Hence if $r^n S_n(\theta,\lambda)$ is a solution, $r^{-n-1} S_n(\theta,\lambda)$ is another, and conversely. Such solutions are called *solid harmonics* of degree n or $-n-1$ as the case may be, and S_n is called a *surface harmonic*.

Derivatives of $1/r$ as solutions

This fact leads to another way of developing the standard solutions. For since $1/r$ is a solution of Laplace's equation, of degree -1, any derivative $\dfrac{\partial^{l+m+n}}{\partial x^l \partial y^m \partial z^n} \dfrac{1}{r}$ is another, of degree $-l-m-n-1$. If we multiply it by $r^{2l+2m+2n+1}$ we shall therefore get another, of degree $l+m+n$. Hence the functions

$$r^{2l+2m+2n+1} \frac{\partial^{l+m+n}}{\partial x^l \partial y^m \partial z^n} \left(\frac{1}{r}\right) \tag{1}$$

constitute a set of solid harmonics of degree $l+m+n$. Those of given degree are not independent, since, for instance,

$$r^5 \frac{\partial^2}{\partial x^2}\frac{1}{r} + r^5 \frac{\partial^2}{\partial y^2}\frac{1}{r} + r^5 \frac{\partial^2}{\partial z^2}\frac{1}{r} = r^5 \nabla^2 \frac{1}{r} = 0. \tag{2}$$

It is easy, however, to obtain an independent set in terms of them. We take

$$K^s_{-n-1} = \left(\frac{\partial}{\partial x} + i\frac{\partial}{\partial y}\right)^s \left(\frac{\partial}{\partial z}\right)^{n-s} \frac{1}{r}. \tag{3}$$

This is a solid harmonic, being a linear combination of derivatives of $1/r$. Suppose that $z^2 > x^2 + y^2$. Then

$$\frac{1}{r} = \frac{1}{z}\left(1 + \frac{x^2+y^2}{z^2}\right)^{-1/2} = \frac{1}{z} + \sum_{m=1}^{\infty}(-1)^m \frac{\frac{1}{2}\cdot\frac{3}{2}\ldots m-\frac{1}{2}}{m!}\frac{(x^2+y^2)^m}{z^{2m+1}}. \tag{4}$$

Now
$$\left(\frac{\partial}{\partial x} + i\frac{\partial}{\partial y}\right) F(\varpi^2) = 2(x+iy)\frac{dF}{d\varpi^2}, \tag{5}$$

$$\left(\frac{\partial}{\partial x} + i\frac{\partial}{\partial y}\right) g(x+iy) = g'(x+iy) - g'(x+iy) = 0. \tag{6}$$

Successive operations $\dfrac{\partial}{\partial x} + i\dfrac{\partial}{\partial y}$ therefore introduce powers of $x+iy$, but further differentiation of these gives nothing; and

$$\left(\frac{\partial}{\partial z}\right)^n \frac{1}{r} = (-1)^n \frac{n!}{z^{n+1}} + \sum_{m=1}^{\infty}(-1)^{m+n}\frac{1.3\ldots 2m-1}{2^m m!}\frac{(2m+n)!}{(2m)!}\frac{(x^2+y^2)^m}{z^{2m+n+1}}, \tag{7}$$

$$\left(\frac{\partial}{\partial x} + i\frac{\partial}{\partial y}\right)^s \left(\frac{\partial}{\partial z}\right)^{n-s}\frac{1}{r}$$
$$= \Sigma(-1)^{n-s+m}\frac{1.3\ldots 2m-1}{2^m m!}\frac{2^s m!}{(m-s)!}\frac{(2m+n-s)!}{(2m)!}\frac{(x+iy)^s(x^2+y^2)^{m-s}}{z^{2m+n-s+1}} \quad (s>0). \tag{8}$$

Every term contains the factor $(x+iy)^s$ and therefore $e^{is\lambda}$. The lowest non-zero term has $m = s$, and reduces to

$$(-1)^n \frac{(n+s)!}{2^s s!}\frac{(x+iy)^s}{z^{n+s+1}}. \tag{9}$$

Hence
$$K^s_{-n-1} = (-1)^n\frac{(n+s)!}{2^s s!}\frac{\sin^s\theta\, e^{is\lambda}}{r^{n+1}}(1 + O(\sin^2\theta)). \tag{10}$$

Since this is a solid harmonic of degree $-n-1$ proportional to $e^{is\lambda}$, and of order $\sin^s\theta$ for θ small, $r^{n+1}K^s_{-n-1}$ must be a constant multiple of $\sin^s\theta\, e^{is\lambda}\dfrac{d^{n+s}}{d\mu^{n+s}}(\mu^2-1)^n$.

Differentiating by Leibniz's theorem and picking out the only term that does not vanish at $\mu = 1$ we have

$$\frac{d^{n+s}}{d\mu^{n+s}}\{(\mu-1)^n(\mu+1)^n\} = \frac{(n+s)!}{n!s!}n!\frac{n!}{(n-s)!}(\mu+1)^{n-s}+\dots$$

$$= \frac{2^{n-s}n!(n+s)!}{(n-s)!s!} + O(\sin^2\theta), \qquad (11)$$

and

$$r^{n+1}K^s_{-n-1} = \frac{(-1)^n(n-s)!}{2^n n!}\sin^s\theta\, e^{is\lambda}\frac{d^{n+s}}{d\mu^{n+s}}(\mu^2-1)^n. \qquad (12)$$

This form suggests the most convenient way of assigning the constant factors in the standard solutions. We shall take

$$p^s_n(\mu) = \frac{(n-s)!}{2^n(n!)^2}\sin^s\theta\,\frac{d^{n+s}}{d\mu^{n+s}}(\mu^2-1)^n \qquad (13)$$

The usual definition is to take the constant factor as $1/2^n n!$ and call the resulting function $P^s_n(\mu)$. The present form has one considerable advantage in symmetry. With it, let us see what happens if we replace s by $-s$. Since

$$\frac{\partial^2}{\partial z^2}\left(\frac{1}{r}\right) = -\left(\frac{\partial}{\partial x}+i\frac{\partial}{\partial y}\right)\left(\frac{\partial}{\partial x}-i\frac{\partial}{\partial y}\right)\frac{1}{r}, \qquad (14)$$

we can make the interpretation

$$\left(\frac{\partial}{\partial x}+i\frac{\partial}{\partial y}\right)^{-s}\frac{\partial^{n+s}}{\partial z^{n+s}}\left(\frac{1}{r}\right) = (-1)^s\left(\frac{\partial}{\partial x}-i\frac{\partial}{\partial y}\right)^s\frac{\partial^{n-s}}{\partial z^{n-s}}\left(\frac{1}{r}\right)$$

$$= (-1)^s K^{*s}_{-n-1}, \qquad (15)$$

where the asterisk denotes that we replace $i\lambda$ by $-i\lambda$. Now see whether we get the same relation by taking $-s$ for s in (13); we have

$$2^n(n!)^2 p^{-s}_n(\mu) = (n+s)!(1-\mu^2)^{-\frac{1}{2}s}\frac{d^{n-s}}{d\mu^{n-s}}\{(\mu-1)(\mu+1)\}^n$$

$$= (n+s)!(1-\mu^2)^{-\frac{1}{2}s}\sum_{m=0}^{n-s}\frac{(n-s)!}{m!(n-s-m)!}\frac{n!(\mu-1)^{n-m}}{(n-m)!}\frac{n!(\mu+1)^{s+m}}{(s+m)!},$$

where terms with $m > n-s$ vanish. Hence all terms contain $(\mu^2-1)^s$ as a factor, and the sum is

$$(-1)^s(n+s)!(1-\mu^2)^{\frac{1}{2}s}\sum_{m=0}^{n-s}\frac{(n-s)!n!n!(\mu-1)^{n-m-s}(\mu+1)^m}{m!(n-s-m)!(n-m)!(s+m)!}. \qquad (16)$$

But $\quad 2^n(n!)^2 p^s_n(\mu) = (n-s)!(1-\mu^2)^{\frac{1}{2}s}\sum_{m=s}^{n+s}\frac{(n+s)!}{m!(n+s-m)!}\frac{n!(\mu-1)^{n-m}}{(n-m)!}\frac{n!(\mu+1)^{-s+m}}{(-s+m)!} \qquad (17)$

and terms with $m < s$ vanish; putting $m = s+u$ we have

$$(n-s)!(1-\mu^2)^{\frac{1}{2}s}\sum_{u=0}^{n}\frac{(n+s)!}{(s+u)!(n-u)!}\frac{n!(\mu-1)^{n-s-u}}{(n-s-u)!}\frac{n!(\mu+1)^u}{u!} \qquad (18)$$

which is the same series as (16), terms with $u > n-s$ vanishing. Hence

$$p^{-s}_n(\mu) = (-1)^s p^s_n(\mu). \qquad (19)$$

In our original derivation we were not concerned with negative values of s, but we should naturally like $p_n^s \exp i(\gamma t - s\lambda)$ and $p_n^{-s} \exp i(\gamma t + s\lambda)$ to represent two waves of equal amplitude, passing around a sphere in opposite directions. With the present definition this condition is satisfied, and apart from it there appears to be no reason for considering negative s at all. With the usual definition, which omits the factor $(n-s)!/n!$, the amplitudes expressed by the usual solutions are very different. For $n = s = 4$ the ratio is $8! = 40320$. The factor $(-1)^s$ presents a minor difficulty in securing symmetry, and could be absorbed by including a factor i^s, but this would make the functions imaginary for odd s and does not seem worth while. C. G. Darwin* has already introduced the factor $(n-s)!$ into the definition for the sake of symmetry, but it seems best at the same time to divide by $n!$ so as to retain the usual standard solutions when $s = 0$.

Hobson associated the factor $(-1)^s$ with $P_n^s(\mu)$, not with $P_n^{-s}(\mu)$. In this respect he is followed by Condon and Shortley,[†] who use normalized functions.

A common modern procedure is to *normalize* the functions, that is, to introduce a constant factor so that the integral of the square of the function over the range used is 1. This device simplifies the writing of general proofs in, for instance, the theory of integral equations. But in a simple application it would mean that we must not use $\cos x$ and $\sin x$; we must use $(2/\pi)^{1/2} \cos x$ and $(2/\pi)^{1/2} \sin x$ if the range used is π, and $\pi^{-1/2} \cos x$ and $\pi^{-1/2} \sin x$ if the range is 2π. Presumably separate tables would be wanted in the two cases. For more complicated functions the normalizing factor introduces square roots everywhere, and especially it needlessly complicates the recurrence relations. When the range is infinite the normalizing integral may diverge $\left(\text{e.g.} \int_0^\infty x J_n^2(x)\, dx\right)$ and other devices are needed. We shall therefore take as the standard functions of the first kind (i.e., behaving like $\sin^s \theta$ near $\theta = 0$ and π)

$$p_n^s(\mu) = \frac{(n-s)!}{n!} P_n^s(\mu) = \frac{(n-s)!}{2^n (n!)^2} \sin^s \theta \frac{d^{n+s}}{d\mu^{n+s}} (\mu^2 - 1)^n, \tag{20}$$

and the corresponding solutions of Laplace's equation are

$$(r^n, r^{-n-1}) p_n^s(\mu) (\cos s\lambda, \sin s\lambda) \tag{21}$$

related to the solid harmonics of degree $-n-1$ by

$$K_{-n-1}^s = \left(\frac{\partial}{\partial x} + i\frac{\partial}{\partial y}\right)^s \left(\frac{\partial}{\partial z}\right)^{n-s} \left(\frac{1}{r}\right) = (-1)^n n! r^{-n-1} p_n^s(\mu) e^{is\lambda}. \tag{22}$$

The functions $p_n^0(\mu)$ are usually called the *Legendre functions or polynomials*, or *zonal harmonics*; it is usual to suppress the explicit mention of s when it is zero. p_n^s is called an *associated Legendre function*, and $p_n^s(\cos s\lambda, \sin s\lambda)$ *tesseral harmonics*, apparently after a kind of dice known to the Romans. If $s = n$ the tesseral harmonic is called a *sectorial harmonic*.

It is important to have a general idea of the appearance of the functions. By the general principle that the zeros of a derivative of a continuous function separate those of the function, since $(\mu^2 - 1)^n$ has n zeros at $+1$ and n at -1, its first derivative has $n-1$ at each of these values and one between, its second two between -1 and $+1$, and p_n has n,

---

* *Proc. Roy. Soc.* A, 118, 1928, 668.
† *The Theory of Atomic Spectra*, 1935, 52.

all real and between $+1$ and -1. Powers of $\sin\theta$ never vanish except at $\mu = \pm 1$ in this range, and for $s > 0$, if we do not count the zeros at $\theta = 0$ and π, p_n^s will have $n-s$ zeros, all real. Thus the zonal harmonics keep the same sign each over $n+1$ belts, as between parallels of latitude, counting each polar cap as a belt; each increase of s by 1 reduces the number of parallels where the harmonic vanishes by 1, but increases by 2 the number of meridians where it does so. We see easily from

$$p_n(\mu) = \left[\frac{1}{2^n n!}\frac{d^n}{d\mu^n}\{(\mu-1)^n(\mu+1)^n\}\right]$$

on differentiating by Leibniz's theorem that all terms but one vanish at $\mu = \pm 1$, giving

$$p_n(1) = 1, \quad p_n(-1) = (-1)^n. \tag{23}$$

Actually $|p_n(\mu)|$ never exceeds 1. For μ near 1 the lowest term in p_n^s is got by taking $u = n-s$ in (18); we get

$$p_n^s = \frac{(n+s)!}{2^s n! s!}\sin^s\theta(1+O(\sin^2\theta)). \tag{24}$$

24·05. Expansion of $(r^2-2rh\cos\theta+h^2)^{-1/2}$: Green's function for a sphere. The harmonics with $s = 0$ are particularly important, since many disturbances are symmetrical about an axis. We have

$$\left(\frac{\partial}{\partial z}\right)^n \frac{1}{r} = (-1)^n n! r^{-n-1} p_n(\mu). \tag{1}$$

Consider the function

$$\frac{1}{R} = \frac{1}{\{x^2+y^2+(z-h)^2\}^{1/2}} = \frac{1}{(r^2-2rh\cos\theta+h^2)^{1/2}}, \tag{2}$$

which has a convergent expansion in negative powers of r if $h < r$. But since R involves z and h only through $z-h$,

$$\left(\frac{\partial}{\partial h}\right)^n \frac{1}{R} = \left(-\frac{\partial}{\partial z}\right)^n \frac{1}{R}, \tag{3}$$

and by Taylor's theorem

$$\frac{1}{R} = \Sigma \frac{h^n}{n!}\left\{\left(-\frac{\partial}{\partial z}\right)^n \frac{1}{R}\right\}_{h=0} = \Sigma \frac{h^n}{r^{n+1}}p_n(\mu). \tag{4}$$

This expansion is often taken as providing the definition of $P_n(\mu)$ ($= p_n(\mu)$). The explicit form of $P_n(\mu)$ can be found from it quite easily by Lagrange's expansion.* But there are few practical cases where this expansion yields explicit expressions for the general term without great difficulty, and we prefer to regard it as an existence theorem, in spite of this solitary instance of its use. In practice the associated functions are extremely important, and it seems best to have a definition that can deal with them from the start.

If $0 < \alpha < 1$,

$$(1-2\alpha\cos\theta+\alpha^2)^{-1/2} = (1-\alpha e^{i\theta})^{-1/2}(1-\alpha e^{-i\theta})^{-1/2}$$

$$= \left(1+\tfrac{1}{2}\alpha e^{i\theta}+\frac{\tfrac{1}{2}\cdot\tfrac{3}{2}}{2!}\alpha^2 e^{2i\theta}+\ldots\right)\left(1+\tfrac{1}{2}\alpha e^{-i\theta}+\frac{\tfrac{1}{2}\cdot\tfrac{3}{2}}{2!}\alpha^2 e^{-2i\theta}+\ldots\right), \tag{5}$$

and the coefficient of α^n is a sum of cosines, all with positive coefficients. It is therefore greatest numerically if they are all ± 1; hence

$$|p_n(\cos\theta)| \leq p_n(1) = 1. \tag{6}$$

---

* Cf. Jeans, *Electricity and Magnetism*, 1908, 215.

A related series of much importance is

$$S = \sum_{n=0}^{\infty} (2n+1) \frac{h^n}{r^{n+1}} P_n(\mu). \tag{7}$$

Differentiating (4) and multiplying by h we have

$$h \frac{\partial}{\partial h} \frac{1}{R} = \Sigma n \frac{h^n}{r^{n+1}} P_n(\mu) \quad (h < r),$$

and therefore

$$S = \left(2h \frac{\partial}{\partial h} + 1\right) \frac{1}{(r^2 - 2rh\cos\theta + h^2)^{1/2}} = \frac{r^2 - h^2}{(r^2 - 2rh\cos\theta + h^2)^{3/2}}, \tag{8}$$

which is the function that arises in the determination of a potential function given its values on a sphere (6·091, 6·092). If $h > r$ the same expansions hold with the exception that h and r must be interchanged.

24·06. Potential outside matter. To the theorem about expansibility of a potential function in a sphere within a cavity corresponds one about expansibility outside a sphere that contains the whole of the matter whose potential is being considered. Let $P(x_i)$ be a point outside such a sphere, the origin being the centre, and $Q(\xi_i)$ the position of a mass dm; then the potential is $\gamma \int dm/R$, where

$$R^2 = (x_i - \xi_i)^2, \quad \rho^2 = \xi_i^2, \tag{1}$$

$$\frac{1}{R} = \Sigma \frac{\rho^n}{r^{n+1}} P_n(\cos\vartheta), \tag{2}$$

which is uniformly convergent on and outside the sphere, and therefore can be multiplied by dm and integrated term by term. Thus the potential is developed in a series of negative powers of r, which can be differentiated term by term as often as we like provided that r is greater than every value of ρ. Hence $\nabla^2 \phi$ is another convergent series, and is identically zero, and the terms in ϕ of every degree separately must satisfy $\nabla^2 \phi = 0$. But if $r^{-n-1} f(\theta, \lambda)$ satisfies Laplace's equation so does $r^n f(\theta, \lambda)$, and therefore $f(\theta, \lambda)$ is linearly expressible in terms of our solutions. Hence ϕ can be expressed by a series of the form

$$\phi = \frac{a_0}{r} + \sum_{n=1}^{\infty} \sum_{s=0}^{n} \frac{1}{r^{n+1}} P_n^s(\mu)(a_{ns}\cos s\lambda + b_{ns}\sin s\lambda). \tag{3}$$

The condition that r is greater than every value of ρ is sufficient for the existence of this expansion, but not necessary. Consider any distribution of matter within a surface S, the maximum of ρ on which is c. Take a further surface S' outside it. The field outside S' is the same, by the theorem of the equivalent stratum, as that of a suitable distribution of sources and doublets over S', which does not need to be a sphere. Let a be the maximum, b the minimum, of ρ on S'. If the condition that r must be greater than every value of ρ was necessary to the existence of an expansion of the form (3), the expansion of the potential due to the distribution on S' would exist only for $r > a$. But it is the same as the potential due to the distribution within S, which has an expansion for all $r > c$. If then $b > c$ the expansion will exist right down to S' even though S' is not a sphere. It is therefore possible for the potential outside a surface with matter on it to have an expansion in negative powers of r without the surface being a sphere, so that the expansion is being applied at some places where r is less than the largest value of ρ. This feature in potential theory is the analogue of analytic continuation in the theory of the complex variable; the expansion will exist if the potential outside S' is the same as that due to some distribution of matter within a surface inside S' such that the largest value of ρ on this surface is less than the smallest on S'. It is particularly important in problems relating to boundaries that are not exact spheres, in particular in the theory of the figure of the Earth. The external potential can then often be continued into the body, but the result will not be the actual potential in the body, since the continuation of the external potential, if it exists, will satisfy Laplace's equation and the actual potential within the body will not. There is also an analogue of singularities. In the case of a charged sphere with a projecting point on it there will in general be a local concentration of charge on the projection, and the potential due to this cannot be represented by that due to any internal distribution.

24·07. Orthogonality relations: expansion theorem. Given that an expansion exists, it can be determined by considering the values of the function over a sphere. *All our standard solutions are mutually orthogonal in the sense that the product of any two of them, multiplied by the surface element dS, and integrated over a sphere, gives* 0. First, let S_m and S_n be any two surface harmonics of different degrees. Then $\phi_m = r^m S_m$ and $\phi_n = r^n S_n$ satisfy Laplace's equation. Therefore if we apply Green's theorem to a sphere of radius a

$$\iint \left(\phi_m \frac{\partial \phi_n}{\partial r} - \phi_n \frac{\partial \phi_m}{\partial r} \right) dS = 0. \tag{1}$$

But this is the same as

$$(m-n) a^{m+n-1} \iint S_m S_n dS, \tag{2}$$

and the first factor cannot vanish if $m \neq n$. Hence if $m \neq n$

$$\iint S_m S_n dS = 0. \tag{3}$$

Also, for harmonics of the same degree, any pair of $p_n^s \cos s\lambda$, $p_n^s \sin s\lambda$, $p_n^t \cos t\lambda$, $p_n^t \sin t\lambda$ are orthogonal since the integral with regard to λ vanishes, except in the case where $s = t$ and we take either the cosine factor in both cases or the sine factor in both and are therefore integrating the square of a harmonic.

Since we can take $S_m = p_m^s \cos s\lambda$, $S_n = p_n^s \cos s\lambda$, it follows that if $m \neq n$

$$\iint p_m^s p_n^s \cos^2 s\lambda \sin\theta \, d\theta \, d\lambda = 0, \tag{4}$$

and therefore

$$\int_{-1}^{1} p_m^s p_n^s d\mu = 0 \quad (m \neq n). \tag{5}$$

Linear independence between the standard harmonics follows immediately, though we have already verified it by another method. For if we denote any of our harmonics by Y_p and there was a general relation

$$\Sigma a_p Y_p = 0,$$

we could multiply by any Y_q with a non-zero coefficient and integrate over the sphere, and the result would be 0. But every term separately gives 0 by the orthogonality relations except Y_q, which gives $a_q \iint Y_q^2 dS$, and this cannot vanish since by hypothesis $a_q \neq 0$. Hence the assumption of any linear relation between the harmonics leads to a contradiction.

24·07 Expansion theorem

It also follows that *every zonal harmonic p_n is orthogonal to every polynomial in μ of lower degree*. For if $f(\mu)$ is a polynomial in μ of degree $m < n$, the term of highest degree being $a_m \mu^m$, we can subtract such a multiple of p_m as will remove this term. We can then subtract such a multiple of p_{m-1} as will remove the term in μ^{m-1} in the remainder, and so proceed. The process ends in $m+1$ steps and a sum of multiples of zonal harmonics of degrees $\leq m$ is found that is identically equal to $f(\mu)$. But each of these harmonics has degree different from n and therefore is orthogonal to p_n. Hence

$$\int_{-1}^{1} f(\mu)\, p_n(\mu) = 0.$$

Now suppose that a function $f(\theta, \lambda)$ has an expansion in spherical harmonics, so that

$$f(\theta, \lambda) = \sum_{n=0}^{\infty} \sum_{s=0}^{n} (a_{ns} p_n^s \cos s\lambda + b_{ns} p_n^s \sin s\lambda), \tag{6}$$

where we suppose that $f(\theta, \lambda)$ is known, and require to determine the coefficients a_{ns}, b_{ns}, assuming that the expansion exists. Multiplying by the respective harmonics and integrating with respect to $\sin\theta\, d\theta\, d\lambda$ we have, by the orthogonality relations,

$$a_{n0} \int_{-1}^{1} \int_{0}^{2\pi} (p_n)^2 d\mu\, d\lambda = \int_{-1}^{1} \int_{0}^{2\pi} f(\theta, \lambda)\, d\mu\, d\lambda, \tag{7}$$

$$a_{ns} \int_{-1}^{1} \int_{0}^{2\pi} (p_n^s)^2 \cos^2 s\lambda\, d\mu\, d\lambda = \int_{-1}^{1} \int_{0}^{2\pi} f(\theta, \lambda)\, p_n^s \cos s\lambda\, d\mu\, d\lambda, \tag{8}$$

$$b_{ns} \int_{-1}^{1} \int_{0}^{2\pi} (p_n^s)^2 \sin^2 s\lambda\, d\mu\, d\lambda = \int_{-1}^{1} \int_{0}^{2\pi} f(\theta, \lambda)\, p_n^s \sin s\lambda\, d\mu\, d\lambda. \tag{9}$$

The coefficients are therefore determined in terms of definite integrals. For those on the left, integration with regard to λ gives 2π or π. Also

$$\int_{-1}^{1} (p_n)^2 d\mu = \frac{1}{2^{2n}(n!)^2} \int_{-1}^{1} \left\{ \frac{d^n}{d\mu^n} (\mu^2 - 1)^n \right\}^2 d\mu. \tag{10}$$

Integrate by parts n times; the integrated parts all vanish at both limits and we are left with

$$\frac{1}{2^{2n}(n!)^2} (-1)^n \int_{-1}^{1} (\mu^2 - 1)^n \frac{d^{2n}}{d\mu^{2n}} (\mu^2 - 1)^n d\mu = \frac{(2n)!}{2^{2n}(n!)^2} (-1)^n \int_{-1}^{1} (\mu^2 - 1)^n d\mu$$

$$= \frac{(2n)!}{2^{2n}(n!)^2} 2 \int_{0}^{\frac{1}{2}\pi} \sin^{2n+1}\theta\, d\theta$$

$$= \frac{2(2n)!}{2^{2n}(n!)^2} \frac{2n \cdot 2n-2 \ldots 2}{2n+1 \cdot 2n-1 \ldots 1} = \frac{2}{2n+1}. \tag{11}$$

The corresponding integration for $(p_n^s)^2$ can be simplified a little by using p_n^{-s}. We have

$$\int_{-1}^{1} (p_n^s)^2 d\mu = (-1)^s \int_{-1}^{1} p_n^s p_n^{-s} d\mu$$

$$= \frac{(-1)^s}{2^{2n}(n!)^4} \int_{-1}^{1} (n-s)! \sin^s\theta \frac{d^{n+s}}{d\mu^{n+s}} (\mu^2-1)^n (n+s)! \sin^{-s}\theta \frac{d^{n-s}}{d\mu^{n-s}} (\mu^2-1)^n d\mu$$

$$= \frac{(-1)^s (n-s)!(n+s)!}{2^{2n}(n!)^4} \int_{-1}^{1} \frac{d^{n+s}}{d\mu^{n+s}} (\mu^2-1)^n \frac{d^{n-s}}{d\mu^{n-s}} (\mu^2-1)^n d\mu.$$

Integrate by parts s times; we get

$$\frac{(n-s)!\,(n+s)!}{2^{2n}(n!)^4}\int_{-1}^{1}\left\{\frac{d^n}{d\mu^n}(\mu^2-1)^n\right\}^2 d\mu = \frac{(n-s)!\,(n+s)!}{(n!)^2}\int_{-1}^{1}(p_n)^2 d\mu = \frac{2}{2n+1}\frac{(n-s)!\,(n+s)!}{(n!)^2}. \tag{12}$$

Hence
$$\int_{0}^{\pi}\int_{0}^{2\pi}(p_n)^2 \sin\theta\,d\theta\,d\lambda = \frac{4\pi}{2n+1}, \tag{13}$$

$$\int_{0}^{\pi}\int_{0}^{2\pi}(p_n^s)^2\,(\cos^2 s\lambda,\sin^2 s\lambda)\sin\theta\,d\theta\,d\lambda = \frac{2\pi}{2n+1}\frac{(n-s)!\,(n+s)!}{(n!)^2}, \tag{14}$$

whence the coefficients can be found from (7), (8), (9).

It will be noticed that this expansion has the property, like the Fourier expansion and all other expansions in orthogonal functions, that if Σ is the sum of any finite number of terms of a series of harmonics, with arbitrary coefficients, and we adjust the coefficients so as to make $\iint (f(\theta,\lambda)-\Sigma)^2 d\mu d\lambda$ a minimum, the resulting coefficients are the coefficients a_{ns}, b_{ns}. There is an immediate analogue of Parseval's theorem,

$$\iint\{f(\theta,\lambda)\}^2 d\mu d\lambda = \sum_{n=0}^{\infty}\sum_{s=0}^{n} a_{ns}^2 \iint (p_n^s)^2 \cos^2 s\lambda\,d\mu d\lambda + \sum_{n=0}^{\infty}\sum_{s=0}^{n} b_{ns}^2 \iint (p_n^s)^2 \sin^2 s\lambda\,d\mu d\lambda,$$

expressing that the mean square of f over the sphere is the sum of the mean squares of its harmonic components.

24·071. The above argument assumes that the expansion exists. We have proved this only for the potential over a sphere such that matter is either all exterior or all interior to it. Extensions to more general forms of $f(\theta,\lambda)$ can be made in various ways, as for Fourier series. A proof, on the supposition that $f(\theta,\lambda)$ has continuous second derivatives, is given by Courant and Hilbert.* If $f(\theta,\lambda)$ does not satisfy this condition, but nevertheless can be uniformly approximated to over the sphere, except possibly in a set of points capable of being enclosed within an arbitrarily small total area, by functions that do satisfy it, it will follow immediately that a series of the form (6) exists that will agree with $f(\theta,\phi)$ to any assignable accuracy, except in the region excluded. Such a set of functions can be assigned in many ways; one is by an extension of the argument of 14·08, but perhaps the simplest is to note that if $f(\theta,\lambda)$ is the potential on a sphere of radius a, and $\iint f(\theta,\lambda)\sin\theta\,d\theta\,d\lambda$ over the sphere exists, we can take a set of interior concentric spheres of radii $a-\delta_n$, where $\delta_n \to 0$, and the potentials over these spheres have derivatives of all orders. Further, by a similar argument to that of 14·05, we can show that as $\delta_n \to 0$ the potentials on these spheres tend uniformly to $f(\theta,\lambda)$ in any closed region of θ, λ such that $f(\theta,\lambda)$ is continuous. Consequently a sufficient condition that $f(\theta,\lambda)$ can be approximated to by a series of surface harmonics almost everywhere is that it shall be integrable over the sphere.

As for Fourier series, this type of approximation is possible in some cases where there is no expansion of the form 24·07 (6). Conditions that 24·07 (6), with coefficients given by (7), (8), (9), may converge to $f(\theta,\lambda)$ are more difficult to state than for Fourier series if $f(\theta,\lambda)$ has discontinuities.

* *Methoden der Mathematischen Physik*, 1, 1924, 421–22.

24·08. Explicit forms of the functions p_n^s, up to $n = 4$, are as follows. The polynomials obtained by multiplying by $r^n(\cos s\lambda, \sin s\lambda)$, and the mean square values of the functions, associated with their factors in λ, over a sphere are also given.

n	s	p_n^s	Mean square	Polynomials
0	0	1	1	1
1	0	$\cos\theta$	$\frac{1}{3}$	z
	1	$\sin\theta$	$\frac{1}{3}$	x, y
2	0	$\frac{3}{2}\cos^2\theta - \frac{1}{2}$	$\frac{1}{5}$	$\frac{1}{2}(2z^2 - x^2 - y^2)$
	1	$\frac{3}{2}\cos\theta\sin\theta$	$\frac{3}{20}$	$\frac{3}{2}zx, \frac{3}{2}zy$
	2	$\frac{3}{2}\sin^2\theta$	$\frac{3}{5}$	$\frac{3}{2}(x^2-y^2), 3xy$
3	0	$\frac{5}{2}\cos^3\theta - \frac{3}{2}\cos\theta$	$\frac{1}{7}$	$\frac{1}{2}z(2z^2 - 3x^2 - 3y^2)$
	1	$\frac{1}{2}\sin\theta(5\cos^2\theta - 1)$	$\frac{3}{21}$	$\frac{1}{2}x(4z^2 - x^2 - y^2), \frac{1}{2}y(4z^2 - x^2 - y^2)$
	2	$\frac{5}{2}\sin^2\theta\cos\theta$	$\frac{5}{21}$	$\frac{5}{2}z(x^2-y^2), 5xyz$
	3	$\frac{5}{2}\sin^3\theta$	$\frac{10}{7}$	$\frac{5}{2}(x^3 - 3xy^2), \frac{5}{2}(3x^2y - y^3)$
4	0	$\frac{1}{8}(35\cos^4\theta - 30\cos^2\theta + 3)$	$\frac{1}{9}$	$\frac{1}{8}\{8z^4 - 24z^2(x^2+y^2) + 3(x^2+y^2)^2\}$
	1	$\frac{5}{8}\sin\theta(7\cos^3\theta - 3\cos\theta)$	$\frac{5}{72}$	$\frac{5}{8}x\{4z^3 - 3z(x^2+y^2)\}, \frac{5}{8}y\{(4z^3 - 3z(x^2+y^2)\}$
	2	$\frac{5}{8}\sin^2\theta(7\cos^2\theta - 1)$	$\frac{5}{36}$	$\frac{5}{8}(x^2-y^2)(6z^2 - x^2 - y^2), \frac{5}{4}xy(6z^2 - x^2 - y^2)$
	3	$\frac{35}{8}\sin^3\theta\cos\theta$	$\frac{35}{72}$	$\frac{35}{8}z(x^3 - 3xy^2), \frac{35}{8}z(3x^2y - y^3)$
	4	$\frac{35}{8}\sin^4\theta$	$\frac{35}{9}$	$\frac{35}{8}(x^4 - 6x^2y^2 + y^4), \frac{35}{2}xy(x^2 - y^2)$

In the mean square values, when $s \neq 0$, the factor $\frac{1}{2}$ obtained by averaging $\cos^2 s\lambda$ or $\sin^2 s\lambda$ has been taken into account.

24·09. Analogue of Laurent's theorem. The theorems of 24·03 and 24·06, relating to the expansions of the potential in a spherical cavity or outside a sphere, can be extended immediately to the case where ϕ satisfies Laplace's equation in the region between two spheres, one inside the other. We can apply the theorem of the equivalent stratum to the region in question: the potential will be the sum of those due to distributions over both the inner and the outer spheres, and can be represented by a series of solid harmonics, but these will now include both positive and negative powers of r. This is the spherical analogue of Laurent's theorem, and has been much used in terrestrial magnetism. Part of the variable part of the magnetic field at the Earth's surface is due to electric (ionization) currents in the upper atmosphere, part to currents in the Earth. The former will give a potential at the surface expressible by a series of solid harmonics of positive degrees, the latter a series of negative degrees. The variation of the potential over the surface can be found by integrating the horizontal intensity of magnetic force, and the vertical intensity can be measured directly. Now if the potential is

$$\phi = \Sigma (A_n r^n + B_n r^{-n-1}) S_n,$$

the vertical intensity is

$$-\frac{\partial\phi}{\partial r} = \Sigma\{-nA_n r^{n-1} + (n+1) B_n r^{-n-2}\} S_n.$$

The terms in each harmonic for r equal to the radius of the Earth being found from observation, the coefficients give a pair of equations for A_n and B_n, from which it can be determined how much of the field is due to external and how much to internal currents.

24·10. Recurrence formulae. From

$$(1 - 2\mu\alpha + \alpha^2)^{-1/2} = \Sigma p_n(\mu) \alpha^n, \tag{1}$$

by differentiation with respect to α, we get

$$\frac{\mu-\alpha}{(1-2\mu\alpha+\alpha^2)^{3/2}} = \Sigma n p_n \alpha^{n-1}. \tag{2}$$

Multiply (1) by $(\mu-\alpha)$ and (2) by $1-2\mu\alpha+\alpha^2$, and compare coefficients of α^n. We find

$$(n+1)p_{n+1} - (2n+1)\mu p_n + n p_{n-1} = 0, \tag{3}$$

a recurrence relation connecting three consecutive zonal harmonics.

Differentiating (1) with respect to μ we have

$$\frac{\alpha}{(1-2\mu\alpha+\alpha^2)^{3/2}} = \Sigma \frac{dp_n}{d\mu} \alpha^n, \tag{4}$$

and comparison with (2) leads to

$$\mu \frac{dp_n}{d\mu} - \frac{dp_{n-1}}{d\mu} = n p_n. \tag{5}$$

If we now differentiate (3) and eliminate $\mu\, dp_n/d\mu$ we get

$$(2n+1)p_n = \frac{dp_{n+1}}{d\mu} - \frac{dp_{n-1}}{d\mu}, \tag{6}$$

and hence

$$(2n+1)\int_1^\mu p_n d\mu = p_{n+1} - p_{n-1}. \tag{7}$$

These can be generalized by differentiation to give recurrence relations between the sth derivatives, and hence between the p_n^s.*

$$(2n+1)\frac{d^{s-1}p_n}{d\mu^{s-1}} = \frac{d^s p_{n+1}}{d\mu^s} - \frac{d^s p_{n-1}}{d\mu^s}, \tag{8}$$

$$(n-s+1)\frac{d^s p_{n+1}}{d\mu^s} = (2n+1)\mu\frac{d^s p_n}{d\mu^s} - (n+s)\frac{d^s p_{n-1}}{d\mu^s}. \tag{9}$$

Direct relations between the p_n^s are probably less convenient, since it is desirable to keep the $\sin^s\theta$ factor outside the differentiation. The following formula, however, is easily proved from 24·04 (3) and has the peculiarity that s does not appear in the coefficients:

$$p_{n+1}^s = \mu p_n^s - \frac{1-\mu^2}{n+1}\frac{dp_n^s}{d\mu}. \tag{10}$$

Other recurrence relations, the proofs of which present no difficulty, are as follows:

$$(1-\mu^2)\frac{dp_n}{d\mu} = \int_\mu^1 n(n+1)p_n d\mu = \frac{n(n+1)}{2n+1}(p_{n-1}-p_{n+1})$$
$$= (n+1)(\mu p_n - p_{n+1}) = n(p_{n-1}-\mu p_n).$$

If $f(\mu) = 0$ for $\mu < \alpha$, and $= 1$ for $\mu > \alpha$, where $|\alpha| < 1$,

$$\int_{-1}^1 f(\mu) p_n(\mu) d\mu = \frac{1}{2n+1}\{p_{n-1}(\alpha) - p_{n+1}(\alpha)\} \quad (n \geq 1),$$

$$\int_{-1}^1 f(\mu) d\mu = 1 - \alpha,$$

* Adams, *Collected Scientific Papers*, 2, 243–96.

and the expansion of $f(\mu)$ in Legendre polynomials is

$$\tfrac{1}{2}(1-\alpha) + \tfrac{1}{2}\sum_{n=1}^{\infty}\{p_{n-1}(\alpha) - p_{n+1}(\alpha)\}p_n(\mu).$$

We can now give a few examples of the use of spherical harmonics.*

24·11. Potential due to a uniform circular disk. We have seen (6·032) that the potential on the axis is

$$\phi = 2\pi\gamma\sigma\{\sqrt{(b^2+x^2)} - |x|\}, \tag{1}$$

where b is the radius of the disk and x is the distance from the plane of the disk. Replace x by r and expand in descending powers of r; this is valid provided $r > b$.

$$\begin{aligned}\phi &= 2\pi\gamma\sigma r\left\{\frac{1}{2}\frac{b^2}{r^2} + \frac{\tfrac{1}{2}(-\tfrac{1}{2})}{2!}\frac{b^4}{r^4} + \ldots + \frac{\tfrac{1}{2}(-\tfrac{1}{2})\ldots(\tfrac{3}{2}-m)}{m!}\frac{b^{2m}}{r^{2m}} + \ldots\right\}\\ &= \pi\gamma\sigma b^2\left\{\frac{1}{r} - \frac{\tfrac{1}{2}}{2!}\frac{b^2}{r^3} + \ldots + \frac{(-\tfrac{1}{2})\ldots(\tfrac{3}{2}-m)}{m!}\frac{b^{2m-2}}{r^{2m-1}} + \ldots\right\}.\end{aligned} \tag{2}$$

We now interpret r as the distance from the centre of the disk and introduce the polar coordinate θ measured from the axis of the disk. The above form of ϕ is correct only for $\theta = 0$. But we know that the potential for $r > b$ is expansible in a series of the form $\Sigma A_m r^{-m-1}S_m$, and by symmetry it must depend on r, θ only. Hence for every value of r the expression of S_m in terms of the standard functions can contain only p_m. Further, $p_m = 1$ for $\theta = 0$, and therefore the only form of ϕ that (1) satisfies Laplace's equation, (2) has symmetry about the axis, (3) reduces to (2) on the axis, is

$$\phi = \pi\gamma\sigma b^2\left(\frac{1}{r} - \frac{1}{2.2!}\frac{b^2}{r^3}p_2 + \ldots + \frac{(-\tfrac{1}{2})\ldots(\tfrac{3}{2}-m)}{m!}\frac{b^{2m-2}}{r^{2m-1}}p_{2m-2} + \ldots\right). \tag{3}$$

If $r < b$ a similar expansion in *ascending* powers of r is possible.

24·111. Potential due to given surface density over a sphere. By the expansion theorem we can express σ as a series of spherical harmonics

$$\sigma = \sum_{n=0}^{\infty}\sum_{s=0}^{n} a_{ns}p_n^s\cos s\lambda + \sum_{n=0}^{\infty}\sum_{s=1}^{n} b_{ns}p_n^s\sin s\lambda, \tag{1}$$

which we can write shortly as $\Sigma c_{ns}S_{ns}$. The potentials inside and outside the sphere are expressible by series

$$\phi_1 = \Sigma A_{ns}\left(\frac{r}{a}\right)^n S_{ns}, \quad \phi_0 = \Sigma B_{ns}\left(\frac{r}{a}\right)^{-n-1} S_{ns}. \tag{2}$$

The potential is continuous on crossing the sphere. Hence when $r = a$ both ϕ_0 and ϕ_1 must reduce to the potential on the sphere, and $A_{ns} = B_{ns}$ by the expansion theorem. Again, the discontinuity in $\partial\phi/\partial n$ is $-4\pi\gamma\sigma$; that is,

$$\left(\frac{\partial\phi_0}{\partial r}\right)_{r\to a} - \left(\frac{\partial\phi_1}{\partial r}\right)_{r\to a} = -4\pi\gamma\sigma.$$

---

* Integrals of products of three spherical harmonics are given by Adams, loc. cit. pp. 343–400; J. A. Gaunt, *Phil. Trans.* A, **228**, 1929, 192–6.

Again using the expansion theorem, we can equate coefficients of all the harmonics, and

$$-(2n+1)\frac{A_{ns}}{a} = -4\pi\gamma c_{ns},$$

$$A_{ns} = B_{ns} = \frac{4\pi\gamma a c_{ns}}{2n+1}.$$

Thus ϕ_0 and ϕ_1 are completely determined. The treatment of all harmonics with the same n is exactly similar; S_{ns} is therefore usually replaced simply by S_n. But if we are given the values of σ over a sphere it will be necessary in any case to perform the expansion with regard to s as well as n before we have the answer.

24·112. Potential due to given density within a sphere. We take

$$\rho = \Sigma \rho_{ns} S_{ns},$$

where ρ_{ns} will now be a function of r', the distance from the centre. Also

$$\phi_1 = \Sigma A_{ns}\left(\frac{r}{a}\right)^n S_{ns}, \quad \phi_0 = B_{ns}\left(\frac{a}{r}\right)^{n+1} S_{ns},$$

where A_{ns} will be a function of r, but B_{ns} will be a constant since ϕ_0 must satisfy Laplace's equation. Two methods are available. We can use the conditions that ϕ_1 must satisfy Poisson's equation and ϕ and $\partial\phi/\partial r$ must be continuous at $r = a$. Alternatively, the shell between r' and $r'+dr'$ can be regarded as a surface distribution of density $\rho dr'$. The potential due to this, for $r < r'$, is

$$\Sigma \frac{4\pi\gamma}{2n+1}\rho_{ns} S_{ns} r' dr' \left(\frac{r}{r'}\right)^n,$$

and for $r > r'$, is

$$\Sigma \frac{4\pi\gamma}{2n+1}\rho_{ns} S_{ns} r' dr' \left(\frac{r'}{r}\right)^{n+1}.$$

Adding up for all shells,

$$\phi_0 = 4\pi\gamma \int_0^a \Sigma \frac{\rho_{ns} S_{ns}}{2n+1} \frac{r'^{n+2}}{r^{n+1}} dr',$$

$$\phi_1 = 4\pi\gamma \int_0^r \Sigma \frac{\rho_{ns} S_{ns}}{2n+1} \frac{r'^{n+2}}{r^{n+1}} dr' + 4\pi\gamma \int_r^a \Sigma \frac{\rho_{ns} S_{ns}}{2n+1} \frac{r^n}{r'^{n-1}} dr'.$$

24·113. Potential of a nearly spherical conductor. We take the equation of the conductor to be

$$r = a\left(1 + \sum_{n=1}^{\infty} \sum_{s=0}^{n} \epsilon_{ns} S_{ns}\right), \tag{1}$$

where the ϵ_{ns} are constants, small enough for their squares to be neglected. We assume also

$$\phi_0 = A_{00}\frac{a}{r} + \Sigma A_{ns}\left(\frac{a}{r}\right)^{n+1} S_{ns}. \tag{2}$$

ϕ_1 will of course be a constant since the surface is supposed to be at uniform potential v. Since ϕ_0 would reduce to its first term if all the ϵ_{ns} were zero we can suppose that all the A_{ns} ($n > 0$) are small of the same order of magnitude as the ϵ_{ns}. There is a difficulty here about the substitution of (1) into (2) directly, because (2) is not necessarily true if r is less than *some* value of r on the conductor. But we can take a sphere $r = a+h$, where h is positive and just large enough for the sphere to enclose the whole of the conductor. (2) holds outside this. Outside the conductor ϕ is continuous, and so are all its derivatives. Hence on the conductor

$$\phi = (\phi_0)_{r=a+h} - \left(\frac{\partial \phi_0}{\partial r}\right)_{r=a+h}(a+h-r) + o(h), \qquad (3)$$

$$= \Sigma A_{ns}\left(\frac{a}{a+h}\right)^{n+1} S_{ns} + \Sigma (n+1) A_{ns} \frac{a^{n+1}}{(a+h)^{n+2}}(a+h-r) S_{ns} + o(h)$$

$$= \Sigma A_{ns}\left(\frac{a}{r}\right)^{n+1} S_{ns} + o(h), \qquad (4)$$

and we *can* apply (2) down to the conductor with a negligible error. Hence

$$v = \Sigma A_{ns}\{1 - \Sigma(m+1)\epsilon_{ml}S_{ml}\}S_{ns}$$
$$= A_{00}\{1 - \Sigma(n+1)\epsilon_{ns}S_{ns}\} + \Sigma A_{ns}S_{ns} + o(\Sigma \epsilon^2), \qquad (5)$$

for all θ, λ. Therefore $A_{ns} = (n+1)\epsilon_{ns}A_{00}$, $A_{00} = v$,

$$\phi_0 = v\left\{\frac{a}{r} + \Sigma (n+1)\left(\frac{a}{r}\right)^{n+1}\epsilon_{ns}S_{ns}\right\}, \qquad (6)$$

to the first order in the departures from a sphere.

24·114. The figure of the Earth. In the problems just considered the form of the surface and the values of ϕ or $\partial \phi/\partial r$ over it are enough to determine ϕ at external points. We now come to a problem where the form of the surface itself is to be found, but we know a great deal already about *both* ϕ and $\partial \phi/\partial r$ over it. The Earth is not quite a sphere, the chief departures being the ellipticity and the elevations and depressions of the solid surface above and below sea level. The distribution of density inside is not known directly, but a great deal can be found out about the gravitational field outside the earth and about its external form from observations of gravity at the solid surface. The departures of the outer surface, gravity, and the gravitation potential from symmetry about the centre are small enough for their squares to be neglected in a first approximation, the range of each being of the order of 1/200 or 1/300 of the mean value.

The external gravitational potential can be written

$$U = \frac{fM}{r} + U', \qquad (1)$$

where f is the constant of gravitation, M the mass of the Earth, and U' satisfies Laplace's equation and tends to zero like r^{-3} for large r provided that the centre of mass is taken as origin. The acceleration of a free particle is grad U. But the solid earth is rotating with angular velocity ω, and each particle of it has component accelerations relative to non-rotating axes

$$(-\omega^2 x, -\omega^2 y, 0) = -\operatorname{grad} \tfrac{1}{2}\omega^2(x^2+y^2) = -\operatorname{grad} \tfrac{1}{2}\omega^2 r^2 \sin^2\theta. \qquad (2)$$

Hence the difference between the accelerations of a free particle and the ground is grad Ψ, where
$$\Psi = U + \tfrac{1}{2}\omega^2 r^2 \sin^2\theta. \qquad (3)$$

The function Ψ is called the geopotential. Observed gravity is its gradient, and the surfaces of constant Ψ are the level surfaces.

On the ocean surface Ψ is a constant, which we shall denote by C. Take a standard value of r, which we shall denote by a and define precisely later; at present we need only say that it is such that r over the Earth's surface differs from it by small quantities of the first order. Then since $\Psi - fM/r$ is of the first order, it and its derivatives vary only by quantities of the second order when we change r from a to $a+r'$, where r' is of the first order. If

$$\Psi = \frac{fM}{r} + \tfrac{1}{2}\omega^2 r^2 \sin^2\theta + U' = C - gh, \tag{4}$$

where h is small, and we choose a, g_0 so that

$$C = \frac{fM}{a}, \quad g_0 = \frac{fM}{a^2}, \tag{5}$$

then to the first order

$$\frac{fM}{r} = C - g_0 r', \tag{6}$$

and

$$r' = h + \frac{1}{g_0}(\tfrac{1}{2}\omega^2 r^2 \sin^2\theta + U'). \tag{7}$$

We shall show that to the first order h is the measured height above sea level. We shall denote the second term by h'; it represents the departure of the level surfaces from spheres due to rotation and to the higher harmonics in the potential.

In triangulation differences of level dh are measured upwards, normally to a level surface at the point of observation; therefore along a survey route the change of Ψ is $-\int g\,dh$ (the negative sign because g is the downward gradient of Ψ). Ψ is a single-valued function of position, but g varies, and therefore the measured height of a given place will depend somewhat on the route taken from sea level. The difference, however, is of the second order in departures from a sphere and will be neglected. Hence on the surface

$$\Psi = C - gh, \tag{8}$$

where g is the local gravity and h the measured height.

Gravity at the outer surface is given by

$$g^2 = \left(\frac{\partial \Psi}{\partial r}\right)^2 + \left(\frac{\partial \Psi}{r\partial\theta}\right)^2 + \left(\frac{\partial \Psi}{r\sin\theta\, d\lambda}\right)^2, \tag{9}$$

and the second and third terms are of the second order. Hence to the first order

$$g = -\frac{\partial \Psi}{\partial r} = \frac{fM}{r^2} - \frac{\partial U'}{\partial r} - \omega^2 r \sin^2\theta. \tag{10}$$

The relations between g_0 and mean gravity, and between a and the mean radius, will have to be found. Then from (4)

$$\frac{fM}{r} = g_0\left(a - \frac{gh}{g_0}\right) - U' - \tfrac{1}{2}\omega^2 a^2 \sin^2\theta, \tag{11}$$

and from (10), (11)

$$g = \frac{1}{fM}\left(\frac{fM}{r}\right)^2 - \frac{\partial U'}{\partial r} - \omega^2 a \sin^2\theta$$

$$= g_0\left(1 - \frac{2gh}{ag_0}\right) - \frac{\partial U'}{\partial r} - \frac{2U'}{a} - 2\omega^2 a \sin^2\theta, \tag{12}$$

$$\frac{\partial U'}{\partial r} + \frac{2U'}{a} = g_0 - g\left(1 + \frac{2h}{a}\right) - \tfrac{4}{3}\omega^2 a - 2\omega^2 a(\tfrac{1}{3} - \cos^2\theta). \tag{13}$$

This determines the left side in terms of g at the outer surface; but as it is small and the outer surface is itself nearly a sphere we commit only a second order error in supposing it to hold at $r = a$. The first order corrections have been taken into account by the terms in h and ω^2. The observed value of gravity appears only in the expression $g(1+2h/a)$, which is practically the result of multiplying it by $(a+h)^2/a^2$, and would be the change required if gravity had been observed at the same height over the ocean surface and then the value at sea level was calculated from it according to the inverse square law. This rule, given by Stokes and later by Helmert, is known as the free air reduction.

To solve, we first take the last term. If

$$U_1' = \frac{2\omega^2 a^5}{r^3}(\tfrac{1}{3} - \cos^2\theta), \tag{14}$$

$$\frac{\partial U_1'}{\partial r} + \frac{2U_1'}{a} = -2\omega^2 a(\tfrac{1}{3} - \cos^2\theta) \quad (r=a), \tag{15}$$

and since

$$\tfrac{1}{3} - \cos^2\theta = -\tfrac{2}{3}(\tfrac{3}{2}\cos^2\theta - \tfrac{1}{2}) = -\tfrac{2}{3}p_2, \tag{16}$$

U_1' is a solution of Laplace's equation. Then putting

$$U' = U_1' + U_2', \tag{17}$$

$$g\left(1 + \frac{2h}{a}\right) = \gamma_0 + \sum_{n=2}^{\infty} g_{ns} S_{ns}, \tag{18}$$

$$U_2' = \sum_{n=2}^{\infty} \frac{A_{ns}}{r^{n+1}} S_{ns}, \tag{19}$$

we find by equating coefficients

$$g_0 = \gamma_0 + \tfrac{4}{3}\omega^2 a, \tag{20}$$

$$A_{ns} = \frac{a^{n+2}}{n-1} g_{ns}, \tag{21}$$

$$U' = \frac{2\omega^2 a^5}{r^3}(\tfrac{1}{3} - \cos^2\theta) + \sum_{n=2}^{\infty} \frac{a^{n+2}}{(n-1)r^{n+1}} g_{ns} S_{ns}. \tag{22}$$

Thus U' is determined. The elevation of sea level above the standard sphere is h'. If in the interior of the land we take a point at a depth below the visible surface equal to the measured height h, this point also will be at a height h' above the standard sphere. The locus of such points is called the *geoid*.

Special interest is attached to the main ellipticity term. Returning to (4) we have on the geoid

$$\frac{g_0 a^2}{r} + \tfrac{1}{3}\omega^2 a^2 + \tfrac{1}{2}\omega^2 a^2(\tfrac{1}{3} - \cos^2\theta) + U' = g_0 a, \tag{23}$$

whence

$$r - a = \frac{1}{3}\frac{\omega^2 a^2}{\gamma_0} + \frac{5}{2}\frac{\omega^2 a^2}{\gamma_0}(\tfrac{1}{3} - \cos^2\theta) + \sum_{n=2}^{\infty} \frac{a g_{ns}}{(n-1)\gamma_0} S_{ns}. \tag{24}$$

The constant terms in r give the mean radius, in terms of which our a is therefore determined. Denoting this by a_0 and ignoring now all terms but p_2, we can denote the equatorial and polar radii by $a_0(1 + \tfrac{1}{3}e)$ and $a_0(1 - \tfrac{2}{3}e)$, and for a general latitude

$$r = a_0(1 + \tfrac{1}{3}e)\sin^2\theta + a_0(1 - \tfrac{2}{3}e)\cos^2\theta + O(e^2)$$
$$= a_0\{1 + e(\tfrac{1}{3} - \cos^2\theta)\} + O(e^2). \tag{25}$$

e is the *ellipticity*. Comparing terms in $\tfrac{1}{3} - \cos^2\theta$ we have

$$e = \frac{5}{2}\frac{\omega^2 a_0}{\gamma_0} + \frac{g_{20}}{\gamma_0}, \tag{26}$$

and gravity, including the main ellipticity term, is to the first order

$$\gamma_0\{1 - (\tfrac{5}{2}m - e)(\tfrac{1}{3} - \cos^2\theta)\}, \tag{27}$$

where $m = \omega^2 a_0/\gamma_0$, to the present order the ratio of the acceleration at the equator due to the Earth's rotation to gravity. This is *Clairaut's formula*. Actually the analysis of gravity leads to a better determination of e than survey does. The extension to higher harmonics is due to Stokes.

The analysis leads also to a determination of the difference between the Earth's principal moments of inertia. In MacCullagh's formula (18·09) we know that the neglected terms are of order r^{-4} and therefore cannot contain any terms in the S_{2s}. Hence the term in $\tfrac{1}{3} - \cos^2\theta$ in U' has an exact relation to the moments of inertia. If we take $A = B$

$$I = A\sin^2\theta + C\cos^2\theta, \quad A + B + C - 3I = (C - A)(1 - 3\cos^2\theta), \tag{28}$$

the term in question is

$$Jf\frac{Ma^2}{r^3}(\tfrac{1}{3} - \cos^2\theta) = J\frac{\gamma_0 a^4}{r^3}(\tfrac{1}{3} - \cos^2\theta), \tag{29}$$

where

$$J = \frac{3}{2}\frac{C - A}{Ma^2}. \tag{30}$$

But by (22) it is

$$\left(\frac{2ma^4\gamma_0}{r^3} + \frac{a^4}{r^3}g_{20}\right)(\tfrac{1}{3} - \cos^2\theta). \tag{31}$$

Therefore

$$J = \frac{g_{20}}{\gamma_0} - 2m = e - \tfrac{1}{2}m. \tag{32}$$

J is about $1/600$. From the theory of precession of the equinoxes it is known that $(C - A)/C$ is about $1/300$, whence C/Ma^2 is about $\tfrac{1}{3}$. This ratio is clear evidence of the increase of density of the Earth towards the centre.

The neglect of second order terms makes it likely that the quantities calculated by the first order theory will be inaccurate by about 1 part in 300. Modern observational determinations are capable of giving them with a higher accuracy, and for this purpose it has become necessary to extend the theory to the second order of small quantities.

24·12. Value of $\iint p_n(\cos\theta) S_n(\theta, \lambda) d\omega$ over a sphere. We know that there is an expansion

$$S_n = a_{n0}p_n + \sum_{s=1}^{n} p_n^s(a_{ns}\cos s\lambda + b_{ns}\sin s\lambda), \tag{1}$$

and therefore

$$\iint p_n(\cos\theta) S_n d\omega = a_{n0}\iint (p_n)^2 d\omega = \frac{4\pi a_{n0}}{2n+1}. \tag{2}$$

Also since all p_n^s vanish at $\theta = 0$ for $s \geqslant 1$, $S_n(0, \lambda) = a_{n0}$ for all λ; and a_{n0} is the value of $S_n(\theta, \lambda)$ on the axis $\theta = 0$.

24·13. Change of axes of a zonal harmonic. Let ϑ be the angular distance from a fixed direction $OP(\theta, \lambda)$. The angular coordinates of a general point on the sphere are (θ', λ'), and by a fundamental formula of spherical trigonometry

$$\cos\vartheta = \cos\theta\cos\theta' + \sin\theta\sin\theta'\cos(\lambda' - \lambda). \tag{1}$$

24·131   Change of axes

Consider $p_n(\cos\vartheta)$. This is a surface harmonic of degree n, since we could have taken OP as axis of reference. Hence it can be expressed in terms of the harmonics of degree n in θ', λ'. Take

$$p_n(\cos\vartheta) = a_0 p_n(\cos\theta') + \sum_{s=1}^{n} p_n^s(\cos\theta')(a_s \cos s\lambda' + b_s \sin s\lambda'). \qquad (2)$$

Integrating over a unit sphere we have

$$\iint p_n(\cos\vartheta) p_n(\cos\theta')\, d\omega = \frac{4\pi a_0}{2n+1}, \qquad (3)$$

$$\iint p_n(\cos\vartheta) p_n^s(\cos\theta')(\cos s\lambda', \sin s\lambda')\, d\omega = \frac{2\pi}{2n+1} \frac{(n-s)!(n+s)!}{n!n!} (a_s, b_s). \qquad (4)$$

But if in the result of the last paragraph we replace $S_n(\theta, \lambda)$ by $p_n^s(\cos\theta')\cos s\lambda'$, and θ by ϑ, we have

$$\iint p_n(\cos\vartheta) p_n^s(\cos\theta') \cos s\lambda'\, d\omega = \frac{4\pi}{2n+1} p_n^s(\cos\theta) \cos s\lambda, \qquad (5)$$

with analogous relations; whence

$$(a_s, b_s) = 2 \frac{n!n!}{(n-s)!(n+s)!} p_n^s(\cos\theta)(\cos s\lambda, \sin s\lambda), \qquad (6)$$

except for a_0, for which the factor 2 does not occur; and

$$p_n(\cos\vartheta) = p_n(\cos\theta) p_n(\cos\theta') + 2 \sum_{s=1}^{n} \frac{n!n!}{(n-s)!(n+s)!} p_n^s(\cos\theta) p_n^s(\cos\theta') \cos s(\lambda' - \lambda).$$

This result, due to Legendre, is often called the *addition theorem* for spherical harmonics, and $p_n(\cos\vartheta)$ a *biaxial harmonic*.

24·131. Derivation from two-dimensional transformation. In the wave mechanics of complex atoms it is necessary to study the transformation properties of $p_n^s(\cos\theta) e^{is\lambda}$ under a rotation. This can be done by applying the method of 4·102, relating a rotation in three dimensions to a unitary transformation of two variables.

Since $\dfrac{\partial^{l+m+n}}{\partial x^l\, \partial y^m\, \partial z^n} \dfrac{1}{r}$ transforms like $a^l b^m c^n$, where a, b, c may be components of any vector, and $\nabla^2(1/r) = 0$, we may impose on (a, b, c) the further restriction that it is a complex null vector. Then we have the following correspondence of transformation properties for s positive or zero:

$$K_{-n-1}^s = \left(\frac{\partial}{\partial x} + i\frac{\partial}{\partial y}\right)^s \left(\frac{\partial}{\partial z}\right)^{n-s} \frac{1}{r} \text{ like } (a+ib)^s c^{n-s} \text{ like } x_1^{n-s} x_2^{n+s}, \qquad (1)$$

$$(-1)^s K_{-n-1}^{-s} = K_{-n-1}^{s*} = \left(\frac{\partial}{\partial x} - i\frac{\partial}{\partial y}\right)^s \left(\frac{\partial}{\partial z}\right)^{n-s} \frac{1}{r} \text{ like } (a-ib)^s c^{n-s} \text{ like } (-1)^s x_1^{n+s} x_2^{n-s}. \qquad (2)$$

Therefore for all s irrespective of sign the functions K_{-n-1}^s transform like $x_1^{n-s} x_2^{n+s}$. Now $x_1^* x_1 + x_2^* x_2$ is invariant and so therefore is $(x_1^* x_1 + x_2^* x_2)^{2n}$, that is,

$$\sum_{s=-n}^{n} \frac{(2n)!}{(n-s)!(n+s)!} (x_1^* x_1)^{n-s} (x_2^* x_2)^{n+s}. \qquad (3)$$

If we write therefore

$$X_s = \frac{x_1^{n-s} x_2^{n+s}}{\{(n-s)!(n+s)!\}^{1/2}} \qquad (4)$$

$\Sigma_s X_s^* X_s$ is invariant under our unitary transformations of x_1, x_2 and the transformation of X_s is also unitary. Therefore

$$(n!)^2 \sum_{s=-n}^{n} \frac{p_n^s(\mu) \, p_n^s(\mu')}{(n-s)!(n+s)!} e^{is(\lambda'-\lambda)} \tag{5}$$

is invariant. But if we rotate the axes so that θ' becomes 0, $\theta = \vartheta$ and all the $p_n^s(\mu')$ vanish for $s \neq 0$, and $p_n(\mu') = 1$. Then we have again

$$p_n(\cos\vartheta) = p_n(\cos\theta) p_n(\cos\theta') + 2 \sum_{s=1}^{n} \frac{n!\,n!}{(n-s)!(n+s)!} p_n^s(\cos\theta) p_n^s(\cos\theta') \cos s(\lambda'-\lambda). \tag{6}$$

For transformation formulae of associated Legendre functions under a rotation with given Euler angles, see B. Jeffreys, *Geophys. J.*, 10, 141–6, 1965.

24·14. Relation to Bessel functions. Consider the neighbourhood of a pole of a sphere of large radius a. The curvature of the surface is small and we should expect that the suitable potential functions will approximate to those useful with cylindrical coordinates, z corresponding to $r-a$, ϖ to $a\sin\theta$, and λ to λ. The factor in r can be written approximately, for $(r-a)/a$ small

$$(r/a)^{-n-1} \doteqdot e^{-nz/a} = e^{-\kappa z}, \tag{1}$$

with $n = \kappa a$. Neglecting the difference between θ and $\sin\theta$ we have from 18·06(6) for the θ factor

$$\varpi \frac{d}{d\varpi}\left(\varpi \frac{dy}{d\varpi}\right) + \left(\frac{n^2\varpi^2}{a^2} - s^2\right) y = 0, \tag{2}$$

and the solution finite at $\varpi = 0$ is $J_s(\kappa\varpi)$. Hence the solutions $r^{-n-1} p_n^s(\cos s\lambda, \sin s\lambda)$ correspond to the solutions in cylindrical coordinates $e^{-\kappa z} J_s(\kappa\varpi) (\cos s\lambda, \sin s\lambda)$. The usual n of Bessel functions, however, corresponds to the s of Legendre functions. An appreciable error will accumulate if θ is large enough for the difference between $\sin\theta$ and θ no longer to be neglected. The constant factor follows at once from 24·04(21); the first term in p_n^s is

$$\frac{1}{2^s} \frac{(n+s)!}{s!\,n!} \sin^s\theta \doteqdot \frac{(\tfrac{1}{2}n\sin\theta)^s}{s!}, \tag{3}$$

if n is large compared with s, and this is the first term of $J_s(n\sin\theta)$ as it stands. Hence for given $n\theta$, as $n \to \infty$,

$$p_n^s(\mu) \to J_s(n\sin\theta). \tag{4}$$

With the same approximation the factors $(n!)^2/(n-s)!(n+s)!$ in the formula for change of axes tend to 1, and $a\sin\vartheta$ tends to κR, where R is the distance between two points in the plane. Hence

$$J_0[\kappa\sqrt{\{(\varpi^2+\rho^2-2\varpi\rho)\cos(\lambda-\lambda')\}}] = J_0(\kappa\varpi) J_0(\kappa\rho) + \sum_{s=1}^{\infty} J_s(\kappa\varpi) J_s(\kappa\rho) \cos s(\lambda-\lambda'), \tag{5}$$

which is the addition formula for Bessel functions, originally found by Heine by this limiting process. We notice also that 24·04(19) becomes

$$J_{-s}(\kappa\varpi) = (-1)^s J_s(\kappa\varpi), \tag{6}$$

which is the relation already found for Bessel functions when s is an integer.

24·15. Asymptotic formulae for n large and θ not small.

In the differential equation for p_n^s

$$\frac{d^2y}{d\theta^2} + \cot\theta \frac{dy}{d\theta} - s^2 \csc^2\theta \, y + n(n+1)y = 0$$

we remove the term in y' by the substitution

$$y = \sin^{-1/2}\theta \, z. \tag{1}$$

We get

$$\frac{d^2z}{d\theta^2} + \{(n+\tfrac{1}{2})^2 - (s^2 - \tfrac{1}{4})\csc^2\theta\} z = 0. \tag{2}$$

When n is large and s not large asymptotic solutions are therefore simply

$$y \sim \sin^{-1/2}\theta \, e^{\pm(n+1/2)i\theta}, \tag{3}$$

so that the relation to $J_s\{(n+\tfrac{1}{2})\theta\}$ is much closer than that to $J_s(n\sin\theta)$ found by considering only small values of θ. The constants can be found by noting that we can take θ so that $n\theta$ is large and $n\theta^3$ small, and then comparing with Stokes's asymptotic expansion for J_s in this range, namely (cf. 21·05 (5))

$$J_s\{(n+\tfrac{1}{2})\theta\} \sim \left\{\frac{2}{\pi(n+\tfrac{1}{2})\theta}\right\}^{1/2} \cos\{(n+\tfrac{1}{2})\theta - \tfrac{1}{2}s\pi - \tfrac{1}{4}\pi\}. \tag{4}$$

Hence

$$p_n^s(\mu) \sim \left\{\frac{2}{\pi(n+\tfrac{1}{2})\sin\theta}\right\}^{1/2} \cos\{(n+\tfrac{1}{2})\theta - \tfrac{1}{2}s\pi - \tfrac{1}{4}\pi\}, \tag{5}$$

where θ is not near 0 or π. The full expansion of this type is given by Hobson, of course with a different coefficient on account of his definition of the function. Unfortunately in practice s, if not zero, is usually comparable with n, and expansions of Stokes's type proceed in powers of s^2/n and consequently are not often useful. In fact the approximation obviously breaks down completely when $s = n$, when the function has no zeros between 0 and π.

24·16. Definite integral representations.

We have

$$f^{(n)}(z) = \frac{n!}{2\pi i} \int_C \frac{f(t)}{(t-z)^{n+1}} dt, \tag{1}$$

where $f(t)$ is analytic within C and z is within C. Hence

$$p_n^s(\mu) = (1-\mu^2)^{1/2s} \frac{(n-s)!}{2^n(n!)^2} \frac{d^{n+s}}{d\mu^{n+s}}(\mu^2-1)^n = \frac{(n-s)!(n+s)!(1-\mu^2)^{1/2s}}{2^n(n!)^2} \frac{1}{2\pi i} \int_C \frac{(t^2-1)^n}{(t-\mu)^{n+s+1}} dt, \tag{2}$$

provided C encloses μ. Since n has so far been taken as a positive integer there is no singularity at $t = \pm 1$, and C can be taken as large as we like. But even if n and s are not integers it is easy to verify that (2) satisfies the differential equation for $p_n^s(\mu)$, and also holds for unrestricted μ, provided the path C is such that the integrand returns to its original value on describing it.

The integral (2) is due to Schläfli, and is related to his integral for the Bessel functions. If we put $t - \mu = \lambda$,

$$p_n^s(\mu) = \frac{(n-s)!(n+s)!(1-\mu^2)^{1/2s}}{2^n n! n!} \frac{1}{2\pi i} \int_C \left(2\mu + \lambda + \frac{\mu^2-1}{\lambda}\right)^n \frac{d\lambda}{\lambda^{s+1}}, \tag{3}$$

and with $\lambda = (1-\mu^2)^{1/2} u$

$$p_n^s(\mu) = \frac{(n-s)!\,(n+s)!}{2^n n!\,n!} \frac{1}{2\pi i} \int \left\{2\mu + (1-\mu^2)^{1/2}\left(u - \frac{1}{u}\right)\right\}^n \frac{du}{u^{s+1}}. \quad (4)$$

If $(1-\mu^2)^{1/2}$ is small, the contours fixed, and n large, this approximates to

$$\frac{1}{2\pi i}\int_C \exp\left\{\tfrac{1}{2}n(1-\mu^2)^{1/2}\left(u - \frac{1}{u}\right)\right\} \frac{du}{u^{s+1}}, \quad (5)$$

which is Schläfli's integral for $J_s(n \sin \theta)$. A factor $e^{n(1-\mu)}$ has been replaced by 1, so that the approximation assumes $n\theta$ moderate and $n\theta^2$ small.

With integral n and s but μ complex we can take C in the t-plane to be a circle with centre μ and radius $|(\mu^2-1)|^{1/2}$. In general one of $t = \pm 1$ is inside C and the other outside. Then with

$$t = \mu + (\mu^2 - 1)^{1/2} e^{i\phi}, \quad (6)$$

we have

$$p_n^s(\mu) = \frac{(n-s)!\,(n+s)!\,(1-\mu^2)^{1/2 s}}{2^n (n!)^2} \frac{1}{2\pi} \int_{-\pi}^{\pi} \frac{\{\mu - 1 + (\mu^2-1)^{1/2} e^{i\phi}\}^n \{\mu + 1 + (\mu^2-1)^{1/2} e^{i\phi}\}^n}{(\mu^2-1)^{1/2(n+s)} e^{(n+s)i\phi}} d\phi. \quad (7)$$

But $\{(\mu-1)e^{-1/2 i\phi} + (\mu^2-1)^{1/2} e^{1/2 i\phi}\}\{(\mu+1)e^{-1/2 i\phi} + (\mu^2-1)^{1/2} e^{1/2 i\phi}\}$

$$= (\mu^2-1)(e^{i\phi} + e^{-i\phi}) + 2\mu(\mu^2-1)^{1/2} = 2(\mu^2-1)^{1/2}\{\mu + (\mu^2-1)^{1/2} \cos\phi\}, \quad (8)$$

and

$$p_n^s(\mu) = \frac{(n-s)!\,(n+s)!}{(n!)^2} \frac{i^{\pm s}}{2\pi} \int_{-\pi}^{\pi} \{\mu + (\mu^2-1)^{1/2} \cos\phi\}^n e^{-si\phi} d\phi$$

$$= \frac{(n-s)!\,(n+s)!}{(n!)^2} \frac{i^{\pm s}}{\pi} \int_0^{\pi} \{\mu + (\mu^2-1)^{1/2} \cos\phi\}^n \cos s\phi\, d\phi. \quad (9)$$

Powers of $\cos\phi$ up to the $(s-1)$th will give 0 on integration, and if we take $(\mu^2-1)^{1/2}$ to mean $i(1-\mu^2)^{1/2}$ we must take i^{-s} outside. This is *Laplace's integral*. It will be noticed that the integrand always has modulus $\leqslant 1$ for $-1 \leqslant \mu \leqslant 1$ and therefore $|p_n(\mu)| \leqslant 1$. This integral yields only one solution of the differential equation. If we reverse the sign of i and then put $\phi = \pi - \psi$ we clearly recover the same integral except possibly for a change of sign. Similarly in (2), either the path C encloses μ and gives p_n^s, or does not enclose μ and gives 0 if n and s are integers. It is useless to take a path going to infinity since the integral diverges. Other solutions can be obtained if n is not a positive integer, but the specification of the paths to make the integrand single-valued becomes difficult. A full treatment of this case is given by Hobson. Unlike what has happened in several other cases, we get no fundamentally new solution by varying the path for positive integral n, s.

24·161. Another method is to begin with

$$(1 - 2h\mu + h^2)^{-1/2} = \Sigma h^n p_n \quad (h < 1), \quad (10)$$

$$\frac{d^s}{d\mu^s}(1 - 2h\mu + h^2)^{-1/2} = \Sigma h^n (1-\mu^2)^{-1/2 s} p_n^s \quad (11)$$

$$= 2^s h^s \tfrac{1}{2} \cdot \tfrac{3}{2} \ldots (s - \tfrac{1}{2})(1 - 2h\mu + h^2)^{-s - 1/2}, \quad (12)$$

$$(1-\mu^2)^{1/2 s} \frac{1 \cdot 3 \ldots 2s - 1}{(1 - 2h\mu + h^2)^{s + 1/2}} = \Sigma h^{n-s} \frac{n!}{(n-s)!} p_n^s, \quad (13)$$

24·161 *Definite integrals*

and
$$p_n^s = \frac{(n-s)!}{n!} 1.3\ldots(2s-1) \frac{(1-\mu^2)^{1/2 s}}{2\pi i} \int_C \frac{dh}{h^{n-s+1}(1-2h\mu+h^2)^{s+1/2}}, \quad (14)$$

where C surrounds the origin but neither of the points $h = e^{\pm i\theta}$.

To verify that this satisfies the differential equation, put
$$p_n^s = A(1-\mu^2)^{1/2 s} \Theta, \quad (15)$$

where A is the constant factor, and
$$1 - 2h\mu + h^2 = X; \quad (16)$$

$$(1-\mu^2)\Theta'' - 2(s+1)\mu\Theta' + (n-s)(n+s+1)\Theta$$
$$= \int_C \left\{ \frac{(2s+1)(2s+3)(1-\mu^2)}{h^{n-s-1} X^{s+5/2}} - \frac{2(2s+1)(2s+1)\mu}{h^{n-s} X^{s+3/2}} + \frac{(n-s)(n+s+1)}{h^{n-s+1} X^{s+1/2}} \right\} dh. \quad (17)$$

Now
$$1-\mu^2 = X - (\mu-h)^2, \quad \frac{\partial X}{\partial h} = -2(\mu-h), \quad (18)$$

and we proceed to eliminate μ by partial integration. We arrange (17) as follows:

$$\int_C \left[(2s+1)(2s+3)\left(\frac{1}{h^{n-s-1} X^{s+3/2}} - \frac{(\mu-h)^2}{h^{n-s-1} X^{s+5/2}}\right) \right.$$
$$\left. - 2(s+1)(2s+1)\left(\frac{\mu-h}{h^{n-s} X^{s+3/2}} + \frac{1}{h^{n-s-1} X^{s+3/2}}\right) + \frac{(n-s)(n+s+1)}{h^{n-s+1} X^{s+1/2}} \right] dh. \quad (19)$$

Now
$$-\int_C \left(\frac{(2s+1)(2s+3)(\mu-h)^2}{h^{n-s-1} X^{s+5/2}} + \frac{2(s+1)(2s+1)(\mu-h)}{h^{n-s} X^{s+3/2}} \right) dh$$
$$= -\left[\frac{(2s+1)(\mu-h)}{h^{n-s-1} X^{s+3/2}}\right] - (2s+1) \int_C \left(\frac{(n-s-1)(\mu-h)}{h^{n-s} X^{s+3/2}} + \frac{1}{h^{n-s-1} X^{s+3/2}} + \frac{2(s+1)(\mu-h)}{h^{n-s} X^{s+3/2}} \right) dh$$
$$= -[\,] - (2s+1) \int_C \left(\frac{(n+s+1)(\mu-h)}{h^{n-s} X^{s+3/2}} + \frac{1}{h^{n-s-1} X^{s+3/2}} \right) dh$$
$$= -[\,] - \left[\frac{n+s+1}{h^{n-s} X^{s+1/2}}\right] - (n+s+1)(n-s) \int_C \frac{dh}{h^{n-s+1} X^{s+1/2}} - (2s+1) \int_C \frac{dh}{h^{n-s-1} X^{s+3/2}}. \quad (20)$$

The first integral remaining cancels the last term of (19); the second, taken with the remaining terms of (19), gives an expression containing the factor

$$(2s+1)(2s+3) - 2(s+1)(2s+1) - (2s+1) = 0.$$

Hence
$$(1-\mu^2)\Theta'' - 2(s+1)\mu\Theta' + (n-s)(n+s+1)\Theta = -\left[\frac{(2s+1)(\mu-h)}{h^{n-s-1} X^{s+3/2}} + \frac{n+s+1}{h^{n-s} X^{s+1/2}}\right], \quad (21)$$

and therefore the integral will satisfy the differential equation provided that the expression in [ ] returns to its original value on describing the path; and this is true not only for integral n and s. In particular it is true if C is any loop from infinity around one of the possible singularities at $h = 0$ and $e^{\pm i\theta}$, provided $n+s+1 > 0$. If $n-s$ is an integer (21) vanishes if C is a closed path about the origin, not including the other singularities, and the integrand will be single-valued if we break the path and complete it by two lines to infinity, so that the infinite lines of the path will cancel. Hence if we take a loop from infinity, passing about O, it will afford a definition of p_n^s when n and s are fractional, $(n+s+1>0)$, reducing when they are integers to the previous definition.

If n and s are integers the integral around a large circle will be zero, and we can replace C by a path about the points $h = \exp(\pm i\theta)$. It is usual to reduce the path to a circular arc $|h| = 1$ about the origin connecting the points $h = \exp(\pm i\theta)$. The result is *Mehler's integral*; but clearly it will diverge if $s \geq \frac{1}{2}$. On the other hand we could replace s by $-s$, but then the factor outside the integral is proportional to $(1-\mu^2)^{-1/2s}$, which is not convenient in a function that behaves like $(1-\mu^2)^{1/2s}$ when μ is near ± 1. It seems that integrals of this type can be useful in practice only for $s = 0$; then it is found that

$$p_n(\cos\theta) = \frac{2}{\pi}\int_0^\theta \frac{\cos(n+\frac{1}{2})\psi}{\{2(\cos\psi-\cos\theta)\}^{1/2}}d\psi, \tag{22}$$

$$= \frac{2}{\pi}\int_\theta^\pi \frac{\sin(n+\frac{1}{2})\psi}{\{2(\cos\theta-\cos\psi)\}^{1/2}}d\psi. \tag{23}$$

These formulae are due to Mehler.

24·162. Bateman's integral. A different type of definite integral solution is given by Bateman. We have, with some modifications of his method, if

$$R^2 = x^2+y^2+(z-a)^2, \tag{24}$$

and a and b are small constants,

$$\frac{1}{R} = \frac{1}{\pi}\int_{-\infty}^\infty \frac{dw}{x^2+y^2+(z-a)^2+(w-b)^2}. \tag{25}$$

Then

$$\left(\frac{\partial}{\partial x}+i\frac{\partial}{\partial y}\right)^s \left(\frac{\partial}{\partial a}+i\frac{\partial}{\partial b}\right)^{n-s}\frac{1}{R} = (-1)^s\frac{2^n n!}{\pi}\int \frac{(x+iy)^s\{z-a+i(w-b)\}^{n-s}}{\{x^2+y^2+(z-a)^2+(w-b)^2\}^{n+1}}dw. \tag{26}$$

But $1/R$ is a function of $z-a$ and is independent of b. Hence the operation $\partial/\partial b$ gives 0, and $\partial/\partial a$ is equivalent to $-\partial/\partial z$, and if we make a and b tend to 0 we have

$$\left(\frac{\partial}{\partial x}+i\frac{\partial}{\partial y}\right)^s\left(\frac{\partial}{\partial z}\right)^{n-s}\frac{1}{r} = (-1)^n\frac{2^n n!}{\pi}\int_{-\infty}^\infty \frac{(x+iy)^s(z+iw)^{n-s}}{(x^2+y^2+z^2+w^2)^{n+1}}dw. \tag{27}$$

But the left side is

$$(-1)^n n!\frac{p_n^s e^{is\lambda}}{r^{n+1}},$$

and therefore

$$p_n^s = \frac{2^n}{\pi}\sin^s\theta \int_{-\infty}^\infty \frac{(\mu+iw/r)^{n-s}}{(1+w^2/r^2)^{n+1}}\frac{dw}{r}$$

$$= \frac{2^n}{\pi}\sin^s\theta \int_{-\infty}^\infty \frac{(\mu+it)^{n-s}}{(1+t^2)^{n+1}}dt. \tag{28}$$

This converges for s positive if $n+s>0$ and μ is not purely imaginary; even if μ is purely imaginary the point $t=-i\mu$ gives no trouble if $n>s-1$. It therefore provides a definition of the function in all practical circumstances. If we put $\lambda = it$ we get

$$p_n^s = \frac{2^n}{\pi i}\sin^s\theta \int_{-i\infty}^{i\infty}\frac{(\lambda+\mu)^{n-s}}{(1-\lambda^2)^{n+1}}d\lambda. \tag{29}$$

If we use a different path the integral may still satisfy the differential equation; for if we take
$$\Theta = \int \frac{(\mu+it)^{n-s}}{(1+t^2)^{n+1}} dt, \qquad (30)$$

$(1-\mu^2)\Theta'' - 2(s+1)\mu\Theta' + (n-s)(n+s+1)\Theta$
$$= \int \frac{(n-s)(n-s-1)(1-\mu^2)(\mu+it)^{n-s-2} - 2(n-s)(s+1)\mu(\mu+it)^{n-s-1} + (n-s)(n+s+1)(\mu+it)^{n-s}}{(1+t^2)^{n+1}} dt \qquad (31)$$

$n-s$ is a factor, and
$$(n-s-1)(1-\mu^2) - 2(s+1)\mu(\mu+it) + (n+s+1)(\mu+it)^2$$
$$= n-s-1 + 2n\mu it - (n+s+1)t^2$$
$$= n-s-1 + 2int(\mu+it) + 2nt^2 - (n+s+1)t^2$$
$$= (n-s-1)(1+t^2) + 2int(\mu+it),$$

$$\frac{\{(n-s-1)(1+t^2) + 2int(\mu+it)\}(\mu+it)^{n-s-2}}{(1+t^2)^{n+1}} = -i\frac{d}{dt}\frac{(\mu+it)^{n-s-1}}{(1+t^2)^n}.$$

Hence
$$(1-\mu^2)\Theta'' - 2(s+1)\mu\Theta' + (n-s)(n+s+1)\Theta = \left[-i(n-s)\frac{(\mu+it)^{n-s-1}}{(1+t^2)^n}\right], \qquad (32)$$

and vanishes for any path with infinite ends if $n+s+1>0$. The path chosen for p_n^s passes between the two singularities at $t = \pm i$; but we shall be able to obtain another solution by taking a path from $t = i\mu$ to infinity. The terms obtained from differentiating the limits vanish if $s < n-1$. For $s = n$ they vanish, for $s = n-1$ they cancel the right side of (32). Hence (30), with termini $i\mu$ and ∞, is a solution for $n+s+1>0$, $n-s+1>0$.

24·17. Solutions when μ is not real and between -1 and 1. Far the greater number of applications of Legendre's equation require only the solutions p_n^s. The other solution may arise, possibly with non-integral n, in problems relating to a spherical boundary when the poles are excluded, and also in external problems for spheroids. μ is real and greater than 1 for the region outside a prolate spheroid, purely imaginary outside an oblate spheroid, and its modulus may be arbitrarily large. Hence it is convenient to replace the factor $(1-\mu^2)^{1/2s}$ by $(\mu^2-1)^{1/2s}$ and to take the first solution as

$$t_n^s(\mu) = \frac{(n-s)!}{2^n(n!)^2}(\mu^2-1)^{1/2s}\frac{d^{n+s}}{d\mu^{n+s}}(\mu^2-1)^n. \qquad (1)$$

This will be real and positive for all real $\mu > 1$. By 24·162 (29) we have also

$$t_n^s(\mu) = \frac{2^n}{\pi i}(\mu^2-1)^{1/2s}\int_{-i\infty}^{i\infty}\frac{(\lambda+\mu)^{n-s}}{(1-\lambda^2)^{n+1}}d\lambda \qquad (2)$$

which can be extended to all n, s such that $n+s+1>0$. If we put $\lambda = -\mu - v$ and then $v = -(\mu^2-1)^{1/2}w$ we get

$$t_n^s(\mu) = \frac{2^n}{\pi i}(\mu^2-1)^{1/2s}\int_{-i\infty}^{i\infty}\frac{(-v)^{n-s}dv}{(1-\mu^2-v^2-2\mu v)^{n+1}}$$
$$= \frac{1}{2\pi i}(\mu^2-1)^{1/2s}\int_{-i\infty}^{i\infty}(-v)^{-s-1}\left(\mu + \frac{\mu^2-1}{2v} + \frac{v}{2}\right)^{-(n+1)}dv$$
$$= \frac{1}{2\pi i}\int_{-i\infty}^{i\infty}w^{-s-1}\left\{\mu - \tfrac{1}{2}(\mu^2-1)^{1/2}\left(w + \frac{1}{w}\right)\right\}^{-n-1}dw. \qquad (3)$$

As in 24·16 (5), using 21·01 (16), if μ is real and > 1, this tends to $I_s\{n(\mu^2-1)^{1/2}\}$ when $n(\mu^2-1)^{1/2}$ is fixed and $n \to \infty$. The path must always be taken so as to pass between the poles of the integrand; these tend to 0 and $+\infty$.

The most convenient form for the second solution is, for $n-s+1$ and $n+s+1$ positive,

$$q_n^s(\mu) = \frac{2^{n+1}}{\pi}(\mu^2-1)^{1/2 s}\int_\mu^\infty \frac{(u-\mu)^{n-s}}{(u^2-1)^{n+1}}du, \tag{4}$$

where the path does not intersect the real axis between ± 1. If we take a cut in the μ plane from -1 to $+1$, $q_n^s(\mu)$ will be analytic and single-valued except possibly at points on the cut. If we put $u-\mu = \lambda$, we get

$$q_n^s(\mu) = \frac{1}{\pi}(\mu^2-1)^{1/2 s}\int_0^\infty \lambda^{-s-1}\left\{\mu + \frac{1}{2}\left(\frac{\mu^2-1}{\lambda}+\lambda\right)\right\}^{-(n+1)}d\lambda. \tag{5}$$

If $n(\mu^2-1)^{1/2} = x$, and $n \to \infty$, $\mu \to 1+$ we find

$$q_n^s(\mu) \to \mathrm{Kh}_{-s}(x) = \mathrm{Kh}_s(x) \tag{6}$$

by a similar argument to that leading to 24·16 (5), using 21·022 (60). Hence $q_n^s(\mu)$ defined by (4) is related to $\mathrm{Kh}_s\{n\sqrt{(\mu^2-1)}\}$ in the same way as t_n^s to I_s and p_n^s to J_s.

In the expression (4) put $u = \dfrac{\mu}{1-v}$

then

$$q_n^s(\mu) = \frac{2^{n+1}}{\pi}(\mu^2-1)^{1/2 s}\int_0^1 \mu^{-n-s-1} v^{n-s}(1-v)^{n+s}\left\{1-\frac{(1-v)^2}{\mu^2}\right\}^{-n-1}dv.$$

Expand in powers of μ^{-2} and integrate term by term; we have

$$q_n^s(\mu) = \frac{2^{n+1}}{\pi}(\mu^2-1)^{1/2 s}\mu^{-n-s-1}\frac{(n-s)!}{n!}\sum_{m=0}^\infty \frac{(n+m)!}{m!}\frac{(n+s+2m)!}{(2n+1+2m)!}\mu^{-2m}. \tag{7}$$

This converges for $|\mu^2| > 1$.

It is possible to express q_n^s in finite terms. Take first $s = 0$; then $t_n(\mu) = p_n(\mu)$. The roots of the indicial equation at $\mu = \pm 1$ are both 0, and the differential equation has no other singularities. Hence any solution of Legendre's equation has the form

$$\Theta = At_n(\mu)\log(\mu-1) + Bt_n(\mu)\log(\mu+1) + Ct_n(\mu) - f_n(\mu), \tag{8}$$

where $f_n(\mu)$ is analytic at $\mu = \pm 1$ and is therefore an integral function. But $t_n(\mu)$ and $q_n(\mu)$ are single-valued for $|\mu| > 1$; hence $A = -B$. If $\Theta = q_n(\mu)$, $\Theta = O(\mu^{-n-1})$ for $|\mu|$ large, and $t_n(\mu)$ is a polynomial of degree n. Then the first two terms are together of order μ^{n-1}. Take C zero. Then Θ will tend to zero as $|\mu| \to \infty$ if and only if

$$q_n(\mu) = B\left\{t_n(\mu)\log\frac{\mu+1}{\mu-1} - f_n(\mu)\right\}, \tag{9}$$

where $f_n(\mu)$ must be the expansion of the first term in the brackets { } in descending powers of μ, as far as the constant term. Evidently it is of degree $n-1$. (Further the coefficients of powers from μ^{-1} to μ^{-n} must evidently be zero.)

To determine B, return to (7) with $s = 0$, and notice that if m is large

$$\frac{2^{n+1}}{\pi}\frac{(n+m)!}{m!}\frac{(n+2m)!}{(2n+1+2m)!}\mu^{-2m} = \frac{1}{\pi}\mu^{-2m}\left\{\frac{1}{m}+O\left(\frac{1}{m^2}\right)\right\}$$

$$= \frac{2}{\pi}\mu^{-2m}\left\{\frac{1}{2m+1}+O\left(\frac{1}{m^2}\right)\right\} \tag{10}$$

and therefore for $|\mu| > 1$

$$q_n(\mu) = \mu^{-n}\left\{\frac{1}{\pi}\log\frac{\mu+1}{\mu-1}+O(1)\right\}. \tag{11}$$

By comparison with (9), since $t_n(1) = 1$, $B = 1/\pi$, and finally

$$q_n(\mu) = \frac{1}{\pi}\left\{t_n(\mu)\log\frac{\mu+1}{\mu-1} - f_n(\mu)\right\}, \tag{12}$$

also

$$q_n^s(\mu) = (-1)^s\frac{(n-s)!}{n!}(\mu^2-1)^{1/2 s}\frac{d^s}{d\mu^s}q_n(\mu). \tag{13}$$

This expresses $q_n^s(\mu)$ in finite terms. It will be noticed that the differentiations will give terms in $(\mu-1)^{-s}$ and $(\mu+1)^{-s}$, while the extra variable factor needed is $(\mu^2-1)^{1/2 s}$. Hence near $\mu = \pm 1$, $q_n^s(\mu) \to \infty$ like $(\mu^2-1)^{-1/2 s}$.

Another definite integral for $q_n(\mu)$ can be found as follows. We start with

$$S = \sum_{n=0}^{\infty} h^n q_n(\mu) = \frac{2}{\pi} \int_{\mu}^{\infty} \sum \frac{2^n (u-\mu)^n h^n}{(u^2-1)^{n+1}} du$$

$$= \frac{2}{\pi} \int_{\mu}^{\infty} \frac{du}{u^2-1-2uh+2\mu h} = \frac{2}{\pi} \int_{\mu}^{\infty} \frac{du}{(u-h)^2-(1-2\mu h+h^2)}$$

on summation under the integral sign, which is possible for all real $u > \mu > 1$ if $|h| < 1$; then

$$S = \frac{1}{\pi(1-2\mu h+h^2)^{1/2}} \log \frac{\mu-h+(1-2\mu h+h^2)^{1/2}}{\mu-h-(1-2\mu h+h^2)^{1/2}}, \tag{14}$$

which can be regarded as the generating function of $q_n(\mu)$.

On the other hand consider the series

$$T = \sum_{n=0}^{\infty} h^n \int_{-1}^{1} \frac{p_n(\nu) d\nu}{\mu-\nu} = \int_{-1}^{1} \frac{d\nu}{(1-2\nu h+h^2)^{1/2}(\mu-\nu)}. \tag{15}$$

With the substitution

$$1-2\nu h+h^2 = v^2, \tag{16}$$

$$T = -2 \int_{1-h}^{1+h} \frac{dv}{1-2\mu h+h^2-v^2} = -\frac{1}{(1-2\mu h+h^2)^{1/2}} \left[\log \frac{(1-2\mu h+h^2)^{1/2}+v}{(1-2\mu h+h^2)^{1/2}-v} \right]_{1-h}^{1+h}$$

$$= \frac{1}{(1-2\mu h+h^2)^{1/2}} \log \frac{\mu-h+(1-2\mu h+h^2)^{1/2}}{\mu-h-(1-2\mu h+h^2)^{1/2}}$$

$$= \pi S. \tag{17}$$

Hence by equating coefficients of h^n,
$$q_n(\mu) = \frac{1}{\pi} \int_{-1}^{1} \frac{p_n(\nu)}{\mu-\nu} d\nu. \tag{18}$$

Note that if n is even, $t_n(\mu)$ is an even function, $q_n(\mu)$ an odd one, and conversely. The following are specimen values of q_n.

$$\left. \begin{aligned} q_0 &= \frac{1}{\pi} \log \frac{\mu+1}{\mu-1}, \\ q_1 &= \frac{\mu}{\pi} \log \frac{\mu+1}{\mu-1} - \frac{2}{\pi}, \\ q_2 &= \frac{p_2}{\pi} \log \frac{\mu+1}{\mu-1} - \frac{3\mu}{\pi}, \\ q_3 &= \frac{p_3}{\pi} \log \frac{\mu+1}{\mu-1} - \frac{1}{\pi} (5\mu^2 - \tfrac{4}{3}). \end{aligned} \right\} \tag{19}$$

24·171. Asymptotic approximation for n large. The series 24·17 (7) is convergent, so that the need for approximations of Stokes's type for given n, s and large $|\mu|$ does not arise. If we want the behaviour for given μ and increasing n, s we can apply the method of steepest descents to 24·17(4). Put

$$\phi(u) = (n-s) \log (u-\mu) - n \log (u^2-1),$$

$$\phi'(u) = \frac{n-s}{u-\mu} - \frac{2nu}{u^2-1},$$

$$\phi''(u) = -\frac{n-s}{(u-\mu)^2} - \frac{2n}{u^2-1} + \frac{4nu^2}{(u^2-1)^2}.$$

The condition for a saddle-point gives

$$(n+s) u = n\mu + \{n^2(\mu^2-1)+s^2\}^{1/2} = n\mu + M,$$

since we want the root greater than μ for μ real and > 1; then

$$q_n^s(\mu) \sim \left(\frac{2}{\pi}\right)^{\frac{1}{2}} (\mu^2-1)^{\frac{1}{2}s} \left(\frac{n-s}{n}\right)^{n+\frac{1}{2}} M^{-\frac{1}{2}} \frac{(n+s)^{n+s+\frac{1}{2}}}{(M-s\mu)^s (M+n\mu)^{n+\frac{1}{2}}} \tag{20}$$

valid if $u - \mu$ at the saddle-point is several times $|\phi''(u)|^{-\frac{1}{2}}$; this is satisfied if

$$\frac{n^2-s^2}{n} \frac{M}{n\mu+M} \gg 0. \tag{21}$$

It may be verified easily that if n is large but s is not, the approximation reduces to

$$\left(\frac{2}{\pi}\right)^{\frac{1}{2}} \frac{1}{n^{\frac{1}{2}}(\mu^2-1)^{\frac{1}{4}}} \{\mu + \sqrt{(\mu^2-1)}\}^{-(n+\frac{1}{2})}, \tag{22}$$

and if we put

$$n(\mu^2-1)^{\frac{1}{2}} = x, \tag{23}$$

this is approximately $(2/\pi x)^{\frac{1}{2}} e^{-x}$, which is the first term of the asymptotic expansion of $\mathrm{Kh}_s(x)$.
For $t_n^s(\mu)$ it is found that the relevant saddle-point is where

$$(n+s)u = n\mu - M$$

and we get

$$t_n^s(\mu) \sim \frac{1}{\sqrt{(2\pi)}} (\mu^2-1)^{\frac{1}{2}s} \left(\frac{n-s}{n}\right)^{n+\frac{1}{2}} \frac{(n+s)^{n+s+\frac{1}{2}}}{(s\mu+M)^s} \frac{M^{-\frac{1}{2}}}{(n\mu-M)^{n+\frac{1}{2}}}. \tag{24}$$

(20) and (24) are valid for all arg μ if M is defined by continuity.

24·18. Second solution when $-1 < \mu < 1$. We have from 21·022 (67)

$$\mathrm{Hs}_s(nu) = -ie^{-\frac{1}{2}s\pi i}\mathrm{Kh}_s(nu\,e^{-\frac{1}{2}\pi i}), \tag{1}$$

$$\mathrm{Hi}_s(nu) = ie^{\frac{1}{2}s\pi i}\mathrm{Kh}_s(nu\,e^{\frac{1}{2}\pi i}), \tag{2}$$

$$2Y_s(nu) = -e^{-\frac{1}{2}s\pi i}\mathrm{Kh}_s(nu\,e^{-\frac{1}{2}\pi i}) - e^{\frac{1}{2}s\pi i}\mathrm{Kh}_s(nu\,e^{\frac{1}{2}\pi i}). \tag{3}$$

We take u to be positive.

We see from 21·02 (42), 21·022 (69) that the coefficients in the expansions of $Y_s(x)$ and $\mathrm{Kh}_s(x)$ are equal in magnitude; but as $x \to 0$ through positive values $\mathrm{Kh}_s(x) \to \infty$, $Y_s(x) \to -\infty$. It is convenient therefore to take our solution as corresponding to $-Y_s(nu)$ instead of $Y_s(nu)$. We know that when n is large the function $q_n^s(\mu)$ behaves like $\mathrm{Kh}_s\{n(\mu^2-1)^{\frac{1}{2}}\}$. If μ is moved clockwise about $+1$ so as to reach a point between -1 and 1,

$$q_n^s(\mu) \to \frac{2^{n+1}}{\pi} (\mu^2-1)^{\frac{1}{2}s} \int_{\mu,1-i}^{\infty} \frac{(u-\mu)^{n-s}}{(u^2-1)^{n+1}} du$$

$$= \frac{2^{n+1}}{\pi} e^{-\frac{1}{2}is\pi}(1-\mu^2)^{\frac{1}{2}s} \int_{\mu,1-i}^{\infty} \frac{(u-\mu)^{n-s}}{(u^2-1)^{n+1}} du = q_n^s(\mu - 0i) \tag{4}$$

say. If μ moves counter-clockwise to the same point

$$q_n^s(\mu) \to \frac{2^{n+1}}{\pi} e^{\frac{1}{2}is\pi}(1-\mu^2)^{\frac{1}{2}s} \int_{\mu,1+i}^{\infty} \frac{(u-\mu)^{n-s}}{(u^2-1)^{n+1}} du = q_n^s(\mu + 0i). \tag{5}$$

Hence we can define a second solution, real in $(-1, 1)$ and corresponding to $-Y_s\{n(1-\mu^2)^{\frac{1}{2}}\}$ by taking

$$q_n^s(\mu) = \frac{2^n}{\pi} (1-\mu^2)^{\frac{1}{2}s} \left\{ e^{is\pi} \int_{\mu,1+i}^{\infty} \frac{(u-\mu)^{n-s}}{(u^2-1)^{n+1}} du + e^{-is\pi} \int_{\mu,1-i}^{\infty} \frac{(u-\mu)^{n-s}}{(u^2-1)^{n+1}} du \right\}, \tag{6}$$

$$= \frac{(n-s)!}{n!} (1-\mu^2)^{\frac{1}{2}s} \frac{d^s}{d\mu^s} q_n(\mu), \tag{7}$$

also

$$q_n^s(\mu) = \tfrac{1}{2} e^{\frac{1}{2}is\pi} q_n^s(\mu+0i) + \tfrac{1}{2} e^{-\frac{1}{2}is\pi} q_n^s(\mu-0i). \tag{8}$$

If n is even, $q_n(\mu)$ is an odd function when μ is not on the cut; it follows from (8) that it is also an odd function when defined by (6) on the cut. Similarly if n is odd, $q_n(\mu)$ will be an even function on the cut. Hence $q_n(\mu)$ on the cut is always a constant multiple of the non-terminating series solution found in 16·04.

24·18 Second solution for $-1 < \mu < 1$

To identify the constant, notice first that, from 24·17 (12),

$$q_n(\mu+0i) = \frac{1}{\pi}\left\{p_n(\mu)\log\frac{1+\mu}{1-\mu} - i\pi p_n(\mu) - f_n(\mu)\right\}, \tag{9}$$

$$q_n(\mu-0i) = \frac{1}{\pi}\left\{p_n(\mu)\log\frac{1+\mu}{1-\mu} + i\pi p_n(\mu) - f_n(\mu)\right\}, \tag{10}$$

and therefore

$$q_n(\mu) = \frac{1}{\pi}\left\{p_n(\mu)\log\frac{1+\mu}{1-\mu} - f_n(\mu)\right\} \tag{11}$$

$f_n(\mu)$ being the same polynomial as in 24·17 (9) and (12). Now if n is even the odd solution 16·04 (5) is

$$\Theta_2 = \mu - \frac{(n-1)(n+2)}{3!}\mu^3 + \frac{(n-3)(n-1)(n+2)(n+4)}{5!}\mu^5 \ldots \tag{12}$$

The coefficient of μ in q_n (n even) can be obtained from (19) below by finding the first derivative with respect to μ for $\mu = 0$, and expanding it in powers of h.
If n is odd, the even solution, with the sign changed, is found to be

$$\Theta_2 = -1 + \frac{n(n+1)}{2!}\mu^2 - \frac{(n-2)n(n+1)(n+3)}{4!}\mu^4 + \ldots \tag{13}$$

The constant term in q_n (n odd) can be obtained by putting $\mu = 0$ in (19) and expanding in powers of h. Then

$$q_n(\mu) = k_n \frac{2}{\pi}\Theta_2, \tag{14}$$

where, in particular,

$$k_0 = 1, \quad k_1 = 1, \quad k_2 = -2, \quad k_3 = -\tfrac{2}{3}. \tag{15}$$

The integral 24·17 (18) can also be adapted to give a form for $q_n(\mu)$ on the cut. If $\mu \to \mu_0$, where μ_0 is on the cut, we can indent the path from -1 to 1 by a small semicircle below μ_0; hence

$$q_n(\mu+0i) = \frac{1}{\pi}\int_{-1,\,\mu-i\epsilon}^{1}\frac{p_n(\nu)}{\mu-\nu}d\nu, \tag{16}$$

and similarly

$$q_n(\mu-0i) = \frac{1}{\pi}\int_{-1,\,\mu+i\epsilon}^{1}\frac{p_n(\nu)}{\mu-\nu}d\nu, \tag{17}$$

$$q_n(\mu) = \frac{1}{\pi}P\int_{-1}^{1}\frac{p_n(\nu)}{\mu-\nu}d\nu, \tag{18}$$

where P denotes the principal value.

We find also, for h small,

$$\Sigma h^n q_n(\mu) = \frac{1}{\pi\sqrt{(1-2\mu h + h^2)}}\log\frac{\sqrt{(1-2\mu h + h^2)}+\mu-h}{\sqrt{(1-2\mu h + h^2)}-\mu+h}. \tag{19}$$

Now if

$$r = \sqrt{(x^2+y^2+z^2)}, \quad h = a/r, \quad z = \mu r, \quad R = \sqrt{\{x^2+y^2+(z-a)^2\}}, \tag{20}$$

this leads to

$$\frac{1}{r}\Sigma\left(\frac{a}{r}\right)^n q_n(\mu) = \frac{1}{\pi R}\log\frac{R+z-a}{R-z+a}, \tag{21}$$

whence

$$r^{-n-1}q_n(\mu) = \lim_{a\to 0}\frac{1}{\pi n!}\frac{\partial^n}{\partial a^n}\left(\frac{1}{R}\log\frac{R+z-a}{R-z+a}\right) \tag{22}$$

$$= \frac{(-1)^n}{\pi n!}\frac{\partial^n}{\partial z^n}\left(\frac{1}{r}\log\frac{r+z}{r-z}\right). \tag{23}$$

Also by comparison of the terms containing $\log\frac{r+z}{r-z}$, using 24·04 (22)

$$r^{-n-1}q_n^s(\mu)e^{is\lambda} = \frac{(-1)^n}{\pi n!}\left(\frac{\partial}{\partial x}+i\frac{\partial}{\partial y}\right)^s\frac{\partial^{n-s}}{\partial z^{n-s}}\left(\frac{1}{r}\log\frac{r+z}{r-z}\right). \tag{24}$$

24·19. Asymptotic approximation to $p_n^s(\mu)$ for n and s both large.* By the method of steepest descents we find, using 24·162 (29), for $0 < \mu < 1$,

$$p_n^s(\mu) \sim \frac{1}{\sqrt{(2\pi)}} \sin^s\theta \left(\frac{n-s}{n}\right)^{n+\frac{1}{2}} \frac{(n+s)^{n+s+\frac{1}{2}} M^{-\frac{1}{2}}}{(M+\mu s)^s (n\mu - M)^{n+\frac{1}{2}}}, \qquad (1)$$

where $M = \sqrt{(s^2 - n^2 \sin^2\theta)} > 0$.

If $M^2 < 0$, put $N = \sqrt{(n^2 \sin^2\theta - s^2)} > 0$, $\arg(s\mu + iN) = \alpha$, $\arg(n\mu - iN) = -\beta$;

then $\qquad p_n^s(\mu) \sim \sqrt{\left(\frac{2}{\pi}\right) \frac{(n-s)^{\frac{1}{2}n-\frac{1}{2}s+\frac{1}{4}}(n+s)^{\frac{1}{2}n+\frac{1}{2}s+\frac{1}{4}}}{n^{n+\frac{1}{2}}}} N^{-\frac{1}{2}} \sin\{(n+\tfrac{1}{2})\beta - s\alpha + \tfrac{1}{4}\pi\}.$ (2)

24·20. Other solutions of Laplace's equation in spherical polar coordinates. In the simplest possible case, $n = s = 0$, the solution of Legendre's equation is

$$\Theta = A + B\log\frac{1+\mu}{1-\mu} = A + B\log\frac{r+z}{r-z}.$$

The term A represents that in p_0, which is a constant. The second has branch points at $\mu = \pm 1$, and is a multiple of $q_0(\mu)$. We have virtually had another harmonic of degree 0 already, since $\phi = \lambda$ is an obvious solution of Laplace's equation in spherical polar coordinates, excluded for a complete sphere by the condition that the solution must be a single-valued function of x, y, z. Another is $\log \varpi$; and since any functions of $z + ix$ and $z + iy$ are other solutions of Laplace's equation, further harmonics of degree 0 are the real and imaginary parts of their logarithms, namely

$$\log(z^2 + x^2)^{\frac{1}{2}} = \log r + \tfrac{1}{2}\log(1 - \sin^2\theta \sin^2\lambda), \quad \log(z^2 + y^2)^{\frac{1}{2}} = \log r + \tfrac{1}{2}\log(1 - \sin^2\theta \cos^2\lambda), \quad (1)$$

$$\tan^{-1}(\tan\theta \cos\lambda), \quad \tan^{-1}(\tan\theta \sin\lambda). \qquad (2)$$

In fact if we assume a function of θ and λ only as a solution of Laplace's equation we get

$$\sin\theta \frac{\partial}{\partial\theta}\left(\sin\theta \frac{\partial\phi}{\partial\theta}\right) + \frac{\partial^2\phi}{\partial\lambda^2} = 0, \qquad (3)$$

which is satisfied by the real and imaginary parts of any function of $\log\tan\tfrac{1}{2}\theta + i\lambda$, i.e. of $\tan\tfrac{1}{2}\theta\, e^{i\lambda}$. This does not include solutions containing $\log r$. The multiplicity of solutions of Laplace's equation in spherical polar coordinates is therefore endless. Apart from the constant solution, however, the new solutions all either become infinite like $\log r$ at the centre, are infinite for some θ, λ, or do not return to their original values when θ or λ is increased continuously by 2π. By differentiation with regard to x, y, z and multiplying by appropriate powers of r we can build up harmonics of any other degree, just as we built up the functions of the first kind from derivatives of $1/r$. But none of these solutions satisfy our fundamental rule, that they must be expansible in a sphere in a triple Taylor series in x, y, z, and they are therefore excluded by physical principles at the outset from any solution intended to hold within a complete sphere or outside one. If we make the restriction that the solutions are to be finite and periodic in λ they become limited to the solutions of Legendre's associated equation. This condition arises in the problems of spheroids. For the prolate spheroid, the condition that ϕ over a surface of constant ξ is finite and continuous ensures that it is expansible in a series in $p_n^s(\cos\eta)(\cos s\lambda, \sin s\lambda)$, the coefficients depending on ξ. But then the condition that the terms must satisfy Laplace's equation ensures that the factors in ξ must be $p_n^s(\cosh\xi)$ and $q_n^s(\cosh\xi)$. The modifications for an oblate spheroid are obvious. The physical conditions therefore show that the solutions that we have obtained are those actually required. In practice $\cosh\xi$ will be real and greater than 1 for a prolate spheroid on the surface and at all external points, so that the singularity of q_n^s is excluded from the region considered and the q_n^s solution becomes admissible. For an oblate spheroid the functions that occur are $p_n^s(i\sinh\xi)$, $q_n^s(i\sinh\xi)$, with ξ real; and $i\sinh\xi$ cannot be ± 1, so that the q_n^s is again admissible in external problems. (We have seen that it is inadmissible in internal problems for a different reason.)

---

* Other asymptotic approximations are given by Watson, *Mess. Math.* 47, 1917, 151–60; *Camb. Phil. Soc. Mems.* 22, 1918, 277–308.

24·21. Expansions of $e^{-i\kappa R}/R$.

In many applications of the wave equation we need an expansion of this function, analogous to that of $1/R$ in potential theory, where $R^2 = (x_i - x_i')^2$ and the expansion is to be in a series of terms in $p_n(\cos\vartheta)$ multiplied by the product of two functions, one of r and one of r'. The expansion can be determined as follows. The function satisfies

$$\nabla^2 \psi = \kappa^2 \psi, \qquad (1)$$

and can therefore, if $r < r'$, be expanded in a series

$$\frac{e^{-i\kappa R}}{R} = \sum_{n=0}^{\infty} A_n(r') \frac{J_{n+1/2}(\kappa r)}{r^{1/2}} p_n(\cos\vartheta). \qquad (2)$$

We write

$$\cos\vartheta = \mu, \qquad (3)$$

$$R^2 = r^2 + r'^2 - 2rr'\mu. \qquad (4)$$

Then with r and r' kept constant

$$R\, dR = -rr'\, d\mu, \qquad (5)$$

$$-\frac{\partial}{\kappa R \partial(\kappa R)} = \frac{\partial}{\kappa^2 rr' \partial\mu}. \qquad (6)$$

Hence

$$\left(-\frac{\partial}{\kappa R \partial(\kappa R)}\right)^m \frac{e^{-i\kappa R}}{R} = \sum A_n(r') \frac{J_{n+1/2}(\kappa r)}{r^{1/2}} \left(\frac{1}{\kappa^2 rr'}\right)^m \frac{d^m}{d\mu^m} p_n(\mu). \qquad (7)$$

Now if $m > n$, $d^m p_n(\mu)/d\mu^m = 0$. If $m < n$,

$$\lim_{r \to 0} \frac{J_{n+1/2}(\kappa r)}{r^{m+1/2}} = 0. \qquad (8)$$

So if we make r tend to zero the sum on the right of (7) reduces to the limit of the term with $n = m$, namely

$$\lim_{r \to 0} A_m(r') \frac{J_{m+1/2}(\kappa r)}{r^{1/2}(\kappa^2 rr')^m} \frac{(2m)!}{2^m m!} = \frac{A_m(r')}{(\kappa r')^m} \frac{\kappa^{1/2}}{2^{m+1/2}(m+\tfrac{1}{2})!} \frac{(2m)!}{2^m m!}. \qquad (9)$$

But

$$(2m)! = 2^{2m} m!(m-\tfrac{1}{2})!/\sqrt{\pi}, \qquad (10)$$

and hence (9) reduces to

$$\frac{A_m(r')}{(\kappa r')^m} \frac{\kappa^{1/2}}{(m+\tfrac{1}{2})\sqrt{(2\pi)}}. \qquad (11)$$

Putting $r = 0$ on the left of (7) we have therefore

$$\left(-\frac{\partial}{\kappa r' \partial(\kappa r')}\right)^m \frac{e^{-i\kappa r'}}{r'} = \frac{A_m(r')}{(\kappa r')^m} \frac{\kappa^{1/2}}{(m+\tfrac{1}{2})\sqrt{(2\pi)}}.$$

But

$$\text{Hi}_{m+1/2}(x) = i\sqrt{\left(\frac{2}{\pi}\right)} x^{m+1/2} \left(-\frac{\partial}{x\partial x}\right)^m \frac{e^{-ix}}{x},$$

and therefore

$$A_m(r') = -(m+\tfrac{1}{2})\pi i \frac{\text{Hi}_{m+1/2}(\kappa r')}{\sqrt{r'}}.$$

Hence for $r < r'$

$$\frac{e^{-i\kappa R}}{R} = -\pi i \sum_{n=0}^{\infty} (n+\tfrac{1}{2}) \frac{\text{Hi}_{n+1/2}(\kappa r')}{\sqrt{r'}} \frac{J_{n+1/2}(\kappa r)}{\sqrt{r}} p_n(\cos\vartheta). \qquad (12)$$

For $\kappa = 0$ this reduces to 24·05 (4). For $r > r'$ we must interchange r and r'.

24·22. The classification of multipole radiation.

A familiar solution of the equations of propagation of electromagnetic waves corresponds to radiation from an oscillating dipole. Other solutions are, however, important physically, and we consider now how they may be classified.

We have seen that the $2n+1$ quantities K^s_{-n-1} ($s = -n, -n+1, ..., n$) transform like the set of quantities $x_1^{n-s} x_2^{n+s}$ when x_1, x_2 undergo a unitary transformation of determinant unity. We shall speak of such a set of quantities as transforming according to R_n and as giving a representation of a real rotation of order n. The three quantities $ib - a = x_1^2$, $c = x_1 x_2$, $ib + a = x_2^2$, where a, b, c are the components of a null vector in three dimensions, transform according to R_1.

We denote the electromagnetic vector potential by $A(A_1, A_2, A_3)$ and the scalar potential by ϕ. Then we have in free space

$$\nabla^2 A - \frac{1}{c^2} \frac{\partial^2 A}{\partial t^2} = 0, \quad \nabla^2 \phi - \frac{1}{c^2} \frac{\partial^2 \phi}{\partial t^2} = 0. \tag{1}$$

It will be convenient to take as components of A, not A_1, A_2, A_3 but $(iA_2 - A_1, A_3, iA_2 + A_1)$, which transform according to R_1. We call these A_μ ($\mu = -1, 0, 1$). They satisfy

$$\nabla^2 A_\mu - \frac{1}{c^2} \frac{\partial^2 A_\mu}{\partial t^2} = 0. \tag{2}$$

These equations have solutions of the form

$$u^{s'}_{n'} = e^{i\kappa ct} f_{n'}(\kappa r) p^{s'}_{n'} e^{is'\lambda} = \frac{(-1)^{n'}}{n'!} e^{i\kappa ct} f_{n'}(\kappa r) r^{n'+1} K^{s'}_{-n'-1}, \tag{3}$$

from 24·04(22); where $(\kappa r)^{1/2} f_{n'}(\kappa r)$ is a Bessel function of κr of order $n' + \frac{1}{2}$. We consider the case where the Bessel function is the function Hi; then

$$\text{Hi}_{n'+1/2}(x) = \left(\frac{2}{\pi}\right)^{1/2} (-1)^{n'} x^{n'+1/2} \left(\frac{d}{x\,dx}\right)^{n'} \left(\frac{ie^{-ix}}{x}\right). \tag{4}$$

Let c^s_n be a set of quantities that transform according to R_n. Then it may be shown* that if and only if n' has one of the values $n, n \pm 1$, three linear combinations of $c^s_n u^{s'}_{n'}$ can be formed which transform according to R_1. We write these as

$$W^S_1 = \sum_{s+s'=S} \begin{pmatrix} n & n' & | & 1 \\ s & s' & | & S \end{pmatrix} c^s_n u^{s'}_{n'} \quad (S = -1, 0, 1), \tag{5}$$

where the coefficients $\begin{pmatrix} n & n' & | & 1 \\ s & s' & | & S \end{pmatrix}$ can be determined.

The three components A_μ transform like W^S_1. We have three cases to consider. To distinguish them we write in turn h, a and b for c.

(1) $n' = n$. $\quad A_\mu = \sum_s \begin{pmatrix} n & n & | & 1 \\ s & \mu-s & | & \mu \end{pmatrix} h^s_n u^{\mu-s}_n.$ (6)

(2) $n' = n+1$. $\quad A_\mu = \sum_s \begin{pmatrix} n & n+1 & | & 1 \\ s & \mu-s & | & \mu \end{pmatrix} a^s_n u^{\mu-s}_{n+1}.$ (7)

(3) $n' = n-1$. $\quad A_\mu = \sum_s \begin{pmatrix} n & n-1 & | & 1 \\ s & \mu-s & | & \mu \end{pmatrix} b^s_n u^{\mu-s}_{n-1}.$ (8)

---

* B. L. v. d. Waerden, *Die Gruppentheoretische Methode in der Quantenmechanik*, §18.

The h_n^s solutions differ fundamentally from the others with respect to reflexion of the axes in the origin. This reverses all coordinates and is a change from a right-handed to a left-handed set of axes. Some vectors, for instance displacements, have all their components reversed under such a transformation, and are called polar vectors. Others, including all vector products of polar vectors, are unchanged and are called axial vectors. We may speak of any set of c_n^s as an n-vector, and again we have two types. Those which are multiplied by $(-1)^n$ on reflexion we shall call polar n-vectors, and those multiplied by $(-1)^{n+1}$ we shall call axial n-vectors. Now A_μ and $u_n^{\mu-s}$ are polar, and hence h_n^s is an axial n-vector and a_n^s and b_n^s are polar n-vectors.

It is convenient to define (6) as a vector potential for the electromagnetic field of a magnetic 2^n-pole and either (7) or (8) as a vector potential for the electromagnetic field of an electric 2^n-pole.

The scalar potential ϕ also satisfies

$$\nabla^2 \phi - \frac{1}{c^2}\frac{\partial^2 \phi}{\partial t^2} = 0; \quad \operatorname{div} \mathbf{A} + \frac{1}{c}\frac{\partial \phi}{\partial t} = 0. \tag{9}$$

ϕ is a true scalar, that is, it does not change sign on reflexion in the origin. It is of the form

$$\Sigma \begin{pmatrix} n & n & 0 \\ s & -s & 0 \end{pmatrix} c_n^s u_n^{-s}, \tag{10}$$

where the c_n^s are components of a polar n-vector. It follows that the scalar potential corresponding to (6) must be zero, since h_n^s is an axial n-vector, and that the c_n^s for cases (7) (8) are related to a_n^s, b_n^s respectively.

From the relations $\quad \mathbf{H} = \operatorname{curl} \mathbf{A}, \quad \mathbf{E} = -\operatorname{grad} \phi - \frac{1}{c}\frac{\partial \mathbf{A}}{\partial t}$

the properties of $\mathbf{E}$ and $\mathbf{H}$ corresponding to the three cases can be deduced. They are summarized in the following tables, in which it is understood that linear combinations are to be formed of the quantities given:

	A	ϕ	E	H
2^n electric pole	$a_n^s u_{n+1}^{s'}$	$a_n^s u_n^{s'}$	$a_n^s \langle u_{n+1}^{s'}, u_{n-1}^{s'} \rangle$	$a_n^s u_n^{s'}$
	$b_n^s u_{n-1}^{s'}$	$b_n^s u_n^{s'}$	$b_n^s \langle u_{n+1}^{s'}, u_{n-1}^{s'} \rangle$	$b_n^s u_n^{s'}$
2^n magnetic pole	$h_n^s u_n^{s'}$	0	$h_n^s u_n^{s'}$	$h_n^s \langle u_{n+1}^{s'}, u_{n-1}^{s'} \rangle$

It will be noticed that whichever form is taken for the potentials for the 2^n electric pole the form of $\mathbf{E}$ and $\mathbf{H}$ is the same, as it should be. In comparing the fields for magnetic and electric poles of the same order we see that the roles of $\mathbf{E}$ and $\mathbf{H}$ are interchanged.

We give some simple cases.

Electric dipole. $n = 1$, $n' = 0$:
$$A_\mu = b_1^\mu u_0^0 = b_1^\mu e^{i\kappa ct}\left(\frac{2}{\kappa\pi}\right)^{1/2}\frac{ie^{-i\kappa r}}{r} = i\kappa d_1^\mu \frac{e^{i\kappa(ct-r)}}{r}$$

say. In the special case $d_1^0 = 1$, $d_1^{\pm 1} = 0$,
$$\phi = -\frac{\partial}{\partial z}\frac{e^{i\kappa(ct-r)}}{r}.$$

For $\kappa = 0$ this reduces to the scalar potential for a dipole of unit strength with its axis along the z axis.

Electric quadripole. $n = 2, n' = 1$:

Suppose $b_2^s = 0$ unless $s = 0$. Then we obtain a solution
$$A_x = \frac{Cx}{r^2}\left(1 - \frac{i}{\kappa r}\right)e^{i\kappa(ct-r)},$$
$$A_y = \frac{Cy}{r^2}\left(1 - \frac{i}{\kappa r}\right)e^{i\kappa(ct-r)},$$
$$A_z = -\frac{2Cz}{r^2}\left(1 - \frac{i}{\kappa r}\right)e^{i\kappa(ct-r)},$$
$$\phi = -2C\frac{p_2(\cos\theta)}{r}\left(1 - \frac{3i}{\kappa r} - \frac{3}{\kappa^2 r^2}\right)e^{i\kappa(ct-r)}.$$

Magnetic dipole. $n = 1$, $n' = 1$:

Suppose $h_1^0 \neq 0$, $h_1^{\pm 1} = 0$. Then we obtain a solution
$$A_x = \frac{Cy}{r^2}\left(1 - \frac{i}{\kappa r}\right)e^{i\kappa(ct-r)}; \quad A_y = -\frac{Cx}{r^2}\left(1 - \frac{i}{\kappa r}\right)e^{i\kappa(ct-r)}; \quad A_z = \phi = 0.$$

For further discussion the reader is referred to H. C. Brinkman, Diss. Utrecht, 1932, 1–59. W. Heitler, *Proc. Camb. Phil. Soc.* 32, 1936, 112–26. H. A. Kramers, *Physica*, 10, 1943, 261–72. V. Berestetzky, *J. Phys. U.S.S.R.* 11, 1947, 85–90. B. Jeffreys, *Proc. Camb. Phil. Soc.* 48, 1952, 470–81.

24·23. Multipole expansion of scalar and vector potentials. In the theory of electromagnetic radiation we sometimes need an expansion that takes together all terms containing the same power of $\kappa r'$. This can be obtained from 24·21 (12), but also directly in the following manner. Suppose that we have a finite distribution of charge density $\rho(x_i')$ and current density $j(x_i')$, both containing a time factor $\exp(i\kappa ct)$, and satisfying the equation of continuity
$$\frac{\partial j_i}{\partial x_i'} + \frac{1}{c}\frac{\partial \rho}{\partial t} = 0, \tag{1}$$

that is,
$$\frac{\partial j_i}{\partial x_i'} + i\kappa\rho = 0. \tag{2}$$

Then the scalar and vector potentials (ϕ, A) at $P(x_i)$ are given by
$$\phi = \iiint \rho\frac{e^{-i\kappa R}}{R}dx_1'dx_2'dx_3', \quad A_i = \iiint j_i\frac{e^{-i\kappa R}}{R}dx_1'dx_2'dx_3', \tag{3}$$

24·23 Multipole expansions

ρ and j_i being regarded as functions of the position of $Q(x_i')$. These satisfy the equation

$$\frac{\partial A_i}{\partial x_i} + \frac{1}{c}\frac{\partial \phi}{\partial t} = 0, \tag{4}$$

for if f is any function of x_i, x_i', writing $dx_1' dx_2' dx_3' = d\tau$, we have

$$i\kappa \iiint \rho f d\tau = -\iiint \frac{\partial j_i}{\partial x_i'} f d\tau = -\iiint \frac{\partial}{\partial x_i'}(fj_i) d\tau + \iiint j_i \frac{\partial f}{\partial x_i'} d\tau. \tag{5}$$

We can find a bounding surface such that $j_i = 0$ at every point of it; hence the first integral vanishes by Green's lemma. Now if f is a function of $x_i - x_i'$ only,

$$\frac{\partial f}{\partial x_i} = -\frac{\partial f}{\partial x_i'}, \tag{6}$$

and if in particular

$$f(x_i, x_i') = \frac{e^{-i\kappa R}}{R}, \tag{7}$$

$$i\kappa \iiint \rho \frac{e^{-i\kappa R}}{R} d\tau + \iiint j_i \frac{\partial}{\partial x_i}\left(\frac{e^{-i\kappa R}}{R}\right) d\tau = 0, \tag{8}$$

that is,

$$\frac{1}{c}\frac{\partial \phi}{\partial t} + \frac{\partial A_i}{\partial x_i} = 0. \tag{9}$$

In (5) put successively $f = 1, x_k', x_k' x_m'$. Then if $\kappa \neq 0$,

$$\iiint \rho d\tau = 0, \tag{10}$$

$$i\kappa \iiint \rho x_k' d\tau = \iiint j_i \delta_{ik} d\tau = \iiint j_k d\tau, \tag{11}$$

$$i\kappa \iiint \rho x_k' x_m' d\tau = \iiint j_i(x_m' \delta_{ik} + x_k' \delta_{im}) d\tau = \iiint (j_k x_m' + j_m x_k') d\tau. \tag{12}$$

Now the Taylor expansion of ϕ for $r' < r$ may be written

$$\phi = \iiint \rho \exp\left(-x_i' \frac{\partial}{\partial x_i}\right)\left(\frac{e^{-i\kappa r}}{r}\right) d\tau$$

$$= \sum_{n=0}^{\infty} \frac{(-1)^n}{n!} \iiint \rho x_i' \frac{\partial}{\partial x_i}\left(x_k' \frac{\partial}{\partial x_k}\right)(\ldots) \text{(to } n \text{ operations)}\left(\frac{e^{-i\kappa r}}{r}\right) d\tau$$

$$= 0 - \iiint \rho x_i' \frac{\partial}{\partial x_i}\left(\frac{e^{-i\kappa r}}{r}\right) d\tau + \frac{1}{2}\iiint \rho x_i' x_k' \frac{\partial^2}{\partial x_i \partial x_k}\left(\frac{e^{-i\kappa r}}{r}\right) d\tau - \ldots$$

$$\phi e^{i\kappa r} = \kappa \iiint \rho \frac{x_i' x_i}{r^2}\left(i + \frac{1}{\kappa r}\right) d\tau + \tfrac{1}{2}\kappa^2 \iiint \rho x_i' x_k'\left\{-\frac{x_i x_k}{r^3} + \frac{1}{\kappa r^2}\left(i + \frac{1}{\kappa r}\right)\left(\frac{3x_i x_k}{r^2} - \delta_{ik}\right)\right\} d\tau + \ldots. \tag{13}$$

We put:

$$P_i = \iiint \rho x_i' d\tau, \quad P_{ik} = \iiint \rho(x_i' x_k' - \tfrac{1}{3}\delta_{ik} x_j' x_j') d\tau; \quad P = \iint \rho x_j' x_j' d\tau. \tag{14}$$

Then P_i is a vector which can be expressed as a linear combination of three b_1^a. Also $P_{ii} = 0$ and the tensor P_{ik} has five independent components, which can be expressed as linear combinations of five b_2^a. P is a scalar. Then

$$\phi = e^{-i\kappa r}\left[\kappa P_i \frac{x_i}{r^2}\left(i + \frac{1}{\kappa r}\right) + \tfrac{1}{2}\kappa^2 P_{ik}\left\{-\frac{x_i x_k}{r^3} + \frac{1}{\kappa r^2}\left(i + \frac{1}{\kappa r}\right)\left(\frac{3x_i x_k}{r^2} - \delta_{ik}\right)\right\} - \frac{1}{6}\frac{\kappa^2 P}{r} + \ldots\right]. \tag{15}$$

The term in P_i gives the scalar potential for an electric dipole, that in P_{ik} for an electric quadripole and that in P for an electric single pole. The corresponding expansion for A_i is

$$A_i = \iiint j_i \exp\left(-x'_k \frac{\partial}{\partial x_k}\right)\left(\frac{e^{-i\kappa r}}{r}\right) d\tau$$

$$= \iiint j_i \frac{e^{-i\kappa r}}{r} - \iiint j_i \frac{x_k x'_k}{r} \frac{\partial}{\partial r}\left(\frac{e^{-i\kappa r}}{r}\right) d\tau$$

$$= i\kappa \frac{e^{-i\kappa r}}{r} P_i - \frac{1}{2}\iiint (j_i x'_k + j_k x'_i) \frac{x_k}{r} \frac{d}{dr}\left(\frac{e^{-i\kappa r}}{r}\right) d\tau$$

$$- \frac{1}{2}\iiint (j_i x'_k - j_k x'_i) \frac{x_k}{r} \frac{d}{dr}\left(\frac{e^{-i\kappa r}}{r}\right) d\tau, \tag{17}$$

by (11) and (14). But by (12) the second term is

$$-\frac{1}{2}\frac{i\kappa}{r} x_k P_{ik} \frac{d}{dr}\left(\frac{e^{-i\kappa r}}{r}\right) - \frac{i\kappa}{6} P \frac{x_i}{r} \frac{d}{dr}\left(\frac{e^{-i\kappa r}}{r}\right); \tag{18}$$

these two terms give vector potentials for an electric quadripole and electric single pole respectively, corresponding to those in the scalar potential. The field (E, H) of the electric single pole vanishes. (In the conditions stated after (5) there cannot be an oscillating electric single pole.)

In the last term of (17) we write

$$M_{ik} = \tfrac{1}{2}\iiint (x'_i j_k - x'_k j_i) d\tau. \tag{19}$$

This is an antisymmetrical tensor and can be related to an axial h_1^s. The term containing it is a vector potential for a magnetic dipole. We have then as far as terms in $(\kappa r')^2$

$$A_i = e^{-i\kappa r}\left\{i\kappa \frac{P_i}{r} - \tfrac{1}{2}\kappa^2 P_{ik} \frac{x_k}{r^2}\left(1 - \frac{i}{\kappa r}\right) - i\kappa M_{ik} \frac{x_k}{r^2}\left(1 - \frac{i}{\kappa r}\right) - \tfrac{1}{6}\kappa^2 P \frac{x_i}{r^2}\left(1 - \frac{i}{\kappa r}\right)\right\}. \tag{20}$$

From 24·21 (12) we can write A_i as

$$\iiint j_i \sum_{n=0}^{\infty} \sum_{s=-n}^{n} C_{n,s} \frac{\mathrm{H}i_{n+1/2}(\kappa r)}{\sqrt{r}} \frac{J_{n+1/2}(\kappa r')}{\sqrt{r'}} p_n^s(\cos\theta') p_n^s(\cos\theta) e^{is(\lambda-\lambda')} d\tau, \tag{21}$$

where $C_{n,s}$ are constants. It can be shown that the terms

$$\iiint j_i \frac{J_{n+1/2}(\kappa r')}{\sqrt{r'}} p_n^s(\cos\theta') e^{-is\lambda'} d\tau, \tag{22}$$

can be expressed as linear combinations of quantities $c^{s'}_{n\pm 1}, c^{s'}_n$. Sets of linear combinations of products of these with the $p_n^s(\cos\theta) e^{is\lambda}$ that give a polar 1-vector can be formed in the following ways.

$c^{s'}_{n+1} P_n^s(\cos\theta) e^{is\lambda}$ $c^{s'}_{n+1}$ polar 2^{n+1} electric pole

$c^{s'}_{n-1} p_n^s(\cos\theta) e^{is\lambda}$ $c^{s'}_{n-1}$ polar 2^{n-1} electric pole

$c^{s'}_n p_n^s(\cos\theta) e^{is\lambda}$ $c^{s'}_n$ axial 2^n magnetic pole.

The terms in ϕ and A_i containing

$$\frac{\partial^n}{\partial x_i \partial x_k \ldots} \frac{e^{-i\kappa r}}{r},$$

give rise in general to mixed multipole radiation, as we have seen from the case of

$$(\partial^2/\partial x_i \partial x_k)(e^{-i\kappa r}/r).$$

EXAMPLES

1. An electric dipole of moment M is at a distance a from the centre of a sphere of dielectric constant K and radius b ($b<a$) and the centre of the sphere is on the axis of the dipole. Prove that the force of attraction between the sphere and the dipole is

$$\frac{K-1}{a^4} M^2 \sum_0^\infty \frac{n(n+1)^2(n+2)}{Kn+n+1} \left(\frac{b}{a}\right)^{2n+1}. \tag{M.T. 1939.}$$

2. Obtain the gravitational potential of a thin spherical layer of matter, the surface density of which has axial symmetry.

A uniform nearly spherical solid of density ρ has the surface $r = a(1+\epsilon P_2)$ as its boundary. It is surrounded by liquid of volume $\frac{4}{3}\pi(b^3-a^3)$ and uniform density σ. Show that, provided the solid is completely covered with liquid, the equation of the free surface is $r = b(1+\eta P_2)$, where

$$\eta = \frac{3(\rho-\sigma)a^5\epsilon}{b^2\{5(\rho-\sigma)a^3+2\sigma b^3\}}. \tag{M.T. 1936.}$$

3. A nearly spherical conductor is bounded by the surface

$$r = a\{1+\epsilon P_n(\cos\theta)\} \quad (n>1),$$

where ϵ is small. It is insulated and placed in a uniform field F with its axis in the direction of the field. Find the disturbance in the field due to the presence of the conductor, and show that the surface density of the induced charge at a point of the conductor is

$$\frac{3F}{4\pi}\cos\theta + \frac{3F}{4\pi}\frac{n}{2n+1}\epsilon\{(n-2)P_{n-1}(\cos\theta)+(n+1)P_{n+1}(\cos\theta)\}. \tag{M.T. 1939.}$$

4. A small magnet is placed at the centre of a spherical shell of iron of radii a and b and permeability μ. Show that the field outside the shell is reduced by the presence of the iron in the ratio

$$\left\{1+\frac{2}{9}\frac{(\mu-1)^2}{\mu}\left(1-\frac{a^3}{b^3}\right)\right\}^{-1}.$$

5. A condenser is formed from two conducting spheres of radii a, b ($a<b$) with centres at A, B, where AB is of length c and $(c/a)^2$ may be neglected. Find the capacity of the condenser when the outer sphere is earthed, and show that if Q is the charge on the inner sphere, the surface density at a point P of that sphere is

$$\frac{Q}{4\pi a^2}\left\{1-\frac{3a^2c\cos PAB}{b^3-a^3}\right\}. \tag{Prelim. 1936.}$$

6. An electric charge e is distributed uniformly along the segment BC of a straight line OBC. P is any point such that $OP<OB$. Writing

$$OP=r, \quad OB=b, \quad OC=c, \quad \cos BOP=\mu,$$

obtain an expression in terms of r, P_0, P_1, ... for the potential at P due to the charge.

An earthed conducting sphere of radius a ($<b$) is placed with its centre at O. Obtain an expression for the potential at a point P in the region $a \leqslant r < b$, and show that the charge induced on the sphere is

$$-ea\frac{\log c - \log b}{c-b}.$$

666   *Examples*

7. If K_n is a solid harmonic of degree n, show that

$$\nabla^2(r^m K_n) = m(m+2n+1)r^{m-2}K_n.$$

If U is a homogeneous polynomial in x, y, z of degree $2n$, show that

$$\iint U\,dS = \frac{4\pi a^{2n+2}}{(2n+1)!}(\nabla^2)^n U,$$

integration being over the sphere $r = a$.

8. Prove from the integral definition that

$$(n+1)q_{n+1}(\mu) - (2n+1)\mu q_n(\mu) + n q_{n-1}(\mu) = 0.$$

Evaluate $q_0(\mu)$, $q_1(\mu)$ by integration; and hence prove 24·17 (9) and 24·17 (18).

9. Prove directly that $\displaystyle\int_{-1}^1 \frac{p_n(\nu)\,d\nu}{\mu-\nu}$ is a solution of Legendre's equation and hence that it is a multiple of $q_n(\mu)$.

10. Prove that if μ is not real and between -1 and 1,

$$q_n(\mu) = \frac{1}{2^n \pi}\int_{-1}^1 \frac{(1-\nu^2)^n}{(\mu-\nu)^{n+1}}\,d\nu.$$

11. Prove that if μ is not real and between -1 and 1, and ν is in this interval,

$$\frac{1}{\mu-\nu} = \tfrac{1}{2}\pi \sum_{n=0}^\infty (2n+1)p_n(\nu)q_n(\mu). \qquad \text{(Heine.)}$$

12. Express $\sin^2\theta(1+\cos\theta)\sin 3\lambda$ as a sum of tesseral harmonics.

13. Toroidal coordinates σ, ψ, λ are related to cylindrical coordinates ϖ, z, λ by the equation

$$z + i\varpi = a \cot \tfrac{1}{2}(\psi + i\sigma).$$

Prove that
$$ds^2 = \frac{a^2(d\sigma^2 + d\psi^2 + \sinh^2\sigma\, d\lambda^2)}{(\cosh\sigma - \cos\psi)^2},$$

and hence express Laplace's equation in terms of σ, ψ, λ.

Show that there exist simple toroidal harmonics of the type

$$(\cosh\sigma - \cos\psi)^{1/2}\cos n\psi \cos m\lambda\, f(\cosh\sigma),$$

and determine f. (I.C. 1937.)

14. Prove that $\displaystyle e^{-i\kappa z} = (\tfrac{1}{2}\pi)^{1/2}\sum_{n=0}^\infty (2n+1)(-i)^n \frac{J_{n+\frac12}(\kappa r)}{(\kappa r)^{1/2}} p_n(\mu).$

15. If $u = \tfrac{1}{2}(x+iy)$, $v = \tfrac{1}{2}(-x+iy)$, $a+b+c = n$, $a-b = s$, where a, b, c are positive integers or zero, show that

$$\sum_c \frac{u^a v^b z^c}{a!\,b!\,c!}$$

satisfies Laplace's equation.

If $n - s$ is even show that all c must be even and that (24·03) (5) with

$$g(x,y) = \frac{u^{\frac12(n+s)}v^{\frac12(n-s)}}{[\tfrac12(n+s)]!\,[\tfrac12(n-s)]!}, \quad h(x,y) = 0,$$

is of the given form. Find the corresponding solution when $n - s$ is odd.

The terms in ϕ and A_i containing

$$\frac{\partial^n}{\partial x_i \partial x_k \ldots} \frac{e^{-i\kappa r}}{r},$$

give rise in general to mixed multipole radiation, as we have seen from the case of

$$(\partial^2/\partial x_i \partial x_k)(e^{-i\kappa r}/r).$$

EXAMPLES

1. An electric dipole of moment M is at a distance a from the centre of a sphere of dielectric constant K and radius b ($b<a$) and the centre of the sphere is on the axis of the dipole. Prove that the force of attraction between the sphere and the dipole is

$$\frac{K-1}{a^4} M^2 \sum_0^\infty \frac{n(n+1)^2(n+2)}{Kn+n+1} \left(\frac{b}{a}\right)^{2n+1}. \tag{M.T. 1939.}$$

2. Obtain the gravitational potential of a thin spherical layer of matter, the surface density of which has axial symmetry.

A uniform nearly spherical solid of density ρ has the surface $r = a(1+\epsilon P_2)$ as its boundary. It is surrounded by liquid of volume $\tfrac{4}{3}\pi(b^3-a^3)$ and uniform density σ. Show that, provided the solid is completely covered with liquid, the equation of the free surface is $r = b(1+\eta P_2)$, where

$$\eta = \frac{3(\rho-\sigma) a^5 \epsilon}{b^2\{5(\rho-\sigma) a^3 + 2\sigma b^3\}}. \tag{M.T. 1936.}$$

3. A nearly spherical conductor is bounded by the surface

$$r = a\{1+\epsilon P_n(\cos\theta)\} \quad (n>1),$$

where ϵ is small. It is insulated and placed in a uniform field F with its axis in the direction of the field. Find the disturbance in the field due to the presence of the conductor, and show that the surface density of the induced charge at a point of the conductor is

$$\frac{3F}{4\pi}\cos\theta + \frac{3F}{4\pi} \frac{n}{2n+1} \epsilon\{(n-2) P_{n-1}(\cos\theta) + (n+1) P_{n+1}(\cos\theta)\}. \tag{M.T. 1939.}$$

4. A small magnet is placed at the centre of a spherical shell of iron of radii a and b and permeability μ. Show that the field outside the shell is reduced by the presence of the iron in the ratio

$$\left\{1 + \frac{2}{9}\frac{(\mu-1)^2}{\mu}\left(1-\frac{a^3}{b^3}\right)\right\}^{-1}.$$

5. A condenser is formed from two conducting spheres of radii a, b ($a<b$) with centres at A, B, where AB is of length c and $(c/a)^3$ may be neglected. Find the capacity of the condenser when the outer sphere is earthed, and show that if Q is the charge on the inner sphere, the surface density at a point P of that sphere is

$$\frac{Q}{4\pi a^2}\left\{1 - \frac{3a^2 c \cos PAB}{b^3 - a^3}\right\}. \tag{Prelim. 1936.}$$

6. An electric charge e is distributed uniformly along the segment BC of a straight line OBC. P is any point such that $OP < OB$. Writing

$$OP = r, \quad OB = b, \quad OC = c, \quad \cos BOP = \mu,$$

obtain an expression in terms of r, P_0, P_1, ... for the potential at P due to the charge.

An earthed conducting sphere of radius a ($<b$) is placed with its centre at O. Obtain an expression for the potential at a point P in the region $a \leqslant r < b$, and show that the charge induced on the sphere is

$$-ea\frac{\log c - \log b}{c-b}.$$

Examples

7. If K_n is a solid harmonic of degree n, show that

$$\nabla^2(r^m K_n) = m(m+2n+1)\, r^{m-2} K_n.$$

If U is a homogeneous polynomial in x, y, z of degree $2n$, show that

$$\iint U\, dS = \frac{4\pi a^{2n+2}}{(2n+1)!}\, (\nabla^2)^n U,$$

integration being over the sphere $r = a$.

8. Prove from the integral definition that

$$(n+1)\, q_{n+1}(\mu) - (2n+1)\, \mu q_n(\mu) + n q_{n-1}(\mu) = 0.$$

Evaluate $q_0(\mu)$, $q_1(\mu)$ by integration; and hence prove 24·17 (9) and 24·17 (18).

9. Prove directly that $\displaystyle\int_{-1}^{1} \frac{p_n(\nu)\, d\nu}{\mu - \nu}$ is a solution of Legendre's equation and hence that it is a multiple of $q_n(\mu)$.

10. Prove that if μ is not real and between -1 and 1,

$$q_n(\mu) = \frac{1}{2^n \pi}\int_{-1}^{1} \frac{(1-\nu^2)^n}{(\mu-\nu)^{n+1}}\, d\nu.$$

11. Prove that if μ is not real and between -1 and 1, and ν is in this interval,

$$\frac{1}{\mu - \nu} = \tfrac{1}{2}\pi \sum_{n=0}^{\infty} (2n+1)\, p_n(\nu)\, q_n(\mu). \qquad \text{(Heine.)}$$

12. Express $\sin^2\theta\, (1 + \cos\theta) \sin 3\lambda$ as a sum of tesseral harmonics.

13. Toroidal coordinates σ, ψ, λ are related to cylindrical coordinates ϖ, z, λ by the equation

$$z + i\varpi = a \cot \tfrac{1}{2}(\psi + i\sigma).$$

Prove that
$$ds^2 = \frac{a^2(d\sigma^2 + d\psi^2 + \sinh^2\sigma\, d\lambda^2)}{(\cosh\sigma - \cos\psi)^2},$$

and hence express Laplace's equation in terms of σ, ψ, λ.

Show that there exist simple toroidal harmonics of the type

$$(\cosh\sigma - \cos\psi)^{1/2} \cos n\psi \cos m\lambda\, f(\cosh\sigma),$$

and determine f. (I.C. 1937.)

14. Prove that $\displaystyle e^{-i\kappa z} = (\tfrac{1}{2}\pi)^{1/2} \sum_{n=0}^{\infty} (2n+1)(-i)^n \frac{J_{n+\frac{1}{2}}(\kappa r)}{(\kappa r)^{1/2}}\, p_n(\mu).$

15. If $u = \tfrac{1}{2}(x+iy)$, $v = \tfrac{1}{2}(-x+iy)$, $a+b+c = n$, $a-b = s$, where a, b, c are positive integers or zero, show that

$$\sum_c \frac{u^a v^b z^c}{a!\, b!\, c!}$$

satisfies Laplace's equation.

If $n - s$ is even show that all c must be even and that (24·03) (5) with

$$g(x, y) = \frac{u^{\frac{1}{2}(n+s)} v^{\frac{1}{2}(n-s)}}{[\tfrac{1}{2}(n+s)]!\, [\tfrac{1}{2}(n-s)]!}, \quad h(x, y) = 0,$$

is of the given form. Find the corresponding solution when $n-s$ is odd.

Chapter 25

ELLIPTIC FUNCTIONS

'Double, double, toil and trouble.'
SHAKESPEARE, *Macbeth*

25·01. Definition: illustrations. The elliptic functions are characterized by the following properties. (1) They are single-valued analytic functions over the whole plane, except at isolated points where they have poles. (2) Two numbers ω and ω' exist, whose ratio is not real, such that for all values of z

$$f(z+\omega) = f(z+\omega') = f(z)$$

and therefore

$$f(z+m\omega+n\omega') = f(z),$$

for all positive and negative integral values of m and n.

The name 'elliptic function' arose first from the relation of the functions to the integral that arises in the determination of the perimeter of an ellipse. They present themselves in physics in numerous ways. They have an extensive purely mathematical theory, possibly more extensive than any other family of transcendental functions, and consisting mainly of three-volume works.* Unlike most functions treated in this book they satisfy no linear differential equation; they satisfy non-linear differential equations of the first order. While most of the functions of mathematical physics, in one form or another, arise from the equations of waves and heat conduction, elliptic functions turn up in all sorts of places, usually unexpectedly.

They occur naturally in potential problems concerning the interior of a rectangle, the potential being kept zero over the boundary, while there may be charges inside. The boundary conditions are satisfied if a suitable doubly-infinite series of images is built up by successive reflexions in the sides, and the total complex potential is easily seen to have the essential properties of an elliptic function. An important series of problems relating to lines of equally spaced vortices between rigid barriers has been solved in this way by Rosenhead,† and have applications to the resistance of solids in wind tunnels.

25·02. Periodicity of solutions of a type of differential equation: pendulum and central orbits. A common mode of occurrence of periodicities is through a differential equation of the form

$$\left(\frac{dx}{dt}\right)^2 = f(x). \tag{1}$$

which is to be regarded as a first integral of a second-order equation $\ddot{x} = \tfrac{1}{2}f'(x)$; where $\dot{x}$ is real and not zero at $x = x_0$, and $f(x)$ has simple zeros at $x = a$ and $x = b$, where $a < x_0 < b$. The solution is

$$t - t_0 = \int_{x_0}^{x} \frac{dx}{\sqrt{\{f(x)\}}}. \tag{2}$$

---

* Standard works are those of Tannery and Molk, Enneper, and Halphen. A recent one (in one volume) by E. H. Neville achieves greater symmetry in the treatment.
† *Phil. Trans.* A, 228, 1929, 275–329.

If $\dot{x}$ is positive at $t = t_0$, it remains positive till x reaches b. A formally possible solution henceforward would be that x remains equal to b; for then we should have $\dot{x} = 0$, $f(x) = f(b) = 0$.* But this solution is excluded by the condition that $\ddot{x} = \tfrac{1}{2}f'(b)$, which is not zero because by hypothesis b is a simple zero of $f(x)$. (An equivalent condition, if complex values are admitted, is that x is an analytic function of t.) Hence $\dot{x}$ must reverse its sign and x decreases steadily to a, where $\dot{x}$ again reverses. It follows that in the conditions stated x is a periodic function of t, with a real period

$$\omega = 2 \int_a^b \frac{dx}{\sqrt{\{f(x)\}}}. \tag{3}$$

In one of its simplest forms this type of solution occurs in the motion of a simple pendulum given that at $t = 0$, $\theta = \theta_0$, $\dot\theta = \dot\theta_0$. The energy equation is

$$\dot\theta^2 = \dot\theta_0^2 + \frac{2g}{l}(\cos\theta - \cos\theta_0) = \dot\theta_0^2 + \frac{4g}{l}(\sin^2 \tfrac{1}{2}\theta_0 - \sin^2 \tfrac{1}{2}\theta).$$

This is positive at $\theta = \theta_0$ and at $\theta = 0$, and has no zero for $0 < \theta < \theta_0$. It is negative at $\theta = \pi$ provided that

$$\dot\theta_0^2 < \frac{2g}{l}(1 + \cos\theta_0),$$

and then has a simple zero between $\theta = \theta_0$ and $\theta = \pi$. Similarly it has one between $-\theta_0$ and $-\pi$. Hence θ is a function of t with a real period.

Again, consider motion in a central orbit under an acceleration Ar^m towards the centre; the work function per unit mass is

$$U = -\frac{Ar^{m+1}}{m+1} \quad (A > 0).$$

The energy and angular momentum equations are

$$\dot r^2 + r^2\dot\theta^2 = \dot r_0^2 + r_0^2\dot\theta_0^2 + 2(U - U_0) = V^2 + 2(U - U_0),$$

$$r^2\dot\theta = r_0^2\dot\theta_0,$$

whence
$$\dot r^2 = V^2 - \frac{r_0^4\dot\theta_0^2}{r^2} + 2(U - U_0) = f(r).$$

The signs of $f(r)$ are seen to run as follows ($f(r_0) = \dot r_0^2 > 0$)

	$r \to 0$	$r = r_0$	$r \to \infty$
$m < -3$	$+\infty$	$+$	$V^2 - 2U_0$
$-3 < m < -1$	$-\infty$	$+$	$V^2 - 2U_0$
$-1 < m$	$-\infty$	$+$	$-\infty$

Hence $f(r)$ always has a pair of zeros on opposite sides of r_0 if $m > -1$, and if $-3 < m < -1$ provided $V^2 < 2U_0$. They are easily seen to be simple. Hence in motion under a central acceleration r is periodic unconditionally if $m > -1$ (including as an important special case the law of the direct distance), and if $-3 < m < -1$ provided $V^2 < 2U_0$. The latter covers as a special case the inverse square law $m = -2$, and the critical case $V^2 = 2U_0$ is that of parabolic motion.

* Professor D. R. Hartree informs us that this solution is actually often given by mechanical integrating machines unless special precautions are taken.

25·03. Differential equation satisfied by sn t: inversion of the integral. This is a powerful type of argument because it enables us to demonstrate exact periodicity in suitable conditions by mere inspection of the differential equation, without the need to solve it first. Let us apply it to the differential equation

$$\left(\frac{dx}{dt}\right)^2 = (1-x^2)(1-k^2x^2) = X^2, \tag{1}$$

where k is real and less than 1, and $\ddot{x}$ is assumed always to exist. When $x = 0$, we take $dx/dt = +1$, and the right side has simple zeros at $x = \pm 1$. Then $\dfrac{d^2x}{dt^2} = \dfrac{1}{2}\dfrac{dX^2}{dx}$. Hence there is a period

$$\omega = 4\int_0^1 \frac{dx}{\sqrt{\{(1-x^2)(1-k^2x^2)\}}} = 4K, \tag{2}$$

say. Also, if $t = 0$ when $x = 0$,

$$t = \int_0^x \frac{dx}{\sqrt{\{(1-x^2)(1-k^2x^2)\}}}. \tag{3}$$

This is known as the first elliptic integral and K as the first complete elliptic integral. Integrals of these forms were tabulated and much studied by Euler and Legendre, but the whole theory was revolutionized by a remark of Abel. Clearly if $k = 0$, $t = \sin^{-1}x$; and the study of the integrals as such is analogous to studying the properties of $\sin^{-1}x$ instead of the much more manageable function $\sin x$. The method initiated by Abel and developed by Jacobi is to write sn $(t; k)$ for x in the upper limit of (3) and to regard the equation as specifying the upper limit as a function of t, which will have period $4K$ analogous to the period 2π of the circular functions. Then sn $(t; k)$ is to be regarded as a generalization of $\sin t$. It is usual to suppress explicit statement of k when the same value is to be understood through the work.

K can be expressed as a power series in k by putting $x = \sin\phi$; then

$$K = \int_0^{\frac{1}{2}\pi} \frac{d\phi}{\sqrt{(1-k^2\sin^2\phi)}} = \int_0^{\frac{1}{2}\pi} \left\{1 + \sum_{n=1}^{\infty} \frac{1.3.5\ldots 2n-1}{2^n n!} k^{2n}\sin^{2n}\phi\right\}d\phi$$

$$= \tfrac{1}{2}\pi\left\{1 + \sum_{n=1}^{\infty}\left(\frac{1.3.5\ldots 2n-1}{2.4.6\ldots 2n}\right)^2 k^{2n}\right\}. \tag{4}$$

This is a hypergeometric series with radius of convergence 1.
Evidently

$$\operatorname{sn} 0 = 0, \quad \operatorname{sn} K = 1; \tag{5}$$

$$\begin{aligned}
&\operatorname{sn}(-t) = -\operatorname{sn} t, \quad \operatorname{sn}(2K-t) = \operatorname{sn} t; \\
&\operatorname{sn}(2K+t) = -\operatorname{sn} t, \\
&\frac{d}{dt}\operatorname{sn}(2K-t) = -\frac{d}{dt}\operatorname{sn} t = \frac{d}{dt}\operatorname{sn}(2K+t).
\end{aligned} \tag{6}$$

So far we have considered only real values of x and t. But if sn t possesses an analytic continuation to complex values of t this can be taken as completing the definition of sn t. Further, since the equalities (6) hold for all real values in certain intervals they will also hold for all values accessible by continuation.

25·031. We have therefore to consider whether (3) can be inverted when x is complex, to give x as an analytic function of t, and whether this function will be single-valued. We still take k real and $0 < k < 1$. Other values will be considered later. The differential equation shows that t, considered as a function of x, is analytic in a region that does not enclose a branch-point of X, where

$$X^2 = (1-x^2)(1-k^2x^2). \tag{1}$$

Also
$$\frac{d^2x}{dt^2} = \frac{1}{2}\frac{d}{dx}\left(\frac{dx}{dt}\right)^2 = \frac{1}{2}\frac{dX^2}{dx}, \tag{2}$$

so that the extension to complex values introduces just the extra condition that the differential equation is a first integral of a second order equation of the form $\ddot{x} = \tfrac{1}{2}f'(x)$. The integral for t converges when x tends to infinity, so that we must also consider its behaviour near the corresponding values of t (which will differ according to the path chosen), and there will be infinities of x considered as a function of t.

If a is $\neq \alpha$, where α is any of ± 1, $\pm 1/k$, then $t(x) - t(a)$ will begin with a linear term and the series for t can be inverted, giving x as a single-valued analytic function of t within any circle about $t(a)$ such that no point $x = \alpha$ lies within it and such that x does not tend to infinity. Near any point $x = \alpha$, we have

$$t - t(\alpha) = (x-\alpha)^{1/2}\phi(x), \tag{3}$$

where $\phi(x)$ is analytic and not zero at $x = \alpha$. This is of a form considered in 12·052, and x is single-valued near $t = t(\alpha)$, while $dx/dt = 0$ there.

For x large on any path going to infinity

$$t = \left[\int_0^\infty - \int_x^\infty\right]\frac{dx}{X} = M \pm \int_x^\infty \frac{dx}{kx^2}(1+\psi)$$
$$= M \pm \frac{1}{kx}(1+\chi), \tag{4}$$

where M is finite and ψ, χ are analytic and of order $1/x$. Hence

$$x = \pm\frac{1}{k(t-M)} + g(t), \tag{5}$$

where $g(t)$ is analytic at $t = M$. The points $t = M$ are therefore simple poles of x, of residue $\pm k$.

Hence x is a single-valued analytic function of t over the set of values of t taken by the integral for all modifications of the path, with simple poles as its only singularities.

If X^2 was a quintic or some polynomial of higher degree in x, with no triple or higher zeros, x considered as a function of t would still be analytic at values of t that make X^2 zero. But for x large the term in $1/x$ in (4) would be replaced by one in $x^{-\gamma}$, with $\gamma > 1$. Hence every infinity of x would also be a branch-point, and x would not be single-valued. In other words we could take x along a large arc, not closed, and arrive at the same value of t as we started from.

If X^2 was a cubic we should get a term in $x^{-1/2}$ instead of $1/x$, and x would still be single-valued as a function of t, but with double poles instead of simple poles. The student may examine for himself what happens if X^2 is linear or quadratic. The result is that inversion of the integral makes x a single-valued function of t, provided that X^2 is of degree not higher than 4.

It follows at once that 25·03 (6) can be extended to all accessible t, and that $\operatorname{sn} t$ has period $4K$.

Now consider the continuation when $1 < x < 1/k$. To give X a definite sign we take x to pass 1 by a semicircle above it, so that in the new range

$$\sqrt{(1-x^2)} = -i\sqrt{(x^2-1)}. \tag{6}$$

Then
$$t = K + i\int_1^x \frac{dx}{\sqrt{\{(x^2-1)(1-k^2x^2)\}}} = K + iv, \tag{7}$$

say; and when $x \to 1/k$,
$$t \to K + iK', \tag{8}$$

where
$$K' = \int_1^{1/k} \frac{dx}{\sqrt{\{(x^2-1)(1-k^2x^2)\}}}. \tag{9}$$

The quartic in the denominator is real and positive in the range, with simple zeros at $1, 1/k$. Then x is a function of v with period $2K'$. Therefore by (7) it is a function of t with period $2iK'$, and therefore

$$\operatorname{sn}(t + 2iK') = \operatorname{sn} t, \tag{10}$$

for all t. Also
$$\operatorname{sn}(2iK' - t) = -\operatorname{sn} t. \tag{11}$$

Now take x purely imaginary and the path along the imaginary axis. Then if $x = iy$,

$$t = i\int_0^y \frac{dy}{\sqrt{(1+y^2)(1+k^2y^2)}}. \tag{12}$$

Take
$$\frac{1+y^2}{1+k^2y^2} = z^2, \tag{13}$$

and take the upper limit as $x = i\infty$. Then

$$t = i\int_1^{1/k} \frac{dz}{\sqrt{\{(z^2-1)(1-k^2z^2)\}}} = iK'. \tag{14}$$

Hence iK' is a pole of x considered as a function of t. Also for large y

$$t \doteqdot iK' - i\int_y^\infty \frac{dy}{ky^2} = iK' - \frac{i}{ky}, \tag{15}$$

$$x = iy \doteqdot \frac{1}{k(t-iK')}, \tag{16}$$

and the residue of $\operatorname{sn} t$ at iK' is $+1/k$. Since $2iK'$ is a period there is another pole at $-iK'$ with residue $1/k$. Since $\operatorname{sn}(2K+t) = -\operatorname{sn} t$, there are poles at $2K \pm iK'$ with residue $-1/k$.

K' can be put in another form; in (9) put

$$z^2 = \frac{1-k^2x^2}{1-k^2}, \quad k'^2 = 1-k^2, \tag{17}$$

$$K' = \int_0^1 \frac{dz}{\sqrt{\{(1-z^2)(1-k'^2z^2)\}}}, \tag{18}$$

so that K' is the same function of k' as K is of k.

Now if we use only paths starting from $x = 0$ and such that $\Im(x) \geq 0$ for $\Re(x)$ between $\pm 1/k$, t will tend to iK' when x tends to infinity in any direction. On the real axis it increases from 0 to K as x goes from 0 to 1; as x increases further to $1/k$, the real part of t remains constant and the imaginary part increases to K'; as $x \to \infty$, the imaginary part remains constant and the real part decreases to 0. As x goes from 0 to -1, $-1/k$, $-\infty$ (passing -1 and $-1/k$ on the positive side), t goes to $-K$, $-K+iK'$, iK'. Thus t is bounded in the half-plane of x, and both its real and imaginary parts must take their greatest and least values on the boundary. Hence for all x in the upper half-plane

$$-K \leq \Re(t) \leq K, \quad 0 \leq \Im(t) \leq K'.$$

If we take paths confined to the lower half-plane, we get similar results except for the reversal of all imaginary parts. For any t satisfying these inequalities it follows that an x can be found, and is unique since the region contains no branch-point of t. But the integral provides no definition of t for $|\Im(t)| > K'$ unless the path crosses the real axis where $\Re(x) > 1$ or $\Re(x) < -1$. We can however make the real and imaginary parts of t as large as we like by including circuits about *two* branch-points. In the figure the integrals about C and C', in the directions shown, are $4K$ and $2iK'$; and by including a sufficient number of

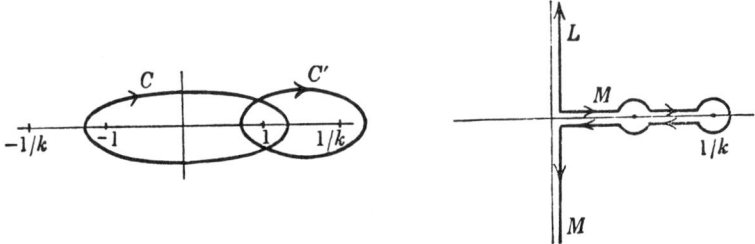

circuits of either type before proceeding from 0 to x we can make the real and imaginary parts of the integral as large as we like, positive or negative, without altering dx/dt. By modification of the contour we can get a simple proof of the equivalence of our expressions (9), (12) for K'. For L can be deformed into M without crossing a branch-point so as to pass to $-i\infty$ as shown. The loop around part of the real axis contributes $2iK'$ as defined in (9), the two parts between 0 and 1 cancel, and the path from 0 to $-i\infty$ makes a contribution equal and opposite to the integral along L. Since the integrals along L and M must be equal the equivalence of (9) and (12) follows.[a]

We can take the path to include a single loop about $x = +1$; this will contribute $2K$, but X is reversed in sign when we get back to 0. By taking the integral to $-x$ we therefore get

$$\operatorname{sn}(2K+t) = -\operatorname{sn} t, \quad \frac{d}{dt}\operatorname{sn}(2K+t) = -\frac{d}{dt}\operatorname{sn} t. \tag{19}$$

Taking it to x we get
$$\operatorname{sn}(2K-t) = \operatorname{sn} t. \tag{20}$$

By suitable combinations of paths C and C' in either sense, with or without a loop about $+1$, we can therefore attach a meaning to $\operatorname{sn} t$ by inversion of the integral for any value of t. Hence $\operatorname{sn} t$ as defined by continuation is a single-valued analytic function over the whole t plane, with simple poles of residue $1/k$ at all points $4mK + (2n+1)iK'$, and simple poles of residue $-1/k$ at all points $(4m+2)K + (2n+1)iK'$, where m and n are positive or negative integers, and is a doubly periodic function with periods $4K$ and $2iK'$.

25·04. Impossibility of three independent periods. If an analytic function $f(z)$ is not constant, and has a set of periods $\omega_1, \omega_2, \ldots$ there must be one of them with the smallest modulus. For if not, take any regular point z of the function; then $f(z) = f(z+\omega_r)$ for all r. If there is no smallest $|\omega_r|$, there are infinitely many points $t = \omega_r$ with 0 as a limit point such that $f(z+t) = f(z)$. Hence $f(z+t) - f(z) = 0$ for all t, and therefore $f(z)$ is a constant, contrary to hypothesis. Denote the ω_r with least modulus by ω. If possible let Ω be a period with the same argument as ω but not an integral multiple of ω. Let m be an integer such that $m|\omega| < |\Omega| < (m+1)|\omega|$. Then $\Omega - m\omega$ is a period with smaller modulus than ω, and we have a contradiction. Hence all periods with the same argument as ω are integral multiples of ω. Let ω' be the period with smallest modulus that is not an integral multiple of ω. Consider the plane marked out into parallelograms whose corners are the points $m\omega + n\omega'$, m and n being integers. If possible let ω'' be a third period, not expressible in the form $m\omega + n\omega'$. Then any expression $\omega'' - m\omega - n\omega'$ is also a period, and we can choose m and n so as to make it lie within the parallelogram whose corners are $0, \omega, \omega', \omega+\omega'$. Denote the result by Ω, and draw the diagonal AC connecting the two obtuse angles. Then if Ω lies within the triangle OAC, the length of $O\Omega$ is less than the larger of $|\omega|$ and $|\omega'|$, and therefore $< |\omega'|$, contrary to our hypothesis that ω' is the period of smallest modulus that is not a multiple of ω. Similarly if Ω lies within ABC, the length $B\Omega$ is less than $|\omega'|$. Hence an analytic function, not a constant, cannot have more than two independent periods.

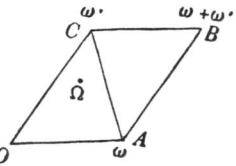

This result is relevant to the inversion of an integral $\int dx/X$, where X^2 is of higher degree than 4. For a contour about any two simple zeros of X^2 should define a period. For X^2 of degree ≤ 4 it can be shown that the integrals about such paths are connected by relations of the form $\epsilon\omega + \epsilon'\omega' + \epsilon''\omega'' = 0$, where $\epsilon, \epsilon', \epsilon''$ are 0 or ± 1, and determine at most two independent periods. For higher degrees there are more independent periods, which are possible *only* because the function is no longer single-valued. Such functions are called *Abelian functions*.

We shall call the pair of periods ω, ω' defined as in the last proposition the *fundamental periods*. Any parallelogram whose corners are at $z_0, z_0+\omega, z_0+\omega', z_0+\omega+\omega'$ will be called a fundamental parallelogram. In general z_0 will be taken to be such that there is no pole on any side. The greater part of the theory of elliptic functions rests on two simple theorems.

25·05. *The integral of an elliptic function about a fundamental parallelogram is zero.* For if $ABCD$ is a fundamental parallelogram the integrals along opposite sides AB, DC, are equal and opposite on account of the property of periodicity and the fact that the sides are traversed in opposite directions; similarly those along BC, DA cancel.

An elliptic function with no singularities in a fundamental parallelogram is a constant. For if it is bounded in a fundamental parallelogram it is bounded over the whole plane on account of the periodicity; and therefore it is a constant by Liouville's theorem.

If $f(z)$ is an elliptic function, $f'(z)/f(z)$ is another. Applying the first theorem to $f(z)$, we see that *the sum of the residues at all poles in a fundamental parallelogram is* 0. Applying it to $f'(z)/f(z)$, we see that *the number of poles of $f(z)$ is equal to the number of zeros, multiple poles and zeros being taken multiply*. Applying it to $f'(z)/\{f(z)-c\}$, where c is any constant, the same holds. But the poles of $f(z) - c$ are those of $f(z)$; hence $f(z) - c$ *has the same number of zeros in a fundamental parallelogram as $f(z)$, whatever c may be*.

Now take round a fundamental parallelogram

$$\frac{1}{2\pi i}\int \frac{zf'(z)}{f(z)}dz.$$

If the poles of $f(z)$ are α_r, the zeros β_r, the integral is equal to $\Sigma \beta_r - \Sigma \alpha_r$, multiple poles and zeros being taken multiply. But

$$\left[\int_{z_0}^{z_0+\omega} + \int_{z_0+\omega+\omega'}^{z_0+\omega'}\right]\frac{zf'(z)}{f(z)}dz = -\omega'\int_{z_0}^{z_0+\omega}\frac{f'(z)}{f(z)}dz = 2\pi i p'\omega',$$

where p' is an integer, and similarly the integrals along the other two sides give a multiple of $2\pi i \omega$. Hence *the sums of the values of z at the zeros and poles of f(z) in a fundamental parallelogram differ by an expression of the form $p\omega + p'\omega'$, where p and p' are integers (possibly zero)*.

25·06. Other Jacobian elliptic functions: cn, dn, etc. The function sn z has poles in every fundamental parallelogram. This is surprising in a generalization of sin z, but we recall that sin z has an essential singularity at infinity. When $k \to 0$ the imaginary period of sn z tends to infinity, and the poles merge into an essential singularity at infinity.

There are four poles in a parallelogram of sides $4K$, $4iK'$; hence the function takes every other value four times in such a parallelogram.

Associated with sn t there are functions corresponding to the other trigonometric functions. The first two are defined by

$$\operatorname{cn} t = \sqrt{(1-\operatorname{sn}^2 t)}, \qquad \operatorname{cn} 0 = 1,$$
$$\operatorname{dn} t = \sqrt{(1-k^2\operatorname{sn}^2 t)}, \qquad \operatorname{dn} 0 = 1.$$

The zeros and poles of $1 - \operatorname{sn}^2 t$, $1 - k^2\operatorname{sn}^2 t$ are double when these functions are considered as functions of t; hence cn t and dn t are single-valued when continued analytically. The other functions are defined by division:

$$\operatorname{sc} t = \operatorname{sn} t/\operatorname{cn} t, \qquad \operatorname{cs} t = \operatorname{cn} t/\operatorname{sn} t,$$
$$\operatorname{sd} t = \operatorname{sn} t/\operatorname{dn} t, \qquad \operatorname{ds} t = \operatorname{dn} t/\operatorname{sn} t,$$
$$\operatorname{cd} t = \operatorname{cn} t/\operatorname{dn} t, \qquad \operatorname{dc} t = \operatorname{dn} t/\operatorname{cn} t,$$
$$\operatorname{ns} t = 1/\operatorname{sn} t, \qquad \operatorname{nc} t = 1/\operatorname{cn} t.$$
$$\operatorname{nd} t = 1/\operatorname{dn} t,$$

Then the fundamental differential equation can be written

$$\frac{d}{dt}\operatorname{sn} t = \operatorname{cn} t \operatorname{dn} t. \qquad (1)$$

Hence
$$\frac{d}{dt}\operatorname{cn}^2 t = -\frac{d}{dt}\operatorname{sn}^2 t = -2\operatorname{sn} t \operatorname{cn} t \operatorname{dn} t,$$
$$\frac{d}{dt}\operatorname{cn} t = -\operatorname{sn} t \operatorname{dn} t, \qquad (2)$$

and similarly
$$\frac{d}{dt}\operatorname{dn} t = -k^2 \operatorname{sn} t \operatorname{cn} t. \qquad (3)$$

Since
$$\operatorname{sn} K = 1, \quad \operatorname{cn} K = 0, \quad \operatorname{dn} K = k'; \qquad (4)$$
$$\operatorname{sn}(K+iK') = 1/k, \quad \operatorname{cn}(K+iK') = -ik'/k, \quad \operatorname{dn}(K+iK') = 0. \qquad (5)$$

To reach values of t between K and $K+iK'$, x must pass $+1$ on the upper side; then $\sqrt{(1-x^2)}$ becomes $-i\sqrt{(x^2-1)}$ and cn t is negative imaginary in this range. When $k = 0$, cn t reduces to $\cos t$, dn t to 1, sc t to $\tan t$, and so on. The functions are also reducible to elementary functions in the other extreme case $k = 1$, when

$$\operatorname{sn} t = \tanh t, \quad \operatorname{cn} t = \operatorname{dn} t = \operatorname{sech} t.$$

25·07. Differentiation. By direct transformation we find the following differential equations satisfied by the functions:

$$y = \operatorname{cn} t \quad \frac{dy}{dt} = -\sqrt{\{(1-y^2)(k'^2+k^2y^2)\}}, \qquad y = \operatorname{nc} t \quad \frac{dy}{dt} = \sqrt{\{(y^2-1)(k'^2y^2+k^2)\}},$$

$$y = \operatorname{dn} t \quad \frac{dy}{dt} = -\sqrt{\{(1-y^2)(y^2-k'^2)\}}, \qquad y = \operatorname{nd} t \quad \frac{dy}{dt} = \sqrt{\{(y^2-1)(1-k'^2y^2)\}},$$

$$y = \operatorname{sc} t \quad \frac{dy}{dt} = \sqrt{\{(1+y^2)(1+k'^2y^2)\}}, \qquad y = \operatorname{ns} t \quad \frac{dy}{dt} = -\sqrt{\{(y^2-1)(y^2-k^2)\}},$$

$$y = \operatorname{cs} t \quad \frac{dy}{dt} = -\sqrt{\{(y^2+1)(y^2+k'^2)\}}, \qquad y = \operatorname{cd} t \quad \frac{dy}{dt} = -\sqrt{\{(1-y^2)(1-k^2y^2)\}},$$

$$y = \operatorname{sd} t \quad \frac{dy}{dt} = \sqrt{\{(1-k'^2y^2)(1+k^2y^2)\}}, \qquad y = \operatorname{dc} t \quad \frac{dy}{dt} = \sqrt{\{(y^2-1)(y^2-k^2)\}},$$

$$y = \operatorname{ds} t \quad \frac{dy}{dt} = -\sqrt{\{(y^2-k'^2)(y^2+k^2)\}}.$$

The roots are to be taken positive for $0 \leqslant t \leqslant K$.

We notice that the functions ns t, cs t, ds t have poles of residue 1 at the origin and that the signs of the constants in the roots are respectively both negative, both positive, and different. E. H. Neville finds it convenient to take these as the fundamental functions.

25·08. Residues at poles. We have seen that all poles of sn t are simple poles of residue $\pm 1/k$; hence all those of cn t are simple with residue $\pm i/k$, and of dn t with residue $\pm i$. When $x = \operatorname{sn} t \to +\infty$, passing $+1$ and $+1/k$ on the positive side, $t \to iK'$, cn t behaves like $-i \operatorname{sn} t$, dn t like $-ik \operatorname{sn} t$. But the residue of sn t at iK' is $1/k$; therefore those of cn t and dn t are $-i/k$ and $-i$. Since cn t and dn t are even functions their residues at $-iK'$ are i/k and i; for

$$\frac{A}{t-\beta} - \frac{A}{t+\beta} = \frac{2A\beta}{t^2-\beta^2},$$

which is an even function of t.

Apart from a sign, cd t and sn t satisfy the same differential equation. But if $K < t < 2K$, sn t is a decreasing function. Since the equation is of the first order and does not contain t explicitly, it follows that

$$\operatorname{sn}(t+\gamma) = \operatorname{cd} t,$$

where γ is some constant. With $\gamma = K$ this holds for $t = 0$, and therefore universally. Hence
$$\operatorname{sn}(t+K) = \operatorname{cn} t / \operatorname{dn} t. \tag{1}$$

By transformation, attending to the signs of the functions between K and $2K$, we get

$$\operatorname{cn}(t+K) = -k' \operatorname{sn} t / \operatorname{dn} t, \quad \operatorname{dn}(t+K) = k'/\operatorname{dn} t. \tag{2}$$

It follows that
$$\operatorname{sn}(t+2K) = -\operatorname{sn} t, \quad \operatorname{cn}(t+2K) = -\operatorname{cn} t, \quad \operatorname{dn}(t+2K) = \operatorname{dn} t. \tag{3}$$

Hence $\operatorname{sn} t$ and $\operatorname{cn} t$ have period $4K$; but $\operatorname{dn} t$ has period $2K$. This permits us to say that the residues of $\operatorname{cn} t$ at $2K + iK'$ and $2K - iK'$ are $+i/k$ and $-i/k$; those of $\operatorname{dn} t$ are $-i$ and $+i$. We therefore have the following set of residues.

	iK'	$-iK'$	$2K+iK'$	$2K-iK'$
$\operatorname{sn} t$	$1/k$	$1/k$	$-1/k$	$-1/k$
$\operatorname{cn} t$	$-i/k$	$+i/k$	$+i/k$	$-i/k$
$\operatorname{dn} t$	$-i$	$+i$	$-i$	$+i$

If we put $\operatorname{ns} t = kz$, z also satisfies the differential equation for $\operatorname{sn} t$ except for a sign. Hence there is a γ such that

$$\operatorname{sn}(t+\gamma) = \pm \frac{1}{k \operatorname{sn} t}.$$

To make the poles correspond we take $\gamma = iK'$, and taking t small we fix the sign as positive since the residue of $\operatorname{sn} t$ is $+1/k$ at $t = iK'$. Hence

$$\operatorname{sn}(t+iK') = \frac{1}{k \operatorname{sn} t}, \qquad \operatorname{sn}(t+2iK') = \operatorname{sn} t. \tag{4}$$

Further, fixing the constants suitably, we get

$$\operatorname{cn}(t+iK') = -\frac{i \operatorname{dn} t}{k \operatorname{sn} t}, \tag{5}$$

$$\operatorname{dn}(t+iK') = -\frac{i \operatorname{cn} t}{\operatorname{sn} t}, \tag{6}$$

and therefore

$$\operatorname{cn}(t+2iK') = -\operatorname{cn} t, \quad \operatorname{dn}(t+2iK') = -\operatorname{dn} t. \tag{7}$$

Also

$$\operatorname{sn}(t+K+iK') = \frac{\operatorname{dn} t}{k \operatorname{cn} t}, \quad \operatorname{cn}(t+K+iK') = -\frac{ik'}{k \operatorname{cn} t}, \quad \operatorname{dn}(t+K+iK') = ik' \frac{\operatorname{sn} t}{\operatorname{cn} t}. \tag{8}$$

$$\operatorname{sn}(t+2K+2iK') = -\operatorname{sn} t, \quad \operatorname{cn}(t+2K+2iK') = \operatorname{cn} t, \quad \operatorname{dn}(t+2K+2iK') = -\operatorname{dn} t. \tag{9}$$

Each of the functions is therefore periodic in a parallelogram whose area is *half* that of the parallelogram of sides $4K$ and $4iK'$. All are periodic in the latter parallelogram. It is usually convenient to use the parallelogram of sides $4K$ and $4iK'$ for this reason.

25·09. Partial fraction and trigonometric expansions. One of the most prolific sources of formulae in elliptic functions is the comparison of the principal parts at poles. The method is illustrated most simply by the functions $\operatorname{ns} t$, $\operatorname{cs} t$, $\operatorname{ds} t$. These are all odd functions with residue 1 at $t = 0$. The effects of adding $2K$, $2iK'$, $2K + 2iK'$ to t are shown in the following table of signs.

	$2K$	$2iK'$	$2K+2iK'$
sn	−	+	−
cn	−	−	+
dn	+	−	−
ns	−	+	−
cs	+	−	−
ds	−	−	+

Hence by Mittag-Leffler's theorem (the functions being bounded on a suitably chosen set of parallelograms tending to infinity)

$$\operatorname{ns} t = \sum_{m=-\infty}^{\infty} \sum_{n=-\infty}^{\infty} \left(\frac{1}{t-4mK-4niK'} - \frac{1}{t-(4m+2)K-4niK'} \right.$$
$$\left. + \frac{1}{t-4mK-(4n+2)iK'} - \frac{1}{t-(4m+2)K-(4n+2)iK'} \right),$$

$$\operatorname{cs} t = \Sigma\Sigma \left(\frac{1}{t-4mK-4niK'} + \frac{1}{t-(4m+2)K-4niK'} \right.$$
$$\left. - \frac{1}{t-4mK-(4n+2)iK'} - \frac{1}{t-(4m+2)K-(4n+2)iK'} \right),$$

$$\operatorname{ds} t = \Sigma\Sigma \left(\frac{1}{t-4mK-4niK'} - \frac{1}{t-(4m+2)K-4niK'} \right.$$
$$\left. - \frac{1}{t-4mK-(4n+2)iK'} + \frac{1}{t-(4m+2)K-(4n+2)iK'} \right).$$

The integrals 25·03 (2), 25·031 (9) or (14) would define K and K' if k was complex (K having singularities at $k = \pm 1$ and K' at $k = 0$), and the series just given will still converge and be analytic functions both of t and of k. Hence they lead to definitions of the elliptic functions, satisfying the same differential equations, even if k is complex, and we can now remove the restriction that k is real.

Now
$$\frac{\pi}{2K} \cot \frac{\pi v}{2K} = \frac{1}{v} + \Sigma' \left(\frac{1}{v-2nK} + \frac{1}{2nK} \right),$$
$$\frac{\pi}{2K} \operatorname{cosec} \frac{\pi v}{2K} = \frac{1}{v} + \Sigma' (-1)^n \left(\frac{1}{v-2nK} + \frac{1}{2nK} \right),$$

and therefore
$$\operatorname{ns} t = \frac{\pi}{2K} \left[\operatorname{cosec} \frac{\pi t}{2K} + \sum_1^{\infty} \left\{ \operatorname{cosec} \frac{\pi}{2K}(t-2niK') + \operatorname{cosec} \frac{\pi}{2K}(t+2niK') \right\} \right]$$
$$= \frac{\pi}{2K} \left\{ \operatorname{cosec} \frac{\pi t}{2K} + \sum_1^{\infty} \frac{4 \sin \frac{\pi t}{2K} \cosh \frac{n\pi K'}{K}}{\cosh \frac{2n\pi K'}{K} - \cos \frac{\pi t}{K}} \right\},$$

$$\operatorname{cs} t = \frac{\pi}{2K} \left\{ \cot \frac{\pi t}{2K} + \sum_1^{\infty} \frac{\sin \frac{\pi t}{K}}{\cosh \frac{4n\pi K'}{K} - \cos \frac{\pi t}{K}} - \sum_0^{\infty} \frac{\sin \frac{\pi t}{K}}{\cosh \frac{(4n+2)\pi K'}{K} - \cos \frac{\pi t}{K}} \right\},$$

$$\operatorname{ds} t = \frac{\pi}{2K} \left\{ \operatorname{cosec} \frac{\pi t}{2K} + \sum_1^{\infty} (-1)^n \frac{4 \sin \frac{\pi t}{2K} \cosh \frac{n\pi K'}{K}}{\cosh \frac{2n\pi K'}{K} - \cos \frac{\pi t}{K}} \right\}.$$

The series are rapidly convergent if K'/K is not very small. A physical illustration is given by a lattice of charged wires arranged regularly in planes, in such a way that every set of four taken at the corners of a cell contains two positively and two negatively charged wires. In such conditions the potential at a point is determined mainly by the charges in the neighbouring planes, the contributions from more distant charges nearly cancelling.

25·10. The addition formulae.

If v is constant, $\operatorname{sn}(u+v)$ is an elliptic function of u, with the same periods as $\operatorname{sn} u$. For $k = 0$ and $k = 1$ it can be expressed in terms of functions of u and v, as follows:

$$\sin(u+v) = \sin u \cos v + \cos u \sin v,$$

$$\tanh(u+v) = \frac{\tanh u + \tanh v}{1 + \tanh u \tanh v}.$$

There is no obvious similarity between these formulae to suggest an analogous expression for general k. But for small v and any k

$$\operatorname{sn}(u+v) = \operatorname{sn} u + \operatorname{cn} u \operatorname{dn} u \operatorname{sn} v + O(v^2)$$

$$= \operatorname{sn} u \operatorname{cn} v \operatorname{dn} v + \operatorname{cn} u \operatorname{dn} u \operatorname{sn} v + O(u^2 v^2),$$

by symmetry. For $k = 0$ the error term vanishes. For $k = 1$ the right side is

$$\tanh u \operatorname{sech}^2 v + \tanh v \operatorname{sech}^2 u + O(u^2 v^2)$$

$$= (\tanh u + \tanh v)(1 - \tanh u \tanh v) + O(u^2 v^2)$$

$$= (1 - \tanh^2 u \tanh^2 v) \tanh(u+v) + O(u^2 v^2).$$

Thus both extreme cases are included in the formula

$$\operatorname{sn}(u+v) = \frac{\operatorname{sn} u \operatorname{cn} v \operatorname{dn} v + \operatorname{cn} u \operatorname{dn} u \operatorname{sn} v}{1 - f(k) \operatorname{sn}^2 u \operatorname{sn}^2 v},$$

where $f(0) = 0, f(1) = 1$; and the formula is right to $O(u^2 v^2)$ for any k.

Now $\operatorname{sn}(u+v)$, considered as a function of u, has a pole at $iK' - v$. But the numerator is finite unless u or v itself is of the form $iK' + 2mK + 2niK'$; hence the pole can arise only from the vanishing of the denominator. But

$$\operatorname{sn}(iK' - v) = -\frac{1}{k \operatorname{sn} v},$$

and therefore this condition is satisfied for all v if, and only if,

$$f(k) = k^2,$$

and then

$$\operatorname{sn}(u+v) = \frac{\operatorname{sn} u \operatorname{cn} v \operatorname{dn} v + \operatorname{cn} u \operatorname{dn} u \operatorname{sn} v}{1 - k^2 \operatorname{sn}^2 u \operatorname{sn}^2 v}. \tag{1}$$

If there is any formula expressing $\operatorname{sn}(u+v)$ in terms of elliptic functions of u and v, it must therefore be (1). It remains to show that (1) is true.

The easiest method is by comparison of residues. The left side has poles of residue $1/k$ at $u = -v \pm iK'$, and of residue $-1/k$ at $u = -v + 2K + iK'$. Adding $2K$ to u reverses both sides of (1), and both sides have period $2iK'$. We need therefore consider only $u = -v + iK'$.

At any pole of $\operatorname{sn} u$ the right side is analytic. Hence poles of the right side can arise only from the vanishing of the denominator, and will be simple. The denominator vanishes if $u = -v \pm iK'$, but also if $u = v \pm iK'$. We must therefore also consider $u = v + iK'$. But

$$\operatorname{sn}(v+iK') \operatorname{cn} v \operatorname{dn} v + \operatorname{sn} v \operatorname{cn}(v+iK') \operatorname{dn}(v+iK')$$

$$= \frac{1}{k \operatorname{sn} v} \operatorname{cn} v \operatorname{dn} v + \operatorname{sn} v \left(-\frac{i \operatorname{dn} v}{k \operatorname{sn} v}\right)\left(-i \frac{\operatorname{cn} v}{\operatorname{sn} v}\right) = 0,$$

and therefore $v+iK'$ is not a pole of the right side. Hence the left and right sides have the same poles. Again, if $u+v-iK'$ is small, say θ,

$$\operatorname{sn}^2(iK'-v+\theta) = \frac{1}{k^2\operatorname{sn}^2(v-\theta)} = \frac{1}{k^2\operatorname{sn}^2 v} + \frac{2\operatorname{cn} v\operatorname{dn} v}{k^2\operatorname{sn}^3 v}\theta + O(\theta^2),$$

and the right of (1) is

$$\frac{2\operatorname{cn} v\operatorname{dn} v}{k\operatorname{sn} v} \bigg/ \frac{2\operatorname{cn} v\operatorname{dn} v}{\operatorname{sn} v}\theta = \frac{1}{k\theta}.$$

Hence both functions have the same poles and the same residues, and can therefore differ only by a constant, which we identify as 0 by taking $u=0$.

Alternatively, knowing that both sides of (1) have the same poles and zeros we infer that their ratio is a constant; and we show that this constant is 1 by taking u small.

The corresponding formulae for $\operatorname{cn}(u+v)$ and $\operatorname{dn}(u+v)$ are

$$\operatorname{cn}(u+v) = \frac{\operatorname{cn} u\operatorname{cn} v - \operatorname{sn} u\operatorname{sn} v\operatorname{dn} u\operatorname{dn} v}{1-k^2\operatorname{sn}^2 u\operatorname{sn}^2 v}, \qquad (2)$$

$$\operatorname{dn}(u+v) = \frac{\operatorname{dn} u\operatorname{dn} v - k^2\operatorname{sn} u\operatorname{sn} v\operatorname{cn} u\operatorname{cn} v}{1-k^2\operatorname{sn}^2 u\operatorname{sn}^2 v}. \qquad (3)$$

These are easily verified by direct transformation.

We have thought it interesting to show how these formulae *might* have been discovered by study of extreme cases. They were actually found in a totally different way. Euler found a complicated identity connecting elliptic *integrals*, which became (1) when translated into Jacobi's notation.

Another method of verification, quite straightforward but best suited for a long spell in a railway waiting room, is to differentiate the right sides and show that the derivatives of each with regard to u and v are equal. They are therefore functions of $u+v$, and are identified by taking $u=0$.

25·11. Infinite products for $\operatorname{sn} t$, $\operatorname{cn} t$, $\operatorname{dn} t$. The function $d(\log\operatorname{sn} t)/dt$ has simple poles of residue $+1$ at all zeros of $\operatorname{sn} t$, and of residue -1 at all poles of $\operatorname{sn} t$. Mittag-Leffler's theorem therefore shows that

$$\frac{d}{dt}\log\operatorname{sn} t - \frac{1}{t} =$$

$$\sum_{-\infty}^{\infty}\Sigma'\left(\frac{1}{t-2niK'-2mK}+\frac{1}{2niK'+2mK}-\frac{1}{t-(2n+1)iK'-2mK}-\frac{1}{(2n+1)iK'+2mK}\right),$$

$m=0$, $n=0$ being excluded from the first two terms. If we take equal numbers of positive and negative values of m and n the constant terms will cancel. Hence

$$\frac{d}{dt}\log\operatorname{sn} t = \frac{\pi}{2K}\left\{\cot\frac{\pi t}{2K} + \Sigma'\cot\frac{\pi(t-2niK')}{2K} - \Sigma\cot\frac{\pi\{t-(2n+1)iK'\}}{2K}\right\},$$

$$\operatorname{sn} t = A\sin\frac{\pi t}{2K}\frac{\prod_{-\infty}^{\infty}{}'\sin\frac{\pi(t-2niK')}{2K}}{\prod_{-\infty}^{\infty}\sin\frac{\pi\{t-(2n+1)iK'\}}{2K}}$$

$$= A\sin\frac{\pi t}{2K}\frac{\prod_{1}^{\infty}\left(\cosh\frac{2n\pi K'}{K}-\cos\frac{\pi t}{K}\right)}{\prod_{1}^{\infty}\left(\cosh\frac{(2n-1)\pi K'}{K}-\cos\frac{\pi t}{K}\right)}.$$

Put $iK' = K\tau$; $q = e^{i\pi\tau} = e^{-\pi K'/K}$. We assume $\Re(K'/K) > 0$. Then $|q| < 1$, and

$$\operatorname{sn} t = A' \sin\frac{\pi t}{2K} \prod_1^\infty \frac{1 - 2q^{2n}\cos\frac{\pi t}{K} + q^{4n}}{1 - 2q^{2n-1}\cos\frac{\pi t}{K} + q^{4n-2}}$$

$$= \frac{2K}{\pi} \sin\frac{\pi t}{2K} \prod_1^\infty \frac{(1-q^{2n-1})^2}{(1-q^{2n})^2} \cdot \frac{1 - 2q^{2n}\cos\frac{\pi t}{K} + q^{4n}}{1 - 2q^{2n-1}\cos\frac{\pi t}{K} + q^{4n-2}},$$

the constant being adjusted to give the correct derivative at $t = 0$. The products are all absolutely convergent since $|q| < 1$. Similarly

$$\frac{d}{dt}\log\operatorname{cn} t = \sum_{-\infty}^\infty \sum_{-\infty}^\infty \left(\frac{1}{t-(2m+1)K-2niK'} - \frac{1}{t-(2n+1)iK'-2mK} \right)$$

$$= -\frac{\pi}{2K} \left\{ \sum_0^\infty \tan\frac{\pi(t-2niK')}{2K} + \sum_0^\infty \cot\frac{\pi\{t-(2n+1)iK'\}}{2K} \right\},$$

whence
$$\operatorname{cn} t = \cos\frac{\pi t}{2K} \prod_1^\infty \frac{(1-q^{2n-1})^2(1+2q^{2n}\cos\frac{\pi t}{K} + q^{4n})}{(1+q^{2n})^2(1-2q^{2n-1}\cos\frac{\pi t}{K} + q^{4n-2})}.$$

Also
$$\operatorname{dn} t = \prod_1^\infty \frac{(1-q^{2n-1})^2(1+2q^{2n-1}\cos\frac{\pi t}{K} + q^{4n-2})}{(1+q^{2n-1})^2(1-2q^{2n-1}\cos\frac{\pi t}{K} + q^{4n-2})}.$$

25·111. ϑ functions. These formulae express the elliptic functions as ratios of four integral functions all possessing the period $4K$. These functions can also be expressed as series. It is convenient to put

$$z = \frac{\pi t}{2K}, \tag{1}$$

so that the period becomes 2π. Then any of the four integral functions, for z with constant imaginary part, can be expressed as a Fourier series, which can be extended to other values of the imaginary part by continuation (if it remains convergent). Take first $q = e^{\pi i\tau}$,

$$\phi(z) = \prod_1^\infty (1 - 2q^{2n-1}\cos 2z + q^{4n-2})$$

$$= \prod_1^\infty (1 - q^{2n-1}e^{2iz})(1 - q^{2n-1}e^{-2iz}).$$

Then
$$\phi(z+\pi\tau) = \prod_1^\infty (1 - q^{2n+1}e^{2iz})(1 - q^{2n-3}e^{-2iz})$$

$$= \frac{1 - q^{-1}e^{-2iz}}{1 - qe^{2iz}} \phi(z) = -q^{-1}e^{-2iz}\phi(z). \tag{2}$$

Also if
$$\phi(z) = A_0 + 2A_2\cos 2z + 2A_4\cos 4z + \ldots$$
$$= A_0 + A_2(e^{2iz} + e^{-2iz}) + A_4(e^{4iz} + e^{-4iz}) + \ldots,$$

and also
$$\phi(z+\pi\tau) = A_0 + A_2(q^2e^{2iz} + q^{-2}e^{-2iz}) + A_4(q^4e^{4iz} + q^{-4}e^{-4iz}) + \ldots$$
$$= -q^{-1}e^{-2iz}\{A_0 + A_2(e^{2iz} + e^{-2iz}) + \ldots\}.$$

On equating coefficients we find

$$\phi(z) = A_0\{1 + 2\sum_1^\infty (-1)^n q^{n^2} \cos 2nz\}, \tag{3}$$

which we define as

$$A_0 \vartheta_0(z). \tag{4}$$

Then by substituting in turn the values $z + \tfrac{1}{2}\pi$, $z + \tfrac{1}{2}\pi\tau$, $z + \tfrac{1}{2}\pi + \tfrac{1}{2}\pi\tau$ for z we find

$$\phi(z + \tfrac{1}{2}\pi) = \prod_1^\infty (1 + 2q^{2n-1} \cos 2z + q^{4n-2}), \tag{5}$$

$$\phi(z + \tfrac{1}{2}\pi\tau) = 2ie^{-iz} \sin z \prod_1^\infty (1 - 2q^{2n} \cos 2z + q^{4n}), \tag{6}$$

$$\phi(z + \tfrac{1}{2}\pi + \tfrac{1}{2}\pi\tau) = 2e^{-iz} \cos z \prod_1^\infty (1 + 2q^{2n} \cos 2z + q^{4n}), \tag{7}$$

Three functions $\vartheta_1(z)$, $\vartheta_2(z)$, $\vartheta_3(z)$ are defined in terms of $\vartheta_0(z)$ and expressed in series as follows:

$$\vartheta_0(z + \tfrac{1}{2}\pi) = \vartheta_3(z), \quad \vartheta_3(z) = 1 + 2\sum_1^\infty q^{n^2} \cos 2nz. \tag{8}$$

$$\vartheta_0(z + \tfrac{1}{2}\pi\tau) = iq^{-1/4} e^{-iz} \vartheta_1(z), \quad \vartheta_1(z) = 2\sum_{n=0}^\infty (-1)^n q^{(n+1/2)^2} \sin(2n+1)z, \tag{9}$$

$$\vartheta_0(z + \tfrac{1}{2}\pi + \tfrac{1}{2}\pi\tau) = q^{-1/4} e^{-iz} \vartheta_2(z), \quad \vartheta_2(z) = 2\sum_{n=0}^\infty q^{(n+1/2)^2} \cos(2n+1)z, \tag{10}$$

The four functions $\vartheta_0, \vartheta_1, \vartheta_2, \vartheta_3$ are Jacobi's theta-functions.* They are seen to be directly related to the four infinite products that we have found in the expression of sn, cn, dn. In fact if we put $1/A_0 = G$,

$$\vartheta_0 = G \prod_1^\infty (1 - 2q^{2n-1} \cos 2z + q^{4n-2}), \tag{11}$$

$$\vartheta_1 = 2Gq^{1/4} \sin z \prod_1^\infty (1 - 2q^{2n} \cos 2z + q^{4n}), \tag{12}$$

$$\vartheta_2 = 2Gq^{1/4} \cos z \prod_1^\infty (1 + 2q^{2n} \cos 2z + q^{4n}), \tag{13}$$

$$\vartheta_3 = G \prod_1^\infty (1 + 2q^{2n-1} \cos 2z + q^{4n-2}), \tag{14}$$

and then

$$\operatorname{sn}\frac{2Kz}{\pi} = \frac{K}{\pi} q^{-1/4} \frac{\vartheta_1(z)}{\vartheta_0(z)} \prod \frac{(1 - q^{2n-1})^2}{(1 - q^{2n})^2}, \tag{15}$$

$$\operatorname{cn}\frac{2Kz}{\pi} = \tfrac{1}{2} q^{-1/4} \frac{\vartheta_2(z)}{\vartheta_0(z)} \prod \frac{(1 - q^{2n-1})^2}{(1 + q^{2n})^2} = \frac{\vartheta_0(0)}{\vartheta_2(0)} \frac{\vartheta_2(z)}{\vartheta_0(z)}, \tag{16}$$

$$\operatorname{dn}\frac{2Kz}{\pi} = \frac{\vartheta_3(z)}{\vartheta_0(z)} \prod \frac{(1 - q^{2n-1})^2}{(1 + q^{2n-1})^2} = \frac{\vartheta_0(0)}{\vartheta_3(0)} \frac{\vartheta_3(z)}{\vartheta_0(z)}. \tag{17}$$

It is readily verified that all the ϑ functions satisfy the equation of heat conduction in the form

$$\frac{\partial^2 \vartheta}{\partial z^2} = \frac{4i}{\pi} \frac{\partial \vartheta}{\partial \tau}. \tag{18}$$

* *Fundamenta Nova*, 1829. Whittaker and Watson denote by ϑ_4 the function here called ϑ_0. Their $\vartheta_r(\pi z)$ is our $\vartheta_r(z)$.

Since $\operatorname{cn} 0 = \operatorname{dn} 0 = 1$,

$$\frac{\vartheta_2(0)}{2q^{1/4}\Pi(1+q^{2n})^2} = \frac{\vartheta_3(0)}{\Pi(1+q^{2n-1})^2} = \frac{\vartheta_0(0)}{\Pi(1-q^{2n-1})^2} = G. \tag{19}$$

When $z = \tfrac{1}{2}\pi$, $\operatorname{dn} 2Kz/\pi = k'$; hence

$$\sqrt{k'} = \frac{\vartheta_0(0)}{\vartheta_3(0)}. \tag{20}$$

Also when $z = \tfrac{1}{2}\pi + \tfrac{1}{2}\pi\tau$, $\operatorname{cn} 2Kz/\pi = -ik'/k$; then

$$\sqrt{\frac{k'}{k}} = \frac{\vartheta_0(0)}{\vartheta_2(0)}, \quad \text{whence} \quad \sqrt{k} = \frac{\vartheta_2(0)}{\vartheta_3(0)}, \tag{21}$$

and

$$\operatorname{sn}\frac{2Kz}{\pi} = \frac{1}{\sqrt{k}}\frac{\vartheta_1(z)}{\vartheta_0(z)}, \tag{22}$$

by choosing the constant factor so that $\operatorname{sn} 2Kz/\pi = 1/k$ when $z = \tfrac{1}{2}\pi + \tfrac{1}{2}\pi\tau$. Hence also

$$\operatorname{cn}\frac{2Kz}{\pi} = \left(\frac{k'}{k}\right)^{1/2}\frac{\vartheta_2(z)}{\vartheta_0(z)}, \quad \operatorname{dn}\frac{2Kz}{\pi} = k'^{1/2}\frac{\vartheta_3(z)}{\vartheta_0(z)}. \tag{23}$$

Now we can write (22) as

$$\vartheta_1(z) = \sqrt{k}\,\vartheta_0(z)\operatorname{sn}\frac{2Kz}{\pi}, \tag{24}$$

whence

$$\vartheta_1'(z) = \sqrt{k}\left\{\vartheta_0'(z)\operatorname{sn}\frac{2Kz}{\pi} + \frac{2K}{\pi}\vartheta_0(z)\operatorname{cn}\frac{2Kz}{\pi}\operatorname{dn}\frac{2Kz}{\pi}\right\}, \tag{25}$$

and

$$\vartheta_1'''(0) = \sqrt{k}\left\{\frac{6K}{\pi}\vartheta_0''(0) - \left(\frac{2K}{\pi}\right)^3\vartheta_0(0)(1+k^2)\right\},$$

$$\frac{\vartheta_1'''(0)}{\vartheta_1'(0)} = \frac{3\vartheta_0''(0)}{\vartheta_0(0)} - \left(\frac{2K}{\pi}\right)^2(1+k^2). \tag{26}$$

But from (23)

$$\frac{\vartheta_2''(0)}{\vartheta_2(0)} = \frac{\vartheta_0''(0)}{\vartheta_0(0)} - \left(\frac{2K}{\pi}\right)^2, \quad \frac{\vartheta_3''(0)}{\vartheta_3(0)} = \frac{\vartheta_0''(0)}{\vartheta_0(0)} - k^2\left(\frac{2K}{\pi}\right)^2, \tag{27}$$

whence

$$\frac{\vartheta_1'''(0)}{\vartheta_1'(0)} = \frac{\vartheta_0''(0)}{\vartheta_0(0)} + \frac{\vartheta_2''(0)}{\vartheta_2(0)} + \frac{\vartheta_3''(0)}{\vartheta_3(0)}. \tag{28}$$

But by the partial differential equation (18) this is equivalent to

$$\frac{d}{d\tau}\log\vartheta_1'(0) = \frac{d}{d\tau}\log[\vartheta_0(0)\vartheta_2(0)\vartheta_3(0)],$$

and

$$\vartheta_1'(0) = c\,\vartheta_0(0)\,\vartheta_2(0)\,\vartheta_3(0),$$

where c is independent of τ and z. Taking the lowest powers of q in each series we find that

$$c = 1,$$

whence

$$\vartheta_1'(0) = \vartheta_0(0)\,\vartheta_2(0)\,\vartheta_3(0). \tag{29}$$

Now

$$\left(\frac{d}{dz}\operatorname{sn}\frac{2Kz}{\pi}\right)_{z=0} = \frac{2K}{\pi},$$

whence from (13), (14) and (15)

$$1 = \frac{\vartheta_1'(0)}{\vartheta_0(0)\,\vartheta_2(0)\,\vartheta_3(0)} = \frac{\Pi(1-q^{2n})^2}{G^2\,\Pi(1+q^{2n})^2(1+q^{2n-1})^2(1-q^{2n-1})^2}. \tag{30}$$

The infinite product in the denominator is equal to 1; for

$$\Pi (1+q^{2n})(1+q^{2n-1})(1-q^{2n-1}) = \Pi (1+q^n)(1-q^{2n-1})$$
$$= \Pi \frac{(1-q^{2n})(1-q^{2n-1})}{1-q^n}$$
$$= 1. \tag{31}$$

Hence
$$G = \prod_1^\infty (1-q^{2n}). \tag{32}$$

Also from (22)
$$\frac{2K}{\pi} = \frac{1}{\sqrt{k}} \frac{\vartheta_1'(0)}{\vartheta_0(0)} = \frac{1}{\sqrt{k}} \vartheta_2(0)\vartheta_3(0) = \vartheta_3^2(0), \tag{33}$$

and
$$iK' = \tau K = \tfrac{1}{2}\pi\tau\vartheta_3^2(0). \tag{34}$$

The ϑ functions have one period equal to π or 2π, which is an advantage in problems relating to fixed boundaries. The series are extremely rapidly convergent. If $K'/K = 1$, $q = e^{-\pi}$, and q^4 is practically always negligible. (We can always arrange that $K'/K \geq 1$; for we see from the differential equations that if we put $t = iu$ the differential equation for sc t is converted into that for sn t with modulus k'. This is known as Jacobi's imaginary transformation.) On the other hand k is often among the data, and both periods are to be found. The sn, cn, dn notation is then more convenient.

The transformation for the ϑ functions is closely related to an identity that we have had in relation to heat conduction. From 20·02 (7), (12)

$$\frac{x}{l} + \sum_{n=1}^\infty \frac{2}{n\pi} \sin\frac{n\pi x}{l}\exp\left(-\frac{n^2\pi^2h^2t}{l^2}\right) = 1 - \sum_{r=0}^\infty \left(1-\operatorname{erf}\frac{x+2rl}{2ht^{1/2}}\right) + \sum_{r=0}^\infty \left(1-\operatorname{erf}\frac{2(r+1)l-x}{2ht^{1/2}}\right),$$

for all positive t and $0 < x < l$. By continuation it can be extended to all x. Differentiate with regard to x and multiply by l. Then

$$1 + 2\sum_{n=1}^\infty \cos\frac{n\pi x}{l}\exp\left(-\frac{n^2\pi^2h^2t}{l^2}\right) = \frac{l}{h\sqrt{(\pi t)}} e^{-x^2/4h^2t}\left(1 + 2\sum_{r=1}^\infty e^{-r^2l^2/h^2t}\cosh\frac{rlx}{h^2t}\right).$$

Put
$$\frac{\pi x}{l} = 2z, \quad \sigma = \frac{\pi h^2 t}{l^2} = -i\tau = \frac{K'}{K},$$

then
$$1 + 2\sum_{n=1}^\infty \cos 2nz \exp(-\pi n^2\sigma) = \sigma^{-1/2}e^{-z^2/\pi\sigma}\left(1 + 2\sum_{r=1}^\infty e^{-\pi r^2/\sigma}\cosh\frac{2rz}{\sigma}\right),$$

Hence
$$\vartheta_3(z:\sigma) = \sigma^{-1/2}e^{-z^2/\pi\sigma}\vartheta_3\left(\frac{iz}{\sigma}:\frac{1}{\sigma}\right).$$

The transformation is usually stated in terms of τ, but is more convenient in terms of σ, which is real in most applications. The notation adopted makes the use of σ explicit. If σ is small the terms on the left diminish slowly at first, those on the right rapidly so long as $\Re(z) < \pi$. The form of the expression on the right does not make the periodicity in $\Re(z)$ explicit, but for larger values of $\Re(z)$ the values of the function can be inferred from the periodicity.

By using the relations (8), (9), (10), we derive the corresponding transformations of the other ϑ functions.

$$\vartheta_0(z:\sigma) = \sigma^{-1/2} e^{-z^2/\pi\sigma} \vartheta_2\left(\frac{iz}{\sigma}:\frac{1}{\sigma}\right),$$

$$\vartheta_2(z:\sigma) = \sigma^{-1/2} e^{-z^2/\pi\sigma} \vartheta_0\left(\frac{iz}{\sigma}:\frac{1}{\sigma}\right),$$

$$\vartheta_1(z:\sigma) = -i\sigma^{-1/2} e^{-z^2/\pi\sigma} \vartheta_1\left(\frac{iz}{\sigma}:\frac{1}{\sigma}\right).$$

25·112. Change of argument. When the argument of a ϑ function is increased by $\tfrac{1}{2}\pi$, $\tfrac{1}{2}\pi\tau$, or $\tfrac{1}{2}\pi+\tfrac{1}{2}\pi\tau$ the result is another ϑ function multiplied by a simple factor; the following table gives the results, several of which have already been used. We write

$$M = q^{-1/4} e^{-iz}, \quad M' = q^{-1/2} M^2 = q^{-1} e^{-2iz},$$

z	$z+\tfrac{1}{2}\pi$	$z+\tfrac{1}{2}\pi\tau$	$z+\tfrac{1}{2}\pi+\tfrac{1}{2}\pi\tau$	$z+\pi$	$z+\pi\tau$	$z+\pi+\pi\tau$
ϑ_0	ϑ_3	$iM\vartheta_1$	$M\vartheta_2$	ϑ_0	$-M'\vartheta_0$	$-M'\vartheta_0$
ϑ_1	ϑ_2	$iM\vartheta_0$	$M\vartheta_3$	$-\vartheta_1$	$-M'\vartheta_1$	$M'\vartheta_1$
ϑ_2	$-\vartheta_1$	$M\vartheta_3$	$-iM\vartheta_0$	$-\vartheta_2$	$M'\vartheta_2$	$-M'\vartheta_2$
ϑ_3	ϑ_0	$M\vartheta_2$	$iM\vartheta_1$	ϑ_3	$M'\vartheta_3$	$M'\vartheta_3$

25·113. Expression of elliptic function with given periods in terms of ϑ functions. Since

$$\vartheta_1(z+\pi) = -\vartheta_1(z), \quad \vartheta_1(z+\pi\tau) = -q^{-1} e^{-2iz} \vartheta_1(z),$$

$$\frac{\vartheta_1'(z+\pi)}{\vartheta_1(z+\pi)} = \frac{\vartheta_1'(z)}{\vartheta_1(z)}, \quad \frac{\vartheta_1'(z+\pi\tau)}{\vartheta_1(z+\pi\tau)} = -2i + \frac{\vartheta_1'(z)}{\vartheta_1(z)}.$$

Hence the function $\vartheta_1'(z)/\vartheta_1(z)$ has period π. It has simple poles of residue 1 at $z = 0$, $\pi\tau$, $\pi+\pi\tau$, ... and its derivative has also the period $\pi\tau$.

This property can be used to express any elliptic function $\phi(u)$ of periods ω and ω' in terms of ϑ_1'/ϑ_1. Put

$$\phi(u) = f\left(\frac{\pi u}{\omega}\right) = f(z), \quad \omega'/\omega = \tau,$$

where ω', ω are taken so that $\Im(\omega'/\omega) > 0$.

Then $f(z)$ has the periods π, $\pi\tau$. Let the poles of $f(z)$ in a fundamental parallelogram be simple poles at $\alpha_1, \alpha_2, \ldots$ of residues $A_1, A_2, \ldots$. Then

$$F(z) = \Sigma A_i \frac{\vartheta_1'(z-\alpha_i)}{\vartheta_1(z-\alpha_i)}$$

differs from $f(z)$ by at most an integral function. Further,

$$F(z+\pi) = F(z),$$

$$F(z+\pi\tau) = \Sigma A_i \frac{\vartheta_1'(z-\alpha_i)}{\vartheta_1(z-\alpha_i)} - 2i \Sigma A_i$$

$$= F(z),$$

since ΣA_i is the sum of the residues of $f(z)$ in a fundamental parallelogram and therefore is zero. Hence

$$f(z) - F(z) = \text{constant}.$$

It follows that
$$\int f(z)\,dz = \Sigma A_i \log \vartheta_1(z-\alpha_i) + Cz,$$
so that any elliptic function with only simple poles can be integrated in terms of ϑ_1.
This can be extended at once to multiple poles by making use of the derivatives of ϑ_1'/ϑ_1.
Again, if $f(z)$ has n poles α_i, and n zeros β_i we can take
$$G(z) = \frac{\Pi \vartheta_1(z-\beta_i)}{\Pi \vartheta_1(z-\alpha_i)},$$
multiple poles and zeros being repeated in the products. Then
$$G(z+\pi) = G(z),$$
since the numbers of poles and zeros are equal; and
$$G(z+\pi\tau) = \frac{\exp(2i\Sigma\beta_i)}{\exp(2i\Sigma\alpha_i)} G(z).$$
But $\Sigma\beta_i - \Sigma\alpha_i$ is of the form $p\pi + q\pi\tau$, where p and q are integers; and this can be made zero by a suitable choice of the β_i and α_i, if necessary going outside the original parallelogram. Then $f(z)/G(z)$ is a constant.

25·12. Reduction of elliptic integrals to the standard form. Let $R(x)$ be a quartic with real coefficients and no repeated factor. If we make a bilinear transformation
$$x = \frac{\alpha t + \beta}{\gamma t + \delta}, \tag{1}$$
$R(x)$ takes the form $R_1(t)/(\gamma t+\delta)^4$, $R_1(t)$ being another quartic, and dx/dt is proportional to $(\gamma t+\delta)^{-2}$. Hence $\int dx/\sqrt{R(x)}$ is transformed into the form $\int dt/\sqrt{R_1(t)}$. We want to choose the transformation so that $R_1(t)$ will be an even function. Let the roots of $R(x)$ be a, b, c, d, those of $R_1(t)$ be $-h, -g, g, h$. Try
$$\frac{x-b}{x-c} = l\frac{t+g}{t-g}. \tag{2}$$
Then b corresponds to $-g$ and c to $+g$. For the other roots to correspond we must have
$$\frac{a-b}{a-c} = l\frac{-h+g}{-h-g} = l\frac{h-g}{g+h}, \quad \frac{d-b}{d-c} = l\frac{h+g}{h-g}, \tag{3}$$
whence
$$\left(\frac{g+h}{g-h}\right)^2 = \frac{d-b}{d-c}\frac{a-c}{a-b}, \tag{4}$$
the cross ratio of the roots d, a with respect to b, c; also
$$l = \frac{a-b}{a-c}\frac{g+h}{h-g}. \tag{5}$$
Then if g is taken arbitrarily (4) determines h and then (5) determines l.

Case 1. Let all the roots be real and $a<b<c<d$. Then if g is real so is h. Take $g=1$. Then taking the positive root
$$\frac{h+1}{h-1} > 1, \quad h > 1, \tag{6}$$
$$R_1(t) = C(t^2-1)(t^2-h^2) = C'(1-t^2)(1-k^2t^2), \tag{7}$$
and the transformation leads to $t = A\,\text{sn}\,u$ with $k = 1/h < 1$.

Case 2. Let b and c be real, a and d conjugate complexes. Take $g = 1$. Then

$$\left|\frac{g+h}{g-h}\right| = 1, \tag{8}$$

and h is purely imaginary $= ij$. Then

$$R_1(t) = C(1-t^2)(1+j^2t^2), \tag{9}$$

and integration can be carried out by putting $t = \operatorname{cn} u$, $\operatorname{sd} u$, or $\operatorname{ds} u$.

Case 3. Let a and d be conjugate complexes, b and c also. Take $g = i$. h will be purely imaginary since (4) is again real; and

$$R_1(t) = C(1+t^2)(1+\nu^2 t^2). \tag{10}$$

Integration can be carried out by putting $t = \operatorname{sc} u$ or $\operatorname{cs} u$.

Integrals of the form $\int dt/\sqrt{R_1(t)}$ are called elliptic integrals of the first kind, and can be evaluated in terms of the functions sn, cn, dn. More complicated integrals involving the square root of a quartic can be reduced to the form

$$\int \frac{tf_1(t^2) + f_2(t^2)}{\sqrt{R_1(t)}} dt, \tag{11}$$

where f_1 and f_2 are rational functions. Then this can be broken up by partial fractions into terms of the types

$$\int \frac{tf(t^2) dt}{\sqrt{R_1(t)}}, \quad \int \frac{t^{2p} dt}{\sqrt{R_1(t)}}, \quad \int \frac{dt}{(1+nt^2)^p \sqrt{R_1(t)}}. \tag{12}$$

The first form gives an elementary function. For definiteness we take the case when $R_1(t) = (1-t^2)(1-k^2t^2)$. In the second, if we put $t = \operatorname{sn} u$, we have

$$u_p = \int \operatorname{sn}^{2p} u \, du.$$

But
$$\frac{d}{du}(\operatorname{sn}^{2p-3} u \operatorname{cn} u \operatorname{dn} u) = (2p-3)\operatorname{sn}^{2p-4} u \operatorname{cn}^2 u \operatorname{dn}^2 u - \operatorname{sn}^{2p-2} u \operatorname{dn}^2 u - k^2 \operatorname{sn}^{2p-2} u \operatorname{cn}^2 u$$
$$= \alpha \operatorname{sn}^{2p} u + \beta \operatorname{sn}^{2p-2} u + \gamma \operatorname{sn}^{2p-4} u,$$

where α, β, γ are constants. Hence by successive reduction u_p can be reduced to u and $\int \operatorname{sn}^2 u \, du$. The usual standard form is

$$E(u) = \int_0^u \operatorname{dn}^2 u \, du = \int_0^{\operatorname{sn} u} \sqrt{\left(\frac{1-k^2 x^2}{1-x^2}\right)} dx, \tag{13}$$

and is called the second elliptic integral.

The second complete elliptic integral is

$$E = E(K) = \int_0^K \operatorname{dn}^2 u \, du. \tag{14}$$

$E(u)$ is not periodic; but the function

$$Z(u) = E(u) - \frac{E}{K} u \tag{15}$$

has period $2K$. It is called the elliptic Zeta function, not to be confused with the Riemann ζ function

$$\zeta(s) = \sum_{n=1}^{\infty} n^{-s}.$$

The third form in (12) is called the third elliptic integral. It can be evaluated in terms of ϑ functions at the cost of a good deal of algebra; direct recourse to arithmetic is probably usually the best method.

Integration of elliptic functions can always be carried out in terms of ϑ functions. For the necessary transformations the special treatises should be consulted.

25·13. Complete elliptic integrals. The complete elliptic integrals often arise by themselves as definite integrals. They are fully tabulated, but care should be taken in using the tables; k is sometimes denoted by $\sin\alpha$, sometimes by $\sin\tfrac{1}{2}\alpha$, the latter being a survival from the time when the pendulum was the only application of K. x is often denoted by $\sin\phi$, $\sqrt{(1-k^2x^2)}$ by $\Delta\phi$. Approximations to K' and E' when k is small are of interest. We have

$$K' = \int_0^{\tfrac{1}{2}\pi} \frac{d\phi}{\Delta'(\phi)} \quad \text{where} \quad \Delta' = (1-k'^2\sin^2\phi)^{\tfrac{1}{2}}$$

$$= \int_0^{\tfrac{1}{2}\pi} \frac{1-k'\sin\phi}{\Delta'(\phi)} d\phi + \int_0^{\tfrac{1}{2}\pi} \frac{k'\sin\phi}{\Delta'(\phi)} d\phi = I_1 + I_2, \text{ say,}$$

$$I_1 = \int_0^{\tfrac{1}{2}\pi} \sqrt{\left(\frac{1-k'\sin\phi}{1+k'\sin\phi}\right)} d\phi \to \int_0^{\tfrac{1}{2}\pi} \sqrt{\left(\frac{1-\sin\phi}{1+\sin\phi}\right)} d\phi$$

$$= \log 2.$$

I_2 is integrable exactly;

$$I_2 = -\int_{\phi=0}^{\tfrac{1}{2}\pi} \frac{k'd(\cos\phi)}{\sqrt{(k^2+k'^2\cos^2\phi)}} = -\left[\log\{k'\cos\phi + \sqrt{(k^2+k'^2\cos^2\phi)}\}\right]_0^{\tfrac{1}{2}\pi}$$

$$= \log\frac{1+k'}{k} = \log\frac{2}{k} + O(k),$$

and

$$K' = \log\frac{4}{k} + O(k).$$

Similarly if k is small $\quad E' = \int_0^{\tfrac{1}{2}\pi} \Delta'(\phi) d\phi \doteqdot \int_0^{\tfrac{1}{2}\pi} \cos\phi\, d\phi = 1,$

while

$$E = \int_0^{\tfrac{1}{2}\pi} \Delta(\phi) d\phi = \tfrac{1}{2}\pi + O(k^2).$$

25·14. Reduction of integrals containing a cubic. If

$$t = \int_0^x \frac{dx}{\sqrt{X}}, \quad X^2 = (x-a)(b-x)(c-x),$$

we can put $x - a = \xi^2$, and then

$$= \int \frac{2d\xi}{\{(b-a-\xi^2)(c-a-\xi^2)\}^{\tfrac{1}{2}}},$$

which is in one or other of the standard forms according to the signs of $b-a$ and $c-a$. These forms are more generally useful than the Weierstrass one, which takes the standard function as

$$\wp(u) = \frac{1}{u^2} + \Sigma\Sigma'\left\{\frac{1}{(u-m\omega-n\omega')^2} - \frac{1}{(m\omega+n\omega')^2}\right\}.$$

This satisfies
$$\wp'^2(u) = 4\wp^3(u) - g_2\wp(u) - g_3,$$
where g_2 and g_3 are known functions of ω, ω'. This function has the property that for u small
$$\wp(u) - \frac{1}{u^2} = O(u^2).$$

Had Abel been alive he would probably have remarked that this property corresponds to taking the fundamental trigonometric function as $\operatorname{cosec}^2 x - \tfrac{1}{3}$.

The square of any of sn, cn, dn and their reciprocals and ratios is a function with only double poles, and having the periods $2K$ and $2iK'$. These functions can therefore be used in the same way as $\wp(u)$ in Weierstrass's method.

25·15. Change of modulus: Landen's transformation. This important transformation is most symmetrically expressed if we take the integral

$$I = \int_0^\phi \frac{d\phi}{\{a^2\cos^2\phi + b^2\sin^2\phi\}^{1/2}} = \int_0^u \frac{du}{\sqrt{\{(1+u^2)(a^2+b^2u^2)\}}} \quad (u = \tan\phi), \tag{1}$$

where $a \geqslant b > 0$. We put
$$u = \frac{v}{A - Bv^2}; \tag{2}$$

then
$$I = \int_0^v \frac{(A + Bv^2)\,dv}{[\{(A-Bv^2)^2 + v^2\}\{a^2(A-Bv^2)^2 + b^2v^2\}]^{1/2}}. \tag{3}$$

With a suitable choice of A and B we can arrange that the second square root is proportional to the numerator; this is seen to impose only one condition, and we can add the further condition that the first factor in the denominator is to vanish when $v^2 = -1$. The result will be a new integral of the same form as that in u. The conditions are

$$(A+B)^2 = 1, \quad 4AB = b^2/a^2, \tag{4}$$

and A and B can be taken as $\tfrac{1}{2} \pm \tfrac{1}{2}\sqrt{(1-b^2/a^2)}$. Then

$$I = \int_0^v \frac{dv}{\{(1+v^2)(a^2A^2 + a^2B^2v^2)\}^{1/2}}, \tag{5}$$

and
$$\alpha^2 = (A+B)^2 a^2, \quad \beta^2 = 4ABa^2;$$

thus
$$\int_0^v \frac{dv}{\{(1+v^2)(\alpha^2 + \beta^2 v^2)\}^{1/2}} = \frac{1}{2}\int_0^u \frac{du}{[(1+u^2)\{\tfrac{1}{4}(\alpha+\beta)^2 + \alpha\beta u^2\}]^{1/2}}, \tag{6}$$

with
$$v = \frac{\alpha+\beta}{2\beta u}\left\{\left(1 + \frac{4\alpha\beta u^2}{(\alpha+\beta)^2}\right)^{1/2} - 1\right\}, \tag{7}$$

or
$$\int_0^\psi \frac{d\psi}{\{\alpha^2\cos^2\psi + \beta^2\sin^2\psi\}^{1/2}} = \frac{1}{2}\int_0^\phi \frac{d\phi}{\{\tfrac{1}{4}(\alpha+\beta)^2\cos^2\phi + \alpha\beta\sin^2\phi\}^{1/2}}, \tag{8}$$

with
$$\tan\phi = \frac{(\alpha+\beta)\tan\psi}{\alpha - \beta\tan^2\psi}. \tag{9}$$

If $\psi = \tfrac{1}{2}\pi$, $\phi = \pi$, and we have a simple relation between the complete elliptic integrals. If we denote the integrals by t we have

$$\sin\psi = \operatorname{sn}\left\{\alpha t\left|\left(1 - \frac{\beta^2}{\alpha^2}\right)^{1/2}\right.\right\}, \quad \sin\phi = \operatorname{sn}\left(\tfrac{1}{2}(\alpha+\beta)t\left|\frac{\alpha-\beta}{\alpha+\beta}\right.\right), \tag{10}$$

with (9) holding between the corresponding sc functions.

In (8) the constants α, β are replaced by their arithmetic and geometric means. Thus if $\alpha = 0\cdot 9$, $\beta = 0\cdot 1$, the arithmetic and geometric means are $0\cdot 5$ and $0\cdot 3$; repeating the transformation we get $0\cdot 40$ and $0\cdot 387$. By successive applications of the transformation we can therefore reduce the elliptic integral as closely as we like to a linear function of the new argument.

The complete integrals of the form (8) are symmetrical in α and β; but

$$\int_0^{\frac{1}{2}\pi} \frac{d\psi}{(\alpha^2 \cos^2 \psi + \beta^2 \sin^2 \psi)^{1/2}} = \int_0^{\frac{1}{2}\pi} \frac{d\psi}{\{\alpha^2 - (\alpha^2 - \beta^2)\sin^2 \psi\}^{1/2}} = \frac{1}{\alpha} K\left\{\left(1 - \frac{\beta^2}{\alpha^2}\right)^{1/2}\right\}, \quad (11)$$

in which the symmetry is no longer obvious. But it is also equal to

$$\frac{2}{\alpha + \beta} K\left(\frac{\alpha - \beta}{\alpha + \beta}\right), \quad (12)$$

in which the symmetry is obvious since K is an even function of k.

The second elliptic integral can be treated similarly. We take

$$J = \int_0^\phi \{\tfrac{1}{4}(\alpha + \beta)^2 \cos^2 \phi + \alpha\beta \sin^2 \phi\}^{1/2} d\phi$$
$$= \tfrac{1}{2}(\alpha + \beta)^2 \int_0^\psi \frac{(\alpha \cos^2 \psi + \beta \sin^2 \psi)^2}{(\alpha^2 \cos^2 \psi + \beta^2 \sin^2 \psi)^{3/2}} d\psi, \quad (13)$$

by the same transformation. We have

$$\frac{d}{d\psi} \frac{\cos \psi \sin \psi}{(\alpha^2 \cos^2 \psi + \beta^2 \sin^2 \psi)^{1/2}} = \frac{\alpha^2 \cos^4 \psi - \beta^2 \sin^4 \psi}{(\alpha^2 \cos^2 \psi + \beta^2 \sin^2 \psi)^{3/2}},$$

and

$$(\alpha \cos^2 \psi + \beta \sin^2 \psi)^2 = -\frac{\alpha - \beta}{\alpha + \beta}(\alpha^2 \cos^4 \psi - \beta^2 \sin^4 \psi)$$
$$+ \frac{2\alpha\beta}{(\alpha + \beta)^2}(\alpha^2 \cos^2 \psi + \beta^2 \sin^2 \psi) + \frac{2}{(\alpha + \beta)^2}(\alpha^2 \cos^2 \psi + \beta^2 \sin^2 \psi)^2.$$

The easiest way to find the coefficients is to insert a factor $\cos^2 \psi + \sin^2 \psi$ in the second term on the right and then equate coefficients. Then

$$J = -\tfrac{1}{2}(\alpha^2 - \beta^2)\left[\frac{\cos \psi \sin \psi}{(\alpha^2 \cos^2 \psi + \beta^2 \sin^2 \psi)^{1/2}}\right]_0^\psi$$
$$+ \alpha\beta \int_0^\psi \frac{d\psi}{(\alpha^2 \cos^2 \psi + \beta^2 \sin^2 \psi)^{1/2}} + \int_0^\psi (\alpha^2 \cos^2 \psi + \beta^2 \sin^2 \psi)^{1/2} d\psi, \quad (14)$$

which can be written

$$\int_0^\psi (\alpha^2 \cos^2 \psi + \beta^2 \sin^2 \psi)^{1/2} d\psi = \int_0^\phi \{\tfrac{1}{4}(\alpha + \beta)^2 \cos^2 \phi + \alpha\beta \sin^2 \phi\}^{1/2} d\phi$$
$$- \tfrac{1}{2}\alpha\beta \int_0^\phi \frac{d\phi}{\{\tfrac{1}{4}(\alpha + \beta)^2 \cos^2 \phi + \alpha\beta \sin^2 \phi\}^{1/2}} + \tfrac{1}{2}(\alpha^2 - \beta^2)\frac{\cos \psi \sin \psi}{(\alpha^2 \cos^2 \psi + \beta^2 \sin^2 \psi)^{1/2}}. \quad (15)$$

Equivalent relations were found by Landen in 1775 in determining the length of an arc of a hyperbola.

For the complete integrals we have

$$\int_0^{1/2\pi} (\alpha^2 \cos^2\psi + \beta^2 \sin^2\psi)^{1/2} d\psi = 2\int_0^{1/2\pi} \{\tfrac{1}{4}(\alpha+\beta)^2 \cos^2\phi + \alpha\beta \sin^2\phi\}^{1/2} d\phi$$
$$-\alpha\beta \int_0^{1/2\pi} \frac{d\phi}{\{\tfrac{1}{4}(\alpha+\beta)^2 \cos^2\phi + \alpha\beta \sin^2\phi\}^{1/2}}. \quad (16)$$

These relations were used by Legendre in his numerical calculation of the elliptic integrals.

EXAMPLES

1. Prove that
$$\operatorname{sn} \tfrac{1}{2}K = \frac{1}{k\sqrt{2}}\{\sqrt{(1+k)} - \sqrt{(1-k)}\},$$
and find what values of z make $\operatorname{sn} z$ equal to the four expressions obtained by reversing the signs of the roots. (Use $\operatorname{cn} K = 0$.)

2. Prove that
$$\operatorname{sn} \tfrac{1}{2}iK' = i/\sqrt{k},$$
and find what values of z make $\operatorname{sn} z = 1/\sqrt{k}$.

3. If
$$u = \int_0^x \frac{dx}{\sqrt{(x^4 - 7x^2 + 10)}},$$
express x as a single-valued function of u.

4. Solve the pendulum equation
$$\dot{\theta}^2 = \dot{\theta}_0^2 + \frac{4g}{l}(\sin^2 \tfrac{1}{2}\theta_0 - \sin^2 \tfrac{1}{2}\theta)$$
in terms of elliptic functions.

5. Prove that
$$k\operatorname{sn} t = \lim_{m\to\infty} \lim_{n\to\infty} \sum_{\mu=-m}^{m} \sum_{\nu=-n}^{n} \left\{\frac{1}{t+4\mu K+(2\nu-1)iK'} - \frac{1}{t+(4\mu+2)K+(2\nu-1)iK'}\right\},$$
$$k\operatorname{cn} t = \lim_{m\to\infty} \lim_{n\to\infty} \sum_{\mu=-m}^{m} \sum_{\nu=-n}^{n} \left\{\frac{2K'}{(t+4\mu K+4\nu iK')^2 + K'^2} - \frac{2K'}{(t+(4\mu+2)K+4\nu iK')^2 + K'^2}\right\},$$
$$\operatorname{dn} t = \lim_{m\to\infty} \lim_{n\to\infty} \sum_{\mu=-m}^{m} \sum_{\nu=-n}^{n} \left\{\frac{2K'}{(t+4\mu K+4\nu iK')^2 + K'^2} + \frac{2K'}{(t+(4\mu+2)K+4\nu iK')^2 + K'^2}\right\},$$
and express the double series as series of trigonometric functions of $2Kt/\pi$ and of hyperbolic functions of $2K't/\pi$.

6. Prove that
$$k\int \operatorname{sn} u\, du = \log\frac{\vartheta_1}{\vartheta_2}\left\{\frac{\pi}{4K}(u-iK')\right\} + \log\frac{\vartheta_1}{\vartheta_2}\left\{\frac{\pi}{4K}(u+iK')\right\}$$
$$= \tfrac{1}{2}\log\frac{\operatorname{dn} u - k\operatorname{cn} u}{\operatorname{dn} u + k\operatorname{cn} u} + \text{constant}.$$

NOTES

1·116 a. Theorem of bounded convergence. We need first a few definitions. For any finite set I of non-overlapping intervals we define lI as the sum of their lengths. For an infinite set I within (a,b) we define lI as the upper bound of the sums of lengths of finite sets included in I. If E is a set of points of (a,b) we define the complementary set CE as the set of points of (a,b) that are not members of E. The complementary set of a finite set of closed non-overlapping intervals is a finite set of non-overlapping intervals. If all points of E_1 are members of E_2 we write $E_1 \subset E_2$ (read, E_1 is included in E_2).

Lemma 1. *If I is a set of non-overlapping intervals included in (a,b), and δ is any positive quantity, however small, there exists a finite set J of the intervals of I such that $lI \geq lJ > lI - \delta$.* This follows at once from the fact that lI is the upper bound of lJ for all finite sets.

Lemma 2. *In the same conditions there exists a finite set of closed intervals K included in I such that $lK > lI - 2\delta$.* For let the J of Lemma 1 be m in number. For any interval of J, say (α_i, β_i) such that $\beta_i - \alpha_i > \delta/m$, define an interval of K as the closed interval $(\alpha_i + \delta/2m, \beta_i - \delta/2m)$. Then the set of intervals K have the property required.

I consists of K together with another set of non-overlapping intervals K' such that $lI = lK + lK'$.

Lemma 3. *If $\{I_n\}$ is a sequence of sets of non-overlapping intervals included in (a,b) such that $I_n \subset I_{n-1}$, and such that no x of (a,b) belongs to all I_n, then $lI_n \to 0$.* Given $\delta > 0$, we can take a finite closed set K_n included in I_n such that $lK_n > lI_n - 2^{-n}\delta$. Denote the set of points contained in I_n and not in K_n by K'_n. Take L_n to be the set of points common to $K_1, K_2, ..., K_n$. Then a point of I_n is a point of $I_1, I_2, ..., I_n$ and therefore is either a point of L_n or of at least one of $K'_1, ..., K'_n$; and $lI_n \leq lL_n + lK'_1 + ... + lK'_n < lL_n + \delta$. Hence L_n is a finite set of non-overlapping closed intervals (and therefore closed) satisfying $L_n \subset L_{n-1}$ and $lL_n > lI_n - \delta$.

Since no point of (a,b) belongs to all I_n, every point belongs to some CI_n and therefore to some CL_n since $CI_n \subset CL_n$. It is not an end-point of any interval of this CL_n because end-points of CL_n are end-points of L_n, which belong to L_n since L_n is closed. Hence for every point x of (a,b) there is an n such that x is interior to some interval of CL_n. Consider the set of all such intervals. Then, by the Heine-Borel theorem, there is a finite set of intervals $d_1, d_2, ..., d_k$, each part of some CL_n, that covers (a,b). Let d_r be a member of CL_{n_r}, and let n_0 be the greatest of $n_1, ..., n_k$. Then since $CL_n \subset CL_{n+1}$, CL_{n_0} includes all d_r and therefore CL_{n_0} is the whole interval (a,b), and L_n is empty for all $n \geq n_0$. Hence $lL_n = 0$, $lI_n < \delta$ for all $n \geq n_0$. Since δ is arbitrary, $lI_n \to 0$.

Lemma 4. *If $f_n(x)$ is non-negative and integrable in (a,b), $f_n(x) < M$ for all x and n, and $f_n(x) \to 0$ for all x, then $\int_a^b f_n(x)\,dx \to 0$.* For every n define a subdivision such that the corresponding lower sum as in 1·101 differs from $\int_a^b f_n(x)\,dx$ by less than $1/n$. Take $g_n(x) = 0$

at every point of subdivision, and inside any subinterval take $g_n(x)$ equal to the lower bound of $f_n(x)$ at points of that subinterval. Then

$$f_n(x) \geqslant g_n(x) \geqslant 0; \quad 0 \leqslant \int_a^b \{f_n(x) - g_n(x)\} dx < 1/n; \quad g_n(x) \to 0.$$

Given $\epsilon > 0$, denote by I_n the set of all x where $g_p(x) > \epsilon$ for at least one $p \geqslant n$. Interior to any interval chosen for $f_p(x)$, $g_p(x)$ is constant; hence $g_p(x) > \epsilon$ throughout a subinterval if it is $> \epsilon$ at any point. Thus I_n is a set of non-overlapping intervals, and $I_n \subset I_{n-1}$. No x belongs to all I_n, for if it did we should have $f_p(x) \geqslant g_p(x) > \epsilon$ for an infinite sequence of values of p, and $f_n(x)$ would not tend to 0. Hence I_n satisfies the conditions of Lemma 3, and $lI_n \to 0$. Take n_0 such that for $n \geqslant n_0$, $lI_n < \delta$. Then for $n \geqslant n_0$

$$\int_a^b g_n(x) dx \leqslant \epsilon(b-a-\delta) + M\delta,$$

$$\int_a^b f_n(x) dx \leqslant \epsilon(b-a-\delta) + M\delta + 1/n.$$

But ϵ, δ and $1/n$ can all be taken arbitrarily small. Hence the lemma is proved.

Theorem. If $f_n(x)$ is integrable in (a,b), $|f_n(x)| < M$, and $f_n(x) \to f(x)$, where $f(x)$ is integrable, then $\int_a^b f_n(x) dx \to \int_a^b f(x) dx$. Take $h_n(x) = |f_n(x) - f(x)|$. Then $h_n(x)$ satisfies the conditions of Lemma 4 and therefore $\int_a^b h_n(x) dx \to 0$. But

$$\left| \int_a^b f_n(x) dx - \int_a^b f(x) dx \right| \leqslant \int_a^b |f_n(x) - f(x)| dx,$$

whence
$$\int_a^b f_n(x) dx \to \int_a^b f(x) dx.$$

It will be noticed that $\int_a^x f_n(u) du \to \int_a^x f(u) du$ uniformly in (a,b).

The above proof is one of three (somewhat overlapping) given to us by Professor Besicovitch. Dr Smithies has given an independent proof.

As an illustration, take in $(0,1)$ $f_n(x) = 0$ for all irrational x. If x is rational and equal to m/k in its lowest terms, take $f_n(x) = 1$ for $n = k$ and otherwise $= 0$. Then $f_n(x) \to 0$ everywhere, and

$$\lim \int_0^1 f_n(x) dx = \int_0^1 \lim f_n(x) dx.$$

But in any interval of x, for any n, there are rationals whose denominators exceed n, and therefore $f_n(x)$ does not tend to 0 uniformly in any interval of x. The reason for using the lower sum, instead of, as usual, the upper sum, is that the upper sum leads to no definition of a set of intervals I_n with the properties required.

1·134a. The most important applications are to integrals of the forms

$$\int_a^b \{e^{-tx}, \cos tx, \sin tx\} v(x) dx \quad (b > a),$$

when t is positive and large. Note first that if $v(x)$ is given only to be bounded, say $|v(x)| < A$, we can assert (without using Abel's lemma) that

$$\left|\int_a^b e^{-tx} v(x)\,dx\right| < Ae^{-ta}/t. \tag{1}$$

Without further restrictions we cannot say that $\int_a^b \{\cos tx, \sin tx\} v(x)\,dx$ are $O(1/t)$. But with a further condition on $v(x)$ we can prove the latter statements, and we can find a result to replace (1) in the case where the integral is improper.

(a) If t is positive, and $\left|\int_a^\xi f(x)\,dx\right| < M$ in $a \leqslant \xi \leqslant b$, then e^{-tx} satisfies the conditions imposed on $v(x)$ in Abel's lemma, and we have

$$\left|\int_a^b e^{-tx} f(x)\,dx\right| < Me^{-ta}. \tag{2}$$

(b) If t is positive, and $v(x)$ is non-negative, bounded, and non-increasing in $a \leqslant x \leqslant b$, then

$$\left|\int_a^\xi \cos tx\,dx\right| \leqslant \frac{2}{t}, \quad \left|\int_a^\xi \sin tx\,dx\right| \leqslant \frac{2}{t},$$

whence, by taking $f(x) = \cos tx$ or $\sin tx$ in Abel's lemma, we have

$$\left|\int_a^b \cos tx\, v(x)\,dx\right| \leqslant \frac{2}{t} v(a), \quad \left|\int_a^b \sin tx\, v(x)\,dx\right| \leqslant \frac{2}{t} v(a). \tag{3}$$

(c) If t is positive, and $v(x)$ has total variation V in $a \leqslant x \leqslant b$, let its positive and negative variations in (a, x) be $P(x)$ and $N(x)$. Then

$$v(x) = v(a) + P(x) - N(x)$$
$$= v(a) + \{N(b) - N(x)\} - \{P(b) - P(x)\} + P(b) - N(b)$$
$$= v(b) + \{N(b) - N(x)\} - \{P(b) - P(x)\}. \tag{4}$$

Here $v(b)$ is constant. $N(b) - N(x)$, $P(b) - P(x)$ satisfy the conditions imposed on $v(x)$ in Abel's lemma, and reduce to $N(b), P(b)$ when $x = a$. Hence

$$\left|\int_a^b \cos tx\, v(x)\,dx\right| \leqslant \frac{2}{t}\{|v(b)| + N(b) + P(b)\} = \frac{2}{t}\{|v(b)| + V\}, \tag{5}$$

$$\left|\int_a^b \sin tx\, v(x)\,dx\right| \leqslant \frac{2}{t}\{|v(b)| + V\}. \tag{6}$$

3·03a. The statement that the only isotropic tensors of orders 2 and 3 are scalar multiples of δ_{ik} and ϵ_{ikm} respectively is easily proved by the method of 3·031.

5·04a. If

$$f(x, y) = (x^2 + y^2)^{1/2} \frac{x^2 y}{x^4 + y^2}, \quad f(0, 0) = 0,$$

$\partial f/\partial x = \partial f/\partial y = 0$ at $(0, 0)$. Also if $x = r\cos\theta$, $y = r\sin\theta$, $f/r \to 0$ when $r \to 0$ for any fixed θ. Hence $\partial f/\partial r = 0$, and the gradient of f in any direction satisfies the vector rule.

But f is not differentiable, because for $r < \delta$ we could always take $y = x^2$, which makes $f/r = \frac{1}{2}$. Hence differentiability of a function is not a necessary condition for the gradient to be a vector as defined in 3·06. The property (5) is, however, so important that the vector property of the derivative tells us little if (5) is not satisfied. For instance, if f is differentiable, and $r \to 0$, $\theta \to \phi$ in any manner,

$$\frac{1}{r}\{f(x,y) - f(0,0)\} \to \frac{\partial f}{\partial x}\cos\phi + \frac{\partial f}{\partial y}\sin\phi;$$

but this is false for the above example.

5·051 a. Differentiation under the integral sign.

We show that, under certain conditions,

$$\frac{d}{dc}\int_a^b f(x,c)\,dx = \int_a^b \frac{\partial f(x,c)}{\partial c}\,dx,$$

where a and b are independent of c.

We assume that $\int_a^b f(x,y)\,dx$ exists and that $\dfrac{\partial f(x,y)}{\partial y}$ is a continuous function of the two variables x and y for $a \leqslant x \leqslant b$ and $|y-c| \leqslant \delta$, $\delta > 0$. Then by 5·051, since the double integral

$$\int_a^b dx \int_c^{c+h} \frac{\partial f(x,y)}{\partial y}\,dy$$

exists for $|h| < \delta$,

$$\int_a^b dx \int_c^{c+h} \frac{\partial f(x,y)}{\partial y}\,dy = \int_c^{c+h} dy \int_a^b \frac{\partial f(x,y)}{\partial y}\,dx.$$

The integral on the left is

$$\int_a^b [f(x,c+h) - f(x,c)]\,dx.$$

That on the right, since the single integral is a continuous function of y, is equal to

$$h\int_a^b \left(\frac{\partial f(x,y)}{\partial y}\right)_{y=c} dx + o(h).$$

Hence, dividing by h and making $h \to 0$ we have

$$\frac{d}{dc}\int_a^b f(x,c)\,dx = \int_a^b \frac{\partial f(x,c)}{\partial c}\,dx.$$

If a, b depend on c, terms $f(b,c)\dfrac{db}{dc} - f(a,c)\dfrac{da}{dc}$ must be added.

The extension for a or b tending to infinity follows from 1·12 if $\int_a^b \dfrac{\partial f}{\partial y}\,dx$ is uniformly convergent.

A proof under different conditions is as follows:

Let $\int_a^b f(x,y)\,dx$ exist for all y such that $|y-c| \leqslant \delta$ and let $\dfrac{\partial f(x,y)}{\partial y}$ exist for $a \leqslant x \leqslant b$,

$|y-c| \leqslant \delta$. Also let $\dfrac{\partial^2 f(x,y)}{\partial y^2}$ exist in these ranges and be bounded with upper bound M but not necessarily continuous. Then

$$\frac{1}{h}\left\{\int_a^b f(x, c+h)\,dx - \int_a^b f(x, c)\,dx\right\} = \int_a^b f_y(x, c+\theta h)\,dx,$$

where $0 < \theta < 1$, $|h| \leqslant \delta$. Also

$$f_y(x, c+\theta h) - f_y(x, c) = f_{yy}(x, c+\phi h)\,\theta h,$$

where $0 < \phi < \theta < 1$. Hence

$$\left|\int_a^b f_y(x, c+\theta h) - \int_a^b f_y(x, c)\right| < M(b-a)\,|h|$$

so that, proceeding to the limit as $h \to 0$ we have

$$\frac{d}{dc}\int_a^b f(x, c)\,dx = \int_a^b \frac{\partial f(x, c)}{\partial c}\,dx.$$

5·07 a. A surface satisfying the conditions of 5·07 can be enclosed in an arbitrarily small volume. Since $l^2 + m^2 + n^2 = 1$, at every point of the surface at least one of l^2, m^2, $n^2 \geqslant 1/3$. Take the points where $n^2 \geqslant 1/3$; these give a region or regions of x, y since n is continuous, and

$$1 + F_x^2 + F_y^2 = 1/n^2 \leqslant 3.$$

Then $\qquad F_x^2 \leqslant 2,\ F_y^2 \leqslant 2.$

Hence $\qquad |F(x+h, y) - F(x, y)| \leqslant h\sqrt{2},\quad |F(x, y+k) - F(x, y)| \leqslant k\sqrt{2}.$

Take $k \leqslant h$. Then a parallelepiped of sides $2h$, $2k$, $2h\sqrt{2}$ centred at (x, y, z) will include all points of S where $|\xi - x| < h$, $|\eta - y| < k$, and overlap similar parallelepipeds, centred at points of S corresponding to adjacent points of the lattice; and such a set of parallelepipeds about points corresponding to all points of an (h, k) lattice in (x, y) will therefore include the whole of S. Let the extents of x, y be H, K. Then the number of lattice points is

$$\leqslant \left(\frac{H}{h}+1,\ \frac{K}{k}+1\right),$$

and the total volume of the parallelepipeds is

$$\leqslant \left(\frac{H}{h}+1\right)\left(\frac{K}{k}+1\right) 8h^2 k\sqrt{2} = 8\sqrt{2}\,(Hh + h^2)(K + k)$$

which tends to 0 with h since $k \leqslant h$.

Apply a similar argument to the points where l^2 or $m^2 \geqslant 1/3$ and the result follows by addition.

5·08 a. Green derived the theorem known by his name* by separating the terms and integrating by parts. M. V. Ostrogradsky† gave the divergence theorem explicitly, but of course all principles used in it are included in Green's argument.

---

* Collected papers, p. 23; Essay on the application of mathematical analysis to electricity and magnetism, Nottingham, 1828.

† *Mem. Acad. Imp. Sci.* St Petersburg (6) **1**, 1831, 130.

6·043a. Clearly if ρ satisfies the condition everywhere, and we altered it by a finite amount at an isolated point, ϕ and therefore $\nabla^2\phi$ would not be altered; but Poisson's equation would be false at this point. Integrability of ρ is therefore not a sufficient condition; these considerations suggest that continuity might be a necessary and sufficient condition. It is in fact necessary, but not sufficient. A weaker sufficient condition than the one we have assumed was given by Hölder, namely that for any point Q different from $P |\rho_Q - \rho_P| < Ar^\alpha$, where A and α are fixed and α is positive. It is an extension to three dimensions of the Lipschitz condition for one.* Continuity alone is not sufficient. If the density in a sphere of radius a is, in polar coordinates,

$$\rho = -\frac{3\cos^2\theta - 1}{\log(b/r)}, \quad b > a,$$

ρ is continuous but has unbounded derivatives near $r = 0$ and does not satisfy a Hölder condition. It can then be shown that for $0 < r \leq a$

$$\nabla^2\phi = -4\pi\gamma\rho$$

has a solution containing a term

$$\phi_0 = \tfrac{4}{5}\pi\gamma r^2 \log\log(b/r)(3\cos^2\theta - 1)$$

but the second derivatives of ϕ_0 do not exist at $r = 0$. Modified definitions of $\nabla^2\phi$ have been proposed that make Poisson's theorem true for all continuous ρ.† But even with the ordinary definition continuous distributions of ρ that make Poisson's equation false are very rare.

9·04a. The throw-back can also be used with Bessel's formula, as has been pointed out by Comrie. The coefficient of the fourth difference in this formula is $\frac{1}{12}(\theta + 1)(\theta - 2)$ times that of the second, and this ratio varies from $-\frac{1}{6}$ to $-\frac{3}{16}$; the variation is even less than that of the corresponding ratio in Everett's formula. The ratio of the coefficient of the fifth difference to that of the third is $\frac{1}{20}(\theta + 1)(\theta - 2)$. Consequently it is advantageous, if fourth and fifth differences cannot be neglected, to take

$$f(x_0 + \theta h) = f_0 + \theta \delta f_{1/2} + \frac{\theta(\theta - 1)}{2}(\mu\delta^2 f_{1/2} - 0{\cdot}184 \mu\delta^4 f_{1/2})$$
$$+ \frac{\theta(\theta - \tfrac{1}{2})(\theta - 1)}{6}(\delta^3 f_{1/2} - 0{\cdot}108 \delta^5 f_{1/2}).$$

9·05a. This method of transforming the equation is given by Newton in *De Analysi*, 1669, and illustrated by the equation $x^3 - 2x - 5 = 0$. Synthetic division is not used in Horner's original paper;‡ he used another method of transformation. Synthetic division was introduced and applied to this problem by Horner in a further series of papers.§ He did not multiply the roots by 10 at each stage. He emphasizes the importance of proceeding one figure at a time in the early stages of the work, and this is perhaps his most important contribution. As it happened, the real root of the equation used by Newton for

---
* O. D. Kellogg, *Foundations of Potential Theory*, 1929, 1525–56.
† H. Petrini, *Acta Math.* 31, 1908, 127–332; G. Birkhoff and L. Burton, *Canadian J. Math. L.*, 1949, 199–208.
‡ *Phil. Trans.* 109, 1819, 306–35. § Leybourne's *Mathematical Repository*, 5, 1820.

illustration is very close to $+2\cdot 10$. Consequently it is impossible to say from this example alone whether Newton habitually tried to obtain several figures at a time or not. He does so in the later stages of the work, when higher powers are becoming negligible, but so does Horner.

It appears from Horner's papers that Newton's method had been completely forgotten. When he speaks of 'Newton's method' he means the iterative method stated in geometrical form in the *Principia* (Lib. 1, Prop. 23) for solving $\phi - e \sin \phi = N$ (e, N given), and still usually known as Newton's method. It seems to have been first applied to algebraic equations by Raphson. It contains no provision for making the determination of early figures facilitate that of later ones. For non-algebraic equations such provision is best made in the method of inverse interpolation.

9·09a. The comparison of the Simpson and three-eighths rules takes the total range the same for both, so that Simpson's rule uses one intermediate value and the three-eights rule two. If the lengths of the intervals are the same for both the advantage is the other way. Thus

$$\int_{-3}^{3} x^4 dx = \tfrac{2}{5} \times 243 = 97\cdot 2.$$

Using unit intervals we have

x	-3	-2	-1	0	1	2	3
x^4	81	16	1	0	1	16	81

Simpson's rule gives $\quad \tfrac{1}{3}\{162 + 4 \times 32 + 2 \times 2\} = \tfrac{1}{3} \times 294 = 98\cdot 0.$

The three-eighths rule gives

$$\tfrac{3}{8}(81 + 3 \times 16 + 3 \times 1 + 0) + \tfrac{3}{8}(0 + 3 \times 1 + 3 \times 16 + 81) = \tfrac{3}{8} \times 132 \times 2 = 99\cdot 0.$$

The possibility of this comparison arises only when the number of intervals is a multiple of 6.

9·09b. These rules have been used with success by S. Chandrasekhar in the solution of integral equations.* When a method analogous to that of 4·17 is used, and 10 equally spaced values would be needed to give the accuracy needed, this accuracy may be achieved with 5 points suitably spaced. The smaller number of equations to be solved compensates for the inconvenience of interpolation.

9·10a. With any method of numerical solution of differential equations, rounding-off errors tend to accumulate, and as each is carried on to the next step they cannot be detected by differencing. This is particularly serious for a differential equation of the form

$$y'' = f(x)\,y,$$

where $f(x)$ is positive. One solution, y_1, increases with x, another, y_2, decreases. Then the first two values of a solution y can be represented exactly by a function of the form $Ay_1 + By_2$. It we start from 0 and try to compute y_2, the first two values actually chosen will have rounding-off errors, which can be represented by a term in y_1, and the latter

* *Radiative Transfer*, 1950, Chapter II.

will increase steadily throughout the work, while y_2 itself diminishes. Thus the proportional error will increase for both reasons. Consequently it is desirable, in solution of equations of this type, to work in the direction of increasing x if we want y_1, but in that of decreasing x if we want y_2.

9·11a. Since
$$f(a+h) = f(a) + \nabla f(a+h), \quad \nabla^n f(a) = (\nabla^n - \nabla^{n+1}) f(a+h)$$
we can write the Adams-Bashforth formula as
$$\frac{1}{h}\int_a^{a+h} f(x)\,dx = f(a+h) - \nabla f(a+h) + (\tfrac{1}{2}\nabla + \tfrac{5}{12}\nabla^2 + \ldots)(1-\nabla)f(a+h)$$
$$= f(a+h) - (\tfrac{1}{2}\nabla + \tfrac{1}{12}\nabla^2 + \tfrac{1}{24}\nabla^3 + \tfrac{19}{720}\nabla^4 + \tfrac{3}{160}\nabla^5 \ldots)f(a+h).$$

In this formula the coefficients of the second and higher differences are much smaller than in the original one, and it is correspondingly more accurate. As for the central-difference formulae, the procedure is first to extrapolate a value of $f(a+h)$ and then improve it by successive approximation.

9·14a. The Gauss-Jackson method can be adapted to the solution of
$$y'' = f\left(x, y, \frac{dy}{dx}\right)$$
if we have a means of calculating dy/dx at the tabular values of x. We have
$$h\frac{dy}{dx} = (\mu\delta - \tfrac{1}{6}\mu\delta^3 + \tfrac{1}{30}\mu\delta^5 - \tfrac{1}{140}\mu\delta^7 + \ldots)y.$$
Substituting y from (10) we find
$$\frac{dy}{dx} = h(\mu\delta^{-1} - \tfrac{1}{12}\mu\delta + \tfrac{11}{720}\mu\delta^3 - \tfrac{191}{60480}\mu\delta^5 \ldots)f.$$

The coefficients are the same as in 9·084 (8). The extra trouble of forming
$$(\delta^{-1} - \tfrac{1}{12}\delta + \ldots)(f_{n-1/2}, f_{n+1/2})$$
and taking the mean is not prohibitive.

The Euler-Maclaurin formula leads at once to the integration formula
$$y_1 - y_0 = \tfrac{1}{2}h\{(y_0' + y_1') - \tfrac{1}{6}h\delta y_{\frac{1}{2}}'' + \tfrac{1}{360}h\delta^3 y_{\frac{1}{2}}''\} + O(h^7).$$

This resembles the central-difference formulae for double integration in the small factor associated with the third term. Consequently, if y'' is easily calculable, this formula can be used for solution of first-order equations as easily as the central-difference formulae for second-order equations with the first derivative absent.*

9·16a. Southwell's method depends on the same principles as Seidel's. Its distinctive features are: (1) at each stage a record is made of the outstanding residuals of all equations; (2) the next step is to reduce (*liquidate*, in Southwell's language) the largest residual; (3) no attempt is made to obtain more than one figure at a time in the next approximation.

* D. R. Hartree, *Proc. Camb. Phil. Soc.* 46, 1950, 523–4.

Thus, with the same equations as before, the largest term on the right is in the third equation. Take $x = y = 0$, $z = +2$ as the first approximation. The left sides are $+2\cdot 0$, $-5\cdot 2$, $+11\cdot 4$. Subtract these from the right sides, leaving the residuals $+5\cdot 8$, $+2\cdot 9$, $-2\cdot 8$. The largest is the first; take $x = +1$ and proceed. The values given in later approximations are, of course, corrections to the approximations already found:

	$z = +2$		$x = +1$		$y = +1$		$x = +0.5$		$z = -0.3$	
$6\cdot 3x - 3\cdot 2y + 1\cdot 0z = +7\cdot 8$	$+2\cdot 0$	$+5\cdot 8$	$+6\cdot 3$	$-0\cdot 5$	$-3\cdot 2$	$+2\cdot 7$	$+3\cdot 15$	$-0\cdot 45$	$-0\cdot 30$	$-0\cdot 15$
$-3\cdot 2x + 8\cdot 4y - 2\cdot 6z = -2\cdot 3$	$-5\cdot 2$	$+2\cdot 9$	$-3\cdot 2$	$+6\cdot 1$	$+8\cdot 4$	$-2\cdot 3$	$-1\cdot 6$	$-0\cdot 7$	$+0\cdot 78$	$-1\cdot 48$
$+1\cdot 0x - 2\cdot 6y + 5\cdot 7z = +8\cdot 6$	$+11\cdot 4$	$-2\cdot 8$	$+1\cdot 0$	$-3\cdot 8$	$-2\cdot 6$	$-1\cdot 2$	$+0\cdot 5$	$-1\cdot 7$	$-1\cdot 71$	$+0\cdot 01$
	$y = -0.2$		$x = -0.1$		$z = -0.1$		$y = -0.05$		$x = -0.04$	
$6\cdot 3x - 3\cdot 2y + 1\cdot 0z = -0\cdot 15$	$+0\cdot 64$	$-0\cdot 79$	$-0\cdot 63$	$-0\cdot 16$	$-0\cdot 10$	$-0\cdot 06$	$+0\cdot 16$	$-0\cdot 22$	$-0\cdot 25$	$+0\cdot 03$
$-3\cdot 2x + 8\cdot 4y - 2\cdot 6z = -1\cdot 48$	$-1\cdot 68$	$+0\cdot 20$	$+0\cdot 32$	$-0\cdot 12$	$+0\cdot 26$	$-0\cdot 38$	$-0\cdot 42$	$+0\cdot 04$	$+0\cdot 13$	$-0\cdot 09$
$+1\cdot 0x - 2\cdot 6y + 5\cdot 7z = +0\cdot 01$	$+0\cdot 52$	$-0\cdot 51$	$-0\cdot 10$	$-0\cdot 41$	$-0\cdot 57$	$+0\cdot 16$	$+0\cdot 13$	$+0\cdot 03$	$-0\cdot 04$	$+0\cdot 07$
	$y = -0.01$		$z = +0.01$		$y = +0.002$		$z = -0.001$			
$6\cdot 3x - 3\cdot 2y + 1\cdot 0z = +0\cdot 03$	$+0\cdot 03$	$0\cdot 00$	$+0\cdot 01$	$-0\cdot 01$	$-0\cdot 01$	$0\cdot 00$	$-0\cdot 00$	$0\cdot 00$		
$-3\cdot 2x + 8\cdot 4y - 2\cdot 6z = -0\cdot 09$	$-0\cdot 08$	$-0\cdot 01$	$-0\cdot 03$	$+0\cdot 02$	$+0\cdot 02$	$+0\cdot 00$	$+0\cdot 00$	$0\cdot 00$		
$+1\cdot 0x - 2\cdot 6y + 5\cdot 7z = +0\cdot 07$	$+0\cdot 03$	$+0\cdot 04$	$+0\cdot 06$	$-0\cdot 02$	$-0\cdot 01$	$-0\cdot 01$	$-0\cdot 01$	$0\cdot 00$		

The solution is
$$x = +1\cdot 0 + 0\cdot 5 - 0\cdot 1 \ -0\cdot 04 = +1\cdot 36,$$
$$y = +1\cdot 0 - 0\cdot 2 - 0\cdot 05 - 0\cdot 01 + 0\cdot 002 = +0\cdot 742,$$
$$z = +2\cdot 0 - 0\cdot 3 - 0\cdot 1 \ +0\cdot 01 - 0\cdot 001 = +1\cdot 609.$$

It is, in general, worth while to overcorrect at each stage in this method (and in Seidel's). If, for instance, we increase x in one approximation to remove the residual in the first equation exactly, then it will increase the residual of the second. This will be compensated by an increase in y. But this will again increase the residual of the first equation, and x will need a further increase. For this reason, especially if the non-diagonal coefficients are not small, convergence can be made more rapid by overcorrecting.

If, for instance, the correction to x needed to remove the residual in the x equation at some stage is δ_x, and we actually increase x by anything between δ_x and $2\delta_x$, it is easy to see that we shall always decrease S. In the relaxation method applied to differential equations it is often worth while to take the correction as $\frac{4}{3}\delta_x$ or even $\frac{3}{2}\delta_x$.

10·11 a. The transformation of the form (37) has been put in a more general form by Brown and Shook, their form being itself an extension of one due to H. von Zeipel. The following is a further extension.

If
$$H = H_0(p) + m \sum_s H_s(p) \cos(l_{rs} q_r + n_s t), \tag{1}$$

$m = 0$ gives
$$\frac{dp_r}{dt} = -\frac{\partial H_0}{\partial q_r} = 0; \quad \frac{dq_r}{dt} = \frac{\partial H_0}{\partial p_r} = a_r(p) = \text{constant}. \tag{2}$$

Take
$$J = p'_r q_r - m\Sigma' H_s(p') \frac{\sin(l_{rs} q_r + n_s t)}{\nu_s}, \tag{3}$$
$$\nu_s = l_{rs} a_r + n_s, \tag{4}$$

where Σ' is a sum over any set of terms with $\nu_s \neq 0$ and usually not with ν_s small. Then

$$p_r = p_r' - m\Sigma' H_s(p') \frac{l_{rs}}{\nu_s} \cos(l_{rs}q_r + n_s t), \tag{5}$$

$$q_r' = q_r - m\Sigma' \frac{\partial H_s}{\partial p_r'} \frac{\sin(l_{rs}q_r + n_s t)}{\nu_s}, \tag{6}$$

$$H' = H_0(p) + m\Sigma H_s(p) \cos(l_{rs}q_r + n_s t) - m\Sigma' H_s(p') \frac{n_s}{\nu_s} \cos(l_{rs}q_r + n_s t). \tag{7}$$

But
$$H_0(p) = H_0(p') + (p_r - p_r') \frac{\partial H_0}{\partial p_r} + O(m^2)$$

$$= H_0(p') + a_r(p_r - p_r') + O(m^2)$$

$$= H_0(p') - m\Sigma' H_s(p') \frac{a_r l_{rs}}{\nu_s} \cos(l_{rs}q_r + n_s t) + O(m^2). \tag{8}$$

Then by (4) and (7), since $p' - p$ is $O(m)$, all terms of order m in H' cancel except those not included in Σ', and

$$H' = H_0(p') + m\Sigma'' H_s(p) \cos(l_{rs}q_r + n_s t) + O(m^2), \tag{9}$$

where Σ'' is a sum over all s not included in Σ'.

Then (9) must be expressed in terms of the p' and q' to give a fresh Hamiltonian, which can be expanded in the form of (1). The $O(m^2)$ terms may contain some with $\nu_s = 0$ or small. Terms with $\nu_s = 0$ are called secular, those with ν_s small long-period. Secular terms can be included in H_0'. If there are no terms in Σ'' and we neglect m^2 the solution follows immediately as in (2). This may not, however, give sufficient accuracy. If there are long period terms, $p_r' - p_r$ and $q_r' - q_r$ contain coefficients m/ν_s, which may not be small, and the $O(m^2)$ terms may not be negligible.

In the theory of the motion of the planets m is the ratio of the mass of the disturbing planet to that of the Sun and some long-period terms are needed to the third order to complete the calculations within the errors of observation. The method can be repeated to give corrections of these higher orders.

A special case which gives essentially von Zeipel's form is the following. If H_0 depends on p_1 only, and all $n_s = 0$ we have

$$a_r = 0 \quad (r \neq 1) \tag{10}$$

and
$$J = p_r' q_r - m\Sigma' \frac{H_s(p') \sin(l_{rs}q_r)}{l_{1s}a_1}, \tag{11}$$

$$\nu_s = l_{1s}a_1. \tag{12}$$

Thus
$$J = p_r' q_r - \frac{m}{a_1} \int \Sigma H_s(p') \cos(l_{rs}q_r) \, dq_1, \tag{13}$$

where $q_2, q_3, \ldots$ are kept constant in the integration.

The following method is applicable when some ν_s is small. Suppose that H_0 depends on p_1 only, but $n_s \neq 0$ and some n_s, say $s = k$, is small. We suppose H to have been reduced to the form

$$H = H_0(p_1) + mH_k(p) \cos(l_{rk}q_r + n_k t). \tag{14}$$

We write
$$N = l_{rk}q_r + n_k t. \tag{15}$$

Then
$$\frac{dp_r}{dt} = mH_k(p)\, l_{rk} \sin N, \tag{16}$$

$$\frac{dq_1}{dt} = \frac{dH_0}{dp_1} + m\frac{\partial H_k(p)}{\partial p_1}\cos N, \tag{17}$$

$$\frac{dq_r}{dt} = m\frac{\partial H_k(p)}{\partial p_r}\cos N \quad (r \neq 1). \tag{18}$$

Differentiating (17) we have
$$\frac{d^2q_1}{dt^2} = \frac{d^2H_0}{dp_1^2}\frac{dp_1}{dt} + m\frac{\partial^2 H_k}{\partial p_1 \partial p_r}\frac{dp_r}{dt}\cos N - m\frac{\partial H_k}{\partial p_1}\sin N\,\frac{dN}{dt}. \tag{19}$$

The first and third terms are $O(m)$, the second $O(m^2)$. The expressions for q_r'' ($r \neq 1$) are similar except that the first term is absent. Then to order m
$$\ddot{N} = A\sin N - B\sin N\,.\,\dot{N}, \tag{20}$$
where A and B can be calculated and are $O(m)$. Except for the term in B this is a pendulum equation, and it may be shown that N either varies continually in one direction or oscillates about 0 or π, with a speed $O(m^{1/2})$. The latter case is known as libration and occurs for several satellites in the Solar System.

11·171 a. Another way of stating the theorem for $m = 0$ is: If $f(z)$ is bounded and analytic in a neighbourhood of $z = 0$, except possibly at $z = 0$, then a function $g(z)$ exists that is equal to $f(z)$ except possibly at $z = 0$ and is analytic also at $z = 0$. For the Laurent expansion of $f(z)$ has the property required.

13·05 a. The statement that the failure of a moving liquid to turn a sharp corner smoothly is due to the formation of a negative pressure is still to be found in text-books of hydrodynamics. As Rayleigh pointed out many years ago, the same phenomenon occurs in a gas, in which there is no question of negative pressure.*

13·091 a. If in Osgood's function (11·18) we put $z' = z + 1/z$ we get a function of z' bounded over the whole z' plane, but not constant. This does not contradict Liouville's theorem because there is a line of singularities from -2 to 2.

14·08 a. Lebesgue† has given a direct proof of the theorem for polynomial approximations, not depending on the use of integration. Other proofs independent of the use of Fourier series exist. One due to Weierstrass, applicable to any number of dimensions, is to take, for instance,
$$g(x,y,z) = \frac{\iiint_D f(\xi,\eta,\zeta)\exp[-k\{(\xi-x)^2+(\eta-y)^2+(\zeta-z)^2\}]\,d\xi\,d\eta\,d\zeta}{\iiint_D \exp[-k\{(\xi-x)^2+(\eta-y)^2+(\zeta-z)^2\}]\,d\xi\,d\eta\,d\zeta},$$
where f is continuous in D. k can be taken large enough for $|g - f|$ to be uniformly $< \omega$,

* See also H. Jeffreys, *Proc. Roy. Soc.* A, 128, 1930, 376–93.
† *Bull. des Sci. Math.* 22, part 1, 1898, 278–87.

in any D' interior to D; and then by expanding the exponentials in D we get the required approximation. See also Courant and Hilbert, 1, 69–72; Littlewood, *A Mathematician's Miscellany*, 30–4.

14·13a. A method extensively recommended in recent years for the solution of linear differential equations is (1) to apply the Laplace transformation to the whole of the equation, (2) hence determine the Laplace transform of the solution, (3) identify this by reference to a list of Laplace transforms of known functions. In criticism of (1), without preliminary study of the properties of solutions of the equation in general, there is no reason to suppose that the Laplace transform exists; and of (3), that the Bromwich integral actually gives the answer whether the function has already appeared in the various lists or not. Even if it has, a special theorem is still needed to establish uniqueness of the solution. It is remarkable that such treatment is advocated even for finite sets of ordinary differential equations, for which the direct operational treatment leads to a straightforward proof of existence and uniqueness of the solution. For partial differential equations substitution of complex integrals in Bromwich's manner proves existence; uniqueness would require use of the general theory of such equations. The Laplace transform method proves neither.*

If a solution is wanted for $0 \leqslant t \leqslant T$, it is possible to apply the Laplace transform method using integration from 0 to $T' \geqslant T$ instead of from 0 to ∞. The resulting transform naturally depends on T', but its interpretation by the Bromwich integral can be proved without much difficulty to be the same for all $t \leqslant T$ so long as $T' \geqslant T$.

17·07a. Ai(x) and Bi(x) are tabulated in the British Association Mathematical Tables.

For information about existing tables of numerical values of functions the *Index of Mathematical Tables* by A. Fletcher, J. C. P. Miller and L. Rosenhead (1946) should be consulted.

18·02a. The gradient of a scalar function ϕ in general orthogonal coordinates has components, in the direction of ξ_1, ξ_2, ξ_3 increasing,

$$u_1 = \frac{\partial \phi}{h_1 \partial \xi_1}, \quad u_2 = \frac{\partial \phi}{h_2 \partial \xi_2}, \quad u_3 = \frac{\partial \phi}{h_3 \partial \xi_3}.$$

The divergence of a general vector function u is found by considering the flux out of an element $d\xi_1 d\xi_2 d\xi_3$ as in 18·02; it is

$$\text{div } u = \frac{1}{h_1 h_2 h_3} \left\{ \frac{\partial}{\partial \xi_1} (h_2 h_3 u_1) + \frac{\partial}{\partial \xi_2} (h_3 h_1 u_2) + \frac{\partial}{\partial \xi_3} (h_1 h_2 u_3) \right\}.$$

The curl of a general vector is found by considering the integral of its normal component over a surface of constant ξ_1, with ξ_2 constant over one pair of edges and ξ_3 constant over the other pair; the values of ξ_2, ξ_3 over opposite edges differ by $\delta\xi_2$, $\delta\xi_3$. Then by Stokes's theorem this integral is $\int u_i dx_i$ around the element. Expressing this in terms of the components and taking $\delta\xi_2$, $\delta\xi_3$ small we find

$$(\text{curl } u)_1 = \frac{1}{h_2 h_3} \left[\frac{\partial}{\partial \xi_2} (h_3 u_3) - \frac{\partial}{\partial \xi_3} (h_2 u_2) \right].$$

---

* See also H. Jeffreys, *Proc. Inst. Elec. Eng.* (Heaviside Centenary Volume, 1950).

Cylindrical coordinates (ϖ, λ, z):

$$\operatorname{grad} \phi = \left(\frac{\partial \phi}{\partial \varpi}, \frac{\partial \phi}{\varpi \partial \lambda}, \frac{\partial \phi}{\partial z}\right),$$

$$\operatorname{div} u = \frac{1}{\varpi}\frac{\partial}{\partial \varpi}(\varpi u_\varpi) + \frac{\partial u_\lambda}{\varpi \partial \lambda} + \frac{\partial u_z}{\partial z},$$

$$(\operatorname{curl} u)_\varpi = \frac{1}{\varpi}\frac{\partial u_z}{\partial \lambda} - \frac{\partial u_\lambda}{\partial z},$$

$$(\operatorname{curl} u)_\lambda = \frac{\partial u_\varpi}{\partial z} - \frac{\partial u_z}{\partial \varpi},$$

$$(\operatorname{curl} u)_z = \frac{1}{\varpi}\frac{\partial}{\partial \varpi}(\varpi u_\lambda) - \frac{\partial u_\varpi}{\varpi \partial \lambda}.$$

Spherical polar coordinates (r, θ, λ):

$$\operatorname{grad} \phi = \frac{\partial \phi}{\partial r}, \frac{\partial \phi}{r \partial \theta}, \frac{\partial \phi}{r \sin\theta \partial \lambda},$$

$$\operatorname{div} u = \frac{1}{r^2 \sin\theta}\left\{\frac{\partial}{\partial r}(r^2 \sin\theta\, u_r) + \frac{\partial}{\partial \theta}(r \sin\theta\, u_\theta) + \frac{\partial}{\partial \lambda}(r u_\lambda)\right\},$$

$$(\operatorname{curl} u)_r = \frac{1}{r \sin\theta}\left\{\frac{\partial}{\partial \theta}(\sin\theta\, u_\lambda) - \frac{\partial u_\theta}{\partial \lambda}\right\},$$

$$(\operatorname{curl} u)_\theta = \frac{1}{r \sin\theta}\left\{\frac{\partial u_r}{\partial \lambda} - \frac{\partial}{\partial r}(r \sin\theta\, u_\lambda)\right\},$$

$$(\operatorname{curl} u)_\lambda = \frac{1}{r}\left\{\frac{\partial}{\partial r}(r u_\theta) - \frac{\partial u_r}{\partial \theta}\right\}.$$

Components of strain in these coordinates are given in Love's *Elasticity*, 1905, p. 56; but his h_i are the present $1/h_i$; and his components e_{23}, e_{31}, e_{12} are twice those adopted here.

18·05a. There are many papers on Mathieu's and related equations, especially by E. L. Ince and S. Goldstein, who first produced adequate general methods for computation. References are given by W. G. Bickley.[*] The fullest account is by N. W. McLachlan.[†]

23·07a. A summary of standard solutions of this equation and their asymptotic expansions is given by M. J. Seaton, *M.N.R.A.S.* **118**, 1958, 504–518.

Tables also are available:

(i) *O.N.R. Technical Reports*, Nos. 204 and 260. Cruft Laboratory, Harvard University, Cambridge, Massachusetts.

(ii) *Royal Society Mathematical Tables*, Vol. 11, *Coulomb Wave Functions*, A. R. Curtis, 1964.

For negative E an alternative treatment to that given in 23·07 is as follows: the solution for large ρ is
$$P = e^{-\frac{1}{2}\rho}\rho^{l+1} U(l+1-\nu, 2l+2, \rho).$$

From 23·051 we have, using the formulae,

$$(x-1)!\,(-x)! = \frac{\pi}{\sin x\pi}, \quad F(-x) = F(x-1) + \pi \cot x\pi,$$

---

[*] *Phil. Mag.* (7), **30**, 1940, 310–22.   [†] *Theory and Application of Mathieu Functions*, Oxford, 1947.

$$U(l+1-\nu, 2l+2, \rho) = \frac{(-1)^{l+1}(\nu+l)!}{(2l+1)!} \Big\{ {}_1F_1(l+1-\nu, 2l+2, \rho)\Big[\cos\nu\pi + \frac{\sin\nu\pi}{\pi}$$

$$\times\Big(\log\rho + F(\nu-l-1)\Big)\Big] + \frac{\sin\nu\pi}{\pi}\sum_{r=0}^{\infty}\frac{\rho^r}{r!}\frac{(l+r-\nu)!}{(l-\nu)!}\frac{(2l+1)!}{(2l+r+1)!}$$

$$\times\Big(-F(2l+1+r) - F(r) + \sum_{p=1}^{r}\frac{1}{l+p-\nu}\Big)\Big\} + \frac{(2l)!}{(l-\nu)!}\rho^{-2l-1}G(\rho).$$

The summation over p is omitted for $r = 0$.

For P to vanish at $\rho = 0$ we must have ν equal to an integer n greater than l.

23·08 a. Wave problems. Transmission and reflexion. Solutions of

$$\frac{d^2y}{dz^2} + h^2(z^2 - c^2)y = 0 \tag{1}$$

are $\qquad \exp(-\tfrac{1}{2}ihz^2)(U, V)(\alpha, \tfrac{1}{2}, ihz^2), \quad \alpha = \tfrac{1}{4}(1 - ihc^2). \tag{2}$

If the time factor for $\gamma > 0$ is $e^{i\gamma t}$, $U\exp(i\gamma t - \tfrac{1}{2}ihz^2)$ represents an advancing wave for $z > 0$. For $z < 0$ this solution becomes

$$\exp(-\tfrac{1}{2}ihz^2) U(\alpha, \tfrac{1}{2}, ihz^2 e^{2i\pi}). \tag{3}$$

From 23·04 (25) and 23·05 (23) we find that for $|\arg x| < \pi$

$$U(xe^{2i\pi}) = \frac{2\pi i e^{-i\gamma\pi}}{(\alpha-1)!(\alpha-\gamma)!} V(x) + 2e^{(-i\gamma\pi}\cos\gamma\pi - e^{2i(\alpha-\gamma)\pi}) U(x), \tag{4}$$

from which we find that (3) is equal to

$$\exp(-\tfrac{1}{2}ihz^2)\Big[\frac{2\pi i e^{-\tfrac{1}{2}i\pi}}{(\alpha-1)!(\alpha-\tfrac{1}{2})!} V(\alpha, \tfrac{1}{2}, ihz^2) - e^{2i(\alpha-\tfrac{1}{2})\pi}U(\alpha, \tfrac{1}{2}, ihz^2)\Big] \tag{5}$$

$$= \exp(-\tfrac{1}{2}ihz^2)\Big[\frac{2\pi}{(\alpha-1)!(\alpha-\tfrac{1}{2})!} V + ie^{\tfrac{1}{2}hc^2\pi}U\Big]. \tag{6}$$

The moduli of the coefficients are found to be $(1 + e^{hc^2\pi})^{\frac{1}{2}}, e^{\tfrac{1}{2}hc^2\pi}$. The term in V represents the incident wave for $z < 0$, since its phase decreases with $|z|$. Thus the amplitudes of the incident, reflected and transmitted waves are in the ratios*

$$(1 + e^{hc^2\pi})^{\frac{1}{2}}, \quad e^{\tfrac{1}{2}hc^2\pi}, \quad 1.$$

25·031 a.

When x goes from -1 to 1 by any path in the upper half-plane, $\Re(t)$ goes from $-K$ to K; hence every value of $\Re(t)$ in this interval specifies a curve going from a point between -1 and 1 on the real axis to the boundary; on this curve $\Im(t)$ goes from 0 to K'. Hence for any t satisfying $-K < \Re(t) \leq K$, $0 \leq \Im(t) \leq K'$, excluding $\Re(t) = 0$, $\Im(t) = K'$, where x has a pole, there is a value of x specified by a path from 0 with $\Im(x) \geq 0$. Similarly if $\qquad -K < \Re(t) \leq K, \quad -K' < \Im(t) \leq 0,$

there is a suitable path with $\Im(x) \leq 0$. Hence, by including suitable numbers of circuits about the paths C and C', we can make t take any assigned finite value. This is the important property of *ubiquity*: that the integral can take any value of t and hence that its inversion defines x over the whole t plane.

* Dr J. Heading gave us this result, found by a different method.

ADDENDA

3·031a. For further results on isotropic tensors in n dimensions the reader is referred to papers by M. Pastori, *Atti Acc. Naz. Lincei: Rendiconti* CL. VI: 12, 1930, 374–9; 12, 1930, 499–502; 13, 1931, 109–14; 17, 1933, 439–43 and also in *Scritti matematici offerti a Luigi Berzolari*, Pavia, 1936, 223–38, and *Boll. dell'Unione Matematica Italiana*, ser. II, anno I, 1939.

9·041a. Interpolation when first derivatives are given. If $f(0)$, $f(h)$, $f'(0)$, $f'(h)$ are given, we can construct a table of divided differences by using the device at the end of 9·013 and derive the formula

$$F(x) = f(0) + \frac{f(h)-f(0)}{h} x + \left[\frac{1}{h^2}\{f(h)-f(0)\} - \frac{1}{h}f'(0)\right] x(x-h)$$

$$+ \left[\frac{1}{h^2}\{f'(h)+f'(0)\} - \frac{2}{h^3}\{f(h)-f(0)\}\right] x^2(x-h),$$

$$= f(0) + \{f(h)-f(0)\}\left(\frac{3x^2}{h^2} - \frac{2x^3}{h^3}\right) + f'(0)\frac{x(x-h)^2}{h^2} - f'(h)\frac{x^2(h-x)}{h^2}.$$

It can be verified that this gives the correct results when $f = 1, x, x^2$ and x^3. For $f = x^4$ the maximum error in $(0, 1)$ is $-0·0625$. The maximum error for Bessel's formula with third differences (which uses the same number of data) is $-0·5625$.

When values of f' are available at the tabular values of x, the method has the following advantages.

(1) It is more accurate than any formula based on values of f alone.

(2) It uses no values outside the range of interpolation and therefore can be used in the terminal intervals, unlike the central difference formulae.

(3) If f, f' are given at equal intervals of x, and integrals of the form $\int x^r f(x)\, dx$ are wanted, the intervals may be too wide for accurate integration by the usual formulae, but when the intervals are halved by the above formulae they may be accurate enough.

(4) f and f' may be given at unequal intervals of x. Then the formula may be used to interpolate to equal intervals of x with much less trouble and loss of accuracy than by divided differences.

9·181a The great development in automatic computation has taken place since the first edition of this book. We have not tried to describe it but there are many books on the subject. Many of the methods treated here are, however, still used as they stand, and the principles of others are still applicable. Relaxation methods as described in 9·16 and 9·18 are now made more systematic.* The application in note 9·16a would not be considered worth while on a high-speed machine, which would carry 8 or 9 figures automatically. A problem that would take a day by hand can now be carried out in a few seconds. However, it takes much longer than this to write out the instructions for the machine, and there are still many problems that can be solved on a hand machine in a shorter time than it takes to write the programme and get access to a high-speed machine.

* Cf. G. E. Forsythe and W. R. Wasow, Finite difference methods for partial differential equations, 1960.

APPENDIX ON NOTATION

The difficulty of learning mathematical physics is much increased by confusion of notation, especially the overworking of certain letters and the introduction of awkward sign conventions. The only criterion usually recognized is conformity with 'standard practice'. Unfortunately standard practice is not unique and students are put to much unnecessary trouble by having to accustom themselves to work with different conventions in rapid succession. Research workers in border-line subjects are also inconvenienced by finding the usual symbols in one subject pre-empted for different meanings in another.

The following principles are important in choosing conventions:

(1) Complications should be reduced to a minimum. Negative signs should not be introduced without good reason.

(2) Genuine physical differences should be recognized as such and not disguised as conventions; attempts to disguise them always lead to later difficulties that should have been forestalled.

(3) Where a mathematical theory has applications in several subjects the notation should be such that it can be carried over into those subjects unchanged; so far as possible it should not use symbols already used with other meanings in those subjects.

The outstanding difficulty of notation at present is the ambiguous use of V and ϕ. V is used for potential energy, which is a property of a complete system, but also for the various potential functions, which are functions of position within the system. It is also used for Hamilton's characteristic function, and, in hydrodynamics, for a component of the velocity at a great distance. There is a tendency at present for potential functions to be denoted by ϕ, a usage long established in hydrodynamics. This would remove most of the difficulty. The characteristic function, which has a rather special field of application, could be denoted by A, and the velocity component by U_2 if tensor notation is adopted. It would therefore be easy to remove the ambiguity of V. The trouble is now that ϕ is also used for one spherical polar coordinate, corresponding to the longitude, and there are many potential problems that require this coordinate. One alternative would be to use capital Φ for potential in such problems, but this is difficult to write. The other is to find a new symbol for longitude and, with it, Euler's second angle. Lamb here uses ω in problems where a velocity potential exists and ϕ where none does. The disadvantage of ω as a regular notation for this purpose is obvious. ϕ has the disadvantage that in geodesy, which also depends greatly on the theory of the gravitational potential, ϕ is used for the *latitude*, and the longitude is denoted by λ. The same notation is used in meteorology. Again, in classical hydrodynamics we often require to use the velocity potential and the gravitational potential in the same problem, and it is therefore impossible to use ϕ for both. If we maintain its original use for the velocity potential, therefore, we must for this reason alone find another symbol for the gravitational potential. A universal rule that potential functions are to be denoted by ϕ is out of the question. The exceptional treatment of the gravitational potential would remove the difficulty in geodesy and dynamical meteorology, where velocity potentials do not occur, but not that of electrical and hydrodynamical systems with axial symmetry, where we need another symbol for the azimuthal coordinate anyhow. The geodetic practice suggests λ as a suitable one. Something could be said for ψ; its use as the allied function to ϕ in two-dimensional theory would lead to no confusion because the corresponding angle is there usually denoted by θ, and ψ was used in the works of Routh for the second Eulerian angle (χ would be available for the third). The replacement of ψ by ϕ was made only in recent works. Lamb's *Higher Mechanics* uses θ, ψ, χ. Either change would lead to changes in the notation of spherical polar coordinates, but there seems to be no escape, and in fact the notation given in mathematical textbooks is not used in a large fraction, perhaps the majority, of problems where position on a sphere has to be specified.

In dynamics it is generally convenient to use the work function, the work done on a system in transporting it from some standard state to its state at time t. This avoids a negative sign in the formation of generalized force-components. The potential energy, which is the work function with its sign changed, is convenient in the general theory of small oscillations about equilibrium, since it is then a positive quadratic form. This is a case where a little extra complication in notation is justified. The interest of potential energy in its own right really arises in relation to stable systems and becomes dominant in electricity and thermodynamics. In the treatment of large motions it is a nuisance. There is therefore a definite advantage in having both the work function and the potential energy as part of our equipment. In electrical problems potential energy is often denoted by W when V has been preoccupied by the potential function; but if we denote the latter by ϕ we can use V for the potential energy and release W. The position therefore is that we need symbols for work-function and gravitational potential, and U and W are available. U is at present widely used for both. It is suggested that W should be used for the work-function and U for the gravitational potential.

With regard to the choice of signs, the following usages have become common, for the sake of formal similarity:

Generalized force: $\qquad Q_r = -\dfrac{\partial V}{\partial q_r},$ (1)

Electric intensity: $\qquad X_i = -\dfrac{\partial \phi}{\partial x_i},$ (2)

Velocity in fluid: $\qquad u_i = -\dfrac{\partial \phi}{\partial x_i},$ (3)

Gravitational acceleration: $\qquad \ddot{x}_i = -\dfrac{\partial \Omega}{\partial x_i}.$ (4)

The first arises if we use potential energy; the sign is reversed if we use the work function, which is, for instance, the easier in all problems of orbits. The second usage has definite recommendations. In electrostatics the potential energy is a minimum and is convenient to write down, since that of two charges is ee'/r in electrostatic units. We must either have the negative sign in the function to be differentiated or insert it after the differentiation. The usual potential is the change of potential energy per unit change of charge at the point considered. If we used the work function per unit charge instead we should have to put a negative sign into the definition of ϕ, which would mean reversing the signs of all recorded potential differences. Accordingly (2) must be kept. (4) was introduced by Lamb to make it analogous with (2). But this is a false analogy. There is a fundamental physical difference between gravitation and electrostatics: two masses attract, two like charges repel, and a difference of sign somewhere is inevitable. What Lamb's convention does is to make the gravitational potential always negative and reverse the sign in Poisson's equation, a heavy price to pay for a thin analogy. The obvious course here is to call the work function per unit mass U, and replace (4) by the form in use before Lamb's convention

$$\ddot{x}_i = +\frac{\partial U}{\partial x_i}.$$

The negative sign in (3) was also introduced by Lamb. It has never been used outside Britain, but several other British writers have copied it on the basis either of Lamb's authority or of a belief that the usage was general in this country. The latter belief is mistaken; the chief users of the velocity potential in this country are the workers on aerofoil theory, who have continued to use the positive sign as in Glauert's book. Further, Lamb's book is as generally recognized as the chief authority abroad as here, but his convention is not adopted; and Love's *Elasticity*, an equally authoritative work, uses the positive sign when irrotational displacements occur. The negative sign in (3) can therefore be regarded only as an annoying and useless complication. Accordingly all considerations of convenient expression indicate that the best relations to take are:

$$Q_r = \frac{\partial W}{\partial q_r} \text{ (general case)} = -\frac{\partial V}{\partial q_r} \text{ (small oscillations)}, \tag{1'}$$

$$X_i = -\frac{\partial \phi}{\partial x_i}, \tag{2'}$$

$$u_i = \frac{\partial \phi}{\partial x_i}, \tag{3'}$$

$$\ddot{x}_i = \frac{\partial U}{\partial x_i}. \tag{4'}$$

The difference of sign in (2') and (3') is of little importance because electrostatic and hydrodynamical aspects seldom occur in the same problem. Gravitation is often important in hydrodynamics and it is convenient to have the same sign in (3') and (4').

A few words are also desirable about the constant γ. This has different values in different subjects and special symbols may be introduced. It is desirable not to suppress it in electrostatics and magnetism even though its numerical measure has been made 1 by a choice of units. The choice of a unit does not make it into a number, and inadequate analysis of the nature of physical measurement has led to the assertion, which still appears in textbooks intended for mathematical students, that the ratio of the electrostatic and electromagnetic units is the velocity of light. Further, the theoretical absolute electromagnetic units are never used, and with the practical units these constants are far from having numerical measure 1. The omission of the constants is possible only if the teaching of electromagnetic theory is completely separated from application to concrete systems.

In Tisserand's *Mécanique Céleste*, and in many other works on the subject where the law of gravitation is most used, the constant is denoted by f (roman or italic). f appears to be quite free from objection. It has been suggested that it might be mistaken for a function, but one of us has been using it since 1914 without finding a case where some letter other than f was not indicated for any function considered. The alternative G seems to lead to more difficulty in avoiding ambiguity.

The choice of units recommended by Lorentz and Heaviside, so as to absorb the 4π of Poisson's equation into γ, seems to have about as much to be said for it as against it. The theorem of Green's equivalent stratum asserts nothing but relations between values of ϕ, and contains the 4π, and there is no way of removing it from one place without putting it in in another.

Several different notations are used to indicate that two quantities are not very different: O, $\sim$, $\asymp$, and $\doteqdot$ are all used. The first two have precisely defined mathematical senses. But in physics we often want to say, without detailed calculation, that two quantities are unlikely to differ by more than a factor of 10, or by more than, say, 10 per cent. We suggest that the former statement, usually read as 'a and b are of the same order of magnitude', should be expressed by '$a \asymp b$'; and that the second, which can be read 'a is a rough estimate of b' or 'a is approximately equal to b' can be denoted by '$a \doteqdot b$', meaning something more precise than $a \asymp b$ and something less precise than a statement of extreme possible values of the difference. The degree of approximation that is interesting naturally depends on the problem. A statistical estimate $a = b \pm \sigma$ has a meaning defined in works on probability theory and needs no change.

The expression 'very approximately' literally means 'very closely'; its use to mean 'very roughly' is to be condemned.

Index

Abel, N. H., 405, 473, 669, 688
Abelian functions, 673
Abel's lemma (series), 41
   integrals, 45
Abel's test (integrals), 46, (series), 42, 370
Abel summation, 370, 436, 455
Abraham, M. and Becker, R., 106
Absolute convergence, *see* Convergence
Adams, J. C., 283, 293, 640, 641
Adams-Bashforth formula, *see* Differential equations
Airey, J. R., 290, 502, 616, 626
Airy integral, 476, 508, 517, 525, 702
Ai (x), *see* Airy integral
Algebra, complex numbers, 333
   infinity, 12
   matrices, 114
   operators, 228
   rules of, 1
Algebraic equations, linear, 118
   higher degrees, 379
   numerical solution, 274, 304, 696, 698
Allied Fourier integral, 453, 456
   series, 431
Almost everywhere, 29
Amplitude (complex number), *see* Argument
Analytic continuation, 362
Analytic function, 333, 339
   integral of, 343, 348
Andress, W. R., 408
Angular velocity, 79, 81, 97
Appell, P., 323
Area of surface, 188, 695
Argand diagram, 245, 341
Argument of complex number, 341
Arithmetic, rules of, 1
Associative law, 1, 66, 115, 116, 334
Asymptotic expansions, 498, 624
   formulae, 282, 466, 649
   solutions of differential equations, 519
      Stokes's type, 520, 582, 609
      Green's type, 523, 586
      uniformity, 523, 526

Baber, W. G., 488
Baker, H. F., 21, 229
Barnes, E. W., 612
Bashforth, F., 293
Bateman, H., 652
Bayes, T., 467
Bell, E. T., 341
Berestetzky, V., 662
Bernoulli, Daniel, 436
Bernoulli, James, numbers and polynomials, 280, 387, 431, 439
Bernoulli, John, 316
Besicovitch, A. S., 53, 188, 192, 692
Bessel functions, 574
   applications, 595
   ber, bei, kher, khei, 588
   complex integrals, 490, 574, 580
   definite integrals containing, 589
   expansions containing, 591, 593

Hankel functions, 575, 591, 660
I_n, Kh_n, 579
$J_{n+\frac{1}{2}}$, 539, 588, 659
   large order, 586
   operational forms, 574
   of operators, 585, 612
   recurrence formulae, 581
Y_n, 577
Bessel's equation, 489, 521, 527, 534
Bessel's interpolation formula, *see* Interpolation
Beta function, 463
Bethe, H. A., 619
Biaxial harmonics, 647
Bickley, W. G., 144, 289, 312, 383, 425, 495, 703
Bicycle, 323
Birkhoff, G., 696
Bi (x), *see* Airy integral
Bleieck, W. E., 383
Block matrices, 129
Bôcher, M., 136
Bolzano, N., 10, 21
Bonnet, O., 52
Boole, G., 229
Born, M., 202
Bounded, 11
Bounded convergence, 43, 691
Bounded variation, 24
Bounds, 13, 21
Bourguet, L., 467
Bowman, F., 312
Brachistochrone, 316
Branch points, 355
Briggs, L. J., 273
Brinkman, H. C., 662
Bromwich, T. J. I'A., 48, 142, 246, 392, 557, 561, 571
Bromwich integral, 392, 394, 491
Brown, E. W., 329, 488, 699
Bückner, H., 168
Burkill, J. C., 29

Cable, submarine, 602
Calculus of variations, 314
   restricted variation, 319
   variation of limits, 317
Campbell, N. R., 3
Cantor, G., 6, 12
Capacity, 310, 418
Caqué, J., 229, 476
Carathéodory, 192
Carlini, G., 525
Carnap, R., 2
Carter, G. W., 260
Cartesian coordinates, 57
Cauchy, A. L., 363, 476
Cauchy's inequality, 54, 336
   inequality (power series), 361
   integral, 360
   theorem, 344, 347
Cauchy-Riemann relations, 333, 337, 338, 346, 361
Cavalieri, B., 286
Centre of mass, 83, 543

Césaro, E., 436
Chandrasekhar, S., 697
Change of variable in single integral, 32
  in multiple integral, 182
Chappell, E., 273
Characteristic values, 127
Charge function, 412
Checking, 275
Circle of convergence, 350
Circulation, 196
Ci (x), ci (x), 471
Clairaut's formula, 646
Closed interval, 19
  region, 174
Coaxal circles, 415
Collineatory transformation, 128
Commutative laws, 1, 62, 66, 106, 115, 334
Commuting matrices, 115
Comparability, 2
Comparison series, 43
Complex numbers, 333
Complex potential, 412
Comrie, L. J., 270, 273, 274, 696
Condenser, charging of, 244
  discharge of, 245
Condon, E. J., and Shortley, G. H., 133, 633
Conduction of heat, 529, 563, 600
Confluent hypergeometric function, 607
  integrals for, 608
  Whittaker's form, 616
Confocal conics, 419; *see also* Mathieu's equation
Conformal mapping, 409
Consistent units, 4
Contact transformation, 328
Continuation, analytic, 362
Continuity, 17, 176, 342
  sectional, 19
  uniform, 23, 342
Continuous distributions (mass or charge), 202
Contour, 173, 344
  integration, 335
Contraction of tensor, 87
Contravariant components, 158
Convergence, 11, 52
  absolute, 16, 36, 350
  bounded, 43, 691
  conditional, 16
  of integrals, 33
  radius of, 350
  uniform, 37, 44, 48, 351, 371
Convergence factors, 502
Cooling of bars, 563
Cooling of Earth, 569
  thermometer bulb, 571
Cosine transform, 457
Courant, R., and Hilbert, D., 127, 495, 619, 638, 702
Courant, R., and Robbins, H., 189
Covariant components, 158
Covering theorems, 19, 175
Cowell, P. H., 300
Crommelin, A. C. D., 300
Crossley, A. F., 259
Crystals, 155
Curl, 90, 224
Curves, 172
  length of, 173
Curvilinear coordinates, 157, 532, 702

Cuts (complex), 341, 356
Cylinder, lift on, 414
Cylindrical coordinates, 534, 703
  pulse, 595

δ_{ik}, 59; δ-function, 393
$D_n(x)$, 623
D'Alembert, J. le R., 436, 547
D'Alembert's principle, 83, 322
Dalzell, D. P., 395, 396
Darwin, C. G., 633
Darwin, G. H., 450
Debye, P., 501, 503
Decreasing functions, 22
Dedekind, R., 1
Dedekind section, 6, 13
Definite integration, as linear operator, 229
Derived magnitudes, 4
Determinants, infinite, 488
  multiplication of, 119
Determinantal equations, 127
  root-separation theorem, 140, 146, 254
Diagonal block matrices, 130
Diagonal matrices, 94, 117, 128
Diameter of region, 174
Differentiability of function of several variables, 178, 180, 337, 693
  on one side, 50
Differential equations
  asymptotic solutions, 519
  existence of solution, 475
  indicial equation, 481
  numerical solution, 290
    Adams-Bashforth, 292, 698
    central differences, 293
    Gauss-Jackson, 300, 698
    jury problems, 306
    Taylor's series, 290
  operational method, first order, 232
    higher orders, 239
  regular singularities, 478
  singular points, 474, 477
  singularities at infinity, 480
  solution by complex integrals, 489
  three-term recurrence relations, 485
Differentiation, as linear operator, 229, 398
  non-commutative property, 229
  numerical, 277
  of integral, 30
  of power series, 352
  under integral sign, 45, 694
Diffraction, 621
Diffusion, 600
Digamma function, 465
Dimensions, 4
Dipole, 206
Dirac, P. A. M., 153, 393
Direction vector, 64
Dirichlet, P. G. L., 436
Dirichlet integrals, 468
Dirichlet-Hardy test (series), 42
  integrals, 46
Discontinuity, 18, 26
  and non-uniform convergence, 39, 43, 47
  of arbitrary constants, 511
  removable, 18, 25, 357
  simple, 17, 25

Index

Dispersion, 511, in string, 597
  of water waves, 515
Dissipative systems, 254
Distance, 57, 171
  from set, 177
Distributive law, 1, 66, 116, 334
Divergence (sequences), 14
Divergence of vector, 90
  theorem, 193, 345
Divided differences, 262
Division of vectors, 73
Donkin, W. F., 329
Doodson, A. T., 450
Double integrals, 180
Doublet, 206
  shell, 207, 215
Doubly-periodic functions, see Elliptic functions
Durrell, C. V., and Robson, A., 73
Dyadics, 89

ϵ, 8
ϵ_{ikm}, 69, 73, 75
Earth, cooling of, 569
  figure of, 221, 643
Earthquakes, 253
Eddington, A. S., 108, 152
Edser, E., 568
ei (x), 470; Ei (x), 471
Eigenvalues, 127
Einstein, A., 332
Elasticity, 99
Electromagnetic stress tensor, 105
  radiation, 660
  theory, 160
Ellipsoidal coordinates, 541
Elliptic coordinates, 419, 536
Elliptic equations, 531
Elliptic functions, 667
  addition formulae, 678
  infinite products, 679
  residues, 675
  trigonometric expansions, 677
  ubiquity, 704
Elliptic integral, complete, 687
  first, 669
  second, 686
  third, 687
  standard form, 685
Empirical periodicities, 450
Enneper, A., 667
Enumerable sets, 10
erf x, 403, 569
Ergodic theorem, 166
Essential singularity, 357, 367
  isolated, 359, 367
Euclid, 5, 57
Euler, L., 261, 370, 436, 472, 669, 679
Euler pendulum, 248
Euler's angles, 108
Euler's constant, 282, 402, 470
Euler's equations, 110
Euler-Maclaurin formula, 279, 280, 406, 466, 499
Everett's formula, 269
Ewald, P. P., 157
Explosion wave, cylindrical, 595
  spherical, 560

Factorial function, 187, 401, 462
Fermat's principle, 318
Ferrar, W. L., 122, 137
Figure of Earth, 643
Fletcher, A., 702
Fluctuation, 24
Fluid, heated below, 442
  motion of, 97, 701
Flux, 196
Fourier, J., 436, 566
Fourier series, 368, 430
  cosine series, 437
  differentiation of, 441
  integration of, 440
  sine series, 437
Fourier-Bessel expansion, 591
Fourier-Mellin theorem, 458
Fourier's integral theorem, 452
Fox, L., 312
Fresnel integrals, 473
Frobenius, method of, 482
Frullani integrals, 406
Fuchs, L., 229
Functions of complex variable, 337
  real variable, 17
  several variables, 176
  two variables, 36
Fundamental magnitudes, 3

Galitzin seismograph, 250
Gamma function, see Factorial function
Gans, R., 526
Gaunt, J. A., 641
Gauss, C. F., 312, 464, 472
Gauss-Jackson, (J.), method, 300, 698
Gauss's Π function, see Factorial function
Gauss's theorem, 201, 211
Generalized coordinates, 321
Germane units, 4
Gibbs (Willard) phenomenon, 445
Gibson, G. A., 196
Gilles, D. C., 312
Glauert, H., 423, 707
Gödel, K., 2
Goldstein, S., 304, 404, 405, 444, 487, 527, 604, 612, 703
Goursat, E., 21, 342, 345, 347
Graphical methods, 290
Grassmann, H. G., 155
Gray, A., Matthews, G. B., and MacRobert, T. M., 577
Green, G., 525
Green's equivalent stratum, 218
Green's function, 220, 495, 543, 634
Green's lemma, 193, 345
Green's theorem, 195, 695
Gregory, James, 265, 284, 286
Group velocity, 512
Guggenheim, E. A., 4
Gyroscopic systems, 145, 254, 331

Hall, P., 133
Halphen, G. H., 667
Hamilton-Jacobi equation, 325
Hamilton's equations, 326
Hamilton's principle, 318, 320
Hardy, G. H., 15, 42, 436, 461, 524
Harkness, J., and Morley, F., 363

Harmonic analysis, 429, 449
Harmonic oscillation of finite duration, 459
Harmonic wave train, interrupted, 518
Hartree, D. R., 300, 570, 616, 622, 668, 698
Hassé, H. R., 488
Heading, J., 704
Heat conduction, 529, 563, 600
Heat, line source of, 604
　periodic supply of, 572
Heaviside, O., 18, 229, 238, 254, 266, 565, 568, 577, 579, 602, 708
Heaviside's unit function, 18, 243, 393
Heine, E., 21, 648
Heine-Borel theorem, 20, 175, 365
Helmert, F. R., 645
Henderson, J. B., 4
Hermite functions, 620
Hermitian forms, 136
　matrices, 117, 133, 137
$Hh_n(x)$, 622
$Hi_n(x)$, 575
Hill, G. W., 487, 488
Hobson, E. W., 22, 173, 181, 541, 650
Hölder, O., 696
Horner, W. G., 275, 696
$Hs_n(x)$, 575
Hurwitz, A., and Courant, R., 363
Hydrodynamics, sources and sinks, 201
　complex potential, 412
Hydrogen-like atom, 618
Hydrogen molecular ion, 488
Hyperbolic equations, 531
Hypergeometric function, 606

$I_n(x)$, 574
Idelson, N., 221
Imaginary numbers, see Complex numbers
Improper integrals, 33
Ince, E. L., 520, 536, 543, 703
Incomplete factorial function, 498
Increasing functions, 22
Indicial equation, 481
Indirect proof, 8
Induction, mathematical, 9
Inertia tensor, 95
Infinite determinants, 488
Infinite instability, 392
Infinite integrals, 33
Infinite products, 52
Infinity, 12
Ingen-Hausz's experiment, 568
Ingersoll, L. R., 571
Inner product, 115
Instrument, response, of, 460
Integral equations, 167, 405, 457, 473, 495
Integral function, 352, 362
Integrals, double, 180
　change of variable, 32, 182
　principal value, 376
　repeated, 180
Integration, 26
　by parts, 32
　change of variable, 33, 182
　complex, 343, 348
Integration, numerical, 278
　central difference, 284
　Euler-Maclaurin, 279

Gauss, 288, 698
Gregory, 283
　of $x^{\frac{1}{2}}f(x)$, $x^{-\frac{1}{2}}f(x)$, 289
Simpson, 286, 697
three-eighths rule, 287, 697
Weddle, 287
Interpolation, 261
　Everett, 269, 272
　Gregory, 265
　inverse, 274
　Lagrange, 261
　Newton, 265
　Newton-Bessel, 269, 272, 696
　Newton-Gauss, 268
　Newton-Stirling, 268
Intervals, closed, 19
　nests of, 6
　open, 19
Inverse functions, 22, 379
Irrotational vector, 196
Isotropic tensors, 87

$J_n(x)$, 574
Jackson, D., 431, 449
Jackson, J., 300
Jacobi, C. G. J., 382, 520, 681
Jacobian elliptic functions, 669
Jacobians, 184
Jacobi's imaginary transformation, 683
Jacobi's theorem (determinants), 135
　(dynamics), 327
Jahnke, E., and Emde, F., 577
Jeans, J. H., 634
Jeffreys, B., 527, 662
Jeffreys, H., 57, 84, 87, 229, 275, 304, 399, 444, 451, 460, 490, 498, 525, 526, 527, 563, 570, 600, 702
Jentzsch, R., 372
Jordan's lemma, 392
Joukowsky aerofoils, 423
　transformation, 422

Kellogg, O. D., 191, 696
Kelvin, 501, 507
$Kh_n(x)$, 579
$Khei_n(x)$, $Kher_n(x)$, 588
Kinetic theory of gases, 167, 202
Klein bottle, 188
Kneser, A., 543
Knopp, K., 10, 12
Kramers, H. A., 662
Kronecker δ, 59

Lagrange's equations, 321, 324
Lagrange's expansion, 382
Lagrange's interpolation formula, 261
Laguerre polynomials, 619
Lamb, H., 142, 453, 525, 597, 604, 621, 706
Lamé functions, 541
Lamé's constants, 102
Landau, E., 368
Landen's transformation, 688
Langer, R. E., 526
Language, mathematics as a, 2
Laplace, P. S., 254, 457, 485, 650
Laplace equation, 197, 199, 339, 437, 529, 658
　uniqueness theorems, 215, 216, 217
Laplace transform, 458, 702

Index

Larmor, J., 160
Latent roots, 127
Laurent's theorem, 366, 701
  three-dimensional analogue of, 639
Leap, 22, 26
Least action, 330
Leathem, J. G., 11, 427
Lebesgue, H., 21, 192, 701
Lebesgue integral, 29
Legendre, A. M., 467, 628, 633, 647, 669, 690
Legendre functions, 538, 540, 579, 628
  asymptotic formulae, 649, 655
  definite integrals, 649
  explicit forms, 639
  recurrence formulae, 639
  relation to Bessel functions, 648, 656
Legendre's equation, 477, 484, 538
Lennard-Jones, J. E., 202
Levi-Civita, T., 159
Libration, 701
Light, velocity of, 518
Lighthill, M. J., 527
Limiting processes, inversion of, 36, 48
Limit-points, 9, 172, 353
Limits of functions, 17
  of operators, 390
  of sequences, 11
Line density, 205
Linear algebraic equations, 118
  numerical solution, 304
Linear relation between functions, condition for, 492
Liouville, J., 525
Liouville's theorem, 362
Lipschitz condition, 53, 431, 434, 435, 696
Littlewood, D. E., 148
Littlewood, J. E., 348, 436, 702
Livens, G. H., 327
Localized vector, 64
Lodge, A., 473
Logarithm, 358, 376
Lommel, E., 594
Lorentz, H. A., 160, 708
Love, A. E. H., 253, 555, 561, 707
Lowan, A. N., 273

M test, integrals, 45
  series, 40
MacColl, J. W., 531
McConnell, A. J., 108, 159
MacCullagh's formula, 543
Macdonald, H. M., 408, 557, 579
McLachlan, N. W., 459, 703
Maclaurin, C., 267, 280
McLeod, A. R., 571
Magnitude, 57
Magnitude, derived, 4
  elementary, 3
  fundamental, 3
Malkin, N., 221
Mallock's machine, 168
Mathieu's equation, 486, 488, 527, 536, 541, 703
Matrices, 114
  adjugate, 117
  block, 129
  commuting, 115, 132
  conjugate complex, 117
  diagonal, 117
  division, 118
  hermitian, 117, 133
  null, 116
  orthogonal, 118, 120, 121, 122, 148
  rank, 124
  reciprocal, 117
  roots, 127
  symmetric, 117, 123
  transformation, 120
  transposed, 116
  unit, 116
  unitary, 118, 148
Maximum modulus principle, 365
Mean value theorems, 49
Measurement, 3, 57
Measure zero, 29
Mechanics, 60, 108
Mehler, F. G., 652
Meteorology, 572
Method of steepest descents, 400, 503
  stationary phase, 400, 506
Miller, J. C. P., 291, 383, 528, 594, 702
Miller, W. H., 155
Milne-Thomson, L. M., 270, 280
Milne, W. E., 522
Minimal conditions, 10
Minors, 125
  of matrix of cofactors, 135
Mittag-Leffler's theorem, 383
Möbius strip, 189
Modulus, complex numbers, 335
  real numbers, 9
  vectors, 63
Moivre, A. de, 467
Monotonic, 22
Moore, E. H., 48
Mordell, L. J., 275, 469
Morera's theorem, 371
Morris, R. M., 425
Motion, under gravity, 77
  of charged particle, 78
Motz, H., 312
Multiple integrals, 180
  change of variables in, 182
Multipole expansions, 660
Murnaghan, F. D., 105
Mutual induction, 245

Natural boundary, 369
Necessary condition, 10
Neighbourhood, 172, 342
Nests of intervals, 6
Neumann, C., 591
Neville, E. H., 667, 675
Newman, M. H. A., 22, 175, 347
Newton, I., 203, 275, 696; see also Interpolation
Nicholson, J. W., 594
Nielsen, 576
Non-concentric circles, 418
Non-holonomic systems, 322
Normal coordinates, 145
Normal modes, 138
Notation, 203, 705
Null vector, 64
Numbers, 1
  complex, 333
  real, 5

$O(x)$, $o(x)$, 24
Oblate spheroidal coordinates, 539
Oblique axes, 153
Ocean, heating of, 572
Offord, A. C., 431
Open interval, 19
   region, 174
Operational methods, 228, 388, 473, 548, 585, 595
Operators, composition, 231
   division by, 233
   interpretation, 229, 233, 237, 392
   limits of, 390
   series of, 230
Orbits, 668
Ordering relation, 2
Orders of magnitude, 24, 707
Orthogonal property of normal modes, 139, 142
Orthogonality relations, 541, 636
Oscillation (of function), 22
Oscillations, small, 144, 252
Oscillatory, 11
Osgood, W. F., 369, 701
Osgood-Vitali theorem, 372, 396
Ostrogradsky, M. V., 695
Outer product, 115

p (Heaviside symbol), 237, 398
p_n^s, 632
Pairman, E., 466
Parabolic cylinder coordinates, 535, 620
Parabolic equations, 531
Parallax, 111
Parallelogram law, 58
Parapet function, 405
Parseval's theorem, 448, 457, 638
Partial differential equations, types of, 531
Partial fraction rule, 237, 389
Pathology, 5, 17
Pauli, W., 151
Peano, G., 173, 229
Pearson, K., 261, 269
Pellew, A., 444
Pendulum, 668
Pendulum, inverted, 488
Periodic disturbance at internal point, 557
Periodicity, of solutions of differential equation, 667
   empirical, 450
Periodogram, 450
Petrini, H., 696
Physical magnitudes, 3, 57
Picard, E., 229, 367, 476
Planets, theory of motion of, 700
Planetary orbits, 329, 494, 668
Poincaré, H., 221, 499
Poisson, S. D., 405, 437
Poisson's equation, 204, 211, 696
Poisson's integral equation, 405
Poisson's ratio, 102
Pol, B. van der, 459, 491, 581
Poles, 356, 358, 368
Pollard, S., 26, 30, 338, 347
Polygonal boundaries, 424
Potential, 199, 412, 453, 707
   at external points, 543, 635, 641
   in cavity, 210, 629
   in polarized medium, 222
   inside continuous distribution, 208
   of disk, 641
   of line density, 205
   of non-uniform sphere, 642
   of sphere, 204
   of spherical cap, 204
   of spherical shell, 203
   theory, 199
Power series, 349, 476
   differentiation, 352
   integration, 352
   multiplication, 353
Powers, non-integral, 358, 359
Pressure, fluid, 103
Principal axes, 93
Principal part at pole, 356
Principal value of argument, 341
   of integral, 376
   of logarithm, 359
Principia Mathematica, 2
Principle of superposition, 239, 404
Probabilities in chains, 163
Progressive wave, 548
Prolate spheroidal coordinates, 540
Proudman, J., 450
Pulse, 561, 595
   refraction of, 519

q_n^s, 579, 654, 656
Quadratic forms, 136
Quantum theory, 133, 152, 167, 202
Quaternions, 74, 169

Radioactivity, 256
Radius of convergence, 350
Raphson, J., 697
Rayleigh, 253, 526, 553, 701
Rayleigh-Ritz method, 218
Rayleigh's principle, 144, 148, 302
Real numbers, 5
Reciprocal lattice, 155
Rectifiable curves, 173, 188
Reductio ad absurdum, 8
Reflexion, 71, 120, 121, 150, 170, 661
Refraction of pulse, 519
Region, 172, 174, 339
   multiply-connected, 175, 348
Regular function, 339
   singularity, 478
Relaxation methods, 305, 307, 698
Removable discontinuity, 18, 357
Residue, 359
Resonance, 251
Response of recording instrument, 460
Reymond, P. du Bois, 28, 52
Richardson, L. F., 218, 288, 306, 312
Riemann, G. F. B., 26, 332, 345, 363
Riemann's lemma, 431
Riemann-Weber, 567
Riesz, F., 488
Rigidity, 102
Rimington, E. C., 246
Ritz, W., 218, 304
Robson, A., 73
Rolle's theorem, 49
Rolling, 322
Rosenhead, L., 667, 702

Index

Rotating axes, 106
Rotation, large, 62, 96, 122, 149
  small, 79
Rouché's theorem, 378
Routh, E. J., 142, 145, 254, 330
Russell, B., 2
Rybner, J., 251

Saddle-point, 504
Sadler, D. H., 291
Satellites, 701
Sawtooth function, 405
Scalar product, 65
Scalars, 61
Schläfli, 575, 649
Schlicht functions, 380
Schmidt, E., 168, 543
Schrödinger's equation, 300, 331, 488, 618
Schwarz, H. A., 54, 190, 192, 366
Schwarz's inequality, 54
Schwarz-Christoffel transformation, 426
Scientific laws, 57
Seaton, M. J., 703
Sectionally continuous, 19
Second mean-value theorem, 52, 692
Secular instability, 256
Secular terms, 700
Sedgwick, W. F., 259
Seidel, P. L. von, 305
Seismograph, 248
Self-induction, 246
Separation of variables, 530
Sequences, 10
Series, 14
  double, 16
  expressed by integrals, 491
Sets, 9, 172
Sgn (z), 341, 399, 506
Shook, C. A., 329, 699
si (x), Si (x), 471
Simple discontinuity, 17
Simple functions, 380
Simpson's rule, 286, 697
Sine transform, 457
Singular points of differential equation, *see* Differential equations
Singularities, 355
  essential, 357, 367
  at infinity, 357
  lines of, 357, 369
Singularities of inverse functions, 381
Slater, L. J., 627
Small oscillations, 144, 252
Smithies, F., 168, 543, 692
Soddy, F., 258
Solar system, 701
Solenoidal vector, 196
Solid harmonics, 630
Solution of algebraic equations, numerical, 274
  Horner, 275
  linear, 304
  Newton, 275, 696
  Raphson, 697
Southwell, R. V., 218, 309, 312, 444, 698
Sphere, Green's function for, 220, 634
Spherical harmonics, *see* Legendre functions
Spherical polar coordinates, 537, 703

Spherical waves, 559, 659
Spheroidal coordinates, 537
Spin matrices, 151
Stability, 144, 255, 488
Staircase function, 405
Standing waves, 549
Stationary phase, method of, 506
Steepest descents, method of, 503, 625
Stieltjes, 26, 467
Stieltjes integral, 26, 30, 32, 239
Stirling's formula, 467, 507
Stokes, G. G., 40, 221, 507, 511, 561, 645
Stokes's theorem, 195
Stolz, O., 178, 366
Stoneley, R., 253, 519, 616
Strain, 97, 99
Stream function, 412
Stress, 99
String, vibrating, 546
  loaded, 553
  numerous loads, 597
Stroud, W., 4
Sturm-Liouville method, 543
Submarine cable, 602
Substitution tensor, 59
Sufficient condition, 10
Suffix notation, 58
Summation convention, 59
Sunspots, 451
Superposition, principle of, 239, 404
Surface density, 204, 212
Surface integrals, 188

t_n^a, 653
Tannery, J., and Molk, J., 667
Tannery's theorem, 48
Tauber, A., 436
Taylor, G. I., 531
Taylor's theorem, 50, 266, 354, 361, 362
Temple, G., 144, 304, 495
Tensors, 86
  antisymmetrical, 91, 93, 97
  isotropic, 87, 693
  quotient rule, 89
  symmetrical, 91, 97
  in two dimensions, 110
Termini, 26
Terrestrial magnetism, 639
Tesseral harmonics, 633
Thermodynamics, 563
Theta functions, 680
Thompson, A. J., 273
Three-eighths rule, 287, 697
Throwback, 270, 696
Tidal equation, 485
Tides, prediction of, 450
Tisserand, F., 707
Titchmarsh, E. C., 29, 368, 453, 522
Top, 110, 146, 324
Topology, 175
Toroidal coordinates, 666
Transformation of rectangular coordinates, 59
Transformation theory (dynamics), 328
Trigonometrical functions, 384
Triple products of vectors, 74
Turnbull, H. W., 267
Turner, H. H., 253, 430

Unbounded, 11
Uniformity of continuity, 23
 of convergence (series), 37, 48, 351, 371
 (integrals), 44, 48
Uniqueness theorems, 215
Unit function, 18, 243, 393
Unit vector, 64
Units, 3, 4, 707

Variation of function, 24
Variation of parameters, 493, 551
Variations, calculus of, 314
Vector, 60
 area, 69
 potential, 224, 660
 of point charge and doublet, 226
 product, 67, 92
 single letter notation, 62
 of tensor, 92
Vector diagram, 245
Velocity potential, 412
Vibrating string, 546
Viscosity, 103
Vitali, G., 372
Vortex, line, 206, 604

Waerden, B. L. van der, 660
Wagner, K. W., 392
Wallis, J., 245, 341, 468
Waring, E., 261
Watson, G. N., 525, 576, 579, 658
Watson's lemma, 501
Wave equation, 529, 546
Wave mechanics, 331, 527
 harmonic oscillator, 621
 hydrogen-like atom, 618
 hydrogen molecular ion, **488**
 potential barrier, 527, 700
 rotation of axes, 647
Wave velocity, 512
Weatherburn, C. E., 84
Webb, H. A., 616
Weber, H., 1
Webster, A. G., 466, 495, **531**, 541
Weddle's rule, 287
Weierstrass, 10, 20, 41, 142, 363
Weierstrass's approximation, 446, **701**
Weierstrass's elliptic functions, **687**
Wessel, C., 245, 341
Whipple, F. J. W., 323
Whitehead, A. N., 2
Whittaker, E. T., 461, 532, **616**
Whittaker and Robinson, G., 286
Whittaker and Watson, 481, 488, **532, 681**
Widder, D. V., 26
Wiechert, E., 248, 332
Wirtinger, W., 279
Wood-Anderson seismograph, **248**
Work function, 321, 707
Wronskian, 492, 587

$Y_n(x)$, 576
Young, G. C., 53
Young, W. H., 21, 53, 178, **366, 431**
Young's modulus, 102

Zeipel, H. von, 699
Zeta function (elliptic), **686**
 (Riemann), 15
Zobel, O. J., 571
Zonal harmonics, **633**